Abridged Table of Atomic and Molar Masses of the Elements*

Element	Symbol	Atomic Mass (u) Molar Mass ($g \cdot mol^{-1}$)	Element	Symbol	Atomic Mass (u) Molar Mass ($g \cdot mol^{-1}$)
Aluminum	Al	26.98	Lithium	Li	6.941
Argon	Ar	39.95	Magnesium	Mg	24.30
Arsenic	As	74.92	Manganese	Mn	54.94
Barium	Ba	137.3	Mercury	Hg	200.6
Beryllium	Be	9.012	Neon	Ne	20.18
Boron	B	10.81	Nickel	Ni	58.69
Bromine	Br	79.90	Nitrogen	N	14.01
Calcium	Ca	40.08	Oxygen	O	16.00
Carbon	C	12.01	Phosphorus	P	30.97
Cesium	Cs	132.9	Potassium	K	39.10
Chlorine	Cl	35.45	Rubidium	Rb	85.47
Chromium	Cr	52.00	Silicon	Si	28.09
Cobalt	Co	58.93	Silver	Ag	107.9
Copper	Cu	63.55	Sodium	Na	22.99
Fluorine	F	19.00	Strontium	Sr	87.62
Gold	Au	197.0	Sulfur	S	32.07
Helium	He	4.003	Tin	Sn	118.7
Hydrogen	H	1.008	Titanium	Ti	47.88
Iodine	I	126.9	Tungsten	W	183.9
Iron	Fe	55.85	Vanadium	V	50.94
Krypton	Kr	83.80	Xenon	Xe	131.3
Lead	Pb	207.2	Zinc	Zn	65.39

*The values in this table are given to four significant figures.
See Appendix C for a complete table of atomic and molar masses.

Atoms, Molecules, and Reactions

An Introduction to Chemistry

Ronald J. Gillespie
McMaster University

Donald R. Eaton
McMaster University

David A. Humphreys
McMaster University

Edward A. Robinson
University of Toronto

Prentice-Hall International, Inc.
Englewood Cliffs, New Jersey 07632

Library of Congress Cataloging-in-Publication Data

Atoms, molecules, and reactions : an introduction to chemistry /
Ronald J. Gillespie . . . [et al.].
p. cm.
ISBN 0-13-088790-0
1. Chemistry. I. Gillespie, Ronald J. (Ronald James)
QD33.A78 1994
540--dc20 93-42571
CIP

Acquisition Editor: Paul Banks
Editor in Chief: Tim Bozik
Development Editors: Karen Karlin/Robin Fox
Production Project Manager: Barbara DeVries
Marketing Manager: Kelly McDonald
Assistant Editor: Mary Hornby
Product Manager: Trudy Pisciotti
Design Director: Florence Dara Silverman
Text Designer: Judith A. Matz-Coniglio
Page Layout: Natasha Sylvester
Cover Designer: Jeannette Jacobs
Cover Photographs: Richard Megna, Fundamental Photographs
Photo Editor: Lorinda Morris-Nantz
Photo Researcher: Diane Austin
Editorial Assistant: Veronica A. Wade
Copy Editor: Barbara Ligouri
Art Studio: Academy ArtWorks, Inc.
Text Composition: Better Graphics, Inc.

A Paramount Communications Company
Englewood Cliffs, New Jersey 07632

Printed in the United States of America
10 9 8 7 6 5 4 3 2 1

ISBN 0-13-088790-0

Prentice-Hall International (UK) Limited, *London*
Prentice-Hall of Australia Pty. Limited, *Sydney*
Prentice-Hall Canada Inc., *Toronto*
Prentice-Hall Hispanoamericana, S.A., *Mexico*
Prentice-Hall of India Private Limited, *New Delhi*
Prentice-Hall of Japan, Inc., *Tokyo*
Simon & Schuster Asia Pte. Ltd., *Singapore*
Editors Prentice-Hall do Brasil, Ltds., *Rio de Janeiro*

Brief Contents

1 Atoms and Molecules: The Building Blocks of Substances *1*
2 The Atmosphere, Gases, and the Gas Laws *42*
3 The Periodic Table and Chemical Bonds *80*
4 Chemical Reactions and the Halogens *118*
5 Stoichiometry *156*
6 The Electronic Structure of Atoms and Molecules *180*
7 Sulfur, Phosphorus, and Chlorine: Period 3 Nonmetals *222*
8 Carbon and the Hydrocarbons *263*
9 Thermochemistry and Thermodynamics *301*
10 Metals: Properties, Structures, and Reactions *342*
11 Solids, Liquids, and Intermolecular Forces *385*
12 Chemical Equilibrium: Quantitative Aspects *423*
13 Electrochemistry *464*
14 Organic Chemistry *498*
15 The Rates and Mechanisms of Reactions *549*
16 Chemistry of the Environment *582*
17 Biochemistry: The Chemistry of Life *614*
18 Nuclear Chemistry *652*
19 Cosmochemistry and Geochemistry *684*
20 Polymers and Materials Science *715*
Appendix C Chemistry: Important Units and Tables *A-1*
Appendix P Physics Review *A-9*
Appendix M Mathematics Review *A-14*
Answers to Exercises *E-1*
Answers to Selected Problems *P*
Glossary *G-1*
Index *I-1*

CHAPTER

Contents

Preface *xvii*

About the Authors *xxv*

1 Atoms and Molecules: The Building Blocks of Substances *1*

1.1 Atoms and Molecules 1
1.2 Elements, Compounds, and Formulas 2
Elements: Symbols and Molecular Formulas 2
Compounds and Their Molecular Formulas 4
Representing Bulk Substances 5 / Empirical Formulas 6
Structural Formulas 7
1.3 Substances: Properties and Purity 9
Substances and Chemicals 9 / Physical and Chemical Properties 9
■ ***Demonstration 1.1*** Properties of Water and Hydrogen Peroxide 10
Purity 11
1.4 Mixtures 12
Homogeneous Mixtures (Solutions) 12 / Heterogeneous Mixtures 12
■ ***Demonstration 1.2*** Mixtures and Compounds 13
The Separation of Mixtures 14
■ ***Demonstration 1.3*** Paper Chromatography of Ink 17
1.5 Chemical Reactions and Chemical Equations 18
Balancing Chemical Equations 18
1.6 The Structure of Atoms 21
Electrons, Protons, and Neutrons 21
Atomic Number and Mass Number 24 / Isotopes 24
Atomic Mass 25
1.7 The Mole: Counting by Weighing 29
1.8 Unstable Atoms: Radioactivity 33
Radioactivity 33 / The Stability of Nuclei 34
1.9 Conservation of Mass and Energy 35
Summary 36 / Important Terms 37
Review Questions 37 / Problems 38

2 The Atmosphere, Gases, and the Gas Laws *42*

2.1 The Atmosphere 43
2.2 Oxygen 46
Oxides 46
■ ***Demonstration 2.1*** Decomposition of Water by an Electric Current 46
■ ***Demonstration 2.2*** Reactions of Metals and Nonmetals with Oxygen 47
Oxidation and Reduction 48
■ ***Demonstration 2.3*** Reduction of Copper Oxide with Carbon 49
2.3 Nitrogen 50
2.4 Hydrogen 51
2.5 Solids, Liquids, and Gases 53
Gas Pressure and Atmospheric Pressure 55
2.6 The Gas Laws 56
Pressure and Volume; Boyle's Law 56 / Temperature and Volume: Charles' Law and the Kelvin Temperature Scale 57
■ ***Demonstration 2.4*** Liquefaction of Air and Charles' Law 58
Molecules and Volume: Avogadro's Law 59
The Ideal Gas Law and Molar Volume 60
2.7 The Kinetic Theory of Gases 61
Kinetic Theory and Ideal Gas Behavior 61
Kinetic Energy, Heat, and Temperature 64
2.8 Using the Ideal Gas Law 64
Number of Moles 65 / Molar Mass 65
Molecular Formula from Molecular Mass 66
Gas Density and Molar Mass 66
Dependence of Volume on Temperature and Pressure 67
Dalton's Law of Partial Pressures 68
2.9 Diffusion and Effusion 69
■ ***Demonstration 2.5*** Diffusion of Bromine Vapor 70
Molecular Speeds 70 / Rates of Diffusion and Effusion 72
■ ***Demonstration 2.6*** Rates of Diffusion 73
2.10 Real Gases and Intermolecular Forces 73
Summary 74 / Important Terms 76
Review Questions 76 / Problems 77

3 The Periodic Table and Chemical Bonds *80*

3.1 The Periodic Table 81
Organization of the Periodic Table 83 / Metals and Nonmetals 84
The Groups 86 / Valence and the Periodic Table 88
3.2 The Shell Model of the Atom 90
3.3 Chemical Bonds and Lewis Structures 92
The Atomic Core and Valence Electrons 92 / The Octet Rule 93
Lewis (Electron-Dot) Symbols 94
Ions, Ionic Bonds, and Ionic Compounds 94
Understanding the Octet Rule 97
Covalent Bonds and Lewis Structures 98
3.4 Electronegativity 101
Bond Polarity and Bond Type 103

3.5 Molecular Shape and the VSEPR Model 104
■ ***Demonstration 3.1*** Electron-Pair Arrangements: Styrofoam Sphere Models 107
Shapes of Molecules with Only Single Bonds 108
Shapes of Molecules with Multiple Bonds 110
Shapes of Molecules with More than One "Central Atom" 111
Predicting the Shapes of Molecules by Using the VSEPR Model 112
Summary 112 / Important Terms 114
Review Questions 114 / Problems 115

4 Chemical Reactions and the Halogens *118*

4.1 Chemical Reactions 119
Reaction Rates 120 / Equilibrium 121 / Types of Reactions 122
■ ***Demonstration 4.1*** Decomposition Reactions 123
4.2 The Halogens 123
■ ***Demonstration 4.2*** Physical Properties of the Halogens 124
Sources and Uses 124 / Reactions with Nonmetals 126
■ ***Demonstration 4.3*** Solubility of Hydrogen Chloride in Water 126
Reactions with Metals: Metal Halides 128
■ ***Demonstration 4.4*** Reactions of Metals with Chlorine 128
Electrical Conductivity 129
■ ***Demonstration 4.5*** Electrical Conductivity of Molten Lithium Chloride 130
4.3 Oxidation-Reduction (Redox) Reactions 131
The Halogens as Oxidizing Agents 133
■ ***Demonstration 4.6*** Oxidation–Reduction Reactions of the Halogens 134
Industrial Preparation of the Halogens 135
4.4 Acid-Base Reactions 138
Aqueous Solutions of Hydrogen Halides: The Hydronium Ion 138
Acids 138
■ ***Demonstration 4.7*** Effect of Acids and Bases on Some Natural Indicators 139
■ ***Demonstration 4.8*** Electrical Conductivity of Acid Solutions 141
Bases 142 / Acid-Base Reactions in Aqueous Solution 144
Polyatomic Ions: Lewis Structures 145
4.5 Precipitation Reactions 146
■ ***Demonstration 4.9*** Precipitation Reactions 146
Solubility Rules 147
4.6 Reaction Types 149
Summary 151 / Important Terms 152
Review Questions 152 / Problems 153

5 Stoichiometry *156*

5.1 Amounts of Reactants and Products 157
5.2 Limiting Reactants 160
5.3 Theoretical Yield and Percent Yield 162

5.4 Determination of Empirical and Molecular Formulas 163
Empirical Formula Determination by Synthesis 165 / Empirical Formula Determination by Combustion of an Organic Compound 166
Molecular Formulas 168
5.5 Stoichiometry of Gas Reactions 168
5.6 Stoichiometry of Reactions in Solution 169
Molarity 169 / Concentration of Trace Constituents 172
Acid-Base Reactions: Titrations 173
Summary 175 / Important Terms 176
Review Questions 177 / Problems 177

6 The Electronic Structure of Atoms and Molecules *180*

6.1 Light and Other Electromagnetic Radiation 181
Wavelike Properties of Electromagnetic Radiation 182
Particlelike Properties of Electromagnetic Radiation 185
6.2 Atomic Spectra 189
Line Spectra and the Quantization of Energy in the Atom 189
■ ***Demonstration 6.1*** Colored Flames and Atomic Spectra 190
The Spectrum of the Hydrogen Atom 192
Spectra of Other Atoms 194
6.3 Quantum Mechanics 195
6.4 Electron Configurations 197
Ionization Energies from Photoelectron Spectroscopy 198
Energy Levels and Electron Configurations 198 / Electron Spin 204
Atomic Orbitals and the Pauli Exclusion Principle 205
6.5 The Hydrogen Molecule and the Covalent Bond 208
The Hydrogen Molecule 208
Valence and Electron Configurations 209
Hybrid Orbitals and Molecular Geometry 210
Summary 216 / Important Terms 217
Review Questions 218 / Problems 218

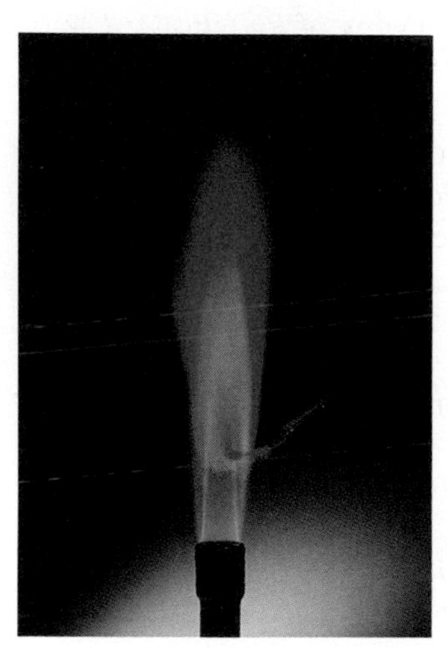

7 Sulfur, Phosphorus, and Chlorine: Period 3 Nonmetals *222*

7.1 Sulfur: Oxidation Numbers and Oxidation States 224
Occurrence and Production 224 / Properties of the Element: The Allotropes of Sulfur 224
■ ***Demonstration 7.1*** Plastic Sulfur 226
Oxides of Sulfur 227 / Sulfuric Acid 227
Concentrated Sulfuric Acid 229
■ ***Demonstration 7.2*** Sulfuric Acid as a Dehydrating Agent 231
■ ***Demonstration 7.3*** Sulfuric Acid as an Oxidizing Agent 231
Redox Reactions of Sulfur Compounds: Oxidation Numbers and Oxidation States 232 / Oxidation States of Sulfur 235
Sulfurous Acid and Sulfites 236
■ ***Demonstration 7.4*** Sulfur Dioxide as a Reducing Agent 237
Hydrogen Sulfide and Metal Sulfides 238

■ ***Demonstration 7.5*** Properties of Hydrogen Sulfide 238
Sulfur Halides 239
7.2 Phosphorus 240
Occurrence and Properties 240
■ ***Demonstration 7.6*** Oxidation of White Phosphorus: Chemiluminescence 241
Oxides and Oxoacids 241
■ ***Demonstration 7.7*** Reaction of White Phosphorus with Oxygen 243
■ ***Demonstration 7.8*** The Reaction of P_4O_{10} with Water 244
Phosphides and Phosphine 247 / Phosphorus Halides 247
7.3 Oxoacids 247
Naming Oxoacids and Their Anions 247 / Strengths of Oxoacids 248
7.4 Chlorine 250
Chlorine Oxoacids 250
■ ***Demonstration 7.9*** Properties of Sodium Hypochlorite 251
Chlorine Oxides and Fluorides 252
7.5 Lewis Structures 252
Summary 258 / Important Terms 259
Review Questions 259 / Problems 259

8 Carbon and the Hydrocarbons *263*

8.1 Allotropes of Carbon 265
■ ***Demonstration 8.1*** Adsorption by Activated Charcoal 268
8.2 Inorganic Compounds of Carbon 268
Carbon Dioxide and Carbonic Acid 268
■ ***Demonstration 8.2*** Carbon Dioxide Behaves as an Acid in Water 269
Carbon Monoxide 270 / Carbon Disulfide 271
Hydrogen Cyanide 271 / Carbides 271
Multiple Bonding in the Compounds of Carbon, Nitrogen, and Oxygen 272
8.3 Alkanes 274
Structure of Alkanes 274 / Structural Isomers 275
Naming Alkanes 277 / Conformations 280
Reactions of Alkanes 281 / Cycloalkanes 281
Petroleum and Natural Gas 283
8.4 Alkenes and Alkynes 284
Ethene 284 / Other Alkenes 286 / Naming Alkenes 286
Reactions of Alkenes 288
■ ***Demonstration 8.3*** Testing for Unsaturated Hydrocarbons with Bromine 289
Alkynes 290
8.5 Benzene and the Arenes: Aromatic Hydrocarbons 290
The Structure of Benzene 291 / Reactions of Benzene 293
Naming Aromatic Compounds 293 / Polycyclic Arenes 294
8.6 Resonance Structures and Bond Order 294
Summary 296 / Important Terms 297
Review Questions 297 / Problems 298

9 Thermochemistry and Thermodynamics *301*

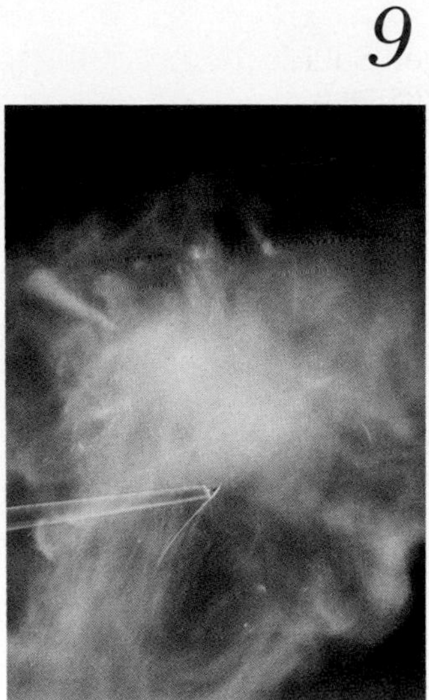

9.1 Thermochemistry and the First Law of Thermodynamics 303
Heat 303 / Internal Energy 304
The First Law of Thermodynamics and Enthalpy 304
Standard Reaction Enthalpy 306 / Calorimetry 307
Hess's Law 310 / Standard Enthalpy of Formation 312
9.2 Bond Energies 314
Bond Energy and Bond Strength 316
■ ***Demonstration 9.1*** The Decomposition of Nitrogen Triiodide 318
Estimating Reaction Enthalpies from Bond Energies 318
9.3 Alternative Energy Sources 319
Coal Conversion 319 / Hydrogen 320 / Biomass 320
Solar Energy 321 / Nuclear Energy 321
9.4 Entropy and the Second Law of Thermodynamics 321
Spontaneous Reactions 321
■ ***Demonstration 9.2*** A Spontaneous Reaction 323
Disorder and Entropy 323
■ ***Demonstration 9.3*** Disorder and Entropy 325
Entropy Changes in a Reacting System 326
Finding the Total Entropy Change 329
9.5 Gibbs Free Energy 330
Gibbs Free Energy and Equilibrium 333
Gibbs Free Energy and Work 333 / Coupled Reactions 334
Postscript 335
Summary 336 / Important Terms 337
Review Questions 337 / Problems 338

10 Metals: Properties, Structures, and Reactions *342*

10.1 Physical Properties and Uses 344
The Alkali and Alkaline Earth Metals 344
■ ***Demonstration 10.1*** Physical Properties of the Alkali Metals 344
Aluminum 345 / Iron 345 / Copper 346
10.2 Structure and Bonding 346
Hexagonal and Cubic Close-Packed Structures 346
■ ***Demonstration 10.2*** Close Packing of Spheres 347
Metallic Bonding 348 / Properties of Metals 348
10.3 Reactions and Compounds of Group I and II Metals 350
Reactions with the Halogens and Oxygen 351
Reactions with Water 352
■ ***Demonstration 10.3*** Reactions of Sodium, Potassium, and Calcium with Water 353
Reactions with Acids 354 / Salts 355 / Flame Tests 359
10.4 Reactions and Compounds of Aluminum 360
Aluminum Oxide 361
Aluminum Hydroxide: An Amphoteric Hydroxide 361
Aluminum Chloride 362
■ ***Demonstration 10.4*** The Preparation of Aluminum Bromide 363
Lewis Acids and Bases 363

10.5 The Transition Elements 365
Oxidation States 365 / Magnetic Properties 366
Complex Ions 367 / Reactions and Compounds of Iron 367
■ ***Demonstration 10.5*** Iron(II) and Iron(III) Hydroxides 369
Reactions and Compounds of Copper 371
Other Transition Metals 374
10.6 Metallurgy: Extraction of Metals 375
Manufacture of Iron and Steel 375
■ ***Demonstration 10.6*** The Thermite Process 376
Production of Copper 377
10.7 Summary of the Reactions of Metals and Metal Ions 378
Reactions of Metals 378 / Reactions of Metal Ions 378
Summary 380 / Important Terms 381
Review Questions 381 / Problems 382

11 Solids, Liquids, and Intermolecular Forces *385*

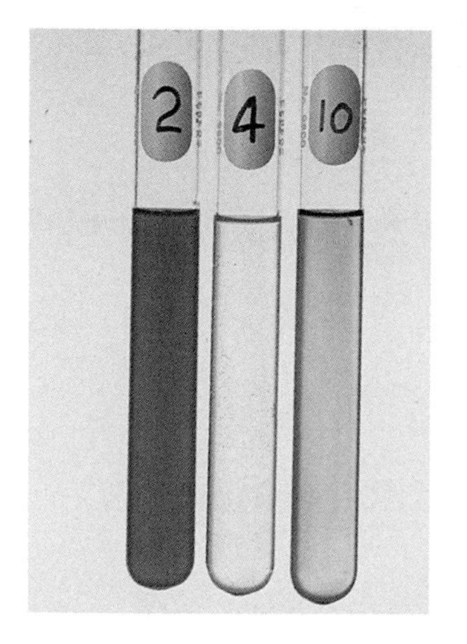

11.1 Solids and Liquids 387
Types of Solids 390 / Lattices and Unit Cells 391
Metal Crystals 391 / Covalent Network Crystals 394
Covalent Molecular Crystals 395
The Sizes of Atoms: Metallic, Covalent, and Ionic Radii 395
Ionic Crystals 397 / Amorphous Solids 399 / Liquids 400
11.2 Phase Changes 401
Melting and Freezing 401 / Evaporation and Condensation 402
Vapor Pressure 403 / Entropy and Phase Changes 406
11.3 Intermolecular Forces 407
Dipole-Dipole Forces 407 / London (Dispersion) Forces 409
11.4 Water and the Hydrogen Bond 411
Properties of Water 411 / The Hydrogen Bond 413
11.5 Solutions 415
■ ***Demonstration 11.1*** Immiscible Liquids and Solubility 416
Summary 417 / Important Terms 418
Review Questions 419 / Problems 419

12 Chemical Equilibrium: Quantitative Aspects *423*

12.1 The Equilibrium Constant 425
12.2 Acid-Base Equilibria in Aqueous Solution 425
The Acid Ionization Constant 426
The Base Ionization Constant 429
The Autoionization of Water 431 / The pH Scale 431
Conjugate Acid-Base Pairs 434
Acid-Base Properties of Anions, Cations, and Salts 435
Anions 435 / Cations 436
Buffer Solutions 438 / Le Châtelier's Principle 441
■ ***Demonstration 12.1*** Le Châtelier's Principle 442
The Reaction Quotient, Q 443 / The Measurement of pH 443
■ ***Demonstration 12.2*** A Natural Indicator 444

■ ***Demonstration 12.3*** The pH of Some Common Solutions 447
Titrations 447 / Determination of K_a and K_b 448
12.3 Gas-Phase Equilibria 449
Pressure Changes 450 / Temperature Changes 452
12.4 Heterogeneous Equilibria 453
12.5 Gibbs Free Energy and the Equilibrium Constant 455
Summary 458 / Important Terms 459
Review Questions 459 / Problems 460

13 Electrochemistry *464*

13.1 Electrochemical Cells 465
Cells and Cell Reactions 465
■ ***Demonstration 13.1*** The Reaction of Zinc with an Aqueous Solution of Copper Sulfate 466
Cell Potentials 467 / Standard Reduction Potentials 469
Direction of Oxidation-Reduction Reactions 472
■ ***Demonstration 13.2*** Metal Displacement Reactions 473
Calculation of Cell Potentials 475
Gibbs Free Energy and Cell Potential 475
Effect of Concentration on Cell Voltage: The Nernst Equation 477
13.2 Applications of Electrochemical Cells 480
Batteries 480 / Corrosion 483
13.3 Electrolysis 484
Electrolysis of Molten Sodium Chloride 485
Preparation of Aluminum by Electrolysis 486
Quantitative Aspects of Electrolysis 487
Electrolysis of Aqueous Solutions 489
Electrolytic Preparation of Chlorine and Sodium Hydroxide 490
Electrolytic Refining of Copper 491
■ ***Demonstration 13.3*** Electroplating 492
Electroplating 492 / Electrolysis and Gibbs Free Energy 492
Summary 492 / Important Terms 494
Review Questions 494 / Problems 494

14 Organic Chemistry *498*

14.1 Functional Groups 500
14.2 Haloalkanes (Alkyl Halides) 501
Preparation of Haloalkanes 501 / Reactions of Haloalkanes 503
14.3 Alcohols and Ethers 504
Naming Alcohols 506 / Diols and Triols 507
Properties of Alcohols 508 / Reactions of Alcohols 508
Phenols 510 / Ethers 511
14.4 Thiols and Disulfides 512
14.5 Aldehydes and Ketones 514
Preparation of Aldehydes and Ketones 515
Properties of Aldehydes and Ketones 516
Reactions of Aldehydes and Ketones 517

■ ***Demonstration 14.1*** Silver Mirror Test for Aldehydes 518
14.6 Carboxylic Acids and Esters 518
Preparation of Carboxylic Acids 520 / Some Common Carboxylic Acids 520 / Properties of Carboxylic Acids 521 / Esters 522
14.7 Amines, Amides, and Amino Acids 528
Amines 528 / Amides 530 / Amino Acids 531
Chirality and Amino Acids 531
14.8 Determining the Structure of Organic Compounds 534
Determining Molecular Mass by Mass Spectrometry 534
Determining Molecular Structure 536 / Infrared Spectroscopy 536
Nuclear Magnetic Resonance Spectroscopy 538
X-ray Crystallography 541 / Gas Chromatography 541
Summary 542 / Important Terms 543
Review Questions 543 / Problems 544

15 The Rates and Mechanisms of Reactions *549*

15.1 Reaction Rate 550
15.2 Effect of Reactant Concentration on Reaction Rate 554
Rate Laws and Reaction Mechanisms 555
Reaction Intermediates 557
Integrated Form of the First-Order Rate Law 558
Half-Life of a First-Order Reaction 559
Equilibrium Constant and Reaction Mechanism 560
15.3 Activation Energy and the Effect of Temperature on Reaction Rate 561
Activation Energy 562 / The Collision Model 563
The Arrhenius Equation 565 / Unimolecular Reactions 568
15.4 Catalysis 568
■ ***Demonstration 15.1*** Catalysis 569
Homogeneous Catalysis 570 / Heterogeneous Catalysis 570
Enzymes 572
15.5 Chain Reactions 573
Summary 575 / Important Terms 576
Review Questions 576 / Problems 577

16 Chemistry of the Environment *582*

16.1 Oxides and Oxoacids of Nitrogen 584
Nitrogen Monoxide 584
Nitrogen Dioxide and Dinitrogen Tetraoxide 585
■ ***Demonstration 16.1*** Nitrogen Monoxide, NO 586
The Structures of NO and NO_2: Free Radicals 586
■ ***Demonstration 16.2*** Nitrogen Dioxide and Dinitrogen Tetraoxide 587
Other Oxides of Nitrogen 588
Nitric Acid and Nitrous Acid 588
■ ***Demonstration 16.3*** Preparation of NO by the Catalytic Oxidation of NH_3 589

16.2 Ozone and Peroxides 590
Ozone 590
Peroxides and the Hydroxyl and Hydroperoxyl Radicals 591
16.3 Pollution of the Stratosphere 592
Solar Radiation and Atmospheric Photochemistry 592
The Ozone Layer 594
Chlorofluorocarbons and the Depletion of the Ozone Layer 596
Ozone Depletion by Nitrogen Oxides 599
Solutions to the Problem of Ozone Depletion 599
16.4 Photochemical Smog 599
Ozone and Other Constituents of Photochemical Smog 600
Toxic Effects of Photochemical Smog 601
Solutions to the Problem of Photochemical Smog 602
16.5 Acid Rain 602
Sources of Acid Rain 602 / Effects of Acid Rain 603
Solutions to the Problem of Acid Rain 605
16.6 The Greenhouse Effect 606
Postscript 608
Summary 608 / Important Terms 609
Review Questions 609 / Problems 610

17 Biochemistry: The Chemistry of Life *614*

17.1 Proteins 616
Primary Structure 617 / Secondary Structure 617
Tertiary Structure 622 / Quaternary Structure 625
Globular and Fibrous Proteins 625
17.2 Enzymes 626
17.3 Carbohydrates 627
Monosaccharides, Disaccharides, and Polysaccharides 627
17.4 Energy in Biochemical Reactions: ATP and ADP 629
Carbohydrate Metabolism 629 / Adenosine Triphosphate, ATP 629
17.5 DNA and RNA 631
Nucleic Acids 632 / Replication and Protein Synthesis 635
The Genetic Code 637
17.6 Lipids 641
Triacylglycerols 641 / Glycerophospholipids and Membranes 642
17.7 Nutrition 644
Carbohydrates 644 / Fats 644 / Proteins 645 / Vitamins 645
Minerals 645
Summary 648 / Important Terms 649
Review Questions 649 / Problems 650

18 Nuclear Chemistry *652*

18.1 Radioactivity 653
Nuclear Structure and Stability 653
Radioactive Decay Processes 654
The Measurement of Radioactivity 658
The Biological Effects of Radiation 658

18.2 Radioactive Decay Rates 662
Nuclear Half-Life 662 / Using Radioisotopes for Dating 664
18.3 Artificial Radioisotopes 667
Synthesis of Radioisotopes 667 / Uses of Radioisotopes 668
Transuranium Elements 670
18.4 Nuclear Energy 671
Mass Defect and Binding Energy 671 / Nuclear Fission 673
Nuclear Reactors 674 / Chemical Aspects of Nuclear Power 676
Nuclear Fusion 677 / Nuclear Power: Risks and Benefits 678
Summary 679 / Important Terms 680
Review Questions 680 / Problems 680

19 Cosmochemistry and Geochemistry *684*

19.1 The Origin of Atoms and Molecules 685
Nucleosynthesis in the Stars 686 / Interstellar Molecules 688
19.2 The Formation and Composition of the Solar System 690
The Planets 691 / Comets, Asteroids, and Meteorites 692
19.3 The Structure and Composition of the Earth 693
The Structure of the Earth 693 / Silicon and Its Compounds 694
Minerals and Other Geological Resources 701
The Atmosphere 702 / The Hydrosphere 704
19.4 The Biosphere 705
The Formation of the Biosphere 705
Maintaining the Biosphere 708
Postscript: Chemistry, The Central Science 710
Summary 711 / Important Terms 712
Review Questions 712 / Problems 713

20 Polymers and Materials Science *715*

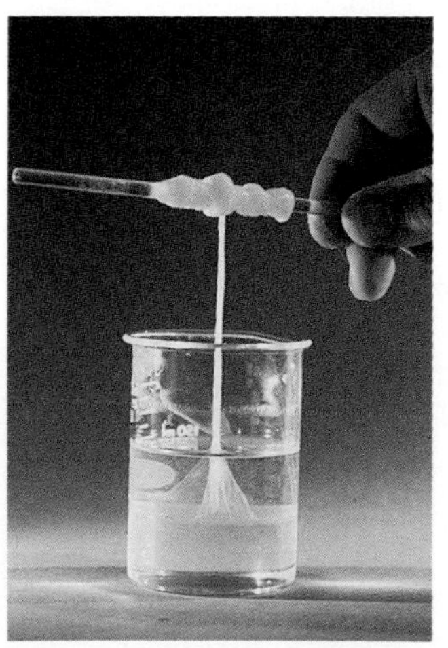

20.1 Polymer Chemistry 716
Addition Polymers 717 / Condensation Polymers 725
- ***Demonstration 20.1*** Synthesis of Nylon 610 727
- ***Demonstration 20.2*** The Formation of Phenol-Formaldehyde Polymer 729

20.2 Materials Science 730
Zeolites 731 / Semiconductors 734 / Electrical Properties of Metal Oxides 741 / Superconductivity 742
Summary 743 / Important Terms 745
Review Questions 745 / Problems 745

Appendix C Chemistry: Important Units and Tables *A-1*

1. **SI Units A-1**
2. **Derived Units A-2**
3. **Conversion of Units: Unit Factors A-3**
4. **Temperature Conversions A-4**
5. **Atomic and Molar Masses of the Elements A-4**

6. Electron Configurations of the Elements A-6
7. Thermodynamic Data A-7

Appendix P Physics Review *A-9*

1. The Laws of Motion A-9
2. Momentum A-10
3. Work and Energy A-11
4. Electricity and Magnetism A-12

Appendix M Mathematics Review *A-14*

1. Scientific (Exponential) Notation A-14
2. Significant Figures A-15
3. Logarithms A-17
4. Linear Equations and Straight-Line Graphs A-19
5. Quadratic Equations A-19

Answers to Exercises *E-1*

Answers to Selected Problems *P*

Glossary *G-1*

Index *I*

Photo Credits *C-1*

Boxes

Box 1.1 Atoms, Dalton, and the Scanning Tunneling Microscope *2*
1.2 Rutherford and the Nuclear Atom *22*
1.3 The Mass Spectrometer: Measuring Masses of Atoms and Abundances of Isotopes *26*
1.4 Mass and Weight *29*
2.1 Hydrogen as a Fuel *52*
2.2 Balloons *59*
2.3 Laws, Models, and Theories *62*
3.1 Dimitri Mendeleev and the Periodic Table *82*
3.2 Linus Pauling *101*
3.3 Molecular Models *106*
5.1 Analytical Chemistry *164*
6.1 Albert Einstein *186*
6.2 Interference and Diffraction *196*
6.3 Experimental Evidence for Electron Spin *205*
7.1 The Chemistry of Matches *242*
7.2 Fertilizers *245*
7.3 More on Lewis Structures *257*
8.1 Buckyball: The Third Allotrope of Carbon *267*
8.2 The Internal Combustion Engine *285*
9.1 Explosives and the Nobel Prize *317*
10.1 Hard Water and Limestone Caves *357*
10.2 Fireworks *359*
10.3 Hemoglobin *371*
11.1 X-ray Diffraction and the Structures of Crystals *388*
11.2 Liquid Crystals *402*
12.1 Body Fluids: pH and Buffers *439*
13.1 Michael Faraday *476*
13.2 Charles Martin Hall *487*
14.1 Inhalation Anesthetics *513*
14.2 Detergents and Soaps *526*
16.1 James Lovelock and the Gaia Hypothesis *597*
17.1 Molecular Biology *623*
17.2 Chemistry in Medicine *638*
17.3 Computer Modeling of Biomolecules *640*
17.4 Vitamins *646*
18.1 The Curie Family and Radioactivity *660*
18.2 Calibrating the ^{14}C Clock *667*
19.1 Chemical Analysis and the Extinction of the Dinosaurs *710*
20.1 Polymers in Medicine *722*
20.2 Polytetrafluoroethylene *723*

Preface

TO THE STUDENT

As educators in the field of chemistry, we recognize that we are teaching not only chemistry, but more importantly that we are teaching *students*. To help you, the student, is the motivation behind this book.

WHY STUDY CHEMISTRY?

Some students take a college chemistry course because they have already discovered the fascination of chemistry. Others do so because chemistry is a requirement in their chosen field. Some of you may have studied chemistry before, whereas others have not. Whatever your background and whatever your goals, we have written this book for you. Our aim is to help you succeed in the course and to provide you with the background knowledge and understanding of chemistry that is needed in many other science courses and for any other courses in chemistry. Chemistry may truly be called the central science. Chemistry deals with matter—that is, with substances and their properties. Physics deals with the basic principles of the universe, but because the universe consists of matter and energy, an understanding of chemistry is needed in many areas of physics. Geology is concerned with the rocks and other substances of which the earth is composed, and biology is concerned with the properties of living matter; both involve chemistry. Indeed, some of the most exciting areas of chemistry today are those that are on the borderlines between chemistry and other sciences such as biochemistry, medicine, geochemistry, materials science, and cosmochemistry.

UNDERSTANDING CHEMISTRY

Sometimes students have the impression that chemistry is a particularly difficult subject, perhaps because it deals with so many different substances and the reactions between them. But the basic ideas of chemistry are very simple: Matter consists of atoms and molecules, which are composed of groups of atoms; these atoms and molecules are in constant random motion and have energies that increase with increasing temperature. It is in terms of these two basic ideas that chemists think about the properties and reactions of substances. You need to learn to think about chemistry in this way, too—the relation of the observed properties and reactions of substances to the atoms and molecules of which they are composed. Although there are very many substances and many different reactions, there are far fewer *types* of substances and *types* of reactions. You will learn how to recognize the different types of substances, such as metals and nonmetals, and the different types of reactions, such as acid–base and oxidation–reduction reactions. This basic knowledge provides the foundation for understanding any substances and reactions that you might meet later in your studies in other sciences or engineering.

HOW TO STUDY

Authors of textbooks such as ourselves have been learning and teaching for many years. We have to study the work of others in our own areas of research, and we have to study a broader range of topics to write a textbook. Hence we can offer you some advice based on our experience. We suggest that you

- Read the appropriate material in the text *before* you attend the lecture. This will help you follow the lecture and take meaningful notes as necessary. If you attend the lecture unprepared, you may miss important points in the lecture.
- Understanding is the key to success. You may be able to memorize passages from the text, equations, or typical problem solutions, but that does not necessarily mean that you understand them. If you *understand*, you may not need to memorize, or at least any memorization you must do will be easier.
- Understanding is enhanced by efficient study habits, by discussion with your instructors and with other students, and by answering review questions and working out exercises and problems.

HOW TO USE THIS BOOK

We recommend that you first read each chapter or section quickly to get an overall impression. Then reread the chapter or section slowly and carefully, making sure you understand everything. Keep in mind that important new words are defined and appear in **boldface** type the first time they are used. **Margin Notes** are offered to provide you with additional explanations and reminders, and **Problem-Solving Margin Notes**, in blue, are aimed at helping you understand and solve problems within the text. Use the numerous illustrations and their accompanying captions to enhance not only your visual appreciation of the text, but also your understanding of the concepts and principles presented in the text. Highlight the text sparingly, or make notes in the margins and/or in a separate notebook. Study the worked **Examples**, and then try the **Exercises** that follow an example or are at the end of a section. The answers to the exercises are given in the back of the book. Then read the **Summary** at the end of the chapter to be sure you understand all the key concepts and relationships covered in the chapter.

When you have gone through a chapter in this way, try to answer the **Review Questions** in a few words or a sentence or two. Look through the list of **Important Terms** to make sure that you can explain or define each one. If you cannot, refer back to the text or ahead to the **Glossary**, which is a complete list of important terms, with a definition or explanation of each.

It is then time to solve some of the **Problems** given at the end of each chapter. Answers to about half these problems are supplied at the back of the book. Your instructor will discuss the solutions with you. Note that problems marked with an asterisk are more difficult than the others, for students who want a challenge.

Color photos of nearly 60 **Demonstrations** of the properties of substances and their reactions are an important feature of this book. Your instructor may perform some of these demonstrations in class or show them to you on videotape. Watching the demonstration carefully and studying the photos in the text will help you remember and understand many important facts about the properties and reactions of substances. The book also offers **Boxes** about some of the major contributors to the field of chemistry and some important applications. These features are intended to bring the ''faces of chemistry'' to life and to show how chemistry affects your daily life.

Understanding the material in this book will take you into a new and fascinating world—the world of atoms, molecules, and reactions. All the new polymers, drugs, and solid-state devices were developed by people who completed the journey upon which you are about to embark. As you travel into the challenging new world of chemistry, you are preparing yourself to live in a society of significant challenges, such as achieving a clean and healthy environment and improving living conditions for all living things. The solutions will come from a new generation of people who approach these problems with knowledge and understanding. Working hard to understand the principles and relationships given in this book is one of the best ways to prepare yourself to play a key role in meeting the challenges that will face all of us tomorrow.

TO THE INSTRUCTOR

PHILOSOPHY

There has been a growing concern in recent years that something is wrong with the general chemistry course. This concern has been expressed in many articles in the *Journal of Chemical Education* and in the symposia organized by the Task Force of the Chemical Education Division of the American Chemical Society (of which Ron Gillespie has been a long-standing member). Among students the general chemistry course has the reputation of being one of the most difficult of the first-year college courses. To many students it seems uninspiring, uninteresting, and irrelevant. And yet chemistry is all around us. Chemistry is involved in almost every aspect of the material world, and it is certainly not uninteresting or irrelevant to our daily lives or to many other sciences. It seems clear that there must be something wrong with the course as it has traditionally been taught and with the books on which they have been based

In our opinion there are five major problems with the traditional course:

1. Despite its name, the course is not a truly general introduction to chemistry. There is too much emphasis on physical chemistry and not enough on inorganic and organic chemistry, and these different aspects of chemistry are not well integrated.
2. The course contains too much material. New material has been constantly added to the course, but very little has been deleted.
3. The course contains too much theoretical and abstract material. A fair amount of the new material is theory, and much of it is difficult and abstract theory, such as ligand field theory. We feel that some of this information is not appropriate for a first introduction to chemistry.
4. The course content is determined too much by the perceived need to prepare students for the chemistry majors course and not enough by the needs of the

majority of the students in the course. In particular, there is very little mention of new frontier areas in which many of the most exciting developments are being made.

5. There is too much emphasis on the solving of numerical problems and not enough on the understanding of concepts.

A NEW APPROACH

Our response to the concerns about the course that we and many others have expressed is *Atoms, Molecules, and Reactions: An Introduction to Chemistry.* Our new approach has the following distinguishing features.

An Integrated, More Balanced Course

The traditional course is heavily biased toward a physical-chemistry-based treatment of principles and theories. Inorganic and organic chemistry—so-called descriptive chemistry—is usually treated separately near the end of the course, where it often gets a cursory treatment due to time limitations.

We believe that a good general introduction to chemistry should include some basic factual material of inorganic and organic chemistry and a discussion of these facts in terms of simple concepts and theories. This basic factual knowledge is required for understanding the chemistry that we all meet in our daily lives and the chemistry involved in today's frontier areas, such as environmental chemistry, biochemistry, cosmochemistry, and materials science. It is good pedagogy as well as good science to put *facts before theories*—to show how theories and principles are needed to understand and explain observations. An important feature of our treatment is therefore to discuss some simple descriptive chemistry and to use these facts, and *only* these facts, to introduce the relevant theory and principles needed to explain them. In Chapter 4, for example, we use some of the chemistry of the halogens to introduce three important reaction types: acid–base, redox, and precipitation reactions. In Chapter 8 we discuss the chemistry of carbon and of the hydrocarbons before we use the combustion reactions of hydrocarbons to discuss thermochemistry in Chapter 9. In this way we integrate the descriptive chemistry with the physical chemistry and make it easier and more interesting for students to understand and learn both. Thus, students are dealing with new concepts and principles in terms of substances they have already encountered and are beginning to feel comfortable with.

Although there is too much material in most courses, the amount of organic chemistry is typically inadequate and certainly not commensurate with the importance of this subdiscipline. A knowledge and understanding of some simple organic compounds such as methane, ethane, benzene, acetone, acetic acid, and methylamine and their reactions is just as important as a knowledge and understanding of sulfur dioxide, sulfur trioxide, and sulfuric acid, or of nitrogen monoxide, nitrogen dioxide, and nitric acid and their reactions. We have therefore introduced some organic chemistry rather early—namely, in Chapter 8, where we discuss the hydrocarbons. We follow up in Chapter 14 with some important classes of organic compounds in terms of functional groups and in Chapter 17 with biochemistry.

Less—But Carefully Selected—Material

Attempting to cover too much, particularly when the material is purely theoretical or when it is simply fact, overwhelms students and destroys their interest. We believe that the amount of material in the course should be reduced, and we have attempted to do this. We have omitted such topics as the molecular orbital treatment of diatomic molecules and ligand field theory because they are too difficult and abstract at this level and very little use is made of them in the rest of the course. We have omitted the quantitative treatment of solubility in terms of the solubility product because oversimplified calculations neglecting activities and competing equilibria tend to give grossly inaccurate values. And we have omitted colligative properties such as boiling point elevation and freezing point depression because they are rarely used today for the determination of molecular mass.

We treat inorganic chemistry by discussing the chemistry of only a few selected elements rather than attempting to give a necessarily very cursory treatment of each group of the periodic table. We have simplified and shortened the discussions of atomic orbitals, atomic structure, and bonding while keeping them theoretically sound and emphasizing the experimental evidence for the concepts we use. For example, we introduce electron configurations by way of atomic ionization energies rather than in terms of quantum numbers, for which no experimental or theoretical justification can be given at the introductory level.

Addressing the Needs of the Audience

The major aim of the introductory course should be to provide a reasonably rigorous and intellectually challenging introduction to chemistry that emphasizes the central role of chemistry in other sciences and features its importance in everyday life from polymers to the environment. The introductory course should stand on its own as a general introduction to chemistry. It should provide students with the background knowledge of chemical substances and an understanding of their properties and reactions that they will need in their future careers.

The purpose of the introductory chemistry course should not be only, or even primarily, to prepare students for subsequent courses in chemistry. The majority of students in the introductory course will not become chemistry majors or even take other courses in chemistry. The content of the introductory course should therefore not be dictated by the needs of other courses in chemistry. Instead, the general chemistry course should provide students with an understanding of chemistry that will be useful to them in whatever branch of science, medicine, engineering, business, politics, or other field they choose for their future careers. Subsequent courses in chemistry and in other disciplines should build on the introductory chemistry course but should not dictate its content. We believe that the type of introduction to chemistry provided by *Atoms, Molecules, and Reactions* is needed for all students whether or not they continue in chemistry courses and no matter what career path they follow. The particular interests of special groups of students can be addressed by a suitable choice of material from the last five chapters.

A Conceptual Approach

It is important that students learn to appreciate how chemists think about the material world and to understand the more important concepts of chemistry. The heavy emphasis on solving numerical problems that is common in the traditional course may enable students to become adept at plugging numbers into formulas and using their calculators, but it does not necessarily improve their understanding of the concepts. We believe it is important for students to gain a qualitative but sound understanding of concepts and principles before they attempt to solve numerical problems based on those concepts and principles. We cover some common elements—hydrogen, nitrogen, oxygen, and the halogens—and their compounds in

Chapters 2 and 4 before we explain stoichiometry in Chapter 5, our treatment of which is based largely on the reactions discussed previously. We discuss the kinetic theory of gases very early (Chapter 2) for three reasons: (1) The concept of random chaotic molecular motion is second in importance only to the fundamental concept of atoms and molecules themselves; (2) it provides a simple example of a model or theory; and (3) some of the most familiar elements are gases. We introduce early (Chapter 4) the ideas of equilibrium, reaction rate, acids and bases, and oxidation and reduction qualitatively and use these ideas to discuss the properties and reactions of substances before we treat them quantitatively in Chapters 12, 13, and 15.

About 20 **Review Questions** at the end of each chapter are designed to help students review the chapter and to test their understanding of concepts and principles. The end-of-chapter **Problems** contain many qualitative problems in addition to numerical problems. These problems are grouped according to the sections within the chapter, with the more difficult problems toward the end of each group and with those that are clearly most challenging denoted by an asterisk. There is also a set of General Problems that test students' ability to synthesize information they learned in various sections within a chapter or in earlier chapters. The answers to these problems have been checked by two independent chemistry instructors as well as by the authors.

It is important for students to observe for themselves the properties and reactions that form the basis for theories and principles. So we illustrate the properties and reactions of substances in many photographs of **Demonstrations**. Many of these demonstrations, which are also available on videotape and videodisc, were a popular feature of our earlier book *Chemistry*, Allyn and Bacon, 1986, 1989. The new text also includes many other photographs illustrating the properties of substances and their reactions and applications as well as much new line art that enhances the text discussions.

Other Pedagogical Features

Every chapter opens with a full-page photograph and an explanatory caption. These **Chapter-Opening Photographs** are intended to attract the attention of students by illustrating some particularly relevant, interesting, and important aspect of the material in the chapter. Each chapter includes many worked **Examples**, which are typically followed by one or two **Exercises** on the same topic. Answers to the exercises are provided at the back of the book. Important new words are defined and appear in **boldface** type the first time they are used. These **Important Terms** are listed at the end of each chapter, and a definition or explanation of each one is given in the **Glossary** at the back of the book. Most chapters have **Boxes** that are designed to show students the human side of chemistry and to show the relevance of chemistry to other sciences and to life in general. **Margin Notes** provide additional explanations and reminders to students. **Problem-Solving Margin Notes**, in blue, are directed at helping students understand and solve problems.

EXPLORING THE FRONTIERS OF CHEMISTRY

Students should realize that chemistry is an active, challenging field that is making vital contributions to our understanding of the world around us and is improving our health and material well being. We feel therefore it is important to include in an introduction to chemistry some of the areas in which exciting new developments are taking place. Because our new integrated approach to the fundamentals has allowed us to discuss them in the first fifteen chapters, we have been able to devote five chapters to the chemistry of the environment, biochemistry, nuclear chemistry, cosmochemistry, geochemistry, and polymers and materials science. We use these topics to illustrate the wide-ranging importance of chemistry, to review facts and theories, and to show some applications of the fundamentals discussed in the earlier chapters. This material also provides students with the background needed to understand and judge critically the related information that often appears in the media, such as air pollution and the origin of the universe. In other words, these additional topics supply the knowledge students need to become chemically literate citizens.

SUPPLEMENTS

The following supplements are available from Prentice Hall:

Instructor's Manual—emphasizes the relationships among the principles, theories, and descriptive chemistry in each chapter; provides instructions and tips for carrying out the Demonstration experiments; and offers sample syllabuses and tips on how to make the transition from a principles-based curricula.
Chemistry Toolkit (Gillespie version)—offers students reinforcement of the skills necessary to succeed in chemistry.
Study Guide—guides students through the text by means of chapter overviews, learning objectives, key terms lists, progressive reviews, and self-tests.
Student Solutions Manual—offers solutions to all problems in the text; written by the text authors.
Prentice Hall Testmanager 2.0 (in IBM 3.5″ and MAC formats)—offers full control over printing (including a print preview); complete mouse support; on-screen VGA graphics with import capabilities for .TIFF and PCX file formats; the ability to export files to WordPerfect, Word, and ASCII; and context-sensitive help as well as toll-free technical support.
Test Item File—provides a menu of more than 1200 test items in multiple-choice, true/false, and fill-in formats.
Prentice Hall Telephone Testing—allows you to create tests by phone, toll-free, and to pick the questions you want, even in multiple versions.
Prentice Hall/*The New York Times* Program—THE NEW YORK TIMES and PRENTICE HALL are sponsoring a Themes of the Times: a program designed to enhance student access to current information of relevance in the classroom.

Through this program, the core subject matter provided in the text is supplemented by a collection of time-sensitive articles from one of the world's most distinguished newspapers, THE NEW YORK TIMES. These articles demonstrate the vital, ongoing connection between what is learned in the classroom and what is happening in the world around us.

To enjoy the wealth of information of THE NEW YORK TIMES daily, a reduced subscription rate is available in deliverable areas. For information, call toll-free: 1-800-631-1222.

PRENTICE HALL and THE NEW YORK TIMES are proud to co-sponsor Themes of the Times. We hope it will make the reading of both textbooks and newspapers a more dynamic, involving process.
Transparencies—consists of 150 full-color acetates taken from the text.
Lab Demonstration Videos by David Humphreys—contains two videocassettes with approximately 100 brief chemistry laboratory demonstrations; parallels the photo Demonstrations in the text.
Chemistry Explorer Software (in WIN, MAC, and IBM 3.5″ formats)—

interactively simulates selected worked Examples and Problems from the text; allows users to manipulate variables and physical parameters to observe how they affect results; includes data analysis tools, such as spreadsheets and graphs; can be used as a lecture demonstration tool in the laboratory or in a tutorial setting.

PH Multimedia Chemistry Laserdisc—combines lecture demonstrations, images from both the text and outside sources, flow diagrams, and other problem-solving frames.

PH Multimedia Chemistry—drives the use of the PH Multimedia Chemistry Laserdisc and can be customized to individual needs; offers modular presentation with minimal start-up time.

ACKNOWLEDGMENTS

The following chemistry instructors reviewed all or part of the manuscript at various stages during its preparation, and we are most grateful to all of them: Nigel Bunce, University of Guelph; Joseph Laposa, McMaster University; Lyman Rickard, Millersville University; Dennis Shaw, McMaster University; and Jim Spencer, Franklin and Marshall College.

We are also indebted to the following instructors, who provided invaluable assistance in working examples and solving problems and/or critiquing the manuscript:

Margaret R. Asirvatham, University of Colorado
George Baldwin, University of Manitoba
Joseph F. Bieron, Canisius College
James E. Davis, Harvard University
Norman Duffy, Kent State University
Natalie Foster, Lehigh University
R. H. Gibson, University of North Carolina–Charlotte
Fred Goellner, South Campus Community College of Allegheny County
David B. Green, Pepperdine University
Paul W. W. Hunter, Michigan State University
Albert W. Jache, Marquette University
Louis J. Kirschenbaum, University of Rhode Island
David Miller, California State University–Northridge
Mark Palmer, Rensselaer Polytechnic Institute
Jack Passmore, University of New Brunswick
Cyndi Wilson Porter, University of Akron
Bernard L. Powell, University of Texas at San Antonio
Ruben Puentedura, Bennington College
Joseph Sneddon, McNeese State University
Satya P. Sood, University of Hawaii at Monoa
James H. Weber, University of New Hampshire
Gary Wnek, Rensselaer Polytechnic Institute

It is a pleasure to acknowledge the enthusiastic support and assistance given to us by the staff at Prentice Hall. We especially wish to thank Chemistry Editors Diana Farrell, who guided and encouraged us during the early stages of the project, and Paul Banks, under whose direction the project was brought to a successful conclusion; and Editor in Chief Tim Bozik. Our sincere thanks also go to Development Editors Robin Fox and Karen Karlin, who made invaluable contributions to the readability and accuracy of our book, and to Production Editor Barbara DeVries, who so competently looked after all the intricacies of production and kept the authors on schedule. It has been a pleasure to work with all of them. We also much appreciate the enthusiastic support of Jim Smith, who guided and encouraged us in our earlier efforts at textbook writing.

We owe a debt of gratitude to Tom Bochsler, who took the many outstanding color photographs that illustrate our Demonstrations. His patience and fortitude in sometimes trying circumstances in the unfamiliar surroundings of the laboratory were greatly appreciated. The cheerful assistance of Josie Petrie, who retyped parts of the manuscript and reliably dispatched many parcels of proofs, is gratefully acknowledged.

About the Authors

Ronald J. Gillespie is renowned for his work in chemical education. He has received the MCA College Chemistry Teacher Award, the Union Carbide Award of the Chemical Institute of Canada for Chemical Education, and the McMaster Students Union Award for Excellence in Teaching. His VSEPR model has done much to simplify and improve the teaching of molecular geometry. An advocate of change in the general chemistry course, he has worked tirelessly for reform as a member of the Task Force on General Chemistry of the Chemical Education Division of the ACS and the Committee on the Teaching of Chemistry of IUPAC. Dr. Gillespie received his Ph.D. in chemistry from University College, London. He is Professor of Chemistry at McMaster University, where he has taught inorganic chemistry and general chemistry for more than 30 years. He is internationally known for his work in fluorine and superacid chemistries and has received such honors as the ACS Awards for Creative Work in Fluorine Chemistry and for Distinguished Service to Inorganic Chemistry. He is a Fellow of the Royal Societies of London and of Canada, the Royal Society of Chemistry, and the Chemical Institute of Canada. Dr. Gillespie has published more than 300 papers in research journals and two texts on his VSEPR model. *Atoms, Molecules, and Reactions* fully reflects the ideas he has long advocated for general chemistry reform.

Donald R. Eaton received his undergraduate degree and D.Phil. from Oxford University and spent two years as a post-doctoral fellow at the Division of Pure Physics, National Research Council, Ottawa. After spending several years at the Central Research Department of E. I. DuPont Co. in Wilmington, Delaware, he joined the Department of Chemistry at McMaster University in 1968 and has been there ever since. His research interests are in transition metal chemistry, catalysis, and magnetic resonance, and he has published extensively in these areas. Dr. Eaton is particularly interested in new approaches to the teaching of chemistry, including a chemistry course for nonscientists.

David A. Humphreys received his Ph.D. from McMaster University, where he is Professor of Chemistry. He was the founding director of the Instructional Developmental Center and chairman of the M.Sc. (Teaching) Program at McMaster. He has introduced chemistry to more than twenty thousand students and has been recognized for his efforts in using laboratory demonstrations to integrate reactions and reality with theory and principles. Dr. Humphreys has won such teaching awards as the Catalyst Medal from the MCA, the 3–M National Teaching Award, the Union Carbide Award of the Chemical Institute of Canada for Chemical Education, and the Distinguished Educator Award of the Ontario Institute for Studies in Education. He has been the HERDSA Fellow in Australia and New Zealand. His current interests include the development of puzzle experiments to encourage thinking in large classes and new ways of showing reactions via videodiscs and CD Rom. His credo is that the main measure of an effective teacher is student learning.

E. A. (Peter) Robinson did his Ph.D. work with Dr. Ronald Gillespie at University College, London. Afterward in 1958, Dr. Robinson was the first member of the Gillespie research team in Canada. He has been at the University of Toronto since 1961, where he is Professor of Chemistry. In 1965 he became a founding member of Toronto's suburban campus in Mississauga. As the first Associate Dean, second Dean, and third Principal of Erindale College, he played a central role in the organization of a college now equally recognized for the quality of its teaching and research. His current research interests include relating force constants to bond lengths and molecular geometry, attempts to systematize structures with an expanded octet, and C–H hydrogen bonding. He was awarded a D.Sc. by the University of London in 1969.

Atoms and Molecules: The Building Blocks of Substances

1.1 Atoms and Molecules
1.2 Elements, Compounds, and Formulas
1.3 Substances: Properties and Purity
1.4 Mixtures
1.5 Chemical Reactions and Chemical Equations
1.6 The Structure of Atoms
1.7 The Mole: Counting by Weighing
1.8 Unstable Atoms: Radioactivity
1.9 Conservation of Mass and Energy

Chemistry is the science of the properties and transformations of substances. *Substances are composed of* atoms, *which may combine to form* molecules. *Here the* element *iron (in the form of steel wool), made up of iron atoms, burns vigorously in the element oxygen, made up of oxygen molecules. The result is the* compound *iron oxide.*

Chemistry is the science that studies substances. It is concerned with all the substances of which the earth is composed: the water in the oceans, the oxygen and nitrogen in the air, all the substances in rocks and minerals, and all those found in the plants and animals that inhabit the earth. It also deals with all the substances that chemists have made, many of which are not found in nature. Indeed, the realm of chemistry extends beyond our planet to the substances that compose the other planets and that are found in interstellar space.

Chemistry is concerned with the composition and properties of all these substances and with the changes, called chemical reactions, in which one substance is converted into another. If an iron tool is left outside in moist air, the iron combines with oxygen to form rust, a substance composed of both iron and oxygen. When lightning streaks through the sky, nitrogen and oxygen in the air combine to form a new substance called nitrogen monoxide. Using the energy of light, plants convert carbon dioxide and water from the air and soil into carbohydrates and oxygen in a series of chemical reactions called photosynthesis. Chemical reactions within our cells break carbohydrates down into simpler substances, liberating energy for movement and for our bodies' growth and maintenance.

The O_2 molecule has two identical atoms.

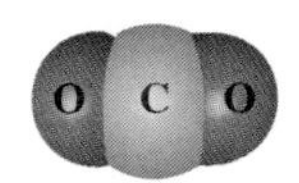

The CO_2 molecule, made of two kinds of atoms, is linear.

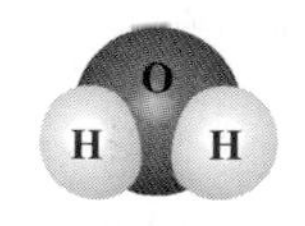

The H_2O molecule, made of two kinds of atoms, is bent or angular.

The eight identical atoms in the S_8 molecule are arranged in a ring.

FIGURE 1.1 Models of Some Molecules Oxygen (O_2) and sulfur (S_8) have molecules made up of only one kind of atom, so these two substances are elements. A carbon dioxide molecule (CO_2) contains more than one kind of atom, as does a water molecule (H_2O), so these substances are compounds.

1.1 ATOMS AND MOLECULES

All substances consist of **atoms**, and in most substances atoms are joined in groups called **molecules**. So chemistry may be said to be the science of atoms and molecules. A major aim of chemistry is to explain the properties and reactions of substances in terms of the atoms and molecules of which they are composed.

Atoms and molecules are extremely small—it is difficult to imagine just how small they are. One hundred million (10^8) carbon atoms placed side by side would form a row slightly less than 2 cm long. Even molecules that contain thousands of atoms are so small we cannot detect them with our ordinary senses. Thus, although substances are composed of atoms and molecules, they appear to us to be continuous. Because we cannot see them directly, we construct models of atoms and molecules that help us visualize and think about them. A simple model that has been used ever since atoms were first postulated represents an atom as a tiny sphere and a molecule as two or more such spheres joined together (Figure 1.1).

Methods for observing atoms and molecules indirectly have been devised that allow us to measure their size and shape. (One such method is described in Box 1.1, and we shall describe some other methods later.) We can also determine the numbers, kinds, and arrangements of atoms that make up a particular molecule (Figure 1.1). We know, for example, that the oxygen in air consists of oxygen molecules, each of which is made up of two oxygen atoms. A water molecule is made up of an oxygen atom and two hydrogen atoms. Carbon dioxide, another gas found in air, consists of molecules composed of one carbon atom and two oxygen atoms. Although we typically represent atoms as featureless spheres, they have an internal structure that is very important for understanding their behavior and why and how they combine to form molecules. We will describe this structure later in the chapter.

1.2 ELEMENTS, COMPOUNDS, AND FORMULAS

Long before chemists knew about atoms, they knew that the simplest substances, which they called *elements*, can react to form more complex substances, which they called *compounds*. Chemists found that a compound can be decomposed into the elements from which it was formed, but an element cannot be further decomposed into other substances. Since the early nineteenth century, we have been able to explain these observations in terms of the combining of atoms into molecules (Box 1.1).

Elements: Symbols and Molecular Formulas

Substances that are composed of only one kind of atom are called **elements**. For example, each nitrogen molecule in the air consists of two nitrogen atoms. Nitrogen

BOX 1.1

Atoms, Dalton, and the Scanning Tunneling Microscope

The idea that matter consists of atoms originated with the Greek philosophers well over two thousand years ago. Although Plato and Aristotle believed that matter is continuous, Democritus (c. 460–370 B.C.) had earlier argued that matter is composed of very small, indivisible particles. However, chemists did not begin to take this idea seriously until the nineteenth century, when John Dalton showed that atoms could form the basis for understanding the nature of elements and compounds and their reactions.

John Dalton (1766–1844) was born in Cumberland, England. He attended a small country school, and he made such rapid progress that by the age of 12 he had become a teacher. In 1793 he moved to Manchester, where he remained the rest of his life, teaching chemistry and mathematics to private students. Dalton made many contributions to chemistry and other sciences, but he is best known for the **atomic theory,** which he proposed in 1803. The essentials of the theory were that

- The atoms of any given element are identical;
- The atoms of an element are different in mass from those of every other element;
- Atoms of two or more elements may combine in definite ratios to form compounds;
- Atoms remain unchanged in chemical reactions.

Dalton had no direct evidence for the existence of atoms. However, by making these assumptions he could explain the observation that the masses of elements that combine to form a given compound have a fixed ratio. Since Dalton's time a great deal of more direct evidence for the existence of atoms has been obtained, and his assumptions about atoms have been fully confirmed. Although it is not true that atoms are indivisible, as both the Greeks and Dalton believed, this does not affect the validity of Dalton's four basic assumptions. Dalton's atomic theory was an important breakthrough that led to a much better understanding of the nature of elements and compounds and a rapid growth of the science of chemistry.

In the 1980s, an instrument that for the first time enabled us to "see" individual atoms, the **scanning tunneling microscope** (STM), was developed. A *microscope* is any instrument with which we can obtain an enlarged image of an object. The familiar optical microscope uses light to produce this enlarged image, whereas electron microscopes use electrons. An electron microscope provides much greater magnification than does a light microscope, but not nearly great enough to reveal individual atoms. In the STM a very sharp metal tip (only a few atoms in diameter) is held very close to the surface of a solid and moved slowly across the surface; it is said to be *scanning* the surface (Figure A).

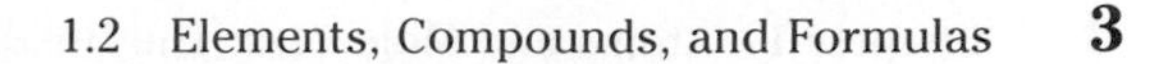

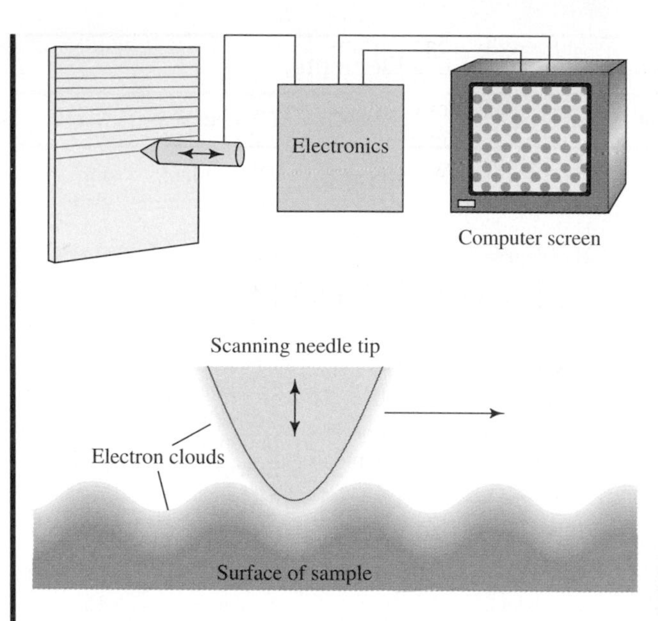

Figure A Electron Tunneling *An electron cloud occupies the space between the surface of the sample and the needle tip. An electric current (a flow of electrons) can pass through this cloud if there is a voltage difference between the tip and the surface. Because the density of the electron cloud decreases exponentially with distance from the surface of the sample, the current is very sensitive to the distance between the tip and the surface. As the tip is swept across the surface, a feedback mechanism senses the flow (called the tunneling current) and holds constant the height of the tip above the surface. In this way the tip traces out the contours of the surface.*

The electron clouds (Chapter 6) of the atoms in the metal tip and the solid surface extend out a very short distance, and when the tip is just a few atom diameters from the solid, these electron clouds overlap. When a voltage is applied between the tip and the solid surface, electrons from the tip pass through the electron clouds to the solid; they are said to be *tunneling* from the tip to the surface. In other words, an electric current flows between the tip and the solid. Because the density of the electron cloud of an atom decreases very rapidly with increasing distance from the nucleus, the magnitude of the current is extremely sensitive to the distance between the atoms in the tip and those in the solid surface. Changing this distance by an amount equal to the diameter of a single atom causes the current to change by a factor of as much as 1000. As the tip sweeps across the surface, a feedback mechanism senses the current and continually adjusts the height of the tip above the surface to keep the current constant. The resulting motion of the tip is read and processed by a computer and displayed on a screen or a plotter. Sweeping the tip through a pattern of parallel lines produces a three-dimensional image with such fine detail that atoms can be seen as individual bumps on the surface of the solid (Figure B).

Figure B Surface of Silicon *The surface of silicon as disclosed by the scanning tunneling microscope. The bumps on the surface correspond to individual silicon atoms.*

is therefore an element. Sulfur contains only sulfur atoms, and iron only iron atoms, so both of these substances are elements. One hundred and nine different kinds of atoms are known, so of the many millions of known substances, only 109 are elements. Each element is given its own symbol, which is generally based on the first letter or two of the first few letters of its English name, such as H for hydrogen and Cl for chlorine. In some cases the symbol is based on the name of the element in another language. For example, the symbol for iron is Fe, from the Latin *ferrum* (Table 1.1).

Although a few elements such as neon (Ne) and argon (Ar) consist of single, isolated atoms, in most elements the atoms are combined in some way. Both the oxygen and nitrogen in air consist of molecules, each of which contains two atoms. Using the element symbol to represent one atom, we represent these molecules as N_2 and O_2. These representations are called the **molecular formulas** of nitrogen and oxygen, respectively. Chlorine similarly consists of Cl_2 molecules, but sulfur consists of S_8 molecules, each of which contains eight sulfur atoms (Figure 1.1).

TABLE 1.1 Names and Symbols of the Elements

Element	Symbol	Element	Symbol	Element	Symbol
Actinium	Ac	Hafnium	Hf	Promethium	Pm
Aluminum	Al	**Helium**	He	Protactinium	Pa
Americium	Am	Holmium	Ho	Radium	Ra
Antimony	Sb	**Hydrogen**	H	**Radon**	Rn
Argon	Ar	Indium	In	Rhenium	Re
Arsenic	As	**Iodine**	I	Rhodium	Rh
Astatine	At	Iridium	Ir	**Rubidium**	Rb
Barium	Ba	**Iron**	Fe	Ruthenium	Ru
Berkelium	Bk	**Krypton**	Kr	Samarium	Sm
Beryllium	Be	Lanthanum	La	Scandium	Sc
Bismuth	Bi	Lawrencium	Lr	Selenium	Se
Boron	B	**Lead**	Pb	**Silicon**	Si
Bromine	Br	**Lithium**	Li	**Silver**	Ag
Cadmium	Cd	Lutetium	Lu	**Sodium**	Na
Calcium	Ca	**Magnesium**	Mg	**Strontium**	Sr
Californium	Cf	**Manganese**	Mn	**Sulfur**	S
Carbon	C	Mendelevium	Md	Tantalum	Ta
Cerium	Ce	**Mercury**	Hg	Technetium	Tc
Cesium	Cs	Molybdenum	Mo	Tellurium	Te
Chlorine	Cl	Neodymium	Nd	Terbium	Tb
Chromium	Cr	**Neon**	Ne	Thallium	Tl
Cobalt	Co	Neptunium	Np	Thorium	Th
Copper	Cu	**Nickel**	Ni	Thulium	Tm
Curium	Cm	Niobium	Nb	**Tin**	Sn
Dysprosium	Dy	**Nitrogen**	N	**Titanium**	Ti
Einsteinium	Es	Nobelium	No	Tungsten	W
Erbium	Er	Osmium	Os	**Uranium**	U
Europium	Eu	**Oxygen**	O	**Vanadium**	V
Fermium	Fm	Palladium	Pd	**Xenon**	Xe
Fluorine	F	**Phosphorus**	P	Ytterbium	Yb
Francium	Fr	**Platinum**	Pt	Yttrium	Y
Gadolinium	Gd	**Plutonium**	Pu	**Zinc**	**Zn**
Gallium	Ga	Polonium	Po	Zirconium	Zr
Germanium	Ge	**Potassium**	K		
Gold	Au	Praseodymium	Pr		

Note: The elements discussed in this book are in bold type. Elements 104–109 have been omitted because no international agreement has yet been reached on their names and symbols.

Compounds and Their Molecular Formulas

A substance composed of identical molecules that contain more than one kind of atom is called a **compound**. Water is a compound: It consists of identical water molecules, each of which is composed of 1 oxygen atom and 2 hydrogen atoms. We represent water by the molecular formula H_2O. Nitrogen monoxide, formed when lightning discharges in the atmosphere, is a compound of nitrogen and oxygen. Each of its molecules contains 1 nitrogen atom and 1 oxygen atom, so it has the molecular formula NO. A solution of ammonia in water is a common household cleaner; ammonia is a compound made of molecules containing 1 nitrogen atom and 3 hydrogen atoms each. Thus, ammonia has the molecular formula NH_3. Methane, the main component of natural gas, has the molecular formula CH_4, because each of its molecules contains 1 carbon atom and 4 hydrogen atoms.

All the compounds we have mentioned so far have small molecules that contain only a few atoms, but there are many molecules that contain larger numbers of atoms—in some cases, very large numbers of atoms. The molecules of sucrose (ordinary cane or beet sugar) contain 12 carbon atoms, 22 hydrogen atoms, and 11 oxygen atoms, so the molecular formula of sucrose is $C_{12}H_{22}O_{11}$. Many other compounds found in living organisms have large molecules. For example, cholesterol has the molecular formula $C_{27}H_{46}O$, adenosine triphosphate (ATP) has the molecular formula $C_{10}H_{16}N_5O_{13}P_3$, and enzyme molecules may have thousands of atoms. Just as every molecule in a sample of water has 2 hydrogen atoms and 1 oxygen atom, every molecule in a sample of cholesterol has 27 carbon atoms, 46 hydrogen atoms, and 1 oxygen atom.

If every molecule in a compound is identical in composition, then every sample of that compound, regardless of how many molecules it contains, must contain the same kinds of atoms in the same fixed ratio:

A compound has a constant, characteristic composition.

Every sample of nitrogen monoxide, whatever its size, contains *exactly* equal numbers of oxygen and nitrogen atoms, because each of its molecules contains 1 nitrogen atom and 1 oxygen atom. Likewise, both a drop of water and a swimming pool contain *exactly* twice as many hydrogen atoms as oxygen atoms, and in every single crystal and every cupful of sucrose (sugar), the ratio of carbon atoms to hydrogen atoms to oxygen atoms is *exactly* 12:22:11. Constant composition is also a characteristic of compounds that have no discrete molecules, as we will see shortly.

Representing Bulk Substances

The molecular formula of an element or compound indicates the numbers of each kind of atom in one molecule of the substance. It also represents the *relative* numbers of each kind of atom in any *bulk sample* (large quantity) of the substance—a sample that may contain many billions of molecules. This dual usage can lead to confusion if we are not careful, because a bulk sample has many properties that are not possessed by a single atom or molecule.

One of the properties of a bulk sample of any substance is its ability to exist in different physical states: liquid, solid, or gas. Water, for example, is a solid below 0°C, a gas above 100°C, and a liquid at temperatures in between. Because these states (which we will discuss in Chapter 2) depend on the attractions among molecules, it is meaningless to talk about the physical state of a single water molecule. The physical state of bulk substances gives us a convenient way to distinguish them from their component atoms or molecules. When we use an atomic symbol or molecular formula to refer to a bulk sample, we will add a letter designating the sample as solid(s), liquid(l), or gas(g). Thus, a sample of water at room temperatures is $H_2O(l)$, frozen water (ice) is $H_2O(s)$, and gaseous water is $H_2O(g)$. Gaseous water is commonly called *water vapor.* When we use an atomic symbol or molecular formula alone, it will refer to a single atom or molecule. Thus, O represents one atom of oxygen, O_2 represents one molecule of oxygen, H_2O represents one molecule of water, and $CO_2(g)$ represents a bulk sample of carbon dioxide gas.

Empirical Formulas

We have repeatedly mentioned water, which has the molecular formula H_2O. Another compound of hydrogen and oxygen is hydrogen peroxide, which is commonly used as a bleach, particularly for hair. Hydrogen peroxide has the molecular formula H_2O_2, so in any sample of hydrogen peroxide, the ratio of hydrogen atoms to oxygen atoms is 1:1. We express this ratio by saying that the *empirical formula* of hydrogen peroxide is HO.

The simplest formula that tells us the ratios of the different kinds of atoms in a substance is called its empirical formula.

Silicon dioxide, better known as quartz, is the main component of beach sand.

The empirical formula does *not* tell us how many atoms are in a molecule. The molecular formula does. The molecular formula of hydrogen peroxide, H_2O_2, which tells us that there are two hydrogen and two oxygen atoms in each hydrogen peroxide molecule, is twice the empirical formula, HO.

We also use empirical formulas for the many compounds that have no discrete molecules. One such substance is silicon dioxide, which is known also as silica or quartz and is the major component of most types of sand. Silicon dioxide consists of a giant network of silicon and oxygen atoms in the ratio 1:2. Thus, the empirical formula of silicon dioxide is SiO_2. Each silicon atom is joined to four oxygen atoms, and each oxygen atom is joined to two silicon atoms (Figure 1.2a). This arrangement of the atoms extends throughout any solid sample of silicon dioxide, and no individual molecules can be recognized. We might think of a grain of pure sand as one enormous molecule with the molecular formula Si_nO_{2n} or $(SiO_2)_n$, where n is a very large number. However, it is much more convenient to use the empirical formula.

Silicon dioxide is a compound because it has a definite composition—it is *not* a random mix of silicon and oxygen atoms. Another example of a nonmolecular

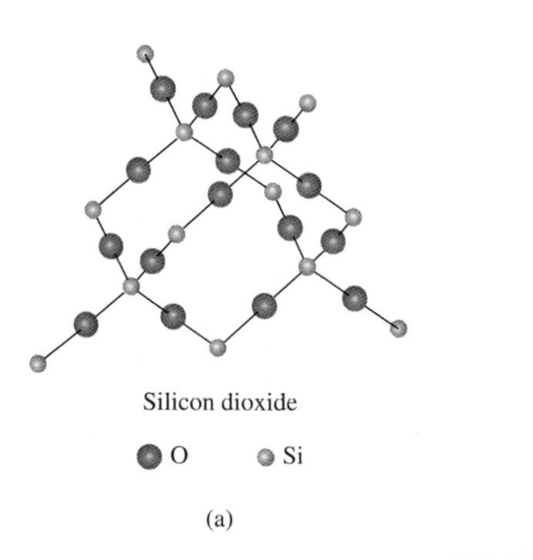

Sodium chloride

Na^+ Cl^-

(b)

FIGURE 1.2 Silica and Common Salt (a) Silica, or silicon dioxide $(SiO_2)_n$, has a network structure, in which no individual molecules can be recognized. Its empirical formula is SiO_2. (b) Salt, or sodium chloride $(NaCl)_n$, also has a network structure. Its empirical formula is NaCl.

compound is sodium chloride (common table salt), which has the empirical formula NaCl. As we will see in later chapters, sodium (Na) and chlorine atoms occur in sodium chloride as charged particles called *ions*. These two ions always occur in NaCl in a 1:1 ratio, alternating with one another in the same pattern throughout every sodium chloride crystal (Figure 1.2b). In this way, sodium chloride has a definite composition but no discrete molecules. A grain of salt, or for that matter a large salt crystal, is another example of a single "giant molecule."

Atoms are electrically neutral because they are made of a positively charged nucleus and enough (negatively charged) electrons to balance the charge on the nucleus. An ion is an atom or a molecule that has a positive or negative charge as a result of losing or gaining electrons.

Structural Formulas

Once we know the molecular formula of an element or molecular compound, the next question that we can ask is, How are the atoms arranged in the molecule? We saw some possible arrangements in Figure 1.1. In the carbon dioxide molecule, the three atoms are arranged in a straight line with the carbon atom in the middle; we say that this molecule is *linear*. In the water molecule, the oxygen is in the middle, and the molecule is not linear but bent, or *angular*. When we are sitting in a bathtub or swimming in the ocean, it is hard to imagine that the water surrounding us consists of many billions of these tiny, angular molecules. But that is what chemists have shown water to be. By thinking about water in this way, we are able, as we shall see, to understand its properties and reactions.

Sodium chloride (salt) forms giant molecules rather than the small, discrete molecules that molecular compounds form. Each crystal shown here is one giant molecule.

Many years ago, chemists started to draw a line between the symbols of the atoms that they believed were attached to each other, so the water molecule was written as H–O–H. This representation is called a **structural formula**. The lines that show how atoms are connected together are called *chemical bonds*. Structural formulas are often written in such a way as to give an idea of the three-dimensional shape of the molecule. The standard structural formulas and three-dimensional structural formulas of water, ammonia, and methane are given in Figure 1.3. The

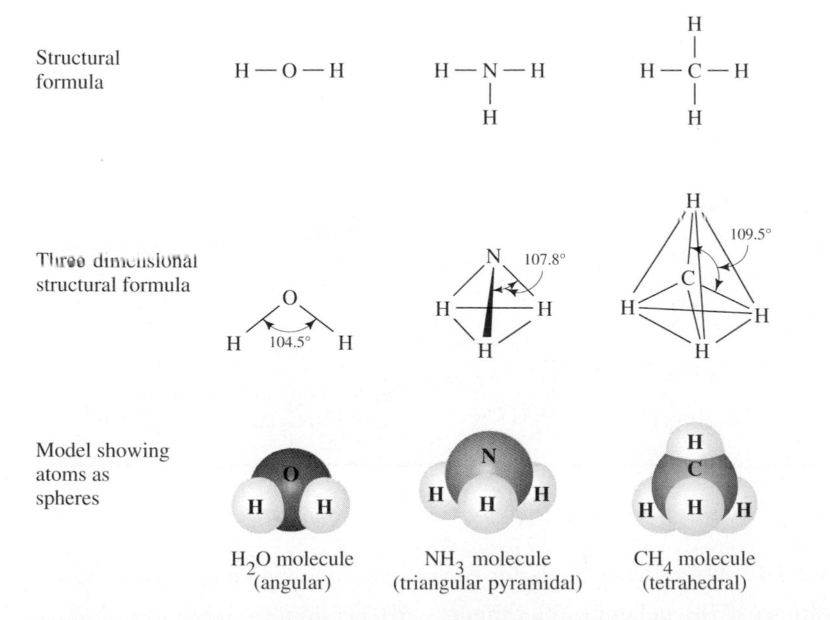

FIGURE 1.3 Structural Formulas and Shapes of Some Molecules

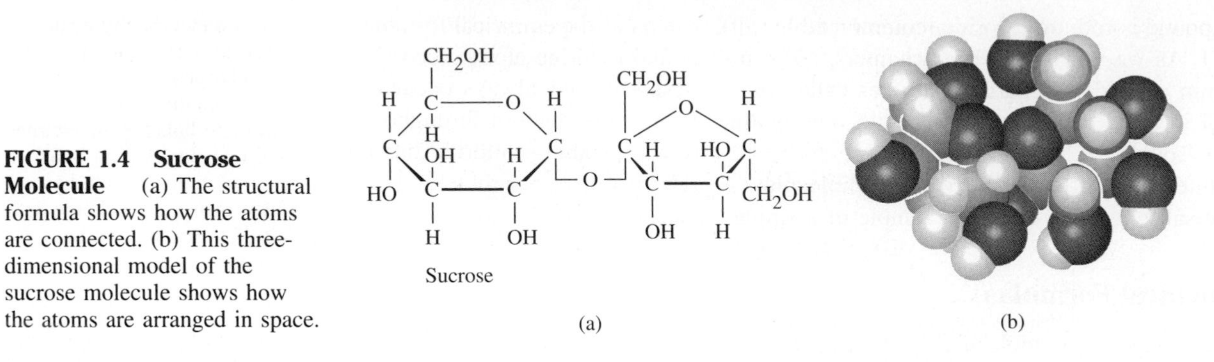

FIGURE 1.4 Sucrose Molecule (a) The structural formula shows how the atoms are connected. (b) This three-dimensional model of the sucrose molecule shows how the atoms are arranged in space.

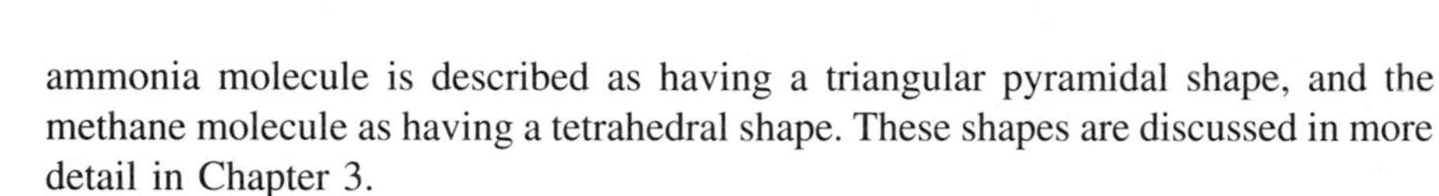

ammonia molecule is described as having a triangular pyramidal shape, and the methane molecule as having a tetrahedral shape. These shapes are discussed in more detail in Chapter 3.

The structures of large molecules such as sucrose and cholesterol are more complex than the structures we have just described. But the principles that determine the structures and reactions of these molecules are the same as those that apply to small molecules. The structural formula of sucrose and a model of the sucrose molecule are shown in Figure 1.4. The relationships of the different types of formulas are summarized in Table 1.2.

TABLE 1.2 Types of Formulas

Formula	Examples: Hydrogen Peroxide	Examples: Ethane	Information Provided
Empirical formula	HO	CH_3	Relative number of atoms in the molecule and in the bulk sample
Molecular formula	H_2O_2	C_2H_6	Number of atoms of each kind in the molecule
Structural formula	H—O—O—H	H H / | | / H—C—C—H / | | / H H	Connections among the atoms within the molecule. Each line represents a chemical bond.
Three-dimensional structural formula	H, O—O, H; 94.5°; 97 pm; 147 pm	H, C—C, H, H, H, H; 109°; 153 pm; 111 pm	Distances and angles. Wedges show bonds in front of the page; dotted lines show bonds behind.

Exercise 1.1 Write the symbols for each of the following elements.
(a) hydrogen **(b)** helium **(c)** lithium **(d)** beryllium **(e)** boron **(f)** carbon **(g)** nitrogen **(h)** oxygen **(i)** fluorine **(j)** neon

Exercise 1.2 Name the elements with each of the following symbols.

(a) H (b) He (c) B (d) Be (e) C (f) Ca (g) N (h) Na

(i) F (j) Fe (k) K (l) Kr

Exercise 1.3 Name each of the substances with the following molecular formulas, and classify each as a constituent of an element (E) or a compound (C).

(a) H_2 (b) N_2 (c) O_2 (d) NO (e) S_8 (f) Cl_2 (g) H_2O

(h) NH_3 (i) CH_4 (j) CO_2

Exercise 1.4 Write the empirical formulas for the substances with each of the following molecular formulas.

(a) H_2O_2 (b) CH_4 (c) C_2H_6 (d) S_8 (e) CO_2 (f) N_2

(g) $C_{12}H_{22}O_{11}$ (h) C_4H_8

1.3 SUBSTANCES: PROPERTIES AND PURITY

Substances and Chemicals

In the preceding sections we have used the word ''substance'' many times, but we have not explained exactly what we mean by a substance. In everyday speech the word may be used to describe almost any material. However, we must be careful not to confuse the everyday usage of a term with its precise meaning in chemistry. In chemistry, a **substance** is either an element or a compound. A compound is composed of two or more kinds of atoms, and the numbers of each kind of atom are always in exactly the same ratio to each other. We say that a compound has a constant characteristic composition. The same must be true of an element, because it consists of only one type of atom.

Every substance (element or compound) has a constant, characteristic composition.

Substances are often referred to as chemicals. In the news media and in everyday speech, the word ''chemical'' is all too often associated with a material that is harmful or dangerous. ''Chemical'' is also frequently used to mean a synthetic substance, as opposed to a naturally occurring substance, often with the mistaken implication that the former is necessarily dangerous and the latter necessarily safe. Thus, there is considerable confusion and misunderstanding about the meaning of the word ''chemical'' when it is used as a noun. To a chemist, a chemical is simply a substance, that is, a material with a fixed characteristic composition. We will avoid the use of ''chemical'' as a noun, because it is unnecessary and misleading. It is useful, however, as an adjective in terms such as ''chemical industry'' and ''chemical reaction.''

Physical and Chemical Properties

Every substance has a unique set of properties that enable us to distinguish it from all other substances. We recognize two different types of properties: **physical properties** and **chemical properties**.

The *physical properties* of a substance are those properties that can be observed and measured without changing the substance into other substances.

Physical properties include color, physical state at room temperature (solid, liquid, or gas), freezing point and boiling point, density, and electrical conductivity. Some physical properties of water are that it is a liquid at room temperature, that it freezes at 0°C and boils at 100°C, and that it is colorless.

The *chemical properties* of a substance are those properties related to its participation in chemical reactions.

Iron has the chemical property of rusting (combining with oxygen) when it is exposed to air and water. Nitrogen has the chemical property of combining with oxygen to form nitrogen monoxide when heated very strongly or when a spark or electric discharge is passed through a mixture of the two gases. Although both water (H_2O) and hydrogen peroxide (H_2O_2) are composed of hydrogen and oxygen atoms and are colorless liquids, they have very different chemical properties. As we see in Demonstration 1.1, the hydrogen peroxide in a solution of hydrogen peroxide in water decomposes rapidly to oxygen and water when solid manganese dioxide is added, whereas water remains unchanged. A drop of blood contains an enzyme that decomposes hydrogen peroxide even more rapidly but does not decompose water.

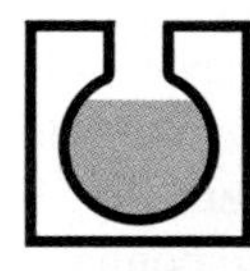

DEMONSTRATION 1.1 Properties of Water and Hydrogen Peroxide

Both water (left) and a solution of hydrogen peroxide in water (right) are colorless liquids. Both hydrogen peroxide and water are compounds of the elements hydrogen and oxygen. When black solid manganese dioxide is added to water, no reaction is observed. When manganese dioxide is added to the hydrogen peroxide solution, it causes the hydrogen peroxide to decompose, producing bubbles of oxygen that carry the manganese dioxide to the surface. The oxygen that is evolved ignites a glowing splint.

When a drop of blood is added to water, it slowly mixes, but no reaction is observed. When a few drops of blood are added to a hydrogen peroxide solution, the hydrogen peroxide decomposes rapidly, producing bubbles of oxygen that form a thick foam that fills the beaker and overflows the top.

Purity

The terms ''pure'' and ''impure'' are typically used to describe substances. An **impurity** is a substance that is mixed with or contaminates other substances. In other words, it is a mixture of two or more substances. Sea water is not pure water, as it contains dissolved salt (sodium chloride) and many other substances. Substances that are present in small amounts in a mixture are called *impurities.* Truly pure water would consist entirely of water molecules. However, a truly pure substance is an ideal concept that is impossible to obtain. No matter how much water or any other substance is purified, there will always be some other molecules present. Chemists use ''**pure substance**'' to mean that impurities are present in such small amounts that they cannot be detected or are of no practical significance.

Thus, purity is a relative term to chemists. It is usually expressed as a **mass percentage**. If a sample of 100 grams (g) of a substance has 0.1 g of impurities, it is said to be 99.9% pure. A substance that is 99.999% pure has only 0.001% of impurities. The labels on containers of substances found in the chemistry laboratory typically list both the names and the amounts of the impurities (Figure 1.5).

In everyday speech the term ''pure'' is often used to mean that a substance contains nothing injurious to health. For example, bottled water labeled ''pure'' normally contains dissolved oxygen and carbon dioxide and small amounts of other substances, all of which a chemist would call impurities. In this case, pure means only that harmful or toxic impurities, such as bacteria, are not present in significant amounts.

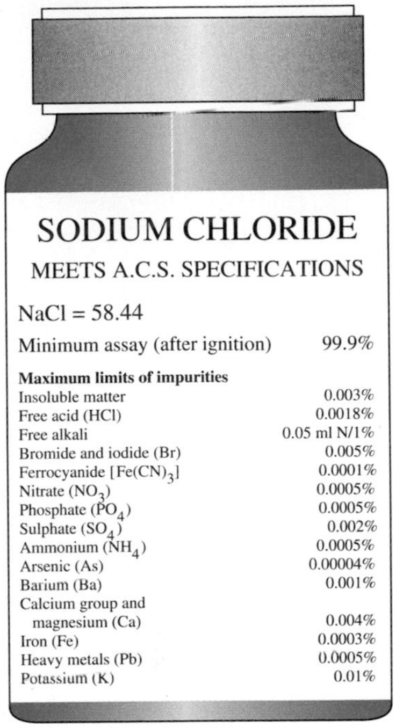

SODIUM CHLORIDE

MEETS A.C.S. SPECIFICATIONS

NaCl = 58.44

Minimum assay (after ignition)	99.9%
Maximum limits of impurities	
Insoluble matter	0.003%
Free acid (HCl)	0.0018%
Free alkali	0.05 ml N/1%
Bromide and iodide (Br)	0.005%
Ferrocyanide [$Fe(CN)_3$]	0.0001%
Nitrate (NO_3)	0.0005%
Phosphate (PO_4)	0.0005%
Sulphate (SO_4)	0.002%
Ammonium (NH_4)	0.0005%
Arsenic (As)	0.00004%
Barium (Ba)	0.001%
Calcium group and magnesium (Ca)	0.004%
Iron (Fe)	0.0003%
Heavy metals (Pb)	0.0005%
Potassium (K)	0.01%

FIGURE 1.5 Label on Bottle of Sodium Chloride Indicates Impurities This sodium chloride is described as 99.9% pure. It therefore contains a total of 0.1% of impurities, which are listed on the label. A.C.S. is the abbreviation for the American Chemical Society.

EXAMPLE 1.1 Chemical and Physical Properties

Classify each of the following properties of water as a chemical property or a physical property: Water is **(a)** a colorless liquid, **(b)** forms ice at 0°C, and **(c)** boils at 100°C. It has **(d)** a high surface tension and **(e)** a very small electrical conductivity. **(f)** Its density is 1.000 $g \cdot cm^{-3}$ at 4°C. **(g)** When water is passed over hot coke (carbon), hydrogen, $H_2(g)$, forms. **(h)** Water vapor strongly absorbs infrared light but is transparent to visible and ultraviolet light. **(i)** When magnesium, $Mg(s)$, is heated in steam, the white powder magnesium oxide, $MgO(s)$, forms. **(j)** When an electric current is passed through water, $H_2(g)$ and $O_2(g)$ form.

Solution: Physical properties of a substance are those properties that can be observed and measured without the substance changing into other substances. Chemical properties of a substance are those properties that relate to its participation in chemical reactions. Thus, properties **(a)** through **(f)** are physical properties of water, as is **(h)**. However, **(g)**, **(i)**, and **(j)** are chemical properties of water, because they involve chemical changes.

Exercise 1.5 Classify each of the underlined terms in the following statement as a physical property (P) or a chemical property (C).

''Hydrogen peroxide is **(a)** a colorless, **(b)** viscous **(c)** liquid that readily **(d)** decomposes to water and oxygen. It has **(e)** a density of 1.44 $g \cdot mL^{-1}$, **(f)** a melting point of -0.89°C, and **(g)** a boiling point of 151°C. A solution of hydrogen peroxide in water is used as **(h)** a mild antiseptic and as **(i)** a bleaching agent.''

Exercise 1.6 A 10.000-g sample of table salt, $NaCl(s)$, contains 0.0030 g of impurities. What is the purity of the sample, expressed as the mass percentage of NaCl?

mass percentage purity = $100\% \times \dfrac{(\text{mass of sample} - \text{mass of impurities})}{(\text{mass of sample})}$

1.4 MIXTURES

A material containing two or more substances is called a **mixture**. There are two types of mixtures: *homogeneous* (uniform) and *heterogeneous* (nonuniform).

Homogeneous Mixtures (Solutions)

A *homogeneous* mixture has uniform properties throughout.

It is uniform in composition, color and other physical properties, and chemical properties. A more common name for a homogeneous mixture is a **solution**. If we add sugar to water and stir, the sugar dissolves, forming a clear solution (homogeneous mixture) of sugar in water. The individual water molecules and sucrose molecules are uniformly mixed, so every part of the solution has the same numbers of water and sucrose molecules as every other part. The solution has a uniform composition and therefore uniform properties: We cannot distinguish the sugar from the water. A homogeneous mixture is not a substance, however, because it does not have a constant characteristic composition. Very different amounts of sugar can be dissolved in a given amount of water. We cannot write an empirical formula for a sugar–water mixture, because it has no fixed ratio of sugar molecules to water molecules.

Although we commonly use the term ''solution'' for a homogeneous mixture of a solid and a liquid (such as sugar and water) or two liquids (such as alcohol and water), there are other types of solution. Most gold jewelry is made from a homogeneous mixture of silver and gold that is a **solid solution**. A solid solution of two or more metals is called an **alloy**. The atmosphere is a good example of a **gaseous solution**, but it is usually called a mixture. All gases mix with each other, forming homogeneous mixtures.

We will frequently be concerned with solutions in water, which are called **aqueous solutions**, so we indicate that a substance is in aqueous solution by (*aq*) following its formula. Thus NaCl(*aq*) means a solution of NaCl in water, just as NaCl(*s*) means solid sodium chloride.

Heterogeneous Mixtures

If we add powdered sulfur to water and stir, the sulfur does not dissolve. Rather, it forms a **suspension**, a mixture in which individual particles (in this case, sulfur particles) are still visible. In other words, the sulfur molecules do not mix with the water molecules to form a solution but remain together as small particles, each of which, nevertheless, contains many millions of sulfur molecules. The mixture of sulfur and water is heterogeneous.

A *heterogeneous* mixture does not have uniform properties.

Some parts of the mixture consist just of sulfur, and other parts just of water. If the mixture is left to stand, the sulfur particles will gradually settle to the bottom of the container to form a distinct layer (Figure 1.6).

Two powdered solids mixed together form a heterogeneous mixture, unless they react with each other chemically. Thus, if we mix powdered sulfur with iron powder, we get a heterogeneous mixture. We can separate this mixture by using a magnet, to which only the iron powder is attracted. If we heat the mixture, however, a chemical reaction occurs, producing the compound iron sulfide, FeS(*s*). Iron cannot be separated from this compound by using a magnet (Demonstration 1.2).

(a)

(b)

FIGURE 1.6 A Heterogeneous Mixture (a) Sulfur and water form a heterogeneous mixture when powdered sulfur is stirred in water. (b) After the mixture stands for some time, the sulfur particles settle to the bottom.

DEMONSTRATION 1.2 Mixtures and Compounds

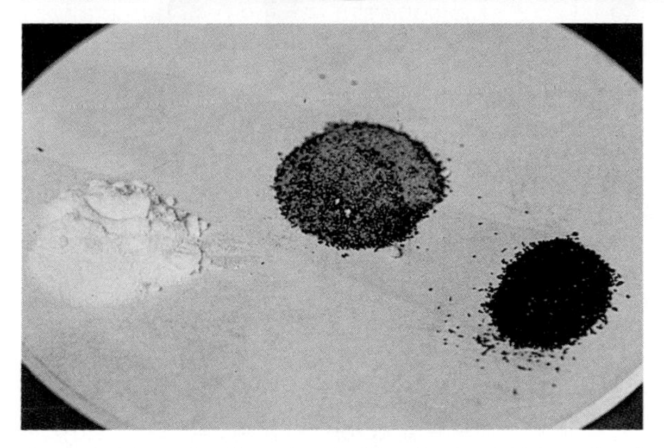

On the left is a pile of powdered sulfur. On the right is a pile of iron filings. In the middle is a mixture of the iron and sulfur.

The iron filings can be separated from the sulfur by using a magnet.

When the mixture is heated, it glows brightly as the iron and the sulfur combine to form the compound iron sulfide.

Iron sulfide is gray-black and brittle, unlike either sulfur or iron. The iron in the compound iron sulfide cannot be separated from the sulfur by means of a magnet.

FIGURE 1.7 Granite Granite is a heterogeneous solid. Different parts of this sample have different properties.

Heterogeneous mixtures are common in the natural world around us. Soil and many rocks, such as granite, are visibly heterogeneous mixtures (Figure 1.7).

The relationships between homogeneous and heterogeneous mixtures and among substances, compounds, and elements are shown in Figure 1.8.

The Separation of Mixtures

In general, mixtures can be separated into their components by making use of their different physical properties, that is, by carrying out some physical change. In a **physical change** the composition and amount of each substance present does not

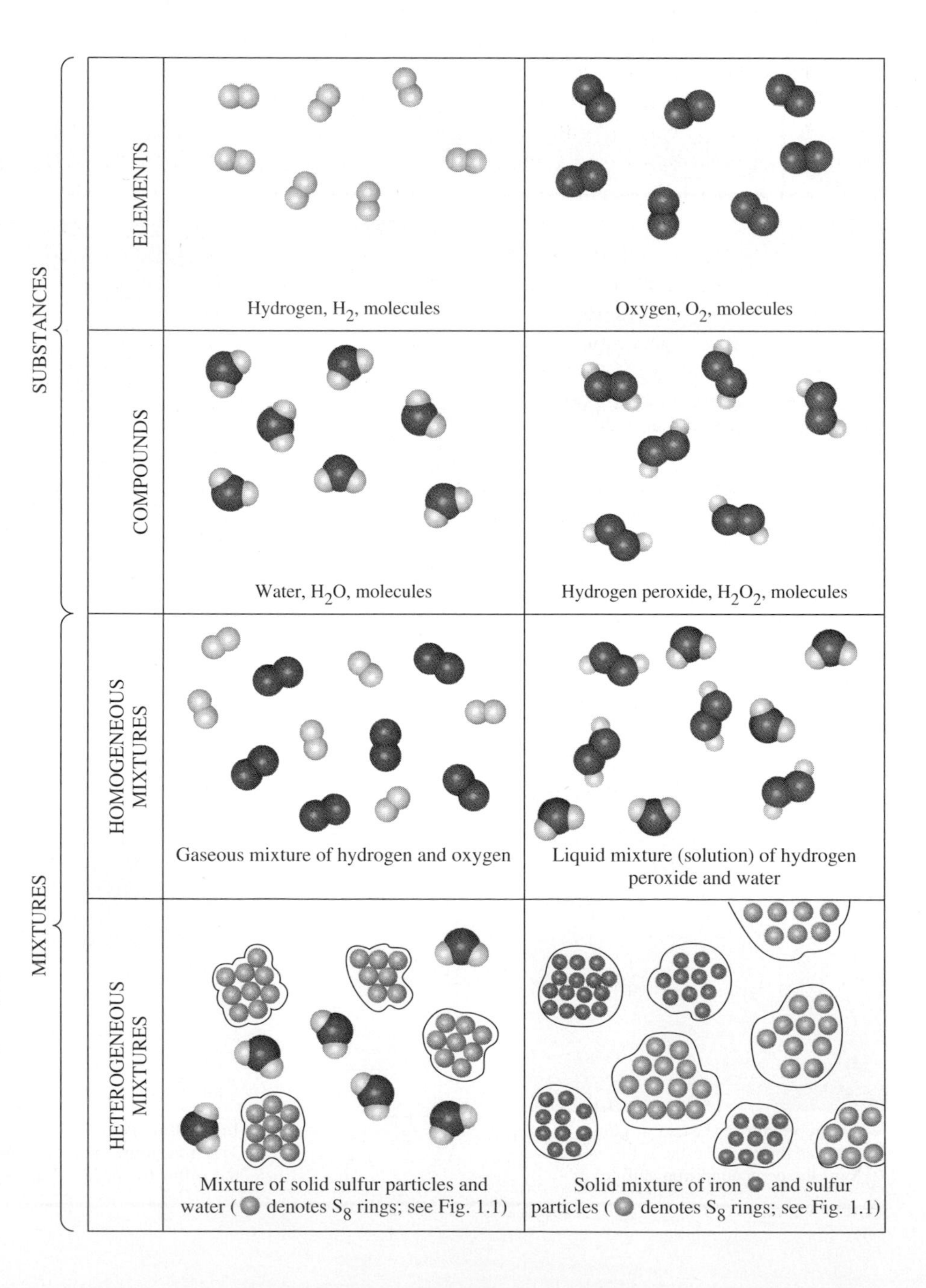

FIGURE 1.8 Elements, Compounds, and Mixtures Hydrogen, oxygen, water, and hydrogen peroxide are all substances. Hydrogen and oxygen are elements; water and hydrogen peroxide are compounds. A mixture of hydrogen peroxide and water is homogeneous; it is a solution. A mixture of hydrogen and oxygen is also homogeneous; it is a (gaseous) solution. A mixture of sulfur and water is heterogeneous, as is a mixture of the solid elements iron and sulfur.

change. In contrast, in a **chemical change** (*chemical reaction*), new substances with different compositions are formed.

The separation of mixtures into their component substances is an essential technique both in the laboratory and in industry. We usually need purer substances than we find in nature, so we must separate naturally occurring substances from their impurities. Many chemical reactions give several products, of which only one is needed, so the desired substance must be separated from the other products. Chemists use many procedures for separating mixtures. Three of the most common methods are filtration, distillation, and chromatography.

Filtration is used to separate a heterogeneous mixture of a solid and a liquid; the liquid may be a pure substance or a solution. For example, if a mixture of salt and sand is stirred with water, the salt dissolves, leaving a solution of salt mixed with solid sand. To separate this mixture, we pour it through a *filter funnel* containing either a porous paper known as *filter paper* or a *sintered glass disk* (Figure 1.9). The filter allows the solution to pass through but retains the solid sand, which may then be washed with water to remove any adhering solution. Another common method of separating a solid from a liquid is to use a *centrifuge*. In this device, a tube containing the mixture is spun rapidly, and centrifugal force causes solid particles to accumulate at one end of the tube (Figure 1.10).

Distillation is used to separate substances that have different boiling points (bp). For example, we can distill pure water from a solution of solids in water, using the apparatus shown in Figure 1.11. When sea water is heated, the water boils when it reaches a certain temperature. But at this temperature sodium chloride and other dissolved substances are *nonvolatile*, that is, they cannot vaporize (form a gas). The water vapor passes into the *condenser*, where it is cooled and converted back into liquid. The liquid emerging from the condenser is called the **distillate**. After all the water has been removed, the solid remaining in the distillation flask contains salt (sodium chloride) and the other nonvolatile substances present in sea water. Water that has been purified by distillation is commonly called **distilled water**.

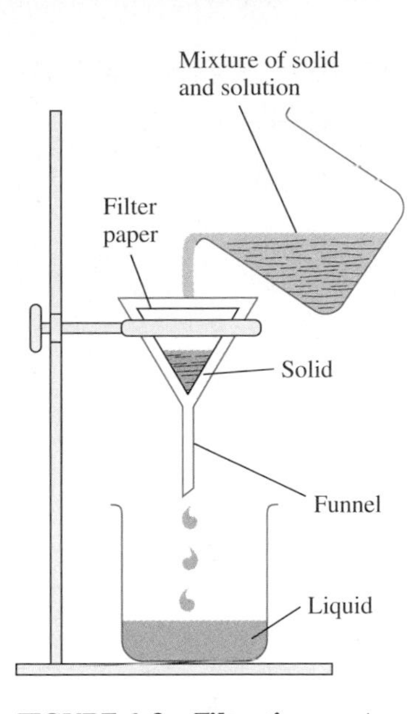

FIGURE 1.9 Filtration A filter allows the liquid to pass through but retains the solid and thus separates the heterogeneous mixture.

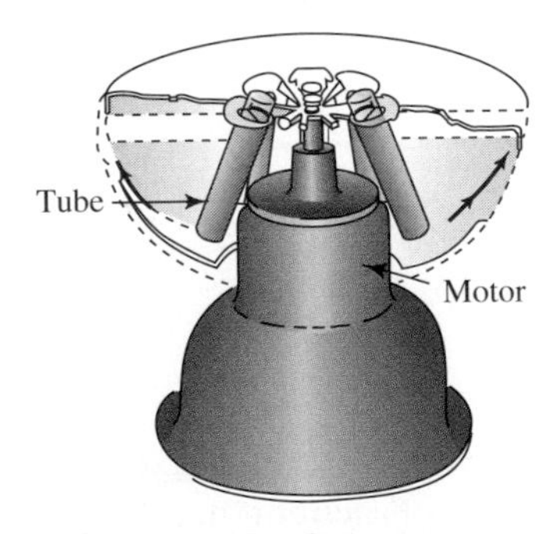

FIGURE 1.10 Centrifuge A heterogeneous mixture to be centrifuged is placed in a tube and inserted in one of the aluminum holders. When the centrifuge is operating, the tubes swing into the dashed positions, and centrifugal force packs the solid particles tightly in the bottom of the tube. After the tube has been removed from the centrifuge, the liquid above the solid can be poured off.

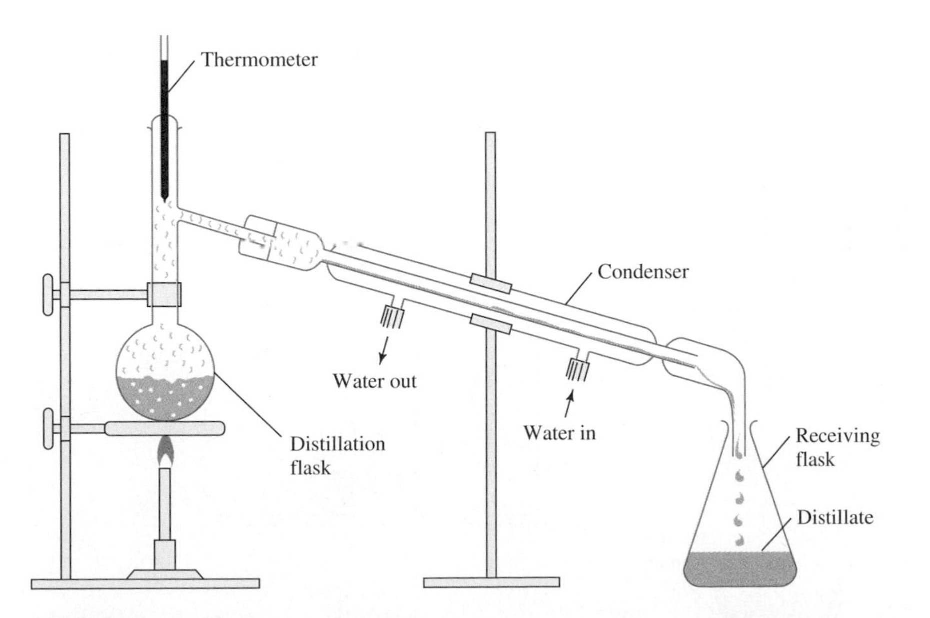

FIGURE 1.11 Distillation Apparatus The nonvolatile component of the mixture remains in the distillation flask. The volatile component is converted to vapor, which passes into the water-cooled condenser. Here the vapor is converted back to a liquid, which flows into the receiving flask.

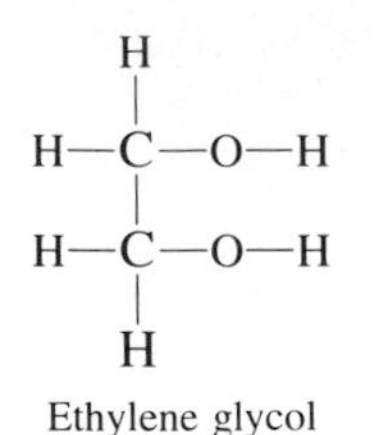

Ethylene glycol

Distillation is used also to separate a mixture of two or more liquids, although separation is not complete unless the liquids have very different boiling points. Ethylene glycol, $C_2H_6O_2(l)$, is a major component of antifreeze, which boils at a much higher temperature than water. In the distillation of a mixture of water (bp 100°C) and ethylene glycol (bp 198°C), the vapor entering the condenser is richer in water vapor than in glycol. When it condenses, the distillate has a higher ratio of water to glycol than does the original mixture. The liquid remaining in the flask has a higher ratio of glycol to water. Redistillation of the distillate further enriches it in the lower-boiling component, water. By repeated distillation a sample of pure water is eventually obtained.

The tedious and time-consuming procedure just described is normally avoided by using a **fractionating column** between the distillation flask and the condenser, as shown in Figure 1.12. This process is known as **fractional distillation**. Fractional distillation is important in many industrial processes. Petroleum, a mixture of many hydrocarbons, is separated by fractional distillation into gasoline, heating oil, and other fractions, each of which contains only a small number of hydrocarbons of similar boiling points (Chapter 8). Pure oxygen and nitrogen are obtained from air by first liquefying air and then carrying out a fractional distillation.

Chromatography (from the Greek word *chroma*, meaning ''color'') was first used for the separation of colored substances found in plants. There are several types of chromatography, but they are all based on the same principle. Demonstration 1.3 illustrates the use of **paper chromatography** to separate the different colored substances in ink. The solvent dissolves these substances and moves up the paper, carrying them with it. However, some of the substances move more slowly than

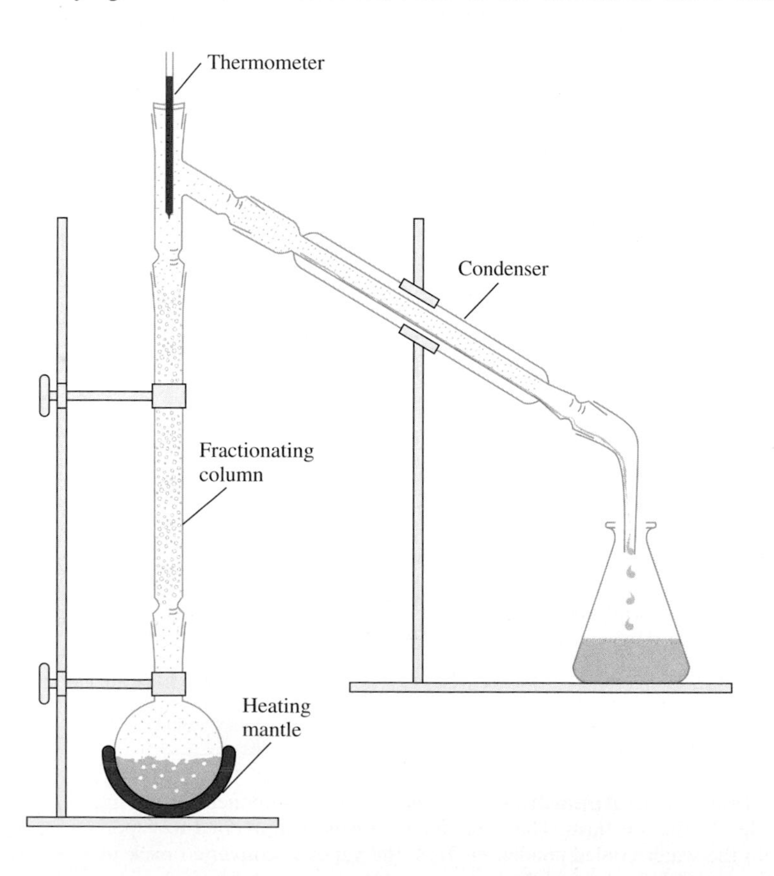

FIGURE 1.12 Distillation with a Fractionating Column The column is packed with an inert material, such as glass beads. Vapor rises up the column and condenses on the beads. The condensed liquid is revaporized by the hot vapor rising up the column, and it recondenses farther up in the column, and so on. The vapor becomes successively richer in the component of lower boiling point, and the pure, lower-boiling component leaves the top of the column and is reconverted to liquid in the condenser. The process of condensation and reevaporation is equivalent to repeated distillations.

DEMONSTRATION 1.3
Paper Chromatography of Ink

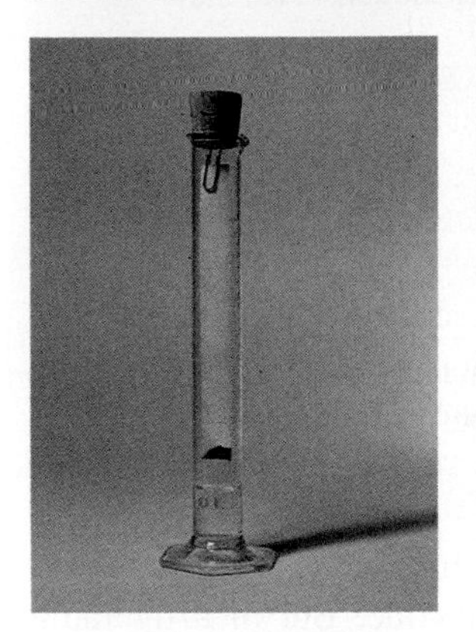

A black ink line is drawn at the bottom of a long strip of absorbent paper. The paper is suspended in a mixture of ethanol and water.

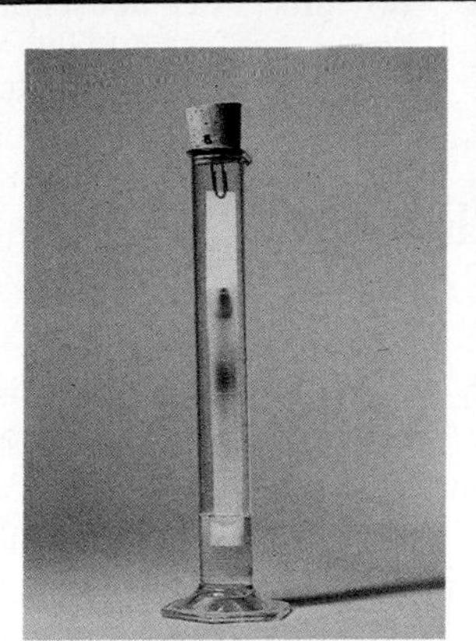

As the solvent is drawn up the paper, the ink separates into colored bands. Each color corresponds to a compound in the ink.

This shows the progress of the separation with time. The strips were removed from the solvent at different times during the separation.

others because they adsorb onto (stick to) the paper more strongly. Eventually several colored bands appear on the paper, corresponding to the various substances in the ink.

Paper chromatography is useful for identifying substances, but it is not very useful for the separation of large amounts of substances in a mixture. For this purpose, **liquid-column chromatography** can be used (Figure 1.13). Another form of chromatography is gas chromatography (Chapter 14).

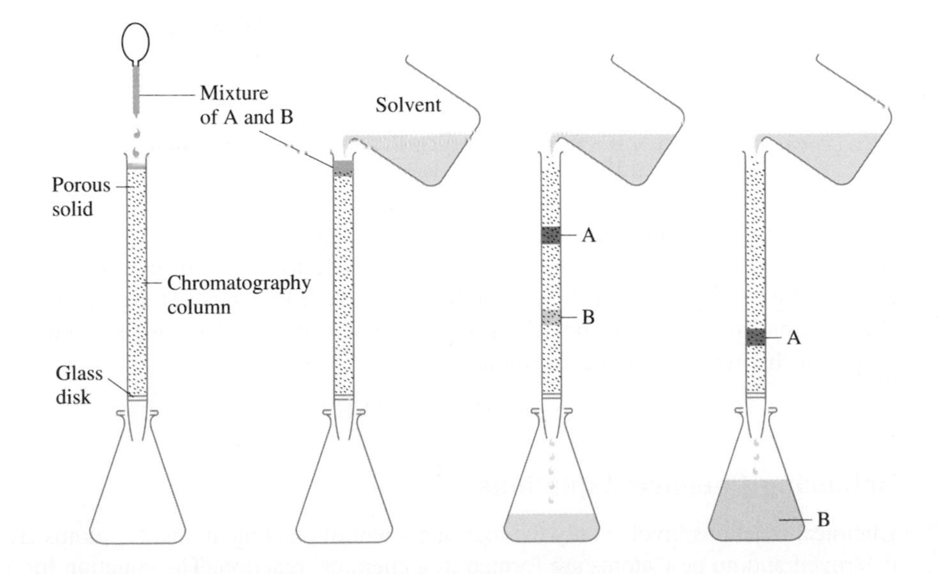

FIGURE 1.13 Liquid Column Chromatography The column is packed with an adsorbent material such as aluminum oxide, Al_2O_3. A solution of the mixture to be separated is poured into the top of the column. A suitable solvent is then poured slowly through the column. The solvent carries the different substances in the mixture down the column at different rates, so they separate into bands that can be readily identified if they are colored. Each band can then be separately washed out of the column into the receiving flask.

Exercise 1.7 Classify each of the following as a heterogeneous mixture (H) or as a homogeneous mixture (solution) (S).

(a) gasoline **(b)** smog **(c)** milk **(d)** household bleach **(e)** soil **(f)** vinegar **(g)** natural gas

1.5 CHEMICAL REACTIONS AND CHEMICAL EQUATIONS

When nitrogen reacts with oxygen in a lightning discharge in the atmosphere, nitrogen monoxide, NO, is formed. This reaction can be represented as

$$N_2 + O_2 \rightarrow 2NO$$

This equation represents the reaction of one molecule of nitrogen with one molecule of oxygen to produce two molecules of nitrogen monoxide. But in a lightning discharge, many millions of nitrogen molecules react with *exactly* the same number of oxygen molecules to give *exactly* twice this number of NO molecules. We therefore write the equation as

$$N_2(g) + O_2(g) \rightarrow 2NO(g)$$

In this reaction, the gaseous element nitrogen combines with the gaseous element oxygen and produces the gaseous compound nitrogen monoxide.

When methane, CH_4, burns in air, it reacts with oxygen to form carbon dioxide and water. This reaction can be represented as

$$CH_4 + 2O_2 \rightarrow CO_2 + 2H_2O$$

or as

$$CH_4(g) + 2O_2(g) \rightarrow CO_2(g) + 2H_2O(g)$$

The reaction occurs at a temperature above 100°C, so water is in the gaseous state.

These shorthand descriptions of chemical reactions are called **chemical equations**. The substances that we start with, called the **reactants**, are on the left side of the equation; the substances that are produced, called the **products**, are on the right side.

A *chemical reaction* is a process in which the atoms in the reactant molecules are rearranged to form the molecules of the products.

Making new molecules is a very important part of chemistry. Ammonia, $NH_3(g)$, is used as a fertilizer and for making other fertilizers. This compound can be made by heating nitrogen, $N_2(g)$, and hydrogen, $H_2(g)$, together at high pressure. We say that ammonia can be *synthesized* from nitrogen and hydrogen, and we represent this reaction by the equation

$$N_2(g) + 3H_2(g) \rightarrow 2NH_3(g)$$

Balancing Chemical Equations

Chemical reactions involve only the rearrangement of existing atoms. No atoms are destroyed and no new atoms are formed in a chemical reaction. The equation for a reaction must therefore have exactly the same number of atoms on the right side as on the left side—we say that the equation must *balance*. By counting the atoms on each side of the equations that we have given above, we see that they do indeed balance. In the equation for the burning of methane (natural gas) in air, there are one carbon atom, four hydrogen atoms, and four oxygen atoms on the left side and exactly the same numbers of each kind of atom on the right side. We have given examples of simple reactions, but no matter how complicated the reaction and no matter how large the molecules involved, there are always exactly the same numbers of each kind of atom at the end of the reaction as at the beginning.

Balancing chemical equations is not difficult, but a little practice is required. Consider the equation for the burning of methane in oxygen to give carbon dioxide and water. We first write the reactants and products as

$$\underbrace{CH_4 + O_2}_{\text{Reactants}} \rightarrow \underbrace{CO_2 + H_2O}_{\text{Products}} \quad \text{(Unbalanced)}$$

This equation is called an unbalanced equation because it correctly indicates the reactants and products but not the relative numbers of molecules of each. *Before any equation can be balanced, all the reactants and products and their correct formulas must be known.* If all this information is not available, then we cannot write a correct, balanced equation for the reaction.

The simplest procedure for balancing an equation is to consider first those elements that appear the least frequently in the equation. In this example hydrogen and carbon appear in only two formulas each, whereas oxygen appears three times. So we begin by balancing the numbers of carbon and hydrogen atoms. All the carbon in methane, CH_4, must be converted to carbon dioxide, CO_2, because no other product contains carbon. Thus one molecule of CH_4 must give one molecule of CO_2, as written. Each molecule of CH_4, however, contains four hydrogen atoms; and because all the hydrogen ends up in water molecules, two water molecules must be produced for each methane molecule. Hence we must place a 2 in front of the formula for water, to give

$$CH_4 + O_2 \rightarrow CO_2 + 2H_2O \quad \text{(Unbalanced)}$$

Now we can balance the remaining element, oxygen. We note that there are four oxygen atoms on the right-hand side of the equation: two in the CO_2 molecule and two in the two H_2O molecules. We must therefore place a 2 in front of the formula for oxygen, O_2, so that we have four oxygen atoms on both sides of the equation:

$$CH_4 + 2O_2 \rightarrow CO_2 + 2H_2O \quad \text{(Balanced)}$$

Generally the smallest whole numbers are used in writing balanced chemical equations, although any multiple of a balanced equation is correct.

We can now make a final check of the numbers of each kind of atom on both sides of the equation. We have one C atom, four H atoms, and four O atoms on both sides of the equation. The equation is therefore balanced.

Some remarks on incorrect ways of balancing equations might be helpful. The equation for the reaction between hydrogen and oxygen to give water,

$$H_2 + O_2 \rightarrow H_2O \quad \text{(Unbalanced)}$$

cannot be balanced by writing a 2 after the O in H_2O,

$$H_2 + O_2 \rightarrow H_2O_2 \quad \text{(Incorrect)}$$

This equation is balanced, but it is *not* the equation for the reaction between hydrogen and oxygen to give *water*. The product has been changed to hydrogen peroxide, and this equation now represents a *different reaction*.

Nor should the equation be balanced as

$$H_2 + O_2 \rightarrow H_2O + O \quad \text{(Incorrect)}$$

Although this equation is also balanced, an additional product that was not originally specified, O atoms, has been introduced. In balancing an equation, we cannot change the nature of either the reactants or the products. Coefficients must be inserted to balance the equation without introducing new products or reactants. The balanced equation for this reaction is

$$2H_2 + O_2 \rightarrow 2H_2O \quad \text{(Balanced)}$$

Finally, note that equations can be balanced only after all the products are known. Incomplete equations such as

$$H_2 + O_2 \rightarrow \quad \text{(Incomplete)}$$

cannot be balanced in an unambiguous way because no product is specified, and we have no way of choosing among the three possibilities discussed above. Do not be tempted to invent products simply to balance an equation. You *must* know all the reactants and products before you can write a balanced equation. The only way that we can be certain what the products of a reaction are is by carrying out appropriate experiments. After completing this course you should know the products of a variety of important reactions. You should also be able to make reasonable predictions about the products of many others by analogy with the reactions that you know. For the moment we are concerned primarily with balancing an equation once the products are known. In Exercise 1.8 the products of both reactions are given.

EXAMPLE 1.2 Balancing Chemical Equations

Balance each of the following equations.
(a) $Fe(s) + O_2(g) \rightarrow Fe_2O_3(s)$
(b) $C_3H_8(g) + O_2(g) \rightarrow CO_2(g) + H_2O(g)$

We usually avoid fractional coefficients in chemical equations, because $1\frac{1}{2}$ O_2, for example, could be taken to mean $1\frac{1}{2}$ oxygen molecules, but half an oxygen molecule is an oxygen atom, not an oxygen molecule.

Solution: **(a)** We first balance the Fe atoms,

$$2Fe(s) + O_2(g) \rightarrow Fe_2O_3(s) \text{ (Unbalanced)}$$

The O atoms can then be balanced by writing

$$2Fe(s) + \mathit{1\frac{1}{2}}O_2(g) \rightarrow Fe_2O_3(s) \text{ (Unbalanced)}$$

Finally, to remove the fractional number of O_2 molecules, we multiply both sides by 2 to give

$$4Fe(s) + 3O_2(g) \rightarrow 2Fe_2O_3(s) \text{ (Balanced)}$$

(b) We first balance the C atoms,

$$C_3H_8(g) + O_2(g) \rightarrow \mathit{3}CO_2(g) + H_2O(g) \text{ (Unbalanced)}$$

next the H atoms,

$$C_3H_8(g) + O_2(g) \rightarrow 3CO_2(g) + \mathit{4}H_2O(g) \text{ (Unbalanced)}$$

and then the O atoms,

$$C_3H_8(g) + \mathit{5}O_2(g) \rightarrow 3CO_2(g) + 4H_2O(g) \text{ (Balanced)}$$

Finally, to check our answers, we could count up the number of atoms on each side of each equation to ensure that the numbers of atoms are the same, which is indeed the case.

Exercise 1.8 Write balanced equations for the following.

(a) the reaction of carbon monoxide, $CO(g)$, and oxygen, $O_2(g)$, to give carbon dioxide, $CO_2(g)$

(b) the reaction of methane, $CH_4(g)$, and water vapor, $H_2O(g)$, at high temperature to give a mixture of carbon monoxide, $CO(g)$, and hydrogen gas, $H_2(g)$

1.6 THE STRUCTURE OF ATOMS

We have said that all the atoms of an element are of the same kind. In fact, they are not absolutely identical, as we will see shortly, but they are similar enough that we can speak of them all as the same kind: hydrogen atoms, for example, or iron atoms. But how do the atoms of one element differ from those of another? To answer this question, we must look at the internal structure of atoms.

Electrons, Protons, and Neutrons

The smallest and simplest atoms are those of hydrogen. The simplest type of hydrogen atom, shown in Figure 1.14, consists of two electrically charged particles. One of these is a positively charged particle called a **proton**, and the other is a negatively charged particle called an **electron**. The mass of the hydrogen atom is concentrated in the proton, which is 1836 times as massive as the electron. A proton has a mass of 1.6726×10^{-27} kg, whereas an electron has a mass of only 9.1096×10^{-31} kg (Table 1.3).

Distance in chemistry is measured in meters (m), mass is measured in kilograms (kg), and charge is measured in coulombs (C). These units are three of the internationally accepted units of measurement known as the SI system (Appendix C).

The electron moves around the proton, held by the electrostatic attraction of their opposite charges. The charge on the electron is -1.6022×10^{-19} C, and the charge on the proton is equal in magnitude but opposite in sign. Charges within atoms are generally expressed in terms of the charge on one electron, e:

$$e = 1.6022 \times 10^{-19} \text{ C}$$

Very large and very small numbers are conveniently represented by scientific notation, in which a large number such as 10,000 is written as 10^4, and a small number such as $\frac{1}{1000}$ is written as 10^{-3} (Appendix M).

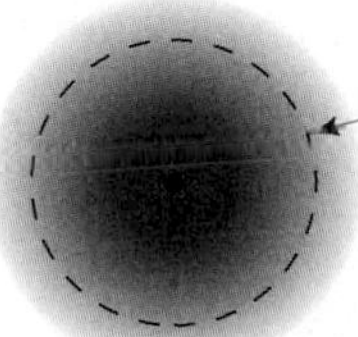

FIGURE 1.14 The Hydrogen Atom The simplest hydrogen atom consists of one (positively charged) proton and one (negatively charged) electron. The electron moves at high speed around the proton, but its exact path cannot be determined. If we could take a time-exposure photograph of a hydrogen atom, its electron would appear, because of its rapid motion, as a cloud of negative charge surrounding the proton. This cloud has no precise boundary but is much larger than the proton. A radius of 1.4×10^{-10} m would include 90% of the electron cloud. Atoms of other elements have more electrons and protons than hydrogen has. But in these atoms, too, the mass is concentrated in the nucleus, whereas the electrons account for most of the volume.

TABLE 1.3 Masses and Charges of Some Fundamental Particles

Particle	Mass (u)*	Charge (e)†
Proton	1.007 28	+1
Neutron	1.008 66	0
Electron	$5.485\,80 \times 10^{-4}$	−1

*1 u = $1.660\,54 \times 10^{-27}$ kg
†1 e = 1.6022×10^{-19} C

Common SI Prefixes*

Factor	*Prefix*	*Symbol*
10^3	kilo-	k
10^{-2}	centi-	c
10^{-3}	milli-	m
10^{-9}	nano-	n
10^{-12}	pico-	p

* See Appendix C for other SI prefixes.

An electron is said to have a charge of $-1e$, or -1. A proton is said to have charge of $+1e$, or $+1$.

As we shall see in Chapter 6, we cannot determine the path of the electron as it moves around the proton. We can say only that, as a result of its motion, the electron of a hydrogen atom effectively occupies a sphere of approximate radius 0.000 000 000 14 m = 1.4×10^{-10} m = 140 pm. Small as this sphere is, it is much larger than the proton, which has a radius of only 10^{-14} m. Instead of thinking of the electron as a moving point charge, it is often more convenient to picture it as a cloud of negative charge (Figure 1.14).

All atoms are similar to the one just described in that they consist of a positively charged particle surrounded by negatively charged electrons. The central particle of all atoms is called the **nucleus**. The nucleus of the simplest hydrogen atom consists of a single proton, but other nuclei are more complex. In every case, however, the nucleus is much smaller than the space occupied by the electrons: If an atom were magnified to the size of a hot-air balloon about 10 m in diameter, the nucleus would be a tiny speck of dust less than 1 mm in diameter. The positive charge of the nucleus is always equal to the negative charge of the surrounding electrons, so the atom is electrically neutral (has no net charge). Thus, the number of electrons surrounding the nucleus always equals the number of protons inside the nucleus. The famous experiment that led to this *nuclear model* of the atom is described in Box 1.2.

Protons and neutrons are not truly fundamental particles, in that they are composed of still smaller particles, such as quarks. This subject is still under active investigation by physicists.

An atom of the element helium has 2 electrons (total charge = −2) surrounding a nucleus with a charge of +2. From what we have said so far, we might expect the nucleus of this atom to consist of 2 protons and to have a mass twice that of a single proton. But the mass of the nucleus of the commonest type of helium atom is nearly *four times* the mass of a single proton. In addition to 2 protons, this nucleus contains 2 particles called **neutrons**. A neutron has very nearly the same mass as a proton but zero charge (Table 1.3).

BOX 1.2 Rutherford and the Nuclear Atom

The presently accepted model of the atom was first proposed by Ernest Rutherford (Figure A) in 1910 as a result of investigations into the newly discovered phenomenon of radioactivity. In that year, two of Rutherford's collaborators, Ernest Geiger (after whom the Geiger counter is named; Chapter 18) and his student Ernest Marsden, carried out an experiment on the scattering of α particles by thin metal foil. As we will see in Section 1.8, α particles are

Figure A Ernest Rutherford (1871–1937) *Rutherford, the grandson of Scottish immigrants and the second of 12 children, was born on a farm in New Zealand in 1871. Upon graduating from the University of New Zealand, he obtained a scholarship to Cambridge University, England, and began his research in the then-new field of radioactivity. At the age of 27 Rutherford was appointed professor of physics at McGill University, Montreal. In 1907 he moved to the University of Manchester, England, and in 1908 he was awarded the Nobel Prize in chemistry. Among his many discoveries were showing that α particles are the nuclei of helium atoms and finding the evidence that led to our present model of the atom. In 1919 Rutherford made another discovery of fundamental importance: He was the first to observe a nuclear reaction. By bombarding nuclei with subatomic particles, he changed the atoms of one element (nitrogen-14) into atoms of another element (oxygen-17), thus achieving the ancient dream of the alchemists. In 1919 Rutherford returned to Cambridge to head the internationally renowned Cavendish Laboratory.*

helium nuclei emitted by unstable (radioactive) nuclei of other atoms. Positively charged, they are emitted at very high speeds ($\simeq 10^7\ m \cdot s^{-1}$) and travel several centimeters before they are stopped by collisions with the molecules of the air. Alpha particles can pass right through very thin metal foil, but in so doing they are deflected from their original path by encounters with atoms in the foil (Figure B).

At the time of Geiger and Marsden's experiment, the model of the atom most favored by scientists was one proposed by J. J. Thomson in 1898. The atom was thought to consist of (negatively charged) electrons distributed through a uniform mass of positively charged matter. Because of the uniform distribution of the positive charge, atoms were expected to deflect each (positively charged) α particle by approximately the same small amount (Figure C). However, Geiger and Marsden found that although most of the α particles were indeed deflected through very small angles, a few were deflected through very large angles, and some even bounced right back toward the source. This result was so unexpected that Rutherford later remarked, "It was quite the most incredible event. . . . It was almost as if a gunner were to fire a shell at a piece of tissue and the shell bounced right back." Rutherford concluded that all the positive charge and almost all the mass of the atom must be concentrated in a very small particle situated at the center of the atom, which he called the nucleus. The rest of the space in the atom is occupied by the electrons, which, because of their very small mass, have little effect on the heavy α particles. Most of the α particles pass through this space at some distance from the nucleus, so they are deflected only slightly by the (positively charged) nucleus. However, the few α particles that pass close to the nucleus are repelled strongly and therefore deflected through large angles, and the very few that make a direct hit on the nucleus bounce straight back (Figure D).

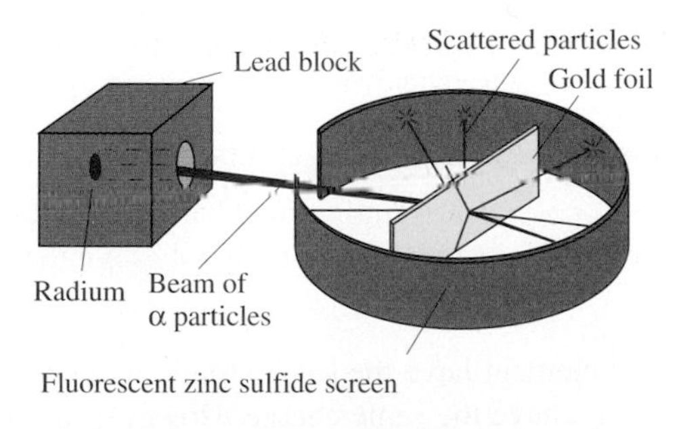

Figure B Rutherford's Apparatus for Studying Scattering of α Particles *Alpha particles emitted by radium are impeded by a lead block, but a small hole in the block allows a narrow beam of particles to pass through. This beam then passes through a very thin gold foil and strikes a screen coated with zinc sulfide. A momentary flash is observed on the screen whenever it is struck by an α particle. (The screen in a television set works on the same principle.)*

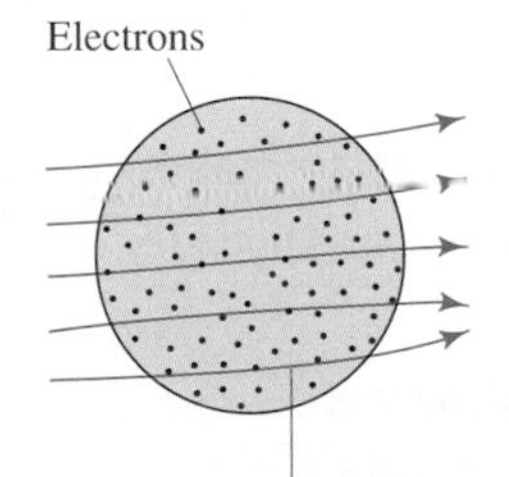

Figure C Scattering of α Particles according to Thomson's Model of the Atom *The α particles are deflected only slightly.*

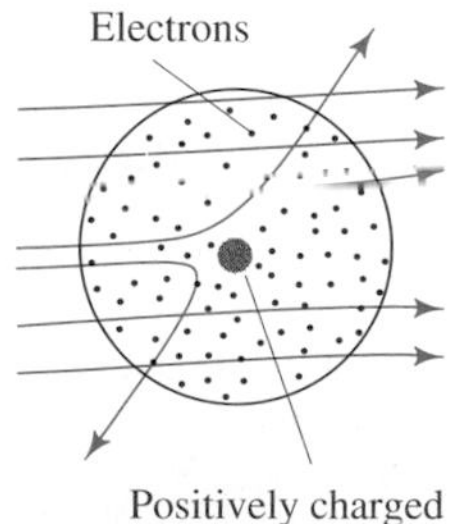

Figure D Scattering of α Particles according to Rutherford's Model of the Atom *A few α particles are deflected through large angles.*

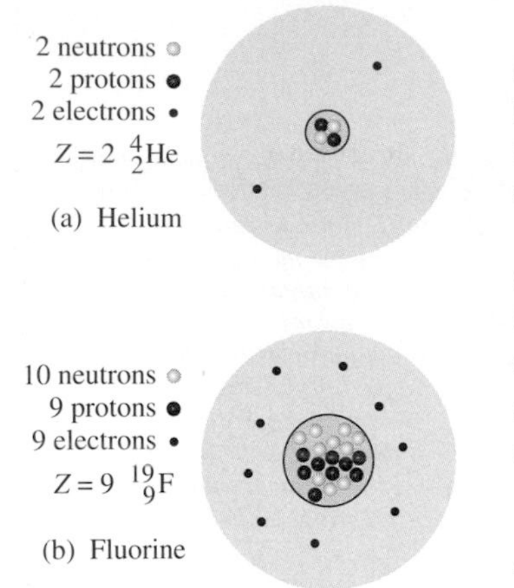

FIGURE 1.15 Structures of Helium and Fluorine Atoms These diagrams do not accurately represent the relative sizes of the electron cloud and the nucleus. They show only the numbers of particles of each kind.

With the exception of the simplest hydrogen nucleus, all nuclei are made up of protons and neutrons. The charge of a nucleus is determined by the number of protons it contains, but its mass is determined by the total number of protons and neutrons, collectively called **nucleons**. The helium nucleus shown in Figure 1.15a consists of four nucleons, giving it a mass approximately four times that of a proton.

The nucleus of a fluorine atom is shown in Figure 1.15b. This nucleus consists of 9 protons and 10 neutrons, so it has a charge of +9 and a mass approximately 19 times the mass of a proton. The fluorine atom has 9 electrons, which balance the charge of the 9 protons in the nucleus. *The atoms of every element are characterized by a specific number of electrons and an equal number of protons.* Only the number of neutrons varies, as we shall see.

Atomic Number and Mass Number

We can summarize the composition of any nucleus by just two numbers:

- The number of protons in the nucleus, called the **atomic number** Z.
- The total number of nucleons (protons plus neutrons), called the **mass number** A.

The difference between the mass number and the atomic number, $A - Z$, is equal to the number of neutrons in the nucleus.

The atomic number is written at the bottom left corner of the symbol for an atom, and the mass number is written at the top left corner: $^{A}_{Z}X$. Examples are $^{1}_{1}H$, $^{4}_{2}He$, $^{7}_{3}Li$, and $^{12}_{6}C$. The symbol $^{12}_{6}C$ indicates that there are 12 nucleons in the nucleus of a carbon atom, of which 6 are protons. From this we know that the carbon atom has $12 - 6 = 6$ neutrons.

Because each atom has a zero charge overall, its **nuclear charge**, $+Z$, must be balanced by the total charge of the surrounding electrons, each of which has a charge of -1. For example, a $^{7}_{3}Li$ atom ($Z = 3$) has a nuclear charge of +3 and 3 surrounding electrons. A $^{12}_{6}C$ atom ($Z = 6$) has 6 electrons surrounding the nucleus, and a $^{16}_{8}O$ atom has 8 electrons. It is the nuclear charge, represented by the atomic number (Z), that differentiates the atoms of one element from those of another.

An *element* may be defined as a substance whose atoms all have the same atomic number.

The 109 known elements are represented by the atomic numbers from 1 to 109 (Table 1, Appendix C).

Isotopes

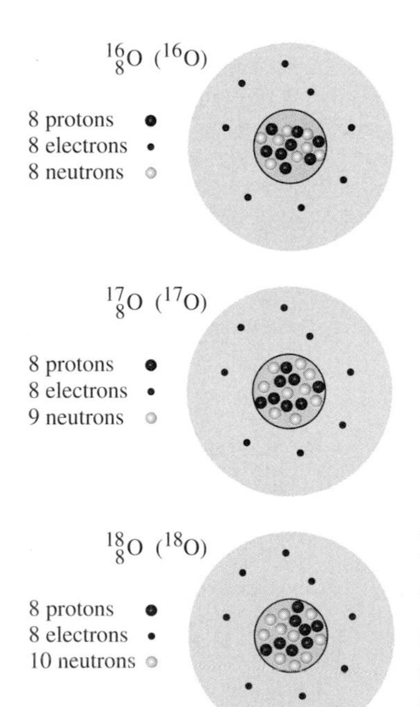

FIGURE 1.16 The Isotopes of Oxygen A diagrammatic representation of the three isotopes of oxygen. They differ only in the number of neutrons in the nucleus.

The nuclei of all the atoms of a given element have the same atomic number; they contain the same number of protons and have the same charge. However, all these nuclei do not necessarily have the same number of neutrons. For example, all oxygen atoms ($Z = 8$) have 8 protons in the nucleus and 8 electrons surrounding the nucleus. But three different types of stable oxygen atoms are known: $^{16}_{8}O$ atoms, with 8 protons and 8 neutrons in the nucleus; $^{17}_{8}O$ atoms, with 8 protons and 9 neutrons in the nucleus; and $^{18}_{8}O$ atoms, with 8 protons and 10 neutrons in the nucleus (Figure 1.16). Different oxygen atoms therefore have different masses, depending on the number of neutrons in their nuclei.

Atoms of an element that have different masses are called *isotopes*.

Elements occur in nature as mixtures of their isotopes. Some elements, such as $^{19}_{9}F$ and $^{31}_{15}P$, have just a single naturally occurring isotope. But many elements have two or more naturally occurring isotopes. Even the simplest element, hydrogen, has two stable isotopes, $^{1}_{1}H$ and $^{2}_{1}H$. The nucleus of the lighter isotope, which we described earlier, is a proton; the nucleus of the heavier isotope, $^{2}_{1}H$, contains a proton and a neutron. Consequently, the mass of the $^{2}_{1}H$ isotope is very nearly twice the mass of $^{1}_{1}H$. Helium also has two stable isotopes: $^{4}_{2}He$, which we described earlier, and a lighter isotope, $^{3}_{2}He$. Isotopes are distinguished by the mass number in their symbol, and they are normally not given different names. However, the $^{2}_{1}H$ isotope of hydrogen is an exception. It is called *deuterium* or heavy hydrogen and is sometimes given the symbol D.

Hydrogen also has an unstable (radioactive) isotope $^{3}_{1}H$, called tritium, T. Unstable isotopes are discussed in Section 1.8.

Because the atomic number of an element is implied by the symbol of that element, it is often omitted. The isotopes of oxygen, for example, are often represented simply by the symbols ^{16}O, ^{17}O, and ^{18}O, which are read as oxygen-16, oxygen-17, and oxygen-18. All these isotopes have the atomic number 8.

Atomic Mass

The mass of a single atom is much too small to be conveniently expressed in kilograms. For example, the mass of a hydrogen atom is only 1.6735×10^{-27} kg = 1.6735×10^{-24} g. The unit that is used to express **atomic mass** is the **atomic mass unit** (u).

The atomic mass unit (u) is equal to $\frac{1}{12}$ the mass of a single carbon-12 atom.

In biochemistry the atomic mass unit is generally called the **dalton** and given the symbol **Da**. This is a convenient and suitable name, but it has not been internationally adopted by chemists.

In other words, the mass of one $^{12}_{6}C$ atom is defined as *exactly* 12 u. Because the mass of one $^{12}_{6}C$ atom is 1.9926×10^{-23}g,

$$1 \text{ u} = \frac{1.9926 \times 10^{-23} \text{ g}}{12} = 1.6605 \times 10^{-24} \text{ g}$$

You may wonder how we can find the mass of a single atom, which is much too small to weigh. An important method for finding the masses of atoms uses a **mass spectrometer**, as described in Box 1.3.

Strictly speaking, atomic mass refers to the mass of an atom of a single isotope. However, most elements occur naturally as mixtures of two or more isotopes. The mass of a sample of an element depends on the relative abundance of each isotope in the sample. The mass of a hundred atoms of hydrogen could be anywhere between 100.783 u and 201.210 u, depending on the relative numbers of ^{1}H and ^{2}H atoms in the sample. During the earth's long geological history, however, the isotopes of most elements have become thoroughly mixed with one another, so the isotopic composition of most elements is constant throughout the surface of the earth. For example, the carbon in any sample of coal is 98.90% carbon-12 and 1.10% carbon-13, as is the carbon in CO_2 in a sample of air. Carbon-12 is said to have an abundance of 98.90% and carbon-13 an abundance of 1.10%. This means that out of every 1000 carbon atoms in any carbon-containing substance on the earth's surface, 989 are carbon-12 atoms and 11 are carbon-13 atoms.

The abundances of isotopes can be measured with a mass spectrometer (Box 1.3). Table 1.4 lists the masses and abundances of the isotopes of the elements with atomic numbers 1 to 19.

Because the isotopic composition of most elements is constant, most elements have a constant **average atomic mass**. This average value is often called simply the

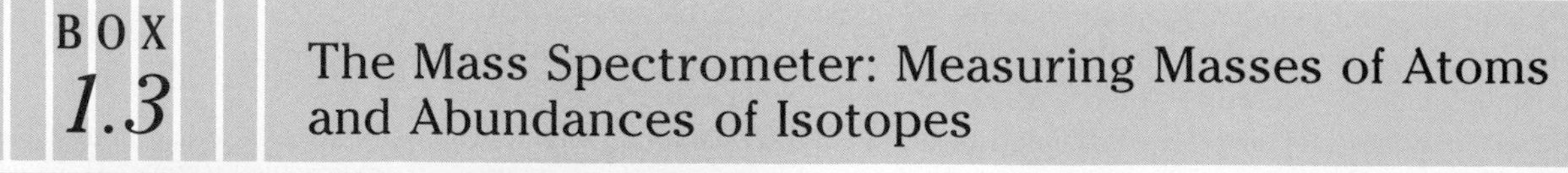

BOX 1.3 The Mass Spectrometer: Measuring Masses of Atoms and Abundances of Isotopes

Figure A J. J. Thomson (1856–1940)

In 1913 the British physicist J. J. Thomson (Figure A) was working at the Cavendish Laboratory in Cambridge, England, on methods to determine the masses of individual atoms. For this work he developed an instrument known as the *mass spectrometer.* It is based on the principle that if a force is applied at right angles to the path of a moving object, the force will change the object's direction of motion. A light object will be deflected from its original path more than a heavy object (Figure B).

Thomson found, to his surprise, that a sample of pure neon gas gave two deflected beams rather than just one. This discovery showed for the first time that neon contains atoms with two different masses; it was thus the first demonstration of the existence of isotopes.

The simple instrument invented by Thomson has been developed into the modern mass spectrometer, which is capable of measuring the abundances and masses of isotopes with great precision. Almost all the values given in a table of atomic masses have been determined by this method. Today the mass spectrometer is used primarily for the identification of substances and the analysis of mixtures (see Chapter 14).

In Figure C, a diagram of a mass spectrometer, the substance to be studied is introduced in the form of a gas at A. It passes into the region between the two metal plates B and C, which have a large voltage across them. High-energy electrons, which are injected at D, collide with the atoms or molecules of the gas. If the gas is a simple monatomic (single-atom) gas, such as neon or argon, one or more electrons are knocked out of the atoms and leave behind positively charged ions, such as Ne^+, Ar^+, and Ar^{2+}. These ions are attracted to the negatively charged plate C and are accelerated to high speeds. Some of the ions pass through the slit in that plate and form a narrow beam that then passes between the poles of a powerful magnet. The magnet deflects the beam into a circular path whose curvature depends on the mass of the ions and on their charge. Lighter ions are deflected more than heavier ions, and ions with greater charge are deflected more than ions with a single positive charge.

With a given strength of the magnetic field, only ions with a given value of Q/m, the ratio of charge to mass, pass through the slit E and reach the plate F, where they produce a measurable electric current. When the magnetic field is varied, ions with different values of Q/m reach plate F. Knowing the strength of the magnetic field needed to deflect the ions sufficiently to pass through slit E, we can determine their mass. And knowing the magnitude of the current produced at F by each ion of different mass, we can determine the relative abundances of the different ions.

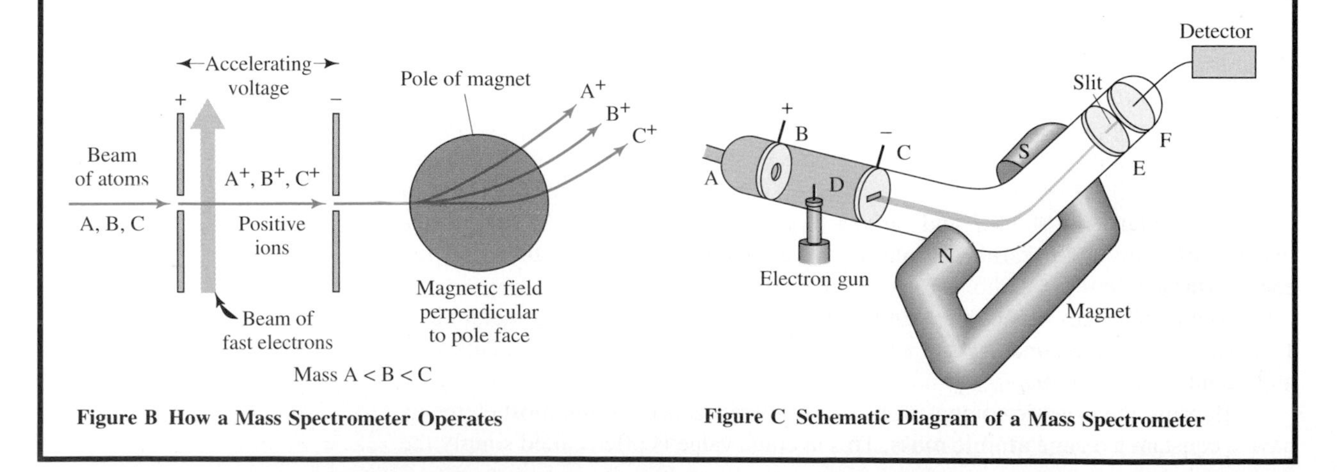

Figure B How a Mass Spectrometer Operates

Figure C Schematic Diagram of a Mass Spectrometer

TABLE 1.4 Mass and Natural Abundance of Some Isotopes

Isotope	Mass (u)	Percent Natural Abundance	Isotope	Mass (u)	Percent Natural Abundance
$^{1}_{1}H$	1.007 83	99.985	$^{23}_{11}Na$	22.989 77	100
$^{2}_{1}H$	2.014 10	0.015	$^{24}_{12}Mg$	23.985 04	78.99
$^{3}_{2}He$	3.016 03	0.000 14	$^{25}_{12}Mg$	24.985 84	10.00
$^{4}_{2}He$	4.002 60	99.999 86	$^{26}_{12}Mg$	25.982 59	11.01
$^{6}_{3}Li$	6.015 12	7.5	$^{27}_{13}Al$	26.981 53	100
$^{7}_{3}Li$	7.016 00	92.5	$^{28}_{14}Si$	27.976 93	92.23
$^{9}_{4}Be$	9.012 18	100	$^{29}_{14}Si$	28.976 50	4.67
$^{10}_{5}B$	10.012 94	19.9	$^{30}_{14}Si$	29.973 77	3.10
$^{11}_{5}B$	11.009 31	80.1	$^{31}_{15}P$	30.973 76	100
$^{12}_{6}C$	12.000 00	98.90	$^{32}_{16}S$	31.972 07	95.02
$^{13}_{6}C$	13.003 35	1.10	$^{33}_{16}S$	32.971 46	0.75
$^{14}_{7}N$	14.003 07	99.634	$^{34}_{16}S$	33.967 87	4.21
$^{15}_{7}N$	15.000 11	0.366	$^{36}_{16}S$	35.967 08	0.02
$^{16}_{8}O$	15.994 91	99.762	$^{35}_{17}Cl$	34.968 85	75.77
$^{17}_{8}O$	16.999 13	0.038	$^{37}_{17}Cl$	36.965 90	24.23
$^{18}_{8}O$	17.999 16	0.200	$^{36}_{18}Ar$	35.967 55	0.337
$^{19}_{9}F$	18.998 40	100	$^{38}_{18}Ar$	37.962 73	0.063
$^{20}_{10}Ne$	19.992 44	90.48	$^{40}_{18}Ar$	39.962 38	99.60
$^{21}_{10}Ne$	20.993 84	0.27	$^{39}_{19}K$	38.963 71	93.26
$^{22}_{10}Ne$	21.991 38	9.25	$^{40}_{19}K$	39.964 00	0.012
			$^{41}_{19}K$	40.961 83	6.73

atomic mass of the element, but it is not the mass of any one atom of that element (unless the element has only one isotope). For example, in a sample of chlorine, 75.77% of the atoms are chlorine-35 atoms with a mass of 34.97 u, and 24.23% are chlorine-37 atoms with a mass of 36.97 u. Expressing the percent abundances as fractions, we have the *fractional abundances* ^{35}Cl 0.7577 and ^{37}Cl 0.2423. We can now find the average mass of an atom of chlorine as follows:

Contribution of ^{35}Cl atoms	0.7577×34.97 u =	26.497 u
Contribution of ^{37}Cl atoms	0.2423×36.97 u =	8.958 u
Sum of contributions from ^{35}Cl and ^{37}Cl atoms	=	35.46 u

Thus, the average mass of one chlorine atom is 35.46 u.

In general, we calculate the average mass of an atom of an element by

1. Converting the percent abundance of each isotope to a fractional abundance (by dividing by 100);
2. Multiplying the fractional abundance of each isotope by its mass to give its contribution to the average mass; and
3. Finding the sum of the fractional contributions to the mass obtained in step 2.

Our calculated value differs slightly from the accepted value for the average atomic mass of chlorine, 35.453, because the values we use for the masses and abundances were rounded to 4 significant figures.

For elements that have a single isotope or a very constant isotopic composition, the average atomic mass can be given with great precision (that is, to a large number of significant figures). But for elements such as sulfur that have a slightly variable isotopic composition, the average atomic mass must be stated less precisely. For most purposes, including most of the calculations in this book, a value of the average atomic mass rounded to four significant figures is adequate. Values of the average atomic mass to four significant figures for the most common elements are given on the inside front cover. A complete table of average atomic masses is given in Appendix C.

The average mass of an atom of an element is often expressed *relative* to $\frac{1}{12}$ the mass of a ^{12}C atom rather than in atomic mass units, u. This **relative average atomic mass** is numerically the same as the average atomic mass, but it has no units. The average mass of a Cl atom is 35.45 times $\frac{1}{12}$ the mass of a ^{12}C atom, so the relative average mass of a Cl atom is 35.45. An element's relative average atomic mass is often called the *atomic weight*, but this is, strictly speaking, an incorrect usage, because weight is different from mass, as explained in Box 1.4. However, the term ''atomic weight'' has a long history and is still in common use.

EXAMPLE 1.3 Atomic Structure

For both $^{7}_{3}Li$ and $^{32}_{16}S$ atoms, give **(a)** the mass number, **(b)** the number of protons in the nucleus, **(c)** the number of neutrons in the nucleus, and **(d)** the number of electrons in a neutral atom.

Solution: For an isotope $^{A}_{Z}X$, A is the mass number, the total number of nucleons (protons plus neutrons), and Z is the atomic number (the number of protons). The number of neutrons is $A - Z$, and the number of electrons is equal to the number of protons, Z.

	$^{7}_{3}Li$	$^{32}_{16}S$
(a) *mass number, A*	7	32
(b) *number of protons, Z*	3	16
(c) *number of neutrons, A − Z*	4	16
(d) *number of electrons, Z*	3	16

For any neutral atom, it is always true that Z = number of protons = number of electrons. In later chapters we will see that for charged ions, Z = number of protons ≠ number of electrons.

Exercise 1.9 The nucleus of an atom of phosphorus contains 15 protons and 16 neutrons. What are **(a)** the charge on the nucleus, **(b)** the approximate mass relative to that of the proton, and **(c)** the number of electrons?

EXAMPLE 1.4 Average Atomic Mass from Isotopic Abundance

Naturally occurring oxygen consists of a mixture of the isotopes ^{16}O, ^{17}O, and ^{18}O in the relative abundances 99.762%, 0.038%, and 0.200%, respectively. Their masses are 15.994 91, 16.999 13, and 17.999 16 u. What is the average mass of an oxygen atom?

Solution: To find the average mass, we calculate the mass contributed by each isotope and then add these amounts. It is convenient first to express the percentage abundances as fractional abundances. For example, 99.762% abundance = 0.997 62 fractional abundance.

Mass number	Mass	Fractional abundance	Mass × Abundance
^{16}O	15.994 91 u	× 0.997 62	= 15.956 84 u
^{17}O	16.999 13 u	× 0.000 38	= 0.006 46 u
^{18}O	17.999 16 u	× 0.002 00	= 0.035 60 u
	Average mass per O atom		= 15.999 u

BOX 1.4 Mass and Weight

The terms *"mass"* and *"weight"* are often used interchangeably, but they are not the same. We can describe mass very loosely as the amount of matter in an object. More strictly speaking, we say that an object's **mass** is the quantity that measures its resistance to a change in its state of rest or motion. The greater an object's resistance to such change (that is, the greater its mass), the greater the force that must be applied to bring about the change. A rocket needs a much greater applied force than a pebble to set it in motion if it is at rest or to change its motion if it is already moving.

Mass can be measured by the force necessary to give an object a certain acceleration. On earth we use the force of gravitational attraction between the earth and another object to measure the object's mass. We call this gravitational force the **weight** of the object. Weight (on earth) depends on the mass of the object, the mass of the earth, and the distance of the object from the center of the earth. In SI, *mass* is measured in kilograms, but weight, which is a force, is measured in units called *newtons*.

A given object has the same mass everywhere. But if that object were on a planet with a mass different from that of the earth, it would be subject to a different gravitational force and would therefore have a different weight. Astronauts were able to bound around with ease on the surface of the moon because, although their mass had not changed, their weight was considerably less than on earth. Even on earth, an object can have different weights in different locations on the earth's surface. Although the mass of the object and the mass of the earth are constant, the object's distance from the center of the earth may vary slightly—from the top of a mountain to the bottom of a valley, for example—so the object's weight will vary slightly, but its mass does not change.

The mass of an object is determined by comparison with a set of standard masses. Because the weights of two objects of equal mass are the same at any one place on the earth's surface, these objects balance each other when placed on the pans of a scale (balance) with equal arms. We call this procedure weighing, that is, the determination of weight, but it is really the determination of mass.

Figure A *The astronaut on the moon has the same mass as on earth, but he weighs much less. Thus, he can move easily, despite his bulky clothing and equipment.*

The terms "mass" and "weight" are frequently confused in everyday conversation, so we speak most often of a weight of 1 kg rather than a mass of 1 kg (Figure A). It is unfortunate, however, that scientists continue to use the inaccurate term "atomic weight" in place of the more accurate "relative average atomic mass." Scientists, like everyone else, tend to resist change, and "atomic weight" is short and familiar, whereas "relative average atomic mass" is long and less familiar. In this book we have tried to keep things simple by expressing the masses of atoms in atomic mass units (u) so that we do not need the term "relative." Moreover, we assume that when we are talking about a naturally occurring element, we mean the natural mixture of isotopes—thus, we do not need to use the term "average" each time.

Exercise 1.10 Boron has isotopes ^{10}B and ^{11}B with relative abundances of 19.9% and 80.1% and masses of 10.012 94 u and 11.009 31 u, respectively. What is the average mass of a boron atom?

1.7 THE MOLE: COUNTING BY WEIGHING

When carbon is strongly heated with oxygen, it gives carbon dioxide. When carbon is strongly heated with sulfur, it gives carbon disulfide, CS_2, in a similar reaction:

$$C(s) + 2S(s) \rightarrow CS_2(l)$$

Suppose that we wish to know how much carbon and sulfur to heat together and how much product to expect. We have

C	+	2S	→ CS_2
1 atom		2 atoms	1 molecule

From the table of average atomic masses, we find that carbon has an atomic mass of 12.01 u and that sulfur has an atomic mass of 32.07 u. The mass of the carbon disulfide molecule, its **molecular mass**, is the sum of the masses of its component atoms: 12.01 u + 2(32.07) u = 76.15 u. Thus, we know that 12.01 u of carbon combines with 64.14 u of sulfur to give 76.15 u of carbon disulfide:

1 atomic mass C (12.01 u) + 2 atomic masses S (64.14 u)
→ 1 molecular mass CS_2 (76.15 u)

Of course, we cannot weigh out individual atoms and molecules. Suppose that we start with one hundred molecules of each reactant:

100 C atoms + 200 S atoms → 100 CS_2 molecules

100 atomic masses C (1201 u) + 200 atomic masses S (6414 u)
→ 100 molecular masses CS_2 (7615 u)

Because 1 u = 1.6605×10^{-24} g, these are still exceedingly small masses; we will never encounter samples this small in the chemistry laboratory. The mass of a single atom is so tiny that the smallest sample we ever work with contains vastly more than 100 atoms. If we want to mix together sulfur atoms and carbon atoms in a 2:1 ratio in the laboratory, we need a simple way of ''counting out'' an extremely large number of atoms.

The number 6.022×10^{23} is awe-inspiringly large. A pile of 6.022×10^{23} pages (1 mol of sheets of paper) would form a column 6.1×10^{16} km high, about 1000 times the distance to the nearest star. Avogadro's number of seconds, 6.022×10^{23} s, is 2×10^{14} centuries, roughly a million times the best estimates of the age of the universe!

The SI symbol for mole is mol.

Chemists have devised a unit for measuring large numbers of atoms or molecules, based on *the number of atoms in an exactly 12-g sample of carbon-12.* Let us calculate how many atoms that is. Each carbon-12 atom has a mass of $12(1.6605 \times 10^{-24}$ g) = 1.9926×10^{-23}g, so there are 12 g/1.9926×10^{-23} g = 6.022×10^{23} atoms in 12 g of carbon-12. This enormous number is called **Avogadro's number, N_A**, and the work of the chemist for whom it was named, Amedeo Avogadro, is described in Chapter 2. The amount of carbon-12 that contains Avogadro's number of atoms is called a **mole** of carbon-12. A mole of carbon-12 therefore has a mass of exactly 12 g. Although the mole is defined in terms of ^{12}C atoms, it applies to all atoms and molecules and indeed, to any entity.

A mole of any substance is the amount of that substance that contains Avogadro's number of entities (such as atoms, molecules, or ions).

Thus, a mole is a counting unit, like a pair or a dozen:

1 pair of shoes is 2 shoes;
1 dozen eggs is 12 eggs;
1 dozen oranges is 12 oranges;
1 mol of C atoms is 6.022×10^{23} C atoms;
1 mol of S atoms is 6.022×10^{23} S atoms;
1 mol of CS_2 molecules is 6.022×10^{23} CS_2 molecules.

FIGURE 1.17 The Mole Clockwise from bottom: One mole each of mercury atoms (200.6 g), copper atoms (63.55 g), water (H_2O) molecules (18.01 g), sulfur (S_8) molecules (256.5 g), and sugar (sucrose, $C_{12}H_{22}O_{11}$) molecules (342.3 g).

If one egg has a mass of 30.0 g, then a dozen eggs has a mass of 12 × 30.0 g. One atom of ^{12}C has a mass of 12 u, and Avogadro's number of ^{12}C atoms—that is 1 mol—has a mass of $12 \times 1.6605 \times 10^{-24}$ g $\times 6.022 \times 10^{23} = 12$ g. Similarly, one S atom has a mass of 32.07 u, so Avogadro's number of S atoms—1 mol of S atoms—has a mass of $32.07 \times 1.6605 \times 10^{-24}$ g $\times 6.022 \times 10^{23} = 32.07$ g (Figure 1.17). The average mass of one carbon atom (that is, of the ^{12}C and ^{13}C atoms in naturally occurring carbon) is 12.01 u, so 1 mol of carbon atoms has a mass of 12.01 g.

The mass of 1 mol of a substance is called the *molar mass* of the substance.

It is the mass of Avogadro's number of the entities (such as atoms or molecules) of which the substance is composed. Molar mass is expressed in grams per mole. The molar mass of carbon-12 atoms is 12 $g \cdot mol^{-1}$, the molar mass of naturally occurring carbon is 12.01 $g \cdot mol^{-1}$, and the molar mass of CS_2 molecules is 76.15 $g \cdot mol^{-1}$.

Because we know that 1 atom of C reacts with 2 atoms of S to produce 1 molecule of CS_2 and that 100 atoms of C react with 200 atoms of S to produce 100 molecules of CS_2, we know also that Avogadro's number of C atoms react with 2 × Avogadro's number of S atoms to produce Avogadro's number of CS_2 molecules. We express this more briefly by saying that 1 mol of C atoms reacts with 2 mol of S atoms to produce 1 mol of CS_2 molecules:

$$C(s) \quad + \quad 2S(s) \quad \rightarrow \quad CS_2(l)$$

$$N_A \text{ C atoms} + 2N_A \text{ S atoms} \rightarrow N_A \text{ CS}_2 \text{ molecules}$$

$$1 \text{ mol C atoms} + 2 \text{ mol S atoms} \rightarrow 1 \text{ mol CS}_2 \text{ molecules}$$

▮▮▮ A balanced chemical equation describes not only the relationship between *atoms* (or *molecules*) of reactants and products, but also that between moles of reactants and products. Knowing the molar mass of each reactant and product, we can calculate the grams of reactants and products.

We know the mass of 1 mol each of carbon atoms, sulfur atoms, and CS_2 molecules, so we can now write

$$12.01 \text{ g C}(s) + 64.14 \text{ g S}(s) \rightarrow 76.15 \text{ g CS}_2\ (l)$$

If we heat a mixture of 12.01 g of carbon and 64.14 g of sulfur, we expect to obtain 76.15 g of CS_2.

When we use moles and molar masses, we should always specify the entities that we are talking about, that is, whether they are atoms, molecules, or some other entity. The molar mass of hydrogen *atoms* is the mass of 1 mol of H atoms, or

1.008 g · mol^{-1}. The molar mass of hydrogen *molecules* is the mass of 1 mol of H_2, or 2.016 g · mol^{-1}. However, chemists often speak of the molar mass of water rather than the more specific molar mass of water molecules (18.02 g · mol^{-1}), because it is well known that water consists of H_2O molecules. We might speak of the molar mass of oxygen as 32.00 g · mol^{-1}, because oxygen normally consists of O_2 molecules. However, we must be careful to specify the entities if there can be *any* doubt about the particular entity with which we are concerned. It is essential to specify the entities if they are not those of which the substance is composed under ordinary conditions: for example, if we are concerned with O or H atoms rather than O_2 or H_2 molecules.

When we are dealing with a network substance (a giant molecule, or nonmolecular compound) such as silicon dioxide, we use the empirical formula as the entity for expressing the molar mass. The empirical formula of silicon dioxide is SiO_2. One empirical formula of SiO_2 has a mass of 60.09 u, so the molar mass of SiO_2 is 60.09 g · mol^{-1}. A mole of silicon dioxide is a 60.09-g piece of a giant SiO_2 network or 60.09 g of a number of smaller SiO_2 crystals.

Iron sulfide, FeS (Demonstration 1.2), is a nonmolecular compound that consists of an infinite array of iron and sulfur ions arranged in a regular pattern. The empirical formula of iron sulfide is FeS, and its molar mass is (molar mass of Fe) + (molar mass of S) = 55.85 g · mol^{-1} + 32.07 g · mol^{-1} = 87.92 g · mol^{-1}.

It is very easy to find the molar mass of the atoms of any element:

If the average atomic mass of an atom of an element is x u, then the molar mass of the atoms of that element is x g · mol^{-1}.

Similarly,

If the mass of a molecule or of an empirical formula is x u, then the molar mass of a substance represented by the molecular or empirical formula (usually a network substance) is x g · mol^{-1}.

EXAMPLE 1.5 Conversion of Grams to Moles

How many moles of aluminum atoms are in 50.00 g of aluminum?

Solution: We can summarize the problem as follows:

$$50.00 \text{ g Al} = x \text{ mol Al atoms}$$

Because the atomic mass of aluminum is 26.98 u, 1 mol Al atoms has a mass of 26.98 g, so we can use the unit conversion factor

$$\frac{1 \text{ mol Al atoms}}{26.98 \text{ g Al}} = 1$$

This conversion factor enables us to convert grams to moles:

$$x \text{ mol Al atoms} = (50.00 \cancel{\text{g Al}})\left(\frac{1 \text{ mol Al atoms}}{26.98 \cancel{\text{g Al}}}\right) = 1.853 \text{ mol Al atoms}$$

The numerator of the unit conversion factor contains the desired units (here, mol Al), whereas the denominator contains the *units to be removed* (here, g Al). Desired units = units to be removed × unit conversion factor.

EXAMPLE 1.6 Empirical Formula Mass, Molecular Mass, and Molar Mass

What are the empirical formula mass, the molecular mass, and the molar mass of butane, $C_4H_{10}(g)$?

Solution: The empirical formula of butane is C_2H_5, and its molecular formula is C_4H_{10}. Thus, the empirical formula mass is {2(12.01) + 5(1.008)} u = 29.06 u. The molecular mass is {4(12.01) + 10(1.008)} u = 58.12 u, which could equally well have been calculated by using the relationship molecular mass = 2(empirical formula mass). The molar mass is 58.12 g · mol^{-1}.

A molecular mass is never smaller than an empirical formula mass.

Exercise 1.11 How many moles of water molecules, and how many molecules, are there in 1.00 g of water?

Exercise 1.12 What are the molar masses of each of the following molecules?

(a) CH_4 **(b)** H_2O_2 **(c)** S_8 **(d)** $C_{12}H_{22}O_{11}$

Exercise 1.13 What are the molar masses of silicon dioxide and sodium chloride?

1.8 UNSTABLE ATOMS: RADIOACTIVITY

So far we have dealt only with stable atoms. By *stable* we mean that the nuclei of these atoms do not change spontaneously into nuclei of other kinds of atoms. All but two of the elements with atomic numbers of 83 (bismuth) or less have one or more stable isotopes. But the nuclei of all the isotopes of elements with atomic numbers greater than 83 are *unstable*, that is, they decompose spontaneously into nuclei of other elements.

The two elements with $Z < 83$ that have no stable isotopes are technetium ($Z = 43$) and promethium ($Z = 61$).

Radioactivity

The spontaneous disintegration of a nucleus is called *radioactivity*, and an isotope with atoms containing an unstable nucleus is said to be *radioactive*.

An example of a radioactive isotope is uranium-238. Uranium-238 nuclei emit helium nuclei, which for historical reasons are called **α particles**. A uranium nucleus is thereby transformed into a thorium nucleus:

α is the Greek letter alpha.

$$^{238}_{92}U \rightarrow ^{234}_{90}Th + ^{4}_{2}He \text{ (α particle)}$$

Such a reaction, in which one type of nucleus is transformed into another, is an example of a **nuclear reaction**. It is quite different from a chemical reaction, in which atoms are rearranged but not changed. In a nuclear reaction, the nucleons (protons and neutrons) comprising the nuclei of one or more atoms are rearranged into new nuclei, but their total number remains constant.

Not all radioactive nuclei emit α particles—some emit electrons. This may seem strange, because there are no electrons in nuclei. An electron emitted in a nuclear reaction comes from the disintegration of a neutron to give a proton (which is retained in the nucleus) and an electron. Thorium-234, which is produced by radioactive disintegration of uranium-238, emits electrons and is transformed into protactinium-234:

$$^{234}_{90}Th \rightarrow ^{234}_{91}Pa + ^{0}_{-1}e$$

In equations for nuclear reactions, the electron is written as $^{0}_{-1}e$. The subscript is the charge of the electron, and the superscript is its mass number. The mass of the electron is not zero, but it is much smaller than that of the proton, which has a mass number of 1, so the electron is given a mass number of zero. Emitted electrons were originally called **β particles**, and the electron emission is still often called β emission.

In nuclear reactions, the sum of the mass numbers of all reactants equals the sum of the mass numbers of all products.

β is the Greek letter beta.

There are still other types of radioactivity that we discuss in more detail in Chapter 18.

The Stability of Nuclei

What makes some nuclei stable and others unstable? To answer this question, we must first consider why any nucleus with more than one proton should hold together at all. Like charges repel one another: Electrons repel electrons, and protons repel protons. The distances between the particles in nuclei are very small—less than 10^{-15} m—so the electrostatic repulsion among the protons is extremely large. Most nuclei do not fly apart, however, because there are also very strong attractive forces, called **nuclear forces**, between nucleons. As the number of protons increases, the electrostatic repulsion between them increases, so a higher proportion of neutrons is needed to provide additional attractive forces to hold the nucleus together.

Let us plot the number of neutrons ($A - Z$) against the number of protons (Z) for all the known stable nuclei (Figure 1.18). We see that all these nuclei fall in a narrow band, which we call the **band of stability**. For stable nuclei with low atomic numbers, the number of protons is equal to the number of neutrons $(A - Z)/Z = 1$, but as Z increases, the number of neutrons exceeds the number of protons. The ratio of neutrons to protons reaches a value of approximately 1.5 for the heaviest stable nuclei. For elements heavier than bismuth (that is, $Z > 83$), no increase in the number of neutrons is sufficient to hold the nucleus together. Thus, all nuclei with more protons than bismuth are unstable, although some disintegrate very slowly and others very rapidly (Chapter 18).

Exercise 1.14 The unstable hydrogen-3 isotope (tritium) disintegrates with the emission of an electron (β particle) to give an isotope of helium. Write the equation for this nuclear reaction, and name the helium isotope.

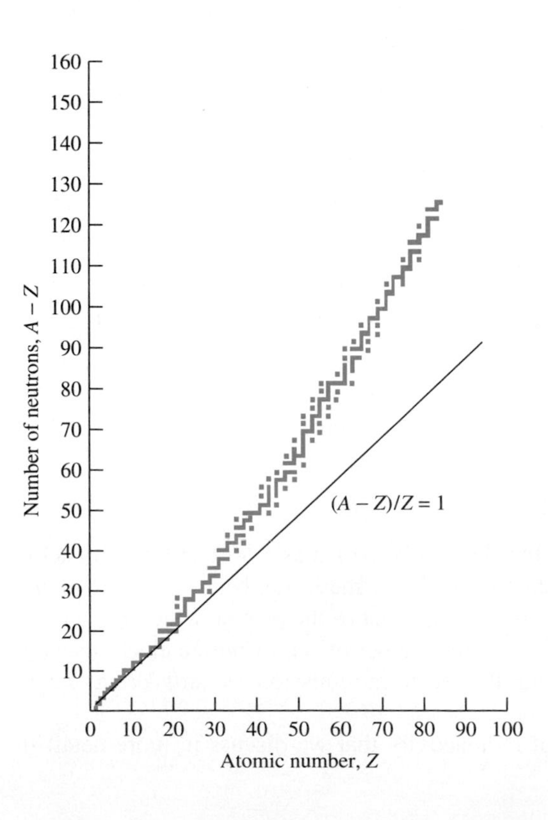

FIGURE 1.18 Plot Showing the Band of Stability for Nuclei All the stable nuclei lie in a narrow band called the band of stability. For the light nuclei $A - Z/Z = 1$, but heavier stable nuclei contain more neutrons than protons.

1.9 CONSERVATION OF MASS AND ENERGY

We might think that the mass of an atom would be equal to the sum of the masses of its electrons, protons, and neutrons. But it is not. Consider, for example, a ^4_2He atom. The sum of the masses of its electrons, protons, and neutrons is

$$\begin{aligned} 2 \text{ protons} &= 2 \times 1.00728 \text{ u} = 2.01456 \text{ u} \\ 2 \text{ neutrons} &= 2 \times 1.008\,66 \text{ u} = 2.017\,32 \text{ u} \\ 2 \text{ electrons} &= 2 \times 0.000\,55 \text{ u} = \underline{0.001\,10 \text{ u}} \\ & \qquad\qquad\qquad\quad 4.032\,98 \text{ u} \end{aligned}$$

This sum is greater than the experimentally determined mass of a helium-4 atom, which is 4.002 60 u. The difference of 0.030 38 u is called the **mass defect**. A similar mass defect is found for all atoms: The sum of the masses of the protons, neutrons, and electrons of which an atom is composed is always greater than the experimentally determined mass.

Because the forces holding the protons and the neutrons together in a nucleus are very strong, a very large amount of energy is released when protons and neutrons combine to form a nucleus. Conversely, this amount of energy must be supplied to pull the nucleus apart into its constituent particles. This amount of energy is so large that it has a significant **mass equivalent**. In other words, when a new nucleus is formed, mass is converted into energy, and the amount of mass lost is large enough to be detectable. The mass equivalent is given by the equation

The relationship between force and energy is discussed in Appendix P.

$$\Delta E = \Delta mc^2$$

where ΔE is the energy evolved (released), Δm is the mass lost (the mass defect), and c is the speed of light. Nuclear reactions, in which nuclei are either built up from or broken down into simpler nuclei, are the origin of the enormous quantities of energy produced by nuclear reactors and nuclear explosions (Chapter 18).

The important equation $\Delta E = \Delta mc^2$, proposed by Albert Einstein in 1905 and subsequently verified in many experiments, gives the relationship between the two fundamental concepts by which we describe the universe, mass and energy.

The forces holding atoms together in molecules are much weaker than those holding protons and neutrons together in nuclei. Therefore, the energy changes accompanying the rearrangement of atoms in chemical reactions are too small for the accompanying mass changes to be detectable. For example, when 1 metric ton (10^3 kg) of gasoline is burned, the large amount of energy released is equivalent only to 5×10^{-9} kg, which is much too small a mass to be detected experimentally. For all practical purposes, the total mass of products and reactants is conserved (remains the same) in a chemical reaction. One of the earliest-discovered laws of chemistry was the **law of conservation of mass**. Formulated more than 200 years ago by the French chemist Antoine Lavoisier (Figure 1.19), this law states that

In a chemical reaction, mass is conserved.

FIGURE 1.19 Antoine Laurent Lavoisier (1743–1794) Lavoisier, the son of a wealthy Parisian lawyer, qualified as a lawyer in 1764 but became interested in science and devoted much of his life to research in chemistry. He is shown here with his wife, Marie-Anne Pierrette, who assisted him in his experiments. Fascinated by the phenomenon of burning, he was the first to recognize clearly that air consists of two gases, one that does not support combustion (nitrogen) and the other that does. Lavoisier called this latter gas oxygen and categorized it as an element. One of the first chemists to recognize the importance of accurate quantitative measurements, and his accurate determination of the masses of substances formed in reactions led him to the law of conservation of mass.

But in a nuclear reaction, as we have seen, measurable amounts of mass are transformed into energy (which is transferred to the surroundings), or energy from the surroundings is transformed into measurable amounts of mass. In these situations, mass is not conserved; only the total mass plus energy is conserved. This statement expresses the **law of conservation of mass and energy**.

SUMMARY

Chemistry is the study of the composition, properties, and chemical reactions of all known substances. Elements are substances composed of only one kind of atom. Each element has its own symbol, and 109 elements are known. A few elements exist as single, isolated atoms, but in most atoms the elements are combined to form molecules. A compound, a large collection of identical molecules, has molecules containing two or more different elements (kinds of atoms). A molecular formula gives the number of atoms of each element in one molecule. A structural formula shows how the atoms are joined (bonded) and is sometimes written to indicate the molecular geometry (three-dimensional shape). A compound has a fixed constant composition. H_2O is written for *one* water molecule; any sample of *bulk* water contains immense numbers of molecules and is represented by $H_2O(l)$, $H_2O(s)$, or $H_2O(g)$ depending on whether it is liquid, solid (ice), or gas (water vapor). An empirical formula is the simplest formula that depicts the ratios of the different kinds of atoms in a substance. Substances with a giant network of atoms in which no individual molecules can be recognized, such as silica, SiO_2, are customarily represented by their empirical formulas.

Every substance has a distinct set of chemical and physical properties that distinguish it from all others. Physical properties are properties that can be observed and measured without the substance changing into other substances. Chemical properties describe the participation of a substance in chemical reactions where it is changed into other substances. A substance contaminated by others is impure. Absolutely pure substances are unobtainable in practice; in chemistry "pure" is taken to mean that other substances are present only in insignificant amounts. Purity is expressed in terms of mass percentage.

Mixtures of two or more substances are heterogeneous (with diverse properties in different parts) or homogeneous (with uniform properties throughout but with no required fixed composition). Homogeneous mixtures are solutions. The separation of mixtures utilizes one or more different physical properties of their components. Separation methods include filtration, simple distillation, fractional distillation, and various kinds of chromatography.

A chemical equation is a shorthand representation of a chemical reaction that gives the reactants on the left side and the products on the right side. In a balanced equation, the total numbers of different atoms in the formulas appearing on both sides are conserved.

The three important constituents of atoms are the proton, with a charge of $+e$ $(+1)$; the neutron, with almost the same mass as the proton and a charge of 0; and the electron, with a negligible mass compared to the proton or neutron and a charge of $-e$ (-1). Virtually all the mass of the atoms of each element is concentrated in a very small, central nucleus. A nucleus with mass number A (the total number of nucleons—that is, protons plus neutrons) contains Z protons and $A - Z$ neutrons. A neutral atom has Z electrons surrounding its nucleus; these electrons determine the chemical behavior. An element has atoms all with the same atomic number, Z. Atoms with the same Z but different mass (nucleon) numbers A differ only in the numbers of nuclear neutrons and are isotopes of the same element. In the symbols of isotopes, such as ${}^{35}_{17}Cl$ and ${}^{37}_{17}Cl$, the numerical subscript is the atomic number, and the superscript is the mass number. The masses of atoms are measured in atomic mass units, u ($\frac{1}{12}$ the mass of one carbon-12 atom). The average atomic mass of an element depends on the natural abundances of its isotopes, which are usually invariable in samples of elements or compounds on earth, and thus the atoms of most elements have constant average masses (at least to 4 significant figures).

The mass of a molecule, its molecular mass, is given by the sum of the masses of its atoms. A mole (mol), the SI unit used for measuring the amount of a substance, is Avogadro's number (6.022×10^{23}) of entities (atoms, molecules, empirical formula units, or ions). A mole of entities has the same numerical mass in grams as the mass of one entity in atomic mass units, u. The mass of 1 mol of a substance is its molar mass.

Radioactivity is the spontaneous disintegration of unstable (radioactive) nuclei. Two important processes are $^{4}_{2}He$-nucleus (α-particle) emission and electron, $^{0}_{-1}e$, (β-particle) emission. There are no stable isotopes of the elements with $Z > 83$.

The difference between the mass of the constituent electrons, protons, and neutrons of an atom and its actual mass is its mass defect, Δm, which equals the very large amount of energy, ΔE, released when the protons and neutrons combine to form the nucleus, given by Einstein's equation $\Delta E = \Delta mc^2$. In nuclear reactions, the sum of mass and energy is conserved, but in chemical reactions the mass equivalent of any energy change is negligible, and mass is conserved; in other words, the atoms are conserved.

IMPORTANT TERMS

alpha (α) particle (page 33)
atom (page 1)
atomic mass (page 25)
atomic mass unit (page 25)
atomic number (page 24)
Avogadro's number (page 30)
band of stability (page 34)
beta (β) particle (page 33)
chemical equation (page 18)
chemical property (page 10)
chemical reaction (page 18)
chromatography (page 16)
compound (page 4)
distillation (page 15)
electron (page 21)
element (pages 2, 24)
empirical formula (page 6)
filtration (page 15)
heterogeneous mixture (page 12)
homogeneous mixture (solution) (page 12)
impurity (page 11)
isotope (page 24)
law of conservation of mass (page 35)
law of conservation of mass and energy (page 36)
mass defect (page 35)
mass number (page 24)
mass spectrometer (page 25)
mixture (page 12)
molar mass (page 31)
mole (page 30)
molecular formula (page 3)
molecular mass (page 30)
molecule (page 1)
neutron (page 22)
nuclear charge (page 24)
nuclear force (page 34)
nuclear reaction (page 33)
nucleon (page 24)
nucleus (page 22)
physical property (page 10)
proton (page 21)
product (page 18)
pure substance (page 11)
radioactivity (page 33)
reactant (page 18)
solution (page 12)
structural formula (page 7)
substance (page 9)

REVIEW QUESTIONS

The review questions are designed to define the principal objectives of each chapter and to test the extent of your understanding and knowledge of the material covered. After studying each chapter, you should be able to give brief answers to the questions without referring to the text.

1. Define each of the following.
(a) substance **(b)** element **(c)** compound
(d) atom **(e)** molecule

2. Distinguish among the terms "empirical formula," "molecular formula," and "structural formula."

3. Name each of the substances with the following molecular formulas.
(a) CH_4 **(b)** NH_3 **(c)** H_2O
(d) H_2O_2 **(e)** NaCl **(f)** SiO_2

4. Using water to exemplify your answer, explain the terms "physical property" and "chemical property."

5. What is meant by the purity of a substance?

6. Define "mass percentage."

7. Why are solutions and gaseous mixtures, such as air, described as homogeneous mixtures?

8. What is a chemical reaction?

9. **(a)** For the reaction in which ammonia, $NH_3(g)$, is synthesized from $H_2(g)$ and $N_2(g)$, name the reactants and the products. **(b)** Write the balanced equation for this reaction.

10. What principle regarding the atoms involved must be observed in arriving at the balanced equation for a reaction?

11. **(a)** What are the three particles that chemists use to describe the structures of atoms? **(b)** What are their electrical charges and approximate relative masses?

12. **(a)** What are the constituents of the nucleus of an atom? **(b)** What is the size of the nucleus relative to that of the atom?

13. Explain the terms ''atomic number,'' *Z*, and ''mass number,'' *A*.

14. What information does an atomic symbol such as $^{12}_{6}C$ convey?

15. How is an element defined in terms of atomic number?

16. Using oxygen as an example, explain what is meant by the term ''isotope.''

17. Define the quantity ''atomic mass unit,'' u.

18. Why is the atomic mass of most elements an average value?

19. What is the definition of ''mole''?

20. What is meant by the terms ''molecular mass'' and ''molar mass''?

21. What is the relationship between the mass of one molecule in atomic mass units, u, and the mass of 1 mol of molecules in grams?

22. What is meant by the term ''unstable'' or (''radioactive'') atom?

23. Why is the sum of the masses of an atom's constituent particles always greater than the experimentally observed mass?

24. Why, within the accuracy of experimental measurements, is mass conserved in chemical reactions?

PROBLEMS

Elements, Compounds, and Formulas

1. Give the symbol of each of the following elements.

(a) sodium **(b)** potassium **(c)** mercury **(d)** gold
(e) tin **(f)** lead **(g)** iron

2. Give the symbol of each of the following elements.

(a) hydrogen **(b)** oxygen **(c)** carbon
(d) nitrogen **(e)** chlorine **(f)** magnesium
(g) calcium **(h)** barium **(i)** sulfur
(j) phosphorus

3. Give the names of the elements with each of the following symbols.

(a) H **(b)** He **(c)** Ne **(d)** F **(e)** Mg
(f) Al **(g)** P **(h)** S **(i)** K **(j)** Na

4. What formulas represent each of the following?

(a) a molecule of solid sulfur containing 8 S atoms

(b) a molecule of sulfuric acid containing 2 H, 1 S, and 4 O atoms

(c) liquid sulfuric acid

(d) a solution of sulfuric acid in water

(e) solid silicon carbide (carborundum) containing an infinite array of Si and C atoms in the ratio 1:1

5. Write molecular and empirical formulas, where possible, and the formula for a sample of the bulk substance for each of the following.

(a) solid white phosphorus with molecules containing 4 P atoms

(b) solid red phosphorus consisting of infinite sheets of P atoms

(c) a gaseous hydrocarbon with molecules containing 3 C atoms and 6 H atoms

(d) a gaseous fluorocarbon with molecules containing 2 C, 2 F, and 4 H atoms

6. What is the empirical formula of each of the substances with the following molecular formulas?

(a) N_2O_4 **(b)** N_2O_5 **(c)** S_8 **(d)** C_2H_6 **(e)** Al_2Cl_6
(f) C_3H_8 **(g)** C_4H_{10} **(h)** $C_6H_{12}O_6$ **(i)** C_6H_6

7. Give the molecular formula and the formula for a bulk sample of each of the following substances.

(a) hydrogen **(b)** oxygen **(c)** nitrogen
(d) water **(e)** hydrogen peroxide **(f)** methane
(g) carbon dioxide **(h)** magnesium oxide **(i)** silicon dioxide

8. Give the molecular formula and bulk formula of each of the following substances.

(a) nitrogen monoxide **(b)** carbon monoxide
(c) sulfur **(d)** ammonia **(e)** sucrose

Substances: Properties and Purity; Mixtures

9. What problems are associated with the use of the term ''chemical''?

10. **(a)** Explain the terms ''mixture'' and ''impure.'' **(b)** Although it is commonly used, why is ''pure'' strictly speaking a redundant term? **(c)** What is meant when a substance is described as 99.98% pure?

11. Classify each of the following as an element, a compound, or a mixture.

(a) water **(b)** iron **(c)** beer **(d)** sugar

(e) wine (f) silicon dioxide (g) sulfur (h) cement
(i) air (j) magnesium oxide

12. Classify each of the following as a pure substance or as a mixture. Divide the pure substances into elements and compounds and the mixtures into homogeneous or heterogeneous mixtures.

(a) nitrogen (b) gasoline (c) sterling silver
(d) sodium chloride (e) carbon dioxide (f) black coffee
(g) diamond (h) distilled water (i) filtered sea water
(j) carbon (k) vegetable soup (l) concrete

13. Classify each of the following substances as an element, a compound, or a mixture.

(a) tin (b) lemon juice (c) natural gas
(d) uranium (e) 3% aqueous hydrogen peroxide
(f) sodium nitrate (g) hydrogen peroxide (h) soda water

The Separation of Mixtures

14. Briefly explain each of the following.

(a) filtration (b) distillation (c) homogeneous mixture
(d) solution (e) heterogeneous mixture

15. On what differences in physical properties do each of the following depend?

(a) filtration (b) distillation (c) paper chromatography

16. Suggest possible ways to separate the components of each of the following mixtures.

(a) sugar and water (b) water and gasoline
(c) iron filings and wood sawdust
(d) sugar and powdered glass (e) food coloring and water

Chemical Reactions and Chemical Equations

17. (a) What are meant by the *reactants* and the *products* of a chemical reaction? (b) What information is needed before a balanced chemical equation can be written to describe a reaction between carbon and oxygen?

18. In terms of atoms and energy changes, why is mass conserved in a chemical reaction?

19. Balance those among the following equations that are not already balanced.

(a) $2SO_2 + H_2O + O_2 \rightarrow 2H_2SO_4$
(b) $CH_3OH + 2O_2 \rightarrow CO_2 + 2H_2O$
(c) $H_2O_2 \rightarrow H_2O + O_2$
(d) $H_2SO_4 + KOH \rightarrow KHSO_4 + H_2O$

20. Balance each of the following equations.

(a) $Al + HCl \rightarrow AlCl_3 + H_2$
(b) $C_5H_{12} + O_2 \rightarrow CO_2 + H_2O$
(c) $C_3H_8 + H_2O \rightarrow CO + H_2$
(d) $Na_2CO_3 + HCl \rightarrow NaCl + H_2O + CO_2$
(e) $Al + O_2 \rightarrow Al_2O_3$ (f) $Zn + HCl \rightarrow ZnCl_2 + H_2$
(g) $Al_2O_3 + H_2SO_4 \rightarrow Al_2(SO_4)_3 + H_2O$

21. Balance each of the following equations.

(a) $S + O_2 \rightarrow SO_3$ (b) $C_2H_2 + O_2 \rightarrow CO_2 + H_2O$
(c) $Cu_2S + O_2 \rightarrow Cu_2O + SO_2$
(d) $Na_2SO_4 + H_2 \rightarrow Na_2S + H_2O$
(e) $Na_2CO_3 + Ca(OH)_2 \rightarrow NaOH + CaCO_3$

22. Write balanced equations for the reactions between the following reactants to give the products indicated.

Reactants	*Products*
(a) carbon, oxygen	carbon monoxide
(b) methane, oxygen	carbon monoxide, water
(c) methane, oxygen	carbon dioxide, water
(d) magnesium, steam	magnesium oxide, hydrogen
(e) methane, steam	carbon monoxide, hydrogen

The Structure of Atoms

23. What are the chemical symbols and the atomic numbers of one atom of each of the following?

(a) hydrogen (b) oxygen (c) fluorine
(d) neon (e) magnesium (f) phosphorus
(g) chlorine (h) calcium (i) zinc

24. The nuclei of which elements have atoms with each of the following numbers of protons?

(a) 5 (b) 9 (c) 32 (d) 54 (e) 92

25. How many protons are there in the nucleus of one atom of each of the following?

(a) boron (b) nitrogen (c) hydrogen
(d) neon (e) chlorine (f) oxygen
(g) sulfur (h) potassium (i) iron

26. Identify each of the following neutral atoms, and give the symbol of the particular isotope: an atom with (a) 1 proton and 1 neutron; (b) 8 protons and 9 neutrons; (c) 6 protons and 8 neutrons; (d) 19 nucleons and 9 electrons; (e) 32 nucleons and 16 electrons; (f) 12 neutrons and 12 electrons; (g) 45 neutrons and 35 electrons.

27. Give the symbols of the isotopes with each of the following atomic numbers (Z) and mass numbers (A).

(a) $Z = 19, A = 40$ (b) $Z = 14, A = 30$
(c) $Z = 18, A = 40$ (d) $Z = 7, A = 15$
(e) $Z = 16, A = 32$ (f) $Z = 11, A = 23$
(g) $Z = 13, A = 27$

28. Complete each of the following symbols, and give the number of protons, neutrons, and electrons, respectively, in each atom.

(a) ^{4}He (b) ^{27}Al (c) ^{14}C (d) ^{31}P
(e) ^{37}Cl (f) ^{85}Rb (g) ^{108}Ag (h) ^{131}Xe

29. Give the numbers of protons, neutrons, and electrons in each of the following atoms.

(a) ^{2}H (b) ^{19}F (c) ^{40}Ca
(d) ^{112}Cd (e) ^{117}Sn (f) ^{235}U

30. Complete the following table.

Atomic symbol	^{9}Be	^{15}N	^{18}O	—	—
Mass number	9	—	—	—	23
Atomic number	4	—	—	—	11
Number of protons	4	—	—	6	—
Number of electrons	4	—	—	6	—
Number of neutrons	5	—	—	6	—

31. Complete the following table.

Atomic symbol	^{24}Mg	^{106}Ag	^{137}Ba
Mass number	—	—	—
Atomic number	—	—	—
Number of protons	—	—	—
Number of electrons	—	—	—
Number of neutrons	—	—	—

32. Write the symbols for each of the isotopically different molecules of water, given that the isotopes of hydrogen and oxygen are ^{1}H, ^{2}H, ^{16}O, ^{17}O, and ^{18}O.

33. **(a)** What information is given by the symbol ^{24}Mg? **(b)** From the data in the accompanying table, calculate the average atomic mass of magnesium.

Mass Number	*Abundance (%)*	*Atomic mass (u)*
24	78.99	23.986
25	10.00	24.986
26	11.01	25.983

34. Naturally occurring copper has an average atomic mass of 63.55 u and contains isotopes with masses 62.9298 and 64.9278 u. What are the symbol and relative abundance of each isotope of copper?

35. Uranium has an average atomic mass of 238.03 and consists of ^{235}U, mass 235.044 u, and ^{238}U, mass 238.051 u. The ^{235}U isotope is the fuel used in nuclear power reactors. What is the percentage abundance (by mass) of ^{235}U in natural uranium?

36. Naturally occurring chlorine has an average atomic mass of 35.45 u and consists of isotopes with mass numbers 35 and 37, respectively.
(a) What is the approximate abundance of each isotope?
(b) What is the average mass of a Cl_2 molecule?

37.* Bromine atoms and chlorine atoms combine to give BrCl molecules. BrCl(*g*) consists of molecules with approximate masses 114, 116, and 118 u, and chlorine has just two isotopes, with mass numbers 35 and 37. **(a)** Deduce the possible isotopes of bromine, and write their symbols. **(b)** Give the formulas of each of the isotopically different BrCl molecules.

The Mole: Counting by Weighing

38. Explain why Avogadro's number of atoms or molecules is a convenient quantity in considering the quantitative aspects of the masses of substances involved in chemical reactions.

39. If scientists had selected 1 lb (453.6 g) as the basic unit for measuring mass, what difference would this have had on **(a)** the definition of the mole and **(b)** the value of Avogadro's number?

40. The charge on the electron is $1.602\,19 \times 10^{-19}$ C. What is the charge on **(a)** 1 mol of electrons in coulombs (C) and **(b)** 1 mol of protons?

41. An atom of element X has a mass of 3.155×10^{-23} g. What are **(a)** the atomic mass of X compared to 12 u for one ^{12}C atom, **(b)** the molar mass of X, and **(c)** the symbol for element X?

42. Using the atomic masses given inside the front cover, calculate the molar mass of each of the following compounds.

(a) $H_2O(l)$ **(b)** $H_2O_2(l)$ **(c)** $NaCl(s)$ **(d)** $MgBr_2(s)$
(e) $CO(g)$ **(f)** $CO_2(g)$ **(g)** $CH_4(g)$ **(h)** $C_2H_6(g)$
(i) $NH_3(g)$ **(j)** $HCl(g)$

43. Using atomic masses inside the front cover, calculate the molar mass of each of the following compounds.

(a) $NH_4Cl(s)$ **(b)** $Ca_3(PO_4)_2(s)$ **(c)** $KI(s)$
(d) $PCl_5(s)$ **(e)** $C_4H_{10}(g)$

44. What are the empirical formula mass and the molecular mass of each of the following?

(a) acetic (ethanoic) acid, $C_2H_4O_2(l)$
(b) formic (methanoic) acid, $H_2CO_2(l)$
(c) cane sugar, $C_{12}H_{22}O_{11}(s)$
(d) hexane, $C_6H_{14}(l)$
(e) hydrogen peroxide, $H_2O_2(l)$

45. The artificial sweetener Nutrasweet is the compound aspartamine, with the molecular formula $C_{14}H_{18}N_2O_5$.
(a) What is the mass of 1.000 mol of aspartamine?
(b) How many moles of aspartamine are in 6.22 g of the substance?
(c) What is the mass, in grams, of 0.245 mol of aspartamine?
(d) How many molecules are in 4.28 mg of aspartamine?

46. What is the mass in grams of each of the following?
(a) 7.1×10^{-3} mol of oxygen gas
(b) 5.43 mol of iron
(c) 3.14×10^{-2} mol of phosphorus, $P_4(s)$
(d) 9.6×10^{-4} mol of sulfur, $S_8(s)$
(e) 0.452 mol of magnesium phosphate, $Mg_3(PO_4)_2(s)$

47. **(a)** How many moles of sulfur dioxide are in 0.028 g $SO_2(g)$? **(b)** What mass contains exactly 3 mol of $SO_2(g)$?

48. Express 10.00 g of each of the following in moles.
(a) water, $H_2O(l)$ **(b)** sulfur dioxide, $SO_2(g)$
(c) acetic acid, $C_2H_4O_2(l)$ **(d)** sulfuric acid, $H_2SO_4(l)$

49. The pain reliever aspirin is the compound acetylsalicylic acid, $C_9H_8O_4(s)$. **(a)** What is the molecular mass of aspirin? **(b)** How many moles, and how many molecules, of $C_9H_8O_4$ are there in a 500-mg aspirin tablet?

50. **(a)** How many molecules of oxygen, O_2, are there in 2.00 g of oxygen gas?
(b) If the oxygen molecules were split into atoms, how many moles of oxygen atoms would result?
(c) If the oxygen atoms were combined to give ozone molecules, $O_3(g)$, what would be the mass of ozone produced?

51. **(a)** Assuming that the human body contains 6×10^{13} cells and that the earth's population is 4×10^{9} persons, approximately how many moles of human body cells are there on the earth?

(b) Assuming that the human body is 80% water by mass, how many water molecules are there in the body of a person with a mass of 65 kg?

52. Carbon monoxide, $CO(g)$, taken into the lungs reduces the ability of the blood to transport oxygen. An amount of CO as high as 2.38×10^{-4} g · L^{-1} is fatal. Calculate the number of CO molecules that must be emitted from an automobile exhaust to produce a fatal concentration of CO in a garage of volume 150 m^3.

Unstable Atoms: Radioactivity; Conservation of Mass and Energy

53. Explain **(a)** why some nuclei are unstable and **(b)** why the mass of any atom is not exactly the same as the mass of its constituent protons, neutrons, and electrons.

54. Explain why in nuclear reactions it is the sum of mass and energy that is conserved, whereas in chemical reactions we can assume that mass alone is conserved.

55.* Explain why the product of the decay of a radioactive atom by **(a)** the emission of an α particle gives an atom in which the mass number has decreased by 4 and the atomic number has decreased by 2 and **(b)** the emission of a β particle gives an atom in which the mass number is unchanged but the atomic number has increased by 1.

56.* Radioactive ^{14}C atoms and stable ^{1}H atoms form in the atmosphere by the bombardment of ^{14}N atoms with cosmic rays (neutrons), and ^{14}C atoms decay to ^{14}N atoms by the emission of a β particle. Write balanced equations for **(a)** the formation of ^{14}C and **(b)** the decay of ^{14}C.

General Problems

57. Ammonium sulfate, $(NH_4)_2SO_4(s)$ ("sulfate of ammonia"), is an important commercial fertilizer produced by the reaction of hydrogen, $H_2(g)$, with nitrogen, $N_2(g)$, to give ammonia, $NH_3(g)$. The ammonia is passed into aqueous sulfuric acid, $H_2SO_4(aq)$, until reaction is complete, and the ammonium sulfate formed is then crystallized from the solution. Write balanced equations for these reactions.

58. Nitric acid, $HNO_3(aq)$, is manufactured by the reaction of ammonia, $NH_3(g)$, with oxygen at high temperature to give water and nitrogen monoxide, $NO(g)$. These products react with more oxygen to give nitrogen dioxide, $NO_2(g)$, which is subsequently dissolved in water to give an aqueous solution of nitric acid, $HNO_3(aq)$, and nitrogen monoxide, $NO(g)$. Write balanced equations for these reactions.

59. **(a)** Write the symbol for the isotope that contains as many protons as neutrons and has a mass number of 28.

(b) The element of the isotope in (a) also has isotopes of mass numbers 29 and 30. Write their symbols.

(c) The isotopes of mass numbers 28, 29, and 30 have atomic masses of 27.976 93 u, 28.976 50 u, and 29.973 77 u, with relative abundances of 92.23%, 4.67%, and 3.10%, respectively. What is the average atomic mass of the element?

60. Heavy water, $^2H_2O(l)$, or $D_2O(l)$, occurs to the extent of 0.003% by mass in natural water. Pure heavy water, which is used to reduce the energies of fast neutrons in the CANDU nuclear reactor, is a colorless liquid that forms hexagonal crystals upon freezing and has a density of 1.105 g · cm^{-3}, a melting point of 3.8°C, and a boiling point of 101.4°C. It is separated from ordinary water by fractional distillation or by electrolysis and reacts with $SO_3(g)$ to give deuterosulfuric acid, $D_2SO_4(l)$. From this description list **(a)** the physical properties and **(b)** the chemical properties of heavy water that are mentioned.

61. **(a)** Explain why the molecular mass of water in grams is the same as the mass of one water molecule in atomic units.

(b) How many molecules are there in 1.000 g of water?

(c) Given that the density of water is 0.997 g · cm^{-3} at 25°C, what is the volume of one water molecule in liquid water at this temperature?

62. Explain the following law proposed in 1913: 1. In α-particle emission, the nucleus produced falls into a group in the periodic table *two places lower* than that containing the parent substance. 2. In β-particle emission, the product falls into a group in the periodic table *one group higher* than that containing the parent substance.

63.* Gunpowder is a mixture of potassium nitrate, $KNO_3(s)$, sulfur, and charcoal (carbon). **(a)** How could a sample of gunpowder be separated into its components? **(b)** Write the balanced equation for the reaction between the components of gunpowder to give potassium nitrite, $KNO_2(s)$, sulfur dioxide, $SO_2(g)$, and carbon dioxide, $CO_2(g)$, when heated.

* Asterisks denote more difficult problems.

CHAPTER 2 The Atmosphere, Gases, and the Gas Laws

2.1 The Atmosphere
2.2 Oxygen
2.3 Nitrogen
2.4 Hydrogen
2.5 Solids, Liquids, and Gases
2.6 The Gas Laws
2.7 The Kinetic Theory of Gases
2.8 Using the Ideal Gas Law
2.9 Diffusion and Effusion
2.10 Real Gases and Intermolecular Forces

The tank on this scuba diver's back contains a compressed mixture of the gases oxygen and helium. Ordinary air is not used because it contains the gas nitrogen, which would form bubbles in the diver's blood when she ascends to the surface. This condition, known as the "bends," is very painful and can lead to death. Hence this diver must know something about the chemistry of gases just to enjoy herself at the Grand Cayman Islands!

The atmosphere surrounding the earth is a mixture of gases called air. In addition to containing the oxygen that sustains the biosphere, the atmosphere protects us from harmful ultraviolet radiation and controls the temperature of the earth's surface. In the first part of this chapter we describe the atmosphere and some of the chemistry of its two principal components, the gaseous elements nitrogen and oxygen. We will also consider some of the chemistry of hydrogen, the most abundant element in the universe. There is almost no hydrogen, $H_2(g)$, in the air, but there is a great deal of combined hydrogen on earth, most of it in molecules of water.

The atmosphere has played an important role in the history of chemistry. When chemists first studied the properties of gases, they began with air, and this work produced some of the first scientific laws. In the second part of the chapter we discuss the laws that describe the behavior of gases. Unlike solids and liquids, all gases behave in much the same way with changing conditions such as temperature and pressure, so it is easier to describe their behavior than that of substances in other physical states.

Two fundamental concepts underlie all of chemistry. The first of these, which we introduced in Chapter 1, is the idea that substances consist of atoms and molecules. The second fundamental concept, which we will begin to discuss in this chapter, is that

Atoms and molecules are in constant random motion and move more rapidly with increasing temperature.

The energy of motion is called **kinetic energy**. We will see that the characteristic properties of gases can be understood in terms of a model called the *kinetic theory of gases*, according to which a gas consists of widely separated atoms or molecules moving randomly at very high speeds.

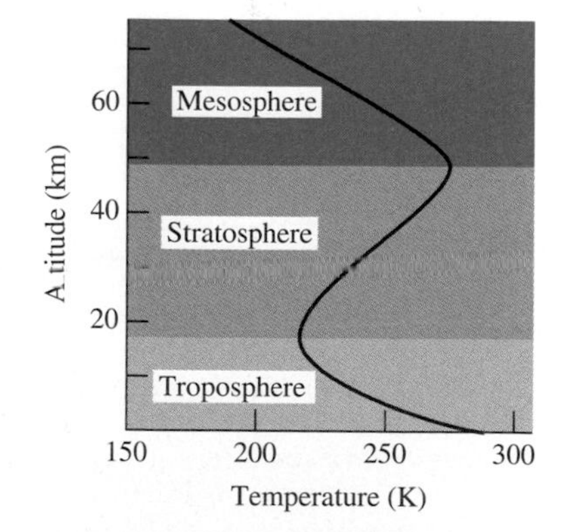

FIGURE 2.1 Variation of Temperature with Altitude in the Earth's Atmosphere The temperature decreases in the troposphere with increasing distance from the warm surface. It then increases in the stratosphere due to the absorption of ultraviolet light by oxygen molecules, and it decreases again in the mesosphere.

2.1 THE ATMOSPHERE

The **atmosphere** is held on the surface of the earth by gravity. It is most dense at sea level and thins rapidly with increasing altitude. Almost all the atmosphere (99%) lies within 30 km of the earth's surface.

The atmosphere can be divided into several layers on the basis of the variation of temperature with altitude (Figure 2.1). In the lowest layer, the **troposphere**, the average temperature decreases with increasing altitude as the distance from the warm earth increases. The troposphere has a nearly uniform and constant composition, except for small amounts of water vapor and certain industrial pollutants. In the troposphere winds blow, and the temperature and pressure vary from day to day, producing the changes we call weather. These changes continually mix the different gases, giving the troposphere its nearly uniform composition.

At an altitude of about 15 km, where the average temperature is about −60°C, the temperature begins to increase again, marking the lower limit of the **stratosphere**. In this layer, which extends up to about 50 km, the temperature increases to about 0°C. This temperature increase is the result of the absorption of ultraviolet radiation from the sun by oxygen molecules. These molecules are decomposed to oxygen atoms, which combine with O_2 molecules to form ozone molecules, O_3, which also absorb ultraviolet radiation. In this way, living organisms are protected from the harmful effects of ultraviolet radiation, as we shall discuss in more detail in Chapter 16. Above about 50 km there are so few molecules to absorb the ultraviolet radiation that the temperature decreases again, marking the lower limit of a third layer, called the *mesosphere*.

The mixture of gases in the troposphere that we call air consists mostly of oxygen and nitrogen, which make up 99% of its volume (Table 2.1). Among the other components are the noble gases helium, neon, argon, krypton, and xenon, of which argon is the most abundant. Carbon dioxide is present only in a very small amount (0.03%), but it is nevertheless essential to life, because it is the source of the carbon in biological molecules.

The process of incorporating carbon from the air into the molecules of living organisms is called *carbon fixation*. The dominant form of carbon fixation on earth today is *photosynthesis*, in which green plants use solar energy in synthesizing carbohydrates from carbon dioxide and water, releasing oxygen into the atmosphere as a waste product. Most present-day organisms obtain energy by combining oxygen from the air with carbohydrates in a process called *respiration*. This energy is used to synthesize a variety of other compounds that the organism needs to live and grow.

There are many different carbohydrates with molecules of different sizes. The general formula for a carbohydrate is $C_xH_{2y}O_y$, where x and y have many possible values.

The earth is unique among the planets in the solar system in that its atmosphere contains a large amount of oxygen. But the earth has not always had the oxygen-rich atmosphere that it has today. When the first life forms appeared on earth more than 3 billion years ago, the atmosphere contained little, if any, oxygen. The earliest organisms did not use O_2 for respiration; they relied instead on chemical reactions that produce much smaller amounts of energy. But as life evolved and photosynthesis became the dominant form of carbon fixation, enor-

TABLE 2.1 Composition of Dry Air

Component	Formula	Percent by Volume
Nitrogen	N_2	78.084
Oxygen	O_2	20.948
Argon	Ar	0.934
Carbon dioxide	CO_2	0.0314
Neon	Ne	0.00182
Helium	He	0.00052
Methane	CH_4	0.0002
Krypton	Kr	0.00011
Hydrogen	H_2	0.00005
Dinitrogen monoxide	N_2O	0.00005
Xenon	Xe	0.000008

mous amounts of oxygen began to accumulate in the atmosphere. This has allowed the evolution of millions of plant and animal species, all of which use the oxygen in air (or dissolved in water) for respiration. Our neighboring planets, Venus and Mars, have atmospheres composed mainly of carbon dioxide, with very little oxygen. It seems reasonable to conclude that there has been no form of life on these planets that engaged in photosynthesis.

The earth's original atmosphere probably contained considerable quantities of hydrogen and helium. However, the earth's gravity is not strong enough to hold these very light atoms and molecules, so they have gradually diffused away from our planet's surface. Considerably larger amounts of hydrogen and helium are found in the atmospheres of the four larger and more massive outer planets of the Solar System (Jupiter, Saturn, Uranus, and Neptune).

Human activities are producing small but important changes in the atmosphere, which does not therefore have a completely constant or uniform composition. In industrial regions, the atmosphere contains variable amounts of sulfur dioxide, SO_2, nitrogen monoxide, NO, and nitrogen dioxide, NO_2, formed primarily during the combustion of oil and coal and in the production of iron and other metals. These undesirable impurities are the cause of acid rain and photochemical smog, as we discuss in more detail in Chapter 16. Moreover, the combustion of fuels and the burning of forests are contributing to a gradual increase in the concentration of carbon dioxide in the atmosphere (Figure 2.2). This small but definite increase could have significant long-term effects on the earth's climate as a result of what is called the *greenhouse effect* (Chapter 16).

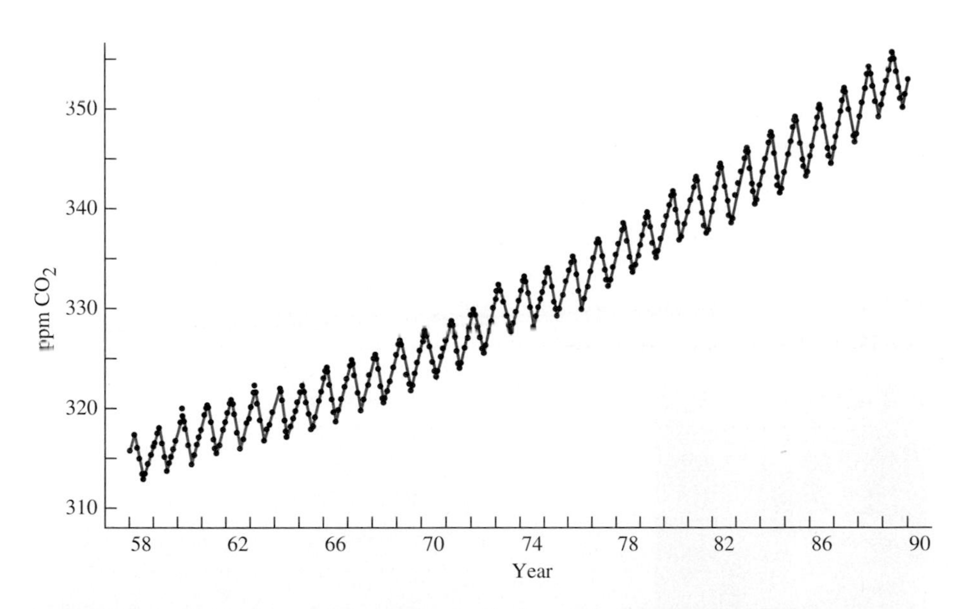

FIGURE 2.2 Variation of Atmospheric Carbon Dioxide Concentrations
Atmospheric carbon dioxide concentrations measured at the Mauna Loa Observatory, Hawaii. There has been a steady increase from a concentration of 315 ppm (parts per million) in 1958 to 350 ppm in 1988, and this increase is continuing. Superimposed on the general trend are annual oscillations caused by seasonal changes in the photosynthetic activity of plants.

2.2 OXYGEN

The diatomic oxygen, O_2, nitrogen, N_2, and hydrogen, H_2, molecules should strictly be called dioxygen, dinitrogen, and dihydrogen, respectively. But it is common to call them simply the oxygen, nitrogen, and hydrogen molecules. This is not likely to cause any confusion: Nitrogen and hydrogen have no other stable forms, and although there is another oxygen molecule, trioxygen, O_3, this is always called ozone.

The element oxygen consists of diatomic O_2 molecules. It is a colorless gas at ordinary temperatures, condensing to a blue liquid at −183°C and freezing to a pale blue solid at −218°C. Oxygen is obtained on a large scale by liquefying air and then distilling it to separate the components. Very pure oxygen can be obtained by *electrolysis* of water, that is, by passing an electric current through water to split the molecules into oxygen and hydrogen (Demonstration 2.1):

$$2H_2O(l) \xrightarrow{\text{electric current}} 2H_2(g) + O_2(g)$$

Oxygen reacts with almost all other elements and with numerous compounds. Many of these reactions are very slow at ordinary temperatures, but as we will see in Chapters 4 and 15, the rates of reactions are normally increased upon raising the temperature. Most of the reactions between oxygen and other elements proceed at a significant rate only at relatively high temperatures.

Oxides

Sulfur does not react with oxygen at a measurable rate at room temperature. But if we heat sulfur until it melts and place it in oxygen, it burns with a bright blue flame (Demonstration 2.2). The product of this reaction is sulfur dioxide, SO_2, which is a colorless gas with a pungent odor:

$$S(l) + O_2(g) \rightarrow SO_2(g)$$

Magnesium reacts very slowly with oxygen at ordinary temperatures, but it reacts very rapidly when heated, emitting a brilliant white light and forming magnesium oxide, MgO, a white solid:

$$2Mg(s) + O_2(g) \rightarrow 2MgO(s)$$

This reaction, and those of some other elements with oxygen, are also shown in Demonstration 2.2.

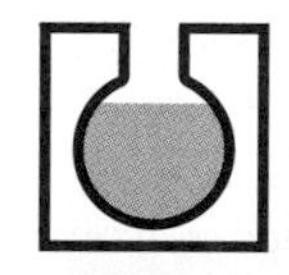

DEMONSTRATION 2.1 Decomposition of Water by an Electric Current

Water can be decomposed into the elements hydrogen and oxygen by passing an electric current through it. Bubbles of hydrogen can be seen rising from the wire at the left. The pure elements are collected in separate tubes. The volume of hydrogen produced is twice the volume of oxygen. A few drops of sulfuric acid have been added to the water to increase its electrical conductivity (its ability to pass electric current).

DEMONSTRATION 2.2 Reactions of Metals and Nonmetals with Oxygen

If sulfur is warmed until it melts, it burns in oxygen with a bright blue flame, producing the pungent-smelling gas sulfur dioxide, SO_2.

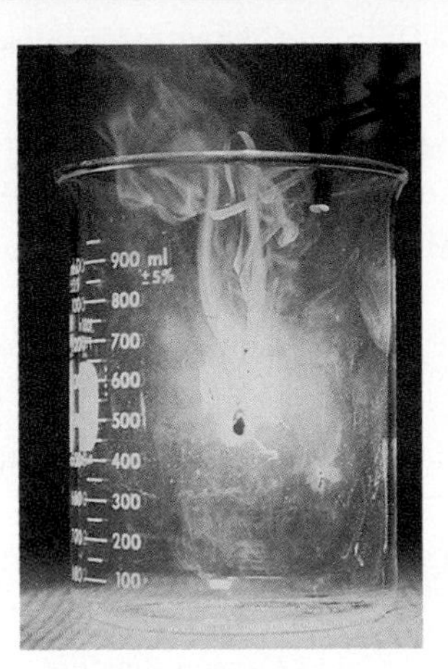

Magnesium burns even more violently in oxygen than in air, forming a white smoke of solid particles of magnesium oxide, MgO.

If steel wool is heated in a flame and then plunged into an atmosphere of oxygen, it burns vigorously, giving a shower of sparks and forming brown, solid iron oxide, Fe_2O_3. (Steel is an alloy of iron; see Chapter 10.)

White phosphorus ignites spontaneously in oxygen and burns with a very bright flame, forming a dense white smoke of solid P_4O_{10} particles.

The products of the reactions of elements with oxygen are called **oxides**, such as magnesium oxide. Some elements form more than one oxide; carbon combines with oxygen to form both carbon monoxide, CO, and carbon dioxide, CO_2, and sulfur forms both sulfur dioxide, SO_2, and sulfur trioxide, SO_3. Nitrogen forms several oxides, including nitrogen monoxide, NO(*g*), and dinitrogen tetraoxide, N_2O_4(*l*). In such cases the numerical prefixes in the name indicate the numbers of each kind of atom in the molecular formula of the compound (or the empirical formula of a network solid, such as silicon dioxide, SiO_2). The prefix ''mono-'' is omitted when there is only one atom of an element other than oxygen—carbon dioxide, *not* monocarbon dioxide. Table 2.2 shows some typical reactions in which elements are converted to their oxides.

The prefixes used to denote one to six atoms of the same element in a molecular or empirical formula are mono-, di-, tri-, tetra-, penta-, and hexa-.

Combustion The reactions of many elements with oxygen give off heat. We saw in Demonstration 2.2 that once elements such as sulfur and magnesium are heated to a temperature at which they react at an appreciable speed, they continue to react. The heat given off in these reactions maintains or increases the temperature, so the reaction rate is maintained or even increased without the need for further heating. An element undergoing such a self-sustaining reaction with oxygen is said to be *burning*, or to be undergoing **combustion**.

Many compounds undergo combustion reactions. For example, methane, $CH_4(g)$ (natural gas), burns when ignited in oxygen, giving carbon dioxide and water:

$$CH_4(g) + 2O_2(g) \rightarrow CO_2(g) + 2H_2O(g)$$

TABLE 2.2 Reactions of Some Elements with Oxygen

Element	Reaction with Oxygen	Oxide
Copper, a reddish metal	$2Cu + O_2 \rightarrow 2CuO$	**Copper oxide**, a black solid, insoluble in water
Mercury, a silvery liquid metal	$2Hg + O_2 \rightarrow 2HgO$	**Mercury oxide**, a red solid, insoluble in water
Magnesium, a silvery metal	$2Mg + O_2 \rightarrow 2MgO$	**Magnesium oxide**, a white solid, insoluble in water
Sulfur, a yellow solid	$S + O_2 \rightarrow SO_2$	**Sulfur dioxide**, a colorless and pungent-smelling gas, soluble in water.
Phosphorus, a red solid	$4P + 5O_2 \rightarrow P_4O_{10}$	**Tetraphosphorus decaoxide**, a white solid, soluble in water
Hydrogen, a colorless gas	$2H_2 + O_2 \rightarrow 2H_2O$	**Water**, a colorless liquid
Carbon (graphite), a black solid	$C + O_2 \rightarrow CO_2$	**Carbon dioxide**, a colorless gas, slightly soluble in water

The reactions of elements and compounds with oxygen also occur in air, but less rapidly than in pure oxygen. A simple **test for oxygen** is to hold a glowing splint of wood in a tube of the gas in question. If the gas is oxygen, the wood splint will burst into flame, because the combustion of wood, which is mainly a mixture of carbohydrates, is much faster in oxygen than it is in air. The flames that we see in many combustion reactions are the hot gaseous products emitting light as they are given off from the burning substance.

Exothermic and Endothermic Reactions Reactions, such as combustion reactions, that give off heat to their surroundings are called **exothermic reactions**. Other reactions, such as the reaction between nitrogen and oxygen to give nitrogen monoxide,

$$N_2(g) + O_2(g) \rightarrow 2NO(g)$$

take in heat from their surroundings, such as the heat provided by a lightning discharge in the atmosphere. Reactions that take in heat from their surroundings are called **endothermic reactions**. Some reactions also emit light, another form of energy. In Chapter 9 we shall discuss the origin and magnitude of the heat and other energy changes that accompany chemical reactions.

Oxidation and Reduction

When an element combines with oxygen to form an oxide, the element is said to have been **oxidized**, and the process is called **oxidation**. Thus, when copper combines with oxygen, it is said to be oxidized to copper oxide, CuO. When methane burns to form carbon dioxide and water, the carbon in methane is oxidized to CO_2.

Because most elements combine with oxygen and because oxygen is very abundant on the surface of the earth, many elements, particularly metals, are found in the form of their oxides. Some common oxides are $H_2O(l)$, $Al_2O_3(s)$, $Fe_2O_3(s)$ and $CuO(s)$. Many oxides are major sources of the element, which can be obtained by removing the oxygen in a process called **reduction**. Iron is made by the reduction of the oxide Fe_2O_3 with carbon monoxide:

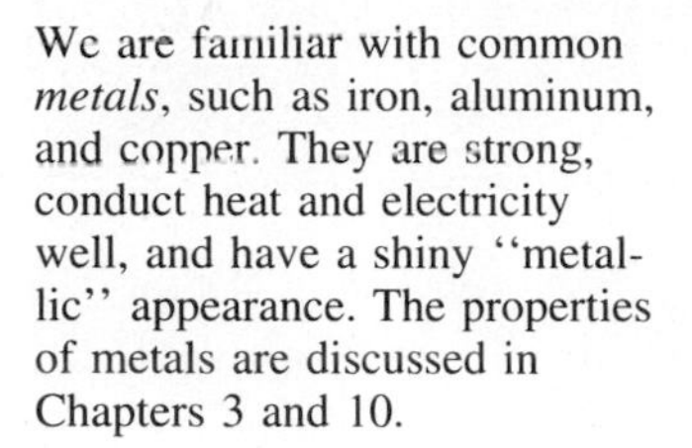

We are familiar with common *metals*, such as iron, aluminum, and copper. They are strong, conduct heat and electricity well, and have a shiny "metallic" appearance. The properties of metals are discussed in Chapters 3 and 10.

$$Fe_2O_3(s) + 3CO(g) \rightarrow 2Fe(s) + 3CO_2(g)$$

In this reaction, carbon monoxide removes the oxygen from the iron oxide and is thereby converted to carbon dioxide. The iron oxide is said to be **reduced** to iron, and at the same time the carbon monoxide is oxidized to carbon dioxide. Other substances can be used to reduce oxides to their elements. For example, copper oxide is reduced to copper when it is heated with hydrogen or with carbon (Demonstration 2.3):

$$CuO(s) + H_2(g) \rightarrow Cu(s) + H_2O(g)$$

$$CuO(s) + C(s) \rightarrow Cu(s) + CO(g)$$

In all these reactions, as the oxide is being reduced, the other substance (carbon monoxide, hydrogen, or carbon) is being oxidized.

Oxidation is always accompanied by reduction, and vice versa.

Early chemists used the term "oxidation" to mean the combination of an element with oxygen to give an oxide and the term "reduction" to mean the reverse process, that is, the removal of oxygen to give back the original element. Today we use more general definitions of these terms, as we will discuss in Chapter 4.

DEMONSTRATION 2.3 Reduction of Copper Oxide with Carbon

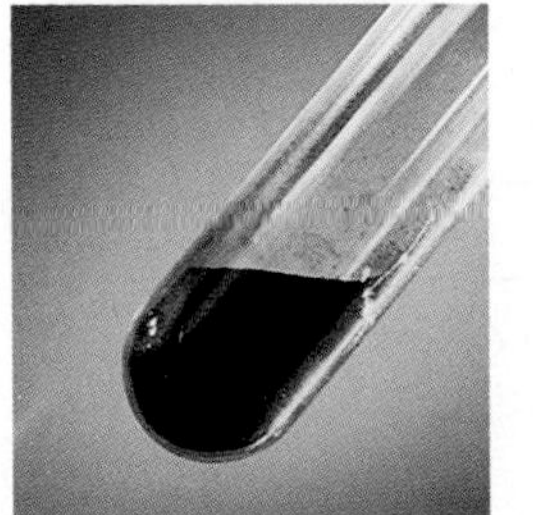

A mixture of black copper oxide and black carbon

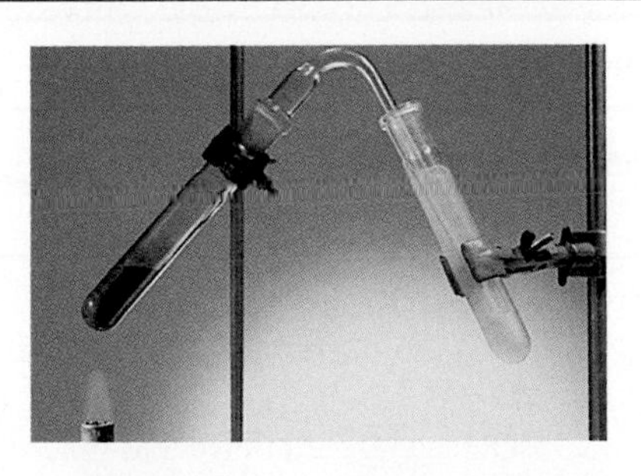

When the mixture is heated, the copper oxide is reduced to copper and the carbon is oxidized to carbon dioxide. The formation of carbon dioxide is shown by passing the gases given off through limewater—an aqueous solution of calcium hydroxide, $Ca(OH)_2$. A white solid, calcium carbonate, is formed.

$$Ca(OH)_2(aq) + CO_2(g) \rightarrow CaCO_3(s) + H_2O(l)$$

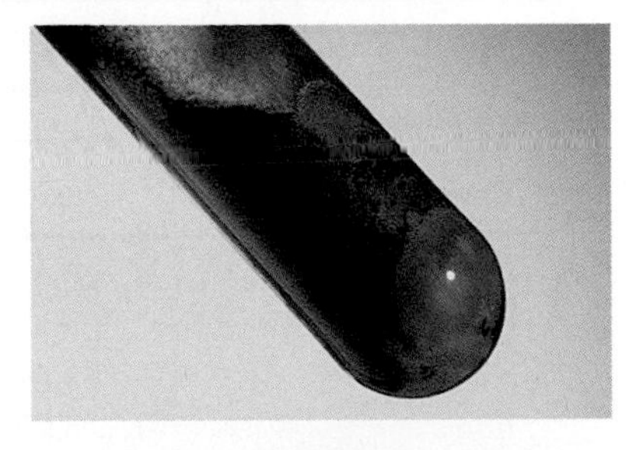

When the tube is cooled after the reaction, the red-brown copper that has been produced is clearly visible.

FIGURE 2.3 Liquid Nitrogen The white mist is a cloud of fine water droplets condensed from the air as it is cooled by the liquid nitrogen.

2.3 NITROGEN

The element nitrogen is a colorless, odorless, tasteless gas consisting of diatomic N_2 molecules. It boils at −196°C and freezes at −210°C (Figure 2.3). Like oxygen, it can be obtained from the air by liquefaction and distillation. Although nitrogen is the major component of the atmosphere and is one of the essential elements of living organisms, it is not very abundant on the earth as a whole, and few nitrogen-containing compounds are found in the earth's crust (Table 2.3).

As we might suspect from its scarcity in compounds, nitrogen is a rather unreactive element. It reacts with very few substances at ordinary temperatures, and even at higher temperatures it is much less reactive than oxygen. It does not react with oxygen at ordinary temperatures, and it takes the very high temperatures generated in a lightning discharge or in the cylinders of an automobile engine to cause this reaction to occur at a significant rate to give nitrogen monoxide:

$$N_2(g) + O_2(g) \rightarrow 2NO(g)$$

The NO(*g*) emitted in car exhaust is a serious atmospheric pollutant (Chapter 16).

Because the only widely available source of nitrogen is the atmosphere, all nitrogen compounds are ultimately obtained from this source. The process of converting the nitrogen of air into useful compounds is called **nitrogen fixation**. The only practical method of large-scale nitrogen fixation at present is the preparation of ammonia, NH_3 in the **Haber process** by the reaction of nitrogen with hydrogen:

$$N_2(g) + 3H_2(g) \rightarrow 2NH_3(g)$$

Because nitrogen is relatively unreactive, this reaction is carried out at a high temperature of about 400°C and at a high pressure in the presence of a *catalyst*. A

TABLE 2.3 Abundance of Elements in the Earth's Crust

		Mass Percent		Atom Percent		
Element	**Z**	**Crust**	**Surface***	**Crust**	**Surface***	**Occurrence**
Oxygen	8	46.9	49.1	59.3	62.5	Rocks, water, air
Silicon	14	26.9	25.4	21.1	18.5	Granite, sand
Aluminum	13	8.08	7.65	6.04	5.78	Rocks, clay
Iron	26	5.09	4.82	1.84	1.75	Oxides, granite
Calcium	20	4.99	4.72	2.52	2.40	Chalk, limestone
Potassium	19	2.55	2.41	1.31	1.25	Rocks
Magnesium	12	2.33	2.21	1.94	1.84	Rocks
Sodium	11	2.13	2.02	1.88	1.79	Salt, rocks
Carbon	6	0.52	0.49	0.86	0.83	Limestone
Titanium	22	0.50	0.47	0.22	0.20	Rocks
Hydrogen	1	0.15	0.15	3.3	3.3	Hydrosphere (water)
Others		<0.10	<0.10			

*Crust, hydrosphere, and atmosphere

catalyst is a substance that increases the rate of a reaction without being used up in the reaction. We consider catalysts in more detail in Chapters 4 and 15.

Ammonia is a colorless gas with a unique, penetrating odor that condenses to a liquid at −33.4°C. It is very soluble in water. Ammonia solutions are extensively used as household cleaning agents. The major use of ammonia, however, is in the manufacture of *fertilizers*. Plants require nitrogen-containing compounds for growth, but they cannot make these compounds from the N_2 in the atmosphere, as it is too unreactive. Instead, they make use of nitrogen compounds present in the soil. When the same land is used repeatedly for growing the same crops, natural processes do not restore the nitrogen compounds rapidly enough, and it becomes necessary to add nitrogen-containing fertilizers. Liquid ammonia stored at low temperature can be added directly to the soil, but more commonly it is converted to solid nitrogen compounds that are easier to handle (Chapter 7).

Certain bacteria in the soil and in the root nodules of legumes (plants in the pea family) carry out nitrogen fixation at ordinary temperatures. In other words, these bacteria convert the nitrogen of the air into nitrogen compounds that plants can use. Much research has been carried out on the fixation of nitrogen, and the reactions used by nitrogen-fixing bacteria are now largely understood. However, these or similar reactions do not yet provide an economical process by which to replace the Haber process for fixing nitrogen.

2.4 HYDROGEN

Hydrogen is the most abundant element in the universe. The stars consist mainly of hydrogen atoms, which also make up most of the matter found in interstellar space. Under ordinary conditions on earth, hydrogen is a colorless, odorless gas composed of H_2 molecules. It has a very low boiling point (−252.8°C) and melting point (−259.1°C).

Because of the very high temperatures in stars, the hydrogen atoms are almost entirely ionized into protons and separate electrons, except in the cooler outer regions of a star.

Hydrogen is not found in any significant amount in the atmosphere, but compounds of hydrogen are very common. A very large portion of the hydrogen on the earth's surface is present as water, but hydrogen is also found in hydrocarbons, in proteins and carbohydrates, and in almost all other substances in living organisms.

Water is by far the cheapest and most abundant source of hydrogen. Hydrogen is made industrially by the reaction of water with natural gas (methane) at a high temperature in the presence of a nickel catalyst:

$$CH_4(g) + H_2O(g) \xrightarrow{Ni} CO(g) + 3H_2(g)$$

In this reaction, methane is oxidized to carbon monoxide, and water is reduced to hydrogen. The mixture of carbon monoxide and hydrogen obtained in this reaction is called **synthesis gas** because it is often used directly for the industrial synthesis of other compounds of carbon (organic compounds), as discussed in Chapter 14. We can obtain reasonably pure hydrogen from these mixtures by removing the carbon monoxide, but the purest hydrogen is obtained by the electrolysis of water. When an electric current is passed through water, it decomposes into hydrogen and oxygen, which can be collected separately (Demonstration 2.1).

Under suitable conditions, hydrogen, $H_2(g)$, combines with most elements to form compounds. Compounds of hydrogen with metals are called **hydrides**. Examples include sodium hydride, $NaH(s)$, calcium dihydride, $CaH_2(s)$, and aluminum trihydride, $AlH_3(s)$. Most compounds of nonmetals with hydrogen have special names, such as methane, $CH_4(g)$, ammonia, $NH_3(g)$, and water $H_2O(l)$.

The reaction of hydrogen with oxygen to form water is an exothermic reaction that liberates a great deal of heat:

$$2H_2(g) + O_2(g) \rightarrow 2H_2O(g) + \text{heat}$$

This reaction is used to fuel rockets, such as those that power the space shuttle (Box 2.1).

BOX 2.1 Hydrogen as a Fuel

Hydrogen is an important rocket fuel. A primary consideration for a rocket fuel is that its mass be as small as possible for a given amount of energy produced. Liquid hydrogen was used in the *Saturn V* rocket that enabled the first astronauts to land on the moon, and it is the main fuel in the space shuttle rockets (Figure A). Both the hydrogen and the oxygen needed to burn the hydrogen are carried on the rocket in liquid form.

Hydrogen shows promise as a fuel for more general use as well. But its use at present is hampered by the difficulty of safely storing, transporting, and distributing such a highly flammable, potentially explosive material. In addition, because there are no natural sources of hydrogen, energy from another source must be expended to obtain H_2 from hydrogen compounds such as water. Therefore, hydrogen is said to be an energy carrier rather than an energy source. Both sunlight and excess electric energy from nuclear reactors are being studied as possible sources of energy for producing hydrogen from water. Widespread use of hydrogen as a fuel will be economical only when hydrogen can be produced cheaply.

If the problems of safety and economics are resolved, hydrogen may eventually be delivered to homes and industry by pipelines, as natural gas is today. Hydrogen could even be used to fuel vehicles (Figure B). For this purpose it might be stored as a solid metal hydride that, when heated, decomposes to hydrogen and metal. Hydrogen has already been used on a trial basis as a fuel for jet airplanes. An important advantage of hydrogen as a jet fuel or automobile fuel is that the only product of its combustion is water: There is no emission of CO, SO_2, or hydrocarbons, which pollute the atmosphere.

Figure A Columbia Space Shuttle *The rocket that launches the shuttle uses liquid hydrogen as a fuel. It is stored in a tank 40 m long and 8.4 m in diameter with a capacity of 385,000 gal. The oxygen for burning the hydrogen is stored in a similar tank containing 143,000 gal of liquid.*

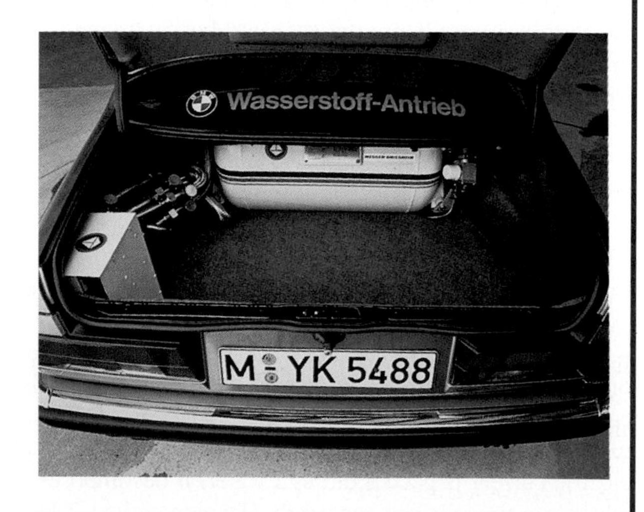

Figure B Hydrogen as a Fuel *Vacuum-insulated fuel tank for liquid hydrogen in a German car modified to run on hydrogen.* Wasserstoff antreib *means "hydrogen powered".*

Exercise 2.1 Name the oxides with the following empirical formulas.
(a) ZnO **(b)** Na_2O **(c)** SO_3 **(d)** CaO **(e)** Cl_2O **(f)** Al_2O_3 **(g)** P_4O_6

Exercise 2.2 Write balanced equations for both of the following reactions, and state which reactant is oxidized and which is reduced.
(a) the reaction of methane with water to give carbon monoxide and hydrogen
(b) the reaction of diiron trioxide with carbon to give carbon dioxide and iron

2.5 SOLIDS, LIQUIDS, AND GASES

One of the most obvious physical properties of a substance is its **physical state**: solid, liquid, or gas.

Solids A **solid** is rigid and has a shape that is independent of the shape of its container: If a rock is placed in a jar, it does not change shape to fill the jar. The atoms or molecules of a solid are packed together closely in a regular pattern and can move only small distances about their average positions. As a result of this packing, a solid sample cannot be compressed very much.

Liquids A **liquid** is fluid, that is, it flows. A liquid takes the shape of the part of the container that it fills, but it does not expand to fill the whole container. A small amount of water in a deep jar covers the bottom of the jar, but it has an upper surface beyond which it does not expand. The molecules of a liquid move farther from their average positions than those of a solid and continually change places (mix), so they lack the regular arrangement found in solids. However, the molecules remain close to one another, so a liquid is almost as difficult to compress as is a solid.

Gases A **gas** is also fluid, but it expands until it fills its container—that is, it has no surface of its own. In a gas, the molecules are no longer in contact: They are much farther apart than those of a solid or liquid and move rapidly with a random, chaotic motion. Because there is so much space between the molecules, a gas can readily be compressed to a smaller volume.

Because molecules become more loosely packed as a substance changes from a solid to a liquid to a gas, the density of most substances decreases in the same sequence (Figure 2.4). The change of a substance from one of its physical states to another is called a **change of state**. Changes of state include freezing, as when water changes to ice, and vaporization, as when water boils. The water molecules remain unchanged through these transitions: The changes of state of water, and of many other substances, are physical, not chemical, changes. The alternative name **vapor** is often used for a gas, particularly when the liquid state is also present. Hence we call the change from liquid water to gaseous water (water vapor) vaporization. The names for other changes of state are given in Figure 2.4. They are discussed in more detail in Chapter 11.

$$\text{Density} = \text{mass per unit volume} = \frac{\text{mass}}{\text{volume}}$$

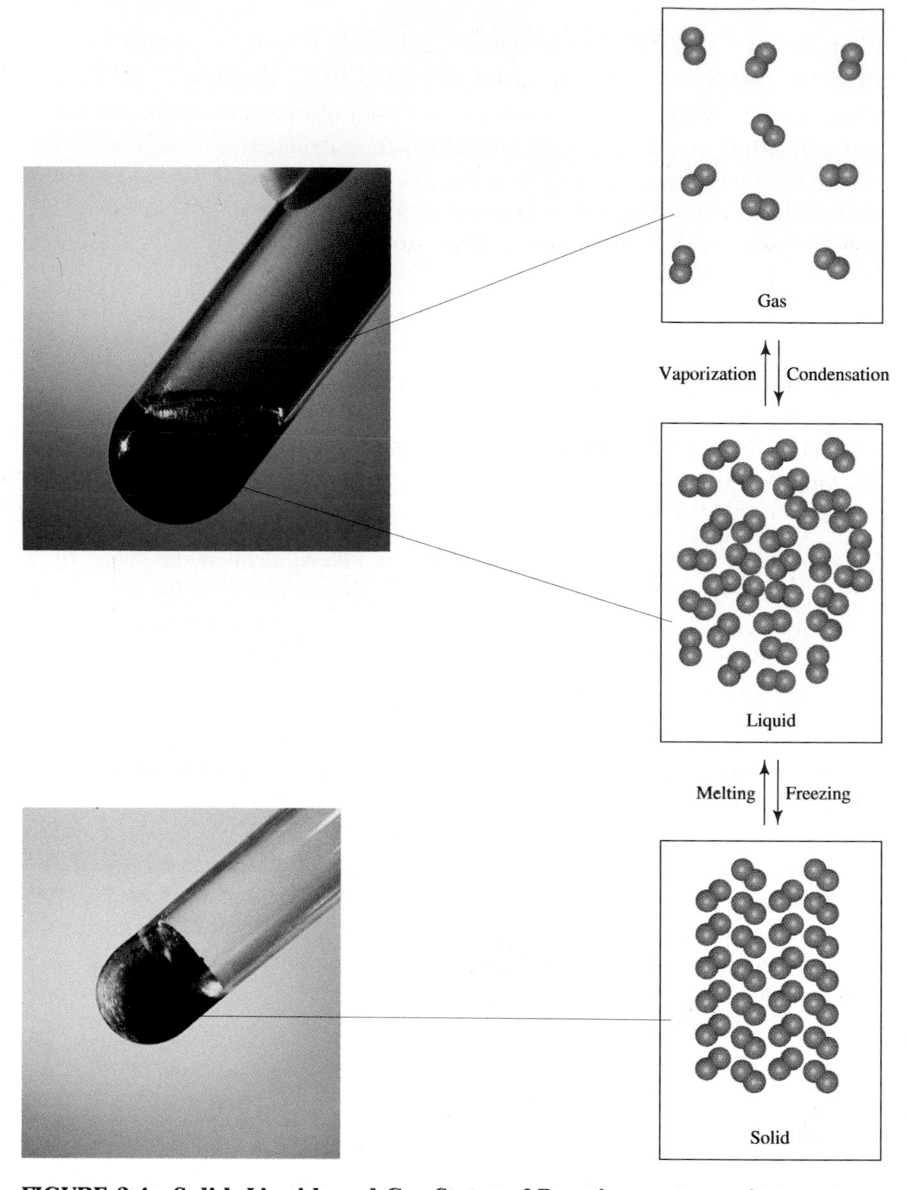

FIGURE 2.4 Solid, Liquid, and Gas States of Bromine At −15°C bromine is a dark brown solid. The bromine molecules are packed together closely in a regular arrangement. At room temperature bromine is a dark brown liquid. The molecules are still packed closely, but their arrangement is less regular, and they are constantly moving around one another. Brown bromine gas is also seen clearly in the tube above the liquid. The molecules in bromine gas are far apart and are moving rapidly so that they fill the whole tube. Because the molecules are far apart, the color of the gas is much less intense than the color of the liquid.

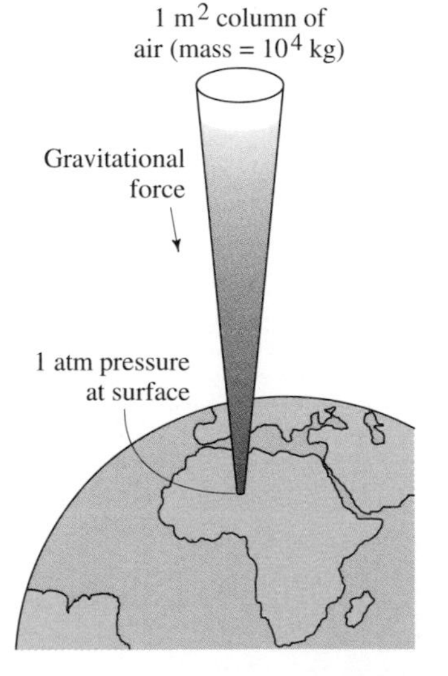

FIGURE 2.5 Atmospheric Pressure A column of the atmosphere 1 m^2 in cross-sectional area and extending to the top of the atmosphere has a mass of approximately 10 kg. It exerts a force $F = mg = (10 \text{ kg})(9.81 \text{ m} \cdot \text{s}^{-1}) = 1 \times 10^5$ N, where g is the acceleration due to gravity. Thus, the pressure P exerted by the atmosphere is $P = F/A = 1 \times 10^5 \text{ N} \cdot \text{m}^{-2} = 1 \times 10^5 \text{ Pa} = 100$ kPa.

Gas Pressure and the Atmospheric Pressure

When we compress a gas into a smaller volume, as when we pump up a bicycle tire or blow up a balloon, we can feel the pressure exerted by the gas on the piston of the pump or the skin of the balloon. **Pressure** is defined as force per unit area (Appendix P). In a car tire we measure this pressure in pounds per square inch (psi). For scientific measurements we use SI units (Appendix C). In SI the unit of force is the newton, and the unit of pressure is newtons per square meter ($N \cdot m^{-2}$), which is called a **pascal** (Pa):

One newton is the force required to give a 1-kg mass an acceleration of $1\ m \cdot s^{-2}$.

$$1\ Pa = 1\ N \cdot m^{-2}$$

The gravitational attraction between the earth and the gases of the atmosphere causes the atmosphere to exert a force on the earth's surface. **Atmospheric pressure** is the force exerted by the atmosphere on a unit area of the earth's surface (Figure 2.5). Atmospheric pressure was first measured by the Italian physicist Evangelista Torricelli (1608–1647) in 1643, using what we now call a Torricellian **barometer**, or mercury barometer (Figure 2.6). A long glass tube closed at one end is filled with mercury, and its open end is plugged. With the tube vertical, the plugged end is inserted under the surface of mercury in an open dish, and the plug is then removed. The mercury in the tube exerts a downward pressure due to gravitational force, so it begins to flow out into the dish, leaving a vacuum at the top of the tube. The flow stops, however, when the top of the mercury column in the tube is approximately 76 cm higher than the surface of the mercury in the dish. At this point, the downward pressure due to gravity is exactly balanced by the pressure of the atmosphere on the surface of the mercury in the dish, pressure that is pushing mercury *up* into the tube. The height of the mercury column supported by the atmosphere varies somewhat with the atmospheric conditions (the weather) and decreases with altitude, but it is normally about 76 cm = 760 mm at sea level.

Any liquid could in principle be used in a barometer. However, the denser the liquid, the shorter the column required to exert a given pressure. Mercury, because of its high density ($13.6\ g \cdot cm^{-3}$), gives a column of convenient height. Water, which has a much lower density ($1.00\ g \cdot cm^{-3}$), would give a column 13.6 times as high (10.3 m)! Because the mercury barometer is widely used, pressures are frequently expressed in terms of the height of a mercury column, that is, in millimeters of mercury (mm Hg). One millimeter of mercury is often called 1 *torr* in honor of Torricelli. A pressure of 760 mm Hg is called a *standard atmosphere* (atm):

$$1\ \textbf{standard atmosphere} = 1\ atm = 760\ mm\ Hg = 760\ torr$$

In SI, as we mentioned above, pressure is measured in pascals (newtons per square meter):

$$1\ Pa = 1\ N \cdot m^{-2}$$

The pascal is a very small unit, so kilopascals (kPa) are used more frequently. The relationship between kilopascals and the standard atmosphere is

$$1\ atm = 101.325\ kPa$$

Although the atmosphere is not an SI unit, it is a convenient unit for many purposes; we will use it often in this book.

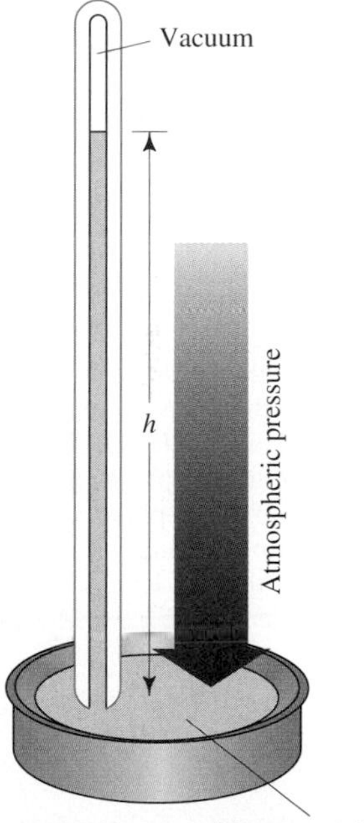

FIGURE 2.6 The Mercury Barometer The pressure of the atmosphere on the surface of the mercury pushes mercury up into the tube. Thus atmospheric pressure may be measured by the height, h, of the column of mercury. Average atmospheric pressure at sea level supports a column of mercury 760 mm high.

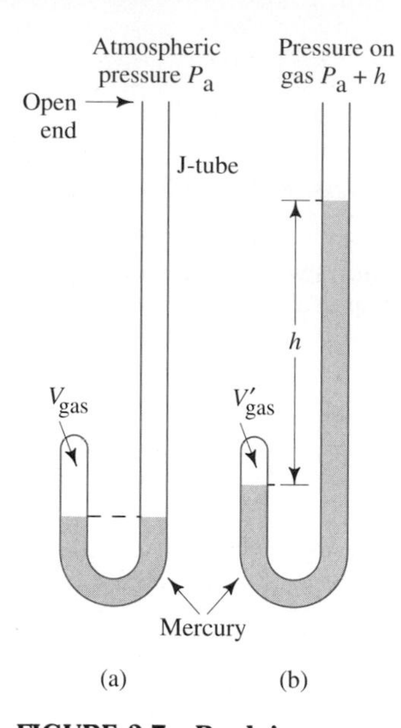

FIGURE 2.7 Boyle's Apparatus Boyle used this apparatus to study the volume and pressure of a gas sample. Mercury is poured into the J-tube, trapping air at the closed end of the tube. (a) When the height of the mercury is the same in the open and the closed parts of the tube, the pressure exerted on the gas equals the atmospheric pressure. (b) Mercury is added to the tube, increasing the pressure of the gas. Then the pressure exerted on the gas equals h (the difference in heights of the two mercury surfaces) plus the atmospheric pressure, and the volume of the gas is less.

2.6 THE GAS LAWS

Within ordinary ranges of temperature and pressure, most gases behave in very similar ways. The behavior of gases is expressed by a set of relationships among the pressure, the volume, and the temperature known as the gas laws.

Pressure and Volume: Boyle's Law

When we compress a gas into a smaller volume, as when we pump up a bicycle tire, the pressure exerted by the gas on its container increases. We can feel this pressure as we try to force more air in or if we squeeze the tire. Pumping up a bicycle tire shows us that every force is opposed by an equal and opposite force. The air in the tire exerts a pressure on the walls of the tire and on the piston of the pump; the walls of the tire and the piston of the pump exert an equal and opposite force on the air in the tire. So when we speak of the pressure of a gas, we mean equally the pressure exerted by the gas on its container and the pressure exerted by the container on the gas. Forces always come in equal and opposite pairs. This is a brief approximate statement of Newton's third law of motion (Appendix P).

In 1660 Irish chemist Robert Boyle (1627–1691) studied the effect of pressure on the volume of air, using the apparatus shown in Figure 2.7. He found that if he doubled the pressure, the volume of the air was halved; if he increased the pressure fourfold, the volume decreased to one-quarter of its original value. In general, the volume, V, of a given mass of air is inversely proportional to its pressure, P, if the temperature, T, is held constant:

$$V \propto \frac{1}{P} \quad \text{(if } T \text{ is held constant)}$$

This can also be written as $V = a \times 1/P$, where a is a constant. This relationship is true not just for air but for any gas. Still another way to state it is

$$PV = \text{constant} \quad \text{(if } T \text{ is held constant)}$$

Pressure times volume is constant for a given amount of gas at a constant temperature.

This relationship is known as **Boyle's law**. Figure 2.8 shows the relationship between P and V graphically.

Boyle's law expresses quantitatively the fact that a gas is compressible. The more a gas is compressed, the greater the pressure it exerts, and the greater the pressure exerted on the gas by its container. The atmosphere is compressed by the mass of gas above it. Thus the higher the altitude, the less the air is compressed, and the lower its pressure. At 2500 m (8000 ft) in the Rocky Mountains, the pressure is only 0.75 atm, and at 8000 m (26,000 ft) in the Himalayas, the world's highest mountains, the atmospheric pressure is only 0.47 atm (Figure 2.9). In contrast, the changes in pressure at sea level on earth are quite small. A high-pressure region such as a summer anticyclone has a pressure of about 1.03 atm, whereas a low-pressure region typically has a pressure of about 0.98 atm. The pressure in the eye of a hurricane may, however, fall as low as 0.90 atm.

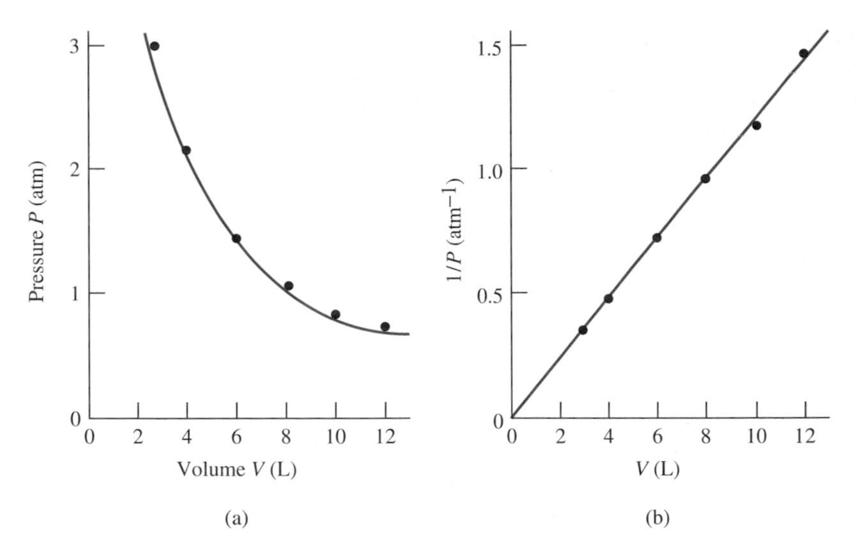

FIGURE 2.8 Boyle's Law Boyle's law gives the relationship between pressure and volume at constant temperature. If the temperature is held constant, the volume of a given amount of gas is inversely proportional to its pressure. (a) A plot of P versus V gives a curve (a hyperbola). (b) A plot of $1/P$ versus V (or $1/V$ versus P) gives a straight line.

FIGURE 2.9 Pressure and Altitude These climbers on the summit of Mt. Everest (8,848 m, or 29,948 ft) must carry oxygen tanks to assist in breathing the "thin" (low-pressure) air at that altitude.

Temperature and Volume: Charles' Law and the Kelvin Temperature Scale

In 1787 the French scientist Jacques Charles (1746–1823) found that the volume of a given amount of a gas increases linearly with increasing temperature when the pressure is kept constant. If we plot volume against temperature (Figure 2.10a), different amounts of gas at different temperatures give different straight lines, but all the lines extrapolate to a volume of zero at $-273°C$. At a still lower temperature, a gas would apparently have a negative volume. We can attach no meaning to a negative volume, so we must assume that we cannot obtain a temperature lower than $-273°C$, which is therefore called the **absolute zero of temperature**.

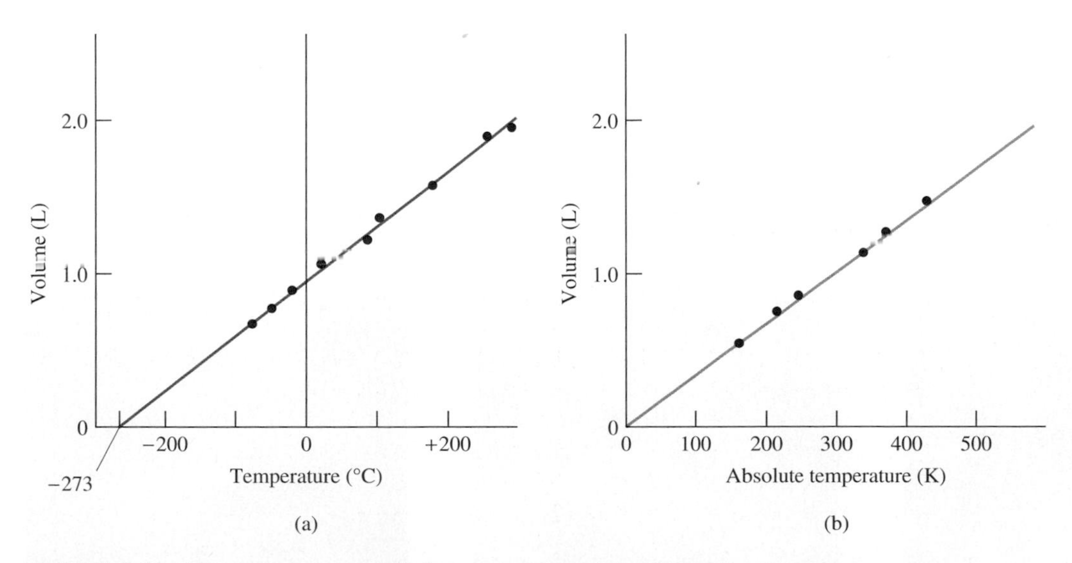

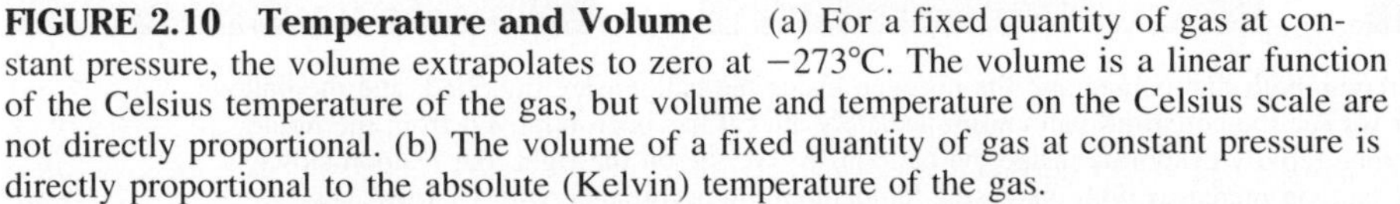
FIGURE 2.10 Temperature and Volume (a) For a fixed quantity of gas at constant pressure, the volume extrapolates to zero at $-273°C$. The volume is a linear function of the Celsius temperature of the gas, but volume and temperature on the Celsius scale are not directly proportional. (b) The volume of a fixed quantity of gas at constant pressure is directly proportional to the absolute (Kelvin) temperature of the gas.

In practice, all gases condense to liquids and solids at temperatures above −273°C, so no gas can actually be cooled until it has zero volume. Nevertheless, the idea of a lowest possible temperature, the absolute zero, is very important. It is logical to use the absolute zero rather than an arbitrary temperature as the zero of a temperature scale—for example, the melting point of ice, as for the Celsius scale. The use of the absolute zero is the basis of the **Kelvin temperature scale**, first suggested by the British scientist Lord Kelvin (1824–1907). One unit of the Kelvin scale represents the same interval as one unit of the Celsius scale, but the zero point of the Kelvin scale is the absolute zero. According to accurate measurements, the absolute zero of temperature is −273.15°C. Thus 0 K = −273.15°C, and the Kelvin scale is related to the Celsius scale by the expression

$$T = t + 273.15$$

where T is the temperature on the Kelvin scale and t is the temperature on the Celsius scale. In other words, temperatures on the Celsius scale are converted to temperatures on the Kelvin scale by adding 273.15. By convention, the degree sign (°) is not used when we express temperatures on the Kelvin scale. The unit of the Kelvin scale is the *kelvin* (K), and a temperature such as 100 K is read as "one hundred kelvins."

If the temperature is expressed on the Kelvin scale, the volume of a gas held at constant pressure is directly proportional to the temperature (Figure 2.10b):

$$V \propto T, \qquad \text{or} \qquad V = b \times T \quad (\text{if } P \text{ is held constant})$$

where T is the temperature on the Kelvin scale and b is a constant. These expressions summarize **Charles' law**, which may be stated as follows:

The volume of a given mass of gas at constant pressure is proportional to its temperature on the Kelvin scale.

The variation of the volume of a gas with temperature is shown in Demonstration 2.4. Hot-air balloons are an interesting application of Charles' law (see Box 2.2).

DEMONSTRATION 2.4 Liquefaction of Air and Charles' Law

When a balloon is plunged into liquid nitrogen, both the oxygen and the nitrogen inside the balloon are liquefied, and the balloon collapses almost completely. In the center we see the collapsed balloon immediately after it has been removed from the liquid nitrogen. The liquid oxygen and nitrogen then rapidly evaporate inside the balloon. As we see on the right, the balloon slowly expands as the volume of the gases in the balloon increases with increasing temperature, in accordance with Charles' law.

BOX 2.2 Balloons

You may have seen a brightly colored hot-air balloon floating across the sky on a summer day (Figure A). The sport of hot-air ballooning has revived our earliest means of getting into the air.

A balloon filled with any gas less dense than air rises in the atmosphere, because the mass of the air displaced by the balloon is greater than the mass of the balloon. The density of the atmosphere decreases with increasing altitude, so the balloon rises until its mass and its load equal the mass of the air it displaces, at which point the balloon floats in the atmosphere. Two French scientists, Jacques Charles and Joseph Louis Gay-Lussac, who made some of the first quantitative studies of the properties of gases, were pioneers in the use of balloons. The first balloon flight, with a hot-air balloon, was made in France by the brothers Joseph and Jacques Montgolfier in June 1783. A few months later, Charles filled a balloon with hydrogen, which he made by the reaction of about 250 kg of acid with 500 kg of iron! The balloon remained aloft 45 min and traveled a distance of about 25 km. In 1804 Gay-Lussac set a record by ascending to 7 km (23,000 ft) in a hydrogen-filled balloon. He used balloon ascents to carry out experiments that included studies of the composition of the atmosphere and variations in the earth's magnetic field. Hydrogen-filled balloons are still used for weather observations today. Because of the very low density of hydrogen, they can reach heights of approximately 40 km.

By the 1930s a logical development of the hydrogen balloon, the airship, was providing regular transportation across the Atlantic. Instead of just drifting in the wind, airships were driven by engine-powered propellers and included cabins for passengers. The disastrous fire that destroyed the German airship *Hindenburg* in 1937 marked the end of the airship era, but the possibility of building airships that use helium, which is nonflammable but expensive, has attracted new interest. Although they would be too slow to compete with jets, helium airships might have other uses, such as transporting timber in forestry operations.

Figure A *Hot-air balloons.*

Figure B *A nineteenth-century representation of Gay-Lussac's August 1804 ascent (with physicist Jean-Baptiste Biot) in a hydrogen-filled balloon. This ascent reached 4 km (13,000 ft). Gay-Lussac (alone) reached 7 km (23,000 ft) in a September 1804 balloon flight.*

Molecules and Volume: Avogadro's Law

The Italian scientist Amedeo Avogadro (1776–1856) was the first to suggest that the volume occupied by a gas at a given temperature and pressure is proportional to the number of molecules present and is independent of the type of gas. One mole of hydrogen molecules, H_2, occupies the same volume as 1 mol of oxygen molecules, O_2, or 1 mol of methane molecules, CH_4, if the gases are at the same temperature and pressure. Stated in terms of the number of moles of molecules, n,

$$V \propto n, \qquad \text{or} \qquad V = c \times n \quad (\text{if } P \text{ and } T \text{ are held constant})$$

where c is a constant.

The volume of a sample of a gas at a given temperature and pressure is proportional to the number of moles of molecules in the sample.

This statement is known as **Avogadro's law**.

The Ideal Gas Law and Molar Volume

If we combine the three relationships expressing Boyle's law, Charles' law, and Avogadro's law,

$$V \propto \frac{1}{P}, \qquad V \propto T, \qquad \text{and} \qquad V \propto n$$

we obtain the relationship

$$V \propto \frac{nT}{P}, \qquad \text{or} \qquad V = \frac{RnT}{P}$$

where R is a constant called the **gas constant**. This equation is usually written in the form

$$PV = nRT$$

and is known as the **ideal gas equation**, or the **ideal gas law**. If pressure is expressed in atmospheres, volume in liters, and temperature in kelvins, R has the value 0.08206 atm · L · mol^{-1} · K^{-1}. If the pressure, volume, and temperature are expressed in SI units, the value of R is 8.314 kPa · dm^3 · mol^{-1} · K^{-1}.

An **ideal gas** is a hypothetical gas that obeys the ideal gas law exactly, under all conditions of pressure and temperature. No such gas exists, but the ideal gas law describes the behavior of nearly all real gases quite closely, particularly if the pressure is 1 atm or less and the temperature is well above the boiling point. Boyle's, Charles', and Avogadro's laws are special cases of the ideal gas law: the first at constant n and T, the second at constant n and P, and the third at constant P and T. Like the ideal gas law, they are only approximately true for real gases.

We can use the ideal gas law to calculate the **molar volume** (the volume of 1 mole) for an ideal gas at a given temperature and pressure. It is convenient to choose 1 atm (101.33 kPa) and 0°C (273.15 K) as a *standard pressure and temperature* for specifying volumes of gases. These conditions are abbreviated as STP. The molar volume of an ideal gas at STP is

$$V = \frac{nRT}{P} = \frac{1 \text{ mol} \times 0.08206 \text{ atm} \cdot \text{L} \cdot \text{mol}^{-1} \cdot \text{K}^{-1} \times 273.15 \text{ K}}{1 \text{ atm}} = 22.41 \text{ L}.$$

The molar volume of most real gases at STP is fairly close to 22.41 L. Some examples are given in Table 2.4.

TABLE 2.4 Molar Volumes of Some Gases

Gas	Molar Volume at STP (L · mol^{-1})
Hydrogen, H_2	22.43
Neon, Ne	22.44
Oxygen, O_2	22.39
Nitrogen, N_2	22.40
Carbon dioxide, CO_2	22.26

Exercise 2.3 Express the following pressures in atmospheres and temperatures in kelvins: 745 mm Hg, 783 torr, 100 kPa, 25.0°C, 100.0°C, − 150°C.

Exercise 2.4 Consider an experiment in which water is electrolyzed. The resulting hydrogen and oxygen are collected separately at 25°C and 1 atm pressure, and the volume of oxygen collected is 158.9 mL. What volume of hydrogen is collected?

Exercise 2.5 What would be the molar volume of an ideal gas at 1 atm and 25.0°C?

2.7 THE KINETIC THEORY OF GASES

We have described a gas as a collection of randomly moving molecules with rather large distances between them. This description is the basis of the kinetic theory of gases, which enables us to explain the gas laws. The **kinetic theory of gases** is a *model* (Box 2.3) of a gas based on the following assumptions:

1. A gas consists of a collection of molecules in continuous random motion.
2. The molecules are regarded as infinitely small particles with zero volume.
3. Each molecule moves in a straight line unless it collides with another molecule or with the walls of the container.
4. The molecules exert no force on one another or on the container, except when they collide with one another or with the walls of the container. When molecules collide, one molecule may gain energy and the other may lose energy, but their total energy remains constant. As a result, the total energy of all the molecules in the container remains constant.
5. The average kinetic energy of the molecules of a gas is proportional to the absolute temperature. *The kinetic energy of a molecule is* $\frac{1}{2}mv^2$*, where m is its mass and v its speed* (Appendix P).

On the basis of these assumptions, the behavior of an ideal gas can be calculated by using the laws of mechanics. We will not show how this can be done, but we will see how the theory enables us to understand the gas laws qualitatively.

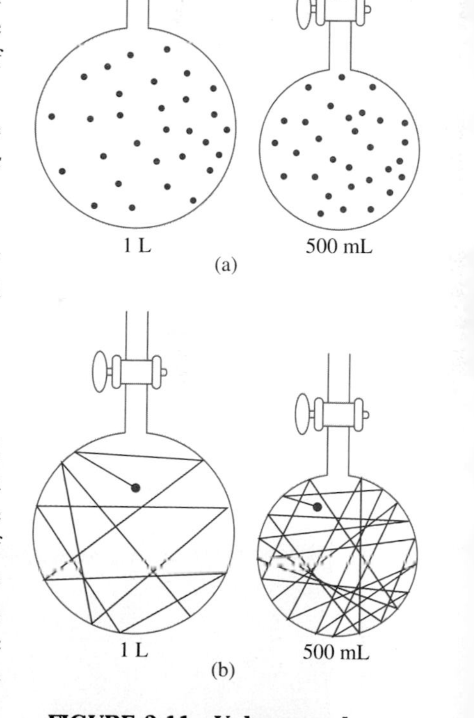

FIGURE 2.11 Volume and Pressure (a) If the volume of gas is halved while the temperature and number of molecules are held constant, the molecules pack closer together and the density doubles. (b) On average, a molecule now hits the container wall twice as often. The total number of impacts with the wall is therefore doubled, which doubles the pressure.

Kinetic Theory and Ideal Gas Behavior

The pressure exerted by a gas on the walls of its container results from the continual bombardment of the walls by the molecules. Every time a molecule collides with a wall, it exerts a force on the wall. Because pressure is force per unit area, the pressure is proportional to the number of collisions per unit area in a given time. If we decrease the total volume of a gas sample to half its original volume, the molecules will undergo twice as many collisions per unit area with the walls of the container (Figure 2.11). Consequently, the pressure will double. Thus the kinetic theory provides a simple explanation of Boyle's law, that pressure and volume are inversely proportional:

$$P \propto \frac{1}{V}$$

If we double the number of molecules in a given volume, the number of collisions that the molecules undergo with the container's walls will also double and the pressure will double correspondingly. Or if we keep the pressure (and therefore the number of collisions) constant, the volume will double. Thus the kinetic theory also explains Avogadro's law, that the pressure is proportional to the number of molecules and hence to the number of moles, n:

$$P \propto n$$

BOX 2.3 Laws, Models, and Theories

A fundamental activity of science is making **observations** of the world around us. Carefully made observations that can be repeated by other persons constitute scientific **facts**. When a large number of observations have been made and certain facts established, it may be possible to make a concise statement or give a mathematical equation that summarizes these observations. Such a statement is known as a scientific **law**. Boyle's law, Charles' law, and the law of conservation of mass are statements that in each case summarize a large number of observations. Another familiar example is the law of gravity, which summarizes the numerous observations that show that separate masses attract one another.

Once a law has been established, scientists ask, Why does nature behave in the way that is summarized by the law? Why for example, is the volume of a gas inversely proportional to the pressure, as stated by Boyle's law? To try to answer such a question, we propose a **model**, a set of assumptions proposed to explain certain observations. For example, the model that has been proposed to explain the gas laws is that a gas consists of widely separated molecules moving at high speed. This model enables the behavior of gases to be understood both qualitatively and quantitatively. To be useful, a model should suggest new experiments, so we can see whether new observations are consistent with it. In other words, a model should result in **predictions** that can be tested. A model that has not been adequately tested is called a **hypothesis**. A model that withstands such testing usually comes to be called a **theory**.

A theory is always subject to change. It is useful only as long as no observations are made for which it cannot account. A single observation that a theory cannot explain will eventually cause the theory to be modified or replaced by a new model or theory. Dalton's atomic theory, for instance, is still very useful today, but it has been modified as a result of experiments carried out since Dalton's time. As we have seen, it no longer states that an atom is indivisible or that all the atoms of an element have the same mass. It is a mistake to believe that if we know all the current theories, we need not know the experimental facts. A new theory can be developed only by those who have a wide knowledge of the facts, particularly those facts that have not been satisfactorily accounted for by existing theories.

Some theories become so well established that we may begin to call them facts. For example, now that we can see individual atoms and molecules with the scanning tun-

Figure A *The mid-1920s saw the development of the quantum theory, which had a profound effect on chemistry. Many theories in science are first presented at international meetings. This photograph of well-known scientists was taken at the international Solvay Conference in 1927. Among those present are many whose names are mentioned in this book. Front row, left to right: I. Langmuir, M. Planck, M. Curie, H. A. Lorentz, A. Einstein, P. Langevin, C. E. Guye, C. T. R. Wilson, O. W. Richardson. Second row, left to right: P. Debye, M. Knudsen, W. L. Bragg, H. A. Kramers, P. A. M. Dirac, A. H. Compton, L. V. de Broglie, M. Born, N. Bohr. Standing, left to right: A. Piccard, E. Henriot, P. Ehrenfest, E. Herzen, T. De Donder, E. Schrödinger, E. Verschaffelt, W. Pauli, W. Heisenberg, R. H. Fowler, L. Brillouin.*

neling microscope, for example (Box 1.1) and can draw detailed pictures of molecules on computer screens, it is inconceivable that the atomic theory will ever be abandoned or even modified to any significant extent (Figure A). So we are now entitled to say that it is a fact that matter consists of atoms and molecules. Similarly, although no one has ever seen an intact dinosaur, the theory that dinosaurs once existed on earth but became extinct about 65 million years ago is now so well established by the fossil record that we accept this theory as a fact. But the reason for the extinction of the dinosaurs is still a matter open to discussion and to the proposal of new hypotheses.

The deduction of laws from observations and the development of models and theories that lead to predictions—which in turn must be tested by experiment, leading to new observations—is a continuous and never-ending process. This process is known as the **scientific method** (Figure B). Although this might appear to be a logical process, science does not advance in a completely organized and logical manner. Success in scientific research often depends on the ability to observe and interpret the unexpected. (Box 8.1 describes one particularly interesting example of an unexpected result for which a new model had to be developed.) The experiment that "does not work" can be the clue to an important discovery. New theories are generally not developed in an entirely logical and planned manner but through a long process of trial and error in which many incorrect ideas and models may be proposed before an adequate theory is formulated. Slow and tortuous though this process may be, it is often exciting and in the end very satisfying. The development of a new model or theory may depend as much on flashes of insight and on inspiration and imagination as on logical argument. One brilliant inspiration may bring to an end years of groping in the dark for understanding.

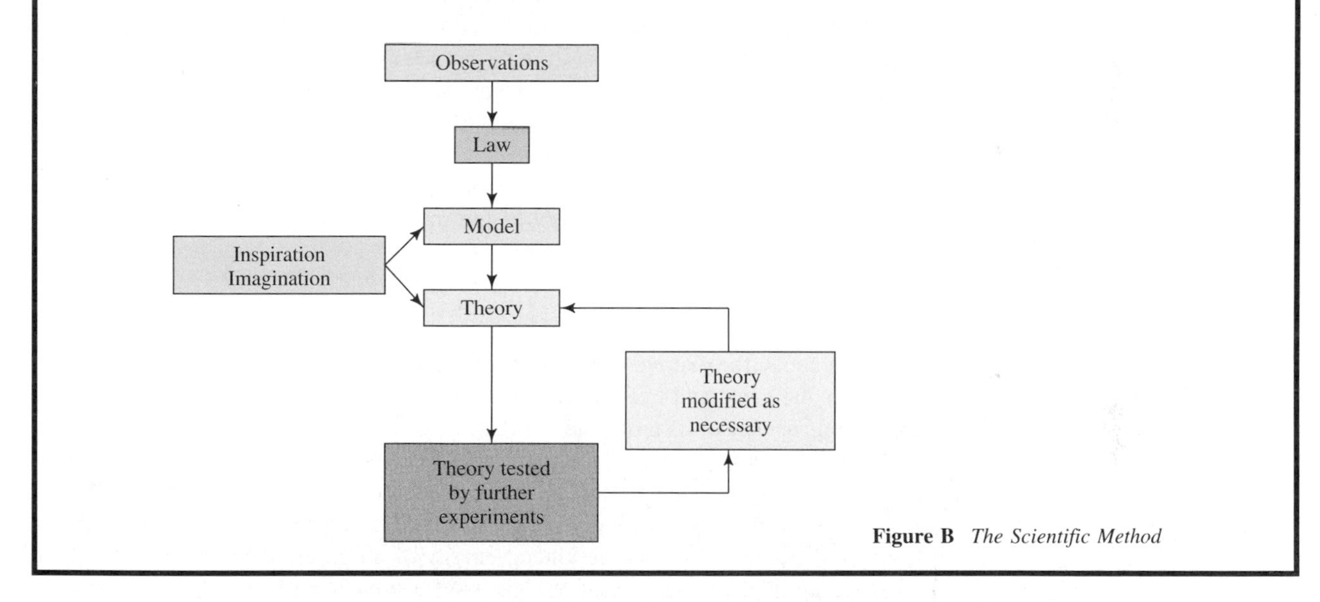

Figure B *The Scientific Method*

If we increase the temperature of a given volume of gas, the molecules move faster and therefore have greater kinetic energies. As a result, they hit the container walls more frequently and with higher kinetic energy, exerting a greater force on the walls. Therefore, the pressure of the sample increases as its temperature rises. A detailed calculation of the force exerted on the walls shows that the pressure is proportional to the temperature, as stated by Charles' law:

$$P \propto T$$

Combining the preceding three relationships, we have

$$P \propto \frac{nT}{V} \qquad \text{or} \qquad P = k\frac{nT}{V}$$

If we then write R for the constant k, we obtain the ideal gas law,

$$PV = nRT$$

Kinetic Energy, Heat, and Temperature

An assumption we made in presenting the kinetic theory was that the average kinetic energy of the molecules of a gas is proportional to the temperature. This statement is not so much an assumption as a definition of temperature. Suppose that a cool gas sample is allowed to mix with a warm gas sample. Each molecule has a kinetic energy of $\frac{1}{2}mv^2$, where m is the molecule's mass and v is its speed. A molecule with a higher speed has more kinetic energy than one with a lower speed. In the collisions between the slower molecules of the cooler gas and the faster molecules of the warmer gas, the slower molecules are speeded up, and faster molecules are slowed down. Similarly, in a rear-end collision between two cars, the slower car in front is speeded up while the faster car at the back is slowed down. The faster car loses some of its energy, and the slower car gains this energy. This is a consequence of the laws of conservation of energy and of momentum (Appendix P). In the same way, the average kinetic energy of the molecules of the warmer gas decreases, whereas the average kinetic energy of the molecules of the cooler gas increases. Kinetic energy is transferred from the warmer gas to the cooler gas. We refer to this energy that is transferred as a result of a temperature difference as **heat**. Although we say that heat flows from the hot gas to the cooler gas, we really mean that *kinetic energy* is transferred from the hot gas to the cooler gas. We will discuss heat further in Chapter 9.

Temperature, which we perceive subjectively as how hot or cold something is, measures the average kinetic energy of the molecules of a substance:

The temperature of a substance is proportional to the average kinetic energy of its molecules.

As a gas cools, the average speed and kinetic energy of its molecules decrease. In principle, the gas could eventually reach a temperature at which the average speed and kinetic energy of its molecules were zero. Because the speed and kinetic energy could not decrease further, this temperature must be the lowest possible temperature, that is, the absolute zero, 0 K.

The average kinetic energy ($\frac{1}{2}mv^2$) of the molecules of a gas depends only on the temperature, so the average kinetic energy of all gases is the same at the same temperature. If the molecules of one gas have a greater mass than those of another gas, they move more slowly and therefore have the same kinetic energy. If two gases at the same temperature had *different* average kinetic energies and we mixed the gases, collisions between the molecules of the two gases would increase the average kinetic energy of one gas and decrease the average kinetic energy of the other. In other words, kinetic energy (heat) would flow from one gas to the other: One gas would get warmer and the other would get cooler. We know from experience that no heat flows if two gases are at the same temperature, so two gases at the same temperature must have the same average kinetic energy:

$$\tfrac{1}{2}m_1{v_1}^2 = \tfrac{1}{2}m_2{v_2}^2$$

2.8 USING THE IDEAL GAS LAW

If we know any three of the variables P, V, n, and T that describe the physical state of a gas, we can calculate the fourth by using the ideal gas equation. The ideal gas equation describes the behavior of real gases only approximately, but for many purposes this approximation is sufficiently accurate. For the calculations in this book we assume that real gases behave like an ideal gas.

Number of Moles

We can find n, the number of moles, in a gas sample of known volume, pressure, and temperature from the relationship

$$n = \frac{PV}{RT}$$

EXAMPLE 2.1 Calculation of the Number of Moles of a Gas

How many moles of gas are there in a 2.00-L sample of the hydrocarbon ethane at 740 mm Hg pressure and 25.0°C?

Solution: Using $R = 0.0821\ \text{atm} \cdot \text{L} \cdot \text{mol}^{-1} \cdot \text{K}^{-1}$, remember that P must be in atmospheres, V in liters, and T in kelvins:

$$P = (740\ \text{mm Hg})\left(\frac{1\ \text{atm}}{760\ \text{mm Hg}}\right)$$

and $T = (25.0 + 273.1) = 298.1$ K. Then, using $PV = nRT$,

$$\text{moles ethane} = n = \frac{PV}{RT} = \frac{(740\ \text{mm Hg})\left(\frac{1\ \text{atm}}{760\ \text{mm Hg}}\right)(2.00\ \text{L})}{(0.0821\ \text{atm} \cdot \text{L} \cdot \text{mol}^{-1} \cdot \text{K}^{-1})(298.1\ \text{K})}$$

$$= 0.0796\ \text{mol} \qquad \text{(3 significant figures)}$$

We have assumed ideal gas behavior, so this result will apply to any gas under the same conditions.

Because 1 atm = 760 mm Hg, we form a unit conversion factor by dividing both sides by 760 mm Hg to get

$$\frac{1\ \text{atm}}{760\ \text{mm Hg}} = 1.$$

This factor converts pressures in mm Hg to pressures in atm.

Exercise 2.6 How many moles and how many molecules are there in a 1.00-L sample of $H_2O(g)$ at 100°C and exactly 1 atm pressure?

Molar Mass

We saw in Example 2.1 that we can use the ideal gas law to find the number of moles of a gas if we can measure V, P, and T. If we also know the mass of the gas, we can calculate its molar mass, M, from the expression

$$M = \frac{\text{mass of gas}}{\text{number of moles}} = \frac{m}{n}$$

EXAMPLE 2.2 Calculation of the Molar Mass of a Gas

If the ethane sample of Example 2.1 has a mass of 2.393 g, what is the molar mass of ethane?

Solution: The number of moles of ethane is 0.0796 mol; the mass is 2.393 g:

$$1\ \text{molar mass of ethane} = \frac{2.393\ \text{g}}{0.0796\ \text{mol}} = 30.1\ \text{g} \cdot \text{mol}^{-1}.$$

Exercise 2.7 A 3.525-g sample of $CO_2(g)$ occupies a volume of 2.00 L at a pressure of 745 mm Hg and a temperature of 25.0°C. How many moles of CO_2 are present, and what is the molar mass of $CO_2(g)$?

Molecular Formula from Molecular Mass

If we know the empirical formula of a compound and if we determine its molar mass, we can find its molecular formula from the relationships

The integer n in this equation is not related to the number of moles n in the ideal gas law $PV = nRT$.

$$\frac{\text{molecular mass}}{\text{empirical formula mass}} = n$$

and molecular formula = $n \times$ (empirical formula).

EXAMPLE 2.3 Calculation of a Molecular Formula from an Empirical Formula

If the empirical formula of ethane is CH_3, what is its molecular formula?

Solution: From the solution to Example 2.2, the molecular mass of ethane is 30.1 u. From the formula CH_3, the empirical formula mass of CH_3 is (12.01 u + 3.02 u) = 15.03 u, so

$$n = \frac{30.1 \text{ u}}{15.03 \text{ u}} = 2.00$$

Thus the molecular formula of ethane is $2(CH_3)$, or C_2H_6.

Exercise 2.8 The empirical formula of benzene is CH. What is its molecular formula, if 0.638 g of benzene occupies a volume of 250 mL at 100°C and a pressure of 1.00 atm?

Gas Density and Molar Mass

Another property that we can calculate from the ideal gas law is the density of a gas at any given temperature and pressure. Using the relationship

$$\text{number of moles} = \frac{\text{mass}}{\text{molar mass}} \quad \text{or} \quad n = \frac{m}{M}$$

we may express the ideal gas law, $PV = nRT$, in the form

$$PV = \frac{mRT}{M} \quad \text{or} \quad \frac{m}{V} = \frac{PM}{RT}$$

Because

$$\text{density} = \frac{\text{mass}}{\text{volume}} \quad \text{or} \quad d = \frac{m}{V}$$

we have

If a molecule's molar mass is not stated in a problem, it can be calculated from the sum of the molar masses of its atoms.

$$d = \frac{PM}{RT}$$

From this equation we can calculate the density of a gas at any pressure and temperature if we know its molar mass.

EXAMPLE 2.4 Calculation of the Density of a Gas

What is the density of sulfur dioxide, $SO_2(g)$, of molar mass 64.07 $g \cdot mol^{-1}$, at STP?

Solution: STP is 0°C and a pressure of 1 atm. Therefore, $T = 273.1$ K and $P = 1$ atm, and

$$d = \frac{MP}{RT} = \frac{(64.07 \text{ g} \cdot \text{mol}^{-1})\ (1 \text{ atm})}{0.0821 \text{ atm} \cdot \text{L} \cdot \text{mol}^{-1} \cdot \text{K}^{-1})\ (273.1 \text{ K})} = 2.86 \text{ g} \cdot \text{L}^{-1}$$

We can use the same equation in the form $M = dRT/P$ to find the molar mass of a gas from the known density. This calculation is simply a variation of the calculation of the molar mass from the known mass of a measured volume of a gas that we described in Example 2.2.

EXAMPLE 2.5 Calculation of Molar Mass from Gas Density

The density of a gas at 25.0°C and a pressure of 2.00 atm is $3.596\ g \cdot L^{-1}$. What is its molar mass? The gas is an oxide of carbon. What is its molecular formula?

Solution: Rearranging the equation $d = PM/RT$, we have

$$M = \frac{dRT}{P}$$

$$= \frac{(3.596\ g \cdot L^{-1})(0.0821\ atm \cdot L \cdot mol^{-1} \cdot K^{-1})(298.1\ K)}{2.00\ atm} = 44.01\ g \cdot mol^{-1}$$

The gas is CO_2, molar mass $= 12.01\ g \cdot mol^{-1} + 2(16.00\ g \cdot mol^{-1}) = 44.01\ g \cdot mol^{-1}$.

If the gas is an oxide of carbon, its molecular mass must be $x(12.01) + y(16.00)$. By trial and error, we find that $x = 1$ and $y = 2$.

Exercise 2.9 What is the density of water vapor, molar mass 18.02 $g \cdot mol^{-1}$, at 100°C and a pressure of 1.00 atm?

Exercise 2.10 The density of a hydrocarbon found in natural gas at 25.0°C and a pressure of 1.00 atm is $2.375\ g \cdot L^{-1}$. What are its molar mass and molecular formula?

Dependence of Volume on Temperature and Pressure

We can use the ideal gas law to find the change in volume of a given sample of a gas when the temperature and/or the pressure are changed. Suppose that a given mass of gas has a volume V_1 at a pressure P_1 and a temperature T_1. From the ideal gas law we have

$$\frac{P_1V_1}{T_1} = nR$$

Now suppose that the pressure is changed to P_2 and the temperature to T_2, so the volume is now V_2. We now have

$$\frac{P_2V_2}{T_2} = nR$$

Because the number of moles, n, remains unchanged,

$$\frac{P_1V_1}{T_1} = \frac{P_2V_2}{T_2}$$

so

$$V_2 = \frac{V_1P_1T_2}{P_2T_1}$$

This expression can be used to obtain either a new volume V_2, a new temperature T_2, or a new pressure P_2, depending on which properties are given in a problem.

From this relationship we can calculate the new volume V_2 for any new conditions P_2 and T_2.

EXAMPLE 2.6 Calculation of the Dependence of Volume on Temperature and Pressure

A weather balloon on the ground is charged with 50.0 L of hydrogen at a pressure of 1.00 atm at 25.0°C. What is the volume of the balloon when it ascends to an altitude where the temperature is −15.0°C and the pressure inside it is 0.300 atm?

Solution: For a given mass of gas,

$$\frac{PV}{T} = \text{constant}$$

$$\frac{P_1V_1}{T_1} = \frac{P_2V_2}{T_2};\ V_2 = \frac{P_1V_1T_2}{P_2T_1} = \frac{(1.00\ \text{atm})\ (50.0\ \text{L})\ (258.1\ \text{K})}{(0.300\ \text{atm})\ (298.1\ \text{K})} = 144\ \text{L}$$

Exercise 2.11 The volume of gas inside an automobile cylinder is 0.50 L at a pressure of 1.00 atm and a temperature of 20.0°C. What is the volume when the gas is compressed to a pressure of 5.00 atm at a temperature of 200°C?

Dalton's Law of Partial Pressures

We have seen that all gases behave in approximately the same way, so it is not surprising that a mixture of gases behaves very much like a single gas. Each molecule in a mixture moves independently, just as it would in a single pure gas. Each gas distributes itself uniformly throughout the container. The molecules of each gas strike the container walls with the same frequency and force and therefore exert the same pressure as they do when no other gas is present. In other words, the pressure exerted by any one gas in a mixture is the same as it would be if the gas occupied the container by itself. This pressure is called the **partial pressure** of the gas. The total pressure is the sum of the partial pressures of all the gases in the mixture. **Dalton's law of partial pressures**, formulated by John Dalton (1766–1844) in 1803, states that

> **The total pressure of a mixture of gases is the sum of the partial pressures of its components:**

$$P_{\text{total}} = p_1 + p_2 + p_3 + \ldots$$

where p_1, p_2, and so on represent the partial pressures of each gas in the mixture.

We can apply the ideal gas equation to any gas in the mixture. For gas 2, for example, $p_2V = n_2RT$, where V is the total volume of the gas mixture. Generalizing, we can say that, for gas i,

$$p_iV = n_iRT.$$

We see that the partial pressure of a gas is proportional to the number of molecules of that gas in the mixture, or $p_i \propto n_i$, because the number of molecules is proportional to the number of moles.

EXAMPLE 2.7 Calculation of Partial Pressures

The composition of dry air at sea level is 78.08% nitrogen, 20.95% oxygen, 0.93% argon, and 0.03% carbon dioxide by volume, with traces of other gases. What are the partial pressures of nitrogen, oxygen, argon, and carbon dioxide at 21°C if the barometer reading is 754.1 torr?

Solution: The number of moles of each gas is proportional to the volume at a constant temperature, and the partial pressures are proportional to the numbers of moles. Thus

$$p_{N_2} = \left(\frac{78.08}{100}\right)(754.1 \text{ torr}) = 588.8 \text{ torr}$$

$$p_{O_2} = \left(\frac{20.95}{100}\right)(754.1 \text{ torr}) = 158.0 \text{ torr}$$

$$p_{Ar} = \left(\frac{0.93}{100}\right)(754.1 \text{ torr}) = 7.0 \text{ torr}$$

$$p_{CO_2} = \left(\frac{0.03}{100}\right)(754.1 \text{ torr}) = 0.2 \text{ torr}$$

for a total pressure of 754.0 torr. The total pressure is not quite equal to the barometer reading (754.1 torr) because we have neglected the traces of other gases.

Check for errors by summing the calculated partial pressures to get the total pressure.

Exercise 2.12 In the electrolysis of water, a dry (water-free) sample of the mixture of $H_2(g)$ and $O_2(g)$ produced is collected in a single vessel at a total pressure of 736.2 mm Hg. What is the partial pressure of each gas in the mixture?

2.9 DIFFUSION AND EFFUSION

The mixing of gases is an everyday occurrence. We smell substances such as perfume or coffee because molecules of the perfume or coffee mix with the O_2 and N_2 molecules of the air and eventually reach our nose. In the case of a colored gas such as bromine vapor, we can actually see the mixing (Demonstration 2.5). This mixing of one gas with another is called **diffusion**, and it illustrates that the molecules of a gas are in motion, as assumed by kinetic theory of gases. A closely related phenomenon is called **effusion**, the process by which a gas escapes from a container through a very small opening.

Both diffusion and effusion result from the chaotic, random motion of gas molecules (Figure 2.12). An important feature of these two processes is that they are irreversible. Two gases mix by diffusing into each other, but they never "unmix":

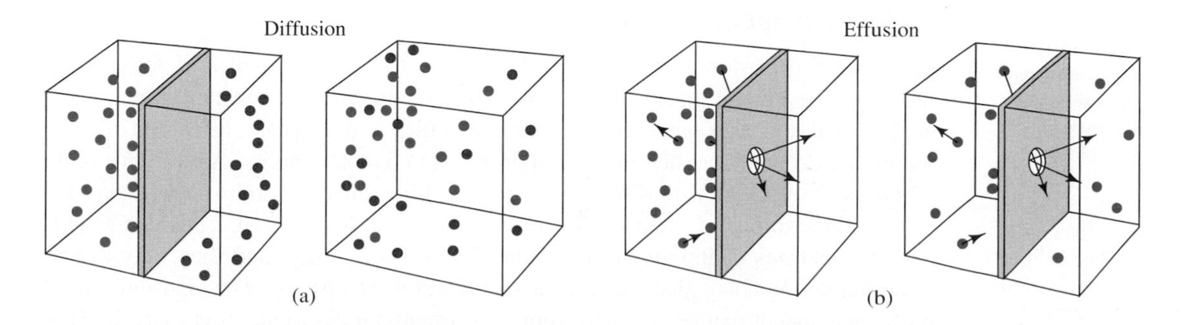

FIGURE 2.12 Diffusion and Effusion (a) Diffusion: When the partition is removed, the molecules spread out and fill the container. (b) Effusion: The molecules escape through the very small hole in the partition.

DEMONSTRATION 2.5
Diffusion of Bromine Vapor

A few drops of bromine are placed at the bottom of a tube that contains air at atmospheric pressure.

After 12 s, the bromine molecules have diffused a short distance up the tube. Although the bromine molecules are moving at speeds of several hundred meters per second, their constant collisions with nitrogen and oxygen molecules of the air cause them constantly to change direction. Thus their overall movement up the tube is rather slow.

After 1 min 30 s, some of the bromine molecules have traveled about halfway up the visible portion of the tube.

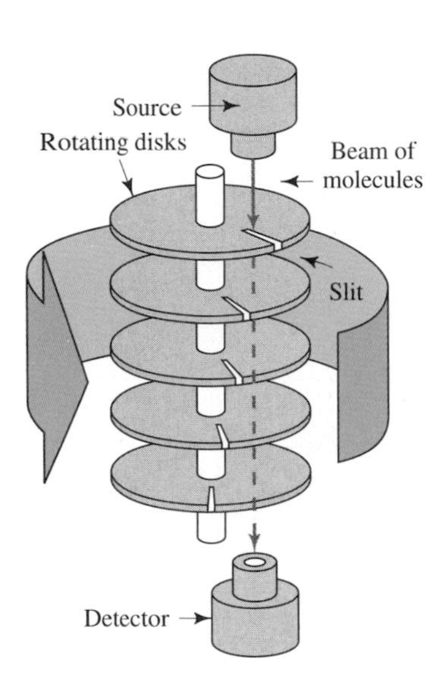

FIGURE 2.13 Determination of Molecular Speeds
Molecular speeds can be measured by passing a beam of molecules through a *velocity selector*, a series of disks mounted on a common axis. Each disk has a slit displaced by the same angle from the preceding slit. When the disks are rotated, the slits arrive successively in the same angular position. Any molecule that moves the distance between the disks in the same time it takes the next disk to arrive in the same angular position will pass through each slit and reach the detector. Molecules moving at other speeds will be blocked. From the rotation speed of the disks, the speed of the molecules arriving at the detector can be calculated. Then if the disk rotation speed is varied, the number of molecules arriving at the detector for each different speed can be found. We can then plot curves such as Figures 2.14 and 2.15.

The oxygen and the nitrogen of air never separate into pure oxygen and pure nitrogen. Likewise, the random motion of molecules results in gas effusing out of a pressurized container but never into it. The same random molecular motions that cause gases to mix but never ''unmix'' control the direction of all processes involving molecules, including chemical reactions. We will return to this important idea in Chapter 9.

Molecular Speeds

The rate of diffusion of a gas depends on how fast the molecules of the gas are moving. The kinetic theory tells us that if the temperature is constant, the average energy and the average speed of the molecules are constant also. But all the molecules do not have the same speed or the same kinetic energy. As we have seen, when two molecules with different energies collide, energy is transferred from one to the other, so one of them speeds up and the other slows down. Thus, the speeds and the energies of individual molecules change constantly, and they vary over a wide range. We say that there is a distribution of speeds. The distribution of molecular speeds can be obtained from experimental measurements (Figure 2.13) or from a detailed treatment of the kinetic theory. The distributions of molecular speeds in oxygen at 25°C and at 1000°C are shown in Figure 2.14. At higher temperatures, the average speed is greater, and the speed distribution is broader.

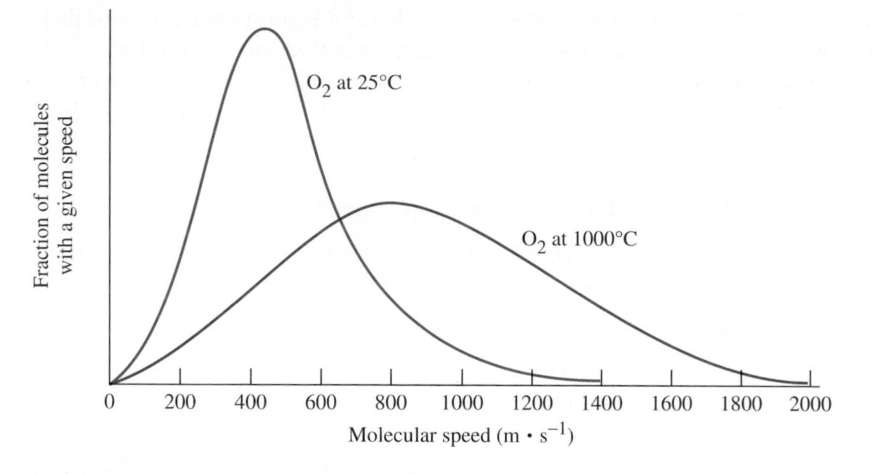

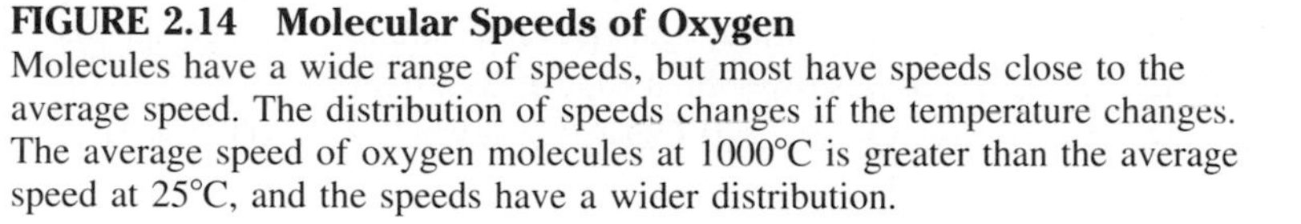
FIGURE 2.14 Molecular Speeds of Oxygen
Molecules have a wide range of speeds, but most have speeds close to the average speed. The distribution of speeds changes if the temperature changes. The average speed of oxygen molecules at 1000°C is greater than the average speed at 25°C, and the speeds have a wider distribution.

What is the relationship of average molecular speed to the mass of the molecules in a gas? We saw in our discussion of temperature (Section 2.7) that two gases at the same temperature have the same average kinetic energy,

$$\tfrac{1}{2}m_1v_1^2 = \tfrac{1}{2}m_2v_2^2$$

Rearranging, we see that the average speed of the molecules of a gas is inversely proportional to the square root of its molecular mass:

$$\frac{v_1}{v_2} = \sqrt{\frac{m_2}{m_1}}$$

Molecules with a low molecular mass have higher average speeds than those with a high molecular mass (Figure 2.15). This relationship accounts for the absence of the

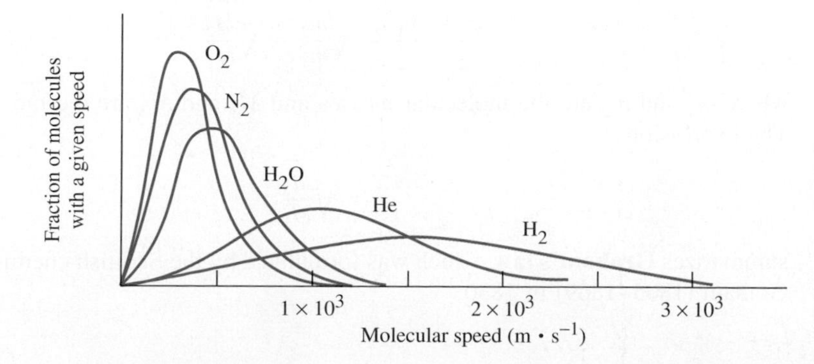

FIGURE 2.15 Distribution of Molecular Speeds for Some Gases at 25°C Light gases, such as hydrogen and helium, have higher average speeds and a wider distribution of speeds than do heavier gases, such as oxygen and nitrogen.

light gases hydrogen and helium from the atmosphere. A significant fraction of their molecules have a high enough speed to escape from the gravitational attraction of earth. The moon has no atmosphere, because the moon's weaker gravitational attraction and correspondingly lower escape speed have allowed all the gaseous molecules to escape.

EXAMPLE 2.8 Calculation of Average Speed

The average speed of hydrogen molecules at 25°C is 1793 $m \cdot s^{-1}$. What is the average speed of ammonia, $NH_3(g)$, molecules at the same temperature?

Solution: The molecular mass of $H_2(g)$ is 2.016 u, and that of $NH_3(g)$ is 17.03 u:

$$\frac{v_1}{v_2} = \sqrt{\frac{m_2}{m_1}}; \qquad v_1 = v_2 \times \sqrt{\frac{m_2}{m_1}} = 1793 \text{ m} \cdot \text{s}^{-1} \times \sqrt{\frac{2.016 \text{ u}}{17.03 \text{ u}}} = 617 \text{ m} \cdot \text{s}^{-1}$$

||| At constant temperature, heavier molecules travel more slowly than lighter molecules.

Exercise 2.13 At a given temperature, by what factor are the average speeds of each of the following gases slower than that of hydrogen?
(a) $He(g)$ **(b)** $N_2(g)$ **(c)** $O_2(g)$ **(d)** $CO_2(g)$ **(e)** $SO_2(g)$ **(f)** $Br_2(g)$

Rates of Diffusion and Effusion

We just saw in Example 2.8 that the average speed of ammonia molecules at 25°C is 617 $m \cdot s^{-1}$. Why then, does it take several seconds after you open a bottle of ammonia or perfume before you can detect the odor, even if your nose is only 1 m away from the bottle? In other words, why do molecules diffuse so slowly, even though they are moving at very high speeds? The answer is that molecules are constantly colliding with one another.

A detailed calculation based on the kinetic theory shows that, at 1 atm pressure and at ordinary temperatures, a molecule makes about 10^{10} collisions per second with other molecules. Any one molecule therefore has a very tortuous path, because its direction is changed in every collision. The average distance a molecule travels between collisions is known as its **mean free path**. The mean free path of oxygen molecules at 0°C and 1 atm is only 60 nm, or 60×10^{-9} m. Thus although an oxygen molecule is moving at high speed, its motion in any particular direction is slow (Figure 2.16).

FIGURE 2.16 Molecular Speeds and Rates of Diffusion In traveling from A to B, a molecule undergoes an enormous number of collisions, so the actual distance it travels is very much greater than the direct distance between A and B.

Although the diffusion (or effusion) of a gas is much slower than the speeds of the individual molecules, the diffusion (or effusion) rate is *proportional* to the average speed of the molecules. If r_1 is the rate at which one gas diffuses into another, and if r_2 is the rate at which the second gas diffuses, we may write

$$\frac{r_1}{r_2} = \frac{v_1}{v_2} = \sqrt{\frac{m_2}{m_1}} = \sqrt{\frac{M_2}{M_1}}$$

where m_1 and m_2 are the molecular masses and M_1 and M_2 are the molar masses. The expression

$$\frac{r_1}{r_2} = \sqrt{\frac{M_2}{M_1}}$$

summarizes **Graham's law**, which was formulated by the Scottish chemist Thomas Graham (1805–1869) in 1830:

DEMONSTRATION 2.6 Rates of Diffusion

A large beaker containing hydrogen is inverted over a porous clay cylinder. The hydrogen molecules diffuse through the holes in the porous cylinder faster than the oxygen and nitrogen molecules inside diffuse out, creating an excess pressure in the cylinder and in the flask to which it is connected. The excess pressure forces the colored water in the flask out through a narrow tube, forming a jet of water that collects in the dish on the right.

The rate of diffusion or effusion of a gas is inversely proportional to the square root of its molar mass.

Hydrogen, therefore, diffuses more rapidly than other gases, as we see in Demonstration 2.6.

Exercise 2.14 Nitrogen diffuses through a porous barrier about 50% faster than a gaseous oxide of sulfur diffuses under the same conditions. What are the approximate molar mass of the oxide of sulfur and its likely molecular formula?

2.10 REAL GASES AND INTERMOLECULAR FORCES

The kinetic theory makes the simplifying assumption that the total volume of the molecules of a gas is negligibly small compared with the volume that the molecules occupy. The theory assumes also that the molecules of a gas exert no forces on one another, because of the large distances between them. The derivation of the ideal gas law is based on these assumptions. Real gases do not obey the ideal gas law exactly for two reasons:

1. The volume of the gas molecules may *not* be negligible compared with the space that they occupy. If this is the case, the volume of a real gas is greater than the volume of an ideal gas.
2. There are attractive forces among all molecules. These **intermolecular forces** pull the molecules together, so the pressure exerted by a real gas is less than it would be if there were no intermolecular forces—that is, less than the pressure of an ideal gas.

The volume of the molecules of a gas becomes more important at high pressures, as the volume occupied by the gas decreases. The effect of intermolecular forces becomes more important at low temperatures, when the molecules have a lower kinetic energy and are therefore less able to overcome the intermolecular forces pulling them together. Thus, gases deviate most from the ideal gas law at low temperatures and at high pressures.

For most common gases at ordinary temperatures and at pressures of 1 atm or less, we can neglect deviations from the ideal gas law for most purposes. However, many industrial processes involving gases, such as the Haber process for the synthesis of ammonia, use pressures of several hundred atmospheres. In such cases, the behavior of the gases is not well represented by the ideal gas law, and we must use modified forms of this law that take into account intermolecular forces and the size of the molecules.

At sufficiently low temperatures, intermolecular forces pull the molecules of a gas close enough together that the gas condenses to a liquid. The molecules of a liquid are packed rather closely, so a liquid has a relatively high density and is not very compressible. But the molecules still have enough kinetic energy to move around one another. When the temperature is lowered still further, the molecules no longer have sufficient energy to be able to jostle past one another. Each one becomes trapped in a hole formed by the surrounding molecules, and it can only rotate and oscillate about a fixed mean position. The molecules then usually take up the regular ordered arrangement that is characteristic of most solids.

Intermolecular forces are electrostatic in origin. They are a consequence of the attractions and repulsions between the negative electrons and the positive nuclei of adjacent molecules. Except when the molecules are very close together—essentially touching each other—the overall force between two molecules is attractive. In other words, all molecules attract one another. We will consider the nature and strength of intermolecular forces in more detail in Chapter 11. Because the molecules of liquids and solids are very close together, the properties of liquids and solids depend very much on the sizes and shapes of the molecules and on the strengths of the intermolecular forces. The relationship of volume and density to temperature and pressure is therefore much more complex for liquids and solids than it is for gases. For liquids and solids, this relationship cannot be described even approximately by a single equation like the ideal gas law.

SUMMARY

The atmosphere is a mixture of gases surrounding the earth that can be divided into several layers. In the lowest layer, the troposphere, temperature decreases with increasing altitude. In the layer above that, the stratosphere, temperature increases with increasing altitude. The troposphere is a nearly uniform mixture consisting of 99% nitrogen, N_2, and oxygen, O_2, in the ratio 4:1; almost 1% argon; about 0.03% $CO_2(g)$; and very small amounts of the other noble gases. Earth is the only planet in the solar system with an atmosphere with a high $O_2(g)$ content, a result of photosynthesis.

Oxygen, $O_2(g)$, obtained by the distillation of liquid air, is a colorless, reactive gas. Almost all other elements and many compounds react with oxygen, usually at elevated temperatures. Most of these reactions generate sufficient heat that they are self-sustaining—they are then said to be undergoing combustion. Elements give their oxides, such as $MgO(s)$ and $CO_2(g)$, and combustible compounds, such as methane, give oxides of their constituent elements. Many metal oxides are major sources of the corresponding metals. Reactions in which an element or a compound combines with oxygen are examples of oxidation. Reactions in which oxygen is removed from an oxide or other oxygen-containing compound are examples of reduction. When oxygen is removed from $Fe_2O_3(s)$, it is reduced to the element $Fe(s)$.

Reactions, such as oxidation reactions, that give off heat to the surroundings are called exothermic reactions. Reactions that take in heat from the surroundings are called endothermic reactions.

Nitrogen, $N_2(g)$, obtained by the distillation of liquid air, is a gas that reacts with few other elements, usually at rather high temperatures. An important industrial process is the reaction of $N_2(g)$ with $H_2(g)$ to give $NH_3(g)$ at high pressures and 400°C in the presence of a catalyst (the Haber process). $N_2(g)$ from air is the only abundant source from which nitrogen compounds can be made. Conversion of the elemental form to useful compounds is called nitrogen fixation. At present the only economical method for nitrogen fixation is the Haber process. Much of the ammonia made by the Haber process is used for making fertilizers.

Hydrogen, $H_2(g)$, is obtained with $CO(g)$ in the high-temperature catalyzed reaction of water vapor with methane, $CH_4(g)$. The most abundant element in the universe, hydrogen is a colorless gas that combines with most elements. Metal hydrides include NaH, CaH_2, and AlH_3.

Atoms and molecules are in constant random motion, and their speeds increase with increasing temperature. In a solid the atoms or molecules are packed closely and move only small distances about their average positions. In a liquid the molecules are packed slightly less closely, move larger distances, and continually change places, so they do not have the regular, ordered arrangement of a solid. In a gas the molecules are far apart and move rapidly with a chaotic random motion. Accordingly, the density of most substances decreases in the sequence solid > liquid > gas. Changes of state between solid, liquid, and gas in which the constituent molecules of a substance are unchanged are physical, not chemical, changes.

Pressure is the force per unit area. In the mercury barometer, a mercury column about 760 mm long supports the pressure of the atmosphere; 1 standard atmosphere = 1 atm = 760 mm Hg = 760 torr = 101.325 kPa. Here Pa is the pascal; 1 Pa = $N \cdot m^{-2}$.

The gas laws are relations involving the pressure P, volume V, absolute temperature T, and amount (number of moles, n) of a gas:

Boyle's law: $PV = \text{constant}$ (n, T held constant)

Charles' law: $\frac{V}{T} = \text{constant}$ (n, P held constant)

Avogadro's law: $\frac{V}{n} = \text{constant}$ (P, T held constant)

The absolute zero of temperature is −273.15°C (0 K), at which all the molecules of a substance are at rest. The relationship between t°C on the Celsius scale and T K on the Kelvin temperature scale is $T = t + 273.15$. Combination of the three gas laws gives the ideal gas law, $PV = nRT$, where R is the gas constant (R = 0.08206 $atm \cdot L \cdot mol^{-1} \cdot K^{-1}$, with P in atmospheres, V in liters, n in moles, and T in kelvins). STP (standard temperature and pressure) is 1 atm and 0°C. The molar volume of an ideal gas at STP is 22.41 $L \cdot mol^{-1}$.

The kinetic theory of gases is used to explain the gas laws. For an ideal gas, molecules are assumed to be infinitely small particles that have zero volume, are in continuous random motion, and have no forces acting on them, except when they collide with each other or with the container walls. The pressure exerted by a gas results from the collisions of its molecules with the walls of the container.

Using the ideal gas law, we can calculate any of P, V, n, or T if the other three are known or if some are kept constant. If the mass of a sample of gas is known and we find n from the ideal gas law, we can then find the molar mass of the gas.

Dalton's law of partial pressures states that the total pressure of a mixture of gases is the sum of the partial pressures of the components: $P_{total} = p_1 + p_2 + p_3 + \ldots + p_i$. The partial pressure of any gas in a mixture is proportional to the number of moles of that gas in the mixture: $p_i \propto n_i$.

Diffusion is the mixing of two gases, and effusion is the escape of a gas through a tiny hole in its container. Both are irreversible processes resulting from the chaotic random motion of gas molecules. Their rates depend on how fast the molecules are moving. Although the speeds and the energies of individual molecules change constantly and there is a distribution of speeds, the average kinetic energy ($\frac{1}{2}mv^2$) of the molecules of all gases is the same at a given temperature. Graham's law states that the rate of diffusion (or effusion) of a gas is inversely proportional to the square root of its molar mass:

$$\frac{r_1}{r_2} = \sqrt{\frac{M_2}{M_1}}$$

In real gases neither the size of the molecules nor the intermolecular forces between them are negligible. Thus real gases deviate from ideal behavior, particularly at low temperatures and high pressures. Intermolecular forces become increasingly important as molecules get closer together, so a gas eventually condenses to a liquid and then a solid as the temperature decreases.

IMPORTANT TERMS

absolute zero of temperature (page 57)
atmosphere (page 43)
atmospheric pressure (page 55)
Avogadro's law (page 59)
barometer (page 55)
Boyle's law (page 56)
catalyst (page 51)
change of state (page 53)
Charlcs' law (page 58)
combustion (page 47)
Dalton's law of partial pressures (page 68)
diffusion (page 69)
effusion (page 69)
endothermic reaction (page 48)
exothermic reaction (page 48)
gas (page 53)
gas constant (page 60)
Graham's law (page 72)
Haber process (page 50)
hydride (page 51)
ideal gas (page 60)
idcal gas law (page 60)
intermolecular forces (page 73)
kelvin (page 58)
Kelvin temperature scale (page 58)
kinetic energy (page 43)
kinetic theory of gases (page 61)
liquid (page 53)
mean free path (page 72)
molar volume (page 60)
nitrogen fixation (page 50)
oxidation (page 48)
oxide (page 47)
partial pressure (page 68)
pascal (page 55)
physical state (page 53)
pressure (page 55)
reduction (page 49)
solid (page 53)
standard atmosphere (page 55)
standard temperature and pressure (STP) (page 60)
stratosphere (page 44)
temperature (page 64)
torr (page 55)
troposphere (page 43)
vapor (page 53)

REVIEW QUESTIONS

1. How does the composition of the earth's atmosphere differ substantially from that of all the other planets?

2. Give two physical and two chemical properties of oxygen. (Describe the physical properties qualitatively.)

3. **(a)** What is meant by the term "combustion"?

(b) Give two examples of combustion reactions.

4. Name the oxides with each of the following molecular formulas.

(a) CO **(b)** CO_2 **(c)** SO_2 **(d)** SO_3 **(e)** N_2O **(f)** NO **(g)** MgO

5. What is a simple laboratory test for oxygen gas?

6. How is pure nitrogen obtained commercially?

7. Write balanced equations for the reactions between **(a)** nitrogen and hydrogen; **(b)** nitrogen and oxygen.

8. By what large-scale industrial process is the fixation of nitrogen achieved?

9. What are the names and formulas of the hydrides of calcium, carbon, and nitrogen?

10. Explain the terms "gas," "vapor," and "vaporization."

11. **(a)** Why is the pressure of the atmosphere often expressed in millimeters of mercury?

(b) What is the standard pressure of the atmosphere in mm Hg?

12. Explain how Charles' law leads to the concept of an absolute zero of temperature.

13. State Avogadro's law.

14. State the ideal gas equation.

15. How does the kinetic theory of gases account for the existence of an absolute zero of temperature?

16. What equation relates the number of moles of a gas, n, to pressure, volume, and temperature?

17. State Dalton's law of partial pressures, and explain it on the basis of the kinetic theory of gases.

18. Describe diffusion and effusion.

19. Why are the diffusion and effusion of gases described as irreversible processes?

20. Give two reasons why real gases do not obey the ideal gas law.

PROBLEMS

The Atmosphere; Oxygen; Nitrogen; Hydrogen

1. Describe the earth's atmosphere in terms of its present composition. In what important way does it differ from its composition before life appeared on earth?

2. **(a)** What are the three most abundant gases in the atmosphere?

(b) Name three gases in the air that are chemically reactive.

(c) Explain why carbon dioxide is an essential constituent of the atmosphere.

3. From the elements mentioned in this chapter, name **(a)** two that are monatomic gases; **(b)** two that are diatomic gases.

4. With the aid of two examples in each case, explain what is meant by the terms **(a)** ''oxidation'' and **(b)** ''reduction.''

5. Write balanced equations to describe the reactions that occur when **(a)** sulfur burns in air; **(b)** magnesium burns in oxygen; **(c)** methane burns in air; **(d)** hydrogen burns in oxygen.

6. Write balanced equations to describe each of the following.

(a) the reduction of $Fe_2O_3(s)$ to $Fe(s)$ with $CO(g)$

(b) the reduction of $Fe_2O_3(s)$ to $Fe(s)$ with $H_2(g)$

(c) the reduction of $CuO(s)$ to $Cu(s)$ with $CO(g)$

(d) the oxidation of $Mg(s)$ with $H_2O(g)$

7. Write a balanced equation for the reaction of steam at high temperatures with the following.

(a) iron **(b)** magnesium **(c)** methane

8. Write a balanced equation to describe the reaction of nitrogen at high temperatures with the following, and name the products.

(a) hydrogen **(b)** oxygen

9. **(a)** Explain what is meant by the term ''nitrogen fixation.''

(b) Give two examples of processes involving nitrogen fixation.

10. Write the balanced equations for two reactions by which hydrogen may be formed from water.

11. From the substances discussed in this chapter, name and give the formula of each of the following, and write a balanced equation for the reaction described.

(a) a liquid used to power the space shuttle

(b) a gas used to make fertilizers

(c) a gas that burns when ignited in air

12. Write a balanced equation for each of the following.

(a) the reaction of carbon when heated in air to give carbon monoxide

(b) the reaction of calcium with oxygen to give calcium oxide, $CaO(s)$

(c) the combustion of propane, $C_3H_8(g)$, in excess oxygen

(d) the combustion of ethanol, $C_2H_6O(l)$, in excess oxygen

13. Write a balanced equation for each of the following.

(a) the reaction of copper oxide, CuO, with carbon to give copper metal and carbon dioxide

(b) the decomposition of lead dioxide, $PbO_2(s)$, upon heating to give lead monoxide, $PbO(s)$, and oxygen

(c) the reaction of magnesium with nitrogen to give magnesium nitride, $Mg_3N_2(s)$

(d) the reaction of calcium with hydrogen to give calcium hydride, $CaH_2(s)$

Solids, Liquids, and Gases

14. Give the physical state (gas, liquid, or solid) at room temperature of each of the following elements.

(a) magnesium **(b)** nitrogen **(c)** oxygen **(d)** sulfur

(e) copper **(f)** hydrogen **(g)** bromine

15. In molecular terms, describe the principal differences between the solid, liquid, and gaseous states, and the way in which these differences influence the bulk properties of each physical state.

The Gas Laws

16. **(a)** State Boyle's law and **(b)** Charles' law, and explain how these laws were established experimentally.

17. An automobile tire of volume 28 L is filled with air at a pressure of 2.3 atm. What volume does this air occupy at atmospheric pressure (1.00 atm) at the same temperature?

18. Oxygen gas is commonly sold in 50.0-L steel containers at a pressure of 150 atm at room temperature. What volume would the gas occupy at a pressure of 1.00 atm and the same temperature?

19. A cylinder in a gasoline engine initially has a volume of 0.50 L and contains gases that exert a pressure of 1.00 atm. What is the pressure of the gases when the volume of the cylinder is reduced to 0.20 L, assuming that there is no change in the temperature?

20. A meteorological balloon has a volume of 150 L when filled with hydrogen at a pressure of 1.00 atm. To what volume will the balloon expand when it rises in the atmosphere to a height of 2500 m (8000 ft), where the atmospheric pressure is only 0.75 atm, assuming that the temperature remains constant?

21. A sample of hydrogen gas has a pressure of 0.98 atm when it is confined in a bulb of volume 2.00 L. A tap connects the bulb to another bulb that has a volume of 5.00 L and has been completely evacuated. What will be the pressure in the two bulbs after the tap is opened?

22. A sample of nitrogen has a volume of 400 mL at 100°C. At what temperature will it have a volume of 200 mL if the pressure does not change?

23. To what temperature must 30.0 L of helium at 25°C be cooled at 1 atm pressure for its volume to be reduced to 1.00 L at the same pressure?

24. A balloon filled with helium has a volume of 1.60 L at a pressure of 1.00 atm at 25°C. What will be the volume of the balloon when it is cooled to −196°C, the temperature of liquid nitrogen, assuming that the pressure remains constant at 1.00 atm?

Molecules and Volume: Avogadro's Law

25. **(a)** Which sample contains more molecules: 1.00 L of $O_2(g)$ at STP or 1.00 L of $H_2(g)$ at STP? **(b)** Which sample has the greater mass?

26. How many liters of oxygen are needed to burn 1.00 L of **(a)** methane and **(b)** hexane, C_6H_{14}, to carbon dioxide and water, if initially all the gases are at the same temperature and pressure?

27. The anesthetic cyclopropane is a gas containing only carbon and hydrogen. Suppose 0.550 L of cyclopropane at 120°C and 0.900 atm reacts with oxygen to give 1.65 L of $CO_2(g)$ and 1.65 L of $H_2O(g)$ at the same temperature and pressure. What are **(a)** the molecular formula of cyclopropane and **(b)** the balanced equation for its reaction with oxygen?

The Kinetic Theory of Gases

28. What are the five assumptions of the kinetic theory of gases?

29. In terms of the kinetic theory of gases, explain why the gas laws are the same for all gases and for mixtures of gases, such as air.

30. In terms of the kinetic theory of gases, explain why the pressure of a gas **(a)** doubles when the volume is halved and **(b)** increases with increasing temperature.

Using the Ideal Gas Law

31. How many moles of methane, $CH_4(g)$, occupy a volume of 4.00 L at 1.00 atm pressure and 25°C?

32. What volume is occupied by 0.200 mol of oxygen gas at 20°C and a pressure of 740 mm Hg?

33. The temperature of a closed vessel containing air at a pressure of 1.02 atm is raised from 20 to 200°C. What is the pressure inside the vessel if its volume increases by 10% as the temperature is increased from 20 to 200°C?

34. A metal cylinder with a safety valve that opens at a pressure of 100 atm is to be filled with nitrogen gas and heated to 300°C. What is the maximum pressure to which it can be filled at 25°C?

35. An automobile tire of fixed volume is inflated to a pressure of 2.50 atm at 20°C. What is the pressure inside the tire at **(a)** 30°C; **(b)** −10°C?

36. A 1.00-L evacuated vessel is to be filled with carbon dioxide gas at 300 K and a pressure of 500 mm Hg by placing a piece of dry ice, $CO_2(s)$, inside the flask and allowing it to vaporize. What mass of dry ice should be used?

37. What mass of gaseous hydrogen chloride, $HCl(g)$, is needed to provide a pressure of 0.240 atm in a container of volume 250 mL at 37°C?

38. The atmosphere of Venus is mainly carbon dioxide. At the surface of Venus, the temperature is about 800°C and the pressure is about 75 atm. In the event that an inhabitant of Venus (if any!) took these values as STP for Venus, what value would the Venusian find for the molar volume of an ideal gas?

39. As a publicity stunt, a water-bed retailer filled a water bed with helium and floated it above his store. Calculate the mass of helium required to fill a water-bed bag at a pressure of 1.03 atm at 23°C, if its dimensions are 2.00 m × 1.50 m × 0.20 m.

40. An anesthetic used to relax patients contains oxygen mixed with dinitrogen monoxide, $N_2O(g)$. The mixture has a density of 1.482 $g \cdot L^{-1}$ at 25°C and 0.980 atm. What is the mass percentage of $N_2O(g)$ in the gas mixture?

41. What is the density of the chlorofluorocarbon (CFC) of molecular formula CF_2Cl_2 at 1.00 atm and 20°C?

Molar Mass

42. A sample of a noble gas has a mass of 0.20 g and exerts a pressure of 0.48 atm in a container of volume 0.26 L at 27°C. Is the gas helium, neon, argon, krypton, or xenon?

43. Oxygen can exist not only as diatomic O_2 molecules but also as ozone. What is the molecular formula of ozone, given that at the same pressure and temperature it has a density 1.50 times that of O_2?

44. A gas at a pressure of 740 mm Hg at 20°C occupies a volume of 1.00 L and has a mass of 1.134 g.

(a) What is its molar mass?

(b) If the empirical formula of the gas is CH_2, what is its molecular formula?

45. The density of a gaseous chlorofluorocarbon (CFC) at 23.8°C and 432 mm Hg is 3.23 $g \cdot L^{-1}$.

(a) What is its molar mass?

(b) If it contains only one carbon atom per molecule, what is its molecular formula?

Dalton's Law of Partial Pressures

46. Air is composed of 78% nitrogen, 21% oxygen, and 1% argon by volume. What are the partial pressures of each gas when the atmospheric pressure is 758 mm Hg?

47. The atmospheric pressure at the summit of Mt. Whitney is 0.58 atm. The partial pressure of oxygen at sea level is 160 mm Hg. What is the partial pressure of oxygen at the summit of Mt. Whitney?

48. A mixture of 0.200 g of helium and 0.200 g of hydrogen is confined in a vessel of volume 225 mL at 27°C.

(a) What are the partial pressures of each gas?

(b) What is the total pressure exerted by the gaseous mixture?

49. One day in Los Angeles, the concentration of nitrogen monoxide (nitric oxide), $NO(g)$, in the atmosphere was 0.94 ppm (parts per million) by volume. The atmospheric pressure was 750 mm Hg, and the temperature was 30°C.

(a) What was the partial pressure of $NO(g)$?

(b) What was the number of NO molecules per cubic meter of air?

50. The partial pressure of oxygen in air at 37°C is 159 mm Hg. When a human exhales a breath, the partial pressure of oxygen is only 115 mm Hg. How many oxygen molecules per liter of inhaled air are used by the lungs? (Take body temperature to be 37°C.)

51. Is the density of moist air greater than, less than, or the same as the density of dry air at the same temperature and pressure? Explain your answer.

52. The mass percentage composition of the atmosphere of Mars is 95% CO_2, 3% N_2, and 2% other gases, principally Ar. What are the partial pressures of each gas at the surface, where the total pressure is 5 torr?

Diffusion and Effusion

53. The average speed of oxygen molecules at 25°C is 450 $m \cdot s^{-1}$. What are the average speeds of each of the following gaseous molecules at 25°C?

(a) hydrogen **(b)** chlorine, Cl_2 **(c)** carbon monoxide

(d) water **(e)** carbon dioxide

54. **(a)** Which diffuses faster, molecular nitrogen or molecular oxygen? **(b)** What is the ratio of their rates of diffusion?

55. $H_2(g)$ and $D_2(g)$ are separated by utilizing their different rates of diffusion. What is the expected ratio of these rates?

56. The original separation of ^{235}U (required for nuclear reactors) from ^{238}U was achieved by exploiting the slight difference in the diffusion rates of $^{235}UF_6$ and $^{238}UF_6$. Calculate the ratio of the rates of diffusion of these two gases, assuming that the mass of each isotope is exactly the same as its mass number.

57. The average speed of helium atoms is 0.707 $mi \cdot s^{-1}$ at 25°C. What is the average speed at the same temperature of **(a)** hydrogen, **(b)** nitrogen, and **(c)** oxygen molecules?

58. A gaseous element that exists as a diatomic gas at STP effuses through a small hole at 0.324 times the rate of effusion of helium gas under the same conditions. Identify the element.

Real Gases

59. Explain why a real gas obeys the ideal gas law best at low pressures and high temperatures.

General Problems

60. **(a)** Calculate the average volume effectively occupied by each molecule in an ideal gas at 27°C and 1 atm pressure. Assume the molecule to be spherical with a radius of 100 pm.

(b) Calculate the actual volume of a molecule.

(c) Compare the actual volume with the volume effectively occupied by a molecule in an ideal gas.

61. By what factor does water expand when converted from liquid at 100°C to vapor at 100°C at 1 atm pressure, given that the density of water at 100°C is 0.96 $g \cdot cm^{-3}$?

62. In terms of the kinetic theory of gases, account for each of the following.

(a) A gas exerts a pressure on the walls of any vessel in which it is confined.

(b) The pressure of a given mass of gas increases when its volume is decreased.

(c) Two gases readily mix, and their total pressure is the sum of the pressures that each gas exerts alone in the same volume at the same temperature.

(d) The absolute zero of temperature is −273.15°C (0 K).

(e) Real gases obey the ideal gas law to a good approximation but not exactly.

63. Gases collected over water contain water vapor, $H_2O(g)$, the partial pressure (vapor pressure) of which depends only on temperature. The pressure in a 250-mL collecting bulb filled with wet $Cl_2(g)$ is 754.3 mm Hg at 25.0°C, and the pressure falls to 644.7 mm Hg when the bulb is cooled to −10.0°C, at which temperature the water vapor has all condensed to ice on the walls of the bulb. What is the vapor pressure of water at 25.0°C?

64. Igniting ammonium dichromate, $(NH_4)_2Cr_2O_7(s)$, results in a spectacular "chemical volcano" due to its decomposition to give $Cr_2O_3(s)$, $N_2(g)$, and water.

(a) Write the balanced equation for the reaction. If 2.000 g (7.93×10^{-3} mol) of ammonium dichromate is ignited in an evacuated 1.00-L flask, when the temperature of the flask is 150°C, what will be **(b)** the total pressure in the flask after the reaction; **(c)** the partial pressures of $N_2(g)$ and $H_2O(g)$?

65. A pearl diver inhales air at 1.00 atm, holds her breath, and dives into water. Given that the densities of water and mercury are 1.00 $g \cdot cm^{-3}$ and 13.6 $g \cdot cm^{-3}$, respectively, what would be the pressure of air in her lungs at depths of **(a)** 9.1 m (about 30 ft); **(b)** 15.2 m (about 50 ft)?

CHAPTER 3 The Periodic Table and Chemical Bonds

3.1 The Periodic Table
3.2 The Shell Model of the Atom
3.3 Chemical Bonds and Lewis Structures
3.4 Electronegativity
3.5 Molecular Shape and the VSEPR Model

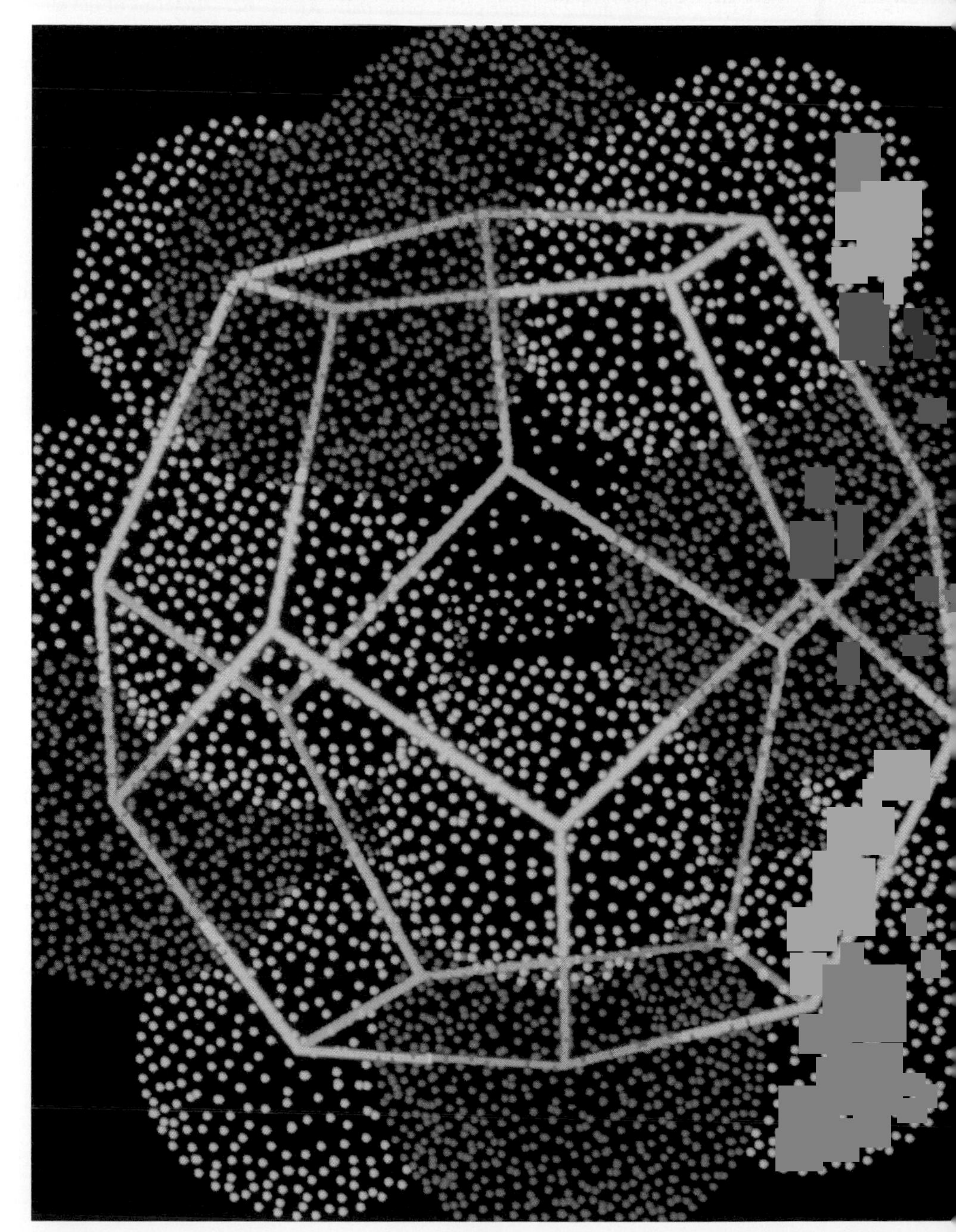

Computers are now a common, useful tool for generating molecular models. Such models can be viewed from different angles and can show a molecule's three-dimensional geometry, bonding between atoms, or even how the replacement of one atom with another affects the molecule. This computer-generated model of a molecule has an unusual dodecahedral arrangement of 12 carbon atoms (in green) and 8 titanium atoms (in blue).

In Chapter 2 we caught a glimpse of the fascinating variety in the behavior of the elements. We saw that oxygen, $O_2(g)$, reacts with almost all the other elements, whereas nitrogen, $N_2(g)$, reacts with rather few. We saw also that although a nitrogen atom combines with three hydrogen atoms to form an ammonia molecule, NH_3, an oxygen atom combines with two hydrogen atoms to form a water molecule, H_2O, and a chlorine atom combines with only one hydrogen atom to form a hydrogen chloride molecule, HCl. But argon forms no compounds with hydrogen, or indeed, with any other element. The various elements have different combining powers.

The foremost challenge for early chemists was to explore this great variety of behaviors and to try to bring some order to it. A major advance was made when it was found that if the elements are listed in order of their atomic masses, they can be arranged in the form of a table, called the *periodic table of the elements,* in which elements with similar properties occupy the same column of the table. We introduce the periodic table in this chapter, and we will use it extensively thereafter. It is one of the simplest and most useful tools that we have for understanding and systematizing the great variety of chemical behaviors.

To understand why the properties of the elements vary in the regular way that is summarized by the periodic table, we need to consider the internal structure of atoms—in particular, how the electrons are arranged. In this chapter we introduce a simple model of the atom, the *shell model*, that we will elaborate in later chapters. We will see that the shell model provides the basis for understanding the patterns observed in the periodic table, the different combining powers of the elements, and how atoms are joined together in molecules by chemical bonds.

Finally, we will see that when atoms are linked together in molecules, they adopt definite geometric arrangements. We will describe a convenient model, the *VSEPR model*, that enables us to understand and to predict the geometric arrangement of atoms in molecules.

3.1 THE PERIODIC TABLE

Chemists have long recognized that some elements, such as sodium and potassium, have very similar properties, whereas other elements, such as sodium and chlorine, have very different properties. Attempts to classify the elements in terms of these similarities and differences culminated in the formulation of the **periodic table of the elements** by Dimitri Mendeleev in 1869 (Box 3.1). When Mendeleev listed the elements in order of atomic mass, he found that elements with similar properties recur in a periodic manner. This recurrence of similar properties enabled him to arrange the elements in table form, with similar elements in the same column.

BOX 3.1 Dimitri Mendeleev and the Periodic Table

Dimitri Mendeleev was born in Tobolsk, Siberia, and was the youngest of a family of 17. After his father died, his mother, determined that Dimitri should have the best possible education, moved the family to St. Petersburg. In 1856 Mendeleev obtained a master's degree in chemistry and then taught at the University of St. Petersburg. In 1867 he was appointed professor of inorganic chemistry.

While giving his lecture course, Mendeleev felt the need for a new textbook, and he began writing what was to become a famous and widely used textbook, *Principles of Chemistry* (1868–1870). He realized that the order and system that was then becoming apparent in organic chemistry was lacking in inorganic chemistry. In attempting to remedy this situation, he was led to formulate the periodic table in 1869.

To consider that the properties of the elements are in some way related to their atomic masses was an imaginative idea, because the structures of atoms were unknown at the time, and it seemed to most chemists that there was no reason why chemical properties should be related to atomic mass. Moreover, to bring certain elements into the correct group according to their chemical properties, Mendeleev was bold enough to reverse the order of some pairs of elements and to predict that their atomic masses were incorrect. Only some of these predictions were correct, because we now know that the fundamental basis of the periodic table is atomic number rather than atomic mass, and it is the number and arrangement of the electrons that determines properties. Also, to keep elements in the correct groups according to their chemical properties, Mendeleev left gaps in his table, predicting that they would be filled by elements that were still undiscovered at that time. From the trends that he observed among the properties of related elements, he predicted the properties of these undiscovered elements.

At first the periodic table attracted little attention. Then, when Mendeleev's predictions concerning undiscovered elements were fulfilled by the successive discoveries of gallium (1874), scandium (1879), and germanium (1885), chemists began to realize that in the periodic table they had a tool of great value. From that time on Mendeleev was recognized as one of the foremost scientists of his day.

Mendeleev was a versatile genius who was interested in many fields of science, both pure and applied. He worked on many problems associated with Russia's natural resources, such as coal, salt, and petroleum, and in 1876 he visited the United States to study the Pennsylvania oil fields. He invented an accurate barometer; he made an ascent in a balloon to study a total eclipse of the sun; and he was interested in the possibility of air travel.

Dimitri Mendeleev (1834–1907)

As a professor at the University of St. Petersburg, Mendeleev was unavoidably involved in the political turmoil that affected all nineteenth-century Russia. Although in the middle of the century universities had been left comparatively free to carry out their research and teaching as they saw fit, the government came to suspect that academic freedom encouraged political unrest. Many repressive measures were taken against the universities, and the students reacted with demonstrations and riots. The universities were frequently closed for long periods, and many students were exiled to Siberia. In 1890 Mendeleev resigned from the university because of the government's oppressive treatment of students and the lack of academic freedom. Fortunately, he still had friends at the czar's court, and he was appointed director of the Bureau of Weights and Measures, where he continued to carry out important research.

Mendeleev was a colorful character as well as a genius. Portraits of Mendeleev reveal his characteristic enormous head of hair. Biographers say that he had it cut only once a year, in the spring, and that he would not deviate from this custom even when he had an audience with the czar.

Many different forms of the periodic table have been proposed. It is important to understand that there is no single *correct* form; we can use whatever form is most useful for us. The table given in Figure 3.1 is a simple and popular version. As in all modern versions, *the elements are arranged in order of atomic number, Z,* rather than atomic mass. This gives the same order as Mendeleev's table in all but a few cases. The atomic number is the more fundamental property, because it is equal to

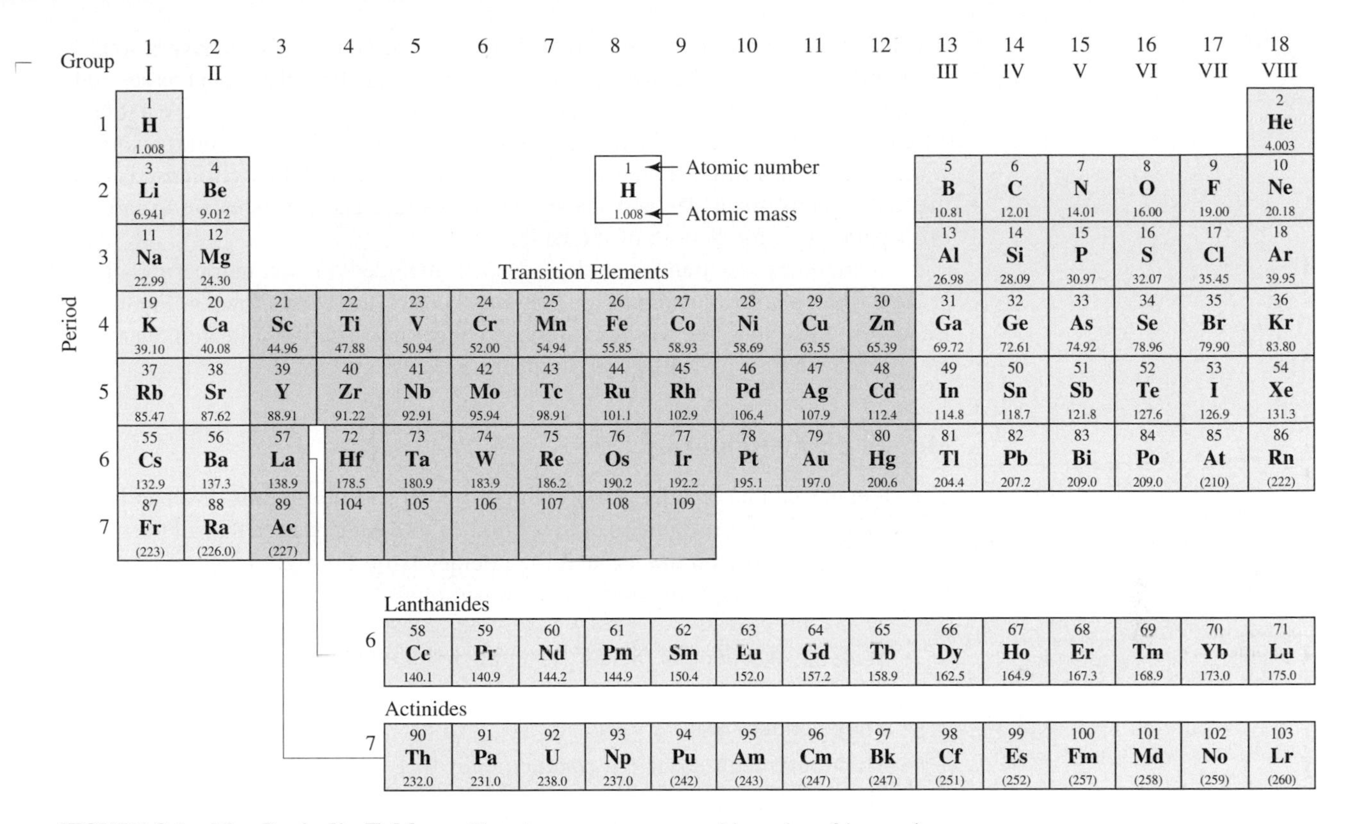

FIGURE 3.1 The Periodic Table The elements are arranged in order of increasing atomic number, which usually (but not always) matches the order of increasing average atomic mass. This arrangement places elements with similar properties in the same column. The atomic numbers are noted above the symbols of the elements; average atomic masses appear below each symbol. The horizontal rows, or *periods*, are numbered from 1 to 7. The vertical columns, or *groups*, are numbered in two different ways. If only the main groups are numbered, the Roman numerals I to VIII are used. If the transition element groups are also numbered, then the Arabic numerals 1 to 18 are used. Because in this book we have no need to number the transition metal groups, we use the simpler I-to-VIII numbering scheme.

the number of protons in the nucleus and therefore to the number of electrons surrounding the nucleus. As we will see, the number of electrons is what determines the properties of an element.

The formulation of the periodic table marked the beginning of a new era in chemistry. It created order out of apparent chaos and led to a much greater understanding of the properties of the elements. It remains today, more than 120 years later, a most important working tool for the chemist. It is therefore essential to know the general form of the table, and it is extremely useful to memorize the symbols and positions of at least the first 20 elements. It is also wise to note the position of every element that you meet for the first time.

Organization of the Periodic Table

The periodic table has eight principal vertical columns, called **groups**, and seven horizontal rows, known as **periods**. The groups are numbered I to VIII from left to right, and the periods are numbered 1 to 7 from top to bottom. The 10 elements that fall between Groups II and III in Periods 4, 5, and 6 (which have a total of 18 elements each) are known as **transition elements**. Period 6 contains an additional 14 elements between La (lanthanum) and Hf (hafnium), so there are a total of 32 elements in this period. These 14 elements are called the **lanthanides**, and the 14 elements that form a similar series in Period 7 are called the **actinides**. To give the table a convenient size and shape, the lanthanides and actinides are usually set off separately at the bottom of the table.

Sometimes the transition elements are included in the numbering of the groups. The 18 vertical columns are then numbered 1 to 18, as shown in Figure 3.1. Because in this book we will not be much concerned with the transition elements, we use the simpler I—VIII group numbering.

Metals and Nonmetals

The elements can be broadly classified into **metals** and **nonmetals**. We can immediately appreciate the usefulness of the periodic table when we see that all the metals are grouped together on the left and the nonmetals on the right (Figure 3.2).

Metals have the following characteristic physical properties:

- With the exception of mercury (melting point −39°C), they are solids at room temperature.
- They conduct heat very well.
- They have high *electrical conductivities* that increase with decreasing temperature.
- They have a high reflectivity and a shiny metallic luster.
- They are *malleable* (can be beaten out into sheets or foil) and *ductile* (can be pulled into thin wires without breaking).
- They emit electrons when they are exposed to radiation of sufficiently high energy or when they are heated. These two effects are known as the *photoelectric effect* and the *thermionic effect*, respectively.

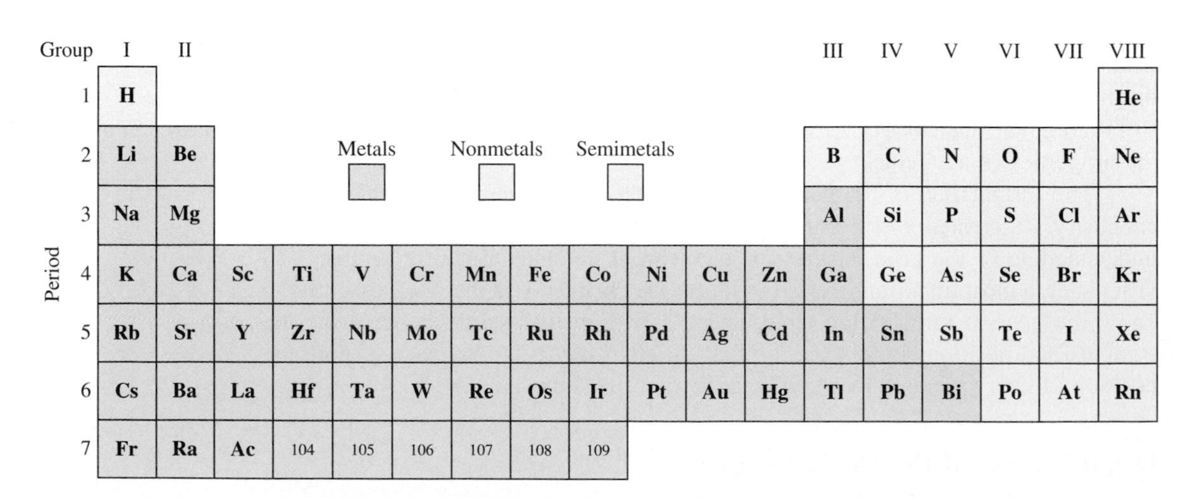

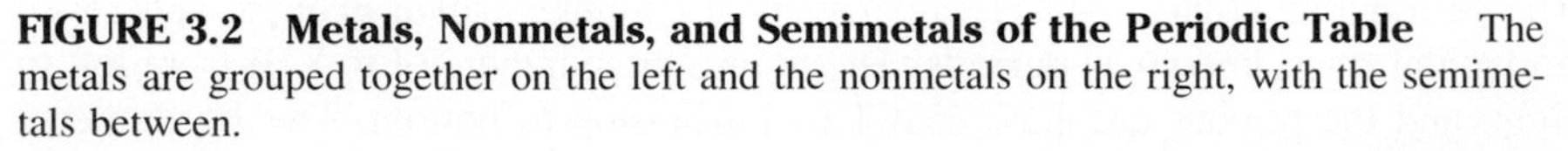

FIGURE 3.2 Metals, Nonmetals, and Semimetals of the Periodic Table The metals are grouped together on the left and the nonmetals on the right, with the semimetals between.

FIGURE 3.3 Metals Some familiar metals: aluminum, copper, iron, and gold

Some typical metals are shown in Figure 3.3.

Nonmetallic elements have the following characteristic physical properties:

- They may be gases, liquids, or solids at room temperature.
- Most nonmetals are poor conductors of heat.
- Most nonmetals are insulators (very poor conductors of electricity).
- They do not have a high reflectivity or a shiny metallic appearance.
- In the solid form they are generally brittle and fracture easily under stress rather than being malleable and ductile.
- They do not exhibit the photoelectric or thermionic effects.

Some typical nonmetals are shown in Figure 3.4.

Not every metal or nonmetal has all these characteristic properties, and there are exceptions to these generalizations. For example, the metal chromium is quite brittle; diamond, a form of the nonmetal carbon, is a very good conductor of heat; graphite, another form of carbon, is a fairly good conductor of electricity. Furthermore, the transition from nonmetal to metal in the periodic table is not sharp. Several elements, such as germanium and antimony, have intermediate properties. These intermediate elements are usually called **semimetals** or **metalloids** (Figure 3.2). There are only 17 nonmetals, but this small number does not represent their relative importance. Most nonmetals are rather common and have many important compounds. In contrast, many metals are rare and have few important compounds.

FIGURE 3.4 Some Nonmetals *Clockwise from bottom left:* yellow sulfur, black carbon (charcoal and graphite), white phosphorus, red phosphorus, and iodine

FIGURE 3.5 Sodium
Sodium is a metal that is soft enough to cut with a knife.

Exercise 3.1 Name and classify each of the following elements as a metal or a nonmetal.
(a) Li **(b)** Mg **(c)** C **(d)** P **(e)** F **(f)** Na **(g)** Cl **(h)** N **(i)** Ca

The Groups

Each group of the periodic table is a family of elements that have many similar properties, although within each group there is a gradual change in properties from the top of the column to the bottom.

Group I metals are called alkali metals because, before the metals were themselves isolated, their hydroxides and carbonates were known to give alkaline (basic) solutions in water.

Group I: Alkali Metals—Li, Na, K, Rb, Cs, Fr The Group I elements are metals known as the **alkali metals**. They have most of the typical physical properties of metals. For example, they are shiny solids and good conductors of heat and electricity. However, they are much softer than more familiar metals such as iron and aluminum; in fact, they can be cut with a knife (Figure 3.5). Sodium and potassium are among the ten most abundant elements (Table 2.3), and in this book we will be concerned mainly with these two alkali metals.

The alkali metals are the most reactive of the metals. They react with all the nonmetals except the noble gases (Group VIII). For example, alkali metals react with hydrogen to form compounds called *hydrides* and with chlorine to form chlorides:

$$2K(s) + H_2(g) \rightarrow 2KH(s) \qquad \text{potassium hydride}$$

$$2Na(s) + Cl_2(g) \rightarrow 2NaCl(s) \qquad \text{sodium chloride (common salt)}$$

Hydrogen is usually placed in Group I, although it is a nonmetal and has little in common with the alkali metals. The difficulty of finding a satisfactory place for hydrogen in the periodic table reflects the fact that it is in many ways a unique element, sharing few properties with other elements.

Early chemists gave the name "earth" to many nonmetallic substances, such as metal oxides, that have no metallic properties. Calcium oxide, CaO, and the other oxides of Group II metals give alkaline (basic) solutions in water. Thus when these metals were isolated, they became known as the alkaline earth metals.

The full systematic name for CaH_2 and $MgCl_2$ are calcium dihydride and magnesium dichloride, respectively. But calcium forms only one hydride, CaH_2, and magnesium only one chloride, $MgCl_2$, so the prefix *di-* is usually omitted.

Group II: Alkaline Earth Metals—Be, Mg, Ca, Sr, Ba, Ra All the Group II elements, usually called the **alkaline earth metals**, are solids at room temperature and have typical metallic properties. They are harder than the alkali metals and have higher melting points. Of the Group II elements, we will be concerned mostly with calcium and magnesium, which are among the 10 most abundant elements in the earth's crust.

Group II elements react with nonmetals, although more slowly than Group I metals. Among the products of these reactions are hydrides, chlorides, and oxides:

$$Ca(s) + H_2(g) \rightarrow CaH_2(s) \qquad \text{calcium hydride}$$

$$Mg(s) + Cl_2(g) \rightarrow MgCl_2(s) \qquad \text{magnesium chloride}$$

$$2Mg(s) + O_2(g) \rightarrow 2MgO(s) \qquad \text{magnesium oxide}$$

Group III: B, Al, Ga, In, Tl Of the Group III elements, boron is a semimetal and all the others are metals. Aluminum is by far the most abundant and important of the elements in this group and the only one we shall discuss in detail. The third most abundant element in the earth's crust, aluminum is widely distributed in the form of aluminosilicates, which are compounds of metals such as sodium and potassium with aluminum, silicon, and oxygen. Aluminosilicates are the major component of many rocks and clays (Chapter 19). Like most other metals, aluminum reacts with nonmetals such as chlorine and oxygen:

The common rock granite, for example, contains the mineral albite, which is an aluminosilicate with the formula $NaAlSi_3O_8$.

$$2Al(s) + 3Cl_2(g) \rightarrow 2AlCl_3(s) \qquad \text{aluminum chloride}$$

$$4Al(s) + 3O_2(g) \rightarrow 2Al_2O_3(s) \qquad \text{aluminum oxide}$$

Group IV: C, Si, Ge, Sn, Pb Among the Group IV elements, carbon is a nonmetal, silicon and germanium are semimetals, and tin and lead are metals. Thus there is a larger, although still regular, change in properties in this group than in Groups I and II. Both carbon and silicon are abundant elements; in fact, silicon is the most abundant element on earth except oxygen. The other elements of this group are much less abundant, although tin and lead are well known and useful elements.

The Group IV elements react with chlorine and oxygen to give chlorides and oxides. For example,

$$C(s) + O_2(g) \rightarrow CO_2(g) \qquad \text{carbon dioxide}$$

$$Si(s) + 2Cl_2(g) \rightarrow SiCl_4(l) \qquad \text{silicon tetrachloride}$$

They also form hydrides such as CH_4 and SiH_4. These hydrides cannot easily be made by direct combination of the elements, although they form in other reactions.

Group V: N, P, As, Sb, Bi As is the case in Group IV, the properties of Group V elements change from top to bottom from those typical of nonmetals to those typical of metals. Nitrogen and phosphorus are nonmetals; arsenic and antimony are semimetals; and bismuth is a metal. Once again it is the first two members of the group that are most abundant and important. Nitrogen, $N_2(g)$, is the major component of the atmosphere. Phosphorus compounds are found in all living cells. An example is ATP (adenosine triphosphate), a compound necessary for the transfer of energy in cellular reactions.

Both nitrogen and phosphorus react with oxygen to give oxides and with the halogens to give halides; for example,

$$2P(s) + 3Cl_2(g) \rightarrow 2PCl_3(l) \qquad \text{phosphorus trichloride}$$

The hydride of nitrogen is ammonia, $NH_3(g)$, and there is a corresponding gaseous hydride of phosphorus, $PH_3(g)$, called phosphine.

Group VI: O, S, Se, Te, Po As we proceed from left to right across the periodic table, the elements become progressively more nonmetallic. In Group VI, oxygen, sulfur, and selenium are nonmetals, whereas tellurium and polonium are semimetals. We will be concerned primarily with the first two elements of Group VI, oxygen and sulfur, which are the most abundant.

As we mentioned in Chapter 2, oxygen combines with almost all the other elements to form oxides, and sulfur forms analogous sulfides:

$$2Mg(s) + O_2(g) \rightarrow 2MgO(s) \qquad \text{magnesium oxide}$$

$$Mg(s) + S(s) \rightarrow MgS(s) \qquad \text{magnesium sulfide}$$

Oxygen and sulfur form the hydrides water, $H_2O(l)$, and hydrogen sulfide, $H_2S(g)$.

The name "halogen" comes from two Greek words: *hals*, meaning "sea salt," and *gennao*, meaning "I produce." The halogen elements react with metals to form NaCl and other salts found in the sea.

Group VII: Halogens—F, Cl, Br, I, At The first two members of Group VII, fluorine and chlorine, are again among the most abundant elements in the earth's crust. Bromine and iodine are much less abundant but nevertheless have important uses. Astatine is radioactive and very rare. The **halogens** exist as diatomic molecules, for example, Cl_2 and I_2. We will consider the chemistry of the halogens in some detail in Chapter 4. We note here only that they are very reactive nonmetals (fluorine is more reactive than *any* other element). They all react with hydrogen to form the hydrogen halides HF, HCl, HBr, and HI, which are gases at room temperature; for example,

$$H_2(g) + Cl_2(g) \rightarrow 2HCl(g) \qquad \text{hydrogen chloride}$$

Before 1962 Group VIII elements were called the inert gases. ("Inert" means "unreactive" or "inactive.") But after it was discovered that compounds can be made from some of these elements, they became more commonly known as the noble gases, implying that they are much less reactive than other elements but not completely inert.

Group VIII: Noble Gases—He, Ne, Ar, Kr, Xe, Rn Among the Group VIII elements, known as the **noble gases**, only argon is relatively abundant on earth, constituting 1% of the atmosphere. All the others are present in the atmosphere but in very small amounts. The noble gases are very unreactive. No compounds of any of these elements were prepared until 1962. Today there are still no known compounds of helium, neon, or argon, and only a few compounds of krypton, xenon, and radon with fluorine and oxygen have been prepared.

Exercise 3.2 Name each of the following elements. Without consulting the periodic table, place each in its appropriate group and period.
(a) C **(b)** Na **(c)** N **(d)** Cl **(e)** Mg **(f)** Si **(g)** Al **(h)** O **(i)** F **(j)** Ca

Valence and the Periodic Table

A very important property of an element is its capacity for combining with other elements, which we call its **valence**. For example, with very few exceptions, hydrogen combines with no more than one atom of any other element. Hydrogen is therefore said to have a valence of 1.

One oxygen atom combines with *two* hydrogen atoms to give the water molecule, H_2O. One nitrogen atom combines with *three* hydrogen atoms to give the ammonia molecule, NH_3. One carbon atom combines with *four* hydrogen atoms to give the methane molecule, CH_4. Oxygen, nitrogen, and carbon are said to have valences of 2, 3 and 4, respectively. Some elements have more than one valence, but here we will consider only the **principal valence**, which may be defined as follows:

The principal valence of an element is equal to the number of hydrogen atoms that will combine with one atom of the element.

The principal valence of an element is given by its position in the periodic table:

- **All the elements in a group have the same principal valence.**
- **The principal valence of an element is equal to the group number for Groups I to IV. For Groups V to VIII the principal valence is equal to 8 − *n*, where *n* is the group number.**

For example, sodium, potassium, and all the other elements in Group I have a valence of 1; all the elements in Group II have a valence of 2; and all the elements in

TABLE 3.1 Valences of the Main Group* Elements

Group	I	II	III	IV	V	VI	VII	VIII
Element	H							He
	Li	Be	B	C	N	O	F	Ne
	Na	Mg	Al	Si	P	S	Cl	Ar
Valence	1	2	3	4	3	2	1	0

* Groups I through VIII

Group V have a valence of 8 − 5 = 3. The principal valences of the elements of Periods 2 and 3 are shown in Table 3.1. From these valences we can immediately write the empirical formulas for the compounds of these elements with hydrogen:

Period 2:	LiH	BeH_2	BH_3	CH_4	NH_3	H_2O	HF
Period 3:	NaH	MgH_2	AlH_3	SiH_4	PH_3	H_2S	HCl

Neon and argon (Group VIII) have a principal valence of zero; they do not form compounds with hydrogen or with any other element.

Like hydrogen, fluorine and chlorine (Group VII) have a valence of 1, so we can write the empirical formulas of their compounds with the Period 2 and 3 elements as follows:

Period 2:	*Fluorides*	LiF	BeF_2	BF_3	CF_4	NF_3	OF_2	F_2
	Chlorides	LiCl	$BeCl_2$	BCl_3	CCl_4	NCl_3	OCl_2	FCl
Period 3:	*Fluorides*	NaF	MgF_2	AlF_3	SiF_4	PF_3	SF_2	ClF
	Chlorides	NaCl	$MgCl_2$	$AlCl_3$	$SiCl_4$	PCl_3	SCl_2	Cl_2

We can also write empirical formulas for compounds of these elements with elements of a higher valence than 1. Oxygen, with a valence of 2, combines with *two* atoms of an element with a valence of 1, such as hydrogen, and with *one* atom of an element with a valence of 2, such as magnesium. Thus, one oxygen atom can combine with two hydrogen atoms to form H_2O, or with one magnesium atom to form MgO. The empirical formulas of the oxides of the elements in Periods 2 and 3, based on the principal valences of these elements, are predicted to be

In writing an empirical formula for the compound formed from elements A and B, use (principal valence of atom A) × (number of atoms of A) = (principal valence of atom B) × (number of atoms of B)

Period 2:	*Oxides*	Li_2O	BeO	B_2O_3	CO_2	N_2O_3	O_2	F_2O
Period 3:	*Oxides*	Na_2O	MgO	Al_2O_3	SiO_2	P_2O_3	SO	Cl_2O

We can now see one reason it is useful to know the positions of the first 20 elements of the periodic table. Knowing the group that an element is in gives us its principal valence and enables us to write the empirical formulas of many of its compounds.

Exercise 3.3 Write the formulas of the compounds formed between each of the following pairs of atoms.
(a) Ca and H **(b)** K and O **(c)** Mg and S **(d)** Na and N **(e)** Al and S

Exercise 3.4 Predict the following properties of the halogen astatine, At: its physical state and the molecular and empirical formulas of the compounds it forms with H, O, Na, and Mg.

3.2 THE SHELL MODEL OF THE ATOM

To explain the properties of the elements, we must consider the structure of an atom and how it differs from one element to another. The **shell model** is a simple model of an atom that we can use to begin to understand how atoms combine to form molecules and network solids and why each group of elements in the periodic table has a characteristic valence and forms similar compounds. According to the shell model, the electrons surrounding the nucleus of an atom are arranged in successive spherical layers called **shells**, each of which can contain only a limited number of electrons.

Figure 3.6 shows a simple representation of the shell structures of the first 18 elements. The successive shells are designated by $n = 1$ for the first (innermost) shell, $n = 2$ for the second shell, and so on. The first shell can contain up to 2 electrons, the second up to 8, and the third up to 18. In general, the maximum allowable number of electrons in a shell is $2n^2$. The shells overlap each other to some extent and are not as well defined as shown in Figure 3.6. But on average, the electrons in the $n = 1$ shell are considerably closer to the nucleus than are those in the $n = 2$ shell, those in the $n = 3$ shell are still farther from the nucleus, and so on.

As Figure 3.6 shows, the hydrogen atom ($Z = 1$) has 1 electron in its first ($n = 1$) shell, whereas the He atom ($Z = 2$) has 2 electrons in its $n = 1$ shell. Because the maximum number of electrons the $n = 1$ shell can have is 2 ($2n^2 = 2$), the $n = 1$ shell of helium is full. Lithium ($Z = 3$), the first element in Period 2, has 3 electrons, so 2 electrons occupy the $n = 1$ shell, and the third electron must go into the next ($n = 2$) shell. In the succeeding elements of Period 2, Be, B, C, N, O, F, and Ne, electrons continue to enter the outer ($n = 2$) shell until it contains a maximum of 8 electrons. Neon ($Z = 10$) has 2 electrons in the $n = 1$ shell and 8 electrons in the $n = 2$ shell. The maximum number of electrons the $n = 2$ shell can have is 8

Period 1	H $Z = 1$	He $Z = 2$
	$1e^-$ • +1	$2e^-$ • +2
Electrons in shell $n = 1$	1	2

Period 2	Li $Z = 3$	Be $Z = 4$	B $Z = 5$	C $Z = 6$	N $Z = 7$	O $Z = 8$	F $Z = 9$	Ne $Z = 10$
	$2e^-$ • +3 $1e^-$	$2e^-$ • +4 $2e^-$	$2e^-$ • +5 $3e^-$	$2e^-$ • +6 $4e^-$	$2e^-$ • +7 $5e^-$	$2e^-$ • +8 $6e^-$	$2e^-$ • +9 $7e^-$	$2e^-$ • +10 $8e^-$
Electrons in shell $n = 1$	2	2	2	2	2	2	2	2
$n = 2$	1	2	3	4	5	6	7	8

Period 3	Na $Z = 11$	Mg $Z = 12$	Al $Z = 13$	Si $Z = 14$	P $Z = 15$	S $Z = 16$	Cl $Z = 17$	Ar $Z = 18$
	$2e^-$ • +11 $8e^-$ $1e^-$	$2e^-$ • +12 $8e^-$ $2e^-$	$2e^-$ • +13 $8e^-$ $3e^-$	$2e^-$ • +14 $8e^-$ $4e^-$	$2e^-$ • +15 $8e^-$ $5e^-$	$2e^-$ • +16 $8e^-$ $6e^-$	$2e^-$ • +17 $8e^-$ $7e^-$	$2e^-$ • +18 $8e^-$ $8e^-$
Electrons in shell $n = 1$	2	2	2	2	2	2	2	2
$n = 2$	8	8	8	8	8	8	8	8
$n = 3$	1	2	3	4	5	6	7	8

FIGURE 3.6 Shell Models of the First 18 Elements These diagrams show how electrons are arranged in shells; they do not indicate the relative sizes of shells or atoms. Starting with the innermost shell, we designate shells as $n = 1$, $n = 2$, and so on. Within a group, each element has a different number of shells but the same number of electrons in its outermost shell. Helium, with only two electrons in its outermost shell, is an exception.

TABLE 3.2 Shell Structure of Atoms of the First 20 Elements

			Number of Electrons in Each Shell			
Period	Z	**Element**	$n = 1$	2	3	4
1	1	H	1			
	2	He	2			
2	3	Li	2	1		
	4	Be	2	2		
	5	B	2	3		
	6	C	2	4		
	7	N	2	5		
	8	O	2	6		
	9	F	2	7		
	10	Ne	2	8		
3	11	Na	2	8	1	
	12	Mg	2	8	2	
	13	Al	2	8	3	
	14	Si	2	8	4	
	15	P	2	8	5	
	16	S	2	8	6	
	17	Cl	2	8	7	
	18	Ar	2	8	8	
4	19	K	2	8	8	1
	20	Ca	2	8	8	2

Note: Color shading indicates the valence shell.

($2n^2 = 8$), so the $n = 2$ shell is now full. We can write the electron arrangements of the elements in the first two periods as follows: H, 1; He, 2; Li, 2, 1; . . . , Ne, 2, 8, where the first number is the number of electrons in the $n = 1$ shell, the second number is the number of electrons in the $n = 2$ shell, and so on.

In the elements of Period 3, electrons begin to fill the $n = 3$ shell. In sodium ($Z = 11$), 10 electrons fill the $n = 1$ and $n = 2$ shells, and there is 1 electron in the $n = 3$ shell. In other words, sodium has the electron arrangement 2, 8, 1. In the succeeding elements of Period 3, electrons continue to fill this shell, so argon ($Z = 18$) has the electron arrangement 2, 8, 8. A fourth shell ($n = 4$) is commenced in the next period. Potassium ($Z = 19$) has the electron arrangement 2, 8, 8, 1, and calcium ($Z = 20$) has the electron arrangement 2, 8, 8, 2. We will discuss in Chapter 6 the electron arrangements of the elements after potassium. The electron arrangements for the first 20 elements are shown in Table 3.2.

The outer shell of an atom is called its *valence shell*, because it is the electrons in this shell that are involved when an atom combines with another atom and that determine the valence of the atom. We see from Table 3.2 and Figure 3.6 that the valence shells of the elements of Group I contain only 1 electron. The valence shells of all the Group II elements have 2 electrons; the valence shells of the Group III elements have 3 and so on.

- **Elements in the same group of the periodic table have the same number of electrons in their valence shell (except for helium, which has only 2 electrons in its valence shell although the other members of its group have 8).**
- **For a main group element (Groups I through VIII), the number of valence-shell electrons is equal to the group number (except in the case of helium).**

III The number of valence electrons for an atom with atomic number Z is equal to Z for a Period 1 element, $Z - 2$ for a Period 2 element, and $Z - 10$ for a Period 3 element.

In the rest of this chapter we will see how the shell model of the atom accounts for the existence of families of elements with the same principal valence and with similar properties and, thus, for the general form of the periodic table. In Chapter 6 we will give some experimental evidence for the shell model and study it in more detail.

Exercise 3.5 Give the shell-model electron arrangements for each of the following atoms.
(a) C **(b)** O **(c)** F **(d)** Mg **(e)** K **(f)** P **(g)** S **(h)** Cl

Exercise 3.6 Give the number of the group to which each of the following elements belongs. How many valence electrons does an atom of each of these elements have?
(a) F **(b)** N **(c)** Cl **(d)** B **(e)** Na **(f)** S **(g)** Ca **(h)** P **(i)** C **(j)** Al

3.3 CHEMICAL BONDS AND LEWIS STRUCTURES

Whenever two atoms are held together in a molecule, we say that there is a **chemical bond** between them. How many bonds an atom forms (its valence) and the types of bonds that it forms are determined by the number of electrons in its valence shell and by the strength of the force with which these electrons are held by the atom.

The Atomic Core and Valence Electrons

It is convenient to think of the nucleus plus the completed inner shells of an atom as the **core** of the atom. The core electrons are closer to the nucleus and are therefore held to it more strongly than the valence electrons in the outer shell. For this reason, the core electrons are not generally involved when atoms combine, and the core remains unchanged.

The overall positive charge on the core of an atom is called the **core charge**:

Core charge = nuclear charge + total charge of core electrons

For example, magnesium ($Z = 12$), has a nuclear charge of 12, 2 valence electrons, and 10 core electrons, so the core charge of magnesium is $12 - 10 = 2$. In an electrically neutral atom, the number of valence-shell electrons is equal to the magnitude of the core charge, which increases from $+1$ for Group I to $+8$ for Group VIII:

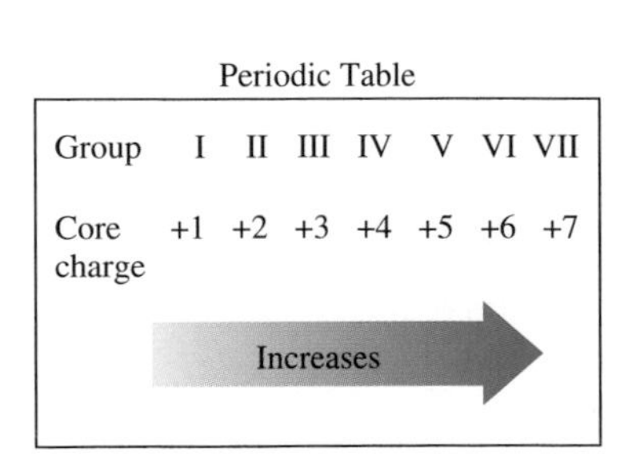

Magnitude of core charge = group number

As the core charge increases, so does the electrostatic force exerted by the core on the valence electrons. In other words, the strength with which an atom holds on to its outermost electrons increases from Group I to Group VIII.

Because the electron shells overlap somewhat and because the electrons in the valence shell repel each other, the net force attracting a valence-shell electron is somewhat less than the force due to the core charge. Nevertheless, the core charge is a useful but approximate guide to the force acting on the valence electrons, which increases markedly left to right across the periodic table. Although core electrons do not participate in the formation of compounds the magnitude of the core charge is very important in determining how readily the valence electrons will do so.

Exercise 3.7 What is the core charge of each of the following atoms?
(a) F **(b)** N **(c)** Cl **(d)** B **(e)** Na **(f)** S **(g)** Ca **(h)** P **(i)** C **(j)** Al

The Octet Rule

The valences of neon and argon, both of which have eight valence-shell electrons, are 0: These elements are not known to form any compounds and are themselves monatomic. Thus, a completely filled valence shell containing eight electrons appears to be an especially stable, or inert, arrangement. This observation led the American chemist Gilbert Lewis (Figure 3.7) to propose that

In compound formation an atom gains or loses electrons, or shares pairs of electrons, until its valence shell has eight electrons.

A valence shell containing eight electrons is called an **octet**, and the rule just stated is known as the **octet rule**. We will now see how the octet rule enables us to understand the number and types of bonds an atom will form.

FIGURE 3.7 Gilbert Newton Lewis (1875–1946) Lewis was born in Massachusetts, but in 1884 his family moved to Lincoln, Nebraska, where he received little formal schooling. He began his college career at the University of Nebraska but transferred to Harvard, where he obtained his Ph.D in 1899. After a period of advanced studies in Germany, he joined the faculty of the Massachusetts Institute of Technology in 1905. In 1912 he became professor of chemistry at the University of California, Berkeley. Under his guidance the chemistry department at Berkeley gained international recognition.

Lewis's inquisitive, imaginative mind led him to make important contributions in several areas of chemistry. He was the first to propose that atoms could be held together by sharing an electron pair, and he proposed the electron-dot (Lewis) structures that we use. He made outstanding contributions to thermodynamics (the study of energy changes) and was the co-author of a textbook that profoundly influenced the teaching of thermodynamics. In addition, he proposed a new definition of acids and bases (see Chapter 10) and was the first to prepare and study pure heavy water, $^{2}H_2O$ (D_2O).

TABLE 3.3 Lewis Symbols and Valences of the Main Group Elements

Group	I	II	III	IV	V	VI	VII	VIII
Number of valence-shell electrons	1	2	3	4	5	6	7	8
Valence	1	2	3	4	3	2	1	0
Lewis symbols:								
Period 2	Li·	·Be·	·B·	·C·	·N·	:O·	:F·	:Ne:
Period 3	Na·	·Mg·	·Al·	·Si·	·P·	:S·	:Cl·	:Ar:

Lewis (Electron-Dot) Symbols

In using the octet rule it is very convenient to use symbols that were introduced by Lewis, called **Lewis symbols** or **electron-dot symbols**. The electrons in the valence shell of an atom are indicated by dots surrounding the symbol of the element (Table 3.3). For the elements of Groups I through IV, these electrons are arranged as single dots. For the elements of Groups V through VIII, the electrons in excess of four are arranged in pairs. For example, the Lewis symbol for a carbon atom (Group IV) has four single dots, but that for a nitrogen atom (Group V) has three single dots and one pair of dots. The Lewis symbol for a chlorine atom (Group VII) has one single dot and three pairs of dots. We will see later why it is convenient to arrange the dots in this way.

Exercise 3.8 Write the Lewis symbol for each of the following atoms.
(a) H **(b)** F **(c)** C **(d)** Al **(e)** N **(f)** Ne **(g)** Ca **(h)** O

Ions, Ionic Bonds, and Ionic Compounds

When an atom gains or loses electrons, a charged atom called an **ion** is formed.

Positive ions are called *cations*, and negative ions are called *anions*.

Cations formed from metals are given the name of the metal: for example, Na^+ is called the sodium ion; K^+, the potassium ion; Mg^{2+}, the magnesium ion; and Ca^{2+}, the calcium ion. Anions formed from nonmetals have names related to that of the nonmetal, with the ending *-ide*. For instance, H^- is called the hydride ion; F^-, the fluoride ion; Cl^-, the chloride ion; O^{2-}, the oxide ion; S^{2-}, the sulfide ion; and N^{3-}, the nitride ion.

Sodium has the electron arrangement 2, 8, 1, with only one electron in its valence shell (Table 3.2, Figure 3.6). If a sodium atom loses this valence electron, a cation is formed. This sodium ion Na^+, has the electron arrangement 2, 8. Thus the $n = 2$ shell, with its eight electrons, has become the outer, or valence, shell. In this way sodium can acquire an octet by losing a single electron.

Chlorine has the electron arrangement 2, 8, 7. If a chlorine atom gains an electron, the chloride ion, Cl^-, an anion, is formed; Cl^- has the electron arrangement 2, 8, 8, with eight electrons in the valence shell. In this way chlorine can acquire an octet by gaining a single electron.

When sodium reacts with chlorine to form sodium chloride, NaCl, each sodium atom loses one electron and thereby forms a sodium ion, Na^+, and each chlorine atom gains one electron and thereby forms a chloride ion, Cl^-. We can summarize this reaction very conveniently in terms of Lewis symbols:

$$Na\cdot + \cdot\ddot{\underset{\cdot\cdot}{Cl}}: \rightarrow (Na^+)(:\ddot{\underset{\cdot\cdot}{Cl}}:^-)$$

Because opposite charges attract, the sodium and chloride ions are attracted to each other and form solid sodium chloride, NaCl. This compound consists of equal numbers of sodium ions, Na^+, and chloride ions, Cl^-, held together by electrostatic attraction (Figure 3.8).

The electrostatic attraction between oppositely charged ions is called an *ionic bond*.

Compounds that consist of ions held together by ionic bonds are called **ionic compounds**.

In Demonstration 2.2, we saw that magnesium oxide can be made by burning magnesium in air or oxygen. Magnesium has two electrons in its valence shell. When it loses these two electrons, it becomes the magnesium cation Mg^{2+}, with a valence shell containing eight electrons. Oxygen has six electrons is its valence shell. When it gains two electrons, it becomes the oxide anion, O^{2-}, which has a valence shell containing eight electrons. Magnesium oxide consists of Mg^{2+} and O^{2-} ions held together by electrostatic attraction, that is, by ionic bonds:

$$\cdot Mg\cdot + \cdot\ddot{\underset{\cdot}{O}}: \rightarrow (Mg^{2+})(:\ddot{\underset{\cdot\cdot}{O}}:^{2-})$$

In a reaction between a metal and a nonmetal, electrons are removed from the metal and transferred to the nonmetal. This reaction forms positive metal ions (cations) and negative nonmetal ions (anions) held together by ionic bonds.

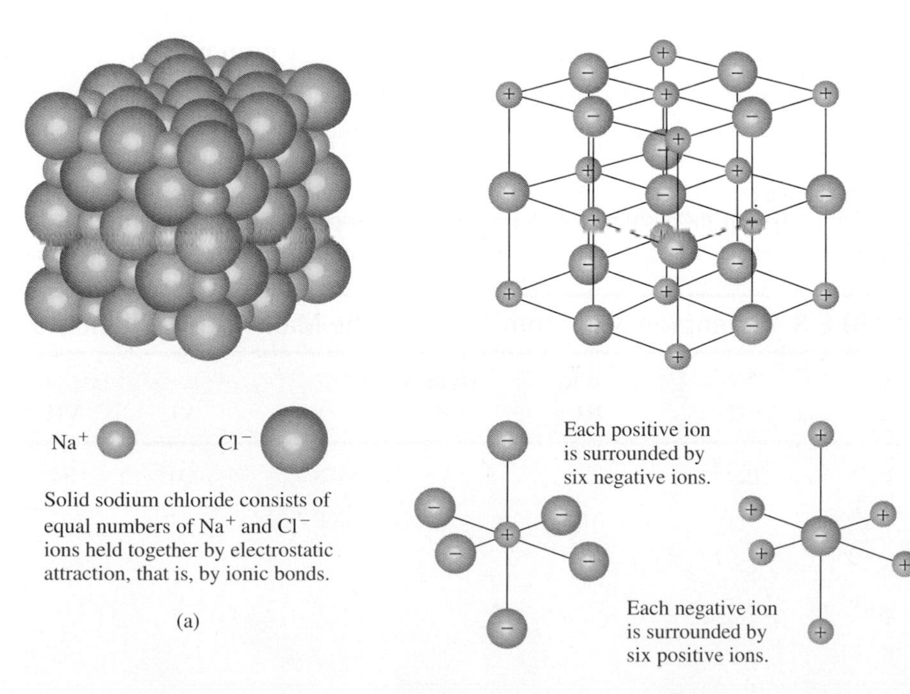

FIGURE 3.8 The Structure of Sodium Chloride
(a) The regular array of sodium ions and chloride ions extends indefinitely throughout the crystal. The crystal can therefore be regarded as one huge molecule. No individual NaCl molecules can be recognized in the structure. (b) An expanded view in which the ions are shown separated from each other, but their arrangement has been maintained. Each sodium ion is surrounded by six chloride ions, which have an octahedral arrangement. Each chloride ion is surrounded by six sodium ions, which also have an octahedral arrangement.

FIGURE 3.9 Crystals of Common Salt The crystals of common salt (sodium chloride) are cubes. Their shape is a result of the cubic structure of the crystal lattice.

Structure of Ionic Compounds An ionic compound has no overall charge, so the total positive charge on its cations must equal the total negative charge on its anions. Thus, in Na^+Cl^- and $Mg^{2+}O^{2-}$ there are equal numbers of positive and negative ions, but in $Mg^{2+}(Cl^-)_2$, magnesium chloride, there are twice as many negative Cl^- ions as positive Mg^{2+} ions.

All ionic compounds have *network*, or giant molecule, *structures* (such that individual molecules cannot be identified) and are solids at room temperature. In sodium chloride each sodium ion is surrounded by six chloride ions that form an octahedron around it, and each chloride ion is likewise surrounded by six octahedrally arranged sodium ions (Figure 3.8). Overall, the arrangement of ions has a cubic shape. Because this cubic arrangement persists throughout the solid, sodium chloride forms crystals that have this same cubic shape (Figure 3.9). Potassium chloride, magnesium oxide, and many other ionic compounds with the general formula MX have the same cubic structure. Other ionic compounds may have different structures, depending on the relative numbers of the ions of opposite charge and their relative sizes (Chapter 10).

Formulas of Ionic Compounds The formula of an ionic compound such as sodium chloride may be written as Na^+Cl^-, but it is more often written simply as NaCl, in which its ionic structure is not explicitly shown. We always use empirical formulas for ionic compounds: Because they have no individual molecules, they can have no molecular formula.

The empirical formulas of simple ionic compounds may be derived from the charges of the common ions of the main group elements, which are listed in Table 3.4. The metals of Groups I, II, and III form M^+, M^{2+}, and M^{3+} ions, respectively, by losing 1, 2, or 3 electrons. The nonmetals of Groups V, VI, and VII form X^{3-}, X^{2-}, or X^- ions, respectively, by gaining 3, 2, or 1 electron. To obtain the empirical formula of an ionic compound, we combine the appropriate number of ions of opposite charge so that the sum of the charges is zero.

EXAMPLE 3.1 Empirical Formulas for Ionic Compounds

Give the empirical formulas for the ionic compounds formed by each of the following pairs of elements.

(a) Na, O **(b)** Mg, Br **(c)** K, S **(d)** Al, F **(e)** Na, P

Because an ionic compound $(A^{m+})_x(B^{n-})_y$ must be neutral, $m \times x = n \times y$.

Solution: We can find the charge on each of the ions formed by these elements from Table 3.4 or directly from the position of each element in the periodic table. Remembering that the

TABLE 3.4 Common Monatomic Ions of the Main Group Elements

Group I	II	III	IV	V	VI	VII
Li^+	Be^{2+}			$:\ddot{N}:^{3-}$	$:\ddot{O}:^{2-}$	$:\ddot{F}:^-$
Na^+	Mg^{2+}	Al^{3+}		$:\ddot{P}:^{3-}$	$:\ddot{S}:^{2-}$	$:\ddot{Cl}:^-$
K^+	Ca^{2+}					$:\ddot{Br}:^-$
Rb^+	Sr^{2+}					$:\ddot{I}:^-$
Cs^+	Ba^{2+}					

Note: All these ions except Li^+ and Be^{2+} have eight electrons in their outer shell. Both Li^+ and Be^{2+} have only two electrons in the $n = 1$ shell, as does the noble gas He. All the other ions have the same valence-shell electron arrangements as the noble gases Ne to Xe.

total charge of an ionic compound is zero, we can deduce the ratio of the numbers of cations and anions in each compound and hence write the empirical formula. The results:

Elements	Ions	Empirical Formula	
Na, O	$Na^+ \ :\ddot{\underset{..}{O}}:^{2-}$	Na_2O	$(Na^+)_2\,(O^{2-})$
Mg, Br	$Mg^{2+} \ :\ddot{\underset{..}{Br}}:^-$	$MgBr_2$	$(Mg^{2+})\,(Br^-)_2$
K, S	$K^+ \ :\ddot{\underset{..}{S}}:^{2-}$	K_2S	$(K^+)_2\,(S^{2-})$
Na, P	$Na^+ \ :\ddot{\underset{..}{P}}:^{3-}$	Na_3P	$(Na^+)_3\,(P^{3-})$

Understanding the Octet Rule

We can understand the inertness of neon and the tendency of atoms such as sodium and fluorine to attain the electron configuration of neon in terms of the core charge and the number of electrons in the valence shell. A fluorine atom, $:\ddot{\underset{..}{F}}\cdot$, has a core charge of $+7$ that strongly attracts the valence-shell electrons, and it has space in its valence shell ($n = 2$) for one additional electron. Therefore, the F atom has a strong tendency to add one additional electron to its valence shell to form a F^- ion. A neon atom, $:\ddot{\underset{..}{Ne}}:$, has a core charge of $+8$, and it therefore attracts the electrons in its valence shell still more strongly than fluorine, but it has no space in its valence shell for an additional electron. Therefore, neon cannot gain an electron and form a Ne^- ion. Moreover, because neon attracts its valence-shell electrons so strongly, it is very difficult to remove an electron from neon to form a Ne^+ ion. Neon does not form either positive or negative ions. Sodium, the element following neon in the periodic table, has one electron in its valence shell ($n = 3$). This electron is attracted by a core charge of only $+1$, so it is not strongly held and can easily be removed to form a Na^+ ion. Thus, when sodium reacts with fluorine, an electron is removed from a sodium atom, and an electron is added to a fluorine atom—both sodium and fluorine attain the same stable electron arrangement as neon.

The atoms of the other noble gases also have eight electrons in their valence shells, so, like neon, they also do not form either a positive ion or a negative ion. In contrast, the atoms of many other elements tend to form ionic compounds, gaining or losing electrons until they have the same stable electronic arrangement as a noble gas. In other words, they obey the octet rule. Electrons can rather easily be removed from metals to form cations with the electron configuration of the noble gas that precedes them in the table. Nonmetals attract electrons to form anions with the electron arrangement of the noble gas that follows them. Metals frequently react with nonmetals to form ionic compounds because electrons can be removed from metals rather easily, whereas nonmetals strongly attract electrons.

Helium is an obvious exception to the octet rule, because only two electrons are needed to fill its $n = 1$ shell and give the atom a stable inert electron arrangement. The ions H^-, Li^+, and Be^{2+} have just two electrons in their $n = 1$ shell. Thus H, Li, and Be also do not obey the octet rule in forming ionic compounds, but they may be said to obey a **duet rule**. As we shall see later, there are some other exceptions to the octet rule, particularly in Periods 3 to 7.

Isoelectronic Ions The ions N^{3-}, O^{2-}, F^-, Na^+, Mg^{2+}, and Al^{3+} (Table 3.4) differ in their nuclear charges, which range from $+7$ for N to $+13$ for Al, but all of them have the same electron arrangement. This arrangement, 2, 8, is the same as that of an atom of neon. We say that these ions are **isoelectronic** with neon:

Atoms and ions that have the same electron arrangement are said to be isoelectronic.

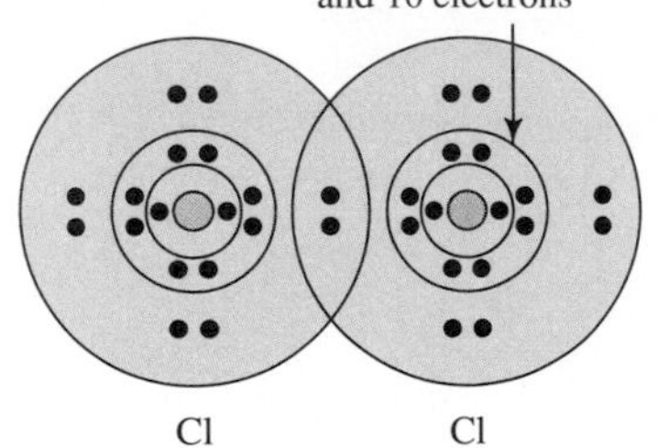

FIGURE 3.10 Formation of a Covalent Bond by Sharing of an Electron Pair By sharing a pair of electrons, each Cl atom in the Cl_2 molecule effectively acquires a valence shell of eight electrons.

Covalent Bonds and Lewis Structures

We have seen that when a metal such as magnesium reacts with a nonmetal such as chlorine, an ionic compound is formed in which ions are held together in a network structure by electrostatic attraction. But how do we explain the existence of molecules such as H_2O, NH_3, or Cl_2 in which nonmetal atoms are held together? If one chlorine atom in a Cl_2 molecule were to gain an electron and become Cl^-, the other chlorine atom would have to lose an electron and become Cl^+ and would have only six, rather than the expected eight, electrons in its valence shell.

In 1916 Lewis proposed that in Cl_2 and similar molecules, atoms are held together by a pair of shared electrons. Thus, he viewed the formation of the Cl_2 molecule in the following way:

$$:\ddot{\underset{..}{Cl}}\cdot + \cdot\ddot{\underset{..}{Cl}}: \rightarrow :\ddot{\underset{..}{Cl}}:\ddot{\underset{..}{Cl}}:$$

shared pair of electrons

By sharing a pair of electrons, both atoms effectively acquire eight electrons in their valence shells (Figure 3.10). This type of bond is called a **covalent bond**.

Whenever two atoms are held together by a shared pair of electrons, we say that there is a covalent bond between them. The shared pair of electrons is called a *bonding pair*.

Many compounds of the nonmetals consist of molecules held together by covalent bonds. For example, nitrogen, in Group V, has a valence of 3 and forms the fluoride NF_3. Nitrogen needs three electrons to complete its octet; it can obtain these electrons by sharing a pair of electrons with each of three fluorine atoms:

$$\cdot\ddot{\underset{\cdot}{N}}\cdot + 3\cdot\ddot{\underset{..}{F}}: \rightarrow :\ddot{\underset{..}{F}}:\ddot{\underset{..}{N}}:\ddot{\underset{..}{F}}: \atop :\underset{..}{F}:$$

The diagrams

$$:\ddot{\underset{..}{Cl}}:\ddot{\underset{..}{Cl}}: \quad \text{and} \quad :\ddot{\underset{..}{F}}:\ddot{\underset{..}{N}}:\ddot{\underset{..}{F}}: \atop :\underset{..}{F}:$$

which show the arrangement of the valence-shell electrons in the Cl_2 and NF_3 molecules, are called **Lewis structures** or **electron-dot diagrams**. (Note that Lewis *symbols* illustrate elements, whereas Lewis *structures* illustrate molecules.)

We can describe the bonds in the hydrogen molecule, H_2, and in many molecules formed by hydrogen with other nonmetallic elements in the same way. Recall, however, that hydrogen is an exception to the octet rule and needs only two electrons to fill its $n = 1$ shell. The Lewis structures for the H_2 molecule is

$$H:H$$

The Lewis structures for CH_4, NH_3, H_2O, and HF are

$\begin{matrix} & H & \\ & .. & \\ H: & C & :H \\ & .. & \\ & H & \end{matrix}$	$\begin{matrix} & .. & \\ H: & N & :H \\ & .. & \\ & H & \end{matrix}$	$\begin{matrix} & .. & \\ H: & O & : \\ & .. & \\ & H & \end{matrix}$	$H:\ddot{\underset{..}{F}}:$
Methane	Ammonia	Water	Hydrogen fluoride

A carbon atom needs four electrons, nitrogen needs three, oxygen needs two, and fluorine needs one electron to complete a valence shell of eight electrons, an octet. Therefore, carbon can form four covalent bonds, nitrogen three, oxygen two, and fluorine one covalent bond.

Lewis structures of molecules are typically simplified by representing a bonding electron pair by a line drawn between the two atoms. The Lewis structures of the methane, ammonia, water, and hydrogen fluoride molecules are commonly drawn in the following manner:

```
     H
     |           ..          ..        ..
H—C—H      H—N—H      H—O:     H—F:
     |           |           |         ..
     H           H           H
```

Depending on the number of bonds that it forms, an atom with an octet of electrons may have one or more pairs of electrons that are not forming bonds. Such pairs are called **unshared pairs**, **nonbonding pairs**, or **lone pairs**. In the ammonia molecule the nitrogen atom has one unshared pair of electrons:

```
unshared pair
       ..↙
    H—N—H
       |
       H
```

In the water molecule the oxygen atom has two unshared pairs of electrons. The fluorine atom in a hydrogen fluoride molecule, HF, has three unshared pairs of electrons.

To draw the Lewis structure of any molecule that contains three or more atoms, we need to know which atoms are bonded together. The arrangement of the atoms in a molecule can be found with complete certainty only by experiment, using one of the methods that we describe later. For now we will be concerned only with simple molecules in which atoms of H, F, Cl, or O are attached to a central atom. In these cases, the atom with the highest valence—that is, the atom that needs the most electrons to complete its valence shell—is normally the central atom of the molecule. Nitrogen is the central atom in the NH_3 molecule, carbon in the CH_4 molecule, oxygen in the H_2O molecule, and so on. To draw the Lewis structure, we write the Lewis symbol of the central atom, then form bonds from the central atom to the other atoms until each atom has an octet (or a single electron pair, in the case of hydrogen).

In Chapter 7 we describe how to draw the Lewis structures of a greater variety of molecules.

EXAMPLE 3.2 Drawing Lewis Structures for Molecules

Draw the Lewis structures for the covalent compounds formed between the following pairs of elements.

(a) C and F **(b)** S and H **(c)** P and Cl

Solution: (a) C (in Group IV) has 4 valence electrons, and F (in Group VII) has 7 valence electrons; C needs 4 electrons and F needs 1 electron to complete an octet of electrons:

Nonbonding electrons must be included in any Lewis structure.

```
                           ..
                          :F:
     ..       .         .. .. ..
  4:F· + ·C· → :F:C:F:
     ..       .         .. .. ..
                          :F:
                           ..
```

(b) S (in Group VI) has 6 electrons and needs 2 more to complete an octet; H (in Group I) has 1 electron and needs 1 more to complete a duet:

```
 ..                ..
:S· + 2·H → H:S:H
 .                 ..
```

(c) P (in Group V) has 5 electrons and needs 3 more, and Cl (in Group VII) has 7 electrons and needs 1 more, to complete an octet:

$$:\dot{\underset{\cdot}{P}}\cdot + 3:\ddot{\underset{\cdot\cdot}{Cl}}: \rightarrow \begin{matrix} & :\ddot{Cl}: & \\ & :\ddot{P}:\ddot{\underset{\cdot\cdot}{Cl}}: & \\ & :\underset{\cdot\cdot}{Cl}: & \end{matrix}$$

Exercise 3.9
Draw the Lewis structure of each of the following molecules.
(a) F_2 **(b)** CCl_4 **(c)** OF_2 **(d)** SiF_4 **(e)** SCl_2

Multiple Bonds Covalent bonds formed by *one* electron pair shared between two atoms are called **single bonds. Double bonds**, covalent bonds in which two pairs of electrons are shared between two atoms, and **triple bonds**, covalent bonds in which three pairs of electrons are shared between two atoms, are also common. Double and triple bonds collectively are called **multiple bonds**.

The carbon dioxide molecule has double bonds. Carbon needs four electrons to complete its octet, and oxygen needs two. Each bond results from the sharing of *two* electrons from *each* atom:

$$\cdot\dot{\underset{\cdot}{C}}\cdot + 2\cdot\dot{\underset{\cdot\cdot}{O}}: \rightarrow :\underset{\cdot\cdot}{O}::C::\underset{\cdot\cdot}{O}: \quad or \quad :\underset{\cdot\cdot}{O}{=}C{=}\underset{\cdot\cdot}{O}:$$

Each CO bond consists of two pairs of shared electrons.

Carbon–carbon double bonds also are common. The hydrocarbon ethene, C_2H_4 (ethylene), for example, has the Lewis structure

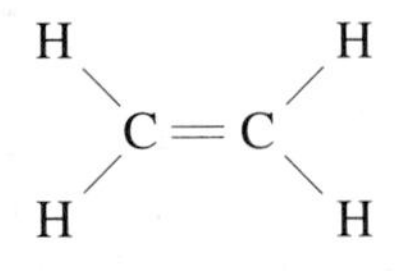

where each C atom forms single bonds with each of two H atoms and a double bond with the other C atom, giving each C atom an octet of electrons in its valence shell.

In the nitrogen molecule, N_2, each N atom shares its three unpaired valence electrons with the other N atom to form a triple bond, $:N:::N:$ or $:N{\equiv}N:$, which completes the octet of electrons around each nitrogen atom. The hydrocarbon ethyne, C_2H_2 (acetylene), has the Lewis structure $H:C:::C:H$ or $H{-}C{\equiv}C{-}H$, with a triple carbon–carbon bond.

EXAMPLE 3.3 Drawing Lewis Structures for Molecules with Multiple Bonds

Draw the Lewis structures of **(a)** O_2 and **(b)** P_2.

Solution: (a) Both $:\dot{\underset{\cdot\cdot}{O}}\cdot$ atoms (in Group VI) need two electrons to complete their octets, so they share two pairs of electrons:

$$:\dot{\underset{\cdot\cdot}{O}}\cdot + \cdot\dot{\underset{\cdot\cdot}{O}}: \rightarrow :\underset{\cdot\cdot}{O}{=}\underset{\cdot\cdot}{O}:$$

(b) Each P atom (in Group V) has the Lewis symbol $:\dot{\underset{\cdot}{P}}\cdot$ and can share its three unpaired electrons with the other P atom to form a triple bond, exactly as in N_2:

$$:\dot{\underset{\cdot}{P}}\cdot + \cdot\dot{\underset{\cdot}{P}}: \rightarrow :P{\equiv}P:$$

Exercise 3.10 Draw the Lewis structure of each of the following molecules. In each case the carbon atom or atoms are in the center of the molecule.
(a) H_2CO **(b)** CS_2 **(c)** C_2F_4 **(d)** FCCF **(e)** HCN

3.4 ELECTRONEGATIVITY

In a molecule such as H_2, F_2, or Cl_2, the electron pair of the covalent bond is shared equally between two atoms, because identical atoms attract the electrons of the bond equally strongly. However, in molecules such as HF or HCl, in which the two atoms are not the same, one atom attracts electrons more strongly than the other. So the electron pair forming the bond is shared unequally between the two atoms.

The ability of an atom in a molecule to attract the electrons of a covalent bond to itself is called its *electronegativity*, χ.

χ is the Greek letter chi (pronounced kye).

The greater the electronegativity of an atom, the more strongly the atom attracts the electrons of a bond. This important concept was proposed by Linus Pauling in 1937 (Box 3.2).

BOX 3.2 Linus Pauling

Linus Pauling was born in Portland, Oregon, in 1901. He graduated from Oregon State College in 1922 and obtained his Ph.D. in chemistry in 1925 from the California Institute of Technology. After studying in Europe, he became a professor at Cal Tech and remained there until 1963. Pauling was one of the first chemists to use quantum mechanics to understand the chemical bond. In 1939 he published a book entitled *The Nature of the Chemical Bond*, which proved to be one of the most influential chemistry texts of the twentieth century. He introduced the concepts of electronegativity and resonance structures (Chapter 7). He was among the few chemists who, as early as the 1930s, considered the possibility that the noble gases might form compounds, although the first compound, XeF_4, was not made until 1962. In 1954, Pauling was awarded the Nobel Prize in chemistry for his work on molecular structure.

In the 1950s Pauling began to study the structures of biomolecules. He was the first to propose the helical structure of many protein molecules. This led Pauling and others to consider that a similar helix might also be important in the structure of DNA. (Soon after, James Watson and Francis Crick discovered the now famous double-helix structure; Chapter 17). Pauling introduced the concept of molecular diseases, such as sickle-cell anemia, which is caused by an abnormal structure of the protein part of the hemoglobin molecule.

Linus Pauling (1901–)

After World War II Pauling became a passionate supporter of nuclear weapons disarmament. In 1962 he was awarded the Nobel Peace Prize for his work in this area, thus becoming the first person in history to receive two full Nobel Prizes. In 1970 he made the headlines again with his contention that large doses of vitamin C are effective in the prevention of the common cold and other ailments, including cancer. In his nineties, Pauling is still active in research in chemistry and biochemistry.

The net electrostatic force of attraction exerted by an atom on the electrons of a bonding pair results from the attraction exerted by the atom's nucleus and from the repulsion due to all the other electrons. This force cannot be calculated accurately or measured directly. However, we would expect the electronegativity of an atom to depend on the magnitude of the core charge and on the distance of the bonding pair of electrons from the nucleus. Thus, there are two important trends in electronegativity within the periodic table:

- Electronegativity increases across a period as the core charge increases.
- Electronegativity generally decreases from top to bottom in a group, because, with each successive shell, the bonding electrons are farther from the nucleus.

Because the electronegativity of an atom cannot be defined quantitatively, it cannot be given a precise value, but approximate values have been estimated by several methods. Figure 3.11 gives values for the electronegativities of the elements of the main groups. Values are not given for helium, neon, or argon because there are no known compounds of these elements.

The nonmetals on the right side of the periodic table have high electronegativities, and the metals on the left side have low electronegativities. To a good approximation, elements with an electronegativity of 2.0 or greater are nonmetals. Elements with an electronegativity of less than 2.0 are metals. The halogens are among the most electronegative elements—indeed, fluorine is the most electronegative element ($\chi = 4.1$). The second most electronegative element is oxygen ($\chi = 3.5$), followed by krypton and nitrogen ($\chi = 3.1$), and chlorine ($\chi = 2.8$).

Because electronegativities cannot be determined in a quantitative manner and because different methods of obtaining electronegativities give somewhat different values, the electronegativity values in Figure 3.11 have only a limited quantitative significance. Moreover, the electronegativity of a given element does not have a truly constant value. It varies slightly from one molecule to another, depending on the number and the nature of the other atoms that are bonded to it. Nevertheless, electronegativity is an important and useful concept, as we shall see in the next section and on many other occasions.

	Group							
Period	I	II	III	IV	V	VI	VII	VIII
1	H 2.2							He —
2	Li 1.0	Be 1.5	B 2.0	C 2.5	N 3.1	O 3.5	F 4.1	Ne —
3	Na 1.0	Mg 1.2	Al 1.3	Si 1.7	P 2.1	S 2.4	Cl 2.8	Ar —
4	K 0.9	Ca 1.0	Ga 1.8	Ge 2.0	As 2.2	Se 2.5	Br 2.7	Kr 3.1
5	Rb 0.9	Sr 1.0	In 1.5	Sn 1.7	Sb 1.8	Te 2.0	I 2.2	Xe 2.4

Electronegativity increases

Electronegativity decreases

FIGURE 3.11 Electronegativities of the Main Group Elements Across any period the electronegativity increases from left to right. Generally, the electronegativity decreases down any group from top to bottom, although among the heavier elements there are exceptions to this trend.

Bond Polarity and Bond Type

If two atoms forming a diatomic molecule have exactly the same electronegativity, as do the atoms in H_2 and F_2, the bonding electron pair is shared equally between the two atoms (Figure 3.12a). Each atom in effect acquires one electron of the pair and thereby gets just the number of electrons needed to balance its core charge, one for hydrogen and seven for fluorine. In contrast, if the atoms forming a bond have different electronegativities, the bonding pair is shared unequally. An example is the C—F bond. Fluorine has a greater electronegativity than carbon, so the fluorine atom attracts the electrons of the bonding pair more strongly than does the carbon atom (Figure 3.12b). As a result the fluorine atom in effect acquires slightly *more* than one electron in its valence shell and thereby gets a total of slightly more than the seven electrons needed to balance its core charge of +7. The fluorine atom therefore has a small negative charge, and the carbon atom acquires a corresponding small positive charge. These small charges are equal in magnitude but opposite in sign, and the overall charge of the molecule is zero. Such small charges are denoted by the symbol $\delta+$ (delta plus) and $\delta-$ (delta minus). When the difference in the electronegativites of the two atoms forming a bond is very large, as it is for Li and F, the ''bond pair'' is transferred almost completely to the more electronegative atom—here, the fluorine (Figure 3.12c). So there is now an ionic bond rather than a covalent bond.

Here we are thinking of an electron as a charge cloud rather than as a particle. The charge cloud of the two bonding electrons is unequally shared, so more than half this charge cloud is in the valence shell of the fluorine atom—it has in effect acquired more than one electron but less than two electrons.

A bond in which the atoms carry small charges of equal magnitude but

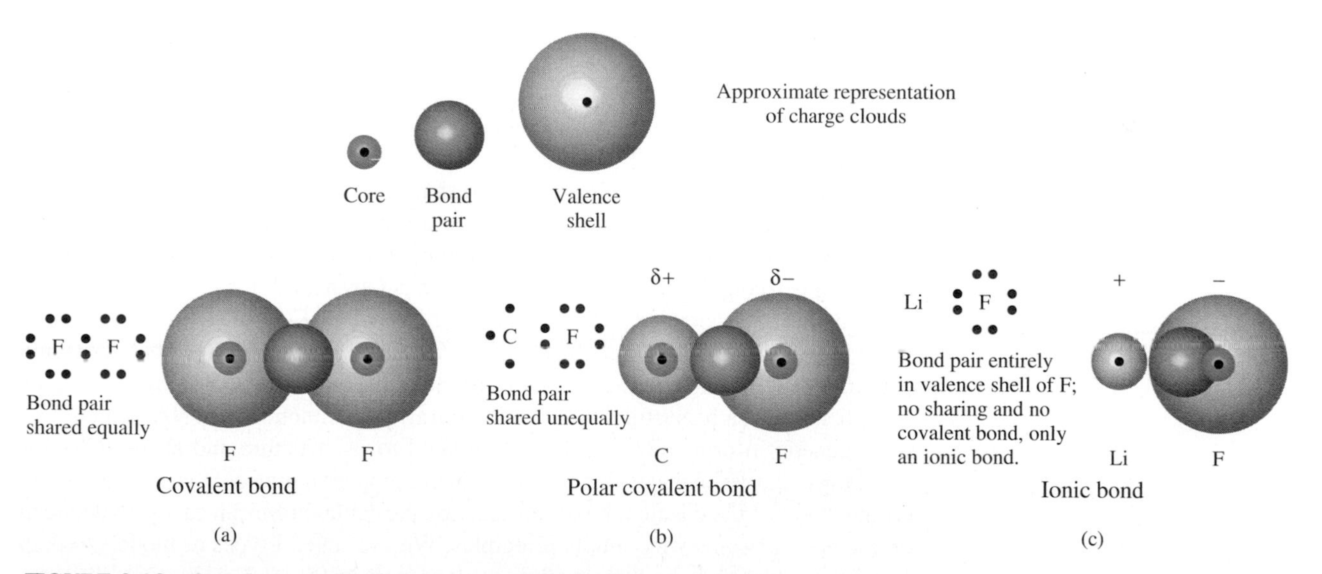

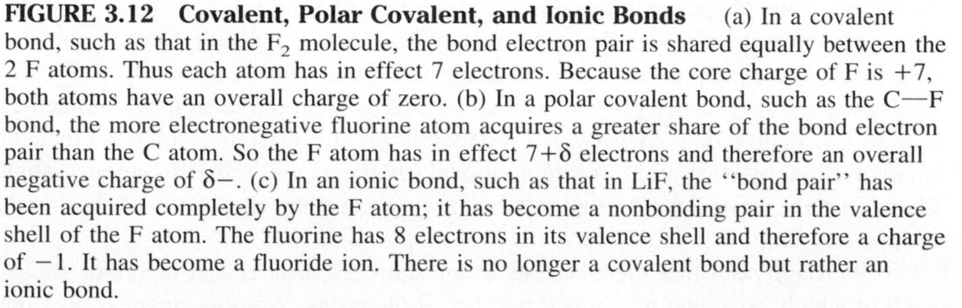
FIGURE 3.12 Covalent, Polar Covalent, and Ionic Bonds (a) In a covalent bond, such as that in the F_2 molecule, the bond electron pair is shared equally between the 2 F atoms. Thus each atom has in effect 7 electrons. Because the core charge of F is +7, both atoms have an overall charge of zero. (b) In a polar covalent bond, such as the C—F bond, the more electronegative fluorine atom acquires a greater share of the bond electron pair than the C atom. So the F atom has in effect $7+\delta$ electrons and therefore an overall negative charge of $\delta-$. (c) In an ionic bond, such as that in LiF, the ''bond pair'' has been acquired completely by the F atom; it has become a nonbonding pair in the valence shell of the F atom. The fluorine has 8 electrons in its valence shell and therefore a charge of −1. It has become a fluoride ion. There is no longer a covalent bond but rather an ionic bond.

Δ is the capital Greek letter delta. It is used to denote "the difference of" a quantity.

opposite sign is called a **polar covalent bond**. The electronegativity values in Figure 3.11 can be used to a limited extent to predict whether a particular bond is covalent (F–F), polar covalent ($C^{\delta+}$–$F^{\delta-}$), or ionic (Li^+F^-). If the electronegativity difference between the two atoms that form the bond is zero, then the bond is certainly covalent. If the electronegativity difference, $\Delta\chi$, is between zero and 1.0, the bond is polar covalent. Examples include the bonds in PCl_3 ($\Delta\chi = 2.8 - 2.1 = 0.7$) and NF_3 ($\Delta\chi = 4.1 - 3.1 = 1.0$). If the electronegativity difference is greater than 2.0, the bond is almost certainly ionic. Examples include the bonds in NaF ($\Delta\chi = 3.1$) and K_2O ($\Delta\chi = 2.6$). For electronegativity differences between 1.0 and 2.0, however, reliable predictions cannot be made. For example, AsF_3 ($\Delta\chi = 1.9$) is a molecular compound with polar covalent bonds, but NaCl ($\Delta\chi = 1.8$) is an ionic compound composed of Na^+ and Cl^- ions.

A simpler and more reliable guide to the nature of the bonds between two atoms of the main group elements is given by the following generalizations:

- Bonds between a Group I or a Group II metal atom and a nonmetal atom are ionic.
- Bonds between two different nonmetal atoms are polar covalent.
- Bonds between two identical nonmetal atoms are covalent.

Exercise 3.11 Classify the bonds in the following substances as covalent, polar covalent, or ionic.
(a) Br_2 **(b)** ClF **(c)** BF_3 **(d)** CaF_2 **(e)** O_2 **(f)** K_2S **(g)** $SiCl_4$ **(h)** PF_3

3.5 MOLECULAR SHAPE AND THE VSEPR MODEL

Molecules occur in a fascinating variety of shapes. Some are long and thin, some are spherical, some are flat, some are rings, and others are spirals. We have already mentioned the shapes of several simple molecules. Water is an angular molecule, ammonia is triangular pyramidal, and methane is tetrahedral. Larger molecules may have much more complex shapes. A sucrose molecule (Figure 1.4) is considerably more complex than a water molecule but is simple compared with a DNA molecule (Figure 3.13). Although the detailed structure of DNA is very complicated, its overall shape is a beautiful double helix (spiral). The function of DNA, which is to store genetic information, is intimately related to its structure and shape (Chapter 17). However, we shall see that no matter how complex a molecule is, the geometric arrangement of the covalent bonds around any particular atom can easily be deduced on the basis of some very simple principles. We use several types of models to help us visualize the shapes of molecules, as described in Box 3.3.

We have already seen that the covalent bonds in a molecule form in specific directions. For example, the water molecule is angular, with a bond angle of 105°. Molecular shape (geometry)—the geometric arrangement of the atoms in molecules and in covalently bonded networks—depends on the directions of the covalent bonds formed by each atom. A simple model known as the **VSEPR (valence-shell electron-pair repulsion) model** enables us to predict the directions of the bonds around any atom in a molecule.

We have seen that the electrons in an atom are arranged in layers or shells, each of which can contain a limited number of electrons. We now consider how the

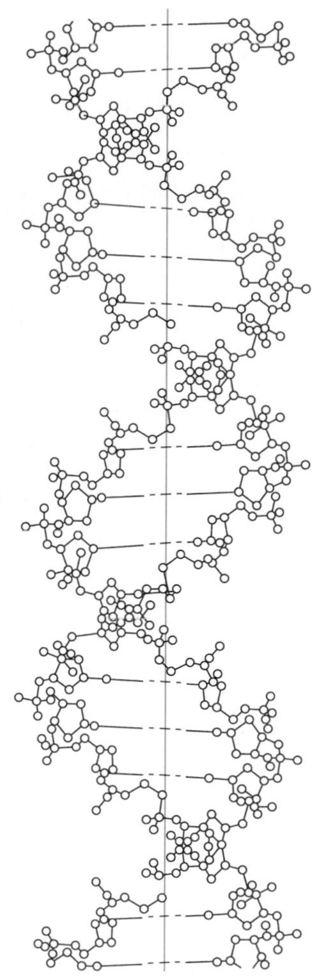

FIGURE 3.13 The DNA Molecule The DNA molecule consists of two long chains that spiral around each other and form a double helix. Only a small portion of this very long molecule is shown here.

electrons are arranged in a shell, in particular in the valence shell. According to the VSEPR model,

> **The electrons in the valence shell of an atom occur in pairs arranged so that the pairs are as far apart from each other as possible. That is, electron pairs behave as if they repel each other and are arranged so as to minimize this repulsion.**

These arrangements are

Number of electron pairs	*Arrangement*
2	**Linear**
3	**Triangular**
4	**Tetrahedral**

Just as we think of the electrons in each shell as occupying a region of space that is largely outside the space occupied by the electrons in underlying shells, so we may think of each pair of electrons in the valence shell as occupying its own region of space. We call this space the **domain** of the electron pair.

> **An electron-pair domain is the approximate region of space occupied by an electron pair.**

BOX 3.3 Molecular Models

To visualize the shapes of molecules, you will find it very helpful to use a set of molecular models. *Stick models* use plastic or metal forms consisting of 3 to 6 spokes radiating from a central point (Figure A). Plastic tubes can be pushed onto these spokes to represent the bonds to other atoms and unshared electron pairs. The atoms are not directly represented in these models. Alternatively, plastic or wooden balls drilled with holes at appropriate angles can be used to represent atoms, and wooden or plastic rods or stiff springs can be used to represent bonds. This type of model is often called a *ball-and-stick model* (Figure B). It shows the arrangement of the atoms in space but does not correctly show their relative sizes.

Space-filling models are more realistic but are somewhat more elaborate (Figure C). Because the charge clouds of two atoms overlap when they form a bond, atoms have a smaller radius in this direction than in directions in which they are not forming a bond, so the atoms in a molecule are not spherical. In space-filling models they are represented by spheres with one or more slices cut off at suitable angles. Each of the flat faces thus formed has a fastener that enables it to be connected to other spheres. These models give no direct representation of bonds, but they give the best representation of the shape of a molecule and of the relative sizes of its atoms.

Computer programs are now available that enable molecules to be modeled on the screen of a personal computer (Figure D), so *computer models* are now for many purposes replacing the model sets. Once the appropriate information has been stored in a computer, models can be generated very rapidly. They can be rotated on the screen and viewed from many angles and can be instantly modified by replacing atoms or groups with other atoms or groups. Very large molecules containing hundreds of atoms can be modeled almost as easily as very simple molecules. Computer modeling is fast becoming a new and important tool for the chemist. For example, it is already playing an important role in the development of new drug molecules (Chapter 17).

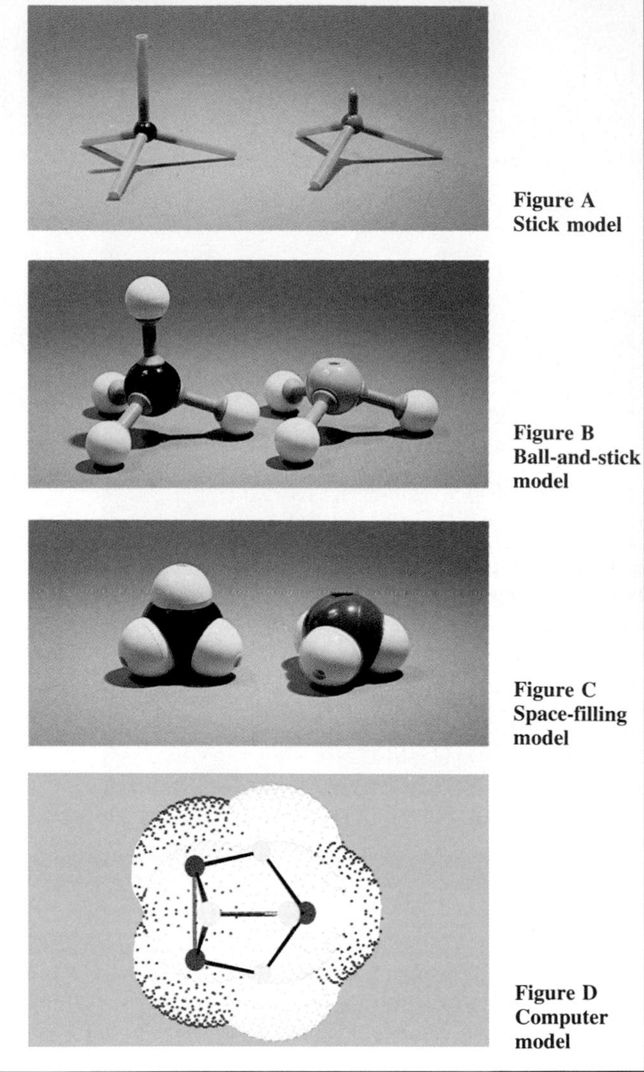

Figure A Stick model

Figure B Ball-and-stick model

Figure C Space-filling model

Figure D Computer model

It is a convenient, although a very rough, approximation to represent each electron-pair domain by a sphere. Then, using a model with Styrofoam spheres as shown in Demonstration 3.1, we see that two, three, and four electron-pair domains surrounding a central core naturally take up the linear, triangular, and tetrahedral arrangements, respectively. The same arrangements can also be easily demonstrated by using toy balloons (Figure 3.14).

Electron-pair domains are not as well defined as we have illustrated them. Like electron shells, electron-pair domains do not have sharp boundaries, and they

DEMONSTRATION 3.1 Electron-Pair Arrangements: Styrofoam Sphere Models

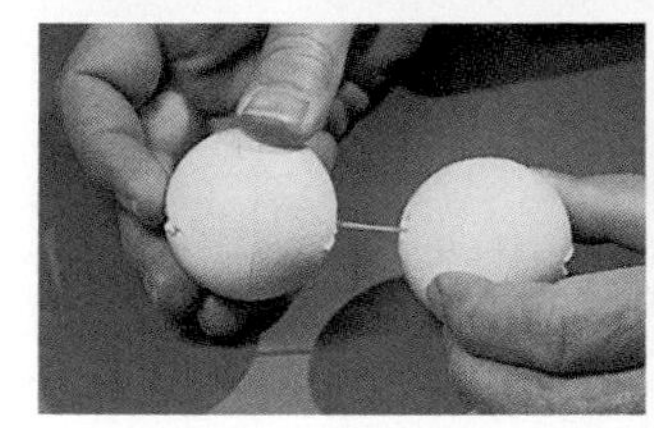
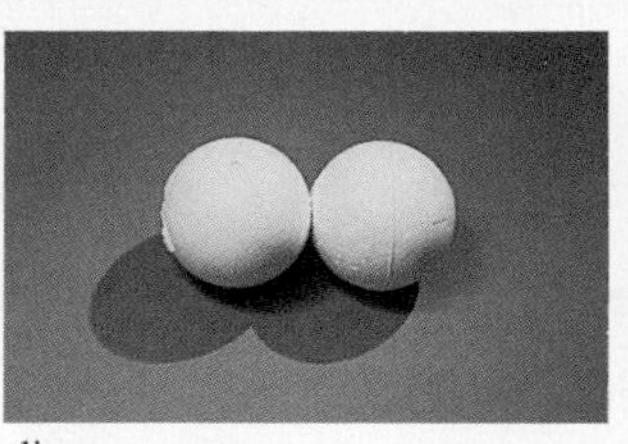
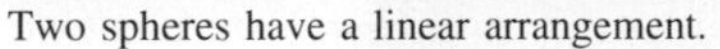

Two spheres have a linear arrangement.

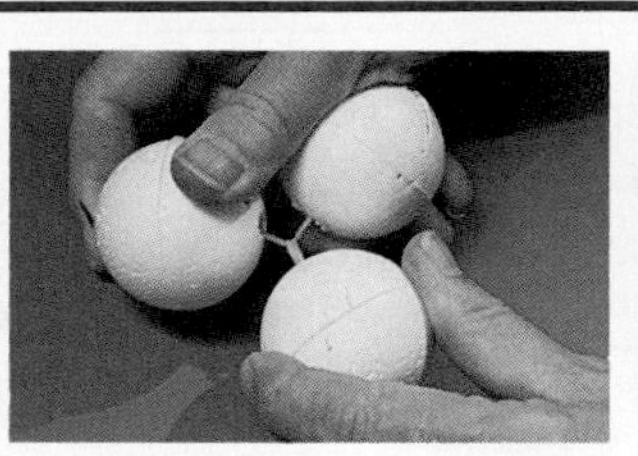

Three spheres adopt a triangular arrangement.

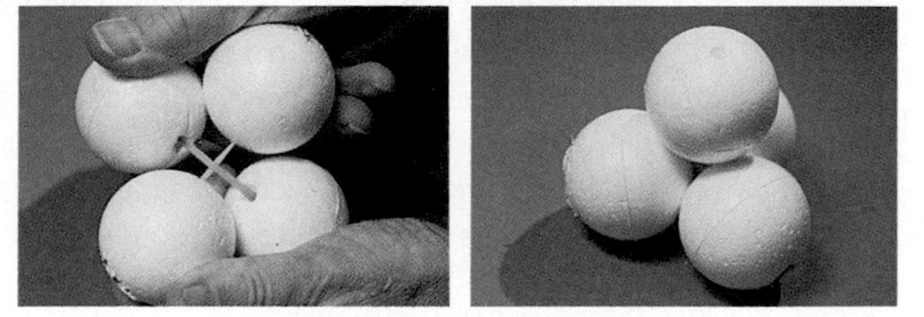

Four spheres adopt a tetrahedral arrangement.

Electron-pair arrangements can be demonstrated by joining pairs or threes of Styrofoam spheres by elastic bands held in place by small nails or toothpicks. Each sphere represents a pair of electrons or, more exactly, the space occupied by a pair of electrons. The elastic band provides a force of attraction between the spheres that corresponds to the electrostatic attraction between the electrons and a positive core situated at the midpoint of the elastic band. The spheres naturally adopt the arrangements shown. If they are forced into some other arrangement—such as the square planar arrangement for four spheres—they immediately adopt the tetrahedral arrangement when the restraining force is removed.

overlap somewhat. But we can regard each domain as a region in which an electron pair has a high probability of being found.

In later chapters we will discuss many applications of the VSEPR model to predict molecular shape. In this section we will consider the geometric arrangement of the bonds in some molecules in which two, three, or four atoms are attached to a central atom. First we consider molecules in which all the bonds are single bonds.

(a) (b) (c)

FIGURE 3.14 Electron-Pair Domain Arrangements: Balloon Models
(a) Linear arrangement of two domains (b) Triangular arrangement of three domains
(c) Tetrahedral arrangement of four domains

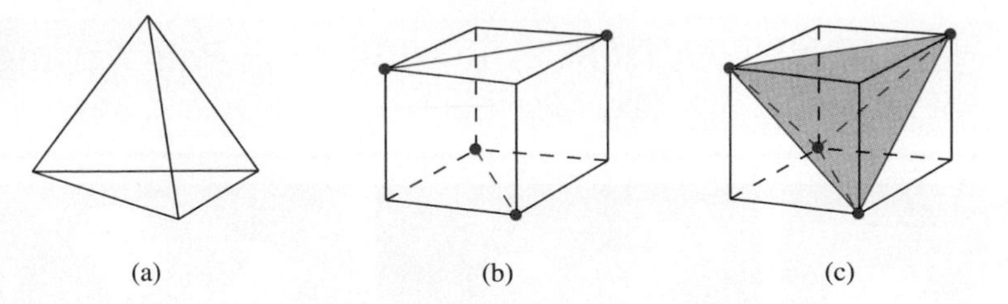

FIGURE 3.15 The Tetrahedron (a) The tetrahedron is a regular solid with four equivalent equilateral triangular faces and six equal edges. (b) The shape of the tetrahedron is closely related to the shape of a cube. This relation can be seen by drawing a diagonal across one face of the cube and another diagonal at right angles across the opposite face of the cube, (c) then joining the ends of the diagonals to form the tetrahedron.

Shapes of Molecules with Only Single Bonds

In most molecules the central atom obeys the octet rule and has eight electrons (four pairs) in its valence shell, so we consider this case first. Four electron pairs are kept as far apart as possible when they are arranged at the four corners of a tetrahedron. A **tetrahedron** is a regular solid with four equivalent vertices and four equilateral triangular faces (Figure 3.15). The tetrahedral arrangement of four electron pairs in a valence shell can give rise to four different shapes for molecules in which the bonds are all single bonds: tetrahedral, triangular pyramidal, angular, and linear, depending on how many of the four electron pairs are bonding pairs and how many are lone pairs.

The X in AX_4 may denote different atoms—both CF_4 and CF_2Cl_2 are AX_4 molecules. However, all the bond angles in CF_2Cl_2 are not exactly 109.5°. Similarly, BF_2Cl is an AX_3 molecule, but all its bond angles are not exactly 120°.

Tetrahedral AX_4 Molecules (Figure 3.16) In the methane molecule, CH_4, all four electron pairs in the valence shell of the carbon atom are bonding pairs. The CH_4 molecule has a tetrahedral shape, with the C atom in the center of the tetrahedron and one H atom at each of the four corners. We describe this molecule as an AX_4 type, where A represents the central atom and X is an attached atom. The angle between any pair of bonds in such a molecule is 109.5°. All AX_4 molecules formed by the elements of Group IV, such as CCl_4 and SiF_4, have the same tetrahedral shape.

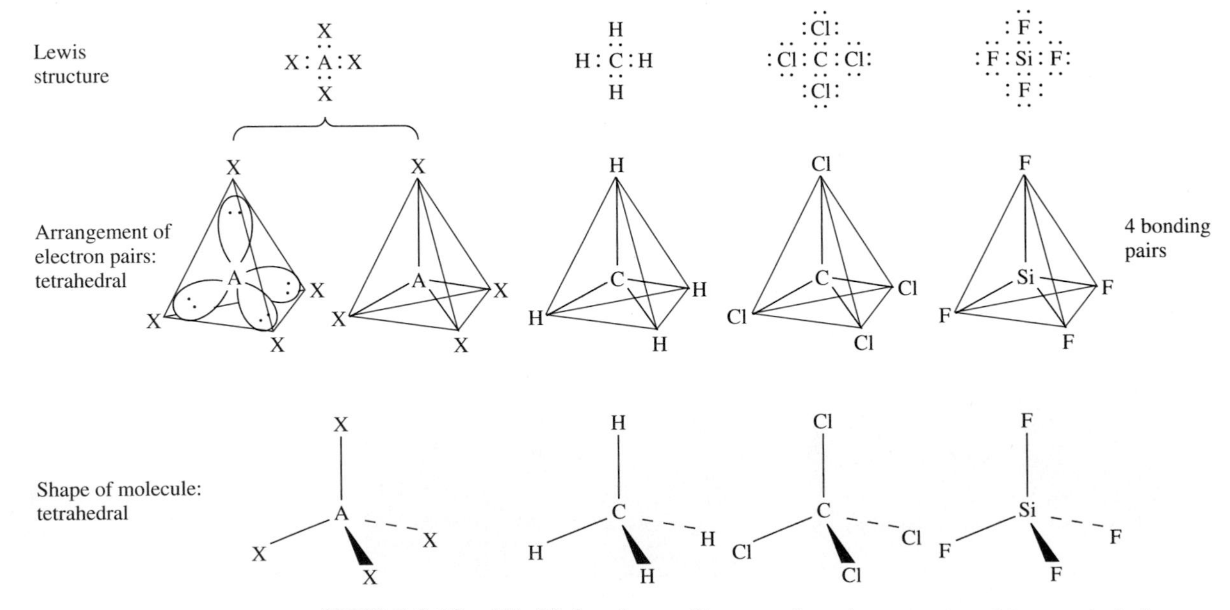

FIGURE 3.16 AX_4 Molecules There are four electron pairs with a tetrahedral arrangement in the valence shell of the central atom A. All four electron pairs are bonding pairs, so the molecules have a tetrahedral shape. In the left-hand electron-pair diagram, the electron pairs are shown as domains (the approximate volume of space occupied by the electron pairs), ☉. In the remaining diagrams the bonding electron pairs are shown as bond lines.

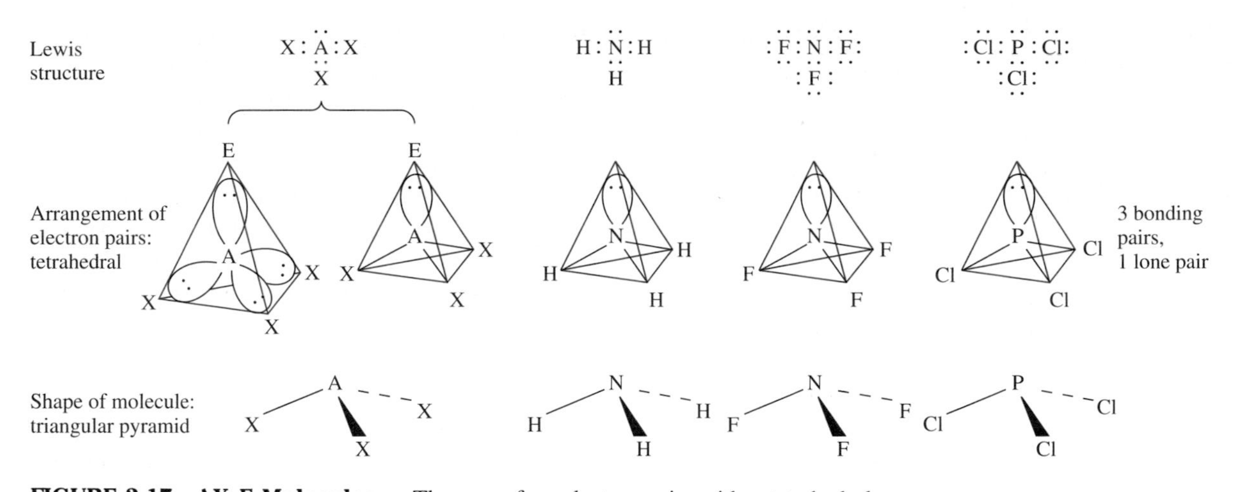

FIGURE 3.17 AX_3E Molecules There are four electron pairs with a tetrahedral arrangement in the valence shell of the central atom A. Only three of the electron pairs are bonding pairs, so these molecules have a triangular pyramid shape. In the left-hand electron-pair diagram, all the electron pairs are shown as domains. In the remaining diagrams, the bonding pairs are shown as bond lines and the lone pairs are shown as domains, (:).

Triangular Pyramidal AX_3E Molecules (Figure 3.17) The valence shell of nitrogen in the ammonia molecule, NH_3, has four electron pairs. These electron pairs have a tetrahedral arrangement, but one of them is a nonbonding, or lone, pair; the other three are bonding pairs. The resulting molecule has a triangular pyramidal shape. We describe such a molecule as an AX_3E molecule, where E represents the lone pair. Its shape is *not* tetrahedral, because it is the positions of the atomic nuclei A and X that determine molecular shape. In NH_3 the nitrogen atom is at the apex of a triangular pyramid whose base is formed by three hydrogen atoms. All AX_3E molecules formed by the elements of Group V, such as NF_3 and PH_3, have the same triangular pyramidal shape.

Distinguish carefully between the arrangements of the *electron domains* around central atom A and the arrangements of the *atoms* X bonded to A.

Angular AX_2E_2 Molecules (Figure 3.18) There are four electron pairs in the valence shell of the oxygen atom in H_2O; two of them are nonbonding pairs, and two are bonding pairs. We describe such a molecule as an AX_2E_2 molecule. The four electron pairs have a tetrahedral arrangement, but the molecule has an angular shape. All AX_2E_2 molecules formed by the elements of Group VI, such as OF_2 and H_2S, have the same angular shape.

FIGURE 3.18 AX_2E_2 Molecules There are four electron pairs with a tetrahedral arrangement in the valence shell of the central atom A. Only two of the electron pairs are bonding pairs, so the molecule has an angular shape.

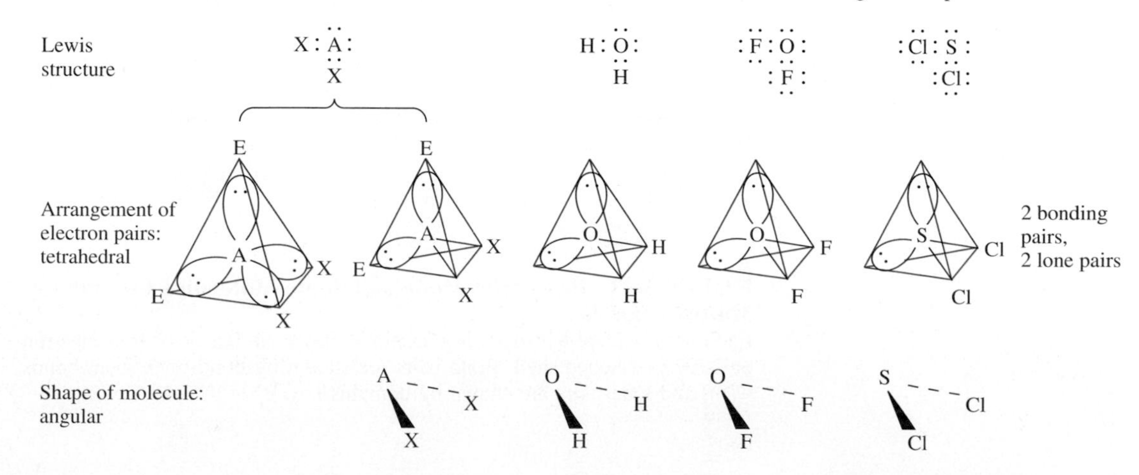

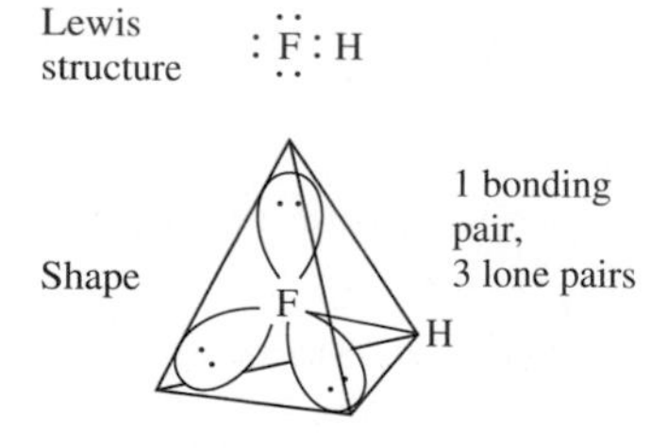

FIGURE 3.19 AXE_3 Molecules There are four electron pairs with a tetrahedral arrangement in the valence shell of the central atom A. Only one of the electron pairs is a bonding pair, so the (diatomic) molecule has a linear shape.

Linear (Diatomic) AXE_3 Molecules (Figure 3.19) Fluorine in the HF molecule has three nonbonding pairs and one bonding pair in its valence shell. It is described as an AXE_3 molecule. Because they are diatomic, AXE_3 molecules are necessarily linear.

Triangular AX_3 (Figure 3.20) Group III elements form halides such as BF_3 and $AlCl_3$ with three polar covalent bonds. There are only three bonding pairs of electrons in the valence shell of the central atom, which in these molecules does not obey the octet rule. Because three electron pairs have a triangular arrangement, AX_3 molecules, such as BF_3 and $AlCl_3$, have a planar triangular shape.

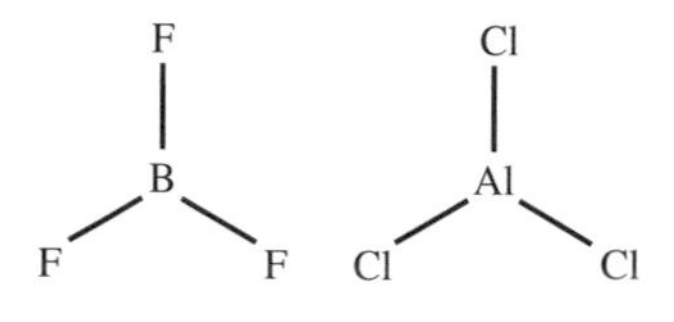

FIGURE 3.20 AX_3 Triangular Molecules Group III AX_3 molecules, such as BF_3, and $AlCl_3$, have a planar triangular shape. There are only three bonding pairs and no lone pairs in the valence shell of the central atom (B, Al).

Linear AX_2 Molecules (Figure 3.21) A few covalent molecules of Group II elements, such as $BeCl_2$, have only two electron pairs in the valence shell of beryllium. Because two electron pairs have a linear arrangement, these AX_2 molecules have linear shapes.

Cl —— Be —— Cl

FIGURE 3.21 AX_2 Linear Molecules Group II AX_2 molecules, such as $BeCl_2$, are linear. There are only two bonding pairs and no lone pairs in the valence shell of the central atom (Be).

Shapes of Molecules with Multiple Bonds

The Bent-Bond Model If we represent the four tetrahedrally arranged electron pairs in the valence shell of a carbon atom by four bond lines but then connect two or three of these lines to the same atom to form a double bond or triple bond, each line must be curved or bent. Such a bond is called a **bent bond** (Figure 3.22). The shapes that can arise in this way are as follows:

AX_3 Triangular Molecules An example is methanal, H_2CO (formaldehyde), in which the carbon atom forms two single bonds and one double bond (two bent bonds) (Figure 3.22a). The oxygen atom also has four tetrahedrally arranged electron pairs in its valence shell—two bonding pairs and two lone pairs. All four atoms are in the same plane with the two hydrogen atoms and the oxygen atom, forming a triangle around the carbon.

AX_2 Linear Molecules Examples include carbon dioxide, CO_2, in which there are two double bonds, and hydrogen cyanide, HCN, in which there are a single bond and a triple bond (Figure 3.22b). There are four tetrahedrally arranged electron pairs in the valence shell of the nitrogen atom—three bonding pairs and a lone pair.

The Multiple Bond Domain Model The AX_3 and AX_2 molecules containing multiple bonds have the same shapes, triangular and linear, respectively, as the

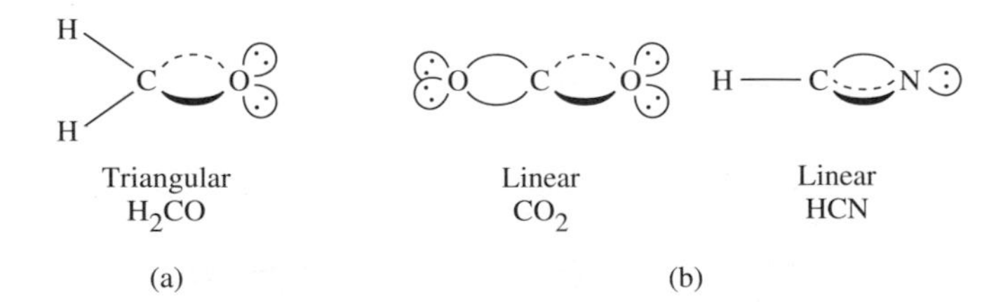

FIGURE 3.22 Bent-Bond Models of Some Molecules Containing Multiple Bonds
Each atom except hydrogen has a tetrahedral arrangement of four electron pairs in its valence shell. Bond pairs are shown by bond lines (bent bonds ⌒), and lone pairs are shown by domains (⊙).

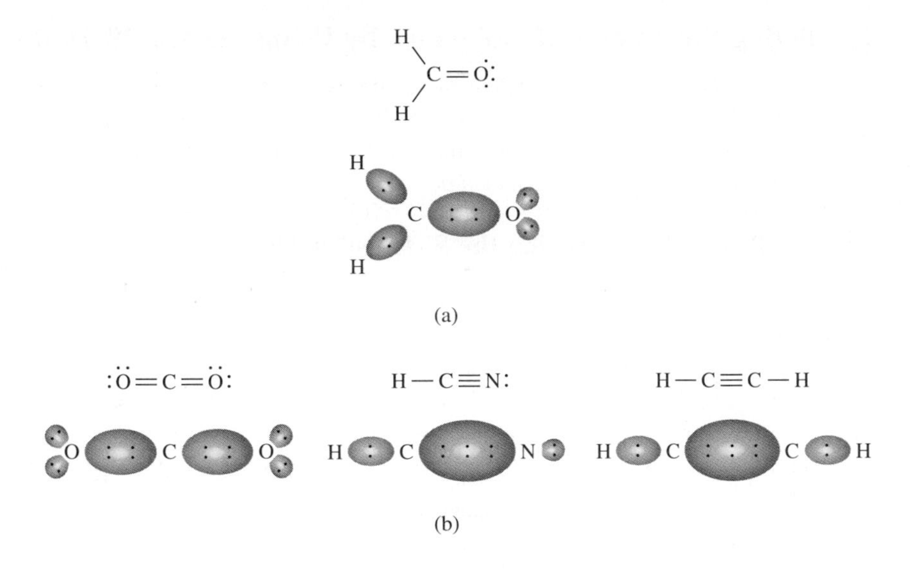

FIGURE 3.23 Domain Models of Molecules with Multiple Bonds Single bond domain (:); double bond domain (: :); triple bond domain (: : :); lone pair domains. (a) Triangular AX_3 geometry: three bonding domains in the valence shell of the carbon atom. (b) Linear AX_2 geometry: two bonding domains in the valence shell of each carbon atom.

corresponding molecules with single bonds. This gives us a simple alternative method for predicting the shapes of molecules that contain multiple bonds. Because the two electron pairs of a double bond or the three electron pairs of a triple bond are in the same bonding region, we represent a double bond or a triple bond by just one bond domain—a double bond domain containing 4 electrons, and a triple bond domain containing 6 electrons. Then the valence shell of carbon in methanal has just three bond domains and no lone pairs. Three bond domains have a triangular arrangement, so methanal is a planar triangular AX_3 molecule (Figure 3.23a). The valence shell of the carbon atom in the CO_2 and HCN molecules has just two bond domains (two double bond domains in CO_2, and one single and one triple bond domain in HCN). The two bond domains have a linear arrangement, so both molecules are linear AX_2 molecules (Figure 3.23b).

Shapes of Molecules with More than One "Central Atom"

The arrangements of the bonds around an atom apply to every atom in a molecule, so we can easily extend our discussion to larger molecules. For example, each of the carbon atoms in ethene, C_2H_4, has a triangular AX_3 geometry, and the molecule as a whole is planar (Figure 3.24a). In ethane, C_2H_6, both carbon atoms have a tetrahedral geometry (Figure 3.24b).

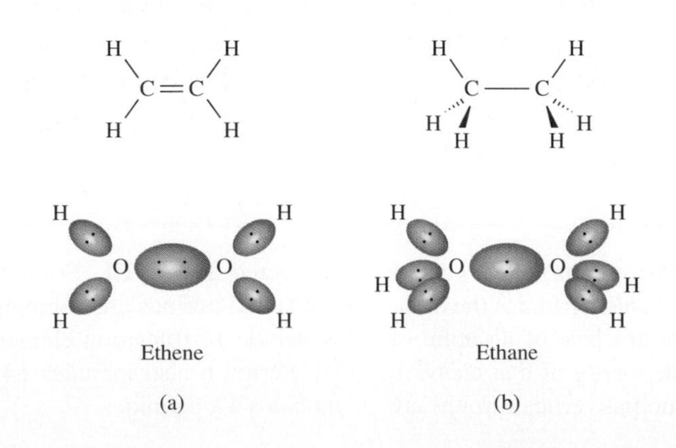

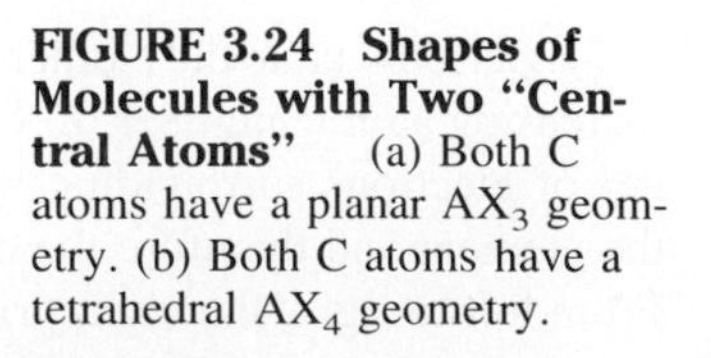
FIGURE 3.24 Shapes of Molecules with Two "Central Atoms" (a) Both C atoms have a planar AX_3 geometry. (b) Both C atoms have a tetrahedral AX_4 geometry.

Predicting the Shapes of Molecules by Using the VSEPR Model

To determine the shape of a covalent molecule, we first draw its Lewis structure, then count the number n of bonding pairs in the valence shell of the central atom A and the number m of unshared (lone) pairs in the formula AX_nE_m, which gives us the type of molecule and hence its shape.

EXAMPLE 3.4 Predicting the Shape of a Molecule

Predict the shape of the PH_3 molecule.

Solution: PH_3 has the Lewis structure

$$\mathrm{H{-}\ddot{P}{-}H}$$
$$\mathrm{|}$$
$$\mathrm{H}$$

There are three bonding pairs and a lone pair in the valence shell of the P atom. It is therefore an AX_3E molecule with a triangular pyramidal shape.

Exercise 3.12 Classify each of the following molecules in terms of the AX_nE_m nomenclature, and give the geometric shape for each.
(a) CF_4 **(b)** NF_3 **(c)** BCl_3 **(d)** F_2O **(e)** HCl

EXAMPLE 3.5 Shapes of Molecules with Multiple Bonds

What are the molecular shapes of **(a)** CS_2 and **(b)** F_2CO?

Solution: **(a)** We first deduce the Lewis structure

$$:\!\ddot{\underset{\cdot}{S}}\cdot + \cdot\dot{\underset{\cdot}{C}}\cdot + \cdot\ddot{\underset{\cdot}{S}}\!: \rightarrow :\!\ddot{S}\!=\!C\!=\!\ddot{S}\!:$$

Because the valence shell of the central C atom contains two bond domains, it is a linear AX_2-type molecule.
(b) We first deduce the Lewis structure, which is analogous to that of methanal:

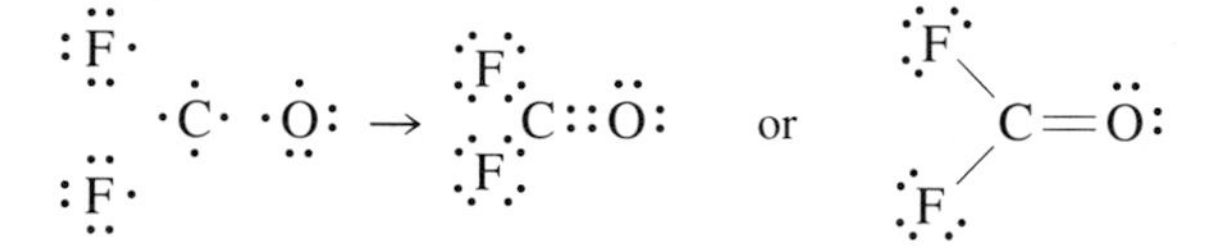

Because the valence shell of carbon contains three bond domains, this is a planar triangular AX_3 molecule.

Exercise 3.13 Classify each of the following molecules in terms of the AX_nE_m nomenclature, and give the geometric shape for each.
(a) OCS **(b)** Cl_2CO **(c)** ClCN **(d)** FCCF **(e)** ONCl

SUMMARY

In the periodic table the elements are arranged in order of increasing atomic number, Z, which determines the number of electrons surrounding the nucleus of an atom of the element and therefore the chemistry of that element. From left to right, the eight principal vertical groups are numbered I to VIII. From top to bottom, the seven horizontal periods are numbered 1 to 7. Periods 4, 5, and 6 include 10 transition elements between Groups II and III. Period 6 also includes 14 lanthanides, and Period 7 includes 14 actinides.

The elements may be classified as metals, semimetals (metalloids), and nonmetals. With the exception of H in Group I, metals are on the left and nonmetals are on the right of the periodic table. Semimetals are on the border between metals and nonmetals. Elements in each group have many similarities but increase in metallic character from top to bottom. Group I elements are commonly known as the alkali metals, Group II elements as the alkaline earth metals, Group VII elements as the halogens, and Group VIII elements as the noble gases.

The combining power of an element for other elements is its valence. The principal valence of hydrogen is 1; for any other element, it is the number of H atoms that combine with one atom of the element. All the elements in a group have the same principal valence. From Group I to Group VIII the principal valences are 1, 2, 3, 4, 3, 2, 1, 0.

According to the shell model, the electrons in an atom are arranged in spherical shells, labeled $n = 1, 2, 3, \ldots$ from the nucleus out. Each shell can accommodate a maximum of $2n^2$ electrons (2 for $n = 1$, 8 for $n = 2$, 18 for $n = 3$). The outermost shell in an atom is the valence shell. Atoms of elements in the same group have a number of valence electrons equal to the group number. Helium is an exception.

The core charge is the overall charge of the nucleus plus the inner completed shells of electrons. It is equal in magnitude to the group number and to the number of electrons in the valence shell in a neutral atom.

According to the Lewis octet rule, in forming compounds an atom gains or loses electrons or shares pairs of electrons, until it has eight electrons in its valence shell (and therefore the same electron arrangement as in a noble gas). Lewis symbols use dots to represent valence electrons. For Group I to IV atoms, they are arranged singly; for Group V to VIII atoms, the electrons in excess of four are arranged in pairs.

Electrons are easily removed from the valence shell of a metal, because metal atoms have a low core charge. The removal of electrons gives cations such as Na^+, Mg^{2+}, and Al^{3+}, which have an octet of electrons in a new outer shell. Nonmetal atoms, which have a high core charge, attract electrons to give anions, such as O^{2-} and Cl^-, which also have an octet of electrons. An ionic bond is the electrostatic attraction between oppositely charged ions in an ionic compound, such as Na^+Cl^- and $Ca^{2+}O^{2-}$. Ions with the same electron arrangement (O^{2-}, F^-, Na^+, and Mg^{2+}) are isoelectronic.

Compounds between nonmetals, such as $CF_4(g)$, contain molecules in which the atoms achieve octets by sharing electron pairs to form covalent bonds. The number of covalent bonds formed by a neutral atom equals the number of single (unpaired) electrons in its Lewis symbol. Pairs of electrons forming covalent bonds are bonding pairs; pairs of electrons not used for bonding are called unshared pairs, nonbonding pairs, or lone pairs. Lewis structures are diagrams of molecules showing all the bonding and unshared pairs of electrons in the valence shells of their atoms. Covalent bonds may be single (one shared electron pair), double (two shared electron pairs), or triple (three shared pairs).

Electronegativity, χ, is the ability of an atom in a molecule to attract the electrons of a covalent bond. It depends on the magnitude of the core charge and the distance of the bonding pair of electrons from the nucleus. Electronegativity increases across a period from left to right with increasing core charge and decreases down any group with increasing distance of the valence electrons from the nucleus. Two atoms with the same electronegativity form a covalent bond, in which the bonding pair is shared equally. Two atoms with a small electronegativity difference form a polar covalent bond, in which the bonding pair is shared unequally. Two atoms with an electronegativity difference $\Delta\chi$ greater than approximately 2.0 form an ionic bond, in which the bonding pair is transferred completely to the atom of higher electronegativity and becomes a nonbonding pair in the valence shell of the anion. In general, bonds between a Group I or II metal atom and a nonmetal atom are ionic; bonds between two different nonmetal atoms are polar covalent; and bonds between two identical nonmetal atoms are covalent.

According to the VSEPR model, pairs of electrons in the valence shell of an atom (bonding or nonbonding) are arranged as far apart as possible. Two pairs have a linear arrangement; three pairs, triangular; and four pairs, tetrahedral. For a single-bonded molecule AX_nE_m, where n is the number of bonding pairs and m is the number of unshared electron pairs E in the valence shell of A, the predicted geometries (shapes) are AX_4, tetrahedral; AX_3E, triangular pyramidal; AX_2E_2, angular; AXE_3, linear; AX_3, triangular planar; and AX_2, linear. In the bent-bond model for molecules containing multiple bonds, double bonds are represented by two bent single bonds, and triple bonds by three bent single bonds. In the electron-domain model, a double bond is represented by a two-electron-pair domain, and a triple bond by a three-electron-pair domain. In an AX_nE_m molecule, X represents a single, double, or triple bond domain and E an unshared-pair domain so, for example, $O{=}CH_2$ has a triangular planar AX_3 geometry and $O{=}C{=}O$ and $H{-}C{\equiv}N$ have a linear AX_2 geometry.

IMPORTANT TERMS

actinide (page 84)
alkali metal (page 86)
alkaline earth metal (page 86)
anion (page 94)
bonding electron pair (page 98)
cation (page 94)
chemical bond (page 92)
core (atomic) (page 92)
core charge (page 92)
covalent bond (page 98)
double bond (page 100)
electron-dot symbol (page 94)
electron-pair domain (page 105)
electronegativity (page 101)
group (page 83)
halogen (page 88)
ionic bond (page 95)
ionic compound (page 95)
ion (page 94)
isoelectronic (page 97)
lanthanide (page 84)
Lewis structure (page 98)
Lewis symbol (page 94)
lone pair (page 99)
metal (page 84)
metalloid (page 85)
molecular shape (geometry) (page 104)
multiple bond (page 100)
noble gas (page 88)
nonbonding pair (page 99)
nonmetal (page 85)
octet (page 93)
octet rule (page 93)
period (page 83)
periodic table (page 81)
polar covalent bond (page 103)
principal valence (page 88)
semimetal (page 85)
shell model (page 90)
single bond (page 100)
transition elements (page 84)
triple bond (page 100)
unshared pair (page 99)
valence (page 88)
valence shell (page 91)
VSEPR model (page 104)

REVIEW QUESTIONS

1. What property is used to arrange the elements in the form of the modern periodic table?

2. Among the first 20 elements, which are metals and which are nonmetals?

3. Give three characteristic physical properties each of **(a)** metals and **(b)** nonmetals.

4. How is the principal valence of an element related to its position in the periodic table?

5. What are the empirical formulas of the hydrides and the chlorides of the elements in Period 3?

6. What is the maximum number of electrons that could be accommodated in each of the shells designated $n = 1$, $n = 2$, and $n = 3$?

7. Why is the outer shell of an atom called its valence shell?

8. For the main group elements (except He), how is the number of electrons in the valence shell related to the element's group number?

9. **(a)** How is the core of an atom defined?

(b) How is core charge related to the position of an element in the periodic table?

10. How is the strength of the attraction of an atom for its valence electrons related to the position of the element in the periodic table (group and period)?

11. What is an ionic bond?

12. What is meant by the term "isoelectronic"?

13. What is a covalent bond?

14. What is an unshared pair (a lone or nonbonding pair)?

15. What is meant by the term "electronegativity"?

16. How do the electronegativities of the elements change **(a)** in going across any period from left to right and **(b)** in descending any group of the periodic table? Explain.

17. What generalization can be made about the electronegativities of **(a)** metals and **(b)** nonmetals?

18. What generalizations can be made about the types of bonds formed between **(a)** a metal atom and a nonmetal atom, **(b)** identical nonmetal atoms, and **(c)** two different nonmetal atoms?

19. What is the basic postulate of the VSEPR model of molecular geometry?

20. What are the arrangement of electron pairs in the valence shell of atom A and the molecular shape for molecules formulated as AX_4, AX_3E, AX_2E_2, and AXE_3?

PROBLEMS

The Periodic Table

1. Write the first 20 elements in the form that they appear in the periodic table.

2. Locate each of the following elements in the periodic table by group and by period. Name the element, and classify it as a metal, a semimetal (metalloid), or a nonmetal.

(a) He **(b)** P **(c)** K **(d)** Ca **(e)** S **(f)** Br **(g)** Al **(h)** F

3. **(a)** To what period and group of the periodic table do the elements with each of the following atomic numbers belong? (i) 2 (ii) 7 (iii) 9 (iv) 11 (v) 6 (vi) 19

(b) Give the symbol of each of the elements in (a). From their positions in the periodic table, classify each as a metal or a nonmetal.

4. The element with atomic number 22 forms crystals that melt at 1668°C to give a liquid that boils at 3313°C. The crystals are hard, conduct heat and electricity, can be drawn into fine wires, and emit electrons when exposed to ultraviolet light. From these properties, classify the element as metal or a nonmetal. Which element is it?

5. By referring to the periodic table, classify each of the following elements as a main group element or a transition metal. For the main group elements, indicate the group to which the element belongs and whether it is a metal or a nonmetal.

(a) Se **(b)** P **(c)** Mn **(d)** Kr **(e)** W **(f)** Al **(g)** Pb

6. Repeat Problem 5 for the following elements.

(a) Ar **(b)** Rb **(c)** V **(d)** Br **(e)** Ba **(f)** Fe **(g)** Au

7. For each of the following elements, state whether it is a metal or a nonmetal. Indicate the group to which it belongs, and give its number of valence electrons and principal valence.

(a) Li **(b)** Mg **(c)** S **(d)** P **(e)** Br **(f)** Ne **(g)** As **(h)** Se **(i)** Cl **(j)** Ba

8. Repeat Problem 7 for the following elements.

(a) Ca **(b)** Al **(c)** F **(d)** Cs **(e)** N **(f)** I **(g)** K **(h)** Ar **(i)** O **(j)** Si

9. Predict the empirical formulas of the chlorides formed by the elements in Group III of the periodic table.

10. Francium, Fr, is an alkali metal; germanium, Ge, is in the same group of the periodic table as carbon; astatine, At, is a halogen, and radon, Rn, is a noble gas. Write formulas for their expected **(a)** hydrides and **(b)** the products of their reactions with chlorine.

11. Assign each of the following elements to its appropriate group in the periodic table. Classify each as a metal or a nonmetal, and write the empirical formula of its hydride.

(a) C **(b)** Ca **(c)** He **(d)** B **(e)** Cl **(f)** Li **(g)** O **(h)** F **(i)** P **(j)** Mg

12. Complete and balance each of the following equations.

(a) $Mg(s) + Br_2(l) \rightarrow$ **(b)** $Ca(s) + O_2(g) \rightarrow$
(c) $Na(s) + I_2(s) \rightarrow$ **(d)** $Mg(s) + N_2(g) \rightarrow$

13. Complete and balance each of the following equations for those cases in which a reaction occurs.

(a) $Li(s) + S(s) \rightarrow$ **(b)** $Mg(s) + H_2(g) \rightarrow$
(c) $Ne(g) + H_2O(l) \rightarrow$ **(d)** $Sr(s) + Br_2(l) \rightarrow$

14. Arrange the following elements in pairs in terms of the greatest similarity of chemical and physical properties: Na, Mg, C, Cl, Ca, Si, K, and F.

15. **(a)** To which group and period of the periodic table does each of the following belong?
(i) Mg (ii) K (iii) Br (iv) P (v) Si (vi) Al (vii) S (viii) C

(b) What are the empirical formulas of the fluorides and the sulfides of the elements in (a)?

The Shell Model of the Atom

16. **(a)** Which among the electron shells of an atom is its valence shell?

(b) What is the common feature of the valence shells of elements in the same group of the periodic table?

(c) Give the electron shell structures for atoms of each of the elements in Periods 2 and 3.

17. The atomic numbers of phosphorus, carbon, and potassium are 15, 6, and 19, respectively. For atoms of each of these elements, without reference to any other information, deduce **(a)** the electron shell structure and **(b)** the core charge.

18. Give the electron shell structures of each of the following atoms, and give their core charges.

(a) Li **(b)** Mg **(c)** S **(d)** P

19. Repeat Problem 18 for each of the following atoms.

(a) Ca **(b)** Al **(c)** F **(d)** N

20. **(a)** Give the electron shell structures of each of the following: (i) N (ii) N^{3-} (iii) Be (iv) Be^{2+} (v) O (vi) O^{2-}

(b) Which of the ions in (a) are isoelectronic?

21. Repeat Problem 20 for each of the following.

(a) Na **(b)** Na^+ **(c)** Mg **(d)** Mg^{2+} **(e)** Cl **(f)** Cl^-

22. What is the core charge of the atoms of each of the following?

(a) fluorine **(b)** nitrogen **(c)** sulfur **(d)** lithium **(e)** oxygen **(f)** magnesium **(g)** cesium **(h)** silicon

23. What is the core charge of each of the following atoms and ions?

(a) C **(b)** Mg **(c)** Mg^{2+} **(d)** Si **(e)** O **(f)** O^{2-} **(g)** S^{2-} **(h)** Br

Chemical Bonds and Lewis Structures

24. For the Lewis electron-dot symbol of an atom of an element, what relationship exists between **(a)** the *number* of valence electrons (dots) and the position of the element in the periodic table; **(b)** the number of unpaired dots (electrons) and its principal valence?

25. Draw Lewis symbols for atoms of each of the following elements.

(a) K **(b)** Ca **(c)** B **(d)** Sn **(e)** Sb
(f) Te **(g)** Br **(h)** Xe **(i)** As **(j)** Ge

26. To attain a structure in which all the electron shells are filled, how many electrons must be **(a)** *lost* by each of the atoms Li, Al, Na, Ca, Mg, Rb, and Sr; **(b)** *gained* by each of the atoms O, H, Cl, N, P, S, F, and I?

27. On the basis of the octet rule and the group in the periodic table to which each of the following elements belongs, predict the charges on the ions formed by each element.

(a) magnesium **(b)** aluminum **(c)** bromine
(d) sulfur **(e)** phosphorus **(f)** potassium

28. What are the empirical formulas of the ionic solids composed of each of the following pairs of ions?

(a) NH_4^+, S^{2-} **(b)** Fe^{3+}, O^{2-} **(c)** Cu^+, O^{2-}
(d) Al^{3+}, Cl^-

29. Predict the empirical formula of the ionic compound formed by each of the following pairs of elements.

(a) Li, S **(b)** Ba, O **(c)** Mg, Br **(d)** Na, H **(e)** Al, I

30. Repeat Problem 29 for the following pairs of elements.

(a) Ca, I **(b)** Ca, O **(c)** Al, S **(d)** Ca, Br
(e) Rb, Se

31. Write the empirical formula of each of the following compounds, and give the balanced equation for its formation from its elements.

(a) barium iodide **(b)** aluminum chloride
(c) calcium oxide **(d)** sodium sulfide
(e) aluminum oxide

32. Select from the following as many pairs of elements as possible that would be expected to form *ionic* binary compounds (containing two elements), and give the empirical formulas of the compounds.

(a) H **(b)** O **(c)** F **(d)** Mg **(e)** Al **(f)** Ca

33. Using the molecules F_2, O_2, and N_2 as appropriate to illustrate your answers, define **(a)** single covalent bond; **(b)** double covalent bond; **(c)** triple covalent bond; **(d)** shared (bonding) electron pair; **(e)** unshared (nonbonding or lone) electron pair.

34. Draw the Lewis structure for each of the following molecules, for which all the bonds are single covalent bonds.

(a) H_2 **(b)** HCl **(c)** PH_3 **(d)** SiF_4 **(e)** F_2O **(f)** Cl_2

35.* Draw the Lewis structure for each of the following molecules and ions.

(a) H_2CO **(b)** P_2 **(c)** CN^- **(d)** H_2NNH_2 **(e)** HOOH

36.* Use the positions of their elements in the periodic table to predict the empirical formulas of the hydrides of antimony, bromine, tellurium, and tin, and draw their Lewis structures.

37.* In which of the following molecules is the central atom an exception to the octet rule?

(a) Cl_2O **(b)** $BeCl_2$ **(c)** $AlCl_3$ **(d)** PCl_3

Electronegativity

38. **(a)** What are the two principal properties of an atom that determine its electronegativity?

(b) How do the electronegativities of the elements change in going across any period from left to right and in descending any group? Explain these variations.

39. Without reference to Figure 3.11, select the element of higher electronegativity in each of the following pairs.

(a) F, Cl **(b)** F, O **(c)** P, S **(d)** C, Si
(e) O, P **(f)** Br, Se **(g)** P, Al

Ionic, Polar Covalent, and Covalent Bonds

40. Classify the bonds in each of the following as ionic, covalent, or polar covalent.

(a) Cl_2 **(b)** PCl_3 **(c)** LiCl **(d)** ClF **(e)** $MgCl_2$

41. Repeat Problem 40 for each of the following.

(a) Li_2O **(b)** MgO **(c)** O_2 **(d)** SO_2
(e) Cl_2O **(f)** NO

42. Repeat Problem 40 for each of the following.

(a) HBr **(b)** I_2 **(c)** BrCl **(d)** H_2 **(e)** LiH

43. Name the compounds with each of the following formulas. State which are ionic and which are covalent substances. For the covalent substances, draw the Lewis structure and indicate the polarity of the bonds by writing $\delta+$ and $\delta-$ on the appropriate atoms.

(a) $MgCl_2$ **(b)** SCl_2 **(c)** PCl_3 **(d)** HF
(e) OCl_2 **(f)** CS_2 **(g)** LiH **(h)** NF_3

44. Classify the bonds in the oxides of the Period 3 elements as ionic, polar covalent, or covalent.

45. Explain why sodium chloride at room temperature forms ionic crystals rather than a gas consisting of Na^+Cl^- molecules (ion pairs).

Molecular Shape and the VSEPR Model

46. Draw diagrams to illustrate the geometry of each of the following types of molecules, and give the value of the XAX bond angle in each case.

(a) a linear AX_2 molecule
(b) an equilateral triangular AX_3 molecule
(c) a tetrahedral AX_4 molecule

47. Draw diagrams to illustrate the geometry of each of the following types of molecules. In each case name the geometric shape.

(a) AX_3E (b) AX_2E_2 (c) AXE_3 (d) AX_3 (e) AX_2

48. BCl_3 is a planar molecule with 120° bond angles, whereas PCl_3 is triangular pyramidal with 100° bond angles. Explain why these two tetratomic molecules have such different shapes.

49. Deduce the AX_nE_m type to which each of the following molecules belongs, and hence predict the shape.

(a) H_2O (b) PCl_3 (c) BCl_3 (d) SiH_4

50. Repeat Problem 49 for the following molecules.

(a) H_2S (b) NF_3 (c) BeH_2 (d) GeH_4

51. How are each of the following domains arranged in the valence shell of a central atom?

(a) 4 single bond domains

(b) 3 single bond domains and 1 lone pair domain

(c) 2 single bond domains and 2 lone pair domains

(d) 1 single bond domain and 3 lone pair domains

(e) 2 single bond domains and 1 double bond domain

(f) 2 double bond domains

(g) 1 single bond domain and 1 triple bond domain

(h) 1 triple bond domain and 1 lone pair domain

52. Use the electron-pair domain model to predict the geometric shape of each of the following molecules.

(a) CF_4 (b) CO_2 (c) H_2CO (d) ClCN (e) F_2CCF_2

53. Draw Lewis structures for each of the covalent chlorides of the elements of Period 2 from Be to F. Describe each by the AX_nE_m terminology, name the geometric shape, and illustrate the shape.

General Problems

54.* Draw a cube and connect the four appropriate corners to form a tetrahedron. Join the midpoint of the cube (the midpoint of the tetrahedron) to each of the four corners of the tetrahedron. These lines represent the bonds in a tetrahedral AX_4 molecule, such as CH_4. Using trigonometry, calculate the angle between the bonds.

55.* When Mendeleev proposed his periodic table in 1869, he predicted that an element unknown at the time would be discovered between calcium and titanium in Period 4, and he foretold some of its expected properties. When the element was discovered in 1879, it was called scandium, Sc. By comparison with the following properties of calcium and titanium, make your own predictions about some of the properties of scandium.

Element	Melting Point (K)	Boiling Point (K)	Density ($g \cdot cm^{-3}$)
Ca metal	1110	1757	1.55
Ti metal	1941	3560	4.50
Sc ?	?	?	?

56.* Indium oxide contains 82.7 mass percent indium. It occurs naturally in ores containing zinc oxide, $ZnO(s)$, and it was therefore originally assumed to have the empirical formula InO. On this basis, calculate the atomic mass of indium, and predict its location in the periodic table. Explain why this location is unsuitable for indium. Mendeleev suggested that the empirical formula of the oxide must be In_2O_3. Calculate the atomic mass of indium on this basis, and find its location in the periodic table. Is this location reasonable?

57.* Examine the periodic table to identify three pairs of elements that are exceptions to the general rule that the average atomic masses of the elements are in the same sequence as their atomic numbers. Explain these exceptions.

Chemical Reactions and the Halogens

4.1 Chemical Reactions
4.2 The Halogens
4.3 Oxidation–Reduction (Redox) Reactions
4.4 Acid–Base Reactions
4.5 Precipitation Reactions
4.6 Reaction Types

Bromine, Br_2, a deep brown liquid that forms a red-brown vapor, is one of the five elements known as the halogens. (Fluorine, chlorine, iodine, and astatine are the others.) Bromine is produced from sea water at an industrial plant such as this when bromide ions are reacted with chlorine gas. Electrons are transferred in this type of chemical reaction, called an oxidation-reduction (or redox*) reaction.*

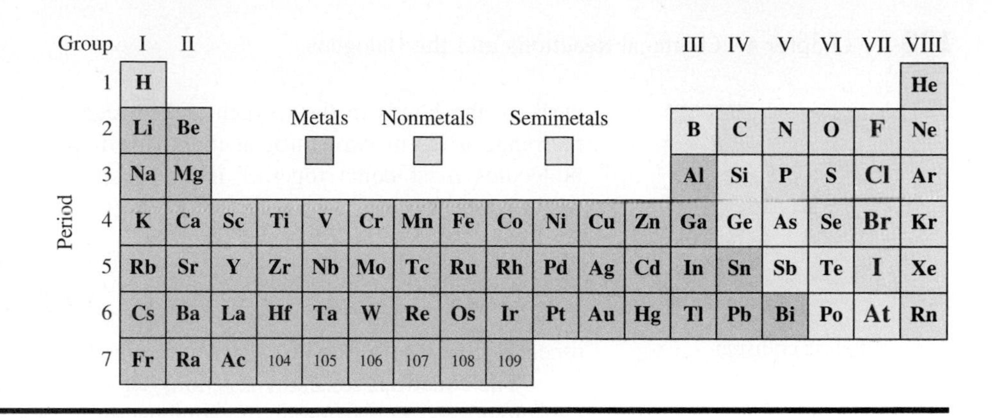

The **halogens** are a family of nonmetals in Group VII of the periodic table. They have very similar properties that change in a regular way from the top to the bottom of the group. They provide an excellent illustration of the similarities and trends within a group. In addition, the halogens provide us with good examples of several important classes of chemical reactions. Although there are many different chemical reactions, most of them can be classified into a few important types. A knowledge of these reaction types gives us a useful basis for understanding many different reactions. In this chapter we will use some of the reactions of the halogens to illustrate the three most important reaction types, namely, oxidation–reduction reactions, acid–base reactions, and precipitation reactions.

Group VII, the halogens, consists of the elements fluorine, chlorine, bromine, iodine, and astatine. Astatine is an extremely rare, radioactive element, but the other elements have many practical uses and form many important compounds. For example, chlorine is used to purify water supplies and as a bleach. Fluoride ion is added to the water supply of many cities to help prevent tooth decay. Iodine is present in mammals in a molecule called thyroxine, which regulates the body's metabolic rate. The halogens and their compounds provide many examples of both ionic and covalent substances, so we will be able to extend our knowledge and understanding of the properties of these two important types of substances, too. We begin by considering some fundamental ideas about reactions.

4.1 CHEMICAL REACTIONS

Although we discuss a very limited number of reactions in this text, even this modest selection may at first seem overwhelming and difficult to remember. But there are only a few common and important *types* of reactions, and the first step toward understanding reactions is to learn to recognize these key types. We can often then understand why a given reaction gives certain products and can even predict the probable products. However, first we will look more closely at how reactions take place and at the factors that influence how fast and to what extent a reaction occurs.

For most reactions to occur, molecules must come into contact so that atoms may be exchanged between them and form new molecules. Molecules come into contact when they collide as a consequence of their constant random motion. For example, the reaction

We shall see later that a reaction can also occur when a molecule absorbs energy, for example, from light.

$$N_2 + O_2 \rightarrow 2NO$$

may occur when an oxygen molecule collides with a nitrogen molecule, as shown in Figure 4.1. This collision must occur with sufficient energy to break, or at least

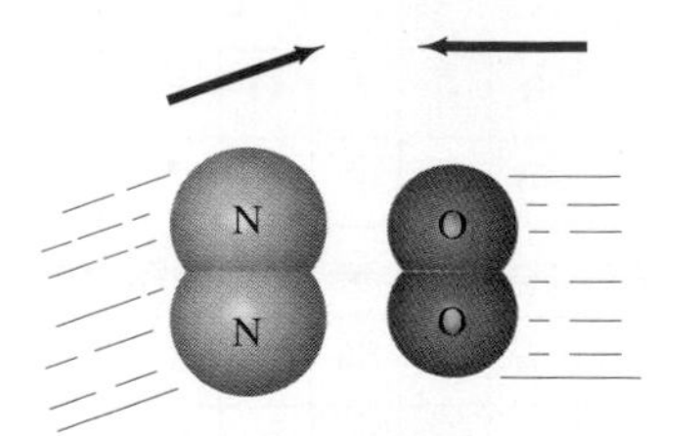

Before collision

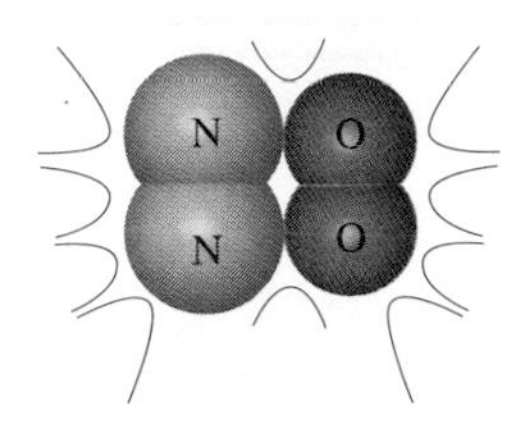

Collision

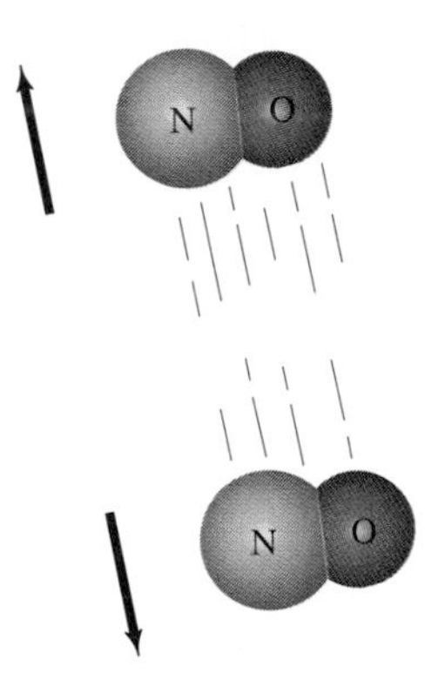

After collision

FIGURE 4.1 Collision between a Nitrogen Molecule and an Oxygen Molecule
If the molecules collide with sufficient energy, the atoms rearrange to form two NO molecules.

weaken, the bonds in the oxygen and nitrogen molecules, so that the atoms can rearrange to form two nitrogen monoxide molecules. Moreover, the N_2 and O_2 molecules must come together in the proper orientation so that the necessary rearrangement of the atoms can occur. In general,

For two molecules to react, they must collide with the correct relative orientation and with sufficient energy to break or weaken some of the bonds;

otherwise they simply bounce apart unchanged. So not all collisions lead to reaction.

The equations for many reactions suggest that they occur between more than two molecules. For example, the combustion of ethene

$$C_2H_4 + 3O_2 \rightarrow 2CO_2 + 2H_2O$$

involves the reaction of one ethene molecule with three oxygen molecules. The chances of the simultaneous collision of one ethene molecule with three oxygen molecules are so small as to be essentially zero. This reaction, like many others, must proceed in more than one step, each of which involves the collision of only two, or occasionally three, molecules. This series of steps is called the **mechanism** of the reaction. We discuss the mechanisms of some reactions in later chapters and particularly in Chapter 15.

Reaction Rates

Chemical reactions vary greatly in the speed with which the reactants are converted into products. The explosive reaction between a hydrocarbon and oxygen in the cylinders of an automobile engine occurs in a fraction of a second, whereas the rusting of iron takes many months or years. The rate of a given reaction is affected by the pressure of the reactants in the gas phase or the concentration of the reactants in solution, the temperature, and the presence of catalysts.

For a reaction between two reactants, the greater the number of molecules of each reactant in a given volume, the more collisions there will be in a given time, and therefore the faster the reaction occurs. For a reaction in the gas phase, the greater the number of molecules of a substance in a given volume, the greater the partial pressure of that substance. Thus

The rate of a reaction increases with increasing partial pressures of the reactants.

For a reaction in solution, the number of molecules of a reactant in a given volume of solution is expressed as the *concentration* of the reactant—the greater the number of molecules of a substance in a given volume, the greater the concentration of that substance. We discuss concentration in a quantitative way in Chapter 5. Thus for a reaction in solution,

The rate of a reaction increases with increasing concentrations of the reactants.

We have seen that, according to the kinetic theory of gases, the higher the temperature, the faster molecules move and the greater their kinetic energy. So the number of collisions that occur in a given time and the energy involved in a collision increase with increasing temperature. It follows that

The rate of a reaction increases with increasing temperature.

Another way to increase the rate of a reaction is to use a catalyst.

A *catalyst* is a substance that increases the rate of a reaction without being used up in the reaction.

Even though a catalyst is not used up in a reaction, it must take part in the reaction if it is to influence the rate. Catalysts operate in several different ways. Solids often act as catalysts in reactions between gases. A nickel catalyst is used, for example, in the Haber process for the synthesis of ammonia (Section 2.2). In such cases the solid catalyst provides a surface to which reactant molecules tend to stick when they collide with it, thus bringing them into close contact and thereby increasing the rate of the reaction. The many different catalysts that function in the reactions that occur in living organisms are called **enzymes**. Enzymes are very large molecules that strongly attract the reacting molecules to some part of their surface and bring them together in just the right way for the reaction to occur (Chapter 17). We will meet several other types of catalysts in the following chapters.

In Chapter 15 we discuss quantitatively the dependence of the reaction rate on the concentrations of the reactants and the temperature. For now it is sufficient to have just a qualitative understanding of the effects of concentration and temperature on reaction rates.

Equilibrium

In the reaction between N_2 and O_2,

$$N_2(g) + O_2(g) \rightarrow NO(g) + NO(g)$$

the concentration of the product NO increases and the concentrations of the reactants N_2 and O_2 decrease as the reaction proceeds. As the concentration of NO builds up, the probability increases that two NO molecules will collide, react, and give back N_2 and O_2 if they collide with sufficient energy and with the correct orientation:

$$NO(g) + NO(g) \rightarrow N_2(g) + O_2(g)$$

We usually write the equation for this reaction with a double arrow to show that it can occur in both the forward direction (left to right) and in the reverse direction (right to left):

$$N_2(g) + O_2(g) \rightleftharpoons 2NO(g)$$

As the concentrations of N_2 and O_2 decrease, the rate of the forward reaction decreases; as the concentration of NO increases, the rate of the reverse reaction increases. Eventually the two rates become equal. And although both reactions are still proceeding, there is no further change in the concentrations of either the reactants or the products, so it appears that the reaction has stopped. We say that the system has reached **equilibrium**.

A reaction reaches equilibrium when the rates of the forward and reverse reactions are equal.

To emphasize that at equilibrium a reaction has not stopped but is still occurring at equal rates in both directions, we call such an equilibrium a **dynamic equilibrium**. All reactions can in general proceed in both directions.

All reactions are in principle reversible and eventually reach a state of dynamic equilibrium.

However, if the reverse reaction is very slow, the reaction will not reach equilibrium until the concentrations of the products are very large and the concentrations of reactants are very small, so it appears that the reaction has gone to completion. In such a case we say that the position of equilibrium lies far to the right and that the reaction has gone essentially to completion.

A reversible reaction may also go to completion if the products are continually removed so that they never reach a concentration large enough for the reverse reaction to become significant. A reaction will reach equilibrium only if it is carried out in a closed reaction vessel to which nothing is added, or from which nothing is removed, as the reaction proceeds.

Let us summarize some important properties of reactions:

- Rates of reactions increase with increasing concentration of the reactants.
- Rates of reactions increase with increasing temperature.
- Catalysts increase the rate of a reaction.
- If left undisturbed, all reactions eventually reach a state of equilibrium.

Types of Reactions

Reactions may be classified in several different ways. Many reactions fall into one of two very broad classes, **synthesis** or **decomposition**. We have seen that two elements often combine to give a compound. In particular, metals react with nonmetals to produce a compound that is generally ionic. Magnesium reacts with oxygen to give magnesium oxide:

$$2Mg(s) + O_2(g) \rightarrow 2MgO(s)$$

Sodium reacts with chlorine to give sodium chloride

$$2Na(s) + Cl_2(g) \rightarrow NaCl(s)$$

These are **synthesis reactions**, reactions in which a substance is produced by the combination of two or more simpler substances (in these cases, elements).

If we pass an electric current through water, we decompose the water into hydrogen and oxygen (Demonstration 2.1):

$$2H_2O(l) \xrightarrow{\text{electric current}} 2H_2(g) + O_2(g)$$

When we heat the red solid mercury oxide, HgO, it decomposes into silvery liquid mercury and oxygen (Demonstration 4.1a):

$$2HgO(s) \rightarrow 2Hg(l) + O_2(g)$$

If we add a drop of blood, which contains the enzyme catalase, to an aqueous solution of hydrogen peroxide, the hydrogen peroxide rapidly decomposes into water and oxygen (Demonstration 4.1b). The preceding three reactions are **decomposition reactions**. The reverse of each of these reactions is a synthesis reaction.

There are other more specific and often more useful ways of classifying reactions. Three very important types are **oxidation–reduction** reactions, which we introduced in Chapter 3; **acid–base** reactions; and **precipitation** reactions. We will discuss these three reaction types in this chapter, taking our examples from the chemistry of the halogens.

DEMONSTRATION 4.1 Decomposition Reactions

(a) The Thermal Decomposition of Mercury Oxide

When red mercury oxide is heated, it darkens and decomposes to give oxygen and mercury, which condenses as silver droplets on the cooler part of the tube.

When the tube is allowed to cool, the remaining mercury oxide returns to its original red color.

(b) Properties of Water and Hydrogen Peroxide

Water (left) and hydrogen peroxide (right) are both colorless liquids. Both are compounds of the elements hydrogen and oxygen. When black solid manganese dioxide is added to water, no reaction is observed. When manganese dioxide is added to hydrogen peroxide, it causes hydrogen peroxide to decompose, producing bubbles of oxygen that carry the manganese dioxide to the surface. The oxygen that is evolved ignites a glowing splint.

When a drop of blood, which contains the enzyme catalase, is added to water, it slowly mixes, but no reaction is observed. When a few drops of blood are added to an aqueous solution of hydrogen peroxide, the hydrogen peroxide decomposes rapidly, producing bubbles of oxygen that form a thick foam that fills the beaker and overflows the top.

Exercise 4.1 Classify each of the following reactions as a synthesis reaction (S) or a decomposition reaction (D).

(a) $2KClO_3(s) \rightarrow 2KCl(s) + 3O_2(g)$

(b) $N_2(g) + 3H_2(g) \rightarrow 2NH_3(g)$

(c) $C_2H_4(g) + H_2(g) \rightarrow C_2H_6(g)$

(d) $CaCO_3(s) \rightarrow CaO(s) + CO_2(g)$

4.2 THE HALOGENS

The halogens are composed of diatomic molecules in which the two atoms are held together by a single covalent bond. They all have similar Lewis structures:

$$:\ddot{\underset{\cdot\cdot}{F}}-\ddot{\underset{\cdot\cdot}{F}}: \qquad :\ddot{\underset{\cdot\cdot}{Cl}}-\ddot{\underset{\cdot\cdot}{Cl}}: \qquad :\ddot{\underset{\cdot\cdot}{Br}}-\ddot{\underset{\cdot\cdot}{Br}}: \qquad :\ddot{\underset{\cdot\cdot}{I}}-\ddot{\underset{\cdot\cdot}{I}}:$$

The halogens are among the most reactive of the elements, but their reactivity decreases as we move down the group from fluorine to iodine. Compounds of the halogens with another element are called fluorides, chlorides, bromides, and iodides, or in general, **halides**; examples are sodium chloride, NaCl, and phosphorus trifluoride, PF_3.

Note the difference between an element in the free state—say, fluor*ine*—and the same element as part of a compound—say, sodium fluor*ide*. It is not the free element fluorine, F_2, that some toothpastes contain or that is added to some water supplies, but rather fluorine in the combined form as a fluoride, such as sodium fluoride.

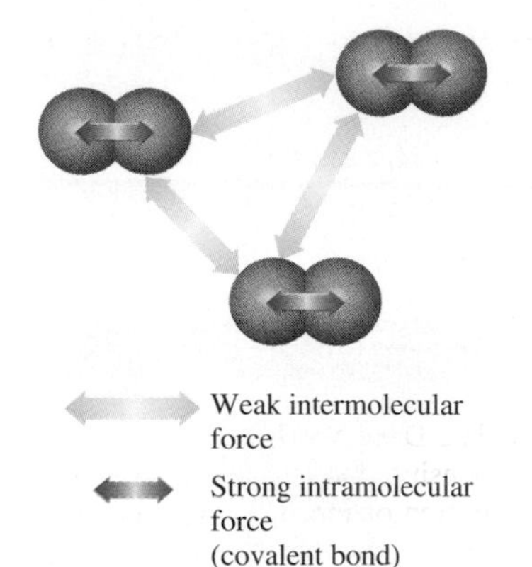

It is important to distinguish clearly between *inter*molecular forces and *intra*molecular forces. *Inter*molecular forces are the forces *between* molecules. *Intra*molecular are the forces *inside* molecules—forces that hold the atoms together in a molecule, usually covalent bonds.

TABLE 4.1 Some Properties of the Halogens

	Melting Point (°C)	Boiling Point (°C)	Color
F_2	−219	−188	Pale yellow
Cl_2	−101	−34	Yellow-green
Br_2	−7	59	Red-brown
I_2	113	185	Black

Table 4.1 summarizes some physical properties of the halogens. Their melting points and boiling points increase as we move down the group. At room temperature fluorine and chlorine are gases, whereas bromine is a red-brown liquid, and iodine is a black solid (Demonstration 4.2). This change in physical state indicates that the strength of the forces between the halogen molecules (intermolecular forces) increases in the order $F_2 < Cl_2 < Br_2 < I_2$ with increasing molecular size. This relationship between molecular size and the strength of intermolecular forces is typical of many substances, for reasons that we discuss in Chapter 9.

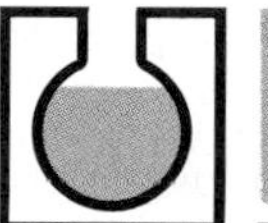

DEMONSTRATION 4.2
Physical Properties of the Halogens

Chlorine is a pale yellow-green gas at room temperature. The tube inside the flask contains dry ice (solid carbon dioxide, −78°C). Yellow liquid chlorine (bp −34°C) is condensing on the cold tube.

Bromine is a red-brown liquid that is sufficiently volatile to fill the flask with vapor.

When solid iodine is gently heated, it forms a violet vapor, which can be seen recrystallizing on the cold flask on top of the beaker.

Sources and Uses

Chlorine is the most abundant of the halogens, making up about 0.20% of the earth's crust (Table 2.3). Fluorine comprises only 0.03% of the earth's crust; bromine and iodine are far less abundant. Because the halogens are very reactive, combining

The Dead Sea has such a high density, due to the high concentration of metal halides and other salts, that the human body floats easily on the surface.

directly with many other elements and reacting with many compounds, they are not found in the free (uncombined) state in nature, but only as compounds, particularly halides of metals.

Common halides are those of sodium, potassium, magnesium, and calcium. Except for calcium fluoride, all these halides are soluble in water and are therefore found dissolved in the oceans and inland seas. They are found also as underground solid salt deposits or as concentrated aqueous solutions (subterranean brines) formed by the evaporation of ancient seas. Calcium fluoride, CaF_2, is insoluble in water. It occurs as the mineral fluorspar in many parts of the world.

Sodium chloride and other metal halides are obtained by mining the solid deposits or by pumping water down to dissolve them and bring them to the surface as a concentrated aqueous solution. Metal halides can be obtained also by evaporating sea water.

Chlorine and Its Compounds Chlorine, Cl_2, is used for bleaching wood pulp in the manufacture of paper, for bleaching textiles, for destroying bacteria in water for domestic use, and for making plastics such as polyvinylchloride (PVC), insecticides, and dry-cleaning solvents.

Sodium chloride, familiar as common salt, is by far the most important of the alkali metal halides. It is a major raw material of the chemical industry from which many other compounds of chlorine and of sodium are made. Sodium and potassium ions are present throughout the human body, and they play several essential roles (Chapters 13 and 17). The total amount of sodium chloride in the body is about 200 g. Approximately 2 g a day is needed to maintain this level.

Bromine and Its Compounds Important uses for bromine include the manufacture of bromine compounds used as gasoline additives, pesticides, and fireproofing agents. Silver bromide, $AgBr(s)$, and silver iodide, $AgI(s)$, are used in the manufacture of photographic film.

Iodine and Its Compounds Although only very small amounts are required, iodine is an essential element in our diet because the growth-regulating hormone thyroxine, which is produced by the thyroid gland, is a molecule containing iodine. Lack of iodine in the diet can cause goiter, a disease in which the thyroid gland becomes much enlarged so as to capture as much iodine as possible. The necessary iodine can be obtained from fish or from "iodized" table salt, which is sodium chloride containing 0.01% sodium iodide or potassium iodide.

Fluorine and Its Compounds An important use of fluorine is to make uranium hexafluoride, $UF_6(g)$. The controlled diffusion of this gas (Chapter 3) is employed to enrich uranium in the ^{235}U isotope for use in the fuel elements of nuclear power stations (Chapter 18).

Sodium fluoride, $NaF(s)$, is added to water supplies and to toothpaste to help prevent tooth decay (Chapter 7). The polymer Teflon, polytetrafluoroethene $(C_2F_4)_n$ (Chapter 20), is a very inert white solid with a slippery surface. It is widely used in chemical plants and for ball bearings that require no lubrication. It is familiar in the kitchen as a nonstick surface on cooking utensils. CFCs (chlorofluorocarbons) such as CCl_2F_2 have been used as the working fluid in refrigerators and air conditioners. However, CFCs are now being replaced with other substances, because they react with and destroy the ozone in the upper atmosphere that protects us from exposure to harmful ultraviolet radiation from the sun (Chapter 16).

FIGURE 4.2 Hydrogen Burning in Chlorine
Hydrogen burns with a blue-white flame in an atmosphere of chlorine to form HCl(*g*). The remaining yellow chlorine can be seen in the lower part of the beaker.

Reactions with Nonmetals

The halogens combine readily with almost all the other nonmetals and with the semimetals to form compounds that consist of covalent molecules.

Hydrogen Halides Hydrogen reacts with each of the halogens, producing the **hydrogen halides**—HF, HCl, HBr, and HI. For example, hydrogen chloride is produced when hydrogen reacts with chlorine:

$$H_2(g) + Cl_2(g) \rightarrow 2HCl(g)$$

The reactivity of the halogens toward hydrogen decreases down the group from fluorine to iodine. The reaction between hydrogen and fluorine is extremely rapid under all conditions. The reaction of hydrogen with chlorine is slow in the dark but becomes explosive in sunlight or other bright light or with strong heating. When a stream of hydrogen is ignited in air, the flame continues to burn in an atmosphere of $Cl_2(g)$, producing HCl(*g*) (Figure 4.2). Bromine and iodine react very slowly with hydrogen. High temperatures and a catalyst are needed to obtain a reasonably rapid reaction.

The hydrogen halides are colorless gases that have pungent odors and are very soluble in water (Demonstration 4.3). Some of their properties are summarized in Table 4.2. The melting points and boiling points increase from HCl to HI as the intermolecular forces become stronger with increasing atomic number and increasing size of the halogen atom. Hydrogen fluoride, however, has unexpectedly high

TABLE 4.2 Some Properties of the Hydrogen Halides

	Melting Point (°C)	Boiling Point (°C)
HF	−83	20
HCl	−115	−85
HBr	−89	−67
HI	−51	−35

DEMONSTRATION 4.3 Solubility of Hydrogen Chloride in Water

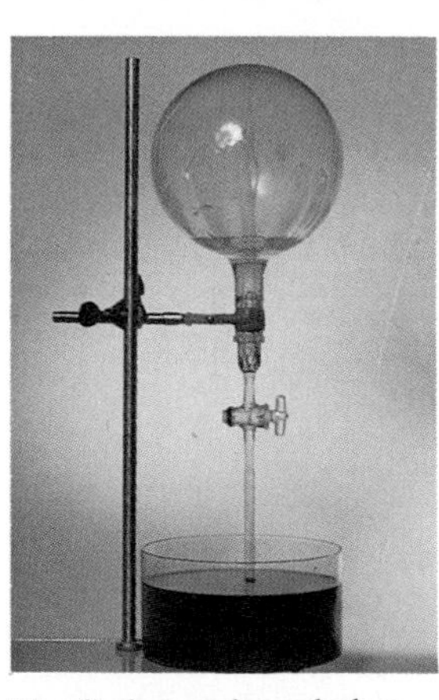

The flask contains colorless hydrogen chloride gas. When the tap is opened, hydrogen chloride dissolves in the water. This reduces the pressure in the flask, so water is drawn into the flask from the dish below and creates a fountain.

Water is drawn into the flask so rapidly that it forms a jet that hits the top of the flask.

A few drops of the indicator bromothymol blue have been added to the water in the dish. The color changes to yellow as the water enters the flask, because hydrogen chloride forms an acid solution in water. Hydrogen chloride is so soluble in water that the flask fills almost completely.

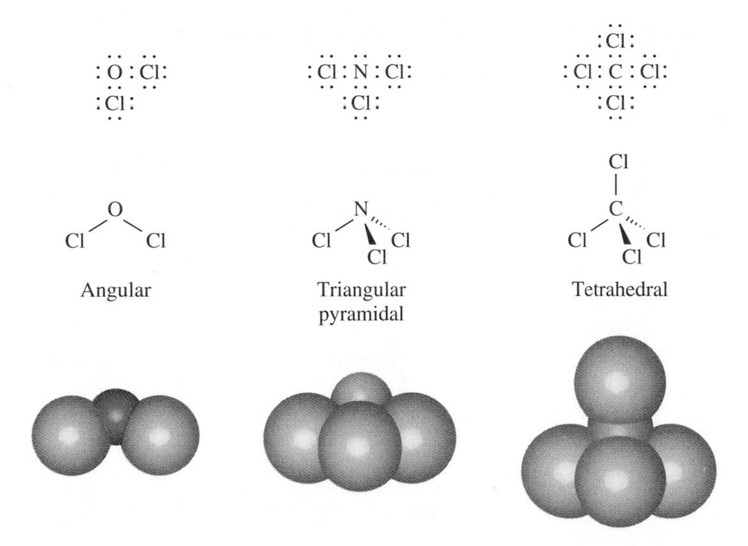

FIGURE 4.3 Lewis Structures and Molecular Shapes of Nonmetal Halides
All halides of Group VI elements, such as Cl_2O, are angular molecules like H_2O. All the "trihalides" of Group V, such as PCl_3 and NCl_3, are triangular pyramidal AX_3E molecules; they have the same shape as NH_3. The halides of Group IV are tetrahedral AX_4 molecules with the Group IV atom at the center and a halogen atom at each of the four corners of the tetrahedron; CCl_4 is an example.

melting and boiling points compared with the other halogen halides. These unusual properties are discussed in Chapter 11.

Nonmetal Halides Many other nonmetal halides, such as SCl_2, PCl_3, NCl_3, CCl_4, CBr_4, and OF_2, can be made by direct combination of the elements or by other methods. For example, carbon tetrachloride, $CCl_4(l)$, can be made by heating carbon in chlorine:

$$C(s) + 2Cl_2(g) \rightarrow CCl_4(l)$$

Phosphorus trichloride, $PCl_3(l)$, can be made by heating phosphorus in chlorine:

$$2P(s) + 3Cl_2(g) \rightarrow 2PCl_3(l)$$

Nonmetal and semimetal halides are covalent molecular compounds.

The shapes of some nonmetal halide molecules and their Lewis structures are given in Figure 4.3. Table 4.3 summarizes the properties of some nonmetal chlorides. Most of them are gases or liquids with low boiling points. The boiling points and melting points increase with increasing atomic mass as we move down each group, so some of the halides of the heavier elements are solids.

TABLE 4.3 Properties of Some Chlorides of Nonmetals

Period	Chloride	Melting Point (°C)	Boiling Point (°C)
2	CCl_4	−23	77
	NCl_3	−40	71
	OCl_2	−20	4
	FCl	−154	−101
3	$SiCl_4$	−68	57
	PCl_3	−91	74
	SCl_2	−122	59
	Cl_2	−102	−35

Reactions with Metals: Metal Halides

The halogens react with the Group I and II metals to form halides (Demonstration 4.4). Some typical reactions are the following:

$$2Na(s) + Cl_2(g) \rightarrow 2NaCl(s)$$

$$Ca(s) + F_2(g) \rightarrow CaF_2(s)$$

$$Mg(s) + Cl_2(g) \rightarrow MgCl_2(s)$$

As we saw in Chapter 3,

The halides of the Group I and II metals are ionic compounds

consisting of oppositely charged ions held together by electrostatic attraction. The formation of these ionic compounds can be represented as follows:

$$Na\cdot + \cdot\ddot{\underset{..}{Cl}}: \rightarrow (Na^+)(:\ddot{\underset{..}{Cl}}:^-)$$

$$\cdot Ca\cdot + 2\cdot\ddot{\underset{..}{F}}: \rightarrow (Ca^{2+})(:\ddot{\underset{..}{F}}:^-)_2$$

We do not usually write the formulas of ionic compounds in terms of the Lewis symbols for the ions $(Na^+)(:\ddot{\underset{..}{Cl}}:^-)$ as we do here. Sodium chloride does not consist of pairs of ions, as this formula might suggest, but an infinite array of sodium ions and chloride ions. We normally write its formula as NaCl(s) without showing the charges on the ions.

Because metal halides of Group I and II metals have ionic bonds, their physical and chemical properties are quite different from the covalent molecular nonmetal halides. Some of the physical properties of the fluorides and chlorides of the alkali and alkaline earth metals are summarized in Table 4.4. In contrast to the nonmetal halides, they are all colorless crystalline solids (Figure 4.4) with high melting points.

DEMONSTRATION 4.4 Reactions of Metals with Chlorine

Sodium metal heated until it melts burns with an intense yellow flame when inserted into a jar of chlorine. A white smoke of sodium chloride particles, NaCl, is formed.

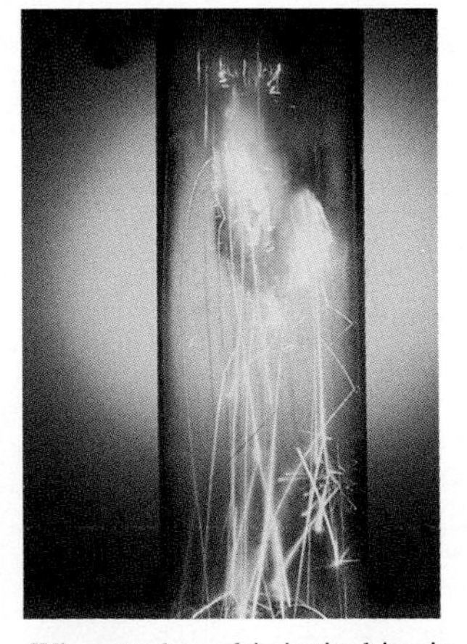

When steel wool is ignited in air and then inserted into a jar of chlorine, it continues to burn, forming dense red-brown fumes of iron chloride, $FeCl_3$, which deposit on the sides of the jar. The sparks are white-hot iron.

Small pieces of antimony react immediately, becoming white hot when dropped into a jar of chlorine. They can be seen here bouncing from the bottom of the jar. A white smoke of antimony trichloride, $SbCl_3$, is formed.

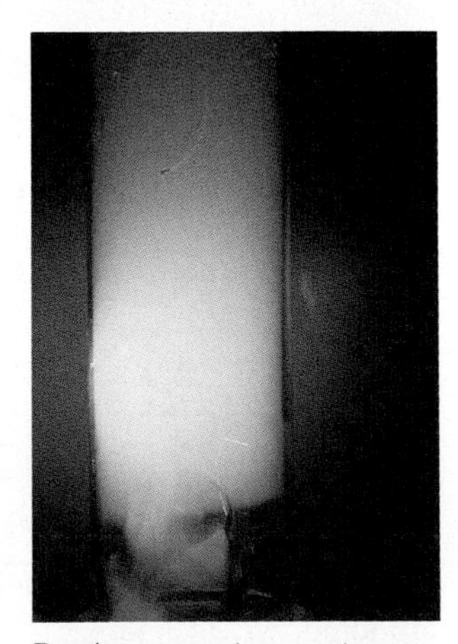

Burning magnesium continues to burn with a bright white flame when inserted into a jar of chlorine gas. White solid magnesium chloride, $MgCl_2$, is formed.

TABLE 4.4 Some Properties of the Alkali and Alkaline Earth Metal Fluorides and Chlorides

Compound	Melting Point (°C)	Boiling Point (°C)	Solubility in Water at 25°C ($g \cdot L^{-1}$)
LiF	845	1681	1.3
NaF	995	1704	40.1
KF	856	1501	1020*
RbF	775	1408	1310
CsF	682	1250	3700*
LiCl	610	1382	850*
NaCl	808	1465	360
KCl	772	1407	350
RbCl	777	1381	940
CsCl	645	1300	1900
BeF_2	—	800†	5500
MgF_2	1263	2227	0.13
CaF_2	1418	2500	0.016
SrF_2	1400	2460	0.12
BaF_2	1320	2260	1.6
$BeCl_2$	405	488	720*
$MgCl_2$	714	1418	550
$CaCl_2$	772	>1600	830*
$SrCl_2$	875	1250	560*
$BaCl_2$	1350	—	370*

* Solubilities of hydrates $KF \cdot 2H_2O$, $CsF \cdot H_2O$, $LiCl \cdot H_2O$, $BeCl_2 \cdot 4H_2O$, $CaCl_2 \cdot 6H_2O$, $SrCl_2 \cdot 6H_2O$, $BaCl_2 \cdot 2H_2O$

† Sublimes (changes directly from solid to gas without forming a liquid)

FIGURE 4.4 Sodium Chloride Crystals Sodium chloride forms colorless cubic crystals. We can see these crystals by looking at table salt under a microscope. Large crystals are sometimes found in salt deposits; they are called halite or rock-salt crystals.

Electrical Conductivity

How do we know that metal halides such as potassium fluoride and lithium chloride are ionic, whereas nonmetal halides, such as carbon tetrachloride and phosphorus trichloride, are covalent? One important piece of evidence is that ionic substances conduct an electric current when melted, whereas covalent substances are nonconductors.

If graphite rods connected by copper wires to a current source and a light bulb are dipped into molten lithium chloride, the bulb lights up (Demonstration 4.5). This observation shows that the molten (liquid) lithium chloride conducts electricity. An

Graphite, a form of carbon, is one of the few exceptions to the rule that nonmetals are nonconductors of electricity. Some of the valence electrons in graphite are free to move from atom to atom as in a metal (Chapter 10).

DEMONSTRATION 4.5 Electrical Conductivity of Molten Lithium Chloride

The crucible contains lithium chloride, which is heated to its melting point (610°C) to give a colorless liquid. The light bulb connected to a circuit with graphite electrodes shines brightly, showing that molten lithium chloride is an electrical conductor.

If the heating is stopped and the electrodes are removed from the molten lithium chloride, a plug of solid lithium chloride forms between the electrodes. The bulb does not light, showing that solid lithium chloride is not an electrical conductor.

electric current is composed of electric charges moving under the influence of an electric potential (Appendix P). In the copper wires and the graphite rods, the current consists of moving electrons (Chapter 10). In the molten lithium chloride, the current consists of positive lithium ions moving in one direction and negative chloride ions moving in the opposite direction. In contrast, liquid nonmetal halides such as carbon tetrachloride and phosphorus trichloride, which are not composed of ions, are nonconductors, or insulators.

As we saw in Demonstration 4.5, a solid metal halide does not conduct an electric current. The ions in the solid are fixed in position and cannot move about as they can in the liquid state. A solid metal does conduct an electric current, however, because electrons in a metal can move from one atom to another.

Water is a covalent molecular substance that, when pure, has an extremely small electrical conductivity. But when a solid metal halide is dissolved in water, it gives a solution that has a much larger conductivity. The ions of which the ionic solid is composed separate from each other in the solution and can then move about, carrying their charges with them. In contrast, substances that dissolve in water but consist of covalent molecules such as sucrose (sugar) or oxygen, O_2, do not give conducting solutions. The molecules can move through the solution, but because they are electrically neutral, they do not conduct a current.

Substances that give conducting solutions when dissolved in water are called **electrolytes**. Substances that give nonconducting solutions when dissolved in water are called **nonelectrolytes**.

EXAMPLE 4.1 Reactions of the Halogens with Metals and Nonmetals

Write a balanced equation for each of the following reactions.

(a) barium with fluorine **(b)** silicon with fluorine

(c) potassium with iodine **(d)** phosphorus with bromine

In each case state whether you expect the product to be (i) a solid or (ii) a liquid or gas at room temperature and atmospheric pressure.

Solution: **(a)** Barium is in Group II, so the formula of its fluoride is BaF_2. Therefore, the equation for the reaction is

$$Ba(s) + F_2(g) \rightarrow BaF_2(s)$$

We expect barium difluoride to be a solid, because it is the halide of a metal and consists of Ba^{2+} and F^- ions.

(b) Silicon is in Group IV, so the formula of its fluoride is SiF_4, and the equation for the reaction is

$$Si(s) + 2F_2(g) \rightarrow SiF_4(g)$$

We expect silicon tetrafluoride to be either a gas or a liquid, because it is the fluoride of a semimetal and consists of covalent SiF_4 molecules. It is in fact a gas at room temperature.

(c) Potassium is an alkali metal in Group I, so the formula of its iodide is KI, and the equation for the reaction is

$$2K(s) + I_2(s) \rightarrow 2KI(s)$$

We expect potassium iodide to be a solid, because it is the halide of a metal and thus consists of K^+ and I^- ions.

(d) Phosphorus is in Group V, so we expect the formula of its bromide to be PBr_3, and the equation for the reaction is

$$2P(s) + 3Br_2(l) \rightarrow 2PBr_3(l)$$

We expect phosphorus tribromide to be a liquid or gas, because it is the halide of a nonmetal and therefore consists of covalent molecules. It is in fact a liquid.

By remembering the division of the periodic table into metals and nonmetals, it is easy to apply the rules "metal + nonmetal → ionic compound" and "nonmetal + nonmetal → covalent compound."

Exercise 4.2 Write balanced equations for the reactions in Demonstration 4.4.

Exercise 4.3 Write balanced equations for each of the following reactions.
(a) strontium with bromine **(b)** arsenic with chlorine
(c) rubidium with fluorine **(d)** bromine with chlorine
In each case state whether you expect the product to be (i) a solid or (ii) a liquid or gas at room temperature and atmospheric pressure.

4.3 OXIDATION–REDUCTION (REDOX) REACTIONS

In the reaction between sodium and chlorine to give sodium chloride,

$$2Na(s) + Cl_2(g) \rightarrow 2(Na^+Cl^-)(s)$$

sodium atoms give up electrons to become sodium ions, and chlorine molecules gain electrons to become chloride ions. We can therefore conveniently imagine that the reaction takes place in two steps, called **half-reactions**:

(1) Electron loss: $Na \rightarrow Na^+ + e^-$

(2) Electron gain: $Cl_2 + 2e^- \rightarrow 2Cl^-$

To obtain the balanced equation for the overall reaction, we must multiply the first equation by 2 and add it to the second equation. In this way we indicate that the electrons produced in the first reaction are used up in the second reaction:

$$\begin{array}{r} 2Na \rightarrow 2Na^+ + 2e^- \\ Cl_2 + 2e^- \rightarrow 2Cl^- \\ \hline 2Na + Cl_2 \rightarrow 2(Na^+Cl^-) \end{array}$$

The overall reaction is an **electron-transfer reaction**: Electrons are transferred from sodium to chlorine. Electron-transfer reactions are commonly called **oxidation–reduction reactions**, or **redox reactions**.

In our discussion of oxides (Chapter 2), we defined oxidation as the addition of oxygen to an element or compound and reduction as the removal of oxygen from an oxide. For example, the oxidation of sodium to sodium oxide is represented by the equation

$$4Na(s) + O_2(g) \rightarrow 2(Na^+)_2O^{2-}(s)$$

We can see that this is also an electron-transfer reaction, like the reaction of sodium with chlorine, by splitting the equation into two halves. Each sodium atom loses an electron to form a sodium ion:

$$(1) \qquad Na \rightarrow Na^+ + e^-$$

Each oxygen molecule gains four electrons to form two oxide ions:

$$(2) \qquad O_2 + 4e^- \rightarrow 2O^{2-}$$

Combining the two equations

$$4(Na \rightarrow Na^+ + e^-) \qquad \textit{Oxidation}$$

$$O_2 + 4e^- \rightarrow 2O^{2-} \qquad \textit{Reduction}$$

gives the overall equation

$$4Na + O_2 \rightarrow 2(Na^+)_2O^{2-}$$

Thus, the oxidation of sodium to sodium oxide is an electron-transfer reaction that involves the same half-reaction as in the reaction of sodium with chlorine. In both cases a sodium atom loses an electron to become a sodium ion, Na^+. We say that sodium is oxidized in both reactions. More generally,

An element is oxidized when it loses electrons,

and

Oxidation is the loss of electrons.

Conversely, the half-reactions for chlorine and oxygen in the sodium–chlorine reaction and in the sodium–oxygen reaction show that they both gain electrons:

$$Cl_2 + 2e^- \rightarrow 2Cl^-$$

$$O_2 + 4e^- \rightarrow 2O^{2-}$$

We say that both chlorine and oxygen are reduced. In general,

An element is reduced when it gains electrons,

and

Reduction is the gain of electrons.

An oxidation half-reaction *must* be accompanied by an appropriate reduction half-reaction that uses up the electrons produced in the oxidation half-reaction. The overall reaction is an oxidation–reduction, or redox reaction. In a redox reaction,

Electrons are transferred from an element that is oxidized to an element that is reduced.

According to this more general definition of oxidation, many substances other than oxygen can oxidize other substances—that is, remove electrons from them. Substances that can oxidize other substances are said to be oxidizing agents. Any substance that tends to gain electrons (for example, nonmetals such as oxygen, sulfur, and the halogens) may behave as an oxidizing agent.

An oxidizing agent is a substance that can gain electrons.

Any substance from which electrons can be readily removed, such as a metal, may reduce other substances—that is, behave as a reducing agent.

A reducing agent is a substance that can give up electrons.

Thus, in a redox reaction the oxidizing agent is reduced, and the reducing agent is oxidized.

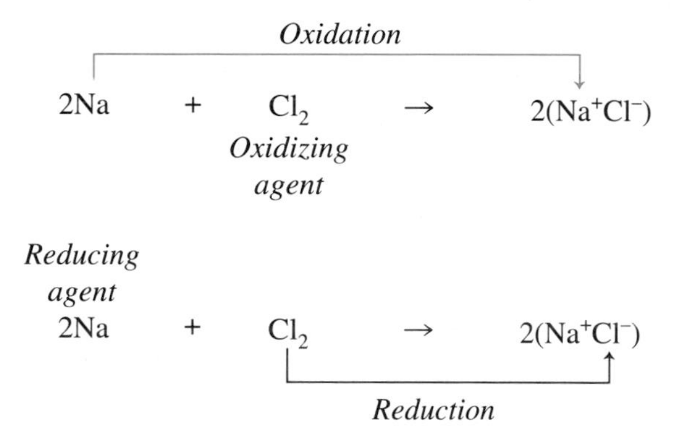

Oxidation–reduction reactions are a very important general class of reactions, and we will encounter many examples in later chapters. In general, the reaction between a metal and a nonmetal is an oxidation–reduction reaction, in which the metal behaves as a reducing agent and is oxidized and the nonmetal behaves as an oxidizing agent and is reduced.

Any oxidation–reduction reaction can be written as the sum of two half-reactions—one involving a loss of electrons (oxidation), and the other a gain of electrons (reduction).

The Halogens as Oxidizing Agents

Oxidizing Strengths of the Halogens The halogens are strong oxidizing agents, but their strength decreases in the series $F_2 > Cl_2 > Br_2 > I_2$. When an aqueous solution of chlorine, Cl_2, is added to a colorless aqueous solution of sodium bromide, NaBr, the solution rapidly becomes orange-red because bromine, Br_2, is formed (Demonstration 4.6):

$$Cl_2(g) + 2Br^-(aq) \rightarrow 2Cl^-(aq) + Br_2(l)$$

This reaction is an oxidation–reduction reaction in which bromide ion is oxidized to bromine,

$$2Br^-(aq) \rightarrow Br_2(l) + 2e^-$$

while chlorine is reduced to chloride ion

$$Cl_2(g) + 2e^- \rightarrow 2Cl^-(aq)$$

DEMONSTRATION 4.6 Oxidation–Reduction Reactions of the Halogens

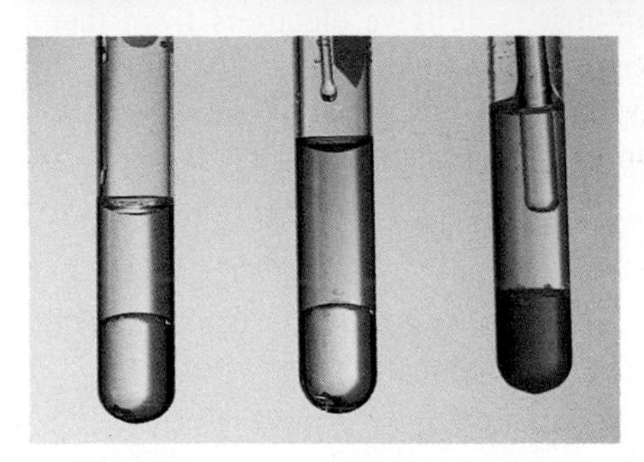

Left: The top layer is an aqueous solution of NaBr. The bottom layer is carbon tetrachloride, CCl_4. Center: When a few drops of a dilute aqueous solution of Cl_2 are added, the NaBr solution becomes yellow, because Br_2 is formed. Right: Bromine is much more soluble in carbon tetrachloride than in water. Upon being stirred, it is extracted into the carbon tetrachloride and forms a brown solution.

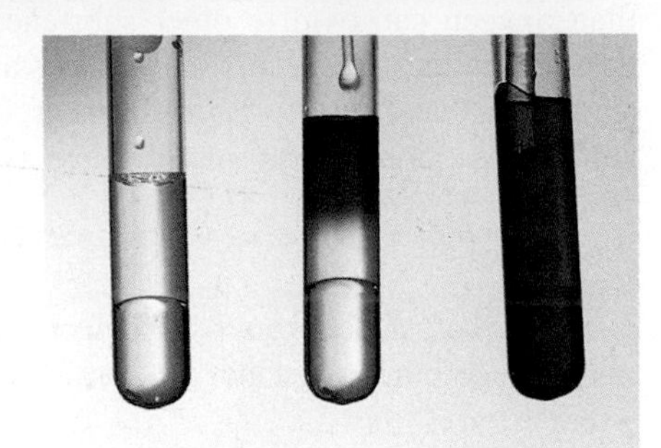

Left: The top layer is an aqueous solution of NaI. The bottom layer is CCl_4. Center: When a few drops of an aqueous solution of Cl_2 are added, iodine, I_2, is formed. Iodine reacts with iodide ion, I^-, to form a brown solution of the triiodide ion, I_3^-. Right: Upon being stirred, I_2 is extracted into the CCl_4 and forms a violet solution.

The overall reaction is

$$Cl_2(g) + 2Br^-(aq) \rightarrow 2Cl^-(aq) + Br_2(l)$$

Chlorine also oxidizes iodide ion to iodine (Demonstration 4.6):

$$Cl_2(g) + 2I^-(aq) \rightarrow 2Cl^-(aq) + I_2(s)$$

Similarly, bromine oxidizes iodide ion to iodine:

$$Br_2(l) + 2I^-(aq) \rightarrow 2Br^-(aq) + I_2(s)$$

However, if bromine is added to an aqueous solution of a chloride, no change in color is observed: Chloride ion is not oxidized to chlorine. This is not surprising, because we have already seen that chlorine oxidizes bromide ion to bromine. In other words, in the reversible reaction

$$Br_2(l) + 2Cl^-(aq) \rightleftharpoons 2Br^-(aq) + Cl_2(g)$$

the position of the equilibrium is far to the left. If we start with Br^- and Cl_2, as we did in Demonstration 4.6, we find that large amounts of Br_2 and Cl^- are formed before the system reaches equilibrium. But if we start with Br_2 and Cl^-, as in the last equation above, equilibrium is reached with the formation of only a very small amount of Br^- and Cl_2. We say that this reaction proceeds essentially completely to the left but almost not at all to the right.

To summarize:

- Chlorine, Cl_2, oxidizes bromide ion, Br^-, and iodide ion, I^-.
- Bromine, Br_2, oxidizes I^- but not Cl^-.
- Iodine, I_2, will not oxidize either Cl^- or Br^-.

So we conclude that the order of oxidizing strengths is $Cl_2 > Br_2 > I_2$.

In view of this order, it should not surprise us that fluorine is an even stronger oxidizing agent than chlorine. Fluorine reacts vigorously with water, oxidizing it to oxygen:

$$2F_2(g) + 2H_2O(l) \rightarrow 4HF(aq) + O_2(g)$$

In contrast, chlorine reacts only slowly to give small equilibrium concentrations of HCl and hypochlorous acid, HOCl (Chapter 7):

$$Cl_2(g) + H_2O(l) \rightleftharpoons HCl(aq) + HOCl(aq)$$

Bromine has only a small solubility in water, and iodine is insoluble (Figure 4.5). (**Solubility** is the amount of a substance that will dissolve in a given quantity of solvent.) Fluorine is such a strong oxidizing agent, reacting vigorously with many substances, that it is difficult and dangerous to handle without special precautions and equipment.

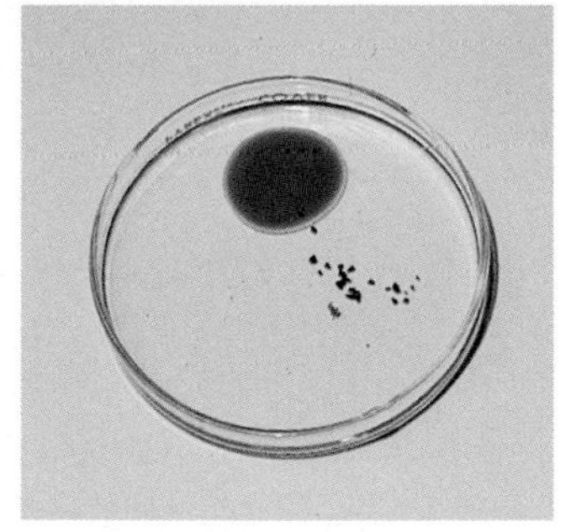

FIGURE 4.5 Carbon Tetrachloride Dissolves Covalent Nonpolar Molecules
Carbon tetrachloride, CCl_4, is not soluble in water. It forms a small pool when added to water. Solid iodine dissolves in the colorless CCl_4 to give a violet solution, but it is insoluble in water.

That F_2 is the strongest oxidizing agent among the halogens means that it shows the greatest tendency to be reduced, so we expect that F^- is very difficult to oxidize to F_2. Thus, the ease with which a halide ion can be oxidized to the corresponding element—in other words, the strength of the halide ions as reducing agents—increases in the series $F^- < Cl^- < Br^- < I^-$.

The order of halide ions, X^-, as reducing agents is the opposite of the order of the halogens, X_2, as oxidizing agents.

The order of oxidizing strengths $F_2 > Cl_2 > Br_2 > I_2$ is not unexpected when we recall that the electronegativities of F, Cl, Br, and I decrease in the same sequence. The electronegativity of a halogen is a measure of the tendency of a halogen atom to attract the electrons of a covalent bond and therefore to acquire a partial negative charge. It is not the same as the oxidizing strength of a halogen, which is a measure of the tendency of the diatomic molecule X_2 to acquire two electrons to become two X^- ions. Nevertheless, both the oxidizing strength and the electronegativity of the halogens are related to the tendency of halogen atoms to acquire electrons. Not surprisingly, as the electronegativity of X decreases, the oxidizing strength of X_2 also decreases in the order $F_2 > Cl_2 > Br_2 > I_2$.

The Triiodide Ion A brown solution forms when chlorine oxidizes an aqueous solution of an iodide (Demonstration 4.6) because the soluble triiodide ion, I_3^-, is formed by the combination of iodine and the iodide ion:

$$\underset{\text{Soluble in } CCl_4\text{, insoluble in water}}{I_2(s)} + I^-(aq) \rightleftharpoons \underset{\text{Soluble in water, insoluble in } CCl_4}{I_3^-(aq)}$$

When this brown solution is shaken with carbon tetrachloride, which is insoluble in water and forms a separate layer below the water, the carbon tetrachloride layer takes on a violet color. This violet color is similar to the color of iodine vapor (Demonstration 4.2) and is due to iodine molecules. Iodine is very soluble in CCl_4 (Figure 4.5), whereas the triiodide ion, I_3^-, has a very low solubility. Thus, the position of equilibrium is far to the left in CCl_4, whereas it is far to the right in aqueous solution.

Industrial Preparation of the Halogens

Because the halogens occur naturally as metal halides, the elements are prepared by oxidation of the corresponding halide ion.

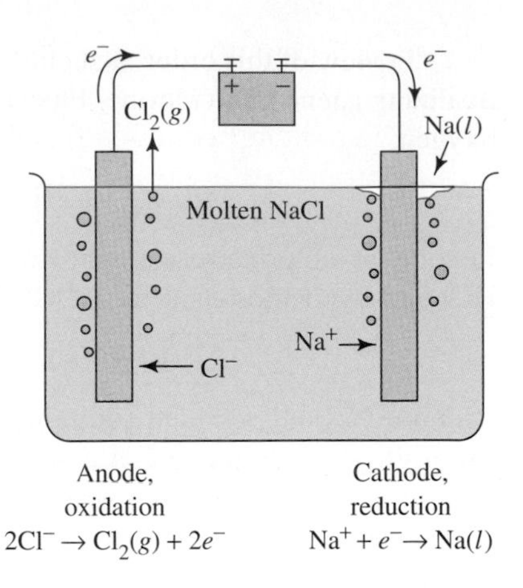

FIGURE 4.6 Electrolysis of Molten Sodium Chloride The electric current outside the electrolytic cell is carried by the electrons, which are pushed around the circuit by the battery. Inside the molten electrolyte, the current is carried by the movement of the positive and negative ions toward the electrodes. At the anode Cl^- ions give up electrons to the electrode to form Cl atoms, which combine to give Cl_2 molecules. At the cathode Na^+ ions accept electrons from the electrode to give Na atoms, which form liquid sodium metal.

Preparation of Chlorine by Electrolysis In principle, chlorine could be made by oxidation of Cl^- with F_2. However, this method is impractical, because fluorine is expensive to produce and difficult to handle. Chlorine, $Cl_2(g)$, is manufactured industrially by oxidizing chloride ion, Cl^-, by removing electrons in a process called **electrolysis**, in which an electric current is passed through molten sodium chloride or through a concentrated aqueous solution of sodium chloride. Here we discuss only the electrolysis of molten sodium chloride; we leave the discussion of aqueous solutions until Chapter 13.

Two graphite or metal rods called *electrodes* are inserted into the molten sodium chloride and connected to a battery or other source of electric current (Figure 4.6). The current source drives electrons into one of the electrodes, called the negative electrode, or **cathode**. At the surface of the cathode, these electrons are transferred to Na^+ ions in the molten NaCl. The product of this electron transfer is sodium, which floats to the top of the molten salt. Thus, Na^+ ions are reduced to Na atoms in a reduction half-reaction that takes place at the cathode:

$$Na^+ + e^- \rightarrow Na$$

Reduction takes place at the cathode, and oxidation occurs at the anode.

The electrons used up in this half-reaction are replaced at the other electrode, called the positive electrode, or **anode**. Chloride ions, Cl^-, in the molten NaCl are attracted to the anode, where they give up electrons to form Cl atoms. The Cl atoms combine to form $Cl_2(g)$, which bubbles off from the anode, and the electrons flow back to the battery. In the oxidation half-reaction that takes place at the anode, Cl^- ions are oxidized to Cl_2:

$$2Cl^- \rightarrow Cl_2 + 2e^-$$

The two half-reactions in an electrolysis are separated in space, but both must occur.

Preparation of the Other Halogens Because chlorine, $Cl_2(g)$, is manufactured in large amounts and is relatively inexpensive, it is economical to use chlorine to prepare bromine from the bromide ion found in the oceans and in subterranean brines:

$$Cl_2(g) + 2Br^-(aq) \rightarrow Br_2(l) + 2Cl^-(aq)$$

Chlorine and steam are passed through the concentrated aqueous bromide solution. The bromine formed is carried off from the solution with steam and excess chlorine. It is condensed from this mixture and purified by distillation. Iodine can be obtained from iodides by the same method.

Because the fluoride ion, F^-, is so difficult to oxidize, the only practical method of making fluorine, F_2, is the electrolysis of molten sodium fluoride or of a solution of sodium fluoride in liquid hydrogen fluoride:

$$2F^- \rightarrow F_2 + 2e^-$$

Exercise 4.4 Among the species (molecules or ions) I_2, I^-, Cl_2, F_2, and F^-, which is the strongest reducing agent?

Exercise 4.5 Among the species Br_2, Br^-, Cl^-, I_2, and F^-, which is the strongest oxidizing agent?

EXAMPLE 4.2 Redox Reactions

In each of the following reactions, identify the oxidizing agent, the reducing agent, the substance that is oxidized, and the substance that is reduced.

(a) $Ca(s) + Br_2(l) \rightarrow Ca^{2+}(Br^-)_2(s)$ **(b)** $4Li(s) + O_2(g) \rightarrow 2(Li^+)_2O^{2-}(s)$

(c) $Fe(s) + S(s) \rightarrow Fe^{2+}S^{2-}(s)$ **(d)** $2Fe^{2+}(aq) + Cl_2(g) \rightarrow 2Fe^{3+}(aq) + 2Cl^-(aq)$

Solution: **(a)** The Ca loses electrons to become Ca^{2+}. It is therefore a reducing agent and is itself oxidized.

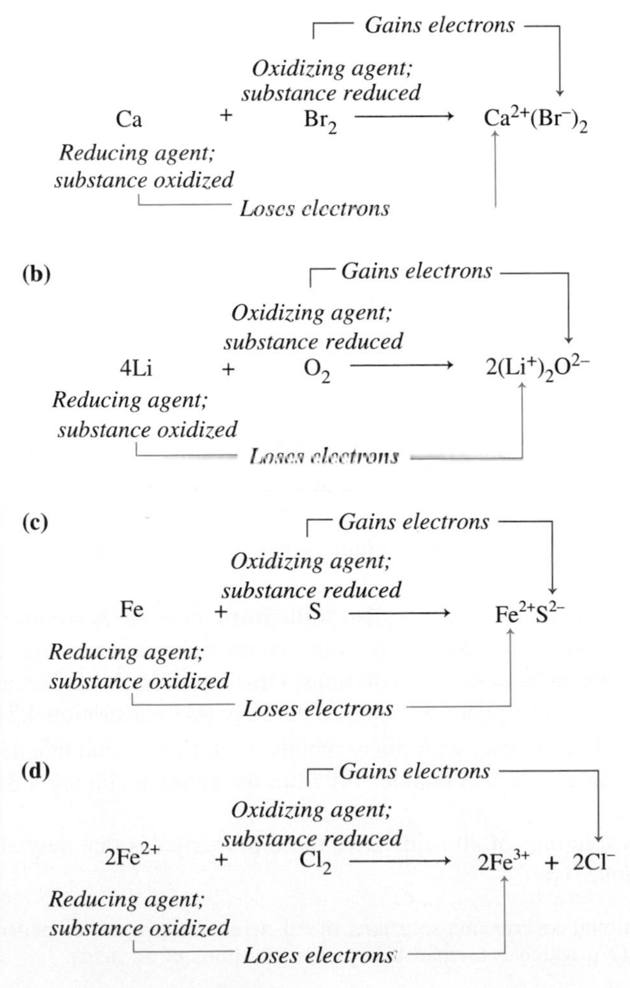

Exercise 4.6 In each of the following reactions, identify the oxidizing agent, the reducing agent, the substance that is oxidized, and the substance that is reduced.

(a) $2Cu(s) + O_2(g) \rightarrow 2CuO(s)$ **(b)** $2Cs(s) + I_2(s) \rightarrow 2CsI(s)$

(c) $Zn(s) + S(s) \rightarrow ZnS(s)$ **(d)** $Zn(s) + Cu^{2+}(aq) \rightarrow Zn^{2+}(aq) + Cu(s)$

4.4 ACID–BASE REACTIONS

Aqueous Solutions of Hydrogen Halides: The Hydronium Ion

Hydrogen chloride and the other hydrogen halides are typical molecular covalent substances. They are gases at room temperature and are nonconductors of electricity in both the liquid and solid states. But unlike many molecular covalent substances, the hydrogen halides are very soluble in water, and they form solutions that are very good conductors of electricity. These solutions must therefore contain ions formed by a reaction between the hydrogen halide and water.

Because the differences in electronegativity between hydrogen ($\chi_H = 2.2$) and either chlorine ($\chi_{Cl} = 2.8$) or oxygen ($\chi_O = 3.5$) are substantial, both HCl and H_2O are polar molecules. When HCl is dissolved in water, the positively charged H atom of H—Cl is attracted to the negatively charged O atom of H_2O. The H atom then leaves behind the electron pair of the H—Cl bond and is transferred as a proton (hydrogen ion, H^+) to a lone pair of electrons on the O atom of the water molecule:

The symbols δ^+ and δ^- are small equal but opposite charges. In the water molecule, each hydrogen atom has a positive charge of δ^+. Because the water molecule has an overall zero charge, the charge on the oxygen atom must be $2\delta^-$.

$$^{\delta-}\!:\ddot{\underset{..}{Cl}}—H^{\delta+} + {}^{2\delta-}\!:\ddot{O}(—H^{\delta+})—H^{\delta+} \rightarrow :\ddot{\underset{..}{Cl}}:^- + H—\ddot{O}^+(—H)—H$$

The products of this proton transfer are a chloride ion, Cl^-, and a **hydronium ion, H_3O^+**. The hydronium ion has the same Lewis structure as the ammonia molecule, NH_3, and both are pyramidal AX_3E molecules (Figure 4.7).

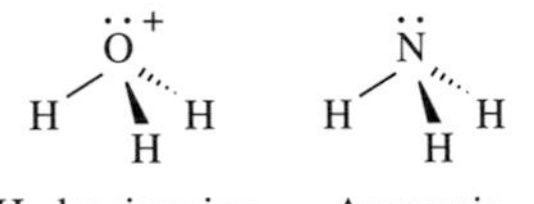

FIGURE 4.7 Triangular Pyramidal AX_3E Molecules

Acids

Aqueous solutions of HCl and the other hydrogen halides have some properties that have long been used to identify them as **acids**:

- Aqueous solutions of acids have a sour taste. Vinegar, for example, is an aqueous solution of acetic acid, and lemon juice is an aqueous solution containing citric acid.
- Acids change the colors of substances called **indicators**, or dyes. A common indicator is litmus, a substance extracted from certain lichens, which has a characteristic red color in aqueous acid solutions. Other substances that act as indicators are found in grape juice, tea, and red cabbage (Demonstration 4.7).
- Aqueous solutions of acids react with many metals, such as zinc and magnesium, producing hydrogen, which bubbles off from the solution (Figure 4.8).

The reason that aqueous solutions of all acids have these properties is that they all contain the hydronium ion, H_3O^+.

The properties common to aqueous solutions of all acids are properties of the hydronium ion, H_3O^+, which is formed in aqueous solutions of all acids.

DEMONSTRATION 4.7 Effect of Acids and Bases on Some Natural Indicators

The middle tube contains purple grape juice. Acid has been added to the left-hand tube, changing the color to red. Base has been added to the right-hand tube, changing the color to olive green.

When lemon juice is added to tea, the color changes from brown to yellow-orange.

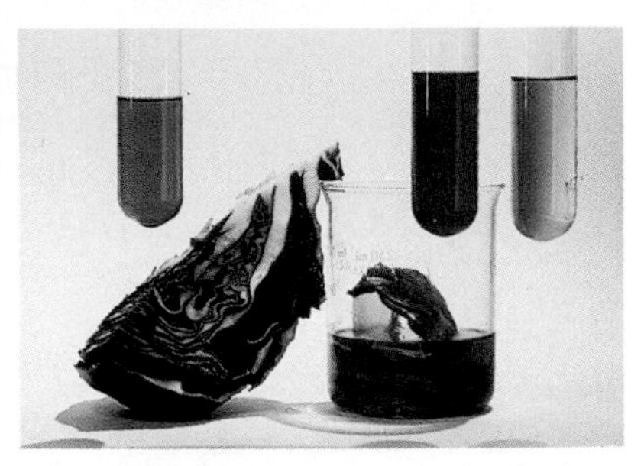

In the beaker the natural color of red cabbage is being extracted into methanol. The middle tube contains the red methanol extract of red cabbage. HCl (aq) has been added to the left-hand tube, changing the color to pink. NaOH(aq) has been added to the right-hand tube, changing the color to green.

We may write the formula of an acid as HA, where A is an electronegative atom or group of atoms (such as Cl or ClO_4). The HA bond is therefore polar:

$$H^{\delta+}—A^{\delta-}$$

In a solution of HA in water, the $H^{\delta+}$ atom of H—A is attracted to the $O^{2\delta-}$ atom of a water molecule and then transferred as H^+ to form H_3O^+ and A^-, as we described above for HCl(aq). In general we can represent the reaction of an acid with water by the equation

$$HA + H_2O \rightarrow H_3O^+ + A^-$$

For example,

$$HCl + H_2O \rightarrow H_3O^+ + Cl^-$$

$$HClO_4 + H_2O \rightarrow H_3O^+ + ClO_4^-$$

A solution of HCl(g) in water, HCl(aq), is called **hydrochloric acid**. Perchloric acid, $HClO_4$, is described in Chapter 7.

All acids react with water to form the hydronium ion, H_3O^+.

Because in the reaction of any acid with water, the acid transfers a proton, H^+, to a water molecule to give H_3O^+, such a reaction is called a **proton-transfer reaction**.

Brønsted-Lowry Definition of Acids and Bases In 1923 the Danish chemist Johannes Brønsted (1879–1947) and the English chemist Thomas Lowry (1874–1936) proposed that an **acid** be defined as a molecule or ion that can transfer a proton to a suitable molecule or ion that is called a **base**. In short,

An acid is a proton donor.
A base is a proton acceptor.

Therefore, a proton-transfer reaction is an **acid–base reaction**.

FIGURE 4.8 Acids React with Metals Magnesium reacts with dilute hydrochloric acid, producing bubbles of $H_2(g)$.

When an acid is dissolved in water, there is an acid–base reaction in which water behaves as a base—that is, it accepts a proton from the acid:

$$\underset{\textit{Acid}}{HA(aq)} + \underset{\textit{Base}}{H_2O(l)} \rightarrow H_3O^+(aq) + A^-(aq)$$

Many other substances can behave as bases, as we shall soon see. An essential feature of a base is that, like the water molecule, it has an unshared pair of electrons to which a proton can be added. The equation for the reaction of an acid with water is sometimes written in the form

$$HA(aq) \rightarrow H^+(aq) + A^-(aq)$$

The acid is commonly said to **ionize** in water. Remember, however, that in this equation $H^+(aq)$ is merely an abbreviation for $H_3O^+(aq)$. There is no evidence for the existence of free hydrogen ions—that is, free protons—in any solutions. Because the proton is a bare nucleus and is therefore extremely small, it can approach an unshared electron pair very closely and is then held very strongly by the electron pair. Consequently, a proton in water becomes attached to an unshared pair of electrons on the oxygen atom of a water molecule, and the hydronium ion, H_3O^+, is formed.

Strong and Weak Acids The reaction of an acid with water, like all reactions, is an equilibrium:

$$HA(aq) + H_2O(l) \rightleftharpoons H_3O^+(aq) + A^-(aq)$$

For hydrochloric acid, HCl(*aq*), the position of equilibrium lies far to the right. Thus, HCl is said to be completely ionized in water:

$$HCl(aq) + H_2O(l) \rightarrow H_3O^+(aq) + Cl^-(aq)$$

Acids that react completely with water in this way are called **strong acids**.

Strong acids are completely ionized in water.

Because solutions of strong acids contain a large concentration of ions, they have a high electrical conductivity (Demonstration 4.8). In contrast, the reaction of hydrofluoric acid, HF, with water,

$$HF(aq) + H_2O(l) \rightleftharpoons H_3O^+(aq) + F^-(aq)$$

is incomplete. Equilibrium is reached with a substantial concentration of un-ionized HF molecules remaining in the solution. The position of equilibrium lies on the left, and there are only small concentrations of H_3O^+ and F^-. The electrical conductivity of an aqueous solution of HF is therefore much smaller than that of an HCl solution of the same concentration. Acids such as HF that react incompletely with water are called **weak acids**.

Weak acids are incompletely ionized in water.

Table 4.5 lists some strong and weak acids with which you should become familiar. Sulfuric, hypochlorous, chloric, and perchloric acids are discussed in Chapter 7. The *only* common strong acids are the acids given in this table; almost all other acids are weak. If you remember the common strong acids, you can assume that any other acid you encounter is weak.

DEMONSTRATION 4.8 Electrical Conductivity of Acid Solutions

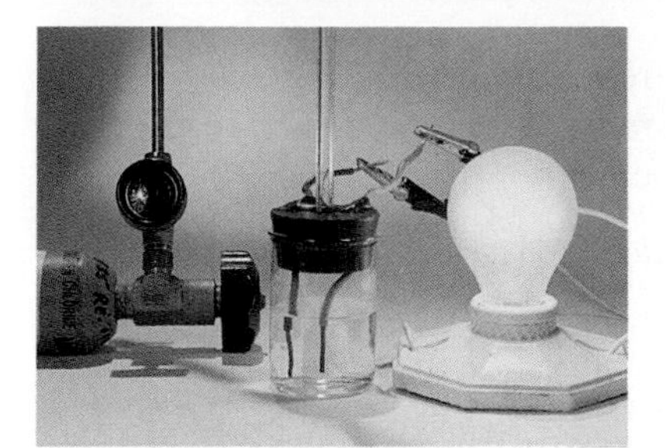

A. Two metal electrodes connected to a power supply are dipped into a beaker of water. When $HCl(g)$ is added to the water, the light bulb in the circuit glows brightly. The large electrical conductivity of the solution shows that many ions are present. Water behaves as a base, accepting protons from the strong acid $HCl(g)$ to give $H_3O^+(aq)$ and $Cl^-(aq)$.

B. The left-hand beaker contains a solution of HF and the right-hand beaker a solution of acetic acid, CH_3CO_2H. The light bulb glows dimly, showing that the solution has only a small electrical conductivity. Acetic acid and hydrofluoric acid are weak acids and are only slightly ionized in water.

Acetic acid, $C_2H_4O_2$, is a common weak acid (Demonstration 4.8). The ionization of acetic acid may be written as

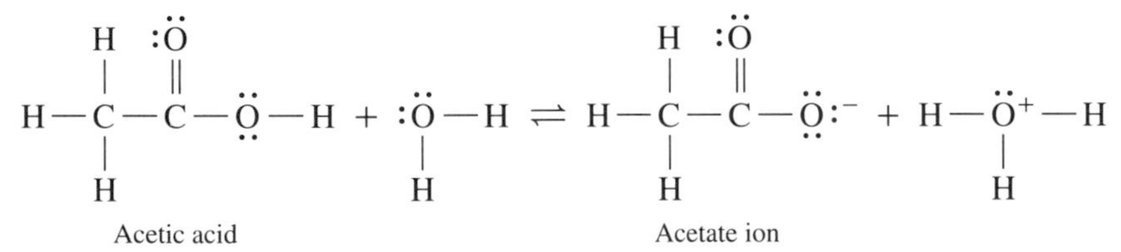

or more simply as

$$CH_3CO_2H(aq) + H_2O(l) \rightleftharpoons CH_3CO_2^-(aq) + H_3O^+(aq)$$

The anion $CH_3CO_2^-$ is called the **acetate ion.**

Acetic acid belongs to a family of acids called *carboxylic acids* (Chapter 14). Like the other carboxylic acids, acetic acid has both C—H bonds and O—H bonds. The O—H bond has a considerable polarity, and it is the hydrogen of this bond that

TABLE 4.5 Some Common Acids

Strong Acids		Weak Acids	
Hydrochloric acid	HCl	Hydrofluoric acid	HF
Hydrobromic acid	HBr	Carbonic acid	H_2CO_3
Hydroiodic acid	HI	Phosphoric acid	H_3PO_4
Sulfuric acid	H_2SO_4	Acetic acid	CH_3CO_2H
Nitric acid	HNO_3	Hypochlorous acid	$HOCl$
Perchloric acid	$HClO_4$	Boric acid	H_3BO_3
Chloric acid	$HClO_3$		

is transferred to a water molecule. Because the electronegativities of C ($\chi_C = 2.5$) and H ($\chi_H = 2.2$) are similar (Figure 3.11), C—H bonds have only a very small polarity. So these hydrogen atoms have only a very small positive charge, and they have no tendency to be transferred to the oxygen atom of a water molecule.

Redox Reactions of Acids Not all the reactions of the hydronium ion are acid–base reactions; this ion also takes part in redox (electron-transfer) reactions. In the reaction of magnesium with an aqueous acid,

$$Mg(s) + 2H_3O^+(aq) \rightarrow Mg^{2+}(aq) + H_2(g) + 2H_2O(l)$$

each magnesium atom loses two electrons to become Mg^{2+}:

$$Mg(s) \rightarrow Mg^{2+}(aq) + 2e^-$$

Magnesium is therefore oxidized. At the same time the hydronium ion is reduced to molecular hydrogen. If we write the equation in the simpler form

$$Mg(s) + 2H^+(aq) \rightarrow Mg^{2+}(aq) + H_2(g)$$

we see that two hydrogen ions both acquire an electron to form a hydrogen molecule:

$$2H^+(aq) + 2e^- \rightarrow H_2(g)$$

The reactions of metals with the hydronium ion in aqueous acid solutions are therefore oxidation–reduction reactions, and $H^+(aq)$ is an oxidizing agent.

Bases

We have seen above that according to the Brønsted–Lowry definition,

A base is a molecule or ion that is a proton acceptor.

We saw also that the water molecule can behave as a base: It can accept a proton from an acid to form the H_3O^+ ion. A common feature of all bases is that, like the water molecule, they have a lone pair of electrons to which a proton can be transferred to form a covalent bond. If B is a base, then

$$B: + H—A \rightarrow B—H^+ + :A^-$$

For example, when HCl(g) and $NH_3(g)$ are allowed to mix, a dense white cloud of the solid ionic compound ammonium chloride, $NH_4Cl(s)$, is formed (Figure 4.9):

$$HCl(g) + NH_3(g) \rightarrow NH_4{}^+Cl^-(s)$$

$$^{\delta-}:\ddot{\underset{..}{Cl}}—H^{\delta+} + {}^{3\delta-}:N(H^{\delta+})_3 \rightarrow :\ddot{\underset{..}{Cl}}:^- + H—N^+(H)_2—H$$

FIGURE 4.9 The Acid-Base Reaction between HCl and NH_3 When a stream of HCl(g) meets a stream of $NH_3(g)$, a white cloud of solid ammonium chloride, NH_4Cl, is formed.

A polar HCl molecule (the acid) transfers a proton to a lone pair on the nitrogen atom of an NH_3 molecule (the base) to give an ammonium ion and a chloride ion.

The acid–base reaction we have just described takes place in the gas phase, but we will be concerned mostly with reactions that take place in solution in water. In water, all bases react to give a *hydroxide ion*, OH^-:

$$B: + H—OH \rightarrow BH^+ + OH^-$$

In this reaction water behaves as an acid: It donates one of its protons to the base.

Water has both acid and base properties.

We have seen that many substances behave as acids in water; they donate a proton to one of the lone pairs on the oxygen atom of the water molecule to form an H_3O^+ ion. But the water molecule has polar O–H bonds and can donate a proton to a lone pair on a base. This reaction produces a BH^+ ion and a hydroxide ion OH^-, so water can also behave as an acid.

Strong and Weak Bases Ammonia, $NH_3(g)$, is very soluble in water, with which it reacts as a base. When a proton is transferred from a polar H_2O molecule (the acid) to the lone pair on the N atom of a polar ammonia molecule (the base), an ammonium ion, NH_4^+, and a hydroxide ion, OH^-, are formed:

$$^{2\delta-}{:}\ddot{\text{O}}(\text{H}^{\delta+})\text{—H}^{\delta+} + {}^{3\delta-}{:}\text{N}(\text{H}^{\delta+})_2\text{—H}^{\delta+} \rightleftharpoons {:}\ddot{\text{O}}{:}^-\text{—H} + \text{H—N}^+(\text{H})_2\text{—H}$$

or more simply

$$H_2O(l) + NH_3(aq) \rightleftharpoons NH_4^+(aq) + OH^-(aq)$$

The reaction of ammonia with water does not go to completion, so ammonia is described as a **weak base**; only small concentrations of NH_4^+ and OH^- are formed at equilibrium, and a large amount of NH_3 remains un-ionized.

When the reaction between a base and water goes to completion, the base is said to be a **strong base**. When the reaction between a base and water reaches equilibrium before it goes to completion, the base is weak.

A strong base is a molecule or ion that is fully ionized in water to give OH^-. A weak base is a molecule or ion that is incompletely ionized in water to give OH^-.

It is not surprising to find that many negatively charged ions (anions) strongly attract protons and behave as bases. Examples of strong bases in water are the oxide ion, O^{2-}, and the hydride ion, H^-. All soluble ionic oxides are strong bases because, in aqueous solution, O^{2-} is completely converted to hydroxide ion:

$$:\ddot{\underset{\cdot\cdot}{\text{O}}}:^{2-} + \text{H—}\ddot{\underset{\cdot\cdot}{\text{O}}}\text{—H} \rightarrow {}^-{:}\ddot{\underset{\cdot\cdot}{\text{O}}}\text{—H} + {}^-{:}\ddot{\underset{\cdot\cdot}{\text{O}}}\text{—H}$$

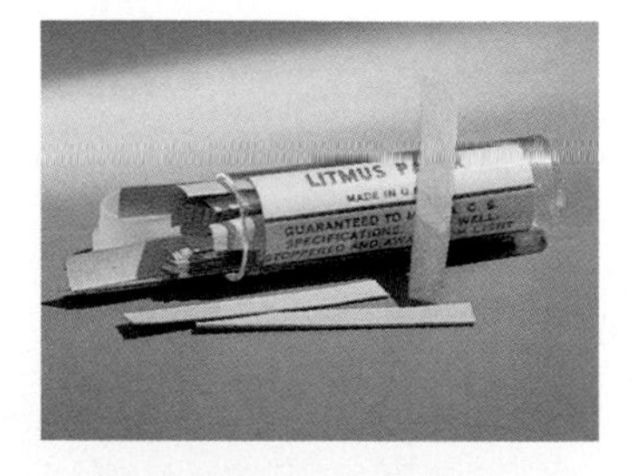

Red litmus paper turns mauve when dipped into basic (alkaline) solution.

All the alkali metal oxides are strong bases in water. They dissolve in water to give solutions of the corresponding alkali metal hydroxide, $M^+OH^-(aq)$, where M stands for any metal. For example,

$$(Na^+)_2O^{2-}(s) + H_2O(l) \rightarrow 2Na^+(aq) + 2OH^-(aq)$$

Solid alkali metal hydroxides, such as NaOH(s) and KOH(s), are ionic compounds composed of alkali metal cations and hydroxide ions. These solids are strong bases in water. For example,

$$K^+OH^-(s) \rightarrow K^+(aq) + OH^-(aq)$$

Metal hydrides are strong bases in water because the hydride ion, $:H^-$, accepts a proton from water completely to give $H_2(g)$ and hydroxide ion:

$$\text{H:}^- + \text{H—}\ddot{\underset{\cdot\cdot}{\text{O}}}\text{—H} \rightarrow \text{H—H} + {}^-{:}\ddot{\underset{\cdot\cdot}{\text{O}}}\text{—H}$$

A base need not contain OH^-; the hydroxide ion can originate from a water molecule. When water acts as an acid and donates H^+, OH^- remains.

For example, sodium hydride reacts completely with water to give $H_2(g)$ and a solution of sodium hydroxide:

$$NaH(s) + H_2O(l) \rightarrow Na^+(aq) + OH^-(aq) + H_2(g)$$

Many other anions also behave as bases in water. For example, the anion of a weak acid such as acetic acid is a weak base:

$$CH_3CO_2^-(aq) + H_2O(l) \rightleftharpoons CH_3CO_2H(aq) + OH^-(aq)$$

We discuss other examples of weak acids and bases in Chapter 12, where we consider these equilibria in a quantitative way.

Acid-Base Reactions in Aqueous Solution

Let us consider two acid–base reactions in aqueous solution:

$$NaOH(aq) + HCl(aq) \rightarrow NaCl(aq) + H_2O(l)$$

$$\underset{\textit{Base}}{KOH(aq)} + \underset{\textit{Acid}}{HBr(aq)} \rightarrow \underset{\textit{Salt}}{KBr(aq)} + \underset{\textit{Water}}{H_2O(l)}$$

The products of these reactions are a salt and water. A **salt** consists of a positive ion (cation) derived from a base and a negative ion (anion) derived from an acid. The salts formed in these two reactions are the alkali metal halides: sodium chloride, Na^+Cl^-, and potassium bromide, K^+Br^-. These salts can be obtained as solids by evaporating the solutions.

The actual reaction that occurs in an acid–base reaction in aqueous solution is the same in every case, as we will now show. Because NaOH consists of Na^+ and OH^- ions and KOH consists of K^+ and OH^- ions, and because HCl and HBr ionize in water to give the H_3O^+, Cl^-, and Br^- ions, we may rewrite the two previous equations as

$$Na^+(aq) + OH^-(aq) + H_3O^+(aq) + Cl^-(aq) \rightarrow Na^+(aq) + Cl^-(aq) + 2H_2O(l)$$

and

$$K^+(aq) + OH^-(aq) + H_3O^+(aq) + Br^-(aq) \rightarrow K^+(aq) + Br^-(aq) + 2H_2O(l)$$

Canceling the ions that occur on both sides of these equations (ions that take no part in the reaction), we obtain the much simpler equation

$$H_3O^+(aq) + OH^-(aq) \rightarrow 2H_2O(l)$$

for both reactions. We can simplify the reaction for any acid–base reaction in aqueous solution in the same way.

The reaction between the hydronium ion, H_3O^+, and the hydroxide ion, OH^-, occurs when any acid reacts with any base in aqueous solution.

The ions that do not take part in this reaction, Na^+ and Cl^- in the first example and K^+ and Cl^- in the second, simply remain in solution unchanged. They are often called **spectator ions**.

The effect of adding a base to an aqueous solution of an acid is therefore to remove the H_3O^+—in other words, to destroy the acid properties of the solution and to reverse the change in the color of an indicator produced by an acid. The base is said to *neutralize* the acid (Demonstration 4.7).

Exercise 4.7 Write an equation for an acid–base reaction by which a solution of each of the following salts could be prepared.

(a) LiI **(b)** $CaBr_2$ **(c)** $Mg(NO_3)_2$

Polyatomic Ions: Lewis Structures

A **polyatomic ion** is an ion that has two or more atoms; in other words, it is a charged molecule. Earlier, we encountered the polyatomic ions NH_4^+, H_3O^+, and OH^-, with the respective Lewis structures

$$\begin{array}{c}H\\|\\H—N^+—H\\|\\H\end{array} \qquad \begin{array}{c}H—\ddot{O}^+—H\\|\\H\end{array} \qquad {}^-:\ddot{\underset{..}{O}}—H$$

We drew the Lewis structures of many neutral molecules in Chapter 3. In Example 4.3 we see how to draw the Lewis structures of polyatomic ions.

EXAMPLE 4.3 Lewis Structures of Polyatomic Ions

Draw the Lewis structures of **(a)** NH_4^+, **(b)** H_3O^+, and **(c)** OH^-. **(d)** Express each structure in terms of the AX_nE_m nomenclature, and give the shape of each.

Solution: **(a)** The Lewis symbol for N is $:\dot{\underset{\cdot}{N}}\cdot$, with three unpaired electrons. For NH_4^+ we need to create a positive charge overall and to form four bonds to nitrogen. To do this, we first remove an electron from the valence shell of $:\dot{\underset{\cdot}{N}}\cdot$ and obtain $\cdot\dot{\underset{\cdot}{N}}\cdot^+$ with four unpaired electrons (Step 1). Then we form single bonds between N^+ and each of four H atoms.

In drawing Lewis structures for simple polyatomic ions, always start with the central atom.

Step 1 $\quad :\dot{\underset{\cdot}{N}}\cdot \rightarrow \cdot\dot{\underset{\cdot}{N}}\cdot^+ + e^-$

Step 2 $\quad \cdot\dot{\underset{\cdot}{N}}\cdot^+ + 4\cdot H \rightarrow \begin{array}{c}H\\|\\H—N^+—H\\|\\H\end{array}$

(b) Similarly, for H_3O^+:

Step 1 $\quad :\ddot{\underset{\cdot}{O}}\cdot \rightarrow :\dot{\underset{\cdot}{O}}\cdot^+ + e^-$

Step 2 $\quad :\dot{\underset{\cdot}{O}}\cdot^+ + 3\cdot H \rightarrow \begin{array}{c}H\\|\\:O^+—H\\|\\H\end{array}$

(c) For OH^-, we first *add* an electron to the valence shell of the O atom and then form one O—H bond:

Step 1 $\quad :\ddot{\underset{\cdot}{O}}\cdot + e^- \rightarrow :\ddot{\underset{..}{O}}\cdot^-$

Step 2 $\quad :\ddot{\underset{..}{O}}\cdot^- + \cdot H \rightarrow {}^-:\ddot{\underset{..}{O}}—H$

(d) NH_4^+: AX_4, tetrahedral; H_3O^+: AX_3E, triangular pyramidal; OH^-: AXE_3, linear.

We see that the Lewis structures of these polyatomic ions are no more difficult to draw than those of uncharged molecules, provided that we first add or remove the appropriate number of electrons corresponding to the charge on the ion.

The charges that appear on the nitrogen and oxygen atoms in the Lewis structures of NH_4^+, H_3O^+, and OH^- are called *formal charges.* A formal charge on an atom in a polyatomic ion is not the real charge on the atom, because real charge depends on the differences in the electronegativities of the atoms in the ion. In assigning formal charges to the atoms in a Lewis structure, we ignore these electronegativity differences. Formal charges should therefore not be confused with the small charges δ^+ and δ^- that we assigned to the atoms in a molecule such as water, in which the bonds are polar due to electronegativity differences. We discuss formal charges more fully in Chapter 7.

4.5 PRECIPITATION REACTIONS

We have seen that the chlorides of the alkali and alkaline earth metals are very soluble in water (Table 4.4). But the chlorides of some other metals, such as silver and lead, have very small solubilities. Thus, if we mix a solution of a soluble silver salt such as silver nitrate, $AgNO_3$, and a solution of sodium chloride, we will have higher concentrations of silver ion and chloride ion than we would expect from the very small solubility of silver chloride. Thus, solid silver chloride must separate from the solution (Demonstration 4.9):

$$AgNO_3(aq) + NaCl(aq) \rightarrow AgCl(s) + NaNO_3(aq)$$

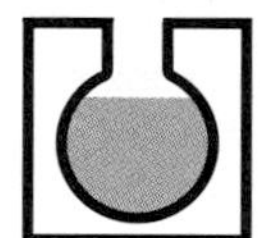

DEMONSTRATION 4.9
Precipitation Reactions

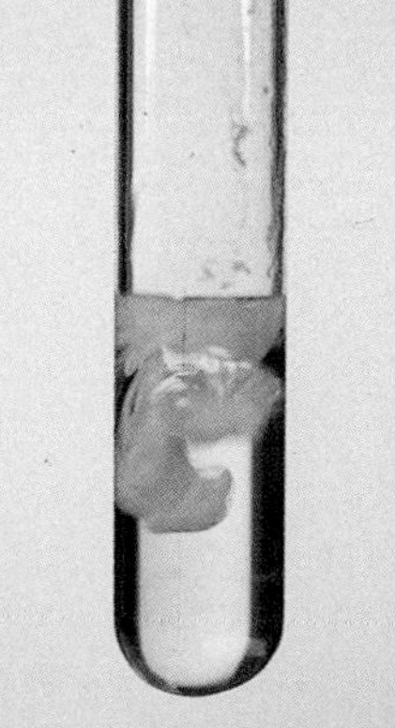

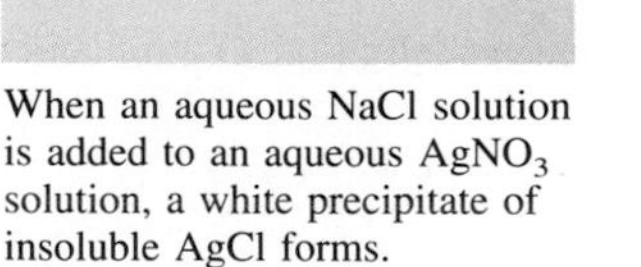

When an aqueous NaCl solution is added to an aqueous $AgNO_3$ solution, a white precipitate of insoluble AgCl forms.

When an aqueous solution of NaI is added to an aqueous solution of $AgNO_3$, a pale yellow precipitate of insoluble AgI forms.

The solid silver chloride formed in this way is called a **precipitate**.

A *precipitation reaction* is a reaction in which a solid compound is formed when solutions of two soluble compounds are mixed.
A precipitate is the insoluble compound formed in a precipitation reaction.

Because all the compounds involved are ionic, we can write the equation for the reaction between solutions of silver nitrate and sodium chloride in terms of their ions:

$$Ag^+(aq) + NO_3^-(aq) + Na^+(aq) + Cl^-(aq) \rightarrow (Ag^+Cl^-)(s) + Na^+(aq) + NO_3^-(aq)$$

The ions Na^+ and NO_3^- appear on both sides of the equation. These ions take no part in the reaction; they are spectator ions. We can cancel them from the equation:

$$Ag^+(aq) + Cl^-(aq) \rightarrow (Ag^+Cl^-)(s)$$

This equation is called a **net ionic equation**. It is usually written as

$$Ag^+(aq) + Cl^-(aq) \rightarrow AgCl(s)$$

where the formula for silver chloride does not show the charges on the ions. Like all reactions, this is an equilibrium, but the position of equilibrium lies far to the right. Only exceedingly small concentrations of silver ion and chloride ion are in equilibrium with solid silver chloride. If more than the very small equilibrium concentrations of these ions are present in solution, a precipitate of solid silver chloride forms immediately. This net ionic equation represents the reaction that occurs when a solution of *any* soluble silver salt is mixed with a solution of *any* soluble chloride. For example, when we mix solutions of silver perchlorate and potassium chloride and a precipitate of silver chloride forms,

$$AgClO_4(aq) + KCl(aq) \rightarrow AgCl(s) + KClO_4(aq)$$

the only reaction that occurs is the reaction represented by the previous net ionic equation.

Like silver chloride, silver bromide and silver iodide are insoluble. We obtain precipitates of silver bromide and silver iodide when we add any aqueous solutions containing silver ion to aqueous solutions containing bromide or iodide ion. Because most other silver salts are soluble, the formation of these precipitates is a good test for chloride, bromide, or iodide ions (Demonstration 4.9).

The equation

$$H_3O^+ + OH^- \rightarrow 2H_2O$$

for the reaction between an acid and a base in aqueous solution is another net ionic equation.

Solubility Rules

There is no general theory that enables us to predict the solubility of a substance. Fortunately, there are some simple rules that enable us to make fairly reliable predictions about the solubilities of ionic substances in aqueous solution (Table 4.6). These rules apply to all compounds composed of the common anions and cations discussed in this book, and you will find it useful to remember them. Although some substances are described in Table 4.6 as being insoluble, no substance is *completely* insoluble. At least a very small amount of any substance will always dissolve. But in the case of substances that are called insoluble, the amount is negligible for most purposes. So the definition of *insoluble* is somewhat arbitrary. We describe any substance that has a solubility less than 0.01 mol · L^{-1} as *insoluble*.

III In memorizing the solubility rules, it is convenient to learn first the generalizations (the first column in Table 4.6) and then the exceptions (the second and third columns in Table 4.6).

TABLE 4.6 Solubilities of Some Common Salts and Hydroxides in Water at 25°C

	Exceptions	
Soluble*	**Insoluble***	**Sparingly Soluble***
Na^+, K^+, and NH_4^+ salts		
Nitrates (NO_3^-)		
Perchlorates (ClO_4^-)		
Fluorides (F^-)	Mg^{2+}, Ca^{2+}, Sr^{2+}, Ba^{2+}, Pb^{2+}	
Chlorides, bromides, and iodides	Ag^+, PbI_2	$PbCl_2$, $PbBr_2$
Sulfates (SO_4^{2-})	Sr^{2+}, Ba^{2+}, Pb^{2+}	Ca^{2+}, Ag^+
Acetates ($CH_3CO_2^-$)		Ag^+

	Exceptions	
Insoluble	**Soluble**	**Sparingly Soluble**
Carbonates (CO_3^{2-})	Na^+, K^+, NH_4^+	
Phosphates (PO_4^{3-})	Na^+, K^+, NH_4^+	
Sulfides (S^{2-})	Na^+, K^+, NH_4^+, Mg^{2+}, Ca^{2+}, Sr^{2+}, Ba^{2+}	
Hydroxides (OH^-)	Na^+, K^+, NH_4^+, Ba^{2+}	Ca^{2+}, Sr^{2+}

* The classification *soluble, sparingly soluble*, and *insoluble* is somewhat arbitrary. The approximate designations are: soluble, $\geqslant 0.1\ mol \cdot L^{-1}$; sparingly soluble, $\leqslant 0.1\ mol \cdot L^{-1}$ and $\geqslant 0.01\ mol \cdot L^{-1}$; insoluble, $\leqslant 0.01\ mol \cdot L^{-1}$.

EXAMPLE 4.4 Solubility Rules

Use the solubility rules in Table 4.6 to predict which of the following substances are insoluble and which are soluble.
(a) KBr **(b)** $NaNO_3$ **(c)** MgF_2 **(d)** $BaSO_4$ **(e)** $FeCl_2$

Solution: **(a)** KBr is soluble, because all K^+ salts are soluble. Br^- salts are also soluble, with a few exceptions that do not include K^+.
(b) $NaNO_3$ is soluble, because all Na^+ salts are soluble. All NO_3^- salts are also soluble.
(c) MgF_2 is insoluble. Although most fluorides are soluble, MgF_2 is one of the exceptions.
(d) $BaSO_4$ is insoluble. Although most sulfates are soluble, $BaSO_4$ is one of the exceptions.
(e) $FeCl_2$ is soluble. Chlorides are soluble, with a few exceptions that do not include Fe^{2+}.

Exercise 4.8 Use the solubility rules to predict which of the following substances are soluble and which are insoluble.
(a) $CaCO_3$ **(b)** PbI_2 **(c)** LiBr **(d)** CuO **(e)** $MgSO_4$

EXAMPLE 4.5 Precipitation Reactions

Use the solubility rules to predict whether a precipitate will form when the following pairs of aqueous solutions are mixed. Explain and write a net ionic equation for those cases in which a precipitate forms.
(a) $MgCl_2$ and Na_3PO_4 **(b)** $Al_2(SO_4)_3$ and $BaCl_2$ **(c)** $Ba(NO_3)_2$ and $MgCl_2$

Solution: **(a)** Possible products are $Mg_3(PO_4)_2$ and NaCl. Because $Mg_3(PO_4)_2$ is insoluble, a precipitation reaction will occur:

$$3Mg^{2+}(aq) + 2PO_4^{3-}(aq) \rightarrow Mg_3(PO_4)_2(s)$$

(b) Possible products are $Al(NO_3)_3$ and $BaSO_4$. Because $BaSO_4$ is insoluble, a precipitation reaction will occur:

$$Ba^{2+}(aq) + SO_4^{2-}(aq) \rightarrow BaSO_4(s)$$

(c) Possible products are $BaCl_2$ and $Mg(NO_3)_2$. Both of these compounds are soluble, so there will be no precipitation reaction. The solution will consist of a mixture of Ba^{2+}, Cl^-, Mg^{2+}, and NO_3^-.

You must know the solubility rules to predict the solubilities of the potential products in a precipitation reaction.

Exercise 4.9 Use the solubility rules to predict whether a precipitate will form when the following pairs of aqueous solutions are mixed.
(a) $MgCl_2$ and NaF **(b)** $MgSO_4$ and NaCl **(c)** $FeCl_2$ and Na_2S

4.6 REACTION TYPES

In this chapter we have discussed oxidation–reduction, acid–base, and precipitation reactions as well as synthesis and decomposition reactions. It is important to be able to identify the main reaction types, so we will first summarize their principal characteristics. Then we will show how to decide the type to which a given reaction belongs.

1. In *oxidation–reduction reactions*, electrons are transferred from one molecule or ion to another molecule or ion. For example, in the reaction

$$2Br^-(aq) + Cl_2(aq) \rightarrow Br_2(aq) + 2Cl^-(aq)$$

electrons are transferred from Br^- ions (the reducing agent) to Cl_2 molecules (the oxidizing agent). Br^- is oxidized to Br_2, and Cl_2 is reduced to Cl^-.

2. In *acid–base reactions*, protons (hydrogen ions), H^+, are transferred from one molecule or ion to another molecule or ion. For example, in the reaction

$$HBr(aq) + H_2O(l) \rightarrow H_3O^+(aq) + Br^-(aq)$$

protons are transferred from HBr (the acid) to H_2O (the base). Or in the reaction between an aqueous acid and an aqueous base, protons are transferred from H_3O^+ to OH^-:

$$H_3O^+(aq) + OH^-(aq) \rightarrow 2H_2O(l)$$

3. In *precipitation reactions*, an insoluble solid is formed upon mixing solutions of soluble substances. For example, in the reaction

$$KBr(aq) + AgNO_3(aq) \rightarrow AgBr(s) + KNO_3(aq)$$

a precipitate of insoluble silver bromide, AgBr, is formed when aqueous solutions of KBr and $AgNO_3$ are mixed.

4. In a *synthesis reaction*, two or more simpler substances combine to give a more complicated substance. An example is a reaction in which two elements combine to form a compound. Chemists frequently talk about synthesizing a compound, by which they mean preparing it from simpler, readily available substances. For example, methanol, CH_3OH, can be synthesized from carbon

monoxide and hydrogen. Plants synthesize carbohydrates from carbon dioxide and water.

5. In a *decomposition reaction*, a substance is converted into two or more simpler substances, such as when a compound is decomposed into its elements. A decomposition reaction is the reverse of a synthesis reaction.

Both synthesis and decomposition cover an enormous number of reactions, so classifying reactions as synthesis or decomposition reactions is less useful than classifying them according to more fundamental characteristics such as acid–base, oxidation–reduction, or precipitation reactions. For example, the combination of a metal with a nonmetal to give an ionic compound such as sodium chloride can be described as a synthesis reaction, but it is usually more useful to describe it as a redox reaction. The latter description gives us information about how the reaction occurs and relates it to other reactions that occur in the same way.

EXAMPLE 4.7 Identifying Reaction Types

Identify each of the following reactions as an acid–base, an oxidation–reduction, or a precipitation reaction. In each case give reasons for your choice.

(a) $Mg(s) + F_2(g) \rightarrow MgF_2(s)$

(b) $2Ca(s) + O_2(g) \rightarrow 2CaO(s)$

(c) $Pb(NO_3)_2(aq) + 2NaCl(aq) \rightarrow PbCl_2(s) + 2NaNO_3(aq)$

(d) $HNO_3(aq) + NH_3(aq) \rightarrow NH_4^+(aq) + NO_3^-(aq)$

(e) $2Na(s) + S(s) \rightarrow Na_2S(s)$

(f) $H_3O^+(aq) + CO_3^{2-}(aq) \rightarrow HCO_3^-(aq) + H_2O(l)$

(g) $Zn(s) + 2HCl(aq) \rightarrow Zn^{2+}(aq) + 2Cl^-(aq) + H_2(g)$

Solution: (a) Oxidation–reduction reaction. The Mg loses electrons to become Mg^{2+}; it is oxidized:

$$Mg \rightarrow Mg^{2+} + 2e^-$$

The F_2 gains electrons to become $2F^-$; it is reduced:

$$F_2 + 2e^- \rightarrow 2F^-$$

(b) Oxidation–reduction reaction. Calcium oxide, CaO, is a compound of a Group II metal with a Group VI nonmetal. It is therefore ionic, $Ca^{2+}O^{2-}$. Thus, Ca loses electrons to become Ca^{2+} and is oxidized. The O_2 gains electrons to become $2O^{2-}$ and is reduced.

(c) Precipitation reaction. An insoluble solid, $PbCl_2$, is formed from an aqueous solution of two soluble substances, $Pb(NO_3)_2$ and NaCl.

(d) Acid–base reaction. Nitric acid, HNO_3, loses a proton to form the nitrate ion, NO_3^-, and NH_3 gains a proton to give the ammonium ion, NH_4^+. Thus it is a proton transfer, or acid–base, reaction.

(e) Oxidation–reduction reaction. Sodium sulfide, Na_2S, is a compound of a Group I metal and a Group VI nonmetal and is therefore ionic. Thus, Na loses an electron to become Na^+, and S gains two electrons to become S^{2-}. It is an electron transfer, or oxidation–reduction, reaction.

(f) Acid–base reaction. The H_3O^+ ion loses a proton to become H_2O; it is an acid. The carbonate ion, CO_3^{2-}, gains a proton to become the hydrogencarbonate ion, HCO_3^-; CO_3^{2-} is a base.

(g) Oxidation–reduction reaction. Although this reaction involves the acid $HCl(aq)$, it is not an acid–base reaction. The proton in HCl is not transferred to a base. In fact, Zn loses two electrons to become Zn^{2+}; it is oxidized. And two H^+ (H_3O^+) ions gain two electrons to become $H_2(g)$:

$$2H^+ + 2e^- \rightarrow H_2$$

Exercise 4.10 Identify each of the following reactions as an oxidation-reduction reaction (OR), an acid–base reaction (AB), or a precipitation reaction (P). In each case give reasons for your choice.

(a) $Mg(s) + H_2(g) \rightarrow MgH_2(s)$

(b) $Cu(s) + S(s) \rightarrow CuS(s)$

(c) $NH_3(g) + HBr(g) \rightarrow NH_4Br(s)$

(d) $CuSO_4(aq) + K_2S(aq) \rightarrow CuS(s) + K_2SO_4(aq)$

(e) $Mg(s) + 2HBr(aq) \rightarrow MgBr_2(aq) + H_2(g)$

(f) $NaH(s) + H_2O(l) \rightarrow NaOH(aq) + H_2(g)$

(g) $Ba(OH)_2(aq) + H_2SO_4(aq) \rightarrow BaSO_4(s) + 2H_2O(l)$

SUMMARY

Most reactions occur when molecules collide with sufficient kinetic energy and with appropriate orientations. Collisions are likely only between two or occasionally three molecules at a time, so many reactions involve several steps, collectively called a reaction mechanism. Reaction rates are increased by increasing the partial pressures or concentrations of the reactants, by increasing the temperature, and by using catalysts. As the concentrations of product molecules increase, so does the rate of the reverse reaction. When the rates of the forward and reverse reactions are equal, equilibrium is achieved. For reactions that appear to have gone to completion, the position of the equilibrium lies far to the right of the balanced equation.

The halogens are the elements F, Cl, Br, I, and At in Group VII of the periodic table. They are nonmetals composed of covalent X_2 molecules. They are not found naturally in elemental form, because they are too reactive. Most of the compounds of the halogens are called halides (fluorides, chlorides, bromides, and iodides). Many halides can be made by the reaction between a halogen and another element. Metal halides are ionic solids and electrical conductors when molten, whereas nonmetal halides are gases, liquids, or solids and are nonconductors (insulators) in the liquid state.

Most reactions belong to one of a few reaction types that are exemplified by reactions of the halogens. Two broad categories are synthesis and decomposition reactions. Important and more fundamental types are oxidation–reduction (redox), acid–base, and precipitation reactions.

Reactions between halogens and metals are oxidation–reduction (redox) reactions. In such reactions, the metal is a reducing agent and is oxidized (loses electrons) to give metal ions, M^{n+}, and the halogen is an oxidizing agent and is reduced (gains electrons) to give halide ions, X^-. For any oxidizing agent we can write an equation for a half-reaction that uses up electrons, such as $Cl_2 + 2e^- \rightarrow 2Cl^-$. For any reducing agent we can write an equation for a half-reaction that provides electrons, such as $Mg \rightarrow Mg^{2+} + 2e^-$. The sum of these two equations is the equation for the overall redox reaction: $Mg + Cl_2 \rightarrow Mg^{2+} + 2Cl^-$.

Group I and II halides—such as $NaX(s)$, $KX(s)$, $MgX_2(s)$, and $CaX_2(s)$, of which NaCl is the most common—occur in salt deposits, oceans, inland seas, and the human body. The halogens are prepared industrially by oxidation of the corresponding halide ion, X^-, by electrolysis or by using a more strongly oxidizing halogen as the oxidizing agent. The oxidizing strength of the halogens decreases in the order $F_2 > Cl_2 > Br_2 > I_2$. Soluble metal halides are electrolytes (give conducting solutions) in aqueous solution.

The hydrogen halides, HX, are gases with molecules that have polar covalent H—X bonds. As liquids they are nonconductors. Very soluble in water, they transfer a proton to the negatively charged oxygen atom of a polar water molecule to give conducting solutions containing hydronium ions, $H_3O^+(aq)$, and halide ions, $X^-(aq)$.

All acids react with water to give $H_3O^+(aq)$ ions, and all bases give $OH^-(aq)$. Strong acids and strong bases in water are completely ionized (the position of the equilibrium lies far to the right), whereas weak acids and weak bases are incompletely ionized. The reaction of an acid and a base gives a salt plus water—for instance, $NaOH(aq) + HBr(aq) \rightarrow NaBr(aq) + H_2O(l)$. All acid–base reactions in aqueous solution can be described by the equation $H_3O^+(aq) + OH^-(aq) \rightarrow 2H_2O(l)$, because the other ions—$Na^+(aq)$ and $Br^-(aq)$ in this example—are spectator ions.

More generally, in the Brønsted–Lowry definitions of acids and bases, an acid is a proton donor, a base is a proton acceptor, and an acid–base reaction is a proton transfer reaction between an acid and a base. Acid–base reactions occur most commonly in aqueous solution, but $NH_3(g) + HCl(g) \rightarrow NH_4Cl(s)$ is an acid–base reaction in the gas phase.

Water, which has both acid and base properties, behaves either as an acid ($H_2O + :B \rightarrow OH^- + BH^+$) or as a base ($H_2O + HA \rightarrow H_3O^+ + A^-$). The hydronium ion, H_3O^+, can behave as an oxidizing agent in redox reactions, where it is reduced to $H_2(g)$ (by, for instance, a reactive metal).

When solutions of soluble salts are mixed, they react only when the cations and anions in solution combine to give an insoluble salt (precipitate) in a precipitation reaction. The solubility rules help us remember which salts are insoluble (solubilities less than 0.01 $mol \cdot L^{-1}$).

IMPORTANT TERMS

acid (page 138)
acid–base reaction (page 139)
base (page 139)
Brønsted–Lowry definition (page 139)
catalyst (page 121)
decomposition reaction (page 122)
electrolyte (page 130)
equilibrium (page 121)
half-reaction (page 131)
halide (page 123)
halogen (page 119)
hydronium ion (page 138)
hydroxide ion (page 142)
indicator (page 138)
net ionic equation (page 147)
nonelectrolyte (page 130)
oxidation (page 132)
oxidation–reduction reaction (page 132)
oxidizing agent (page 133)
polyatomic ion (page 145)
precipitate (page 147)
precipitation reaction (page 147)
proton acceptor (page 139)
proton donor (page 139)
proton–transfer reaction (page 139)
reaction mechanism (page 120)
redox reaction (page 132)
reducing agent (page 133)
reduction (page 132)
salt (page 144)
solubility (page 135)
spectator ion (page 144)
synthesis reaction (page 122)

REVIEW QUESTIONS

1. **(a)** What are the two necessary conditions for two colliding molecules to react?

(b) Why does the speed of a reaction (the reaction rate) increase with increasing concentrations of reactants and with increasing temperature?

2. What can we say about the rates of the forward and reverse reactions of a reaction that has achieved a state of equilibrium?

3. **(a)** Why must many reactions proceed in more than one step?

(b) What is a reaction mechanism?

4. **(a)** What is a halogen? **(b)** What is a halide ion?

5. For each of the halogens, give **(a)** the principal natural source and **(b)** a method of preparation.

6. Why are molten metal halides electrical conductors, whereas liquid nonmetal halides are insulators?

7. What is an oxidation–reduction (redox) reaction?

8. Why does an aqueous solution of sodium iodide turn brown when chlorine is passed into it? Write an equation for the reaction that occurs.

9. When the brown solution obtained in Review Question 8 is shaken with carbon tetrachloride, why does the CCl_4 layer turn violet?

10. What reaction occurs when $HCl(g)$ dissolves in water?

11. What is the hydronium ion?

12. What are the Brønsted–Lowry definitions of an acid and of a base?

13. Explain the terms "strong acid" and "weak acid," and give two examples of each.

14. Give an example of an acid–base reaction that occurs in the gas phase.

15. Give an example of a reaction in which $H_3O^+(aq)$ behaves as an oxidizing agent.

16. **(a)** What ion is formed in aqueous solutions of all bases? **(b)** Why is $Na_2O(s)$ a strong base?

17. Why is ammonia in aqueous solution described as a weak base?

18. **(a)** What is a salt? **(b)** Why are almost all salts solids?

19. Write balanced equations for the preparation of each of the salts $BaCl_2(s)$ and $Na_2SO_4(s)$ from the appropriate acid and base.

20. Give an example of a precipitation reaction.

PROBLEMS

Chemical Reactions

1. Briefly explain three ways by which the rate of a reaction may be increased.

2. **(a)** What are a synthesis reaction and a decomposition reaction?

(b) Write a balanced equation for two examples of each of the reaction types in (a).

The Halogens

3. **(a)** List the halogens in order of increasing atomic number.

(b) Describe the physical states and appearances of the halogens.

(c) Explain why their melting points and boiling points increase, and their electronegativities decrease, with increasing atomic number.

4. **(a)** Explain why the halogens are not found in nature as free elements.

(b) List two compounds of each halogen that occur naturally, and give two important uses for each halogen and/or its compounds.

5. Explain why metal halides are solids with relatively high melting points and boiling points, whereas most nonmetal halides are gases, liquids, or low-melting-point solids.

6. Classify the bonds in each of the following as ionic, covalent, or polar covalent. Draw the Lewis structures for (c), (d), (f), and (i).

(a) NaCl **(b)** CaF_2 **(c)** OF_2 **(d)** PCl_3 **(e)** $MgBr_2$
(f) CCl_4 **(g)** HF **(h)** KI **(i)** ClF

7. **(a)** What experimental proof is there that metal halides are ionic solids, whereas nonmetal halides are covalent compounds?

(b) Define the terms "electrolyte" and "nonelectrolyte," and give two examples of each from among the compounds of the halogens.

8. Write balanced equations for each of the following reactions. Name each of the products.

(a) barium with chlorine **(b)** aluminum with bromine
(c) potassium with iodine **(d)** phosphorus with chlorine
(e) phosphorus with iodine

9. **(a)** Why is tap water a moderately good conductor of electricity? **(b)** How could water be purified until it is almost nonconducting?

10. Why cannot an aqueous solution of sodium fluoride be used to produce fluorine electrolytically?

Oxidation-Reduction (Redox) Reactions

11. The reaction between potassium and bromine to give potassium bromide is classified as an oxidation–reduction reaction.

(a) Write the balanced equation for this reaction.

(b) Show that oxidation–reduction involves the transfer of electrons between the reactants.

(c) In terms of the transfer of electrons, define "oxidation," "reduction," "oxidizing agent," and "reducing agent."

12. **(a)** Describe how the reactions of halogens with halide ions in aqueous solution can be used to demonstrate that the strength of halogens as oxidizing agents decreases in the series $F_2 > Cl_2 > Br_2 > I_2$. **(b)** Why is it reasonable that this order is the same order as that of decreasing electronegativity of the halogens?

13. How are **(a)** chlorine and **(b)** fluorine prepared industrially?

14. What is the most important source of bromine? Briefly describe how bromine is obtained from this natural source.

15. Complete and balance each of the following equations. If no reaction occurs, write NR. Identify the oxidizing agent and the reducing agent in each reaction.

(a) $Br_2(l) + NaCl(aq) \rightarrow$ **(b)** $NaCl(s) + F_2(g) \rightarrow$
(c) $I_2(s) + NaF(aq) \rightarrow$ **(d)** $CaBr_2(s) + F_2(g) \rightarrow$
(e) $KF(aq) + Br_2(l) \rightarrow$ **(f)** $MgI_2(aq) + Cl_2(g) \rightarrow$

16. Would you expect chlorine to react with iodide ion or iodine to react with chloride ion? In the redox reaction that does proceed, identify the oxidizing agent, the reducing agent, the substance oxidized, and the substance reduced.

17. Complete and balance each of the following equations. If no reaction occurs, write NR.

(a) $Cl_2(g) + KI(aq) \rightarrow$ **(b)** $I_2(s) + NaCl(aq) \rightarrow$
(c) $Br_2(l) + NaI(aq) \rightarrow$ **(d)** $F_2(g) + H_2O(l) \rightarrow$

18. Suggest a reason why samples of crystalline sodium iodide are sometimes a pale yellow although both the Na^+ and I^- ions are colorless.

19. For each of the following reactions, identify the oxidizing agent, the reducing agent, the substance that is oxidized, and the substance that is reduced.

(a) $2Rb(s) + I_2(s) \rightarrow 2RbI(s)$

(b) $4Al(s) + 3O_2(g) \rightarrow 2(Al^{3+})_2(O^{2-})_3(s)$

(c) $Zn(s) + S(s) \rightarrow Zn^{2+}S^{2-}(s)$

(d) $Mg(s) + 2HCl(aq) \rightarrow MgCl_2(aq) + H_2(g)$

Acid-Base Reactions

20. Describe two tests that you could make to decide whether a given aqueous solution is a solution of an acid.

21. **(a)** In terms of proton transfer, explain why the reaction between $HBr(g)$ and $NH_3(g)$ to give the salt $NH_4Br(s)$ is described as an acid–base reaction.

(b) Explain why the reaction between magnesium and an aqueous solution of HCl to give $MgCl_2(aq)$ and hydrogen is not an acid–base reaction.

22. For each of the following acids, write a balanced equation for its reaction with water, and classify the acid as a weak acid or a strong acid.

(a) nitric acid **(b)** acetic acid **(c)** hydrobromic acid
(d) perchloric acid **(e)** hydrofluoric acid

23. Write balanced equations to show how each of the following bases is ionized in aqueous solution. Classify each base as a strong base or a weak base.

(a) sodium oxide **(b)** potassium hydroxide
(c) ammonia **(d)** lithium hydride

24. With respect to their behavior in aqueous solution, classify each of the following as a strong acid, a weak acid, a strong base, a weak base, or a salt (as many as apply).

(a) potassium hydroxide **(b)** water **(c)** acetic acid
(d) sulfuric acid **(e)** ammonia **(f)** potassium iodide
(g) calcium chloride **(h)** sodium oxide

25. Write the balanced equation for a suitable acid–base reaction for the preparation of each of the following salts.

(a) calcium sulfate **(b)** lithium fluoride
(c) ammonium nitrate **(d)** magnesium perchlorate

26. **(a)** Classify each of the following as a strong acid, a weak acid, or a molecule or ion with no acidic properties in aqueous solution.

(i) HCl (ii) HF (iii) HNO_3 (iv) CH_4 (v) NH_4^+

(b) Classify each of the following as a strong base, a weak base, or a molecule or ion with negligible basicity in aqueous solution.

(i) Cl^- (ii) F^- (iii) NO_3^- (iv) NH_3 (v) O^{2-}

27. Write the formula, draw the Lewis structure, and describe the shape of each of the following.

(a) hydronium ion **(b)** ammonium ion **(c)** hydroxide ion

28.* **(a)** What is the simplest equation that represents the reaction between a solution of an acid, such as $HCl(aq)$, and a solution of a strong base, such as $NaOH(aq)$? Explain.

(b) When an aqueous solution of a strong base, such as $NaOH(aq)$, is added to a solution of a weak acid, such as $HF(aq)$, why does the reaction go to completion to form a solution of a salt?

29.* Write a balanced equation for the ionization of each of the following in aqueous solution. Indicate which of the reactants and which of the products of these reactions behave as acids and which behave as bases.

(a) HCl **(b)** CH_3CO_2H **(c)** $HClO_4$ **(d)** H_2SO_4
(e) HOCl **(f)** NH_4^+

Polyatomic Ions: Lewis Structures

30.* Draw the Lewis structure of each of the following ions.

(a) NH_2^- **(b)** H_2F^+ **(c)** PH_4^+ **(d)** BF_4^-

31.* Repeat Problem 30 for the following ions.

(a) H_3O^+ **(b)** HO^- **(c)** NF_4^+ **(d)** BH_4^-

Precipitation Reactions

32. **(a)** Write the simplest equation that describes the reaction between $AgNO_3(aq)$ and $NaBr(aq)$ to give a precipitate of $AgBr(s)$.

(b) Explain what is meant by the term ''spectator ion.''

33. Use the solubility rules (Table 4.6) to predict which of the following salts are soluble and which are insoluble or sparingly soluble in water.

(a) MgF_2 **(b)** AgI **(c)** K_2SO_4 **(d)** KF
(e) $BaCl_2$ **(f)** $BaSO_4$ **(g)** NaOH **(h)** $Mg(OH)_2$

34. Repeat Problem 33 for the following salts.

(a) $AlCl_3$ **(b)** $NaNO_3$ **(c)** $CaSO_4$ **(d)** LiOH
(e) $CaCO_3$ **(f)** Na_2S **(g)** MgS

35. Predict whether or not a precipitate will form when aqueous solutions of each of the following pairs of substances are mixed. Where a reaction occurs, write the balanced equation for the reaction and the corresponding net ionic equation.

(a) $FeCl_3 + NaOH \rightarrow$ **(b)** $BaCl_2 + KOH \rightarrow$
(c) $Pb(NO_3)_2 + H_2SO_4 \rightarrow$ **(d)** $AgNO_3 + Na_2S \rightarrow$
(e) $AgNO_3 + HI \rightarrow$ **(f)** $Pb(NO_3)_2 + HCl \rightarrow$

36. Repeat Problem 35 for the following pairs of substances.

(a) $Na_2CO_3 + CaCl_2 \rightarrow$ **(b)** $NaNO_3 + CaBr_2 \rightarrow$
(c) $BaCl_2 + MgSO_4 \rightarrow$ **(d)** $HCl + Pb(NO_3)_2 \rightarrow$
(e) $Ca(NO_3)_2 + KOH \rightarrow$ **(f)** $NaCl + KOH \rightarrow$

Reaction Types

37. Balance each of the following equations for reactions in aqueous solution, and classify each as an acid–base, an oxidation–reduction, or a precipitation reaction.

(a) $AgNO_3 + BaCl_2 \rightarrow Ba(NO_3)_2 + AgCl$

(b) $NH_3 + H_2SO_4 \rightarrow (NH_4)_2SO_4$

(c) $Zn + 2HCl \rightarrow ZnCl_2 + H_2$

(d) $NaHCO_3 + HCl \rightarrow NaCl + CO_2 + H_2O$

38. Using bromine or one of its compounds as a reactant, give at least one example of each of the following reaction types.

(a) acid–base **(b)** oxidation–reduction **(c)** synthesis

39. Which of the following reactions are acid–base reactions, and which are oxidation–reduction reactions? For the acid–base reactions, identify the acid and the base. For the oxidation–reduction reactions, identify the oxidizing agent and the reducing agent.

(a) $Cl_2(aq) + 2I^-(aq) \rightarrow 2Cl^-(aq) + I_2(s)$

(b) $HCl(aq) + H_2O(l) \rightarrow Cl^-(aq) + H_3O^+(aq)$

(c) $Zn(s) + 2HCl(aq) \rightarrow ZnCl_2(aq) + H_2(g)$

(d) $HCO_3^-(aq) + H_3O^+(aq) \rightarrow CO_2(aq) + 2H_2O(l)$

40. Balance each of the following equations. Classify each reaction as an acid–base, oxidation–reduction, or precipitation reaction, and where appropriate as a synthesis or decomposition reaction.

(a) $Na(s) + S(s) \rightarrow Na_2S(s)$

(b) $CaCO_3(s) \rightarrow CaO(s) + CO_2(g)$

(c) $NaNO_3(s) \rightarrow NaNO_2(s) + O_2(g)$

(d) $Fe(s) + Cl_2(g) \rightarrow FeCl_3(s)$

(e) $Ba(NO_3)_2(aq) + K_2SO_4(aq) \rightarrow BaSO_4(s) + KNO_3(aq)$

General Problems

41. **(a)** In terms of its size and core charge, explain why fluorine is the most electronegative element.

(b) Write balanced equations for the reactions of fluorine with (i) water, (ii) potassium, (iii) aluminum, and (iv) phosphorus.

42. Astatine, At, is a member of the halogen family of elements. Predict the following for astatine.

(a) its physical state

(b) its relative strength among the halogens as an oxidizing agent

(c) the balanced equations for its reactions with (i) Na, (ii) Ca, (iii) P, (iv) H_2, and (v) Br_2

(d) the acid strength and reaction of its hydride in water

(e) the type of bond in the molecule it forms with bromine

(f) the type of bond in, and the physical state of, its compound with potassium

43. **(a)** What are the formulas of the simplest hydrides of (i) carbon; (ii) nitrogen; (iii) oxygen; (iv) fluorine?

(b) Write balanced equations for the reactions, if any, of each of the hydrides in (a) with water. Describe the acid–base properties of each of these hydrides as a strong acid, a strong base, a weak acid, a weak base, or neither an acid nor a base (more than one may apply).

44. Describe what is observed when a few drops of an aqueous solution of chlorine (chlorine water) are added to an aqueous solution of potassium iodide and the solution is then shaken with carbon tetrachloride. Explain the color changes, and write balanced equations to describe the reactions that take place.

45.* The salt $MX_2(s)$ is a metal halide that, when dissolved in water, gives a neutral solution. The aqueous solution does not react with bromine but turns brown when $Cl_2(g)$ is bubbled through it. A solution containing 5.35 g of MX_2 requires 710 mL of chlorine, measured at 25°C and 1.00 atm pressure, for complete reaction. Identify the salt MX_2, and explain how each of the preceding observations supports your identification.

46.* Two gaseous elements X and Y have densities of $0.0658\ g \cdot L^{-1}$ and $2.315\ g \cdot L^{-1}$, respectively, at 100°C and 1.00 atm pressure. When 50 mL of X reacted explosively with 100 mL of Y, the resulting volume of gas was 150 mL (all volumes measured at the same temperature and pressure). After the product was shaken with water to give a solution that turned blue litmus red, the volume of the remaining gas was 50 mL. The remaining gas did not support combustion but reacted vigorously with hot sodium to give a white solid. Identify X and Y, and write balanced equations for each of the reactions described.

47.* The density of sodium chloride is $2.17\ g \cdot cm^{-3}$. How many sodium ions and chloride ions are there in a cubic NaCl crystal with sides of length 1 mm?

Stoichiometry

5.1 Amounts of Reactants and Products
5.2 Limiting Reactants
5.3 Theoretical Yield and Percent Yield
5.4 Determination of Empirical and Molecular Formulas
5.5 Stoichiometry of Gas Reactions
5.6 Stoichiometry of Reactions in Solution

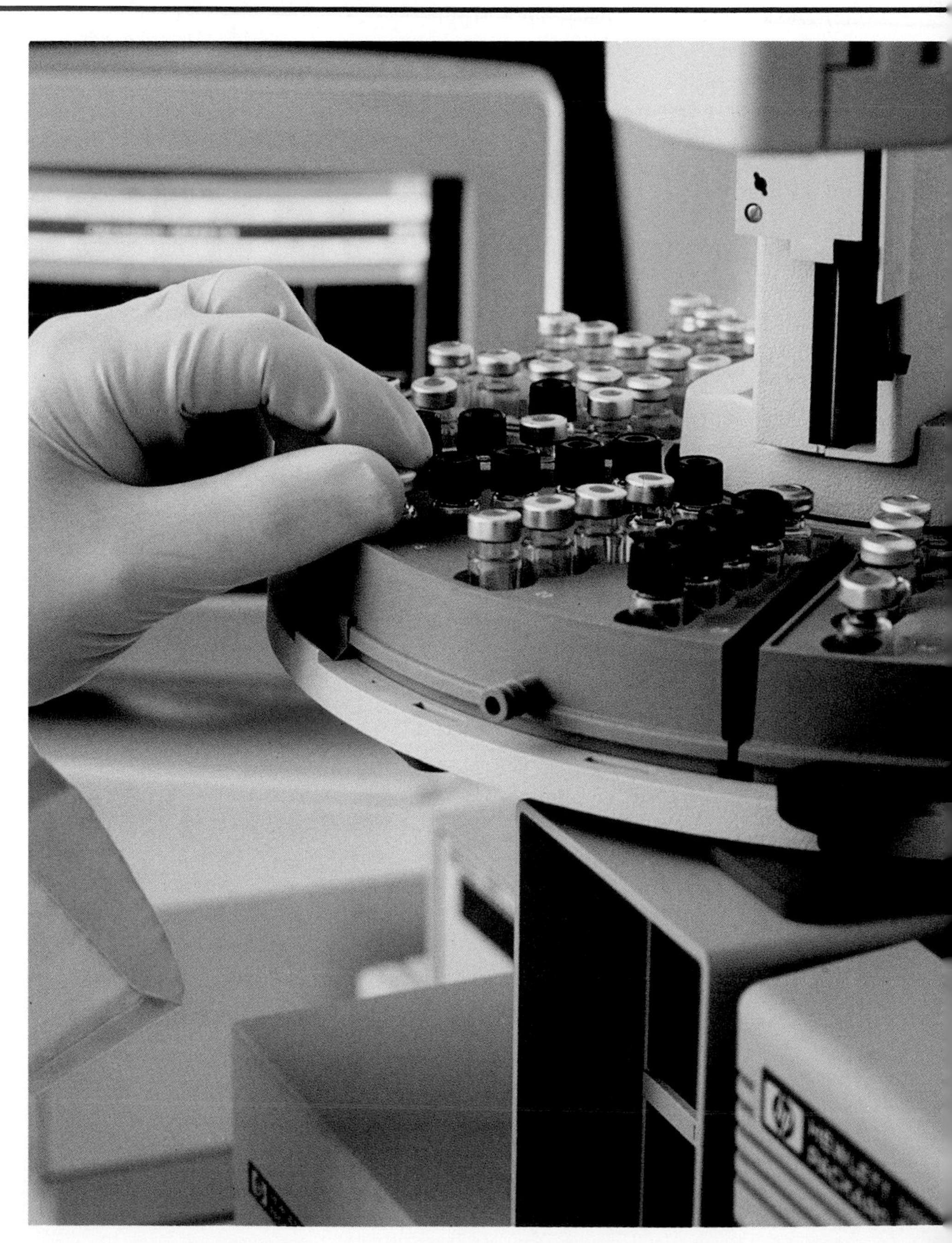

The determination and analysis of quantities of substances comprise the discipline called stoichiometry. Gas chromatography is one of many stoichiometric methods used to analyze substances for industrial or medical needs. The gas chromatograph can be programmed to detect and determine the amounts of many different substances in even the tiniest samples. Here a technician is using a gas chromatograph to test urine samples for drugs.

How much of each element is present in a compound? How much of a substance is present in a mixture? How much of a desired product is produced in a reaction? How pure is a product? These are questions to which chemists and many others always need answers. For example, a police officer may need to know how much alcohol you have in your blood. A mining engineer may need to know how much iron could be obtained from a particular iron ore deposit. An industrial chemist may need to know how much nitrogen and hydrogen are required to make one ton of ammonia. An art restorer may need to find the composition of a paint used by an artist several hundred years ago. A geologist may need to know how much potassium and argon are in a rock to determine its age. A research chemist who synthesizes a new compound needs to know its composition—that is, how much of each element is present—to determine its formula.

The experimental determination of how much of an element is present in a compound, or how much of a compound is present in a mixture, is called *quantitative analysis*. Before we carry out quantitative analysis, we may need to perform *qualitative analysis*—determine which elements are present in a compound or which compounds are present in a mixture. A chemist who carries out such determinations is called an *analytical chemist.* Quantitative analysis may be carried out by weighing the amount of a product formed in a reaction—called *gravimetric analysis*—or by measuring the volumes of solutions used in a reaction—called *volumetric analysis*. These methods have been used by chemists for many years but are being increasingly displaced by *instrumental methods*, which are typically quicker and more accurate. As we shall see in later chapters, many instrumental analytical methods are based on the spectra produced when electromagnetic radiation is absorbed or emitted by atoms and molecules.

Here we are concerned not so much with these analytical techniques but rather with the calculations involved in answering the question, How much? All the quantitative aspects of chemical reactions, chemical composition, and analysis are known collectively as **stoichiometry**, a word derived from the Greek words *stoicheon,* for ''element,'' and *metron,* for ''measure.''

5.1 AMOUNTS OF REACTANTS AND PRODUCTS

We have seen that in chemical reactions, atoms are only rearranged—they are not created or destroyed. Because each kind of atom has a constant characteristic mass, there is no change in the total mass during a chemical reaction. The law of conservation of mass (Chapter 1) states that the total mass of the products formed is equal to the total mass of the reactants used up.

Enormous numbers of atoms and molecules are normally involved in reactions. That is why, as we discussed in Chapter 1, chemists have found it convenient to measure the amounts of substances in terms of moles.

One *mole* of entities (atoms, molecules, ions, or anything else) is 6.022×10^{23} entities.

This number is called Avogadro's number. The mole is the chemist's equivalent of the storekeeper's dozen. Eggs are conveniently counted in dozens. Atoms and molecules are conveniently counted in moles, because

The mass of one mole of atoms or molecules in grams is numerically equal to the mass of one atom or molecule in atomic mass units, u.

Do not confuse molar mass with molecular mass.

The SI unit for mole is *mol.* The balanced equation for a reaction tells how many moles of each product are produced for a given number of moles of reactants that are used up. So if we know the **molar mass** (the mass of 1 mole) of each reactant and product, we can calculate the mass of each product produced by the reaction for a given mass of a reactant.

Suppose that we want to know how many moles of N_2 are needed to produce 6.0 mol of ammonia, NH_3, from the reaction of nitrogen with hydrogen. In all problems of this type, we begin by writing the *balanced* equation for the reaction:

$$N_2(g) + 3H_2(g) \rightarrow 2NH_3(g)$$

From this equation we see that 1 mol N_2 is needed to produce 2 mol NH_3, so 3.0 mol N_2 will be needed to produce 6.0 mol of NH_3.

Let us examine the argument that we have used here in a little more detail. The balanced equation for the reaction gives us the conversion factor

$$\frac{1 \text{ mol } N_2}{2 \text{ mol } NH_3}$$

which means that for every mole of N_2 that we start with, we get 2 mol NH_3. Using this conversion factor, we can find the number of moles of N_2 needed to give 6.0 mol NH_3. The conversion factor converts a known number of moles of NH_3 to the unknown number of moles of N_2:

$$\text{moles } N_2 = (6.0 \cancel{\text{mol } NH_3})\left(\frac{1 \text{ mol } N_2}{2 \cancel{\text{mol } NH_3}}\right) = 3.0 \text{ mol } N_2$$

In general,

The balanced equation gives the relationship between the numbers of moles of reactants and the numbers of moles of products.

In Chapter 1 we saw how to convert moles to masses, so now we know how to find the masses of substances that take part in a reaction. We convert the masses of the reactants to moles, use the balanced equation to convert moles of reactants to moles of products, and then convert moles of products to masses of products.

In Demonstration 2.2 we saw that when magnesium burns in air, it combines with the oxygen of the air to form the white solid magnesium oxide, MgO. Now we are in a position to consider the quantitative aspects of this reaction. For example, what mass of oxygen will combine with 0.187 g of magnesium, and what mass of magnesium oxide will be formed? Let us break down the solution to this problem into steps.

(1) As always, the first step is to *write the balanced equation for the reaction*:

$$2Mg(s) + O_2(g) \rightarrow 2MgO(s)$$

This equation tells us that 2 mol Mg reacts with 1 mol O_2 to give 2 mol MgO.

(2) The next step is to *convert the mass of Mg to moles*:

$$\text{moles Mg} = (0.187 \text{ g Mg})\left(\frac{1 \text{ mol Mg}}{24.3 \text{ g Mg}}\right) = 7.70 \times 10^{-3} \text{ mol Mg}$$

(3) Now we can *set up the conversion factor to convert moles of Mg to moles of O_2*. The required conversion factor is

$$\frac{1 \text{ mol } O_2}{2 \text{ mol Mg}}$$

so we have

$$\text{moles } O_2 = (7.70 \times 10^{-3} \text{ mol Mg})\left(\frac{1 \text{ mol } O_2}{2 \text{ mol Mg}}\right) = 3.85 \times 10^{-3} \text{ mol } O_2$$

(4) Finally, we *convert moles of O_2 to mass of O_2*:

$$\text{mass } O_2 = (3.85 \times 10^{-3} \text{ mol } O_2)\left(\frac{32.00 \text{ g } O_2}{1 \text{ mol } O_2}\right) = 0.123 \text{ g } O_2$$

In steps 2 through 4 we set up the conversion factor so that all the units cancel except the units of the required quantity: step 2, mol Mg; step 3, mol O_2; step 4, g O_2. To obtain the mass of MgO formed, we repeat steps 3 and 4 for MgO instead of O_2:

$$\text{moles MgO} = (7.70 \times 10^{-3} \text{ mol Mg})\left(\frac{2 \text{ mol MgO}}{2 \text{ mol Mg}}\right) = 7.70 \times 10^{-3} \text{ mol MgO}$$

$$\text{mass MgO} = (7.70 \times 10^{-3} \text{ mol MgO})\left(\frac{40.3 \text{ g MgO}}{1 \text{ mol MgO}}\right) = 0.310 \text{ g MgO}$$

Or, more simply, we can make use of the law of conservation of mass:

$$\text{mass MgO} = \text{mass Mg} + \text{mass } O_2 = (0.187 \text{ g} + 0.123 \text{ g}) = 0.310 \text{ g}$$

In the example just given we used four steps, and the same four steps are applicable to all problems in which we need to calculate the masses of substances taking part in chemical reactions:

1. **Write the balanced equation for the reaction.**
2. **Convert the known mass (grams) of one reactant to moles.**
3. **Use the balanced equation to set up the appropriate conversion factor(s) to find the number of moles of another reactant or product(s).**
4. **Convert from moles back to mass (grams).**

We can diagram this procedure, starting with step 2, as follows:

mass of reactant $\xrightarrow{2}$ *moles of reactant* $\xrightarrow{3}$ *moles of product (or other reactant)* $\xrightarrow{4}$ *mass of product (or other reactant)*

EXAMPLE 5.1 Masses of Products from Masses of Reactants

What mass of water results from the complete combustion of 2.000 g of methane, $CH_4(g)$?

Solution: 1. The balanced equation for the reaction is

$$CH_4(g) + 2O_2(g) \rightarrow CO_2(g) + 2H_2O(g)$$

2. We convert the mass of CH_4 to moles of CH_4, using the molar mass of CH_4, 12.01 $g \cdot mol^{-1}$ + 4(1.008 $g \cdot mol^{-1}$) = 16.04 $g \cdot mol^{-1}$:

$$\text{moles } CH_4 = (2.000 \text{ g } CH_4)\left(\frac{1 \text{ mol } CH_4}{16.04 \text{ g } CH_4}\right) = 0.1247 \text{ mol } CH_4$$

3. From the balanced equation,

$$1 \text{ mol } CH_4 \rightarrow 2 \text{ mol } H_2O$$

so the conversion factor is $\dfrac{2 \text{ mol } H_2O}{1 \text{ mol } CH_4}$:

$$\text{moles } H_2O = (0.1247 \text{ mol } CH_4)\left(\frac{2 \text{ mol } H_2O}{1 \text{ mol } CH_4}\right) = 0.2494 \text{ mol } H_2O$$

4. Finally, we convert moles of H_2O to the mass of H_2O, using the molar mass of H_2O, 2(1.008 $g \cdot mol^{-1}$) + 16.00 $g \cdot mol^{-1}$ = 18.02 $g \cdot mol^{-1}$:

$$\text{mass } H_2O = (0.2494 \text{ mol } H_2O)\left(\frac{18.02 \text{ g } H_2O}{1 \text{ mol } H_2O}\right) = 4.494 \text{ g } H_2O$$

With practice, calculations such as this can be performed in *one* step. In this case, we can multiply the mass of CH_4 by the three conversion factors in steps 2, 3, and 4:

$$\text{mass } H_2O = (2.000 \text{ g } CH_4)\left(\frac{1 \text{ mol } CH_4}{16.04 \text{ g } CH_4}\right)\left(\frac{2 \text{ mol } H_2O}{1 \text{ mol } CH_4}\right)\left(\frac{18.02 \text{ g } H_2O}{1 \text{ mol } H_2O}\right) = 4.494 \text{ g } H_2O$$

Include both units and formulas for substances in *all* conversion factors. Then it will be apparent if a conversion factor error has been made, because the *units* and/or *formulas* will not cancel.

Exercise 5.1 What mass of $CaCl_2(s)$ is produced when 0.2500 g of calcium metal is burned completely in $Cl_2(g)$?

5.2 LIMITING REACTANTS

When methane is burned in the atmosphere, an unlimited amount of oxygen is available, so it is the amount of methane that determines the amounts of carbon dioxide and water that form:

$$CH_4(g) + 2O_2(g) \rightarrow CO_2(g) + 2H_2O(g)$$

Methane is called the **limiting reactant**, because it reacts completely and determines the amounts of the products.

The amounts of reactants that are available do not always match the ratio of moles that appear in the balanced equation for a given reaction. For example, suppose we pass a spark through a mixture of 2.5 mol of $H_2(g)$ and 1.0 mol $Cl_2(g)$. The equation

$$\begin{array}{ccccc} H_2(g) & + & Cl_2(g) & \rightarrow & 2HCl(g) \\ 1.0 \text{ mol} & & 1.0 \text{ mol} & & 2.0 \text{ mol} \end{array}$$

tells us that only 1.0 mol $H_2(g)$ can react with 1.0 mol Cl_2, so 1.50 mol $H_2(g)$ is left in excess. Thus, the amount of Cl_2 determines the amount of HCl that forms; Cl_2 is the limiting reactant. In general,

The limiting reactant in any reaction is that reactant whose amount determines

the amount of products that form. Excess amounts of the other reactants remain unreacted.

EXAMPLE 5.2 Calculating the Mass of a Product of a Reaction with a Limiting Reactant

Chlorine is passed through a solution containing 10.00 g of sodium bromide, NaBr, until no more bromine forms. Which is the limiting reactant? What mass of bromine is produced?

Solution: The balanced equation for the reaction (Chapter 4) is

$$Cl_2(g) + 2NaBr(aq) \rightarrow Br_2(l) + 2NaCl(aq)$$

The limiting reactant is sodium bromide, NaBr, because sufficient chlorine was passed to convert all the bromide ion, Br^-, to bromine; the amount of bromine formed is limited by the initial amount of sodium bromide.

The molar mass of NaBr is (22.99 + 79.90) $g \cdot mol^{-1}$ = 102.89 $g \cdot mol^{-1}$, so

$$\text{moles NaBr} = (10.00 \text{ g NaBr}) \left(\frac{1 \text{ mol NaBr}}{102.9 \text{ g NaBr}}\right) = 0.09718 \text{ mol NaBr}$$

From the balanced equation, the conversion factor to convert moles of NaBr to moles of Br_2 is

$$\left(\frac{1 \text{ mol } Br_2}{2 \text{ mol NaBr}}\right)$$

so

$$\text{moles } Br_2 = (0.09718 \text{ mol NaBr})\left(\frac{1 \text{ mol } Br_2}{2 \text{ mol NaBr}}\right) = 0.04859 \text{ mol } Br_2$$

The molar mass of Br_2 is 2(79.90) = 159.8 $g \cdot mol^{-1}$, so

$$\text{mass } Br_2 = (0.04859 \text{ mol } Br_2)\left(\frac{159.8 \text{ g } Br_2}{1 \text{ mol } Br_2}\right) = 7.76 \text{ g } Br_2$$

Thus, 7.76 g of Br_2 can be obtained by passing an excess of chlorine through an aqueous solution of 10.00 g of NaBr.

In Example 5.2 it was clear that sodium bromide was the limiting reactant because we stated that sufficient chlorine was used to react with all the NaBr. But when the masses of all the reactants are given, we cannot immediately tell which is the limiting reactant. We see how to proceed in such a case in Example 5.3.

EXAMPLE 5.3 Finding the Limiting Reactant and Calculating the Mass of a Product

When zinc and sulfur are heated together, zinc sulfide, ZnS, is formed. This is a reaction between a metal and a nonmetal in which zinc is oxidized to Zn^{2+} ion and sulfur is reduced to sulfide ion, S^{2-}. The reaction equation is

$$Zn(s) + S(s) \rightarrow ZnS(s)$$

Suppose that 12.00 g of zinc is heated with 7.50 g of sulfur.

(a) Which is the limiting reactant?
(b) How much ZnS is formed?
(c) How much of the nonlimiting reactant remains unreacted?

Solution: (a) We convert the mass of each reactant to moles:

$$\text{moles Zn} = (12.00 \text{ g Zn})\left(\frac{1 \text{ mol Zn}}{65.39 \text{ g Zn}}\right) = 0.184 \text{ mol Zn}$$

$$\text{moles S} = (7.50 \text{ g S})\left(\frac{1 \text{ mol S}}{32.07 \text{ g S}}\right) = 0.234 \text{ mol S}$$

We see from the equation that 1 mol Zn reacts with 1 mol S. Hence more S is present than is required; and Zn is the limiting reactant.

(b) The amount of product formed depends on the amount of the limiting reactant, Zn, and not on the amount of S. The equation shows that 1 mol Zn gives 1 mol ZnS, so the mass of ZnS formed is

$$\text{mass ZnS} = (0.184 \text{ mol Zn})\left(\frac{1 \text{ mol ZnS}}{1 \text{ mol Zn}}\right)\left(\frac{97.46 \text{ g ZnS}}{1 \text{ mol ZnS}}\right) = 17.9 \text{ g ZnS}$$

We have combined two steps here: moles Zn → moles ZnS → mass ZnS.
(c) The excess amount of S is (0.234 − 0.184) mol = 0.050 mol. We convert this result to grams:

$$\text{mass S} = (0.050 \text{ mol S})\left(\frac{32.07 \text{ g S}}{1 \text{ mol S}}\right) = 1.60 \text{ g excess S}$$

In solving any limited reactant problem, we use the same four steps that we use in solving any problem that involves the amounts of reactants and products in a reaction, except that we add the step of deciding which is the limiting reactant. The steps are as follows:

If the coefficients of all the reactants in the balanced equation are *not* equal, divide the moles of each reactant by its coefficient in the balanced equation. Then the smallest number of moles corresponds to the limiting reactant.

1. **Write the balanced equation for the reaction.**
2. **Convert mass (grams) of reactants to moles of reactants.**
3. **Use the balanced equation to decide which is the limiting reactant.**
4. **Use the balanced equation to convert moles of limiting reactant to moles of products.**
5. **Convert moles of products to mass (grams) of products.**

We may diagram steps 2 through 5 as follows:

$$\textit{masses of reactants} \xrightarrow{2} \textit{moles of reactants} \xrightarrow{3} \textit{moles of limiting reactant} \xrightarrow{4} \textit{moles of products} \xrightarrow{5} \textit{masses of products}$$

Exercise 5.2 When iron oxide is heated with aluminum powder, a very vigorous reaction occurs in which iron oxide is reduced to molten iron (Demonstration 10.6) according to the equation

$$Fe_2O_3(s) + 2Al(s) \rightarrow Al_2O_3(s) + 2Fe(l)$$

A mixture of 30.0 g of aluminum and 100.0 g of Fe_2O_3 is heated.
(a) Which is the limiting reactant?
(b) How much iron will form?
(c) How much of the nonlimiting reactant remains when the reaction is complete?

5.3 THEORETICAL YIELD AND PERCENT YIELD

The amount of a product that is formed when the limiting reactant is completely used up is called the **theoretical yield** of that product. In each case the amount of product that we calculated in Examples 5.1 through 5.3 was the theoretical yield, the maximum amount that could possibly be obtained. In practice, the theoretical yield is rarely obtained. There are several reasons for this. Other reactions may occur between the reactants to give other products, so the yield of the desired product is reduced. Small amounts of the reactants or the products may be lost when they are separated from the reaction mixture. Or equilibrium may be reached before all the reactants have been used up—in other words, the reaction may not go to completion under the prevailing conditions. It is customary to express the *actual yield* as a percentage by mass of the theoretical yield and to call this the **percent yield**:

$$\text{percent yield} = \frac{\text{actual yield}}{\text{theoretical yield}} \times 100\%$$

EXAMPLE 5.4 Percent Yield

Hydrogen is made industrially by heating carbon with steam at 725°C. In this reaction, carbon is oxidized to $CO(g)$, and water is reduced to $H_2(g)$. What is the theoretical yield of H_2 from 10.00 metric tons of carbon and excess steam? If the actual yield in a particular industrial plant is 1.49 tons, what is the percent yield?

Solution: The balanced equation for the reaction is

$$C(s) + H_2O(g) \rightarrow CO(g) + H_2(g)$$

Because 1 metric ton = 1000 kg and the molar mass of C is 12.01 $g \cdot mol^{-1}$,

$$\text{moles C} = (10.00\ \text{ton})\left(\frac{10^3\ \text{kg}}{1\ \text{ton}}\right)\left(\frac{10^3\ \text{g}}{1\ \text{kg}}\right)\left(\frac{1\ \text{mol C}}{12.01\ \text{g C}}\right) = 8.326 \times 10^5\ \text{mol C}$$

The balanced equation shows that 1 mol C gives 1 mol H_2, so the theoretical mass of hydrogen is

$$\text{mass } H_2 = (8.326 \times 10^5\ \text{mol } H_2)\left(\frac{2.016\ \text{g } H_2}{1\ \text{mol } H_2}\right)\left(\frac{1\ \text{ton}}{10^6\ \text{g}}\right) = 1.68\ \text{ton } H_2$$

The theoretical yield is 1.68 tons, and the actual yield is 1.49 tons, so

$$\text{percent yield} = \left(\frac{1.49\ \text{tons}}{1.68\ \text{tons}}\right) \times 100\% = 88.7\%$$

Exercise 5.3 In an experiment where chlorine was passed into a solution containing 1.000 g of $NaI(aq)$, 0.250 g of pure sodium chloride crystals was obtained after the following steps occurred: evaporating the solution, recrystallizing the sodium chloride from water, and drying it in an oven at 120°C. What are **(a)** the theoretical yield and **(b)** the percent yield of $NaCl(s)$?

5.4 DETERMINATION OF EMPIRICAL AND MOLECULAR FORMULAS

One of the first steps in the study of a new compound is to determine its composition: Which elements are present and in what relative amounts (Box 5.1)? From this information we can determine the empirical formula of the compound. We saw in Section 5.3 that if we can write the balanced equation for a reaction, which implies that we know the formulas of the reactants and products, we can calculate the amount of any product that will be formed from given amounts of the reactants. Conversely, if we determine the amounts of the products experimentally, we can find the empirical formula of one of the substances taking part in the reaction. We will consider two simple methods for determining empirical formulas: (1) the synthesis of a compound from its elements and (2) the combustion of a compound of C, H, and O (an *organic compound*) to obtain carbon dioxide and water.

BOX 5.1 Analytical Chemistry

Chemists need a wide variety of analytical techniques to find out what substances are present in a given sample and to measure quantitatively how much of each substance is present. The modern analytical laboratory uses desk-top computers and digital instruments, as well as the conventional buret and pipet, to analyze millions of different substances, some of which may be present in only trace amounts (Figure A).

Figure A *Analytical chemist using automated analytical equipment linked to a computer*

These instruments include those needed for the techniques we have previously discussed, such as chromatography to separate mixtures and mass spectrometry to identify the individual components. The combination of the two techniques is particularly valuable. Gas chromatography separates a mixture, and mass spectrometry determines the molar mass and gives information about the molecular structures. For example, it was the combination of gas chromatography and mass spectrometry that led to the discovery of more than 168 components in cigarette smoke. The 70 components that were identified with certainty included some later shown to cause cancer.

In addition to chromatography and mass spectrometry, chemists use different types of *spectroscopy*, the study of spectra (the range of frequencies of light an atom absorbs or emits). You will learn more about spectroscopic techniques in later chapters. In the trace analysis of metals, a technique called *atomic absorption spectroscopy* (Chapter 6) can detect metal ions in solution with concentrations as low as one part per billion (ppb). A modern atomic absorption spectrometer can analyze as many as 20 samples for up to 20 elements, with excellent specificity. The combination of spectroscopy and telescopes enables us to analyze the substances present on distant stars and galaxies without leaving earth. Placing such instruments on space probes has also given us information about substances on our neighboring planets (Chapter 19).

Chemical analysis is an integral part of environmental investigations. An interesting example of the application of spectroscopy to understanding environmental pollution is the use of infrared spectroscopy (Chapter 14) to look through more than 1 km of polluted city air. Because each substance has its own characteristic infrared absorption spectrum, chemists can identify all the molecules present in an air sample, including their concentrations down to parts per billion. For example, it is possible to measure the nitric acid and formaldehyde in Los Angeles smog and to monitor the increase in concentration of these pollutants throughout the day as photochemical processes are initiated by sunlight (Chapter 16).

The water we drink comes from ground water or from rivers and lakes. Many natural solutes are dissolved in this water. Some of these solutes are beneficial, such as the salts of calcium, magnesium, sodium, and potassium. Others, salts of metal ions such as those of lead, cadmium, mercury, and aluminum may also enter our water supplies but are toxic, typically even at low concentrations. Each jurisdiction has its own standards for the amounts of various substances permissible in water, usually expressed as a maximum concentration in parts per million (ppm). The monitoring and enforcement of water-quality regulations clearly relies on the accurate measurement by analytical chemists of the amounts of these substances present in the water.

To minimize tooth decay, about half of all North Americans drink artificially fluoridated water. A limit of one part per million of fluoride ion in drinking water is recommended, because higher concentrations can cause mottling of the teeth. Here again the techniques of analytical chemistry must be constantly used to monitor the concentration of fluoride ions in the water.

The skills of the analytical chemist also come to the aid of archeologists, geologists, and historians. Archeologists can determine the age of an ancient object by determining the amount of radioactive carbon-14 it contains (Chapter 18). Geologists can determine the age of an ancient rock by determining how much argon, produced by the decay of radioactive potassium, it contains. It was the discovery of trace amounts of arsenic in samples of Napoleon's hair that led to the theory that his death was caused by poisoning. However, it is now thought that a volatile arsenic compound had slowly evolved from pigment in the wallpaper of his rooms while he was in exile on the island of St. Helena. This work used a technique called *neutron activation analysis*, which involves exposing a sample to a neutron beam to produce radioactive iostopes. After exposure, the radiation from the sample is measured, and this information is used to identify the presence of a specific element. The number of radioactive disintegrations can also be used to determine quantitatively the amount of that element present (Chapter 18). Neutron activation analysis is a nondestructive method of analysis that can detect as little as 10^{-15} g of some elements. More recently, neutron activation analysis was involved in the exhumation of the body of Zachary Taylor,

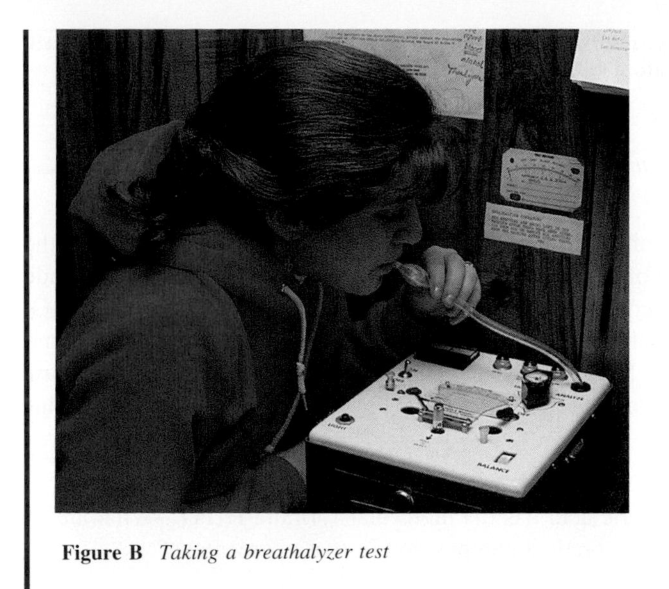

Figure B *Taking a breathalyzer test*

the twelfth president of the United States. Small samples of his hair and fingernails were tested by neutron activation analysis to check for the presence of arsenic. No detectable amounts of arsenic were found, so analytical chemists were able to disprove the theory of some historians that Taylor died of arsenic poisoning.

Many analytical instruments can be made quite portable. For example, the breathalyzer, which measures the amount of alcohol in the breath (Figure B), can be carried in a police car. The breathalyzer utilizes the reaction of alcohol with the orange dichromate ion, which oxidizes ethanol to acetic acid and is reduced to the green Cr^{3+} ion:

$$\underset{\text{Ethanol}}{3CH_3CH_2OH} + \underset{\substack{\text{Dichromate} \\ \text{(orange)}}}{2Cr_2O_7^{2-}} + 16H^+ \rightarrow \underset{\text{(green)}}{4Cr^{3+}} + \underset{\text{Acetic acid}}{3CH_3COOH} + 11H_2O$$

The extent of the color change is proportional to the concentration of alcohol in the breath. The instrument is calibrated to read the alcohol concentration directly.

Empirical Formula Determination by Synthesis

By burning a known mass of phosphorus in excess oxygen and determining the amount of oxide produced, we can find the empirical formula of the oxide. Suppose we find in a particular experiment that when 5.00 g of phosphorus is burned in excess oxygen, the mass of the oxide produced is 11.44 g. The amount of oxygen that combines with 5.00 g of phosphorus must be

$$11.44 \text{ g} - 5.00 \text{ g} = 6.44 \text{ g}$$

Using the molar masses of atomic phosphorus ($30.97 \text{ g} \cdot \text{mol}^{-1}$) and atomic oxygen ($16.00 \text{ g} \cdot \text{mol}^{-1}$), we can express these combining masses in *moles*:

$$\text{moles P} = (5.00 \text{ g P})\left(\frac{1 \text{ mol P}}{30.97 \text{ g P}}\right) = 0.161 \text{ mol P}$$

$$\text{moles O} = (6.44 \text{ g O})\left(\frac{1 \text{ mol O}}{16.00 \text{ g O}}\right) = 0.402 \text{ mol O}$$

We now know the ratio of *moles of O atoms* to *moles of P atoms*. But because 1 mol of atoms of an element is Avogadro's number of atoms of that element, this ratio is the same as the ratio of the *number of O atoms* to the *number of P atoms*:

$$\frac{\text{number of O atoms}}{\text{number of P atoms}} = \frac{0.402 \text{ mol O atoms}}{0.161 \text{ mol P atoms}} = 2.5$$

Because this result is not a whole number, we multiply it by the smallest integer that will convert it to a whole number—in this case $2 \times 2.5 = 5$. Hence the *empirical formula* of the oxide is P_2O_5.

From the mass of each element present in a compound, we can find the empirical formula just as we did for the oxide of phosphorus, P_2O_5. The three steps are

1. **Convert the mass of each element to moles of atoms.**
2. **Find the mole ratios by dividing the number of moles of each element by the number of moles of the element present in the smallest amount. These mole ratios are identical to the atom ratios.**

3. When necessary, multiply the atom ratios by the smallest number that will convert all the atom ratios to whole numbers.

We can summarize these steps as follows:

$$\begin{matrix}\textit{masses of} \\ \textit{elements}\end{matrix} \xrightarrow{1} \begin{matrix}\textit{moles of} \\ \textit{elements}\end{matrix} \xrightarrow{2} \begin{matrix}\textit{mole} \\ \textit{ratios}\end{matrix} = \begin{matrix}\textit{atom} \\ \textit{ratios}\end{matrix} \xrightarrow{3} \begin{matrix}\textit{whole-number} \\ \textit{atom ratios}\end{matrix}$$

Normally when compounds are analyzed, their compositions are given as the mass percentage of each of the elements present. For example, 11.44 g of the oxide of phosphorus with the empirical formula P_2O_5 contains 5.00 g P and 6.44 g O; this composition is usually quoted as $(5.00/11.44) \times 100\% = 43.7\%$ P and $(6.44/11.44) \times 100\% = 56.3\%$ O by mass. A useful way to interpret this **mass percentage**, or **percent composition** by mass, is to say that 100 g of P_2O_5 contains 43.7 g P and 56.3 g O.

Exercise 5.4 Phosphoric acid has the molecular formula H_3PO_4. What are its molar mass and its theoretical composition in mass percentage?

EXAMPLE 5.5 Determining Empirical Formula from Mass Percent Composition

The compound adrenaline, which is released in the human body in times of stress, increases the body's metabolic rate. Like many compounds in living systems, adrenaline is composed of carbon, hydrogen, oxygen, and nitrogen. It has the composition 56.8% C, 6.50% H, 28.4% O, and 8.28% N. What is the empirical formula of adrenaline?

Mass percentage results are independent of the actual sample size. It is convenient to choose a 100-g sample, because then the mass in grams of the desired product is numerically equal to the mass percentage of that product.

Solution: We will assume that we are dealing with 100 g of adrenaline. For convenience we set up the following type of calculation:

	Carbon	*Hydrogen*	*Oxygen*	*Nitrogen*
mass	56.8 g	6.50 g	28.4 g	8.28 g
moles of atoms	$(56.8\text{ g})\left(\frac{1\text{ mol}}{12.01\text{ g}}\right) = 4.73\text{ mol}$	$(6.50\text{ g})\left(\frac{1\text{ mol}}{1.008\text{ g}}\right) = 6.45\text{ mol}$	$(28.4\text{ g})\left(\frac{1\text{ mol}}{16.00\text{ g}}\right) = 1.78\text{ mol}$	$(8.28\text{ g})\left(\frac{1\text{ mol}}{14.01\text{ g}}\right) = 0.591\text{ mol}$
mole (atom) ratio	$\frac{4.73}{0.591} = 8.00$	$\frac{6.45}{0.591} = 10.9$	$\frac{1.78}{0.591} = 3.01$	$\frac{0.591}{0.591} = 1.00$

Thus, the *empirical formula* of adrenaline is $C_8H_{11}O_3N$.

As in this case, the atom ratios in such calculations do not always work out to be exactly whole numbers, because there may be small experimental errors in the percent composition. Provided that the difference from a whole number is small, 0.1 or less, we round off the atom ratio to the nearest whole number.

Empirical Formula Determination by Combustion of an Organic Compound

Many of the compounds of carbon that are known as organic compounds contain only carbon, hydrogen, and oxygen. The composition and empirical formulas of such compounds may be determined by burning a known mass of the compound in sufficient oxygen to convert all the hydrogen to water and all the carbon to carbon dioxide. Figure 5.1 shows how we can determine the masses of the products.

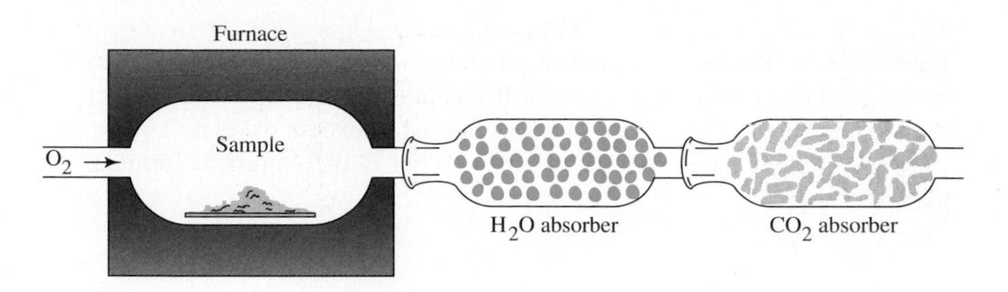

FIGURE 5.1 Combustion Apparatus for Analyzing Organic Compounds A weighed sample of an organic compound is burned in a stream of oxygen. The reaction produces gaseous H_2O and CO_2. These gases then pass through a series of tubes. One tube contains a substance that absorbs H_2O; another tube contains a substance that absorbs CO_2. By comparing the masses of these tubes before and after the reaction, we can determine the masses of hydrogen and of carbon present in the compound that was burned.

EXAMPLE 5.6 Empirical Formula by Combustion of an Organic Compound

Ascorbic acid (vitamin C) is known to contain only C, H, and O. A 6.49-mg sample of ascorbic acid was burned completely in an apparatus like the one in Figure 5.1. The increase in the mass of each absorption tube showed that 9.74 mg of CO_2 and 2.64 mg of H_2O formed. What is the empirical formula of ascorbic acid?

Solution: First, we calculate the mass of C in the original sample from the mass of CO_2, and the mass of H in the sample from the mass of H_2O. For these calculations we need the molar masses of C and CO_2 and of H and H_2O:

$$\text{molar mass C} = 12.01\ \text{g} \cdot \text{mol}^{-1}$$

$$\text{molar mass CO}_2 = [12.01 + 2(16.00)]\ \text{g} \cdot \text{mol}^{-1} = 44.01\ \text{g} \cdot \text{mol}^{-1}$$

$$\text{molar mass H} = 1.008\ \text{g} \cdot \text{mol}^{-1}$$

$$\text{molar mass H}_2\text{O} = [2(1.008) + 16.00]\ \text{g} \cdot \text{mol}^{-1} = 18.02\ \text{g} \cdot \text{mol}^{-1}$$

$$\text{mass C} = (9.74 \times 10^{-3}\ \text{g CO}_2)\left(\frac{12.01\ \text{g C}}{44.01\ \text{g CO}_2}\right) = 2.66 \times 10^{-3}\ \text{g}$$

$$\text{mass H} = (2.64 \times 10^{-3}\ \text{g})\left(\frac{2.016\ \text{g H}}{18.02\ \text{g H}_2\text{O}}\right) = 0.295 \times 10^{-3}\ \text{g}$$

Next we determine the mass of O in the sample. We cannot find this from the masses of CO_2 and H_2O, because some of the oxygen in these products comes from the $O_2(g)$ used to burn the sample. However, we can find the mass of O in the ascorbic acid by subtracting the total mass of C and H from 6.49 mg, the mass of the original sample:

$$\text{mass of O} = (6.49 - 2.66 - 0.30) \times 10^{-3}\ \text{g} = 3.53 \times 10^{-3}\ \text{g}$$

Now that we know the masses of C, H, and O in the sample, we can proceed as follows:

	C	H	O
mass	2.66×10^{-3} g	0.295×10^{-3} g	3.53×10^{-3} g
moles of atoms	$(2.66 \times 10^{-3}\ \text{g})\left(\frac{1\ \text{mol}}{12.01\ \text{g}}\right)$ $= 2.21 \times 10^{-4}$ mol	$(0.295 \times 10^{-3}\ \text{g})\left(\frac{1\ \text{mol}}{1.008\ \text{g}}\right)$ $= 2.93 \times 10^{-4}$ mol	$(3.53 \times 10^{-3}\ \text{g})\left(\frac{1\ \text{mol}}{16.00\ \text{g}}\right)$ $= 2.21 \times 10^{-4}$ mol
mole (atom) ratio	$\frac{2.21 \times 10^{-4}}{2.21 \times 10^{-4}} = 1.00$	$\frac{2.93 \times 10^{-4}}{2.21 \times 10^{-4}} = 1.33$	$\frac{2.21 \times 10^{-4}}{2.21 \times 10^{-4}} = 1.00$

One of these numbers (1.33) is too far from a whole number to be the result of experimental error in determining the composition. We must convert these numbers to whole numbers by multiplying by an appropriate factor, normally a small number such as 2, 3, or 4. In this case, multiplying by 2 does not give integral values, so we try multiplying by 3, which gives

$$\text{C} : \text{H} : \text{O} = 3.00 : 4.00 : 3.00$$

Thus, the *empirical formula* of ascorbic acid is $C_3H_4O_3$.

Exercise 5.5 Phenol, an important product of the chemical industry, is used in the production of certain plastics. It contains only C, H, and O. When a sample of phenol of mass 0.8874 g was burned in excess oxygen, 2.491 g of CO_2 and 0.510 g of H_2O were obtained. What is the empirical formula of phenol?

Molecular Formulas

From the composition of a substance, we can determine only the relative numbers of each kind of atom present—its *empirical formula.* If the compound is a molecular substance, we need to know its *molecular formula,* which is some integral multiple of the empirical formula, and to do this we must determine its *molecular mass.* There are several experimental methods for doing this. As we saw in Chapter 2, we can use the ideal gas equation, $PV = nRT$, to determine the number of moles in a sample of a gas of known mass and volume and thereby find the molar mass. But the most precise method for determining molecular mass is by **mass spectrometry**, as described in Box 1.3 and Chapter 14.

The molecular mass of ascorbic acid is found from mass spectrometry to be 176.1 u. As we saw in Example 5.6, the empirical formula of ascorbic acid is $C_3H_4O_3$, so the empirical formula mass is

$$3(12.01\text{ u}) + 4(1.008\text{ u}) + 3(16.00\text{ u}) = 88.06\text{ u}$$

We can determine the molecular formula of ascorbic acid from the ratio of its molecular mass to its empirical formula mass:

$$\frac{\text{molecular mass}}{\text{empirical formula mass}} = \frac{176.1\text{ u}}{88.06\text{ u}} = 2.00$$

There are twice as many atoms in the molecule as in the empirical formula, so the *molecular formula of ascorbic acid* is $(C_3H_4O_3)_2$, or $C_6H_8O_6$.

$\text{molecular formula} = n \times \text{empirical formula}$ where $n \geq 1$.

Exercise 5.6 The molecular mass of glucose is 180.2 u, and its empirical formula is CH_2O. Deduce its molecular formula.

5.5 STOICHIOMETRY OF GAS REACTIONS

In Chapter 2 we discussed **Avogadro's law**:

> **The volume of a given sample of gas at constant pressure and temperature is proportional to the number of moles of gas in the sample.**

In other words,

$$V \propto n, \quad \textit{or} \quad \frac{V}{n} = \text{constant} \quad (\text{at constant } P \text{ and } T)$$

Using Avogadro's law when we are dealing with gases, we can work with volumes of reactants rather than with moles of reactants. For instance, hydrogen reacts with chlorine according to the equation

$$H_2(g) + Cl_2(g) \rightarrow 2HCl(g)$$
$$1\text{ mol} + 1\text{ mol} \rightarrow 2\text{ mol}$$

or

$$22.4\text{ L} + 22.4\text{ L} \rightarrow 44.8\text{ L} \quad \text{(at STP)}$$

In general,

$$1\text{ volume} + 1\text{ volume} \rightarrow 2\text{ volumes}$$

For example,

$$1.0\text{ L} + 1.0\text{ L} \rightarrow 2.0\text{ L}$$

or

$$30.1\text{ mL} + 30.1\text{ mL} \rightarrow 60.2\text{ mL}$$

Exercise 5.7 The gas propane, C_3H_8, burns in oxygen to give carbon dioxide and water. Write the balanced equation for the reaction. How many liters of carbon dioxide are formed at 25°C and 1.00 atm when 5.00 L of propane at 25°C and 1 atm is burned in oxygen?

In many reactions, we are concerned with solids and liquids as well as gases. In calculations for such reactions, we work in moles, converting volumes of gaseous reactants to moles and moles of gaseous products to volumes as necessary.

EXAMPLE 5.7 Stoichiometry of Reactions Involving Gases

What volume of carbon dioxide, $CO_2(g)$, measured at 20°C and 1.00 atm, could be obtained by reacting 35.0 g of calcium carbonate, $CaCO_3(s)$, with excess hydrochloric acid, $HCl(aq)$?

$$CaCO_3(s) + 2HCl(aq) \rightarrow CaCl_2(aq) + H_2O(l) + CO_2(g)$$

Solution: The molar mass of $CaCO_3$ is $\{40.08 + 12.01 + 3(16.00)\}\ \text{g}\cdot\text{mol}^{-1} = 100\ \text{g}\cdot\text{mol}^{-1}$;

$$\text{moles of } CaCO_3 = \frac{35.0\text{ g } CaCO_3}{100\text{ g}\cdot\text{mol}^{-1}} = 0.350\text{ mol } CaCO_3$$

The equation for the reaction shows that 1 mol $CaCO_3$ gives 1 mol CO_2, so

$$\text{moles of } CO_2 = 0.350\text{ mol } CO_2$$

Then we use the ideal gas equation to calculate the volume of carbon dioxide at 20°C and 1.00 atm pressure:

$$V = \frac{nRT}{P} = \frac{(0.350\text{ mol})(0.0821\text{ atm}\cdot\text{L}\cdot\text{mol}^{-1}\cdot\text{K}^{-1})(293\text{ K})}{1.00\text{ atm}} = 8.42\text{ L}$$

Exercise 5.8 What volume of hydrogen at 26°C and 740 mm Hg would be obtained from the reaction of 2.00 g of magnesium with excess hydrochloric acid, $HCl(aq)$?

5.6 STOICHIOMETRY OF REACTIONS IN SOLUTION

Molarity

Many reactions are carried out in solution. For solutions of solids in liquids, the solid is called the **solute**, and the liquid is called the **solvent**. For reactions in

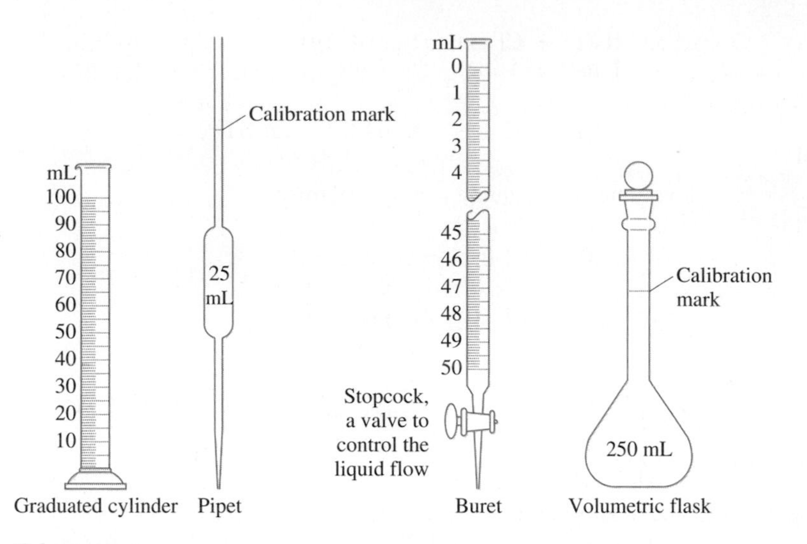

FIGURE 5.2 Apparatus for Measuring Volumes of Solutions

solution, it is convenient to take a known amount of a substance (the solute) by accurately measuring a given volume of a solution of the substance (Figure 5.2) rather than by weighing the pure substance. To do this we need to know how much of the substance is present in a given volume of the solution; that is, we need to know the **concentration** of the solution. There are several ways to express concentrations, but the most convenient for most purposes is in terms of the **molarity** (*molar concentration*) of the solute.

The molarity of a solute in a solution is the number of moles of solute contained in 1 L (1 dm^3 = 10^3 cm^3) of solution.

Thus, the units of molarity (with the symbol M) are $mol \cdot L^{-1}$. Whenever we need to calculate the molarity of a solute in a solution, we calculate the number of moles of solute and divide by the volume of the solution in liters:

$$\text{molarity, } M = \frac{\text{moles of solute}}{\text{liters of solution}} = mol \cdot L^{-1}$$

EXAMPLE 5.8 Solution Molarity

A solution is prepared by dissolving 20.36 g of NaCl in sufficient distilled water to give 1.000 L of solution. What is the molarity of NaCl in the solution?

Solution: To calculate molarity we need to know the number of moles of NaCl and the volume of the solution in liters. Because

$$\text{molar mass NaCl} = (22.99 + 35.45)\ g \cdot mol^{-1} = 58.44\ g \cdot mol^{-1}$$

we can write

$$\text{moles NaCl} = (20.36\ \text{g NaCl})\left(\frac{1\ \text{mol NaCl}}{58.44\ \text{g NaCl}}\right) = 0.3484\ \text{mol NaCl}$$

The volume of solution is 1.000 L; thus

$$\text{molarity NaCl} = \frac{0.3484\ \text{mol}}{1.000\ \text{L}} = 0.3484\ mol \cdot L^{-1} = 0.3484\ M$$

Although we may express the concentration of a solution of NaCl in terms of moles of NaCl, there are no sodium chloride molecules in such a solution. In solution, as in the solid state, NaCl consists of Na^+ and Cl^- ions, so we may write

$$NaCl(aq) \rightarrow Na^+(aq) + Cl^-(aq)$$

Each mole of NaCl gives 1 mol each of Na^+ and Cl^- ions. So for the solution in Example 5.8,

$$\text{molarity } Na^+(aq) = 0.3484\ M; \qquad \text{molarity } Cl^-(aq) = 0.3484\ M$$

Exercise 5.9 Suppose that 5.000 g of $BaCl_2(s)$ is dissolved in water to give 100.0 mL of solution. What are the molar concentrations of $BaCl_2(aq)$, $Ba^{2+}(aq)$, and $Cl^-(aq)$?

EXAMPLE 5.9 Solution Molarity

How many moles and how many grams of sodium hydroxide, NaOH, are in 25.0 mL of a 0.500-*M* NaOH(*aq*) solution?

Solution: A 0.500-*M* NaOH solution contains 0.500 mol NaOH in 1.000 L of solution:

$$0.500\ M\ \text{NaOH} = \frac{0.500 \text{ mol NaOH}}{1 \text{ L solution}}$$

We can use this ratio as a conversion factor to convert volume of solution to moles of solute:

$$\text{mol NaOH} = (25.0 \text{ mL solution})\left(\frac{1 \text{ L}}{1000 \text{ mL}}\right)\left(\frac{0.500 \text{ mol NaOH}}{1 \text{ L solution}}\right) = 0.0125 \text{ mol NaOH}$$

Now we can convert this result to grams (molar mass NaOH = 40.00 g · mol^{-1}):

$$\text{mass NaOH} = (0.0125 \text{ mol NaOH})\left(\frac{40.00 \text{ g NaOH}}{1 \text{ mol NaOH}}\right) = 0.500 \text{ g NaOH}$$

With practice, we can do the calculation in one step:

$$\text{mass NaOH} = (25.00 \text{ mL})\left(\frac{1 \text{ L}}{1000 \text{ mL}}\right)\left(\frac{0.500 \text{ mol}}{1 \text{ L}}\right)\left(\frac{40.00 \text{ g NaOH}}{1 \text{ mol NaOH}}\right) = 0.500 \text{ g}$$

Because 1 L = 1000 mL, dividing both sides by 1000 mL gives

$$\frac{1 \text{ L}}{1000 \text{ mL}} = 1$$

Use this factor when a volume expressed in milliliters must be converted to a volume in liters.

Note that molarity is defined as moles of solute in a given volume of solution, in liters. Thus, we *cannot* prepare a 1-*M* solution of NaOH by adding 1 mol NaOH(*s*) (40.00 g) to 1 L of water. Because of the additional volume occupied by the NaOH, the total volume of the solution would not be exactly 1.000 L, and therefore the concentration of the solution would not be exactly 1.000 *M*. A solution of known molarity is usually prepared by using a *volumetric flask* (Figure 5.2). This flask has a graduation mark on the neck; when the flask is filled with a solution exactly to this point, the volume of the solution is precisely known—for example, 100 mL, 250 mL, 500 mL, or 1 L. The procedure for making up a solution of known concentration by using such a flask is illustrated in Figure 5.3.

To make up a certain volume of a solution of known concentration, we must know what mass of the solute is needed. The required amount may be calculated as shown in Example 5.10.

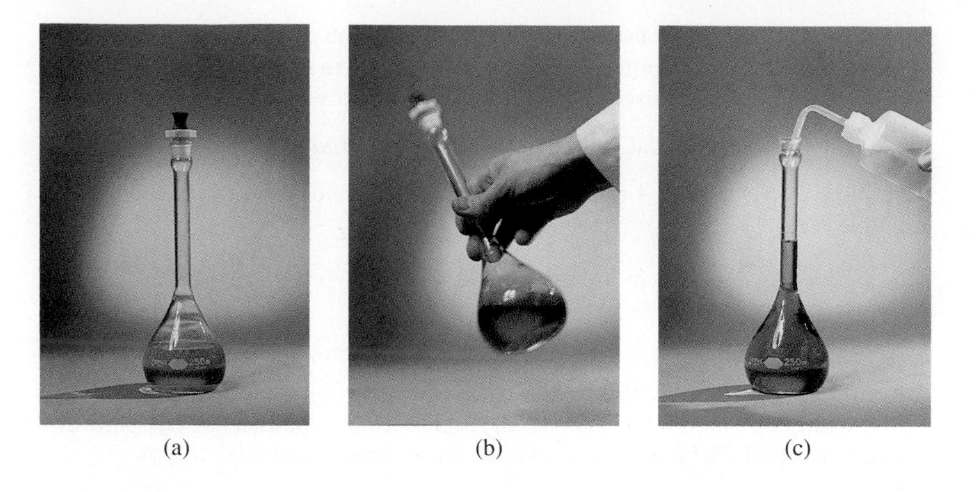

FIGURE 5.3 Using a Volumetric Flask (a) An accurately weighed amount of solute is added to a volumetric flask. (b) Distilled water is added, and the flask is shaken to dissolve the solute. (c) More distilled water is added until the level of the solution reaches the mark on the neck of the flask.

EXAMPLE 5.10 Amount of Solute Needed to Make a Solution of a Given Concentration

How many grams of sodium sulfate, $Na_2SO_4(s)$, are required to prepare 250 mL of 0.500-*M* $Na_2SO_4(aq)$ solution?

Solution: This problem may be restated as follows: How many grams of Na_2SO_4 are in 250 mL of 0.500-*M* $Na_2SO_4(aq)$?

A 0.500-*M* $Na_2SO_4(aq)$ solution contains 0.500 mol in 1 L of solution. To find the number of moles of Na_2SO_4 in 250 mL, we use the conversion factor (0.500 mol)/(1 L), so 250 mL of solution contains

$$(250\ \text{mL})\left(\frac{0.500\ \text{mol}}{1\ \text{L}}\right)\left(\frac{1\ \text{L}}{1000\ \text{mL}}\right) = 0.125\ \text{mol}$$

The molar mass of Na_2SO_4 is 142.0 $g \cdot mol^{-1}$. Therefore,

$$\text{mass } Na_2SO_4 = (0.125\ \text{mol } Na_2SO_4)\left(\frac{142.0\ \text{g } Na_2SO_4}{1\ \text{mol } Na_2SO_4}\right) = 17.8\ \text{g } Na_2SO_4$$

Or we can do the calculation in one step, as follows:

$$\text{mass } Na_2SO_4 = (250\ \text{mL})\left(\frac{0.500\ \text{mol } Na_2SO_4}{1\ \text{L}}\right)\left(\frac{1\ \text{L}}{1000\ \text{mL}}\right)\left(\frac{142.0\ \text{g } Na_2SO_4}{1\ \text{mol } Na_2SO_4}\right)$$
$$= 17.8\ \text{g } Na_2SO_4$$

Exercise 5.10 How many grams of solute are needed to prepare 500 mL of each of the following aqueous solutions?
(a) 0.100-*M* silver nitrate, $AgNO_3(aq)$
(b) 1.00-*M* sodium bromide, $NaBr(aq)$
(c) 0.200-*M* barium chloride, $BaCl_2(aq)$

Concentration of Trace Constituents

Very small concentrations, such as the concentrations of the noble gases (except argon) or of pollutants such as sulfur dioxide in the atmosphere, are often expressed in **parts per million**, ppm. For gases, a part per million refers to one unit by volume in 1 million volume units. A 1 ppm concentration of sulfur dioxide in the atmosphere means that pure sulfur dioxide would occupy 1 liter out of 1 million liters of the atmosphere. Because the volume of a gas is proportional to the number of molecules of that gas, a concentration of 1 ppm sulfur dioxide in the atmosphere means that only one out of 1 million (10^6) molecules is a sulfur dioxide molecule. The present concentration of carbon dioxide in the atmosphere is approximately 350 ppm.

The sensitivity of some modern analytical techniques is such that we can measure concentrations as small as **parts per billion** (ppb), or 1 part in 10^9.

For substances in solution, ''parts per million'' refers to grams of a substance per million grams of solution. For example, a concentration of 1 mg in 1000 g of solution is a concentration of 1 ppm. In general,

$$\text{ppm} = \frac{\text{g solute}}{\text{g soln}} \times 10^6 = \frac{\text{mg solute}}{\text{kg soln}} \simeq \frac{\text{mg solute}}{\text{L soln}}$$

The last expression is approximately true for dilute aqueous solutions, because the density of a dilute aqueous solution is very nearly 1 $\text{kg} \cdot \text{L}^{-1}$. Thus a solution that contains 2 mg of Hg^{2+} in 1 L of solution has a concentration of 2 ppm.

Acid-Base Reactions: Titrations

A common laboratory procedure is the determination of the concentration of a solution of a base by determining the volume of a solution of an acid of known concentration needed to react completely with a known volume of the solution of the base. Or if the concentration of a solution of a base is known, the concentration of a solution of an acid can be determined in the same way. This procedure is called a **titration.** A buret is filled with the solution of the acid of known concentration. The acid solution is added slowly to a known volume of the solution of the base of *unknown* concentration until just sufficient acid solution has been added to react with all the base. This point in the titration is called the **equivalence point**. It is determined by using a suitable *indicator* that changes color when the equivalence point is reached (Figures 5.4 and 5.5). From the amount of acid solution needed to react completely with the solution of the base, the concentration of the base solution can be calculated as shown in Example 5.11. The concentration of a solution of an acid can be determined in the same way by using a solution of a base of known concentration.

The use of indicators in acid–base titrations is discussed in more detail in Chapter 12.

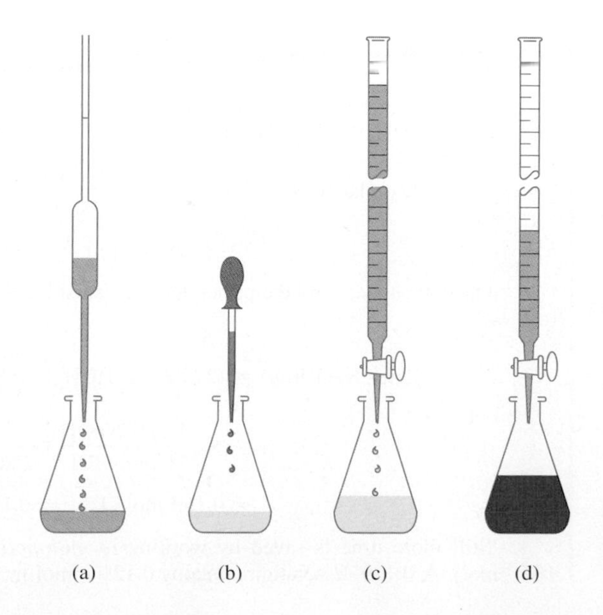

FIGURE 5.4 An Acid-Base Titration (a) A known volume of a solution of a base is placed in a flask, using a pipet. (b) A few drops of a suitable indicator are added. (c) An acid solution of known concentration is placed in a buret, and the level of the solution in the buret is noted. (d) The acid solution is run into the flask until the indicator just changes color. The level of the solution in the buret is then read again.

FIGURE 5.5 An Acid-Base Titration Just sufficient acid is added from a buret to react with all the base in the flask. This point is found by the change in color of the indicator—in this case, methyl red.

EXAMPLE 5.11 Acid–Base Titration

In a titration, 25.00 mL of NaOH(*aq*) reacted completely with 32.72 mL of HCl(*aq*) with a concentration of 0.129 *M*. Find the molarity of the NaOH solution.

Solution: First, we must write the balanced equation for the reaction:

$$\text{NaOH}(aq) + \text{HCl}(aq) \rightarrow \text{NaCl}(aq) + \text{H}_2\text{O}(l)$$

Then we proceed according to the following steps:

$$\text{volume HCl} \xrightarrow[1]{\times \textit{ molarity HCl}} \text{moles HCl} \xrightarrow[2]{\textit{balanced equation}} \text{moles NaOH} \xrightarrow[3]{\div \textit{ volume NaOH}} \text{molarity NaOH}$$

1. We know both the volume and the concentration of the HCl(*aq*), so we can determine the number of moles of HCl that reacted:

$$\text{moles HCl} = (32.72\text{ mL})\left(\frac{0.129\text{ mol HCl}}{1\text{ L}}\right)\left(\frac{1\text{ L}}{1000\text{ mL}}\right) = 4.22 \times 10^{-3}\text{ mol HCl}$$

III The definition of molarity is used twice in this calculation. In step 1, it appears in the form

moles = volume × molarity

In step 3, we find it as

moles/volume = molarity.

2. The balanced equation shows that 1 mol HCl reacts with 1 mol NaOH. Therefore,

$$\text{moles NaOH} = (4.22 \times 10^{-3}\text{ mol HCl})\left(\frac{1\text{ mol NaOH}}{1\text{ mol HCl}}\right) = 4.22 \times 10^{-3}\text{ mol NaOH}$$

3. This amount of NaOH was in 25.00 mL of solution; therefore the molarity (moles per liter) of the NaOH(*aq*) is given by

$$\text{molarity NaOH} = \left(\frac{4.22 \times 10^{-3}\text{ mol NaOH}}{25.00\text{ mL}}\right)\left(\frac{1000\text{ mL}}{1\text{ L}}\right)$$

$$= 0.169\text{ mol} \cdot \text{L}^{-1} = 0.169\ M\text{ NaOH}$$

Once we understand the principles, we can solve this type of problem in one step, as follows:

$$\text{molarity NaOH}(aq) = (37.72\text{ mL HCl})\left(\frac{0.129\text{ mol HCl}}{1\text{ L HCl}}\right)\left(\frac{1\text{ L}}{1000\text{ mL}}\right)$$

$$\times \left(\frac{1\text{ mol NaOH}}{1\text{ mol HCl}}\right)\left(\frac{1}{25.00\text{ mL}}\right)\left(\frac{1000\text{ mL}}{1\text{ L}}\right)$$

$$= 0.169\text{ mol} \cdot \text{L}^{-1} = 0.169\ M\text{ NaOH}$$

Still more time is saved by working in *millimoles* rather than *moles* (1 mmol = 10^{-3} mol). A 0.129-*M* solution contains 0.129 mmol in 1 mL. Therefore,

$$\text{molarity NaOH} = (32.72 \text{ mL HCl})\left(\frac{0.129 \text{ mmol HCl}}{1 \text{ mL}}\right)\left(\frac{1 \text{ mol NaOH}}{1 \text{ mol HCl}}\right)\left(\frac{1}{25.00 \text{ mL}}\right)$$

$$= 0.169 \text{ mmol} \cdot \text{mL}^{-1} = 0.169 \text{ mol} \cdot \text{L}^{-1} = 0.169 \; M \text{ NaOH}$$

We can use the same series of steps in any calculations of this type.

Exercise 5.11 A 25.00-mL sample of KOH(*aq*) reacted completely with 38.60 mL of a 0.0500-*M* solution of hydrobromic acid, HBr(*aq*). What was the concentration of the KOH solution?

Exercise 5.12 A 25.00-mL sample of $Ba(OH)_2(aq)$ reacted completely with 46.25 mL of a 0.0750-*M* solution of perchloric acid, $HClO_4(aq)$. What was the concentration of the $Ba(OH)_2(aq)$?

SUMMARY

Atoms, and therefore mass, are conserved in chemical reactions: the mass of products formed is the same as the mass of reactants consumed. The molar mass (the mass of 1 mole) of a substance in grams is the same as its molecular mass in atomic units, u. For a substance A, an amount in moles is converted to the equivalent mass, or a mass is converted to the equivalent number of moles, by using the appropriate conversion factor:

$$\text{mass of A} = (\text{mol A})\left(\frac{\text{molar mass of A}}{1 \text{ mol A}}\right)$$

$$\text{moles of A} = (\text{mass A})\left(\frac{1 \text{ mol A}}{\text{molar mass of A}}\right)$$

The balanced equation for a reaction gives the relationship between the number of moles of reactants used up and the number of moles of products formed.

In stoichiometric calculations for reactions in which all the reactants are used up and the initial amount of one reactant is known, we follow these four steps: (1) Write the balanced equation for the reaction; (2) convert the known mass of one reactant to moles; (3) use the balanced equation to set up the conversion factor(s) to find the number of moles of another reactant or product(s); (4) convert from moles back to grams.

When the amounts of substances reacting together do not have exactly the same mole ratios as they do in the balanced equation, the limiting reactant is the reactant that reacts completely, leaving excess amounts of the other reactants unreacted. We can find which is the limiting reactant by converting the given masses of reactants to moles and then deciding from the numbers of moles of reactants in the balanced equation which reactant will be used up completely.

The maximum amount of a product that can form when the limiting reactant is completely used up is its theoretical yield. The *actual* yield of a product, expressed as the percentage by mass of the theoretical yield, is its percent yield.

We find the empirical formula of a substance by determining its elemental composition (the elements present and their relative amounts expressed in moles). Among the several methods used are (1) chemical analysis to find the mass percentage of each element and hence the number of moles of each element; (2) synthesis from the elements to determine the combining masses of the elements and hence the number of moles of each element; (3) combustion (or a comparable reaction) to convert completely a sample of known mass to products whose masses can be determined and then converted to moles of each element. By each method we eventually arrive at the ratios of the number of moles of each element in a substance, which is the same as the ratios of the number of atoms of each element. Expressing these ratios in terms of the smallest whole numbers gives the empirical formula.

The composition of a substance gives only its empirical formula. To determine its molecular formula, which is always some whole-number multiple of the empirical formula, we must determine its molecular mass. This is commonly achieved by mass spectrometry. Another method is to calculate the number of moles, n_A, of a substance A in a gaseous sample of known mass, m_A g, from its measured pressure, volume, and temperature by means of the ideal gas law:

$$PV = n_A RT; \quad n_A = \frac{PV}{RT}$$

$$\textit{molar mass} = \frac{\textit{mass of A}}{n_A}$$

The stoichiometry of reactions among gases is governed by Avogadro's law; equal volumes of all gases at the same conditions of temperature and pressure contain equal numbers of moles of molecules. For gaseous reactions, this allows us to work with volumes of reactants and products rather than moles.

Stoichiometric calculations for reactions in solution are based on the volumes of the solutions and the concentrations of the reactants and products. Concentrations are conveniently expressed as molarities. The molarity of a substance in a solution is the number of moles of the substance (the solute) in 1 L of solution:

$$\text{molar concentration} = \frac{\text{moles of substance}}{\text{volume (liters) of solution}} = \frac{\text{mol}}{\text{L}}$$

A solution of known concentration is prepared by weighing out the required mass of solute and dissolving it in sufficient solvent, using a volumetric flask, so that the final volume is exactly known. In an acid–base titration, a solution of an acid (or base) is delivered from a buret into a flask containing a known volume of a base (or acid) until the equivalence point in the titration is reached—when *all* the base (or acid) has reacted. From the balanced equation for the acid–base reaction, the volumes of acid and base used, and the concentration of *either* the acid *or* the base, the concentration of the other can be determined.

Concentrations of very dilute solutions can be expressed in parts per million (ppm) or grams of solute in 10^6 g of solution as well as in moles per liter. The "ppm method" is also useful for gases, such as trace gases in the atmosphere, in which case "parts" refers to volumes.

IMPORTANT TERMS

Avogadro's law (page 168)
concentration (page 170)
equivalence point (page 173)
limiting reactant (page 160)
mass percentage (page 166)
molarity (page 170)
molar mass (page 158)
mole (page 158)
parts per billion (ppb) (page 173)
parts per million (ppm) (page 172)
percent composition (page 166)
percent yield (page 163)
solute (page 169)
solvent (page 169)
stoichiometry (page 157)
theoretical yield (page 162)
titration (page 173)

REVIEW QUESTIONS

1. State the law of conservation of mass.

2. What is Avogadro's number?

3. What is a mole?

4. Why is it convenient to count atoms and molecules in moles?

5. What four steps must be followed to calculate the mass of any product from the given mass of a reactant, assuming that any other reactants are present in excess and that the reaction goes to completion?

6. What is meant by the term "limiting reactant"?

7. What five steps must be followed to determine the amounts of products, given the initial masses of each reactant and assuming that the reaction goes to completion?

8. Define the terms "theoretical yield" and "percent yield."

9. What experimental data are needed to determine the empirical formula of a compound?

10. How can the empirical formula of an organic compound be determined from the results of a combustion experiment?

11. In addition to the empirical formula, what information is needed to determine the molecular formula of a compound?

12. For the complete combustion of ethane, $C_2H_6(g)$, at constant pressure and a temperature at which the products are gases, how is the volume of each of the products related to the initial volume of ethane?

13. How is the molarity of a solute in a solution defined?

14. What expression relates the molarity of a solute to the number of moles of solute and the volume of the solution?

15. Describe how exactly 500 mL of a 0.120-*M* solution of aqueous NaCl might be prepared in the laboratory from NaCl and distilled water.

16. How many grams of silver nitrate, $AgNO_3$, are in 25.20 mL of a 0.110-*M* $AgNO_3$ solution?

17. How is an acid–base titration used in the laboratory to determine the unknown molarity of an aqueous HCl solution if an aqueous solution of NaOH of known molarity is available?

PROBLEMS

Amounts of Reactants and Products

1. What are the molar masses of each of the following?
(a) H_2O (b) CH_4 (c) NH_3 (d) CO_2 (e) NaCl

2. What are the molar masses of each of the following?
(a) HNO_3 (b) H_2SO_4 (c) NaOH (d) $Ba(OH)_2$
(e) Al_2O_3

3. Convert each of the following to moles.
(a) 100.0 g of water (b) 5.000 g of methane
(c) 7.345 g of ammonia (d) 2.367 g of carbon dioxide
(e) 12.50 g of sodium chloride

4. Express each of the following in moles.
(a) 1.00 kg of HNO_3 (b) 25.40 g of H_2SO_4
(c) 10.00 g of NaOH (d) 25.00 g of $Ba(OH)_2$
(e) 0.3654 g of Al_2O_3

5. Balance each of the following equations, and calculate the mass of the second reactant that will react completely with 10.00 g of the first reactant.
(a) $HCl + Ba(OH)_2 \rightarrow BaCl_2 + H_2O$
(b) $Cl_2 + NaBr \rightarrow NaCl + Br_2$
(c) $NaOH + H_2SO_4 \rightarrow Na_2SO_4 + H_2O$
(d) $HNO_3 + CaO \rightarrow Ca(NO_3)_2 + H_2O$

6. Balance each of the following equations, and calculate the mass of product obtained from complete reaction of 1.000 kg of the first reactant.
(a) $NH_3 + H_2SO_4 \rightarrow (NH_4)_2SO_4$
(b) $Na_2O + H_2O \rightarrow NaOH$
(c) $P + Cl_2 \rightarrow PCl_3$
(d) $Na + Br_2 \rightarrow NaBr$

7. Ammonia gas, $NH_3(g)$, and hydrogen chloride gas, $HCl(g)$, react to give the white solid ammonium chloride, $NH_4Cl(s)$. Write the balanced equation for this reaction, and calculate the mass of HCl that reacts completely with 0.200 g of $NH_3(g)$.

8. Write the balanced equation for the reaction of $Cl_2(g)$ with silicon dioxide (silica), $SiO_2(s)$, and carbon, $C(s)$, to give silicon tetrachloride, $SiCl_4(l)$, and carbon monoxide, $CO(g)$. What is the maximum amount of $SiCl_4$ that can be obtained from 15.0 g of SiO_2?

9. Metals, $M(s)$, such as magnesium and zinc, react with dilute sulfuric acid to give hydrogen according to the equation

$$M(s) + H_2SO_4(aq) \rightarrow MSO_4(aq) + H_2(g)$$

When 5.00 g of a mixture of powdered Mg and Zn was dissolved in (excess) sulfuric acid, 0.284 g of hydrogen was collected. What was the composition of the mixture of Mg and Zn, expressed as mass percentages?

10. Determine the maximum amount of carbon dioxide, in liters at STP, that can be obtained by heating 1.00 kg of calcium carbonate, $CaCO_3(s)$, which decomposes according to the equation

$$CaCO_3(s) \rightarrow CaO(s) + CO_2(g)$$

11. A 1.000-kg sample of impure limestone containing 74.2% calcium carbonate, $CaCO_3(s)$, and 25.8% inert and nonvolatile impurities by mass is heated until all the carbonate has decomposed to calcium oxide, $CaO(s)$, and $CO_2(g)$. What mass of CO_2 is produced?

12. What volume of hydrogen at STP is required to (a) reduce 1.00 kg of zinc oxide, $ZnO(s)$, to zinc; (b) form 1.00 kg of water when burned in oxygen; (c) react with lithium to form 1.00 kg of lithium hydride, $LiH(s)$?

13. How many liters of oxygen at 25°C and 740 mm Hg pressure are required to burn 5.00 g of magnesium completely?

14. Hydrogen peroxide decomposes to water and oxygen. How many grams of hydrogen peroxide decompose to give 100 L of oxygen at 25°C and 1 atm?

Limiting Reactants

15. Sulfuric acid is produced when sulfur dioxide reacts with oxygen and water in the presence of a catalyst:

$$2SO_2(g) + O_2(g) + 2H_2O(l) \rightarrow 2H_2SO_4(l)$$

If 5.6 mol of SO_2 react with 4.8 mol of O_2 and a large excess of water, what is the maximum number of moles of H_2SO_4 that can be obtained?

16. Phosphorus trichloride, $PCl_3(l)$, reacts with chlorine, $Cl_2(g)$, to give phosphorus pentachloride, $PCl_5(s)$. What mass of PCl_5 results from the reaction of 5.15 g of PCl_3 with 3.15 g of Cl_2?

17. The unbalanced equation for the reaction of aluminum with hydrogen chloride (gas) is

$$Al(s) + HCl(g) \rightarrow AlCl_3(s) + H_2(g)$$

Determine the maximum mass of $AlCl_3$ produced, and the mass of Al or HCl that remains unreacted, when 2.70 g of Al reacts with 4.00 g of HCl.

18. A strip of zinc metal immersed in an aqueous solution of copper chloride, $CuCl_2(aq)$, reacts to give solid copper and zinc chloride, $ZnCl_2(aq)$:

$$Zn(s) + CuCl_2(aq) \rightarrow ZnCl_2(aq) + Cu(s)$$

If the zinc strip has a mass of 2.00 g and the solution initially contains 2.00 g $CuCl_2$, what mass of copper is produced?

Theoretical Yield and Percent Yield

19. Carbon in the form of graphite was heated strongly with sulfur, and the resulting carbon disulfide, $CS_2(l)$, was distilled off and condensed to liquid. If 2.530 g of graphite gave 12.50 g of CS_2, what was the percent yield?

20. Upon strong heating, potassium nitrate, $KNO_3(s)$, decomposes into potassium nitrite, $KNO_2(s)$, and oxygen. When 2.500 g of $KNO_3(s)$ was heated, the resulting solid mixture of $KNO_3(s)$ and $KNO_2(s)$ had a mass of 2.210 g. What mass of oxygen gas was produced, and what was its percent yield?

Determination of Empirical and Molecular Formulas

21. What are the mass percentage elemental compositions of each of the following?

(a) water, H_2O (b) sodium chloride, NaCl
(c) ethane, C_2H_6 (d) magnesium bromide, $MgBr_2$
(e) carbon dioxide, CO_2.

22. What are the mass percentage elemental compositions of each of the following?

(a) ammonia, NH_3 (b) chlorine, Cl_2
(c) sodium hydroxide, NaOH (d) ethanol, C_2H_6O
(e) nitrobenzene, $C_6H_5NO_2$

23. Calculate the mass percentage of each element in each of the following compounds.

(a) ethene, C_2H_4 (b) ethyne, C_2H_2
(c) magnesium chloride hexahydrate, $MgCl_2 \cdot 6H_2O$
(d) iron sulfate heptahydrate, $FeSO_4 \cdot 7H_2O$

24. What mass percentage of oxygen is contained in **(a)** 3.40 g of nitric acid, HNO_3, and in **(b)** 0.345 mol of sulfuric acid, H_2SO_4?

25. Ammonium sulfate, $(NH_4)_2SO_4$, is a common agricultural fertilizer.

(a) What is the percentage of nitrogen by mass in this compound?

(b) What mass of ammonium sulfate contains 100 g of nitrogen?

26. (a) What is the mass percentage of phosphorus in phosphoric acid, H_3PO_4?

(b) What mass of phosphorus is contained in 28.4 g of this compound?

(c) What mass of the compound contains 64.4 g of phosphorus?

27. The hydrocarbon anthracene has the composition 94.33% C and 5.67% H by mass. What are its empirical formula and empirical formula mass?

28. Caffeine has the composition 49.5% C, 5.2% H, 28.8% N, and 16.6% O by mass. What are the empirical formula and the empirical formula mass of caffeine?

29. When 3.10 g of a compound containing only carbon, hydrogen, and oxygen was completely burned in oxygen, 4.40 g CO_2 and 2.70 g H_2O were produced.

(a) What is the empirical formula of the compound?

(b) If its molecular mass is 62.1 u, what is its molecular formula?

30. When 3.62 g of a compound containing carbon, hydrogen, and oxygen was burned completely in air, 5.19 g of CO_2 and 2.83 g of H_2O were produced. What is the empirical formula of the compound?

31. When 0.100 mol of a compound of carbon, hydrogen, and nitrogen was burned completely in oxygen, 26.4 g of CO_2, 6.30 g H_2O, and 4.60 g of NO_2 were produced. What is the empirical formula of the compound?

32. Cyclopropane is a compound of carbon and hydrogen that is used as a general anesthetic. When 1.00 g of this substance was burned completely in oxygen, 3.14 g CO_2 and 1.29 g of H_2O were produced. What is the empirical formula of cyclopropane?

33. A sample of mass 6.20 g of a compound containing only sulfur, hydrogen, and carbon reacted completely with chlorine and gave 21.9 g of hydrogen chloride, HCl, and 30.8 g of tetrachloromethane, CCl_4. What is the empirical formula of the compound?

34. An oxide of nitrogen contains 3.04 g of nitrogen in a sample of mass 9.99 g. Its molecular mass is determined to be 92 u. What are the empirical formula and the molecular formula of the nitrogen oxide?

35. Upon heating in air, 2.862 g of a red copper oxide gave 3.182 g of a black copper oxide. When the latter was heated strongly in hydrogen, it gave 2.542 g of pure copper. What are the empirical formulas of the two copper oxides?

36. Cesium chloride contains 78.94% Cs by mass.

(a) What is the empirical formula of cesium chloride?

(b) How many grams of cesium metal give 4.34 g of cesium chloride upon complete reaction with chlorine?

37. A volatile compound has the composition 62.04% C, 10.41% H, and 27.55% O by mass. At 100°C and 1.00 atm pressure, 440 mL of the gaseous compound had a mass of 1.673 g. What are the compound's molar mass and molecular formula?

Stoichiometry of Gas Reactions

38. What volumes of $N_2(g)$ and $H_2(g)$, measured at STP, react completely to give 1.00 kg of $NH_3(g)$?

39. What volumes of $H_2(g)$ and $O_2(g)$, measured at 25°C and 754 mm Hg pressure, react completely to give 100 g of water?

40. $CH_4(g)$, methane (natural gas), reacts with steam to give *synthesis gas*, a mixture of $CO(g)$ and $H_2(g)$. What total volume of synthesis gas and what volumes of $CO(g)$ and $H_2(g)$ result from the complete reaction of 100 L of methane (all measured at STP)?

41. Coke (carbon) reacts with steam, $H_2O(g)$, to give *water gas*, a mixture of $CO(g)$ and $H_2(g)$. What volume of steam at 300°C and 1.00 atm pressure reacts completely with 1.00 kg of coke, and what are the resulting volumes of the $CO(g)$ and $H_2O(g)$ produced, in liters?

Stoichiometry of Reactions in Solution

42. How many moles of sodium hydroxide are contained in each of the following?

(a) 1.00 L of 0.0100-*M* NaOH(*aq*)

(b) 250 mL of 0.0100-*M* NaOH(*aq*)

(c) 25.15 mL of 0.0100-*M* NaOH(*aq*)

(d) 25.15 mL of 0.0134-*M* NaOH(*aq*)

43. How many moles of sulfuric acid are contained in each of the following?

(a) 4.00 L of 0.100-*M* $H_2SO_4(aq)$

(b) 125 mL of 0.100-*M* $H_2SO_4(aq)$

(c) 31.46 mL of 0.100-*M* $H_2SO_4(aq)$

(d) 31.46 mL of 0.151-*M* $H_2SO_4(aq)$

44. A 12.00-g sample of potassium permanganate, $KMnO_4(s)$, was dissolved in sufficient distilled water to give 2.00 L of solution. What was the molarity of potassium permanaganate in the solution?

45. **(a)** What mass of glucose, $C_6H_{12}O_6(s)$, must be dissolved in water to give 0.250 L of a 0.100-*M* solution?

(b) What volume of the resulting solution contains 0.0010 mol of glucose?

46. How many milliliters of concentrated sulfuric acid, which is 98.0% H_2SO_4 by mass and has a density of $1.842\ g \cdot mL^{-1}$, are needed to prepare 500 mL of a 0.175-*M* solution of $H_2SO_4(aq)$?

47. What volume of 85.0% H_3PO_4 by mass (density $1.659\ g \cdot mL^{-1}$) is needed to prepare 2.50 L of 1.50-*M* $H_3PO_4(aq)$?

48. How could each of the following be prepared?

(a) 6.30 L of 0.00300-*M* $Ba(OH)_2(aq)$ from a 0.100-*M* $Ba(OH)_2(aq)$ solution

(b) 750 mL of 0.0250-*M* $Cr_2(SO_4)_3(aq)$ from a solution containing 35.0% $Cr_2(SO_4)_3$ by mass (density $1.412\ g \cdot cm^{-3}$)

49. What is the molar concentration of 250 mL of a solution of $AgNO_3(aq)$ that contains 4.630 g of silver nitrate?

50. Write the balanced equation that describes the neutralization of any acid by any base in aqueous solution. Suppose that 250 mL of 1.00-*M* $HCl(aq)$ is neutralized by 500 mL of 0.500-*M* $NaOH(aq)$.

(a) What is the concentration of $NaCl(aq)$ in the resulting solution?

(b) When it is evaporated to dryness, what mass of sodium chloride is obtained?

51. What volume of 0.100-*M* $HCl(aq)$ reacts completely with 5.00 g of calcium hydroxide, $Ca(OH)_2(s)$?

52. **(a)** What volume of 0.124-*M* $HBr(aq)$ would neutralize 25.00 mL of 0.107-*M* $NaOH(aq)$ solution?

(b) What volume of 0.115-*M* $KOH(aq)$ solution would react completely with 100.0 mL of 0.211-*M* $HF(aq)$?

53. **(a)** What volume of 0.200-*M* $H_2SO_4(aq)$ would react completely with 0.100 g of $Al(OH)_3(s)$?

(b) What volume of 0.250-*M* $HCl(aq)$ would react completely with 22.6 g of sodium carbonate, $Na_2CO_3(s)$, according to the equation

$$Na_2CO_3(s) + 2HCl(aq) \rightarrow 2NaCl(aq) + H_2O(l) + CO_2(g)$$

54. **(a)** How much bromine is liberated when 25.00 mL of 0.102-*M* $NaBr(aq)$ reacts with excess $Cl_2(g)$?

(b) How much iodine is liberated when 10.00 mL of 0.120-*M* $Br_2(aq)$ is mixed with 10.00 mL of 0.100-*M* $KI(aq)$?

55. **(a)** What volume of 0.0120-*M* $HCl(aq)$ would react completely with 0.1240 g of magnesium metal?

(b) What is the concentration of the resulting $MgCl_2(aq)$ solution?

General Problems

56. An impure sample of limestone contains calcium carbonate, $CaCO_3(s)$, and inert and nonvolatile impurities. Upon heating, the reaction

$$CaCO_3(s) \rightarrow CaO(s) + CO_2(g)$$

occurs. When 1.000 g of the impure limestone was strongly heated to constant mass, 215 mL of gas was evolved at 25°C and a pressure of 755 torr. What was the mass percentage of $CaCO_3(s)$ in the limestone?

57. A sample of a metal X of mass 4.315 g combined with 0.481 L of $Cl_2(g)$, measured at 1.00 atm pressure and 20°C, to form a metal chloride of empirical formula XCl.

(a) Identify X. **(b)** Is the salt XCl soluble in water?

58. When heated, 0.2800 g of blue hydrated copper sulfate, $CuSO_4 \cdot xH_2O$, gave 0.1789 g of colorless anhydrous copper sulfate, $CuSO_4$. What is the empirical formula of hydrated copper sulfate?

59. Polychlorinated biphenyls, PCBs, known to be dangerous environmental pollutants, are a group of compounds with the general empirical formula $C_{12}H_mCl_{10-m}$, where m is an integer. What is the value of m, and hence the empirical formula, of the PCB that contains 58.9% chlorine by mass?

60. One of the CFC (chlorofluorocarbon) gases with the composition 11.50% C, 54.57% F, and 33.94% Cl by mass has a density of $4.66\ g \cdot L^{-1}$ at STP. What are its empirical formula mass, molecular mass, and molecular formula?

61. Concentrated nitric acid, $HNO_3(aq)$, is 69.0% HNO_3 by mass and has a density of $1.41\ g \cdot mL^{-1}$. What volume of concentrated nitric acid is required to prepare 250 mL of 0.100-*M* nitric acid?

62. The primary active ingredient in a common antacid is the salt $NaAl(OH)_2CO_3(s)$, which reacts with stomach acid according to the equation

$$NaAl(OH)_2CO_3(s) + 4HCl(aq) \rightarrow NaCl(aq) + AlCl_3(aq) + 3H_2O(l) + CO_2(g)$$

What mass of this salt would react completely with 2.00 L of 0.120-*M* $HCl(aq)$?

63. When 0.1573 g of zinc was reacted completely with 25.00 mL of 0.300-*M* $HCl(aq)$, the hydrogen gas evolved was burned to give 0.0434 g of water.

(a) What is the balanced equation for the reaction of zinc with dilute hydrochloric acid?

(b) What were the concentration of zinc chloride and the concentration of unreacted $HCl(aq)$ in the solution after the reaction was complete?

The Electronic Structure of Atoms and Molecules

6.1 Light and Other Electromagnetic Radiation
6.2 Atomic Spectra
6.3 Quantum Mechanics
6.4 Electron Configurations
6.5 The Hydrogen Molecule and the Covalent Bond

The light absorbed or emitted by atoms and molecules provides evidence that the energy of atoms and molecules is quantized—that is, restricted to distinct energy values. The beautiful, shimmering colors of the Aurora Borealis, or Northern Lights, of the northern sky are the result of molecules in the upper atmosphere that have been "excited" to higher energy levels by cosmic radiation from space and then emit light as the molecules return to their "ground states," or lowest energy levels.

In Chapters 3 and 4 we discussed two important types of substances, ionic and covalent. Ionic substances, which are solids under ordinary conditions, consist of oppositely charged ions held together in large three-dimensional networks (giant molecules) by the electrostatic attraction between the ions. Examples of ionic substances are sodium chloride, $NaCl(s)$, magnesium chloride, $MgCl_2(s)$, and magnesium oxide, $MgO(s)$. Covalent substances may be gases, liquids, or solids. They are composed of atoms held together by covalent bonds (shared electron pairs). Examples include the molecular substances water, $H_2O(l)$, chlorine, $Cl_2(g)$, methane, $CH_4(g)$, and the three-dimensional network (giant molecule) substance silicon dioxide, $SiO_2(s)$.

We have described how oppositely charged ions are held together by electrostatic attraction, but we have not yet answered the questions, Why does a covalent bond consist of a pair of shared electrons, and how does a pair of shared electrons hold two atoms together? To answer these questions and extend our knowledge and understanding of chemical bonds, we need to know more about the properties of electrons and their arrangements in atoms.

Most of our information on the structures of atoms and molecules comes from the study of the interaction of light and other forms of electromagnetic radiation, such as X-rays and infrared radiation, with atoms and molecules. These studies have shown that the laws of classical mechanics used to describe the behavior of ordinary-sized objects do not apply to very small particles such as atoms and electrons. To account for the behavior of electrons and other very small particles, the theory of **quantum mechanics** was developed in the 1920s. Because quantum mechanics had its origins in attempts to explain the nature of light and its interactions with atoms and molecules, we begin with a discussion of light and other forms of electromagnetic radiation. We then show how the light emitted or absorbed by an atom—the atomic spectrum—gives us important information about the structure of the atom. We show how more information about the arrangement of electrons in atoms can be obtained from ionization energies; *ionization energy* is the energy needed to remove an electron from an atom. Finally, we answer the two questions posed above and discuss the nature of the covalent bond.

From his experiments with light, Newton believed that light is made of particles. Here he is measuring the transmission of light through a prism.

6.1 LIGHT AND OTHER ELECTROMAGNETIC RADIATION

Isaac Newton (1642–1727), best known for his theory of gravitation and for his laws of motion, believed that light consists of a stream of particles. However, in the nineteenth century, the Scottish physicist James Clerk Maxwell (1831–1879) developed a theory in which light is described as an electromagnetic wave that travels

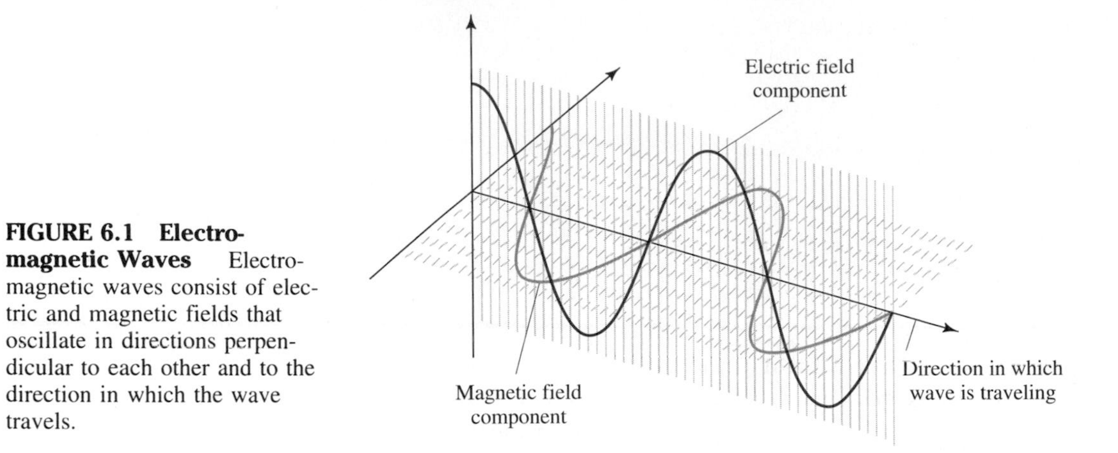

FIGURE 6.1 Electromagnetic Waves Electromagnetic waves consist of electric and magnetic fields that oscillate in directions perpendicular to each other and to the direction in which the wave travels.

through space. Using this theory, he was able to explain all the properties of light known at that time. But, as we shall see, other properties discovered later, particularly those relating to the interaction of light with atoms, could not be understood by Maxwell's theory. It became necessary to revive a particle theory of light. We have to accept the idea, strange as it may seem, that electromagnetic radiation has a dual nature, showing both wavelike and particlelike behaviors. Which behavior we observe depends on the type of experiment being carried out.

Wavelike Properties of Electromagnetic Radiation

There are many different types of waves, including electromagnetic waves, water waves, and sound waves. An **electromagnetic wave** is produced by an oscillating electric charge, such as the oscillating electrons in the antenna of a radio transmitter. An electromagnetic wave consists of oscillating electric and magnetic fields that increase and decrease in strength periodically in directions perpendicular to each other and perpendicular to the direction in which the wave is traveling (Figure 6.1). Water waves are easier to visualize. As a wave travels across the surface of water, the water level at any fixed position oscillates up and down (Figure 6.2). All waves can be described in terms of their frequency, wavelength, and amplitude.

Frequency and Wavelength One complete oscillation of the magnetic and electric fields in an electromagnetic wave or one complete oscillation of the surface of water from its highest level (crest) to its lowest level (trough) and back to its highest level in a water wave is called a *cycle* (Figure 6.2). The number of cycles

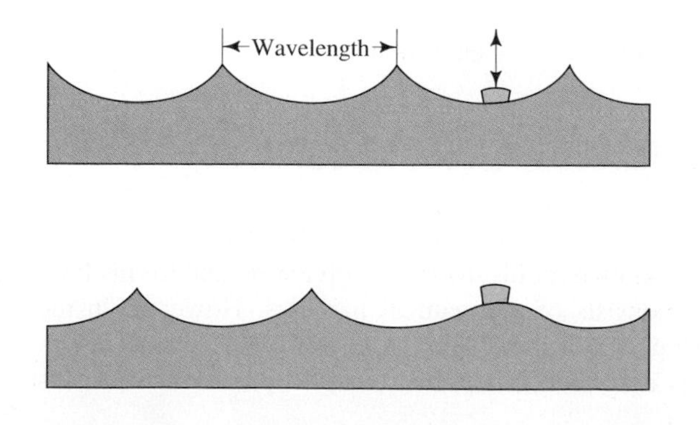

FIGURE 6.2 Water Waves As waves travel across the water surface, the water level at any point oscillates up and down, as shown by the movement of a cork floating on the surface. The crest-to-crest (or trough-to-trough) distance is one *wavelength*. The movement of the cork from a crest to a trough and back to a crest constitutes one *cycle*. The number of cycles per second is the *frequency* of the wave.

occurring per second is called the **frequency**, ν. The unit of frequency is defined as 1 cycle per second and is given the name **hertz (Hz)**:

ν is the Greek letter nu.

$$1 \text{ Hz} = 1 \text{ cycle per second}$$

In SI we omit "cycle," so

$$1 \text{ Hz} = 1 \text{ s}^{-1}$$

The frequency, ν, of a wave is the number of wave crests that pass a given point in 1 s. A frequency of 10 Hz means that there are 10 oscillations per second or that 10 wave crests pass a given point in 1 s.

The **wavelength**, λ, is the distance between points of equal displacement, such as the crests of successive waves (Figures 6.2 and 6.3). Suppose that ν waves per second move past a given point and that the length of each wave is λ. Then the distance traveled by the wave in 1 s is $\lambda\nu$, which is the wave's **speed,** v:

λ is the Greek letter lambda.

$$\text{speed} = (\text{frequency})(\text{wavelength})$$

$$v = \nu\lambda$$

For example, if the waves on a lake have a wavelength of 1.5 m and if they arrive at the shore at a frequency of 2 per second, the speed of the waves, v, is

$$v = (2 \text{ s}^{-1})(1.5 \text{ m}) = 3 \text{ m} \cdot \text{s}^{-1}$$

Light and all other types of electromagnetic radiation have the same constant speed in a vacuum. This speed, the **speed of light**, is $3.00 \times 10^8 \text{ m} \cdot \text{s}^{-1}$. It is given the symbol c:

The current best value of c is $2.997\,924\,6 \times 10^8 \text{ m} \cdot \text{s}^{-1}$. The value $3.00 \times 10^8 \text{ m} \cdot \text{s}$ is accurate enough for our purposes.

$$(\text{frequency})(\text{wavelength}) = \nu\lambda = c = 3.00 \times 10^8 \text{ m} \cdot \text{s}^{-1}$$

The higher the frequency of electromagnetic radiation, the shorter the wavelength; the lower the frequency, the longer the wavelength.

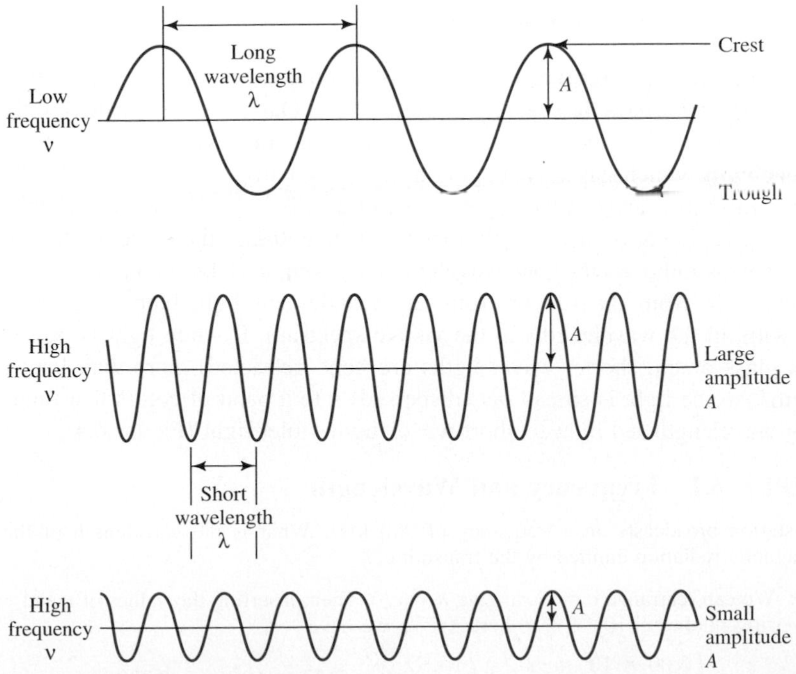

FIGURE 6.3 Properties of Waves Any wave can be described by its wavelength, frequency, speed, and amplitude. The speed of a wave is the horizontal distance traveled by a wave crest in 1 s; it is equal to the product of the wavelength and frequency. The energy of a wave is related to its amplitude, the vertical distance from the undisturbed position of the wave to its crest (or trough). For example, the greater the height of an ocean wave, the greater its destructive power; the greater the amplitude of a light wave, the more intense (brighter) the light.

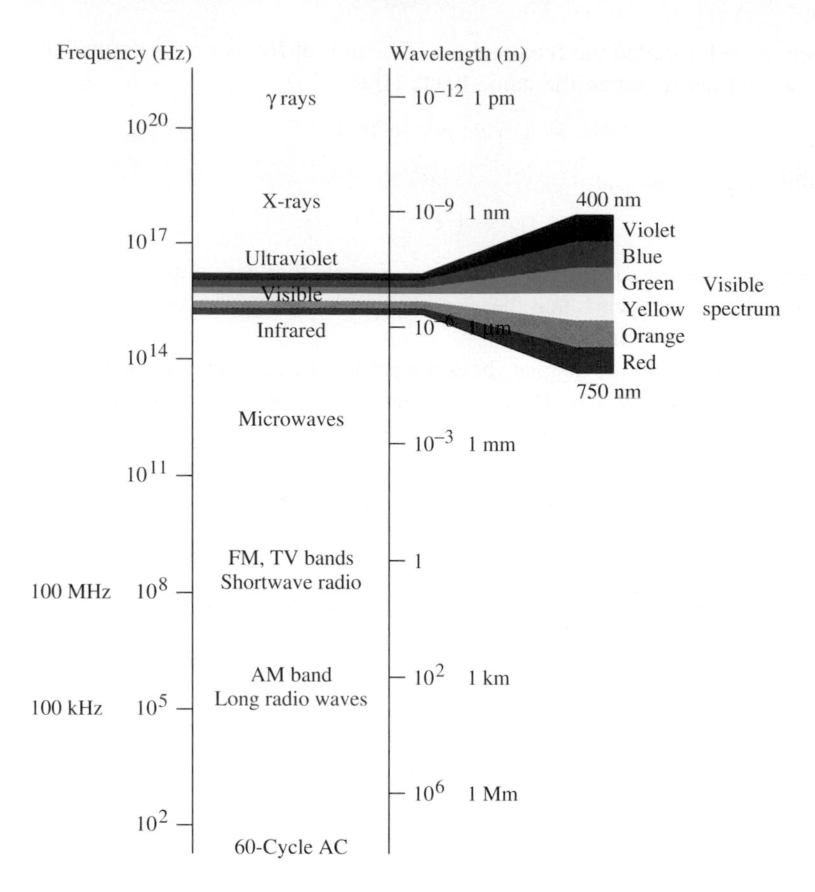

FIGURE 6.4 The Electromagnetic Spectrum All electromagnetic waves have the same speed in a vacuum, but their wavelengths and frequencies vary. Thus, electromagnetic waves with a low frequency have a long wavelength; those with a high frequency have a short wavelength. Electromagnetic radiation with a wavelength of about 400 to 750 nm is detectable by the human eye and thus constitutes visible light.

Amplitude The **amplitude,** A, of a wave is the height of a crest or the depth of a trough as measured from the central point of no displacement (Figure 6.3). The energy stored in a wave is proportional to A^2. In the case of light, the intensity (brightness) of the light is proportional to A^2.

The Electromagnetic Spectrum The complete range of electromagnetic waves is called the **electromagnetic spectrum** (Figure 6.4). Our eyes are sensitive only to the very small part of the electromagnetic spectrum that we call *light*, or the **visible spectrum**, which has wavelengths in the very narrow range 4×10^{-7} to 7.5×10^{-7} m. X-rays have wavelengths as short as 10^{-13} m. Ultraviolet, visible, and infrared radiation have increasingly longer wavelengths, in the range of 10^{-8} to 10^{-4} m, whereas radio waves have wavelengths as long as 1 km or more.

White light from the sun or from an incandescent light bulb consists of radiation with all the wavelengths in the visible spectrum. If white light is passed through a glass prism, shorter wavelengths are bent (refracted) more than longer wavelengths, so the light is spread out (dispersed) into a band of colors that range from long-wavelength red light to short-wavelength violet light (Figure 6.4).

EXAMPLE 6.1 Frequency and Wavelength

A radio station broadcasts on a frequency of 900 kHz. What is the wavelength of the electromagnetic radiation emitted by the transmitter?

Solution: We can rearrange $c = \nu\lambda$ to give $\lambda = c/\nu$. Then, inserting the values of c and ν and converting kilohertz ($10^3\ s^{-1}$) to hertz (s^{-1}), we have

$$\lambda = \frac{c}{\nu} = \left(\frac{3.00 \times 10^8\ \text{m} \cdot \text{s}^{-1}}{900\ \text{kHz}}\right)\left(\frac{1\ \text{kHz}}{10^3\ \text{s}^{-1}}\right) = 3.33 \times 10^2\ \text{m} = 0.333\ \text{km}$$

The equation $c = \nu\lambda$ can be used to solve for the wavelength or the frequency, depending on which quantity is given.

Exercise 6.1 The colors that make up the visible spectrum range in wavelength from 400 nm (violet) to 750 nm (red). What is the corresponding range of frequencies?

$1\ \text{nm} = 1 \times 10^{-9}\ \text{m}$

Particlelike Properties of Electromagnetic Radiation

Photons The results of experiments carried out near the end of the nineteenth century on the absorption and emission of electromagnetic radiation by atoms could be explained only if it was assumed that when light interacts with atoms and molecules, it behaves like a stream of particles. These particles are called **photons**. Each photon has an energy E that is proportional to the frequency of the light:

$$E_{\text{photon}} \propto \nu, \qquad \text{or} \qquad E_{\text{photon}} = h\nu$$

where the proportionality constant h is called the **Planck constant**. The value of the Planck constant has been determined experimentally to be $6.626\,18 \times 10^{-34}\ \text{J} \cdot \text{s}$. Blue light ($\nu = 6.4 \times 10^{14}\ \text{s}^{-1}$) consists of a stream of photons each with the energy

$$E = (6.626 \times 10^{-34}\ \text{J} \cdot \text{s})(6.4 \times 10^{14}\ \text{s}^{-1}) = 4.2 \times 10^{-19}\ \text{J}$$

The photons of yellow light, which has a lower frequency ($\nu = 5.5 \times 10^{14}\ \text{s}^{-1}$) than blue light, have a correspondingly lower energy:

$$E = (6.626 \times 10^{-34}\ \text{J} \cdot \text{s})(5.5 \times 10^{14}\ \text{s}^{-1}) = 3.6 \times 10^{-19}\ \text{J}$$

The more intense (brighter) a beam of light, the more photons arrive at a given point in 1 s. However, each photon carries such a small amount of energy that there is an enormous number of them in even a very feeble beam of light. So we are not conscious of the particle nature of light, just as we are not conscious of the individual molecules in water or any other substance.

Max Planck (1858–1947) was born in Kiel, Germany. His work in theoretical physics laid the foundation for quantum mechanics. In 1918 he won the Nobel Prize for physics. His name has been immortalized by the constant that relates photon energy to the frequency of radiation—the Planck constant.

The SI unit of energy and of work is the joule; $1\ \text{J} = 1\ \text{N} \cdot \text{m}$. See Chapter 9 and Appendix C.

EXAMPLE 6.2 Photon Energy

Calculate the energy of an X-ray photon of wavelength 100 pm and the energy of a radio-wave photon of wavelength 1.00 km.

Solution: When we are given the wavelength, λ, we first combine $E = h\nu$ with $\nu\lambda = c$, which gives $E = hc/\lambda$:

For X-rays

$$E_{\text{photon}} = \frac{hc}{\lambda} = \frac{(6.626 \times 10^{-34}\ \text{J} \cdot \text{s})(3.00 \times 10^{8}\ \text{m} \cdot \text{s}^{-1})}{(100\ \text{pm})\left(\dfrac{1\ \text{m}}{10^{12}\ \text{pm}}\right)} = 1.99 \times 10^{-15}\ \text{J}$$

For radio waves

$$E_{\text{photon}} = \frac{(6.626 \times 10^{-34}\ \text{J} \cdot \text{s})(3.00 \times 10^{8}\ \text{m} \cdot \text{s}^{-1})}{(1.00\ \text{km})\left(\dfrac{10^{3}\ \text{m}}{1\ \text{km}}\right)} = 1.99 \times 10^{-28}\ \text{J}$$

The Photoelectric Effect We mentioned in Chapter 3 that a characteristic property of metals is that they emit electrons when exposed to radiation of a sufficiently high frequency. This property of metals is called the **photoelectric effect**. Studies of the photoelectric effect were among those that led to the idea that light consists of photons. It has been observed that

1. No electrons are emitted unless the frequency of the light is greater than a minimum value that is characteristic for each metal, no matter how intense the light.
2. Electrons are emitted with a kinetic energy that increases with increasing frequency of the light used.
3. When electrons are emitted, the number of electrons is proportional to the intensity of the light.

Albert Einstein (Box 6.1) showed in 1905 that these observations could be explained only by assuming that light consists of photons. This was a revolutionary idea at the time, when it seemed that the wave theory of light had been firmly established.

BOX 6.1 Albert Einstein

Einstein was born in Germany in the old city of Ulm on the Danube. As a child, he was so slow at learning that his parents feared he might be retarded. In high school he disliked the harsh discipline, and when his family emigrated to Milan, Italy, Einstein left school and joined them. He was refused admission to the Swiss Federal Polytechnical School in Zurich because he did not have a high school diploma and failed the entrance examination, although he did very well in mathematics and physics. Einstein spent two years at a small college and finally was able to enter the Polytechnical School. He did not particularly impress his teachers, and after graduation he had difficulty finding employment.

After taking several part-time positions, Einstein went to work as a junior official in the Swiss Patent Office in Berne. The work apparently left him lots of time to think about theoretical physics. In 1905, at age 26, he published three articles, any one of which would have established him as one of the world's leading physicists. In the first, he proposed that light has a particlelike as well as a wavelike nature and explained the photoelectric effect. The second paper explained Brownian motion, the random erratic motion of very small particles suspended in a liquid, as being due to collisions with the rapidly moving molecules of the liquid. This paper provided the first *direct* evidence for the existence of molecules. The third showed that ideas of absolute space and time had to be replaced by the concept that space and time are relative to an independent observer (the theory of relativity). It was in this paper that Einstein derived the famous equation $E = mc^2$.

Einstein reached his revolutionary conclusions by means of rather simple but uncompromising logic based on experimental observations. Remarkably, he did all this work with little contact with other important physicists of the time.

After the publication of these papers, the University of Zurich offered him a position, and Einstein quickly became an important figure in the world of theoretical physics. In 1914 he was persuaded to move to Berlin as the head of the physics department of the world-famous Kaiser Wilhelm Institute. Despite his prestigious position, he was not entirely happy under the militaristic Prussian rulers of Germany, but he continued to work in Germany throughout World War I and the difficult years that followed. Ultimately, Hitler's oppression of Jews forced Einstein to leave. He arrived in New York City in October 1933 and stayed in the United States until his death in 1955.

Albert Einstein (1879–1955)

Albert Einstein is universally recognized as the greatest physicist of our age. Some say that if someone else had discovered the theory of relativity, Einstein's other work would have made him the second greatest physicist of his time. His ideas radically changed our concepts of space and time. He spent the final decades of his life attempting to develop a theory that would embrace both gravitation and electromagnetic phenomena (the unified field theory), but this goal eluded him. Thus far, it has eluded everyone else as well.

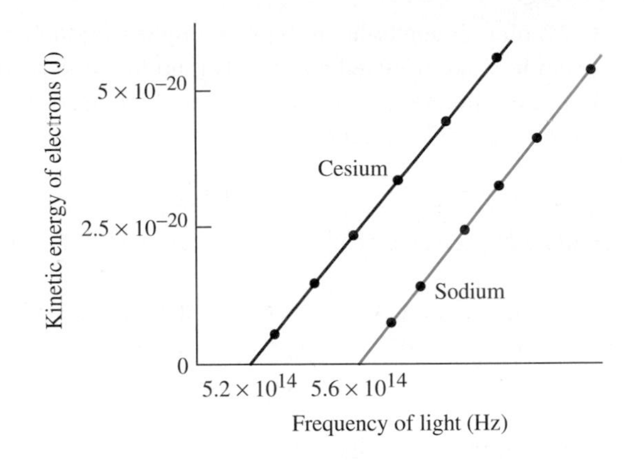

FIGURE 6.5 The Photoelectric Effect When the kinetic energy of the electrons ejected from a metal exposed to radiation is plotted against the frequency of the light, a straight line is obtained. The slope of this line gives the value of the Planck constant. Below a certain frequency, which depends on the metal, no electrons are ejected.

Let us see how Einstein's assumption enables us to understand the photoelectric effect. A photon must have a minimum energy, and therefore a minimum frequency, in order to eject an electron from a metal (observation 1 above), because the forces that hold the electron in the metal must be overcome. If the photon has more than the minimum energy, the excess energy appears as kinetic energy of the emitted electron. So the kinetic energy of the emitted electron is equal to the energy of the photon minus the energy needed to eject the electron from the metal:

$$\begin{pmatrix}\text{kinetic energy of}\\ \text{emitted electron}\end{pmatrix} = \begin{pmatrix}\text{energy}\\ \text{of photon}\end{pmatrix} - \begin{pmatrix}\text{energy needed to}\\ \text{eject electron}\\ \text{from the metal}\end{pmatrix}$$

$$KE_{\text{electron}} = h\nu - \text{constant}$$

This relationship shows that KE_{electron} increases with increasing frequency, ν, of the light (observation 2) and that a plot of KE_{electron} against ν should be a straight line of slope h. Plots for cesium and sodium are indeed straight lines whose slope gives the value for the Planck constant, $h = 6.626 \times 10^{-34}\ \text{J} \cdot \text{s}$ (Figure 6.5). When the photons have at least the required energy, each photon produces one electron. Thus, the number of electrons emitted per second is proportional to the number of photons arriving per second—in other words, to the intensity of the light (observation 3). According to the wave theory of light, the energy of light is related to its amplitude, that is, to its intensity and not to its frequency. So the wave model predicts that light of *any* frequency should be able to eject an electron from an atom provided that the light is intense enough, which is not what is observed (observation 1). The wave theory is therefore not in accord with this experimental observation. The photon model is needed to explain the photoelectric effect.

Photochemical Reactions The description of light in terms of photons is also needed for us to understand chemical reactions caused by the *absorption* of light. These reactions are known as **photochemical reactions**.

We saw in Chapter 4 that no reaction is observed in a mixture of hydrogen and chlorine unless the mixture is exposed to a bright light or is strongly heated. Then the reaction occurs with explosive violence:

$$H_2(g) + Cl_2(g) \rightarrow 2HCl(g)$$

Only blue-green light or light of higher frequency will cause this reaction. The reaction is not initiated by red light, no matter how intense, because a photon of red light does not have enough energy to *dissociate* (split) a chlorine molecule into two chlorine atoms, which is the first step of the reaction:

$$:\ddot{\underset{\cdot\cdot}{Cl}}:\ddot{\underset{\cdot\cdot}{Cl}}: + h\nu \rightarrow 2\cdot\ddot{\underset{\cdot\cdot}{Cl}}:$$

Only a higher-energy photon of blue-green light can dissociate a chlorine molecule in this way.

This first step is followed by the reaction of a chlorine atom with a hydrogen molecule to form a hydrogen chloride molecule and a hydrogen atom:

$$:\ddot{\underset{\cdot\cdot}{Cl}}\cdot + H:H \rightarrow :\ddot{\underset{\cdot\cdot}{Cl}}:H + \cdot H$$

The hydrogen atom then reacts with a chlorine molecule to form another hydrogen chloride molecule and another chlorine atom:

$$H\cdot + :\ddot{\underset{\cdot\cdot}{Cl}}:\ddot{\underset{\cdot\cdot}{Cl}}: \rightarrow H:\ddot{\underset{\cdot\cdot}{Cl}}: + \cdot\ddot{\underset{\cdot\cdot}{Cl}}:$$

This chlorine atom can then react with another hydrogen molecule, and so these two successive reactions can continue indefinitely. The overall reaction is the sum of these two reactions:

$$\begin{array}{c} Cl + H_2 \rightarrow HCl + H \\ H + Cl_2 \rightarrow HCl + Cl \\ \hline H_2 + Cl_2 \rightarrow 2HCl \end{array}$$

The very reactive chlorine and hydrogen atoms are used up as fast as they are formed and do not appear as products of the overall reaction. This reaction is an example of a *chain reaction* (Chapter 15). Although the equation shows only two reactant molecules, H_2 and Cl_2, the reaction does not proceed by collisions between H_2 and Cl_2 molecules but in a series of successive steps that are called the *mechanism* of the reaction (Chapters 4 and 15).

In Chapter 9 we will discuss energy changes in reactions and will see how to measure the energy needed to dissociate a diatomic molecule into two atoms. The energy needed to dissociate chlorine molecules into chlorine atoms is 243 $kJ \cdot mol^{-1}$. Knowing this, we can find the maximum wavelength of light that will initiate the hydrogen–chlorine reaction, as follows.

The energy needed to dissociate one chlorine molecule is

$$E = \left(\frac{243\ kJ}{1\ mol}\right)\left(\frac{1\ mol}{6.022 \times 10^{23}\ molecules\ Cl_2}\right)(1\ molecule\ Cl_2)\left(\frac{1000\ J}{1\ kJ}\right)$$
$$= 4.04 \times 10^{-19}\ J$$

Because one photon is needed to dissociate one chlorine molecule, each photon must have a minimum energy of 4.04×10^{-19} J. From this minimum energy, we could calculate the frequency of the photon from the relation

$$E_{photon} = h\nu$$

However, because

$$\lambda\nu = c$$

we can express the energy of the photon in terms of the wavelength of the light:

$$E_{photon} = h\nu = \frac{hc}{\lambda}$$

III A photon's energy is *directly* proportional to its frequency and *inversely* proportional to its wavelength.

We can then rearrange the equation to find the wavelength of the light:

$$\lambda = \frac{hc}{E_{photon}} = \frac{(6.626 \times 10^{-34}\ \text{J}\cdot\text{s})(3.00 \times 10^{8}\ \text{m}\cdot\text{s}^{-1})}{4.04 \times 10^{-19}\ \text{J}}$$

$$= (4.92 \times 10^{-7}\ \text{m})\left(\frac{10^{9}\ \text{nm}}{1\ \text{m}}\right) = 492\ \text{nm}$$

Any wavelength greater than this maximum wavelength corresponds to a smaller photon energy, which is insufficient to break the Cl—Cl bond.

This is the maximum wavelength of light that will dissociate a chlorine molecule. It is in the blue-green region of the spectrum.

Many other reactions are caused by the absorption of light by molecules. In the upper atmosphere many photochemical reactions involving oxygen and other molecules occur as a result of the absorption of ultraviolet radiation from the sun. One of these reactions is the formation of ozone from oxygen (see Chapters 7 and 16). In each case light of a certain minimum frequency is needed to cause a photochemical reaction.

Exercise 6.2 What is the maximum wavelength of light that will dissociate O_2 molecules into O atoms, given that the dissociation energy of O_2 is 498 $\text{kJ}\cdot\text{mol}^{-1}$?

6.2 ATOMIC SPECTRA

Studies of the emission and absorption of electromagnetic radiation by atoms have given us important information about the structures of atoms. The frequencies of light emitted or absorbed by an atom constitute its **atomic spectrum** (plural, spectra).

Line Spectra and the Quantization of Energy in the Atom

When the alkali metals or their compounds are heated to a high temperature in a flame, they impart distinctive colors to the flame that are characteristic of the particular metal. For example, lithium gives a red color; sodium, yellow; and potassium, lilac (Demonstration 6.1). When the light emitted by the flame is passed first through a slit (to obtain a narrow beam) and then through a glass prism, the light is dispersed into a spectrum. Unlike the spectrum obtained from sunlight, which contains all the colors in the visible region, this spectrum consists of only a few sharp lines, each of which is an image of the slit (Figure 6.6). It is called a **line**

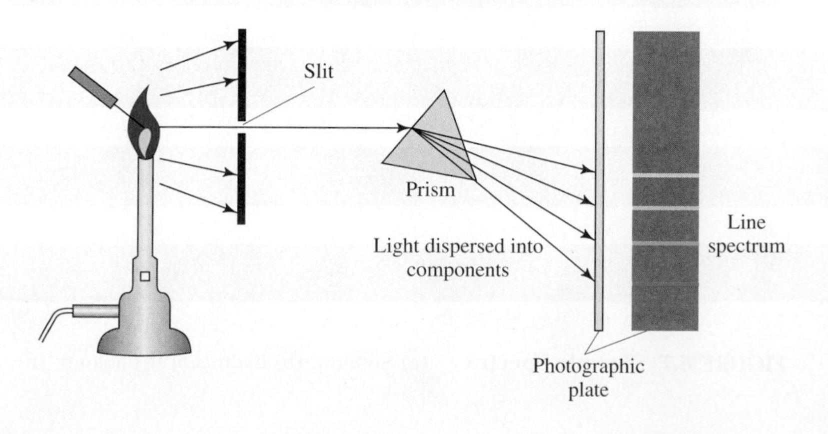

FIGURE 6.6 Obtaining an Atomic Line Spectrum When light from atoms of an alkali metal that has been heated to a high temperature in a flame is passed through a slit and a glass prism, a line spectrum is obtained. The pattern of lines is characteristic of the particular alkali metal.

DEMONSTRATION 6.1 Colored Flames and Atomic Spectra

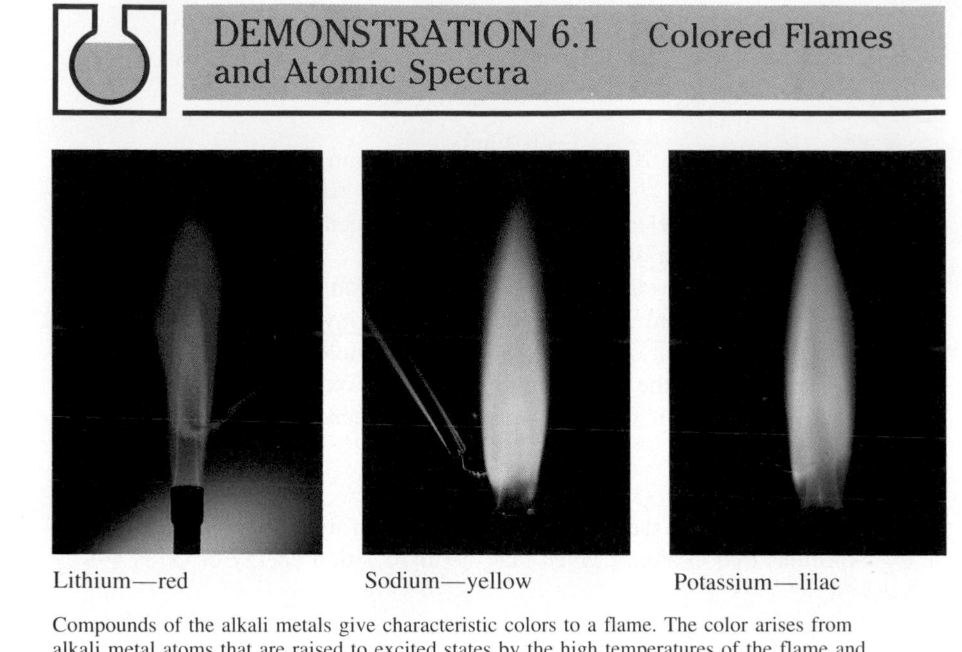

Compounds of the alkali metals give characteristic colors to a flame. The color arises from alkali metal atoms that are raised to excited states by the high temperatures of the flame and then return to the ground state, emitting light of characteristic frequencies. The flames shown here were produced by holding in the flame of a bunsen burner a platinum wire on which a small amount of the alkali metal chloride had been placed.

spectrum. Because this line spectrum is produced by emitted light, it is also known as an **emission spectrum**. Each alkali metal atom has its own characteristic line spectrum (Figure 6.7). Similarly, when an electric discharge is passed through helium, the gas glows with a characteristic blue-violet color. When this light is passed through a prism, the spectrum obtained consists of several sharp lines that are characteristic of helium (Figure 6.7).

What is the origin of these line spectra? When an alkali metal is heated in a flame to a high enough temperature, a sufficiently energetic collision between two fast-moving atoms may cause the electrons in one or both atoms to adopt an

FIGURE 6.7 Atomic Spectra (a) Sodium; (b) hydrogen; (c) helium; (d) neon.

arrangement of higher energy. In other words, some of the kinetic energy (energy of motion) of the atom is converted into potential energy (energy of position) of the electrons. These atoms are said to be raised to higher *energy states*. When an alkali metal compound is heated in a flame, it decomposes to give alkali metal atoms in high-energy states. When an electric discharge is passed through helium gas, the very fast moving electrons of the discharge collide with helium atoms, transferring some of their energy to the helium atoms and thus raising them to higher energy states. The normal, lowest energy state of an atom is called its **ground state**. Atoms that have been raised to a high-energy state are said to be in an **excited state**. An excited atom is unstable, and its electrons tend to rearrange to give a lower-energy arrangement. When this happens, the atom loses some or all of its excess energy by emitting light.

That an excited atom emits only certain frequencies of light and thereby gives a spectrum that consists of a limited number of sharp lines implies that the energy of an atom can change only by certain definite amounts. In other words, an atom cannot have a continuous range of energies but only certain definite energies. We say that

The energy of an atom is quantized.

If an atom has energy E_2 in an excited state and energy E_1 in the ground state, where $E_2 > E_1$, then the energy emitted when the atom returns from the excited state to the ground state is $E_2 - E_1$. This energy is emitted as a single photon of energy $h\nu$, where ν is the frequency of the emitted light:

$$E_{\text{photon}} = \Delta E = E_2 - E_1 = h\nu$$

Each frequency in the emitted light, and thus each line in the spectrum, corresponds to the difference in energy between two different states (Figure 6.8). In Figure 6.8 each energy state is shown as a different level, and we usually call an energy state an **energy level**. An atom can have many different energy levels (E_1, E_2, E_3, and so on), but it cannot have an energy that does not correspond to one of these levels. A set of energy levels is like a set of shelves, each one at a certain height. An object can be placed on any of the shelves but not anywhere in between. The object can have the potential energy that corresponds to the height of any particular shelf, but it cannot have any potential energy between these values. We can determine the energy levels of an atom from its spectrum, as we shall see now for the hydrogen atom.

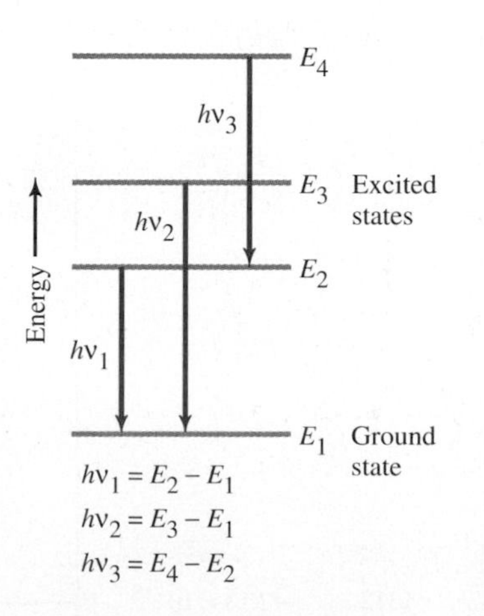

FIGURE 6.8 Energy Levels Each atom has a characteristic set of energy levels. An atom emits light when an electron returns from an excited state to a lower-energy state. Each frequency in the line spectrum therefore corresponds to the difference in energy between two energy levels of the atom.

The Spectrum of the Hydrogen Atom

The spectrum of the hydrogen atom consists of a series of lines in the visible region of the spectrum and other series in the infrared and ultraviolet regions of the spectrum (Figure 6.9). The lines in the ultraviolet and infrared regions of the spectrum cannot be seen by the human eye but can be detected electronically or recorded on suitable photographic film.

Niels Bohr (1885–1962) was born in Copenhagen, Denmark. After attending college there, he studied under Ernest Rutherford, the discoverer of the nucleus, in Cambridge, England. When Bohr was 37, he won the Nobel Prize in physics; he was the youngest winner up to that time. After World War II he was very active in the effort to harness atomic energy for peaceful purposes.

In 1913 Niels Bohr proposed a simple model to account for the spectrum of the hydrogen atom. According to the **Bohr model**, the single electron revolves around the nucleus in one of a limited number of circular orbits. When the electron is in the orbit closest to the nucleus, it has the lowest possible energy, and the hydrogen atom is in its ground state. When the hydrogen atom is in an excited state, the electron is in one of the orbits farther from the nucleus. This electron has a higher energy, because moving the electron away from the nucleus requires that work be done against the electrostatic attraction of the nucleus. When an electron moves from a higher to a lower energy level, that is, when it moves closer to the nucleus, the atom emits a photon of light of energy equal to the difference in energy between the two levels. The movement of an electron from one level to another is called a *transition*.

Bohr was able to calculate from his model that the energy levels of the hydrogen atom are given by the equation

$$E_n = -2.18 \times 10^{-18}\left(\frac{1}{n^2}\right)\text{ J}$$

where $n = 1, 2, 3, \ldots$. The integer n is called a **quantum number**. The ground state of the hydrogen atom (the lowest energy level) corresponds to $n = 1$ and has an energy

$$E_1 = -2.18 \times 10^{-18}\text{ J}$$

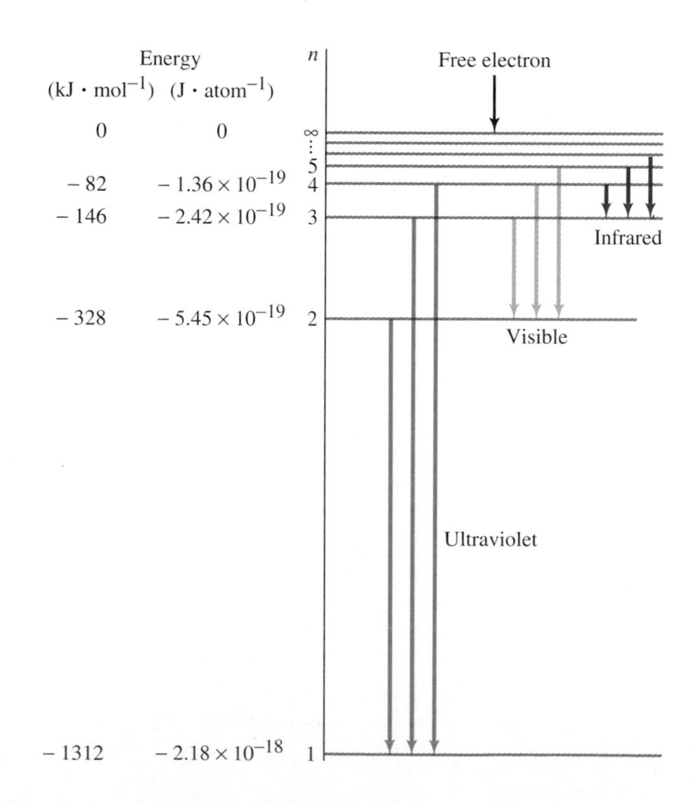

FIGURE 6.9 Energy Levels and the Spectrum of the Hydrogen Atom The series of lines in the ultraviolet region of the spectrum arise when an electron moves from an upper energy level to the ground-state level ($n = 1$). The series of lines in the visible region of the spectrum result from transitions from an upper level to the $n = 2$ level. The lines in the infrared region result from transitions from upper levels to the $n = 3$ level.

The first excited state (the next-lowest energy level) corresponds to $n = 2$ and has an energy

$$E_2 = \frac{-2.18 \times 10^{-18}}{4} \text{ J} = -0.545 \times 10^{-18} \text{ J}$$

The other excited states then correspond to $n = 3, 4, \ldots$ (Figure 6.9).

It is sometimes convenient to express the energy of a given level per mole of H atoms rather than per atom. The energy of 1 mol of H atoms in their ground state is

$$E_1 = (6.022 \times 10^{23} \text{ atoms} \cdot \text{mol}^{-1})(-2.18 \times 10^{-18} \text{ J} \cdot \text{atom}^{-1})$$
$$= -1.31 \times 10^6 \text{ J} \cdot \text{mol}^{-1} = -1.31 \text{ MJ} \cdot \text{mol}^{-1}$$

The energy of 1 mol of H atoms in some of the excited states of hydrogen is given in Figure 6.9.

The negative sign in these expressions for the energy of the electron in the hydrogen atom may seem strange. The energy of the electron in the hydrogen atom is *lower* than the energy of a free electron—an electron at an infinite distance from the nucleus—which is arbitrarily set at zero. This corresponds to setting $n = \infty$ so that $E_\infty = 0$. Thus, an electron in the hydrogen atom always has a negative energy.

When a hydrogen atom in the first excited state, $n = 2$, emits a photon and returns to the ground state, $n = 1$, the energy of the photon is given by

$$E_{\text{photon}} = E_2 - E_1 = -2.18 \times 10^{-18}\left(\frac{1}{2^2} - \frac{1}{1^2}\right) \text{ J} = 1.64 \times 10^{-18} \text{ J}$$

Both E_2 and E_1 are negative, but E_2 is less negative than E_1. Thus the difference $E_2 - E_1$ is positive. An equivalent expression is

$$E_{\text{higher}} - E_{\text{lower}} = E_{\text{photon}}$$

Photons of this energy produce a line in the spectrum of frequency

$$\nu = \frac{E_{\text{photon}}}{h} = \frac{1.64 \times 10^{-18} \text{ J}}{6.626 \times 10^{-34} \text{ J} \cdot \text{s}} = 2.48 \times 10^{15} \text{ s}^{-1}$$

and of wavelength

$$\lambda = \frac{c}{\nu} = \frac{3.00 \times 10^8 \text{ m} \cdot \text{s}^{-1}}{2.47 \times 10^{15} \text{ s}^{-1}} = 1.21 \times 10^{-7} \text{ m} = 121 \text{ nm}$$

In general, the energy of the photon obtained when an electron falls from an initial high-energy orbit with quantum number n_i to a final orbit of lower energy with quantum number n_f is given by

$$E_{\text{photon}} = -2.18 \times 10^{-18}\left(\frac{1}{n_i^2} - \frac{1}{n_f^2}\right) \text{ J}$$

It is convenient to rewrite this equation as

$$E_{\text{photon}} = 2.18 \times 10^{-18}\left(\frac{1}{n_f^2} - \frac{1}{n_i^2}\right) \text{ J}$$

to remove the negative sign.

EXAMPLE 6.3 Calculating the Wavelength of a Hydrogen Spectrum Line

Calculate the wavelength of the longest-wavelength line in the visible region of the spectrum of the hydrogen atom, and identify its color.

Solution: The longest-wavelength line will result from the transition of smallest energy in the visible series of lines, which is the transition from the $n = n_i = 3$ level to the $n = n_f = 2$ level. Substituting in the equation

All the hydrogen emission lines that appear in the *visible* region of the spectrum end in the $n = 2$ level.

$$E_{\text{photon}} = 2.18 \times 10^{-18}\left(\frac{1}{n_f^2} - \frac{1}{n_i^2}\right) \text{J}$$

we have

$$E_{\text{photon}} = 2.18 \times 10^{-18}\left(\frac{1}{2^2} - \frac{1}{3^2}\right) \text{J} = 3.03 \times 10^{-19} \text{J}$$

To find the wavelength of this photon, we use the relationships

$$E_{\text{photon}} = h\nu \quad \text{and} \quad \lambda\nu = c$$

Therefore

$$\lambda = \frac{hc}{E_{\text{photon}}} = \frac{(6.626 \times 10^{-34} \text{ J} \cdot \text{s})(3.00 \times 10^{8} \text{ m} \cdot \text{s}^{-1})}{3.03 \times 10^{-19} \text{ J}} = 656 \text{ nm}$$

This wavelength is in the red region of the spectrum.

Exercise 6.3 **(a)** What is the energy change associated with the transition of a hydrogen atom in the $n = 5$ excited state to the ground state?

(b) What are the frequency and wavelength of the corresponding line in the spectrum?

(c) In what region of the spectrum is this line found?

Spectra of Other Atoms

Because they have more than one electron, other elements have more complicated atomic spectra than does hydrogen, but they can all be accounted for in terms of a set of energy levels characteristic of each element. However, it became clear in the 1920s that Bohr's model of the hydrogen atom could not completely explain the spectra of other atoms and that it was not consistent with the ideas of quantum mechanics, which were being developed at that time. The ''planetary atom'' model, with electrons in fixed orbits, has been replaced by another model, which we discuss in Section 6.3. The lasting contribution of the Bohr model has been the concept of quantized energy states, a concept that is essential to our understanding of atoms.

Because the atomic spectrum of each element is unique, atomic spectra provide a very convenient method for identifying elements. For example, the sodium spectrum has two characteristic intense lines in the yellow region, which enable sodium and its compounds to be recognized very easily.

The emission of light by excited atoms also has some practical uses in everyday life. The yellow light of some street lamps is emitted by excited sodium atoms produced by bombardment of the atoms in sodium vapor by the fast-moving electrons of an electric discharge. This type of lamp produces less heat than does an incandescent light bulb and therefore wastes less energy. Another type of street lamp, the mercury lamp, gives a blue-green light that arises from excited mercury atoms produced in mercury vapor by an electric discharge. The red light emitted by excited neon atoms is familiar from neon signs, which are made of tubes that contain neon gas through which an electric discharge is passed. Other colors are obtained by using other gases, such as helium and mercury vapor.

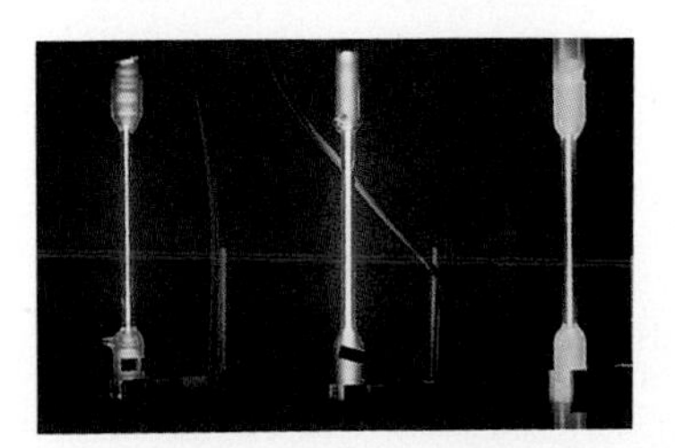

Hydrogen, helium, and mercury discharge tubes

So far we have been discussing the emission spectra produced by the photons emitted when an atom changes from a higher-energy state to a lower-energy state. However, atomic spectra can also be observed when an atom *absorbs* light and is raised to a higher energy level. If we pass white light through a gaseous sample of an element, certain frequencies are absorbed such that these frequencies are missing

FIGURE 6.10 An Absorption Spectrum When white light is passed through a sample of a substance, some frequencies are absorbed, and dark lines are produced in the otherwise continuous spectrum of white light. The absorption spectrum that results is characteristic of the substance—in this case, the protein cytochrome.

from an otherwise continuous spectrum (Figure 6.10). This is called an **absorption spectrum** rather than an emission spectrum. **Atomic absorption spectroscopy** is a very important and sensitive method for analyzing for elements in a sample of almost any material. It is particularly useful in analyzing for trace elements in food and water (Box 5.1).

6.3 QUANTUM MECHANICS

We have seen that light can behave both as a wave and as particles (photons). Likewise, electrons, protons, and neutrons, which were originally thought of as particles, also can behave as waves. That electrons exhibit a wavelike behavior was a revolutionary idea when it was first proposed by the French physicist Louis de Broglie (Figure 6.11) in 1924, but it has been convincingly confirmed by experiment. For example, if a beam of electrons is passed through a thin metal foil, a diffraction pattern is obtained that is quite similar to that obtained by passing X-rays through the same metal foil (Box 6.2).

FIGURE 6.11 Louis de Broglie (1892-1977) DeBroglie was the first to propose that particles such as electrons exhibit wavelike behavior.

The wavelike behavior of electrons helps us to understand why we cannot determine the path that an electron takes around the nucleus in an atom. The wavelength of an electron in an atom is comparable to the size of the atom. Therefore we cannot say precisely where an electron is located in an atom at a particular time, and we cannot, as Bohr did, say that its motion is confined to a precise circular orbit and that this motion can be described by classical mechanics. We must use *quantum mechanics*, according to which, if we think of the electron as a particle, we can find only the *probability* that the electron will be found at some particular location. For the hydrogen atom in its ground state, that is, with the single electron in the lowest energy level, the probability of finding the electron is greatest close to the proton and decreases with increasing distance from the proton. There is a 90% probability that the electron will be found somewhere inside a spherical surface of radius 1.4×10^{-10} m (140 pm) and only a 10% probability that it will be found at a greater distance (Figure 6.12). Alternatively, we may say that in the ground state of the hydrogen atom, the electron spends 90% of its time within a sphere of radius 1.4×10^{-10} m and 10% of its time outside that sphere. We call this description of the electron's probable location a *probability distribution*. If we were able to take a time-exposure photograph of the atom, we would not see the electron

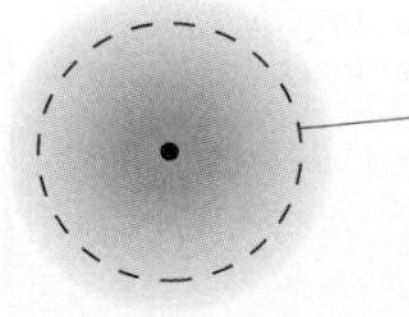

FIGURE 6.12 Electron Probability (or Density) Distribution of the Hydrogen Atom The probability of finding the electron at any point in this model of the hydrogen atom varies with the intensity of color; it is greatest nearest the nucleus. There is only a 10% chance of finding the electron outside the dashed circle. Alternatively, we can think of the electron as a spherical cloud of negative charge around the nucleus. This electron cloud increases in density toward the nucleus, and 90% of it is within a radius of 1.4×10^{-10} m.

BOX 6.2 Interference and Diffraction

An important characteristic of waves is that they exhibit *interference*. This phenomenon is observed when waves from two or more sources meet. We can produce waves on the surface of water by dipping a vibrating object into the water surface. Waves radiate out in all directions from the source (Figure A). In a similar way, vibrating electrons in metal wire, such as a radio transmitting antenna, produce oscillating electric and magnetic fields that radiate out from the wire.

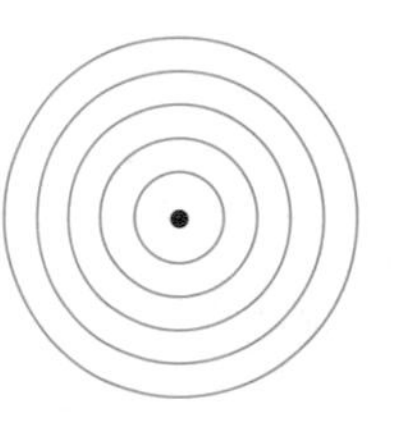

Figure A *A vibrating object sends out a circular wave pattern on the surface of water.*

When waves from two sources meet, they give a resultant wave whose amplitude is the sum of the amplitudes of the individual waves. When a wave crest meets a wave crest, the two waves are said to be *in phase*, and if the two waves have equal amplitude, the resultant amplitude is doubled (Figure B1). The two waves are said to be undergoing *constructive interference*. When a wave crest meets a trough, the waves are said to be *out of phase*, and the resultant amplitude is zero if the two waves have equal amplitude (Figure B2). In general, the two waves are said to be undergoing *destructive interference*. The result of the interacting waves is a pattern consisting of regions of increased amplitude and regions of decreased amplitude called an *interference pattern*. An example of an interference pattern produced by two water waves is shown in Figure C.

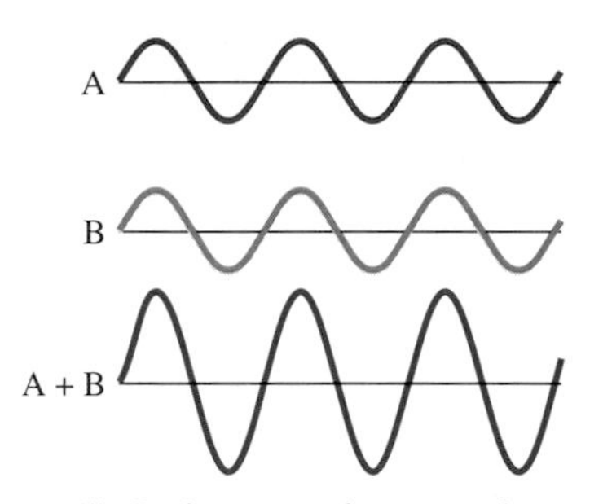

(1) In-phase waves give constructive interference.

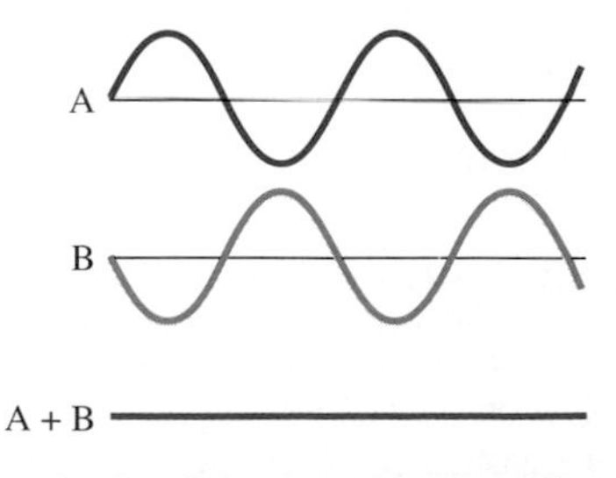

(2) Out-of-phase waves give destructive interference.

Figure B *Constructive and destructive interference*

Figure C *An interference pattern produced on a water surface by two vibrating objects dipping into the surface.*

Interference between light waves can be demonstrated if light of a single wavelength (monochromatic light) from a single source is allowed to pass through two very narrow slits close together (Figure D). If the slits are sufficiently narrow, that is, if the width of each slit is not much greater than the wavelength of the light, light waves spread out from each slit and interfere with each other. When projected onto a screen, the waves produce a series of light

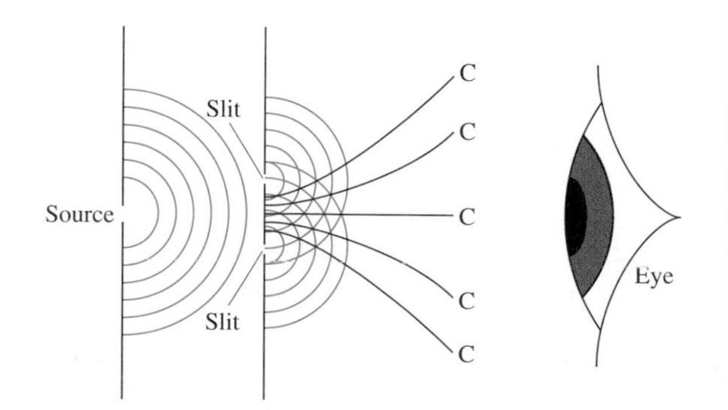

Figure D *Production of an interference pattern by the light emitted through two narrow slits situated close together. The eye observes lines of increased intensity separated by regions of low intensity. The magenta lines C represent lines of constructive interference.*

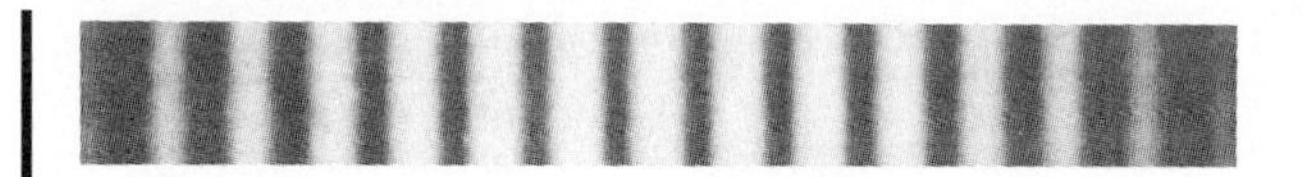

Figure E *Interference pattern produced by the arrangement in Figure D as recorded on a photographic plate.*

and dark lines (Figure E) rather than separate images of each slit.

The spacing of the lines in an interference pattern depends on the wavelength of the light. A *diffraction grating* obtained by ruling many lines (say, 1000 per millimeter) on a sheet of glass is a useful device for measuring the wavelength of light. Conversely, if we know the wavelength of the light, we can find the spacing between the lines of a diffraction grating. The regular arrangement of atoms in a crystal can similarly act as a diffraction grating. However, because the distance between the atoms is smaller than that between the lines in an ordinary diffraction grating, we must use shorter-wavelength radiation, such as X-rays, in order to obtain a diffraction pattern. We can get accurate information about the geometric arrangement of atoms or ions in a crystal and the distances between them from the X-ray diffraction pattern produced by a crystal (Box 11.1).

When a beam of electrons is passed through a thin metal foil, a diffraction pattern consisting of a series of light and dark rings is produced on a photographic plate that is very similar to the pattern produced by X-rays passing through the same metal foil. The independent observations of a diffraction pattern produced by electrons (Figure F) by Clinton Davisson and Lester Germer in the United States and by George Thomson in Britain in 1927 provided the first convincing evidence that electrons exhibit wavelike behavior, as de Broglie had postulated. Although a sheet of metal foil is not a single crystal, it does consist of a random arrangement of many tiny crystals. The combination of the diffraction patterns produced by each crystal gives the concentric rings that are observed.

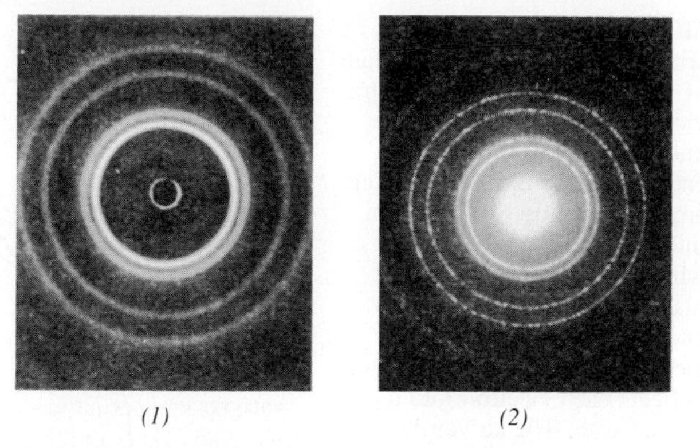

Figure F *(1) Interference pattern produced by X-rays passing through a thin aluminum foil. (2) Interference pattern produced by a beam of electrons passing through the aluminum foil.*

as a point but as a spread-out cloud of negative charge that is densest near the proton and less dense farther away. This picture provides us with another convenient model of the hydrogen atom. We can imagine the electron not as a tiny moving particle but as a cloud of negative charge that decreases in density with increasing distance from the proton. In each excited state of the hydrogen atom, the electron has a different density distribution and is called the **electron density distribution** (Figure 6.12).

We have seen that we must use quantum mechanics, not classical mechanics, to describe the behavior of small particles such as atoms and electrons. Quantum mechanics shows that

1. An electron in an atom can have only certain energies; we say that its energy is quantized.
2. An electron in an atom cannot be precisely located; we can find only the probability that it will be found at any particular location.

Using quantum mechanics, we can calculate the energies of the electrons in an atom or molecule and their probability distributions.

6.4 ELECTRON CONFIGURATIONS

Atomic spectra provide clear evidence that the energies of the electrons in an atom are quantized. We have looked in some detail at the energy levels of the single electron in the hydrogen atom, but what are the energy levels of the electrons in other atoms, which have more than one electron?

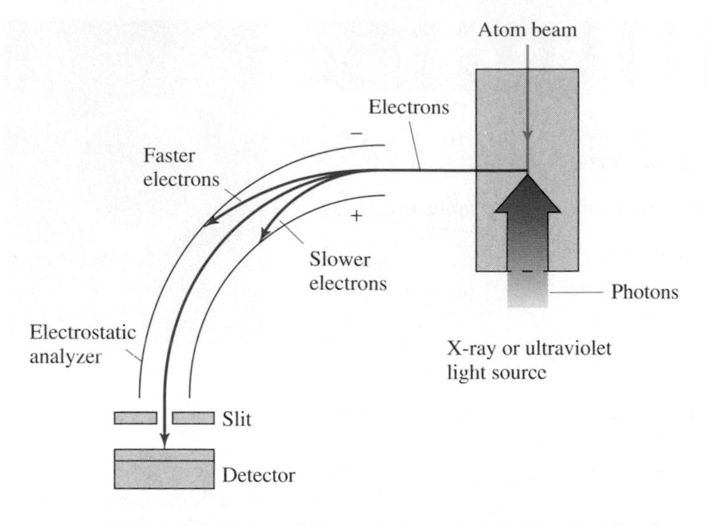

FIGURE 6.13 Photoelectron Spectroscopy A beam of atoms is irradiated with ultraviolet light or X-rays of a known frequency. The kinetic energies of the ejected electrons are measured by passing them into an electrostatic analyzer, which consists of two curved plates, one charged positively and the other negatively. The electric field between the plates deflects each electron into a curved path. The curvature depends on the speed and therefore on the kinetic energy of the electron. Thus, only electrons of one particular speed and therefore one particular kinetic energy will have a path of just the right curvature to pass through the analyzer and through the slit to the detector. The detector counts the electrons as they arrive. By varying the voltage between the plates, electrons of different energies can be detected.

Although it is possible to deduce the energy-level scheme of atoms other than hydrogen from their emission spectra, these spectra are usually complicated. We will find it simpler to use ionization energies to provide us with information about the energy levels of atoms other than hydrogen.

Ionization Energies from Photoelectron Spectroscopy

If an atom absorbs a photon of sufficiently high frequency and therefore of sufficiently high energy, the electron is ejected from the atom, which is converted to a positive ion. The energy needed to do this is the **ionization energy** of the particular electron. Because most of the electrons in other atoms are held more strongly than the single electron in the hydrogen atom, photons of very high energy, such as the photons of very-short-wavelength ultraviolet radiation or even X-rays, are usually needed to provide sufficient energy to remove an electron.

Earlier we discussed the photoelectric effect—the ejection of an electron from the surface of a metal. Here we discuss a related phenomenon, the ejection of an electron from an individual atom. If the energy of an absorbed photon is greater than the ionization energy of the electron, the excess energy appears as the kinetic energy of the ejected electron. The ejected electron is called a **photoelectron**. If *IE* is the ionization energy of the electron and *KE* is the kinetic energy with which it leaves the atom, then

$$E_{photon} = h\nu = IE + KE$$

Rearranging this equation gives

$$IE = h\nu - KE$$

Hence, we can determine the ionization energy *IE* if we know the frequency or wavelength of the photon and if we can measure the kinetic energy of the photoelectron. The measurement of the kinetic energy of the photoelectron is called **photoelectron spectroscopy** (Figure 6.13).

Energy Levels and Electron Configurations

The photoelectrons obtained from helium when we use radiation of a single frequency all have the same kinetic energy. This observation tells us that the two electrons in the $n = 1$ shell in the ground state of the helium atom have the same

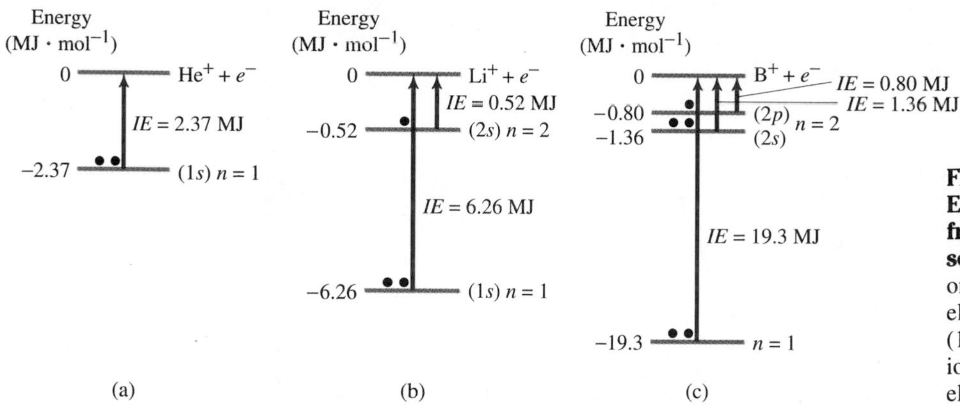

FIGURE 6.14 Ionization Energies and Energy Levels from Photoelectron Spectroscopy (a) Helium has only one ionization energy; its two electrons are both in the $n = 1$ (1s) level. (b) Lithium has two ionization energies; it has two electrons in the $n = 1$ (1s) level and one in the $n = 2$ (2s) level. (c) Boron has three ionization energies; it has two electrons in the $n = 1$ (1s) level and three in the $n = 2$ level, of which two are in the 2s subshell and one is in the 2p subshell.

ionization energy and therefore occupy the same energy level. By measuring the kinetic energy of the photons produced by radiation of a known frequency, we find that the ionization energy of an electron in this level is 2370 kJ · mol^{-1} = 2.37 MJ · mol^{-1}. This value corresponds to an energy level of −2.37 MJ · mol^{-1} if we take the energy of the ionized helium atom, He^+, to be zero (Figure 6.14a). The energy needed to remove either of the two electrons in the helium atom is greater than that needed to remove the single electron of the hydrogen atom, which is 1312 kJ · mol^{-1} = 1.312 MJ · mol^{-1} (Figure 6.9), because the two electrons in helium are attracted by a nuclear charge ($Z = 2$) twice that of the hydrogen atom. But the ionization energy of a helium electron is not *twice* as great as that of a hydrogen electron, because the repulsion between the two electrons reduces the energy needed to remove either one of them.

The ionization energies of the first 21 elements are given in Table 6.1 and

TABLE 6.1 Ionization Energies (MJ · mol^{-1}) of the First 21 Elements

	$n = 1$	$n = 2$		$n = 3$			$n = 4$
Element	**1s**	**2s**	**2p**	**3s**	**3p**	**3d**	**4s**
H	1.31						
He	2.37						
Li	6.26	0.52					
Be	11.5	0.90					
B	19.3	1.36	0.80				
C	28.6	1.72	1.09				
N	39.6	2.45	1.40				
O	52.6	3.04	1.31				
F	67.2	3.88	1.68				
Ne	84.0	4.68	2.08				
Na	104	6.84	3.67	0.50			
Mg	126	9.07	5.31	0.74			
Al	151	12.1	7.19	1.09	0.58		
Si	178	15.1	10.3	1.46	0.79		
P	208	18.7	13.5	1.95	1.06		
S	239	22.7	16.5	2.05	1.00		
Cl	273	26.8	20.2	2.44	1.25		
Ar	309	31.5	24.1	2.82	1.52		
K	347	37.1	29.1	3.93	2.38		0.42
Ca	390	42.7	34.0	4.65	2.90		0.59
Sc	433	48.5	39.2	5.44	3.24	0.77	0.63

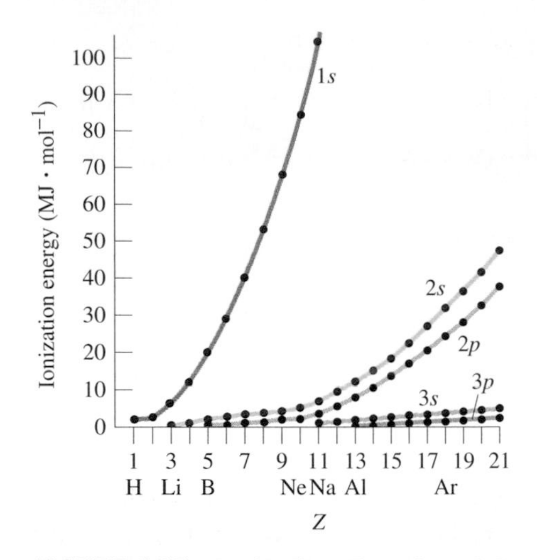

FIGURE 6.15 Ionization Energies of the First 21 Elements The ionization energy of the 1*s* level increases very rapidly with increasing *Z*. The ionization energy of an electron in an *s* level is always slightly greater than that of an electron in the corresponding *p* level.

Figure 6.15. As we shall now see, ionization energies provide experimental evidence for the shell model (Chapter 3). According to that model, the electrons in an atom are arranged in successive layers, or **shells**, designated by $n = 1, 2, 3 \ldots$ in which the electrons have increasingly higher energies.

All the ionization energies in Table 6.1 are for the removal of one electron from an atom, that is, for the process $M \rightarrow M^+$; they are often called first ionization energies. Elsewhere you may find tables of values for the successive removal of electrons from an atom, that is, for the processes $M \rightarrow M^+$, $M^+ \rightarrow M^{2+}$, $M^{2+} \rightarrow M^{3+}$, and so on—the first, second, third, . . . ionization energies.

Table 6.1 shows that lithium has two ionization energies, so that electrons in lithium must be in two energy levels: the $n = 1$ shell, which contains two electrons, and the $n = 2$ shell, which contains the third electron. The electron in the $n = 2$ shell has a much lower ionization energy (0.52 MJ · mol^{-1}) than the two electrons in the $n = 1$ shell (6.26 MJ · mol^{-1}) (Figure 6.14b). This is because the $n = 2$ electron is largely outside the two electrons in the $n = 1$ shell and is subject to a core charge (Chapter 3) of +1, whereas the two electrons in the $n = 1$ shell feel the attraction of the full nuclear charge of +3. Beryllium has two electrons in the $n = 2$ shell, and they have a higher ionization energy (0.90 MJ · mol^{-1}) than the $n = 2$ electron of lithium, due to the stronger attraction of the Be nucleus.

The ionization energies of the succeeding elements show that the $n = 2$ and higher shells have more than one energy level. For example, boron has three ionization energies of 19.3, 1.36, and 0.80 MJ · mol^{-1} (Table 6.1, Figure 6.14c). The electrons with the high ionization energy of 19.3 MJ · mol^{-1} come from the $n = 1$ shell, and those with the much smaller ionization energies of 1.36 and 0.80 MJ · mol^{-1} must come from the $n = 2$ shell. So the $n = 2$ shell must have two energy levels, called **subshells**. These subshells are labeled 2*s* for the lower level (−1.36 MJ · mol^{-1}) and 2*p* for the higher level (−0.80 MJ · mol^{-1}). The $n = 1$ shell has only one energy level, labeled 1*s*. The energy of this level in boron is −19.3 MJ · mol^{-1}. We write the electron arrangement of boron as $1s^22s^22p^1$ to indicate that, of the five electrons of boron, two are in the 1*s* level, two are in the 2*s* level, and one is in the 2*p* level. This representation of the electron arrangement is called an **electron configuration**.

The use of *s*, *p*, *d*, and *f* to name subshells comes from the way in which lines in emission spectra were originally classified. Today they are used simply as labels.

Like boron, the elements carbon through neon have three ionization energies, so we conclude that carbon has two electrons in the 2*p* level, nitrogen three, and so on. The $n = 2$ shell is complete with eight electrons in neon, which has the electron configuration $1s^22s^22p^6$.

The electron configurations of the elements of Periods 1 and 2 are

H $1s^1$
He $1s^2$
Li $1s^2 2s^1$
Be $1s^2 2s^2$
B $1s^2 2s^2 2p^1$
C $1s^2 2s^2 2p^2$
N $1s^2 2s^2 2p^3$
O $1s^2 2s^2 2p^4$
F $1s^2 2s^2 2p^5$
Ne $1s^2 2s^2 2p^6$

The electrons are not all crowded into the lowest energy level—the $1s$ level—in the ground state of these atoms. The capacity of each energy level is limited; the $1s$ and $2s$ levels can hold only two electrons each, and the $2p$ level only six electrons.

The element that follows neon in the periodic table is sodium. Sodium is the first element in Period 3 and has one electron in the $n = 3$ shell. We therefore expect it to have four ionization energies, corresponding to the electron configuration $1s^2 2s^2 2p^6 3s^1$. Table 6.1 shows that this is the case. The electron in the $3s$ energy level is held much less strongly than the electrons in the $n = 2$ and $n = 1$ shells, because it is even farther from the nucleus and is largely outside the electrons in the inner shells. The electron configurations of the Period 3 elements, sodium through argon, follow the same pattern as those of the Period 2 elements:

Na $1s^2 2s^2 2p^6 3s^1$
Mg $1s^2 2s^2 2p^6 3s^2$
Al $1s^2 2s^2 2p^6 3s^2 3p^1$
Si $1s^2 2s^2 2p^6 3s^2 3p^2$
P $1s^2 2s^2 2p^6 3s^2 3p^3$
S $1s^2 2s^2 2p^6 3s^2 3p^4$
Cl $1s^2 2s^2 2p^6 3s^2 3p^5$
Ar $1s^2 2s^2 2p^6 3s^2 3p^6$

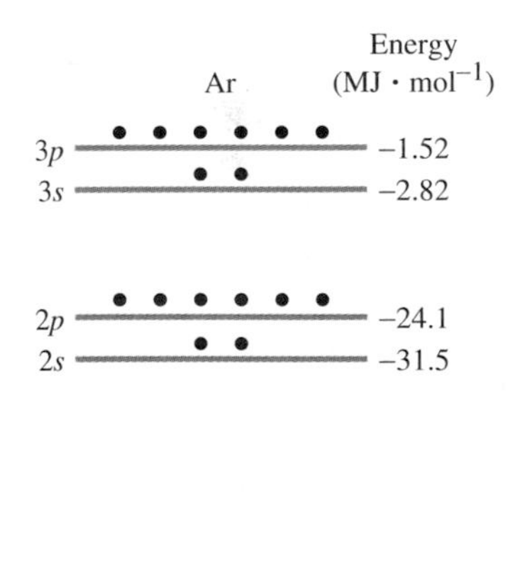

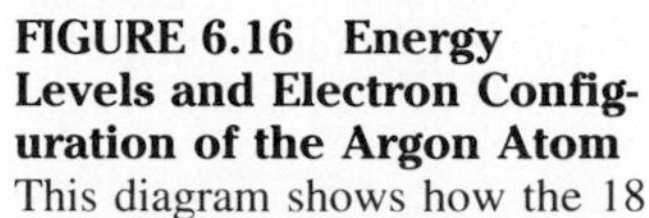

FIGURE 6.16 Energy Levels and Electron Configuration of the Argon Atom This diagram shows how the 18 electrons (•) of the argon atom, with the configuration $1s^2 2s^2 2p^6 3s^2 3p^6$, are distributed among its energy levels. There are 2 electrons in each s level, and 6 in each p level.

The distribution of the 18 electrons of an argon atom among its lowest energy levels is shown in Figure 6.16.

As each electron shell gets larger (that is, as n increases), it can accommodate an increasing number of electrons in an increasing number of subshells. From the experimental ionization energies we have just considered and from those for heavier elements, we can formulate the following rules governing electron configurations. These rules can also be deduced from quantum mechanics.

Rules Governing Electron Configurations We shall be concerned primarily with the electron configurations of the first 36 elements, those of Periods 1 through 4 (Table 6.2). A complete table of the electron configurations of all the elements is given in Appendix C.

The arrangements of electrons in the shells and the subshells (energy levels) of an atom are summarized in Table 6.3. These arrangements are governed by several rules, as follows:

TABLE 6.2 Electron Configurations of the First 36 Elements

Atomic Number	Element	Electron Configuration	Atomic Number	Element	Electron Configuration
1	H	$1s^1$	19	K	$[Ar]4s^1$
2	He	$1s^2$	20	Ca	$[Ar]4s^2$
3	Li	$[He]2s^1$	21	Sc	$[Ar]3d^14s^2$
4	Be	$[He]2s^2$	22	Ti	$[Ar]3d^24s^2$
5	B	$[He]2s^22p^1$	23	V	$[Ar]3d^34s^2$
6	C	$[He]2s^22p^2$	24	Cr	$[Ar]3d^54s^1$
7	N	$[He]2s^22p^3$	25	Mn	$[Ar]3d^54s^2$
8	O	$[He]2s^22p^4$	26	Fe	$[Ar]3d^64s^2$
9	F	$[He]2s^22p^5$	27	Co	$[Ar]3d^74s^2$
10	Ne	$[He]2s^22p^6$	28	Ni	$[Ar]3d^84s^2$
11	Na	$[Ne]3s^1$	29	Cu	$[Ar]3d^{10}4s^1$
12	Mg	$[Ne]3s^2$	30	Zn	$[Ar]3d^{10}4s^2$
13	Al	$[Ne]3s^23p^1$	31	Ga	$[Ar]3d^{10}4s^24p^1$
14	Si	$[Ne]3s^23p^2$	32	Ge	$[Ar]3d^{10}4s^24p^2$
15	P	$[Ne]3s^23p^3$	33	As	$[Ar]3d^{10}4s^24p^3$
16	S	$[Ne]3s^23p^4$	34	Se	$[Ar]3d^{10}4s^24p^4$
17	Cl	$[Ne]3s^23p^5$	35	Br	$[Ar]3d^{10}4s^24p^5$
18	Ar	$[Ne]3s^23p^6$	36	Kr	$[Ar]3d^{10}4s^24p^6$

TABLE 6.3 Arrangements of Electrons in Shells and Subshells

Shell n	Energy Levels	Number of Electrons	Total Number of Electrons in Shell ($2n^2$)
1	$1s$	2	2
2	$2s$	2	8
	$2p$	6	
3	$3s$	2	
	$3p$	6	18
	$3d$	10	
4	$4s$	2	
	$4p$	6	32
	$4d$	10	
	$4f$	14	

Knowing these four rules is very helpful in writing electron configurations for the first 36 elements of the periodic table.

1. The number of subshells (energy levels) in any shell is equal to the quantum number n.

   $n = 1$, $1s$; $n = 2$, $2s$ and $2p$; $n = 3$, $3s$, $3p$, and $3d$; $n = 4$, $4s$, $4p$, $4d$, and $4f$

2. The maximum number of electrons in a shell is $2n^2$.

$$
\begin{aligned}
n &= 1, & (2 \times 1^2) &= 2 \text{ electrons}, & 1s^2 \\
n &= 2, & (2 \times 2^2) &= 8 \text{ electrons}, & 2s^22p^6 \\
n &= 3, & (2 \times 3^2) &= 18 \text{ electrons}, & 3s^23p^63d^{10} \\
n &= 4, & (2 \times 4^2) &= 32 \text{ electrons}, & 4s^24p^64d^{10}4f^{14}
\end{aligned}
$$

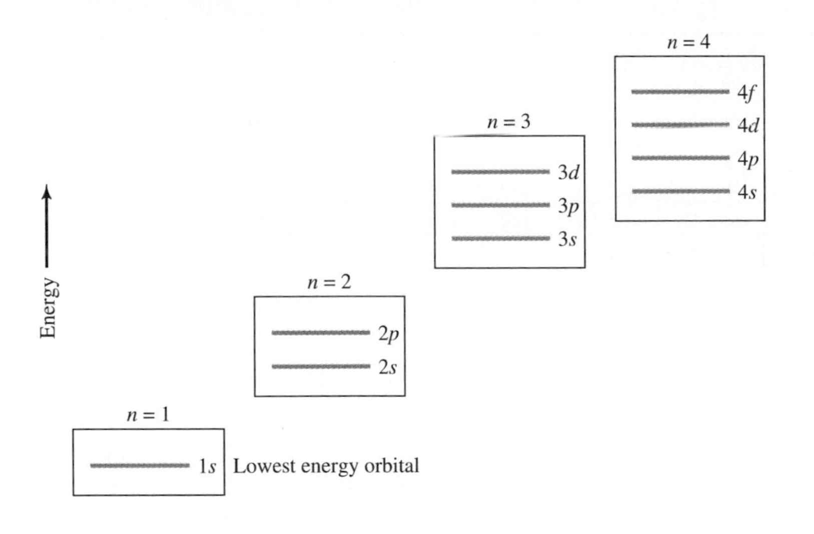

FIGURE 6.17 Atomic Energy Levels This diagram is a very approximate representation of the energy levels of an atom. It is not to scale and does not represent any particular atom, but it illustrates that when $n \geq 3$, there is some overlap between the energy levels of one shell and those of the next highest shell.

3. In subshells, the maximum numbers of electrons are

s	2 electrons
p	6 electrons
d	10 electrons
f	14 electrons

4. For the first 36 elements, the order in which the various energy levels are occupied is 1*s*, 2*s*, 2*p*, 3*s*, 3*p*, 4*s*, 3*d*, 4*p*.

As Figure 6.17 shows, the energies of the levels (subshells) of the $n = 4$ shell overlap slightly with those of the $n = 3$ shell, so the 4*s* level has a slightly lower energy than the 3*d* level and is occupied before electrons enter the 3*d* level. The electron configurations of the two elements after argon are

K	$1s^22s^22p^63s^23p^64s^1$
Ca	$1s^22s^22p^63s^23p^64s^2$

Following calcium, in the series of 10 elements from scandium to zinc, electrons are successively added to the 3*d* level. These elements are the **transition metals** of Period 4. Thus, as we see in Table 6.2, the electron configuration of scandium is $1s^22s^22p^63s^23p^63d^14s^2$; and that of zinc, in which the 3*d* level is filled with 10 electrons, is $1s^22s^22p^63s^23p^63d^{10}4s^2$.

In the Period 4 elements beyond zinc, gallium through krypton, electrons enter the 4*p* level until it is filled in krypton, which has the valence-shell electron configuration $4s^24p^6$. The six elements from Ga through Kr resemble the six elements from Al through Ar ($3s^23p^6$), in which the 3*p* level is progressively filled, and the six elements from B through Ne ($2s^22p^6$), in which the 2*p* level is filled. Figure 6.18 shows the relationship between the periodic table and electron configurations. The alkali and alkaline earth metals (Groups I and II) are called the ***s*-block elements**, because the highest-energy, most easily removed electrons for these elements are *s* electrons. The elements on the right (Groups III through VIII), which are mainly nonmetals, are called the ***p*-block elements**, because the highest-energy electrons for these elements are *p* electrons. The elements in the middle, the transition metals, are called the ***d*-block elements**. Placed at the bottom of the table for convenience are two series of 14 elements that we shall not discuss further. These are the lanthanides and actinides, referred to as the ***f*-block elements**.

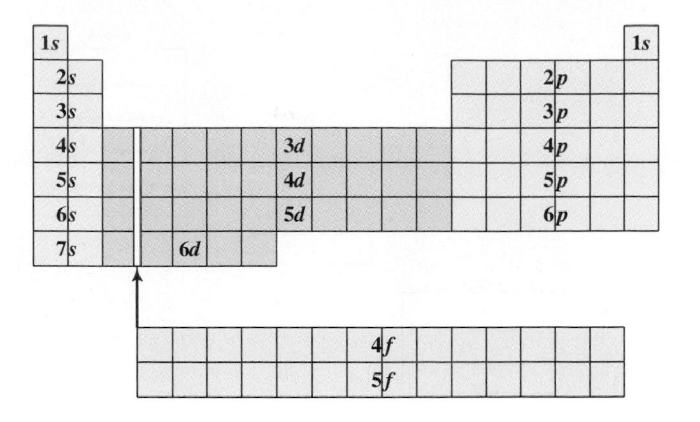

FIGURE 6.18 The Periodic Table and Electron Configurations This diagram of the periodic table shows the energy levels that are being filled with electrons in the different regions of the table. The *s* and *p* regions correspond to the main groups, the *d* region to the transition metals, and the *f* region to the lanthanides and actinides.

We have seen how electrons are distributed among the energy levels of an atom, but how are electrons arranged within each energy level? We need to know this to understand, for example, how the combining power of atoms—their valence—is related to their electron configurations. To understand these arrangements, we must first consider a property of electrons called *spin*.

Electron Spin

Experiments carried out in the 1920s (Box 6.3) showed that an electron behaves like a tiny magnet. To understand this property, it is useful to think of an electron as a rotating sphere of charge, because a rotating charge produces a magnetic field. We call this property of an electron **spin**.

An ordinary bar magnet, such as a compass needle, may be placed in any orientation in a magnetic field, but its energy is lowest when it points in the direction of the field. Work must be done to turn it so that it points in some other direction (Figure 6.19a). Similarly, the energy of an electron depends on the orientation of its magnetic axis relative to the direction of an external magnetic field. In contrast to a

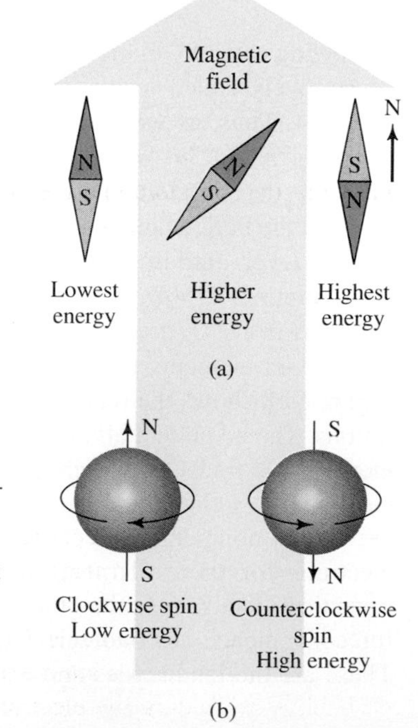

FIGURE 6.19 Electron Spin A spinning electron generates a magnetic field and behaves like a bar magnet. However, unlike a compass needle, it can have only two orientations when placed in a magnetic field. Either its spin axis points in the direction of the magnetic field (clockwise spin), or its spin axis points in the opposite direction (counterclockwise spin). (a) A compass needle can have any orientation in a magnetic field. If left free to rotate, it takes up the lowest-energy orientation on the left. But a force applied to the needle can rotate it into another orientation. Work must be done to rotate the needle, so these other orientations all have a higher energy. (b) In contrast, an electron has only two possible orientations in a magnetic field. Thus, we say that the orientation of the spinning electron is quantized; it can line up only with or against the field.

BOX 6.3

Experimental Evidence for Electron Spin

Experiments carried out by Otto Stern and Walter Gerlach in Germany beginning in 1921 provided the first direct evidence for electron spin. They passed a beam of alkali metal atoms between the poles of a magnet designed to give a very nonuniform field (Figure A1). In each case the beam of atoms was split into two. Because the atoms were not charged, they could be deflected only if they behaved like magnets.

A magnet situated in a nonuniform magnetic field experiences a resultant force that displaces it, because one pole is situated in a stronger magnetic field than the other pole. In a uniform magnetic field, both poles experience the same force, and there is no resultant force to displace the magnet. The amount by which a moving magnet is deflected from its original path in a nonuniform field depends on the orientation of the magnet. When it is lined up along the direction in which the field is changing the most (the greatest field gradient), it experiences the greatest force. When it is at right angles to this direction (zero field gradient), it experiences no deflecting force.

Upon entering a nonuniform magnetic field, a beam of tiny magnets with random orientations would be spread out into a continuous band (Figure A2). The fact that a beam of alkali metal atoms is split into only two beams shows not only that the atoms are magnetic, but also that the atomic magnets can have only two orientations with respect to the magnetic field (Figure A3).

The explanation of the magnetic behavior of these atoms was given by two Dutch physicists, George Uhlenbeck and Sam Goudsmit. To explain certain fine details of atomic spectra, they proposed in 1925 that an electron has spin. Each alkali metal atom has one electron in its valence shell. It is the spin of this single unpaired electron that is responsible for the magnetic properties of these atoms.

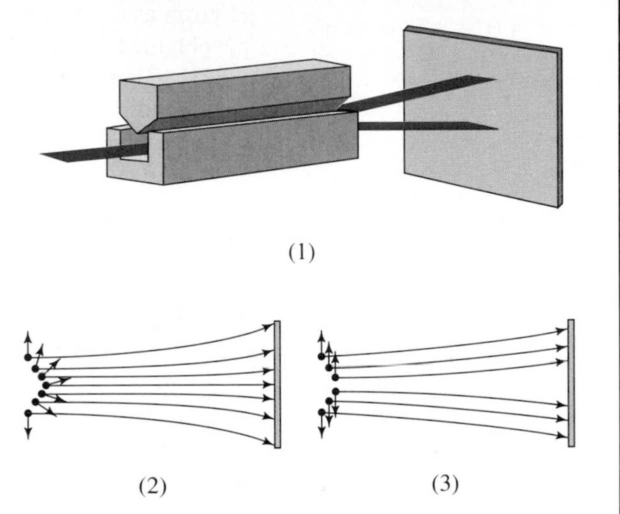

Figure A The Experimental Arrangement for the Stern–Gerlach Experiment (1) *A beam of atoms is passed through a nonuniform magnetic field produced by the specially shaped pole pieces.* (2) *The expected result for magnets that can take up any orientation with respect to the field: The beam of atoms is spread uniformly.* (3) *The experimental result for alkali metal atoms, which have only one unpaired electron in the valence shell: The beam of atoms is split into two distinct beams. This result shows that the magnetic moment due to the single valence-shell electron can take up only two orientations with respect to the field.*

compass needle, however, the energy of an electron is *quantized*, and the electron has only two possible energies in a magnetic field. These two possibilities correspond to rotation in a clockwise or a counterclockwise manner around the direction of an external magnetic field (Figure 6.19b). The energy difference between the two orientations of the spin of an electron is extremely small even in a strong magnetic field and can be ignored for many purposes. Nevertheless, the spin of the electron plays a very important role in determining the arrangement of the electrons in an atom, as we shall now see.

Atomic Orbitals and the Pauli Exclusion Principle

We have seen that an electron in an atom has a probability distribution. The 1*s* electron in the hydrogen atom does not revolve around the nucleus in a circular orbit as proposed by Bohr. It has a spherical distribution in which it has a high probability of being found near the nucleus and a decreased probability of being found farther away, such that within a sphere of radius 140 pm (1.4×10^{-10} m), there is a 90% probability of finding the electron. We can conveniently think of the electron as a cloud of negative charge that increases in density toward the nucleus, 90% of which is within a sphere of radius 140 pm (Figure 6.12). An electron in a 1*s* energy level is said to occupy a **1*s* orbital**. An electron in a 2*s* energy level is said to occupy a **2*s* orbital**, and so on. All electrons in *s* orbitals have a spherical distribution around the nucleus. Strictly speaking, an **atomic orbital** is a mathematical description of the electron as a wave, from which the distribution of the electron density in space can be obtained. However, we commonly use the terms "orbital" and "**electron density distribution**" interchangeably, and we say, for example, that an *s* orbital has a spherical shape, meaning that an electron occupying an *s* orbital has a spherical distribution in space.

In general, for any atom it is found that

> **No orbital can accommodate more than two electrons, and these two electrons must have opposite spins.**

This statement is called the **Pauli exclusion principle**. It was proposed by the Austrian physicist Wolfgang Pauli (1900–1958) in 1925 to explain certain features of the spectra of atoms and to account for the fact that the electrons in an atom are distributed among several energy levels and do not all occupy the lowest (1*s*) energy level. The Pauli exclusion principle expresses a fundamental property of electrons that cannot be explained in terms of classical mechanics. This property is of great importance in chemistry. It determines the electronic configuration of atoms and, as we shall see, it accounts for the importance of the electron pair in the formation of covalent bonds.

Box Diagrams: Hund's Rule An orbital can be conveniently represented by a box in which arrows point either up or down to show the spin of the electron. We can represent a 1*s* orbital or a 2*s* orbital by a single box that may contain one electron or two electrons. If the orbital contains two electrons, they must have opposite spins, so the arrows that represent them point in opposite directions.

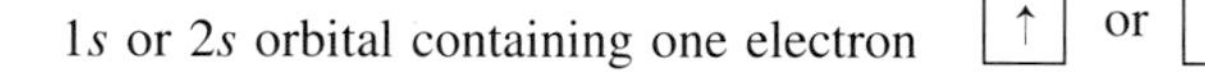

1*s* or 2*s* orbital containing two electrons [↑↓]

The 2*p* level may contain up to six electrons, so there must be three 2*p* orbitals. We represent these by three boxes, each of which may contain up to two electrons:

six electrons in the 2*p* orbitals [↑↓] [↑↓] [↑↓]

Using box diagrams, we can represent the electron configurations of the elements from lithium to neon as shown in Table 6.4. For carbon we appear to have two alternatives for placing the electrons in the 2*p* orbitals. Either the two electrons may be in the same orbital with opposite spins, or they may occupy different orbitals:

	1*s*	2*s*	2*p*
	[↑↓]	[↑↓]	[↑↓] [ ] [ ]
or	[↑↓]	[↑↓]	[↑] [↑] [ ]

TABLE 6.4 Ground-State Electron Configurations (Box Diagrams) and Predicted Valences for Period 2 Elements*

Element	1s	2s	2p	Number of Unpaired Electrons	Predicted Valence	Observed Valence
Li	↑↓	↑		1	1	1
Be	↑↓	↑↓		0	0	2
B	↑↓	↑↓	↑ | |	1	1	3
C	↑↓	↑↓	↑ | ↑ |	2	2	4
N	↑↓	↑↓	↑ | ↑ | ↑	3	3	3
O	↑↓	↑↓	↑↓ | ↑ | ↑	2	2	2
F	↑↓	↑↓	↑↓ | ↑↓ | ↑	1	1	1
Ne	↑↓	↑↓	↑↓ | ↑↓ | ↑↓	0	0	0

*In accordance with Hund's rule, electrons in the same energy level singly occupy as many orbitals as are available before they form pairs. When electrons must occupy the same orbital, they have opposite spins, as the arrows indicate.

However, it has been found that

> **Electrons in the same energy level singly occupy as many orbitals as are available before forming pairs; the electrons in singly occupied orbitals have the same spin.**

This statement is called **Hund's rule** after Fritz Hund, the German spectroscopist who first proposed it. Electrons in different orbitals have different distributions in space (as we shall see in Section 6.5), and their average distance apart is greater than when they are in the same orbital and have the same distribution in space. So the electrostatic repulsion between electrons is minimized when they are in separate orbitals; this is therefore the arrangement of lowest energy. Thus, carbon has two electrons in two different 2*p* orbitals. Using Hund's rule, we can derive the ground-state configurations given in Table 6.4. Many other arrangements of the electrons in the orbitals are possible, but they all represent excited states of the atom rather than the ground states.

All three of the following box diagrams represent a boron atom. Only the first describes the ground state; the others represent excited states.

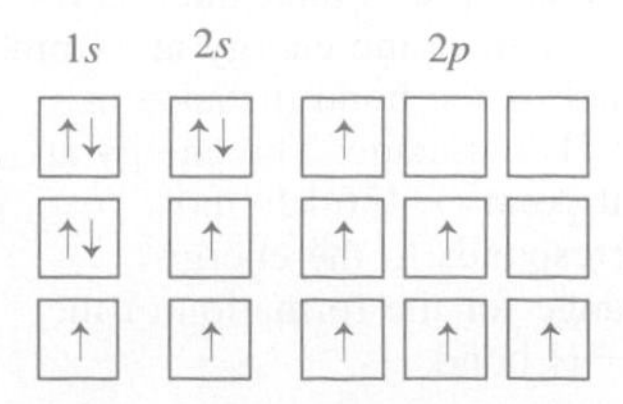

Exercise 6.4 Using only the periodic table and without referring to Figure 6.18, give the electron configurations of the atoms of the following elements. **(a)** Li **(b)** C **(c)** Ne **(d)** Al **(e)** P **(f)** Cl **(g)** Ca

Exercise 6.5 Draw box diagrams for the electron configurations of the atoms in Exercise 6.4.

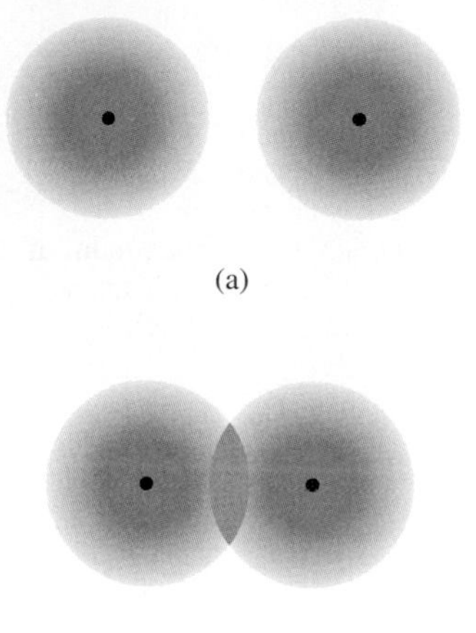

FIGURE 6.20 Electron Density Distribution in the Hydrogen Molecule (a) Two separate hydrogen atoms. (b) In the hydrogen molecule, the electron density distribution near each nucleus is similar to that in the separate hydrogen atoms, but the electron density in the region between the two nuclei is slightly larger than would result from the overlap of the two atomic densities. This increased electron density is responsible for holding the two nuclei together.

6.5 THE HYDROGEN MOLECULE AND THE COVALENT BOND

The Hydrogen Molecule

The simplest molecule is the hydrogen molecule. It has the Lewis structure H:H, with one pair of electrons shared between the hydrogen atoms. To understand better how a pair of electrons can hold two atoms together, let us consider what happens as two hydrogen atoms come together. As the atoms approach one another, each nucleus begins to exert an increasing force on the electron of the other atom. Each H atom can accommodate another electron in its 1*s* orbital. As the two atoms come together, both electrons tend to be drawn into the region between the two nuclei. It is this increased electron density between the two nuclei that pulls them together by electrostatic attraction (Figure 6.20). But as the nuclei come closer together, there is also an increase in the nucleus–nucleus and electron–electron repulsions. At a certain distance between the nuclei, these repulsive forces begin to increase more rapidly than the attractive forces, and the total energy of the system is then a minimum (Figure 6.21). If the nuclei were to come even closer, the repulsive forces would dominate, and the energy would increase. The distance between the nuclei at which the energy is a minimum is the bond length of the hydrogen molecule (74 pm) in its ground state.

A plot of the electron density along the line between the nuclei in the hydrogen molecule is shown in Figure 6.22. Near each nucleus, the electron density is high and similar to that in an isolated hydrogen atom. However, in the region between the nuclei, the electron density is a little greater than would result from simply overlapping the two hydrogen atom densities, because some extra density is drawn into this region. It is the increased electron density in the internuclear region, called the **bonding region**, that holds the two positively charged nuclei together. We can think of the electron charge cloud as acting as a kind of ‘‘electrostatic glue’’ that holds together the two positively charged nuclei.

Bonding Orbitals The two electrons in the hydrogen molecule occupy a **molecular orbital**. Unlike an atomic orbital, a molecular orbital extends over two or more nuclei, but like an atomic orbital, it is occupied by only two electrons with opposite spins. We can represent this molecular orbital very approximately as in

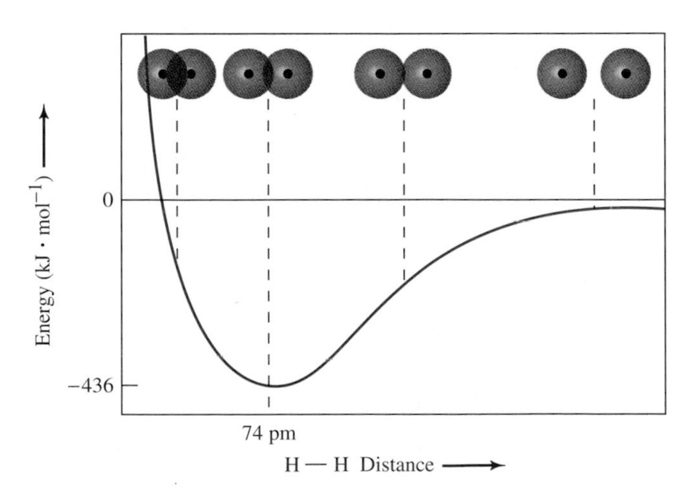

FIGURE 6.21 The Potential Energy Change during the Formation of the H_2 Molecule As two hydrogen atoms approach each other, there is a minimum in the energy at 74 pm, which is the bond distance in the H_2 molecule. The energy at that point ($-436\ kJ \cdot mol^{-1}$) corresponds to the energy change for the formation of the H—H bond.

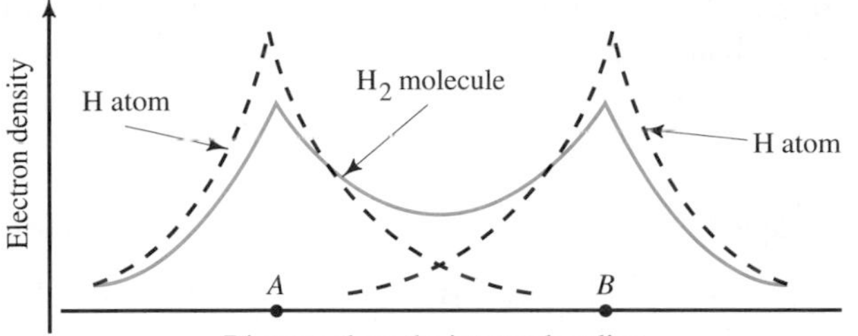

FIGURE 6.22 Electron Density Distribution in the Hydrogen Molecule Variation of the electron density along a line through the two nuclei. There is a maximum at each nucleus and an increased density in the internuclear region.

Figure 6.23, and we can think of it as a combination of the two hydrogen 1*s* orbitals. The two hydrogen 1*s* orbitals merge or *overlap* to form the molecular orbital. Because the two electrons in this orbital attract the two nuclei together, this orbital is called a **bonding molecular orbital** or simply a **bonding orbital**.

Any two atoms that have a singly occupied atomic orbital will be attracted to each other to form a covalent bond in which two electrons with opposite spins occupy a bonding orbital. Thus, quantum mechanics enables us to give answers to the two questions that we posed at the beginning of this chapter. (1) Why does a covalent bond consist of a pair of shared electrons? A covalent bond is formed by a pair of shared electrons because a bonding orbital cannot accommodate more than two electrons. (2) How does a pair of shared electrons hold two atoms together? A shared electron pair leads to an increased electron density in the internuclear region that pulls the two nuclei together by electrostatic attraction. We should not interpret a Lewis diagram such as H:H to mean that two electrons are confined to the region between the two hydrogen nuclei, but only that there is a relatively small increased electron density in the internuclear region.

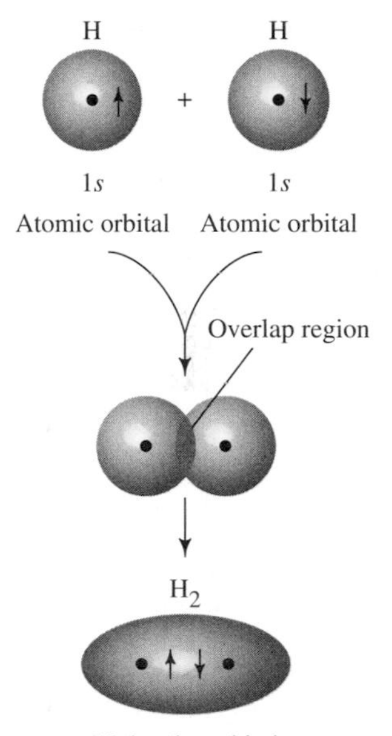

FIGURE 6.23 Formation of the Bonding Molecular Orbital in H_2 The bonding molecular orbital in H_2 results from the overlap of the 1*s* orbitals of the two hydrogen atoms.

Valence and Electron Configurations

We can describe any single bond in terms of a bonding orbital that contains two electrons with opposite spins, one electron from each of the singly occupied orbitals of the two combining atoms. We therefore expect the number of bonds formed by an atom—its valence—to depend on the number of electrons in singly occupied orbitals. Table 6.4 shows the valences we would expect on this basis for the elements of Period 2. These predicted valences agree with the known formulas of their compounds and with the position of the elements in the periodic table for Li, N, O, F, and Ne but not for Be, B, or C. This is because the electron configurations in Table 6.4 are strictly true only for isolated atoms. The energy levels of all the electrons in an atom are changed somewhat when the atom is part of a molecule. In particular, the very small difference in energy between the 2*s* and 2*p* orbitals becomes insignificant, and the electrons spread out among all four orbitals. Alternatively, we say that the energy needed to move an electron from the 2*s* orbital to the 2*p* orbital is more than compensated by the energy obtained from the formation of two additional bonds. These atoms therefore behave *as if* they had the electron configurations

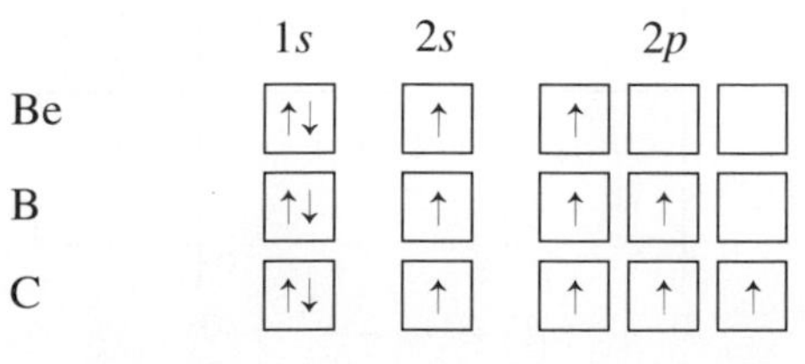

So Be, B, and C use all their valence electrons in forming bonds: They form 2, 3, and 4 bonds, respectively, rather than 0, 1, and 2 bonds.

We saw in Chapter 3 that the number of bonds that a Period 2 element can form follows from its Lewis symbol and the octet rule. In the Lewis symbol for the atom of an element, a single dot can now be interpreted as an electron in a singly occupied orbital, and a pair of dots as a pair of electrons in a filled orbital:

Li· ·Be· ·B· ·C· ·N· :O· :F· :Ne:

Using (·) to signify a singly occupied orbital and (:) to indicate a filled orbital, we can represent the orbitals on each atom as follows:

Li Be B C N O F Ne

The number of bonds that an atom can form is equal to the number of single dots (unpaired electrons) in the Lewis symbol for the element or, in other words, the number of singly occupied orbitals. The doubly occupied orbitals are normally not used to form bonds and are called **nonbonding orbitals**.

Hybrid Orbitals and Molecular Geometry

Let us now look in a little more detail at the electron density distributions in bonding and nonbonding orbitals and how they are related to the electron density distributions in atomic orbitals. So far we have said that electrons in *s* orbitals have a spherical distribution, but we have not described these distributions in detail. Nor have we described the electron density distribution of electrons in *p* orbitals.

The *s* Orbitals By means of quantum mechanics, we can calculate accurate electron density distributions for the single electron in a hydrogen atom. The electron density distributions for an electron in a 1*s*, a 2*s*, and a 3*s* orbital are shown in Figure 6.24. They have an overall spherical shape and get larger with increasing *n*. The average distance of the electron from the nucleus increases with increasing *n*—that is, with increasing energy. The 2*s* and 3*s* orbitals differ from the 1*s* orbital

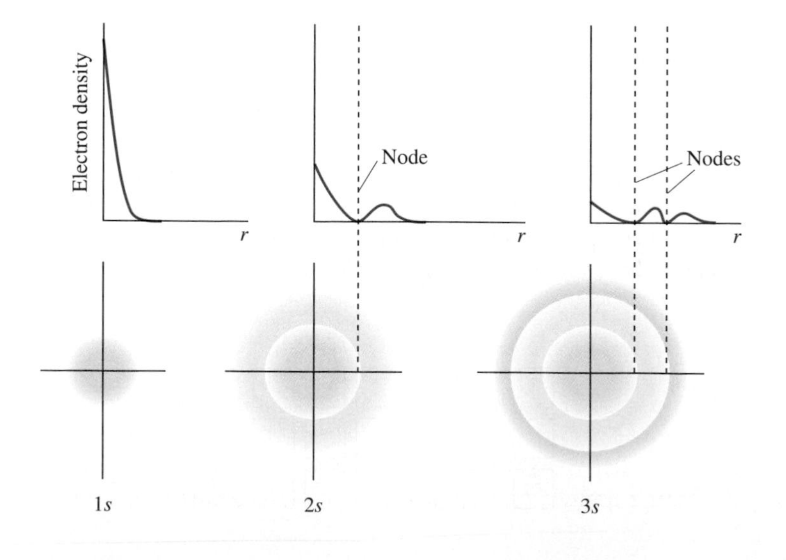

FIGURE 6.24 Electron Density Distribution in 1*s*, 2*s*, and 3*s* Orbitals
The graphs show how electron density varies as a function of distance from the nucleus. In the 2*s* and 3*s* orbitals, the electron density decreases to zero at certain distances from the nucleus. The spherical surfaces at which the electron density is zero are called nodes.

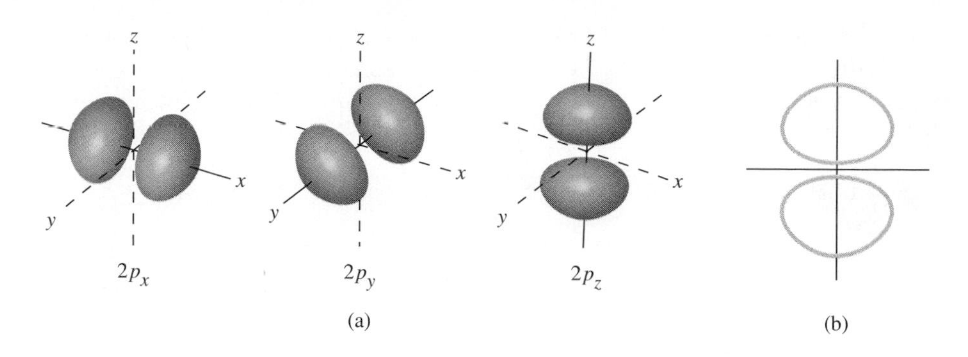

FIGURE 6.25 Electron Density Distributions in the Three 2*p* Orbitals
(a) Approximate shapes of the electron density distribution in the three 2*p* orbitals. The subscript on the orbital label indicates the direction of maximum electron density in the orbital. (b) Cross section through the boundary surface of a single 2*p* orbital.

in that they have internal spherical surfaces on which the electron density is zero. These surfaces are called **nodes**. The 2*s* orbital has one spherical node, and the 3*s* orbital has two spherical nodes. If we plot the electron density in a 1*s* orbital along a radius from the nucleus, the electron density decreases steeply with increasing distance from the nucleus such that, as we said in Section 6.3, a sphere of radius 140 pm encloses 90% of the electron density. The electron density along a radius from the nucleus in the 2*s* orbital decreases steeply to zero at the node, then increases to a maximum and decreases again. Because all *s* orbitals have an overall spherical shape, they are commonly represented simply by a spherical boundary surface.

The *p* Orbitals The distribution of electron density for an electron in the three 2*p* orbitals of a hydrogen atom is shown in Figure 6.25a. The electron density is not distributed in a spherically symmetric fashion as in an *s* orbital. Instead, the electron density is concentrated on the two sides of a planar node that passes through the nucleus. The three 2*p* orbitals have the same size and shape but differ in their orientation in space. There are also three 3*p* orbitals, which differ from the 2*p* orbitals in having an additional spherical node. However, all *p* orbitals are usually represented simply as shown in Figure 6.25b.

Multi-Electron Atoms The *s* and *p* orbitals we just described are valid for the single electron of a hydrogen atom in its ground state—the 1*s* orbital—and in its excited states—the 2*s*, 2*p*, 3*s*, . . . orbitals. The atoms of other elements have two or more electrons that are attracted by the nucleus but are repelled by one another. The resulting electron density distribution and the allowed energy levels are the consequences of a complicated set of forces and are more difficult to calculate than those for the hydrogen atom. However, it is a reasonable approximation for many purposes to assume that the electrons in a multi-electron atom occupy orbitals with shapes similar to those of the hydrogen atom, although they may differ in size depending on the nuclear charge. We call these orbitals **hydrogenlike orbitals**.

Hybrid Orbitals The calculation of the electron density distribution and of energy levels for molecules is more complicated than for atoms, but with the aid of modern computers, accurate results can be obtained for reasonably small molecules. However, for many purposes we can obtain approximate orbitals to describe the bonding in a molecule by merging or overlapping the appropriate atomic orbitals, as we did for the hydrogen molecule (Figure 6.20). The bonding orbital in the hydrogen molecule is cylindrically symmetric around the internuclear axis. An orbital with this shape is called a **σ orbital**, and the bond is called a **σ bond**.

To obtain the bonding orbitals corresponding to the four bonds in the tetra-

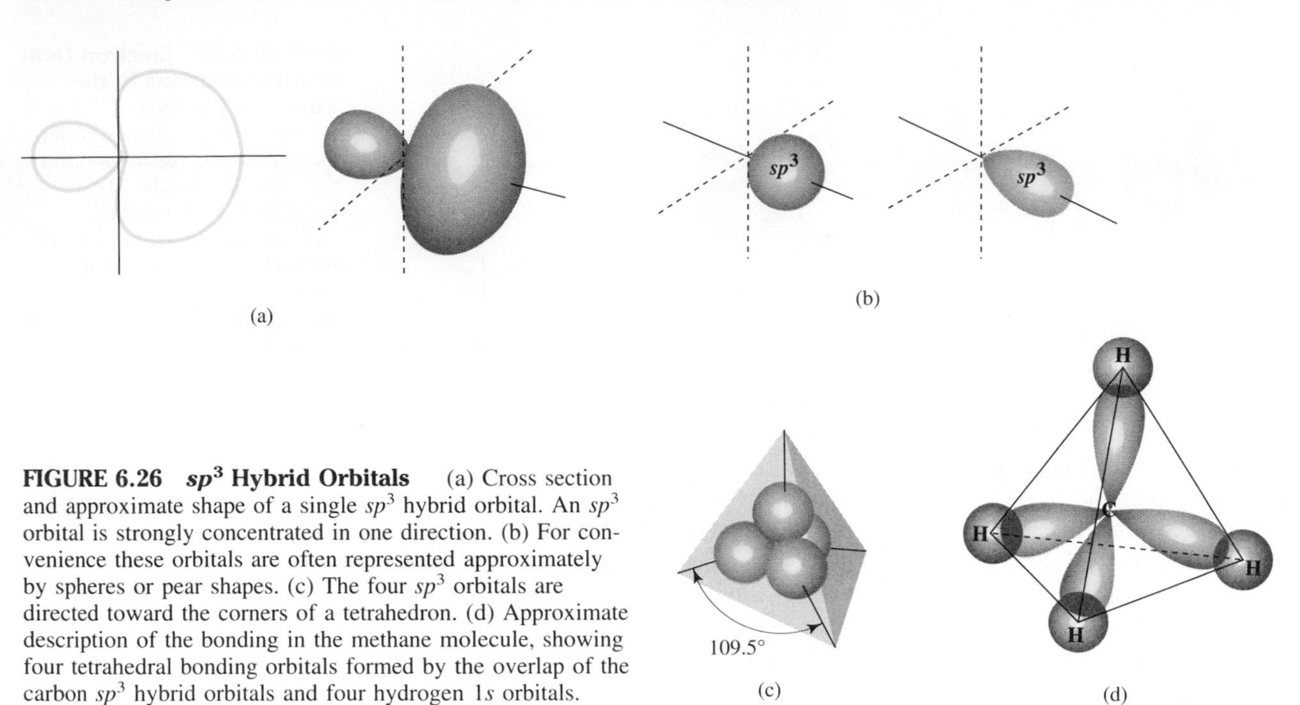

FIGURE 6.26 sp^3 Hybrid Orbitals (a) Cross section and approximate shape of a single sp^3 hybrid orbital. An sp^3 orbital is strongly concentrated in one direction. (b) For convenience these orbitals are often represented approximately by spheres or pear shapes. (c) The four sp^3 orbitals are directed toward the corners of a tetrahedron. (d) Approximate description of the bonding in the methane molecule, showing four tetrahedral bonding orbitals formed by the overlap of the carbon sp^3 hybrid orbitals and four hydrogen $1s$ orbitals.

hedral methane molecule, CH_4, by overlapping the appropriate atomic orbitals, we need four equivalent tetrahedrally oriented orbitals on the carbon atom to combine with four hydrogen $1s$ orbitals. The $2s$ and $2p$ orbitals of the carbon atom have different shapes and do not have a tetrahedral geometry. However, because electrons have wavelike properties and waves can interfere to produce new patterns, the electrons in any atom or molecule with more than one electron can be described by alternative sets of orbitals. These alternative sets of orbitals can be obtained by a mathematical operation called **hybridization**, in which the electron wave patterns are combined. The resulting orbitals are called **hybrid orbitals**. Hybridization of one $2s$ and three $2p$ orbitals produces four equivalent orbitals directed toward the corners of a tetrahedron (Figure 6.26). These orbitals are called **sp^3 hybrid orbitals**, because they are a mixture of one s orbital and three p orbitals. We can then form a set of four *tetrahedral bonding orbitals* by overlapping each sp^3 hybrid orbital with a hydrogen $1s$ orbital. We can use the same set of four orbitals to describe the ammonia and water molecules. But in the ammonia molecule, one of the sp^3 orbitals is a nonbonding orbital containing a lone pair, and in the water molecule, two of the sp^3 orbitals are nonbonding orbitals (Figure 6.27).

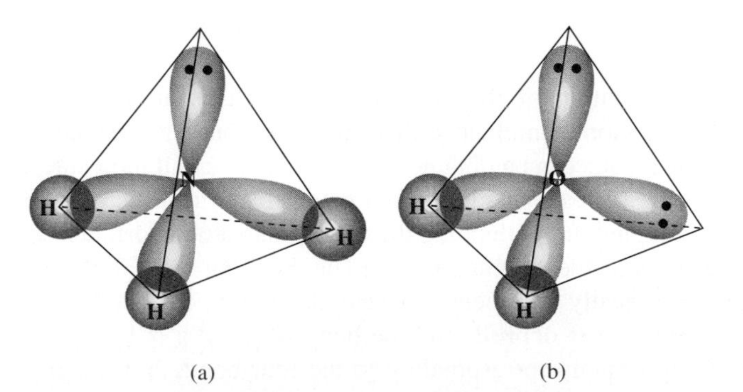

FIGURE 6.27 Hybrid Orbital Descriptions of the Bonding in the Ammonia and Water Molecules (a) In the ammonia molecule, there are three σ-bonding orbitals and one nonbonding (lone-pair) orbital. (b) In the water molecule, there are two σ-bonding orbitals and two nonbonding orbitals.

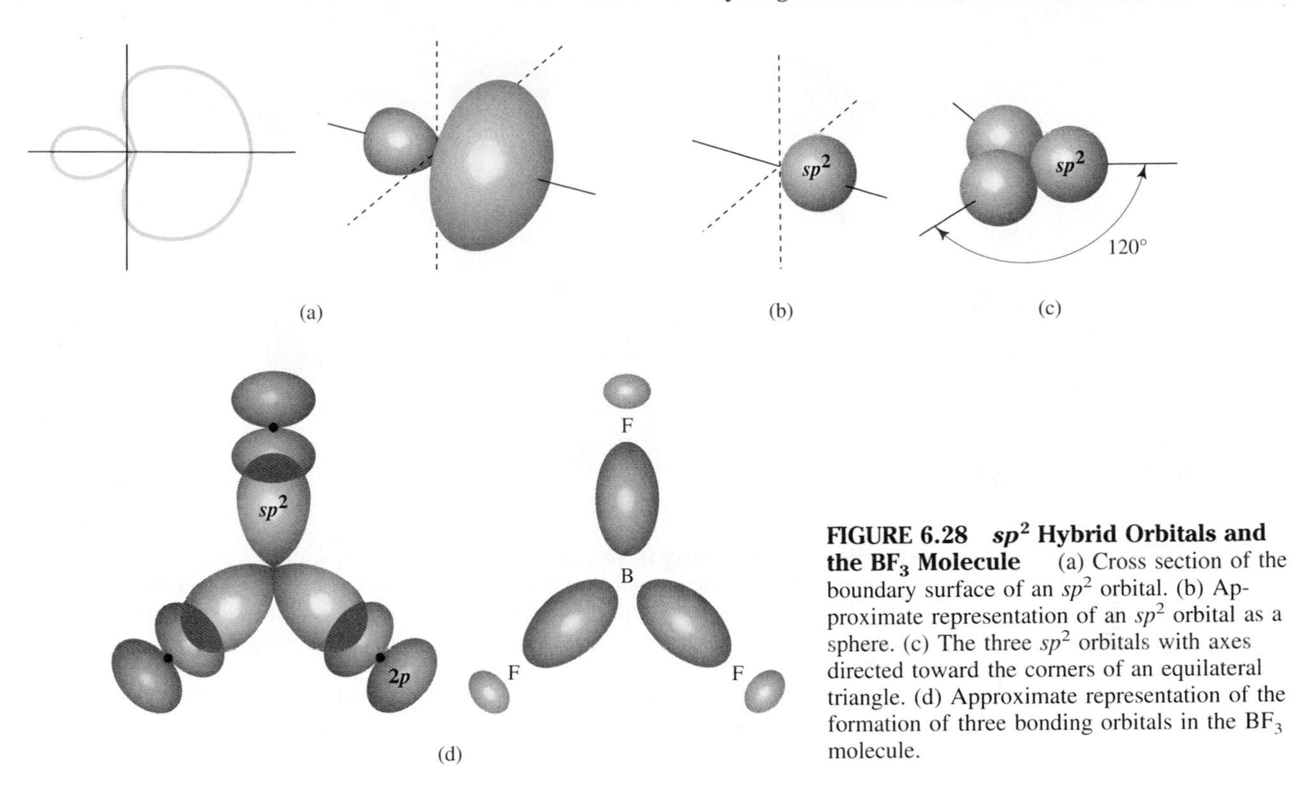

FIGURE 6.28 ***sp*^{2} Hybrid Orbitals and the BF_3 Molecule** (a) Cross section of the boundary surface of an *sp*^{2} orbital. (b) Approximate representation of an *sp*^{2} orbital as a sphere. (c) The three *sp*^{2} orbitals with axes directed toward the corners of an equilateral triangle. (d) Approximate representation of the formation of three bonding orbitals in the BF_3 molecule.

To describe the orbitals in the triangular BF_3 molecule, we need a set of triangular orbitals on boron. Such a set can be obtained by hybridizing one 2*s* and two 2*p* orbitals to form three *sp*^{2} orbitals. They can be overlapped with the singly occupied 2*p* orbital on each of three fluorine atoms to form three bonding orbitals arranged as an equilateral triangle (Figure 6.28). To describe the bonds in a linear molecule such as $BeCl_2$, we use the set of two *sp* hybrid orbitals that can be formed from one 2*s* orbital and one 2*p* orbital (Figure 6.29).

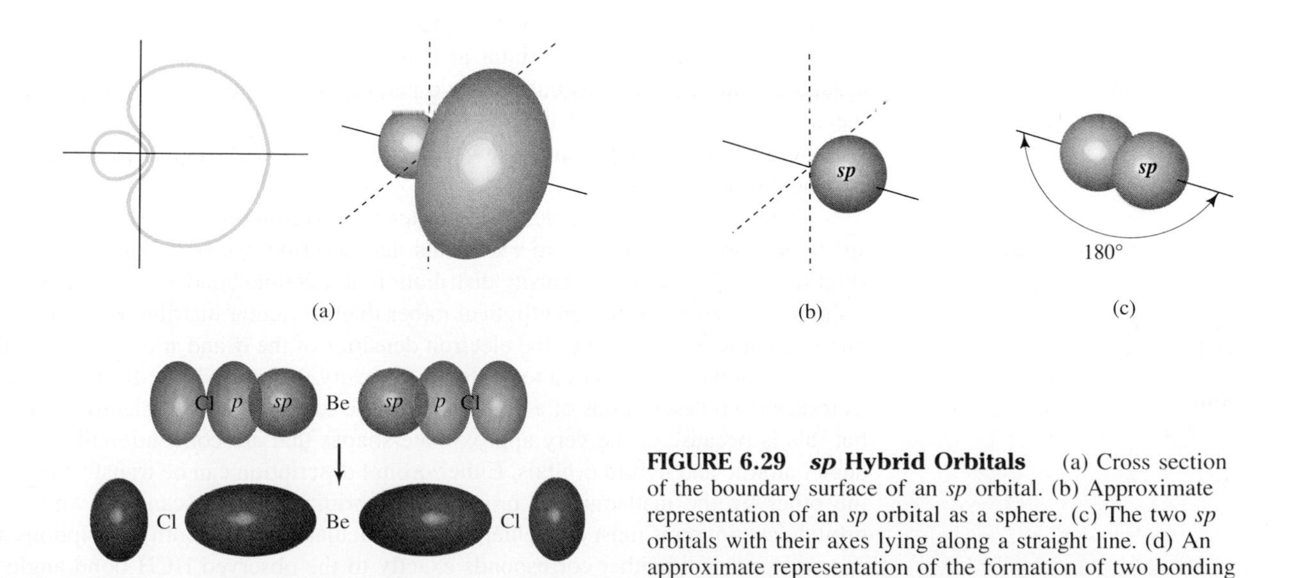

FIGURE 6.29 ***sp* Hybrid Orbitals** (a) Cross section of the boundary surface of an *sp* orbital. (b) Approximate representation of an *sp* orbital as a sphere. (c) The two *sp* orbitals with their axes lying along a straight line. (d) An approximate representation of the formation of two bonding orbitals in the $BeCl_2$ molecule.

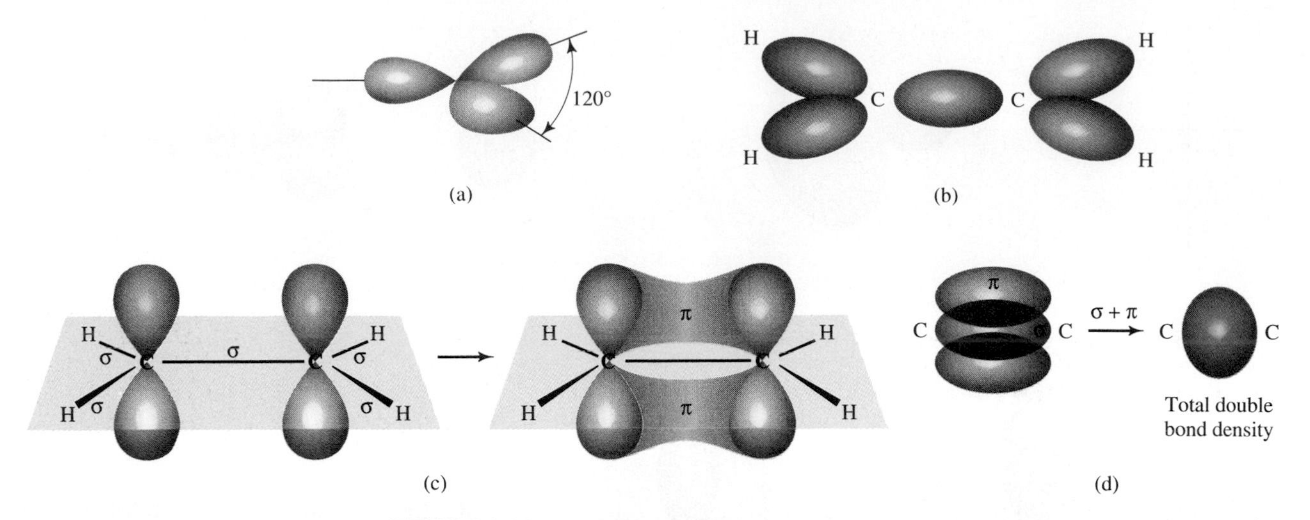

FIGURE 6.30 Bonding in the Ethene Molecule: σ–π model (a) The three sp^2 hybrid orbitals. (b) The σ bonds in the ethene molecule. (c) Overlap of *p* orbitals gives a π bond. The electron density in the π bond is distributed above and below the plane of the molecule. (d) Total electron density (σ + π) of the double bond.

It is important to understand the relationship between the VSEPR model and hybridization. The VSEPR model is a simple method for predicting molecular geometry. Hybridization does not predict geometry. It is a mathematical process for obtaining a set of orbitals (hybrid orbitals) from which we can form bonding orbitals that correspond with the observed and/or predicted geometry of a molecule.

Ethene and Ethyne The sp^2 and *sp* hybrid orbitals are often used to describe the bonding in ethene and ethyne. In ethene, C_2H_4, each carbon atom is bonded to two hydrogen atoms and to the other carbon atom in a planar triangular arrangement. Three corresponding σ-bonding orbitals can be formed from three singly occupied sp^2 hybrid orbitals on each carbon atom (Figure 6.30). This leaves a single electron in the third 2*p* orbital on both carbon atoms. These two 2*p* orbitals can be overlapped to form a second bonding orbital between the two carbon atoms. This orbital is different from a σ orbital in that the electron density is not symmetric around the internuclear axis but is in two parts on either side of a nodal plane, like the 2*p* orbitals from which it is constructed. An orbital of this type is called a **π orbital**. Thus a carbon–carbon double bond may be described as consisting of a σ bond and a **π bond**.

τ is the Greek letter tau.

A double bond can be described instead by overlapping two singly occupied sp^3 hybrid orbitals to form two **τ orbitals** that describe two bent bonds **(τ bonds)** (Figure 6.31). The electron density distribution of a double bond differs from that of a single bond in that it has an elliptical rather than a circular distribution around the internuclear axis. The sum of the electron densities of the σ and π orbitals or of the two bent-bond orbitals gives a total electron density with this shape. It may not look as if these two descriptions of a double bond lead to the same total electron density, but this is because of the very approximate shapes that are conventionally used to depict atomic and hybrid orbitals. Either orbital description can be transformed into the other by the mathematical process of hybridization. We can use whichever orbital description is most convenient for a particular purpose. Both descriptions are approximate, and neither corresponds exactly to the observed HCH bond angle of 116°. If necessary, the orbital model can be refined by constructing hybrid orbitals

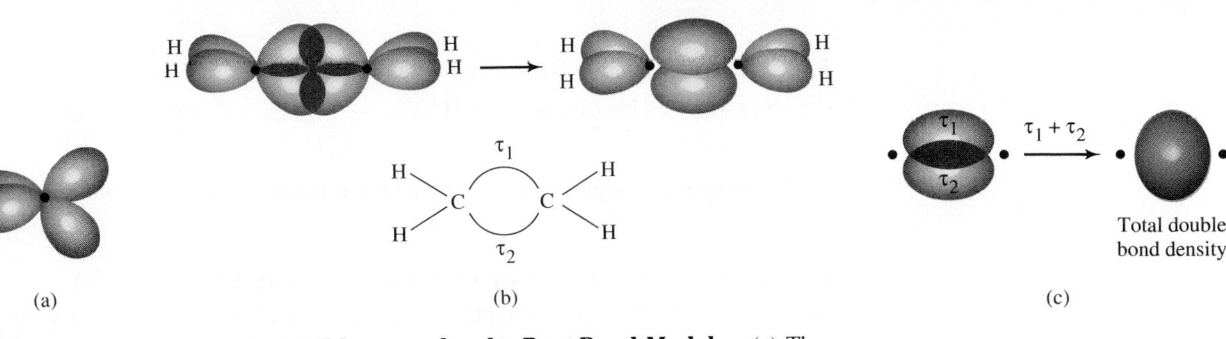

FIGURE 6.31 Bonding in the Ethene molecule: Bent-Bond Model (a) The four tetrahedral sp^3 hybrid orbitals. (b) Overlap of the sp^3 orbitals gives two bent bonds (τ bonds) (c) Total electron density ($\tau_1 + \tau_2$) of the double bond.

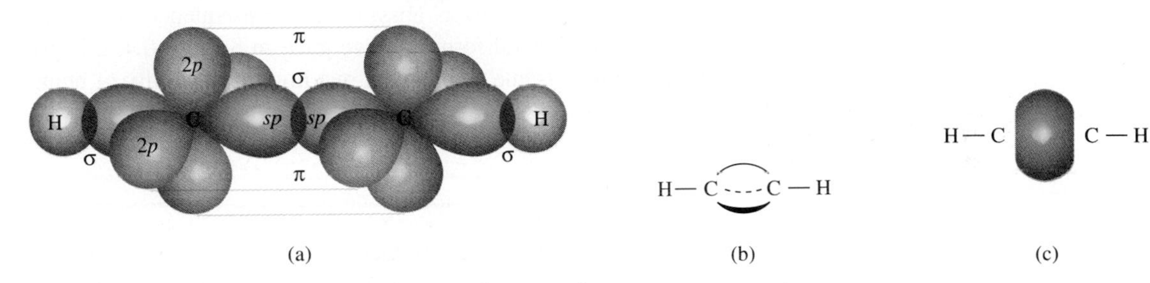

FIGURE 6.32 Models of the Bonding in Ethyne
(a) The σ–π model.
(b) The bent-bond model.
(c) Total electron density in the triple bond.

of intermediate character (for example, between sp^2 and sp^3), that correspond exactly to the bond angle.

In ethyne, C_2H_2, each carbon atom is bonded to one hydrogen and to the other carbon atom in a linear arrangement. The triple bond in the ethyne molecule can be described as consisting of one σ orbital and two π orbitals or as three bent-bond orbitals (Figure 6.32).

EXAMPLE 6.4 Describing the Bonding in a Molecule

Draw the Lewis structure for methanal (formaldehyde), H_2CO. Predict its shape with the VSEPR model, and describe the bonding with the appropriate hybrid orbitals and the σ–π model.

Solution: The Lewis structure is

$$\mathrm{H_2C{=}\ddot{O}:}$$

Carbon is forming three bonds (2 single and 1 double), so methanal is an AX_3 molecule. It therefore has a planar triangular shape with bond angles of approximately 120°. The corresponding hybrid orbitals on carbon are sp^2: Two form the C—H bonds, and one forms the σ component of the C=O double bond. The π component of the double bond is formed from one $2p$ orbital on carbon and one $2p$ orbital on oxygen. The hybrid orbitals on oxygen may also be described as sp^2: Two are nonbonding orbitals, and the other forms the σ component of the double bond. The geometry at oxygen may be described as AXE_2 with angles of approximately 120° between the lone pairs and between each lone pair and the double bond.

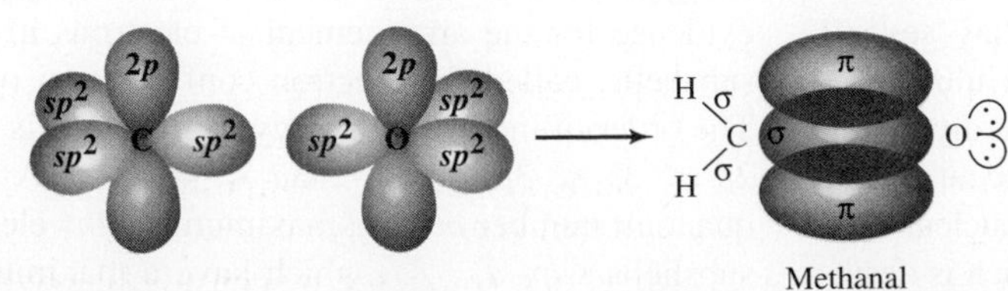

Exercise 6.6 Consider the methyl cyanide molecule, CH_3CN.

(a) Draw the Lewis structure of CH_3CN.

(b) Predict the approximate bond angles at each carbon atom.

(c) Describe the bonding in terms of hybrid orbitals and the σ-π model.

Hybrid Orbitals, Electron-Pair Domains, and the VSEPR Model In discussing the VSEPR model in Chapter 3, we described the region occupied by an electron pair as an electron-pair domain, and as a rough approximation we represented each domain by a sphere. In Section 6.4 we used the same spherical shape as an approximation for sp, sp^2, and sp^3 hybrid orbitals. We see that a domain is just an approximation for an orbital. The basic assumption of the VSEPR model that electron pairs stay as far apart as possible is a consequence of the Pauli exclusion principle. In a more general form than we gave in Section 6.4, this principle states that electrons with the same spin have a maximum probability of being as far apart as possible.

SUMMARY

A wave is described by its frequency, ν, wavelength, λ, speed, $v = \lambda\nu$, and amplitude, A. All electromagnetic waves, waves produced by oscillating charges, have the same speed, $c = 3.00 \times 10^8\ \text{m} \cdot \text{s}^{-1}$. The different kinds of electromagnetic radiation include X-rays, ultraviolet (UV), visible (light), infrared (IR), and radio waves, with wavelengths ranging from 10^{-12} m to 1 km.

Interference and diffraction can be explained in terms of the wave properties of light, but accounting for the photoelectric effect and photochemical reactions requires that light be regarded as a stream of particles called photons. Each photon has energy $E_{\text{photon}} = h\nu$, where h the Planck constant ($6.626 \times 10^{-34}\ \text{J} \cdot \text{s}$).

Light emitted by elements heated in a flame or in an electric discharge give atomic emission spectra consisting of line spectra, sharp lines of only certain frequencies. This phenomenon shows that electrons in atoms have only certain allowed—quantized—energy levels. A spectral line results from the transition (movement) of an electron from a higher energy level E_i to a lower energy level E_f, which gives a photon of energy $E_{\text{photon}} = E_i - E_f$ and frequency $\nu = (E_i - E_f)/h$. When an atom absorbs light, an electron is raised to a higher energy level, giving rise to an absorption spectrum.

The atomic spectrum of hydrogen has several series of lines in the infrared, visible, and ultraviolet regions of the spectrum. The spectrum of the hydrogen atom can be explained by the Bohr model. According to this model, the electron moves around the nucleus in a limited number of circular orbits, each of which is designated by an integer $n = 1, 2, 3, \ldots$ and increases in radius and energy as n increases. The energies of the photons emitted by hydrogen are given by the formula

$$E_{\text{photon}} = 2.18 \times 10^{-18}\left(\frac{1}{n_f^2} - \frac{1}{n_i^2}\right)\ \text{J}$$

The Bohr model is unable to explain completely the spectra of other atoms and has been replaced by a quantum mechanical model.

According to quantum mechanics, the position of an electron in a hydrogen atom cannot be located precisely; we can determine only the probability that an electron will be found at a particular location. The electron is conveniently described as behaving like a spread-out charge cloud or as having a probability distribution.

In photoelectron spectroscopy gaseous atoms are bombarded with high-energy photons of known energy. Electrons are knocked out of each energy level in the atoms and have kinetic energy KE, where $E_{\text{photon}} = IE + KE$ and IE is the ionization energy of the electron in the atom. Each IE can be found by measuring the KE of each emitted electron if we know the energy (frequency) of the photons. Thus photoelectron spectroscopy gives the energies of all the electrons in an atom. It provides evidence for the arrangement of electrons in shells and subshells, called the electron configuration of an atom. The order of increasing energy of subshells is $1s < 2s < 2p < 3s < 3p < 4s < 3d < 4p \ldots$. A shell with quantum number n has a maximum of $2n^2$ electrons in n subshells $s, p, d, \ldots$, which have a maximum of 2, 6, 10, . . . electrons, respectively.

Electrons have the property spin, which can be imagined as either clockwise or counterclockwise with respect to the direction of an external magnetic field. An orbital is a mathematical description of an electron as a wave. The corresponding electron density distribution is usually also called an orbital. The Pauli exclusion principle states that no orbital can accommodate more than two electrons and that these electrons must have opposite spin. Hund's rule states that electrons enter orbitals of the same energy one at a time until each orbital contains one electron before they form electron pairs. Using the exclusion principle and Hund's rule, we can write the electron configuration of any element by filling each orbital in order of increasing energy.

The two electrons in the hydrogen molecule occupy a bonding orbital that we can think of as being formed by the merging or overlap of the two H 1*s* orbitals. The electron density in the region between the nuclei is slightly greater than the sum of the densities in the two atomic orbitals. This additional electron density holds the two nuclei together by electrostatic attraction.

The valence of an atom in its ground state is predicted to be equal to the number of singly occupied orbitals in its valence shell. This is the case for Li, N, O, F, and Ne but not for Be, B, or C. For these latter elements, the perturbation of their energy levels when they form compounds causes the difference in energies of the 2*s* and 2*p* orbitals to become insignificantly small. Thus, they have 2, 3, and 4 singly occupied orbitals, respectively, in agreement with their observed valences.

Electrons in *s* orbitals have spherical electron density distributions. The electron density of electrons in *p* orbitals is concentrated on two sides of a plane through the nucleus called a node, where the electron density is zero. The bonds in molecules can be described in terms of bonding orbitals formed by the overlap of singly occupied atomic orbitals on each of the combining atoms, as in H_2. In a tetrahedral AX_4 molecule, such as CH_4, the 2*s* and 2*p* atomic orbitals are not suitable for this purpose, as they are not all equivalent and do not have a tetrahedral geometry. An alternative set of orbitals called hybrid orbitals can, however, be constructed by mathematically combining the electron waves to form new wave patterns—a process called hybridization. Four equivalent tetrahedrally oriented sp^3 orbitals can be formed in this way from the 2*s* and 2*p* orbitals. They can be used to describe the bonds in CH_4, or in any AX_4 molecule, by overlapping them with the H 1*s* orbital or with a singly occupied orbital of X. Other sets of hybrid orbitals, such as sp^2 and *sp*, can be formed to describe the bonding in molecules in which an atom has a planar triangular or linear geometry, respectively.

The ethene double bond can be approximated in terms of (1) τ orbitals formed by the overlap of two sp^3 orbitals on each carbon atom to give two bent (τ) bonds, or (2) a σ orbital formed by the overlap of an sp^2 hybrid orbital on each carbon atom and a π orbital formed by the "sideways" overlap of a 2*p* orbital on each carbon atom. A σ orbital is symmetric around the bond axis, whereas a π orbital has a nodal plane, like a 2*p* orbital. Bonding in ethyne can be described in terms of three bent bonds or as one σ bond and two π bonds.

IMPORTANT TERMS

absorption spectrum (page 195)
atomic orbital (page 206)
atomic spectrum (page 189)
Bohr model (page 192)
bonding orbital (page 209)
electromagnetic spectrum (page 184)
electromagnetic wave (page 182)
electron configuration (page 200)
electron density distribution (page 197)
emission spectrum (page 190)
energy level (page 191)
excited state (page 191)
frequency (page 183)
ground state (page 191)
Hund's rule (page 207)
hybridization (page 212)
hybrid orbital (page 212)
ionization energy (page 198)
line spectrum (page 189)
molecular orbital (page 208)
Pauli exclusion principle (page 206)
photoelectric effect (page 185)
photoelectron spectroscopy (page 198)
photon (page 185)
π bond (page 214)
π orbital (page 214)
Planck constant (page 185)
quantum mechanics (page 181)
shell (page 200)
σ bond (page 211)
σ orbital (page 211)
spin (page 204)
subshell (page 200)
τ bond (page 214)
τ orbital (page 214)
wavelength (page 183)

REVIEW QUESTIONS

1. Define each of the following properties of a wave, and give the appropriate SI units of each. How are these properties related?

(a) frequency, ν **(b)** wavelength, λ **(c)** speed, v

2. What is the relationship between the energy of a photon and **(a)** its frequency; **(b)** its wavelength?

3. Explain why photons of a certain minimum energy, or maximum wavelength, are needed to eject an electron from a metal.

4. In the photochemical reaction between $Cl_2(g)$ and $H_2(g)$ to give $HCl(g)$, why does blue-green light initiate the reaction, but not red light?

5. What is **(a)** an emission spectrum; **(b)** an absorption spectrum?

6. Why does an emission spectrum consist of a limited number of sharp lines?

7. What are **(a)** the ground state and **(b)** an excited state of an atom?

8. If an electron moves from an energy level E_i to an energy level E_f, what is the frequency of the emitted photon?

9. What evidence is there that electron' have wavelike properties?

10. How can the energy levels of an atom be determined by photoelectron spectroscopy?

11. **(a)** How are the subshells of the $n = 3$ shell designated? **(b)** What is the maximum number of electrons in this shell?

12. In what order are the energy levels (subshells) of the first 36 elements occupied?

13. State the Pauli exclusion principle.

14. State Hund's rule.

15. Explain why carbon has a valence of four although there are only two electrons in singly filled orbitals in the ground state of the carbon atom.

16. How can we obtain an approximate description of a molecular orbital?

17. What are a bonding orbital and a nonbonding orbital?

18. What is a hybrid orbital?

19. Why is it necessary to use hybrid orbitals to describe the bonding in some molecules?

20. What is **(a)** a σ bond; **(b)** a π bond?

PROBLEMS

Light and Other Electromagnetic Radiation

1. The light that is scattered from molecules in the atmosphere and gives the sky its blue color has a frequency of 7.5×10^{14} Hz. What is the corresponding wavelength?

2. Mercury lamps used for street lighting emit the atomic spectrum of mercury. One of the lines in this spectrum is in the blue region and has a wavelength of 435.8 nm. Express this wavelength in **(a)** meters and **(b)** micrometers. What is the frequency of this radiation?

3. Citizens' band (CB) radio operates at a frequency of 27.3 MHz. What is the wavelength of this radio wave?

4. Calculate the range of frequencies associated with each of the regions of the electromagnetic spectrum from the following wavelengths.

(a) radio (1 km to 30 cm)

(b) microwave (30 cm to 2 mm)

(c) infrared (2 mm to 750 nm)

(d) visible (750 nm to 400 nm)

(e) ultraviolet (400 nm to 4 nm)

(f) X-rays (4 nm to 10 pm)

5. A helium-neon laser produces light of wavelength 633 nm. What are the color and frequency of this light?

6. The atomic spectrum of lithium has a strong red line at 670.8 nm.

(a) What is the energy of each photon of this wavelength?

(b) What is the energy of 1 mol of these photons?

7. The yellow color of a sodium-vapor street lamp is due to lines at 589.6 nm and 589.0 nm. What are the frequencies of these transitions, and what is their energy difference?

8. By the use of a suitable filter, a green mercury emission line of wavelength 546.1 nm can be isolated. Calculate the energy of **(a)** one photon of light of this wavelength; **(b)** 1 mol of photons of light of this wavelength.

9. Calculate the energy of single photons and 1 mol of photons corresponding to each of the following.

(a) an X-ray photon of wavelength 20.0 pm

(b) a photon in the ultraviolet of wavelength 100 nm

(c) a photon in the visible of wavelength 500 nm

(d) a photon in the infrared of wavelength 20.0 μm

(e) a microwave photon of wavelength 20.0 cm

10. Photons of minimum energy 496 $kJ \cdot mol^{-1}$ are needed to ionize sodium atoms.

(a) Calculate the lowest frequency of light that will ionize a sodium atom. What is the color of this light?

(b) If light of energy 600 $kJ \cdot mol^{-1}$ is used, what is the kinetic energy of each emitted electron?

11. The longest-wavelength light that causes an electron to be emitted from a gaseous lithium atom is 520 nm. Gaseous lithium atoms are irradiated with light of wavelength 360 nm. What is the kinetic energy of the emitted electrons, in kilojoules per mole?

12. One type of burglar alarm uses the photoelectric effect. Provided that visible light falling on a metal plate causes the emission of photoelectrons, the alarm is inactive. When the light beam is blocked by an intruder, the alarm is set off. Would magnesium metal be a suitable material for the metal plate, given that the lowest frequency that can cause the emission of an electron from magnesium is 8.95×10^{14} Hz?

13. Light of maximum wavelength 493 nm dissociates chlorine molecules into chlorine atoms.

(a) Will light of the same wavelength dissociate bromine molecules into bromine atoms, given that the dissociation energy of $Br_2(g)$ is 224 $kJ \cdot mol^{-1}$?

(b) What is the maximum wavelength of light that will dissociate $Br_2(g)$ molecules into bromine atoms?

14. Nitrogen dioxide, $NO_2(g)$, is one of the components of photochemical smog. The energy required to dissociate $NO_2(g)$ molecules into $NO(g)$ molecules and oxygen atoms is 305 $kJ \cdot mol^{-1}$.

(a) What maximum wavelength of light will cause this dissociation?

(b) What type of radiation is it?

(c) If the minimum wavelength of light that strikes the earth's surface at sea level is 320 nm, will this dissociation occur near the earth's surface?

Atomic Spectra

15. What wavelength of light is emitted when an electron moves from the $n = 6$ to the $n = 2$ energy level of a hydrogen atom? In what region of the electromagnetic spectrum is the corresponding spectral line found?

16. Calculate the energy required to excite an electron from the $n = 2$ to the $n = 4$ energy level of the hydrogen atom. What wavelength of light will cause this excitation?

17. Lines in the ultraviolet region of the emission spectrum of atomic hydrogen arise from transitions to the $n = 1$ level. One of these lines has a wavelength of 103 nm. What is the n quantum number of the electrons in the excited atoms that give rise to this line?

18. How much energy, in megajoules per mole, is needed to ionize hydrogen atoms, starting from **(a)** the ground state; **(b)** the first excited state? In both cases find the maximum wavelength of light that will ionize a hydrogen atom.

19. The visible series of emission lines from atomic hydrogen end in the $n = 2$ energy level. What is the wavelength of the longest-wavelength line in this series?

20. Considering only the $n = 1, 2, 3, 4$, and 5 energy levels of the hydrogen atom, **(a)** how many spectral lines are possible from only this set of levels?

(b) How many are in the ultraviolet region, and how many are in the visible spectrum?

21. The infrared series of emission lines from atomic hydrogen arise from transitions to the $n = 3$ level. One of these lines has a wavelength of 1094 nm. Determine the value of the n quantum number for the upper level involved in this transition.

Electron Configurations

22. **(a)** Explain what is meant by the "ionization energy" of an electron in an atom.

(b) Which electrons in an atom are ionized most easily?

(c) How is the ionization energy of an electron affected by its distance from the nucleus and by other electrons in the same shell?

(d) What information is obtained about an atom by photoelectron spectroscopy?

23. **(a)** Explain briefly how ionization energies are obtained from a photoelectron spectrometer.

(b) Argon has ionization energies of -1.52, -2.82, -24.1, -31.5, and -309 $MJ \cdot mol^{-1}$. Interpret these ionization energies in terms of the electron configuration of argon.

24. When a beam of argon atoms was irradiated with photons, no $2s$ electrons were observed unless the incident photons had a wavelength of at least 3.80 nm. In another experiment, higher-energy photons were used, and $2s$ electrons with kinetic energies of 1.12×10^{-16} J were emitted. What was the wavelength of these higher-energy photons?

25. The ionization energy of helium is 2.37 $MJ \cdot mol^{-1}$. What is the kinetic energy of the electrons produced when helium gas is irradiated with radiation of wavelength 40.0 nm?

26. Explain what information the electron configuration of an atom provides concerning the atom in its ground state.

27. **(a)** What is the maximum number of electrons associated with each of the $n = 1$, $n = 2$, and $n = 3$ energy levels?

(b) How many subshells are associated with each of these levels, and how is each labeled?

(c) What is the maximum number of electrons associated with each subshell?

28. Which of the following energy level designations are not allowed?

(a) $6s$ **(b)** $1p$ **(c)** $4d$ **(d)** $2d$
(e) $3p$ **(f)** $4d$ **(g)** $5p$ **(h)** $2s$

29. Ground-state electronic configurations are determined by the Pauli exclusion principle and Hund's rule.

(a) Explain the Pauli exclusion principle in terms of electron spin.

(b) Give an electrostatic explanation of Hund's rule.

30. In terms of the Pauli exclusion principle and/or Hund's rule, explain the following.

(a) Beryllium cannot have the electron configuration $1s^4$.

(b) The ground state of nitrogen has three unpaired electrons.

31. Without reference to Table 6.2, decide which of the following electron configurations are not allowed by the Pauli exclusion principle, and explain why.

(a) $1s^2 2s^2 2p^4$ **(b)** $1s^2 2s^2 2p^6 3s^3$
(c) $1s^2 3p^1$ **(d)** $1s^2 2s^2 2p^6 3s^2 3p^{10}$

32. Using the orbital box notation, write the ground-state electron configurations of each of the following atoms and ions.

(a) Be (b) N (c) F (d) Mg

(e) Cl^+ (f) Ne^+ (g) Al^{3+}

33. Without reference to Table 6.2, give the ground-state electron configurations of each of the following atoms, using the orbital box notation.

(a) K (b) Al (c) Cl (d) Ti (Z = 22) (e) Zn (Z = 30)

34. Identify the elements with each of the following ground-state configurations.

(a) $1s^22s^1$ (b) $1s^22s^22p^3$ (c) [Ar]$4s^2$ (d) [Ar]$3d^{10}4s^24p^3$

35. How many unpaired electrons are there in the ground states of each of the following?

(a) O (b) O^- (c) O^{2-} (d) S (e) F (f) Ar

36. What electron configuration is associated with **(a)** the main group element with the lowest valence-shell ionization energy in any period; **(b)** the main group element with the highest valence-shell ionization energy in any period?

37. Without reference to Table 6.4, draw orbital box diagrams for the ground-state valence shells of each of the following atoms.

(a) P (b) Ca (c) O (d) Br (Z = 35)

38. Use Hund's rule to decide which of the following electron orbital box diagrams do not correspond to ground states, and identify each element.

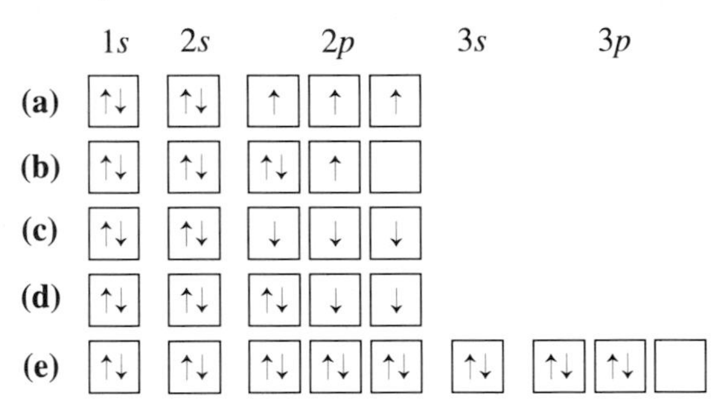

39. By drawing orbital box diagrams, determine how many unpaired electrons there are for the ground-state electron configurations of each of the following atoms.

(a) P (b) Si (c) I (d) Se

40. Categorize each of the following electron configurations as that of a ground state, an excited state, or not possible. For the allowed configurations, identify the element; for those not possible, explain why.

(a) $1s^22s^2$ (b) $1s^23s^1$ (c) $1s^22d^3$

(d) [Ne]$3s^23d^1$ (e) [Ar]$4s^23d^2$ (f) $1s^22s^22p^63s^1$

41. **(a)** To what group and period of the periodic table does arsenic belong?

(b) How many energy levels are occupied in the ground state of arsenic?

(c) How many of the orbitals are only singly occupied?

(d) Name the elements in the same group as arsenic in Periods 2 and 3, and write their ground-state electron configurations.

42. Write the electron configurations of the elements with each of the following atomic numbers. Without reference to any other information, name each element, and classify it as an s-block, p-block, or d-block element.

(a) Z = 5 (b) Z = 11 (c) Z = 19 (d) Z = 22

(e) Z = 23 (f) Z = 29 (g) Z = 33 (h) Z = 36

The Hydrogen Molecule and the Covalent Bond

43. Explain what is understood by the terms **(a)** "atomic orbital" and **(b)** "molecular orbital."

44. Explain why two hydrogen atoms attract each other and combine to form the H_2 molecule, whereas two helium atoms do not form a molecule.

45. **(a)** Write the ground-state electron configuration of silicon.

(b) How many unpaired electrons are there in the ground state?

(c) What valence would you predict for silicon?

(d) How do you explain the fact that silicon forms the compounds $SiCl_4$ and SiH_4?

46. **(a)** Repeat Problem 45a for boron.

(b) How many unpaired electrons are there in this ground state?

(c) How many covalent bonds would you expect boron to form?

(d) How do you explain why boron forms molecules such as BF_3 and ions such as BF_4^-?

47. **(a)** Explain why the bonding in the CH_4 molecule cannot be conveniently described in terms of the atomic 2s and 2p orbitals on carbon.

(b) How are the bonds in this molecule usually described in terms of hybrid orbitals?

48. Describe the bonding in ethene, C_2H_4, in terms of **(a)** the σ–π model and **(b)** the bent-bond model.

49. Give hybrid orbital descriptions of the bonding in the C_2H_2, HCN, and CO_2 molecules.

50. **(a)** What are the geometries of AX_2, AX_3, and AX_4 molecules?

(b) Describe the bonding in these molecules in terms of appropriate hybrid orbitals on the central atom A.

51. For each of the following molecules, use the VSEPR model to find the approximate values of the marked angles.

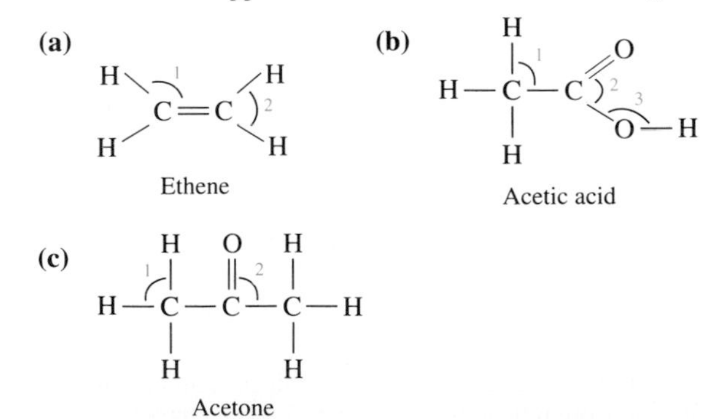

52. Repeat Problem 51 for each of the following molecules.

(a)

Peroxyacetyl nitrate

(b)

Glycine

53. For each of the molecules in Problem 51, describe the bonding in terms of hybrid orbitals and the σ–π model. How many σ bonds and how many π bonds are there in each molecule?

General Problems

54. Some sunglasses have special lenses that darken upon exposure to strong light and become paler in shade. These lenses contain a small amount of silver chloride. Light causes the reaction $AgCl \rightarrow Ag + Cl$ to occur, and the silver that is formed darkens the lens. In the absence of light, the reverse reaction occurs. The energy required for the reaction to take place is $310\ kJ \cdot mol^{-1}$. What is the maximum wavelength of light that can cause this reaction?

55. Excited barium atoms in a bunsen burner flame can return to their ground state by emitting photons of energy 3.62×10^{-19} J. What color is imparted to flame by light of this wavelength?

56. When an excited aluminum atom with an electron in a $4s$ orbital returns to the ground state, it emits light of wavelength 395 nm. When an excited aluminum atom with an electron in a $3d$ orbital returns to the ground state, it emits light of wavelength 319 nm.

(a) Calculate the energy separation between the $3d$ and $4s$ energy levels of aluminum.

(b) Which is higher in energy, the $3d$ energy level or the $4s$ energy level?

57. When light of wavelength 470.0 nm falls on the surface of potassium metal, electrons are emitted with a velocity of $6.4 \times 10^4\ m \cdot s^{-1}$. What is **(a)** the kinetic energy of the emitted electrons; **(b)** the energy of a 470.0-nm photon; **(c)** the minimum energy required to remove an electron from potassium metal?

58. When light of frequency 1.30×10^{15} Hz shines on the surface of cesium metal, photoelectrons are ejected with a kinetic energy of 5.20×10^{-19} J. What is the longest wavelength of light that will cause the removal of electrons from cesium metal?

59. By drawing orbital box diagrams, determine how many unpaired electrons there are for the ground-state electron configurations of each of the following atoms.

(a) Si **(b)** Ge **(c)** As **(d)** Se **(e)** I

60. Explain why two nitrogen atoms attract each other when brought together to form N_2, two oxygen atoms attract each other to form O_2, and two fluorine atoms attract each other to form F_2, whereas two neon atoms do not attract each other to form Ne_2.

Sulfur, Phosphorus, and Chlorine: Period 3 Nonmetals

7.1 Sulfur: Oxidation Numbers and Oxidation States
7.2 Phosphorus
7.3 Oxoacids
7.4 Chlorine
7.5 Lewis Structures

Sulfur, phosphorus, and chlorine are three common and important nonmetallic elements. Many of their compounds have practical uses, and phosphorus and sulfur are essential constituents of biomolecules. In the foreground is powdered red phosphorus and sticks of white phosphorus in a beaker of water, which protects the white phosphorus from reaction with the oxygen in air. In the background is a flask containing phosphorus that is burning in air and is producing a white smoke of the oxide P_4O_{10}. Also in the background is a model of P_4O_{10} (the phosphorus atoms are yellow; the oxygen atoms are red).

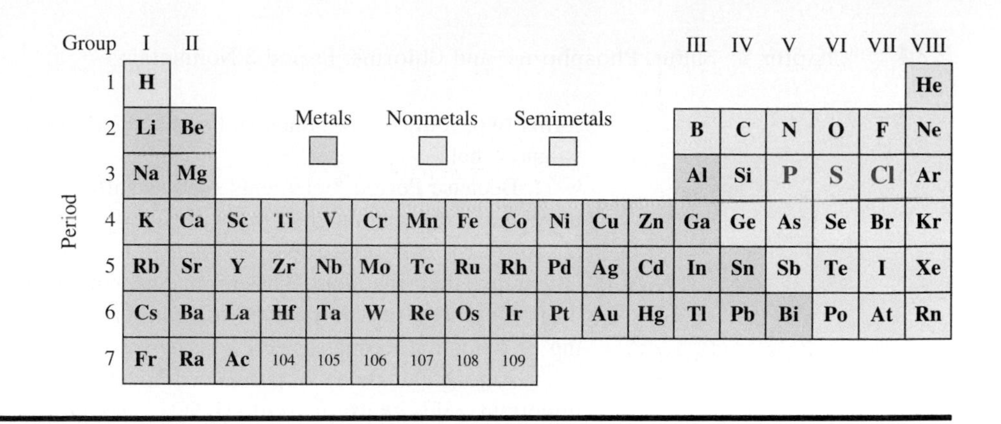

We saw in Chapter 3 that the elements in the upper right part of the periodic table, together with hydrogen, are nonmetals. Nine of these elements (O, Si, H, Cl, P, C, S, N, and F) are among the 20 most abundant elements in the earth's crust, and six of them (H, C, O, N, P, and S) are the most abundant elements in living matter. We described some of the chemistry of oxygen and nitrogen in Chapter 2 and discussed the halogens in Chapter 4. In this chapter we continue our discussion of the nonmetals, focusing on the Period 3 elements sulfur, phosphorus, and chlorine and using their chemistry to illustrate some general rules and concepts. We consider carbon in Chapter 8, some more chemistry of oxygen and nitrogen in Chapter 16, and silicon in Chapter 19.

There are some important differences between the properties of the Period 3 elements and the corresponding elements of Period 2. Most of these differences arise because, as we saw in Chapter 6, the valence (n = 3) shell of the Period 3 elements can accommodate a maximum of 18 electrons, whereas the maximum number of electrons in the valence shell of a Period 2 element is only 8. So the Period 3 elements do not always obey the octet rule in their compounds. For example, the common oxides of sulfur are SO_2 and SO_3, whereas we could predict from the octet rule that the oxide of sulfur would be $:\ddot{S}{=}\ddot{O}:$, in which sulfur forms only two bonds (Chapter 4). However, we find that when sulfur reacts with electronegative elements such as oxygen and fluorine, it may form more than two bonds. For example, in SO_2 there are four bonds (two double bonds) and a total of five pairs of electrons in the valence shell of sulfur. In SO_3 and SF_6 there are six bonds (three double bonds) and six pairs of electrons in the valence shell (Figure 7.1). So when sulfur forms compounds with fluorine or oxygen, it behaves as if it has four or six unpaired electrons in its valence shell. We may therefore conveniently represent a sulfur atom by three different Lewis symbols (Figure 7.1) in which it has two, four, or six unpaired electrons. Thus, in its compounds, sulfur

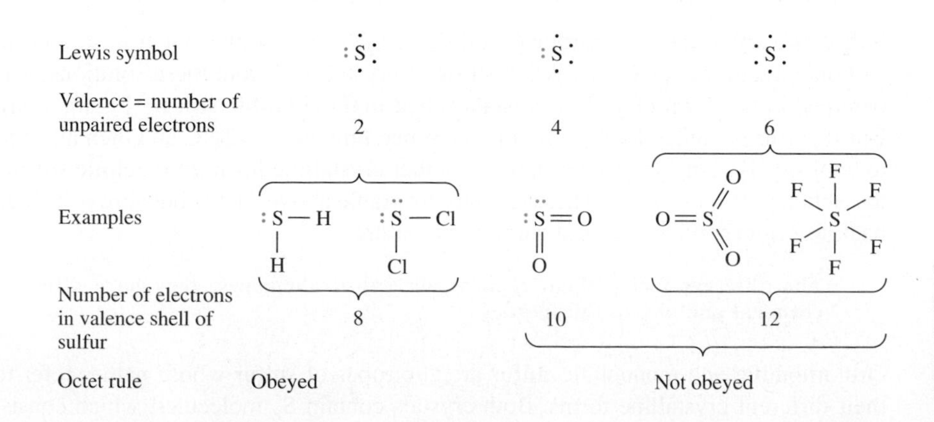

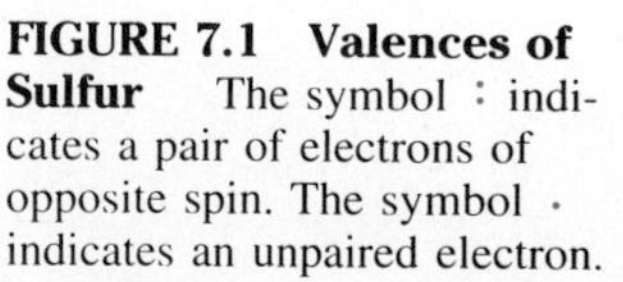

FIGURE 7.1 Valences of Sulfur The symbol : indicates a pair of electrons of opposite spin. The symbol · indicates an unpaired electron.

forms two, four, or six bonds and so has 8, 10, or 12 electrons, respectively, in its valence shell.

Because Period 3 elements such as sulfur form two or more compounds with oxygen and other nonmetals, we will find it convenient to begin using the concepts of *oxidation state* and *oxidation number* to help us classify their compounds and understand their reactions. These concepts, which we will introduce in the following discussion of sulfur compounds, will further extend our knowledge and understanding of oxidation–reduction reactions.

Among the many important compounds of sulfur and phosphorus are two important acids: sulfuric acid, H_2SO_4, and phosphoric acid, H_3PO_4. These are examples of an important class of acids called *oxoacids*. In this chapter we discuss the structures, strengths, and reactions of these acids. We also introduce many new molecules in this chapter. It is important to know how to draw their Lewis structures, so we will give some general rules for doing this.

7.1 SULFUR: OXIDATION NUMBERS AND OXIDATION STATES

Occurrence and Production

Volcanic sulfur deposit in West Java

Sulfur has been known since ancient times, because it is found in small amounts as thc free element in the vicinity of volcanoes. Very large subterranean deposits of sulfur are found in Louisiana, Texas, and other places. These deposits were probably formed by the reduction of calcium sulfate, $CaSO_4$, by bacteria. Sulfur is obtained by boring into the deposit, sending superheated water down to melt the sulfur (melting point 120°C), and forcing it up to the surface, where it is allowed to cool and solidify. The sulfur obtained in this way is usually at least 99% pure and is suitable for most commercial purposes. But today a more important source of sulfur is hydrogen sulfide, H_2S, which is typically present in natural gas to concentrations of 30% or more.

Sulfur occurs widely in the form of many of its compounds. Many metals are found as their sulfides, including FeS_2, PbS, and Cu_2S. These metal sulfides are important sources of the metals (Chapter 10) but not of sulfur, which is more easily obtained from sulfur deposits or from H_2S.

Sulfur awaiting shipment in Vancouver, British Columbia

Properties of the Element: The Allotropes of Sulfur

Sulfur is very soluble in carbon disulfide, $CS_2(l)$, and somewhat less soluble in carbon tetrachloride, $CCl_4(l)$. When sulfur is crystallized from these solutions, it is obtained in the form of yellow crystals called **orthorhombic sulfur** (Figure 7.2a). But if sulfur is melted by heating it to a temperature above 120°C and then allowed to cool slowly, long yellow needles of another crystalline form, **monoclinic sulfur**, are obtained (Figure 7.2b). These crystals are stable above 119°C but slowly change back to orthorhombic sulfur at room temperature.

The different forms of an element are called allotropes; they have different chemical and physical properties.

Orthorhombic and monoclinic sulfur are allotropes of sulfur whose names refer to their different crystalline forms. Both crystals contain S_8 molecules, which consist

(a) (b)

FIGURE 7.2 Crystals of Orthorhombic and Monoclinic Sulfur (a) Orthorhombic sulfur is the stable form of sulfur at room temperature. (b) The needlelike crystals of monoclinic sulfur are stable only above 119°C. They are sometimes found in the vicinity of volcanoes, where they form from the hot volcanic gases.

of a zigzag ring of eight sulfur atoms (Figure 7.3). Each sulfur atom forms two bonds and has an angular AX_2E_2 geometry with a bond angle of 108°C.

A noncrystalline allotrope of sulfur is obtained by quickly cooling molten sulfur in cold water. The sulfur does not react with the water, but the rapid cooling causes it to form a brown, rubbery material known as **plastic sulfur** (Demonstration 7.1). This allotrope is not stable. Within a few hours it is transformed back into crystalline orthorhombic sulfur. When it is heated, sulfur melts to a pale orange liquid at 120°C. As the temperature is raised to about 160°C to 190°C, the liquid becomes dark red-brown in color and increasingly thick and sticky, like maple syrup or molasses. It is described as being very *viscous*. This behavior of molten sulfur is quite unusual; most liquids become *less* viscous upon heating, because the increased thermal motion of the molecules enables them to move past each other more easily. When liquid sulfur is heated, however, S—S bonds in the S_8 rings are broken, and chains of eight sulfur atoms form. The S atoms at the ends of these chains have only seven electrons in their valence shells, so they react rapidly with other S_8 molecules and form chains of hundreds or thousands of atoms. This is an example of a **polymerization reaction**, in which many small molecules join into long-chain molecules called **polymers**. These long-chain molecules become tangled up with each other (Figure 7.4), which makes it more difficult for them to move past each other and thus increases the viscosity. When this viscous liquid sulfur is cooled

A liquid like water, which flows easily, has low viscosity; a liquid like maple syrup or heavy engine oil, which flow slowly, has high viscosity.

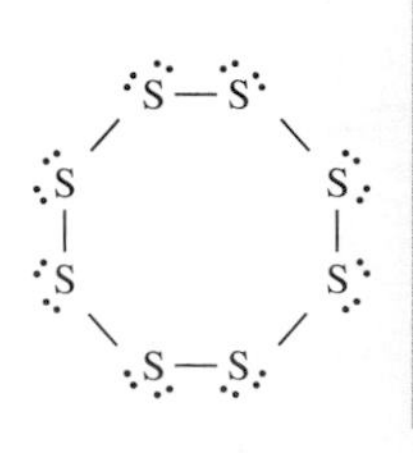

(a)
Structural formula

(b)
Ball-and-stick model

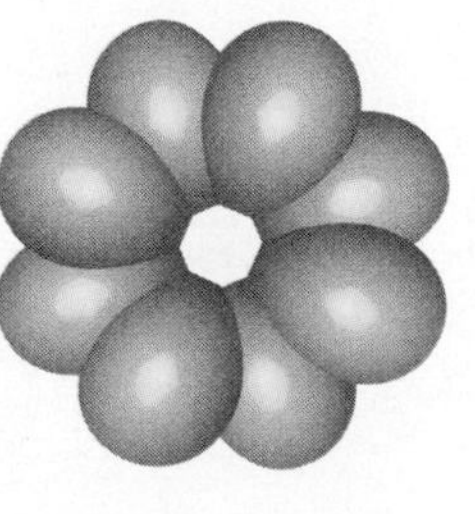

(c)
Space-filling model

FIGURE 7.3 The Crown-Shaped S_8 Molecule Each sulfur atom has an angular AX_2E_2 geometry.

DEMONSTRATION 7.1 Plastic Sulfur

Sulfur is heated until it melts to form an orange liquid.

Upon further heating, the color becomes dark red and the liquid becomes very viscous.

When the liquid is rapidly cooled by pouring it into cold water, a rubbery brown solid—plastic sulfur—is formed.

Plastic sulfur is elastic, like rubber.

rapidly, there is not enough time for the long tangled chains to rearrange to the more stable cyclic S_8 molecules. So the molecules of plastic sulfur do not have the ordered arrangement characteristic of a crystalline solid but retain the random arrangement that is characteristic of liquids. Solids that are not crystalline are called **amorphous solids**.

Plastic sulfur exhibits properties similar to those of rubber, which also consists of long-chain molecules (Chapter 20). Long-chain molecules tend to coil into compact shapes. If plastic sulfur or rubber is pulled, the coiled molecules straighten out a little, and the material stretches. If the stretching force is removed, the molecules resume their coiled structure, and the material contracts again (Figure 7.4). Polymerization and polymer molecules are discussed in more detail in Chapter 20.

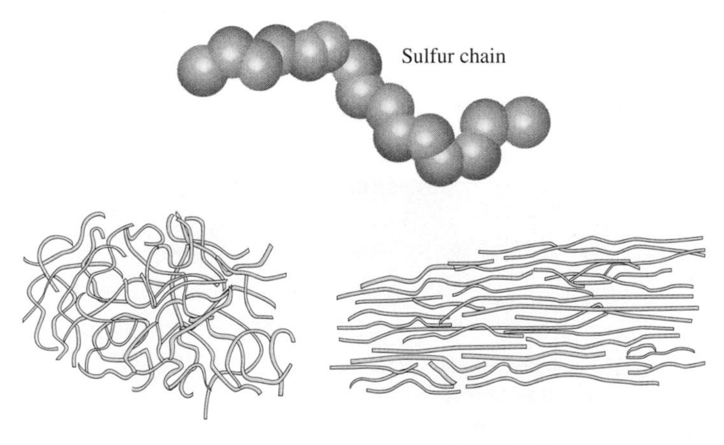

FIGURE 7.4 Plastic Sulfur Plastic sulfur is an amorphous (noncrystalline) allotrope of sulfur consisting of long chains of sulfur atoms. These chains have an irregular, disordered arrangement rather than the regular arrangement characteristic of crystalline substances. The chains tend to coil up and form a tangled mass. When a force is applied to stretch the plastic sulfur, the molecules straighten out a little but coil up again when the force is removed. So the solid can stretch and contract like rubber.

Oxides of Sulfur

From the position of sulfur in Group VI of the periodic table, we expect it to have a valence of 2, as it does in many of its compounds. We would therefore expect it to form the oxide $:\ddot{S}{=}\ddot{O}:$. Although this oxide is known, it is a very unstable and reactive substance. The much more stable and better-known oxides of sulfur are sulfur dioxide, SO_2, and sulfur trioxide, SO_3, in which sulfur has valences of 4 and 6, respectively (Figure 7.1).

Sulfur dioxide reduces the red pigment in a dampened rose to a colorless substance.

When sulfur is heated in air, it ignites and burns with a blue flame to produce **sulfur dioxide**, $SO_2(g)$ (Demonstration 2.2). Sulfur dioxide is a colorless gas with a pungent, choking odor. It is dangerous to breathe because it can damage the respiratory system. It destroys bacteria and is used as a preservative in the storage of fruits, such as apples. Sulfur dioxide condenses to a liquid at $-10°C$ at atmospheric pressure; at a pressure of 3 atm, it can be liquefied at 20°C. It is usually sold in metal cylinders as a liquid under pressure.

Sulfur dioxide molecules have an angular AX_2E structure with a bond angle of 120°.

When sulfur dioxide is heated with oxygen in the presence of a suitable catalyst, **sulfur trioxide**, $SO_3(g)$, forms:

$$2SO_2(g) + O_2(g) \rightarrow 2SO_3(g)$$

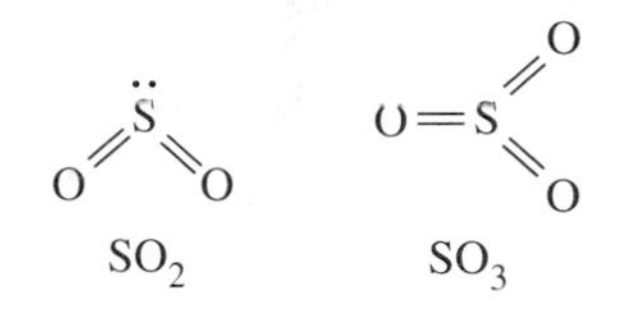

Sulfur trioxide condenses to a colorless liquid at 44.5°C and freezes to transparent crystals at 16.8°C.

Sulfur trioxide reacts with water to form **sulfuric acid**:

$$SO_3(g) + H_2O(l) \rightarrow H_2SO_4(l)$$

An oxide such as SO_3 that gives an acid when it reacts with water is called an **acidic oxide**.

Large quantities of SO_2 are released into the atmosphere by volcanoes and during the burning of coal and other fossil fuels, which often contain appreciable amounts of sulfur. This SO_2 is oxidized to SO_3. The SO_3 then reacts with atmospheric moisture to form a dilute aqueous solution of sulfuric acid, which falls to the ground as acid rain (Chapter 16).

H_2SO_4

Sulfuric Acid

In the sulfuric acid molecule, the sulfur atom forms a total of six bonds (two double bonds and two single bonds). The molecule has a tetrahedral AX_4 geometry. Pure sulfuric acid, $H_2SO_4(l)$, is a colorless liquid with a melting point of 10.4°C and a high boiling point of 338°C.

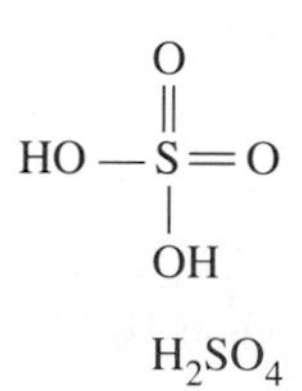

Oxoacids We usually write the formula of sulfuric acid as H_2SO_4, but writing it as $SO_2(OH)_2$ shows more clearly that it contains two OH groups and two oxygen atoms attached to the central sulfur atom. Any acid that has the general formula $XO_m(OH)_n$, where $m = 0, 1, 2, 3, \ldots$ and $n = 1, 2, 3, \ldots$, is called an **oxoacid** (or oxyacid).

An oxoacid has one or more OH groups attached to an electronegative atom. It may also have one or more oxygen atoms attached to the electronegative atom.

We will describe several other important oxoacids of sulfur, phosphorus, and chlorine in this chapter.

In structures that appear in the margin, nonbonding (lone) pairs are shown only on the central atom, where they are important in determining molecular geometry.

Industrial Production of Sulfuric Acid Sulfuric acid may be described as the most important substance produced by the chemical industry; it is produced in a greater amount than any other substance. The annual U.S. production, which is the largest in the world, is approximately 40 million tons. About half this amount is used for the production of phosphate fertilizers. Other uses include the manufacture of paints, dyes, explosives, detergents, and synthetic fibers.

The industrial production of sulfuric acid is based on three reactions:

Although solid sulfur consists of S_8 molecules, it is customary to use its empirical formula, S, in equations for reactions.

1. The burning of sulfur or a metal sulfide in air to give SO_2:

$$S(s) + O_2(g) \rightarrow SO_2(g)$$

$$CuS(s) + O_2(g) \rightarrow Cu(s) + SO_2(g)$$

2. The oxidation of SO_2 to SO_3:

$$2SO_2(g) + O_2(g) \rightarrow 2SO_3(g)$$

3. The combination of SO_3 with water to give H_2SO_4:

$$SO_3(g) + H_2O(g) \rightarrow H_2SO_4(l)$$

The oxidation of SO_2 to SO_3 is very slow at ordinary temperatures. To increase the reaction rate, this oxidation is carried out at about 400°C in the presence of a catalyst, usually vanadium pentaoxide, $V_2O_5(s)$. The process is known as the **contact process**, because the reaction takes place when SO_2 and O_2 molecules come into contact on the surface of the solid catalyst. The SO_3 produced in the contact process is not allowed to react directly with water to produce H_2SO_4, because this is a violent, strongly exothermic, reaction that is difficult to control. Instead, the gaseous SO_3 is absorbed in 98% H_2SO_4, and water is added at a controlled rate to keep the concentration of H_2SO_4 at approximately 98%. This 98% solution is the acid that is normally sold as *concentrated sulfuric acid.* If the amount of water added is suitably adjusted, we can obtain 100% pure sulfuric acid, which contains no water.

Properties of Sulfuric Acid Sulfuric acid behaves as a strong acid—it reacts completely with water to give the hydrogensulfate ion, HSO_4^-:

$$H_2SO_4(l) + H_2O(l) \rightarrow H_3O^+(aq) + HSO_4^-(aq)$$

The HSO_4^- ion is also an acid; it ionizes further to give the **sulfate ion**, SO_4^{2-}. However, it is a weak acid, because its reaction with water is not complete:

$$HSO_4^-(aq) + H_2O(l) \rightleftharpoons H_3O^+(aq) + SO_4^{2-}(aq)$$

An acid with only one ionizable hydrogen atom (such as HCl) is a *monoprotic acid.*

An acid such as sulfuric acid that has two ionizable hydrogen atoms is called a **diprotic acid**. The anions of oxoacids, such as the sulfate ion of the oxoacid sulfuric acid, are known as **oxoanions**.

A solution of sulfuric acid in water contains the H_3O^+, HSO_4^-, and SO_4^{2-} ions. The acidic properties of this solution are those of H_3O^+—the same as those of an aqueous solution of any other strong acid, such as hydrochloric acid.

When we speak of hydrochloric acid, we always mean a solution of hydrogen chloride in water. Hydrogen chloride is a gas that condenses to a liquid only at −85°C, so the pure liquid is not commonly encountered. In contrast, the name *sulfuric acid* is ambiguous: It is used both for the pure substance H_2SO_4 and for aqueous solutions of H_2SO_4 in water.

Because it is a diprotic acid, sulfuric acid reacts with bases to form two series of salts, the hydrogensulfates and the sulfates:

$$H_2SO_4(aq) + KOH(aq) \rightarrow \underset{\text{Potassium hydrogensulfate}}{KHSO_4(aq)} + H_2O(l)$$

$$H_2SO_4(aq) + 2KOH(aq) \rightarrow \underset{\text{Potassium sulfate}}{K_2SO_4(aq)} + 2H_2O(l)$$

Metal hydrogensulfates and sulfates are ionic compounds containing metal ions and HSO_4^- and SO_4^{2-} ions. They are generally soluble in water, but the sulfates of Ca^{2+}, Sr^{2+}, Ba^{2+}, Pb^{2+}, and Ag^+ have low solubilities. Barium sulfate is very insoluble, and white solid barium sulfate is formed when an aqueous solution of barium chloride, $BaCl_2(aq)$, is added to a solution containing even a small concentration of sulfate ion:

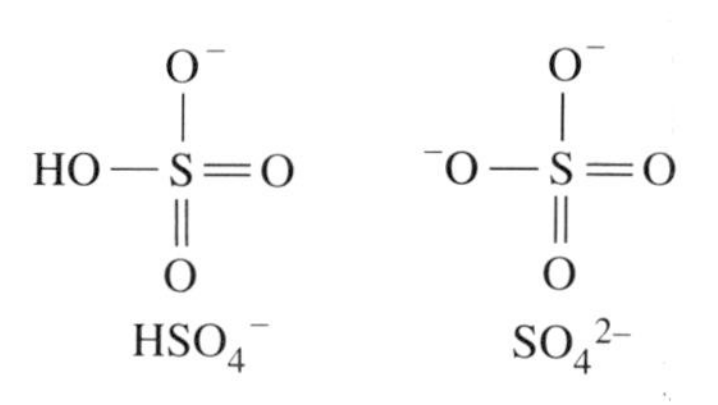

$$Ba^{2+}(aq) + SO_4^{2-}(aq) \rightarrow BaSO_4(s)$$

This precipitation reaction is used as a test for the presence of sulfate ion in an aqueous solution: Barium sulfate is the only common insoluble salt of barium (Table 4.6), so anions other than $SO_4^{2-}(aq)$ do not give a precipitate when mixed with an aqueous barium chloride solution.

The crystals that form when many sulfates crystallize from water contain water molecules as part of their three-dimensional network structure. Examples include $MgSO_4 \cdot 7H_2O$, $ZnSO_4 \cdot 7H_2O$, and $CuSO_4 \cdot 5H_2O$. These salts are said to be **hydrated salts**. The water that is contained in hydrated salts is called **water of crystallization**.

Exercise 7.1 Write balanced equations for the reaction of each of the bases CsOH, NH_3, $Ba(OH)_2$, and MgO with excess sulfuric acid in aqueous solution.

Concentrated Sulfuric Acid

The concentrated (98%) H_2SO_4 that is produced industrially and is commonly used in the laboratory is a mixture of water and sulfuric acid. Because the H_2SO_4 is in a large excess, this mixture should be regarded as a solution of water in sulfuric acid rather than as a solution of sulfuric acid in water. It contains H_2SO_4 molecules and some H_3O^+ and HSO_4^- ions but no H_2O molecules. Its properties are similar to those of 100% H_2SO_4, which consists only of H_2SO_4 molecules. Concentrated sulfuric acid is a strong acid, a strong dehydrating agent, and a strong oxidizing agent.

Strong Acid Concentrated sulfuric acid reacts with sodium chloride to give hydrogen chloride, which is evolved as a gas:

$$NaCl(s) + H_2SO_4(l) \rightarrow HCl(g) + NaHSO_4(s)$$

In this reaction, sulfuric acid *protonates* (transfers a proton to) the chloride ion:

$$Cl^- + H_2SO_4 \rightarrow HCl + HSO_4^-$$

FIGURE 7.5 Copper Sulfate Anhydrous copper sulfate, $CuSO_4$, is white. Hydrated copper sulfate, $CuSO_4 \cdot 5H_2O$, is blue.

We saw in Chapter 4 that chloride ion, Cl^-, is not a base in water. It is not protonated by the hydronium ion, H_3O^+, because it is the anion of HCl, a strong acid in water. But in concentrated sulfuric acid, there are H_2SO_4 molecules as well as H_3O^+ and HSO_4^- ions. Sulfuric acid, H_2SO_4, is a stronger acid than H_3O^+; it can protonate Cl^- to give HCl.

Dehydrating Agent Water is completely ionized in solution in sulfuric acid. It behaves as a strong base:

$$H_2O(l) + H_2SO_4(l) \rightarrow H_3O^+(aq) + HSO_4^-(aq)$$

This property makes sulfuric acid a very good **dehydrating agent**—it removes water by protonating it. If a gas that does not react with sulfuric acid, such as O_2, N_2, CO_2, or SO_2, contains a small amount of water vapor, it can easily be dried by bubbling it through concentrated (98%) sulfuric acid. The water reacts with the acid to produce H_3O^+ and HSO_4^- ions, which remain in solution.

Hydrated salts, such as $CuSO_4 \cdot 5H_2O$, lose their water of crystallization when stored in a *desiccator* (a closed container used for drying) with concentrated sulfuric acid. A small amount of water vapor is normally in equilibrium with blue hydrated copper sulfate, but when this water is absorbed by sulfuric acid, equilibrium is never established. The hydrated salt is completely converted to white *anhydrous* copper sulfate (Figure 7.5). ''Anhydrous'' means ''without water'':

$$\underset{\text{Blue}}{CuSO_4 \cdot 5H_2O} \rightarrow \underset{\text{White}}{CuSO_4} + \underset{\text{Absorbed by } H_2SO_4}{5H_2O}$$

The tendency of sulfuric acid to combine with water is so strong that it will remove hydrogen and oxygen as water from many compounds that do not contain H_2O molecules. For example, it removes hydrogen and oxygen as water from carbohydrates and many other organic compounds, leaving behind a residue of carbon. Wood, paper, starch, cotton, and sugar (sucrose) are all dehydrated in this way (Demonstration 7.2):

Carbohydrates do not contain water molecules, but they contain hydrogen and oxygen atoms in a 2:1 ratio.

$$\underset{\text{Sucrose}}{C_{12}H_{22}O_{11}(s)} + 11H_2SO_4(l) \rightarrow 12C(s) + 11H_3O^+(aq) + 11HSO_4^-(aq)$$

Oxidizing Agent Concentrated sulfuric acid is a strong oxidizing agent. For example, it will oxidize the bromide ion, Br^-, in solid NaBr to bromine, Br_2, the iodide ion in solid NaI to iodine, I_2 (Demonstration 7.3), and copper to the Cu^{2+} ion. In these reactions, the sulfuric acid must be reduced. It is found by experiment that in each case, the other product of the reaction is sulfur dioxide, SO_2. Thus sulfuric acid must be reduced to sulfur dioxide, and we can write the following *unbalanced* equations:

DEMONSTRATION 7.2 Sulfuric Acid as a Dehydrating Agent

Concentrated sulfuric acid (98% H_2SO_4) is added to sugar, $C_{12}H_{22}O_{11}$.

The concentrated sulfuric acid removes water from the sugar to form black carbon and steam.

The steam mixes with the carbon to form a black pillar that rises high above the beaker as the mixture heats and expands.

$$Br^- + H_2SO_4(conc) \rightarrow Br_2 + SO_2$$

$$I^- + H_2SO_4(conc) \rightarrow I_2 + SO_2$$

$$Cu + H_2SO_4(conc) \rightarrow Cu^{2+} + SO_2$$

The abbreviation "(*conc*)" following the symbol of a compound indicates a concentrated aqueous solution of the compound. In the case of sulfuric acid, it means a solution so concentrated that it contains no free H_2O molecules.

We recognize the conversion of copper metal to the copper ion, Cu^{2+}, and the conversion of Br^- to bromine, Br_2, and I^- to iodine, I_2, as oxidations because they involve the loss of electrons. But it is not so clear that the conversion of sulfuric acid

DEMONSTRATION 7.3 Sulfuric Acid as an Oxidizing Agent

Concentrated sulfuric acid (98% H_2SO_4) is added to white solid sodium bromide and white solid sodium iodide.

The sulfuric acid oxidizes the sodium bromide to reddish-brown bromine and the sodium iodide to violet iodine.

to sulfur dioxide involves a gain of electrons, although it clearly involves a loss of oxygen. We will see more clearly in the following subsection that the reduction of sulfuric acid to SO_2 involves a gain of electrons when we introduce the concept of oxidation numbers.

In Chapter 4 we saw that some metals, such as zinc and magnesium, are oxidized by hydrochloric acid to their cations, Zn^{2+} and Mg^{2+}. In these cases the oxidizing agent is the H_3O^+ ion, and the reduction product is hydrogen. Copper is more difficult to oxidize. It is not oxidized by hydrochloric acid, but it is oxidized by concentrated sulfuric acid, which is a stronger oxidizing agent than $H_3O^+(aq)$. In this case the reduction product is $SO_2(g)$, not $H_2(g)$.

Redox Reactions of Sulfur Compounds: Oxidation Numbers and Oxidation States

As we have seen, compounds of sulfur are involved in many oxidation–reduction reactions. To simplify the writing of equations for these reactions, we use the concepts of *oxidation number* and *oxidation state*.

Assigning Oxidation Numbers The **oxidation number** of an atom in a compound is a measure of the extent of its oxidation in that compound. Oxidation numbers are assigned by using the following set of rules.

1. Each atom in an element (in any of its allotropic forms) is assigned an oxidation number of 0.
2. The oxidation number of a monatomic ion is equal to the charge of the ion—$+1$ for Na^+, $+2$ for Cu^{2+}, -1 for Cl^-, and so on.
3. Fluorine, the most electronegative element, has an oxidation number of -1 in all its covalent compounds as well as in the F^- ion.
4. Hydrogen in its compounds usually has an oxidation number of $+1$. The only exceptions occur in metal hydrides, such as Na^+H^-, in which hydrogen has an oxidation number of -1.
5. Oxygen in its compounds usually has the oxidation number -2. There are two important exceptions:
   - When oxygen is combined with fluorine, as in F_2O, its oxidation number is $+2$.
   - In compounds containing oxygen–oxygen bonds, such as hydrogen peroxide, the oxidation number of oxygen is -1.
6. The halogens other than fluorine have an oxidation number of -1 except when they are combined with a more electronegative element, that is, oxygen or a more electronegative halogen.
7. The sum of the oxidation numbers of all the atoms in a neutral compound is 0, and the sum of oxidation numbers in an ion is equal to the charge on the ion.

Exercise 7.2 What is the oxidation number of each of the atoms in the following compounds?
(a) Na_2O **(b)** Al_2O_3 **(c)** BaH_2 **(d)** CaS **(e)** SF_6

What is the basis of these rules? The free element is taken to be the state from which the extent of oxidation or reduction of an atom of an element is measured, so the atoms of an element are given an oxidation number of zero (Rule 1). If an atom of an element loses one electron to give a *monopositive* ion, such as Na^+, we give it an oxidation number of +1. A *dipositive* ion, such as Mg^{2+}, is given an oxidation number of +2, which corresponds to the loss of two electrons. A negative ion, such as Cl^-, is given an oxidation number of −1, which corresponds to the gain of one electron, and so on (Rule 2). But how do we assign oxidation numbers in covalent molecules, such as SO_2, and polyatomic ions, such as SO_4^{2-}?

We saw in Chapter 4 that hydrogen chloride is a polar molecule. The Cl atom has a small negative charge, and the H atom has a small positive charge:

$$H^{\delta+} — \ddot{\underset{..}{Cl}}:^{\delta-}$$

There has been some transfer of electron density from the hydrogen atom to the chlorine atom, so hydrogen has been oxidized and chlorine reduced, but less than one electron has been transferred. However, assigning fractional oxidation numbers to these atoms would be inconvenient and not very useful. So we assign oxidation numbers by assuming that the electron transfer is complete:

The oxidation number of an atom in a covalent molecule or polyatomic ion is defined as the charge that would be associated with that atom if both electrons of each bond that the atom forms were transferred completely to the atom of higher electronegativity.

Thus, the bonding electron pair in hydrogen chloride is assigned to chlorine, because it is more electronegative than hydrogen. The Cl atom is considered to be Cl^- and is assigned an oxidation number of −1. The H atom is considered to be H^+ and is assigned an oxidation number of +1:

$$H:\ddot{\underset{..}{Cl}}: \xrightarrow[\text{to chlorine}]{\text{Transfer bond electrons}} H^{+1}\,:\ddot{\underset{..}{Cl}}:^{-1}$$

Similarly, in PCl_3 the shared electrons are assigned to Cl, which is more electronegative than P. Each Cl atom has an oxidation number of −1, and phosphorus has an oxidation number of +3:

$$:\ddot{\underset{..}{Cl}}:\ddot{P}:\ddot{\underset{..}{Cl}}: \;\; :\ddot{\underset{..}{Cl}}: \xrightarrow[\text{to chlorine}]{\text{Transfer bond electrons}} :\ddot{\underset{..}{Cl}}:^{-1}\;\ddot{P}^{+3}\;:\ddot{\underset{..}{Cl}}:^{-1}\;\; :\ddot{\underset{..}{Cl}}:^{-1}$$

Although in assigning oxidation numbers in covalent compounds we imagine electrons to be transferred from one atom to another, no electrons are added to or removed from the molecule. Therefore, the oxidation numbers of all the atoms in a neutral molecule must add up to zero, as is the case in HCl and PCl_3. In any polyatomic ion, the sum of the oxidation numbers must equal the charge on the ion (Rule 7).

Because fluorine is the most electronegative element in any compound, it always has an oxidation number of −1 (Rule 3). The other halogens also have high electronegativities, so they too have an oxidation number of −1 in most, but not all, of their compounds (Rule 6). Similarly, oxygen in its compounds is normally assigned an oxidation number of −2 (Rule 5). Hydrogen in its covalent compounds is typically less electronegative than the atom to which it is attached, so hydrogen usually has an oxidation number of +1 (Rule 4).

Using these rules, we can assign oxidation numbers to atoms in a great many compounds. For example, each oxygen atom in sulfur dioxide has an oxidation number of −2. The sum of all the oxidation numbers must be zero, so the oxidation number of the sulfur atom is +4:

$$O^{-2} \quad S^{+4} \quad O^{-2}$$

In sulfuric acid, H_2SO_4, the sum of the oxidation numbers of the two hydrogen atoms and the four oxygen atoms is −6:

$$\begin{aligned} 2H &= 2(+1) = +2 \\ 4O &= 4(-2) = \underline{-8} \\ & \qquad\qquad -6 \end{aligned}$$

Because the sum of the oxidation numbers must be zero, sulfur in H_2SO_4 must have an oxidation number of +6:

$$\begin{matrix} H_2 & & S & & O_4 & & \\ (+2) & + & (+6) & + & (-8) & = & 0 \end{matrix}$$

When there is a bond between two like atoms in a molecule, the two atoms share the bond electrons equally. For example, the electrons of the O—O bond in hydrogen peroxide, H—O—O—H, are shared equally by the two oxygen atoms, so one electron is transferred to each oxygen atom,

$$H \quad :\ddot{\underset{\cdot\cdot}{O}}\cdot \quad \cdot\ddot{\underset{\cdot\cdot}{O}}: \quad H$$

which gives the oxidation numbers

$$\begin{matrix} H & & O & & O & & H & & \\ (+1) & + & (-1) & + & (-1) & + & (+1) & = & 0 \end{matrix}$$

Exercise 7.3 Assign oxidation numbers to each of the atoms in the following.
(a) SO_3^{2-} **(b)** S_2^{2-} **(c)** ClO_4^- **(d)** $Al_2(SO_4)_3$

The concept of oxidation number leads to another useful definition of oxidation and reduction:

Oxidation is any change in which the oxidation number of an atom increases. Reduction is any change in which the oxidation number of an atom decreases.

These definitions are equivalent to the definitions that we gave in Chapter 4 in terms of electron loss and gain. For example, for the reaction

$$2Na(s) + Br_2(l) \rightarrow 2Na^+Br^-(s)$$

the oxidation half-reaction is

	Na →	$Na^+ + e^-$	(1 electron lost)
Oxidation numbers	0	+1	increase in oxidation number = 1

and the reduction half-reaction is

	$Br_2 + 2e^- \rightarrow 2Br^-$	(2 electrons gained)
Oxidation numbers	0 $\quad$ 2(−1)	decrease in oxidation number = 2

Thus, we see that

Electron loss (oxidation) is accompanied by an increase in the oxidation number equal to the number of electrons lost. Electron gain (reduction) is accompanied by a decrease in the oxidation number equal to the number of electrons gained.

Br_2 has gained two electrons, one for each Br atom. Each Br atom undergoes a change of oxidation number from 0 in the element to −1 in the ion Br^-.

Using Oxidation Numbers To see how oxidation numbers help us write the equations for oxidation–reduction reactions, let us determine the equation for the half-reaction in which sulfuric acid is reduced to sulfur dioxide. The oxidation number of sulfur decreases from +6 to +4, which shows that two electrons are involved in this reduction:

$$H_2SO_4 + 2e^- \rightarrow SO_2$$

This equation is not balanced, so there must be some other reactants and/or products. H^+ and H_2O are always present in an aqueous acid solution and may be reactants (H^+ in this case is shorthand for H_3O^+). When oxygen is removed from a molecule in a reduction reaction in aqueous solution, water is most commonly a product. Using H^+ to balance for charges and H_2O to balance for atoms, the complete equation for the half-reaction is

$$H_2SO_4 + 2H^+ + 2e^- \rightarrow SO_2 + 2H_2O$$

We can now find the complete equation for any reaction involving the reduction of sulfuric acid to sulfur dioxide. We obtain the equation for the oxidation of Br^- by concentrated sulfuric acid, for example, by combining the equation for the sulfuric acid reduction half-reaction with the equation for the oxidation half-reaction of bromide ion to bromine,

$$2Br^- \rightarrow Br_2 + 2e^-$$

to give

$$2Br^- + H_2SO_4(conc) + 2H^+ \rightarrow Br_2 + SO_2 + 2H_2O$$

Similarly, we can combine the equation for the oxidation of copper to Cu^{2+},

$$Cu \rightarrow Cu^{2+} + 2e^-$$

with the equation for the reduction of sulfuric acid to SO_2 to obtain the complete equation

$$Cu + H_2SO_4(conc) + 2H^+ \rightarrow Cu^{2+} + SO_2 + 2H_2O$$

Oxidation States of Sulfur

The common oxidation numbers found for sulfur in its compounds are +6, +4, +2, 0, and −2. Oxidation numbers, which we use for purposes of calculation, are associated with the different **oxidation states** of an element, which are designated by the corresponding Roman numerals (Table 7.1). When sulfur has the oxidation number +6, as in H_2SO_4, it is said to be in the +VI oxidation state, which is the

TABLE 7.1 Oxidation States of Sulfur

Oxidation state of S	VI	IV	II	0	−II
Examples	H_2SO_4	SO_2	SCl_2	S_n	H_2S
Lewis symbol	⁚Ṣ⁚	:Ṣ⁚	:Ṣ̈·	:Ṣ̈·	:Ṣ̈·

: Pair of electrons of opposite spin
· Unpaired electron

most highly oxidized state of sulfur. When it has an oxidation number of −2, as in H_2S, sulfur is said to be in the −II oxidation state, which is the least highly oxidized, or the most highly reduced, state of sulfur. In its elemental forms, S_n (usually S_8), sulfur is in the 0 oxidation state. Because it can use two, four, or all six of its valence electrons to form bonds when it combines with more electronegative elements, such as fluorine and oxygen, sulfur has three positive oxidation states (II, IV, and VI) but only one negative oxidation state (−II). The positive sign for the positive oxidation states is often not shown, but the negative sign for the negative oxidation states is always shown.

The oxidation state of an element is equivalent to the valence except that the oxidation state also gives the direction of bond polarity. In its valences of 6, 4, and 2, sulfur forms 6, 4, and 2 bonds corresponding to the three Lewis symbols with 6, 4, and 2 unpaired electrons, respectively (Table 7.1). Of the five common oxidation states of sulfur—+VI, +IV, +II, 0, and −II—the last three all correspond to a valence of 2. But a sulfur atom in the +II state forms two bonds to atoms of more electronegative elements; in the 0 state, two bonds to other sulfur atoms; and in the −II state, two bonds to atoms of less electronegative elements. We see that specifying the oxidation state of an element gives us more information than does the valance, so we use the concept of oxidation state more frequently.

Specifying the oxidation state of an atom in a compound gives us an alternative method for naming a compound, as in the following examples:

sulfur dioxide	or	sulfur(IV) oxide
sulfur trioxide	or	sulfur(VI) oxide
phosphorus trichloride	or	phosphorus(III) chloride

This system is used particularly for compounds of metals; for example,

iron dichloride	or	iron(II) chloride
iron trichloride	or	iron(III) chloride

Sulfurous Acid and Sulfites

Sulfur dioxide is very soluble in water, in which it behaves as a weak diprotic acid. The hydronium ion, H_3O^+, the **hydrogensulfite ion**, HSO_3^-, and the **sulfite ion**, SO_3^{2-}, are formed in small amounts in the reactions

DEMONSTRATION 7.4 Sulfur Dioxide as a Reducing Agent

Sulfur dioxide, when passed into an orange-red solution of bromine, reduces the bromine slowly.

After 2 min 10 s the bromine color has noticeably decreased as bromine, Br_2, is reduced to colorless bromide ion, Br^-.

After 2 min 41 s the reduction of bromine to bromide ion is almost complete.

$$SO_2 + 2H_2O \rightleftharpoons H_3O^+ + HSO_3^-$$

$$HSO_3^- + H_2O \rightleftharpoons H_3O^+ + SO_3^{2-}$$

HO—S(=O)—O⁻ (HSO_3^-)   ⁻O—S(=O)—O⁻ (SO_3^{2-})

A solution of SO_2 in water is usually called *sulfurous acid*, but there is no firm evidence that the un-ionized molecule H_2SO_3 exists. Both the HSO_3^- and SO_3^{2-} ions have the expected pyramidal AX_3E geometry.

Metal sulfites, such as $K_2SO_3(s)$ and $MgSO_3(s)$, are soluble in water. Both SO_2 and the sulfite ion, in which sulfur is in the +IV oxidation state, are good reducing agents. They are oxidized by many oxidizing agents to SO_3 and to the sulfate ion, SO_4^{2-}, in which sulfur is in the +VI oxidation state. For example, an orange-red solution of Br_2 in water is decolorized by SO_2 or by a sulfite solution, because red-brown Br_2 is reduced to colorless Br^- (Demonstration 7.4):

$$Br_2(aq) + SO_2(aq) + 2H_2O(l) \rightarrow 2Br^-(aq) + SO_4^{2-}(aq) + 4H^+(aq)$$

This reaction is just the reverse of the reaction that occurs when concentrated sulfuric acid oxidizes bromide ion to bromine:

$$2Br^- + H_2SO_4(conc) + 2H^+ \rightarrow Br_2 + SO_2 + 2H_2O$$

In the presence of a small amount of water, the equilibrium favors the formation of Br_2. In the presence of a large amount of water, the equilibrium favors the formation of Br^-.

Hydrogen Sulfide and Metal Sulfides

Hydrogen Sulfide Hydrogen sulfide, H_2S, is a gas (melting point $-85.5°C$; boiling point $-60.7°C$) (Demonstration 7.5). It has a powerful, unpleasant odor and is very poisonous. The Lewis structure of H_2S (Figure 7.1) shows that it is an AX_2E_2 molecule like H_2O and that it has the same angular shape. Hydrogen sulfide is soluble in water, in which it behaves as a weak diprotic acid:

$$H_2S(aq) + H_2O(l) \rightleftharpoons H_3O^+(aq) + HS^-(aq)$$

$$HS^-(aq) + H_2O(l) \rightleftharpoons H_3O^+(aq) + S^{2-}(aq)$$

In H_2S sulfur is in its lowest oxidation state (−II). H_2S is therefore a good reducing agent and is usually oxidized to sulfur. The half-reaction for this oxidation in aqueous solution is

$$H_2S(aq) \rightarrow S(s) + 2H^+(aq) + 2e^-$$

When an aqueous solution of H_2S is exposed to air, the H_2S is oxidized by oxygen, and a precipitate of sulfur slowly forms:

$$2H_2S(aq) + O_2(g) \rightarrow 2S(s) + 2H_2O(l)$$

When $H_2S(g)$ is passed into an aqueous solution of bromine, Br_2, the red-brown color of the Br_2 disappears as it is reduced to colorless Br^-, and a milky precipitate of sulfur forms:

$$H_2S(g) + Br_2(aq) \rightarrow 2H^+(aq) + 2Br^-(aq) + S(s)$$

Chlorine and iodine are reduced in the same way.

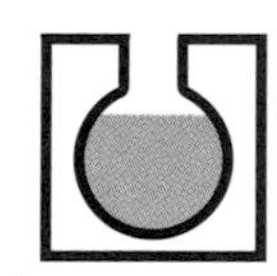

DEMONSTRATION 7.5 Properties of Hydrogen Sulfide

When H_2S is passed into a colorless aqueous solution of zinc sulfate, $ZnSO_4$, a white precipitate of insoluble zinc sulfide, ZnS, forms.

When H_2S is passed into a colorless aqueous solution of lead nitrate, $Pb(NO_3)_2$, a dark brown precipitate of lead sulfide, PbS, forms.

We mentioned at the beginning of this section that the H_2S in natural gas is an important source of sulfur. After it is removed from natural gas, some of the H_2S is burned in a limited supply of air to convert it to sulfur dioxide:

$$2H_2S(g) + 3O_2(g) \rightarrow 2SO_2(g) + 2H_2O(g)$$

The sulfur dioxide is then mixed with more hydrogen sulfide. At about 450°C and in the presence of a catalyst, the H_2S is oxidized to sulfur, and the SO_2 is reduced to sulfur:

$$2H_2S(g) + SO_2(g) \rightarrow 2H_2O(g) + 3S(s)$$

The equation for the net reaction is

$$2H_2S(g) + O_2(g) \rightarrow 2H_2O(g) + 2S(s)$$

||| SO_2 can act as an oxidizing agent (and be reduced from +IV to 0) or as a reducing agent (and be oxidized from +IV to +VI).

Exercise 7.4 By adding the appropriate half equations, derive the equations for the oxidation of $H_2S(aq)$ with $O_2(g)$ and with $Br_2(aq)$.

Metal Sulfides The **sulfides** of the alkali and alkaline earth metals are white solids that are soluble in water. The sulfides of most other metals are insoluble and form when H_2S is passed through a solution of a salt of the metal (Demonstration 7.5). These are further examples of precipitation reactions:

$$Zn^{2+}(aq) + H_2S(aq) \rightarrow ZnS(s) + 2H^+(aq)$$

$$Pb^{2+}(aq) + H_2S(aq) \rightarrow PbS(s) + 2H^+(aq)$$

We can think of these reactions as occurring in two stages, the ionization of H_2S as a weak acid,

$$H_2S(aq) \rightleftharpoons 2H^+(aq) + S^{2-}(aq)$$

followed by the combination of the metal cation with S^{2-} to form the insoluble sulfide:

$$Pb^{2+}(aq) + S^{2-}(aq) \rightarrow PbS(s)$$

Because H_2S is a weak acid, the equilibrium concentration of sulfide ion, S^{2-}, is very small. Nevertheless, all the Pb^{2+} ion (or Zn^{2+}) is precipitated, because sulfide ion is continually removed by combination with the metal ion to form an insoluble precipitate. So the position of the equilibrium for the ionization of H_2S is shifted to the right.

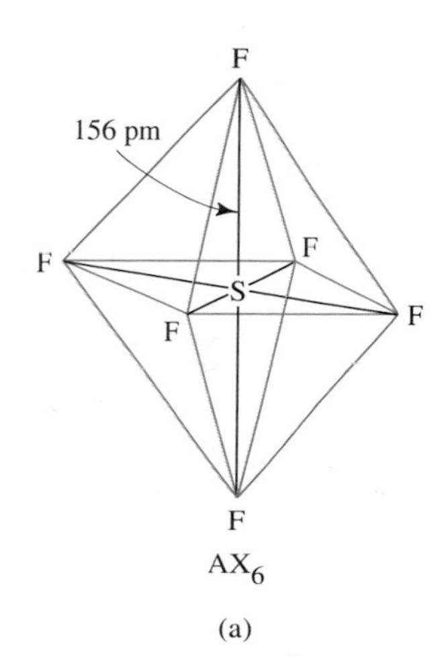

(b)

FIGURE 7.6 The Structure of SF_6, an AX_6 Molecule
(a) Sulfur hexafluoride, SF_6, has an octahedral geometry. (b) Six spheres, each representing a pair of electrons, adopt an octahedral arrangement.

Sulfur Halides

Sulfur reacts readily with all the halogens to form covalent molecular halides, such as SCl_2 and SF_6. Sulfur dichloride, $SCl_2(l)$, in which sulfur is in the +II oxidation state, is a red liquid. The molecule has an angular AX_2E_2 shape. Sulfur hexafluoride, $SF_6(g)$, in which sulfur is in the +VI oxidation state, is very stable and unreactive. It is used as an insulating gas in some high-voltage electrical equipment. According to the VSEPR model, the arrangement that keeps six electron pairs as far apart as possible is the octahedron (Figure 7.6). SF_6 is an example of an octahedral AX_6 molecule.

:S̈—Cl
|
Cl

SCl_2

7.2 PHOSPHORUS

Occurrence and Properties

Unlike sulfur, phosphorus is too reactive to occur naturally as the free element. Most of the phosphorus in the earth's crust is in the form of **phosphate rock**, which consists mainly of calcium phosphate, $Ca_3(PO_4)_2$, and of minerals known as apatites. Apatites are ionic compounds with the formula $Ca_5(PO_4)_3X$, where X represents F, Cl, or OH.

The formula of the apatites is fairly complicated because these compounds contain three ions with different charges (Ca^{2+}, PO_4^{3-}, and X^-), the sum of which must be zero.

Hydroxyapatite, $Ca_5(PO_4)_3OH$, is the main constituent of the bones and teeth of mammals. Fluoridation of water leads to the replacement of the hydroxyapatite in tooth enamel by fluorapatite $Ca_5(PO_4)_3F$. Fluorapatite is less basic than hydroxyapatite and more resistant to attack by acids in food, so it helps to prevent the formation of cavities. Many other compounds that play an essential role in living matter also contain phosphorus. Phosphate groups are an essential part of the structure of DNA (deoxyribonucleic acid), the storehouse of genetic information in animals, and of ATP (adenosine triphosphate), which is used in the transfer and utilization of energy in the body (Chapter 17).

III P has an oxidation number of +5 in PO_4^{3-}, because the *sum* of the oxidation numbers must equal the charge of the ion (−3). Thus the oxidation number of P is $-3 - 4(-2) = +5$.

Phosphorus is manufactured industrially by reducing phosphorus(V) in the **phosphate ion**, PO_4^{3-}, to the element by heating calcium phosphate to a very high temperature with carbon (coke) and silica, SiO_2. The calcium phosphate is converted to phosphorus(V) oxide, P_4O_{10}:

$$2Ca_3(PO_4)_2(s) + 6SiO_2(s) \rightarrow 6CaSiO_3(l) + P_4O_{10}(g)$$

Then the product P_4O_{10} is reduced to phosphorus with carbon:

$$P_4O_{10}(g) + 10C(s) \rightarrow 10CO(g) + P_4(g)$$

The Caribbean flashlight fish is chemiluminescent. Chemical reactions are the source of the light it emits.

The phosphorus vapor is condensed to a liquid and then allowed to solidify. It is typically stored as a liquid at a temperature of 50°C and shipped as a liquid in tank cars.

Like sulfur, phosphorus has several allotropes. **White phosphorus**, obtained when phosphorus vapor is condensed, is a toxic, white solid with a melting point of 44°C. White phosphorus reacts with oxygen at room temperature, emitting a blue-green light (Demonstration 7.6). The emission of light during a reaction is known as **chemiluminescence**. At temperatures above 40°C, the oxidation becomes quite rapid, and phosphorus ignites. White phosphorus is therefore stored out of contact with air—usually under water, in which it is insoluble (Figure 7.7). Both white phosphorus and gaseous phosphorus consist of P_4 molecules, which have a tetra-

FIGURE 7.7 Red and White Phosphorus Red phosphorus is stable in air. White phosphorus rapidly oxidizes in air, so it is usually stored under water, in which it is insoluble.

DEMONSTRATION 7.6 Oxidation of White Phosphorus: Chemiluminescence

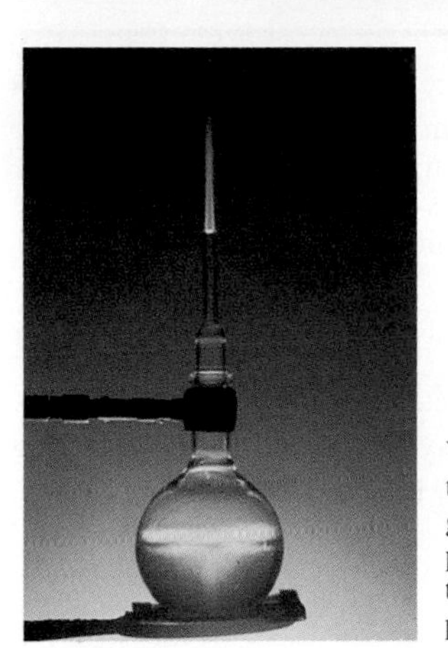

When water containing a piece of white phosphorus is boiled, the jet of steam coming from the tube at the top of the flask glows in a subdued light with a blue-green color as the phosphorus vapor in the steam is oxidized by the oxygen in the air. Some of the energy produced by the oxidation of phosphorus is in the form of light rather than heat.

hedral structure (Figure 7.8). Each of the phosphorus atoms has a pyramidal AX_3E geometry.

Of the other allotropes of phosphorus, the best known is **red phosphorus**, a solid that sublimes (changes directly from solid to vapor) at 417°C and melts only under pressure at 540°C. Red phosphorus can be made by heating white phosphorus at 300°C for several days in the absence of air. It is stable in air and is much less reactive and less poisonous than white phosphorus. These differences between the allotropes result from their different structures. Red phosphorus does not consist of P_4 molecules but has a covalently-bonded three-dimensional network structure. Vaporization of white phosphorus requires only sufficient energy to overcome the relatively weak intermolecular forces between the molecules, whereas much more energy is needed to vaporize red phosphorus and thereby break the strong covalent bonds in the network structure. Red phosphorus is an essential component of the striking surface on a box of safety matches (Box 7.1).

The most important oxidation states of phosphorus are +V, +III, 0, and −III. As we see in Table 7.2, we may conveniently represent the phosphorus atom by two Lewis symbols in which it has either three or five unpaired electrons.

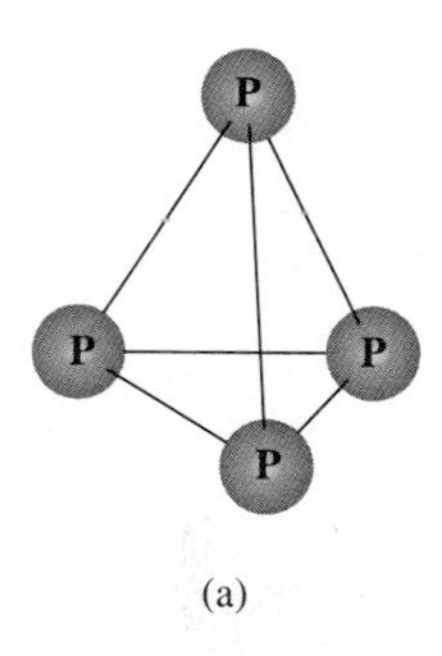

(a)

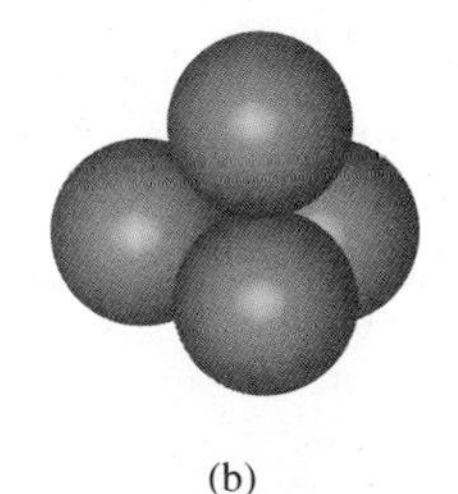

(b)

FIGURE 7.8 The Tetrahedral P_4 Molecule (a) Ball-and-stick model; (b) space-filling model.

TABLE 7.2 Oxidation States of Phosphorus

Oxidation state of P	V	III	0	−III
Examples	P_4O_{10}	P_4O_6	P_4	PH_3
	H_3PO_4	PCl_3		
	PCl_5			
Lewis symbol	$\cdot\!\!\!:\dot{\mathrm{P}}\!:\!\cdot$	$:\dot{\underset{\cdot}{\mathrm{P}}}\cdot$	$:\dot{\underset{\cdot}{\mathrm{P}}}\cdot$	$:\dot{\underset{\cdot}{\mathrm{P}}}\cdot$

BOX 7.1 The Chemistry of Matches

The head of a safety match is made of an oxidizing agent such as potassium chlorate, $KClO_3$, mixed with sulfur, fillers, and glass powder. The striking surface of the matchbox contains red phosphorus, binder, and powdered glass or sand. The heat generated by friction when the match is struck causes a minute amount of red phorphorus to be converted to white phosphorus, which ignites spontaneously in air. This sets off the decomposition of the potassium chlorate to give oxygen. The sulfur then catches fire and ignites the wood.

$$2KClO_3(s) \rightarrow 2KCl(s) + 3O_2(g)$$
$$S(s) + O_2(g) \rightarrow SO_2(g)$$

The head of a "strike-anywhere" match contains an oxidizing agent such as $KClO_3$ or manganese dioxide, MnO_2, together with tetraphosphorus trisulfide, P_4S_3, glass, and binder. The phosphorus sulfide is easily ignited. The $KClO_3$ decomposes to give oxygen, which in turn causes the P_4S_3 to burn more vigorously.

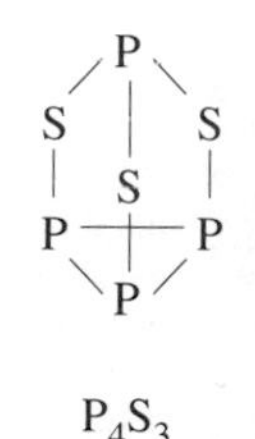

P_4S_3

The red tip on a strike-anywhere match (left) is P_4S_3. The remainder of the tip is composed of $KClO_3$ or MnO_2, sulfur, glue, and a coloring agent. The head of a safety match (right) contains $KClO_3$, sulfur, powdered glass, glue, and a coloring agent.

The striking surface on a box of safety matches consists of red phosphorus, ground glass, and glue.

The striking surface on a box of strike-anywhere matches is simply a rough surface of powdered glass and glue.

Oxides and Oxoacids

When phosphorus burns in air or oxygen, two different oxides may form. If the supply of oxygen is limited, $P_4O_6(s)$, phosphorus(III) oxide (tetraphosphorus hexaoxide), is the major product:

$$P_4(s) + 3O_2(g) \rightarrow P_4O_6(s)$$

Sometimes the oxides of phosphorus are described by their empirical formulas P_2O_3 and P_2O_5 rather than their molecular formulas P_4O_6 and P_4O_{10}.

In excess oxygen, $P_4O_{10}(s)$, phosphorus(V) oxide (tetraphosphorus decaoxide), forms (Demonstration 7.7):

$$P_4(s) + 5O_2(g) \rightarrow P_4O_{10}(s)$$

Figure 7.9 shows the structures of these oxides and how they are related to that of the P_4 tetrahedron. Each phosphorus atom in P_4O_6 forms three bonds and has one unshared pair and so has an octet of electrons. Each phosphorus atom has a pyramidal AX_3E geometry. The structure of P_4O_{10} is similar, except that an

DEMONSTRATION 7.7 Reaction of White Phosphorus with Oxygen

A piece of white phosphorus inserted into a flask of oxygen ignites, burns, and produces a white smoke of P_4O_{10}.

As the phosphorus burns, it produces a brilliant white light.

When the reaction is complete, the inside of the flask is coated with a layer of white P_4O_{10}.

additional oxygen atom is bonded to each phosphorus atom by a double bond. In this case each phosphorus atom uses all five valence electrons to form five covalent bonds (three single bonds and one double bond) and so has a valence shell of 10 electrons. Each phosphorus atom has an AX_4 tetrahedral geometry.

Phosphoric Acid We saw in Section 7.1 that SO_3 is called an acidic oxide because it reacts with water to give the oxoacid H_2SO_4 or $SO_2(OH)_2$. The oxide P_4O_{10} is an acidic oxide that reacts vigorously with water, liberating a considerable

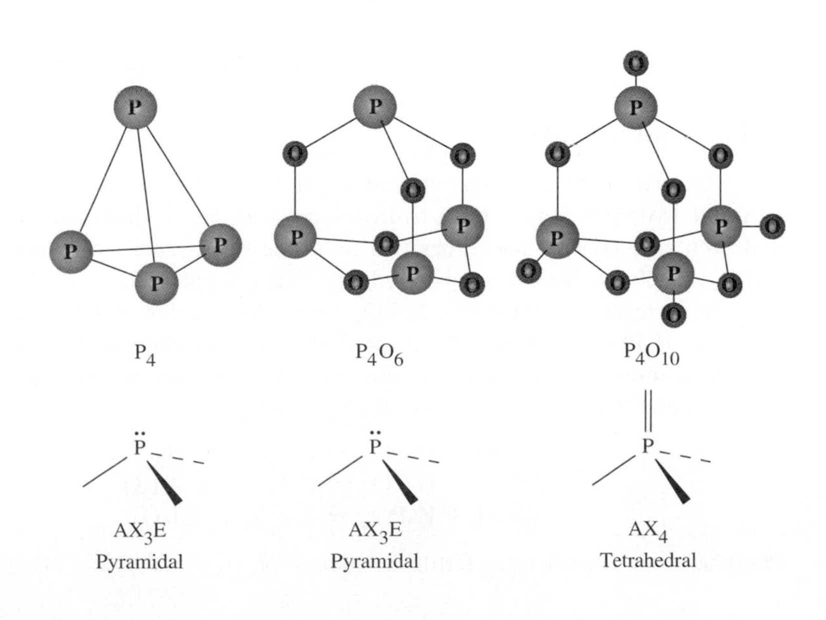

FIGURE 7.9 Structures of P_4, P_4O_6, and P_4O_{10} In P_4O_6 each edge of the P_4 tetrahedron is bridged by an oxygen atom. The structure of P_4O_{10} is the same, except that an additional oxygen atom is bonded to each phosphorus atom by a double bond.

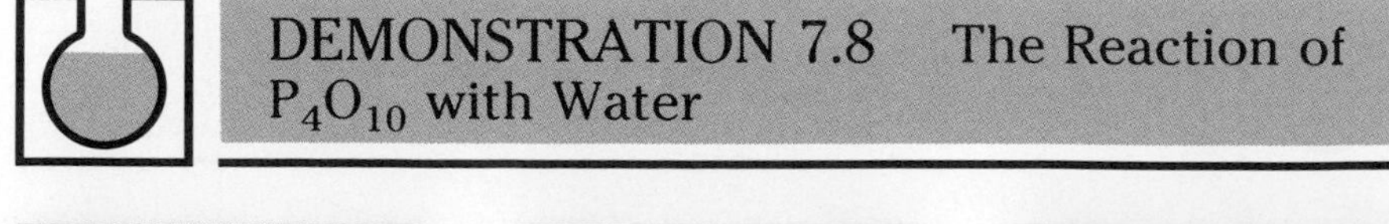

DEMONSTRATION 7.8 The Reaction of P_4O_{10} with Water

The glass spoon contains red phosphorus. The beaker of water is colored yellow because methyl red indicator has been added.

After phosphorus is ignited in a flame, it is placed in the beaker just above the water surface. The white fumes are P_4O_{10}.

The P_4O_{10} dissolves in the water to form phosphoric acid, H_3PO_4, and the methyl red indicator turns red.

amount of heat and forming a solution of the oxoacid H_3PO_4, **phosphoric acid** (Demonstration 7.8):

$$P_4O_{10}(s) + 6H_2O(l) \rightarrow 4H_3PO_4(aq)$$

Phosphoric acid is a colorless solid with a melting point of 42°C. A concentrated (82%) solution in water is usually used in the laboratory.

Phosphoric acid is **triprotic** because it has three OH groups. It ionizes in water in three stages:

$$H_3PO_4 + H_2O \rightleftharpoons H_3O^+ + H_2PO_4^-$$

$$H_2PO_4^- + H_2O \rightleftharpoons H_3O^+ + HPO_4^{2-}$$

$$HPO_4^{2-} + H_2O \rightleftharpoons H_3O^+ + PO_4^{3-}$$

OH
|
OH—P=O
|
OH
H_3PO_4

OH
|
$^-$O—P=O
|
OH
$H_2PO_4^-$

O^-
|
$^-$O—P=O
|
OH
HPO_4^{2-}

O^-
|
$^-$O—P=O
|
O^-
PO_4^{3-}

All three ionizations are incomplete, so phosphoric acid is a weak acid.

There is no relationship between the *strength* of an acid and the *number* of ionizable hydrogen atoms. Both hydrochloric acid, HCl, which has one ionizable hydrogen, and H_2SO_4, which has two ionizable hydrogens, are strong acids. But H_3PO_4, which has three ionizable hydrogens, is a weak acid.

There are only seven common strong acids: sulfuric acid, H_2SO_4; perchloric acid, $HClO_4$; chloric acid; $HClO_3$; nitric acid, HNO_3; hydrochloric acid, HCl; hydrobromic acid, HBr; and hydroiodic acid, HI. All other common acids are weak.

In its reaction with bases, H_3PO_4 forms three series of salts. For example, 1 mol of H_3PO_4 reacts with 1, 2, or 3 mol of potassium hydroxide, KOH, to give potassium dihydrogenphosphate, KH_2PO_4, potassium hydrogenphosphate, K_2HPO_4, and potassium phosphate, K_3PO_4, respectively:

$$KOH + H_3PO_4 \rightarrow KH_2PO_4 + H_2O$$
$$2KOH + H_3PO_4 \rightarrow K_2HPO_4 + 2H_2O$$
$$3KOH + H_3PO_4 \rightarrow K_3PO_4 + 3H_2O$$

Phosphates are important as fertilizers (Box 7.2).

BOX 7.2 Fertilizers

Most of the world's food is produced from plants. The limiting factor in plant growth is the availability of essential inorganic substances containing nitrogen, phosphorus, and potassium. Although most soils contain an appreciable amount of phosphate ion (PO_4^{3-}), this ion is often present in the form of insoluble salts such as $Ca_3(PO_4)_2$, so it is unavailable to plants. Insoluble calcium phosphate (phosphate rock, $Ca_3(PO_4)_2$) can be converted to a more soluble phosphate fertilizer if it is ground up and treated with sulfuric acid. This process produces a mixture of calcium dihydrogen phosphate and gypsum, hydrated calcium sulfate, $CaSO_4 \cdot 2H_2O$. The amount and concentration of the sulfuric acid are adjusted so that all the water is used up in the formation of hydrated salts:

$$Ca_3(PO_4)_2(s) + 2H_2SO_4(aq) + 5H_2O(l) \rightarrow Ca(H_2PO_4)_2 \cdot H_2O(s) + 2(CaSO_4 \cdot 2H_2O)(s)$$

If we write this equation in the ionic form and cancel all the spectator ions, we see that the reaction is simply

$$PO_4^{3-}(aq) + 2H^+(aq) \rightarrow H_2PO_4^-(aq)$$

in which phosphate ion, which is a base, combines with hydrogen ions from the sulfuric acid. The solid mixture of hydrated calcium sulfate (gypsum), $CaSO_4 \cdot 2H_2O$, and hydrated calcium dihydrogen phosphate, $Ca(H_2PO_4)_2 \cdot H_2O$, that results is called *superphosphate of lime.*

A more concentrated phosphorus fertilizer called *triple phosphate* is made by treating phosphate rock with phosphoric acid to give a product that is essentially pure calcium dihydrogen phosphate and contains no calcium sulfate. The phosphoric acid is made by treating phosphate rock with more sulfuric acid than is needed for the production of superphosphate of lime:

$$Ca_3(PO_4)_2(s) + 3H_2SO_4(aq) + 6H_2O(l) \rightarrow 2H_3PO_4(aq) + 3(CaSO_4 \cdot 2H_2O)(s)$$

We can write this equation more simply as

$$PO_4^{3-}(aq) + 3H^+(aq) \rightarrow H_3PO_4(aq)$$

Figure A *Fertilizer composed of 32% N, 3% P, and 5% K*

The insoluble calcium sulfate is removed by filtration. Phosphoric acid is left and then used to treat more phosphate rock to form calcium dihydrogen phosphate (triple phosphate), $Ca(H_2PO_4)_2$:

$$Ca_3(PO_4)_2 + 4H_3PO_4 \rightarrow 3Ca(H_2PO_4)_2$$

Another important fertilizer is made by treating phosphoric acid with ammonia:

$$H_3PO_4 + NH_3 \rightarrow NH_4H_2PO_4$$

The product of this reaction, ammonium dihydrogen phosphate, $NH_4^+H_2PO_4^-$, supplies the soil with both nitrogen and phosphorus.

Because a supply of nitrogen, phosphorus, and potassium is so important for plant growth, bags of fertilizer are usually labeled with their nitrogen, phosphate, and potassium contents (Figure A). A bag of fertilizer labeled 32-3-5, for example, contains 32% nitrogen, 3% phosphorus (expressed as P_2O_5), and 5% potassium (expressed as K_2O).

Polyphosphoric Acids Phosphoric acid readily undergoes condensation reactions.

> **A *condensation reaction* is the reaction of two or more molecules to form a larger molecule by the elimination of small molecules such as water.**

Two hydroxyl (OH) groups on different molecules frequently eliminate a molecule of water in this way. The product is a new molecule containing a bridging oxygen atom:

$$X{-}OH + HO{-}X \rightarrow X{-}O{-}X + H_2O$$

TABLE 7.3 Oxoacids of Phosphorus

Oxidation State	Name	Formula	Structure
V	Phosphor*ic* acid	H_3PO_4	$O{=}P(OH)_2{-}OH$
V	*Di*phosphor*ic* acid	$H_4P_2O_7$	$O{=}P(OH)_2{-}O{-}P(OH)_2{=}O$
V	*Tri*phosphor*ic* acid	$H_5P_3O_{10}$	$O{=}P(OH)_2{-}O{-}P(=O)(OH){-}O{-}P(OH)_2{=}O$
V	*Meta*phosphor*ic* acid	$(HPO_3)_n$	${-}O{-}P(=O)(OH){-}\left(O{-}P(=O)(OH)\right)_n{-}O{-}P(=O)(OH){-}O{-}P(=O)(OH){-}$

For example, when phosphoric acid is heated, it loses water and condenses to give **diphosphoric acid**, $H_4P_2O_7$, also called pyrophosphoric acid:

$$\mathrm{HO{-}P(=O)(OH){-}OH + HO{-}P(=O)(OH){-}OH \rightarrow HO{-}P(=O)(OH){-}O{-}P(=O)(OH){-}OH + H_2O}$$

Continued heating leads to the formation of triphosphoric acid, $H_5P_3O_{10}$ (Table 7.3), and acids whose molecules have still longer chains or ring structures. Complete elimination of water (dehydation) gives $P_4O_{10}(s)$.

Adenosine triphosphate, ATP, is a very important compound found in every living cell (Figure 7.10). The conversion of the triphosphate group of ATP to a

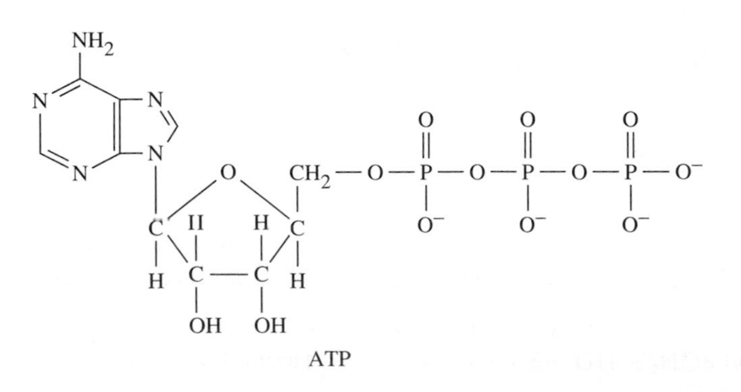

FIGURE 7.10 The Structure of ATP ATP, adenosine triphosphate, is an important energy-transfer molecule in living organisms.

diphosphate group by reaction with water produces adenosine diphosphate, ADP. This reaction provides the energy to drive the many reactions that sustain life (Chapter 17).

Phosphorous Acid Phosphorus(III) oxide, P_4O_6, reacts with water to give phosphorous acid, H_3PO_3, in which phosphorus is in the +III oxidation state:

$$P_4O_6(s) + 6H_2O(l) \rightarrow 4H_3PO_3(aq)$$

Phosphorous acid is a weak acid and a reducing agent.

Phosphides and Phosphine

The lowest oxidation state of phosphorus is the −III state. When phosphorus is heated with some metals, **phosphides**, such as calcium phosphide, $Ca_3P_2(s)$, and aluminum phosphide, $AlP(s)$, are formed. If these phosphides are treated with dilute aqueous acid, **phosphine**, $PH_3(g)$, is formed: For example,

$$Ca_3P_2(s) + 6HCl(aq) \rightarrow 2PH_3(g) + 3CaCl_2(aq)$$

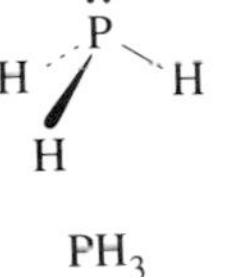

Phosphine is an unpleasant-smelling toxic gas that is a very weak base. The phosphine molecule has a pyramidal AX_3E structure like ammonia.

Phosphorus Halides

Phosphorus reacts readily with the halogens to form covalent halides. The *phosphorus halides* include phosphorus trifluoride, $PF_3(g)$, and phosphorus trichloride, $PCl_3(l)$, in which phosphorus is in the +III oxidation state; and phosphorus pentafluoride, $PF_5(g)$, and phosphorus pentachloride $PCl_5(s)$, in which phosphorus is in the +V oxidation state. According to the VSEPR model, five electron pairs have a trigonal bipyramidal arrangement. Thus, AX_5 molecules, such as PF_5 and PCl_5, are predicted to have a triangular bipyramidal structure (Figure 7.11). PF_3 and PCl_3 have the expected triangular pyramidal AX_3E geometry.

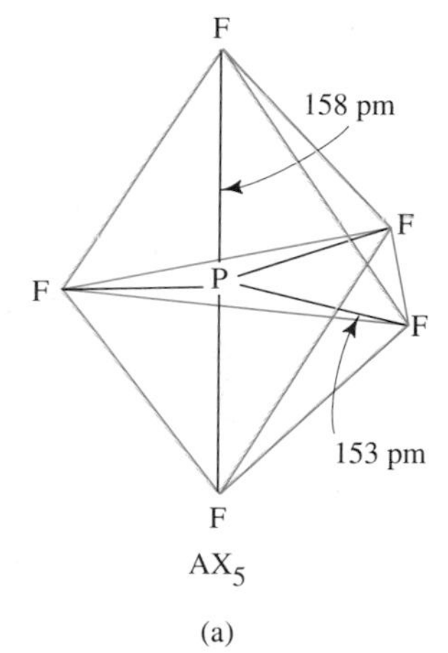

(a)

(b)

FIGURE 7.11 The Structure of PF_5, an AX_5 Molecule
(a) Phosphorus pentafluoride, PF_5, has a trigonal bipyramidal geometry. (b) Five spheres, each representing a pair of electrons, adopt a triangular bipyramidal arrangement.

7.3 OXOACIDS

In Sections 7.1 and 7.2 we discussed several examples of oxoacids, such as sulfuric acid and phosphoric acid. We usually write the formulas of these acids as H_2SO_4 and H_3PO_4, respectively. By writing them as $SO_2(OH)_2$ and $PO(OH)_3$, however, we show more clearly that they contain OH groups and oxygen atoms attached to the central atom. We use the name *oxoacid* for any acid with the general formula $XO_m(OH)_n$, where $m = 0, 1, 2, 3, \ldots$ and $n = 1, 2, 3, \ldots$. In this section we discuss the naming of these acids and their strengths as acids in water.

Naming Oxoacids and Their Anions

If an element has only one oxidation state other than 0, the name of its oxoacid is formed by adding "*-ic acid*" to the stem of the element's name. For example:

$Si(OH)_4$ silico~~n~~ → silic → silic*ic acid*

An element that has two nonzero oxidation states may have two corresponding oxoacids. The acid containing the element in its higher oxidation state is given the suffix *-ic*; and the acid containing the element in its lower oxidation state is given the suffix *-ous*. Thus, we have

		Oxidation state
H_3PO_4	Phosphor*ic acid*	V
H_3PO_3	Phosphor*ous acid*	III

The name of the anion derived from the *-ic* acid ends in *-ate*, whereas the name of the anion derived from the *-ous* acid ends in *-ite*:

				Oxidation state
H_2SO_4	Sulfur*ic* acid	SO_4^{2-}	Sulf*ate*	VI
H_2SO_3	Sulfur*ous* acid	SO_3^{2-}	Sulf*ite*	IV
H_3PO_4	Phosphor*ic* acid	PO_4^{3-}	Phosph*ate*	V
HNO_2	Nitr*ous* acid	NO_2^-	Nitr*ite*	III

A few elements form oxoacids in more than two oxidation states. In this case the prefixes *hypo-* and *per-* are used to designate a lower and a higher oxidation state, respectively. The oxoacids of chlorine provide a good example:

				Oxidation state
$HClO$	*Hypo*chlor*ous acid*	ClO^-	*Hypo*chlor*ite*	I
$HClO_2$	Chlor*ous acid*	ClO_2^-	Chlor*ite*	III
$HClO_3$	Chlor*ic*	ClO_3^-	Chlor*ate*	V
$HClO_4$	*Per*chlor*ic acid*	ClO_4^-	*Per*chlor*ate*	VII

A more logical system has been recommended by the International Union of Pure and Applied Chemistry (IUPAC, pronounced eye-you-pac) in which the names of all acids end in *-ic* and the oxidation state of the nonmetal is indicated by a Roman numeral. For example, HClO would be named chloric(I) acid and $HClO_4$ would be named chloric(VII) acid. This system, however, has not yet been widely adopted.

Many acids are diprotic or triprotic; they have more than one ionizable hydrogen and therefore form two or more anions. The three anions derived from phosphoric acid are dihydrogenphosphate, $H_2PO_4^-$, hydrogenphosphate, HPO_4^{2-}, and phosphate, PO_4^{3-}. When it reacts with sodium hydroxide, phosphoric acid gives the salts sodium dihydrogenphosphate, NaH_2PO_4; disodium hydrogenphosphate (or sodium hydrogenphosphate), Na_2HPO_4; and trisodium phosphate (or sodium phosphate), Na_3PO_4.

Strengths of Oxoacids

The strength of the oxoacids increases with increasing oxidation state of the central atom; sulfuric acid is stronger than sulfurous acid, and phosphoric acid is stronger than phosphorous acid. We can also relate the strength of an oxoacid to the value of m in the general formula $XO_m(OH)_n$. Acids with $m = 0$ or 1 are weak in water, whereas acids with $m = 2$ or 3 are strong in water. For example, sulfuric acid, $SO_2(OH)_2$ ($m = 2$), is a strong acid in aqueous solution, but hypochlorous acid, ClOH ($m = 0$), is a weak acid. Acid strength increases with the number of doubly bonded oxygen atoms on the central atom. Each electronegative oxygen atom draws electrons away from the central atom and thereby increases its electronegativity, which in turn increases the polarity of the O—H bonds.

Exercise 7.5 Write the formula of each of the following acids, and classify each as strong or weak in aqueous solution.
(a) phosphoric acid **(b)** chloric acid **(c)** hypochlorous acid
(d) silicic acid **(e)** boric acid, $B(OH)_3$

The oxoacids, $XO_m(OH)_n$, have one or more hydroxyl, OH, groups. But we know that some compounds containing OH groups, such as NaOH and $Ca(OH)_2$, are ionic hydroxides, Na^+OH^- and $Ca^{2+}(OH^-)_2$, and are therefore strong bases in aqueous solution. What, then, determines whether a molecule containing one or more OH groups is an acid or a base?

The hydroxides of metals are bases, whereas the hydroxides of nonmetals are acids (oxoacids). Acid strength increases and base strength decreases from left to right across any period.

Sodium hydroxide, NaOH, and magnesium hydroxide, $Mg(OH)_2$, are strong bases; $Al(OH)_3$, is **amphoteric** (can behave as both a weak base and a weak acid in aqueous solution); silicic acid, $Si(OH)_4$, and phosphoric acid, $PO(OH)_3$, are weak; and sulfuric acid, $SO_2(OH)_2$, and perchloric acid, $ClO_3(OH)$, are strong. The acid–base properties of the hydroxides of the Period 3 elements are summarized in Figure 7.12.

If the atom X in XOH has a low electronegativity, the X—O bond is ionic or polar. The oxygen has a negative charge, and the atom X a positive charge:

$$X^+\ OH^- \qquad \text{or} \qquad X^{\delta+}\text{—}O^{\delta-}\text{—}H$$

In this case XOH ionizes to X^+ and OH^-, as in NaOH and $Mg(OH)_2$. When X has a high electronegativity, the X—O bond is less polar, so XOH has a decreased tendency to ionize as X^+ and OH^-. Instead, both the oxygen atom and the X atom pull electrons away from the hydrogen, so the O—H bond becomes more polar:

$$X\text{—}O^{\delta-}\text{—}H^{\delta+}$$

There is an increased tendency for the hydrogen to be donated as a proton to a base, B, to give XO^- and BH^+. Thus, acid strength increases with increasing electronegativity of X.

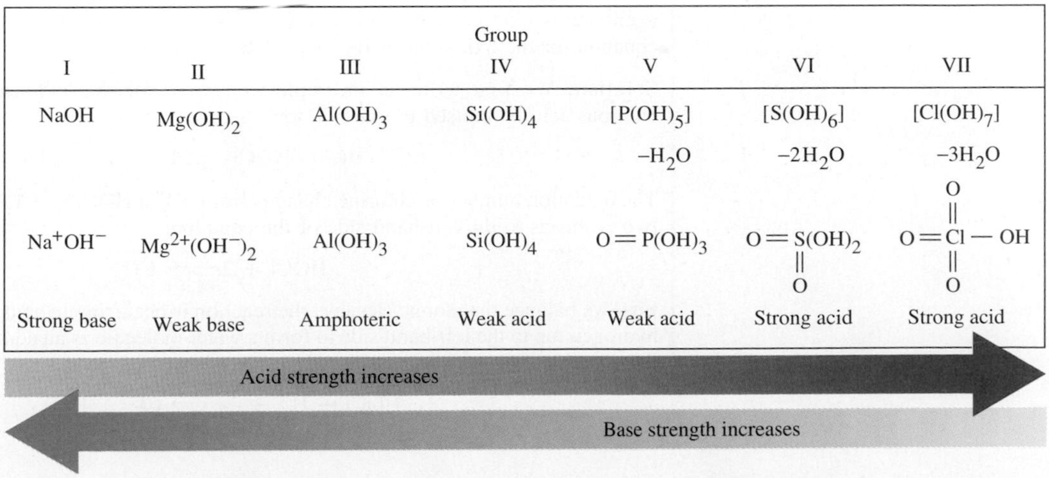

FIGURE 7.12 Acid and Base Strengths of the Hydroxyl Compounds of the Period 3 Elements Acid strength increases from left to right. Base strength increases from right to left.

Group I	II	III	IV	V	VI	VII
NaOH	$Mg(OH)_2$	$Al(OH)_3$	$Si(OH)_4$	$[P(OH)_5]$	$[S(OH)_6]$	$[Cl(OH)_7]$
				$-H_2O$	$-2H_2O$	$-3H_2O$
Na^+OH^-	$Mg^{2+}(OH^-)_2$	$Al(OH)_3$	$Si(OH)_4$	$O{=}P(OH)_3$	$O{=}S(OH)_2$ with $=O$	$O{=}Cl(=O)_2{-}OH$
Strong base	Weak base	Amphoteric	Weak acid	Weak acid	Strong acid	Strong acid

Acid strength increases →

← Base strength increases

TABLE 7.4 Oxidation States of Chlorine

Oxidation state of Cl	VII	V	III	I	0	−I
Examples	$HClO_4$	$HClO_3$	$HClO_2$	HClO	Cl_2	HCl
		ClF_5	ClF_3	ClF		
Lewis symbol	Cl (7 dots)	Cl (5 dots)	Cl (3 dots)	Cl (1 dot)	Cl (7 dots)	Cl (8 dots)

7.4 CHLORINE

We began our discussion of the chemistry of chlorine along with that of the other halogens in Chapter 4, where we restricted our discussion to compounds in which the element was in the −I oxidation state. Here we briefly discuss the higher oxidation states of chlorine. In its compounds with oxygen and fluorine, which have high electronegativities, chlorine may be in the +I, +III, +V, or +VII oxidation states (Table 7.4). As we saw in Section 7.3, there is an oxoacid corresponding to each of these oxidation states. We describe three of these acids next.

Chlorine Oxoacids

Hypochlorous Acid and Hypochlorites Hypochlorous acid, HOCl, is formed when chlorine, Cl_2, is dissolved in water:

$$Cl_2(aq) + 2H_2O(l) \rightleftharpoons HOCl(aq) + H_3O^+(aq) + Cl^-(aq)$$

The reaction is not complete, and the position of the equilibrium lies to the left. Hypochlorous acid is a weak acid and a strong oxidizing agent. The salts of hypochlorous acid are the **hypochlorites**. Calcium hypochlorite, $Ca(OCl)_2(s)$ is known as *bleaching powder*. Solutions of sodium hypochlorite, NaOCl(*aq*) are familiar as *household bleach*. Hypochlorites are used to bleach cotton and linen materials and paper pulp (Demonstration 7.9). They kill microorganisms by their oxidizing action and are therefore used to purify the water in swimming pools.

EXAMPLE 7.1

Write the equation for the half-reaction in which hypochlorous acid acts as an oxidizing agent and is reduced to chloride ion, Cl^-. Use this equation to write a complete balanced equation for the oxidation of $Br^-(aq)$ to $Br_2(aq)$.

Solution: We first write an incomplete equation for the half-reaction in which hypochlorous acid is reduced to chloride ion:

$$HOCl \rightarrow Cl^-$$

The oxidation number of chlorine changes from +1 in HOCl to −1 in Cl^-, so we must add two electrons to the left-hand side of the equation:

$$HOCl + 2e^- \rightarrow Cl^-$$

First we balance the atoms. Because the reaction is occurring in acid solution, we can add a hydrogen ion to the left-hand side to form a water molecule as an additional product. We get the equation

$$HOCl + H^+ + 2e^- \rightarrow Cl^- + H_2O$$

We see that this equation also balances for charges and so is the complete equation for the half-reaction in which HOCl is reduced to Cl^-. Adding this equation to the equation for the half-reaction in which bromide ion is oxidized to bromine,

$$2Br^- \rightarrow Br_2 + 2e^-$$

we obtain the overall balanced equation

$$HOCl + H^+ + 2Br^- \rightarrow Br_2 + Cl^- + H_2O$$

The electrons must cancel when two half-reactions are added to give the overall oxidation-reduction reaction.

Exercise 7.6 Write a complete balanced equation for the oxidation of $SO_3^{2-}(aq)$ to $SO_4^{2-}(aq)$ by hypochlorous acid.

Chloric Acid and Chlorates Chloric acid, $HClO_3$, is a strong acid in aqueous solution, and both chloric acid and the chlorates are powerful oxidizing agents. Chlorates form explosive mixtures with reducing agents such as carbon, sulfur, and organic compounds, which are converted in exothermic reactions to gaseous products such as CO_2 and SO_2. Because of its strength as an oxidizing agent, potassium chlorate is used extensively in matches (Box 7.1), fireworks (Box 10.2), and explosives. Both chloric acid and the chlorate ion have a triangular pyramidal AX_3E structure.

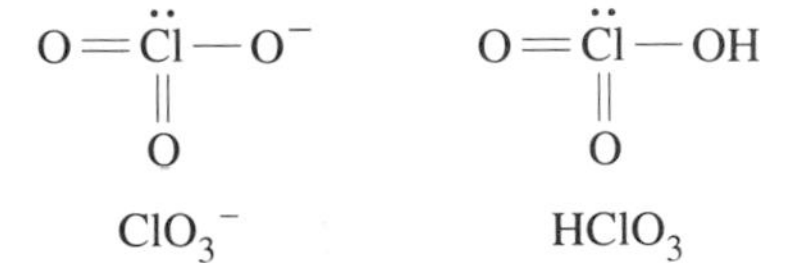

DEMONSTRATION 7.9 Properties of Sodium Hypochlorite

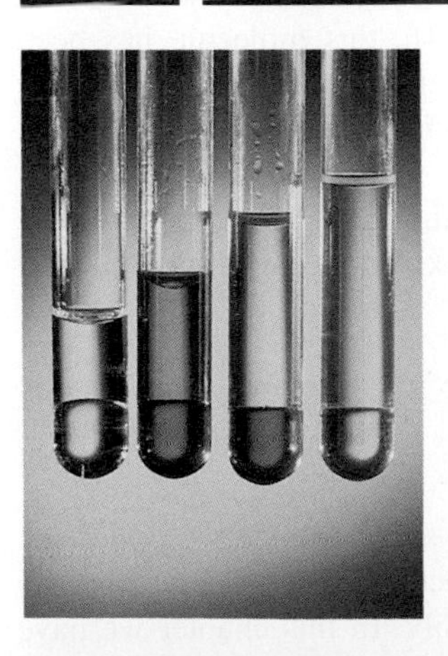

Left to right: (1) Tetrachloroethene, C_2Cl_4, is added to a colorless aqueous solution of potassium iodide. The C_2Cl_4 is immiscible with water and sinks to the bottom of the tube, forming a separate layer. (2) A few drops of a dilute aqueous solution of sodium hypochlorite, NaOCl, are added. This oxidizes iodide ion to iodine, which forms a purple solution in the C_2Cl_4 and a brown solution of I_3^- in the aqueous layer. (3) When the tube is shaken, all the iodine is extracted into the C_2Cl_4. (4) When an excess of sodium hypochlorite solution is added, iodine is further oxidized to iodate ion, IO_3^-. When the tube is shaken, both layers again become colorless.

Household bleach is a solution of sodium hypochlorite. Here its ability to bleach different samples of colored paper is shown. The bleaching action results from the oxidation of the colored dye to a colorless compound.

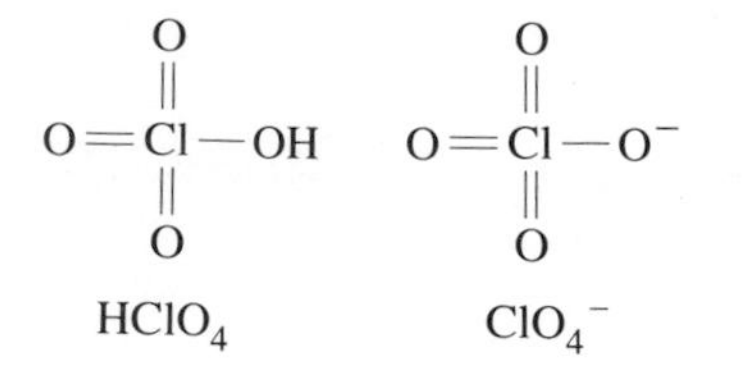

Perchloric Acid and Perchlorates Perchloric acid, $HClO_4$, is a strong acid in aqueous solution, and both the acid and the perchlorate ion are strong oxidizing agents. Both have a tetrahedral AX_4 geometry. Metal perchlorates are used in fireworks and flares and are mixed with carbon and various organic compounds in explosives and rocket fuels. The solid booster rockets of the space shuttle use a solid propellant consisting of 70% ammonium perchlorate, 16% aluminum powder, and 14% organic polymer.

Chlorine Oxides and Fluorides

Chlorine forms several oxides, of which chlorine dioxide, $ClO_2(g)$, is the most important. It is used in the pulp and paper industry as a bleaching agent and in the preparation of white flour. It is an unusual molecule because it has an odd number of valence electrons—namely, 19—and because the chlorine in this molecule is in the unusual +IV oxidation state. The Lewis structure of ClO_2 shows that there is one unpaired electron on the chlorine atom:

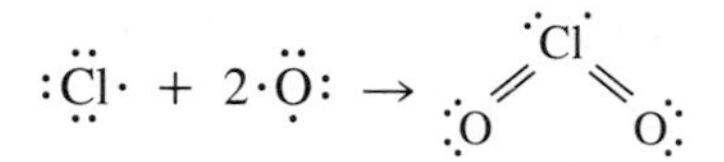

A molecule that has an odd number of electrons and therefore has one unpaired electron is called a *free radical*.

Most free radicals are very reactive and impossible to isolate as pure compounds. They have a strong tendency to react to form compounds in which they have an even number of electrons. ClO_2 is one of a small number of rather stable and relatively unreactive free radicals. As we shall see in Chapter 16, this molecule has been implicated in the destruction of ozone in the upper atmosphere. Another rather stable free radical molecule is nitrogen dioxide, NO_2, which also plays a significant role in the destruction of ozone.

Chlorine reacts readily with fluorine to form the fluorides $ClF(g)$, $ClF_3(g)$, and $ClF_5(g)$ in which chlorine is in the +I, +III, and +V oxidation states, respectively.

7.5 LEWIS STRUCTURES

In Chapter 3 we saw how to draw Lewis structures for simple molecules. Knowing the valences of the elements and taking account of the octet rule, when it is valid, we can often arrive at a correct Lewis structure quite easily. In this chapter we have given more Lewis structures, quite a few of which are for molecules for which the octet rule is not valid. We now describe a general method for drawing Lewis structures for all types of molecules. With practice you will not always need to work systematically through all the steps of this method, but they will be useful when you are uncertain how to proceed.

1. **Draw a diagram of the structure, showing the atoms connected by single bonds**. To draw any Lewis structure, we must first know how the atoms are connected in the molecule. This information can be obtained with complete certainty only by experiment. If we do not know the arrangement of the atoms,

we can often deduce the most probable arrangement as follows: (a) The structure must be consistent with the valences of the elements; (b) if there is a single unique atom, it is normally the central atom; and (c) the central atom is normally the atom with the highest valence or the least electronegative atom.

2. **Find the total number of valence electrons in the molecule**. If the molecule is charged—that is, if it is a polyatomic ion—add one electron for each negative charge and subtract one electron for each positive charge.
3. **Subtract the number of electrons needed to form the single bonds in the structure given by step 1 from the total number of electrons, and use the remainder to complete octets around each atom except hydrogen**. If there are insufficient electrons to complete all the octets, complete those of the more electronegative atoms first. If there are more than enough electrons to complete all the octets, add the remaining electrons to the central atom, which must be from Periods 3 through 7.
4. **If any atom from a group other than Group II or III still has an incomplete octet, convert nonbonding electron pairs to bonding pairs to form double or triple bonds until each atom has an octet**. Calculate formal charges for all atoms.
5. **If the central atom in an AX_nE_m molecule is in Periods 3 through 7 the octet rule may not apply. Form additional multiple bonds to remove as many formal charges as possible.**

To illustrate steps 1 through 4, let us deduce the Lewis structure of PCl_3 (we will soon return to Step 5).

1. The arrangement of the atoms is known to be

```
Cl—P—Cl
   |
   Cl
```

We could have deduced this structure, because (a) the valence of P is 3 and the valence of Cl is 1, (b) P is the single unique atom, and (c) P is the least electronegative atom.

2. The total number of valence electrons in the molecule is 26:

Phosphorus (Group V)	$5e^-$	1P	$5e^-$
Chlorine (Group VII)	$7e^-$	3Cl	$21e^-$
			$26e^-$

3. The three single bonds use up 6 electrons, so 26 − 6 = 20 electrons remain to be allocated. We add 6 electrons to each Cl and 2 electrons to the P atom to complete their octets. This uses up all 20 electrons:

```
 ..   ..   ..
:Cl — P — Cl:
 ..   |    ..
     :Cl:
      ..
```

4. All atoms have an octet.

Next we will deduce the Lewis structure for the polyatomic ion H_3O^+:

1.

```
H—O—H
   |
   H
```

2. The total number of electrons is 3 + 6 − 1 (for the positive charge) = 8.

3. Six of these electrons are used to form the three single bonds, so 2 electrons remain. They are used to complete the octet on oxygen:

```
    ..
H—O—H
    |
    H
```

4. Oxygen has an octet and each hydrogen has a duet, so the structure is complete.

The H_3O^+ ion has a positive charge. We can indicate this charge as follows:

```
⎡    ..    ⎤+
⎢H—O—H⎥
⎢    |     ⎥
⎣    H     ⎦
```

but, it is often convenient to know more precisely where this charge is located in the ion. The actual charge on the atoms can be found only from a detailed calculation of the electron density distribution in the ion. But we can find the approximate charge on each atom by assuming that the bonds are nonpolar, that is, that each bonding pair is shared equally between the two atoms that it bonds. The charge on an atom obtained in this way is called a **formal charge**. To find the formal charges of the atoms in the hydronium ion, we assume that the oxygen atom has an equal share in each of the three bonding pairs (3 electrons) and has a lone pair of its own (2 electrons). This gives oxygen a total of 5 electrons with a total charge of −5. But because oxygen is in Group VI and therefore has a core charge of +6, the oxygen atom has a formal charge of +1. Each hydrogen shares a bonding pair and so has 1 electron. The core charge of hydrogen is +1, so each hydrogen has a formal charge of 0:

```
    ..
H—O+—H
    |
    H
```

The procedure we have used for assigning formal charges can be summarized by the following rule:

$$\begin{matrix}\text{formal}\\ \text{charge}\end{matrix} = \begin{pmatrix}\text{core}\\ \text{charge}\end{pmatrix} - \begin{pmatrix}\text{number of}\\ \text{unshared electrons}\end{pmatrix} - \frac{1}{2}\begin{pmatrix}\text{number of}\\ \text{shared electrons}\end{pmatrix}$$

Exercise 7.7 Assign formal charges to the atoms in **(a)** the ammonium ion, NH_4^+; **(b)** the hydroxide ion, OH^-.

Elements of Periods 3 through 7 do not always obey the octet rule, so we need step 5 for molecules and ions of these elements. As an example that requires step 5, we consider the sulfate ion, SO_4^{2-}:

1. The S atom is in the middle of four O atoms:

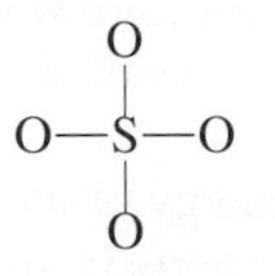

S is the less electronegative atom and the unique atom.

2. The total number of electrons is 32:

Sulfur (Group VI)	$6e^-$	1S	$6e^-$
Oxygen (Group VI)	$6e^-$	4O	$24e^-$
Negative charges			$2e^-$
			$32e^-$

3. The four bonds use 8 of these electrons, leaving 24 with which to complete the octets. We add 6 to each oxygen to complete its octet and get the structure

```
        ..
       :O:
        |
   ..        ..
  :O— S —O:
   ..        ..
        |
       :O:
        ..
```

4. All atoms have an octet. Next we assign the formal charges. Sulfur shares four bonding pairs and so has 4 electrons. It has a core charge of +6, so it has a formal charge of $+6 - 4 = +2$. Each oxygen shares a bonding pair (1 electron) and has three lone pairs (6 electrons) for a total of 7 electrons. Because the core charge of oxygen is +6, each oxygen has a formal charge of $+6 - 7 = -1$. So we write the complete structure, including the formal charges, as follows:

```
          ..
         :O:⁻
          |
 ⁻..              ..
  :O  — S²⁺ —O:⁻
   ..             ..
          |
         :O:⁻
          ..
```

The sum of the formal charges must equal the overall charge on the ion. In this case $+2 + 4(-1) = -2$.

5. This structure is a correct Lewis structure, but we can draw a better structure by using step 5, which recognizes that S is not restricted to 8 electrons in its valence shell. The formal charges on the S atom and on two of the O atoms can be removed by forming double bonds with lone pairs on two of the O atoms. This does not change the number of electrons in the valence shell of each O atom, but it increases the number in the valence shell of S to 12. We then obtain

```
         ..
        :O
         ||
   ..          ..
  :O=S—O:⁻
               ..
         |
        :O:⁻
         ..
```

This is the structure of SO_4^{2-} that we gave in Section 7.1.

In general, when we can draw two or more Lewis structures with different numbers of formal charges, the structure with the fewest formal charges represents more closely the actual electron distribution in the molecule. Removing formal charges transfers electrons from negatively charged atoms to positively charged atoms, thereby decreasing the energy of the structure. It is important to remember that a Lewis structure is only an approximate representation of the electron distribution in a molecule. When it is possible to draw two or more Lewis structures for a molecule, one of them may be a better representation of the electron distribution in the molecule, but none of them is exact (Box 7.3).

EXAMPLE 7.2 Drawing Lewis Structures

Draw the Lewis structure of SO_2.

Solution:

1. The arrangement of atoms is O—S—O.
2. There are 18 valence electrons and 2 bonds. Therefore, 14 electrons are available to complete the octets.
3. Because oxygen is more electronegative than sulfur, we complete the octets on oxygen first. This uses up six pairs of electrons. The remaining pair is placed on sulfur to give

$$:\ddot{\underset{..}{O}}-\ddot{S}-\ddot{\underset{..}{O}}:$$

4. Sulfur still has only six electrons in its valence shell, so we form a double bond to complete an octet on sulfur:

$$:\ddot{O}=\ddot{S}^{+}-\ddot{\underset{..}{O}}:^{-}$$

Assigning formal charges, we see that sulfur has a positive formal charge and oxygen a negative formal charge.

5. The structure we arrived at in step 4 is a correct Lewis structure. However, because sulfur is a Period 3 element, we can obtain a better structure by eliminating the formal charges. To do that, we form another double bond to give the structure

$$:\ddot{O}=\ddot{S}=\ddot{O}:$$

in which sulfur has a valence shell of 10 electrons.

Exercise 7.8 Draw Lewis structures for **(a)** the sulfite ion, SO_3^{2-}; **(b)** the phosphate ion, PO_4^{3-}; **(c)** the phosphorus pentachloride molecule. Assign formal charges where appropriate.

Although the shape of a molecule can be predicted from the Lewis structure by using the VSEPR model, it is not necessary to show this shape when drawing a Lewis structure, but in simple cases it is often convenient to do so. The Lewis structure of SO_2, which is an angular AX_2E molecule, may be correctly drawn as

$$:\ddot{O}=\ddot{S}=\ddot{O}:$$

but it is more common to draw the shape as

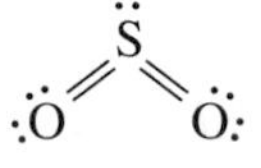

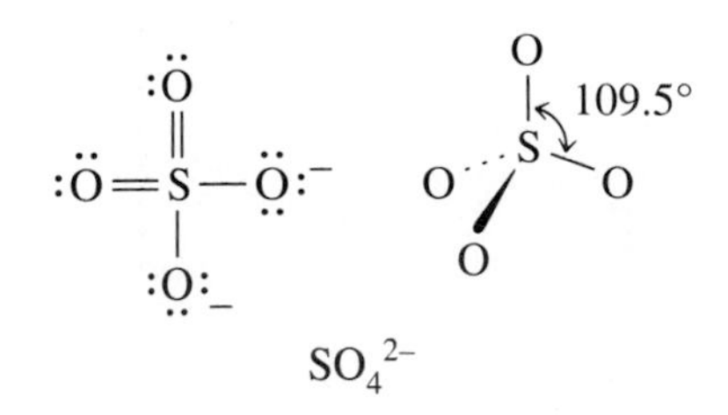

The Lewis structure of a more complicated molecule is often most easily drawn without attempting to show its shape. Thus, without attempting to draw its tetrahedral shape, we draw the Lewis structure of SO_4^{2-} as shown at far left. We can show this shape more conveniently by means of a separate diagram that emphasizes the shape but does not attempt to show all the electrons (near left).

BOX 7.3 More on Lewis Structures

Some of the Lewis structures we have described in this chapter may differ from those that you have learned in a chemistry course in high school or elsewhere. For example, we have given the structures

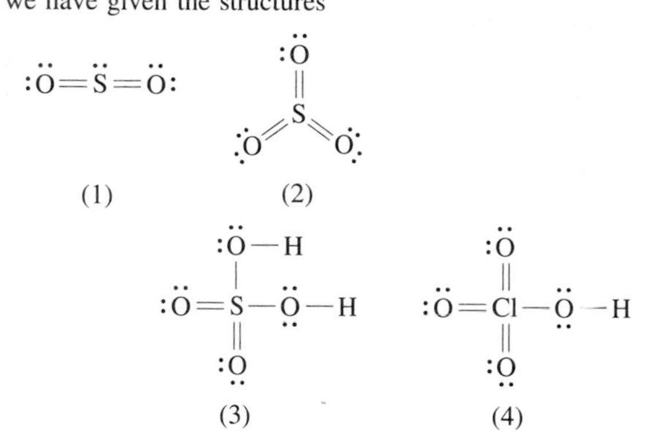

whereas you may have previously learned the structures

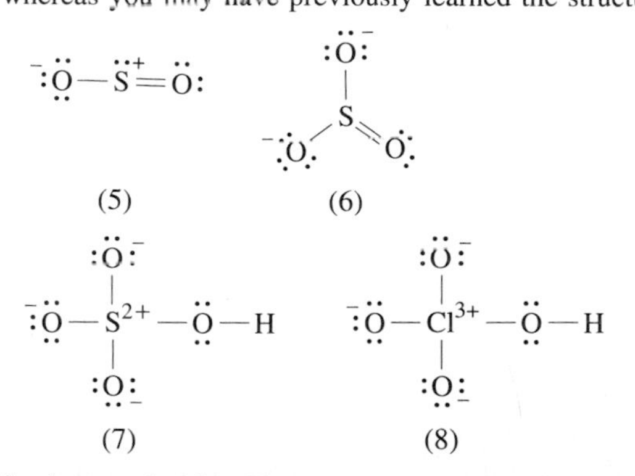

for these molecules. You may be wondering if the structures you had learned are wrong.

They are not wrong, but they are not necessarily the *best* structures for these molecules. Gilbert Lewis first proposed structures (5) through (8) on the basis of his octet rule. But we have since recognized that the octet rule does not always hold for the elements of Periods 3 through 7, which can have more than eight electrons in their valence shell. In combination with electronegative elements such as fluorine and oxygen, these elements may form more than four bonds, as in the molecules PCl_5 and SF_6, which have the Lewis structures.

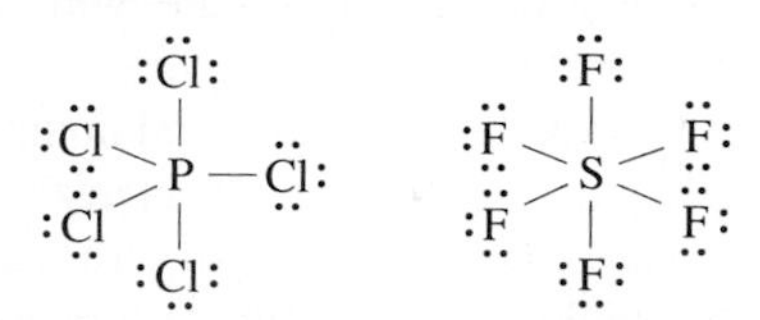

Similarly, we may draw Lewis structures (3) and (4) for H_2SO_4 and $HClO_4$ in which the central atom forms more than four bonds.

How do we decide which structure is the best approximation to the actual electron distribution? We look at the experimental evidence, such as the bond lengths. Measurements of bond lengths show that double bonds are shorter than single bonds and that triple bonds are even shorter. For example, the C—C bond lengths in ethane, ethene, and ethyne are

$H_3C—CH_3$	$H_2C=CH_2$	$HC≡CH$
154 pm	134 pm	120 pm

The bond lengths and angles in SO_2, SO_3, H_2SO_4, and $HClO_4$ have been determined experimentally to be

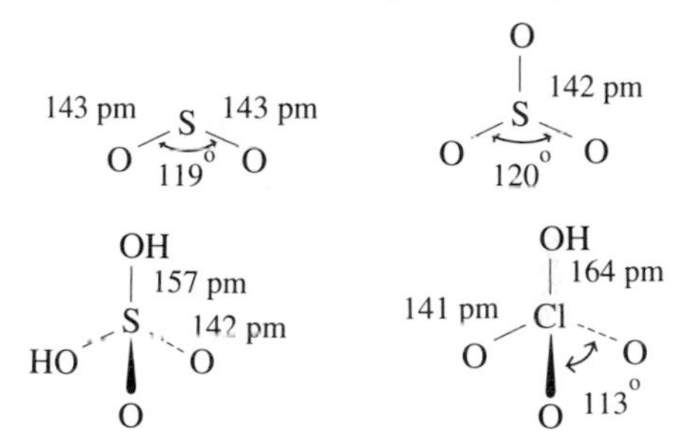

These bond lengths are consistent with Lewis structures containing S=O and Cl=O double bonds with lengths of 142 pm and 141 pm and S—O and Cl—O single bonds with lengths of 157 pm and 164 pm, as we have drawn them in (1) through (4) rather than with single bonds as in (5) through (8).

It is important to remember that a Lewis structure is only a very approximate representation of the distribution of the electrons in a molecule. As we saw in Chapter 6, the electron distribution in a molecule is most accurately represented by a continuous charge cloud that is more dense in some places than others. This electron density distribution can be obtained only by difficult experimental measurements (see Box 11.1) or by lengthy and complex quantum mechanical calculations, so for most purposes we use the approximate electron density distribution given by a Lewis structure. But there has not always been complete agreement as to which Lewis structure is the best approximation. In this book we have chosen to use those structures that

1. Recognize that an atom of an element from any period except 1 and 2 may have more than eight electrons in its valence shell,
2. Are consistent with experimental bond lengths, and
3. Minimize the number of formal charges in the structure.

To summarize, with very few exceptions, in a correct Lewis structure each Period 2 atom (Groups IV–VII) must have an octet of electrons, and each Period 3 atom (Groups IV–VII) must have at least eight electrons in its valence shell. When two or more correct Lewis structures can be drawn, it is to some extent a matter of opinion which one is the best approximation.

SUMMARY

Sulfur is obtained from natural deposits in elemental form or from the hydrogen sulfide, $H_2S(g)$, in natural gas. Sulfur crystallizes from $CS_2(l)$ as orthorhombic sulfur but from the molten state as monoclinic sulfur. These allotropes, or different forms, of sulfur contain S_8 molecules. Another allotrope, plastic sulfur, a rubbery solid formed when molten sulfur is rapidly cooled, is an amorphous mixture of S_n polymers.

Sulfur dioxide, $SO_2(g)$, is formed when sulfur is burned in air, and sulfur trioxide, $SO_3(g)$, results from the catalyzed reaction of SO_2 with O_2. SO_2 and SO_3 are acidic oxides. SO_3 reacts with water to give sulfuric acid, $H_2SO_4(l)$. Sulfuric acid is a diprotic acid, an acid with two ionizable hydrogen atoms. It behaves as a strong acid in water and gives two series of salts, hydrogensulfates and sulfates. Precipitation of insoluble $BaSO_4(s)$ in aqueous solution is a test for the SO_4^{2-} ion. Hydrated salts, such as $MgSO_4 \cdot 7H_2O$, are salts that crystallize from water; their network structures contain water, called water of crystallization. Concentrated (98%) H_2SO_4 is a dehydrating agent and a strong oxidizing agent, stronger than $H_3O^+(aq)$. $H_2SO_4(l)$ is a stronger acid than $H_3O^+(aq)$.

The oxidation number of an atom in a compound is a measure of its extent of oxidation relative to the atom in the element, for which the oxidation number is 0. For a monatomic ion, the oxidation number is equal to the charge (Na^+,+1; Cu^{2+},+2; F^-,−1; O^{2-},−2). In covalent compounds, the oxidation numbers are F, −1; H, +1; O, −2 (but in compounds such as F_2O and H_2O_2, they are +2 and −1, respectively); the other halogens have an oxidation number of −1, except in compounds with F and O. The sum of the oxidation numbers of all the atoms is 0 for a neutral compound and equal to the charge for an ion. In redox reactions, the oxidation number of an atom increases when the atom is oxidized and decreases when it is reduced. In an oxidation the change in oxidation number equals the number of electrons lost. In a reduction the change in oxidation number equals the number of electrons gained.

In compounds in which it has an oxidation number of +6, such as SO_3, sulfur is said to be in the +VI oxidation state. The most important oxidation states of sulfur are +VI, +IV, 0 and −II. Sulfur is in the +IV oxidation state in SO_2 and in the −II oxidation state in H_2S.

In solution in water, SO_2 behaves as the weak diprotic sulfurous acid, $(HO)_2SO$. SO_2, HSO_3^-, and SO_3^{2-} are good reducing agents and are oxidized to $SO_4^{2-}(aq)$. Hydrogen sulfide, $H_2S(g)$, is a weak diprotic acid in water and a good reducing agent. Group I and II metal sulfides are colorless and soluble in water; whereas most other metal sulfides are colored and insoluble and are precipitated when $H_2S(g)$ is passed through a solution of a soluble salt.

Phosphorus is too reactive to be found in the elemental form and occurs mainly as phosphate rock, $Ca_3(PO_4)_2$, and apatites, $Ca_5(PO_4)_3X$ (where X is F, Cl, or OH). Phosphorus has several allotropes. White phosphorus contains tetrahedral P_4 molecules, whereas red phosphorus, P_n, is a three-dimensional network polymer. The most important oxidation states of phosphorus are +V (H_3PO_4, PF_5, and PCl_5), +III (H_3PO_3 and PF_3), 0 (P_4 and P_n), and −III (PH_3). Phosphorus burns in air to give the acidic oxides $P_4O_6(s)$ and $P_4O_{10}(s)$, which react with water to give phosphorous acid, H_3PO_3, and phosphoric acid, H_3PO_4, respectively. Both acids are weak. Phosphoric acid is a triprotic acid, an acid with three ionizable hydrogen atoms. With a base such as $KOH(aq)$, H_3PO_4 forms potassium dihydrogenphosphate, KH_2PO_4, potassium hydrogenphosphate, K_2HPO_4, and potassium phosphate, K_3PO_4. H_3PO_4 readily undergoes condensation reactions, reactions in which water is eliminated, to give polyphosphoric acids, such as diphosphoric acid, $H_4P_2O_7$, and triphosphoric acid, $H_5P_3O_{10}$. Compounds of phosphorus include phosphine, $PH_3(g)$, the trihalides $PF_3(g)$ and $PCl_3(l)$, and the pentahalides $PF_5(g)$ and $PCl_5(s)$.

The most important oxidation states of chlorine are +VII, +V, +III, +I, 0, and −I. Chlorine forms the following oxoacids: +I, hypochlorous acid, HClO; +III, chlorous acid, $HClO_2$; +V, chloric acid, $HClO_3$; and +VII, perchloric acid, $HClO_4$. Their salts contain, respectively, hypochlorite; ClO^-; chlorite, ClO_2^-; chlorate, ClO_3^-; and perchlorate, ClO_4^-, anions. All these acids and anions are strong oxidizing agents and are reduced to Cl^- ion. Chlorine reacts with fluorine to give the fluorides, $ClF(g)$, $ClF_3(g)$, and $ClF_5(g)$. The oxide ClO_2 is a free radical, a molecule with an unpaired electron.

Metal hydroxides are bases, and nonmetal hydroxides are oxoacids. Acid strength increases, and base strength decreases, from left to right across any period of the periodic table. Acids with the general formula $XO_m(OH)_n$, such as sulfuric acid, $SO_2(OH)_2$, and phosphoric acid, $PO(OH)_3$, are called oxoacids. The acid strength of oxoacids increases with increasing value of m (in aqueous solution, $m = 0$ or 1: weak acid; $m = 2$ or 3: strong acid).

The formal charge on an atom in a molecule or ion

is obtained by assuming that it shares equally the electrons of all the bonds in which it takes part. It is given by the expression

$$\begin{pmatrix}\text{formal}\\\text{charge}\end{pmatrix} = \begin{pmatrix}\text{core}\\\text{charge}\end{pmatrix} - \begin{pmatrix}\text{number of}\\\text{unshared}\\\text{electrons}\end{pmatrix} - \frac{1}{2}\begin{pmatrix}\text{number of}\\\text{shared}\\\text{electrons}\end{pmatrix}$$

IMPORTANT TERMS

acidic oxide (page 227)
allotrope (page 224)
amphoteric (page 249)
condensation reaction (page 245)
dehydrating agent (page 230)
diprotic acid (page 228)
formal charge (page 254)
free radical (page 252)
hydrated salt (page 229)
monoprotic acid (page 228)
oxidation number (page 233)
oxidation state (page 235)
oxoacid (page 227)
oxoanion (page 228)
triprotic acid (page 244)
water of crystallization (page 229)

REVIEW QUESTIONS

1. Name the nonmetals in Periods 1, 2, and 3.

2. What is the maximum number of electrons that can occupy the $n = 1$, $n = 2$, and $n = 3$ shells?

3. Give **(a)** the Lewis symbols for P, S, and Cl in each of their possible valence states and **(b)** the corresponding valences.

4. **(a)** State what is meant by the term ''allotrope.'' **(b)** Give examples of two allotropes of sulfur and two of phosphorus.

5. Why are oxides such as SO_2 and SO_3 described as acidic oxides?

6. What molecules and ions are present in each of the following?
(a) an aqueous solution of sulfuric acid, $H_2SO_4(aq)$
(b) 98% sulfuric acid, $H_2SO_4(conc)$
(c) 100% sulfuric acid, $H_2SO_4(l)$

7. What are the three steps in the preparation of $H_2SO_4(conc)$ by the contact process?

8. Give an example of a reaction in which sulfuric acid behaves as a dehydrating agent.

9. What is the common sulfur-containing product in reactions in which $H_2SO_4(conc)$ behaves as an oxidizing agent?

10. What are the oxidation numbers of each of the following in the majority of their compounds?
(a) H **(b)** O **(c)** F **(d)** Cl

11. In terms of **(a)** electron transfer and **(b)** oxidation numbers, how are ''oxidation'' and ''reduction'' defined?

12. Explain why Br_2 is reduced to Br^- ion by SO_2 in aqueous solution, whereas Br^- ion is oxidized to Br_2 in concentrated sulfuric acid.

13. Give an example of a precipitation reaction that involves S^{2-} ions.

14. **(a)** What two oxides can be obtained when phosphorus is burned in air?
(b) What are the oxidation states of phosphorus in these oxides?

15. **(a)** Why is phosphoric acid described as a weak triprotic acid?
(b) Name the possible salts that can be obtained from the reaction of phosphoric acid with $NaOH(aq)$.

16. Why is the ClO_2 molecule described as a free radical?

17. Define the term ''oxoacid.''

18. Classify each of NaOH, $Mg(OH)_2$, $Si(OH)_4$, $PO(OH)_3$, and $ClO_3(OH)$ as a strong acid, weak acid, weak base, or strong base in aqueous solution.

19. How does the strength of an oxoacid of formula $XO_m(OH)_n$ depend on the value of m?

PROBLEMS

Sulfur

1. **(a)** Name and briefly describe the forms and molecular structures of three allotropes of sulfur.
(b) Describe and explain what happens when sulfur is gradually heated to its boiling point.

2. Give an example and write a balanced equation for each of the following types of reactions of sulfuric acid.
(a) acid–base **(b)** oxidation–reduction **(c)** dehydration

3. Compare the boiling points of $H_2S(l)$ and $H_2O(l)$.

(a) Judging from the usual trends in the periodic table, would you have expected these boiling points?

(b) What can you say about the relative strengths of the intermolecular forces in H_2O and H_2S?

4. Write a balanced equation for the reaction of **(a)** zinc with dilute sulfuric acid; **(b)** concentrated sulfuric acid with sodium iodide; **(c)** concentrated sulfuric acid with copper; **(d)** dilute sulfuric acid with solid magnesium hydroxide.

5. A compound containing only iron and sulfur was analyzed by heating in air to convert all the sulfur to sulfur dioxide. At 25°C and a pressure of 750 mm Hg, the volume of SO_2 obtained from 0.4203 g of the compound was 173.6 mL.

(a) What is the empirical formula of the compound?

(b) Is it likely to be an ionic compound or a covalent compound?

6. An aqueous solution is known to contain either $SO_4^{2-}(aq)$ or $SO_3^{2-}(aq)$, but not both. Suggest two tests that could be used to identify the anion in the solution.

7. One method used for the removal of SO_2 from the flue gases of power plants is to react it with an aqueous solution of H_2S. One product of this reaction is sulfur.
(a) Write a balanced equation for the reaction.

(b) What volume of $H_2S(g)$ at 25°C and 750 mm Hg pressure is needed to remove all the SO_2 formed by burning 1 metric ton (10^3 kg) of coal containing 4.00% sulfur by mass?

(c) What mass of elemental sulfur is formed?

8. Water from springs and wells is typically contaminated with small concentrations of hydrogen sulfide, which gives it a bad smell. The H_2S can be removed by treating the water with chlorine, which oxidizes the H_2S to sulfur.
(a) Write a balanced equation for this reaction.

(b) If the H_2S content of the water from a particular source is 2 parts per million (ppm) by mass, how much chlorine will be needed to remove the H_2S from 5000 L of water?

9. **(a)** How could $H_2S(g)$ be prepared from $FeS(s)$?

(b) Hydrogen sulfide burns in air with a blue flame to give sulfur dioxide, but in a limited supply of air, sulfur is formed. Write balanced equations for both reactions.

(c) Write balanced equations to show H_2S acting as a diprotic acid in water.

(d) Give two examples of reactions in which H_2S behaves as a reducing agent.

10. Write the formula for each of the following compounds.

(a) potassium sulfate **(b)** calcium hydrogensulfate

(c) calcium sulfite **(d)** potassium diphosphate

(e) aluminum sulfate

11. Give one example each of a reaction in which sulfuric acid **(a)** behaves as an acid; **(b)** behaves as an oxidizing agent; **(c)** behaves as a dehydrating agent; **(d)** takes part in a precipitation reaction.

12. Name each of the following, and classify each as ionic or covalent. Draw Lewis structures for (c) through (g).

(a) $Na_2S(s)$ (b) $MgS(s)$ (c) $S_8(s)$ (d) $CS_2(l)$

(e) $SO_2(g)$ (f) $SO_3(g)$ (g) $H_2SO_4(l)$

13. Describe an industrial process for converting sulfur to sulfuric acid.

14. Complete and balance each of the following equations.

(a) $H_2SO_4(conc) + Cu(s) \rightarrow Cu^{2+}(aq) +$

(b) $H_2SO_4(aq) + CuS(s) \rightarrow$

(c) $Na_2SO_3(aq) + H_2SO_4(aq) \rightarrow$

(d) $H_2S(aq) + Cl_2(aq) \rightarrow$

(e) $H_2S(aq) + Pb^{2+}(aq) \rightarrow$

(f) $H_2SO_4(conc) + NaI(s) \rightarrow$

15. A convenient laboratory method for preparing sulfur trioxide is to heat tetraphosphorus decaoxide, P_4O_{10}, with concentrated sulfuric acid.

(a) Write a balanced equation for this reaction.

(b) What type of reaction is it?

16.* A compound contains 23.7% S, 52.6% Cl, and 23.7% O by mass and has a boiling point of 69°C. The volume occupied by 0.337 g of the gaseous compound at 100°C and 770 mm Hg pressure is 75.6 mL.

(a) Find the empirical and molecular formulas.

(b) Draw a Lewis structure for the molecule.

17.* A crystalline solid A has the mass percentage composition 14.28% Na, 6.21% H, 9.94% S, and 69.56 O. Upon prolonged heating, it loses 55.9% in mass and leaves a solid residue B with the composition 32.39% Na, 22.53% S, and 45.07% O. What are the empirical formulas of A and B and their likely Lewis structures?

Phosphorus

18. Explain the following properties of the allotropes of phosphorus in terms of their structures.

(a) White phosphorus has a lower melting point than red phosphorus and is more volatile.

(b) White phosphorus is soluble in carbon disulfide, $CS_2(l)$, whereas red phosphorus is insoluble.

19. Write balanced equations for the formation of $P_4O_{10}(s)$ by the combustion in air of **(a)** white phosphorus and **(b)** phosphine $PH_3(g)$, and draw the Lewis structures of all the phosphorus-containing reactants and products.

20. Striking a match involves the combustion of P_4S_3 to produce a white smoke of $P_4O_{10}(s)$ and $SO_2(g)$. Write a balanced equation for this reaction. Calculate the maximum volume of $SO_2(g)$ at 20°C and 772 mm Hg pressure that results from the combustion of 0.157 g of P_4S_3.

21. A 3.064-g sample of phosphorus was burned completely in air to give 7.020 g of an oxide, which reacted with 2.671 g of water to give 9.691 g of an oxoacid of phosphorus. Determine the empirical formulas of the oxide and the oxoacid.

22. **(a)** What reaction is likely between $PH_3(g)$ and $HCl(g)$?

(b) Write an equation for the reaction.

(c) Draw a Lewis structure for the compound formed, and suggest a name for it.

(d) Would you predict it to be a solid, a liquid, or a gas?

23. A sodium salt of a phosphorus oxoacid contains 25.3% P,

43.5% O, and 31.2% Na by mass. What are its empirical formula and the probable Lewis structure of its anion?

24. Write a formula for each of the following compounds.

(a) tetraphosphorus hexaoxide

(b) calcium dihydrogenphosphate

(c) calcium phosphide

(d) phosphorous acid

(e) sodium triphosphate

25. Calcium hydrogenphosphate, $CaHPO_4(s)$, is produced commercially from calcium carbonate and phosphoric acid. Write a balanced equation for the reaction.

(a) How many grams of calcium carbonate are required to react with 48.0 mL of an aqueous solution that is 85.5% phosphoric acid by mass (density $1.70\ g \cdot mL^{-1}$)?

(b) What is the maximum mass of $CaHPO_4(s)$ that can be obtained from this reaction?

26. Starting with white phosphorus, write balanced equations for the preparation of (a) tetraphosphorus hexaoxide; (b) phosphorus pentachloride; (c) phosphoric acid; (d) phosphine.

27. Calculate the volume in milliliters needed for the complete neutralization of 25.00 mL of 0.200-*M* $H_3PO_4(aq)$ by 0.150-*M* $NaOH(aq)$.

28.* Dry hydrogen fluoride, $HF(g)$, reacts with $P_4O_{10}(s)$ to give a gas containing 29.8% P, 54.8% F, and 15.4% O by mass. This gas has a density of $4.64\ g \cdot L^{-1}$ at STP.

(a) What is its molecular formula?

(b) Suggest a possible Lewis structure.

Oxoacids

29. What is the name of each of the oxoacids with the following molecular formulas?

(a) H_2SO_4 (b) H_3PO_4 (c) H_3PO_3 (d) H_2CO_3
(e) H_2SO_3 (f) H_4SiO_4 (g) $HOCl$ (h) $HClO_4$

30. Explain why KOH and $Ca(OH)_2$ are strong bases in aqueous solution whereas $Si(OH)_4$ is a very weak acid.

31. Write the formulas of the following acids, and classify each as a strong acid or a weak acid.

(a) phosphoric acid (b) phosphorous acid
(c) sulfuric acid (d) sulfurous acid
(e) perchloric acid (f) chloric acid
(g) chlorous acid (h) hypochlorous acid

32. Write the formulas of the following ions, and classify each as a strong acid or a weak acid.

(a) hydrogensulfate ion (b) hydrogenphosphate ion
(c) dihydrogenphosphate ion (d) hydrogensulfite ion

33. Explain why the acid strengths of hypochlorous acid, hypobromous acid, and hypoiodous acid decrease in the order HOCl > HOBr > HOI.

34. Write balanced equations for all possible reactions of each of the following oxoacids with sodium hydroxide in aqueous solution, and name each of the possible salts that could be obtained from the solutions.

(a) nitric acid (b) sulfuric acid (c) phosphoric acid
(d) perchloric acid

35. Name each of the compounds with the following formulas.

(a) K_3PO_4 (b) H_3PO_3 (c) $HClO_2$ (d) $Ca(OCl)_2$
(e) $CaSO_3$ (f) $Na_5P_3O_{10}$

36. What molecules or ions are present in aqueous solutions of each of the following? Give both names and formulas.

(a) sulfuric acid (b) sulfur dioxide (c) phosphoric acid
(d) carbonic acid

37.* Name the compounds with the following formulas.

(a) $HOBr$ (b) $Mg(OCl)_2$ (c) $KBrO_3$ (d) $KClO_2$
(e) $Mg(ClO_4)_2$ (f) HIO_3 (g) $HBrO_4$ (h KIO_3

Chlorine

38. Write equations for the half-reactions that describe the behavior of each of ClO_3^-, ClO_2^-, and ClO^- when they are reduced to chloride ion in acidic aqueous solution.

39. Iodine reacts with liquid chlorine at −40°C to give an orange compound containing 54.5% iodine and 45.5% chlorine by mass.

(a) What is the empirical formula of the compound?

(b) If the molar mass is $467\ g \cdot mol^{-1}$, what is the molecular formula?

40.* Bleaching powder is obtained by reacting slaked lime, $Ca(OH)_2(s)$, with chlorine. It has the composition 36.6% Ca, 43.2% Cl, 19.5% O, and 0.6% H by mass. When excess $AgNO_3(aq)$ was added to 1.000 g of bleaching powder dissolved in water, 0.874 g of $AgCl(s)$ precipitated. When an acidified aqueous solution containing 1.000 g of bleaching powder was titrated with 0.100-*M* $KI(aq)$, 121.9 mL of the $KI(aq)$ was needed to reduce all of the hypochlorite ion, $OCl^-(aq)$, to chloride ion, $Cl^-(aq)$.

(a) What is the empirical formula of bleaching powder?

(b) Write a balanced equation for the preparation of bleaching powder.

41.* Hypofluorous acid, HOF, was first prepared in 1971 by the reaction between hydrogen fluoride gas and ice at −40°C. It is a white solid that melts at −11°C to a pale yellow liquid. It is very reactive, decomposing below room temperature into $HF(g)$ and oxygen. It reacts with dilute aqueous acid to give hydrogen peroxide, with dilute aqueous base to give oxygen, and with $HF(g)$ to give fluorine. Write balanced equations for the preparation of HOF and each of the reactions just described.

42.* (a) Using the VSEPR model, predict the geometries of the molecules ClF_3 and ClF_5. How many different possible geometries can you find for ClF_3? How many for ClF_5?

(b) Given that lone pairs always occupy the three equivalent equatorial sites of a triangular bipyramid, draw the predicted structure of ClF_3. Can you explain why this is often called a T-shaped structure?

43.* When 0.3574 g of potassium iodate was heated to constant mass, 0.2772 g of a white solid residue remained. During the reaction, 61.5 mL of dry oxygen at 25°C and a pressure of 755 mm Hg was evolved. What is the balanced equation for this reaction?

Oxidation Numbers and Oxidation States

44. What are the common oxidation states of **(a)** sulfur and **(b)** phosphorus in their compounds? Give an example of a compound for each oxidation state.

45. What is the oxidation number of oxygen in each of the following?

(a) O_2 **(b)** OF_2 **(c)** H_2O_2 **(d)** O_2^{2-} **(e)** O^{2-}

46. What are the oxidation numbers of each of the following?

(a) C in CO **(b)** S in SO_2 **(c)** P in $Ca_5(PO_4)_3(OH)$
(d) S in SF_6 **(e)** S in S_2^{2-} **(f)** P in $H_2PO_3^-$

47. What are the oxidation numbers of each element in each of the following?

(a) Li_2O **(b)** PH_4I **(c)** $NaClO_4$ **(d)** NaOCl
(e) AlP **(f)** S_8 **(g)** $BaSO_3$

48. Assign oxidation numbers to each element in each of the following.

(a) N_2 **(b)** H_2CO_3 **(c)** NH_3 **(d)** PH_3 **(e)** MnO_2
(f) HNO_3

49. What is the oxidation number of the halogen in each of the following molecules and ions?

(a) ClO_3^- **(b)** BrF_3 **(c)** $HClO_4$ **(d)** ClO_2^- **(e)** ClO_2

50. **(a)** If an atom in a molecule is oxidized, does its oxidation number increase or decrease?

(b) If an atom in a molecule is reduced, does its oxidation number increase or decrease?

(c) Write an equation for each of the following half-reactions in aqueous acid.
(i) the oxidation of Cu to Cu^{2+}
(ii) the reduction of S to S^{2-}
(iii) the reduction of NO_3^- to NO

51. Write an equation for each of the following half-reactions in acidic aqueous solution.

(a) the oxidation of H_2S to S
(b) the oxidation of SO_2 to SO_4^{2-}
(c) the oxidation of H_3PO_3 to H_3PO_4

52. **(a)** When an ionic bromide is heated with concentrated sulfuric acid, bromide ion is oxidized to bromine, and the sulfuric acid is reduced to sulfur dioxide. Write the balanced equation for this reaction.

(b) When sulfur dioxide is bubbled through an aqueous solution of bromine, the bromine is reduced to bromide ion, and the sulfur dioxide is oxidized to sulfate ion. Write the balanced equation for this reaction.

Lewis Structures

53. Draw Lewis structures for the following molecules and ions, and then use the VSEPR model to predict the geometry around the S atom.

(a) SO_2 **(b)** SO_3 **(c)** H_2SO_4 **(d)** SO_3^{2-}

54. Draw Lewis structures for each of the following molecules or ions, and use the VSEPR model to predict the molecular geometry.

(a) PH_3 **(b)** H_2S **(c)** PO_4^{3-} **(d)** SO_4^{2-} **(e)** ClO_4^-

55. **(a)** On the basis of the VSEPR model, what are the expected shapes of AX_5 and AX_6 molecules?

(b) Give examples of an AX_5 and an AX_6 molecule.

56. Phosphorus pentachloride is an unusual compound. $PCl_5(g)$ contains covalent PCl_5 molecules, but $PCl_5(s)$ consists of PCl_4^+ and PCl_6^- ions. Draw Lewis structures for PCl_5, PCl_4^+, and PCl_6^-, and predict the shape of each.

57.* Draw Lewis structures for each of the following.

(a) phosphoric acid **(b)** diphosphoric acid
(c) fluorosulfate ion, FSO_3^-
(d) fluorophosphate ion, FPO_3^{2-}

General Problems

58. A sample of coal contains 5.00% sulfur by mass. When burned in a power plant, it gives sulfur dioxide in the stack gas, which is converted in the atmosphere to sulfuric acid and deposited as acid rain.

(a) Write the simplest equations to show how sulfur dioxide is converted to sulfuric acid in the atmosphere.

(b) Calculate the mass of sulfuric acid that could result from burning 1000 kg of this coal.

(c) How much $Ca(OH)_2(s)$ would be required to neutralize this sulfuric acid?

59.* Phosphorous acid, H_3PO_3, forms only two series of salts, and is therefore a diprotic acid. **(a)** Write two balanced equations for the reaction of $H_3PO_3(aq)$ with $NaOH(aq)$.

(b) Suggest a Lewis structure that is consistent with the fact that H_3PO_3 is a diprotic acid.

60.* A yellow solid A was heated in air and gave a colorless gas B, which, when dissolved in water, gave a colorless solution C. Another sample of the yellow solid was mixed with iron powder and heated strongly to produce a black solid D. When dilute sulfuric acid was added to D, a colorless gas E formed. When E was bubbled through C, a yellow precipitate formed. It is identical to A in its chemical and physical properties.

(a) Identify A, B, C, D, and E.

(b) Write balanced equations for all the reactions above.

61.* Excess sulfur was heated with 4.00 g of red phosphorus to give a compound X, which, after excess sulfur was removed, had a mass of 14.35 g. Analysis showed that X contained only phosphorus and sulfur.

(a) Deduce the empirical formula of X.

(b) Suggest a molecular structure.

62.* An element X is a reactive white solid that is insoluble in water but soluble in organic solvents, such as carbon disulfide, from which it can be recovered by evaporating the solvent. It melts at 44°C and boils at 280°C. As a vapor, X has a density of $2.242\ g \cdot L^{-1}$ at 400°C and 1 atm pressure. When 25.0 mL of $X(g)$ reacts completely with 150 mL of chlorine at a given temperature and pressure, it gives 100 mL of a gaseous chloride at the same temperature and pressure. This chloride contains 77.4% Cl by mass, boils below 100°C, is liquid at room temperature, and has a molar mass of $137.3\ g \cdot mol^{-1}$.

(a) Identify X; **(b)** draw its Lewis structure and that of its chloride; and **(c)** write the balanced equation for the reaction of $X(g)$ with $Cl_2(g)$.

Carbon and the Hydrocarbons

8.1 Allotropes of Carbon
8.2 Inorganic Compounds of Carbon
8.3 Alkanes
8.4 Alkenes and Alkynes
8.5 Benzene and the Arenes: Aromatic Hydrocarbons
8.6 Resonance Structures and Bond Order

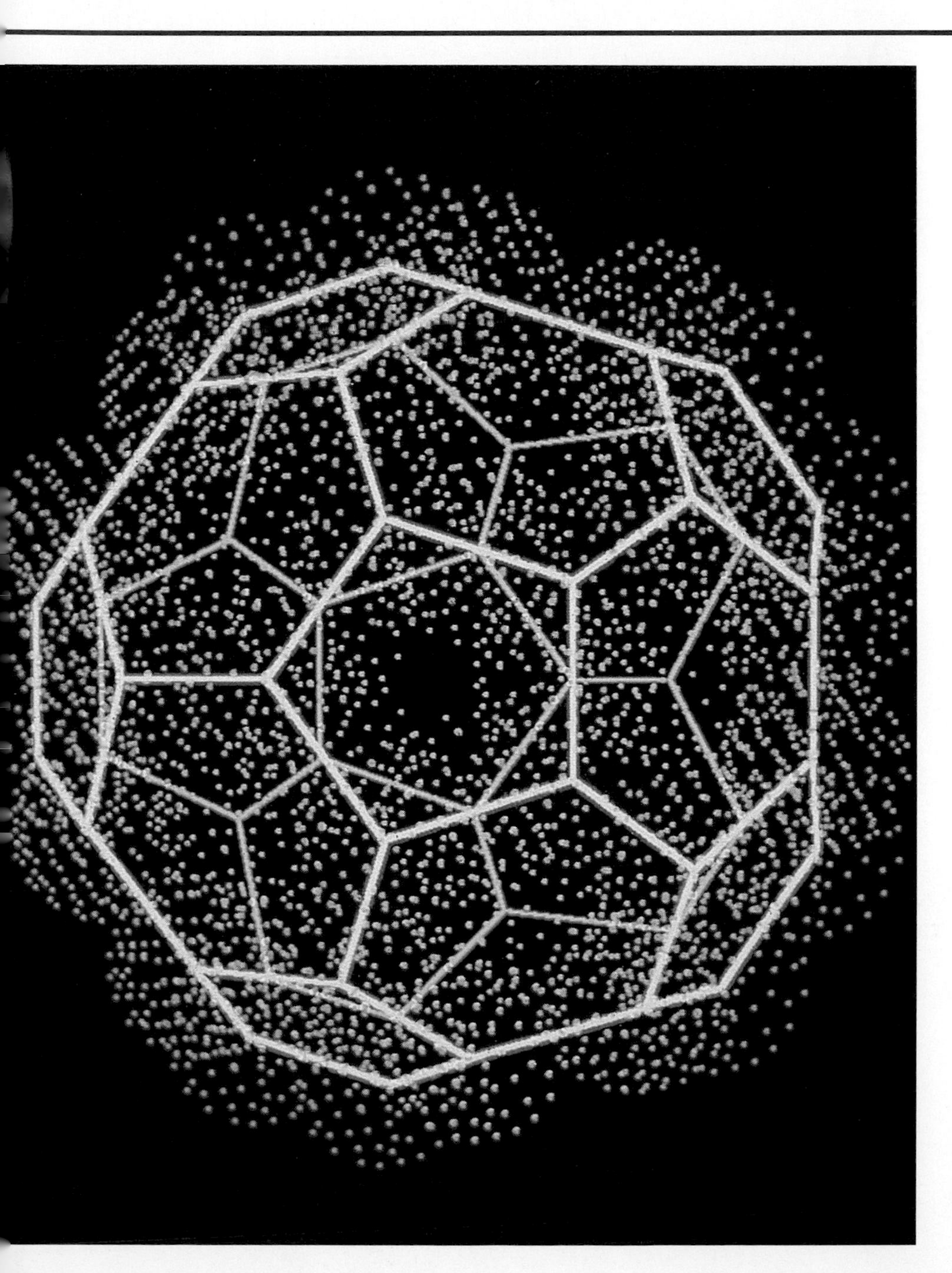

Computer-generated model of buckminsterfullerene (buckyball), C_{60}, a recently discovered form of carbon whose structure was first determined in 1991. The C atoms in this nearly spherical molecule are arranged in a polyhedron with pentagonal and hexagonal faces—the same arrangement as in a soccer ball. This molecule was named after the architect Buckminster Fuller, who designed the "geodesic dome," a dome constructed from pentagons and hexagons to give a spherical shape. Chemists, quick to synthesize many other new molecules by adding other atoms and groups to the C_{60} framework, are now exploring a whole new area of carbon chemistry.

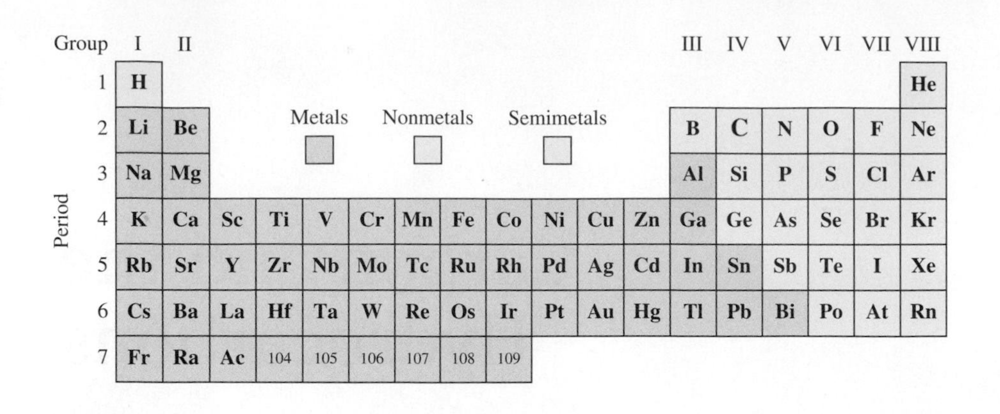

Although we find an enormous variety of carbon compounds in living matter, carbon is not one of the most abundant elements on earth. In fact, it makes up only 0.5% of the earth's crust (Table 2.3). About half this carbon occurs as the carbonate ion, CO_3^{2-}, in calcium carbonate (found mainly in chalk, limestone, and marble) and in other metal carbonates. The rest is present in vegetable and animal matter, in coal and petroleum, and in the atmosphere and oceans as carbon dioxide.

The vast majority of carbon compounds are classified as **organic compounds**. At one time it was believed that many compounds of carbon could be obtained only from organic, or living, matter; hence these compounds were called organic compounds. Those occurring in the nonliving or mineral world, including a few compounds of carbon, such as carbon dioxide, were called **inorganic compounds**. Today we know that carbon compounds found in living matter can be synthesized from inorganic compounds. For example, in the process of photosynthesis, plants convert the inorganic compounds carbon dioxide and water from their environment to organic compounds. Indeed, life itself appears to have originated from simple, mainly inorganic substances such as H_2, H_2O, NH_3, HCN, and CH_4 that were present in the earth's original atmosphere. Chemists have shown that these substances can react under the influence of an electric discharge (lightning) or ultraviolet radiation to form amino acids and other carbon compounds that are the basis of life (Chapter 17). Nevertheless, it is still common to use the name *organic* to denote all carbon compounds except CO_2, carbonates, and a few other compounds that have been traditionally regarded as inorganic. Therefore, with these few exceptions we define **organic chemistry** as the chemistry of carbon compounds, and we define **inorganic chemistry** as the chemistry of all the other elements and their compounds.

The simplest organic compounds are the **hydrocarbons**, compounds that contain only carbon and hydrogen. We can regard all other organic compounds as being derived from hydrocarbons by replacing hydrogen atoms with other atoms or groups of atoms. Modern civilization is almost totally dependent on hydrocarbons, which occur in the earth's crust as *natural gas* and *petroleum.* Gasoline, diesel fuel, and domestic heating oil, all of which are mixtures of hydrocarbons obtained by the distillation of petroleum, provide our major source of energy. Hydrocarbons are also the starting materials for the synthesis of a wide variety of compounds ranging from drugs to plastics. We could be said to be living in the Petroleum Age. In this chapter we consider the structures and properties of the most important types of hydrocarbons. We shall consider these and other organic compounds in Chapters 14, 17, and 20. We begin by discussing the element carbon and some of its inorganic compounds.

FIGURE 8.1 Two Allotropes of Carbon Diamond is transparent and very hard. Graphite is black and soft; it is used in pencils.

8.1 ALLOTROPES OF CARBON

Carbon has several allotropes, the most well known of which are diamond and graphite (Figure 8.1). **Diamond** forms transparent, colorless, very hard crystals that, when cut, are familiar as gemstones. In contrast, **graphite** is a soft, black substance that is used as a lubricant and as the ''lead'' in lead pencils. Diamond has a much higher density ($3.53\ g \cdot cm^{-3}$) than graphite ($2.25\ g \cdot cm^{-3}$) and is an electrical insulator, whereas graphite, unlike most nonmetals, is a fairly good conductor. At ordinary temperatures diamond does not react with any other substance, whereas graphite reacts with concentrated oxoacids, such as nitric and sulfuric acids, and with other strong oxidizing agents. The very different properties of diamond and graphite result from their very different structures.

In diamond, each carbon atom forms four covalent bonds with four neighboring carbon atoms. These four bonds have a tetrahedral arrangement; that is, each carbon atom has a tetrahedral AX_4 geometry. Thus, diamond has a three-dimensional network structure (Figure 8.2). A crystal of diamond may be regarded as one giant covalent molecule.

In contrast, graphite consists of planar sheets of carbon atoms in which each carbon is surrounded by only three other carbon atoms. Thus, it has an overall

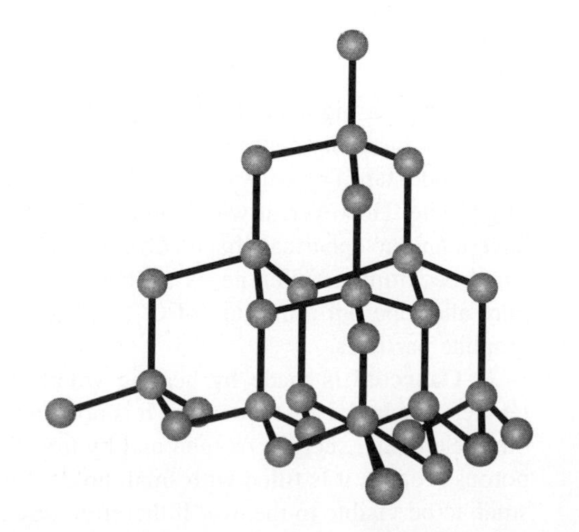

FIGURE 8.2 Structure of Diamond Each carbon atom forms four bonds with neighboring carbon atoms. The four bonds formed by each carbon atom have a tetrahedral AX_4 arrangement. A three-dimensional network of covalent bonds extends throughout the crystal.

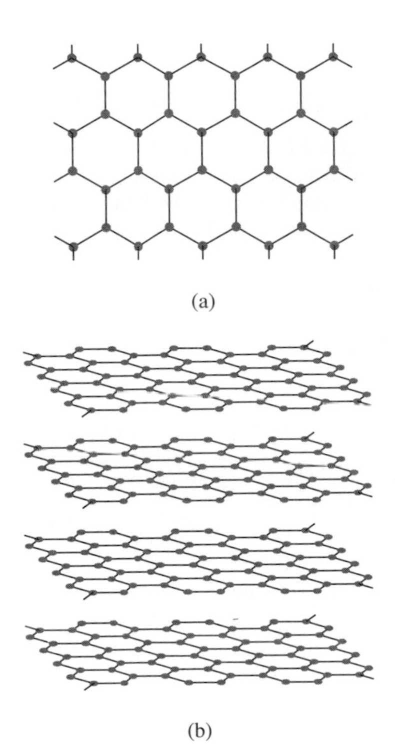

FIGURE 8.3 Structure of Graphite (a) Graphite is composed of flat sheets of carbon atoms in which each carbon atom is bonded to three neighboring carbon atoms in a planar AX_3 arrangement. (b) These sheets of carbon atoms are stacked one over another and held together by relatively weak intermolecular forces.

arrangement of hexagonal rings of atoms similar to chicken wire (Figure 8.3a). The planar sheets are stacked on each other with a separation of 334 pm, which is much greater than the length of a single carbon–carbon bond (154 pm) and shows that there are no covalent bonds between the sheets. Thus, graphite may be considered to be composed of giant planar two-dimensional molecules stacked upon each other and held together by relatively weak intermolecular forces (Figure 8.3b). These layers can slide over each other rather easily, so graphite is soft, feels slippery, and is a good lubricant. When you use a pencil, layers of graphite slide off the point onto the paper. Carbon has four valence electrons and forms four covalent bonds. In graphite this means that the bond to one of the other three carbon atoms to which each carbon is attached must be a double bond (Figure 8.4).

The position of the equilibrium between diamond and graphite

$$\text{diamond} \rightleftharpoons \text{graphite}$$

is to the right at ordinary pressures and temperatures, so diamond is expected to change into graphite. Fortunately for the owners of diamonds, this change is almost infinitely slow at ordinary temperatures; the atoms have far too little energy to enable them to undergo the drastic rearrangement necessary to convert diamond to graphite. At high pressure, however, the position of equilibrium shifts to the left, and graphite will in principle change into diamond. This change, however, is very slow at ordinary temperatures. To increase the rate, it is necessary to use a temperature of 1200°C and a catalyst as well as a very high pressure to shift the position of equilibrium. Synthetic diamonds produced in this way are not normally of gem quality, but they have many practical uses. Because diamonds are so hard, they are used, for example, in oil-well drilling bits, abrasive grinding wheels, and glass- and metal-cutting tools. Diamonds that occur naturally are believed to have been formed in the mantle at least 100 km below the earth's surface where the pressure and temperature are sufficiently high. They then rose to the surface through pipes of molten rock.

Both diamond and graphite have very high melting points, over 3700°C. Graphite is used to make crucibles for melting high-melting-point metals and for the electrodes in electric furnaces.

There are several other forms of carbon, such as carbon black, charcoal, and coke. **Carbon black (soot)** forms when hydrocarbons are burned in a limited supply of air. For example,

$$2C_2H_2(g) + O_2(g) \rightarrow 4C(s) + 2H_2O(g)$$

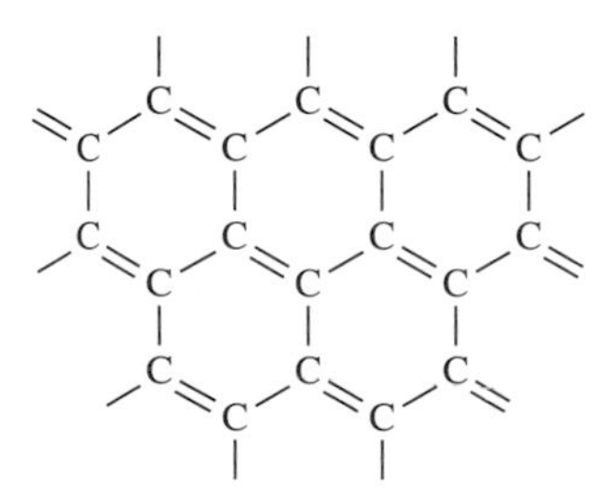

FIGURE 8.4 Bonding in Graphite Each carbon atom forms two single bonds and a double bond and therefore has a planar AX_3 geometry, so each layer in the graphite structure is a very large two-dimensional molecule.

It is used as a pigment for paint, paper, and printer's ink. It is used also in large quantities to strengthen and reinforce automobile tires, an effect that is not fully understood. Until recently carbon black was thought to consist of very small crystals of graphite. However, it was found in the late 1980s to contain C_{60} molecules that have a unique spherical shape (Box 8.1). C_{60}, which has been given the intriguing name buckminsterfullerene, is another allotrope of carbon. Carbon black is not a pure allotrope but a mixture of C_{60} and other similar molecules, such as C_{70}, and graphite particles.

Charcoal is made by heating wood and other organic materials to a high temperature in the absence of air. It is not pure carbon but contains small amounts of other elements, such as oxygen and hydrogen, that are present in wood. It is very porous because it is filled with small holes like a sponge, although the holes are too small to be visible to the eye. It therefore has a very large surface area with respect to its volume, so it can adsorb considerable quantities of other substances. Carbon in

BOX 8.1 Buckyball: The Third Allotrope of Carbon

Carbon has been known from the earliest recorded times, and diamond and graphite have been recognized as allotropes of carbon for more than 200 years. Yet other pure forms of carbon were not discovered until the 1980s. It is even more remarkable that these allotropes are found in ordinary soot (carbon black), such as we obtain from a candle flame. It had been assumed that soot consists of small graphite particles. No one suspected that it might contain quite a different form of carbon.

In the early 1980s Richard Smalley and his colleagues at Rice University in Texas were studying the molecules that could be obtained by subjecting elements to an intense laser beam. They were interested in carbon because unusual molecules that appeared to consist of linear chains with up to 9 or 10 carbon atoms had been detected in outer space. Using a mass spectrometer, they found that their laser technique produced not only these small molecules, but also unexpectedly large amounts of much bigger molecules—in particular, C_{60} and C_{70}. There was considerable speculation about the structures of these molecules and why molecules with these numbers of atoms appeared to be particularly stable.

Smalley noted that if some of the hexagons in a planar sheet were replaced with pentagons, the sheet would no longer be flat and could curl on itself. Indeed, the eighteenth-century Swiss mathematician Leonhard Euler had shown that with at least 12 pentagons, a polyhedron with an almost spherical shape can be formed and that the most symmetrical shape of this type contains 20 hexagons. This polyhedron has 60 corners. Smalley suggested that the C_{60} molecule has just this structure and that its high symmetry is responsible for its exceptional stability (Figure A). He called it buckminsterfullerene after the American architect Buckminster Fuller, who is famous for designing geodesic domes based on the same arrangement of hexagons and pentagons (Figure B). We see exactly the same arrangement of pentagons and hexagons in the panels of a soccer ball (Figure C). Informally, chemists call the C_{60} molecule "buckyball."

Figure B *This geodesic dome at Florida's Epcot Center is based on Buckminster Fuller's design.*

Smalley's method produced only minute amounts of C_{60}, which could be detected in a mass spectrometer but not isolated in a weighable amount. In a 1990 experiment, researchers found that soot deposited from carbon vapor produced by an electric arc between two graphite electrodes dissolved in benzene. Upon evaporation of this solution, they isolated a substantial amount of C_{60} as yellow crystals. This opened up the possibility of determining the structure of C_{60} by X-ray crystallography. In 1991 the structure was shown to be exactly as predicted.

A whole new area of carbon chemistry now awaits investigation. It has already been found, for example, that C_{60} can be fluorinated to give $C_{60}F_{60}$, in which a fluorine atom is attached to each carbon atom. The fluorine atoms form a layer on the outside of the ball, hence the informal names "fuzzyball" or "teflonball." In view of the well-known slippery feel of Teflon (polytetrafluoroethylene, C_nF_{2n}), it has been suggested that these molecules may turn out to be the best lubricants yet developed.

Usable quantities of other allotropes, such as C_{70} (which is expected to have an approximate football shape based on 12 pentagons and 25 hexagons), will certainly be isolated in the future. We can anticipate many exciting developments in this new area of chemistry in the next few years.

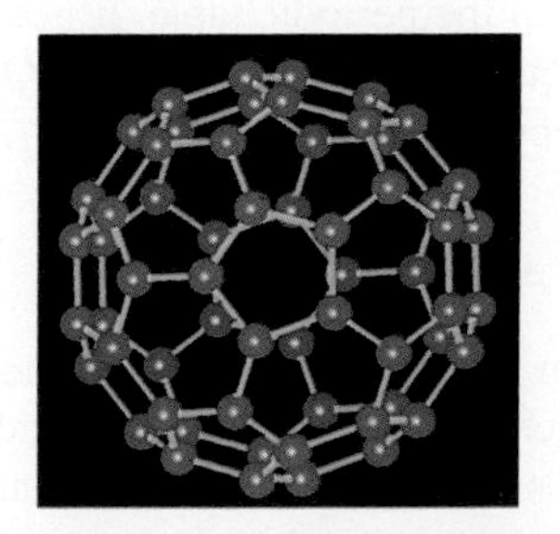

Figure A *Computer-generated model of buckminsterfullerene, the C_{60} molecule, first described in 1990*

Figure C *The patterns of hexagons and pentagons from which the soccer ball and the C_{60} molecule are constructed are the same.*

DEMONSTRATION 8.1 Adsorption by Activated Charcoal

The flask contains bromine vapor. The dish contains charcoal.

When the charcoal is added to the flask, the color of the bromine vapor rapidly diminishes in intensity. After only 20 s, it is substantially reduced.

After 1min 44s no more bromine vapor can be seen, because it has been completely adsorbed by the charcoal.

this form, once cleaned, is referred to as *activated charcoal*. It is used, for example, in removing hydrocarbons from automobile exhaust, purifying gases and water, and removing colored substances in the purification of sugar (Demonstration 8.1).

Coke is made by heating coal in the absence of air. Coal is a complex mixture of many substances that contains 60 to 90% C together with H, O, N, S, Al, Si, and other elements. When heated to a high temperature, it decomposes into a mixture of gases, mainly CH_4 and H_2, known as coal gas; a liquid called coal tar, a mixture of many hydrocarbons and other organic compounds; and coke, a solid residue that contains 90 to 98% C. Coke is used in enormous quantities as a reducing agent, mainly in the manufacture of steel and other metals (Chapter 10). Charcoal and coke are not pure forms of carbon, and their detailed structures are not known.

An egg shell consists mainly of calcium carbonate. When an egg is placed in aqueous hydrochloric acid, bubbles of CO_2 are released.

8.2 INORGANIC COMPOUNDS OF CARBON

Carbon is the first element in Group IV of the periodic table. It is a nonmetal with a valence of 4. Common inorganic compounds include its oxides CO and CO_2, metal carbonates, such as $CaCO_3$, and metal cyanides, such as NaCN.

Carbon Dioxide and Carbonic Acid

The combustion of carbon and carbon-containing compounds in excess oxygen produces **carbon dioxide**, $CO_2(g)$, which is also produced by heating metal carbonates. With the exception of the alkali metal carbonates, metal carbonates decompose to carbon dioxide and the corresponding metal oxide:

$$CaCO_3(s) \xrightarrow{\text{heat}} CaO(s) + CO_2(g)$$

In the laboratory, we can prepare small amounts of carbon dioxide by adding a dilute aqueous acid to a metal carbonate:

$$CaCO_3(s) + 2H^+(aq) \rightarrow Ca^{2+}(aq) + CO_2(g) + H_2O(l)$$

Carbon dioxide is a linear AX_2 molecule with two double bonds:

$$:\ddot{O}=C=\ddot{O}:$$

Carbon dioxide

It is a colorless gas that, when cooled, forms a white solid (dry ice) that sublimes at −78°C at 1 atm pressure (Figure 8.5).

FIGURE 8.5 Solid Carbon Dioxide Carbon dioxide is a white solid that sublimes at −78°C at 1 atm pressure. It is often called "dry ice" and is used for temporary storage of materials at −78°C.

Carbon dioxide dissolves in water to give a solution that contains a small equilibrium concentration of **carbonic acid**, H_2CO_3:

$$CO_2(aq) + H_2O(l) \rightleftharpoons H_2CO_3(aq)$$

This solution is commonly called soda water. Carbonic acid is a weak diprotic acid (Demonstration 8.2),

$$H_2CO_3(aq) + H_2O(l) \rightleftharpoons H_3O^+(aq) + HCO_3^-(aq)$$
$$HCO_3^-(aq) + H_2O(l) \rightleftharpoons H_3O^+(aq) + CO_3^{2-}(aq)$$

||| Water acts as a base (proton acceptor) in these two reactions. There is no oxidation–reduction, because there are no changes in oxidation numbers for H, C, and O.

forming the ions HCO_3^-, hydrogencarbonate (commonly called bicarbonate), and CO_3^{2-}, carbonate. These ions are weak bases and are converted to carbonic acid by hydrogen ion in the reverse of the above reactions. Carbonic acid is unstable and decomposes to give an equilibrium concentration of CO_2. So as we mentioned earlier, carbon dioxide can be prepared in the laboratory by adding an aqueous acid to a metal carbonate.

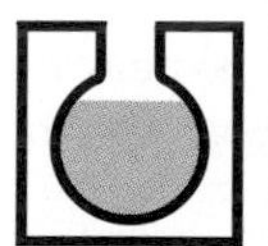

DEMONSTRATION 8.2 Carbon Dioxide Behaves as an Acid in Water

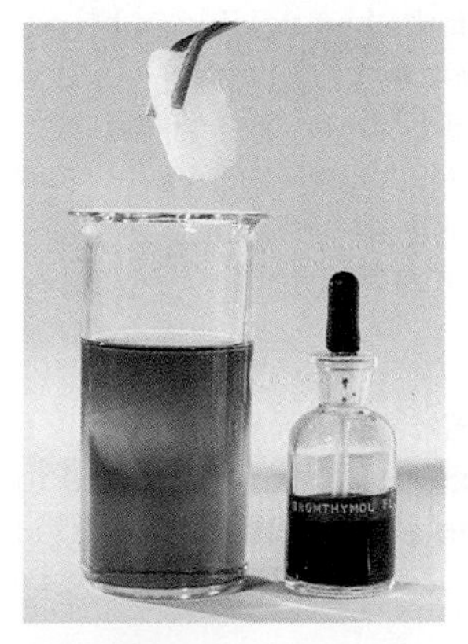

A piece of solid carbon dioxide (dry ice) is dropped into water to which a small amount of bromothymol blue indicator has been added.

As the CO_2 dissolves in the water, it forms carbonic acid, which changes the color of the indicator from blue to yellow. The white smoke is a mist of water droplets condensed from the air by the cold CO_2 gas.

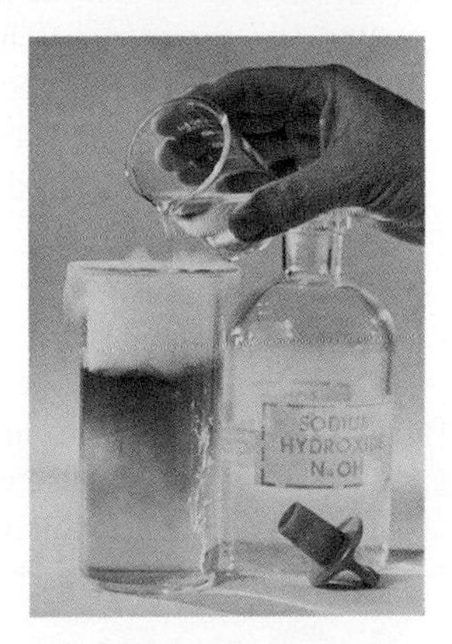

The original blue color of the indicator can be restored by adding sodium hydroxide solution, which converts the carbonic acid to sodium carbonate.

FIGURE 8.6 Carbon Dioxide Fire Extinguisher Extinguishing a Fire

Carbonic acid and its anions, HCO_3^- and CO_3^{2-}, have the following Lewis structures:

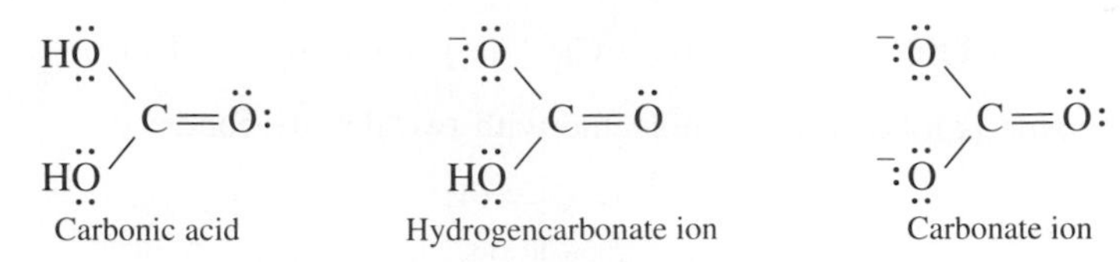

In each case the geometry around the carbon atom is planar AX_3.

Carbon dioxide is used as a fire extinguisher because it cannot be further oxidized and has a higher density than air. It sinks down on a fire, excluding oxygen, and thus extinguishing it (Figure 8.6). In photosynthesis, carbon dioxide combines with water in the presence of light to form carbohydrates.

Carbon Monoxide

When carbon is burned in a limited supply of air, **carbon monoxide** is obtained:

$$2C(s) + O_2(g) \rightarrow 2CO(g)$$

For use in industry, carbon monoxide mixed with hydrogen is made by passing steam over red-hot coke,

$$C(s) + H_2O(g) \xrightarrow{\text{heat}} CO(g) + H_2(g)$$

or by heating methane (natural gas) to a high temperature with steam in the presence of a catalyst

$$CH_4(g) + H_2O(g) \xrightarrow{\text{heat, catalyst}} CO(g) + 3H_2(g)$$

to give a mixture of CO and H_2. This mixture is called **synthesis gas**, or **syngas** for short, because it is used as the starting material for the industrial preparation of many organic compounds (Chapter 14).

Carbon monoxide is a colorless, odorless gas with only a slight solubility in water. It is very toxic to humans because it combines with the hemoglobin in blood, thus preventing the blood from functioning as an oxygen carrier (Chapter 17). It is particularly dangerous because it is odorless and therefore not easily detected. Carbon monoxide is produced when the tobacco in cigarettes burns, and it is present in automobile exhaust.

Carbon monoxide can be represented by the Lewis structure

$$^{-}:C\equiv O:^{+}$$

in which there is a triple bond. It is isoelectronic with N_2. Carbon monoxide is a reducing agent, because carbon is in the +II oxidation state and can be oxidized to carbon dioxide, in which carbon is in the +IV oxidation state. Carbon monoxide burns in air to form carbon dioxide:

$$2CO(g) + O_2(g) \rightarrow 2CO_2(g)$$

It reduces many metal oxides to the corresponding metals. For example, iron oxide, Fe_2O_3, is reduced to iron in the blast furnace (Chapter 10):

$$Fe_2O_3(s) + 3CO(g) \rightarrow 2Fe(l) + 3CO_2(g)$$

Carbon monoxide reacts with hydrogen in the presence of suitable catalysts to give methanal (formaldehyde) and methanol, in which the triple bond is converted first to a double bond and then to a single bond:

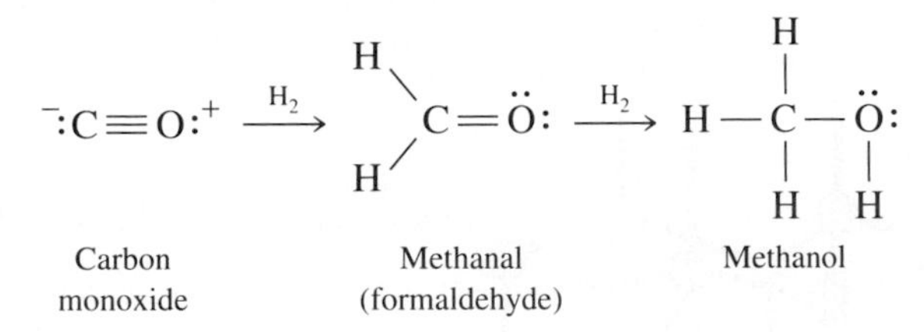

Carbon monoxide   Methanal (formaldehyde)   Methanol

Methanal and methanol are prepared on an industrial scale by these reactions (Chapter 14).

Carbon Disulfide

Carbon disulfide, CS_2, is prepared by the reaction of methane, CH_4, with sulfur at 600°C in the presence of a catalyst:

$$CH_4(g) + 4S(g) \xrightarrow{\text{heat, catalyst}} CS_2(g) + 2H_2S(g)$$

Carbon disulfide is a linear AX_2 molecule with double bonds, just like carbon dioxide:

$$:\ddot{S}=C=\ddot{S}:$$

Carbon disulfide

It is a volatile, flammable liquid with many industrial uses, as in the production of synthetic fibers.

Hydrogen Cyanide

Hydrogen cyanide, HCN, is prepared industrially by the reaction of methane with ammonia at 1200°C in the presence of a platinum catalyst:

$$CH_4(g) + NH_3(g) \xrightarrow{\text{heat, Pt}} HCN(g) + 3H_2(g)$$

Hydrogen cyanide is a colorless liquid that boils just above room temperature (25.6°C). HCN is a linear AX_2 molecule with a carbon–nitrogen triple bond:

$$H—C\equiv N:$$

Hydrogen cyanide

Hydrogen cyanide behaves as a weak acid in water. Its salts are the cyanides, such as sodium cyanide, NaCN. They contain the cyanide ion $^{-}:C\equiv N:$, which is isoelectronic with $^{-}:C\equiv O:^{+}$ and $:N\equiv N:$. Both HCN and the cyanides are very toxic, as is well known to readers of detective stories.

Carbides

When the Group I and II metal oxides are heated with carbon, metal carbides are formed:

$$CaO(s) + 3C(s) \xrightarrow{\text{heat}} CaC_2(s) + CO(g)$$

These *carbides* are ionic compounds containing the ion C_2^{2-}, which is isoelectronic with N_2 and CN^- and, like these molecules, has a triple bond $^{-}:C\equiv C:^{-}$.

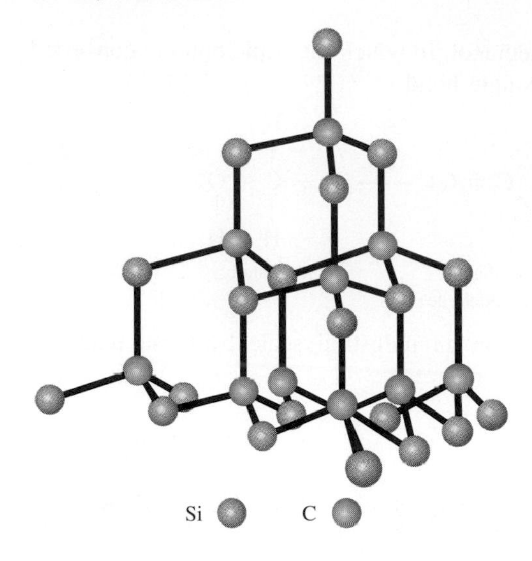

FIGURE 8.7 Silicon Carbide, SiC (Carborundum) SiC has a diamondlike structure in which each C atom is surrounded tetrahedrally by four Si atoms, and each Si atom is surrounded tetrahedrally by four C atoms.

Calcium carbide reacts with water to give *ethyne* (acetylene):

$$CaC_2(s) + 2H_2O(l) \rightarrow Ca(OH)_2(aq) + C_2H_2(g)$$

The carbide ion, C_2^{2-}, is the anion of ethyne, HC≡CH. Ethyne is too weak an acid to ionize in water, so the carbide ion is a strong base. It is protonated in water and completely converted to ethyne:

$$C_2^{2-} + 2H_2O \rightarrow C_2H_2 + 2OH^-$$

With more electronegative elements such as silicon, carbon forms covalent carbides. Silicon carbide, SiC (carborundum), has the diamond structure with alternate carbon atoms replaced by silicon (Figure 8.7). It occurs naturally and is almost as hard as diamond. Because it is so hard and because its crystals fracture in a way that leaves them with sharp edges, silicon carbide is very useful as an abrasive.

Multiple Bonding in the Compounds of Carbon, Nitrogen, and Oxygen

Carbon, nitrogen, and oxygen have a much greater tendency to form multiple bonds than any other elements. For example, the following molecules and ions all have multiple bonds:

$^-$:C≡O:$^+$	$^-$:C≡N:	:N≡N:	$^-$:C≡C:$^-$
Carbon monoxide	Cyanide ion	Dinitrogen	Carbide ion

:Ö=C=Ö:	:S̈=C=S̈:
Carbon dioxide	Carbon disulfide

The Period 3 elements in Groups IV, V, and VI—namely, silicon, phosphorus, and sulfur—form very few analogous molecules or ions containing multiple bonds. For example, although nitrogen consists of N_2 molecules with a triple bond, white phosphorus consists of P_4 molecules with only single bonds. Silicon dioxide, $SiO_2(s)$, has a covalent network structure with single bonds, whereas carbon dioxide, $CO_2(g)$, consists of CO_2 molecules with double bonds. When an element such as sulfur or phosphorus does form a multiple bond, it is almost invariably with carbon, nitrogen, or oxygen, as in SO_2 and SO_3. The great tendency of carbon, nitrogen, and oxygen to form multiple bonds has never been fully explained. But it seems reasonable to suppose that the high electronegativity of these elements enables them to attract two or three bonding pairs into the bonding region despite the increased repulsion between the electron pairs in multiple bonds.

EXAMPLE 8.1 Reactions of Carbon Compounds

Use the information given in this section to predict the products of the following reactions, and write a balanced equation in each case.

(a) $CuO(s) + CO(g) \xrightarrow{heat}$

(b) $BaCO_3(s) \xrightarrow{heat}$

(c) $Na_2CO_3(s) + HCl(aq) \rightarrow$

(d) $HCN(g) + \underset{\text{excess}}{O_2(g)} \xrightarrow{heat} N_2(g) +$

Which reactions are acid–base reactions, and which are redox reactions? What type of reaction is (b)?

Solution: (a) Carbon monoxide is a reducing agent that reduces metal oxides to their corresponding metals. Therefore,

$$CuO(s) + CO(g) \rightarrow Cu(s) + CO_2(g)$$

Oxygen is removed from CuO and added to CO; the oxidation number of Cu changes from +2 to 0. Copper is reduced, and carbon is oxidized. This is a redox reaction.

(b) Metal carbonates, except those of the alkali metals, decompose upon heating to the metal oxide and carbon dioxide. Therefore,

$$BaCO_3(s) \xrightarrow{heat} BaO(s) + CO_2(g)$$

This is a decomposition reaction.

This is not a redox reaction because carbon has the same oxidation number, +4, in $BaCO_3$ and CO_2.

(c) Carbonates react with acids to give carbon dioxide:

$$Na_2CO_3(s) + 2HCl(aq) \rightarrow 2NaCl(aq) + CO_2(g) + H_2O(l)$$

This is an acid–base reaction, as we can see if we write it in the ionic form:

$$CO_3^{2-}(s) + 2H_3O^+(aq) \rightarrow H_2CO_3(aq) + 2H_2O(l)$$
$$H_2CO_3(aq) \rightarrow H_2O(l) + CO_2(g)$$

(d) Here we are told that nitrogen is one of the products. If we make the reasonable assumption that this is the only nitrogen-containing product, we have to worry about only the C and H. We know that hydrocarbons are oxidized to CO_2 and H_2O in excess oxygen, so it is reasonable to assume that this is also the case here. Thus, the balanced equation is

$$4HCN(g) + 5O_2(g) \rightarrow 2N_2(g) + 4CO_2(g) + 2H_2O(g)$$

This is not a completely certain prediction. Another product might have been an oxide of nitrogen, and we cannot be absolutely sure that C and H will be oxidized to CO_2 and H_2O. Nevertheless, it is a reasonable prediction that turns out to be correct. This is a redox reaction.

Exercise 8.1 Predict the products and write the balanced equation for each of the following reactions. Which are redox reactions, and which are acid–base reactions?

(a) $MgCO_3(s) \xrightarrow{heat}$ **(b)** $PbO(s) + CO(g) \xrightarrow{heat}$

(c) $MgO(s) + C(s) \xrightarrow{heat}$ **(d)** $MgC_2(s) + H_2O(l) \rightarrow$

Exercise 8.2 Which of the following anions are strong bases, which are weak bases, and which have no basic properties in water?

(a) CN^- **(b)** CO_3^{2-} **(c)** Cl^- **(d)** HCO_3^- **(e)** C_2^{2-} **(f)** OH^- **(g)** F^-

Exercise 8.3 Draw the Lewis structures and give the names of three molecules or ions that are isoelectronic with the cyanide ion.

8.3 ALKANES

(a)

(b)

(c)

Ball-and-stick models of (a) methane, (b) ethane, (c) propane

Alkanes are a family of hydrocarbons that have single bonds between the carbon atoms and the general formula C_nH_{2n+2}, where n has integral values from 1 up to some very large number. The simplest alkanes are **methane**, CH_4, the major constituent of natural gas; **ethane**, C_2H_6, a minor constituent of natural gas; and **propane**, C_3H_8, and **butane**, C_4H_{10}, both of which are used as fuels. Some of the higher-n members of the family, such as octane, C_8H_{18}, and nonane, C_9H_{20}, are major constituents of gasoline.

The melting points and boiling points of the alkanes increase with increasing molecular size (Table 8.1). The alkanes up to butane are gases at ordinary temperatures and pressures; the higher-n members are liquids and solids. Paraffin wax is a mixture of solid alkanes. All alkanes are insoluble in water and are generally less dense than water.

Structures of Alkanes

The Lewis structures or structural formulas for methane, ethane, and propane are

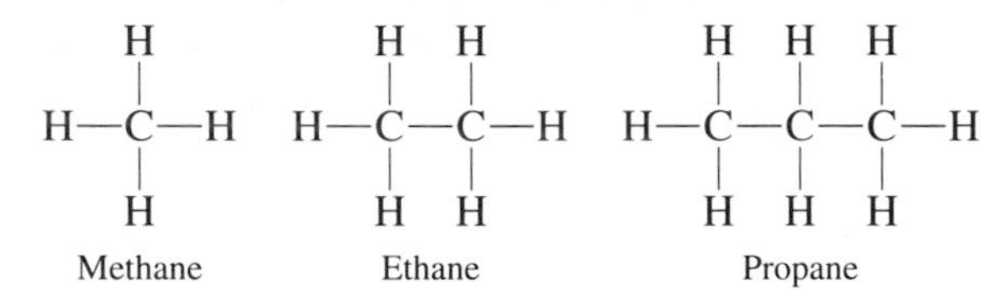

Structural formulas are often simplified by omitting the bond lines to the hydrogen atoms and grouping the hydrogen atoms bonded to a given carbon atom. For example,

$$CH_4 \qquad CH_3—CH_3 \qquad CH_3—CH_2—CH_3$$

Methane Ethane Propane

In these molecules, each carbon atom forms four tetrahedrally arranged bonds, some to hydrogen atoms and some to other carbon atoms (Figure 8.8).

A propane barbeque grill

TABLE 8.1 Boiling Points and Melting Points of n-Alkanes, C_nH_{2n+2}

n	Name	Boiling Point (°C at 1 atm)	Melting Point (°C)	Formula
1	Methane	−162	−183	CH_4
2	Ethane	−89	−172	CH_3CH_3
3	Propane	−42	−188	$CH_3CH_2CH_3$
4	Butane	0	−138	$CH_3(CH_2)_2CH_3$
5	Pentane	36	−130	$CH_3(CH_2)_3CH_3$
6	Hexane	69	−95	$CH_3(CH_2)_4CH_3$
7	Heptane	98	−91	$CH_3(CH_2)_5CH_3$
8	Octane	126	−57	$CH_3(CH_2)_6CH_3$
9	Nonane	151	−54	$CH_3(CH_2)_7CH_3$
10	Decane	174	−30	$CH_3(CH_2)_8CH_3$
20	Eicosane	343	37	$CH_3(CH_2)_{18}CH_3$
30	Triacontane	446	66	$CH_3(CH_2)_{28}CH_3$

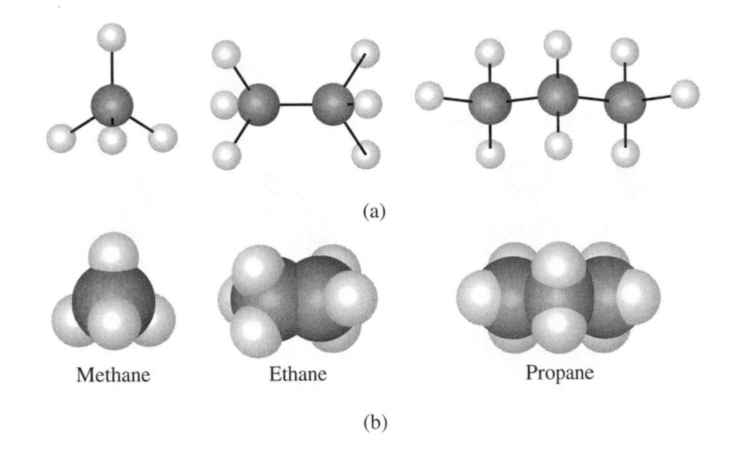

FIGURE 8.8 Structures of Methane, Ethane, and Propane (a) Ball-and-stick models; (b) space-filling models

Continuing the series, we expect the structural formulas for butane, C_4H_{10}, and pentane, C_5H_{12}, to be

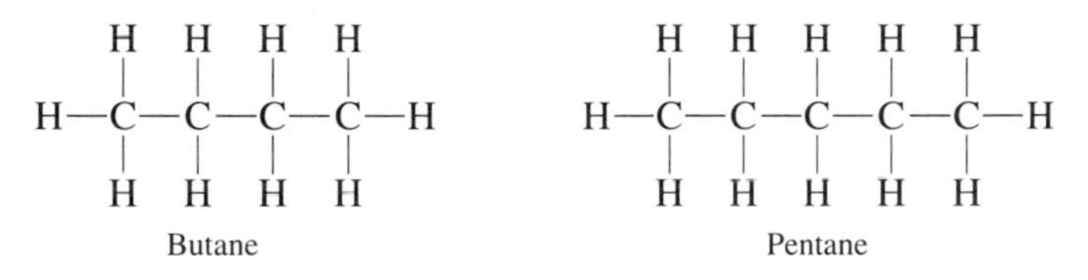

Alkanes in this series are called **continuous-chain alkanes**, because all the carbon atoms are connected in one continuous chain (Figure 8.9).

Lewis structures normally show both bonding and nonbonding electron pairs on all the atoms in a molecule. Nonbonding (lone) pairs are shown as :, and bonding pairs are shown either as : or as a bond line —. A structural formula shows the bonding electron pairs as a bond line, and some or all of the nonbonding pairs may be omitted.

Structural Isomers

There are two substances with the formula C_4H_{10}; one has a boiling point of 0°C, and the other a boiling point of −10°C. It is possible to have two different butanes because in an alkane with four or more carbon atoms, the carbon atoms may be joined in different ways. The four carbon atoms of C_4H_{10} may be joined in one continuous chain, as we have just seen. Alternatively, three carbon atoms may be joined in a chain and the fourth attached to the middle carbon of the chain, forming a **side chain** (Figure 8.10):

```
    H   H   H
    |   |   |
H — C — C — C — H
    |   |   |
    H   |   H
    H — C — H
        |
        H
```

Isobutane

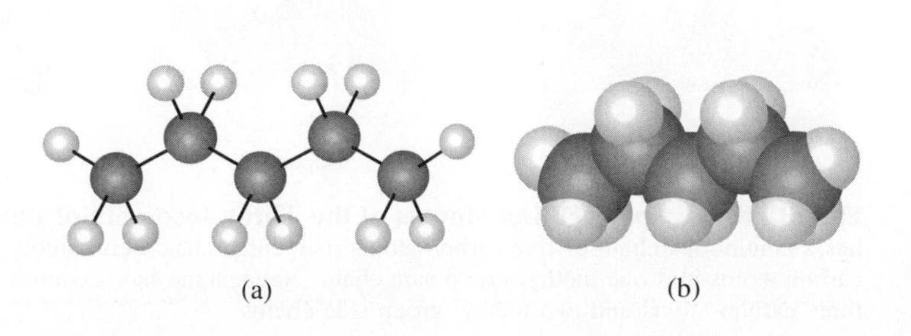

FIGURE 8.9 Structure of Pentane Pentane is called a continuous-chain alkane, because all the carbon atoms are connected in one continuous chain. (a) Ball-and-stick model; (b) space-filling model

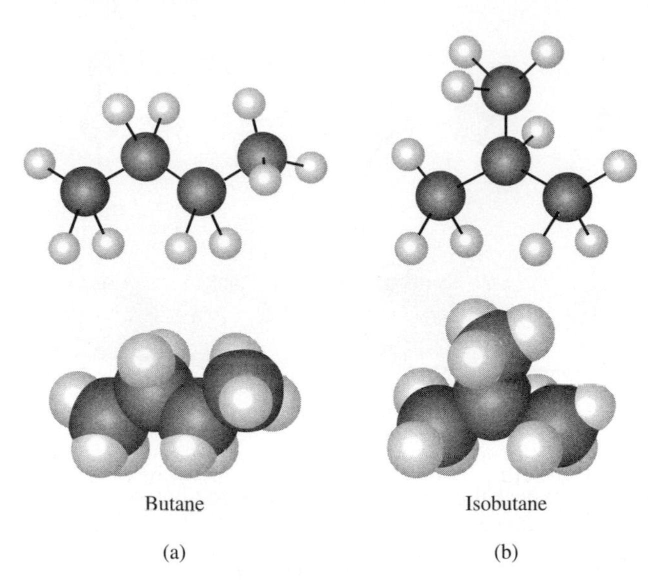

FIGURE 8.10 Butane and Isobutane (a) Butane is a continuous-chain alkane. (b) Isobutane has a side chain.

This structure, an example of a **branched-chain alkane**, is called **isobutane** or, to follow a system that we will describe shortly, **methylpropane**. Both butane and isobutane have the same molecular formula, C_4H_{10}, and the same molar mass. But because they have different structures, they have slightly different properties.

> **Molecules that have the same molecular formula but different structures are called *structural isomers*.**

Structural isomers have the same number of each kind of atom, but the atoms are arranged in different ways.

For pentane, three arrangements of carbon atoms are possible. Therefore, there are three structural isomers of pentane:

Before the introduction of IUPAC nomenclature, the isomer named "pentane" was called "*n*-pentane."

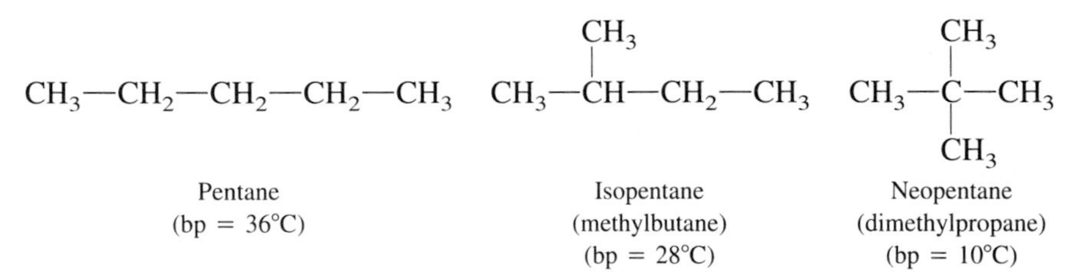

Figure 8.11 shows space-filling models of these three structural isomers. It is important to recognize that a structural formula gives no information about the

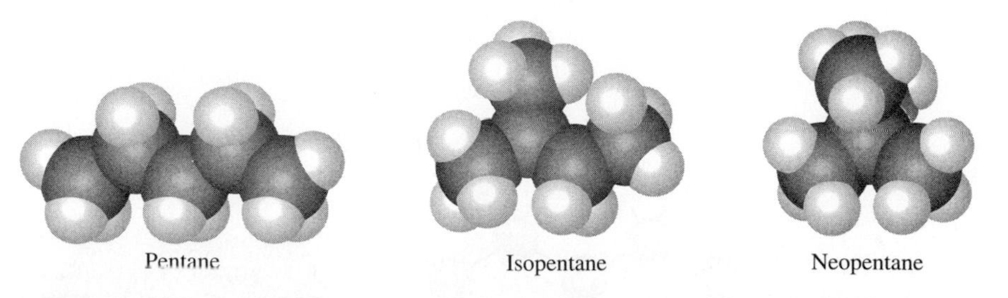

FIGURE 8.11 Space-Filling Models of the Three Iosomers of Pentane Pentane has a continuous chain of five carbon atoms. Isopentane has a continuous chain of four carbon atoms and one methyl-group side chain. Neopentane has a continuous chain of three carbon atoms and two methyl-group side chains.

three-dimensional geometry of a molecule. The structural formula of isopentane can be written in several different ways, such as

$$\begin{array}{lll} & CH_3 & \\ & | & \\ CH_2- & CH- & CH_3 \\ | & & \\ CH_3 & & \end{array} \quad or \quad \begin{array}{llll} & & CH_3 & \\ & & | & \\ CH_3- & CH_2- & CH- & CH_3 \end{array} \quad or \quad \begin{array}{ll} CH_3 & \\ | & \\ CH- & CH_3 \\ | & \\ CH_2- & CH_3 \end{array}$$

but all these formulas represent the same molecule. In each case there is a continuous chain of four carbon atoms and one carbon atom that forms a side chain. It is irrelevant whether the angle between two C—C bonds is shown as 90° or as 180°, because structural formulas are meant not to indicate a molecule's three-dimensional geometry but only how the atoms are connected.

The number of structural isomers increases rapidly as the number of carbon atoms increases. There are 3 structural isomers of pentane, 5 of hexane, 9 of heptane, C_7H_{16}, and 75 of decane, $C_{10}H_{22}$.

Naming Alkanes

In the days when relatively few organic compounds were known, the names of these compounds were chosen in a variety of ways. For example, acetic acid, CH_3CO_2H, was named for the Latin word *acetum,* meaning ''sour wine'' or ''vinegar.'' Urea, $(NH_2)_2CO$, was named because it was first isolated from urine. With the enormous and ever-increasing number of organic compounds, **nomenclature**—a systematic method of naming—became essential. What names should be given to the nine isomers of heptane, for example? The system that was first used to distinguish among the three pentanes by using the prefixes *iso-* and *neo-* cannot easily be extended to the heptanes, as this would require eight different and arbitrary prefixes. The nomenclature used today was devised by IUPAC and is called the IUPAC system.

In this system the original names of the first four alkanes have been retained: methane, ethane, propane, and butane. All the other continuous-chain hydrocarbons are named from Greek numbers according to the number of carbon atoms: pentane, hexane, heptane, and so on. Table 8.1 gives the names, structural formulas, boiling points, and melting points of the first 10 continuous-chain alkanes. The names of other alkanes are based on the names of the continuous-chain alkanes, using the following rules:

1. The name is based on the longest continuous chain of carbon atoms. Single-carbon atoms or shorter chains of carbon atoms may be attached to this longest chain. They are called **substituent groups**, because they substitute for hydrogen atoms of the continuous-chain alkane. The alkane

$$\begin{array}{llll} & & CH_3 & \\ & & | & \\ & CH_3 & CH_2 & \\ & | & | & \\ CH_3- & CH- & CH- & CH_3 \end{array}$$

is named as a substituted pentane and not a substituted butane, because the longest continuous chain has five carbon atoms. It is usually convenient to draw the longest continuous chain in a straight line. But as in the case of the alkane just shown, this is not always done, because structural formulas are not meant to indicate bond angles.

2. The substituent groups attached to the main chain of an alkane are called **alkyl groups**. Each substituent group can be thought of as derived from an alkane by the removal of a hydrogen atom. Their names are obtained by replacing the ending *-ane* of the alkane by *-yl*. The simplest examples are methyl, $—CH_3$, after methane and ethyl, $—C_2H_5$, after ethane.
3. The main chain is numbered from one end, and the substituent groups are assigned numbers corresponding to their position on the chain. The direction of numbering is chosen so as to give the lowest numbers to the side-chain substituents. When there are two or more substituents of the same type, the number is denoted by the prefixes *di-*, *tri-*, *tetra-*, and so on. Thus, the alkane illustrated in rule 1 is 2,3-dimethylpentane, not 3,4-dimethylpentane:

$$_1CH_3—{}_2CH(CH_3)—{}_3CH({}_4CH_2—{}_5CH_3)—CH_3 \quad \textit{not} \quad {}_5CH_3—{}_4CH(CH_3)—{}_3CH({}_2CH_2—{}_1CH_3)—CH_3$$

2,3-Dimethylpentane 3,4-Dimethylpentane

When two different substituents are present on the same carbon atom, they are cited in alphabetical order:

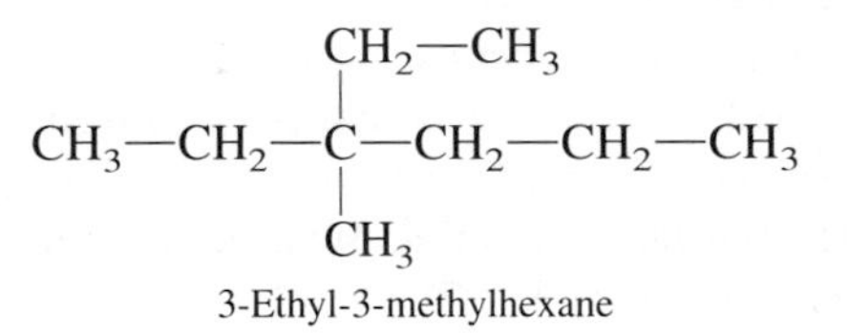

3-Ethyl-3-methylhexane

When there is more than one of the same type of substituent attached to the same carbon atom, we repeat the number of this carbon atom as many times as there are attached groups:

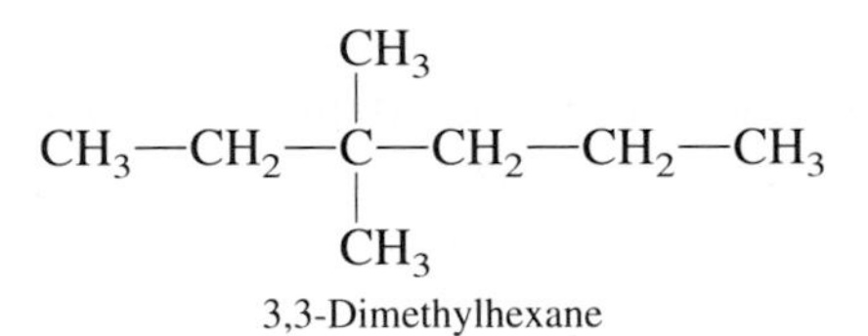

3,3-Dimethylhexane

EXAMPLE 8.2 Naming Alkanes

What are the systematic names of the following alkanes?

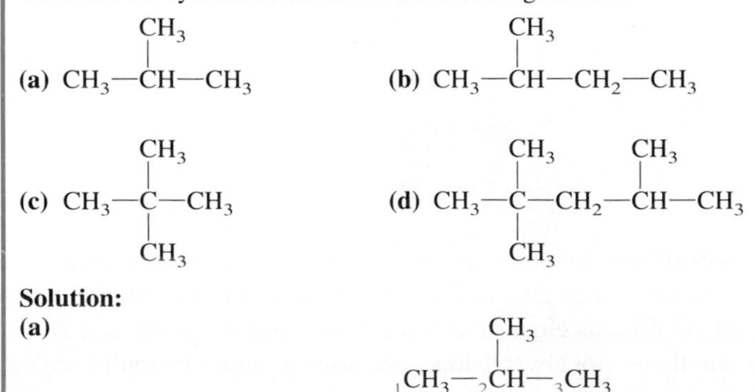

Solution:

(a)

We previously called this compound isobutane. But according to the systematic rules, it is named as a derivative of propane because the longest continuous chain has three carbon atoms. It is 2-methylpropane, or simply methylpropane because the 2 is redundant in this case (both 1- and 3-methylpropane are butane).

(b)

$$\begin{array}{c} \quad CH_3 \\ \quad | \\ {}_1CH_3-{}_2CH-{}_3CH_2-{}_4CH_3 \end{array}$$

This alkane is named as a derivative of butane and numbered from the left, so the substituent $-CH_3$ group is at carbon atom 2. It is 2-methylbutane.

▌If the carbon chain were numbered from right to left here, the incorrect name 3-methylbutane would result.

(c)

$$\begin{array}{c} CH_3 \\ | \\ {}_1CH_3-{}_2C-{}_3CH_3 \\ | \\ CH_3 \end{array}$$

This alkane is a derivative of propane and has two methyl substituents at carbon atom 2. It is named 2,2-dimethylpropane or simply dimethylpropane—the numbers are redundant in this case. Its common name is neopentane.

(d)

$$\begin{array}{c} \quad CH_3 \qquad\qquad CH_3 \\ \quad | \qquad\qquad\quad | \\ {}_1CH_3-{}_2C-{}_3CH_2-{}_4CH-{}_5CH_3 \\ \quad | \\ \quad CH_3 \end{array}$$

Whichever end of the chain we number from, there are substituents at carbon atoms 2 and 4. We choose the numbering that gives the lowest number to the most *highly substituted* carbon—that is, the carbon with the greatest number of substituent groups. This is carbon atom 2. The compound is 2,2,4-trimethylpentane.

Exercise 8.4 What is the IUPAC name of the following alkane?

$$\begin{array}{c} CH_2-CH_3 \qquad\qquad\qquad\qquad CH_3 \\ | \qquad\qquad\qquad\qquad\qquad | \\ CH_3-CH-CH_2-CH_2-CH_2-CH-CH_3 \end{array}$$

EXAMPLE 8.3 Drawing Structures of Alkanes

Draw the structures of the following alkanes.

(a) hexane **(b)** 2-methylpentane **(c)** 3-methylpentane
(d) 2,2-dimethylbutane **(e)** 2,3-dimethylbutane

Solution:

(a) According to systematic nomenclature, hexane has a six-carbon continuous chain:

$$CH_3-CH_2-CH_2-CH_2-CH_2-CH_3$$

(b) The longest continuous chain of 2-methylpentane contains five C atoms, and there is a methyl ($-CH_3$) substituent at C_2:

$$\begin{array}{c} CH_3-CH-CH_2-CH_2-CH_3 \\ | \\ CH_3 \end{array}$$

(c) The longest continuous chain of 3-methylpentane contains five C atoms, and there is a $-CH_3$ substituent at C_3:

$$\begin{array}{c} CH_3-CH_2-CH-CH_2-CH_3 \\ | \\ CH_3 \end{array}$$

(d) The longest continuous chain of 2,2-dimethylbutane contains four C atoms, and there are two methyl substituents at C_2:

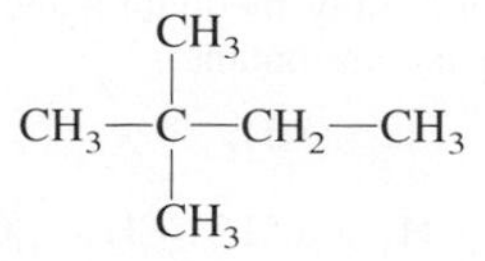

(e) The longest continuous chain of 2,3-dimethylbutane contains four C atoms, and there are methyl substituents at C_2 and C_3:

$$\begin{array}{c} \quad CH_3 \;\; CH_3 \\ \quad | \qquad | \\ CH_3—CH—CH—CH_3 \end{array}$$

These five molecules are the structural isomers of hexane, C_6H_{12}.

Exercise 8.5 Draw the structures of the following alkanes.

(a) 3-methylhexane **(b)** 3,4-dimethyloctane

(c) 2,2,4-trimethylpentane

Conformations

Although we know that the four bonds around each carbon atom in an alkane have a tetrahedral arrangement, this does not tell us the overall shape of the molecule. Because rotation can occur around single bonds, a molecule such as butane can twist into a large number of shapes called **conformations** (Figure 8.12). In a sample of butane, the molecules continually change shape, so the sample will include molecules of many different shapes.

In ethane, two of the infinite number of conformations are given special names. The conformation in which the CH bonds on one carbon atom are directly opposite those on the other carbon atom is called the **eclipsed conformation** (Figures 8.13a, 8.13b, 8.13c). The conformation in which the CH bonds on one carbon atom are rotated through 60° with respect to the CH bonds on the other carbon atom is called the **staggered conformation** (Figures 8.13d, 8.13e, 8.13f). Because it minimizes the interaction between the CH bonds, the staggered conformation has a slightly lower energy than the other conformations, whereas the eclipsed conformation has a slightly higher energy than all other conformations. Consequently, any sample of ethane will have more molecules with conformations closer to the staggered conformation than to the eclipsed conformation.

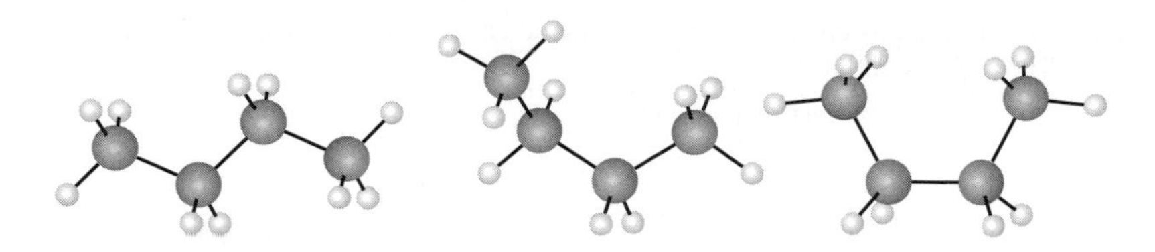

FIGURE 8.12 Some Possible Conformations of Butane Free rotation about single bonds enables a molecule such as butane to twist into many different shapes.

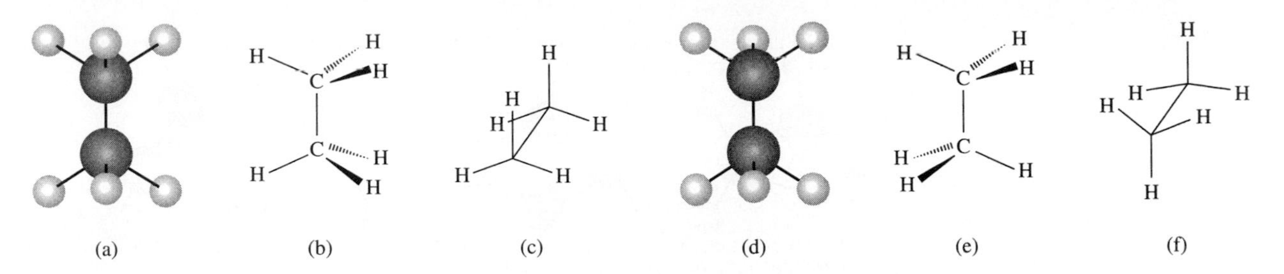

FIGURE 8.13 Eclipsed and Staggered Conformations of Ethane (a), (b), and (c) are representations of the eclipsed conformation; (d), (e), and (f) are representations of the staggered conformation. In (c) and (f) the carbon atoms are not shown specifically; they are represented by the junction of the four bonds.

Reactions of Alkanes

Alkanes are rather unreactive compounds. At ordinary temperatures they do not react with oxidizing agents, such as oxygen, chlorine, bromine, or sulfuric acid, or with reducing agents, such as hydrogen. An important reason for this lack of reactivity is that both the C—H and C—C bonds are strong and therefore not easily broken, as we shall see in Chapter 9. The alkanes are not bases because, unlike nitrogen in ammonia, for example, the carbon atoms have no unshared pairs of electrons to which a hydrogen ion, H^+, can be added. Nor are they acids, because carbon and hydrogen have very similar electronegativities, and so the C—H bonds have very little polarity.

The most important reaction of alkanes is with oxygen at high temperatures. In excess oxygen, complete combustion gives carbon dioxide and water; for example,

$$C_3H_8(g) + 5O_2(g) \rightarrow 3CO_2(g) + 4H_2O(l)$$

Combustion of an alkane is strongly exothermic, which is why we use alkanes as fuels. With insufficient oxygen, carbon monoxide and carbon black (soot) are produced.

Alkanes also react with the halogens at high temperatures or in a photochemical reaction. In the products of such reactions, one or more of the hydrogen atoms are replaced by halogen atoms:

$$CH_4 + Cl_2 \rightarrow \underset{\text{Chloromethane}}{CH_3Cl} + HCl$$

$$CH_4 + 2Cl_2 \rightarrow \underset{\text{Dichloromethane}}{CH_2Cl_2} + 2HCl$$

Cycloalkanes

If the two ends of a continuous-chain alkane are joined into a ring, a **cycloalkane** is formed. The simplest cycloalkanes are as follows:

	Melting Point (°C)	Boiling Point (°C)
Cyclopropane C_3H_6	−127	−33
Cyclobutane C_4H_8	−80	13
Cyclopentane C_5H_{10}	−94	49
Cyclohexane C_6H_{12}	7	81

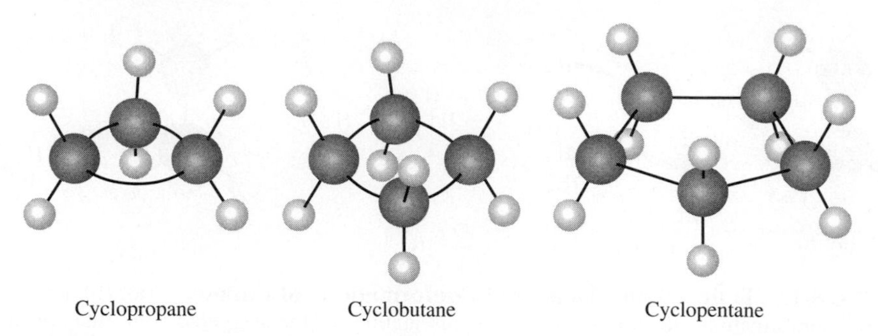

FIGURE 8.14 Structures of Cycloalkanes: Bent Bonds The angle between four single bonds at a carbon atom is approximately 109°C, but the angle of the equilateral triangular shape of the cyclopropane molecule is only 60°. Thus the C—C bonds are described as ''bent.'' The amount of bond bending is less in cyclobutane and is negligible in cyclopentane because the angle of a regular pentagon is 108°, which is very close to the tetrahedral angle of 109°.

An unusual feature of the cyclopropane and cyclobutane molecules is that the angles between the C—C bonds are compressed to 60° and 90°, which are considerably smaller than the tetrahedral angle of 109°. If an approximately tetrahedral angle is maintained between the electron pairs surrounding each carbon atom, the bonds must be regarded as ''bent,'' as shown by ball-and-stick models (Figure 8.14). This unusual feature of the structures of cyclopropane and cyclobutane causes them to be more reactive than the other cycloalkanes. Cyclopentane has an almost planar shape with C—C—C bond angles of 108°. But because the angle in a planar hexagon is 120°, cyclohexane has a puckered nonplanar shape in which bond angles close to the tetrahedral angle are maintained (Figure 8.15). Two of the many possible conformations are the ''boat'' conformation and the ''chair'' conformation; the boat conformation is the less stable of the two.

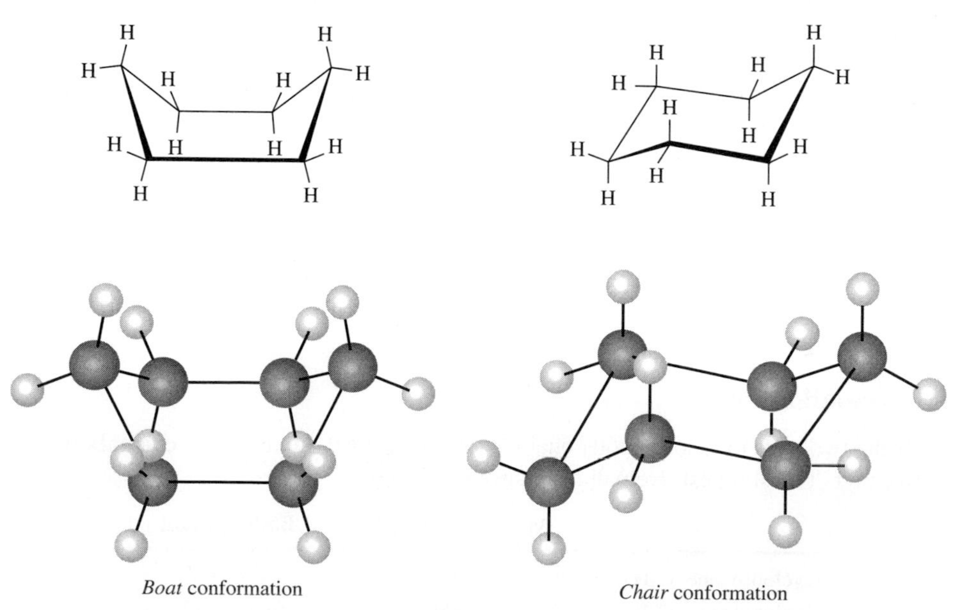

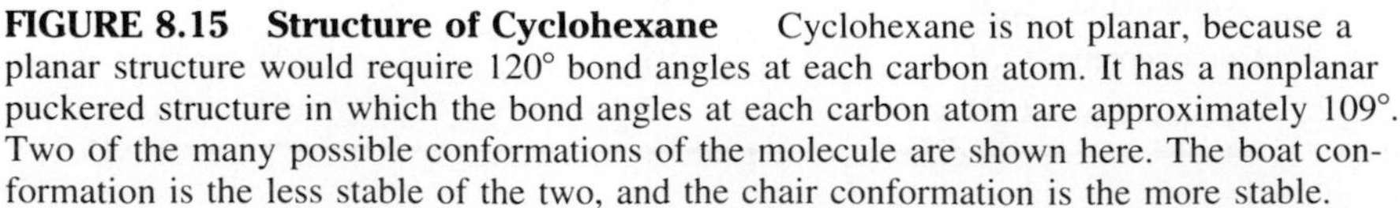
FIGURE 8.15 Structure of Cyclohexane Cyclohexane is not planar, because a planar structure would require 120° bond angles at each carbon atom. It has a nonplanar puckered structure in which the bond angles at each carbon atom are approximately 109°. Two of the many possible conformations of the molecule are shown here. The boat conformation is the less stable of the two, and the chair conformation is the more stable.

Rings, particularly five- and six-membered rings, are very common in organic compounds. To facilitate drawing the structures of cyclic molecules, especially those containing several rings, the rings are represented simply by polygons—a triangle for cyclopropane, a square for cyclobutane, and so on. The carbon and hydrogen atoms are not specifically shown. Each corner of the polygon represents a CH_2 group except where a side chain is attached, and then it represents a CH group. For example, instead of drawing the structure of methyl cyclohexane as

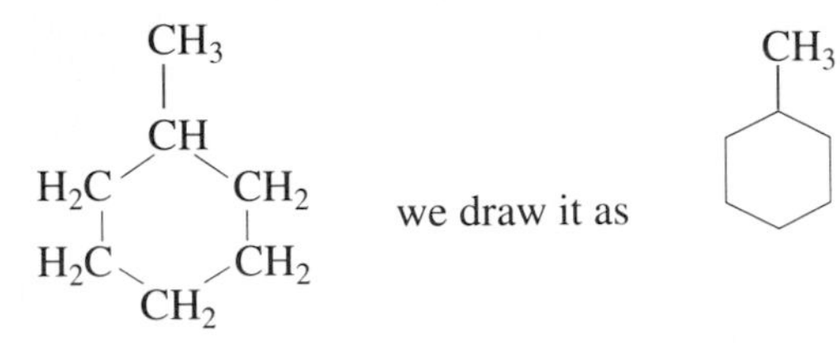

|||Carbon has a valence of four. By noting how many bonds a given carbon atom forms to other carbons, you can deduce how many hydrogen atoms are bonded to that carbon in these simplified structures.

Petroleum and Natural Gas

Ancient deposits of dead marine organisms that were buried under subsequent deposits (and were thereby protected from oxidation) and then subjected to high temperatures and pressures for millions of years have been transformed into a complex mixture of alkanes and other hydrocarbons. The more volatile alkanes in this mixture constitute **natural gas**, and the liquid mixture of all the other alkanes is called **petroleum** or, when it comes straight from the well, **crude oil**.

Natural gas consists mainly of methane and small amounts of ethane, propane, and butane. By cooling and compressing natural gas, propane and butane can be liquefied and separated from the gas. They are sold in liquid form as ''bottled gas.'' Natural gas also contains small amounts of helium formed by the radioactive decay of heavy elements such as uranium (Chapter 18).

Petroleum distillation towers

Petroleum is separated by fractional distillation (Figure 8.16) into fractions, such as gasoline and kerosene, that are mixtures of hydrocarbons of similar boiling

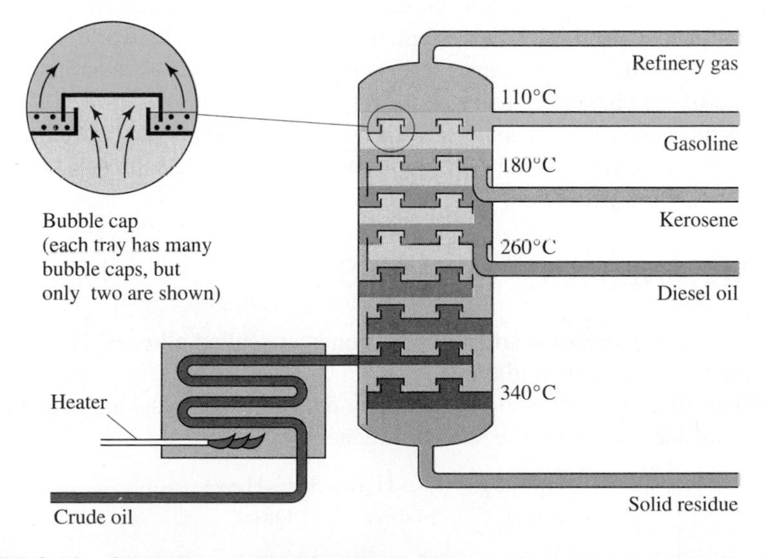

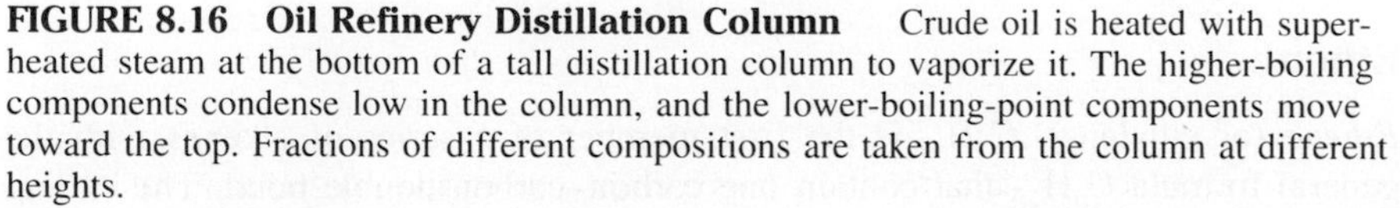

FIGURE 8.16 Oil Refinery Distillation Column Crude oil is heated with superheated steam at the bottom of a tall distillation column to vaporize it. The higher-boiling components condense low in the column, and the lower-boiling-point components move toward the top. Fractions of different compositions are taken from the column at different heights.

TABLE 8.2 Typical Fractions Obtained in the Distillation of Crude Oil

Fraction	Boiling Point (°C)	Composition	Uses
Gas	Up to 20	Alkanes from CH_4 to C_4H_{10}	Synthesis of other carbon compounds; fuel
Petroleum ether	20–70	C_5H_{12}, C_6H_{14}	Solvent; gasoline additive for cold weather
Gasoline	70–180	Alkanes from C_6H_{14} to $C_{10}H_{22}$	Gasoline
Kerosene	180–230	$C_{11}H_{24}$, $C_{12}H_{26}$	Jet engine fuel
Light gas oil	230–305	$C_{13}H_{28}$ to $C_{17}H_{36}$	Heating oil and fuel for diesel engines
Heavy gas oil and light lubricating distillate	305–405	$C_{18}H_{38}$ to $C_{25}H_{52}$	Fuel for generating stations; lubricating oil
Lubricants	405–515	Higher-*n* alkanes	Thick oils, greases, and waxy solids; lubricating grease; petroleum jelly
Solid residue			Pitch or asphalt for roofing and road material

points. Typical fractions are shown in Table 8.2. To meet the high demand for gasoline, a large part of the higher-boiling-point fractions is decomposed by heating with a catalyst to form the shorter-chain alkanes of gasoline. These reactions, called **cracking reactions**, are carried out in reaction vessels called *catalytic converters.*

The industrialized world has come to rely heavily on petroleum as an energy source for the internal combustion engine (Box 8.2) used in cars and other forms of transportation and for generating electricity. Hydrocarbons are also the raw materials for manufacturing many substances, such as plastics, detergents, and drugs, that we have come to think of as essential in modern life. The world's petroleum sources are limited, however, and are expected to last no more than another 100 years. One of the challenges facing humanity is to learn how to utilize efficiently and safely alternative energy sources, such as coal, nuclear power, and solar energy, and to find alternative raw materials from which to make all the substances now manufactured from petroleum.

8.4 ALKENES AND ALKYNES

Hydrocarbons with carbon–carbon double bonds are called **alkenes**. Hydrocarbons with triple bonds are called **alkynes**.

When alkanes are strongly heated, they decompose in cracking reactions into a simpler alkane and an alkene such as ethene, C_2H_4:

$$\underset{\text{Butane}}{C_4H_{10}(g)} \rightarrow \underset{\text{Ethene}}{C_2H_4(g)} + \underset{\text{Ethane}}{C_2H_6(g)}$$

Ethene

Both alkenes and cycloalkanes have the general formula C_nH_{2n}.

Ethene (or ethylene), C_2H_4, is the first member of a series of alkenes with the general formula C_nH_{2n} that contain one carbon–carbon double bond. The ending

BOX 8.2 The Internal Combustion Engine

The old-fashioned steam engine is an *external combustion engine:* The fuel is burned outside the engine, and the heat produced is used to convert water into steam. A wide variety of fuels could be used, of which coal was the most common. But steam engines are inconveniently heavy and require a driving fluid (water) that must be carried. For all forms of transportation, from private automobiles to trains and airplanes, the steam engine has been replaced by the *internal combustion engine,* in which the fuel is burned inside the engine. There are two important types of internal combustion engines, the gasoline engine and the diesel engine.

In the internal combustion engine, a hydrocarbon fuel is burned in air and yields mainly carbon dioxide and water. For example, for the representative alkane nonane, the principal reaction is

$$C_9H_{20}(l) + 14O_2(g) \rightarrow 9CO_2(g) + 10H_2O(g)$$

Some oxides of nitrogen are also formed by reactions that involve nitrogen in the air. These are a source of atmospheric pollution.

In the operation of the gasoline engine, a mixture of gasoline and air is sucked into a cylinder as a piston descends, thereby creating a partial vacuum by increasing the cylinder volume (Figure A). The valve on the cylinder then closes and the piston returns, compressing the air–fuel mixture. An electric spark is then passed through the compressed air–fuel mixture to ignite it. The number of moles of gaseous products exceeds the number of moles of reactants, and the reaction products are much hotter than the reactants because the reaction is exothermic. Therefore, the gas pressure on the piston increases, forcing it to descend and thus delivering power to the engine.

The burning of the air–fuel mixture in the gasoline engine must occur at a rate that delivers a smooth thrust to the descending piston. Too rapid a reaction causes a distinct explosive noise known as engine knocking or pinging, and some of the power is wasted. Most of the hydrocarbons obtained by distillation of petroleum are unbranched alkanes, which tend to explode too rapidly and cause knocking. The highly branched alkane isooctane (2,2,4-trimethylpentane) causes little knocking and is arbitrarily given an octane rating of 100:

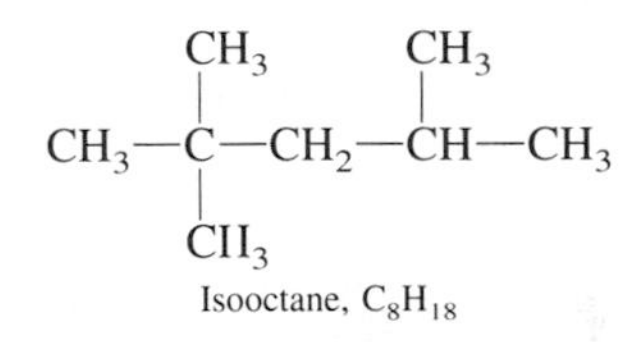

Isooctane, C_8H_{18}

At the other end of the scale is heptane, which causes considerable knocking and is given an octane rating of 0. The octane rating of any fuel is then established by determining the ratio of isooctane and heptane needed to produce the same amount of knocking. For example, a 50-50 mixture has an octane rating of 50. The octane number of gasoline is increased by heating it and passing it over a catalyst (catalytic reforming) to convert some of the less-branched alkanes to more highly branched isomers. For example,

$$CH_3-CH_2-CH_2-CH_2-CH_3 \xrightarrow{\text{catalyst, heat}} CH_3-\underset{}{\overset{CH_3}{\overset{|}{C}H}}-CH_3-CH_3$$

A diesel engine is similar to a gasoline engine, except that the air–fuel mixture is more highly compressed so that its temperature rises sufficiently to ignite the mixture without the use of a spark. Thus, unlike the more familiar gasoline engine, the diesel engine has no spark plugs.

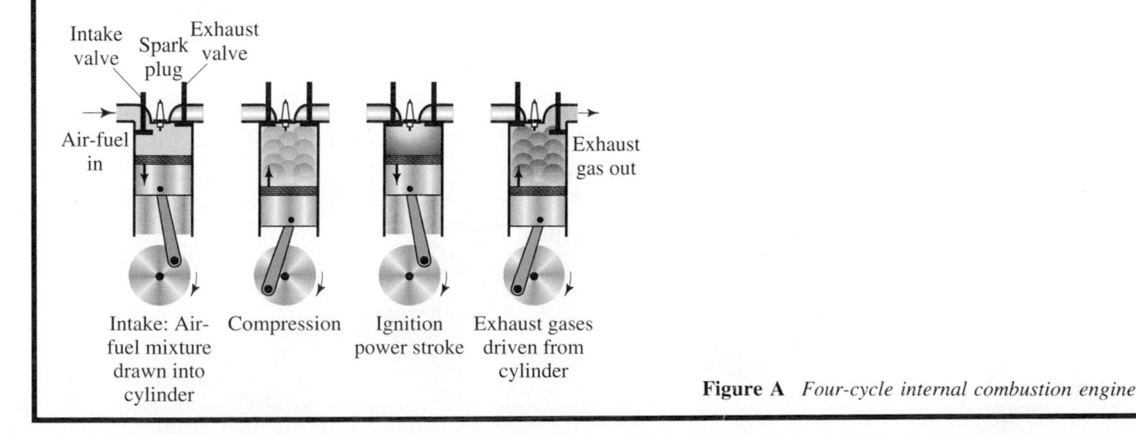

Figure A *Four-cycle internal combustion engine*

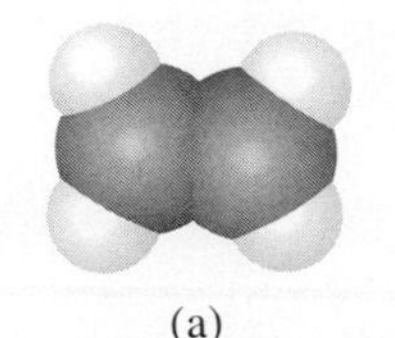

(a)

(b)

(c)

FIGURE 8.17 Structure of the Ethene Molecule (a) Space-filling model showing the planar structure in which both carbon atoms have a planar AX_3 geometry; (b) balloon model of the electron-pair domains; (c) ball-and-stick model with two bent bonds

-ene signifies the presence of a carbon–carbon double bond, C═C, in a hydrocarbon.

In the ethene molecule the two carbon atoms share two pairs of electrons; that is, they are joined by a double bond:

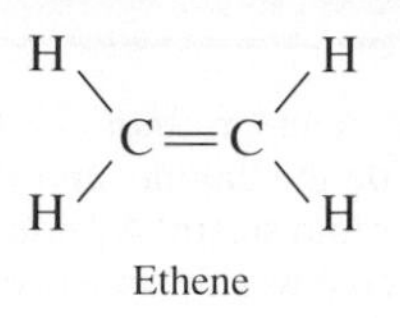

Ethene

As we discussed in Chapter 3, the valence shell of each carbon atom in ethene contains a double bond and two single bonds that have a planar AX_3 arrangement with approximately 120° bond angles (Figures 3.24a, 8.17a, and 8.17b). We saw in Chapter 6 (Figure 6.30) that the bonds at each carbon atom of ethene may be described as being formed by three sp^2 hybrid orbitals and a single electron in a 2*p* orbital on each carbon atom, an arrangement that gives two C—H σ bonds, one C—C σ bond, and one C—C π bond. An alternative model describes the double bond in terms of two bent bonds (Figure 6.32). Both models show that there is a planar geometry at each carbon atom and that the whole molecule is planar. In contrast to alkanes, in which free rotation can occur around a C—C single bond, no rotation is possible around a double bond. Thus, the four attached atoms—the four hydrogen atoms in ethene—always lie in the same plane. To get free rotation around a double bond, one of the two bonds would have to be broken. In the σ–π model the π bond must be broken, because the 2*p* orbitals overlap only when they are in the same plane. In the bent-bond model one of the bent bonds must be broken, as we can easily see with a ball-and-stick model (Figure 8.17c).

Other Alkenes

The names and structures of some simple alkenes are given in Table 8.3. Figure 8.18 shows two models of **propene** (or propylene), $CH_3CH{=}CH_2$, which, like ethene, is a planar molecule (except for the hydrogens in the CH_3 group). The main use of propene is in the production of **polypropylene**, a polymer used, for example, to manufacture synthetic fibers for ropes and carpets (Chapter 20).

Naming Alkenes

According to the IUPAC system, alkenes are named from the longest carbon chain that contains the double bond, and the atoms are numbered from the end nearer the double bond. For example,

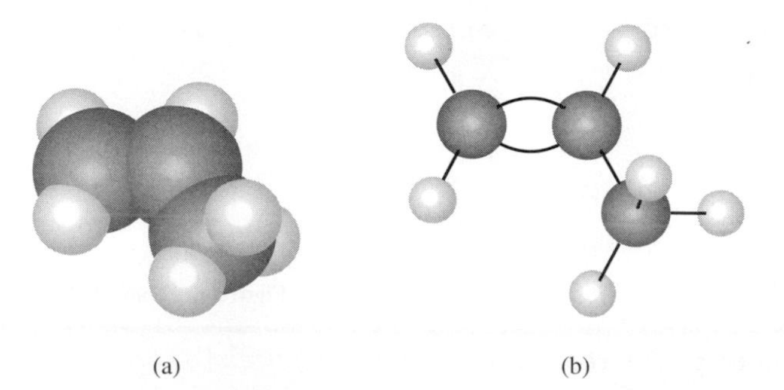

(a) (b)

FIGURE 8.18 Structure of Propene (a) Space-filling model; (b) ball-and-stick model

TABLE 8.3 Some Simple Alkenes

Structure	Name	Boiling Point (°C)
$CH_2{=}CH_2$	Ethene (ethylene)	-104
$CH_3CH{=}CH_2$	Propene (propylene)	-47
$CH_3CH_2CH{=}CH_2$	1-Butene	-6
CH_3 H \ / C=C / \ H CH_3	*trans*-2-Butene	1
CH_3 CH_3 \ / C=C / \ H H	*cis*-2-Butene	4
$CH_3CH_2CH_2CH{=}CH_2$	1-Pentene	30
CH_3CH_2 H \ / C=C / \ H CH_3	*trans*-2-Pentene	36
CH_3CH_2 CH_3 \ / C=C / \ H H	*cis*-2-Pentene	37

$$\underset{\text{1-Butene}}{CH_3CH_2CH{=}CH_2} \quad \textit{and} \quad \underset{\text{2-Butene}}{CH_3CH{=}CHCH_3}$$

which are structural isomers. The alkene 2-butene exhibits another type of isomerism that results from the lack of free rotation around the double bond. In the *cis* isomer the two methyl groups are on the same side of the double bond. In the *trans* isomer the two methyl groups are on opposite sides of the double bond (Figure 8.19):

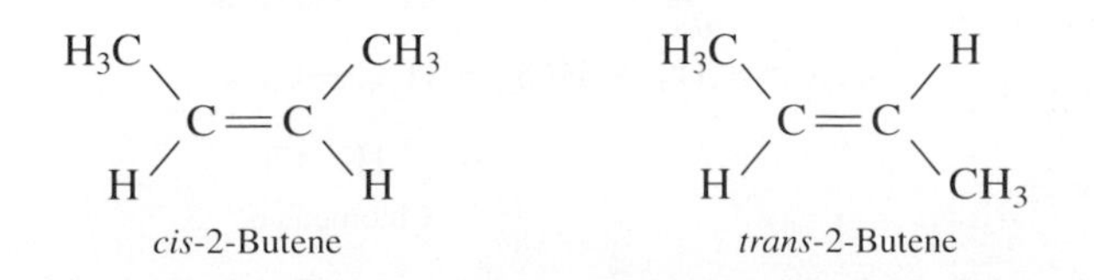

The two 2-butenes are called **geometric isomers** or ***cis-trans*** **isomers**.

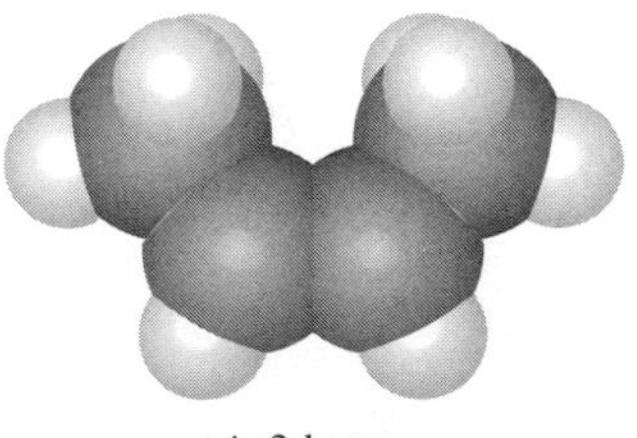

cis-2-butene

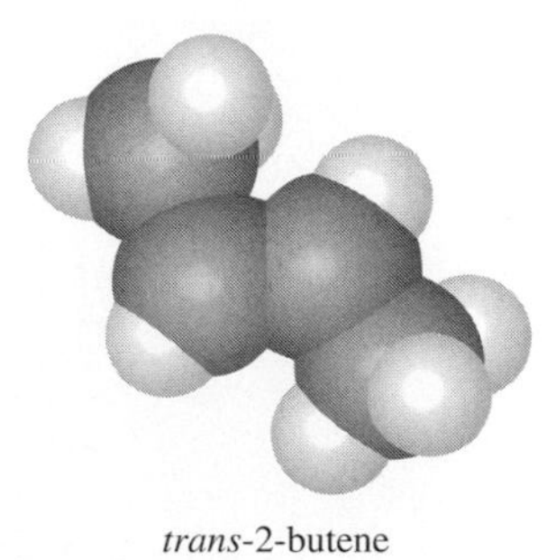

trans-2-butene

FIGURE 8.19 The Geometric Isomers of 2-Butene In *cis*-2-butene the two methyl groups are on the same side of the bond. In *trans*-2-butene the two methyl groups are on opposite sides of the double bond.

There is yet another isomer with the formula C_4H_8. It has the structure

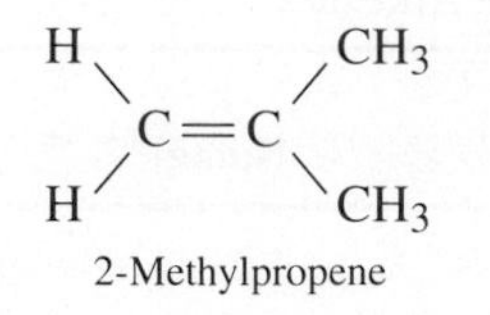

2-Methylpropene

and is called 2-methylpropene (or methylpropene) because the longest carbon chain has only three carbon atoms. Like 1-butene, it is a structural isomer but not a geometric isomer. Thus, there are a total of four alkenes with the formula C_4H_8: 1-butene, *cis*-2-butene, *trans*-2-butene, and 2-methylpropene.

If an alkene has more than one double bond, the endings *-diene, -triene,* and so on are used in their names. The names ethylene for C_2H_4 and propylene for C_3H_6 that were in use long before the IUPAC system was adopted are still widely used, as in polyethylene and polypropylene.

Reactions of Alkenes

Whereas alkanes are relatively unreactive, alkenes react with a variety of substances. In particular, they undergo many **addition reactions**. In this type of reaction, a molecule X—Y, where X and Y are single atoms or groups of atoms, is added to the alkene. X and Y become attached to the carbon atoms at each end of the double bond, and the double bond is converted to a single bond:

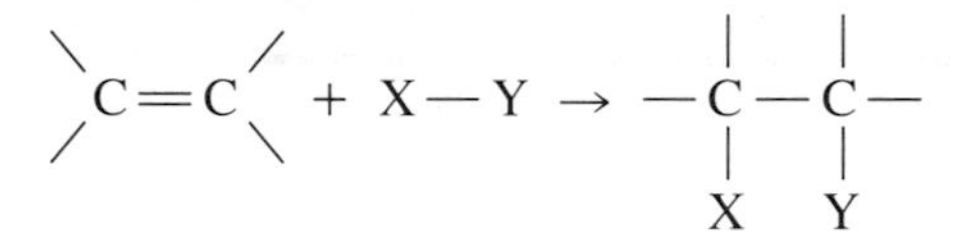

For example, in the presence of suitable catalysts, such as nickel or platinum, hydrogen adds to the C=C double bond in ethene, converting it to a single bond in ethane:

$$H_2C{=}CH_2 + H_2 \rightarrow H_3C{-}CH_3$$

Ethene Ethane

The addition of hydrogen to a double bond is called **hydrogenation**. Alkenes contain less hydrogen than do the corresponding alkanes and are known as **unsaturated hydrocarbons**. By the addition of hydrogen they can be converted to alkanes, which are known as **saturated hydrocarbons**, because they cannot add more hydrogen.

The halogens and the hydrogen halides can also add to a C=C double bond. For example,

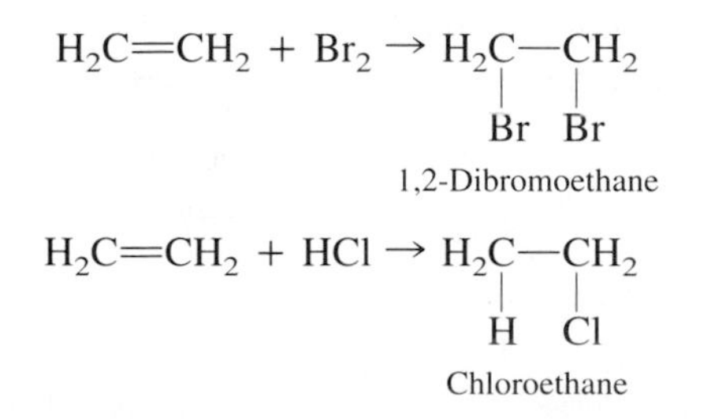

The reaction with bromine provides a convenient test for the presence of a carbon–carbon double bond in a molecule (Demonstration 8.3).

DEMONSTRATION 8.3 Testing for Unsaturated Hydrocarbons with Bromine

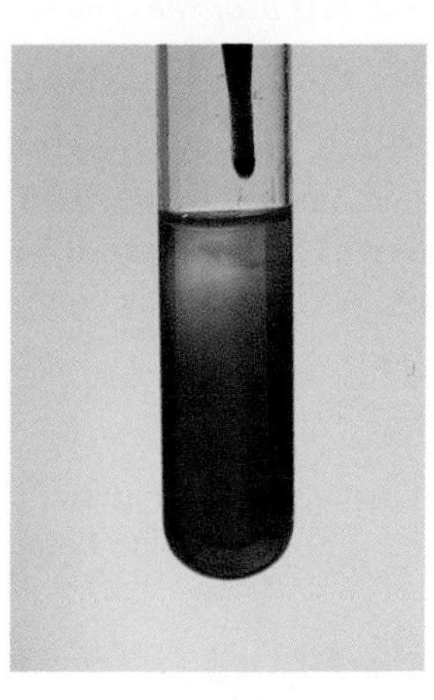

The brown red color of bromine remains when drops of bromine are mixed with the saturated hydrocarbon cyclohexane. The bromine dissolves in cyclohexane but does not react with it.

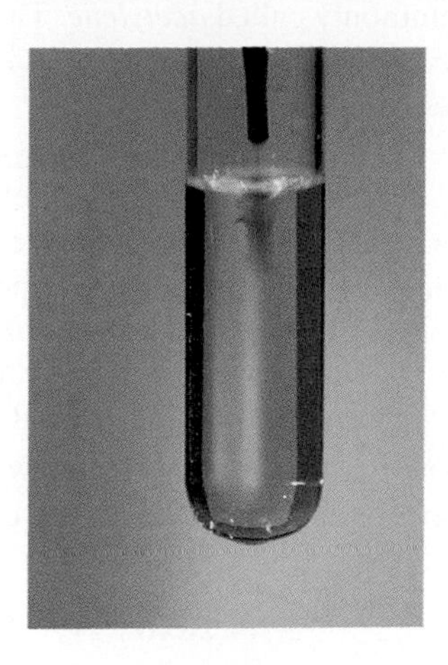

The brown-red color of bromine disappears when drops of bromine are mixed with the unsaturated hydrocarbon cyclohexene. The bromine adds to the double bond in cyclohexene. This reaction makes a convenient test for the presence of a double bond in a molecule.

Although water does not react with ethene under ordinary conditions, at a high temperature in the presence of a strong acid catalyst, such as sulfuric acid, water adds to the double bond to give ethanol;

$$H_2C{=}CH_2 + H_2O \rightarrow \underset{\text{H}}{H_2\underset{|}{C}}{-}\underset{\text{OH}}{\underset{|}{CH_2}}$$

Ethanol

This is an important reaction for the industrial preparation of ethanol (Chapter 14).

Like all hydrocarbons, ethene burns in air or oxygen:

$$C_2H_4 + 3O_2 \rightarrow 2CO_2 + 2H_2O$$

Ethene is used to synthesize ethanol and many other substances and is thus an important product of the chemical industry. Almost half the ethene produced is polymerized to obtain **polyethylene**, $-\!(CH_2{-}CH_2)\!-_n$. Recall from Section 7.1 that in a polymerization reaction, a large number of small molecules combine to form a very large molecule. In the formation of polyethylene, one of the two bonds in the double bond breaks, and new bonds form and join the individual molecules:

$$H_2C{=}CH_2 + H_2C{=}CH_2 + H_2C{=}CH_2 + H_2C{=}CH_2 + H_2C{=}CH_2$$
$$\rightarrow -CH_2{-}CH_2{-}CH_2{-}CH_2{-}CH_2{-}CH_2{-}CH_2{-}CH_2{-}CH_2{-}CH_2{-}$$

Thousands of ethene molecules are joined in this way to form one molecule of polyethylene. Polymers are discussed in more detail in Chapter 20.

Alkynes

Hydrocarbons that contain a carbon–carbon triple bond are called **alkynes**. They have the general formula C_nH_{2n-2}. The simplest alkyne is **ethyne**, HC≡CH, which is commonly called *acetylene*. The alkyne CH_3—C≡CH is *propyne*. The names for larger alkynes (alkynes with $n > 3$) are obtained by using the same numbering system as that for alkenes.

As Figure 8.20 shows, ethyne is a linear molecule in which both carbon atoms have a linear AX_2 geometry. We saw in Chapter 6 that, in terms of the σ–π model, there are two C—H σ bonds, one C—C σ bond, and two C—C π bonds. We saw in Section 8.2 that ethyne can be prepared in the laboratory by the reaction of water with calcium carbide:

$$CaC_2(s) + 2H_2O(l) \rightarrow Ca(OH)_2(aq) + C_2H_2(g)$$

Ethyne burns in air with a yellow, smoky flame that deposits soot (Figure 8.21), but if mixed with a large excess of oxygen, it burns with a very hot blue flame and is fully oxidized to CO_2:

$$2C_2H_2(g) + 5O_2(g) \rightarrow 4CO_2(g) + 2H_2O(g)$$

This very exothermic reaction is used in oxyacetylene torches for welding and cutting metals.

Ethyne is a rather reactive hydrocarbon that, like ethene, undergoes many addition reactions. For example, the halogens and the hydrogen halides readily add to the triple bond of ethyne and convert it to a double bond:

$$\underset{\text{Ethyne}}{\text{H—C≡C—H}} + Cl_2 \rightarrow \underset{\text{1,2-Dichloroethene}}{\text{ClHC=CHCl}}$$

$$\underset{\text{Ethyne}}{\text{H—C≡C—H}} + HCl \rightarrow \underset{\text{Chloroethene}}{H_2C\text{=CHCl}}$$

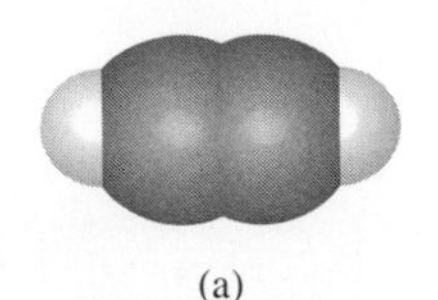

(a)

(b)

(c)

FIGURE 8.20 Structure of the Ethyne Molecule
(a) Space-filling model showing the linear structure, in which both carbon atoms have a linear AX_2 geometry; (b) balloon model of the electron-pair domains; (c) ball-and-stick model with three bent bonds

Exercise 8.6 Draw the structures of **(a)** *cis*-3-hexene, **(b)** *trans*-3-hexene, and **(c)** 2-butyne.

Exercise 8.7 Give the IUPAC names for the compounds that result from the addition of Cl_2 to **(a)** propene; **(b)** propyne.

8.5 BENZENE AND THE ARENES: AROMATIC HYDROCARBONS

Chemists have found it useful to divide all organic compounds into two broad classes: **aliphatic compounds** and **aromatic compounds**. The original meanings of ''aliphatic'' (fatty) and ''aromatic'' (fragrant) no longer have any significance. Aliphatic compounds are the alkanes, alkenes, and alkynes and all the compounds that can be derived from them by replacing the hydrogen atoms with other atoms or groups of atoms. Aromatic compounds include **benzene**, C_6H_6, and other hydrocarbons related to benzene—collectively called **arenes**—and all the compounds that

FIGURE 8.21 Preparation and Combustion of Ethyne Ethyne is being prepared by the reaction of calcium carbide, CaC_2, with water. It burns with a yellow, smoky flame and deposits carbon black (soot) on the surface of the evaporating dish.

may be derived from them. Some benzene is present in petroleum, but most is made by strongly heating cyclohexane in the presence of a catalyst:

$$C_6H_{12}(g) \xrightarrow{\text{heat, catalyst}} C_6H_6(g) + 3H_2(g)$$

Benzene is a colorless liquid that boils at 80°C and freezes at 5°C. It is less dense than and insoluble in water.

The Structure of Benzene

Benzene is a planar cyclic molecule with six carbon atoms arranged in a hexagonal ring. Each carbon atom is bonded to a hydrogen atom and two other carbon atoms and has a planar AX_3 geometry. However, each carbon atom must form four bonds, so it would appear that three of the carbon–carbon bonds must be double bonds. We can draw the following Lewis structure (structural formula):

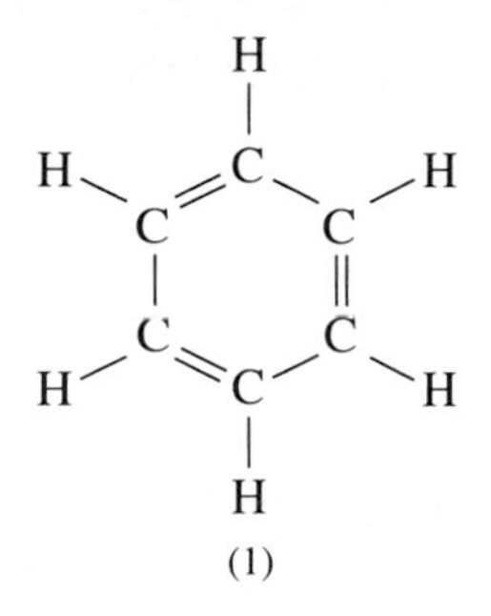

(1)

According to this structure, benzene is an alkene with three double bonds, that is, a cyclic triene. However, benzene does not behave like an alkene. In particular, it does not undergo the typical addition reactions of alkenes. Moreover, a C=C double bond (134 pm) is shorter than a C—C single bond (154 pm), whereas in benzene all the bonds have the same length, 140 pm. It appears therefore that structure (1) is not an adequate representation of benzene. How then should we represent the structure of benzene? Let us first draw a regular hexagonal framework for the molecule with single bonds:

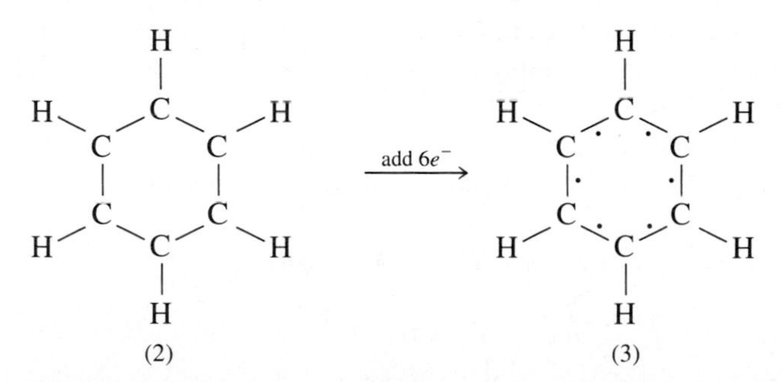

(2) (3)

Each carbon atom in structure (2) forms only three bonds and has only six electrons in its valence shell. We then have six remaining electrons to allocate, and they can be used to form three double bonds, as in structure (1). An alternative way to allocate these six electrons would be to place one electron between each pair of carbon atoms, as in structure (3). In such a structure all six bonds are equivalent and may be described as three-electron bonds or $1\frac{1}{2}$ bonds, intermediate between a two-electron single bond and a four-electron double bond. This description of the benzene molecule is consistent with the actual length of the bonds and with the observation that benzene does not appear to contain double bonds because it does not have the properties of an alkene.

The six electrons distributed around the ring in structure (3) are often referred to as **delocalized electrons**, because they are not localized in three pairs but are more evenly spread around the ring. The structure of benzene is therefore sometimes represented as

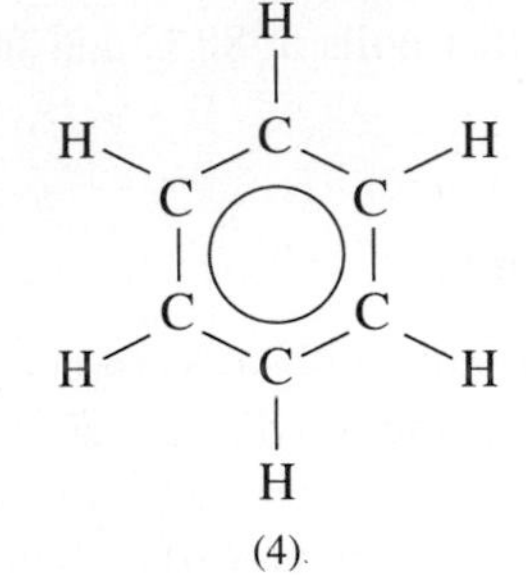

where the circle in the middle represents the six delocalized electrons. Such an arrangement of electrons is more stable than that represented by the Lewis structure (1) because the electrons are farther apart and their electrostatic repulsion is less than in structure (1).

Another way to represent the structure of benzene is as a combination of structures (1) and (5):

▮▮▮ Lewis structures depict bonds as single, double, or triple. Because the carbon–carbon bonds of benzene are neither single bonds nor double bonds, one Lewis structure is insufficient to represent benzene.

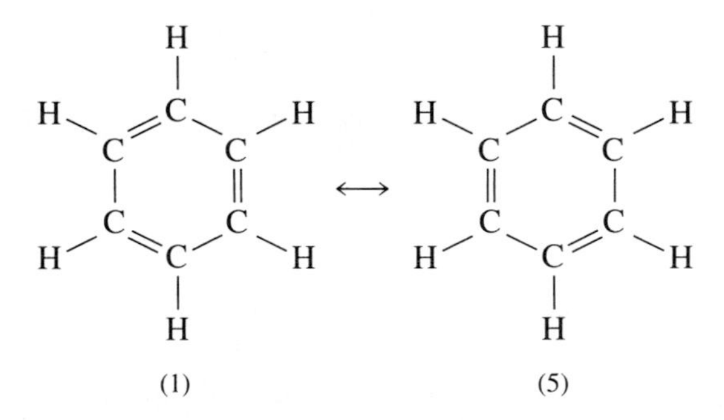

Any given carbon–carbon bond is a double bond in one structure and a single bond in the other structure. This represents the fact that in the real structure, each bond is intermediate between a single and a double bond. When Lewis structures are used in this way, they are called **resonance structures**, and a double-headed arrow is placed between them.

The double-headed arrow ⟷ should not be confused with the double arrow ⇌ that we use to denote an equilibrium. There are not two forms of the benzene molecule in equilibrium. There is only one benzene molecule, but its structure cannot be accurately represented by a single Lewis structure.

It is very convenient to use the abbreviated representations (6) and (7) for structures (1) and (5) and representation (8) for structure (4):

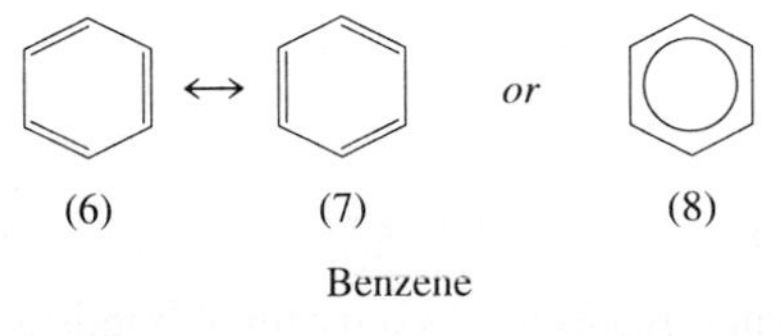

Either of the structures (6) and (7) is typically used alone to represent benzene, with the understanding that benzene behaves quite differently from a hexatriene. As we shall see, benzene and related molecules have their own characteristic properties. These are often referred to as **aromatic properties**, and the benzene ring is called an **aromatic ring**.

Reactions of Benzene

As we mentioned, benzene and other aromatic compounds do not undergo addition reactions as do alkenes. Instead, their characteristic reactions are **aromatic substitution reactions**, in which one of the hydrogen atoms on the ring is replaced by another atom or group of atoms, and the stable aromatic ring remains intact. For example, benzene reacts with nitric acid in the presence of concentrated sulfuric acid as a catalyst to give *nitrobenzene*, $C_6H_5NO_2(l)$. This reaction is called a **nitration reaction**:

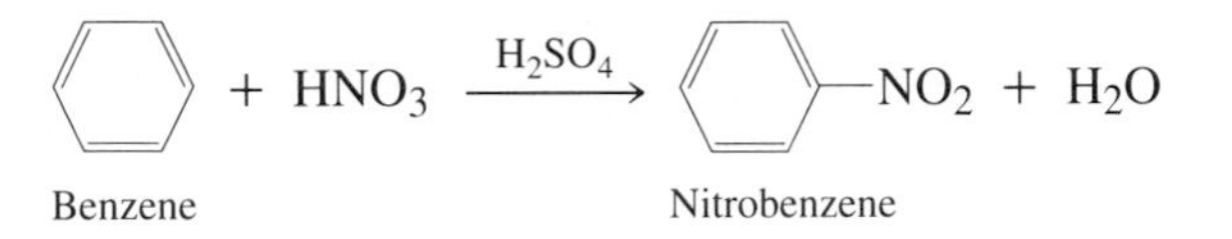

Benzene Nitrobenzene

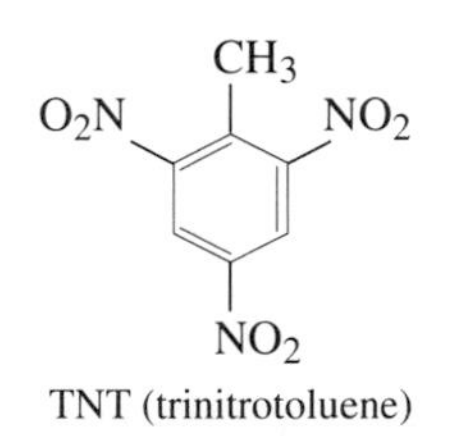

TNT (trinitrotoluene)

The nitration of benzene and other aromatic compounds is used in the preparation of explosives such as TNT (trinitrotoluene) and as a step in the manufacture of many pharmaceuticals and dyes.

When benzene reacts with the halogens, it does not undergo an addition reaction like an alkene does. Instead, one of the hydrogen atoms is replaced by a halogen atom in an aromatic substitution reaction called **halogenation**:

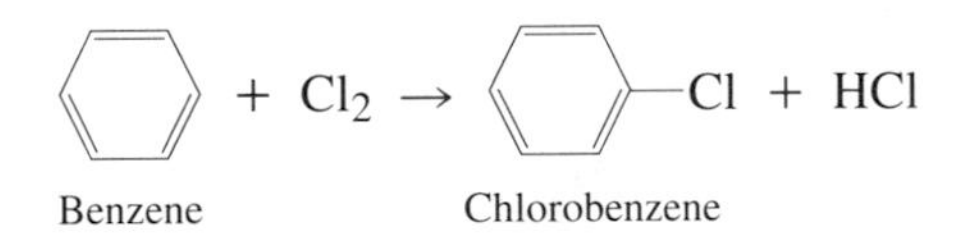

Benzene Chlorobenzene

Naming Aromatic Compounds

Alkyl-substituted benzenes have common names that we always use. For example, methylbenzene is called **toluene**. There are three dimethylbenzenes that are called **xylenes**. They are structural isomers that differ in the relative positions of the methyl groups. If we number the ring in accordance with IUPAC rules, then the methyl groups may be in the 1,2-, the 1,3-, or the 1,4- positions. They are called *ortho-*, *meta-*, and *para-*xylene, respectively:

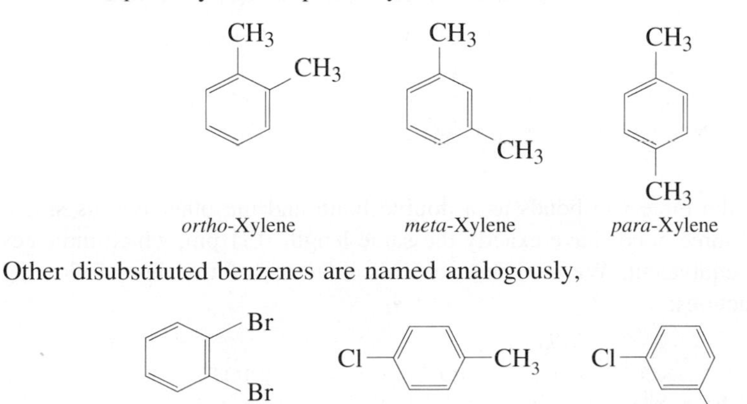

ortho-Xylene *meta*-Xylene *para*-Xylene

Other disubstituted benzenes are named analogously,

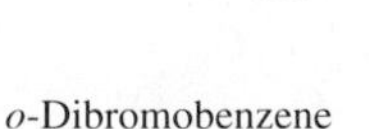

o-Dibromobenzene *p*-Chlorotoluene *m*-Chloronitrobenzene

where *o, m,* and *p* denote *ortho, meta,* and *para,* respectively. If there are three or more substituents, the IUPAC numbering system is used, with the numbers chosen so that the substituents have the lowest possible numbers and are given in alphabetical order (for example, *c*hloro before *n*itro):

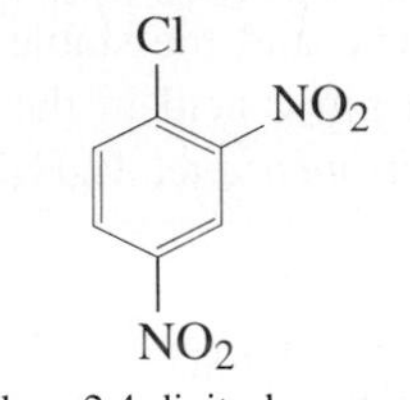

1-Chloro-2,4-dinitrobenzene

Polycyclic Arenes

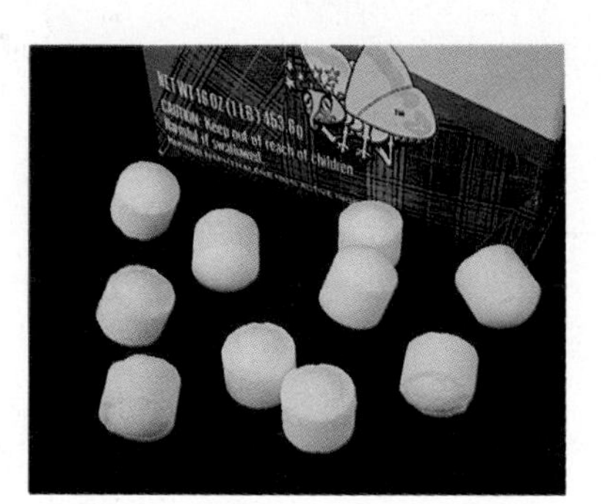

Naphthalene moth balls

Arenes that consist of two or more benzenelike rings that share a common bond are called *polycyclic arenes.* Naphthalene, $C_{10}H_8$, anthracene, $C_{14}H_{10}$, and phenanthrene, also $C_{14}H_{10}$, are examples:

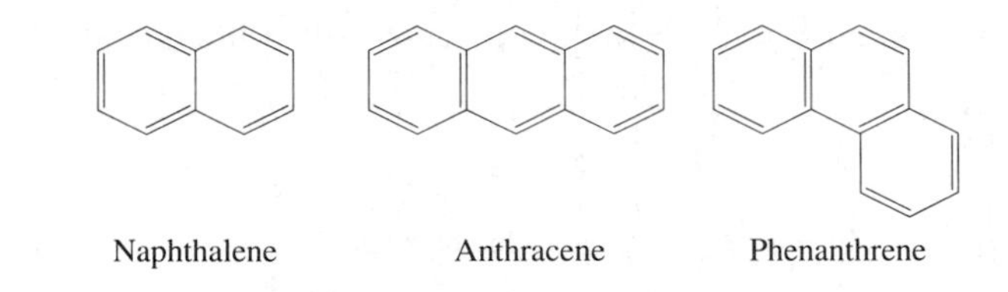

Naphthalene Anthracene Phenanthrene

The sheets of carbon atoms in graphite can be thought of as very large polycyclic arenes.

8.6 RESONANCE STRUCTURES AND BOND ORDER

We can use resonance structures to describe the bonding in any molecule in which some of the electrons are more delocalized than is implied by a single Lewis structure, as in the case of benzene. For example, the Lewis structure of the carbonate ion

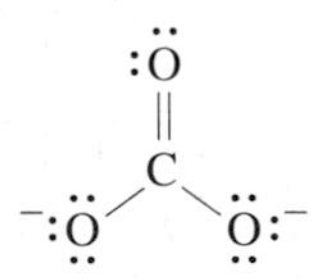

shows one of the three CO bonds as a double bond and the other two as single bonds. But all three bonds have exactly the same length, 131 pm, which indicates that they are equivalent. We can describe this in terms of the three following resonance structures:

To generate resonance structures from one Lewis structure here, convert a bonding electron pair to a nonbonding pair on one oxygen, and convert a nonbonding pair on another oxygen to a bonding pair.

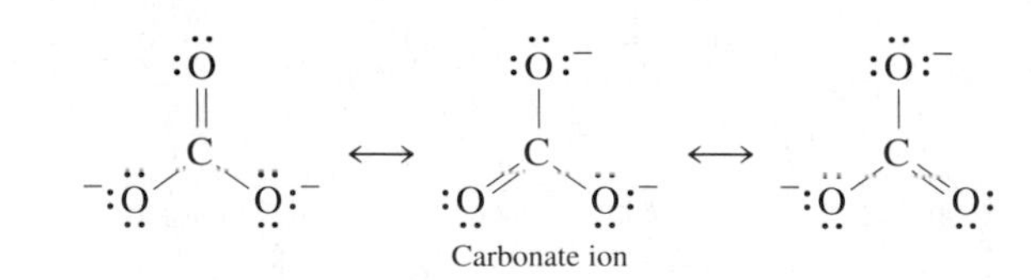

Carbonate ion

In the first structure, the top oxygen atom forms a double bond, whereas in the other two structures, this oxygen forms a single bond. This means that each bond may be described as a $1\frac{1}{3}$ bond: Each is said to have a bond order of $1\frac{1}{3}$.

Bond order **is defined as the total number of electron pairs forming the bond.**

A triple bond has a bond order of 3; a double bond, a bond order of 2; and a single bond, a bond order of 1.

To find the bond order for a bond in a molecule described by a set of resonance structures, we sum the bond orders for this bond in all the structures, and then divide by the number of structures. If we take any of the bonds in the carbonate ion, the sum of the bond orders is $2 + 1 + 1 = 4$. The number of structures is 3, so the bond order is $\frac{4}{3}$ or $1\frac{1}{3}$.

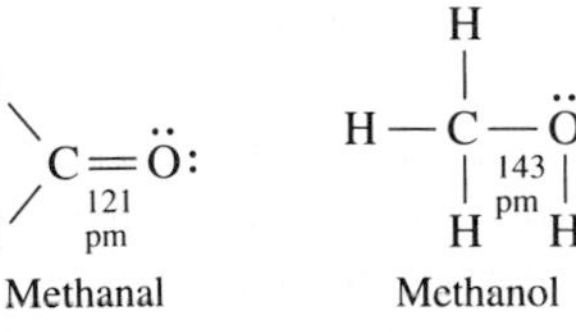

Methanal Methanol

The observed CO bond length of 131 pm in the carbonate ion is intermediate between the length of the C=O double bond in methanal (121 pm) and that for the C—O single bond in methanol (143 pm). It is consistent with a bond order of $1\frac{1}{3}$.

The charge on each oxygen atom in the carbonate ion may be obtained by a similar procedure. The sum of the charges on any given oxygen atom is $0 + (-1) + (-1) = -2$. We then divide by the number of structures, 3, which gives $-\frac{2}{3}$ for the charge on each oxygen.

The bond order of each of the carbon–carbon bonds in the benzene molecule is $1\frac{1}{2}$. The sum of the orders of any particular bond in the two resonance structures is $1 + 2 = 3$. When we divide by the total number of structures, 2, we obtain a bond order of $1\frac{1}{2}$.

As in the case of benzene, for convenience we often use just one of the resonance structures for an ion such as carbonate or phosphate. It is understood that all the bonds are equivalent and are neither double or single but have an intermediate character.

EXAMPLE 8.3 Resonance Structures and Bond Order

Draw four resonance structures for the phosphate ion. Deduce the order of each of the bonds and the charge on each oxygen atom.

Solution: The Lewis structure of the phosphate ion has one double P=O bond and three single P—O bonds. The four resonance structures are

```
      :Ö               :Ö:⁻              :Ö:⁻              :Ö:⁻
       ‖                 |                 |                 |
 ⁻:Ö—P—Ö:⁻  ↔   ⁻:Ö—P=Ö:  ↔   ⁻:Ö—P—Ö:⁻  ↔   :Ö=P—Ö:⁻
       |                 |                 ‖                 |
      :Ö:⁻             :Ö:⁻              :Ö                ⁻:Ö:
```

They show that each bond has an order of $(2 + 1 + 1 + 1)/4 = 1\frac{1}{4}$ and that the charge on each oxygen atom is $(-1 - 1 - 1 + 0)/4 = -\frac{3}{4}$.

Exercise 8.8 The sulfate ion has a tetrahedral structure in which all four SO bonds have the same length (149 pm). Draw six resonance structures for the sulfate ion, and determine the bond order and the charge on each oxygen atom.

SUMMARY

Diamond and graphite are two allotropes of carbon; diamond forms transparent, very hard crystals, does not conduct electricity, and has a three-dimensional covalent network structure. Graphite is a soft, black electrical conductor composed of planar sheets of carbon atoms stacked upon each other. Graphite can be converted to diamond at very high temperatures and pressures. Carbon black (soot), charcoal, and coke are impure forms of carbon. Coke is a useful industrial reducing agent.

In general terms, organic chemistry is the chemistry of carbon compounds, and inorganic chemistry is the chemistry of all other elements and their compounds. However, chemists classify compounds of carbon such as carbon dioxide, carbon monoxide, and metal carbides as inorganic compounds. Carbon dioxide, $CO_2(g)$, is obtained by burning carbon or its compounds in excess $O_2(g)$, by heating metal carbonates (except alkali metal carbonates), or from the reaction of a carbonate with aqueous acid. It is an acidic oxide, and in water it gives weak diprotic carbonic acid, which forms salts such as sodium hydrogencarbonate, $NaHCO_3$, and sodium carbonate, Na_2CO_3. Carbon monoxide, $CO(g)$, a useful industrial reducing agent, is obtained by burning carbon or its compounds in a limited supply of air and by heating methane and steam to high temperatures with a catalyst. The latter process forms a mixture of $CO(g)$ and $H_2(g)$ called synthesis gas. With catalysts, the organic compound methane, $CH_4(g)$, reacts with sulfur at 600°C to give the inorganic compound carbon disulfide, $CS_2(l)$, and with ammonia at 1200°C to give the inorganic compound hydrogen cyanide, $HCN(l)$. Metal carbides, such as $CaC_2(s)$, which contain the carbide ion, $^-:C\equiv C:^-$, react with water to give the organic compound ethyne (acetylene), $C_2H_2(g)$. Silicon carbide (carborundum), $SiC(s)$, has the diamond structure with alternate C atoms replaced by Si atoms. C, N, and O have a much greater tendency to form multiple bonds than do Si, P, and S.

Organic compounds are classified as aliphatic compounds—alkanes, alkenes, and alkynes and their derivatives—and aromatic compounds—including benzene and benzene-containing hydrocarbons. Hydrocarbons, which contain only carbon and hydrogen, are the simplest organic compounds. The systematic names for the hydrocarbons are given by the IUPAC system. Alkanes are hydrocarbons with the general formula C_nH_{2n+2}. Alkanes from C_4H_{10} on exist as two or more structural isomers, molecules with the same molecular formula but with their atoms connected differently. An alkane may have one continuous chain of carbon atoms or a branched chain with one or more alkyl side chains, substituent groups attached to the main chain. Rotation around single bonds allows ethane and higher alkanes to twist into an infinite number of shapes called conformations, such as the eclipsed and staggered conformations of ethane. Alkanes are unreactive, because their strong C—C and C—H bonds are not easily broken, and they behave neither as acids nor as bases. Their only important reactions are combustion and photochemical reactions.

Cycloalkanes, C_nH_{2n}, have rings of C atoms. Cyclopropane, $C_3H_6(g)$, and cyclobutane, $C_4H_8(g)$, are unusually reactive because of their "bent" carbon–carbon bonds.

Alkenes, with one or more C=C double bonds, and alkynes, with one or more C≡C triple bonds, are unsaturated hydrocarbons. Ethene, $H_2C{=}CH_2$, is a planar molecule with an AX_2 planar geometry at each carbon atom. Due to lack of free rotation around the C=C bond, an alkene can have geometric isomers, called *cis*-isomers when similar side chains are on the same side or *trans*-isomers when the side chains are on opposite sides of a C=C bond. Dienes and trienes have two and three C=C bonds, respectively.

Alkenes undergo addition reactions in which a molecule such as hydrogen, H—H, or a halogen, X—X, adds to the C=C bond to give a substituted alkane. Ethene polymerizes to give polyethylene, $-\!(CH_2-CH_2)\!-_n$. Alkynes undergo similar addition reactions.

Subterranean deposits of dead marine organisms subjected to high pressures for millennia are transformed to complex mixtures of alkanes and other hydrocarbons: natural gas and petroleum. Natural gas is methane with small amounts of ethane, propane, and butane. Petroleum, a liquid mixture of other alkanes, is separated into fractions by distillation. In cracking reactions, long-chain alkanes (saturated hydrocarbons) form shorter-chain alkanes and alkenes when heated with a catalyst.

The aromatic compounds consist of the ringed hydrocarbon benzene, C_6H_6, other hydrocarbons with one or more benzene rings, and their derivatives. The benzene molecule has a hexagonal ring structure with six equivalent carbon–carbon bonds of length 140 pm (intermediate between the C—C bond length, 154 pm, and the C=C bond length, 134 pm). A Lewis structure for benzene has alternate single and double bonds, but benzene does not undergo addition reactions as does an alkene. Six of the electrons in benzene are delocalized, not localized in pairs, as is depicted in a Lewis structure. The structure of benzene is conveniently represented as a combination of two equivalent Lewis structures called resonance structures. According to this representation, all the carbon–carbon bonds are equivalent and have a bond order (the total number of electron pairs forming the bond) of $1\frac{1}{2}$. Resonance structures may be used to describe the bonding in any molecule or ion in which some of the electrons are more delocalized than is implied by a Lewis structure, such as the carbonate ion, CO_3^{2-}, and the phosphate ion, PO_4^{3-}.

Aromatic compounds undergo aromatic substitution reactions—such as nitration or halogenation—in which one or more of the H atoms on the ring is replaced by another atom or group of atoms. Polycyclic (multiringed) arenes, such as naphthalene, $C_{10}H_8$, and anthracene, $C_{14}H_{10}$, have two or more benzenelike rings sharing a common bond.

IMPORTANT TERMS

addition reaction (page 288)
aliphatic compound (page 290)
alkane (page 274)
alkene (page 284)
alkyl group (page 278)
alkyne (page 284, 290)
aromatic compound (page 290)
aromatic substitution reaction (page 293)
bond order (page 295)
branched chain (page 276)
cis-isomer (page 287)
conformation (page 280)
continuous chain (page 275)
cracking reaction (page 284)
cycloalkane (page 281)
delocalized electrons (page 292)
geometric isomer (page 287)
halogenation reaction (page 293)
hydrocarbon (page 264)
hydrogenation reaction (page 288)
inorganic chemistry (page 264)
inorganic compound (page 264)
nitration reaction (page 293)
organic chemistry (page 264)
organic compound (page 264)
resonance structure (page 292)
saturated hydrocarbon (page 288)
side chain (page 275)
structural isomer (page 276)
substituent group (page 277)
substitution reaction (page 293)
trans-isomer (page 287)
unsaturated hydrocarbon (page 288)

REVIEW QUESTIONS

1. Define the scope of **(a)** organic chemistry; **(b)** inorganic chemistry.

2. **(a)** What is a hydrocarbon?

(b) What are the principal natural sources of hydrocarbons?

3. **(a)** Define the term "allotrope."

(b) Give the names of two allotropes of carbon, and **(c)** briefly describe their molecular structures.

4. Name four compounds of carbon that are classified as inorganic compounds, and draw their Lewis structures.

5. How are formaldehyde (methanal), $H_2C{=}O$, and methanol, $H_3C{-}OH$, made from synthesis gas?

6. Draw the Lewis structures of CS_2, HCN, and the cyanide ion, CN^-.

7. **(a)** How do the structures of carbon dioxide and silicon dioxide differ?

(b) How do these differences affect their physical properties?

8. Give the formulas, name, and draw the Lewis structures of the alkane, alkene, and alkyne with **(a)** two carbon atoms; **(b)** three carbon atoms.

9. Why are butane and isobutane referred to as "structural isomers"?

10. Draw diagrams to show the two particular conformations of ethane that are referred to as "eclipsed" and "staggered."

11. For ethane, write a balanced equation for **(a)** complete combustion; **(b)** photochemical chlorination.

12. Give two examples of addition reactions involving propene.

13. What is a polymerization reaction?

14. Draw the structures of *cis*- and *trans*-2-butene.

15. Give two examples of addition reactions involving propyne.

16. What are the systematic names of ethylene, propylene, and acetylene?

17. **(a)** What classes of organic compounds are the terms "aliphatic" and "aromatic" used to describe?

(b) What is an arene?

18. Explain why the Lewis structure of benzene is inconsistent with its observed structure and reactions.

19. **(a)** What is a resonance structure?

(b) What resonance structures are used to describe the benzene molecule?

20. **(a)** How is bond order defined?

(b) What are the bond orders of the carbon–carbon bonds in benzene and the carbon–oxygen bonds in the carbonate ion?

PROBLEMS

Allotropes of Carbon

1. Account for the rarity of diamond and the very different physical properties of diamond and graphite.

2. **(a)** How are diamonds made industrially from graphite?

(b) What is known about the composition of carbon black?

(c) How is charcoal made, and why is it useful in the purification of other substances?

3. In terms of their molecular structures, explain why the bond length in diamond is 154 pm, whereas all the bonds in the planar layers of graphite have equal lengths of 142 pm, intermediate in length between that of a single carbon–carbon bond (154 pm) and a double carbon–carbon bond (134 pm).

4.* Diamond is very hard, can be cleaved in many directions to give gemstones, melts above 3700°C, has a density of $3.53\ g \cdot cm^{-3}$, does not conduct electricity, and is chemically very unreactive. Graphite is soft, cleaves easily into flakes, has a density of $2.27\ g \cdot cm^{-3}$, conducts electricity, and is chemically more reactive than diamond. Account for these differences.

Inorganic Compounds of Carbon

5. Write balanced equations for the reaction of carbon monoxide with each of the following, and state the conditions under which each reaction occurs.

(a) $H_2(g)$ **(b)** $O_2(g)$ **(c)** $H_2O(g)$ **(d)** $Fe_2O_3(s)$

6. Write balanced equations for each of the following.

(a) two ways of preparing carbon monoxide

(b) two ways of preparing carbon dioxide

(c) one way of preparing carbon disulfide

7. **(a)** Why is $CO_2(g)$ described as an acidic oxide?

(b) Explain why $CO_2(g)$ can be made by the reaction of a metal carbonate, such as $CaCO_3(s)$ or $Na_2CO_3(s)$, with dilute aqueous acid. *Hint:* Carbonic acid is a weak acid.

8. What is the chemical composition of each of the following?

(a) lime **(b)** soda water **(c)** natural gas

(d) coke **(e)** chalk

9. Write a balanced equation to describe the reaction that occurs when carbon is heated with each of the following to a sufficiently high temperature.

(a) $CuO(s)$ **(b)** $CaO(s)$ **(c)** $O_2(g)$ **(d)** $S(s)$

10. Write a balanced equation to describe the reaction that occurs when methane is strongly heated with each of the following.

(a) $H_2O(g)$ **(b)** $NH_3(g)$ **(c)** $O_2(g)$

11. What is the empirical or molecular formula of each of the following?

(a) limestone **(b)** hydrocyanic acid **(c)** acetylene

(d) calcium carbide **(e)** carborundum

12. **(a)** Write balanced equations for the preparation of ethyne from calcium oxide, coke, and water.

(b) Why is the carbide ion, C_2^{2-}, completely converted to acetylene, $C_2H_2(g)$, in water?

13.* Two Lewis structures can be drawn for the thiocyanate ion, SCN^-.

(a) Draw both Lewis structures.

(b) On the basis that nitrogen forms multiple bonds more readily than does sulfur, which of the structures is the more likely?

14.* When concentrated sulfuric acid is dripped onto carbon tetrabromide at 160°C, a gaseous compound containing 6.4% C, 85.0% Br, and 8.6% O by mass is obtained. At 25°C and a pressure of 1.00 atm, 0.940 g of this compound has a volume of 122 mL.

(a) What is the molecular formula of the compound?

(b) Draw its Lewis structure, and deduce its molecular shape.

15.* When we heat aluminum oxide with coke in an electric furnace, we obtain a yellow carbide of aluminum that is stable up to 1400°C and reacts with water to give methane. We react a sample of the carbide of mass 0.500 g with excess water and collect the methane in a 250-mL bulb at 25°C. The pressure in the bulb is 1.02 atm.

(a) What is the empirical formula of the carbide?

(b) What mass of aluminum carbide is needed to produce 20.0 L of methane at 25°C and 1.00 atm pressure?

Alkanes; Alkenes and Alkynes; Benzene and the Arenes: Aromatic Hydrocarbons

Hydrocarbons: Names, Formulas, and Isomers

16. Give the names and draw the Lewis structures of the alkanes with **(a)** one, **(b)** two, and **(c)** three carbon atoms each.

17. How many carbon atoms are in the longest continuous chain of carbon atoms in each of the following?
(a) butane (b) 2-methylpropane
(c) 2,2-dimethyloctane (d) 2,3-dimethylpentane
(e) 2,2,5,5-tetramethylhexane

18. Give the molecular formula of each of the following.
(a) an alkane with eight carbon atoms
(b) an alkene with six carbon atoms and one double bond
(c) an alkyne with five carbon atoms and one triple bond
(d) a cycloalkane containing a six-membered ring of carbon atoms

19. Classify each of the following as an alkane, alkene, or alkyne. Give the molecular formula, and draw the Lewis structure.
(a) methane (b) ethene (c) propyne
(d) cyclobutane (e) cyclopropene

20. Which of the following names do *not* conform to IUPAC rules for naming organic compounds? Rename those that are incorrect.
(a) 2-ethylbutane (b) 3,3-dimethylbutane
(c) 1-ethylpropane (d) 2,2-dimethylpropane
(e) 1,2-dimethylpropane

21. Draw the structure and give the IUPAC name of each of the isomers with the molecular formula $C_2H_2Cl_2$.

22. Explain why 2-butene exists as *cis–trans* isomers, whereas 2-butyne does not.

23. Draw the structural formula and give the IUPAC name of each of the nine isomers of heptane.

24. Repeat Problem 21 for the six isomers with the molecular formula C_4H_8.

25. Draw a structural formula for each of the following.
(a) 1,3-cyclopentadiene (b) 1,3-cyclohexadiene
(c) 2-methyl-2,4-hexadiene

26. What is the difference between each of the following?
(a) a saturated and an unsaturated hydrocarbon
(b) a continuous-chain and a branched-chain alkane
(c) an aliphatic hydrocarbon and an aromatic hydrocarbon
(d) a molecular conformation and an isomer

27. Give the IUPAC name of each of the following.

(a)
```
      CH3
      |
CH3—C—CH3
      |
      CH3
```

(b)
```
CH3CH2CHCH2CH2CH3
       |
       CH2CH2CH2CH3
```

(c) $CH_3—CH{=}CH_2$ (d) $CH_3CH{=}C(CH_3)_2$
(e) $CH_3CH_2CH_2CH_2CH{=}CHCH_3$

(f)
```
      H   H
      |   |
H3C—C—C—CH3
      |   |
    H2C—C(CH3)2
```

(g) $CH_2{=}CHCH_2CH{=}CHCH_3$

28. Give the systematic name of each of the following.
(a) $CH_3CH_2CH{=}CH_2$ (b) $CH_3CH_2C{\equiv}CH$
(c) $(CH_3)_2C{=}C(CH_3)_2$
(d) $CH_2{=}CHCH_2CH_2CH{=}CH_2$

(e)
```
H3C        CH3
   \      /
    C = C
   /      \
  H        H
```

(f) $CH_3CH{=}CHCH{=}CH_2$

29. Which of the following is *not* an isomer of heptane?
(a) 2-methylhexane (b) 2,2-dimethylpentane
(c) 2,3-dimethylbutane (d) 2,3-dimethylpentane

30. Draw the structures of as many isomers of C_5H_{10} as you can, and give the IUPAC name of each.

31. Write structural formulas for each of the following.
(a) 2,2,4-trimethylpentane (b) 1,3-cyclobutadiene
(c) 4,4-dimethyl-1-pentyne (d) *trans*-2-pentene
(e) 1,3-butadiene (f) 1-methyl-2,3-diethylbenzene
(g) 1,3,5-triethylbenzene

32. Draw the structures and name the isomers of the substituted benzenes with the formulas **(a)** $C_6H_4(CH_3)_2$ and **(b)** $C_6H_3(CH_3)_3$.

33. Why is the tetrafluoroethene molecule, C_2F_4, planar, whereas the difluoroethyne molecule, C_2F_2, is linear?

34. Complete combustion of a sample of a hydrocarbon gave 0.318 g CO_2 and 0.163 g H_2O. The mass of the hydrocarbon that occupied a 250-mL flask at 100°C and 1.00 atm pressure was 0.4743 g. **(a)** Determine the empirical and molecular formulas of the hydrocarbon; **(b)** draw structural formulas for, and name, all isomeric hydrocarbons with this molecular formula.

35. Complete combustion of 0.1540 g of a hydrocarbon gave 0.4832 g of CO_2. The mass of hydrocarbon that filled a 250-mL flask at 100°C and 1.00 atm was 0.4580 g. **(a)** Repeat Problem 34a for this hydrocarbon; **(b)** draw structural formulas for, and name, at least four isomers with this molecular formula.

36. Explain what is meant by a molecular conformation. Draw diagrams to illustrate two important conformations each of **(a)** ethane and **(b)** cyclohexane.

Hydrocarbons: Reactions

37. By writing an appropriate balanced equation, give an example of each of the following reaction types.
(a) the combustion of an alkane
(b) a cracking reaction of an alkane
(c) an addition reaction of an alkene
(d) an addition reaction of an alkyne
(e) a polymerization reaction of an alkene

38. Explain why alkenes and alkynes are very reactive, whereas alkanes are unreactive except at high temperatures.

39. Describe a simple chemical test you might use to distin-

guish between each of the following pairs of gases, and describe what you would observe experimentally.

(a) ethane and ethyne **(b)** carbon dioxide and propane

40. Write the structural formula and name the product of each of the following reactions.

(a) $CO(g) + 2H_2(g) \xrightarrow{\text{heat, catalyst}}$

(b) $(H_3C)_2C{=}C(CH_3)H + Br_2 \rightarrow$

(c) $(H_3C)_2C{=}C(CH_3)H + H_2 \rightarrow$

41. Ethene can be made in the laboratory by heating ethanol, $C_2H_5OH(l)$, with concentrated sulfuric acid.

(a) Write the balanced equation for this reaction.

(b) Suggest how 2-methylpropene might be prepared.

42. Write a balanced equation for the complete combustion of each of the following in oxygen.

(a) pentane **(b)** cyclopropane **(c)** 2-butene **(d)** 1-butyne

43. Each of the following compounds may be synthesized from an alkene, or an alkyne, and another reactant. In each case give the name and structure of the alkene or alkyne and of the other reactant.

(a) 2-propanol **(b)** 2,2,3,3-tetrabromobutane **(c)** 1-butene **(d)** 2-bromopropene

44. Give an example of an addition reaction to **(a)** a C≡O triple bond and **(b)** a C=O double bond.

Resonance Structures and Bond Order

45. Draw Lewis structures for **(a)** CO_2, **(b)** CO, **(c)** CN^- **(d)** C_2^{2-}, and **(e)** HCN, and name each.

46. Using benzene as an example, explain **(a)** what is meant by a resonance structure; **(b)** how resonance structures are used to describe the structure of benzene; and **(c)** the difference between localized and delocalized electrons.

47. Explain why a single Lewis structure is insufficient to describe the observed structure of CO_3^{2-}, which is triangular planar with all the oxygen–carbon–oxygen angles equal to 120° and all the C—O bonds with the same length, 131 pm.

48. Draw Lewis structures, including possible resonance structures where appropriate, for each of the following, and arrange them in order of expected decreasing carbon–oxygen bond length.

(a) carbon monoxide, CO **(b)** carbonate ion, CO_3^{2-}

(c) formaldehyde, H_2CO **(d)** formate ion, HCO_2^-

49. Draw Lewis structures, including possible resonance structures where appropriate, for each of the following. Give the sulfur–oxygen bond orders in each, and use the VSEPR model to predict the expected molecular geometries.

(a) H_2SO_4 **(b)** SO_2 **(c)** SO_4^{2-} **(d)** HSO_3^- **(e)** SO_3

50. Repeat Problem 49 for each of the following, but give the phosphorus–oxygen bond orders instead.

(a) phosphoric acid, $(HO)_3PO$

(b) dihydrogenphosphate ion, $(HO)_2PO_2^-$

(c) hydrogenphosphate ion, $HOPO_3^{2-}$

(d) phosphate ion, PO_4^{3-}

51.* The nitrate ion, NO_3^-, is a triangular planar molecule with all the nitrogen–oxygen bonds of equal length and all the oxygen–nitrogen–oxygen angles equal to 120°. Draw the possible resonance structures, and deduce the expected nitrogen–oxygen bond orders.

General Problems

52. A hydrocarbon contains 82.6% carbon by mass; 0.470 g of it filled a 200-mL flask at 25°C and a pressure of 750 mm Hg.

(a) What are its empirical and molecular formulas?

(b) Can you write a unique structural formula for it?

53.* A small quantity of a liquid hydrocarbon containing 85.6% carbon by mass was placed in a 250-mL flask and heated to 100°C in a boiling-water bath. Under these conditions, all the hydrocarbon vaporized, and excess vapor escaped from the flask into the atmosphere. Upon cooling the liquid in the flask to room temperature, 0.687 g of liquid was in the flask. The atmospheric pressure was 760 mm Hg.

(a) What is the molar mass of the hydrocarbon?

(b) Discuss its possible structure.

54.* A 0.200-g sample of a hydrocarbon containing 85.71% carbon by mass occupies a volume of 95.3 mL at 0.921 atm pressure and 27°C. Draw possible structures for the hydrocarbon, and name them.

55.* An unsaturated hydrocarbon containing 88.8% carbon by mass was reacted with excess hydrogen, using a palladium catalyst. The amount of H_2 used up in complete reaction with 1.00 g of the hydrocarbon was 906 mL at 25°C and 1.00 atm pressure. In a molar mass determination, 0.1200 g of the hydrocarbon had a volume of 67.9 mL at 100°C and 1.00 atm.

(a) What are the empirical and molecular formulas of the hydrocarbon?

(b) Draw the structures of isomers with this molecular formula.

(c) How could the actual structural formula of the hydrocarbon be found by studying the products of its reaction with bromine?

56.* A gaseous mixture of methane and an alkene of volume 1.00 L has a mass of 0.882 g at 25°C and 744 mm Hg pressure. When burned completely in excess oxygen, 2.641 g of carbon dioxide and 1.442 g of water resulted. Identify the alkene, and calculate the mass percentage composition of the mixture.

57.* When mercury(II) cyanide, $Hg(CN)_2(s)$, is heated, a gaseous compound X containing only carbon and nitrogen is obtained. Analysis gives 46.2% C by mass. At 100°C and 0.950 atm pressure, 0.208 g of X has a volume of 126 mL.

(a) What are the empirical and molecular formulas of X?

(b) Draw a possible Lewis structure for X, and deduce its molecular shape.

Thermochemistry and Thermodynamics

9.1 Thermochemistry and the First Law of Thermodynamics
9.2 Bond Energies
9.3 Alternative Energy Sources
9.4 Entropy and the Second Law of Thermodynamics
9.5 Gibbs Free Energy

Thermochemistry and thermodynamics deal with the heat and other forms of energy changes that accompany chemical reactions. The combustion of acetylene in oxygen produces a large amount of heat and so gives a very hot flame. This welder is using the hot flame of an oxyacetylene torch to melt the edges of two metal plates in order to join them.

Chemical reactions supply the energy needed to maintain life. They also supply about 90% of the energy used in industrialized countries for industry and transportation, the rest being produced by hydroelectric power and nuclear reactions. The energy that drives biological reactions comes from the oxidation of carbon compounds, particularly carbohydrates and fats. The chemical reactions that supply industrial energy are the combustion (oxidation) of the hydrocarbons in fossil fuels, in particular petroleum, natural gas, and coal.

Every chemical reaction either produces or absorbs energy. This energy is most frequently produced in the form of heat transferred to or from the environment. The combustion of fossil fuels and many other reactions are **exothermic**: They release energy in the form of heat. Other reactions, such as the combination of N_2 and O_2 to form NO, are **endothermic**: They absorb energy in the form of heat. Energy may also be transferred to or from the surroundings in other forms, such as light and electrical energy. For example, we have seen that when magnesium burns in air, energy is released as both heat and light (Demonstration 2.2).

Whether a reaction is exothermic or endothermic depends on whether the energy stored in the products is less than or greater than the energy stored in the reactants. The energy stored in elements and compounds is called **chemical energy**. We saw in Chapter 6 that when two atoms combine, the resulting molecule has a lower energy than do the two separate atoms. Therefore, the formation of a bond is an exothermic process, and the breaking of a bond is an endothermic process. A reaction is exothermic if forming the bonds of the products releases more energy than is needed to break the bonds of the reactants. But if more energy is needed to break bonds than is released by the formation of new bonds, the reaction is endothermic. Chemical energy is therefore the energy associated with chemical bonds.

It is important to have quantitative information about the energy changes associated with chemical reactions. For example, a scientist designing a rocket must know how much energy is released in the reaction between hydrogen and oxygen in order to calculate how much liquid oxygen and liquid hydrogen the rocket must carry to lift it clear of the earth. An engineer must know how much fuel oil is needed to operate an electricity generating station. A biochemist must know the energy changes associated with cellular reactions to understand how a living organism transforms and utilizes the energy that maintains life. The study of the heat changes accompanying chemical reactions is called **thermochemistry**, the subject of the first part of this chapter. Thermochemistry is part of a subject of much wider scope called **thermodynamics**, which is the study of energy and its transformations.

Learning how to transform heat into mechanical work was one of humankind's greatest achievements. This discovery initiated the Industrial Revolution. The science of thermodynamics grew out of the need to convert efficiently into mechanical work the heat obtained by burning a fuel, using a device called an engine. In engineering the main use of thermodynamics is still in connection with engines. In chemistry, however, thermodynamics has another very important application. It helps us answer the extremely important question (not obviously related to engines), Why do some reactions go to completion, others proceed only partially, and still others proceed hardly at all? We take up this topic in the second half of the chapter.

9.1 THERMOCHEMISTRY AND THE FIRST LAW OF THERMODYNAMICS

Heat

We saw in Chapter 2 that

Heat is energy that is transferred as the result of a temperature difference.

Temperature is a measure of the average kinctic energy of the atoms and molecules of which a substance is composed. The molecules of a warmer object are moving randomly with a higher average speed and have a higher average kinetic energy than the molecules of a cooler object. When the molecules of the warmer object collide with the molecules of the cooler object, they transfer some of their kinetic energy to the molecules of the cooler object, so the slower molecules are speeded up and the faster molecules are slowed down (Figure 9.1). This energy transfer continues until all the molecules have the same average kinetic energy—in other words, until the two objects have the same temperature.

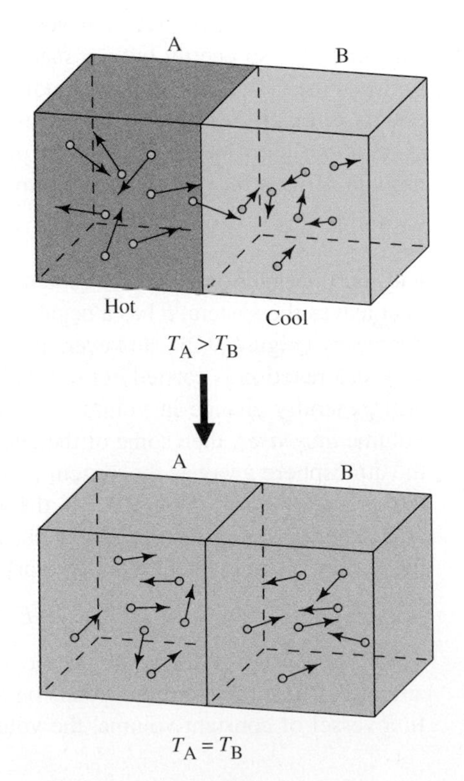

FIGURE 9.1 Heat Flow between Warm and Cold Objects Heat flows from a warmer object to a colder object. The process continues until all the molecules have the same average speed and average kinetic energy. The lengths of the arrows in this figure are proportional to the average speed of the molecules.

Internal Energy

The energy stored in a substance is called its **internal energy**, E. It is the sum of the kinetic and potential energies of all the nuclei and electrons of which the substance is composed. We have no way of measuring the internal energy of a substance in an absolute way. Although in principle we could calculate E by finding the total kinetic and potential energies of all the constituent nuclei and electrons, in practice this is very difficult to do. However, at a given temperature and pressure, a given amount of a substance always has the same internal energy. Thus, the **internal energy change** accompanying a reaction, ΔE, which we *can* measure, has a fixed value for a given set of conditions:

$$\Delta E = E_{\text{products}} - E_{\text{reactants}}$$

The First Law of Thermodynamics and Enthalpy

We can find the energy change accompanying a reaction by carrying out the reaction under such conditions that the change in energy is observed as heat transferred to or from the environment. We call the reaction vessel and its contents the **system**, and the rest of the universe is the **surroundings**. If no energy or matter is transferred to or from a system, we say that the system is isolated from its surroundings. We call such a system an **isolated system**. According to the law of conservation of energy, energy cannot be created or destroyed:

The energy of an isolated system is constant.

In thermodynamics the law of conservation of energy is known as the **first law of thermodynamics**.

Suppose that the reaction vessel and its contents are not isolated from the surroundings, so energy but *not* matter (reactants and products) can be transferred between the system and the surroundings. We call such a system a **closed system**. If energy is transferred only as heat, by measuring the flow of heat into or out of such a system, we can find the change in internal energy that has occurred. If q is the amount of heat that *enters* the system, then

$$\Delta E = q$$

and the internal energy of the system increases. If the reaction is exothermic and heat leaves the system, q has a negative value, and the internal energy of the system decreases (Figure 9.2). However, energy is not always transferred only as heat. Unless a reaction is carried out in a vessel of constant volume, the reacting system will generally change in volume as it proceeds from reactants to products. If the volume *increases,* then some of the energy produced in the reaction is used to push the atmosphere away as the system expands. We say that the system does **work** on the surroundings (Figure 9.3). If the volume of the system *decreases,* then the atmosphere is compressing the system, and we say that the surroundings do work on the system. If w is the amount of work done on the system, then

$$\Delta E = q + w$$

When the system does work on the surroundings, w has a negative value. The amount of heat transferred depends on the way in which the reaction is carried out. In a vessel of constant volume, the volume of the reacting system does not change

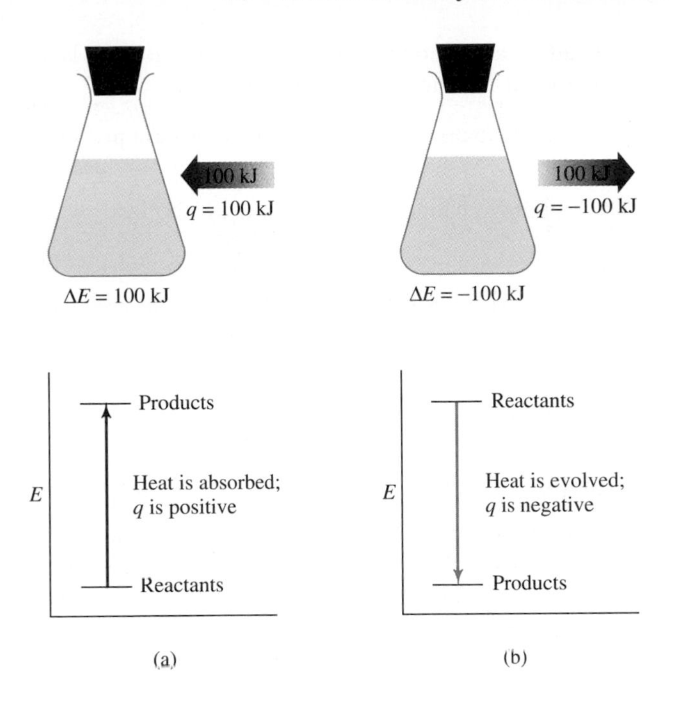

FIGURE 9.2 The Change in E and the Sign of q (a) When an amount of heat q is absorbed, the energy of the system increases, and both ΔE and q are positive. (b) When an amount of heat q is evolved, the energy of the system decreases, and both ΔE and q are negative.

(although its pressure may change). No work is done on or by this system—$w = 0$—and so

$$\Delta E = q \quad \text{(at constant volume)}$$

If instead the reaction is carried out at constant pressure (usually atmospheric pressure), as it is in a vessel whose volume can change (Figure 9.3), then

$$\Delta E = q + w$$

Most reactions, particularly those involving liquids and solutions, are most easily carried out at constant pressure, usually the atmospheric pressure. They may even be carried out in a vessel open to the atmosphere provided no products or reactants are lost to the atmosphere. So it is most convenient to measure the heat absorbed or

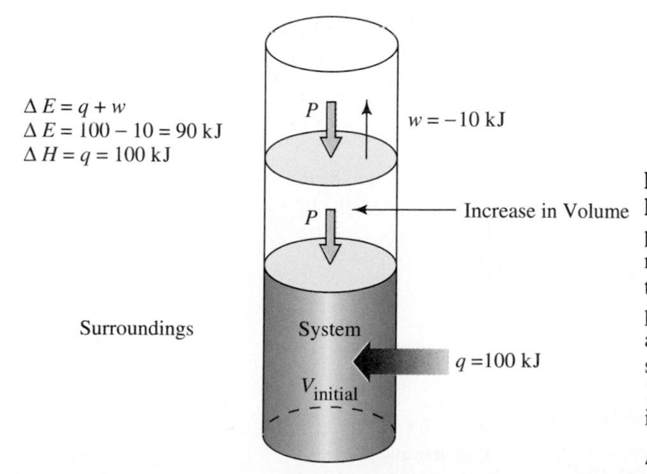

FIGURE 9.3 Expansion at Constant Atmospheric Pressure When volume increases at constant atmospheric pressure, work is done on the surroundings. We imagine the reaction to be occurring in a vessel with a movable piston. If there is an increase in volume during the reaction, the piston is pushed out against the pressure, P, exerted on the piston by the atmosphere. The system does work on the surroundings. If the system does 10 kJ of work on the surroundings, $w = -10$ kJ. If 100 kJ of heat is simultaneously transferred from the surroundings to the system, $q = 100$ kJ.

$\Delta E = q + w = 100 \text{ kJ} - 10 \text{ kJ} = 90 \text{ kJ} \quad \Delta H = q = 100 \text{ kJ}$

"Enthalpy" comes from the Greek word *enthalpein*, meaning "to warm."

evolved by a reaction under the same conditions. It is also convenient, therefore, to define what is called the **enthalpy change**:

> **The heat absorbed in a reaction at constant pressure is called the enthalpy change of the reaction.**

The enthalpy change of a reaction is symbolized by ΔH, and

$$\Delta H = q \quad \text{(at constant pressure)}$$

Because heat is *absorbed* (q is positive) in an endothermic reaction and is *evolved* (q is negative) in an exothermic reaction,

> **ΔH for an endothermic reaction has a positive value; ΔH for an exothermic reaction has a negative value.**

For many reactions the difference between the enthalpy change, ΔH, and the internal energy change, ΔE, is very small. This is particularly true for reactions carried out at pressures of about 1 atm and lower and especially for reactions in solution for which the volume change is usually very small. For many purposes we may regard the enthalpy change as being the same as the internal energy change, although we should be aware that it is not exactly the same.

Standard Reaction Enthalpy

Because the enthalpy change in a reaction depends on the conditions under which it is carried out, we must specify these conditions. When each of the substances in the reaction is at 1 atm pressure and at a specified temperature, usually 25°C, each is said to be in its **standard state**. The enthalpy change under these conditions is denoted by $\Delta H°$ and is called the **standard reaction enthalpy**. For example, the equation and the associated standard reaction enthalpy for the combustion of methane are

$$CH_4(g) + 2O_2(g) \rightarrow CO_2(g) + 2H_2O(l) \qquad \Delta H° = -890.4 \text{ kJ}$$

This reaction is *exothermic* (ΔH is negative), and 890.4 kJ of heat is evolved per mole of CH_4 consumed when the reaction is carried out at a constant pressure of 1 atm and at 25°C (Figure 9.4a). All standard reaction enthalpies in this book are given at 25°C.

It might seem surprising that we can give a value for the standard reaction enthalpy for the combustion of methane at 25°C. When methane is ignited, its temperature rises rapidly as it burns, so the combustion is not in fact occurring at 25°C. But if the reactants are initially at 25°C and if the products are allowed to cool

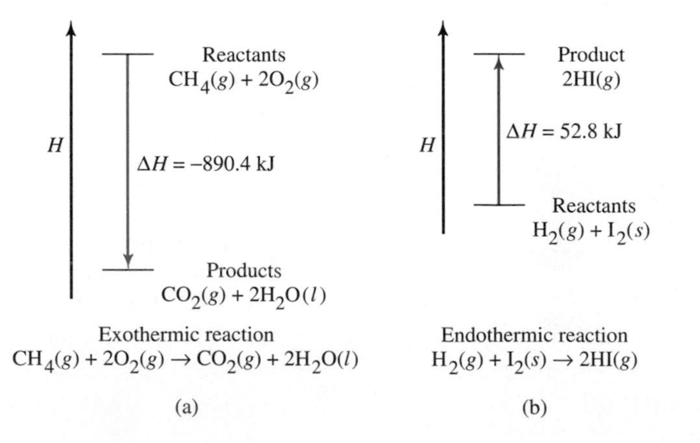

FIGURE 9.4 Enthalpy Changes in Exothermic and Endothermic Reactions (a) Heat is evolved in an exothermic reaction, and the enthalpy change (ΔH) is negative. (b) Heat is absorbed in an endothermic reaction, and the enthalpy change (ΔH) is positive.

down to 25°C at the end of the reaction, all the heat produced by the reaction, including that which was used to raise the temperature of the system, is eventually transferred to the surroundings.

The reaction of $H_2(g)$ with $I_2(s)$ to give $HI(g)$ is *endothermic* (ΔH is positive). In this case 52.8 kJ of heat are absorbed for every mole of H_2 and every mole of I_2 that react at a constant pressure of 1 atm and at 25°C (Figure 9.4b). Thus, the standard reaction enthalpy for this reaction is $\Delta H° = 52.8$ kJ:

$$H_2(g) + I_2(g) \rightarrow 2HI(s) \qquad \Delta H° = 52.8 \text{ kJ}$$

Although we usually avoid fractional coefficients in writing equations for reactions, when we discuss reaction enthalpies, we often write the equation for the reaction of 1 mol of a particular reactant, which can result in fractional coefficients for the other reactants and products. For example, the equation for the combustion of ethane may be written as

$$C_2H_6(g) + \tfrac{7}{2}O_2(g) \rightarrow 2CO_2(g) + 3H_2O(l) \qquad \Delta H° = -1560 \text{ kJ}$$

This equation tells us that the standard reaction enthalpy for the combustion of 1 mol of ethane is -1560 kJ.

Standard reaction enthalpies are expressed for the numbers of moles of reactants and products in the equation as written for a reaction.

If the equation for the combustion of ethane is written in the more usual form, which involves 2 mol of ethane,

$$2C_2H_6(g) + 7O_2(g) \rightarrow 4CO_2(g) + 6H_2O(l)$$

then $\Delta H° = 2(-1560 \text{ kJ}) = -3120$ kJ.

Calorimetry

The measurement of the heat absorbed or evolved in chemical reactions is called **calorimetry**, from *caloric*, the archaic name for heat. The apparatus used for the measurement is known as a **calorimeter**. The amount of heat absorbed or evolved is calculated from the temperature change that it produces in the reacting system or in some other system to which the heat is transferred. To do this calculation we need to know the temperature change produced in a given substance by the addition or the removal of a given amount of heat.

The heat required to raise the temperature of an object or of a given amount of a substance by 1 kelvin (= 1°C) is called the **heat capacity** of the object or of that amount of substance. The heat required to raise the temperature of 1 mol of a substance by 1 kelvin is the **molar heat capacity** of the substance. The molar heat capacity of liquid water is $75.4 \text{ J} \cdot \text{K}^{-1} \cdot \text{mol}^{-1}$. The heat required to raise the temperature of 1 g of a substance is called its **specific heat capacity** or simply its **specific heat**. The specific heat of water is $75.4 \text{ J} \cdot \text{K}^{-1} \cdot \text{mol}^{-1}/18.0 \text{ g} \cdot \text{mol}^{-1} = 4.18 \text{ J} \cdot \text{K}^{-1} \cdot \text{g}^{-1}$.

A very simple calorimeter can be made from two Styrofoam coffee cups (Figure 9.5a). Because Styrofoam is a very good heat insulator, a coffee-cup calorimeter absorbs only very little of the heat produced in a reaction, and very little heat is transferred to the surroundings. So to a reasonable approximation, all the heat produced in the reaction goes into raising the temperature of the contents of the calorimeter. The heat evolved in an acid–base reaction in dilute aqueous solution is conveniently measured in a coffee-cup calorimeter. The results of a typical experiment are given in Example 9.1.

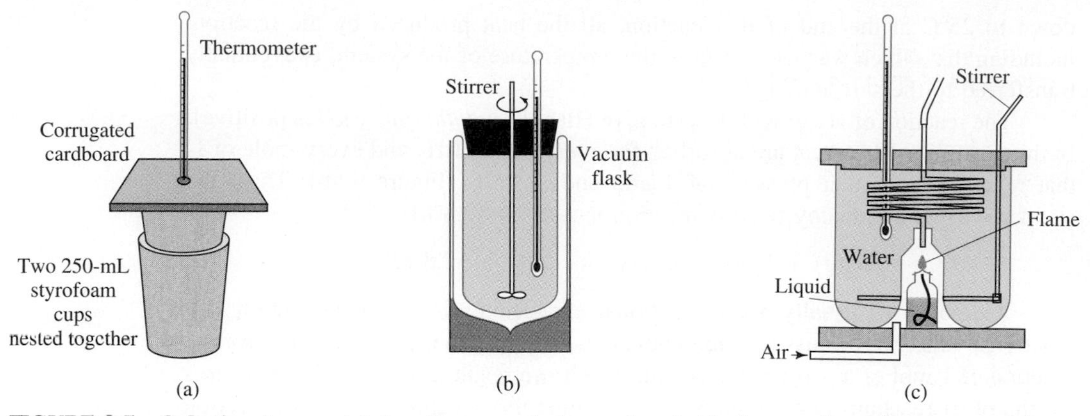

FIGURE 9.5 Calorimeters (a) A coffee-cup calorimeter is suitable for a student laboratory experiment but not for accurate measurements. (b) A calorimeter constructed from a vacuum flask is suitable for making accurate measurements on reactions occurring in solution. (c) A flame calorimeter is used for measuring the enthalpies of combustion of gases and volatile liquids. The heat evolved in the combustion raises the temperature of the water in the calorimeter. From the known heat capacity of water, the quantity of heat released in the reaction can be calculated.

EXAMPLE 9.1 Determination of the Enthalpy Change for a Neutralization Reaction

A sample of 50 mL of a 0.20-*M* solution of HCl was mixed with 50 mL of a 0.20-*M* solution of NaOH in a coffee-cup calorimeter. The initial temperature of both solutions was 22.2°C. After the two solutions were mixed, the temperature rose to 23.5°C. What is the enthalpy change for the neutralization reaction that occurs?

$$H_3O^+(aq) + OH^-(aq) \rightleftharpoons 2H_2O(l)$$

Solution: The total volume of the solution is 100 mL. Because it is a dilute aqueous solution, its density is $1.00\ g \cdot mL^{-1}$. Therefore the mass of the solution is 100 g. The temperature rise was 23.5°C − 22.2°C = 1.3°C = 1.3 K. The specific heat capacity of water is $4.18\ J \cdot K^{-1} \cdot g^{-1}$.

Energy released as heat = specific heat capacity × mass of solution × temperature rise. Thus, the energy released as heat by the reaction is

$$(4.18\ J \cdot K^{-1} \cdot g^{-1})(100\ g)(1.3\ K) = 5.4 \times 10^2\ J$$

Because the heat evolved is 540 J and the sign convention for heat defines heat evolved as negative, $q = \Delta H = -540$ J here.

The HCl solution contained

$$(0.20\ mol \cdot L^{-1})(50\ mL)\left(\frac{1\ L}{1000\ mL}\right) = 0.010\ mol\ HCl$$

The NaOH solution similarly contained 0.010 mol NaOH. Thus, the heat evolved in the reaction of 0.010 mol of NaOH with 0.010 mol of HCl is 5.4×10^2 J. So for the reaction of 1.0 mol of NaOH with 1.0 mol of HCl, we have

$$NaOH(aq) + HCl(aq) \rightarrow NaCl(aq) + H_2O(l) \qquad \Delta H = -54\ kJ$$

There is 100 times as much of the reactants here, so ΔH is 100 times larger:

$$-540\ J \times 100 \times \frac{1\ kJ}{1000\ J} = -54\ kJ$$

Then, writing the equation in terms of the ions present in the solution, we get

$$H_3O^+(aq) + Cl^-(aq) + Na^+(aq) + OH^-(aq) \rightarrow Na^+(aq) + Cl^-(aq) + 2H_2O(l)$$

Canceling the spectator ions, we have

$$H_3O^+(aq) + OH^-(aq) \rightarrow 2H_2O(l) \qquad \Delta H = -54\ kJ$$

Exercise 9.1 A 25-mL sample of a 0.10-*M* aqueous H_2SO_4 solution and 50 mL of an aqueous 0.10-*M* KOH solution were mixed in a coffee-cup calorimeter. The temperature of the solution rose from 21.20°C to 22.10°C as a result of the reaction that occurred upon mixing the solutions. Calculate the enthalpy change for the reaction

$$H_2SO_4(aq) + 2KOH(aq) \rightarrow K_2SO_4(aq) + 2H_2O(l)$$

Compare the value you obtain with that obtained from the neutralization of HCl(*aq*) in Example 9.1, and explain any difference.

Although a coffee-cup calorimeter gives results that are accurate enough for a classroom demonstration, Styrofoam does absorb a small amount of heat, and some is lost to the surroundings. A calorimeter constructed from a vacuum flask, as shown in Figure 9.5b, gives more accurate results. Careful experiments using this type of calorimeter have given the accurate value of $\Delta H = -56.02$ kJ for the enthalpy change for the neutralization reaction in aqueous soution. The equation

$$H_3O^+(aq) + OH^-(aq) \rightarrow 2H_2O(l) \qquad \Delta H = -56.02 \text{ kJ}$$

applies to the reaction of any strong acid with any strong base. For example,

$$HClO_4(aq) + KOH(aq) \rightarrow KClO_4(aq) + H_2O(l) \qquad \Delta H = -56.02 \text{ kJ}$$

$$H_2SO_4(aq) + 2KOH(aq) \rightarrow K_2SO_4(aq) + 2H_2O(l) \qquad \Delta H = 2(-56.02 \text{ kJ}) = -112.04 \text{ kJ}$$

For studying the enthalpies of combustion of gases and liquids, we can use a flame calorimeter, such as the one shown in Figure 9.5c.

EXAMPLE 9.2 Determination of the Enthalpy Change for a Combustion Reaction

When 0.510 g of ethanol was burned in oxygen in a flame calorimeter containing 1200 g of water, the temperature of the water rose from 22.46°C to 25.52°C. What is the enthalpy change, ΔH, for the combustion of 1 mol of ethanol?

$$C_2H_5OH(l) + 3O_2(g) \rightarrow 2CO_2(g) + 3H_2O(l)$$

Solution: The increase in the temperature of the water is

$$25.52°C - 22.46°C = 3.06°C = 3.06 \text{ K}$$

An interval of 1 degree on the Celsius scale is the same as 1 kelvin.

The specific heat capacity of water is $4.18 \text{ J} \cdot \text{K}^{-1} \cdot \text{g}^{-1}$. Thus, the amount of heat added to the water by the combustion of 0.510 g ethanol is

$$(4.18 \text{ J} \cdot \text{K}^{-1} \cdot \text{g}^{-1})(1200 \text{ g})(3.06 \text{ K})\left(\frac{1 \text{ kJ}}{1000 \text{ J}}\right) = 15.3 \text{ kJ}$$

The heat released in the combustion of 1 mol of ethanol (molar mass 46.05 g) is

$$\left(\frac{15.3 \text{ kJ}}{0.510 \text{ g}}\right)\left(\frac{46.05 \text{ g}}{1 \text{ mol}}\right) = 1.38 \times 10^3 \text{ kJ} \cdot \text{mol}^{-1}$$

Therefore, for the combustion of 1 mol of ethanol, $\Delta H = -1.38 \times 10^3$ kJ.

Exercise 9.2 When 1.3 g of butane, C_4H_{10}, was burned in oxygen in a flame calorimeter containing 1800 g of water, the temperature of the water rose from 20.2°C to 28.2°C. What is the enthalpy change for the combustion of 1 mol of butane?

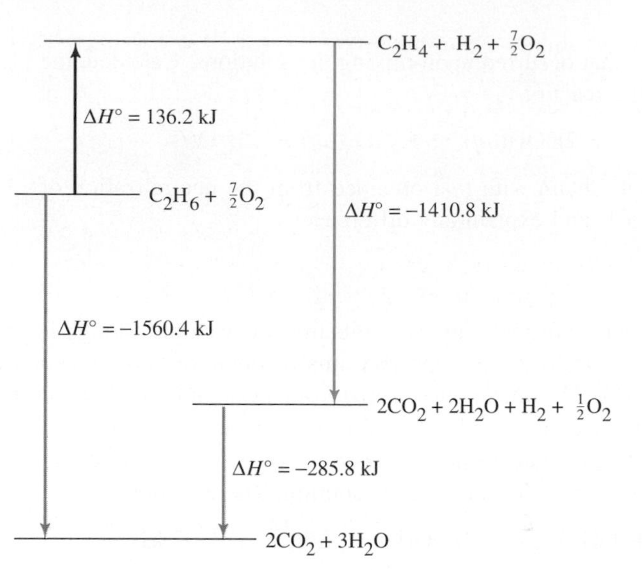

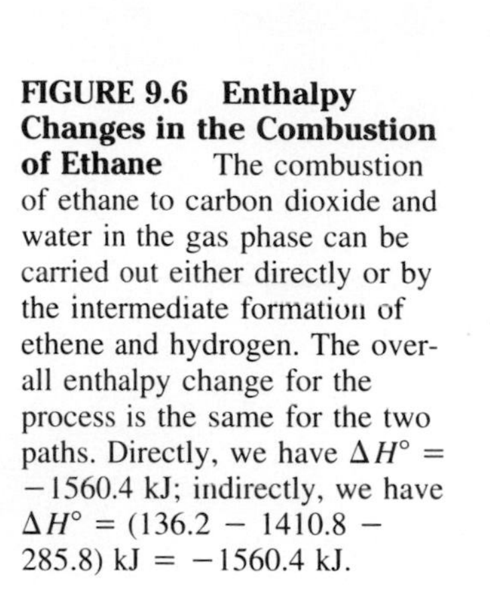
FIGURE 9.6 Enthalpy Changes in the Combustion of Ethane The combustion of ethane to carbon dioxide and water in the gas phase can be carried out either directly or by the intermediate formation of ethene and hydrogen. The overall enthalpy change for the process is the same for the two paths. Directly, we have $\Delta H° = -1560.4$ kJ; indirectly, we have $\Delta H° = (136.2 - 1410.8 - 285.8)$ kJ $= -1560.4$ kJ.

Hess's Law

Many reactions can be carried out by more than one path or series of steps. For example, the combustion of ethane can be carried out directly,

$$C_2H_6(g) + \tfrac{7}{2}O_2(g) \rightarrow 2CO_2(g) + 3H_2O(l) \qquad \Delta H° = -1560.4 \text{ kJ}$$

or it can be carried out in the following three steps (Figure 9.6):

(1) Ethane is decomposed by heating to ethene and hydrogen in the endothermic reaction

$$C_2H_6(g) \rightarrow C_2H_4(g) + H_2(g) \qquad \Delta H° = +136.2 \text{ kJ}$$

(2) The ethene is then burned in oxygen to give CO_2 and H_2O in the exothermic reaction

$$C_2H_4(g) + 3O_2(g) \rightarrow 2CO_2(g) + 2H_2O(l) \qquad \Delta H° = -1410.8 \text{ kJ}$$

(3) The hydrogen is then burned in oxygen to give water in the exothermic reaction

$$H_2(g) + \tfrac{1}{2}O_2(g) \rightarrow H_2O(l) \qquad \Delta H° = -285.8 \text{ kJ}$$

Together these three steps give the same final products as the direct combustion of ethane, as we can see by adding the left-hand and right-hand sides of the equations and canceling terms that appear on each side:

$$\begin{array}{rl} C_2H_6(g) \rightarrow C_2H_4(g) + H_2(g) & \Delta H° = +\ 136.2 \text{ kJ} \\ C_2H_4(g) + 3O_2(g) \rightarrow 2CO_2(g) + 2H_2O(l) & \Delta H° = -1410.8 \text{ kJ} \\ H_2(g) + \tfrac{1}{2}O_2(g) \rightarrow H_2O(l) & \Delta H° = -\ 285.8 \text{ kJ} \\ \hline C_2H_6(g) + \tfrac{7}{2}O_2(g) \rightarrow 2CO_2(g) + 3H_2O(l) & \Delta H° = -1560.4 \text{ kJ} \end{array}$$

Adding the enthalpy changes for the three reactions gives the same overall enthalpy change as for the direct combustion of ethane. This demonstrates that

> **The enthalpy change for a reaction is independent of the path by which the reaction is carried out.**

This statement, which is an important consequence of the first law of thermodynamics, is often called **Hess's law**. It was first stated (in somewhat different form) in 1840 by the Russian chemist Germain Hess (1802–1850) as a result of an extensive series of experiments.

If Hess's law were not true, we could carry out an exothermic reaction to produce a certain amount of energy. Then we could convert the products back to the reactants by a different path that required less energy. As a result, we would have produced energy without making any other change in the system or in the surroundings, contrary to the law of conservation of energy (the first law of thermodynamics).

We can use Hess's law to calculate enthalpy changes that are not easy to measure directly. For example, the enthalpy change for the formation of methane from carbon and hydrogen,

$$C(s) + 2H_2(g) \rightarrow CH_4(g) \qquad \Delta H^\circ = ? \qquad (1)$$

is most easily obtained from the ΔH° values for the combustion of carbon, hydrogen, and methane:

$$C(s) + O_2(g) \rightarrow CO_2(g) \qquad \Delta H^\circ = -393.5 \text{ kJ} \qquad (2)$$
$$H_2(g) + \tfrac{1}{2}O_2(g) \rightarrow H_2O(l) \qquad \Delta H^\circ = -285.8 \text{ kJ} \qquad (3)$$
$$CH_4(g) + 2O_2(g) \rightarrow CO_2(g) + 2H_2O(l) \qquad \Delta H^\circ = -890.4 \text{ kJ} \qquad (4)$$

To combine these three equations we first multiply equation (3) by 2, because in equation (1), the formation of methane, we have $2H_2$ on the left-hand side. And we reverse equation (4) to obtain CH_4 as a product. We can then add the three equations and the ΔH° values to obtain the overall ΔH° value for the formation of methane from carbon and hydrogen:

	ΔH°
$C(s) + O_2(g) \rightarrow CO_2(g)$	-393.5 kJ
$2 \times [H_2(g) + \tfrac{1}{2}O_2(g) \rightarrow H_2O(l)]$	$2(-285.8$ kJ$)$
$CO_2(g) + 2H_2O(l) \rightarrow CH_4(g) + 2O_2(g)$	$+890.4$ kJ
$C(s) + 2H_2(g) + 2O_2(g) + CO_2(g) + 2H_2O(l)$ $\rightarrow CO_2(g) + 2H_2O(l) + CH_4(g) + 2O_2(g)$	-74.7 kJ

Notice that when we reversed equation (4), the sign of ΔH° for that reaction changed. If a reaction is exothermic in one direction, it must be endothermic in the reverse direction. Canceling terms that appear on both sides gives the desired reaction:

$$C(s) + 2H_2(g) \rightarrow CH_4(g) \qquad \Delta H^\circ = -74.7 \text{ kJ}$$

Exercise 9.3 Given the standard enthalpy changes for the two reactions,

$$2P(s) + 3Cl_2(g) \rightarrow 2PCl_3(g) \qquad \Delta H^\circ = -574 \text{ kJ}$$
$$PCl_3(g) + Cl_2(g) \rightarrow PCl_5(s) \qquad \Delta H^\circ = -87.9 \text{ kJ}$$

find the standard enthalpy change for the reaction

$$2P(s) + 5Cl_2(g) \rightarrow 2PCl_5(s)$$

Standard Enthalpy of Formation

By using Hess's law, we can calculate the enthalpy change for any reaction from the enthalpy changes of an appropriate set of related reactions. It is particularly convenient to use the standard enthalpy of formation, ΔH_f°, for this purpose.

> **The standard reaction enthalpy for a reaction in which 1 mol of a substance is formed from its elements in their standard states is called the *standard enthalpy of formation* of the substance, ΔH_f°.**

For example, the standard enthalpies of formation of carbon dioxide, water, and methane are the standard enthalpy changes for the following reactions:

$$\text{C(graphite)} + O_2(g) \rightarrow CO_2(g) \qquad \Delta H_f^\circ(CO_2, g) = -393.5 \text{ kJ} \cdot \text{mol}^{-1}$$
$$H_2(g) + \tfrac{1}{2}O_2(g) \rightarrow H_2O(l) \qquad \Delta H_f^\circ(H_2O, l) = -285.8 \text{ kJ} \cdot \text{mol}^{-1}$$
$$\text{C(graphite)} + 2H_2(g) \rightarrow CH_4(g) \qquad \Delta H_f^\circ(CH_4, g) = -\ 74.8 \text{ kJ} \cdot \text{mol}^{-1}$$

Because an element may exist as a gas, a liquid, or a solid or in several allotropic forms, we must specify the form of the element when we give its standard enthalpy of formation. We choose the most stable form of the element at 1 atm and a specified temperature, usually 25°C. Thus, for bromine we choose $Br_2(l)$, not $Br_2(g)$ or $Br(g)$, and for carbon we choose C (*s*, graphite), not C (*s*, diamond) or C(*g*). Table 9.1 gives the standard enthalpies of formation for some common substances.

TABLE 9.1 Standard Enthalpies of Formation at 25°C

Substance	ΔH_f° (kJ · mol^{-1})	Substance	ΔH_f° (kJ · mol^{-1})
Elements in standard states	0	$CaO(s)$	−635.1
		$CaCO_3(s)$	−1206.9
C(diamond)	+1.9	$Fe_2O_3(s)$	−824.2
H(*g*, atomic)	+218.0	$HCl(g)$	−92.3
C(*g*, atomic)	+716.7	$HBr(g)$	−36.4
O(*g*, atomic)	+249.1	$HI(g)$	+26.4
$AgCl(s)$	−127.1	$HNO_3(l)$	−174.1
$CCl_4(l)$	−135.4	$H_2O(l)$	−285.8
$CH_4(g)$	−74.5	$H_2O(g)$	−241.8
$C_2H_2(g)$ (ethyne)	+226.8	$H_2S(g)$	−20.6
$C_2H_4(g)$ (ethene)	+52.3	$H_2SO_4(l)$	−814.0
$C_2H_6(g)$ (ethane)	−84.7	$NH_3(g)$	−46.2
$C_3H_8(g)$ (propane)	−103.8	$NO(g)$	+90.3
$C_4H_{10}(g)$ (butane)	−126.1	$NO_2(g)$	+33.2
$C_5H_{12}(g)$ (pentane)	−146.4	$NaF(s)$	−569.0
$C_6H_6(l)$ (benzene)	+49.0	$NaCl(s)$	−411.1
$CH_3OH(l)$	−238.7	$NaBr(s)$	−361.1
$C_2H_5OH(l)$	−277.7	$NaI(s)$	−287.8
$C_6H_{12}O_6(s)$ (glucose)	−1268	$NaOH(s)$	−425.6
$CH_3CO_2H(l)$	−484.5	$O_3(g)$	+142.7
$CO(g)$	−110.5	$SO_2(g)$	−296.8
$CO_2(g)$	−393.5	$SO_3(g)$	−395.7

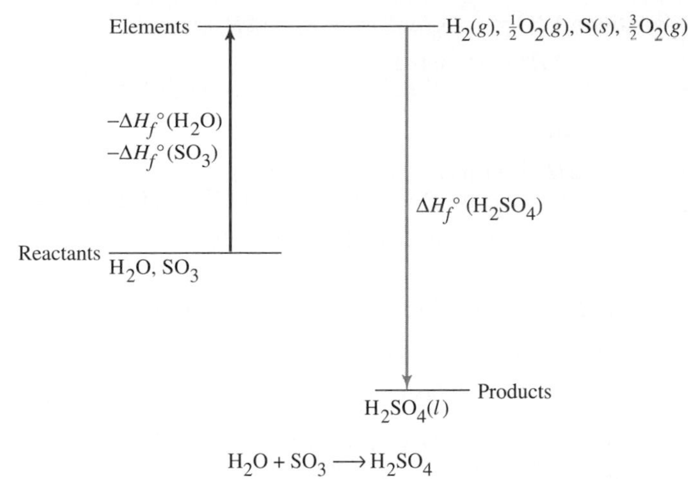

FIGURE 9.7 Hess's Law In principle, any reaction can be carried out by decomposing all the reactants to elements and combining these elements to form products.

It follows from the definition of the standard enthalpy of formation that the standard enthalpy of formation of an element in its most stable form is zero:

$$\text{C(graphite)} \rightarrow \text{C(graphite)} \qquad \Delta H_f^\circ = 0$$

However, an element that is not in its most stable form has a nonzero standard enthalpy of formation. The standard enthalpy of formation of diamond, $\Delta H_f^\circ = +1.9\ \text{kJ} \cdot \text{mol}^{-1}$, is the enthalpy change for the reaction

$$\text{C(graphite)} \rightarrow \text{C(diamond)} \qquad \Delta H^\circ = 1.9\ \text{kJ} \cdot \text{mol}^{-1}$$

Similarly, the standard enthalpy of formation of hydrogen atoms, $H(g)$, in the gas phase is not zero. It is $+218.0\ \text{kJ} \cdot \text{mol}^{-1}$, which is the standard enthalpy change for the reaction

$$\tfrac{1}{2}H_2(g) \rightarrow H(g) \qquad \Delta H^\circ = +218.0\ \text{kJ}$$

We can use standard enthalpies of formation to calculate the reaction enthalpy for any reaction from the following expression:

ΔH° = (sum of standard enthalpies of formation of products) − (sum of standard enthalpies of formation of reactants)

or

$$\Delta H^\circ = \sum[n_p(\Delta H_f^\circ)_p] - \sum[n_r(\Delta H_f^\circ)_r] \qquad (5)$$

$\sum$ is the capital Greek letter sigma, used to denote "the sum of."

where n_p is the number of moles of each product, and n_r is the number of moles of each reactant. Any reaction can in principle be carried out by first decomposing all the reactants to the corresponding elements, for which the overall enthalpy change is $-\sum[n_r(\Delta H_f^\circ)_r]$, and then combining these elements to form the products, for which the overall enthalpy change is $\sum[n_p(\Delta H_f^\circ)_p]$. Hess's law then leads to equation (5) (Figure 9.7).

EXAMPLE 9.3 Using Standard Enthalpies of Formation

Find the standard reaction enthalpy for the reaction

$$H_2O(l) + SO_3(g) \rightarrow H_2SO_4(l)$$

from standard enthalpies of formation (see Figure 9.7).

We must use the ΔH_f° value for liquid water here, because one of the reactants is $H_2O(l)$, not $H_2O(g)$.

Solution: $\Delta H^\circ = \sum[n_p(\Delta H_f^\circ)_p] - \sum[n_r(\Delta H_f^\circ)_r]$

$$\Delta H^\circ = (1\text{ mol})\,\Delta H_f^\circ(H_2SO_4, l) - [(1\text{ mol})\,\Delta H_f^\circ(H_2O, l) + (1\text{ mol})\,\Delta H_f^\circ(SO_3, g)]$$

Using the data in Table 9.1, we have

$$\Delta H^\circ = (1\text{ mol})(-814.0\text{ kJ}\cdot\text{mol}^{-1}) - [(1\text{ mol})(-285.8\text{ kJ}\cdot\text{mol}^{-1}) + (1\text{ mol})(-395.7\text{ kJ}\cdot\text{mol}^{-1})]$$
$$= -814.0\text{ kJ} + (285.8\text{ kJ} + 395.7\text{ kJ}) = -132.5\text{ kJ}$$

EXAMPLE 9.4 Using Standard Enthalpies of Formation

Find the standard enthalpy change for the oxidation of NH_3 according to the equation

$$4NH_3(g) + 5O_2(g) \rightarrow 4NO(g) + 6H_2O(g)$$

from the standard enthalpies of formation given in Table 9.1.

Solution: $\Delta H^\circ = \sum[n_p(\Delta H_f^\circ)_p] - \sum[n_r(\Delta H_f^\circ)_r]$

$$= (4\text{ mol})(+90.3\text{ kJ}\cdot\text{mol}^{-1}) + (6\text{ mol})(-241.8\text{ kJ}\cdot\text{mol}^{-1}) - [(4\text{ mol})(-46.2\text{ kJ}\cdot\text{mol}^{-1}) + (5\text{ mol})(0)]$$
$$= +\,361.2\text{ kJ} - 1450.8\text{ kJ} + 184.8\text{ kJ} - 0 = -\,904.8\text{ kJ}$$

Exercise 9.4 Calculate the standard enthalpy change, ΔH°, for the reaction

$$2SO_2(g) + O_2(g) + 2H_2O(g) \rightarrow 2H_2SO_4(l)$$

from the standard enthalpies of formation in Table 9.1. How much heat would be liberated if 5.20 g of SO_2 was converted to H_2SO_4?

Standard enthalpies of formation provide a useful measure of the energy stored in a compound. If the standard enthalpy of formation of a compound is negative, the compound contains less stored energy than the elements from which it forms. The ΔH_f° for CO_2 is $-393.5\text{ kJ}\cdot\text{mol}^{-1}$, so CO_2 contains less stored energy than C(graphite) and $O_2(g)$. If the standard enthalpy of formation is positive, the compound contains more stored energy than the elements from which it forms. The ΔH_f° for ethyne, $C_2H_2(g)$, is $+226.8\text{ kJ}\cdot\text{mol}^{-1}$; C_2H_2, therefore, contains a large amount of stored energy. If ethyne is decomposed to its elements, this energy is released. As a consequence, when ethyne is burned in oxygen, the reaction is highly exothermic. We can think of this reaction as occurring in three steps: the decomposition of ethyne to its elements, the combustion of carbon to CO_2, and the combustion of hydrogen to H_2O. All three of these steps are exothermic.

Because ethyme is less stable than its elements, it cannot be made by the direct reaction of carbon and hydrogen but is made by other reactions (Chapter 8).

We express the idea that ethyne contains a large amount of stored energy when we say that C_2H_2 is less stable than the elements from which it forms. In contrast, CO_2 is more stable than the elements from which it forms, because it forms in an exothermic reaction.

9.2 BOND ENERGIES

The atoms in a molecule are held together by chemical bonds that result from the electrostatic attraction between the nuclei and the electrons. The energy needed to break a bond is a measure of the strength of the bond. For example, the standard reaction enthalpy for the dissociation of a hydrogen molecule into two hydrogen atoms, the dissociation energy, is a measure of the strength of the bond in the hydrogen molecule:

$$H_2(g) \rightarrow 2H(g) \qquad \Delta H^\circ = 436.0 \text{ kJ}$$

The value of ΔH° is usually called the **bond energy**, although strictly speaking it should be called the *bond enthalpy*. In other words, we ignore the small difference between the internal energy and the enthalpy of a substance. We say that the bond energy *BE* of the H—H bond is 436.0 kJ · mol^{-1}:

$$BE(\text{H—H}) = 436 \text{ kJ} \cdot \text{mol}^{-1}$$

The standard reaction enthalpy for the reaction

$$H_2(g) \rightarrow 2H(g)$$

in which 2 mol of hydrogen atoms form is twice the value given for ΔH_f° in Table 9.1 because the value in the table is the standard enthalpy of formation of 1 mol of H atoms:

$$\tfrac{1}{2}H_2(g) \rightarrow H(g)$$

The bond energy of a diatomic molecule is the energy needed to dissociate 1 mol of molecules in the gas phase into atoms in the gas phase.

The bond energies of diatomic molecules range from 149 kJ · mol^{-1} and 155 kJ · mol^{-1} for the rather weak bonds in I_2 and F_2 to 941 kJ · mol^{-1} and 1070 kJ · mol^{-1} for the very strong bonds in N_2 and CO, respectively (Table 9.2). In Chapter 6 we used the bond energy of Cl_2 (243 kJ · mol^{-1}) to show that light with a maximum wavelength of 493 nm can dissociate the Cl_2 molecule into atoms.

From the energy needed to break all the bonds, we can find the bond energy for a bond in a polyatomic molecule that has only one type of bond. The standard reaction enthalpy for the decomposition of CH_4 into its atoms in the gas phase is

$$CH_4(g) \rightarrow C(g) + 4H(g) \qquad \Delta H^\circ = 1663 \text{ kJ}$$

In this reaction four C—H bonds are broken, so the average energy needed to break one C—H bond is $\frac{1}{4}$(1663 kJ) = 416 kJ. Therefore, the average C—H bond energy in the methane molecule is 416 kJ · mol^{-1}.

TABLE 9.2 Average Bond Energies* (kJ · mol^{-1})

	H	C	N	O	F	Cl	Br	I	Si	P	S
H	436										
C	413	348									
	—	619d									
	—	812t									
N	389	293	159								
	—	616d	418d								
	—	879t	941t								
O	463	335	157	138							
	—	707d	—	498d							
	—	1070t									
F	565	485	270	184	155						
Cl	431	326	200	205	254	243					
Br	364	276	—	—	—	219	190				
I	294	238	—	—	—	210	178	149			
Si	318	—	—	464	—	—	—	—	196		
P	322	—	—	—	—	—	—	—	—	197	
S	338	259	—	—	496	250	212	—	—	—	266

* Values are for single bonds except where otherwise stated; d indicates a double bond and t a triple bond.

In a polyatomic molecule the energy of a bond depends to some extent on the rest of the molecule of which it is a part. However, for most bonds the variations from molecule to molecule are not large, and we can determine a useful average value for each type of bond (Table 9.2). For the C—H bond this average value is $413\ kJ \cdot mol^{-1}$. Many bonds occur only in combination with other bonds. For example, to determine the C—C bond energy in the C_2H_6 molecule, we must first know the C—H bond energy, as we see in Example 9.5.

EXAMPLE 9.5 Bond Energies

From the data in Table 9.1, calculate the enthalpy change for the reaction

$$C_2H_6(g) \rightarrow 2C(g, \text{atomic}) + 6H(g, \text{atomic})$$

Then, assuming that the C—H bond energy is the same as that in CH_4, calculate the C—C bond energy in C_2H_6.

Solution:
$$\Delta H^\circ = (2\ mol)(716.7\ kJ \cdot mol^{-1}) + (6\ mol)(218.0\ kJ \cdot mol^{-1}) - (1\ mol)(-84.7\ kJ \cdot mol^{-1})$$
$$= 2826\ kJ$$

For bond energy calculations, it is useful to draw a Lewis structure for each molecule to see how many bonds of each kind must be broken or formed.

Because C_2H_6 has six C—H bonds and one C—C bond, if *BE*(C—H) is the bond energy of a C—H bond and *BE*(C—C) is the bond energy of a C—C bond, then the energy needed to break all the bonds is

$$BE(C{-}C) + 6BE(C{-}H) = 2826\ kJ \cdot mol^{-1}$$

If we now assume that *BE*(C—H) is the same as it is in CH_4, that is, *BE*(C—H) = $416\ kJ \cdot mol^{-1}$, then

$$BE(C{-}C) = (2826\ kJ \cdot mol^{-1}) - 6(416\ kJ \cdot mol^{-1})$$
$$= 330\ kJ \cdot mol^{-1}$$

Exercise 9.5 From the enthalpy of formation of ethene, C_2H_4, in Table 9.1 and the bond energy of the C—H bond in CH_4 ($416\ kJ \cdot mol^{-1}$), calculate the C=C bond energy.

Bond Energy and Bond Strength

The bond energies in Table 9.2 give a good idea of the relative strengths of different bonds. A stronger bond requires more energy to break than does a weaker one. Molecules that have strong bonds are generally more stable and less reactive than molecules with weaker bonds. The C—C and C—H bonds are among the strongest of the single bonds, which is one important reason organic compounds are so numerous. The corresponding Si—H and Si—Si bonds are much weaker. However, the Si—O bond is one of the strongest single bonds. It is not surprising, therefore, given that silicon is the second most abundant element in the earth's crust (Table 2.3), that SiO_2 and the silicates, which also contain Si—O bonds (Chapter 19), make up about 90% of the solid surface of the earth.

The examples of carbon–carbon, nitrogen–nitrogen, carbon–oxygen, and carbon–nitrogen bonds in Table 9.2 show that bond strength increases in the order single < double < triple bonds. As the number of bonding electrons increases, the atoms are held together more strongly and tightly. The bond lengths of carbon–carbon bonds, for example, decrease as the bond energies increase in the series:

C—C	C=C	C≡C
154 pm	134 pm	120 pm
348 kJ · mol^{-1}	619 kJ · mol^{-1}	812 kJ · mol^{-1}

Because of the great strength of the N≡N triple bond, many reactions in which $N_2(g)$ is formed are highly exothermic. Many nitrogen compounds decompose or burn in explosive reactions. Explosions are simply reactions in which a large amount of energy is released and a large volume of gaseous products is formed very rapidly (Box 9.1; Demonstration 9.1).

BOX 9.1 Explosives and the Nobel Prize

An important application of chemical energy is the use of explosives. Any substance that undergoes a very rapid chemical reaction that is strongly exothermic—or that produces a large volume of gaseous products from a solid or a liquid—is potentially an explosive. The destructive power of an explosion is due to the shock wave caused by the rapid increase in volume from the gases formed or to the rapid expansion of the atmosphere as a consequence of the large amount of heat released in a short time, or to both of these circumstances.

The oldest known explosive is *gunpowder*, which was used in China, Arabia, and India, possibly as early as the tenth century. Gunpowder is a mixture of approximately 75% KNO_3, 12% S, and 13% C. The products include a large volume of gases including CO_2, CO, and N_2, as well as a dense smoke that consists of fine particles of K_2CO_3, K_2SO_4, and K_2S.

For many purposes gunpowder has been replaced by stronger explosives, such as ammonium nitrate. When ammonium nitrate is detonated, it decomposes in a very exothermic reaction to give a large volume of gaseous products:

$$2NH_4NO_3(s) \rightarrow 2N_2(g) + O_2(g) + 4H_2O(g)$$

The oxygen produced can also be used to oxidize other substances, thus increasing the energy released. A commonly used explosive for blasting in mines is composed of 95% NH_4NO_3 and 5% fuel oil. Ammonium nitrate is also used as a fertilizer. Normally, its use as a fertilizer is safe, because ammonium nitrate must be detonated before it will explode. However, the careless handling of large quantities of ammonium nitrate fertilizer can cause a massive explosion. For example, in 1947 a ship carrying ammonium nitrate fertilizer exploded and leveled a huge area of Texas City, Texas, claiming 576 lives.

If an explosive reaction is very rapid, the shock wave may travel at very high speeds, up to 6 km · s^{-1}, and the explosive is classified as a high explosive. Trinitrotoluene (TNT), $C_7H_5O_6N_3$, and nitroglycerin, $C_3H_5O_9N_3$, are examples. The slower combustion that occurs in low explosives, such as gunpowder, produces shock waves that travel at about 100 m · s^{-1}.

Figure A *Alfred Nobel (1833–1896)*

The oxygen required for the very rapid combustion of high explosives cannot come from air, because the oxidation is too rapid. For these high explosives the oxygen comes from the explosive itself. Often these explosives are mixed with other substances, such as NH_4NO_3, to increase the amount of oxygen available. Many common explosives contain nitrogen compounds. Their combustion produces oxides of nitrogen and molecular nitrogen, which is formed in a very exothermic process because of the high bond energy of the nitrogen molecule.

Nitroglycerin is a liquid that explodes 25 times as fast as gunpowder and with three times the energy per gram. Although it was used as an explosive soon after it was first prepared in 1866, nitroglycerin is much too unstable and sensitive to shock to be handled safely. For transportation the containers were packed into a type of clay called *kieselguhr* to cushion them from shocks as much as possible. Alfred Nobel (Figure A), a Swedish inventor, noticed that when nitroglycerin leaked from the containers, it was soaked up by the clay. The nitroglycerin-soaked clay is much more stable and less sensitive to shock than pure nitroglycerin. Nobel called this safer explosive *dynamite*. He left much of the fortune he earned from manufacturing it to found the Nobel prizes in peace, literature, physics, chemistry, and physiology or medicine, which were first distributed in 1901.

DEMONSTRATION 9.1 The Decomposition of Nitrogen Triiodide

The dark brown substance on the filter paper is a compound of nitrogen triiodide and ammonia, $NI_3 \cdot NH_3$.

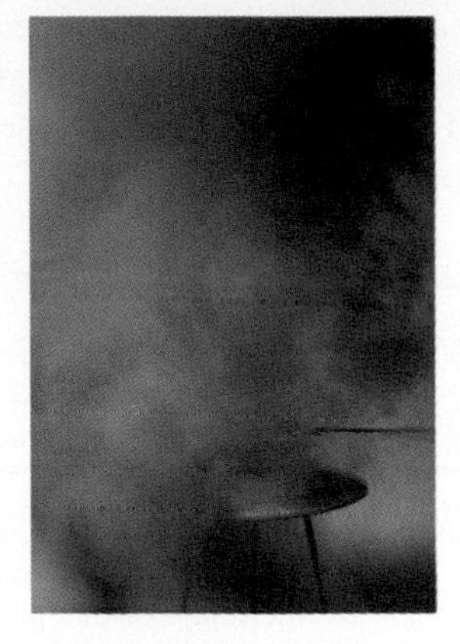

This compound is extremely shock sensitive when dry. When touched lightly with a feather, it explodes violently, producing a cloud of purple-brown iodine vapor.

The explosion is violent enough to punch a hole through the filter papers and the asbestos mat on which they were resting.

Because of the great strength of the triple bond in the N_2 molecule, the decomposition of NI_3 to N_2 and I_2 is a strongly exothermic reaction. It is also very rapid, so a large amount of heat and a large volume of gaseous products form rapidly and produce a violent explosion. Hence it is dangerous to prepare more than very small amounts of this substance.

Estimating Reaction Enthalpies from Bond Energies

Bond energies can be used to obtain an approximate value for a reaction enthalpy of a gas-phase reaction if the appropriate enthalpies of formation are not readily available. From the bond energies we can find the energy needed to break the reactant molecules into atoms and the energy gained by reassembling the atoms into the product molecules (Figure 9.8). Then, from Hess's law, the reaction enthalpy is given by

$$\Delta H^\circ = \sum BE(\text{bonds broken in reactants}) - \sum BE(\text{bonds formed in products})$$

Example 9.6 shows the details of a typical calculation.

EXAMPLE 9.6 Estimating Enthalpy

From the bond energies in Table 9.2, calculate an approximate value for the standard enthalpy of formation of ammonia.

Solution: The equation for the formation of ammonia from its elements in their standard states is

$$N_2(g) + 3H_2(g) \rightarrow 2NH_3(g)$$

First we break the bonds in the reactant molecules to give free atoms, and then we form the new bonds from these atoms.

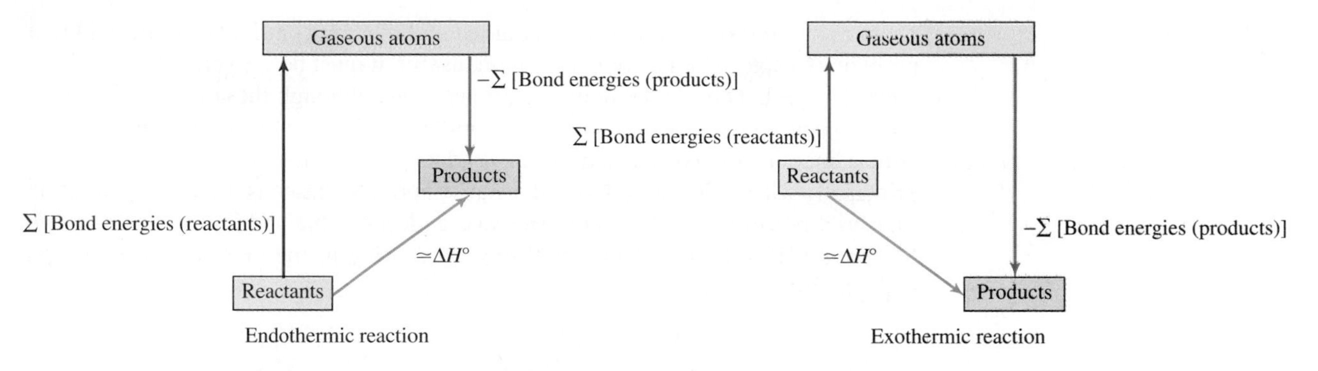

FIGURE 9.8 Approximate Reaction Enthalpies from Bond Energies Any reaction can, in principle, be carried out by first decomposing all the gaseous reactants into free atoms. This process is endothermic, and the heat needed is the sum of the bond energies of the reactants. Then the free atoms may be combined to give the products. This process is exothermic, and the heat evolved is the sum of the bond energies of the products. Hence $\Delta H = \sum[\text{bond energies(reactants)}] - \sum[\text{bond energies(products)}]$. The bond energies are approximate, so this method gives only an approximate value for the enthalpy change of a reaction in the gas phase.

N≡N H—H H—H H—H	→	2 N atoms 6 H atoms	→	H—N—H H—N—H H H

Bonds broken: 1 mol N≡N, 3 mol H—H. Bonds formed: 6 mol N—H

$$\Delta H^\circ = \sum BE(\text{bonds broken}) - \sum BE(\text{bonds formed})$$
$$= 941\ \text{kJ} + 3(436)\ \text{kJ} - 6(389)\ \text{kJ} = -85\ \text{kJ}$$

Exercise 9.6 From the bond energies in Table 9.2, estimate the standard reaction enthalpy for the combustion of 1 mol of ethane. Compare this value with the value calculated from standard enthalpies of formation (Table 9.1).

9.3 ALTERNATIVE ENERGY SOURCES

The main energy source of industrialized nations presently is petroleum. But the world's petroleum resources are limited and, even with the occasional discovery of new deposits, may soon be exhausted—perhaps within as little as 100 years. One of the challenges facing humanity, and in particular scientists and engineers, is to learn how to utilize alternative energy sources efficiently and safely as well as to find alternative raw materials for all the substances now manufactured from petroleum.

Coal Conversion

Coal is the most abundant fossil fuel. It constitutes 80% of the fossil fuel reserves of the United States and 90% of those of the world. However, the use of coal is currently limited by a number of problems. It is difficult to mine and to transport. Coal typically contains large amounts of sulfur, which is released into the atmosphere as sulfur dioxide when coal is burned and thereby contributes to the acid rain problem (Chapter 16). For coal to be more useful, it must be converted to a liquid or a gaseous fuel. This can be done in various ways, although these fuels are usually more expensive than petroleum. For example, heating coal to a high temperature with a mixture of oxygen and steam produces a mixture of gases that contains principally CH_4, CO, and H_2 and is an important gaseous fuel often used in industrial processes. Carbon monoxide and hydrogen obtained in this way can also be converted to the liquid fuel methanol by heating in the presence of a catalyst (Chapter 14):

$$CO(g) + 2H_2(g) \xrightarrow{\text{heat, catalyst}} CH_3OH(g)$$

Hydrogen

We mentioned in Box 2.1 the possibility of using hydrogen as a fuel. Because of its low mass and high enthalpy of combustion, hydrogen is ideal as a rocket fuel. But ways to use it conveniently, safely, and economically as a replacement for gasoline, for example, have not yet been found. Moreover, because hydrogen does not occur naturally, it must be made from one of its compounds. The obvious choice is water, of which there is an essentially inexhaustible supply. However, the decomposition of water to hydrogen and oxygen

$$H_2O(l) \rightarrow H_2(g) + \tfrac{1}{2}O_2(g) \qquad \Delta H^\circ = 286 \text{ kJ}$$

requires 286 kJ per mole of water, and no way has yet been found to carry out this reaction economically. Electrolysis is a possibility if electricity can be provided cheaply enough, or sunlight may be used to decompose water in a photochemical reaction if a catalyst can be found that will increase the rate of the reaction sufficiently. Research is in progress to attempt to modify the photosynthetic process in plants to release hydrogen from water instead of using the hydrogen to produce carbohydrates. Challenging problems await the research chemist in this area of study.

Biomass

The term "**biomass**" refers to animal and plant material—both dead and alive. Wood, leaves, animal excrement, and waste food are all forms of biomass, which is an important source of stored energy. Wood was crucial in the past as an energy source for cooking and heating, and it continues to be used today in less developed countries. Indeed, the destruction of forests during the past 1000 years or so has greatly changed the face of the earth and has led to the disappearance of many animal and plant species. But other forms of biomass are being extensively used today to produce two important fuels, biogas and ethanol.

If animal wastes and plants are allowed to rot in the absence of air, they are broken down by bacteria into **biogas**, which is predominately methane. The methane can be used directly for cooking and heating or to generate electricity. The residue, which has a high nitrogen content, functions as a fertilizer. Biogas production is particularly important in rural areas and in less developed countries.

A methane gas generator in Nepal

The traditional method for the production of **ethanol**, C_2H_5OH, is fermentation, in which sugars and carbohydrates from plants are converted to ethanol by the action of yeast:

$$C_6H_{12}O_6(aq) \rightarrow 2C_2H_5OH(aq) + 2CO_2(g)$$

A car engine needs little modification to enable it to burn ethanol and no modification at all for an ethanol–gasoline mixture called *gasohol.* Indeed, gasohol is already being sold in some parts of the United States, and ethanol made from sugar cane is a widely used fuel for cars in Brazil.

A sign advertising gasohol in New Mexico

Solar Energy

All but a tiny fraction of the energy on earth is derived from the sun, whose energy is produced by nuclear fusion reactions (Chapter 19). It makes sense to look for ways of using this energy directly instead of using it in its stored form as fossil fuels, which will ultimately be depleted. Although an enormous amount of energy reaches the earth from the sun, most of it is radiated back into space, and the remainder is distributed over the whole surface of the earth. The problem, then, is to find a method of trapping and storing this solar energy. On a small scale devices called *photovoltaics,* or solar cells, can convert solar energy directly to electricity. They are used, for example, to provide power in space satellites, but they are at present too expensive to be practical for the large-scale generation of energy—say, for power generation for a small city or even a large factory. No method for large-scale conversion of solar energy into useful energy has yet been developed.

Nuclear Energy

Nuclear reactors based on nuclear fission reactions have become an increasingly important energy source since the 1960s. We discuss nuclear energy in detail in Chapter 18.

9.4 ENTROPY AND THE SECOND LAW OF THERMODYNAMICS

Here and in Section 9.5 we introduce some of the fundamental ideas of thermodynamics. To understand chemical reactions, it is essential to have at least a qualitative understanding of the important thermodynamic concepts of *entropy* and *free energy*, although in an introductory course we can do no more than scratch the surface of these topics. Detailed and rigorous treatments of thermodynamics are presented in more advanced books in chemistry, physics, and engineering.

Spontaneous Reactions

We have seen that some reactions proceed almost to completion before equilibrium is reached, whereas others proceed hardly at all but go almost to completion in the opposite direction. We saw in Demonstration 7.8 that phosphorus combines readily with oxygen to give the oxide P_4O_{10}, but the oxide is never observed to decompose back to phosphorus and oxygen. The reaction proceeds only in one direction and goes essentially to completion. Burning diamond (or graphite) in oxygen gives carbon dioxide, but carbon dioxide does not decompose into diamond and oxygen even when strongly heated (Figure 9.9a). Clearly, something causes some chemical reactions to occur in one direction only. Some physical processes also occur in one direction only. For example, a gas expands to fill the space available to it, but it is never observed to contract to a smaller volume on its own (Figure 9.9b). A hot object cools to the temperature of its environment, but an object at the temperature

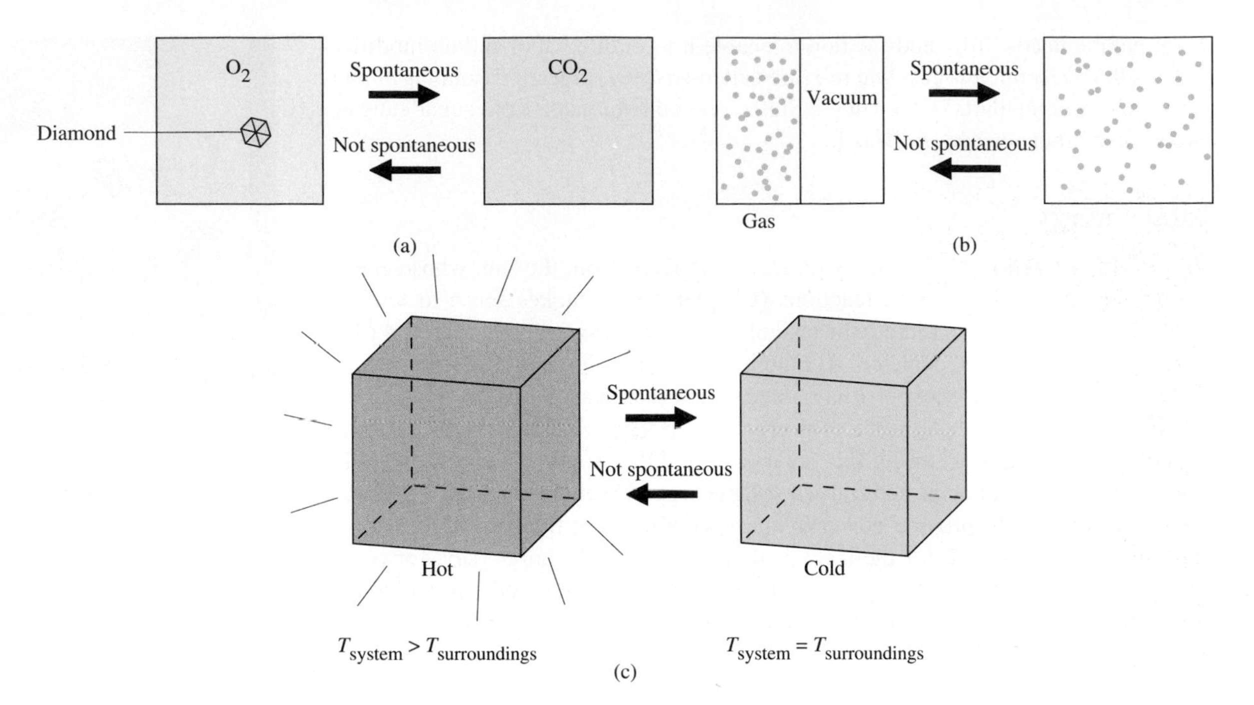

FIGURE 9.9 Spontaneous and Nonspontaneous Processes (a) A diamond burns in oxygen to give carbon dioxide; this process is spontaneous. But carbon dioxide does not decompose into diamond and oxygen; this process is nonspontaneous. (b) A gas expands into a vacuum; this process is spontaneous. But a gas does not collect in one side of a container and leave a vacuum on the other side; this process is nonspontaneous. (c) A hot object cools to the temperature of its surroundings; this process is spontaneous. But an object at the temperature of its surroundings does not become warmer; this process would be nonspontaneous.

of its environment is never observed to get warmer (Figure 9.9c). A change is said to be **spontaneous** if it occurs on its own without any external intervention, and the reverse process is never observed to occur on its own. Demonstration 9.2 shows the spontaneous reaction between hydrogen and oxygen to form water. Of course, these changes can be reversed: A gas *can* be compressed to a smaller volume; and water can be decomposed into hydrogen and oxygen, for example, by passing an electric current through it (Demonstration 2.1). However, in each case work must be done to accomplish these changes. They do not occur without some outside intervention.

We might think at first that a spontaneous change is one in which the energy of a system decreases and in which the final system has a lower energy than the initial system, such as when a hot object cools down. In an exothermic reaction, energy is released into the surroundings in the form of heat, and the products have a lower energy—that is, are more stable—than the reactants. However, there are many spontaneous endothermic reactions, such as the decomposition of calcium carbonate at high temperature,

$$CaCO_3(s) \rightarrow CaO(s) + CO_2(g) \qquad \Delta H^\circ = 178.4 \text{ kJ}$$

and endothermic physical changes, such as the boiling of water at 100°C,

$$H_2O(l) \rightarrow H_2O(g) \qquad \Delta H^\circ = 44.0 \text{ kJ}$$

in which the enthalpy of the system increases. The fact that the difference between internal energy changes and enthalpy changes is generally small means that the internal energy also increases in an endothermic reaction. Moreover, according to

DEMONSTRATION 9.2 A Spontaneous Reaction

The balloon contains hydrogen.

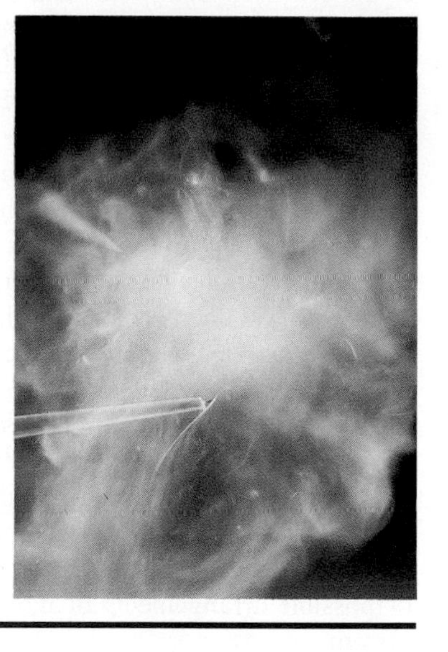

When a flame is touched to the balloon, the hydrogen combines with the oxygen of the air in a rapid, explosive reaction. The position of equilibrium for the reaction

$$2H_2(g) + O_2(g) \rightarrow 2H_2O(g)$$

lies far to the right—it is a spontaneous reaction.

the first law of thermodynamics, the total energy of a system and its surroundings is constant, so when a system *decreases* in energy, its surroundings must *increase* in energy. Energy changes alone cannot account for the direction of either physical changes or chemical reactions. To account for the direction of spontaneous change, we must introduce the concept of entropy.

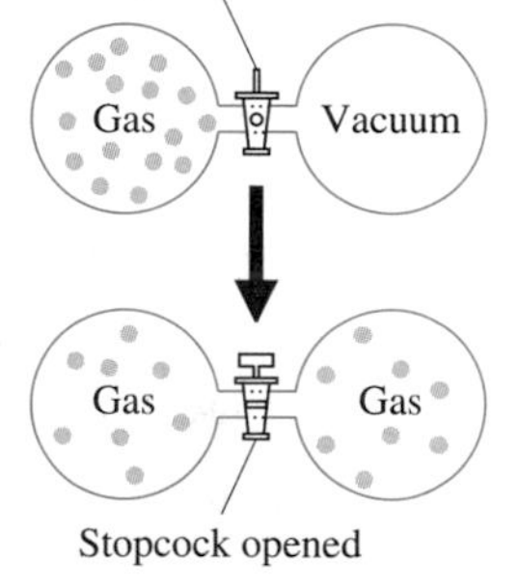

(a)

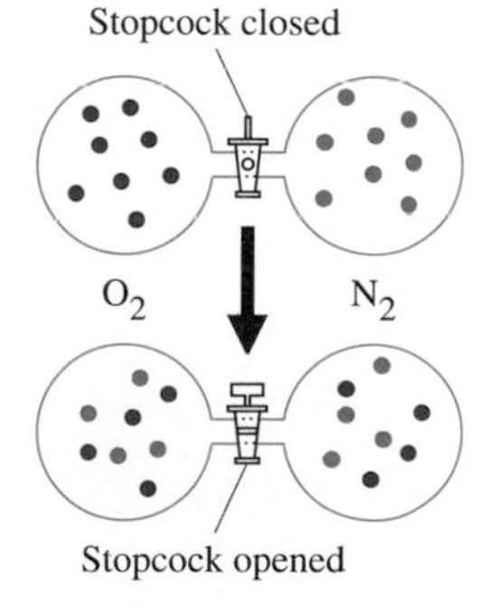

(b)

FIGURE 9.10 Spontaneous Processes (a) The expansion of an ideal gas into a vacuum. (b) The mixing of two ideal gases. In neither process is there any change in the energy of the system or its surroundings.

Disorder and Entropy

Let us first consider a spontaneous process in which there is no change in the energy of the system. An ideal gas expands spontaneously at constant temperature into any space made available to it (Figure 9.10a). Because, as we saw in Chapter 3, the energy of an ideal gas depends only on the temperature, its energy does not change during the expansion. If the gas expands into a vacuum, no work is done by or on the system and therefore no heat is absorbed or evolved. The spontaneous expansion of a gas into a vacuum does not surprise us; indeed, we would be very surprised if it did not occur. But *why* does it occur? It occurs because the molecules of a gas are moving randomly and at high speeds. They move through the valve connecting the two bulbs in Figure 9.10a and into the empty right-hand bulb. Some molecules return to the left-hand bulb, but because there are more molecules in the left-hand bulb than in the right-hand bulb, more molecules move from left to right than from right to left. This net movement to the right continues until there are equal numbers

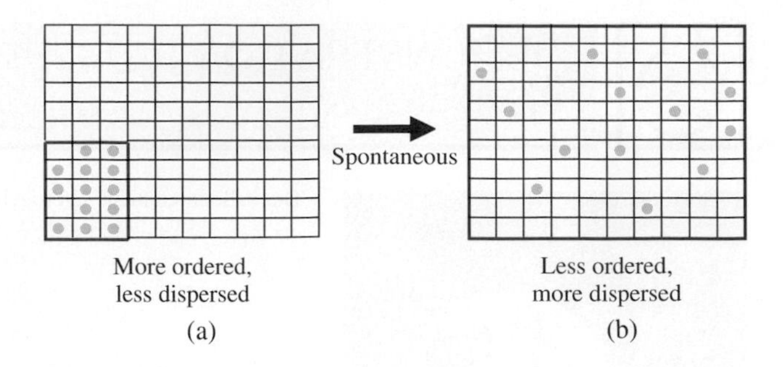

FIGURE 9.11 Spontaneous Processes and Disorder (a) This smaller system is more ordered and less dispersed, because there are fewer possible arrangements of the molecules. (b) This larger system is less ordered and more dispersed, because there are more possible arrangements of the molecules.

of molecules in both bulbs and equilibrium is established. Molecules continue to move from right to left and from left to right between the two bulbs, but there is no further net change. The spontaneity of this change is associated with the random motions of the molecules, which causes them to become more dispersed and eventually to fill all the available space.

If we start with O_2 in one bulb and N_2 in the other bulb (Figure 9.10b) and then open the valve, both gases behave like a gas expanding into a vacuum. The random motions of the molecules produce a net movement of O_2 molecules to the right and N_2 molecules to the left. After a time, equilibrium is reached with equal numbers of O_2 and N_2 molecules in both bulbs. The two gases mix spontaneously in this way, but we never observe the two gases to ''unmix,'' that is, to go back to their initially separated state. Because of their random motions, all molecules have a tendency to become more dispersed in space and, if there is more than one kind, more mixed with each other. When this occurs, we say that the system has become more *disordered.* The concept of dispersion through space is essentially the same as that of disorder if we think of space as being divided into many small boxes (Figure 9.11). The more space the molecules of a substance occupy, the greater the number of possible arrangements of molecules in the boxes, so the more disordered the system.

We are well aware from everyday life that things tend to become more disordered. The shuffling of a pack of cards that was originally ordered according to the four suits mixes the cards, that is, increases the disorder. Continued shuffling is very unlikely to return them to their original order—so unlikely that we can say with considerable confidence that such reordering will never happen. In fact, no matter what order we start with, it is most unlikely that the cards will ever return to that original order. ''Unmixing'' of the cards does not occur because there are many millions of disordered arrangements but only one arrangement with a particular order. So the probability of obtaining a disordered arrangement is very high, whereas the probability of obtaining the ordered arrangement is so small as to be essentially zero. Likewise the probability of an enormous number of molecules of one kind clustering in one part of the available space is infinitesimally small, so the unmixing of the molecules of a mixture of two gases has never been observed. A spontaneous process in a system in which there is no energy change is one in which there is an increase in disorder in the system (Demonstration 9.3).

Now let us consider a spontaneous process in which the energy of the system changes. A hot block of metal spontaneously cools as energy in the form of heat is transferred from the metal to the surroundings. The average kinetic energy of the molecules of the surroundings increases while the average kinetic energy of the molecules of the metal block decreases (Figure 9.12). The energy that was originally concentrated in the metal block *becomes much more dispersed* because of the very large—essentially infinite—size of the surroundings. There is no change in the total

DEMONSTRATION 9.3 Disorder and Entropy

A red liquid and a green liquid and separate layers of red and green balls represent ordered states.

When the red and green liquids are mixed and the red and green balls are mixed, we obtain a more disordered state in each case, that is, a state of higher entropy.

No matter how long the brown mixture of the red and green liquids is stirred, it will never separate into the two original red and green liquids. Similarly, if the red and green balls are stirred or shaken, they will never separate into the two original layers. The two liquids or the two types of balls mix spontaneously, but they will never unmix spontaneously. Spontaneous change in an isolated system is always accompanied by an increase in disorder, randomness, or entropy.

energy of the metal block plus its surroundings, but there has been a dispersal of the energy that was originally concentrated in the block. We never observe this dispersal of energy to reverse itself so that the energy dispersed in the surroundings becomes concentrated in the block again. Dispersed energy is more disordered than concentrated energy, just as dispersed molecules occupying a large space are more disordered than molecules concentrated in a small space. When a hot object cools, there is no change in the total energy of the system and its surroundings, but there is an increase in the disorder of the system and its surroundings.

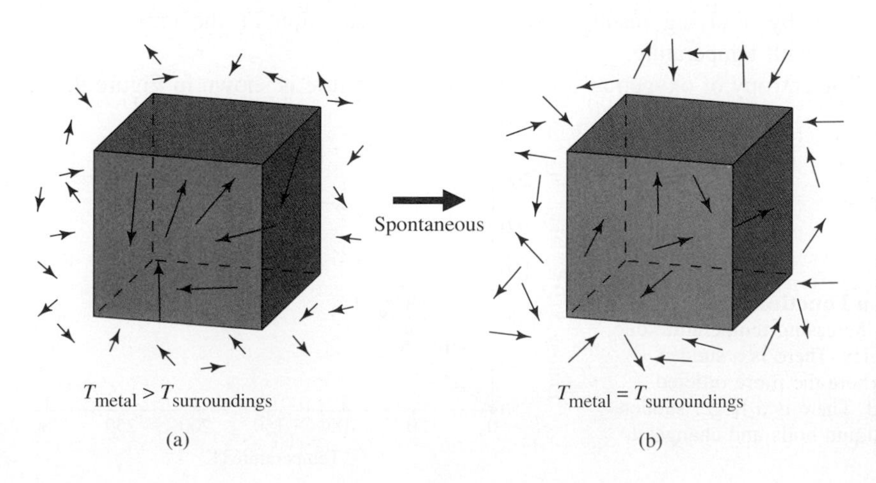

FIGURE 9.12 Spontaneous Cooling A hot block of metal spontaneously cools as heat is transferred to the surroundings. (a) The average kinetic energy of the atoms in the metal is greater than the average kinetic energy of the molecules in the surroundings. (b) The average kinetic energy of the atoms in the metal is the same as the average kinetic energy of the molecules in the surroundings.

Entropy A spontaneous process is a process in which disorder increases due to a dispersal of atoms, molecules, or energy in space as a consequence of the random motions of atoms and molecules. Although we cannot show here how it is done, the extent of disorder in a system can be defined in an exact way and can be calculated or measured experimentally. When it is defined in this way, disorder is called **entropy**, *S*. A spontaneous change is accompanied by an increase in entropy, or more exactly,

In any spontaneous change the total entropy of a system and its surroundings increases.

This statement is called the **second law of thermodynamics**. For any spontaneous process we may write

$$\Delta S_{total} = \Delta S_{system} + \Delta S_{surroundings} > 0$$

In other words, for any spontaneous process, the total entropy change for the system and its surroundings must be positive.

In deciding whether a reaction will occur spontaneously or not, we must take into account the entropy changes in both the system and the surroundings. In an exothermic reaction, the energy transferred to the surroundings as heat causes an increase in the entropy of the surroundings. The entropy of the system may either increase or decrease, but because this change is usually smaller than the change in the entropy of the surroundings, the total entropy of the system and its surroundings usually increases. Most exothermic reactions are therefore spontaneous. In an endothermic reaction, the flow of heat from the surroundings to the reacting system causes a decrease in the entropy of the surroundings, so an endothermic reaction is spontaneous only when there is a larger increase in the entropy of the system. This explains the observation that spontaneous exothermic reactions are more common than spontaneous endothermic reactions.

To determine whether a reaction is spontaneous or not, we must find the total entropy change of the system and its surroundings. We will consider entropy changes in the system first.

Entropy Changes in a Reacting System

It is beyond the scope of this text to discuss how quantitative values of the entropies of substances can be obtained. But we can increase our understanding of the concept of entropy by studying results obtained for the variation of the entropy of a substance with temperature.

The entropy of oxygen as a function of temperature is shown in Figure 9.13.

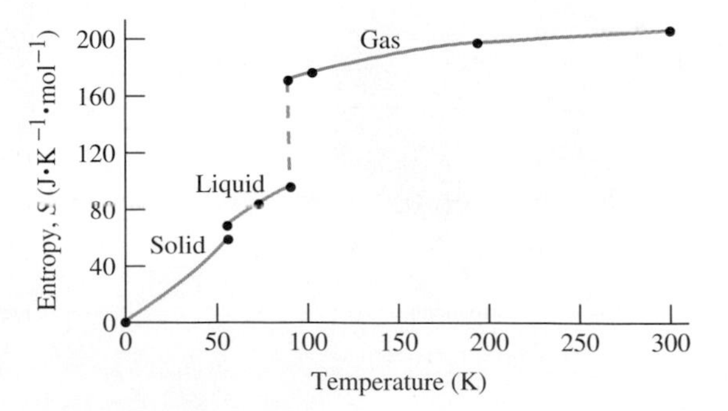

FIGURE 9.13 Entropy of Oxygen as a Function of Temperature Entropy increases with increasing temperature due to the increased random motion of molecules. There is a sudden increase in entropy at the melting point, where the more ordered solid form changes to a less ordered liquid. There is a larger sudden jump in entropy where the more ordered liquid boils and changes to a less ordered vapor.

TABLE 9.3 Standard Molar Entropies, $S°$, at 25°C ($J \cdot K^{-1} \cdot mol^{-1}$)

Solids		Liquids		Gases	
C(diamond)	2.4	H_2O	70.0	H_2	130.6
C(graphite)	5.8	Hg	75.9	CH_4	186.1
Fe	27.3	CH_3OH	126.8	HCl	186.8
S(orthorhombic)	32.0	Br_2	152.2	H_2O	188.7
Cu	33.2	HNO_3	155.6	N_2	191.5
Ag	42.6	H_2SO_4	156.9	NH_3	192.7
I_2	116.1	C_2H_5OH	160.7	CO	197.6
		C_6H_6(benzene)	172.2	O_3	238.8
CaO	38.1	CH_2Cl_2	177.8	O_2	205.0
NaF	51.3	$CHCl_3$	201.7	H_2S	205.6
NaCl	72.4	CCl_4	216.4	HI	206.5
NaBr	87.2	$SiCl_4$	239.7	CO_2	213.7
Fe_2O_3	87.4	C_5H_{12}(pentane)	263.3	Cl_2	223.0
$CaCO_3$	92.9	C_6H_{14}(hexane)	295.9	C_2H_6	229.5
AgCl	96.2			I_2	260.6
NaI	98.5			C_3H_8	269.9
$C_6H_{12}O_6$(glucose)	182.4			C_4H_{10}(butane)	310.1
$C_{12}H_{22}O_{11}$(sucrose)	360			C_5H_{12}(pentane)	348.9

At 0 K the entropy of oxygen or of any pure crystalline substance is zero—the molecules have a perfectly ordered arrangement and no thermal motion, so there is no disorder. As the temperature increases, the molecules begin to vibrate about their mean positions (their random thermal motion begins to increase), and the entropy of oxygen therefore increases. At the melting point, the regular arrangement of molecules in the solid changes to the more random arrangement of molecules in the liquid, so there is an abrupt increase in the entropy. The entropy steadily increases again until the liquid boils. As the oxygen becomes a gas, its volume increases greatly—the molecules become more dispersed, and there is a correspondingly large abrupt increase in the entropy of the oxygen. At high temperatures the entropy of oxygen continues to increase slowly.

The **standard molar entropy**, $S°$, of a substance, is the entropy of 1 mol of the substance in its standard state—that is, at 1 atm pressure—and at 25°C. The units of entropy are joules per kelvin per mole ($J \cdot K^{-1} \cdot mol^{-1}$). Joules, not kilojoules, are normally used in the units of entropy. The values given for standard molar entropies in Table 9.3 show that gases in general have larger entropies than solids. In a gas the molecules have a greater random motion and are more highly dispersed than in a solid. The entropies of liquids are generally between those of gases and solids. And substances consisting of large molecules usually have higher entropies than substances with smaller molecules, because their energy is shared among more atoms and is therefore more dispersed.

The **standard reaction entropy** is the difference between the sum of the entropies of the products in their standard states and the sum of the entropies of the reactants in their standard states:

$$\Delta S° = \sum[n_p(S°)_p] - \sum[n_r(S°)_r]$$

As an example, let us calculate the standard reaction entropy for the rusting of iron:

$$4Fe(s) + 3O_2(g) \rightarrow 2Fe_2O_3(s)$$

For this reaction we may write

Unlike the standard enthalpy of formation, the entropy of the most stable form of an element is not zero.

$$\Delta S^\circ = (2 \text{ mol})\, S^\circ(Fe_2O_3) - [(4 \text{ mol})\, S^\circ(Fe) + (3 \text{ mol})\, S^\circ(O_2)]$$

Using the values given in Table 9.3, we obtain

$$\begin{aligned}\Delta S^\circ &= (2 \text{ mol})(87.4 \text{ J} \cdot \text{K}^{-1} \cdot \text{mol}^{-1}) \\ &\quad - [(4 \text{ mol})(27.3 \text{ J} \cdot \text{K}^{-1} \cdot \text{mol}^{-1}) + (3 \text{ mol})(205.0 \text{ J} \cdot \text{K}^{-1} \cdot \text{mol}^{-1})] \\ &= -549.4 \text{ J} \cdot \text{K}^{-1}\end{aligned}$$

EXAMPLE 9.7 Standard Reaction Entropy

Use the data in Table 9.3 to calculate ΔS° for the reaction

$$CaCO_3(s) \rightarrow CaO(s) + CO_2(g)$$

Solution: $\Delta S^\circ = S^\circ(CaO) + S^\circ(CO_2) - S^\circ(CaCO_3)$

$$\begin{aligned}&= (1 \text{ mol})(38.1 \text{ J} \cdot \text{K}^{-1} \cdot \text{mol}^{-1}) + (1 \text{ mol})(213.7 \text{ J} \cdot \text{K}^{-1} \cdot \text{mol}^{-1}) \\ &\quad - (1 \text{ mol})(92.9 \text{ J} \cdot \text{K}^{-1} \cdot \text{mol}^{-1}) \\ &= 158.9 \text{ J} \cdot \text{K}^{-1}\end{aligned}$$

Exercise 9.7 Use the data in Table 9.3 to find the entropy change for the reduction of iron(III) oxide to iron, using carbon monoxide as the reducing agent:

$$Fe_2O_3(s) + 3CO(g) \rightarrow 2Fe(s) + 3CO_2(g)$$

There is a large decrease in entropy when iron rusts, because the highly dispersed oxygen gas reacts to form a compact, ordered solid. Conversely, when calcium carbonate decomposes, there is a large increase in entropy because a gas is produced from a solid. Whenever a reaction involves one or more gases, we can easily make a qualitative prediction of the entropy change for the reaction.

If the number of moles of gas *increases* during a reaction, the entropy change for the reaction is *positive*.

If the number of moles of gas *decreases* during a reaction, the entropy change for the reaction is *negative*.

EXAMPLE 9.8 Entropy Changes in Reactions Involving Gases

Predict whether the entropy increases or decreases in the following reaction.

$$N_2(g) + 3H_2(g) \rightarrow 2NH_3(g)$$

Solution: Two moles of gas is produced from 4 mol, so we expect a decrease in the entropy in this reaction. In other words, we predict that ΔS will be negative.

Exercise 9.8 Predict the sign of the entropy change for each of the following reactions.

(a) $NH_3(g) + HCl(g) \rightarrow NH_4Cl(s)$ **(b)** $BaO(s) + CO_2(g) \rightarrow BaCO_3(s)$
(c) $CH_4(g) + 2O_2(g) \rightarrow CO_2(g) + 2H_2O(l)$

Finding the Total Entropy Change

We saw earlier that for a spontaneous reaction,

$$\Delta S_{\text{total}} = \Delta S_{\text{system}} + \Delta S_{\text{surroundings}} > 0$$

So to predict whether or not a reaction will occur spontaneously, we must find the total entropy change, that is, the entropy change of the system *and* the surroundings. We can find the entropy change of the system from standard reaction entropies. But how do we find the entropy change of the surroundings? The entropy change of the surroundings results from the energy transferred to the surroundings as heat. If ΔH_{system} is the heat absorbed by the system, then $-\Delta H_{\text{system}}$ is the heat absorbed by the surroundings. It can be shown by thermodynamics that the entropy change of the surroundings is given by the relationship

$$\Delta S_{\text{surroundings}} = \frac{-\Delta H_{\text{system}}}{T}$$

The temperature in this formula must be expressed in kelvins.

Although we have not proved this relationship, it is reasonable that the entropy change in the surroundings should depend on the temperature in this way. A given amount of heat would produce more disorder at a lower temperature, where the surroundings are relatively ordered, than at a higher temperature, where the surroundings are initially more disordered (Figure 9.14).

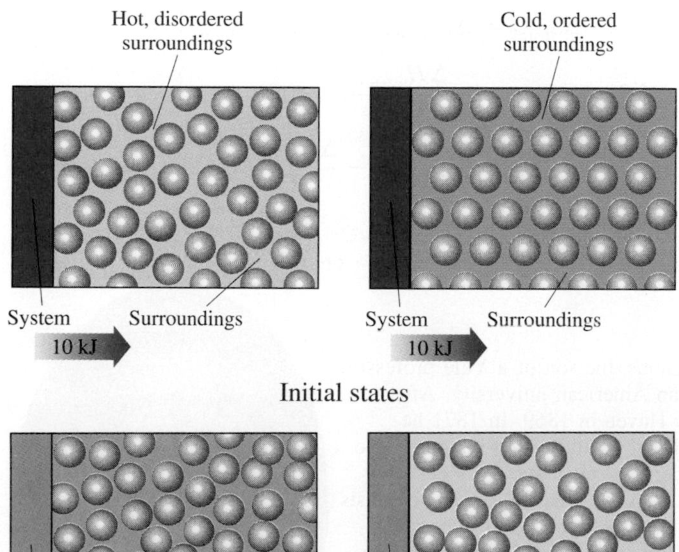

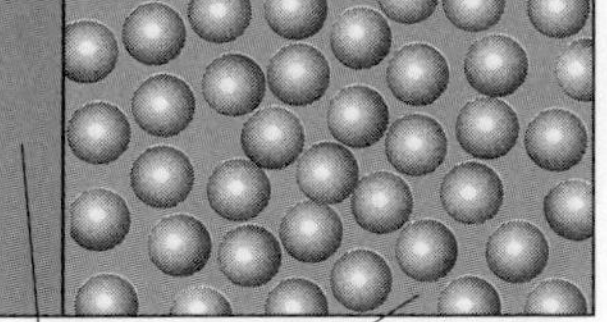

FIGURE 9.14 Effect of Temperature on Change in Entropy (a) When a given amount of heat flows into hot surroundings, it produces very little additional disorder. (b) If the surroundings are cold, however, the same amount of heat produces considerably more disorder.

Let us use this relationship to calculate the entropy change in the surroundings for the rusting of iron:

$$4Fe(s) + 3O_2(g) \rightarrow 2Fe_2O_3(s)$$

The standard reaction enthalpy at 25°C is twice $\Delta H_f^\circ(Fe_2O_3,s)$, which from Table 9.1 is $-824.2\ kJ \cdot mol^{-1}$. Thus,

$$\Delta H^\circ = 2\ \Delta H_f^\circ(Fe_2O_3, s) = (2\ mol)(-824.2\ kJ \cdot mol^{-1}) = -1648.4\ kJ$$

and because 25.00°C = 298.15 K,

$$\Delta S_{surroundings} = -\frac{\Delta H_{system}}{T} = -\left(\frac{-1648.4\ kJ}{298.15\ K}\right)\left(\frac{1000\ J}{1\ kJ}\right) = 5529\ J \cdot K^{-1}$$

We saw in equation (6) that the entropy change for the system is $-549\ J \cdot K^{-1}$, rounded to 3 significant figures. So we have

$$\Delta S_{total} = \Delta S_{system} + \Delta S_{surroundings} = -549\ J \cdot K^{-1} + 5529\ J \cdot K^{-1} = 4980\ J \cdot K^{-1}$$

The rusting of iron is spontaneous because the total entropy increases. The increase in the entropy of the surroundings greatly outweighs the decrease in the entropy of the system.

9.5 GIBBS FREE ENERGY

We have seen that we can determine the spontaneity of a reaction by calculating $\Delta S_{total} = \Delta S_{system} + \Delta S_{surroundings}$. However, it is much more convenient to determine spontaneity by using a quantity that is a property of the system alone. We can derive such a quantity by combining entropy and enthalpy into a single function, as first shown by Josiah Willard Gibbs (Figure 9.15):

$$\Delta S_{total} = \Delta S_{system} + \Delta S_{surroundings} > 0$$

$$\Delta S_{surroundings} = \frac{-\Delta H_{system}}{T}, \quad \text{so}$$

$$\Delta S_{total} = \Delta S_{system} - \frac{\Delta H_{system}}{T} > 0$$

Multiplying through by $-T$ gives

FIGURE 9.15 Josiah Willard Gibbs (1839-1903) Gibbs, the son of a Yale professor, was the first person to be awarded a Ph.D. in science from an American university. After a period of study in France and Germany, he returned to New Haven in 1869. In 1871 he became professor of theoretical physics at Yale and retained that position until his death. He is regarded by many as the most brilliant of native-born American scientists. A modest, reserved person he traveled little and had almost no contact with the great European scientists of the time. Working by himself, he applied the principles of thermodynamics to chemical reactions in a thorough mathematical fashion. He published this work from 1876 to 1878 in the *Transactions of the Connecticut Academy of Sciences*. This journal was not well known in Europe, and he wrote in such a concise, abstract style that most other scientists found his writing difficult to understand. Even Einstein said of a book by Gibbs, "It is a masterpiece, but it is hard to read." Not until the 1890s was Gibbs's work discovered by physical chemists in Europe. He was then universally recognized for his outstanding contribution to thermodynamics. Today his work remains the foundation of modern chemical thermodynamics.

$$-T\,\Delta S_{\text{total}} = -T\,\Delta S_{\text{system}} + \Delta H_{\text{system}} = \Delta H_{\text{system}} - T\,\Delta S_{\text{system}} < 0$$

All the quantities on the right side of the equation refer to the system, so we can drop the subscripts and write

$$-T\,\Delta S_{\text{total}} = \Delta H - T\,\Delta S$$

We call $-T\,\Delta S_{\text{total}}$ the **Gibbs free energy change**, ΔG:

$$\Delta G = \Delta H - T\,\Delta S$$

Although ΔG is composed of two parts, the enthalpy change and the entropy change, it is convenient to think of ΔG as the change in a single property of the system—the **Gibbs free energy**, G. We have seen that for a spontaneous reaction, ΔS_{total} must be positive, so $\Delta G = -T\,\Delta S_{\text{total}}$ must be negative. In other words,

In a spontaneous reaction, the Gibbs free energy of the system decreases: $\Delta G < 0$.

For a reaction in which all the reactants and products are in their standard states, the **standard reaction Gibbs free energy**, $\Delta G°$, is given by

$$\Delta G° = \Delta H° - T\,\Delta S°$$

The **standard Gibbs free energy of formation**, $\Delta G_f°$, is the Gibbs free energy change for the formation of 1 mol of a substance from its elements in their standard states at 25°C and 1 atm pressure. Some standard Gibbs free energies of formation are given in Table 9.4. So we can calculate the standard reaction Gibbs free energy,

TABLE 9.4 Standard Gibbs Free Energies of Formation at 25°C

Substance	$\Delta G_f°$ (kJ · mol^{-1})	Substance	$\Delta G_f°$ (kJ · mol^{-1})
Elements in standard states	0	$Fe_2O_3(s)$	−742.2
C(diamond)	+2.9	$HCl(g)$	−95.3
$AgCl(s)$	−109.8	$HBr(g)$	−53.5
$CCl_4(l)$	−65.3	$HI(g)$	+1.6
$CH_4(g)$	−50.8	$HNO_3(l)$	−80.8
$C_2H_2(g)$ (ethyne)	+209.2	$H_2O(l)$	−237.2
$C_2H_4(g)$ (ethene)	+68.1	$H_2O(g)$	−228.6
$C_2H_6(g)$ (ethane)	−32.9	$H_2S(g)$	−33.4
$C_3H_8(g)$ (propane)	−23.4	$H_2SO_4(l)$	−690.1
$C_4H_{10}(g)$ (butane)	−17.2	$NH_3(g)$	−16.4
$C_5H_{12}(l)$ (pentane)	−9.6	$NO(g)$	+86.6
$C_6H_6(l)$ (benzene)	+124.7	$NO_2(g)$	+51.3
$CH_3OH(l)$	−166.4	$NaF(s)$	−546.3
$C_2H_5OH(l)$	−174.9	$NaCl(s)$	−384.3
$C_6H_{12}O_6(s)$ (glucose)	−919.2	$NaBr(s)$	−349.1
$CH_3CO_2H(l)$	−390.0	$NaI(s)$	−282.4
$CO(g)$	−137.2	$NaOH(s)$	−379.5
$CO_2(g)$	−394.4	$O_3(g)$	+163.2
$CaO(s)$	−603.5	$SO_2(g)$	−300.2
$CaCO_3(s)$	−1128.8	$SO_3(g)$	−371.1

ΔG°, for any reaction for which we know the standard Gibbs free energies of formation, just as we can calculate standard reaction enthalpies from standard enthalpies of formation:

$$\Delta G^\circ = \sum[n_p(\Delta G_f^\circ)_p] - \sum[n_r(\Delta G_f^\circ)_r]$$

EXAMPLE 9.9 Calculating a Standard Reaction Gibbs Free Energy

Use the data in Table 9.4 to calculate ΔG° for the reaction

$$4Fe(s) + 3O_2(g) \rightarrow 2Fe_2O_3(s)$$

Solution:

$$\Delta G^\circ = \sum \Delta G_f^\circ\,(\text{products}) - \sum \Delta G_f^\circ\,(\text{reactants})$$

$$= [(2\ \text{mol})(-742.2\ \text{kJ}\cdot\text{mol}^{-1})] - [(4\ \text{mol})(0\ \text{kJ}\cdot\text{mol}^{-1}) + (3\ \text{mol})(0\ \text{kJ}\cdot\text{mol}^{-1})]$$

$$= -1484\ \text{kJ}\cdot\text{mol}^{-1}$$

ΔG° is negative and, as we concluded previously, the reaction is spontaneous under standard conditions.

Exercise 9.9 Calculate the standard reaction Gibbs free energy for the following reactions, and deduce whether they are spontaneous or not under standard conditions.

(a) $CaCO_3(s) \rightarrow CaO(s) + CO_2(g)$
(b) $CH_3OH(l) + \frac{3}{2}O_2(g) \rightarrow CO_2(g) + 2H_2O(l)$

Examination of the expression $\Delta G = \Delta H - T\Delta S$ enables us to predict when a reaction will be spontaneous, because we simply have to consider the circumstances under which ΔG will be negative. For example, all exothermic reactions (ΔH negative) are spontaneous when the associated entropy change, ΔS, is positive. Moreover, because ΔH is usually much larger than $T\Delta S$, ΔG is usually negative for an exothermic reaction even when ΔS is negative, except at a high temperature. Similarly, most endothermic reactions (ΔH positive) are not spontaneous. They are spontaneous only if ΔS is positive and $T\Delta S$ is greater than ΔH, which is most likely at high temperatures. In total we have to consider four possible combinations of ΔH and ΔS and the magnitude of $T\Delta S$ at high and at low temperatures, as follows:

Enthalpy Change	Entropy Change	Free Energy Change	Spontaneous
1. Exothermic			
$\Delta H < 0$	$\Delta S > 0$	$\Delta G < 0$ at all T	Yes
2. Exothermic			
$\Delta H < 0$	$\Delta S < 0$	$\Delta G < 0$ at low T	Yes
$\Delta H < 0$	$\Delta S < 0$	$\Delta G > 0$ at high T	No
3. Endothermic			
$\Delta H > 0$	$\Delta S < 0$	$\Delta G > 0$ at all T	No
4. Endothermic			
$\Delta H > 0$	$\Delta S > 0$	$\Delta G > 0$ at low T	No
$\Delta H > 0$	$\Delta S > 0$	$\Delta G < 0$ at high T	Yes

Gibbs Free Energy and Equilibrium

A reaction for which ΔG is positive is not spontaneous in the forward direction, but it is spontaneous in the reverse direction, because it then has a negative ΔG. If $\Delta G = 0$, the reaction has no tendency to proceed in either direction; in other words, the system is at equilibrium.

ΔG	Reaction
Negative	Spontaneous
Positive	Not spontaneous
Zero	At equilibrium

The magnitude of ΔG gives us a measure of the extent to which a reacting system is displaced from equilibrium. When ΔG has a large negative value, the system is far from equilibrium. When ΔG has a small negative value, the system is close to equilibrium. When ΔG is zero, the system is at equilibrium.

ΔG tells us whether a reaction is at equilibrium or, if it is not already at equilibrium, the direction in which it will proceed. But it is most important to understand that ΔG tells us nothing about the *rate* at which a reaction proceeds toward equilibrium. For instance, hydrocarbons and oxygen are far from equilibrium at room temperature. The standard reaction Gibbs free energy for their oxidation has a large negative value, but the rate at which these reactants are converted to carbon dioxide and water is almost infinitely slow at ordinary temperatures, and no reaction is observed. We discuss the factors that determine the rates of reactions in Chapter 15.

Although we introduced the Gibbs free energy as a convenient combination of the enthalpy and entropy changes for a reaction that tells the direction in which a reaction will proceed, it has a wider significance and usefulness. For example, we will see in Chapter 13 that it enables us to find the composition of a reacting system that has come to equilibrium. Another very useful application of the Gibbs free energy is to show us how much work we can get from the energy change accompanying a reaction.

Gibbs Free Energy and Work

We might think that it should be possible, at least in principle, to convert all the energy produced by a reaction into any suitable form, such as mechanical energy or work. However, this is not the case. Consider, for example, a reaction that occurs with a negative entropy change, such as the oxidation of iron. For such a reaction there must be some way to increase sufficiently the entropy of the surroundings to compensate for the reduction in the entropy of the system. The entropy of the surroundings can be increased only by heat transferred from the system. Thus, the part of the energy produced by a reaction that is lost as heat cannot be converted to work. Thermodynamics shows that

The maximum amount of energy produced by a reaction that can be converted to work is $-\Delta G$.

Of the total energy $-\Delta H$ produced by a reaction with a negative entropy change, an amount $T\,\Delta S$ is "wasted" as heat transferred to the surroundings. The resulting increase in the entropy of the surroundings compensates for the negative entropy change in the system. We can now see why ΔG is called *free* energy—it is the amount of energy that is "free," or available, to do work.

As an example let us consider the hydrogen–oxygen reaction

$$2H_2(g) + O_2(g) \rightarrow 2H_2O(l)$$

for which

$$\Delta H^\circ = -572 \text{ kJ} \qquad \Delta G^\circ = -474 \text{ kJ} \qquad \Delta S^\circ = -326 \text{ J} \cdot \text{K}^{-1}$$

If we simply burn the hydrogen as a flame in the atmosphere, then 572 kJ are transferred as heat to the surroundings, and no work is done. However, if we use the heat to expand a gas and drive a piston, as in an automobile engine (Box 8.2), we convert to work some of the 572 kJ of energy produced by the reaction. But even if we had a perfect engine, it would convert only a maximum of 474 kJ: An amount $T\,\Delta S = (298 \text{ K}) \times (326 \text{ J} \cdot \text{K}^{-1}) = 98$ kJ must be released to the surroundings to compensate for the decrease in entropy in the reacting system. In practice, even less energy will be converted to work, due to engine inefficiency.

A mechanical engine, such as an automobile engine, is not the most efficient way to transform the energy of a reaction to mechanical energy. A more efficient method—that is, a method that converts more energy to work—is to carry out the reaction in a fuel cell to generate electricity, as we will describe in Chapter 13. But the maximum electrical energy that we can obtain from the reaction is still only 474 kJ or (474/572)(100%) = 83% of the total energy change for the reaction.

Coupled Reactions

Water does not spontaneously decompose into hydrogen and oxygen, because ΔG for this reaction is positive. However, we can make this reaction occur by passing an electric current through the water. The reaction that produces the electric current, either in a battery or in a generating station that burns fossil fuel, has a ΔG that is sufficiently negative that the combination of the two values gives an overall negative free energy change. The two reactions are said to be **coupled** to give an overall reaction with a negative ΔG. Figure 9.16 shows a useful analogy for the coupling of reactions.

The coupling of reactions to bring about a nonspontaneous reaction is very important in biochemical systems. Many of the reactions that are essential for the maintenance of life do not occur spontaneously within the human body. These reactions are made to occur, however, when they are coupled with reactions that are spontaneous and release energy. For example, when glucose is oxidized in the body, a substantial amount of energy is released:

$$C_6H_{12}O_6(s) + 6O_2(g) \rightarrow 6CO_2(g) + 6H_2O(l) \qquad \Delta G^\circ = -2880 \text{ kJ}$$

However, some means is necessary to couple the energy released in this reaction to

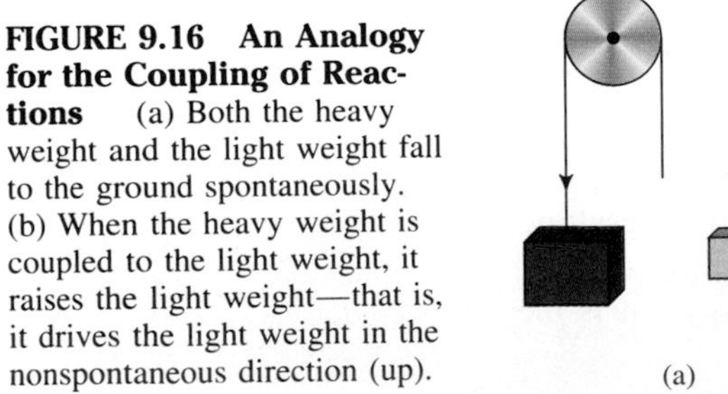

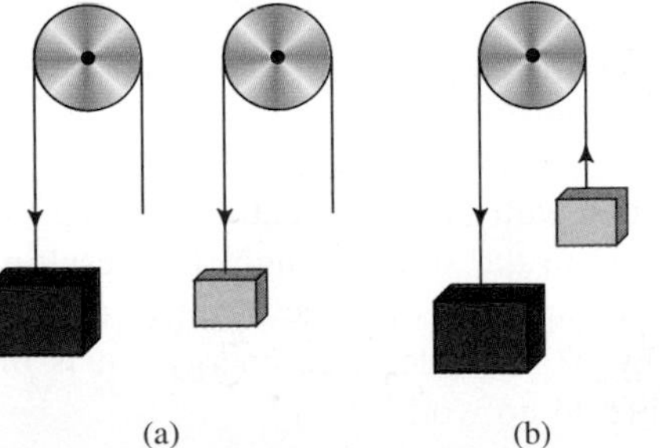

FIGURE 9.16 An Analogy for the Coupling of Reactions (a) Both the heavy weight and the light weight fall to the ground spontaneously. (b) When the heavy weight is coupled to the light weight, it raises the light weight—that is, it drives the light weight in the nonspontaneous direction (up).

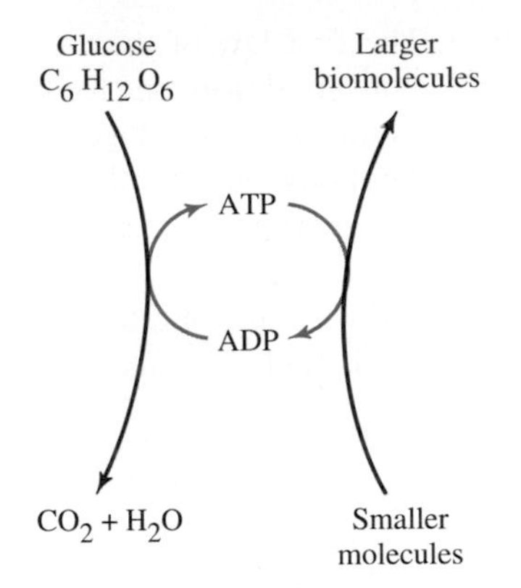

FIGURE 9.17 Schematic Representation of Coupled Reactions The interconversion of ADP and ATP. On the left, the free energy produced by the oxidation of glucose is used to convert ADP to ATP. On the right, the free energy produced by the conversion of ATP back to ADP is used to convert simple molecules to larger biomolecules.

e reactions that *require* energy. This coupling is most commonly accomplished in e body by the interconversion of ATP (Figure 7.10) and ADP,

$$\underset{\text{ATP}^{4-}}{\text{Ad}-\underset{\underset{\text{O}^-}{|}}{\overset{\overset{\text{O}}{\|}}{\text{P}}}-\text{O}-\underset{\underset{\text{O}^-}{|}}{\overset{\overset{\text{O}}{\|}}{\text{P}}}-\text{O}-\underset{\underset{\text{O}^-}{|}}{\overset{\overset{\text{O}}{\|}}{\text{P}}}-\text{O}^-} + \text{H}_2\text{O}$$

$$\rightarrow \underset{\text{ADP}^{3-}}{\text{Ad}-\underset{\underset{\text{O}^-}{|}}{\overset{\overset{\text{O}}{\|}}{\text{P}}}-\text{O}-\underset{\underset{\text{O}^-}{|}}{\overset{\overset{\text{O}}{\|}}{\text{P}}}-\text{O}^-} + \text{HPO}_4^{2-} + \text{H}^+$$

for which, under the conditions in the human body,

$$\Delta G = -30 \text{ kJ} \qquad \Delta H = -20 \text{ kJ} \qquad \Delta S = +34 \text{ J} \cdot \text{K}^{-1}$$

This reaction can drive any reaction for which $\Delta G < 30$ kJ. For example, the biosynthesis of sucrose from glucose and fructose,

$$\underset{\text{Glucose}}{C_6H_{12}O_6} + \underset{\text{Fructose}}{C_6H_{12}O_6} \rightarrow \underset{\text{Sucrose}}{C_{12}H_{22}O_{11}} + H_2O$$

has $\Delta G = 23$ kJ under body conditions, so this reaction can be driven by coupling it to the conversion of ATP to ADP,

$$\text{glucose} + \text{fructose} + \text{ATP}^{4-} \rightarrow \text{sucrose} + \text{ADP}^{3-} + \text{HPO}_4^{2-} + \text{H}^+ \qquad \Delta G^\circ = -7 \text{ kJ}$$

giving an overall reaction with a negative free energy change. The energy obtained from the oxidation of glucose can then be used to convert ADP back to ATP. Thus, as we shall discuss in more detail in Chapter 17, the interconversion of ADP and ATP serves to transfer the energy produced in the oxidation of food (such as glucose) to drive reactions that convert simple molecules to more complex molecules (Figure 9.17).

Postscript

In Section 9.3 we discussed the need to find alternative energy sources to replace petroleum. In light of our discussion of entropy, we are now in a better position to understand the real nature of the problem that we face. The first law of thermodynamics tells us that energy cannot be destroyed but only converted from one form to another, so we need not worry about using it up. There is no energy shortage on earth or in the rest of the universe! The second law of thermodynamics tells us, however, that energy is continually being dispersed because entropy increases in any spontaneous process. It is our supply of highly localized energy that is being depleted. Only this highly localized energy can be utilized to do useful work. So it is alternative sources of highly localized energy that we need to find to replace the energy locked in the chemical bonds of petroleum.

SUMMARY

Thermochemistry is the study of the heat changes in chemical reactions. Thermodynamics is the study of energy and its transformations. Heat is energy that is transferred as a result of a temperature difference. Internal energy, E (units kJ), the energy stored in a substance, is the sum of the kinetic and potential energies of all the nuclei and electrons that comprise the substance. For a reaction, $\Delta E = E_{\text{products}} - E_{\text{reactants}}$. The energy change accompanying a reaction can be observed as heat transferred to or from a system from or to its surroundings.

According to the first law of thermodynamics, energy is conserved. Thus, for an isolated system, in which neither energy nor matter is transferred between a system and its surroundings, the energy is constant. In a closed system, when energy but not matter is transferred between the system and its surroundings, $\Delta E = q + w$, where q is the amount of heat that enters the system and w is the amount of work done on the system. The heat q transferred depends on the way a process is accomplished: At constant pressure, $\Delta E = q + w$, but at constant volume, no work is done on or by the system, so $w = 0$ and $\Delta E = q$.

The heat absorbed in a reaction at constant pressure is called the enthalpy change, $\Delta H = q$. For an endothermic reaction, ΔH has a positive value; for an exothermic reaction, ΔH has a negative value. When all the substances in the reaction are at 1 atm pressure and at a specified temperature, usually 25°C, they are said to be in their standard states. The standard reaction enthalpy, $\Delta H°$ (units kJ), is the enthalpy change for a reaction at 1 atm pressure and 25°C for the numbers of moles of reactants and products specified in the equation for the reaction.

Calorimetry measures the heat absorbed or evolved in a reaction. The molar heat capacity of a substance is the amount of heat required to raise the temperature of 1 mol of the substance 1 K.

According to Hess's law, the reaction enthalpy, ΔH, is independent of the path a reaction follows. The standard enthalpy of formation, ΔH_f° (units kJ · mol^{-1}), of a substance is the reaction enthalpy for the formation of 1 mol of the substance from its elements in their standard states (the most stable form of an element at 1 atm and 25°C). The standard reaction enthalpy, $\Delta H°$, is given by

$$\Delta H^\circ = \sum[n_p(\Delta H_f^\circ)_p] - \sum[n_r(\Delta H_f^\circ)_r]$$

where p refers to the products, r to the reactants, and n to the numbers of moles of each in the balanced equation. A compound with a negative ΔH_f° contains less energy and is more stable, and a compound with a positive ΔH_f° contains more energy and is less stable, than the elements from which they form.

The energy required to break a bond is a measure of the bond's strength. ΔH for the dissociation of a gaseous molecule AX(g) into gaseous atoms is the bond energy, BE(AX). For a polyatomic molecule AX_n, the average BE(AX) is one-nth $\Delta H°$ for the dissociation of an $AX_n(g)$ molecule into its gaseous atoms. Similar bonds in different molecules have similar bond energies. Bond energy increases with increasing bond strength and in the order single bond $<$ double bond $<$ triple bond. Approximate reaction enthalpies are found from bond energies by using the formula

$$\Delta H^\circ = \sum BE \text{ (bonds broken in reactants)} - \sum BE \text{ (bonds formed in products)}$$

Alternative energy sources are important economically and environmentally. Possibilities include coal conversion, which gives gaseous fuels such as CH_4, H_2, and CO; hydrogen from electrolysis or the photosynthetic decomposition of water; the conversion of biomass to biogas or ethanol; and the conversion of solar energy directly to electricity.

Spontaneous processes occur without outside intervention. Nonspontaneous processes never occur without external intervention. To account for the direction of spontaneous change, the concept of entropy, S, which measures the extent of disorder of a system, is needed. In a spontaneous process, disorder increases due to the dispersal of atoms or molecules, or energy, in space. According to the second law of thermodynamics, the total entropy of a system and its surroundings increases in a spontaneous change:

$$\Delta S_{\text{total}} = \Delta S_{\text{system}} + \Delta S_{\text{surroundings}} > 0$$

The standard molar entropy of a substance, S° (units $\text{J} \cdot \text{K}^{-1} \cdot \text{mol}^{-1}$), is the entropy of 1 mol of a substance in its standard state (1 atm and 25°C). The standard reaction entropy, ΔS° (units $\text{J} \cdot \text{K}^{-1}$), for a reaction is given by

$$\Delta S^\circ = \sum[n_p(S^\circ)_p] - \sum[n_r(S^\circ)_r]$$

Writing $-T\,\Delta S_{\text{total}} = \Delta G$, we combine the entropy and enthalpy changes for any process into the function $\Delta G = \Delta H - T\,\Delta S$, where ΔG is the reaction Gibbs free energy change. Because $-T\,\Delta S_{\text{total}}$ decreases in any spontaneous process, ΔG is negative ($\Delta G < 0$) for any spontaneous process. For a nonspontaneous process ($\Delta G > 0$), ΔG is positive.

For a reaction in which all the reactants and products are in their standard states, the standard reaction Gibbs free energy is given by $\Delta G^\circ = \Delta H^\circ - T\,\Delta S^\circ$. The standard Gibbs free energy of formation, ΔG_f°, of a substance is the ΔG° for the formation of 1 mol of the substance from its elements in their standard states (25°C and 1 atm). The standard reaction Gibbs free energy for a reaction is then also given by

$$\Delta G^\circ = \sum[n_p(\Delta G_f^\circ)_p] - \sum[n_r(\Delta G_f^\circ)_r]$$

The larger and more negative ΔG, the farther a reaction is from equilibrium. For a reaction at equilibrium, $\Delta G = 0$. The maximum amount of energy produced by a reaction that can be converted to work is $-\Delta G$. A nonspontaneous reaction, with a positive ΔG, can be driven in the spontaneous direction by coupling it to a spontaneous reaction, with a negative ΔG of greater magnitude.

IMPORTANT TERMS

biomass (page 320)
bond energy (page 315)
calorimetry (page 307)
closed system (page 304)
endothermic reaction (page 302)
enthalpy change, ΔH (page 306)
entropy, S (page 326)
exothermic reaction (page 302)
first law of thermodynamics (page 304)
Gibbs free energy change, ΔG (page 331)
heat, q (page 303)
heat capacity (page 307)
Hess's law (page 310)
internal energy, E (page 304)
internal energy change, ΔE (page 304)
isolated system (page 304)
molar heat capacity (page 307)
second law of thermodynamics (page 326)
specific heat (page 307)
spontaneous process (page 322)
standard enthalpy of formation, ΔH_f° (page 312)
standard Gibbs free energy of formation, ΔG_f° (page 331)
standard molar entropy, S° (page 327)
standard reaction enthalpy, ΔH° (page 306)
standard reaction entropy, ΔS° (page 327)
standard reaction Gibbs free energy, ΔG° (page 331)
standard state (page 306)
surroundings (page 304)
system (page 304)
thermochemistry (page 302)
thermodynamics (page 302)
work, w (page 304)

REVIEW QUESTIONS

1. In terms of the kinetic molecular model, how is heat transferred from a warmer object to a cooler object?

2. What is the internal energy, E, of a substance?

3. In a reaction in which energy is transferred only as heat, how is the change in internal energy, ΔE, related to the amount of heat, q, that flows into or out of the system?

4. **(a)** How is the standard reaction enthalpy, ΔH°, defined? **(b)** Does ΔH° have a positive or a negative value for (i) an exothermic reaction; (ii) an endothermic reaction?

5. Why does ΔH° for a reaction depend only on the initial and final states of the system and not on how the reaction is carried out?

6. Define **(a)** "heat capacity," **(b)** "molar heat capacity," and **(c)** "specific heat."

7. State Hess's law.

8. How is ΔH° for a reaction calculated from the ΔH_f° values for each of its reactants and products?

9. For what reaction is ΔH° the Cl—Cl bond energy?

10. How is the ΔH° of a gas-phase reaction related to the energies of the bonds in the reactant and product molecules of the reaction?

11. Explain what is meant by **(a)** "alternate energy source"; **(b)** "coal conversion"; **(c)** "biogas"; **(d)** "gasohol"; **(e)** "solar energy."

12. What is meant by a spontaneous change?

13. Energy changes alone cannot account for the direction of physical and chemical change. Why not?

14. In any spontaneous change, what can be said about the entropy of a system and its surroundings?

15. If a reaction is spontaneous, what can be said about the reverse reaction under the same conditions?

16. How does the standard molar entropy of a substance, S°, change as its physical state changes from solid to liquid to gas?

17. What equation relates the entropy change of a reaction to the standard molar entropies, S°, of the reactants and products?

18. What equation relates the standard reaction Gibbs free energy, ΔG°, to the standard reaction enthalpy, ΔH°, and to the standard reaction entropy, ΔS°, of a reaction at a given temperature T?

19. What is the sign of ΔG for a spontaneous reaction?

20. What equation relates the standard reaction Gibbs free energy of any reaction to the ΔG_f° values of the reactants and the products?

PROBLEMS

Thermochemistry and the First Law of Thermodynamics

1. When 0.150 g of liquid octane, $C_8H_{18}(l)$, was burned in a flame calorimeter containing 1.500 kg of water, the temperature of the water rose from 25.246 to 26.386°C. What is the standard reaction enthalpy of combustion of octane at 25°C?

2. A 50.0-mL sample of 0.400-M NaOH(aq) was added to 20.0 mL of 0.500-M $H_2SO_4(aq)$ in a calorimeter of heat capacity 39.0 $J \cdot K^{-1}$. The temperature of the resulting solution rose by 3.60°C. What is the standard reaction enthalpy for neutralization of $H_2SO_4(aq)$ with NaOH(aq)?

3. A 25.00-mL sample of HCl(aq) was mixed with 25.00 mL of KOH(aq) of the same concentration in a calorimeter. As a result the temperature rose from 25.00 to 26.60°C. Given that $\Delta H^\circ = -56.02$ kJ for the reaction

$$H_3O^+(aq) + OH^-(aq) \rightarrow 2H_2O(l)$$

and that the heat capacity of dilute aqueous solutions is 75.4 $J \cdot K^{-1} \cdot mol^{-1}$, determine the HCl($aq$) concentration. (Assume the heat capacity of the calorimeter to be negligible.)

4. Calcium oxide (lime) reacts with water in an exothermic reaction to give calcium hydroxide:

$$CaO(s) + H_2O(l) \rightarrow Ca(OH)_2(s)$$

A 5.40-g sample of CaO(s) was added to 500 mL of water in a calorimeter of heat capacity 350 $J \cdot K^{-1}$. The observed temperature increase was 2.60 K. What is the standard reaction enthalpy for the reaction of 1 mol of CaO(s) to give 1 mol of $Ca(OH)_2(s)$?

5. A small, well-insulated hydrogenation apparatus has a heat capacity of 1.500 $kJ \cdot K^{-1}$. When 1.500 g of ethene is hydrogenated completely to ethane in the apparatus, what temperature rise should be observed? (Assume that the heat capacities of the gases are negligible compared to that of the apparatus and that the pressure remains constant.)

6. Calculate ΔH° for the reaction

$$2F_2(g) + 2H_2O(l) \rightarrow 4HF(g) + O_2(g)$$

given that

$$H_2(g) + F_2(g) \rightarrow 2HF(g) \qquad \Delta H^\circ = -542 \text{ kJ}$$
$$2H_2(g) + O_2(g) \rightarrow 2H_2O(l) \qquad \Delta H^\circ = -572 \text{ kJ}$$

7. Calculate ΔH° for the reaction

$$2CO(g) + O_2(g) \rightarrow 2CO_2(g)$$

given that

$$C(s, \text{graphite}) + O_2(g) \rightarrow CO_2(g) \qquad \Delta H^\circ = -393.5 \text{ kJ}$$
$$2C(s, \text{graphite}) + O_2(g) \rightarrow 2CO(g) \qquad \Delta H^\circ = -221.0 \text{ kJ}$$

8. Calculate the standard enthalpy change for the reaction

$$2C(s) + H_2(g) \rightarrow C_2H_2(g)$$

given that

$$2C_2H_2(g) + 5O_2(g) \rightarrow 4CO_2(g) + 2H_2O(l) \quad \Delta H^\circ = -2600 \text{ kJ}$$
$$C(s, \text{graphite}) + O_2(g) \rightarrow CO_2(g) \qquad \Delta H^\circ = -394 \text{ kJ}$$
$$2H_2(g) + O_2(g) \rightarrow 2H_2O(l) \qquad \Delta H^\circ = -572 \text{ kJ}$$

9. Calculate the standard enthalpy change for the reaction

$$2H_2O_2(l) \rightarrow 2H_2O(l) + O_2(g)$$

given that

$$2H_2(g) + O_2(g) \rightarrow 2H_2O(g) \qquad \Delta H^\circ = -483.6 \text{ kJ}$$
$$H_2O(l) \rightarrow H_2O(g) \qquad \Delta H^\circ = 44.0 \text{ kJ}$$
$$H_2(g) + O_2(g) \rightarrow H_2O_2(l) \qquad \Delta H^\circ = -187.6 \text{ kJ}$$

10. It has been proposed that the following reaction occurs in the stratosphere:

$$HO(g) + Cl_2(g) \rightarrow HOCl(g) + Cl(g)$$

Calculate the standard reaction enthalpy from the following data:

$$Cl_2(g) \rightarrow 2Cl(g) \qquad \Delta H^\circ = 242 \text{ kJ}$$
$$H_2O_2(g) \rightarrow 2OH(g) \qquad \Delta H^\circ = 134 \text{ kJ}$$
$$H_2O_2(g) + 2Cl(g) \rightarrow 2HOCl(g) \qquad \Delta H^\circ = -209 \text{ kJ}$$

11. When $PCl_3(g)$ forms from white phosphorus and chlorine, ΔH° for the reaction is $-306 \text{ kJ} \cdot \text{mol}^{-1}$. When $PCl_3(g)$ forms from red phosphorus and chlorine, ΔH° is $-288 \text{ kJ} \cdot \text{mol}^{-1}$. From this information, calculate ΔH° for the conversion of red phosphorus to white phosphorus.

12. What are the enthalpics of formation of each of $H_2O(l)$, $H_2O(g)$, and $NH_3(g)$, given that

$$H_2(g) + \tfrac{1}{2}O_2(g) \rightarrow H_2O(l) \qquad \Delta H^\circ = -285.8 \text{ kJ}$$
$$H_2O(g) \rightarrow H_2O(l) \qquad \Delta H^\circ = -44.0 \text{ kJ}$$
$$2NH_3(g) \rightarrow N_2(g) + 3H_2(g) \qquad \Delta H^\circ = 92.4 \text{ kJ}$$

13. The standard reaction enthalpy of combustion of liquid *n*-heptane, $C_7H_{16}(l)$, is -4816.9 kJ. The products of this combustion are liquid water and carbon dioxide gas. Calculate the standard reaction enthalpy of formation of liquid *n*-heptane.

14. What is the standard reaction enthalpy for the reaction of 1 mol of $SO_3(g)$ with 1 mol of $H_2O(g)$ to give 1 mol of $H_2SO_4(l)$?

15. Calculate ΔH_f° for ethyne, $C_2H_2(g)$, from the standard reaction enthalpy of -312 kJ for the reaction

$$C_2H_2(g) + 2H_2(g) \rightarrow C_2H_6(g)$$

and from the standard enthalpy of formation of ethane, $C_2H_6(g)$, given in Table 9.1.

16. Gas lighters contain liquid butane, $C_4H_{10}(l)$, for which $\Delta H_f^\circ = -126 \text{ kJ} \cdot \text{mol}^{-1}$. Calculate the heat evolved when 1.00 g of liquid butane in a lighter is burned, assuming that the products of combustion are $CO_2(g)$ and $H_2O(g)$ under standard conditions.

17. Calculate ΔH_f° for propane, $C_3H_8(g)$, from its standard reaction enthalpy of combustion ($-2044 \text{ kJ} \cdot \text{mol}^{-1}$) and from other data in Table 9.1.

18. Calculate ΔH° for the conversion of 3 mol of gaseous acetylene, $C_2H_2(g)$, to 1 mol of gaseous benzene, $C_6H_6(g)$, given that it takes 434.5 J to convert 1.00 g of liquid benzene to gaseous benzene and that $\Delta H_f^\circ(C_6H_6, l) = 49.0 \text{ kJ} \cdot \text{mol}^{-1}$.

19. When 2 mol of gaseous hydrogen iodide forms from 1 mol of gaseous hydrogen and 1 mol of solid iodine, 52.8 kJ of heat is absorbed. What is the standard enthalpy of formation of $HI(g)$?

20. Calculate the standard reaction enthalpy for the reaction in a constant-pressure container of 1 mol of octane, $C_8H_{18}(l)$, with oxygen to give $CO_2(g)$ and $H_2O(l)$. ($\Delta H_f^\circ(C_8H_{18}, l) = -224 \text{ kJ} \cdot \text{mol}^{-1}$.)

21. The standard reaction enthalpies of the combustion of graphite and diamond to $CO_2(g)$ are $-393.5 \text{ kJ} \cdot \text{mol}^{-1}$ and $-395.6 \text{ kJ} \cdot \text{mol}^{-1}$, respectively. What is the enthalpy of formation of C(diamond) from C(graphite)? Which is more stable, diamond or graphite?

Bond Energies

22. The standard enthalpies of formation of atomic oxygen gas and atomic nitrogen gas are 249 and 471 $\text{kJ} \cdot \text{mol}^{-1}$, respectively. What is the bond energy of **(a)** the double bond in $O_2(g)$ and **(b)** the triple bond in $N_2(g)$?

23. The standard enthalpy of formation of $ClF(g)$ is $-55.7 \text{ kJ} \cdot \text{mol}^{-1}$, and the bond energies of $F_2(g)$ and $Cl_2(g)$ are 155 $\text{kJ} \cdot \text{mol}^{-1}$ and 242 $\text{kJ} \cdot \text{mol}^{-1}$, respectively. What is the bond energy of $ClF(g)$?

24. Given the ΔH_f° values (in $\text{kJ} \cdot \text{mol}^{-1}$) P(g), 316.2; Cl(g), 121.5; $PCl_3(g)$, -287.0; and $PCl_5(g)$, -374.7, calculate **(a)** the average P—Cl bond energy in PCl_3, **(b)** the average P—Cl bond energy in PCl_5, and **(c)** the average bond energy of the two additional P—Cl bonds in PCl_5.

25. Given the ΔH_f° values (in $\text{kJ} \cdot \text{mol}^{-1}$) H(g), 218.0; O(g), 249.1; $H_2O_2(g)$, -136.4; and $H_2O(g)$, -241.8, **(a)** calculate the average O—H bond energy in water.

(b) Using the result of (a), calculate the O—O bond energy in hydrogen peroxide.

(c) Is the O—O bond in H_2O_2 a strong or a weak bond, compared, for example, to the C—C bond?

26. Calculate the carbon–oxygen bond energies in CO and CO_2, using Table 9.1.

27. Using the average C—C and C—H bond energies given in Table 9.2, estimate the standard enthalpy of formation of ethane, $C_2H_6(g)$.

28. Use the average bond energies given in Table 9.2 to estimate ΔH° for each of the following gas-phase reactions.

(a) $H_2S + Cl_2 \rightarrow SCl_2 + H_2$

(b) $CH_4 + 2F_2 \rightarrow CH_2F_2 + 2HF$

(c) $CH_2Cl_2 + CH_4 \rightarrow 2CH_3Cl$

29. Repeat Problem 28 for each of the following gas-phase reactions.

(a) $C_2H_2 + C_2H_6 \rightarrow 2C_2H_4$

(b) $2H_2O_2 \rightarrow 2H_2O + O_2$

(c) $C_2H_6 \rightarrow C_2H_4 + H_2$

30. From the average bond energies in Table 9.2, estimate the standard enthalpies of formation of each of the following.

(a) $H_2O(g)$ **(b)** $H_2S(g)$ **(c)** $NH_3(g)$ **(d)** $PH_3(g)$

(e) $CH_4(g)$

31. Given that the standard enthalpies of dissociation of $H_2(g)$, $Cl_2(g)$, and $HCl(g)$ are 436, 242, and 431 $\text{kJ} \cdot \text{mol}^{-1}$, respectively, what is the standard enthalpy of formation of $HCl(g)$?

Alternative Energy Sources

32. **(a)** Assuming that coke (graphite) and natural gas (methane) have identical costs per gram, which of these fuels is more economical for heating a home?

(b) If the price per gram of the more economical fuel of (a) is doubled, would that fuel still be more economical?

33. What is the standard enthalpy of formation of propane, $C_3H_8(g)$, given that the combustion of 1.00 g of propane releases 46.3 kJ of heat when it is burned to $CO_2(g)$ and $H_2O(g)$ at 298 K and 1 atm?

34. Using Table 9.1, calculate $\Delta H°$ for the photosynthesis reaction

$$6CO_2(g) + 6H_2O(l) \rightarrow C_6H_{12}O_6(s) + 6O_2(g)$$

(a) per mole and **(b)** per gram of glucose. (In nature, light from the sun rather than heat provides the large energy required for this reaction.)

35. What are some of the problems associated with the use of fossil fuels as energy sources?

36. Discuss the usefulness of **(a)** hydrogen, **(b)** biogas, and **(c)** gasohol as alternative energy sources.

Entropy and the Second Law of Thermodynamics

37. For each of the following, use qualitative reasoning to decide which system will have the greater entropy.

(a) 1 mol of ice at 0°C or 1 mol of water at the same temperature

(b) a pack of cards arranged in suits or a pack of cards randomly shuffled

(c) a collection of jigsaw pieces or a completed puzzle

(d) solid ammonium chloride or an aqueous solution of ammonium chloride

38. Does the degree of disorder increase or decrease as each of the following processes proceeds? Is the entropy change for each positive or negative?

(a) the evaporation of 1 mol of ethanol

(b) $2Mg(s) + O_2(g) \rightarrow 2MgO(s)$

(c) $N_2(g) + 3H_2(g) \rightarrow 2NH_3(g)$

(d) $BaCl_2 \cdot H_2O(s) \rightarrow BaCl_2(s) + H_2O(g)$

39. Predict the sign of the entropy change, ΔS, for each of the following reactions.

(a) $CaCO_3(s) \rightarrow CaO(s) + CO_2(g)$

(b) $NH_3(g) + HCl(g) \rightarrow NH_4Cl(s)$

(c) $BaO(s) + CO_2(g) \rightarrow BaCO_3(s)$

40. Under what conditions is each of the following statements true for a spontaneous process?

(a) The system moves toward a state of lower energy.

(b) The system moves toward a state of greater entropy.

41. Repeat Problem 39 for each of the following reactions.

(a) $2CO(g) + O_2(g) \rightarrow 2CO_2(g)$

(b) $Mg(s) + Cl_2(g) \rightarrow MgCl_2(s)$

(c) $2C_2H_6(g) + 7O_2(g) \rightarrow 4CO_2(g) + 6H_2O(g)$

(d) $CH_4(g) + 2O_2(g) \rightarrow CO_2(g) + 2H_2O(l)$

42. Repeat Problem 39 for each of the following reactions.

(a) $H_2(g) + Br_2(l) \rightarrow 2HBr(g)$

(b) $ZnO(s) + H_2S(g) \rightarrow ZnS(s) + H_2O(l)$

(c) $2H_2(g) + O_2(g) \rightarrow 2H_2O(l)$

(d) $2C_2H_6(g) + 7O_2(g) \rightarrow 4CO_2(g) + 6H_2O(l)$

43. In terms of the changes in the entropy of the system and of the surroundings, explain why endothermic reactions are favored by an increase in temperature.

44. Using Table 9.3, calculate the standard entropy change, $\Delta S°$, associated with each of the following reactions at 298 K.

(a) $C(s, \text{graphite}) + O_2(g) \rightarrow CO_2(g)$

(b) $C_2H_5OH(l) + 3O_2(g) \rightarrow 2CO_2(g) + 3H_2O(l)$

(c) $C_6H_{12}O_6(s) + 6O_2(g) \rightarrow 6CO_2(g) + 6H_2O(l)$

(d) $H_2(g) + I_2(s) \rightarrow 2HI(g)$

45. What is the standard entropy change, $\Delta S°$, associated with each of the following reactions at 298 K? (Use Table 9.3.) Explain qualitatively the sign of each entropy change by comparing the extent of molecular disorder in the reactants and in the products.

(a) $CaCO_3(s) \rightarrow CaO(s) + CO_2(g)$

(b) $2CO(g) + O_2(g) \rightarrow 2CO_2(g)$

(c) $2Na(s) + Cl_2(g) \rightarrow 2NaCl(s)$

46. Using Table 9.3, calculate the standard entropy change, $\Delta S°$, associated with each of the following reactions at 298 K.

(a) $C_2H_2(g) + 2H_2(g) \rightarrow C_2H_6(g)$

(b) $2C(s, \text{graphite}) + O_2(g) \rightarrow 2CO(g)$

(c) $H_2(g) + Br_2(l) \rightarrow 2HBr(g)$

Gibbs Free Energy

47. **(a)** Calculate the standard reaction Gibbs free energy for the reaction

$$H_2(g) + Cl_2(g) \rightarrow 2HCl(g)$$

(b) Is the reaction spontaneous as written?

(c) What are the relative contributions of the enthalpy change and of the entropy change to the spontaneity of the reaction?

(d) Which factor of (c) dominates?

48. Calculate $\Delta G°$ for each of the following reactions, and state whether each is spontaneous or not under standard conditions.

(a) $C_3H_8(g) + 5O_2(g) \rightarrow 3CO_2(g) + 4H_2O(g)$

(b) $2H_2(g) + O_2(g) \rightarrow 2H_2O(g)$

(c) $2SO_2(g) + O_2(g) \rightarrow 2SO_3(g)$

49. Predict the sign of the standard reaction Gibbs free energy change at *low* temperature for reactions where **(a)** ΔH is positive and ΔS is positive; **(b)** ΔH is negative and ΔS is positive; **(c)** ΔH is negative and ΔS is negative; **(d)** ΔH is positive and ΔS is negative.

50. Repeat Problem 49 for *high*-temperature conditions.

51. **(a)** Calculate $\Delta G°$ for the formation of 1 mol of $CO_2(g)$ from $C(s, \text{diamond})$ and $O_2(g)$ at 298 K and 1 atm pressure.

(b) What does the sign of this $\Delta G°$ signify?

(c) Should the owners of diamonds be concerned about the spontaneous conversion of diamond to carbon dioxide? Explain.

52. In terms of the natural tendency for energy to disperse, explain why chemical reactions take place spontaneously in the direction corresponding to a decrease in the Gibbs free energy.

General Problems

53. Define and explain each of the following terms.

(a) "standard reaction enthalpy," $\Delta H°$

(b) "standard enthalpy of formation," $\Delta H_f°$

(c) the "standard state" of an element

(d) "average bond energy"

54. Natural gas typically contains unwanted $H_2S(g)$, which can be removed by reaction with $SO_2(g)$ according to the equation

$$2H_2S(g) + SO_2(g) \rightarrow 3S(s) + 2H_2O(g)$$

(a) Identify the oxidizing agent and the reducing agent in this reaction.

(b) Calculate $\Delta H°$ for the reaction, assuming that the sulfur formed is orthorhombic sulfur consisting of S_8 molecules (its most stable form at room temperature).

55. Calculate the standard reaction Gibbs free energy for each of the following reactions, and use the results to comment on the relative powers of F_2, Cl_2, and Br_2 as oxidizing agents.

(a) $2NaF(s) + Cl_2(g) \rightarrow 2NaCl(s) + F_2(g)$

(b) $2NaBr(s) + Cl_2(g) \rightarrow 2NaCl(s) + Br_2(l)$

56.* **(a)** Define "standard enthalpy of formation" with reference to $MgCO_3(s)$.

(b) When 0.203 g of magnesium were dissolved in an excess of dilute $HCl(aq)$ in a vacuum flask, the temperature rose by 8.61 K. In another experiment, 506 J was required to raise the temperature of the vacuum flask and its contents 1.02. K. Calculate the heat released in the first experiment, and hence find the reaction enthalpy per mole of magnesium.

(c) In a similar experiment with the same apparatus, $MgCO_3(s)$ reacted with excess $HCl(aq)$, and the standard reaction enthalpy was found to be $-90.4\ kJ \cdot mol^{-1}$. Use the result from (b) and the standard enthalpies of formation of $H_2O(l)$ and $CO_2(g)$ to find the standard enthalpy of formation of $MgCO_3(s)$.

57.* Calculate the standard reaction enthalpy, $\Delta H°$, the standard reaction Gibbs free energy, $\Delta G°$, and the standard reaction entropy, $\Delta S°$, for the reaction

$$CH_4(g) + H_2O(g) \rightarrow CO(g) + 3H_2(g)$$

(a) Is this reaction spontaneous at 25°C?

(b) Verify that $\Delta G° = \Delta H° - T\,\Delta S°$ for this reaction.

(c) Discuss whether the enthalpy change and the entropy change, respectively, work for or against the spontaneity of the reaction.

(d) Which factor of (c) dominates?

58.* The standard reaction enthalpies of combustion for C(*s*, graphite), $H_2(g)$, $C_2H_6(g)$, and $C_3H_8(g)$ are -393.5, -285.8, -1559.8, and $-2219.9\ kJ \cdot mol^{-1}$, respectively. **(a)** Calculate the enthalpies of formation of $C_2H_6(g)$ and $C_3H_8(g)$. **(b)** Use bond energies to predict approximate $\Delta H_f°$ and $\Delta H°$ (combustion) values for butane, $C_4H_{10}(g)$.

59.* Consider the formation of acetylene by two processes: (i) the formation of 1 mol of $C_2H_2(g)$ from graphite and $H_2(g)$ and (ii) the formation of $C_2H_2(g)$ from methane gas and oxygen according to the equation

$$4CH_4(g) + 3O_2(g) \rightarrow 2C_2H_2(g) + 6H_2O(g)$$

(a) From a thermodynamic point of view, is either process feasible for the production of acetylene at room temperature? Explain.

(b) Would the use of a suitable catalyst in both reactions change your answer to (a)?

60.* Methanol, $CH_3OH(l)$, is a potentially useful fuel. Although it provides only half as much energy per liter as does gasoline, it burns cleanly and has a high octane number (that is, it has a low tendency to knock in an engine). It is made industrially from synthesis gas at high pressure, using a catalyst, at about 300°C:

$$2H_2(g) + CO(g) \rightarrow CH_3OH(l)$$

Use the following standard reaction enthalpies of combustion to determine $\Delta H°$ for the synthesis of methanol from the previous reaction:

$$CH_3OH(l) + 1\tfrac{1}{2}O_2(g) \rightarrow CO_2(g) + 2H_2O(l) \quad \Delta H° = -726.6\ kJ$$

$$C(s, \text{graphite}) + \tfrac{1}{2}O_2(g) \rightarrow CO(g) \quad \Delta H° = -110.5\ kJ$$

$$C(s, \text{graphite}) + O_2(g) \rightarrow CO_2(g) \quad \Delta H° = -393.5\ kJ$$

$$H_2(g) + \tfrac{1}{2}O_2(g) \rightarrow H_2O(l) \quad \Delta H° = -285.8\ kJ$$

CHAPTER 10 Metals: Properties, Structures, and Reactions

10.1 Physical Properties and Uses

10.2 Structure and Bonding

10.3 Reactions and Compounds of Group I and II Metals

10.4 Reactions and Compounds of Aluminum

10.5 The Transition Elements

10.6 Metallurgy: Extraction of Metals

10.7 Summary of the Reactions of Metals and Metal Ions

Throughout time metals have been vitally important to human civilizations. This decorative Chinese vessel made of bronze, called a "lei," was made between the late eleventh and the tenth centuries B.C. during the "Bronze Age" of China. Bronze is an alloy composed of about 80% copper and 20% tin. It is harder and more durable than copper alone, which explains why bronze proved to be so useful to early peoples as tools and weapons.

The elements may be classified into metals and nonmetals, as we briefly discussed in Chapter 2. In previous chapters we discussed the chemistry of the nonmetals carbon, phosphorus, sulfur, and the halogens. We now turn our attention to the metals. After oxygen and silicon, which are nonmetals, the metals aluminum, iron, calcium, sodium, potassium, and magnesium are the most abundant elements in the earth's crust. All the elements in Groups I, II, and III (apart from hydrogen and boron), all the transition elements, and some of the elements in Groups IV and V are metals.

Metals have played a major role in the development of civilization, as is indicated by the use of terms such as Bronze Age and Iron Age to describe important stages in human history. Although copper is not very abundant, it is sometimes found in elemental form and is very easily obtained from its ores, so it has been known since ancient times. Even before 3000 B.C. it was discovered that copper could be made much harder by forming an alloy with tin called bronze, which could be used to make superior utensils, tools, and weapons. This discovery initiated the period in early history known as the Bronze Age. Iron is more difficult to obtain from its ores, and it did not become important until around 1500 B.C. Steel, an iron–carbon alloy, has been made since around 1000 B.C. However, it was not until the mid-1800s that a cheap process for the manufacture of steel was developed, giving a major impetus to the Industrial Revolution. Modern civilization very much depends on the use of steel for buildings, ships, railways, tools, and machinery. Although aluminum compounds are very abundant in the earth's crust, aluminum is still more difficult to obtain from its compounds than iron. Therefore, only in the twentieth century did aluminum become a widely used metal. Today many other

Metals · Nonmetals · Semimetals

Transition Elements

Period \ Group	I	II											III	IV	V	VI	VII	VIII
1	H																	He
2	Li	Be											B	C	N	O	F	Ne
3	Na	Mg											Al	Si	P	S	Cl	Ar
4	K	Ca	Sc	Ti	V	Cr	Mn	Fe	Co	Ni	Cu	Zn	Ga	Ge	As	Se	Br	Kr
5	Rb	Sr	Y	Zr	Nb	Mo	Tc	Ru	Rh	Pd	Ag	Cd	In	Sn	Sb	Te	I	Xe
6	Cs	Ba	La	Hf	Ta	W	Re	Os	Ir	Pt	Au	Hg	Tl	Pb	Bi	Po	At	Rn
7	Fr	Ra	Ac	104	105	106	107	108	109									

Left to right: the metals magnesium, sodium, and calcium

metals, some of them formerly regarded as quite rare, are becoming increasingly important. For example, a typical jet engine contains approximately 38% Ti, 37% Ni, 12% Cr, 6% Co, and 3% Al, mostly as alloys.

In this chapter we discuss (1) the metals of Groups I and II, the alkali and alkaline earth metals, with particular emphasis on sodium, potassium, magnesium, and calcium; (2) aluminum, the second element in Group III; and (3) a few typical transition metals. We use the chemistry of these elements to discuss some important concepts: (4) amphoteric hydroxides (hydroxides that can behave as both acids and bases); (5) Lewis acids, such as $AlCl_3$, and complex ions such as $Fe(CN)_6^{3-}$; and (6) the activity series of metals, which places the metals in order of their reactivity —for example, toward acids.

10.1 PHYSICAL PROPERTIES AND USES

The Alkali and Alkaline Earth Metals

All the Group I and II elements have properties that are typical of metals: They are good conductors of heat, have high electrical conductivities, are malleable and ductile, and have a shiny metallic luster when freshly cut (Demonstration 10.1). With the exception of magnesium and beryllium, the alkali and alkaline earth metals are softer and mechanically weaker than most other metals. Also, they react rapidly with the oxygen of the atmosphere, so they are of no importance as structural

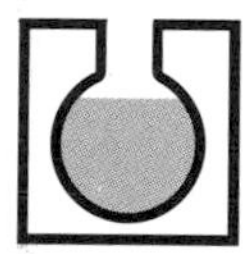

DEMONSTRATION 10.1 Physical Properties of the Alkali Metals

Sodium is a good conductor of electricity.

Sodium is soft enough to be cut with a knife. The bright metallic luster soon dulls as the surface is oxidized.

Sodium melts at 97.5°C, and potassium melts at 62.3°C. The tubes contain sodium (left) and potassium (right) covered with paraffin to protect them from oxidation by the oxygen in air. When the tubes are heated in boiling water, both sodium and potassium melt into a shiny metallic ball that looks like mercury.

TABLE 10.1 Properties of Some Metals

Element	Melting point (°C)	Boiling point (°C)	Density ($g \cdot cm^{-3}$)	Electrical Conductivity Relative to Ag = 100
Lithium	186	1336	0.53	19
Sodium	97.5	880	0.97	36
Potassium	62.3	760	0.86	24
Rubidium	38.5	700	1.53	12
Cesium	28.5	670	1.87	8
Beryllium	1280	2970	1.86	48
Magnesium	650	1100	1.74	36
Calcium	850	1490	1.55	43
Strontium	770	1380	2.6	7
Barium	710	1140	3.6	3
Aluminum	660	2467	2.71	61
Iron	1535	2750	7.86	16
Copper	1083	2567	8.97	97
Silver	960	2210	10.54	100

materials. Magnesium oxidizes less rapidly than do other Group I and II metals and is an important constituent of many useful lightweight alloys.

The alkali metals have several unusual properties. Unlike most other metals, they have low densities and low melting points (Table 10.1). Lithium is the least dense of all metals, and lithium, sodium, and potassium are the only metals that are less dense than water.

Aluminum

Aluminum has a relatively low density and a high electrical conductivity (Table 10.1). It forms several hard, strong alloys, such as duralumin, which contains copper, manganese, and magnesium. These aluminum alloys have a low density and are not subject to corrosion, because the metal surface is protected by a thin, hard, unreactive film of aluminum oxide, Al_2O_3. Hence they have many important uses. For example, aluminum alloys are used in aircraft and space vehicles and for garden furniture, door and window frames, and kitchen utensils. Aluminum is used to manufacture billions of cans each year, and in the form of a thin foil, it is used as a wrapping material. Other applications of aluminum are based on the fact that it is an excellent conductor of electricity. An aluminum wire has only two-thirds the conductivity of a copper wire of the same diameter, and the aluminum wire is much lighter. It is also considerably cheaper than copper, so aluminum is being used increasingly for electrical wiring and is widely used for high-voltage transmission lines.

Iron

Iron has a higher melting point and a much higher density than aluminum but is a poorer electrical conductor (Table 10.1). Pure iron is rather soft and not very strong but is rarely made. Obtaining iron completely free of carbon, oxygen, nitrogen, sulfur, and other metals is difficult but is unnecessary, because its alloys are of much

(a)

(b)

(c)

FIGURE 10.1 Crystals of Metals (a) Gold and (b) silver are sometimes found naturally in crystalline forms. (c) This cross section through a piece of zinc that has been allowed to solidify very slowly shows a mass of interlocking crystals.

greater importance. Almost all iron is produced in the form of steel, an alloy with carbon and transition metals such as chromium, manganese, and vanadium. Steel, which is much harder and stronger than pure iron, is used in buildings, bridges, ships, cars, machinery, and many other familiar items.

Copper

Copper is unusual in having a red color, because most metals have a silvery appearance. Copper is resistant to corrosion and, together with silver and gold, is among the few metals found in the uncombined state. Hence these metals have been used since antiquity for coins; they are often called the *coinage metals*. In recent times, however, these metals have largely been replaced in coins by other, cheaper metals such as aluminum and nickel. Because copper has a very high electrical conductivity, second only to silver (Table 10.1), and because it is rather soft and very ductile (easily drawn out into thin wires), one of its most important uses is for electrical wiring.

10.2 STRUCTURE AND BONDING

Hexagonal and Cubic Close-Packed Structures

All metals except mercury are crystalline solids at room temperature. Usually a piece of metal consists of many tiny crystals packed tightly together. Although they are not normally visible to the eye, these crystals can often be seen under a microscope. Occasionally, small gold and silver crystals are found in nature (Figure 10.1).

As we shall see in more detail in Chapter 11, the atoms in a crystal are arranged in a regular pattern. The structures of many metals are based on the two ways that spheres (which represent atoms) of the same size may pack together as closely as possible. When we push identical spheres together as closely as possible in a tray, we obtain a **close-packed arrangement** (Figure 10.2a; Demonstration

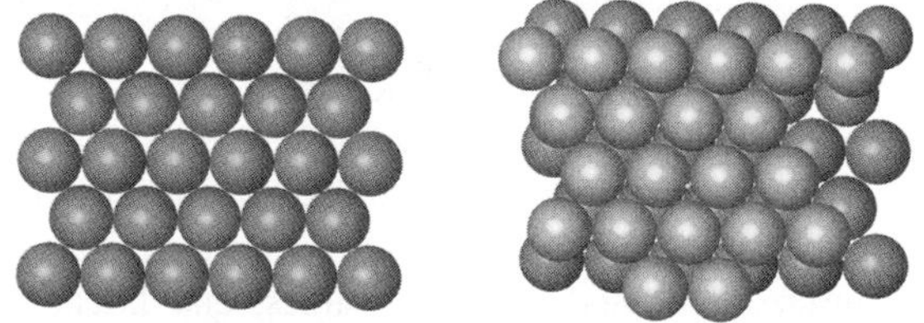

(a) (b)

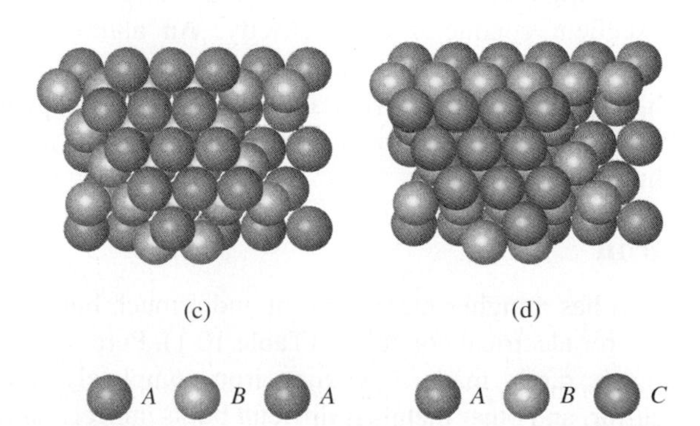

FIGURE 10.2 Close Packing (a) A close-packed layer of spheres. (b) A second close-packed layer can be placed so that it nestles into the depressions in the first layer. A third layer of spheres may be added in two ways; (c) We may place the third layer in the depressions of the second layer so that it lies directly above the spheres in the first layer. In this arrangement we call the bottom layer an *A* layer and the second layer a *B* layer, then the third layer, an exact replica of the first layer, is another *A* layer. Thus, we have an arrangement in which the layers alternate in position—an *ABABAB* . . . arrangement, which is called hexagonal close packing. (d) Or a third layer may be placed so that it does not lie above the spheres in the bottom layer. Only when a fourth layer is added do the spheres lie directly above the spheres in the first layer. This arrangement is therefore an *ABCABCA* . . . arrangement, which is called cubic close packing.

DEMONSTRATION 10.2
Close Packing of Spheres

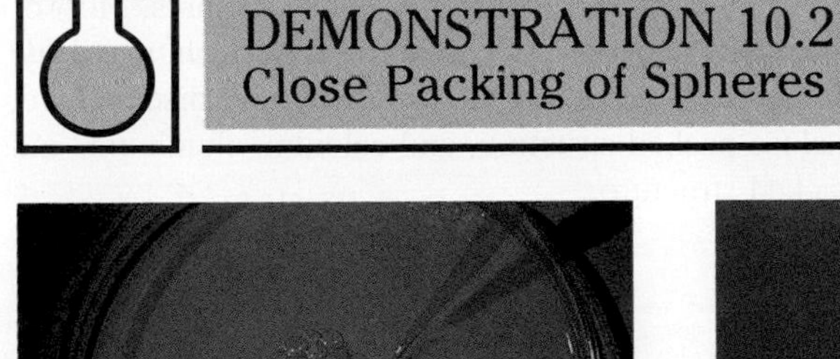

When a fine jet of air is bubbled through a soap solution, a layer of soap bubbles forms on the surface of the solution.

Each bubble is surrounded by a hexagonal arrangement of six more bubbles, which gives a two-dimensional close-packed arrangement.

10.2). The spheres are arranged in straight rows that make angles of 60° with one another. Each sphere is surrounded by a regular hexagonal arrangement of six other spheres.

When layers of close-packed atoms are stacked so that the atoms of one layer nestle into the hollows between the atoms in the adjacent layer, they form a three-dimensional close-packed structure. There are two ways in which close-packed layers can be stacked, giving two closely related but distinct close-packed structures. In the **hexagonal close-packed structure** (Figure 10.2c), successive layers are arranged such that the spheres in alternate layers lie directly above each other in what is described as an *ABABAB* . . . arrangement. In the **cubic close-packed structure** (Figure 10.2d), the layers are arranged such that the spheres of every fourth layer lie directly above each other in what is described as an *ABCABCA* . . . arrangement. In both forms of close packing, each sphere is in contact with 12 neighboring spheres—6 in the same layer, 3 in the layer above, and 3 in the layer below (Figure 10.3).

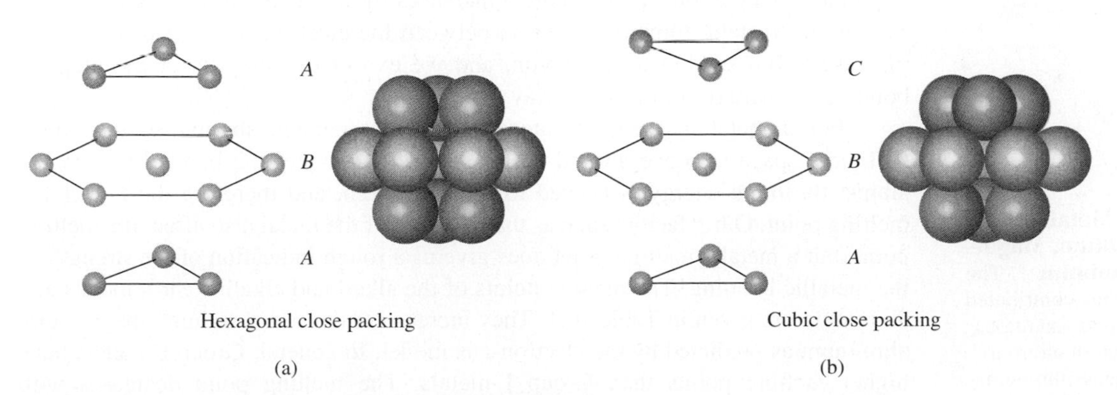

FIGURE 10.3 Hexagonal and Cubic Close Packing
In both forms of close packing, each sphere is in contact with 12 neighboring spheres—6 in its own layer, 3 in the layer above, and 3 in the layer below. No more than 12 spheres all of the same size can be placed so that they all touch a single sphere, and there are two ways of arranging these 12 spheres. (a) The 12 spheres are arranged in hexagonal close packing. (b) They are arranged in cubic close packing.

A majority of the metals have one of the two close-packed structures. Most of the remaining metals have a third type of structure called the *body-centered cubic structure*, which we describe in Chapter 11. Among the metals that we discuss here, magnesium has the hexagonal close-packed structure, and calcium, aluminum, and copper have the cubic close-packed structure.

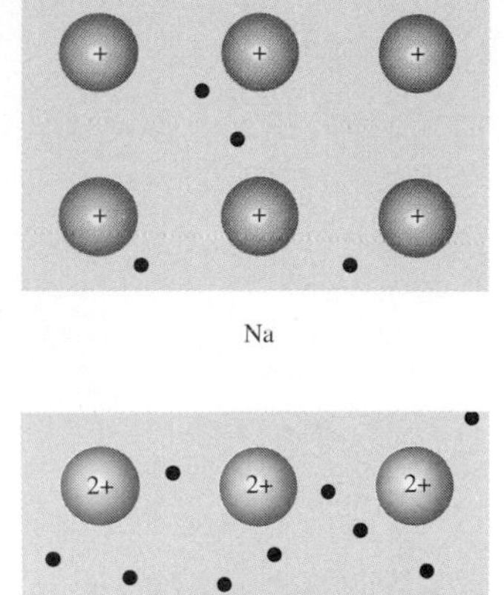

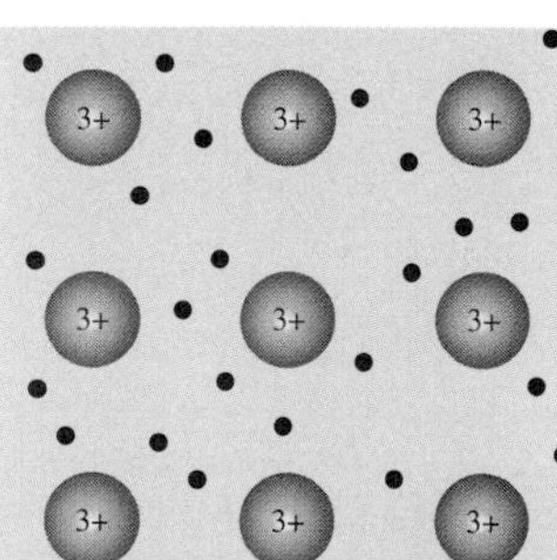

FIGURE 10.4 Metallic Bonding in Sodium, Magnesium, and Aluminum The number of electrons contributed to the metallic bonds increases from one per sodium atom, to two for each magnesium atom, to three for each aluminum atom. Thus, the strength of the metallic bond increases from sodium to magnesium to aluminum.

Metallic Bonding

Metals have a type of bonding that differs from both ionic and covalent bonding, although all three types are electrostatic in origin. A useful model for metallic bonding that accounts for many of the properties of metals is called the **electron-gas model**. This model visualizes a metal as a close-packed or nearly-close-packed arrangement of positive metal ions with the valence electrons free to move among them, forming a kind of electron gas. The electrostatic attraction between the cloud of negative electrons and the positive ions holds them together and constitutes the **metallic bond**. We can alternatively think of the valence-shell charge cloud of each metal atom overlapping that of its neighbors to form a continuous charge cloud. According to this model, sodium consists of Na^+ ions held together by an electron gas or charge cloud formed from one electron per atom, whereas aluminum consists of Al^{3+} ions held together by a charge cloud composed of three electrons per atom (Figure 10.4).

The metallic bond differs in an important way from the covalent bond in that it is nondirectional, whereas, as we have seen, the covalent bond is highly directional. The electron cloud in a metal pulls the metal ions together so that they pack as closely as possible; in most metals, each metal ion has 12 neighbors. In contrast, the carbon atom in a covalent network solid such as diamond can form only four covalent bonds, so diamond has a much more open structure.

Properties of Metals

Melting Point The number of valence electrons per metal atom in the charge cloud increases from one in sodium to two in magnesium and three in aluminum. The charge on the ions (atomic cores) increases from +1 to +3 in the same series. So the electrostatic force of attraction between the electrons and the positive ions increases from sodium to aluminum, and we expect the strength of the metallic bonding to increase in the same way.

For a metal to melt, the atoms must be separated slightly so they have sufficient space to move around each other. The stronger the bonds between the atoms, the more energy is needed to separate them, and therefore the higher the melting point. Other factors such as the structure of the metal also affect the melting point, but a metal's melting point does give us a rough indication of the strength of the metallic bonding. The melting points of the alkali and alkaline earth metals and aluminum are given in Table 10.1. They increase in the series sodium, magnesium, aluminum as predicted by the electron-gas model. In general, Group II metals have higher melting points than Group I metals. The melting point decreases with increasing atomic size down both Groups I and II, although magnesium is an exception. We expect the strength of the metallic bond to decrease with increasing atomic size as the average distance between the electron cloud and the center of each ion increases and the electrostatic force between them decreases correspondingly.

Electrical Conductivity Because the valence electrons in a metal are free to move between the positively charged ions, an applied voltage—from a battery, for example—will cause an electric current to flow through the metal. This type of conduction, which is due to the movement of electrons, is called **metallic conduction**. Silver is the most highly conducting metal, followed by copper and gold and then aluminum (Table 10.1).

In both the solid and liquid states, a metal is said to be a *conductor* of electricity. In contrast, most ionic substances conduct electricity only in the liquid state, and covalent substances do not conduct electricity either in the solid or in the liquid state. The ions in a solid ionic crystal are not free to move; in the molten (liquid) state, the ions can move, and the electric current consists of moving positive and negative ions. Covalent solids or liquids have no ions, and the electrons are not free to move because they are firmly held to the atoms as bonding or nonbonding pairs. These substances are therefore *nonconductors*, or *insulators*.

An important property of metals is that

The conductivity of a metallic conductor increases as the temperature decreases.

This behavior is in contrast to that of an ionic conductor, whose conductivity *decreases* as the temperature decreases. As a molten ionic substance or a solution of an ionic substance cools, the speeds of the ions decrease. Under the influence of an applied voltage, the ions move a shorter distance in a given time with decreasing temperature, and so the conductivity decreases. The conductivity of a metal, however, depends on the speed of the electrons moving between the metal ions. Even at low temperatures, the electrons move at very high speeds, so metals are very good electrical conductors. The electrons are hindered in their motion by collisions with the vibrating ions. But as the temperature decreases, the amplitude of the vibrations of the ions also decreases, so there are fewer such collisions (Figure 10.5). The conductivity of the metal therefore increases as the temperature decreases.

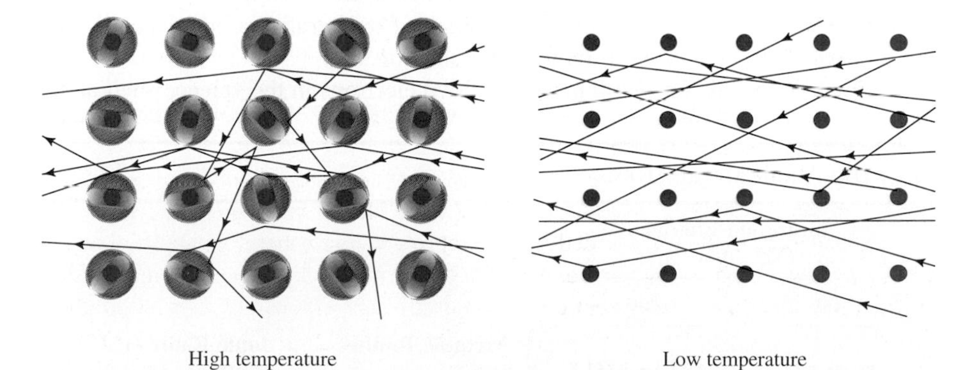

FIGURE 10.5 Effect of Temperature on Electrical Conductivity The movement of electrons through a metal is impeded by their collisions with the atomic cores (metal ions). The atomic cores vibrate, and the amplitude of the vibrations increases with increasing temperature. The greater the amplitude of these vibrations, the greater the effective volume occupied by the core, and the greater chance an electron will collide with an atomic core and that its motion through the solid will be impeded. With decreasing temperature the amplitude of vibration of the atomic cores decreases, as does the chance of an electron colliding with an atomic core.

Mechanical Properties Because the electrons that form the bonds in a metal are not fixed in position between any particular atoms (that is, the bonds are nondirectional), the atoms can move with respect to each other without breaking the bonds between them. For this reason a metal can be distorted in shape relatively easily; that is, it is malleable and ductile. For example, steel can be rolled out into thin sheets for making automobiles and household appliances, and copper can be drawn out into wires for use in electric circuits. Most covalent and ionic solids cannot be distorted in this way. If we try to alter their shape by hammering them, for example, they shatter into pieces, because bonds are broken if the atoms or ions are moved relative to each other.

Thermionic Effect The free electrons in a metal have a distribution of speeds just like the molecules of a gas (Chapter 2). When a metal is heated, the speeds of the electrons increase. At a sufficiently high temperature, some of them acquire enough energy to escape from the metal. This phenomenon is called the **thermionic effect**. In a television cathode-ray tube, the electrons emitted by a heated metal wire are accelerated and focused into a beam. The beam is moved across the coated end of the tube to produce the visible pattern that is the television picture.

10.3 REACTIONS AND COMPOUNDS OF GROUP I AND II METALS

Each alkali metal has only one electron in its valence shell and a core charge of +1. This electron is therefore only weakly attracted to the core, as is shown by its low ionization energy (Table 10.2). It is rather easily removed to give an M^+ ion, so the alkali metals behave as strong reducing agents:

$$M \rightarrow M^+ + e^+$$

For example,

$$Na \rightarrow Na^+ + e^-$$

Because the alkali metals are strong reducing agents, their cations are poor oxidizing agents. Thus $M^+ + e^- \rightarrow M$ does not readily occur.

This half-reaction is the basis of *all* the reactions of the alkali metals. We may say that this is the *only* reaction of the alkali metals. The overall reaction depends only on the oxidizing agent used to remove the electron.

Each of the alkaline earth metals has two electrons in the valence shell and a

TABLE 10.2 Some Properties of the Alkali Metal Atoms

		Ionization Energies ($MJ \cdot mol^{-1}$)			
		1st	**2nd**		
Element	**Electron Configuration**	$M \rightarrow M^+$	$M^+ \rightarrow M^{2+}$	**Metallic Radius (pm)**	**Ionic Radius (M^+) (pm)**
Lithium	$[He]2s^1$	0.52(2*s*)	7.28(1*s*)	157	74
Sodium	$[Ne]3s^1$	0.51(3*s*)	4.56(2*p*)	191	102
Potassium	$[Ar]4s^1$	0.42(4*s*)	3.06(3*p*)	235	138
Rubidium	$[Kr]5s^1$	0.40(5*s*)	2.65(4*p*)	250	149
Cesium	$[Xe]6s^1$	0.38(6*s*)	2.42(5*p*)	272	170

TABLE 10.3 Some Properties of the Alkaline Earth Metal Atoms

		Ionization Energies ($MJ \cdot mol^{-1}$)				
		1st	2nd	3rd		
Element	**Electron Configuration**	$M \rightarrow M^+$	$M^+ \rightarrow M^{2+}$	$M^{2+} \rightarrow M^{3+}$	**Metallic Radius (pm)**	**Ionic Radius (M^{2+}) (pm)**
Beryllium	$[He]2s^2$	0.90(2*s*)	1.75(2*s*)	14.81(1*s*)	112	27
Magnesium	$[Ne]3s^2$	0.74(3*s*)	1.45(3*s*)	7.72(2*p*)	160	72
Calcium	$[Ar]4s^2$	0.59(4*s*)	1.15(4*s*)	4.93(3*p*)	197	100
Strontium	$[Kr]5s^2$	0.55(5*s*)	1.06(5*s*)	4.14(4*p*)	215	126
Barium	$[Xe]6s^2$	0.50(6*s*)	0.96(6*s*)	3.47(5*p*)	224	136

core charge of +2. These two electrons are therefore held somewhat more strongly than the single valence electron of the alkali metals, but they still have quite low ionization energies (Table 10.3) and are relatively easily removed to give an M^{2+} ion. The alkaline earth metals are good reducing agents, although somewhat weaker than the alkali metals:

$$M \rightarrow M^{2+} + 2e^-$$

For example,

$$Ca \rightarrow Ca^{2+} + 2e^-$$

This half-reaction is the basis of *all* the reactions of the alkaline earth metals. The overall reaction depends only on the oxidizing agent used to remove the two electrons.

The atoms increase in size down both groups from Li to Cs and from Be to Ba, so the distance of the valence electrons from the nucleus increases and the ionization energy decreases accordingly. From top to bottom of Groups I and II, the metals become increasingly strong reducing agents.

Reactions with the Halogens and Oxygen

The alkali and alkaline earth metals are readily oxidized by the halogens and oxygen to give the corresponding ionic halides and oxides. For example, potassium reacts with chlorine to give potassium chloride $K^+Cl^-(s)$:

$$2K(s) + Cl_2(g) \rightarrow 2KCl(s)$$

This reaction can be regarded as the sum of the two half-reactions

$$2(K \rightarrow K^+ + e^-) \qquad \text{oxidation of K to } K^+$$
$$Cl_2 + 2e^- \rightarrow 2Cl^- \qquad \text{reduction of } Cl_2 \text{ to } Cl^-$$

Similarly, the reaction of magnesium with oxygen to give magnesium oxide, $Mg^{2+}O^{2-}(s)$,

$$2Mg(s) + O_2(g) \rightarrow 2MgO(s)$$

can be regarded as the sum of the two half-reactions

$$2(Mg \rightarrow Mg^{2+} + 2e^-) \qquad \text{oxidation of Mg to } Mg^{2+}$$
$$O_2 + 4e^- \rightarrow 2O^{2-} \qquad \text{reduction of } O_2 \text{ to } O^{2-}$$

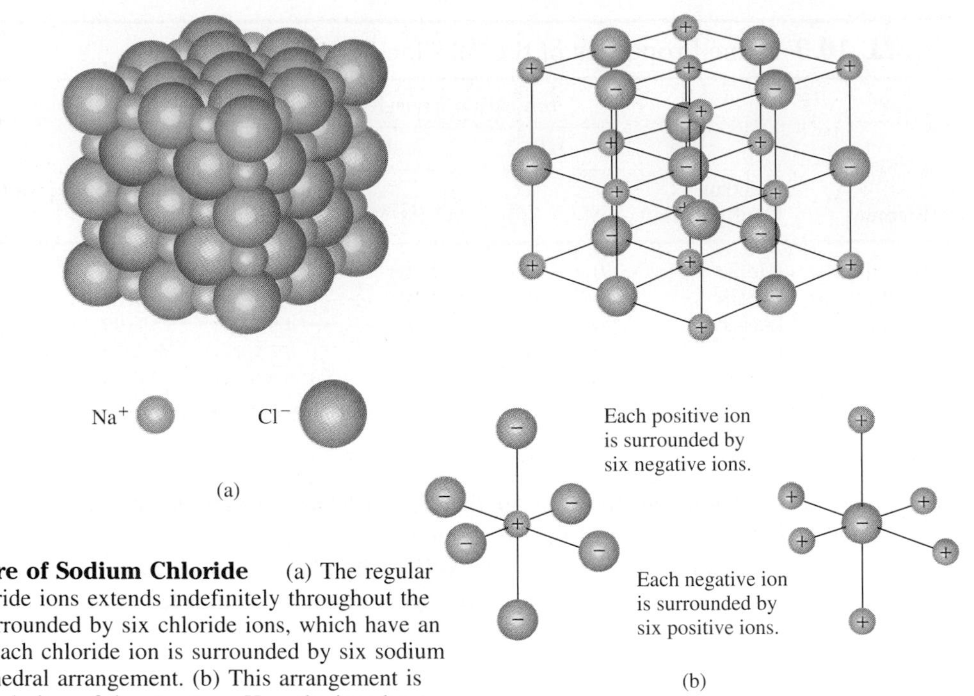

FIGURE 10.6 The Structure of Sodium Chloride (a) The regular array of sodium ions and chloride ions extends indefinitely throughout the crystal. Each sodium ion is surrounded by six chloride ions, which have an octahedral arrangement. And each chloride ion is surrounded by six sodium ions, which also have an octahedral arrangement. (b) This arrangement is most easily seen in an expanded view of the structure. Here the ions have been separated from one another, but their arrangement has been maintained.

Other reactions of this type include

$$2Ca(s) + 2Br_2(l) \rightarrow 2CaBr_2(s)$$

and

$$4Na(s) + O_2(g) \rightarrow 2Na_2O(s)$$

This last reaction must be carried out with a limited supply of oxygen; otherwise the peroxide Na_2O_2, containing the peroxide ion $^{-}\!:\!\underset{..}{\ddot{O}}\!-\!\underset{..}{\ddot{O}}\!:^{-}$, is formed (Chapter 16). With the exception of magnesium, the Group I and II elements react so readily with oxygen, even at room temperature, that they must be stored out of contact with the atmosphere. Magnesium reacts more slowly with oxygen but still burns vigorously when heated in air (Demonstration 2.2).

The halides and oxides of the Group I and II metals all have ionic three-dimensional network structures. In the structure of sodium chloride, Na^+Cl^-, each sodium ion is surrounded by six chloride ions in an octahedral arrangement, and each chloride ion is similarly surrounded by six sodium ions (Figure 10.6). Potassium chloride, K^+Cl^-, and magnesium oxide, $Mg^{2+}O^{2-}$, have the same structure. Some of the other oxides and halides have different structures with different arrangements of their anions and cations (Chapter 11).

Water behaves as an oxidizing agent in this reaction. The hydrogen of H_2O is reduced from oxidation state +I to 0:

$$2H_2O(l) + 2e^- \rightarrow 2OH^-(aq) + H_2(g)$$

Reactions with Water

The alkali metals readily reduce water to hydrogen, $H_2(g)$, forming a solution of a metal hydroxide, such as sodium hydroxide, Na^+OH^-:

$$2Na(s) + 2H_2O(l) \rightarrow 2Na^+(aq) + 2OH^-(aq) + H_2(g)$$

DEMONSTRATION 10.3 Reactions of Sodium, Potassium, and Calcium with Water

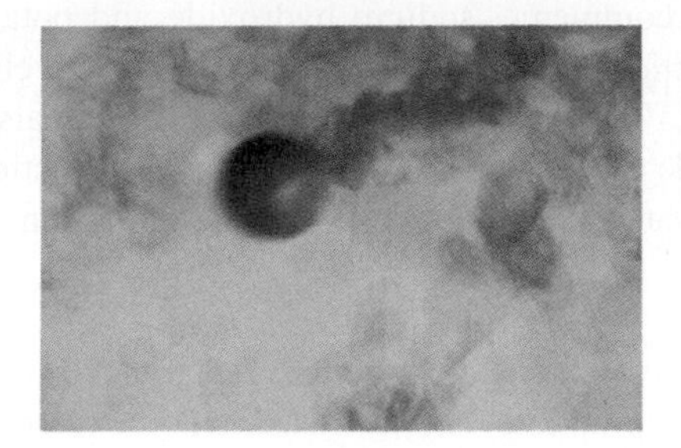

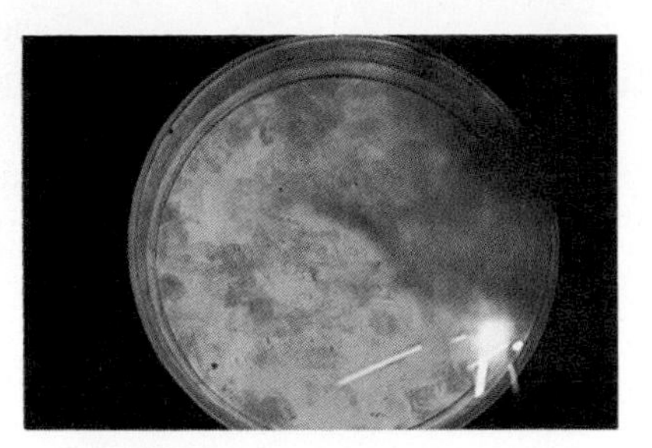

Sodium reacts so vigorously with water that it skates over the surface as hydrogen is evolved. A few drops of the indicator bromothymol blue have been added to the water. The blue trails are due to the sodium hydroxide formed in the reaction, which makes the solution basic, as shown by the change in the color of the indicator.

The heat evolved in the reaction of sodium with water melts the sodium, which forms a small, shiny metallic bead that floats on the water.

Potassium reacts even more violently, skating very rapidly over the surface leaving blue trails as the indicator changes color. The heat of the reaction ignites the hydrogen. The potassium imparts a characteristic lilac color to the flame. Steam can be seen rising from the water as it is heated by the burning potassium.

Calcium is heavier than water and sinks to the bottom of the beaker. A rapid stream of hydrogen bubbles rises to the surface.

Calcium hydroxide, $Ca(OH)_2$, is the other product of the reaction. Because it is not very soluble, a white precipitate soon forms.

These reactions are quite vigorous and become increasingly so from lithium to cesium (Demonstration 10.3). A small piece of sodium floats when dropped onto water. The heat liberated in the reaction melts the sodium, which forms a silvery, spherical ball that moves dramatically on the surface, propelled by the evolved hydrogen. With potassium the reaction is even more spectacular, because the heat liberated is sufficient to ignite the hydrogen, which burns with a flame colored lilac by the potassium (Chapter 6). The reactions of rubidium and cesium are still more violent and are dangerous to demonstrate. These reactions illustrate the increase in reactivity of the Group I metals with decreasing ionization energy. As expected, the Group II metals react similarly but less vigorously with water (Demonstration 10.3):

$$Ca(s) + 2H_2O(l) \rightarrow Ca(OH)_2(s) + H_2(g)$$

In this reaction the calcium hydroxide forms a white precipitate, as it is not very soluble in water (Table 4.6).

Hydroxides The alkali and alkaline earth metal hydroxides are ionic compounds containing a metal ion and the hydroxide ion. Except for $Mg(OH)_2$ and $Ca(OH)_2$, all are soluble in water. They are strong bases. Sodium hydroxide is made industrially by the electrolysis of an aqueous sodium chloride solution (Chapter 13). In industry and commerce, sodium hydroxide and potassium hydroxide are frequently called *caustic soda* and *caustic potash*, respectively.

The hydroxides of these metals can also be made by the reaction of the metal oxide with water. This is an acid–base reaction in which the oxide ion, a strong base in water, is converted to the hydroxide ion:

$$O^{2-} + H_2O \rightarrow 2OH^-$$

The overall equations for the reactions of two typical oxides are obtained by adding the appropriate cation to each side of the equation:

$$(Na^+)_2O^{2-}(s) + H_2O(l) \rightarrow 2Na^+(aq) + 2OH^-(aq)$$
$$Ca^{2+}O^{2-}(s) + H_2O(l) \rightarrow Ca^{2+}(OH^-)_2(s)$$

Just as an acid reacts with a base to give a salt plus water, an acidic oxide reacts with a basic oxide to produce a salt.

Acidic and Basic Oxides The oxides of the Group I and II elements are called **basic oxides**, because they react with water to give basic solutions containing the hydroxide ion. In contrast, the oxides of nonmetals, such as SO_3 and P_4O_{10}, are called **acidic oxides**. They react with water to give solutions of oxoacids, such as sulfuric and phosphoric acids (Chapter 7). Basic oxides react with acidic oxides to give salts. For example, calcium oxide, CaO (lime), can be used to remove SO_2 from the gases discharged from electricity generating stations that use fossil fuels:

$$CaO(s) + SO_2(g) \rightarrow CaSO_3(s) \text{ (calcium sulfite)}$$

Calcium oxide is frequently used in industry whenever an inexpensive, readily available base is required.

Reactions with Acids

A typical reaction of many metals is that they dissolve in aqueous acids to give hydrogen (Chapter 4). This is an oxidation–reduction reaction in which the metal reduces the hydronium (hydrogen) ion to hydrogen:

$$2H_3O^+(aq) + 2e^- \rightarrow H_2(g) + 2H_2O(l)$$

or more simply

$$2H^+(aq) + 2e^- \rightarrow H_2(g)$$

The overall equation for the reaction of magnesium with a dilute aqueous acid is

$$Mg(s) + 2H^+(aq) \rightarrow Mg^{2+}(aq) + H_2(g)$$

Not all the reactions of acids are acid–base reactions. The reaction of hydrochloric acid with magnesium and other metals is a redox reaction.

If the acid is hydrochloric acid the complete equation is

$$Mg(s) + 2HCl(aq) \rightarrow MgCl_2(aq) + H_2(g)$$

Any metal in the left-hand column of Table 10.4 can reduce any ion in the right-hand column that lies below the metal.

Activity Series If we arrange the metals in order of decreasing strength as reducing agents, we obtain what is called the **activity series** (Table 10.4). It is convenient to include hydrogen in this series, although it is not a metal. Metals that can reduce H^+ to hydrogen in dilute aqueous acid lie above hydrogen in the activity series, whereas metals such as copper and silver that are poorer reducing agents are

TABLE 10.4 Activity Series for Common Metals

	Metal		Ion	
Stronger reducing agents, most easily oxidized	Na K Ca	React with water →	Na^+ K^+ Ca^{2+}	Weaker oxidizing agents, less easily reduced
	Mg Zn Al Fe	React with H_3O^+ → in aqueous acids	Mg^{2+} Zn^{2+} Al^{3+} Fe^{2+}	
	H_2		H^+	
Weaker reducing agents, least easily oxidized	Cu Ag Au	Do not react with H_3O^+	Cu^{2+} Ag^+ Au^+	Stronger oxidizing agents, more easily reduced

below hydrogen in the activity series. The most reactive metals (the strongest reducing agents), such as sodium and potassium at the top of the series, reduce not only H_3O^+, but also water, which is a weaker oxidizing agent. Magnesium is on the borderline between the most reactive metals that can reduce water and those that cannot. It reacts with water at an appreciable speed only if the water is boiling or if it is steam (Figure 10.7). The metals below magnesium in the series, such as aluminum, do not react with water.

FIGURE 10.7 Reaction of Magnesium with Boiling Water Magnesium burns in some substances that normally extinguish flames. Here we see it burning in a flask of boiling water. White magnesium oxide and hydroxide are produced.

Salts

Salts of the Group I and II metals, such as the halides, sulfates, carbonates, nitrates, and phosphates, can be made by acid–base reactions between the hydroxides or oxides and the appropriate acid. In each case the reaction is either

$$OH^- + H^+ \rightarrow H_2O$$

or

$$O^{2-} + 2H^+ \rightarrow H_2O$$

Typical balanced equations are

$$2NaOH(aq) + H_2SO_4(aq) \rightarrow Na_2SO_4(aq) + 2H_2O(l)$$

$$Mg(OH)_2(s) + H_2SO_4(aq) \rightarrow MgSO_4(aq) + 2H_2O(l)$$

$$CaO(s) + 2HCl(aq) \rightarrow CaCl_2(aq) + H_2O(l)$$

$$NaOH(aq) + H_2CO_3(aq) \rightarrow NaHCO_3(aq) + H_2O(l)$$

All the salts of the alkali metals are soluble in water. Many of the salts of the Group II metals are also soluble in water, but the carbonates of Mg, Ca, Sr, and Ba are insoluble, as are the sulfates of Ca, Sr, and Ba (Figure 10.8 and Table 4.6). The formation of a precipitate of $BaSO_4$ when a solution of a soluble barium salt, such as

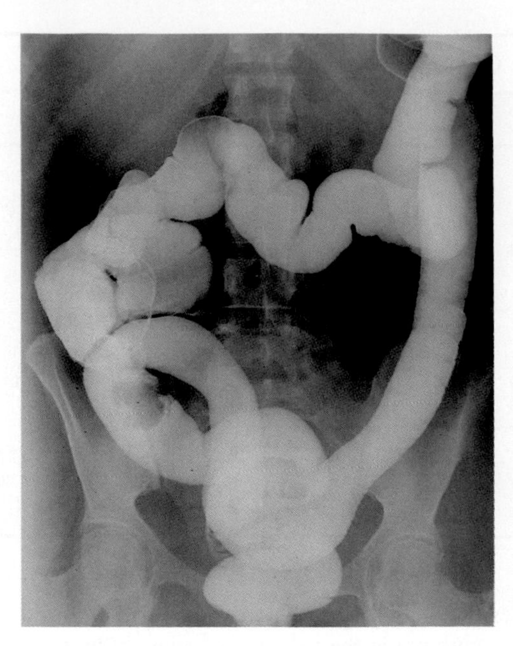

FIGURE 10.8 X-Ray Photograph of the Large Intestine Barium sulfate is opaque to X-rays. Because it has a very low solubility, it passes through the digestive system, and no appreciable amounts of toxic Ba^{2+} are taken up by the human body. The photo shows how finely dispersed barium sulfate particles in the large intestine absorb X-rays and make the intestine visible.

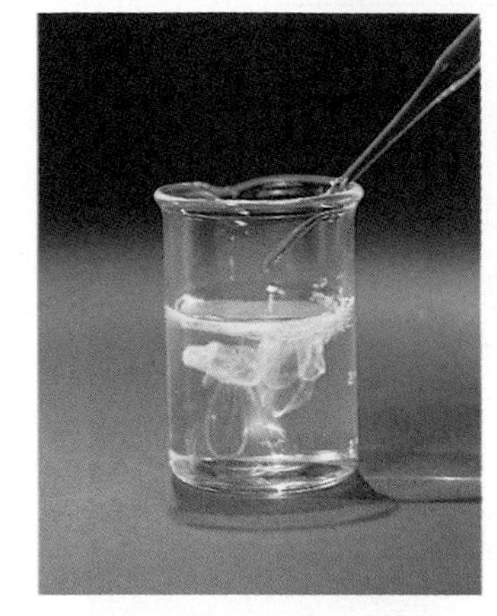

FIGURE 10.9 The Formation of a Barium Sulfate Precipitate A solution of sodium sulfate added to a solution of barium chloride produces a heavy white precipitate of barium sulfate. This reaction can be used as a test for the sulfate ion.

$BaCl_2$, is added to a sulfate solution is used as a test to confirm the presence of the sulfate ion (Figure 10.9):

$$Ba^{2+}(aq) + SO_4^{2-}(aq) \rightarrow BaSO_4(s)$$

Sodium carbonate, Na_2CO_3, is a very important industrial compound. It is used in large quantities in the chemical industry and in the manufacture of glass, paper, detergents, and soap. In the home it is used for washing and cleaning and for softening ''hard'' water (Box 10.1). Most of the sodium carbonate used in North America comes from large deposits of the mineral *trona*, $Na_5(CO_3)_2(HCO_3) \cdot 2H_2O$, found in Wyoming. The ore is crushed and heated to decompose the hydrogen-carbonate ion into carbonate:

$$2HCO_3^- \rightarrow CO_3^{2-} + CO_2 + H_2O$$

Unlike most metal carbonates, such as $CaCO_3$, the alkali metal carbonates do not decompose when heated.

The crude sodium carbonate produced in this way is recrystallized from water to give hydrated sodium carbonate, $Na_2CO_3 \cdot 10H_2O$, which upon heating loses water to give anhydrous sodium carbonate, Na_2CO_3.

Because carbonic acid, H_2CO_3, is a diprotic acid, it forms two series of salts: the carbonates, such as Na_2CO_3, and the hydrogencarbonates, such as sodium hydrogencarbonate, $NaHCO_3$. Sodium hydrogencarbonate can be made, therefore, by the reaction of carbonic acid with sodium hydroxide:

$$H_2CO_3(aq) + NaOH(aq) \rightarrow NaHCO_3(aq) + H_2O(l) \quad (1)$$

Because carbonic acid, H_2CO_3, is in equilibrium with CO_2 and H_2O (Chapter 8),

$$H_2CO_3(aq) \rightleftharpoons CO_2(g) + H_2O(l)$$

we can write equation (1) as

$$CO_2(g) + NaOH(aq) \rightarrow NaHCO_3(aq)$$

which shows that $NaHCO_3$ can be made by passing CO_2 into a solution of sodium hydroxide. Sodium hydrogencarbonate, which is commonly known as *sodium*

BOX 10.1 Hard Water and Limestone Caves

When water flows over limestone, $CaCO_3$, the limestone very slowly dissolves. Although $CaCO_3$ is insoluble in pure water, it is soluble in natural water, which is slightly acidic. This acidity arises from CO_2, which forms carbonic acid:

$$CO_2(g) + H_2O(l) \rightleftharpoons H_2CO_3(aq)$$

Carbonic acid dissolves $CaCO_3$ by converting it to soluble $Ca(HCO_3)_2$:

$$CaCO_3(s) + H_2CO_3(aq) \rightleftharpoons Ca^{2+}(aq) + 2HCO_3^-(aq)$$

Water that contains Ca^{2+} is called ''hard'' water. Mg^{2+} ions are also often present. If these ions are absent, the water is said to be ''soft.''

The use of hard water for domestic purposes and in industry presents several problems. For example, soap consists of the sodium salts of long-chain carboxylic acids such as sodium stearate $C_{17}H_{35}CO_2^-\ Na^+$ (Chapter 14). In hard water, soap forms a scum of insoluble calcium salts of these acids, which has no cleansing power. When hard water is heated the above reactions are reversed, carbon dioxide is driven off, and $CaCO_3$ precipitates. This $CaCO_3$ is the ''scale'' that forms inside teakettles and in hot-water pipes and boilers. Eventually, scale deposits may block a pipe completely.

Hard water can be softened by boiling to precipitate most of the calcium and magnesium as their carbonates. Another method is to dissolve Na_2CO_3 in the water and thereby precipitate the insoluble carbonates:

$$Ca^{2+}(aq) + CO_3^{2-}(aq) \rightarrow CaCO_3(s)$$

$$Mg^{2+}(aq) + CO_3^{2-}(aq) \rightarrow MgCO_3(s)$$

Hard water is commonly softened by ion exchange, as described in Chapter 20.

Caves are found in limestone-rich regions in many parts of the world. They form by water that runs through cracks in the rock and gradually dissolves the $CaCO_3$ by the formation of $Ca(HCO_3)_2$. The cone-shaped deposits found in many limestone caves can grow to be very large and beautifully shaped (Figure A). These structures are formed by the reversal of the reaction that dissolves the $CaCO_3$ and forms the caves. Water containing dissolved CO_2 that has passed through limestone becomes saturated with $Ca(HCO_3)_2$, seeps through the roof of a cave, and forms a droplet hanging from the roof. The water slowly evaporates, and some of the CO_2 is lost from the solution. As a consequence, the position of the equilibrium

Figure A *Stalactites and stalagmites in a limestone cave at Lost Caverns, Pennsylvania*

$$CaCO_3(s) + H_2O(l) + CO_2(g) \rightleftharpoons Ca(HCO_3)_2(aq)$$

shifts to the left, and $CaCO_3$ precipitates. The evaporation of many drops from the same place over many thousands of years gradually leads to the formation of a cone-shaped hanging deposit called a *stalactite*. Alternatively, a drop hanging from the cave roof may fall to the floor, where it evaporates and again deposits a small amount of $CaCO_3$. If drops continue to fall to the same spot, a cone-shaped deposit rising from the cave floor, a *stalagmite*, is built up. A stalagmite may grow tall enough to join a stalactite suspended above it, thus forming a column.

bicarbonate, bicarbonate of soda, or *baking soda*, is used in cooking. Baking powder, a mixture of $NaHCO_3$ and a weak acid such as sodium dihydrogenphosphate, NaH_2PO_4, is used in making biscuits, cakes, and other baked goods. When water is added to baking powder, carbon dioxide is produced in the reaction between the acid $H_2PO_4^-$ and the hydrogencarbonate ion to give carbonic acid:

$$H_2PO_4^- + HCO_3^- \rightarrow H_2CO_3 + HPO_4^{2-}$$

The carbonic acid then decomposes:

$$H_2CO_3(aq) \rightarrow H_2O(l) + CO_2(g)$$

The baked goods ''rise,'' which gives them a light texture, because of the many holes formed by the escaping carbon dioxide.

Calcium carbonate, $CaCO_3$ in the form of **chalk**, **limestone**, or **marble**, is the most abundant compound of calcium in the earth's crust (Box 10.1). Chalk has been subjected to the least pressure during its formation and is the softest $CaCO_3$ rock. Limestone is a more highly compressed form, whereas marble, which is the hardest, has been subjected to such high temperatures and pressures that it has melted and subsequently recrystallized. Marble is an impure form of crystalline calcium carbonate, or *calcite*, and contains many other minerals that affect its color and texture. Pure colorless crystals of calcite also occur naturally.

All the enormous deposits of calcium carbonate found on earth originated in the ocean. The weathering of calcium silicate rocks over millions of years has gradually changed insoluble calcium silicate into soluble calcium salts, which are carried by rivers to the ocean. The dissolved calcium is used by microscopic organisms to form their shells, which are chiefly calcium carbonate. After these organisms die, their shells are deposited on the ocean floor, where they are eventually compressed into sedimentary rock by other deposits above them.

The formation of a white precipitate of calcium carbonate from an aqueous solution of $Ca(OH)_2$ (limewater) is used as a test for carbon dioxide.

Calcium carbonate is obtained as a white precipitate when CO_2 is passed into limewater, an aqueous solution of $Ca(OH)_2$:

$$Ca(OH)_2(aq) + CO_2(g) \rightarrow CaCO_3(s) + H_2O(l)$$

We can think of this reaction as occurring in three stages:

$$CO_2(g) + H_2O(l) \rightarrow H_2CO_3(aq)$$
$$H_2CO_3(aq) + 2OH^-(aq) \rightarrow CO_3^{2-}(aq) + 2H_2O(l)$$
$$CO_3^{2-}(aq) + Ca^{2+}(aq) \rightarrow CaCO_3(s)$$

When calcium carbonate is heated, it decomposes into the oxide CaO and CO_2:

$$CaCO_3(s) \rightarrow CaO(s) + CO_2(g)$$

When the CO_2 is allowed to escape, this reaction goes to completion and the $CaCO_3$ (limestone) is completely converted to calcium oxide, CaO (lime).

EXAMPLE 10.1 Reactions of the Alkali and Alkaline Earth Metals

Starting with the element lithium, explain by suitable balanced equations how you would prepare $LiOH(aq)$ and then convert it to $Li_2SO_4(aq)$. Classify the reactions you use as redox, acid–base, or precipitation.

Solution: Alkali metals react with water to give the corresponding hydroxide:

$$2Li(s) + 2H_2O(l) \rightarrow 2LiOH(aq) + H_2(g) \qquad \textit{Redox}$$

Metal hydroxides react with aqueous acids to give salts:

$$2LiOH(aq) + H_2SO_4(aq) \rightarrow Li_2SO_4(aq) + 2H_2O(l) \qquad \textit{Acid–base}$$

Exercise 10.1 Starting with the element magnesium, explain how you would prepare $MgO(s)$, convert it to $MgCl_2(aq)$, and then to $Mg(OH)_2(s)$. Classify the reactions you use as redox, acid–base, or precipitation.

Exercise 10.2 Write balanced equations for the preparation of the following salts from aqueous solutions of the appropriate acid and base.

(a) K_2SO_4 **(b)** Na_2HPO_4 **(c)** $BaCl_2$

Exercise 10.3 Write balanced equations for the preparation of the following salts from an acidic oxide and a basic oxide.

(a) $MgCO_3$ **(b)** $CaSO_4$ **(c)** Na_2SO_3

Flame Tests

When the alkali and alkaline earth metals and their compounds are strongly heated in a flame, they impart characteristic colors to the flame (Table 10.5). Only beryllium and magnesium do not produce a colored flame. As we saw in Chapter 6, the colors arise from atoms that have been raised to excited states at the high temperature of the flame; atoms in these excited states may lose their excess energy by emitting light of characteristic wavelengths and therefore characteristic colors. These colored flames provide a very convenient qualitative test for these elements in mixtures and compounds (Demonstration 6.1). Many of the brilliant colors of fireworks and flares are provided by salts of the Group I and II metals (Box 10.2).

BOX 10.2 Fireworks

Alkali and alkaline earth metal salts produce many of the color effects in fireworks. The invention of fireworks more than 1000 years ago is usually attributed to the Chinese. The basis of many fireworks is a mixture of potassium nitrate, charcoal, and sulfur that is called *black powder.* The Chinese found that if the mixture was ignited in a sealed container, it produced an explosion and a loud bang. They also used black powder to make the first rockets. The hot gases—mainly CO_2, SO_2, and oxides of nitrogen—produced by the reaction act as the rocket propellant. The potassium nitrate is called the oxidizer because it provides the oxygen needed to convert the carbon and sulfur, which are called the fuels, to CO_2 and SO_2. Potassium chlorate, $KClO_3$, and potassium perchlorate, $KClO_4$, also act as oxidizers. Potassium salts are used rather than sodium salts for two reasons. (1) Many sodium salts are *hygroscopic*, that is, they absorb moisture from the atmosphere. (2) As we saw in Demonstration 6.1, sodium salts impart to a flame a brilliant yellow color so intense that it masks the colors produced by any other salts present. The following are some of the substances used to produce particular colors.

Red: $Sr(NO_3)_2$, $SrCO_3$, and Ca and Li salts

Green: $Ba(NO_3)_2$, $Ba(ClO_4)_2$, and some Cu(II) salts

Blue: $CuCO_3$, $CuSO_4$, CuO, and Cu(I)Cl

Yellow: nonhygroscopic Na salts, such as Na_3AlF_6 and sodium oxalate

Violet/purple: potassium and rubidium salts

White: magnesium and aluminum metals

Figure A *Fireworks above the Manhattan skyline*

Gold sparks: iron filings

White smoke: KNO_3/S mixture

Colored smoke: KNO_3/S/volatile organic dye mixture

Many of the techniques invented by the Chinese are still used today in the manufacture of multistage fireworks involving special effects (Figure A). Despite many improvements in methods of handling the materials used in fireworks, their manufacture is still a hazardous operation that should be left to experts, who take numerous safety precautions. Despite these precautions, we occasionally hear of massive explosions at fireworks factories.

TABLE 10.5 Flame Colors Produced by the Alkali and Alkaline Earth Metals

Metal	Flame Color	Metal	Flame Color
Li	Red	Cs	Blue
Na	Yellow	Ca	Orange red
K	Lilac	Sr	Deep red
Rb	Purple	Ba	Pale green

10.4 REACTIONS AND COMPOUNDS OF ALUMINUM

Aluminum, which is in Group III, has three electrons in its valence shell (Table 10.6). An aluminum atom can lose these three electrons to give the Al^{3+} ion, so aluminum behaves as a reducing agent according to the equation

$$Al \rightarrow Al^{3+} + 3e^-$$

This half-reaction is the basis of all the reactions of aluminum. Examples are the reactions with oxygen, the halogens, and aqueous acids, for which we may write the following equations:

$$4Al(s) + 3O_2(g) \rightarrow 2Al_2O_3(s)$$

$$2Al(s) + 3Cl_2(g) \rightarrow 2AlCl_3(s)$$

$$2Al(s) + 3Br_2(l) \rightarrow 2AlBr_3(s)$$

$$2Al(s) + 6H^+(aq) \rightarrow 2Al^{3+}(aq) + 3H_2(g)$$

The valence electrons of aluminum are less easily removed than those of the Group I and II metals. Thus, aluminum is not as strong a reducing agent as these metals, as is shown by the position of aluminum in the activity series (Table 10.4).

Because of its small size (Table 10.6) and high charge, the aluminum ion attracts electron pairs from any nearby atom into its valence shell. So in most of its compounds, aluminum is not present as a free ion. Instead, it forms bonds that have some covalent character and are better regarded as polar covalent bonds rather than as purely ionic bonds. This behavior of aluminum reflects its position near the border between metals and nonmetals in the periodic table. The Group I and II metals form predominately ionic compounds, whereas the nonmetals of Groups IV, V, VI, and VII form covalent compounds. Aluminum, in Group III, forms compounds in which the bonds have an intermediate character.

Articles made of aluminum

TABLE 10.6 Some Properties of the Aluminum, Iron, and Copper Atoms

Element	Electron Configuration, Atom	Electron Configuration, Ion	Metallic Radius (pm)	Ionic Radius (pm)
Aluminum	$[Ne]3s^23p^1$	Al^{3+}: [Ne]	143	67
Iron	$[Ar]3d^64s^2$	Fe^{2+}: $[Ar]3d^6$	126	75
Copper	$[Ar]3d^{10}4s^1$	Cu^{2+}: $[Ar]3d^9$	128	87

FIGURE 10.10 Ruby and Sapphire Rubies and sapphires are impure forms of corundum (aluminum oxide), Al_2O_3.

Aluminum Oxide

When aluminum is heated strongly in oxygen, it burns with a brilliant flame to give the oxide Al_2O_3, commonly called **alumina**. The oxide can also be prepared by heating the hydroxide:

$$2Al(OH)_3(s) \rightarrow Al_2O_3(s) + 3H_2O(g)$$

Like many other metal oxides, but unlike the alkali metal oxides, Al_2O_3 is insoluble in water. It is very hard and has a very high melting point. These properties can be attributed to the strength of the ionic bonds, which results from the large charges on the Al^{3+} and O^{2-} ions. However, the large charge on the Al^{3+} ion also means that the ion strongly attracts electron pairs on adjacent oxygen atoms, so these electrons are to some extent shared between the two atoms, and the bonds have some covalent character. The nature of the bonds in aluminum oxide is consistent with the position of aluminum in the periodic table between Mg and Si. In MgO the bonds are predominately ionic, whereas in SiO_2 they are predominately covalent. A crystalline form of Al_2O_3 that occurs naturally is called *corundum*. Crystals of corundum are often colored by traces of other metal ions and are familiar as the gems ruby and sapphire (Figure 10.10). An important use of the powdered oxide is as the adsorbing material in chromatographic columns (Chapter 1).

Aluminum metal is normally covered with a strong, thin, continuous layer of Al_2O_3 that protects the metal and makes it appear to be less reactive than its position in the activity series would indicate. Aluminum reacts with acids quite slowly and, unlike iron, does not corrode in moist, damp air.

Aluminum Hydroxide: An Amphoteric Hydroxide

Unlike the alkali metal hydroxides but like most other metal hydroxides, $Al(OH)_3$ is insoluble in water (Table 4.6). When aqueous solutions of NaOH and of an aluminum salt such as $Al_2(SO_4)_3$ are mixed, a white precipitate of $Al(OH)_3$ forms:

$$Al^{3+}(aq) + 3OH^-(aq) \rightarrow Al(OH)_3(s)$$

Aluminum hydroxide is a base that reacts with aqueous acids to give a solution of the aluminum salt of the acid. For example,

$$Al(OH)_3(s) + 3HCl(aq) \rightarrow AlCl_3(aq) + 3H_2O(l)$$

However, a precipitate of aluminum hydroxide, $Al(OH)_3$, formed by the addition of a solution of sodium hydroxide to a solution of an aluminum salt, dissolves if excess hydroxide solution is added. Insoluble $Al(OH)_3$ dissolves because it is converted to

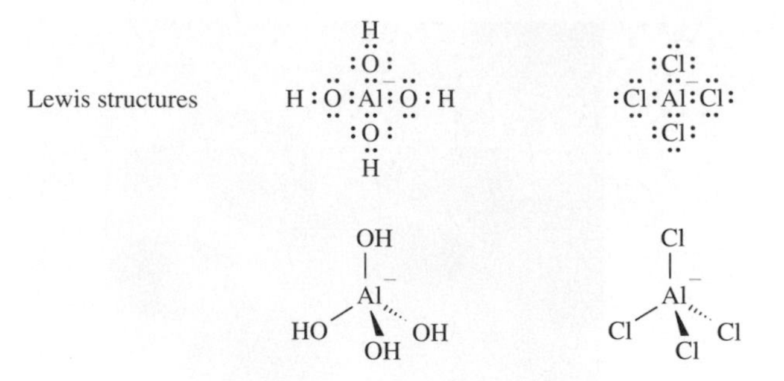

FIGURE 10.11 The Tetrahydroxoaluminate and Tetrachloroaluminate Ions $Al(OH)_4^-$ and $AlCl_4^-$ have a tetrahedral AX_4 geometry.

the soluble compound sodium aluminate, $NaAl(OH)_4$, which contains the AX_4 tetrahedral $Al(OH)_4^-$ ion (Figure 10.11):

$$Al(OH)_3(s) + NaOH(aq) \rightarrow NaAl(OH)_4(aq)$$

By omitting the spectator sodium ion, we can write the equation more simply as

$$Al(OH)_3(s) + OH^-(aq) \rightarrow Al(OH)_4^-(aq)$$

Aluminum hydroxide reacts with acids such as HCl and also with bases such as NaOH. In other words, aluminum hydroxide can behave *both* as a base and as an acid. As we noted in Chapter 7, substances that behave in this way are said to be **amphoteric**.

That aluminum hydroxide is amphoteric is consistent with the position of aluminum in the periodic table near the border between the metals and the nonmetals. The hydroxides of the metals of Groups I and II behave as bases in water, whereas the hydroxides of the nonmetals of periods V, VI, and VII are oxoacids, such as sulfuric acid, which we can write as $O_2S(OH)_2$ (Chapter 7).

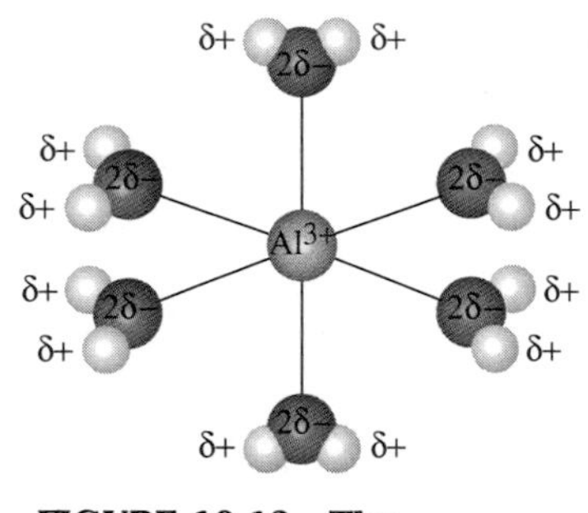

FIGURE 10.12 The Hydrated Aluminum Ion $Al(H_2O)_6^{3+}$ Six water molecules are arranged octahedrally around the aluminum ion. The bonding can best be regarded as intermediate between the two extremes of ionic bonding and covalent bonding.

The formulas of hydrated salts are often written with the number of water molecules separated from the formula of the salt by a dot, as in $AlCl_3 \cdot 6H_2O$ or $CuSO_4 \cdot 5H_2O$. These formulas do not tell us how many water molecules are bonded to the metal ion. The alternative formulas $[Al(H_2O)_6]Cl_3$ and $[Cu(H_2O)_4]SO_4 \cdot H_2O$ show the number of water molecules bonded to the metal ion to form a complex ion.

Aluminum Chloride

Aluminum is oxidized by chlorine upon heating to give the chloride $AlCl_3(s)$ and by bromine to give the bromide $AlBr_3(s)$ (Demonstration 10.4). An aqueous solution of aluminum chloride is obtained when aluminum dissolves in hydrochloric acid,

$$2Al(s) + 6HCl(aq) \rightarrow 2AlCl_3(aq) + 3H_2(g)$$

or by the acid–base reaction between aluminum hydroxide and hydrochloric acid:

$$Al(OH)_3(s) + 3HCl(aq) \rightarrow AlCl_3(aq) + 3H_2O(l)$$

Aluminum chloride is a solid that can to a first approximation be regarded as consisting of Al^{3+} ions and Cl^- ions. Aluminum chloride crystallizes from water as a *hydrated salt* with the formula $AlCl_3 \cdot 6H_2O$. The water that is contained in the crystals is called *water of crystallization*. The negatively charged oxygen atom of a polar water molecule is attracted to the Al^{3+} ion, so six water molecules surround the aluminum ion in an octahedral arrangement (Figure 10.12). It is preferable therefore to write the formula as $[Al(H_2O)_6]^{3+}(Cl^-)_3$. A lone pair on the oxygen atom of each water molecule is partially donated to the valence shell of the Al^{3+} ion, forming six polar covalent bonds. The bonding in the $Al(H_2O)_6^{3+}$ ion can be regarded as intermediate between purely ionic bonding and covalent bonding.

Aluminum chloride reacts with a metal chloride such as NaCl to give $Na(AlCl_4)$:

$$AlCl_3(s) + NaCl(s) \xrightarrow{heat} Na(AlCl_4)(s)$$

DEMONSTRATION 10.4 The Preparation of Aluminum Bromide

Aluminum pellets can be seen on the watch glass on top of the beaker, which contains bromine.

A few minutes after the pellets are tipped into the bromine, the aluminum begins to react vigorously with the bromine. The aluminum pellets ignite, burn with a bright flame, and skate around the surface of the bromine.

The product of the reaction is a white solid, aluminum bromide, which coats the beaker and the watch glass.

$Na(AlCl_4)$ contains the tetrahedral AX_4 tetrachloroaluminate ion, $AlCl_4^-$ (Figure 10.11). This reaction is quite similar to the reaction of $Al(OH)_3$ with OH^- to give $Al(OH)_4^-$.

Lewis Acids and Bases

We have said that $Al(OH)_3$ behaves as an acid in that it reacts with hydroxide ion. However, it does not behave as an acid in the usual Brønsted sense, because it does not give up a proton to the hydroxide ion to form a water molecule. Rather, it removes a hydroxide ion from the solution by combining with it to form the $Al(OH)_4^-$ ion. This ion has four hydroxyl, OH, groups attached to the aluminum by polar covalent bonds (Figure 10.11). The Al atom in $Al(OH)_3$ combines with an OH^- ion by accepting a lone pair on the oxygen atom into its valence shell, thus converting the lone pair to a shared pair:

$$\underset{\text{Lewis acid}}{\mathrm{HO{-}Al(OH)_2}} + \underset{\text{Lewis base}}{:\ddot{\mathrm{O}}\mathrm{H}^-} \rightarrow \mathrm{HO{-}Al^-(OH)_3}$$

We say that $Al(OH)_3$ is behaving as a **Lewis acid**, after Gilbert Lewis (Figure 3.7), who first proposed this extension of the concept of an acid.

> **A substance that can accept one or more lone pairs into the valence shell of one of its atoms to form one or more new bonds is called a *Lewis acid*.**

In short,

> **A Lewis acid is an electron-pair acceptor.**

The ion or molecule that combines with the Lewis acid is called a **Lewis base**. In short,

A Lewis base is an electron-pair donor.

A Lewis base is also a Brønsted base, because the lone pair can, in principle, be used to form a covalent bond to a proton. The hydroxide ion is a base in both the Brønsted and Lewis senses.

A Lewis base must have at least one unshared pair of electrons that can be donated to a Lewis acid to form a covalent bond. The reaction of $Al(OH)_3$ with OH^- is an example of a **Lewis acid–base reaction**.

Another example of a Lewis acid–base reaction is the reaction of $AlCl_3$ with chloride ion:

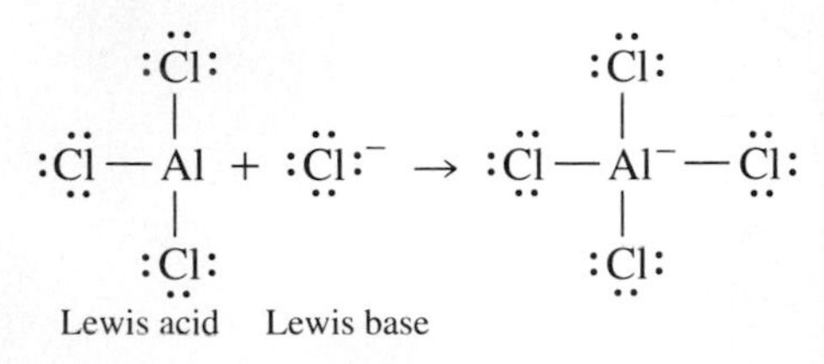

The formation of the hydrated aluminum ion is also a Lewis acid–base reaction, in which the aluminum ion is the Lewis acid (electron-pair acceptor), and the water molecule is the base (electron-pair donor):

$$\underset{\text{Lewis base}}{H_2\ddot{O}:} + \underset{\text{Lewis acid}}{Al^{3+}} \rightarrow H_2\ddot{O}^+\!-\!Al^{2+}$$

$AlCl_3$, $Al(OH)_3$, and Al^{3+} are Lewis acids because the aluminum atom in these molecules has an incomplete valence shell and can accept up to six electron pairs in its valence shell.

Although both Lewis acid–base reactions and redox reactions appear to involve the gain and loss of electrons, it is important not to confuse them. In a Lewis acid–base reaction, there is no oxidation or reduction, as we can see by checking the oxidation numbers. The oxidation number of Al is +3 in $AlCl_4^-$ and $Al(OH)_4^-$, as it is in $AlCl_3$ and $Al(OH)_3$. In forming these ions, aluminum gains a share in an electron pair of a more electronegative atom, so its oxidation number remains unchanged at +3. In contrast, in the reaction of aluminum with aqueous HCl, for example, the oxidation number of aluminum increases from 0 to +3. An Al atom loses electrons, and a hydrogen atom gains electrons. The oxidation number of hydrogen decreases from +1 to 0:

Oxidation numbers

$$\overset{0}{2Al} + 6\overset{+1\,-1}{HCl} \rightarrow 2\overset{+3\,-1}{AlCl_3} + 3\overset{0}{H_2}$$

EXAMPLE 10.2 Reactions of Aluminum and Its Compounds

Complete and balance each of the following equations. In each case, state what type of reaction is occurring.

(a) $Al(s) \rightarrow Al(NO)_3(aq)$

(b) $Al(OH)_3(s) + NaOH(aq) \rightarrow$

(c) $AlCl_3(aq) \rightarrow Al(OH)_3(s)$

Solution: **(a)** Aluminum reacts with aqueous acids to give a salt and hydrogen:

$$2Al(s) + 6HNO_3(aq) \rightarrow 2Al(NO_3)_3(aq) + 3H_2(g) \qquad \textit{Redox}$$

(b) $Al(OH)_3$ reacts with hydroxide ion to give the complex ion $Al(OH)_4^-$. The complete balanced equation is

$$Al(OH)_3(s) + NaOH(aq) \rightarrow NaAl(OH)_4(aq) \qquad \textit{Lewis acid–base}$$

(c) $Al(OH)_3$ is insoluble, so it can be precipitated from the solution of any soluble aluminum compound by the addition of a soluble hydroxide. The complete balanced equation, using NaOH, is

$$AlCl_3(aq) + 3NaOH(aq) \rightarrow Al(OH)_3(s) + 3NaCl(aq) \qquad \textit{Precipitation}$$

Exercise 10.4 Complete and balance the following equations. In each case, state what type of reaction is occurring.
(a) $Al(s) \rightarrow AlBr_3(aq)$
(b) $Al(OH)_3(s) \rightarrow Al_2O_3(s)$
(c) $Al(OH)_3(s) \rightarrow Al_2(SO_4)_3(aq)$

10.5 THE TRANSITION ELEMENTS

In the periodic table the **transition elements** occupy the 10 short columns between main Groups II and III. These elements have several characteristic properties:

1. They are metals.
2. They each have several oxidation states.
3. They form many complex ions.
4. The majority of their compounds are colored.
5. Many of their compounds have magnetic properties.

We consider only the transition elements in Period 4—the first transition series—with particular emphasis on iron and copper.

Oxidation States

The electron configurations of the Period 4 elements are given in Table 10.7. All the transition metals have one or two 4*s* electrons together with one or more 3*d* electrons in their valence shells. The 4*s* electrons and one or more of the 3*d* electrons are rather easily removed to give positive ions, so these elements all behave as metals. Unlike the metals of Groups I and II, the transition metals have several oxidation states depending on the number of electrons lost. The common oxidation states of iron are +II (Fe^{2+}) and +III (Fe^{3+}), and the common oxidation states of copper are +I (Cu^+) and +II (Cu^{2+}).

All the ions of the metals of Groups I and II and the metals of Group III, such as aluminum, have the stable electron configuration of a noble gas (ns^2np^6). The transition metal ions in almost all cases have from 1 to 10 *d* electrons in addition to the noble gas configuration. For example, the Fe^{2+} ion has the electron configuration $1s^22s^22p^63s^23p^63d^6$ with 6 3*d* electrons, and the Cu^{2+} ion has the electron configuration $1s^22s^22p^63s^23p^63d^9$ with 9 3*d* electrons. It is the presence of these *d* electrons that gives the transition metal compounds their characteristic properties.

TABLE 10.7 Electron Configurations of the Period 4 Elements

		$n = 1$	2		3			4		
Period		s	s	p	s	p	d	s	p	d
4	K	$1s^2$	$2s^2$	$2p^6$	$3s^2$	$3p^6$		$4s^1$		
	Ca	$1s^2$	$2s^2$	$2p^6$	$3s^2$	$3p^6$		$4s^2$		
	Sc	$1s^2$	$2s^2$	$2p^6$	$3s^2$	$3p^6$	$3d^1$	$4s^2$		
	Ti	$1s^2$	$2s^2$	$2p^6$	$3s^2$	$3p^6$	$3d^2$	$4s^2$		
	V	$1s^2$	$2s^2$	$2p^6$	$3s^2$	$3p^6$	$3d^3$	$4s^2$		
	Cr	$1s^2$	$2s^2$	$2p^6$	$3s^2$	$3p^6$	$3d^5$	$4s^1$		
	Mn	$1s^2$	$2s^2$	$2p^6$	$3s^2$	$3p^6$	$3d^5$	$4s^2$		
	Fe	$1s^2$	$2s^2$	$2p^6$	$3s^2$	$3p^6$	$3d^6$	$4s^2$		
	Co	$1s^2$	$2s^2$	$2p^6$	$3s^2$	$3p^6$	$3d^7$	$4s^2$		
	Ni	$1s^2$	$2s^2$	$2p^6$	$3s^2$	$3p^6$	$3d^8$	$4s^2$		
	Cu	$1s^2$	$2s^2$	$2p^6$	$3s^2$	$3p^6$	$3d^{10}$	$4s^1$		
	Zn	$1s^2$	$2s^2$	$2p^6$	$3s^2$	$3p^6$	$3d^{10}$	$4s^2$		
	Ga	$1s^2$	$2s^2$	$2p^6$	$3s^2$	$3p^6$	$3d^{10}$	$4s^2$	$4p^1$	
	Ge	$1s^2$	$2s^2$	$2p^6$	$3s^2$	$3p^6$	$3d^{10}$	$4s^2$	$4p^2$	
	As	$1s^2$	$2s^2$	$2p^6$	$3s^2$	$3p^6$	$3d^{10}$	$4s^2$	$4p^3$	
	Se	$1s^2$	$2s^2$	$2p^6$	$3s^2$	$3p^6$	$3d^{10}$	$4s^2$	$4p^4$	
	Br	$1s^2$	$2s^2$	$2p^6$	$3s^2$	$3p^6$	$3d^{10}$	$4s^2$	$4p^5$	
	Kr	$1s^2$	$2s^2$	$2p^6$	$3s^2$	$3p^6$	$3d^{10}$	$4s^2$	$4p^6$	

Magnetic Properties

In Chapter 6 we saw that an electron has an associated magnetic field, which is attributed to its spin. In most molecules and ions, the electrons occur in pairs of opposite spin, so their magnetic fields cancel. Hence, most compounds are nonmagnetic (*diamagnetic*). In many transition metal molecules and ions, however, there are one or more unpaired *d* electrons, so many transition metal compounds have magnetic properties. These compounds are said to be **paramagnetic**. When a paramagnetic substance is placed in a magnetic field, the unpaired electrons tend to line up with their spins in the same direction. This alignment produces a magnetic field in the sample, so it behaves like a small bar magnet and is attracted to the external magnetic field. Paramagnetism can be detected and measured by hanging a sample of a substance just above the poles of a powerful electromagnet. When the electromagnet is switched on, a sample that is paramagnetic is pulled between the magnet's poles.

Some transition metals and some of their compounds—for example, iron and the oxide $Fe_3O_4(s)$—can be permanently magnetized; such substances are said to be **ferromagnetic**. In a ferromagnetic substance the interaction between the unpaired electrons is particularly strong. Hence, even in the absence of an external field, these unpaired electrons become aligned with their spins all in the same direction, producing a strong magnetic field. This alignment is confined to small regions called *magnetic domains*, whose fields are normally oriented in different directions so as to cancel each other (Figure 10.13). But when a ferromagnetic substance is placed in an external magnetic field, the domains are rotated so that their magnetic fields tend to line up in the same direction, producing a strong resultant magnetic field. When the external field is removed, the domains do not all resume their random orientation and the solid remains permanently magnetized.

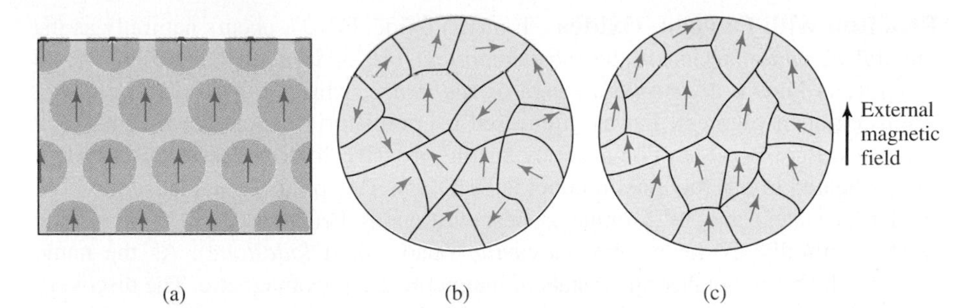

FIGURE 10.13 Ferromagnetism (a) The spins of the electrons in a domain are oriented in the same direction. (b) In the unmagnetized state, the domains have a random orientation. The arrows represent the magnetic field resulting from the spins of all the electrons in the domain. The resultant field of all the domains is zero. (c) In an external magnetic field, the domains rotate and change in size until their magnetic fields are approximately lined up, which produces a strong resultant field. When the external field is removed, all the domains do not resume their random orientation. Thus, the solid remains permanently magnetized.

Complex Ions

We have seen that the aluminum ion, Al^{3+}, because of its small size and high charge, behaves as a Lewis acid. It combines with water molecules to form the $Al(H_2O)_6{}^{3+}$ ion, with hydroxide ions to form the $Al(OH)_4{}^-$ ion, and with chloride ions to form the $AlCl_4{}^-$ ion. Such ions are called **complex ions.**

A complex ion is a polyatomic ion consisting of a central metal ion to which several other ions or molecules are bonded.

The molecules or ions bonded to the metal ion are called **ligands**. A ligand is sometimes said to be *coordinated to* the metal ion, so the number of ligands is called the **coordination number**. Complex ions are sometimes called **coordination compounds**.

Transition metal ions have a particularly strong tendency to form complex ions. The most common coordination numbers for the complex ions of the transition metals are 4 and 6. Iron forms complex ions such as $Fe(H_2O)_6{}^{2+}$ and $Fe(CN)_6{}^{3-}$ (Figure 10.14), and copper forms complex ions such as $CuCl_4{}^{2-}$ and $Cu(NH_3)_4(H_2O)_2{}^{2+}$. In some cases the combination of a central metal atom and ligands gives a neutral molecule, which is then called a **complex** rather than a complex ion. For example, Fe^{2+} combines with water and chloride ions to give the complex $Fe(H_2O)_4Cl_2$, a neutral molecule.

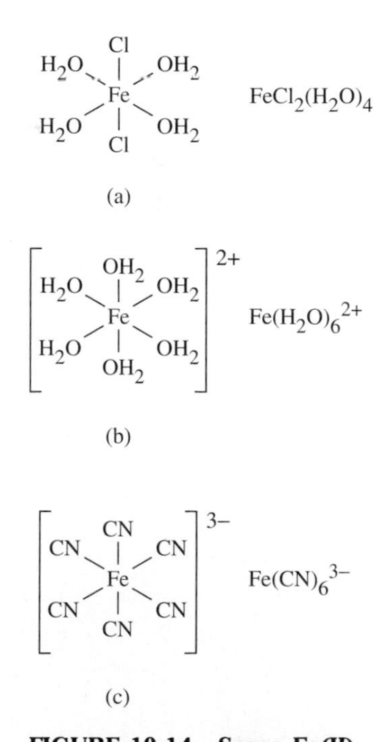

FIGURE 10.14 Some Fe(II) and Fe(III) Complex Ions There are six ligands (Cl^-, H_2O, or CN^-) in each of these complex ions; the metal has a coordination number of six. Each ion has an octahedral AX_6 geometry.

Reactions and Compounds of Iron

Iron can be oxidized to the +II oxidation state, iron(II), or to the +III oxidation state, iron(III). Iron(III) can be reduced to iron(II), and iron(II) oxidized to iron(III). The reactions of iron may be summarized by the three equations

$$Fe \rightarrow Fe^{2+} + 2e^-$$

$$Fe \rightarrow Fe^{3+} + 3e^-$$

$$Fe^{2+} \rightleftarrows Fe^{3+} + e^-$$

Stronger oxidizing agents oxidize iron to iron(III), and weaker oxidizing agents only to iron(II).

Recall that the sum of the oxidation numbers of all the atoms in a neutral compound equals zero. For Fe_3O_4 (that is, Fe(II) $Fe(III)_2O_4$), Fe(+2) + 2Fe(+3) + 4O(−2) = 0.

Australian aboriginal rock paintings made with powdered hematite, Fe_2O_3.

Reaction with Oxygen: Oxides Iron(III) oxide, Fe_2O_3, occurs naturally as the mineral *hematite*, which is the most important ore of iron. In a powdered form hematite is known as *jeweler's rouge* or *red ochre*, which is used as a polishing agent and as a pigment. It was often used by prehistoric peoples in making rock paintings. Pure Fe_2O_3 can be made by heating iron(III) hydroxide, $Fe(OH)_3$. When iron is heated in air, the main product is $Fe_3O_4(s)$. This oxide contains iron in both oxidation states, and its formula is best written as $Fe(II)Fe(III)_2O_4$. This oxide occurs naturally as the mineral *magnetite* (also called *lodestone*). As the name implies, the black octahedral crystals of magnetite are ferromagnetic. The discovery by the Chinese around 2000 years ago that these crystals orient in a particular direction when suspended on a string led eventually to the crude magnetic compasses that were first used on ships in the eleventh century. Iron(II) oxide, FeO, is a reactive oxide that is difficult to prepare because it is very easily oxidized to Fe_2O_3.

Iron(II) Sulfide Sulfur oxidizes iron to the +II oxidation state when the two elements are heated together (Demonstration 1.1):

$$Fe(s) + S(s) \xrightarrow{\text{heat}} FeS(s)$$

The iron(II) sulfide formed is a black solid that to a first approximation can be regarded as an ionic compound composed of Fe^{2+} ions and S^{2-} ions. It may be used for the preparation of hydrogen sulfide by treating it with an aqueous acid (Chapter 7):

$$FeS(s) + H_2SO_4(aq) \rightarrow FeSO_4(aq) + H_2S(g)$$

When $H_2S(g)$ is passed into a solution of an iron(II) salt, the reverse reaction occurs, and a black precipitate of iron(II) sulfide is obtained:

$$Fe^{2+}(aq) + H_2S(aq) \rightarrow FeS(s) + 2H^+(aq)$$

Excess hydrogen sulfide and a very low initial $H^+(aq)$ concentration cause $FeS(s)$ to precipitate. But a high $H^+(aq)$ concentration and a very low initial $H_2S(aq)$ concentration favor the reverse reaction.

Reactions with Acids Iron, like all metals above hydrogen in the activity series, is oxidized by dilute aqueous acids to $Fe^{2+}(aq)$:

$$Fe(s) + 2H^+(aq) \rightarrow Fe^{2+}(aq) + H_2(g)$$

Iron(II) chloride can therefore be made by the reaction of iron with dilute hydrochloric acid:

$$Fe(s) + 2HCl(aq) \rightarrow FeCl_2(aq) + H_2(g)$$

$FeCl_2$ crystallizes from water as the hydrate $FeCl_2 \cdot 4H_2O$. The four water molecules and the two chloride ions are bonded to the Fe^{2+} ion to give a neutral complex (Figure 10.14a).

Similarly, a solution of iron(II) sulfate is obtained from the reaction of iron with dilute sulfuric acid:

$$Fe(s) + H_2SO_4(aq) \rightarrow FeSO_4(aq) + H_2(g)$$

Iron(II) sulfate crystallizes from solution as $FeSO_4 \cdot 7H_2O$. This forms ionic crystals consisting of $Fe(H_2O)_6{}^{2+}$ complex ions (Figure 10.14b), sulfate ions, $SO_4{}^{2-}$, and one water molecule for each sulfate ion. The $Fe(H_2O)_6{}^{2+}$ ion has a pale green color.

Solutions of iron(III) salts can be made by using a stronger oxidizing acid than H_3O^+, such as hot concentrated sulfuric acid or nitric acid.

EXAMPLE 10.3 Reaction of Iron with an Acid

Write the balanced equation for the oxidation of iron to $Fe(NO_3)_3(aq)$ with concentrated nitric acid, which is reduced to $NO_2(g)$.

Solution: $Fe(NO_3)_3(aq)$ contains $Fe^{3+}(aq)$ and $NO_3^-(aq)$ ions, so $Fe(s)$ is oxidized to Fe^{3+}:

$$Fe \rightarrow Fe^{3+} + 3e^- \qquad \textit{Oxidation} \qquad (2)$$

The N (oxidation number +5) in nitric acid, $HNO_3(conc)$, is reduced to N (oxidation number +4) in $NO_2(g)$:

$$HNO_3 + e^- + H^+ \rightarrow NO_2 + H_2O \qquad \textit{Reduction} \qquad (3)$$

Adding equation (2) to three times equation (3) to eliminate the electrons on both sides gives the *balanced* equation

$$Fe + 3HNO_3 + 3H^+ \rightarrow Fe^{3+} + 3NO_2 + 3H_2O$$

Finally, we can add $3NO_3^-$ to each side to give

$$Fe(s) + 6HNO_3(conc) \rightarrow Fe(NO_3)_3(aq) + 3NO_2(g) + 3H_2O(l)$$

Exercise 10.5 Write the balanced equations for the reaction of **(a)** iron with dilute $H_2SO_4(aq)$ and **(b)** iron with hot concentrated $H_2SO_4(aq)$ to give $Fe_2(SO_4)_3(aq)$ and $SO_2(g)$.

Iron Hydroxides When a solution of an alkali metal hydroxide is added to a solution of an iron(II) salt, a precipitate of insoluble iron(II) hydroxide, $Fe(OH)_2$, is formed:

$$Fe^{2+}(aq) + 2OH^-(aq) \rightarrow Fe(OH)_2(s)$$

In this and similar equations we write $Fe^{2+}(aq)$ as short for $Fe(H_2O)_6^{2+}$, just as we frequently write $H^+(aq)$ for $H_3O^+(aq)$. The precipitate is initially white but rapidly changes to a dark green and finally to a red-brown, as air oxidizes $Fe(OH)_2$ to iron(III) hydroxide, $Fe(OH)_3$ (Demonstration 10.5). Iron(III) hydroxide can also be precipitated from a solution of an Fe^{3+} salt by adding hydroxide ion:

$$Fe^{3+}(aq) + 3OH^-(aq) \rightarrow Fe(OH)_3(s)$$

When $Fe(OH)_3$ is heated, it loses water to give iron(III) oxide, Fe_2O_3:

$$2Fe(OH)_3(s) \rightarrow Fe_2O_3(s) + 3H_2O(g)$$

DEMONSTRATION 10.5 Iron(II) and Iron(III) Hydroxides

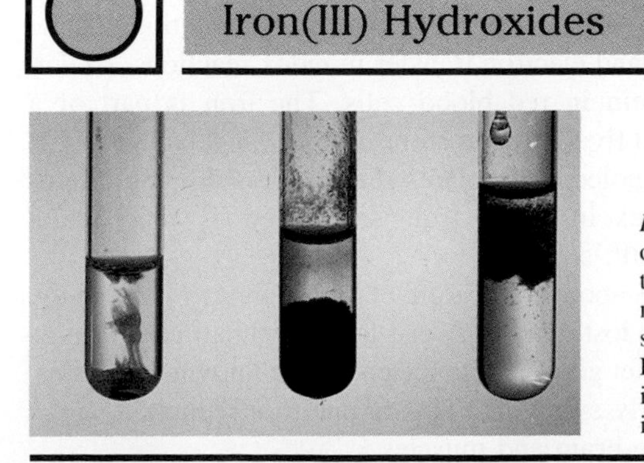

Left: When NaOH(aq) is added to a solution of a soluble Fe^{2+} salt, a dirty white precipitate of $Fe(OH)_2$ is formed. *Center:* In a few minutes this becomes a dark green color as some of the $Fe(OH)_2$ is oxidized by air to $Fe(OH)_3$. *Right:* If aqueous hydrogen peroxide is added to the $Fe(OH)_2$ precipitate, it is immediately oxidized to red-brown $Fe(OH)_3$.

Iron(III) as an Oxidizing Agent Fe^{3+} is a good oxidizing agent and is usually reduced to Fe^{2+}:

$$Fe^{3+} + e^- \rightarrow Fe^{2+}$$

I_2 is insoluble in water but dissolves in excess I^-, forming the brown, soluble I_3^- ion.

For example, Fe^{3+} will oxidize I^- to I_2:

$$2Fe^{3+}(aq) + 2I^-(aq) \rightarrow 2Fe^{2+}(aq) + I_2(s)$$

It is therefore not possible to prepare FeI_3. But Fe^{3+} is not a strong enough oxidizing agent to oxidize Br^- and Cl^- to Br_2 and Cl_2, respectively, so $FeCl_3(s)$ and $FeBr_3(s)$ are stable compounds.

Rusting of Iron Although iron, particularly in the form of steel, is a very useful metal, it has the disadvantage that it corrodes or rusts rather easily when exposed to the atmosphere. Rust is brown hydrated iron(III) oxide, $Fe_2O_3 \cdot H_2O$. Both water and oxygen are required to form rust; iron does not rust in dry air or in water that is free of O_2.

In the first step of rust formation, $Fe(s)$ is oxidized to $Fe^{2+}(aq)$:

$$2Fe(s) + 4H^+(aq) + O_2(g) \rightarrow 2Fe^{2+}(aq) + 2H_2O(l)$$

In the second step Fe^{2+} is oxidized by oxygen to insoluble Fe_2O_3:

$$4Fe^{2+}(aq) + O_2(g) + 4H_2O(l) \rightarrow 2Fe_2O_3(s) + 8H^+(aq)$$

The overall reaction, the sum of these two reactions, is

$$4Fe(s) + 3O_2(g) \rightarrow 2Fe_2O_3(s)$$

Al_2O_3 forms a tough, continuous layer on the surface of aluminum. In contrast, iron(III) oxide does not adhere strongly to the surface of the iron. Rather, it flakes off, exposing more metal to attack by oxygen, and therefore iron continues to rust.

Iron Complexes In addition to the hydrated ions $Fe(H_2O)_6{}^{3+}$ and $Fe(H_2O)_6{}^{2+}$ and the neutral complex $Fe(H_2O)_4Cl_2$, iron(II) and iron(III) form complexes with many other ligands. For example, with the cyanide ion, CN^-, they form the complex ions $Fe(II)(CN)_6{}^{4-}$ and $Fe(III)(CN)_6{}^{3-}$. Like the hydrated ions, these complex ions have an octahedral geometry (Figure 10.14c).

Biochemistry of Iron Iron is the most important transition element involved in living systems, being vital to the whole range of life forms from bacteria to humans. That iron is extremely abundant in the earth's crust and has two easily converted oxidation states are probably important factors in its evolutionary selection for use in many life processes. Proteins containing iron have two major functions in the body: oxygen transport and storage, and electron transfer in redox reactions. *Hemoglobin* is the oxygen-carrying protein in red blood cells. The iron is part of a complex in which it is coordinated to five nitrogen atoms (Box 10.3; Chapter 17). It is interesting that chlorophyll, the molecule in plants that absorbs light in photosynthesis, is a very similar complex in which iron is replaced by magnesium (another abundant element in the earth's crust).

The adult human body contains about 4 g of iron, of which about 3 g are in the form of hemoglobin. Around 1 mg is lost daily in sweat, feces, and hair and must be replaced. An iron deficiency in the diet gives rise to the condition known as anemia, of which chronic tiredness is an early symptom. This is due to a reduction in the amount of oxygen transported to the brain and muscles.

BOX 10.3 Hemoglobin

Among the transition metals that are essential to life, iron is extremely important to humans as an essential component of the oxygen carrier *hemoglobin.* If our diet is deficient in iron, we may suffer from anemia and feel tired and weak. Hemoglobin contains four large protein molecules, each of which has a *heme* molecule as part of its structure. Each heme molecule contains an iron atom coordinated to four nitrogen atoms in a square (Figure A). Octahedral coordination around the iron is completed by a nitrogen atom from another part of the protein molecule and an oxygen molecule or a water molecule (Figure B). Hemoglobin picks up oxygen and forms *oxyhemoglobin* in the lungs. Oxygen is released in the tissues, where it is needed for cell metabolism. It is replaced by a water molecule to form *deoxyhemoglobin.*

Carbon monoxide also behaves as a ligand. It forms a complex with hemoglobin that is about two hundred times as strong as the complex with oxygen and makes the hemoglobin useless as an oxygen carrier. If we breathe carbon monoxide, some of the oxyhemoglobin in our blood is converted to carboxyhemoglobin; breathing air containing only 0.1% CO converts about 60% of the hemoglobin to carboxyhemoglobin in a few hours. If a substantial part of the hemoglobin in the blood is converted to carboxyhemoglobin, the body suffers from acute oxygen deficiency, and death soon results.

Long-term exposure to even small concentrations of carbon monoxide can have serious consequences. For example, air inhaled through a lighted cigarette contains about 400 ppm of carbon monoxide. Consequently, heavy smokers have as much as 6% of the hemoglobin in their blood constantly converted to carboxyhemoglobin. As a result, their blood is not as efficient at carrying oxygen as the blood of nonsmokers, and their hearts must work harder. This is probably a contributing factor in heart disease and heart attacks.

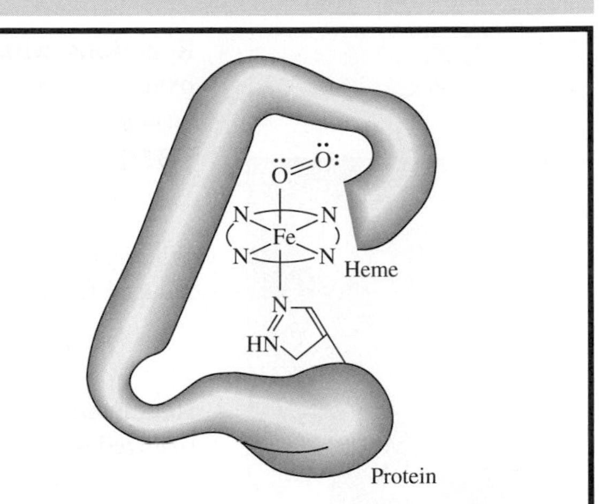

Figure B *One of the four heme–protein complexes in the hemoglobin molecule*

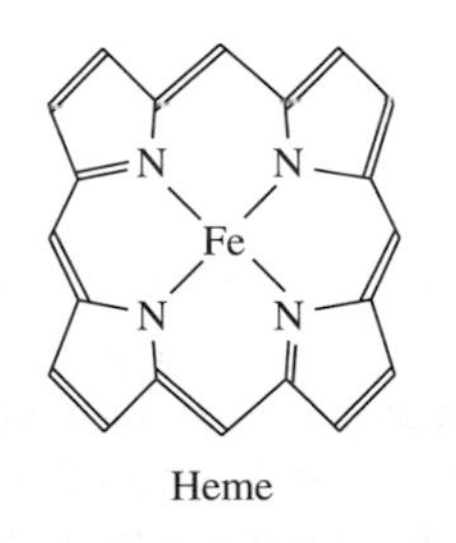

Figure A *Structure of the heme molecule*

Cytochromes are the iron-containing molecules involved in cellular redox reactions. In these molecules iron is in the form of a complex in which it is coordinated to five nitrogen atoms and a sulfur atom. These molecules take part in a series of reactions that lead eventually to the oxidation of glucose to CO_2 and water in which iron is repeatedly converted between its +II and +III oxidation states. Biomolecules of iron are discussed in more detail in Chapter 17, but the role of iron compounds in many life processes is still far from understood, and a large field of research awaits exploration.

Reactions and Compounds of Copper

The important oxidation states of copper are +I and +II. Copper can be oxidized to either of these oxidation states. Copper(I) can be oxidized to copper(II), and copper(II) reduced to copper(I). We can summarize the reactions of copper by the three equations

$$Cu \rightarrow Cu^{+} + e^{-}$$
$$Cu \rightarrow Cu^{2+} + 2e^{-}$$
$$Cu^{+} \rightleftarrows Cu^{2+} + e^{-}$$

Reactions with Oxygen: Oxides When copper is heated strongly in air or oxygen, black copper(II) oxide, CuO, forms. This oxide can also be prepared by heating copper(II) nitrate, which decomposes into CuO, nitrogen dioxide, and oxygen:

$$2Cu(NO_3)_2(s) \rightarrow 2CuO(s) + 4NO_2(g) + O_2(g)$$

When CuO is heated strongly with copper filings or copper powder, copper(II) is reduced to copper(I) to give red copper(I) oxide, $Cu_2O(s)$:

$$CuO(s) + Cu(s) \rightarrow Cu_2O(s)$$

FIGURE 10.15 Precipitation of Copper Sulfide When $H_2S(g)$ is bubbled into a blue solution of a soluble Cu^{2+} salt, a black precipitate of copper(II) sulfide, CuS, is formed.

Reactions with Sulfur: Sulfides When heated with excess sulfur, copper is oxidized to Cu^{2+}, and the sulfur is reduced to S^{2-}. The ionic compound copper(II) sulfide, CuS(*s*), is formed:

$$Cu(s) + S(s) \rightarrow CuS(s)$$

Insoluble black CuS precipitates when $H_2S(g)$ is passed into an aqueous solution of a Cu^{2+} salt (Figure 10.15):

$$Cu^{2+}(aq) + H_2S(g) \rightarrow CuS(s) + 2H^{+}(aq)$$

Black copper(I) sulfide can be prepared by heating the requisite quantities of copper and sulfur according to the equation

$$2Cu(s) + S(s) \rightarrow Cu_2S(s)$$

Reactions with Acids Copper is low in the activity series of metals (Table 10.4). It is not as easily oxidized as iron and cannot be oxidized by $H_3O^{+}(aq)$. It therefore does not react with most dilute aqueous acids, such as HCl(*aq*), in which the oxidizing agent is H_3O^{+}. It is, however, oxidized to Cu^{2+} by hot concentrated sulfuric acid and by nitric acid (Figure 10.16). The equations for these reactions are analogous to those for the oxidation of iron to Fe^{3+}, for example,

$$Cu(s) + 4HNO_3(conc) \rightarrow Cu(NO_3)_2(aq) + 2NO_2(g) + 2H_2O(l)$$

Copper(II) Salts Solutions of copper(II) sulfate and copper(II) nitrate can be obtained, as we have just seen, by the oxidation of copper with concentrated sulfuric acid or nitric acid. An alternative and often more convenient method for preparing

FIGURE 10.16 The Reaction of Copper with Concentrated Nitric Acid A penny contains enough copper to react with concentrated nitric acid, forming a dark green solution of $Cu(NO_3)_2$ and brown NO_2 gas.

copper(II) salts is to use an acid–base reaction between the oxide or hydroxide and the appropriate acid:

$$CuO(s) + H_2SO_4(aq) \rightarrow CuSO_4(aq) + H_2O(l)$$
$$Cu(OH)_2(s) + 2HNO_3(aq) \rightarrow Cu(NO_3)_2(aq) + 2H_2O(l)$$
$$CuO(s) + 2HCl(aq) \rightarrow CuCl_2(aq) + H_2O(l)$$

These salts crystallize from aqueous solution as the hydrates $CuSO_4 \cdot 5H_2O$, $Cu(NO_3)_2 \cdot 6H_2O$, and $CuCl_2 \cdot 2H_2O$. They are typically blue or green in color. When blue $CuSO_4 \cdot 5H_2O$ is heated strongly, it loses water to form white anhydrous $CuSO_4$ (Figure 10.17).

FIGURE 10.17 Copper Sulfate When blue $CuSO_4 \cdot 5H_2O$ is heated, it loses water to give white anhydrous $CuSO_4$.

When an aqueous solution of a soluble hydroxide such as NaOH is added to a solution of a soluble Cu^{2+}salt, a precipitate of pale blue insoluble copper(II) hydroxide, $Cu(OH)_2$, forms:

$$Cu^{2+}(aq) + 2OH^-(aq) \rightarrow Cu(OH)_2(s)$$

Copper(I) Salts Because the Cu^+ ion is not stable in aqueous solution, the only common Cu(I) salts are those that are insoluble in water. For example, insoluble CuI can be obtained by reducing Cu^{2+} with I^- in a reaction analogous to that between Fe^{3+} and I^-:

$$2Cu^{2+}(aq) + 4I^-(aq) \rightarrow 2CuI(s) + I_2(aq)$$

Complex Ions In aqueous solution, Cu^{2+} is always present in the form of the hydrated ion $Cu(H_2O)_6{}^{2+}$ or as some other complex ion. For example, when aqueous ammonia is added to a pale blue solution of $Cu(H_2O)_6{}^{2+}$, the color of the solution changes to an intense deep blue due to the formation of the $Cu(NH_3)_4{}^{2+}$ ion (Figure 10.18). Copper forms many other complex ions, such as $CuCl_4{}^{2-}$ and $Cu(CN)_4{}^{2-}$, and the neutral complex $Cu(H_2O)_2Cl_2$. Following iron and zinc, copper is the third most abundant metal in the human body. Copper-containing biomolecules play important roles in many forms of life.

Exercise 10.6 Write the balanced equations for the reactions needed to carry out the following transformations:

$$Cu(s) \rightarrow CuO(s) \rightarrow CuSO_4(aq) \rightarrow CuS(s) \rightarrow CuCl_2(aq)$$

In each case, state what type of reaction is occurring.

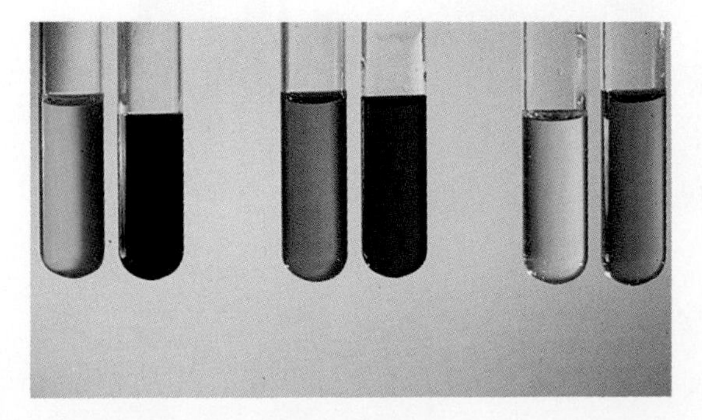

FIGURE 10.18 The Complex Ions Formed by Cu^{2+}, Co^{3+}, and Ni^{2+} *Left:* When an excess of aqueous ammonia is added to a blue solution of $Cu^{2+}(aq)$, the very deep blue $Cu(NH_3)_4{}^{2+}$ complex ion forms. *Middle:* When an excess of ammonia is added to a pink solution of $Co^{2+}(aq)$, it is oxidized to Co^{3+} by the oxygen in air and becomes deep red in color as the $Co(NH_3)_6{}^{3+}$ complex ion forms. *Right:* A green solution of $Ni^{2+}(aq)$ becomes violet in color when an excess of aqueous ammonia is added, because the $Ni(NH_3)_6{}^{2+}$ complex ion forms.

Other Transition Metals

The +II and +III Oxidation States All the Period 4 transition metals except Sc have a +II oxidation state formed by the loss of the two 4*s* electrons to give M^{2+} ions such as Cr^{2+}, Mn^{2+}, Fe^{2+}, Co^{2+}, Ni^{2+}, Cu^{2+}, and Zn^{2+}. The M^{2+} ions form many complex ions. The colors of the complex ions formed by Cu^{2+}, Co^{3+}, and Ni^{2+} with the ligands H_2O and NH_3 are shown in Figure 10.18.

With increasing nuclear charge, the transition metals give up their valence electron with increasing difficulty. Thus Cr^{2+}, Mn^{2+}, Fe^{2+}, and Co^{2+} are good reducing agents that are readily oxidized to the +III state, whereas Ni^{2+} and Cu^{2+} can be oxidized only with difficulty to the +III state, and Zn does not have a +III oxidation state.

Higher Oxidation States The elements up to manganese in the periodic table can be further oxidized up to the oxidation state in which they use all their valence electrons. For example, chromium can use all six of its valence electrons ($3d^5 4s^1$) to form compounds in which it is in the +VI oxidation state, and manganese can use all seven of its valence electrons ($3d^5 4s^2$) to form compounds in which it is in the +VII oxidation state. Examples include the yellow chromate ion, CrO_4^{2-}, which has a Lewis structure analogous to that of the sulfate ion, in which sulfur is in the +VI oxidation state; the orange dichromate ion, $Cr_2O_7^{2-}$; and the deep purple permanganate ion, MnO_4^-, which has a Lewis structure analogous to that of the perchlorate ion, ClO_4^- (Figure 10.19). As would be expected, these oxoanions are strong oxidizing agents.

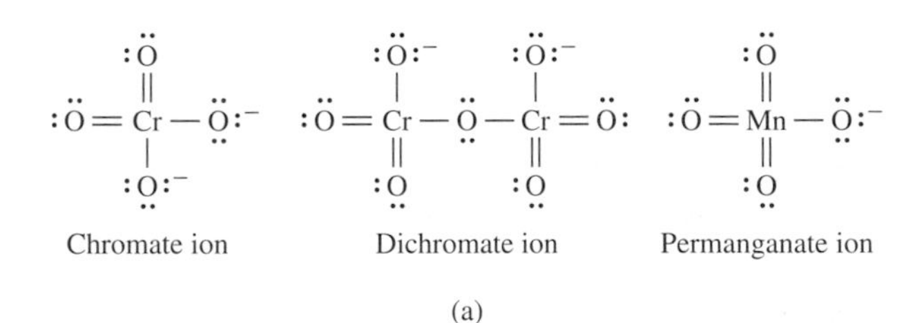

(a)

FIGURE 10.19 The Chromate, Dichromate, and Permanganate Ions (a) Lewis structures. (b) The dish on the left contains a solution of the yellow chromate ion CrO_4^{2-}. HCl(*aq*) has been added to the chromate solution in the middle dish, where we see that the orange-red dichromate ion, $Cr_2O_7^{2-}$, has formed. On the right sufficient HCl(*aq*) has been added to complete the conversion of chromate to dichromate. (c) When an Fe^{2+} solution is added to a purple solution of $KMnO_4$, it is immediately decolorized as purple MnO_4^- is reduced to colorless Mn^{2+}. An aqueous solution of Mn^{2+} is actually pale pink, but here the solution is too dilute for the color to be seen.

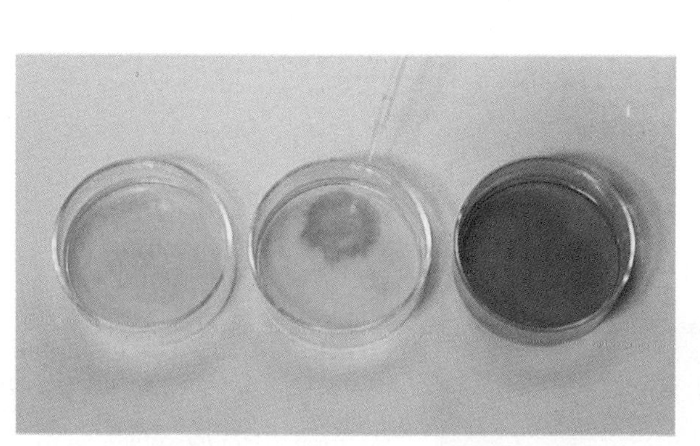

$$2CrO_4^{2-} + 2H^+ \longrightarrow Cr_2O_7^{2-} + H_2O$$

(b)

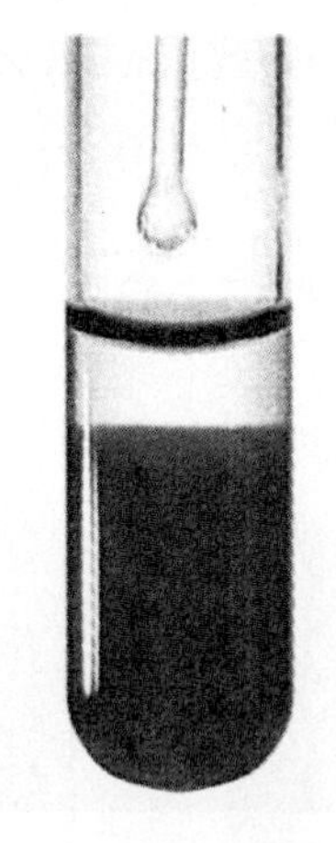

$$\underset{\text{(deep purple)}}{MnO_4^-} \xrightarrow{Fe^{2+}} \underset{\text{(very pale pink)}}{Mn^{2+}}$$

(c)

EXAMPLE 10.4 Transition Metal Ions as Oxidizing Agents

Write the balanced equation for the oxidation in acidic solution of $Fe^{2+}(aq)$ to $Fe^{3+}(aq)$ by $MnO_4^-(aq)$, which is reduced to $Mn^{2+}(aq)$.

Solution: The oxidation half-reaction is

$$Fe^{2+} \rightarrow Fe^{3+} + e^- \qquad \textit{Oxidation} \qquad (4)$$

Mn (oxidation number +7) in MnO_4^- is reduced to Mn^{2+} (oxidation number +2), so the reduction half-reaction is

$$MnO_4^- + 5e^- + 8H^+ \rightarrow Mn^{2+} + 4H_2O \qquad \textit{Reduction} \qquad (5)$$

Adding five times equation (4) to equation (5) gives the balanced equation

$$5Fe^{2+}(aq) + MnO_4^-(aq) + 8H^+(aq) \rightarrow 5Fe^{3+}(aq) + Mn^{2+}(aq) + 4H_2O(l)$$

Exercise 10.7 Write an equation for the half-reaction in which the $Cr_2O_7^{2-}$ ion is reduced to Cr^{3+}. Then write the balanced equation for the oxidation of Fe^{2+} to Fe^{3+} by $Cr_2O_7^{2-}$.

10.6 METALLURGY: EXTRACTION OF METALS

Except for the few metals that are found in the elemental form, metals occur in their minerals as positive ions. The science and technology of the extraction of metals from ores and their preparation for practical use is called **metallurgy**. To obtain the metal, we must reduce the positive metal ion:

$$M^{n+} + ne \rightarrow M$$

Because the ions of Group I and II metals are difficult to reduce, it is necessary to use a very strong reducing agent to prepare them. The most practical method of preparing these metals and aluminum is to use free electrons—that is, an electric current, which is the strongest possible reducing agent. This is done by electrolysis, as will be described in Chapter 13.

To obtain a more easily reduced metal such as iron from its ores, we can use other reducing agents. For example, aluminum can be used for reducing the oxides of some metals. In the **thermite process**, aluminum powder reduces iron(III) oxide, Fe_2O_3 (hematite), to produce small quantities of iron for special purposes such as the welding of railway lines:

$$Fe_2O_3(s) + 2Al(s) \rightarrow Al_2O_3(s) + 2Fe(l)$$

The reaction is strongly exothermic, producing sufficient heat to form molten iron (Demonstration 10.6).

Manufacture of Iron and Steel

Iron is produced on a large scale by using carbon monoxide as the reducing agent:

$$Fe_2O_3(s) + 3CO(g) \xrightarrow{\text{heat}} 2Fe(l) + 3CO_2(g) \qquad (6)$$

Iron ore is impure hematite, $Fe_2O_3(s)$, mixed with silica, silicates, and other impurities. After the mechanical separation of some impurities, the ore is mixed with coke and limestone ($CaCO_3$) and added at the top of a **blast furnace** (Figure 10.20). Preheated oxygen is blown in at the bottom. It reacts with the coke to produce carbon monoxide,

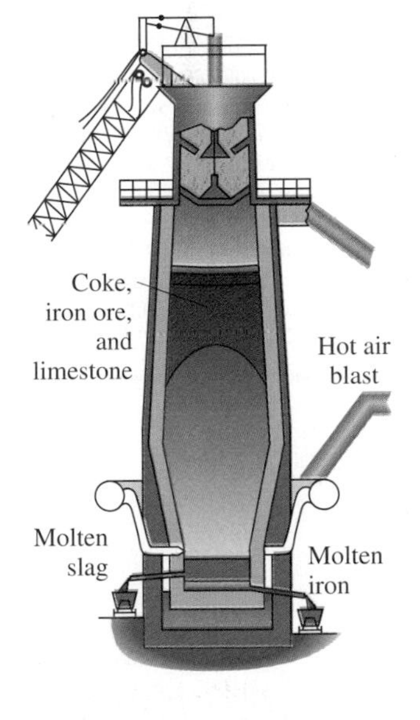

FIGURE 10.20 Blast Furnace Coke, iron ore, and limestone are added at the top of the furnace, and air is blown in at the bottom. The oxygen of the air reacts with the hot coke to form carbon monoxide, which passes up the furnace and reduces the iron(III) oxide as it moves down, first to iron(II) oxide, FeO, and finally to iron. Liquid iron collects at the bottom of the furnace covered by a layer of molten slag, $CaSiO_3$. The iron is tapped off at the bottom of the furnace and allowed to solidify, forming pig iron, and the molten slag is run off separately.

DEMONSTRATION 10.6 The Thermite Process

The crucible contains a mixture of Fe_2O_3 and aluminum powder. For the purpose of demonstration, it is convenient to start the reaction by adding a few drops of concentrated sulfuric acid to a small amount of a mixture of potassium chlorate and sugar placed on top of the Fe_2O_3–Al mixture.

The heat of the strongly exothermic reaction of H_2SO_4 with the $KClO_3$–sugar mixture ignites the Fe_2O_3–Al mixture, which reacts violently in a strongly exothermic reaction. A shower of white-hot sparks is emitted, and the crucible becomes red hot.

A ball of white-hot iron can be seen glowing in the bottom of the crucible.

When the crucible has cooled, a magnet can be used to pick up the ball of iron.

$$2C(s) + O_2(g) \rightarrow 2CO(g)$$

which reduces the Fe_2O_3 to Fe according to equation (6). The limestone removes silica, SiO_2, and silicate impurities as molten calcium silicate, $CaSiO_3$ (called *slag*), which floats on top of the molten iron and can be run off separately. The limestone decomposes at the high temperature of the blast furnace to calcium oxide (lime):

$$CaCO_3(s) \rightarrow CaO(s) + CO_2(g)$$

The calcium oxide reacts with silica and silicates to form molten calcium silicate:

$$CaO(s) + SiO_2(s) \rightarrow CaSiO_3(l)$$

This is a reaction between a basic oxide, CaO, and an acidic oxide, SiO_2, to give a salt, $CaSiO_3$.

The product of the blast furnace is called *pig iron*. It contains about 4% carbon, 2% silicon, and small amounts of other elements, such as phosphorus and sulfur. Pig iron is hard but very brittle. To produce steel, most of the carbon and almost all the other impurities are removed by blowing oxygen through the molten pig iron for a short time in a **basic oxygen furnace** (Figure 10.21). The impurities are converted to their oxides, which either are evolved as gases such as CO_2 and SO_2 or form a slag with added calcium oxide, CaO. The product of the basic oxygen furnace is *carbon steel*, which is strong and malleable and therefore much more useful than pig iron.

FIGURE 10.21 Basic Oxygen Furnace for the Production of Steel A typical basic oxygen furnace is charged with about 200 tons of molten pig iron, 100 tons of scrap iron, and 20 tons of limestone. A stream of hot oxygen gas is blown through the mixture to oxidize the impurities, which either escape as their gaseous oxides or form a slag with the calcium oxide formed from the limestone.

There are many types of steel, which differ primarily in the amount of carbon they contain (0.2 to 1.5%). The small carbon atoms occupy some of the small holes among the iron atoms and, by forming bonds to these atoms, hold them firmly in position, thereby making carbon steel stronger and harder than pure iron. Addition of other metals such as nickel, chromium, manganese, tungsten, and vanadium gives a very large number of alloy steels, of which *stainless steel*, which contains chromium and nickel, is the most common.

Production of Copper

Copper can be obtained relatively easily by heating the ore Cu_2S, *chalcocite*, in air. In the industrial process, oxygen is blown through the molten Cu_2S:

$$Cu_2S(l) + O_2(g) \rightarrow 2Cu(l) + SO_2(g)$$

It may seem odd that oxygen can apparently reduce copper(I) to copper, but this reaction is a little more complicated than it might at first appear. It can be thought to occur in two steps. The first is an equilibrium in which Cu^+ is reduced by S^{2-} to Cu:

$$2Cu^+ + S^{2-} \rightleftharpoons 2Cu + S$$

This is the reverse of the reaction by which Cu_2S is made from the elements, and the position of the equilibrium is to the left. In the second step S is removed as SO_2 by reaction with oxygen,

$$S + O_2 \rightarrow SO_2$$

which causes the above equilibrium to shift to the right. The impure copper obtained in this way is further purified by electrolysis (Chapter 13).

10.7 SUMMARY OF THE REACTIONS OF METALS AND METAL IONS

Although we have described in this chapter what appear to be many different reactions, most of them belong to one of four basic types: redox, Brønsted acid–base, Lewis acid–base, or precipitation. Let us summarize the reactions of metals and of metal ions according to these reaction types.

Reactions of Metals

Metals undergo only one type of reaction, namely *oxidation to a metal ion.*

$$M \rightarrow M^{n+} + ne^-$$

For example,

$$Na \rightarrow Na^+ + e^-$$

Transition metals except Sc can be oxidized to two or more oxidation states. For example,

$$Fe \rightarrow Fe^{2+} + 2e^-$$

$$Fe \rightarrow Fe^{3+} + 3e^-$$

To obtain the overall equation for the oxidation of a metal, we combine the equation for the oxidation half-reaction with the equation for the reduction half-reaction of the oxidizing agent. For example, with chlorine as the oxidizing agent, the half-reaction is

$$Cl_2 + 2e^- \rightarrow 2Cl^-$$

Thus, the overall equations for the oxidation of Na and Fe are

$$2Na(s) + Cl_2(g) \rightarrow 2NaCl(s)$$

$$2Fe(s) + 3Cl_2(g) \rightarrow 2FeCl_3(s)$$

The equations for the half-reactions of the other important oxidizing agents mentioned in this chapter are

$$O_2 + 4e^- \rightarrow 2O^{2-}$$

$$S + 2e^- \rightarrow S^{2-}$$

$$2H_2O + 2e^- \rightarrow H_2 + 2OH^-$$

$$2H^+ + 2e^- \rightarrow H_2$$

Reactions of Metal Ions

Metal ions can undergo three types of reactions: redox, Lewis acid–base (complex-ion formation), and precipitation.

Redox Reactions Reduction to a metal is described by the equation

$$M^{n+} + ne^- \rightarrow M$$

For example,

$$Fe^{3+} + 3e^- \rightarrow Fe$$

In principle, any metal higher than iron in the activity series can be used to reduce iron(II) or iron(III) to the metal. Using Al as the reducing agent gives the overall equation

$$Fe^{3+} + Al \rightarrow Fe + Al^{3+}$$

If a metal has two or more oxidation states, these states may be interconverted by using suitable oxidizing and reducing agents:

$$M^{n+} + e^{-} \rightleftarrows M^{(n-1)+}$$

For example, copper(II) can be reduced to copper(I) with I^{-}:

$$Cu^{2+} + e^{-} \rightarrow Cu^{+}$$
$$2Cu^{2+} + 2I^{-} \rightarrow 2Cu^{+} + I_2$$

Or iron(II) can be oxidized to iron(III) with chlorine:

$$2Fe^{2+} + Cl_2 \rightarrow 2Fe^{3+} + 2Cl^{-}$$

Lewis Acid-Base Reactions (Complex-Ion Formation) Metal ions, particularly small ions with high charge, form complex ions with ligands such as H_2O, NH_3, OH^{-}, and Cl^{-} that have one or more unshared pairs of electrons:

$$M^{n+} + xL \rightarrow ML_x^{n+}$$

For example,

$$Fe^{3+} + 6H_2O \rightarrow Fe(H_2O)_6^{3+}$$
$$Al^{3+} + 4OH^{-} \rightarrow Al(OH)_4^{-}$$

Precipitation Reactions All the salts of the alkali metals are soluble in water, but all other metals have some insoluble salts (Table 4.6). The ions of these metals take part in some precipitation reactions:

$$M^{n+}(aq) + nX^{-} \rightarrow MX_n(s)$$

The most important insoluble compounds of the Group II metals and Al, Cu, and Fe are

1. Group II metals: hydroxides, carbonates, and sulfates (except the sulfate of Mg);
2. Al, Cu, and Fe: oxides, hydroxides, and sulfides.

A typical precipitation reaction is

$$Mg^{2+}(aq) + 2OH^{-}(aq) \rightarrow Mg(OH)_2(s)$$

Exercise 10.8 Predict the products of each of the following reactions. Complete and balance the equations and classify the reaction as a redox, Lewis acid–base, or precipitation reaction.

(a) $FeCl_2(s) + Cl_2(g) \rightarrow$

(b) $FeCl_2(aq) + H_2S(g) \rightarrow$

(c) $CuSO_4(aq) + NaOH(aq) \rightarrow$

(d) $Fe(s) + HClO_4(aq) \rightarrow$

(e) $Fe^{3+}(aq) + CN^{-}(aq)$

SUMMARY

Group I and II metals are unusual in that they have lower densities and melting points than do most other metals. Many metals have structures based on the two types of close packings of spheres: hexagonal close packing and cubic close packing. Metallic bonding can be approximately described by the electron-gas model. The charge clouds of the metal atoms overlap to form a continuous charge cloud (electron gas) between positive metal ions, holding them together by electrostatic attraction. This model accounts for the properties of metals, including their high electrical conductivity and heat conductivity and their malleability and ductility.

Metals undergo only one type of reaction, oxidation to an M^{n+} ion by the loss of valence electrons to give ionic compounds. The activity series arranges the metals and hydrogen in order of their strengths as reducing agents.

Group I and II metals form M^+ and M^{2+} ions, respectively. Reducing strength decreases from Group I to Group II and increases down both groups. These metals are oxidized by the halogens, oxygen, and water to give ionic halides, oxides, and hydroxides, such as NaCl, MgO, and KOH, respectively. Their oxides are basic oxides, whereas nonmetal oxides are acidic oxides. Metal hydroxides or oxides react with acids to give ionic salts. The ions are difficult to reduce back to the corresponding metal, and they undergo few other reactions. The Group II carbonates and Ca, Ba, and Sr sulfates are insoluble and so are formed in precipitation reactions. Calcium carbonate, $CaCO_3(s)$, occurs naturally as chalk, limestone, and marble. Heating limestone gives lime, $CaO(s)$. Most of these metals and their salts give distinctive colors to flames.

Aluminum, a Group III metal, is oxidized to Al^{3+} ion but is a weaker reducing agent than the Group I or II metals. It does not reduce water but will reduce H_3O^+, so aluminum reacts with aqueous acids. It is oxidized by the halogens and oxygen to give halides, such as $AlCl_3$, and the oxide $Al_2O_3(s)$. Aluminum hydroxide is amphoteric: It reacts as a base with aqueous acids (to give Al^{3+} salts) and as an acid with $OH^-(aq)$ (to give $Al(OH)_4^-(aq)$). $Al(OH)_3$ behaves as a Lewis acid (electron-pair acceptor) when it reacts with $:OH^-$, which behaves as a Lewis base (electron-pair donor). The Al^{3+} ion is hydrated in aqueous solution, so aluminum salts often crystallize as hydrated salts with water of crystallization—$AlCl_3 \cdot 6H_2O(s)$ contains octahedral $Al(H_2O)_6^{3+}$ ions.

The transition metals lie between Groups II and III in Periods 4, 5, and 6. Unlike the main group metals, they often have more than one oxidation state. The most common are the +II and +III states, as in the ions Cu^{2+}, Fe^{2+}, and Fe^{3+}. In some of their compounds, the transitional metals have unpaired valence electrons and are therefore paramagnetic (attracted to the field of a magnet) or ferromagnetic (capable of being permanently magnetized). Transition metal ions take part in three types of reactions. (1) In redox reactions, M^{n+} is reduced or oxidized. (2) In Lewis acid–base reactions, an M^{n+} ion behaves as a Lewis acid, forming complex ions (ions that consist of a central metal ion to which other ions or molecules are bonded), such as $Fe(H_2O)_6^{2+}$, $Fe(CN)_6^{3-}$, and $FeCl_4^-$. The coordination numbers (the numbers of ligands, or molecules or ions bonded to the central metal ion) of these complex ions are commonly four or six. (3) Insoluble hydroxides and sulfides are formed by precipitation reactions in aqueous solution.

Iron, $3d^6 4s^2$, has +II and +III oxidation states. $Fe(s)$ is oxidized to Fe^{3+} by strong oxidizing agents, such as $HNO_3(aq)$ and $Cl_2(g)$, and to Fe^{2+} by weaker ones, such as H_3O^+. Salts are formed in reactions of a metal (redox reaction) or of an oxide or hydroxide (acid–base reactions) with acids. These salts are usually hydrated, such as $FeSO_4 \cdot 7H_2O$. The common oxides of iron are $Fe_2O_3(s)$ and $Fe_3O_4(s)$; the hydroxides of iron are $Fe(OH)_2(s)$ and $Fe(OH)_3(s)$; and the sulfide of iron is FeS.

Copper, $3d^{10} 4s^1$, has +I and +II oxidation states. Its oxides and sulfides are copper(II) oxide, $CuO(s)$; copper(I) oxide, $Cu_2O(s)$; copper(II) sulfide, $CuS(s)$; and copper(I) sulfide, Cu_2S. Copper is a weak reducing agent—it is below hydrogen in the activity series, so it is not oxidized by $H_3O^+(aq)$ but is oxidized by $HNO_3(aq)$ or by hot concentrated H_2SO_4 to Cu^{2+}. Copper(II) salts are prepared by the reaction of CuO or $Cu(OH)_2$ with aqueous acids. Copper(II) hydroxide, $Cu(OH)_2(s)$, precipitates when solutions containing $OH^-(aq)$ and $Cu^{2+}(aq)$ are mixed. Common complex ions of copper(II) include $Cu(H_2O)_6^{2+}$, $Cu(NH_3)_4^{2+}$, $CuCl_4^{2-}$, $Cu(CN)_4^{2-}$, and the neutral complex $Cu(H_2O)_2Cl_2$.

All other transition metals form M^{2+} ions. Mn^{2+} and Fe^{2+} are readily oxidized to M^{3+} ions; Co^{2+}, Ni^{2+}, and Cu^{2+} are oxidized to the +III state with increasing difficulty; and Zn has no +III oxidation state. The elements from Ti to Mn can be oxidized to oxidation states higher than +III. Cr, $3d^5 4s^1$, forms CrO_4^{2-} and

$Cr_2O_7^{2-}$ containing chromium(VI), and Mn, $3d^5 4s^2$, forms MnO_4^- containing manganese(VII), all strong oxidizing agents.

Metallurgy is the science and technology of the extraction and subsequent preparation of metals from ores. To prepare metals from ores, M^{n+} ions must be reduced. Group I and II ions require very strong reducing agents. In industry they are prepared by electrolysis. Iron(III) is reduced most easily with easily oxidized metals such as Al (in the thermite process) or with $CO(g)$, as in the production of iron in a blast furnace. Copper is produced by heating $Cu_2S(s)$ with $O_2(g)$.

IMPORTANT TERMS

acidic oxide (page 354)
activity series (page 354)
amphoteric (page 362)
basic oxide (page 354)
cubic close-packed structure (page 347)
complex ion (page 367)
coordination compound (page 367)
coordination number (page 367)
electron-gas model (page 348)
ferromagnetism (page 366)
hexagonal close-packed structure (page 347)
Lewis acid (page 363)
Lewis base (page 364)
ligand (page 367)
metallic bond (page 348)
metallic conduction (page 349)
metallurgy (page 375)
paramagnetism (page 366)
transition element (transition metal) (page 365)

REVIEW QUESTIONS

1. How many metal atoms immediately surround a given atom in a hexagonal close-packed structure in **(a)** the same layer; **(b)** the three-dimensional arrangement?

2. What is the difference between **(a)** a cubic close-packed structure and **(b)** a hexagonal close-packed structure?

3. Briefly describe the electron-gas model of metallic bonding.

4. Why is the melting point of calcium higher than that of potassium?

5. What are the charges on the ions formed by **(a)** the alkali metals; **(b)** the alkaline-earth metals; **(c)** aluminum?

6. Write the equation for the half-reaction of each of the following as a reducing agent.

(a) Na **(b)** Mg **(c)** Al

7. Categorize each of the following as a basic oxide or an acidic oxide, and write the balanced equation for the reaction of each with water.

(a) Na_2O **(b)** BaO **(c)** P_4O_{10}

8. How do their reactions with water show that sodium is a stronger reducing agent than magnesium?

9. Write the empirical formulas for each of the following salts.

(a) sodium sulfate **(b)** potassium carbonate
(c) magnesium phosphate **(d)** aluminum nitrate

10. Write balanced equations for the reactions of aluminum metal with **(a)** oxygen, **(b)** bromine, and **(c)** aqueous sulfuric acid.

11. Why is $Al(OH)_3(s)$ described as an amphoteric substance?

12. Define **(a)** a Lewis acid; **(b)** a Lewis base.

13. Why are the Al^{3+} ion and $Al(OH)_3$ classified as Lewis acids?

14. In terms of their electronic configurations, what distinguishes the transition elements from the other (main group) elements?

15. Define the terms **(a)** "complex ion," **(b)** "coordination compound," and **(c)** "coordination number."

16. Starting with iron, how can **(a)** $FeSO_4(aq)$ and **(b)** $Fe(OH)_3(s)$ be prepared?

17. Starting with copper, how can **(a)** $CuO(s)$ and **(b)** $CuSO_4 \cdot 5H_2O(s)$ be prepared?

18. Write one or more balanced equations to describe the industrial processes for the production from ores of **(a)** copper; **(b)** iron.

PROBLEMS

Physical Properties and Uses

1. Comment on some of the advantages and disadvantages of each of the following uses of metals.

(a) copper as a coinage metal

(b) copper and aluminum for electrical wiring

(c) alloys of magnesium as structural materials

(d) iron and steel as structural materials

2. The metals gold, silver, copper, and iron are mentioned in the Bible, but sodium, potassium, calcium, and aluminum, for example, were not isolated as pure elements until the nineteenth century. What differences in properties between the earliest-known elements and the Group I and II metals and aluminum account for the late discovery of the latter?

3. Explain why **(a)** copper was one of the earliest-known metals; **(b)** pure iron is rarely encountered; **(c)** although aluminum is a rather reactive metal, it has many common uses; **(d)** aluminum was a precious rare metal until the beginning of the twentieth century.

Structure and Bonding

4. Explain **(a)** what a metallic bond is and **(b)** how it differs from a covalent bond.

5. Explain why a metal is malleable and ductile, whereas an ionic solid is brittle and fractures under stress.

6. Explain why **(a)** a metal is a good conductor of heat and electricity as both a liquid and a solid, and **(b)** the electrical conductivity of a metal increases as temperature decreases.

7. **(a)** Explain the following increase in melting point:

$$\text{Na } 98^\circ\text{C} < \text{Mg } 650^\circ\text{C} < \text{Al } 660^\circ\text{C}$$

(b) Suggest a possible reason why iron melts at 1535°C, a temperature much higher than that for the three metals in (a).

8. In terms of their structures, explain why carbon in the form of diamond is very hard and fractures only under a large stress, carbon in the form of graphite splits easily into thin sheets, and copper is relatively soft and deforms relatively easily.

9. **(a)** Draw a diagram of the close packing of spheres in a single layer.

(b) How many neighboring spheres surround each sphere in this layer?

(c) Explain the different ways in which such layers may be stacked, and name these different arrangements.

(d) How many neighboring spheres surround each sphere in each arrangement in **(c)**?

10. **(a)** What is meant by the first ionization energy of an atom and by the second ionization energy of an atom? **(b)** Explain why the first ionization energy of sodium (0.50 $MJ \cdot mol^{-1}$) is lower than that of magnesium (0.74 $MJ \cdot mol^{-1}$), but the second ionization energy of sodium (4.56 $MJ \cdot mol^{-1}$) is much greater than that of magnesium (1.45 $MJ \cdot mol^{-1}$).

Reactions and Compounds of Group I and II Metals; Reactions and Compounds of Aluminum

11. **(a)** Why does the reactivity of the Group I and II metals increase as we descend each group?

(b) Describe the relative reactivities of the alkali metals and of the alkaline earth metals in their reactions with water.

12. Explain each of the following characteristic properties of the Group I alkali metals.

(a) They are reducing agents.

(b) Their compounds are ionic.

(c) Their oxides behave as strong bases in water.

13. **(a)** Place the metals Al, Cu, Mg, and Na in order of decreasing reactivity.

(b) Cite some of the experimental observations that provide the evidence for this order of reactivity.

14. Write balanced equations for the reactions of sodium, calcium, and aluminum, respectively, with **(a)** bromine, **(b)** sulfur, and **(c)** nitrogen. Name each of the products.

15. Write balanced equations, and name the reaction products, for the reactions of potassium and barium with **(a)** water, **(b)** oxygen, and **(c)** bromine.

16. **(a)** The alkali and alkaline earth metals react with hydrogen upon heating to give ionic hydrides. Write balanced equations for the reactions of potassium and calcium with hydrogen. **(b)** Magnesium reacts upon heating with nitrogen to give a nitride containing the N^{3-} ion. Write the balanced equation for this reaction.

17. Complete and balance each of the following equations, and name each product.

(a) $Mg(s) + Cl_2(g) \rightarrow$ **(b)** $Ca(s) + O_2(g) \rightarrow$

(c) $Mg(s) + H_2(g) \rightarrow$ **(d)** $Ca(s) + H_2O(l) \rightarrow$

18. From the metals discussed in this chapter, **(a)** name two metals that react with dilute hydrochloric acid, and write the balanced equations for the reactions that occur; **(b)** name two metals that *do not* react with dilute hydrochloric acid but react with dilute nitric acid, and explain why.

19. Write balanced equations for the reaction of calcium with each of the following reactants. Describe the conditions for reaction, and name each product.

(a) water **(b)** chlorine **(c)** hydrogen **(d)** oxygen

20. **(a)** What volume of hydrogen gas at STP would result from the reaction of 2.00 g of pure calcium hydride with excess water?

(b) What mass of calcium would be required to give the same amount of hydrogen as in (a) in its reaction with water?

21. Write balanced equations for the preparation of each of the following by an acid–base reaction.

(a) $K_2SO_4(aq)$ **(b)** $Mg_2SO_4(aq)$ **(c)** $Mg(ClO_4)_2(aq)$

(d) $Mg_3(PO_4)_2(s)$

22. (a) Write a balanced equation for the production of sodium carbonate by heating the mineral trona.

(b) What is the maximum amount of hydrated sodium carbonate that can be obtained from 1 metric ton (10^3 kg) of trona ore?

23. (a) What reaction occurs when baking soda is used to make baked goods rise?

(b) Write a balanced equation for the reaction of sodium dihydrogenphosphate with sodium hydrogencarbonate when baking powder is used.

24. Explain why magnesium carbonate, which is insoluble in water, dissolves when carbon dioxide is bubbled through a suspension of $MgCO_3(s)$ in water.

25. (a) Explain how you would prepare a solution of sodium nitrate from sodium.

(b) Give a balanced equation for the preparation of sodium perchlorate from an acid–base reaction.

(c) How could you prepare a sample of anhydrous $CaSO_4(s)$ without using aqueous solutions?

26. Write an equation for the reaction of each of the following oxides with water, and classify each as an acidic or basic oxide.

(a) K_2O (b) SrO (c) SO_2 (d) SO_3 (e) CO_2

(f) P_4O_6

27. Write balanced equations for each of the following reactions.

(a) sodium oxide with sulfur dioxide

(b) calcium oxide with sulfur trioxide

(c) sodium oxide with phosphorus(V) oxide

(d) aluminum oxide with sulfur trioxide

28. Write balanced equations for the reaction of calcium oxide with each of the following oxides, and name the products.

(a) water (b) carbon dioxide (c) sulfur dioxide

(d) silica (e) phosphorus(V) oxide

29. Write balanced equations for the reactions that occur when each of the following compounds is heated, and name the products.

(a) calcium carbonate (b) calcium hydroxide

(c) sodium hydrogencarbonate (d) aluminum hydroxide

30. Draw Lewis structures for (a) $AlCl_3$; (b) $AlCl_4^-$; (c) $Al(OH)_4^-$; (d) $Al(H_2O)_6^{3+}$.

31.* Except at very high temperature, both aluminum chloride and aluminum bromide as gases are composed of the dimeric molecules Al_2Cl_6 and Al_2Br_6, respectively, rather than monomeric $AlCl_3$ and $AlBr_3$. They have the following bridged structures. Suggest a reason why dimeric molecules form.

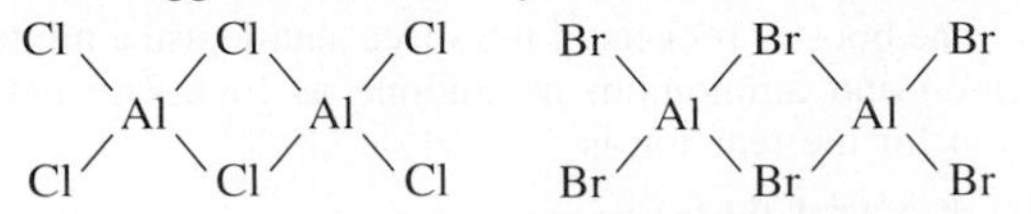

32.* A mixture contains calcium carbonate, calcium hydrogencarbonate, and calcium oxide. When 10.00 g of this mixture is heated to constant mass, 0.200 g of water and 1.500 g of $CO_2(g)$ are obtained. What is the mass percentage composition of the mixture?

33.* Write balanced equations for each of the following reactions.

(a) the reaction of powdered aluminum with $NaOH(aq)$ to give sodium aluminate, $NaAl(OH)_4(aq)$, and hydrogen

(b) the reaction of aluminum oxide with carbon and chlorine to give aluminum trichloride and carbon monoxide

(c) the decomposition, upon heating, of ammonium alum, $NH_4Al(SO_4)_2 \cdot 12H_2O(s)$, to give ammonia, sulfuric acid, aluminum oxide, and water

The Transition Elements

34. Write balanced equations for each of the following reactions.

(a) calcium with water

(b) iron with steam at high temperature

(c) aluminum with dilute sulfuric acid

(d) copper with hot concentrated sulfuric acid

35. Starting with the metal and sulfuric acid, how would you prepare solutions of each of the following?

(a) $Al_2(SO_4)_3(aq)$ (b) $CuSO_4(aq)$ (c) $FeSO_4(aq)$

36. Assign oxidation numbers to each atom in each of the following.

(a) Al_2O_3 (b) $AlCl_4^-$ (c) $Fe_2(SO_4)_3$

(d) $CuSO_4 \cdot 5H_2O$ (e) $[Cu(NH_3)_4]SO_4$

37. Repeat Problem 36 for each of the following.

(a) $Al(OH)_4^-$ (b) $Fe(H_2O)_4Cl_2$ (c) CuCl

(d) $FeCl_4^-$ (e) $KAl(SO_4)_2 \cdot 12H_2O$

38. Classify each of the following as a Lewis acid or a Lewis base, and give a balanced equation for a reaction of each to support your choice.

(a) NH_3 (b) Cu^{2+} (c) Fe^{3+} (d) Cl^- (e) $AlCl_3$

39. Describe and write balanced equations for the reactions that occur when an aqueous solution of sodium hydroxide is added to aqueous solutions of each of the following.

(a) $MgSO_4$ (b) $AlCl_3$ (c) $CuSO_4$

(d) $FeSO_4$ (e) $FeCl_3$

40. (a) Give three examples from this chapter of metal hydroxides that decompose upon heating to give the corresponding oxides.

(b) Write a balanced equation for each reaction.

41. Solutions of (a) $CuSO_4(aq)$, (b) $Al_2(SO_4)_3(aq)$, (c) $Ba(NO_3)_2(aq)$, and (d) $FeSO_4(aq)$ were prepared but not labeled. Describe some simple tests that could be used to identify each solution.

42. Copper(II) ammoniumsulfate was found to contain 27.3% water of crystallization by mass. Upon strongly heating, it gave copper(II) oxide that had a mass 19.9% of the starting mass. What is the empirical formula of hydrated copper(II) ammoniumsulfate?

43. Describe and explain the reactions that could be used to prepare each of the following substances from copper(II) sulfate pentahydrate. (More than one step may be required.) **(a)** copper **(b)** copper(II) chloride **(c)** copper(I) oxide **(d)** copper(II) tetraaminesulfate monohydrate, $Cu(NH_3)_4SO_4 \cdot H_2O(s)$

44. Starting with the oxides, which are insoluble in water, how would you prepare **(a)** $Fe(OH)_3(s)$; **(b)** $Cu(OH)_2(s)$?

45. Of the transition metals in Period 4, what are **(a)** the highest oxidation states of each of the metals from Sc to Mn; **(b)** (i) the commonest and (ii) the highest oxidation states for each of the metals from Co to Zn?

46.* A dark red solid resulted from heating 1.50 g of iron in excess $Cl_2(g)$. When the solid was dissolved in water and excess $NaOH(aq)$ was added, a gelatinous brown precipitate resulted, which when heated formed a red-brown powder.

(a) Write balanced equations for each of the reactions described, and **(b)** identify each product.

(c) What is the maximum mass of red-brown powder that could be obtained?

47.* Aluminum brass contains copper, zinc, and aluminum. When 1.00 g of brass was reacted with 0.100-*M* $H_2SO_4(aq)$, 149.3 mL of $H_2(g)$, measured at 25°C and 1.00 atm pressure, was evolved. When an identical sample was dissolved in hot concentrated sulfuric acid, 411.1 mL of $SO_2(g)$ was obtained at 25°C and 1.00 atm pressure. What is the composition of aluminum brass?

48.* Calculate the empirical formula of hydrated iron(II) ammoniumsulfate from the following experimental data: When heated strongly to constant mass, 0.7840 g of the salt gave 0.1600 g of iron(III) oxide. The addition of an excess of aqueous barium chloride to a solution of 0.7840 g of the salt dissolved in water gave 0.9336 g of barium sulfate. When a solution containing 0.3920 g of the salt was boiled with excess $NaOH(aq)$, ammonia gas was liberated. When absorbed in 50.0 mL of 0.100-*M* $HCl(aq)$, the excess acid that remained after reaction with the ammonia required 30.0 mL of 0.100-*M* $NaOH(aq)$ for neutralization.

Metallurgy: Extraction of Metals

49. A copper ore contains 1.60% Cu_2S by mass. What volume of $SO_2(g)$, measured at 20°C and 1.00 atm pressure, is obtained when 1.00 metric ton (10^3 kg) of the ore is roasted?

50. **(a)** What mass of an ore containing 30% Fe_2O_3 by mass is required to produce 10^6 metric tons of steel?

(b) How many kilograms of iron are present in 1 metric ton of this ore?

(c) What mass of coke (assuming that it is pure carbon and only $CO(g)$ forms) is needed to reduce 1 metric ton of the ore to iron?

51. **(a)** Describe how iron is obtained in a blast furnace.

(b) What function does the limestone added during the smelting of iron play in the process?

52. Calculate $\Delta H°$ for the reduction of $FeO(s)$ to $Fe(s)$ by $CO(g)$ given:

$$Fe_2O_3(s) + CO(g) \rightarrow 2FeO(s) + CO_2(g) \qquad \Delta H° = 38 \text{ kJ}$$
$$Fe_2O_3(s) + 3CO(g) \rightarrow 2Fe(s) + 3CO_2(g) \qquad \Delta H° = -28 \text{ kJ}$$

53.* For the formation of $CO(g)$ and $CO_2(g)$ from their elements at 1500°C, the Gibbs free energies of formation are $-250 \text{ kJ} \cdot \text{mol}^{-1}$ and $-380 \text{ kJ} \cdot \text{mol}^{-1}$, respectively. On the basis of the following information, discuss the feasibility of reducing each of the specified metal oxides to the metal with carbon at 1500°C.

(a) $4Al(s) + 3O_2(g) \rightarrow 2Al_2O_3(s) \qquad \Delta G°(1500°C) = -2250 \text{ kJ}$

(b) $2Fe(s) + O_2(g) \rightarrow 2FeO(s) \qquad \Delta G°(1500°C) = -250 \text{ kJ}$

(c) $2Cu(s) + O_2(g) \rightarrow 2CuO(s) \qquad \Delta G°(1500°C) = 0 \text{ kJ}$

General Problems

54. The mineral *atacamite* has the formula $Cu_2Cl(OH)_3 \cdot xH_2O$, where x is an integer. In a titration experiment, it was found that 21.45 mL of 0.4071-*M* $HCl(aq)$ were required to react completely with 0.6217 g of atacamite. What is x in this empirical formula?

55. A sample of limestone contains both calcium carbonate and magnesium carbonate. Upon heating 2.634 g of this limestone to constant mass, the residue has a mass of 1.288 g. What was the mass percentage of $CaCO_3(s)$ in the sample?

56. Explain qualitatively why the endothermic decomposition of calcium carbonate to calcium oxide and carbon dioxide is spontaneous at high temperatures but not at room temperature.

57. When it acts as an oxidizing agent in aqueous solution, oxygen is reduced to water:

$$O_2 + 4H^+ + 4e^- \rightarrow 2H_2O$$

(a) Write the balanced equation for the oxidation of $Fe(OH)_2(s)$ to $Fe(OH)_3(s)$ by $O_2(g)$.

(b) Describe what is observed experimentally.

58. **(a)** Give the equation for the half-reaction in which dichromate ion, $Cr_2O_7^{2-}$, is reduced to Cr^{3+} in acid solution.

(b) Write the complete equation for the reduction of $Cr_2O_7^{2-}(aq)$ in acidic aqueous acid solution by $SO_2(g)$.

59. One important type of stainless steel contains 18% Ni by mass. What minimum amount of nickel sulfide ore, $NiS(s)$, must be processed to give 1 metric ton (10^3 kg) of this stainless steel?

60.* A sample of hydrated iron(II) sulfate of mass 6.673 g was dissolved in water to give 250 mL of solution. A 25.00-mL sample of this solution required 24.00 mL of 0.0200-*M* $KMnO_4(aq)$ to react with it completely. Calculate **(a)** the mass percentage of $FeSO_4$ in hydrated iron(II) sulfate and hence **(b)** x in its formula, $FeSO_4 \cdot xH_2O$.

61.* The booster rockets of the space shuttle use a mixture of aluminum and ammonium perchlorate as fuel. The balanced equation for the reaction is

$$3Al(s) + 3NH_4ClO_4(s) \rightarrow Al_2O_3(s) + AlCl_3(s) + 3NO(g) + 6H_2O(g)$$

Using data from Table 9.1 and $\Delta H_f°(NH_4Cl, s) = -295$, $\Delta H_f°(Al_2O_3, s) = -1676$, and $\Delta H_f°(AlCl_3, s) = -704$ (all in $\text{kJ} \cdot \text{mol}^{-1}$), calculate the standard reaction enthalpy, $\Delta H°$.

Solids, Liquids, and Intermolecular Forces

11.1 Solids and Liquids
11.2 Phase Changes
11.3 Intermolecular Forces
11.4 Water and the Hydrogen Bond
11.5 Solutions

Water covers three-quarters of the earth's surface and is a common substance in all of its forms. (It is one of only two substances that occur naturally as a liquid; the other is petroleum.) This winter scene of the Oneonta Gorge of the Columbia River, Oregon, includes all three phases of water: the solid phase, as ice and snow; the liquid phase, as the river itself and droplets in the clouds; and the gas phase, as invisible water vapor in the atmosphere. Whereas the properties of every gas are similar, the properties of the solid and liquid forms of water are unusual and are determined by the types of forces that hold the compound's molecules together—the intermolecular forces.

The vast majority of substances are solids under ordinary conditions. The importance in our daily lives of metals, plastics, concrete, glass, and silicon chips, for example, hardly needs emphasizing. Although fewer substances are liquids under ordinary conditions, some of these liquids, in particular water, are of great importance. The earth is unique among the planets in having a large quantity of liquid water on its surface. Life on earth began in the oceans, and water remains essential to life. Water is the solvent in which reactions occur in the human body. In the chemical laboratory and in industry, water is one of the most important of the solvents in which many reactions are carried out.

All gases have similar physical properties. They are all described by the same model—the kinetic theory—and they all obey the ideal gas law approximately. In contrast, different solids and liquids may have very different physical properties. They cannot all be described by the same physical model, and their behavior is not even approximately described by a single equation like the ideal gas law. All gases behave in approximately the same way because they consist of widely separated, rapidly moving molecules with very weak forces between them. But the forces between the atoms, ions, or molecules in solids and liquids are much stronger and vary widely from substance to substance. The arrangement that these atoms, ions, or molecules adopt, and therefore the properties of the solid or liquid, depend very much on their sizes and shapes as well as on the strength of the forces holding them together. That the arrangement of the atoms, ions, or molecules in a solid has a profound influence on its properties is strikingly illustrated by allotropes such as diamond and graphite (Chapter 8) or white and red phosphorus (Chapter 7). These allotropes are composed of exactly the same atoms, but they have very different properties. Their properties depend on how these atoms are arranged in each solid.

In this chapter we first describe the different types of crystalline solids, their structures, and their characteristic properties, and we compare their properties with those of amorphous (noncrystalline) solids. Then we discuss liquids and the changes between the solid, liquid, and vapor (gas) phases. The strength of the forces between molecules (intermolecular forces) is an important factor in determining the properties of molecular solids and liquids. We discuss two important types of intermolecular forces: *dipole–dipole* and *London forces.* We then describe some unusual properties of water (such as the low density of ice) that result from an especially strong intermolecular force called the hydrogen bond. Some substances are soluble in water; others are not, but they may be soluble in another solvent such as a hydrocarbon. Some of the factors determining solubility are discussed in the final section.

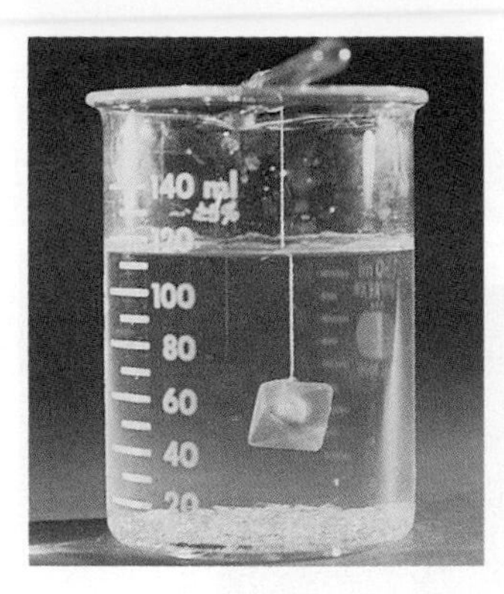

FIGURE 11.1 Growing Crystals An octahedral crystal of potassium aluminum sulfate (alum), $KAl(SO_4)_2 \cdot 12H_2O$, growing in a saturated solution

(a)

(b)

FIGURE 11.2 Crystals (a) Quartz is one of several crystalline forms of silicon dioxide (silica). These crystals are large and well formed, having crystallized very slowly over a long period of time in the earth. (b) Ordinary table salt consists of very small crystals whose cubic shape can be seen under the microscope (about 10× magnification).

11.1 SOLIDS AND LIQUIDS

Crystals

Most pure substances form crystals in the solid state. Crystals may be obtained by cooling a substance in the liquid state to its freezing point or cooling or evaporating a solution of the substance until the solution becomes saturated (Figure 11.1). When crystals are allowed to form very slowly, large, perfectly shaped crystals may be obtained (Figure 11.2a). But the crystals of many substances are often too small to see with the unaided eye (Figure 11.2b). A crystal has flat faces that form characteristic angles with each other. Crystals of the same substance grown under different conditions may have different shapes, because different crystal faces grow at different rates. However, these shapes are all related to one another, because a given pair of faces always forms at the same angle (Figure 11.3). The flat faces of a crystal are a consequence of the regular arrangement of the atoms, ions, or molecules in the crystal. When a crystal breaks, it fractures along the planes formed by the regularly arranged atoms, ions, or molecules. In this way smaller crystals form with new faces that are always at the same angle as the corresponding faces in the original crystal (Figure 11.4). The structure of crystalline solids can be determined with considerable accuracy by **X-ray crystallography**, a technique based on the patterns produced when a beam of X-rays is diffracted by a crystal (Box 11.1).

FIGURE 11.3 Crystal Shapes Crystals of a given substance may have different shapes when grown under different conditions. If some faces grow faster than others (by adding extra layers more quickly), they become smaller, and the shape of the crystal may change as shown here. However, the characteristic angles between each pair of faces are retained.

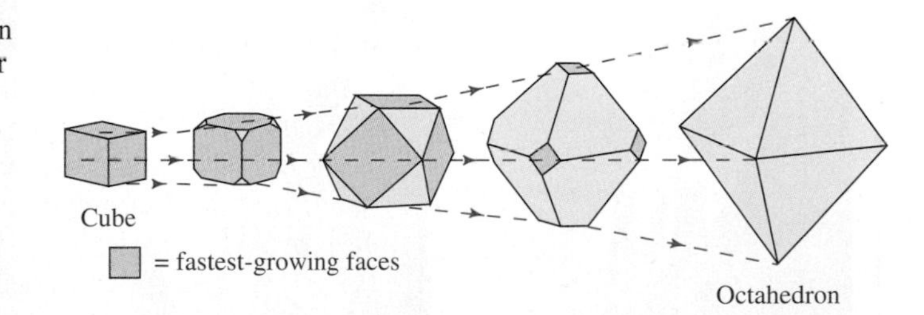

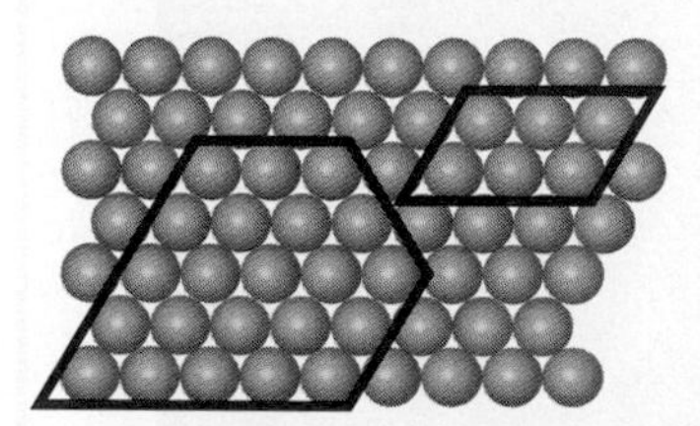

FIGURE 11.4 A Two-Dimensional Representation of a Crystal The thick black lines correspond to the edges of a two-dimensional crystal (the faces of a three-dimensional crystal). A crystal of a particular substance may have several different shapes, as represented by the two different polygons outlined in black, but the angles between corresponding faces of the crystal are always the same. When a crystal is broken, new faces are formed, but the angles between these faces are the same as in the original crystal.

BOX 11.1 X-Ray Diffraction and the Structures of Crystals

Before 1912, chemists had no direct way of determining how the atoms in molecules or network solids are arranged. The discovery by the German physicist Max von Laue, in 1912, that a beam of X-rays is diffracted by a crystalline solid led to the invention and development by William and Lawrence Bragg (Figure A) of *X-ray crystallography*. Since that time this technique has made an enormous contribution to the development of chemistry. By studying the diffraction of X-rays by crystalline solids, we can roughly determine the arrangement of the atoms and molecules in such solids.

Even with the most powerful optical microscope, we cannot see atoms and molecules. Because the wavelength of visible light is much longer than their dimensions, light waves pass over them undisturbed, much as ocean waves are undisturbed by a floating cork. X-rays have wavelengths that are on the same order of magnitude as the distances between atoms in molecules, but X-rays cannot be focused by lenses, as visible light is in a microscope. We can, nevertheless, "see" atoms and molecules in a less direct way by making use of the diffraction pattern produced when X-rays pass through the regularly spaced rows of atoms in a crystal. X-rays interact with the electrons in an atom and scatter in all directions, just as circular ripples form when water waves pass a fixed object, such as a post. The X-rays scattered by different atoms are in phase in some directions and reinforce each other and are out of phase in other directions and cancel each other. Thus a diffraction pattern is produced, just as a diffraction pattern is produced by visible light when it passes through a diffraction grating (Box 6.2). The diffraction pattern produced by X-rays passing through a crystal can be recorded on a photographic plate or in modern instruments by electronic techniques. Figure B shows the diffraction pattern produced on a photographic plate by a sodium chloride crystal. Figure C shows the much more complicated diffraction pattern produced by a crystal of the enzyme lysozyme, which is a large protein molecule with 1950 atoms and the first enzyme whose structure was determined.

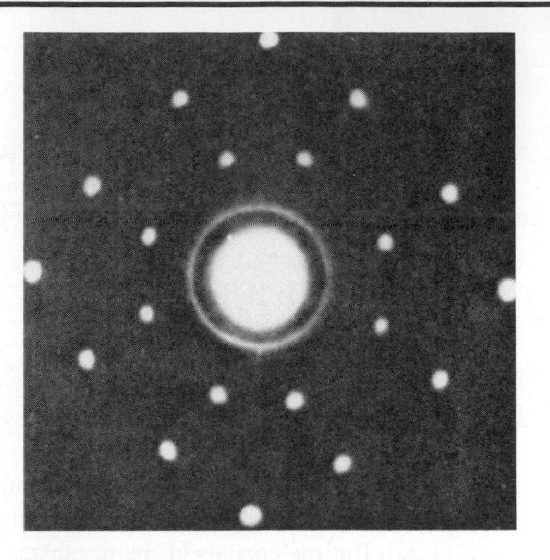

Figure B *Diffraction pattern produced by a sodium chloride crystal*

Figure A W. Lawrence Bragg (1890–1971) and William H. Bragg (1862–1942), *the British son and father who founded the science of X-ray crystallography in 1912. In 1915 they received the Nobel Prize for physics for their work on the determination of the structures of crystals. They are the only father-and-son team to have received this honor, and Lawrence was the youngest recipient.*

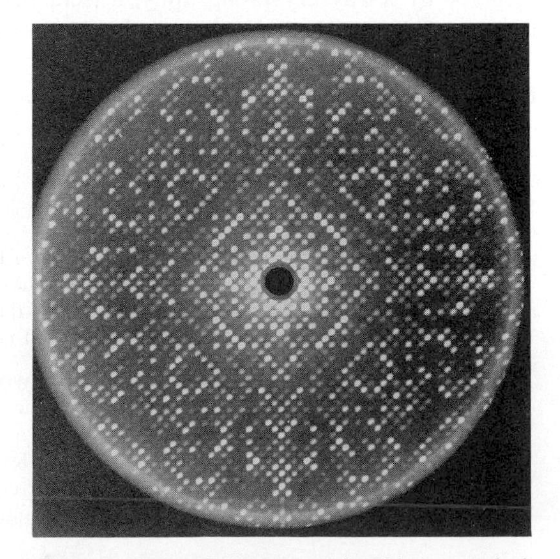

Figure C *Diffraction pattern produced by a crystal of the enzyme lysozyme*

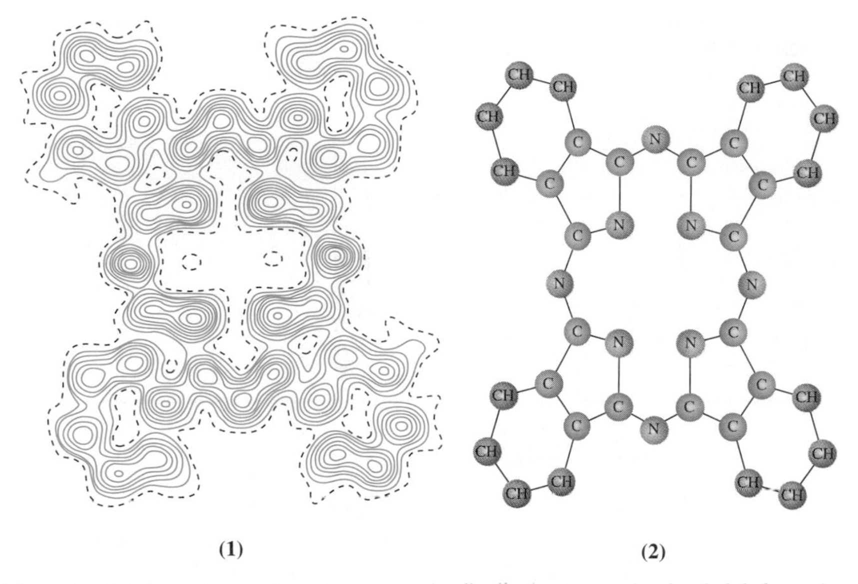

Figure D *(1) Contour map of the electron density distribution in a molecule of phthalocyanine obtained by X-ray crystallography. The peaks of electron density in this map give (2) the positions of the atoms in the molecule. Because hydrogen atoms have a very low electron density, their positions were not separately identified. The structures of chlorophyll and the heme group of the hemoglobin molecule are closely related to that of phthalocyanine.*

The X-ray diffraction pattern depends on the arrangement of the atoms in the crystal. From the arrangement of the spots in the pattern, we can calculate the arrangement of the atoms in the crystal and find the exact distances between the atoms. Not only can the distances between the atoms be determined, but each atom can be identified, because the intensity of the X-rays scattered by an atom depends directly on the number of electrons in the atom. Because X-rays are scattered by electrons, the X-ray diffraction pattern does not give us directly the positions of the nuclei but rather the distribution of electron density in the crystal. This distribution can be plotted as a contour map (Figure D). The electron density has a sharp maximum at each nucleus, so the positions of the nuclei can be found accurately.

Before the advent of computers, the recording of the diffraction pattern, and the measurement of the intensities of all the spots in the pattern, and the distances between them was a tedious and time-consuming process that could take many months of work. But today a modern X-ray diffractometer (Figure E) under the control of a computer automatically records the positions and intensities of all the diffracted beams, of which there may be several thousand or more, and then does the many calculations necessary to determine the structure. A structure can often be completely determined in a few hours or at the most a few days, so the study of increasingly large and complex molecules such as proteins is now routine. Indeed, the biggest problem may be to obtain good crystals, particularly of biomolecules, which often do not readily form crystals.

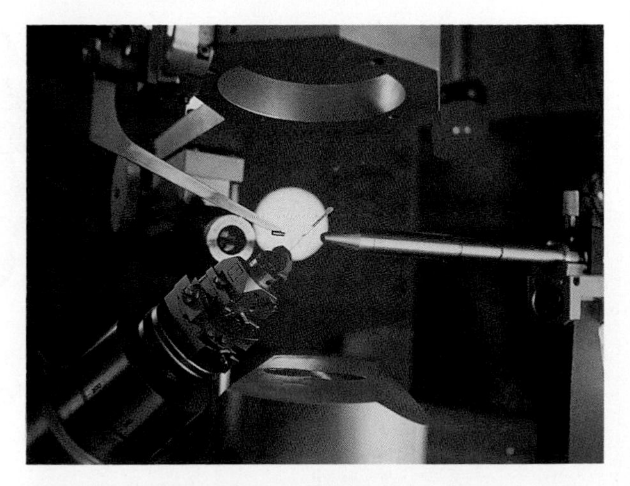

Figure E A Modern X-ray Diffractometer *The tiny crystal being studied (dimensions of about 0.2 mm) is in the middle of the narrow tube in the center.*

TABLE 11.1 Classification of Solids

	Molecular	Network		
Structural units	Molecules	Atoms	Ions of opposite charge	Positive ions and electrons
Type of interaction between structural units	Intermolecular forces	Covalent bonds	Ionic bonds	Metallic bonds
Type of solid	Covalent molecular	Covalent network	Ionic	Metallic
Examples	CH_4, CO_2, HCl, NH_3, I_2, S_8, P_4	C(diamond) SiO_2 SiC	NaCl MgO K_2SO_4	Ca Fe Al

Types of Solids

A classification of the most important types of solids is given in Table 11.1. There are two main types of solids: molecular solids and network solids.

Molecular solids consist of individual covalent molecules held together by relatively weak forces called *intermolecular forces*. We discuss the nature of these forces in Section 11.4. Substances that form molecular solids at ordinary temperatures include I_2, S_8, and P_4O_6. Because intermolecular forces are rather weak, many covalent molecules form solids only at low temperatures and are liquids or gases at room temperature. Examples include Br_2, Cl_2, H_2O, CH_4, and NH_3.

Network solids do not contain individual molecules. Rather, they have a continuous three-dimensional network of atoms or ions, each bound strongly to its neighbors in a definite geometric arrangement that repeats throughout the crystal. There are three types of network solids: covalent, ionic, and metallic. In **covalent network solids**, the atoms are held together by covalent bonds, as in diamond (Figure 11.5a). In **ionic solids**, oppositely charged ions are held together by their

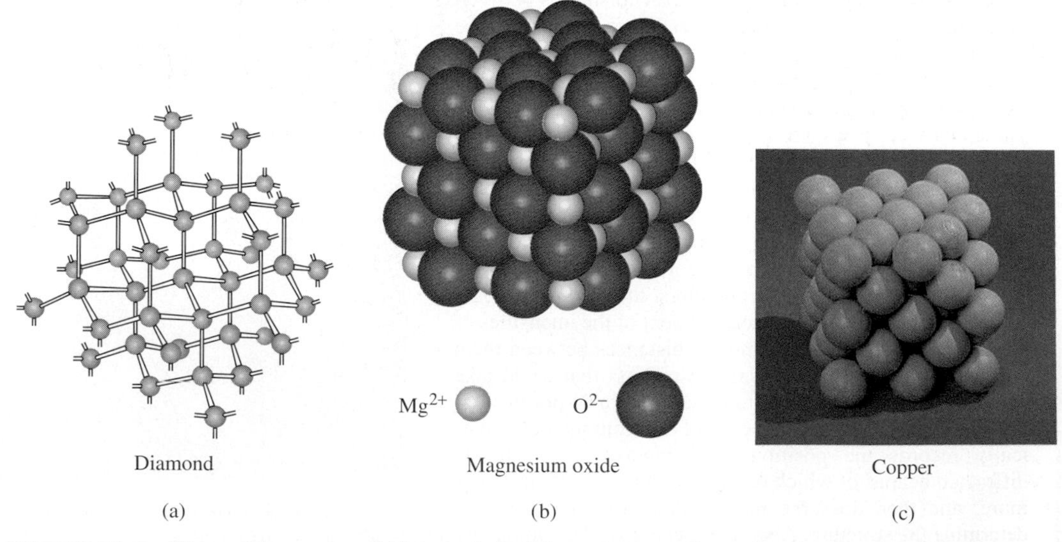

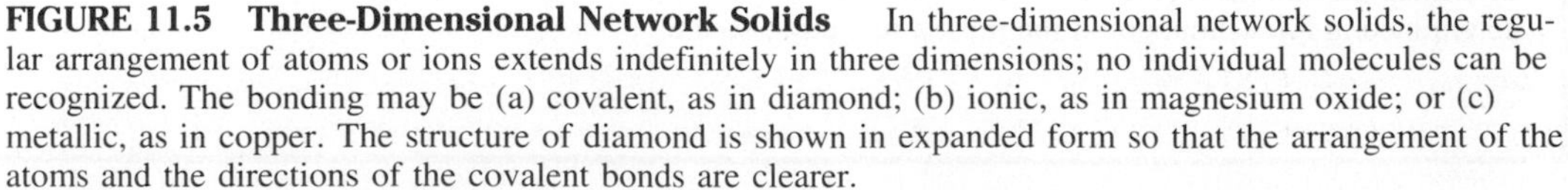

FIGURE 11.5 Three-Dimensional Network Solids In three-dimensional network solids, the regular arrangement of atoms or ions extends indefinitely in three dimensions; no individual molecules can be recognized. The bonding may be (a) covalent, as in diamond; (b) ionic, as in magnesium oxide; or (c) metallic, as in copper. The structure of diamond is shown in expanded form so that the arrangement of the atoms and the directions of the covalent bonds are clearer.

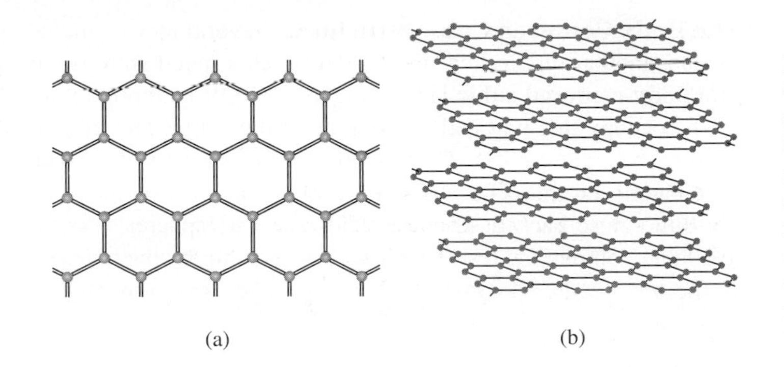

FIGURE 11.6 Structure of Graphite (a) Graphite is composed of flat sheets of carbon atoms in which each atom is bonded to three neighboring atoms in a planar AX_3 arrangement. (b) These sheets of carbon atoms are stacked one on another and held together by relatively weak intermolecular forces.

mutual electrostatic attraction—that is, by ionic bonds (Figure 11.5b). In **metallic solids**, positive ions are held together in a regular arrangement by an electron gas or electron charge cloud (Figure 11.5c)—that is, by a *metallic bond.* Because the bonds in a network solid extend indefinitely in three dimensions, the whole crystal may be regarded as one giant molecule containing many millions of atoms. Network solids typically have high melting points and boiling points because of the strong bonds that must be broken to convert them to a liquid or a gas.

Some solids may be regarded as intermediate between three-dimensional network solids and molecular solids. For example, as we saw in Chapter 8, graphite consists of carbon atoms arranged in sheets or layers. In each sheet the atoms are held together by covalent bonds, forming a giant two-dimensional molecule. These giant two-dimensional molecules are held together by weaker intermolecular forces (Figure 11.6).

Lattices and Unit Cells

It is convenient to think of the structure of any crystalline solid as based on a regular three-dimensional arrangement of points called a **lattice**. The structure of a particular solid is obtained by placing the same atom or group of atoms at each lattice point. Because the same arrangement of points is repeated indefinitely, we need only describe a small part of the lattice called the unit cell.

> **A *unit cell* is the smallest unit of a lattice that, when stacked together repeatedly, reproduces the entire lattice.**

There are several different shapes of unit cells. We shall consider only cubic unit cells, of which there are three types: the simple cubic unit cell, the body-centered cubic unit cell, and the face-centered cubic unit cell (Figure 11.7). The simple cubic cell has a lattice point at each corner of a cube. The body-centered cubic cell has an additional point at the center of the cube, whereas the face-centered cubic cell has an additional point at the center of each face. We will describe the structures of some metals, some ionic crystals, and some covalent network and molecular crystals that are based on these cubic lattices.

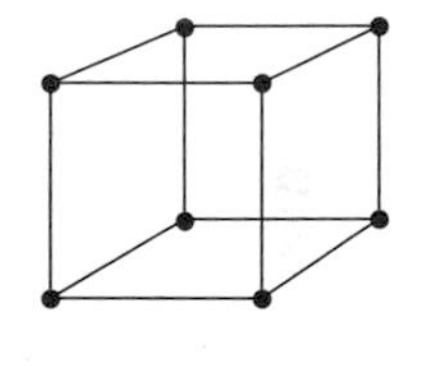

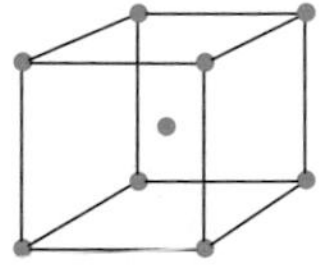

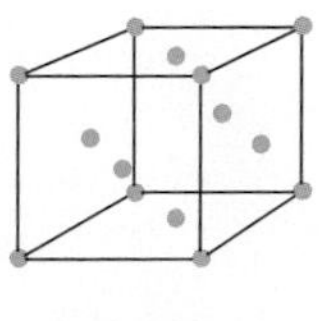

FIGURE 11.7 Unit Cells of the Three Cubic Lattices (a) Simple cubic; (b) body-centered cubic; (c) face-centered cubic

Metal Crystals

Most metals have either a cubic close-packed or a hexagonal close-packed structure, as we discussed in Chapter 10 (Figures 10.2 and 10.3), or a body-centered cubic lattice structure. Cubic close-packed structures are based on the face-centered cubic lattice. Hexagonal close-packed structures do not have a cubic unit cell, so we will not discuss them further here.

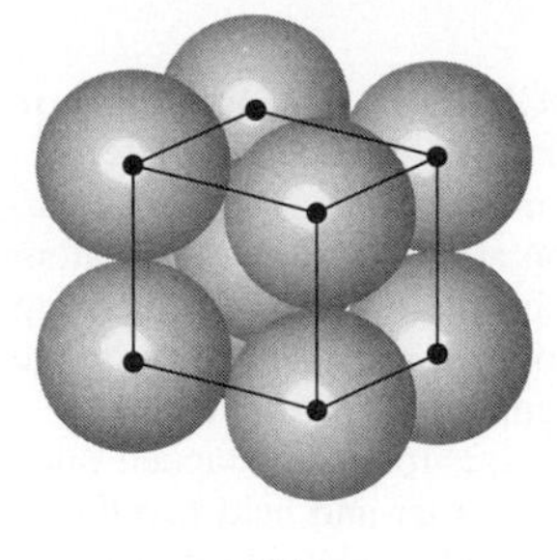

(a)

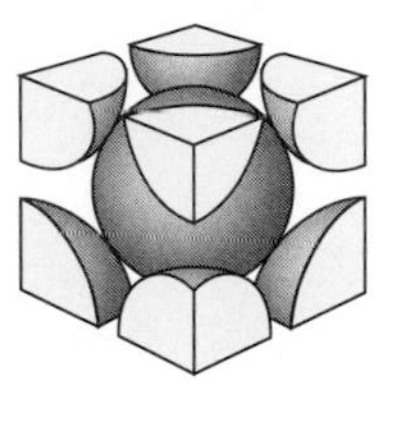

(b)

FIGURE 11.8 The Unit Cell of the Body-Centered Cubic Lattice of Iron (a) One atom is situated at each point of the body-centered cubic lattice. (b) There are two atoms per unit cell: 1 central atom + 8($\frac{1}{8}$) corner atoms = 2 atoms.

The Body-Centered Cubic Structure Several metals, including iron, the alkali metals, and barium, have a structure in which a metal atom is situated at each point of a **body-centered cubic** lattice (Figure 11.8). It is important to note that only two atoms occupy this unit cell—one atom at the center and one-eighth of an atom at each of the eight corners. The remaining seven-eighths of each corner atom is in the seven adjacent unit cells. The body-centered cubic structure may be described as a not-quite-close-packed structure. Close-packed spheres occupy 74% of the total available volume; the rest of the space is taken up by the holes between the spheres. In the body-centered cubic structure, the spheres occupy slightly less, 68%, of the total volume.

The Face-Centered Cubic Structure In Chapter 10 we described copper as having a cubic close-packed structure (Figure 10.3b). Alternatively, we can describe this structure as **face-centered cubic**. We can obtain the copper structure by placing a copper atom at each point of a face-centered cubic lattice (Figure 11.9a). This cell has one-eighth of a copper atom at each of the eight corners and half a copper atom at the middle of each of the six faces, totaling $(8 \times \frac{1}{8}) + (6 \times \frac{1}{2}) = 4$ copper atoms per unit cell (Figure 11.9b). The relationship of this description of the structure to the cubic close-packed description is not immediately obvious, because the close-packed layers are not parallel to the faces of the cubic unit cell. But we see in Figures 11.9c and 11.9d that the close-packed layers are perpendicular to the body diagonal of the cube. (The body diagonal crosses the cube from one corner to an opposite corner, passing through the center of the cube.)

Metals other than copper that have the face-centered cubic structure include aluminum and calcium. The noble gases, which like metals consist of single atoms, also have this structure in the solid state. But the forces holding the atoms together are not metallic bonds. The noble gases form the only solids in which atoms rather than molecules are held together by forces that are exactly the same as those between nonpolar molecules—forces called *intermolecular forces*.

If we find the dimensions of a unit cell by X-ray crystallography (Box 11.1)

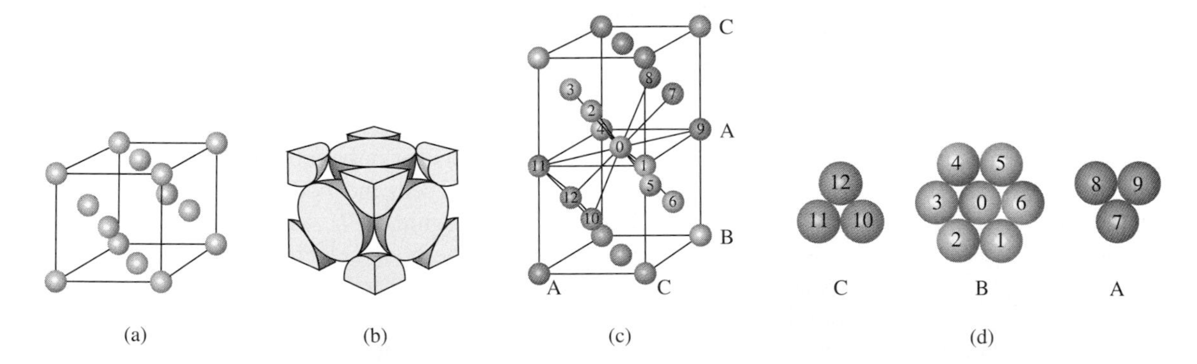

FIGURE 11.9 The Unit Cell of the Face-Centered Cubic Lattice of Copper (a) One atom is situated at each point of the face-centered cubic lattice. (b) There are four atoms per unit cell: 6 ($\frac{1}{2}$) face atoms + 8($\frac{1}{8}$) corner atoms = 4 atoms. (c) This view of two unit cells shows more clearly that the structure may alternatively be described as cubic close-packed. Each atom in a close-packed structure is surrounded by 12 neighbors. The atom 0 is surrounded by atoms 1 through 12 in a close-packed arrangement. (d) The close-packed layers are the layers A, B, and C. Layer B contains the atom 0 surrounded by atoms 1 through 6; layer A contains atoms 7, 8, and 9; and layer C contains atoms 10, 11, and 12.

and we know the number of atoms in the unit cell, we can calculate the density of the metal, as shown in Example 11.1.

EXAMPLE 11.1 Unit Cell Dimensions and Density

Vanadium, V(*s*), has the body-centered cubic structure. The edge length of its unit cell is found by X-ray diffraction to be 305 pm. What is the density of V(*s*)?

Solution: We can find the density (in $g \cdot cm^{-3}$) from the mass and volume of one unit cell:

$$\text{volume} = (\text{edge length})^3 = (305\ \text{pm})^3$$

Because the body-centered cubic structure has two atoms per unit cell, the mass of the unit cell is the mass of two V atoms:

$$\text{mass of two V atoms} = (2\ \text{atoms})\left(\frac{50.94\ \text{g V}}{1\ \text{mol V}}\right)\left(\frac{1\ \text{mol}}{6.022 \times 10^{23}\ \text{atoms}}\right) = 1.692 \times 10^{-22}\ \text{g}$$

$$\text{density} = \frac{\text{mass}}{\text{volume}} = \frac{1.692 \times 10^{-22}\ \text{g}}{(305\ \text{pm})^3\left(\frac{1\ \text{m}}{10^{12}\ \text{pm}}\right)^3\left(\frac{10^2\ \text{cm}}{1\ \text{m}}\right)^3} = 5.96\ \text{g} \cdot \text{cm}^3$$

The density of a unit cell of the metal is the same as the density of the entire metal.

If we know the dimensions of the unit cell of a metal, we can also find the radius of the metal atom, as we see in Example 11.2.

EXAMPLE 11.2 Unit Cell Dimensions and Atomic Radius

The edge length of the unit cell of iron is 286 pm. What is the radius of the iron atom?

Solution: In calculations on structures involving cubic unit cells, it is useful to know the relationships between a cube edge, a face diagonal, and a body diagonal. These relationships can be obtained from trigonometry or from the Pythagorean theorem.

If the edge length of the unit cell is a, if b is a face diagonal, and if c is a body diagonal, then

$$b^2 = a^2 + a^2 = 2a^2 \quad or \quad b = \sqrt{2}a$$

$$c^2 = a^2 + b^2 = 3a^2 \quad or \quad c = \sqrt{3}a$$

We see from Figure 11.8a that for the iron structure, the length of the body diagonal c is equal to $4r$, where r is the radius of the iron atom. Hence

$$r = \frac{c}{4} = \frac{\sqrt{3}}{4}a = \frac{\sqrt{3} \times 286}{4}\ \text{pm} = 124\ \text{pm}$$

The radius of the iron atom is 124 pm.

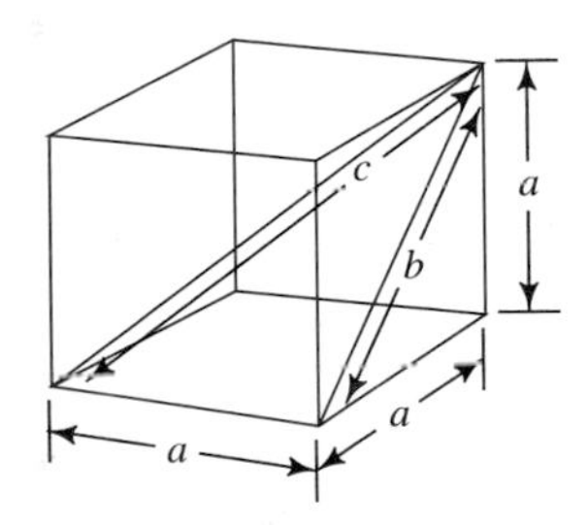

In the body-centered cubic structure, the atoms that touch each other lie on the body diagonal c. For this structure, 4(atomic radius) = c.

Exercise 11.1 Titanium has the body-centered cubic structure, and its density is $4.50\ g \cdot cm^{-3}$. Calculate the edge length of the unit cell and the atomic radius of titanium.

Determining the structure of a metal or other suitable solid and measuring the density of the solid provides a method for finding the value of Avogadro's number, as we see in Example 11.3.

EXAMPLE 11.3 Avogadro's Number from Density and Unit Cell Dimensions

Chromium, Cr(*s*), of density $7.19\ g \cdot cm^{-3}$, has the body-centered cubic structure and a unit cell with edge length 288.4 pm. Use these data to calculate a value of Avogadro's number.

Solution: We first find the mass of the unit cell from its volume and density:

$$\text{mass of unit cell} = \text{volume} \times \text{density} = (288.4\ \text{pm})^3 \left(\frac{7.19\ \text{g}}{1\ \text{cm}^3}\right)\left(\frac{1\ \text{m}}{10^{12}\ \text{pm}}\right)^3\left(\frac{10^2\ \text{cm}}{1\ \text{m}}\right)^3$$
$$= 1.72 \times 10^{-22}\ \text{g}$$

The unit cell of Cr(*s*) contains two Cr atoms, so the mass of the unit cell (1.72×10^{-22}g) is the mass of two Cr atoms. The mass of 2 mol Cr atoms is therefore $N_A(1.72 \times 10^{-22}\ \text{g})$. The molecular mass of Cr is 52.00 $\text{g} \cdot \text{mol}^{-1}$, so the mass of 2 mol Cr atoms is 104.0 g. Thus,

$$N_A(1.72 \times 10^{-22}\ \text{g}) = (2\ \text{mol})\left(\frac{52.00\ \text{g Cr}}{1\ \text{mol Cr}}\right) = 104.0\ \text{g}$$
$$N_A = 6.05 \times 10^{23}$$

This value is not exactly equal to the more accurate value (6.022×10^{23}) that we have been using, because the density was given only to 3 significant figures.

(a)

Exercise 11.2 Iridium has the cubic close-packed structure. The edge length of the unit cell is 383.3 pm. The density of iridium is 22.61 $\text{g} \cdot \text{cm}^{-3}$. Calculate a value for Avogadro's number.

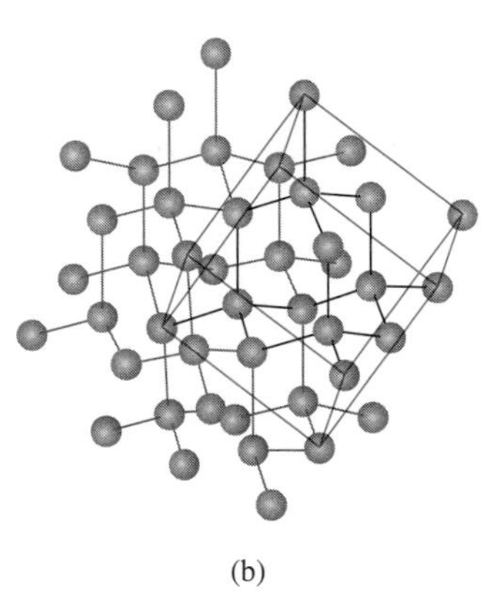

(b)

FIGURE 11.10 The Diamond Structure (a) The unit cell of the diamond structure; (b) another view of the structure, in which a body diagonal of the outlined unit cell is in a vertical position

Covalent Network Crystals

The structures of covalent network crystals are determined primarily by the number and geometry of the bonds formed by each atom. In many simple structures of this type, one atom has a tetrahedral AX_4 arrangement. For example, each carbon atom in the diamond crystal has a tetrahedral AX_4 geometry. The unit cell of the diamond structure (Figure 11.10) is another structure based on the face-centered cubic lattice. Silicon has the same structure.

Silica (silicon dioxide), SiO_2, has many different crystalline forms. The structure of one of these, β-cristobalite, is closely related to that of silicon. Each silicon atom in β-cristobalite has the same tetrahedral AX_4 geometry as in silicon, but in addition an oxygen atom forms two bonds between each pair of silicon atoms (Figure 11.11). However, SiO_2 cannot be regarded as a purely covalent network solid: Its bonds are polar, because oxygen is considerably more electronegative than silicon. No sharp dividing line can, in fact, be drawn between the three types of network solids (covalent, ionic, and metallic). Many solids have bonds of intermediate character. Semiconductors, for example, have bonding that is intermediate between metallic and covalent (Chapter 20).

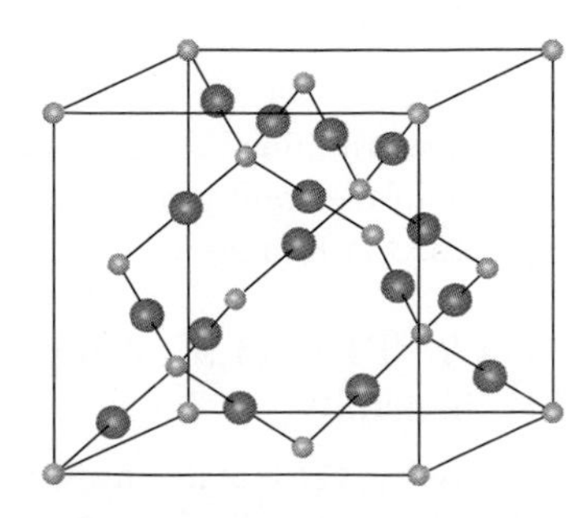

FIGURE 11.11 The Structure of the β-Cristobalite Form of Silicon Dioxide Silicon dioxide (silica), SiO_2, crystallizes in a number of different forms. Each Si atom is surrounded by a tetrahedron of O atoms, each of which in turn is bonded to another Si atom. In the different forms of silica, the SiO_4 tetrahedra are arranged in different ways. The structure shown here is the β-cristobalite form of silica. The O atom of each SiO_4 tetrahedron is shared with another Si atom, so each Si atom has a half share in four O atoms, and the overall composition is SiO_2.

Covalent Molecular Crystals

The packing of molecules in a molecular crystal is determined primarily by the shape of the molecules and, if they are polar, by the charges on their atoms. For example, the structure of carbon dioxide, CO_2, is based on a face-centered cubic lattice with a CO_2 molecule centered on each lattice point (Figure 11.12). The linear molecules are oriented not all in the same direction but in a way that packs them together most efficiently. If the molecules of a substance are large and have a complex shape, they may not pack together easily. Such substances may be difficult to obtain in a crystalline form. Considerable effort was needed to crystallize proteins, which have large complex molecules, so that their structure could be determined by X-ray crystallography. Experiments on substances that are difficult to crystallize have been carried out in spacecraft, because crystals grow more easily in conditions of zero gravity than on earth.

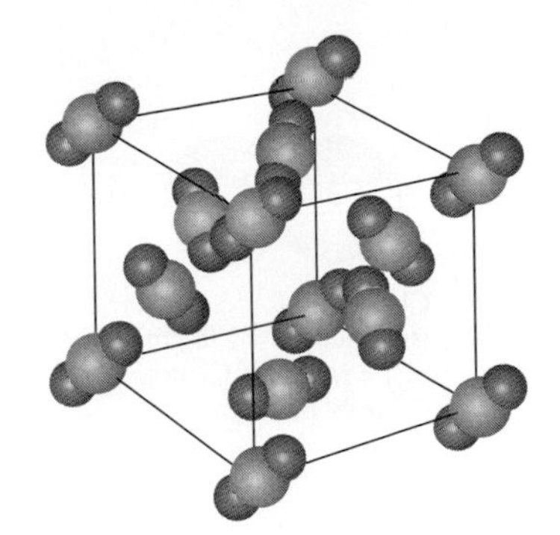

FIGURE 11.12 The Structure of Solid Carbon Dioxide Carbon dioxide molecules, CO_2, form a face-centered cubic lattice.

The Sizes of Atoms: Metallic, Covalent, and Ionic Radii

X-ray crystallography enables us to answer the question, What size are atoms and ions? Because X-rays are scattered by the electrons of an atom, we can find the distribution of electron density in a crystal by X-ray crystallography, as described in Box 11.1. The electron density is very high near the nucleus, so the position of each nucleus is given by the maximum in the electron density that surrounds it. Thus we can determine the distances between the atoms in a crystal with considerable accuracy.

If we measure the distance between two adjacent atoms in a crystal of a metal, we can take half this distance to be the radius of the metal atom, as we did for iron in Example 11.2. This radius, which we call the **metallic radius**, is a measure of the size of the atom in the metal. We gave values of some metallic radii in Tables 10.2 and 10.3, and the values of the Group I and II metals are given in Figure 11.13.

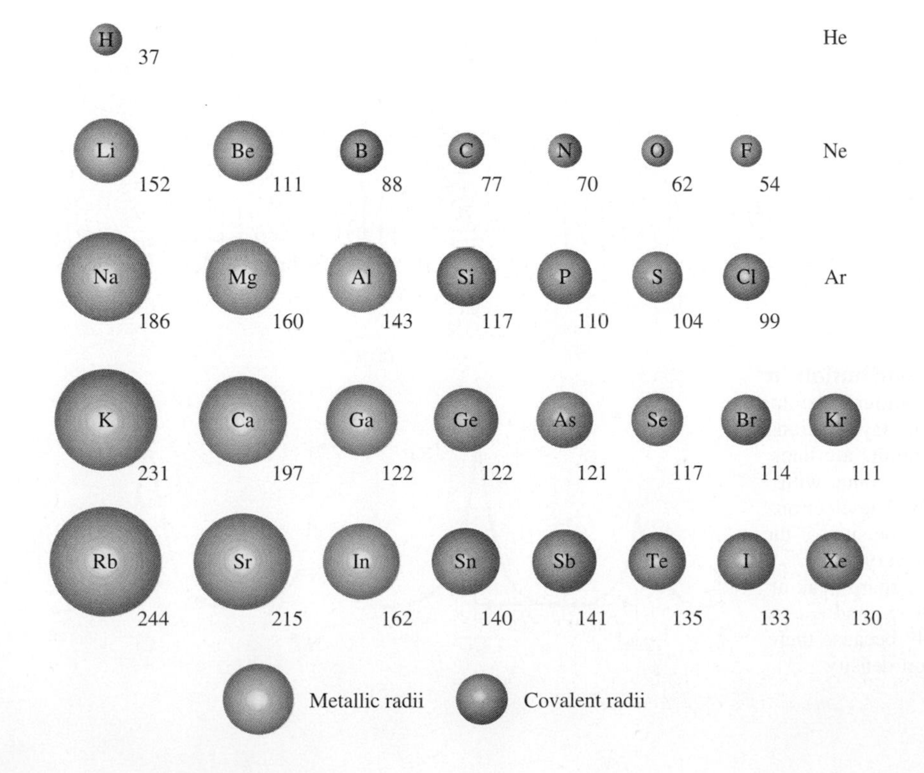

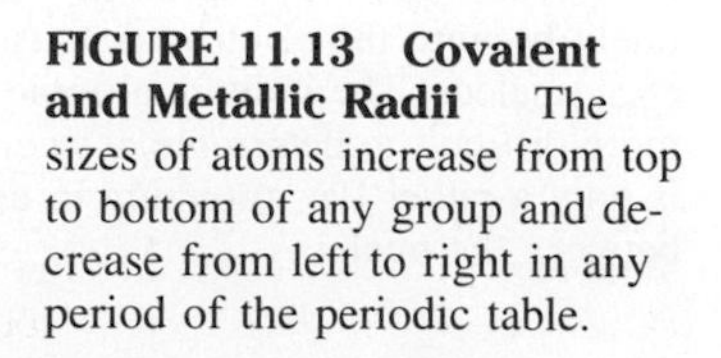

FIGURE 11.13 Covalent and Metallic Radii The sizes of atoms increase from top to bottom of any group and decrease from left to right in any period of the periodic table.

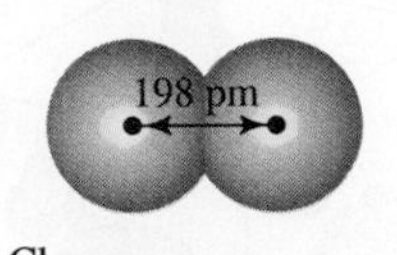

Cl_2
Bond length = 198 pm
rCl = radius of Cl atom
= 99 pm

Similarly, half the distance between two atoms of the same element in a covalent network crystal, such as diamond or silicon, or in a molecule in a molecular crystal, such as $Cl_2(s)$, represents the size of the atom when it is forming a covalent bond. We call this radius the **covalent radius**. Values of the covalent radii of the nonmetals are also given in Figure 11.13. No values are given for the covalent radii of He, Ne, or Ar because no compounds of these elements are known. Because the size of an atom changes only slightly from one molecule to another, covalent radii are roughly additive. This enables us to predict the approximate length of any bond for which we know the appropriate radii. The approximate length of the C—Cl bond in CCl_4, for example, can be obtained by adding the covalent radii of carbon and chlorine:

$$\text{C—Cl bond length} = r_C + r_{Cl} = 99 \text{ pm} + 77 \text{ pm} = 176 \text{ pm}$$

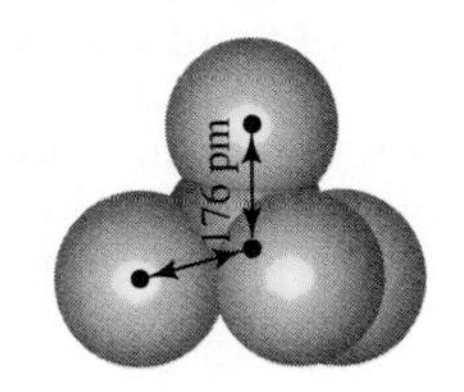

CCl_4
Predicted
bond length = $r_C + r_{Cl}$
= 77 pm + 99 pm
= 176 pm
Observed
bond length = 176 pm

The only sure way to determine the accurate bond length, however, is by experiment, such as by an X-ray crystallographic study of solid CCl_4. The experimental value for the C—Cl bond length in CCl_4 is also 176 pm. But the agreement between observed and calculated bond lengths is not always quite as good.

Finding the radius of an ion, the **ionic radius**, is not so straightforward, because we do not find two ions of the same kind adjacent to each other in an ionic crystal. Positive ions are always surrounded by negative ions, and vice versa. Figure 11.14a shows the electron density distribution in one layer of ions in a crystal of sodium chloride. Each nucleus is surrounded by an approximately spherical distribution of electron density, which is greatest at the nucleus and decreases to a minimum value at some point between the nucleus of the sodium ion and that of the chloride ion. We can measure the distance between the positions of maximum

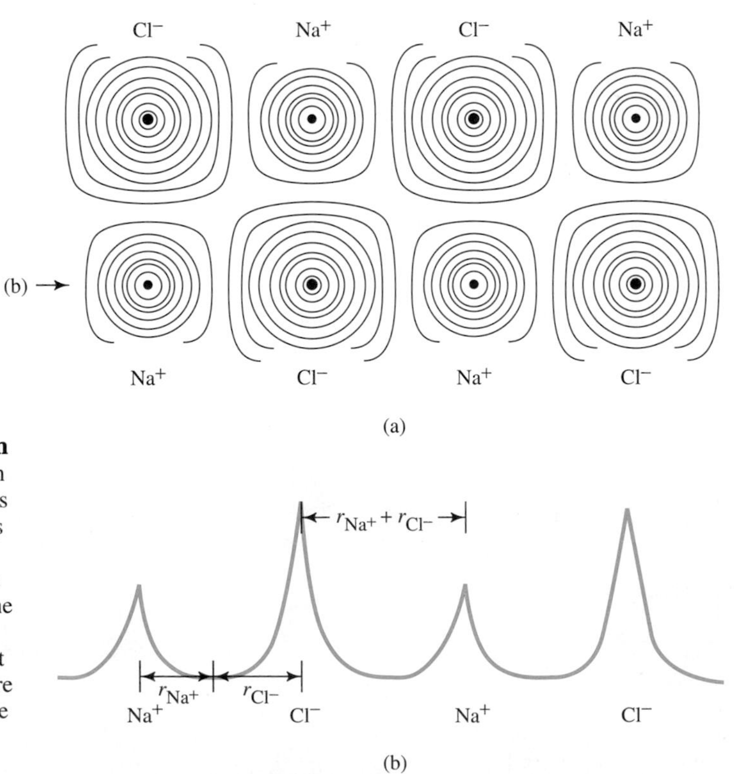

FIGURE 11.14 Electron Density Distribution in a Sodium Chloride Crystal (a) Contour diagram of the electron density distribution in one layer of ions in a crystal of sodium chloride. The contours are lines of equal electron density; they decrease in value with increasing distance from the nucleus. (b) The electron density along a line through the nuclei. The sum of the ionic radii $r_{Na} + r_{Cl}$ can be determined very accurately because the electron density has a sharp peak at each nucleus. The individual ionic radii r_{Na} and r_{Cl} are more difficult to determine as accurately because there is only a rather flat minimum in electron density between the nuclei.

TABLE 11.2 Ionic Radii (pm)

	Group			
Period	**I**	**II**	**VI**	**VII**
2	Li^+ 76	Be^{2+} 27	O^{2-} 138	F^- 133
3	Na^+ 102	Mg^{2+} 72	S^{2-} 184	Cl^- 181
4	K^+ 138	Ca^{2+} 100	Se^{2-} 198	Br^- 196
5	Rb^+ 149	Sr^{2+} 116	Te^{2-} 221	I^- 216
6	Cs^+ 170	Ba^{2+} 136		

electron density with great accuracy. This distance is the sum of the radii of the two ions, but it does not give the radii of the individual ions. Because the electron density is continuous between the ions, it is difficult to decide where the dividing line between the two ions is located. We can reasonably assume it to be the point of minimum electron density. However, this point usually is not well defined, because the minimum is rather flat (Figure 11.14b); this limits the accuracy with which the radii can be determined. Nevertheless, by studying a large number of ionic crystals, we can obtain a reasonably reliable set of average ionic radii (Table 11.2). Like covalent radii, these ionic radii are approximately additive.

As we might expect, ionic radii are not the same as metallic or covalent radii. By comparing the values in Table 11.2 and in Figure 11.13, we see that *positive ions are smaller and negative ions are larger than the corresponding covalently or metallically bound atoms.* Removing the valence electrons from a Group I or II metal to form ions removes the metal's valence shell, so the ionic radius is considerably smaller than the corresponding metallic radius. The addition of one or two electrons to a neutral atom of Group VI or VII to form a negative ion increases the number of repulsions between the valence shell electrons and causes the valence shell to expand. So the ionic radii of the Group VI and VII elements are greater than their covalent radii.

The metallic, covalent, and ionic radii vary in the same way through the periodic table. There are two important trends:

1. **Metallic, covalent, and ionic radii increase down any group** with the increasing number of electron shells.
2. **Metallic, covalent, and ionic radii decrease from left to right in any period.** Although the outer electrons in the atoms of a given period are in the same shell, they are attracted by a core charge that increases from left to right and increasingly contracts the electron cloud.

Ionic Crystals

The structures of ionic crystals are determined by the tendency of each ion to attract as many ions of opposite charge as possible and by the relative sizes of the ions. Cations are generally smaller than anions, so the number of anions that can be

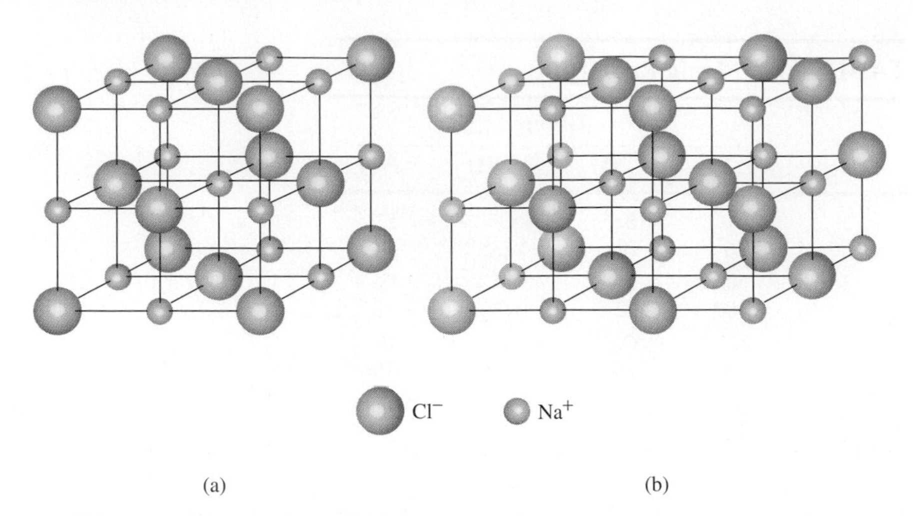

FIGURE 11.15 The Sodium Chloride Structure (a) A unit cell of the sodium chloride structure. The chloride ions form a face-centered cubic lattice. There is a sodium ion midway between each pair of neighboring chloride ions. In this open view of the structure, the ions have been reduced in size so that their arrangement is more easily seen. Each sodium ion is octahedrally coordinated to six chloride ions. (b) An extended view of the structure. The sodium ions also form a face-centered cubic lattice, and each chloride ion is octahedrally coordinated to six sodium ions.

packed around the cation determines the arrangement of the ions. Crystals composed of equal numbers of ions of opposite charge, with the general formula MX(*s*), have the simplest structures. We will consider three cubic structures of this type.

The Sodium Chloride Structure Figure 11.15a shows a unit cell of the sodium chloride structure. There is a chloride ion at each point of a face-centered cubic lattice and a sodium ion situated midway between each pair of chloride ions. The sodium ions also form a face-centered cubic lattice, as we can see if we look at the extended portion of the lattice shown in Figure 11.15b. Each sodium ion is surrounded by six chloride ions in an octahedral arrangement and each chloride ion is surrounded by six sodium ions. The number of ions surrounding an ion of opposite charge in a crystal, or as we saw in Chapter 10, in a complex ion, is called the *coordination number* of the ion. Both the sodium ion and the chloride ion have coordination numbers of 6 in this structure. The other sodium halides, the potassium halides, and the oxides of magnesium and calcium have the same structure.

The Cesium Chloride Structure The cesium ion is larger than the sodium ion (Table 11.2). Thus, it is possible for eight chloride ions to be packed around the cesium ion, as shown in the unit cell of cesium chloride in Figure 11.16a. The chloride ions are at the corners of a simple cubic lattice with a cesium ion at the center of the cube. It is clear that the cesium ion has a coordination number of 8, but the coordination number of a chloride ion is not immediately evident from this diagram. The alternative unit cell in Figure 11.16b, however, shows that the

(a)

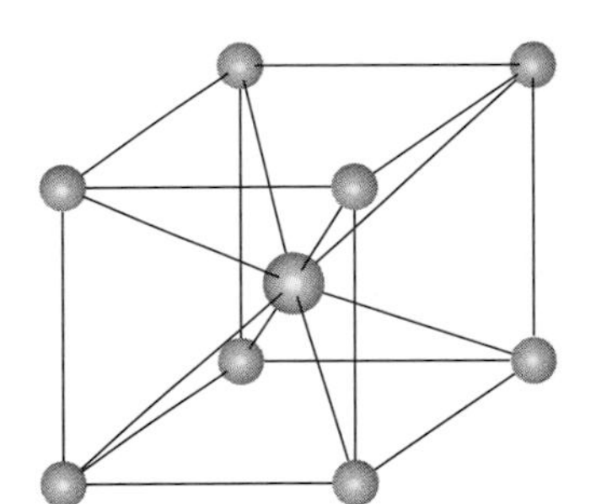

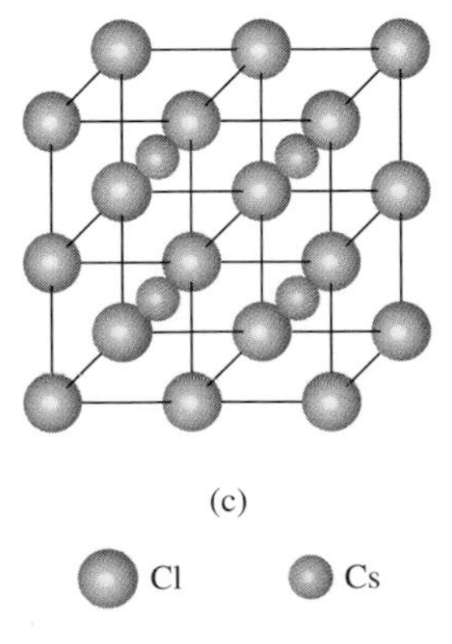

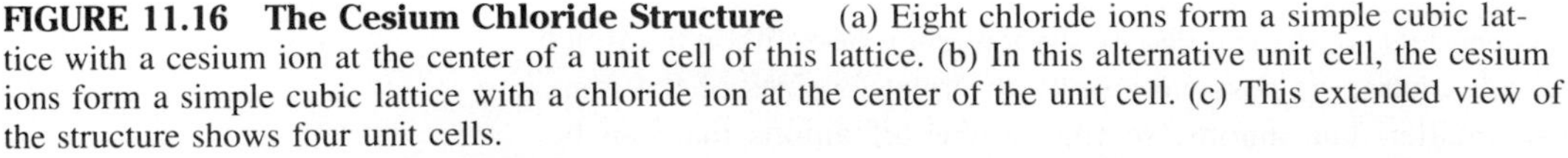

FIGURE 11.16 The Cesium Chloride Structure (a) Eight chloride ions form a simple cubic lattice with a cesium ion at the center of a unit cell of this lattice. (b) In this alternative unit cell, the cesium ions form a simple cubic lattice with a chloride ion at the center of the unit cell. (c) This extended view of the structure shows four unit cells.

coordination number of each chloride ion is also 8 and that the cesium ions also form a simple cubic lattice. Figure 11.16c shows an extended view of the structure with four unit cells. Cesium bromide and cesium iodide also have this structure.

Cesium chloride is not described as a body-centered cubic structure because each point of a lattice must be occupied by the *same* kind of atom.

Exercise 11.3 How many sodium ions and how many chloride ions are there in the unit cell of the sodium chloride structure?

Exercise 11.4 How many cesium ions and how many chloride ions are there in the unit cell of the cesium chloride structure?

The Zinc Sulfide Structure Figure 11.17 shows that the sulfide ions in zinc sulfide form a face-centered cubic lattice with four zinc ions in the interior of the cell. Each zinc ion is tetrahedrally coordinated to four sulfide ions. The structure is just like that of diamond, except that the carbon atoms are replaced by alternate zinc and sulfide ions. Although Figure 11.17 does not show it, the zinc ions also form a face-centered cubic lattice with four sulfide ions in the interior of the unit cell, and each sulfide ion is tetrahedrally coordinated to four zinc ions. Other substances that have this structure include CuCl, CuBr, and CuI.

The description of zinc sulfide as ionic is only a rough approximation. The +2 charge of the zinc ion attracts an unshared electron pair from each neighboring sulfide ion into the bonding region. Thus, each of the four electron pairs on a sulfide ion is to some extent shared by one of the four surrounding zinc ions, so the bonds have partial covalent character and are best described as very polar covalent. They have the expected AX_4 tetrahedral geometry of covalent bonds. Silicon carbide, SiC (Chapter 8), is another substance that has the zinc sulfide structure, but the bonds of silicon carbide are predominately covalent with only a small polarity.

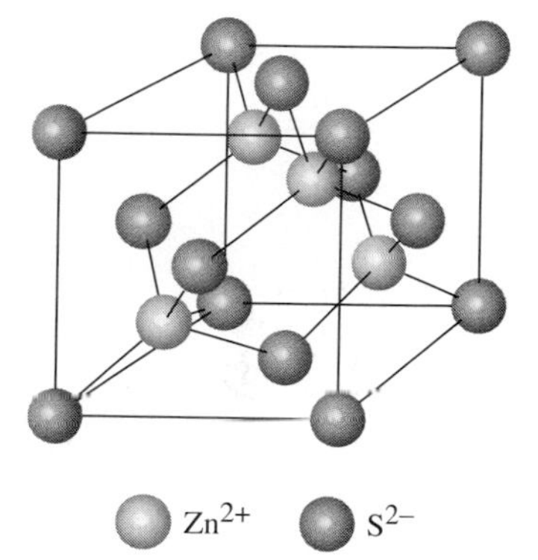

FIGURE 11.17 The Zinc Sulfide Structure This unit cell shows that the sulfide ions form a face-centered cubic lattice with four zinc ions in the interior of the cell. Each zinc ion is in a position of tetrahedral coordination.

Amorphous Solids

Some substances do not form crystals in the solid state. Such a solid is said to be an **amorphous solid**. In an amorphous solid the atoms, ions, or molecules do not have the regular arrangement that is characteristic of crystals. So amorphous solids do not have the plane faces at definite angles that are characteristic of crystals and gives them their regular geometric shape.

Silica (silicon dioxide), SiO_2, crystallizes as the mineral quartz (Figure 11.2a) and in several other forms, such as β-cristobalite (Figure 11.11). But if molten silica cools rapidly, it forms a hard, transparent, noncrystalline (amorphous) solid, because the atoms are strongly bonded together and therefore rearrange only very slowly. Unless cooling is very slow, they do not have time to organize into the regular arrangement of a crystalline solid but retain the random arrangement characteristic of the liquid. Each silicon atom is surrounded by a tetrahedral arrangement of four oxygen atoms, but these tetrahedra are linked in a random manner rather than in a regular pattern. Figure 11.18 illustrates the essential difference between crystalline and amorphous silica.

Amorphous solids are sometimes called *glasses*. A very familiar amorphous solid is the ordinary glass that is used for windows, laboratory apparatus, electric light bulbs, mirrors, and so on. This type of glass is a mixture of silicates of sodium and calcium (Chapter 18); its composition can vary widely depending on its use. Because a given atom does not always have exactly the same environment in the

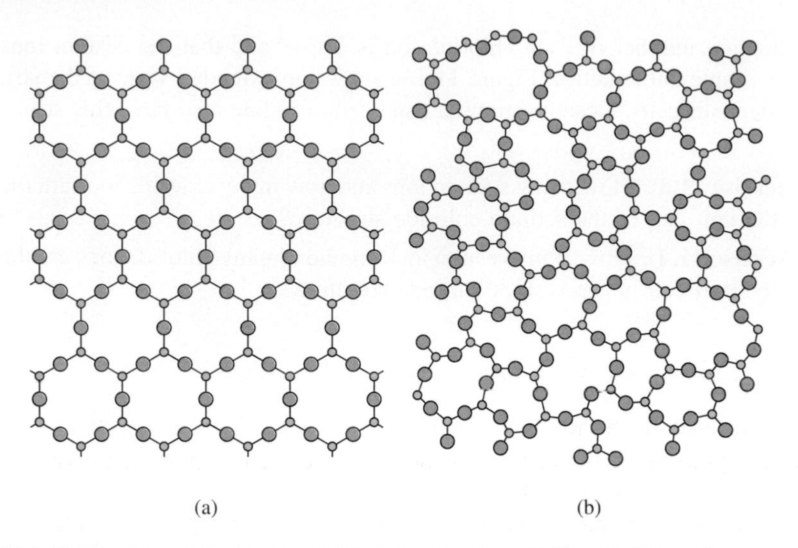

FIGURE 11.18 Two-Dimensional Models of a Crystalline Solid and an Amorphous Solid In both cases there is an equilateral triangular arrangement of the ● atoms around the ∘ atoms. But (a) in a crystalline solid, the arrangement of the atoms is regular throughout the crystal, and (b) in an amorphous solid, the arrangement of the triangular units is irregular.

random structure of an amorphous solid, different amounts of energy are needed to free a given atom from its fixed position. So, unlike crystalline solids, amorphous solids do not have sharp melting points: They gradually soften and become more fluid as the temperature rises. This property enables glass to be blown and molded into a great variety of useful and artistic shapes.

Long-chain polymer molecules, such as polyethylene, $(CH_2)_n$, which can twist and bend into a variety of shapes, become tangled in the liquid state. They do not then readily form crystals, because it is difficult for the molecules to become untangled and lined up with each other in a regular way. That the molecules are generally not all the same length also makes crystallization difficult. Plastics and other substances, such as rubber, that are composed of long-chain polymer molecules are therefore usually amorphous, although some may have crystalline regions (Chapter 20). We saw in Chapter 7 that when molten sulfur is cooled rapidly, we obtain an amorphous allotrope—plastic sulfur—consisting of long-chain S_n molecules.

Liquids

The structures of liquids are not as well understood as those of gases and solids. The assumptions that we make in describing the properties of a gas—namely, that the volume of the molecules is negligible compared to the volume occupied by the gas and that the forces between the molecules are negligible—are not applicable to liquids. The molecules of a liquid are packed closely together, and the forces between them are relatively strong. In contrast to the properties of a gas, the properties of a liquid depend very much on the sizes and shapes of the molecules and the forces between them. Although the molecules of a liquid are held together strongly, they are not in fixed positions, as are the molecules of a crystalline solid. A liquid does, however, have some structure; the arrangement of the molecules, particularly those that are close to one another, is not completely random. The structure at any one point changes with time, but the local structure around any one molecule is retained approximately.

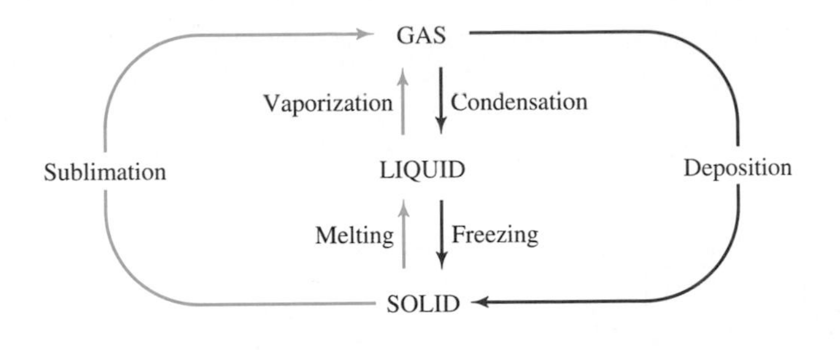

FIGURE 11.19 Phase Changes Phase changes and the terms used to describe them.

11.2 PHASE CHANGES

The transformation of a substance from one state to another is called a **phase change**. Figure 11.19 summarizes the possible phase changes and the terms used to describe them.

Melting and Freezing

The atoms, ions, or molecules in a crystal vibrate and oscillate or rotate about fixed positions, but they are unable to move from one position to another. With increasing temperature the speed and amplitude of their vibrations and oscillations increases until, at a temperature characteristic of the substance, the atoms, ions, or molecules break away from their fixed positions. When this happens the solid melts; that is, its structure collapses, and the solid becomes a liquid.

Melting, also called *fusion*, is an endothermic process, because the energy of the molecules must be increased to break up their regular arrangement in a solid. The **molar enthalpy of fusion**, ΔH_{fus}, is the heat required to melt 1 mol of a substance at 1 atm pressure. Some values of ΔH_{fus} are given in Table 11.3. Molecular solids, in which there are only weak intermolecular forces between the molecules, have lower enthalpies of fusion and lower melting points than do network solids, in which the atoms or ions are held together by strong bonds. Most network solids (such as diamond, sodium chloride, and iron) have very high melting points. In contrast, molecular substances (such as methane, carbon dioxide, water, and iodine) are gases, liquids, or low-melting-point solids at room temperature.

A few substances exhibit an unusual state of matter that is intermediate between a liquid and a crystalline solid. This *liquid crystal* state is described in Box 11.2.

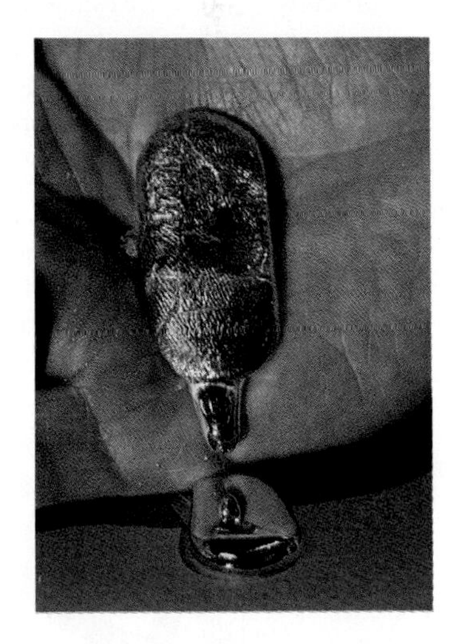

The metal gallium has a melting point of 30°C—so low that it melts when held in the hand.

TABLE 11.3 Molar Enthalpies of Fusion

Substance	Formula	ΔH_{fus} (kJ · mol^{-1})	Melting Point (°C)
Hydrogen	H_2	0.12	−259
Methane	CH_4	0.94	−183
Mercury	Hg	2.3	−40
Tetrachloromethane	CCl_4	2.51	−23
Ethanol	C_2H_5OH	5.01	−115
Water	H_2O	6.01	0
Benzene	C_6H_6	10.6	5.5
Silver	Ag	11.3	961
Sodium chloride	NaCl	27.2	801

BOX 11.2 Liquid Crystals

Figure A *A polarized-light photomicrograph of the long-chain hydrocarbon 4′-ethyl-1,1-bicyclohexyl-4-cyanide*

The electronic displays in calculators, laptop and notebook computers, wrist watches, temperature measuring strips, and numerous other devices use *liquid crystals* (Figure A). In a liquid the molecules have a constantly changing random arrangement (Figure B.1); in a crystalline solid, the molecules have a fixed ordered arrangement. Liquid crystals are substances that retain some of the order characteristic of a solid in (a usually small) temperature range above the melting point. They usually have rather large molecules that are either long and cylindrical—rodlike—or large and flat—platelike. Rodlike molecules can rotate around their own axis and can slide past each other, but they tend to retain their parallel orientation (Figure B.2).

The practical applications of liquid crystals depend on their rather remarkable optical properties. Because the molecules are often arranged in reasonably well-defined layers (Figure B.3), if the spacing between these layers is of the right magnitude, the layers diffract light just as the ordered arrangement of the atoms in a crystal diffract X-rays (Box 11.1). Only one wavelength of white light is diffracted for a given spacing of the layers, so the liquid crystal appears colored. This spacing, however, changes with temperature; hence the color of a liquid crystal is temperature dependent. This property is the basis for their use as temperature measuring devices. One medical application is the location of veins by detecting the color change due to the slightly higher temperature of the skin above a vein.

If the molecules of a liquid crystal are polar, their orientation, and hence their optical properties, may be affected by an electric field. For example, when viewed from a particular direction, a transparent liquid may become opaque when a small voltage is applied. This property is used in the displays of calculators, watches, and laptop and notebook computers.

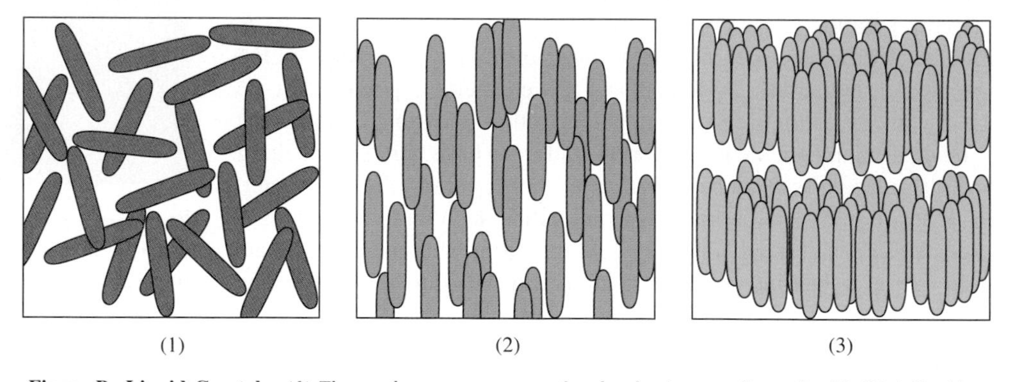

Figure B Liquid Crystals *(1) The random arrangement of molecules in an ordinary liquid. (2) A liquid crystal; the molecules can freely rotate and slip past each other, but they retain their parallel orientation. (3) Another type of liquid crystal; the molecules are arranged in layers. They can rotate and move around each other in the layers, which can slide over each other, but the layers are retained.*

Evaporation and Condensation

The molecules in a liquid have a distribution of energies similar to those in a gas (Figure 11.20). When molecules with a sufficient kinetic energy reach the surface, they may escape the attractions of the other molecules. Such molecules leave the liquid to form a vapor. This process is called **evaporation**. The reverse process is **condensation**.

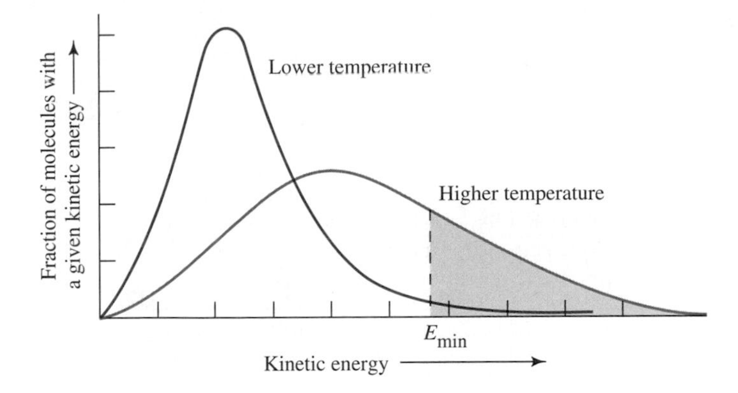

FIGURE 11.20 Distribution of Kinetic Energies of Molecules in a Liquid The average kinetic energy of molecules increases with increasing temperature. With increasing temperature, a rapidly increasing number of molecules of a liquid have more than the minimum energy, E_{min}, needed to escape from the liquid into the vapor phase. The shaded area shows the fraction of the total number of molecules with an energy greater than E_{min}. That fraction is much greater at the higher temperature.

As the more energetic molecules leave a liquid, the average kinetic energy of the remaining molecules must decrease. So if the temperature is to remain constant, heat must be transferred to the liquid from the surroundings. Evaporation is therefore an endothermic process. The **molar enthalpy of vaporization**, ΔH_v, is the heat required to transform a liquid to vapor at 1 atm pressure and at a specified temperature. Values of ΔH_v for some liquids at their boiling points are given in Table 11.4. We see that network substances have higher enthalpies of vaporization and higher boiling points than molecular substances. Although the exact nature of network substances in the liquid state is not entirely clear, their high enthalpies of vaporization indicate that they still contain many unbroken bonds. It appears that bonds are continually being broken and new ones formed as the atoms or ions move around in a liquid. Because a gas consists of widely separated atoms or molecules, all or most of the bonds in the liquid must be broken to form the vapor state.

Vapor Pressure

If water or gasoline is left in an open container, it gradually evaporates until all the liquid has been converted to vapor. But if a sufficient amount of liquid is left in a *closed* container, it does not all evaporate (Figure 11.21). At first the amount of vapor in the container increases, and the pressure it exerts on the container walls increases accordingly. As the amount of vapor increases, the chance that a molecule in the vapor phase will collide with the liquid surface and return to the liquid phase

TABLE 11.4 Molar Enthalpies of Vaporization

Substance	Formula	ΔH_v (kJ · mol^{-1})	Boiling Point (°C)
Hydrogen	H_2	0.9	−253
Methane	CH_4	10.4	−162
Pentane	C_5H_{12}	27.0	36
Tetrachloromethane	CCl_4	30.0	77
Benzene	C_6H_6	30.8	80
Ethanol	C_2H_5OH	38.6	79
Water	H_2O	40.7	100
Mercury	Hg	59.3	357
Sodium chloride	NaCl	207	1465
Carbon (graphite)	C	612	4830

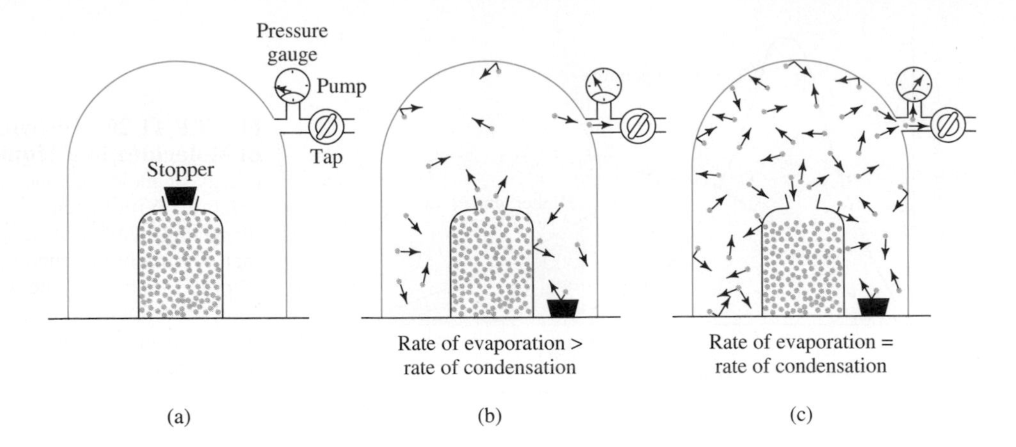

FIGURE 11.21 Vapor Pressure (a) A closed bottle completely filled with water is placed in a container that can be evacuated by pumping out all the air. When all the air has been removed and the pressure gauge registers zero, the tap is turned to close the container, and the stopper of the flask is removed by remote control. (b) The water then begins to evaporate into the container, and the pressure gauge registers the pressure of the water vapor. As the amount of water vapor increases, the chance that a molecule in the vapor phase will collide with the water surface and return to the liquid phase also increases. (c) Eventually, the number of molecules returning to the liquid phase is equal to the number leaving. No further changes in the amounts of liquid and vapor are then observed. The vapor exerts a constant pressure (the vapor pressure of water at the temperature of the experiment), which is registered on the pressure gauge as a constant value.

also increases. Eventually the number of molecules that return to the liquid in a given time interval is equal to the number that leave in the same interval, and a state of dynamic equilibrium exists. For water, this equilibrium can be summarized by the equation

$$H_2O(l) \rightleftharpoons H_2O(g) \qquad \Delta H_v = 40.7\ \text{kJ} \cdot \text{mol}^{-1}$$

No further changes are observed in the amounts of the substance in the liquid phase or in the vapor phase, and the constant amount of vapor in the container exerts a constant pressure.

***Vapor pressure* is the constant pressure exerted by the vapor above a liquid when equilibrium is established.**

If a liquid is not confined in a container, the vapor diffuses into the surroundings, and the concentration of the vapor never reaches the equilibrium value. The rate of condensation is always less than the rate of evaporation, and the liquid continues to evaporate until all of it has been converted to vapor.

Every liquid has a characteristic vapor pressure at a given temperature. This vapor pressure depends on the tendency of the molecules of the particular liquid to escape into the vapor phase. When the intermolecular forces are weak, the molecules escape easily and the vapor pressure is high. When the intermolecular forces are strong, the vapor pressure is low. Thus, the vapor pressure of a liquid at a particular temperature provides a measure of the strength of the intermolecular forces.

The vapor pressure of a liquid increases with increasing temperature (Figure 11.22). As the temperature increases, an increasing fraction of the molecules have sufficient energy to escape from the surface, and the concentration of vapor

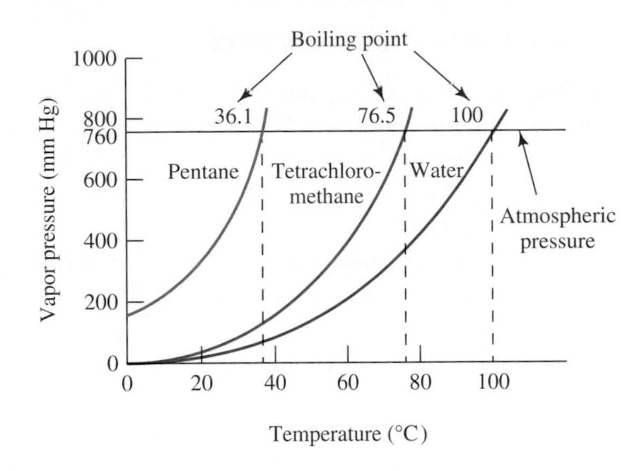

FIGURE 11.22 Variation of Vapor Pressure with Temperature The vapor pressure of a liquid increases with increasing temperature. The temperature at which the vapor pressure equals the pressure exerted by the atmosphere is the boiling point of the liquid.

molecules increases correspondingly. At any given temperature pentane has a higher vapor pressure than tetrachloromethane (carbon tetrachloride), which in turn has a higher vapor pressure than water. Thus, the strength of intermolecular forces decreases from water to tetrachloromethane to pentane.

Boiling Point When the temperature of a liquid is raised to the point at which the vapor pressure is equal to the external pressure, bubbles of vapor form in the liquid, and it is said to boil. A bubble of vapor can form only when the pressure exerted outward by the vapor in the bubble (the vapor pressure) is equal to the pressure exerted by the surrounding liquid on the bubble (usually the atmospheric pressure). The **normal boiling point** of a liquid is defined as the temperature at which the vapor pressure equals 1 atm (760 mm Hg or 101.3 kPa).

The higher the vapor pressure of a liquid, the lower the temperature at which the vapor pressure equals atmospheric pressure—that is, the lower the normal boiling point. Water boils at 100°C at 1 atm pressure; therefore 100°C is its normal boiling point. Tetrachloromethane, which has a higher vapor pressure, boils at 76.5°C at 1 atm pressure, and pentane boils at 36.1°C.

If the atmospheric pressure is lower than 760 mm Hg, as it is at high altitudes, the boiling point of a liquid is correspondingly reduced. In Boulder, Colorado, at an altitude of 1655 m (5430 ft), the atmospheric pressure is 610 mm Hg, and water boils at the temperature at which its vapor pressure reaches 610 mm Hg, 94°C. Because reaction rates decrease with decreasing temperature, spaghetti takes longer to cook in Boulder than it does, for example, in San Francisco. In purifying a liquid with a high normal boiling point, it is often convenient to reduce the boiling point by carrying out the distillation under reduced pressure. This modification is necessary for a liquid that decomposes before it boils at atmospheric pressure.

EXAMPLE 11.4 Boiling Points at Reduced Pressures

Using Figure 11.22, estimate the boiling points of **(a)** C_5H_{12}, **(b)** CCl_4, and **(c)** H_2O, all at a pressure of 400 mm Hg.

Solution: If we draw a horizontal line on Figure 11.22 at a pressure of 400 mm Hg, we can read off the temperature at which this line intersects each vapor pressure curve. Thus, we find the following boiling points: **(a)** C_5H_{12}, 25°C; **(b)** CCl_4, 60°C; **(c)** H_2O, 80°C.

III A liquid can have many different boiling points, depending on the value of the external pressure. We are most familiar with the boiling point at an external pressure of 1 atm, the normal boiling point.

Exercise 11.5 At what pressure will **(a)** tetrachloromethane and **(b)** water boil at a temperature of 60°C?

Entropy and Phase Changes

Melting and evaporation are endothermic processes; they therefore decrease the entropy of the surroundings. These processes are spontaneous, however, because the increase in the entropy of a substance when it melts or evaporates is large enough that the overall entropy change is positive. Let us examine these entropy changes in a little more detail.

The standard molar entropy of water at various temperatures is given in Table 11.5. The standard molar entropy of the fusion of ice at 0°C is

$$\Delta S° = S°(H_2O, l) - S°(H_2O, s) = (65.2 - 43.2)\,J \cdot K^{-1} \cdot mol^{-1} = 22.0\,J \cdot K^{-1} \cdot mol^{-1}$$

The enthalpy of fusion of ice, $\Delta H°_{fus}$, is 6.01 kJ · mol^{-1} (Table 11.3). As we saw in Chapter 9, the change in the entropy of the surroundings is given by $\Delta S_{surr} = -\Delta H_{system}/T$:

$$\Delta S_{surr} = \frac{-\Delta H_{system}}{T} = -\frac{6.01 \times 10^3\,J}{273\,K} = -22.0\,J \cdot K^{-1}$$

Thus, when 1 mol of ice melts, the entropy of the surroundings *decreases* by 22.0 J · K^{-1}. We just saw that the entropy of the ice–water system *increases* by exactly the same amount, so the overall entropy change is zero. Because, as we saw in Chapter 9, all spontaneous changes are accompanied by an increase in entropy, we have to conclude that ice does not melt spontaneously at 0°C! Surprising as this might seem at first, this conclusion is correct. Ice does *not* melt spontaneously at 0°C: Water and ice are *in equilibrium* at that temperature, and their amounts remain constant. Ice may be melting, but the rate at which it melts at 0°C is exactly equal to the rate at which it freezes, and no overall change is observed. The higher the temperature, the smaller the decrease caused in the entropy of the surroundings by a given amount of heat absorbed from the surroundings ($\Delta S_{surr} = -\Delta H_{fus}/T$). Therefore, at any temperature above 0°C, ΔS_{surr} is less negative than $-22.0\,J \cdot K^{-1}$, so the overall entropy change is positive, and the melting of ice is spontaneous. At any temperature below 0°C, ΔS_{surr} is more negative than $-22.0\,J \cdot K^{-1}$, so the overall entropy change is negative, and melting is not spontaneous. Instead, the reverse process of freezing is spontaneous.

We reach the same conclusions if we consider the free energy change for the melting of ice at 273 K. We recall from Chapter 9 that

TABLE 11.5 The Standard Molar Entropy of Water at Various Temperatures

Phase	Temperature (°C)	$S°(J \cdot K^{-1} \cdot mol^{-1})$
Solid	0	43.2
Liquid	0	65.2
	20	69.6
	50	75.3
	100	86.8
Vapor	100	196.9
	200	204.1

for a spontaneous process, $\Delta G = \Delta H - T\,\Delta S < 0$

for a nonspontaneous process, $\Delta G = \Delta H - T\,\Delta S > 0$

for a process at equilibrium, $\Delta G = \Delta H - T\,\Delta S = 0$

The free energy change for the melting of ice at 273 K is

$$\Delta G = \Delta H - T\,\Delta S = (6.01 \times 10^3\ \text{J}) - (273\ \text{K} \times 22.0\ \text{J} \cdot \text{K}^{-1}) = 0$$

There is a zero free energy change, so the system is in equilibrium. If we raise the temperature to 1°C = 274 K,

$$\Delta G = (6.01 \times 10^3\ \text{J}) - (274\ \text{K} \times 22.0\ \text{J} \cdot \text{K}^{-1}) = -22\ \text{J}$$

the free energy change is negative. Therefore, melting is a spontaneous process at 1°C. If we decrease the temperature to −1°C = 272 K,

$$\Delta G = (6.01 \times 10^3\ \text{J}) - (272\ \text{K} \times 22.0\ \text{J} \cdot \text{K}^{-1}) = 22\ \text{J}$$

the free energy change is positive. At this temperature, melting is not spontaneous. But the free energy change for the reverse process, freezing, is −22 J, so freezing is spontaneous at −1°C.

III ΔG for melting (solid → liquid) is the negative of ΔG for freezing (liquid → solid).

The freezing point or the boiling point of a substance is the temperature at which the total entropy change and the free energy change of the system for the phase change are zero. At higher temperatures melting and boiling lead to an increase in the total entropy and a decrease in the free energy of the system, so they are spontaneous processes. At lower temperatures the reverse processes of freezing and condensation lead to an increase in the total entropy and a decrease in the free energy, so these processes are spontaneous.

11.3 INTERMOLECULAR FORCES

We have referred frequently in this chapter to the forces between molecules that cause all substances to condense to liquids and to freeze to solids at sufficiently low temperatures. We will now consider the nature of these **intermolecular forces**. Although it is convenient to distinguish several different types of intermolecular forces, they are all electrostatic. In fact, all the different types of interactions among atoms, molecules, and ions—whether we describe the interactions as covalent, ionic, or metallic bonds or intermolecular forces—result from the electrostatic attractions and repulsions between positive nuclei and negative electrons.

It is important to distinguish clearly between *inter*molecular forces and *intra*molecular forces. Intermolecular forces are the forces acting *between* molecules, such as the force that holds the Br_2 molecules together in liquid bromine. Intramolecular forces are the forces acting *inside* molecules, such as the force holding the two bromine atoms together in each bromine molecule—in this case, a covalent bond.

Dipole-Dipole Forces

When the atoms in a diatomic molecule have different electronegativities, the bond is polar covalent (Chapter 3). For example, Cl has a greater electronegativity than H (Figure 3.11), so the bond electrons in HCl are shared unequally. The Cl atom has a small negative charge, and the H atom has a small positive charge:

$$\overset{\delta+}{\text{H}}\text{—}\overset{\delta-}{\text{Cl}}$$

Two separated equal but opposite charges constitute a **dipole**. A dipole is described quantitatively by its **dipole moment**, μ, which is defined as the product of the magnitude of the charges, Q, and the distance r between the charges. All polar diatomic molecules have a dipole moment.

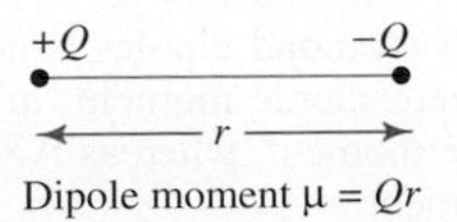

Dipole moment $\mu = Qr$

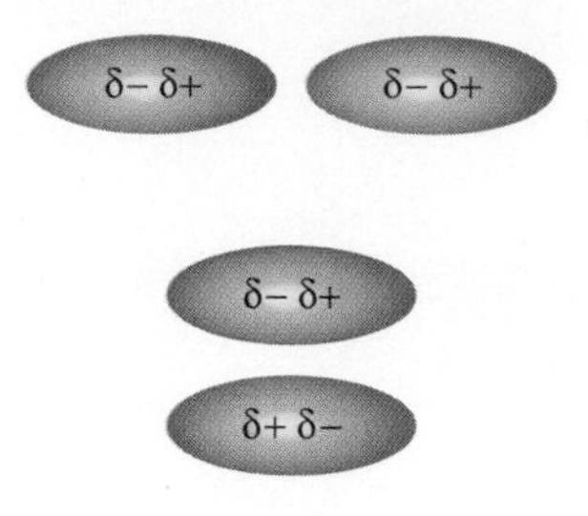

FIGURE 11.23 Intermolecular Forces: Dipole-Dipole Forces Two polar molecules can have many different relative arrangements. Arrangements in which they attract each other, like those shown here, have a lower energy and are therefore more common.

Two dipoles attract each other when they have either of the orientations shown in Figure 11.23, because the charges of opposite sign are closer to each other than are the charges of the same sign. However, the overall force of attraction between two dipoles is much smaller than that between two ions, for several reasons:

The electric force between two charges Q_1 and Q_2 at a distance r apart is given by Coulomb's law:

$$F = k\left(\frac{Q_1Q_2}{r^2}\right)$$

1. The distances between like charges are not much smaller than the distances between opposite charges, so the overall force of attraction is small.
2. The charges on the atoms in a polar molecule are smaller than those on ions.
3. The molecules in a liquid are continually in motion, so they do not always have a favorable orientation such as that shown in Figure 11.21.

The weakness of **dipole–dipole forces** compared with the forces between ions is shown by comparing the boiling point of HCl (−85°C) with that of NaCl (1465°C).

A dipole is associated with each of the polar bonds of polyatomic molecules. Figure 11.24 shows how the overall dipole moment of a molecule depends on its shape. In the linear CO_2 molecule, the two bond dipoles are equal in magnitude but opposite in direction; they cancel each other to give a zero dipole moment. Therefore, carbon dioxide is said to be a **nonpolar molecule**. In the angular water molecule, the two bond dipoles do not point in exactly opposite directions, so they do not cancel each other. The water molecule therefore has an overall dipole moment and is said to be a **polar molecule**. In general, molecules in which the central atom has no unshared pairs of electrons—that is, AX_2, AX_3, AX_4, AX_5, and AX_6 molecules—have a zero dipole moment because of their symmetry and are

III All the X atoms in a molecule AX_n must be the same for the molecule to be nonpolar. CH_3F is an AX_4 molecule, yet it has a nonzero dipole moment; the dipole moment of the C—F bond is not cancelled by the moments of the three C—H bonds.

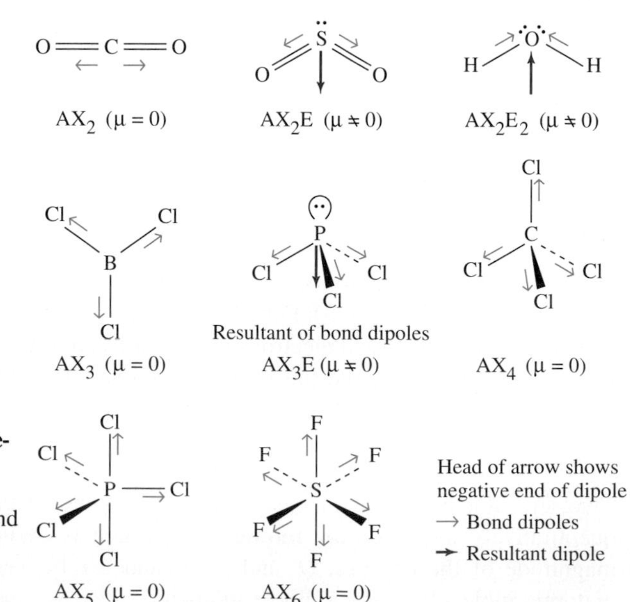

FIGURE 11.24 The Dipole Moments of Polyatomic Molecules Where the resultant of the bond dipoles is zero, the molecule is nonpolar and has a zero dipole moment. Where the resultant of the bond dipoles is not zero, the molecule is polar and has a nonzero dipole moment. In general AX_n molecules have a zero dipole moment, whereas AX_nE_m molecules have a nonzero dipole moment.

nonpolar. Molecules with one or more lone pairs of electrons on the central atom—such as, AX_2E_2 and AX_3E molecules—are less symmetrical. Their bond dipoles do not cancel, and so they have an overall dipole moment and are polar.

Exercise 11.6 Which of the following molecules have a dipole moment?
(a) CS_2 **(b)** $SiCl_4$ **(c)** NH_3 **(d)** PCl_3 **(e)** SO_2 **(f)** SO_3 **(g)** SF_6

London (Dispersion) Forces

Molecules containing atoms all of the same kind, such as O_2, N_2, P_4, and S_8, are nonpolar. Yet P_4 and S_8 are solids at room temperature, and N_2 and O_2 condense to liquids and freeze to solids at a sufficiently low temperature. There must therefore be attractive forces between the molecules of these substances.

The German physicist Fritz London (1900–1954) first proposed an explanation for these attractive forces, which are called **London forces** or **dispersion forces**. The electrons in a molecule are in constant motion. In a nonpolar molecule their *average* distribution is such that the center of negative charge coincides with the center of positive charge, so there is no resultant dipole. However, at any one instant the centers of positive and negative charge do not necessarily coincide. Thus, the molecule has an *instantaneous* dipole moment that is constantly changing in magnitude and direction (Figure 11.25). Such a fluctuating dipole of one molecule induces a fluctuating but opposing dipole in a neighboring molecule. As a result, the two molecules attract each other. All molecules, and even free atoms, such as those of the noble gases, attract each other as a consequence of their fluctuating dipoles.

The magnitude of the dipole moment that can be induced in an atom or molecule by an external electric field is proportional to the polarizability of the atom or molecule. **Polarizability** depends on the number of electrons and the ease with which they can be displaced from their average positions. The polarizability of an atom increases with increasing size; the larger the atom, the more electrons it has and the more easily the outer electrons are displaced from their average positions. As Table 11.6 shows, the polarizability of a molecule is greater the more atoms it contains and the greater the polarizability of these atoms. For example, the polarizabilities of the hydrogen halides from HF to HI increase with increasing size of the halogen atom. The polarizability of nitrogen, N_2, is greater than that of hydrogen, H_2, and that of carbon dioxide, CO_2, is greater than that of carbon monoxide, CO. These differences in polarizabilities—and therefore in the strength of the intermolecular (London) forces—is reflected in the boiling points of these substances, which are also given in Table 11.6.

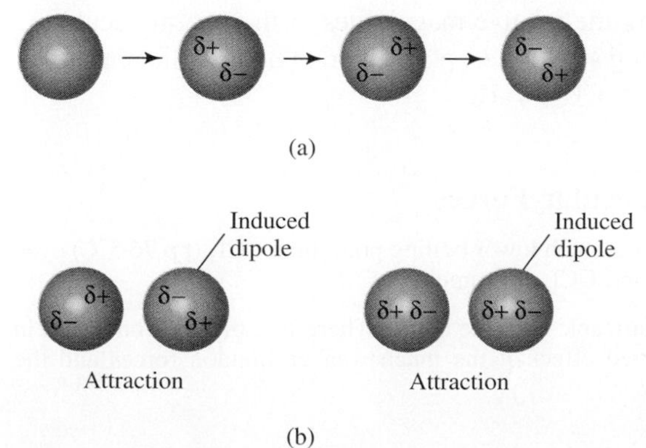

FIGURE 11.25 Intermolecular Forces: London (Dispersion) Forces (a) The movement of the electrons in a nonpolar molecule gives rise to a small instantaneous dipole that continually changes in magnitude and direction. (b) Such a fluctuating dipole can induce a dipole in a neighboring molecule. This induced dipole changes in magnitude and direction, following the changes in the fluctuating dipole in the first molecule. Thus, there is always an attraction between the two molecules.

TABLE 11.6 Dipole Moments and Polarizabilities

Substance	Dipole Moment, μ (10^{-30} C · m)	Polarizability (10^{-31} m^3)	Boiling Point (°C)
He	0	2.0	−269
Ar	0	16.6	−186
H_2	0	8.2	−253
N_2	0	17.7	−196
CO	0.33	19.8	−190
CO_2	0	26.3	−78
HF	6.37	5.1	19
HCl	3.60	26.3	−85
HBr	2.67	30.1	−67
HI	1.40	54.5	−35
H_2O	6.17	14.8	100
NH_3	4.90	22.2	−34
CCl_4	0	105	77
CH_4	0	26.0	−162
SO_2	5.42	43.4	−10

Boiling points increase in the series HCl, HBr, HI owing to the increase in polarizability, despite the fact that the dipole moment and therefore the strength of the dipole–dipole forces decreases in this series. London forces are always present and make an important contribution to the total intermolecular forces even for molecules that have fairly large dipole moments.

EXAMPLE 11.5 Intermolecular Forces

Which of the following substances is most likely to exist as a gas at room temperature and normal atmospheric pressure: PCl_3, Cl_2, $MgCl_2$, or Br_2? Why?

Solution: The weakest intermolecular forces will be found in the substance with the smallest nonpolar molecules. Of the substances listed, $MgCl_2$ is ionic, and PCl_3 consists of polar molecules. Both Cl_2 and Br_2 consist of nonpolar molecules, but because Cl_2 molecules are smaller than Br_2 molecules, they will be less polarizable and therefore have the smaller intermolecular forces. Thus, Cl_2 is the substance most likely to be a gas at room temperature and normal atmospheric pressure. Even if the dipole–dipole forces in PCl_3 were small, PCl_3 is a larger molecule than Cl_2 (three Cl atoms and one P atom rather than two Cl atoms), so we would expect the intermolecular forces in PCl_3 to be larger.

A substance will exist as a gas at room temperature if it has weak intermolecular forces.

Exercise 11.7 By considering the relative magnitudes of the intermolecular forces expected for the following substances, place them in order of increasing boiling point: C_2H_5Cl, C_2H_5Br, CH_4, C_2H_6.

EXAMPLE 11.6 Intermolecular Forces

Explain why CH_3F (bp − 78°C) has a much lower boiling point than CCl_4 (bp 76.5°C) even though CH_3F is a polar molecule and CCl_4 is nonpolar.

Solution: CCl_4 has four large, polarizable chlorine atoms. Therefore, the London forces in CCl_4 are stronger than the combined effect of the much weaker London forces and the dipole–dipole forces in CH_3F.

11.4 WATER AND THE HYDROGEN BOND

Properties of Water

Water is the commonest liquid on earth. Ninety-seven percent of the earth's water is in the oceans. The remainder is fresh water, but 2.1% is in the form of icecaps and glaciers, and only 0.7% is readily available in rivers and lakes and as underground water. Water is essential to life. In photosynthesis in plants, water combines with carbon dioxide to form carbohydrates. Blood, which is 83% water, carries dissolved nutrients and oxygen to, and transports waste products from, the cells of animals. Although water is so common and familiar, it is nevertheless an unusual liquid in that it has several properties that are unlike those of most other liquids.

One unusual property of water is that its density in the solid state is less than that of liquid water at the melting point. For almost all other substances, the solid has a higher density than the liquid. As a liquid cools, the kinetic energy of its molecules decreases, their movements become more restricted, and they pack together more closely. This gradual decrease in volume and increase in density continues until the liquid freezes to a solid, at which point the density increases abruptly. This sudden change in density occurs because the molecules take up still less space when they have the regular arrangement of the solid. However, water behaves in a different manner when it cools (Figure 11.26). Its density increases in the normal way until it reaches a temperature of 3.98°C and a density of 1.000 00 $g \cdot cm^{-3}$. Then the density *decreases* very slightly with decreasing temperature, so water has a density of 0.999 87 $g \cdot cm^{-3}$ at 0°C. Then when water freezes, its density *decreases* abruptly to 0.917 $g \cdot cm^{-3}$ rather than increasing. We will consider why water exhibits this unusual behavior after we look at some of its other unusual properties.

We are used to the fact that water freezes at 0°C and boils at 100°C, and so these properties do not seem strange. But if we compare the melting point and the

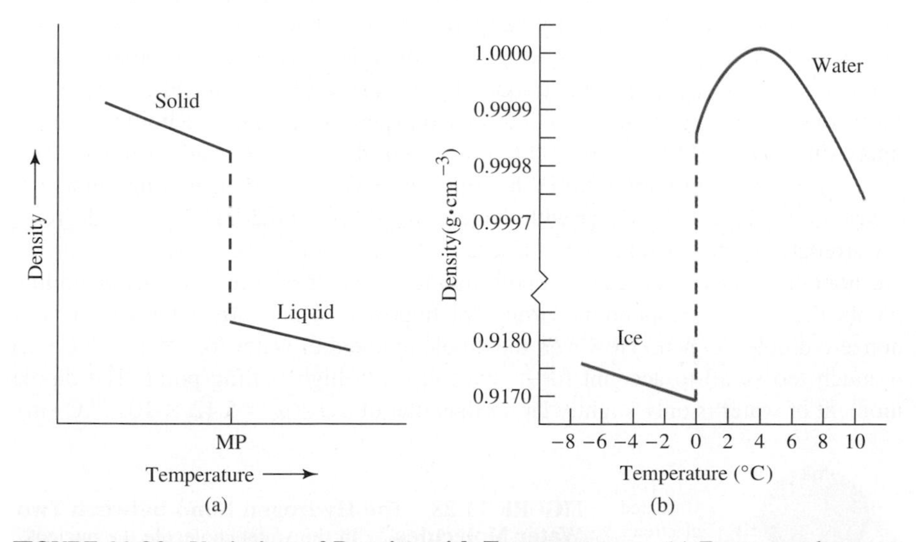

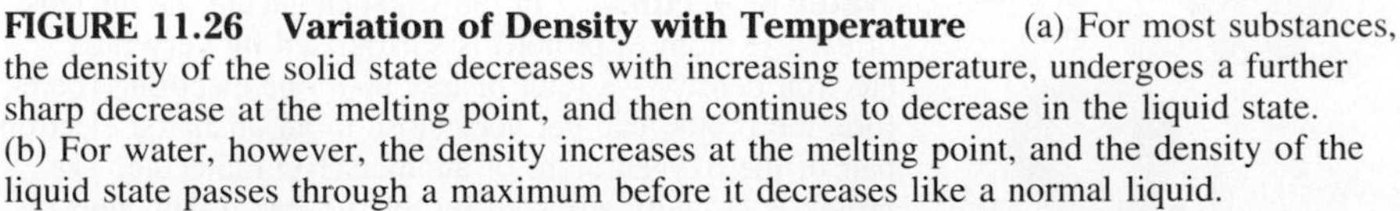
FIGURE 11.26 Variation of Density with Temperature (a) For most substances, the density of the solid state decreases with increasing temperature, undergoes a further sharp decrease at the melting point, and then continues to decrease in the liquid state. (b) For water, however, the density increases at the melting point, and the density of the liquid state passes through a maximum before it decreases like a normal liquid.

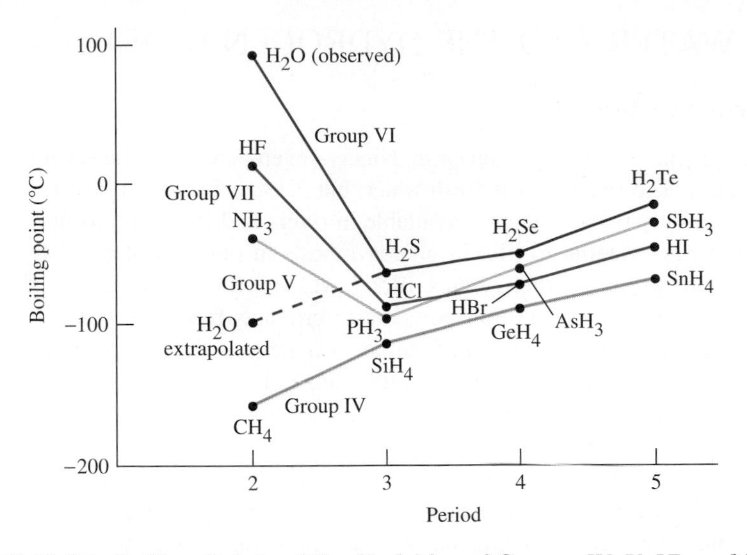

FIGURE 11.27 Boiling Points of the Hydrides of Groups IV, V, VI, and VII The boiling points of the nonpolar Group IV hydrides decrease from SnH_4 with decreasing molecular size and, therefore, decreasing intermolecular (London) forces. Group VI hydrides have higher boiling points than hydrides of Group IV because they are polar molecules, and therefore there are additional dipole–dipole attractions. Nevertheless, London forces make the largest contribution to the overall intermolecular forces. Therefore, the boiling points decrease from H_2Te to H_2S with decreasing molecular size. But the boiling point of H_2O is very much higher than expected by extrapolation from the boiling points of H_2Te, H_2Se, and H_2S. This high boiling point is attributed to a strong additional intermolecular force called the *hydrogen bond*, which is not present in the Group IV hydrides or in H_2S, H_2Se, and H_2Te. Ammonia in Group V and HF in Group VII also have unexpectedly high boiling points, for the same reason.

boiling point of water with those of similar substances, we find that both are far higher than this comparison would lead us to expect. Among the Group VI hydrides, only water is a liquid at ordinary temperatures; the others are gases. The boiling points of the Group VI hydrides decrease from H_2Te to H_2S as their polarizability, and therefore the strength of the London forces, decreases (Figure 11.27). Extrapolation of these boiling points would lead us to expect water to have a boiling point of approximately −100°C, about 200°C lower than the observed boiling point.

In contrast, all the Group IV hydrides are gases, and their boiling points decrease in the expected manner with decreasing size and polarizability and decreasing strength of the London forces. These molecules, unlike the Group VI hydrides, are nonpolar (they have a zero dipole moment) and therefore have lower boiling points than the corresponding Group VI hydrides, which are polar and have a nonzero dipole moment. However, the dipole moment of water (6.7×10^{-30} C · m) is much too small to account for its exceptionally high boiling point. The dipole moment of water is only slightly larger than that of SO_2 ($\mu = 5.42 \times 10^{-30}$ C · m),

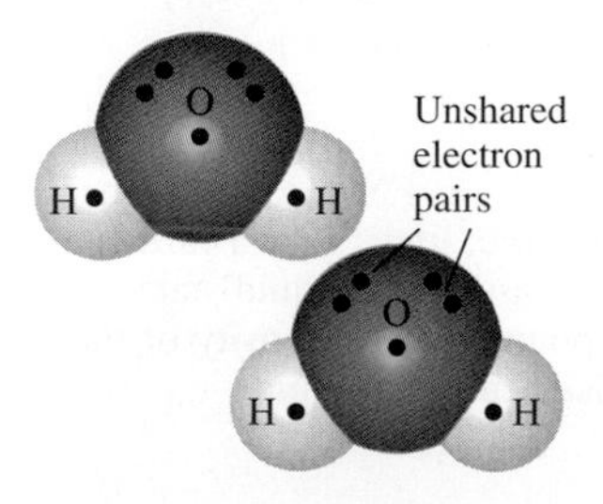

FIGURE 11.28 The Hydrogen Bond between Two Water Molecules In the water molecule the nucleus of each H atom (a proton) is surrounded by very little electron density—a total of less than one electron. Therefore, the proton can get very close to an unshared electron pair of the oxygen atom of another H_2O molecule. The proton is strongly attracted by this unshared pair, thus forming a hydrogen bond.

which also has a much greater polarizability than does water (Table 11.6), yet SO_2 has a boiling point of $-10°C$. The exceptionally high boiling point of water must be due to an additional strong force between water molecules that is not present between SO_2 molecules, nor between H_2S, H_2Se, H_2Te, or CH_4 molecules. This strong force is called the *hydrogen bond.*

The Hydrogen Bond

To understand the hydrogen bond, we need to consider the unique character of the hydrogen atom. It consists of a proton (the nucleus) and a single electron; there are no other electrons forming an inner core, as there are in all other atoms except helium. When one of the positively charged hydrogen atoms of a water molecule is attracted by the negatively charged oxygen atom of another water molecule, the proton can get very close to this oxygen atom, so close that it is attracted very strongly and interacts almost exclusively with just one of the lone pairs (Figure 11.28). This strong interaction is called a **hydrogen bond**. In general, a hydrogen bond forms between

1. A hydrogen atom that is attached to a very electronegative atom (in particular, N, O, or F), so it carries a substantial positive charge, and
2. A small electronegative atom (in particular, N, O, or F) that has small localized lone pairs that strongly attract the nucleus of the hydrogen atom (the proton).

The commonest and strongest hydrogen bonds are therefore the following combinations, in which the hydrogen bond is depicted by a dashed line:

F—H---:F F—H---:O F—H---:N
O—H---:F O—H---:O O—H---:N
N—H---:F N—H---:O N—H---:N

Larger and more diffuse lone pairs on larger atoms such as Cl, P, and S do not interact as strongly with positively charged H atoms and at best form only weak hydrogen bonds. For our purposes, we can consider hydrogen bonds to form only in the nine cases just given.

A hydrogen bond is an intermolecular attraction in which a hydrogen atom that is bonded to an electronegative atom, and therefore has a partial positive charge, is attracted to an unshared electron pair on another small electronegative atom.

The hydrogen bond is a type of dipole–dipole interaction that is much stronger than all other dipole–dipole interactions because of the small size of the hydrogen atom and the absence of any core electrons. It is intermediate in strength between other dipole–dipole interactions and the much stronger covalent and ionic bonds in molecules and network solids. The energies of most hydrogen bonds range from about 5 to 25 $kJ \cdot mol^{-1}$, whereas the energies of most covalent bonds range from 100 to over 500 $kJ \cdot mol^{-1}$ (Table 9.2) and the energies of other dipole–dipole forces are usually less than 1 $kJ \cdot mol^{-1}$. It is the strength of the hydrogen bonds between water molecules that accounts for the exceptionally high boiling point and freezing point of water. Chemists find it convenient to think of the hydrogen bond, as its name implies, as a type of bond rather than as an intermolecular force. But because both bonds and intermolecular forces are electrostatic in origin, there is no fundamental distinction between them.

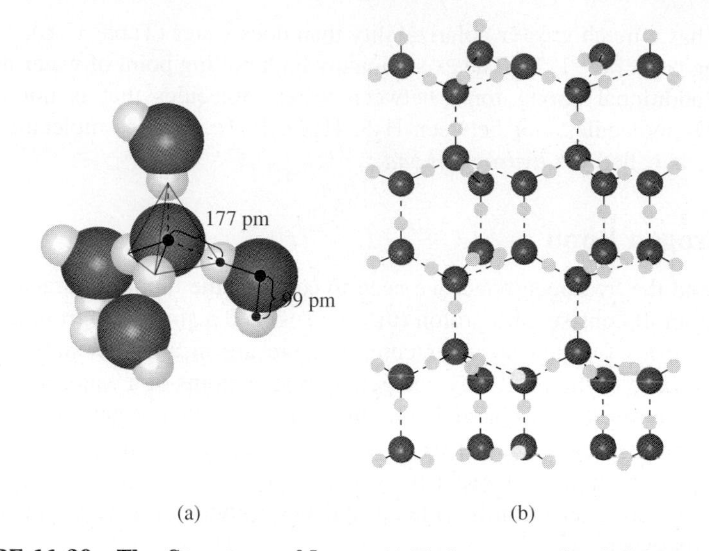

FIGURE 11.29 The Structure of Ice (a) Each water molecule is surrounded by a tetrahedral arrangement of four other water molecules to which it is bound by hydrogen bonds. Each oxygen atom forms covalent bonds to two hydrogen atoms and two hydrogen bonds with hydrogen atoms of neighboring water molecules. (b) The structure is a three-dimensional network similar to the structures of diamond and silica. It is a very open structure. When ice melts and some of the hydrogen bonds are broken, the molecules pack together more closely, and so water has a greater density than ice.

Although we most often describe the water molecule as angular, because the hydrogen atoms are small and their electron density overlaps extensively with that of the larger oxygen atom, the overall shape of the water molecule is not far from spherical.

The Structure of Ice The strong hydrogen bonding between water molecules also accounts for the structure and low density of ice and the expansion of water below 4°C. Ice has a network structure similar to that of diamond, with each oxygen atom surrounded by four other oxygen atoms in a tetrahedral arrangement (Figure 11.29). Each oxygen has two hydrogen atoms covalently bound to it and two lone pairs of electrons, each of which forms a hydrogen bond to a neighboring water molecule. So each water molecule is hydrogen bonded to only four neighboring water molecules, whereas each nearly spherical water molecule would have 12 neighbors if they were close packed. As a result, ice has a very open structure.

When ice melts at 0°C, about 15% of the hydrogen bonds are broken. As a consequence, some of the ''free'' water molecules occupy the holes in the remaining hydrogen-bonded structure, so the density increases. With increasing temperature, more hydrogen bonds are broken and more molecules occupy the holes, so the density continues to increase. Normally, as a liquid is heated, the increased motion of the molecules causes them to take up more space, and therefore the density normally decreases. In the case of water, at 3.98°C the decrease in density due to this increased motion just compensates for the increase in density caused by the collapse of the hydrogen-bonded structure. Above 3.98°C the density of water begins to decrease with increasing temperature, as it does for most other liquids.

Hydrogen bonds are of widespread importance. As we shall see in Chapter 17, they hold long-chain protein molecules into definite shapes that have a profound influence on the properties of proteins, and they hold together the two chains of the double helix of DNA. Hydrogen bonds also play an important role in determining the solvent properties of water, as we see next.

Exercise 11.8 Which of the following substances are expected to form strong hydrogen bonds in the liquid or solid state?

(a) HF **(b)** CH_3SH **(c)** CH_3OH **(d)** $(CH_3)_3N$ **(e)** $(CH_3)_2NH$
(f) CH_3CO_2H

11.5 SOLUTIONS

When two gases are placed in contact, they always mix with each other as a consequence of the rapid random motion of their molecules. As we saw in Chapter 9, the entropy of the mixture is greater than that of the separate gases; the entropy of mixing is always positive. In contrast, when two liquids are placed in contact, they do not necessarily mix. Ethanol, C_2H_5OH, is soluble in water in all proportions: Ethanol and water are said to be **miscible**. But the hydrocarbon hexane is not soluble in water, and water is not soluble in hexane: Hexane and water are **immiscible**. However, hexane and another alkane, such as octane, are miscible. These examples illustrate the often quoted generalization ''Like dissolves like.'' Chemically similar substances are soluble in each other, whereas dissimilar substances are insoluble. Ethanol is like water in that both contain OH groups and are hydrogen bonded. Different alkanes are very similar and are soluble in each other. But water and a hydrocarbon are not at all similar; one is polar and the other is nonpolar, so they are not soluble in each other.

When two gases mix, energy changes are very small (zero in the case of an ideal gas), because the intermolecular forces are very weak. Thus, two gases mix, because the entropy of mixing is positive. However, intermolecular forces are much larger in liquids than in gases, so we cannot ignore the energy changes involved in the mixing of two liquids. Moreover, because liquids—particularly hydrogen-bonded liquids—have some structure, we cannot ignore the contribution of any changes in structure to the entropy change when two liquids are mixed. The large energy and entropy changes that are often involved in the mixing of two liquids means that a detailed discussion of the solubility of one liquid in another is too complicated for us to enter into here. For most purposes we can rely on the generalization ''Like dissolves like.''

For mixing gases,

$$\Delta G_{mix} = \Delta H_{mix} - T\,\Delta S_{mix}$$

ΔH_{mix} is small (zero for ideal gases), ΔS_{mix} is always positive, and T is positive on the Kelvin scale. Thus ΔG_{mix} is always negative, and gases spontaneously mix.

We can understand the solubility of solids in liquids in the same way. Nonpolar substances (such as I_2 and S_8) are soluble in nonpolar liquids (such as carbon tetrachloride, CCl_4, and carbon disulfide, CS_2), but they are insoluble in water, which is polar (Demonstration 11.1). A molecule of a sugar such as glucose contains many OH groups that can form hydrogen bonds with water, so glucose is soluble in water but insoluble in CCl_4.

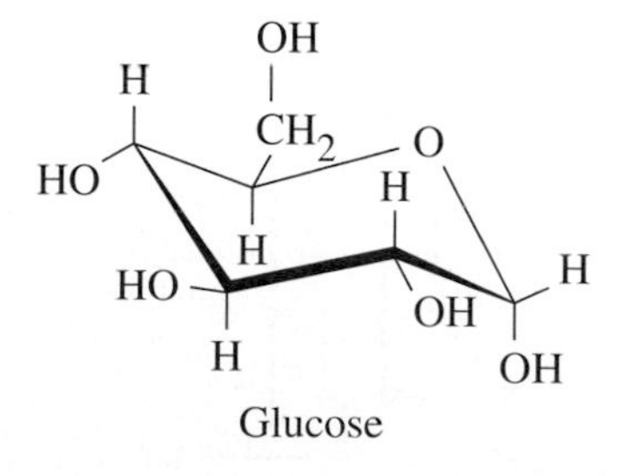

Glucose

Many **ionic solids** such as MgO, AgCl, CuO, and $BaSO_4$ are insoluble in water. Qualitatively, we can say that strong ionic bonds prevent the ions from separating and mixing with the water molecules. The large amount of energy that would be needed to separate the ions would make the overall solution process strongly endothermic. A corresponding large decrease in the entropy of the surroundings would not be compensated by the entropy of mixing of the ions with water, leading to an overall entropy decrease. Thus, the solution process would be nonspontaneous. However, as we saw in Table 4.6, many other ionic solids such as NaCl, $MgSO_4$, $AlCl_3$, and $CuSO_4$ are soluble in water. In these cases, there must be a strong attractive force between the ions and the water molecules that enables the ions to be pulled away from the crystal. We saw in Chapter 10 that many metal ions are hydrated with, commonly, four or six water molecules strongly attached to the metal ion by polar covalent bonds. The hydration of ions, like any process in which

DEMONSTRATION 11.1
Immiscible Liquids and Solubility

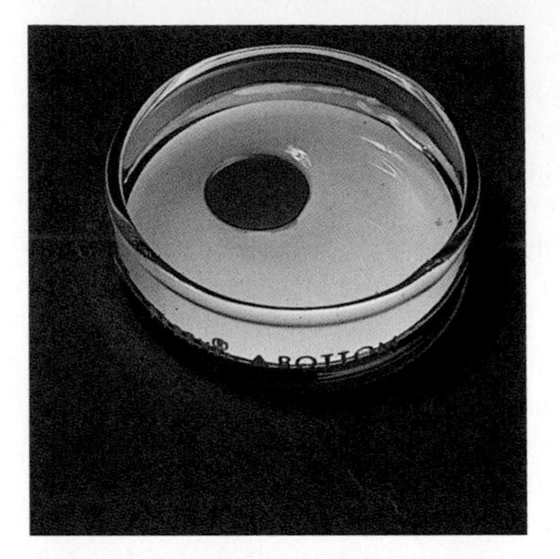

Water and carbon tetrachloride, CCl_4, are immiscible. Carbon tetrachloride forms a small pool when added to water. Solid iodine dissolves in the colorless carbon tetrachloride to give a violet solution, but it is insoluble in water.

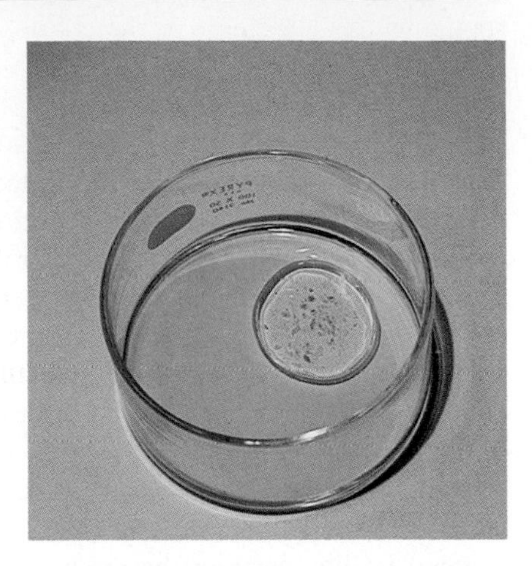

Copper sulfate dissolves in water to give a blue solution, but it is insoluble in carbon tetrachloride, which remains colorless.

bonds are formed, is exothermic and often approximately compensates for the energy needed to break up the crystal, giving a relatively small overall enthalpy change that may be positive or negative. The overall enthalpy change when a solid dissolves in a liquid is called the **enthalpy of solution**. As we just saw, the enthalpy of solution for many ionic solids in water is relatively small and may be either positive or negative. In such cases there is then only a small increase or decrease in the entropy change of the surroundings, so the entropy increase upon mixing the ions with the water leads to an overall entropy increase and is a spontaneous process.

Let us look in a little more detail at the process of solution of an ionic solid, such as NaCl, in water. It is convenient to think of first breaking up the crystal to give gaseous ions (Figure 11.30a). The enthalpy change for this process, ΔH_1, is called the **lattice energy**. For sodium chloride it is 786 kJ · mol^{-1}:

$$NaCl(s) \rightarrow Na^+(g) + Cl^-(g) \qquad \Delta H_1 = 786 \text{ kJ} \cdot \text{mol}^{-1}$$

Then we dissolve the gaseous ions in water. The enthalpy change for this process is called the **enthalpy of hydration**, ΔH_2. For sodium chloride it has been found to have the value -783 kJ · mol^{-1}:

$$Na^+(g) + Cl^-(g) \xrightarrow{H_2O} Na^+(aq) + Cl^-(aq) \qquad \Delta H_2 = -783 \text{ kJ} \cdot \text{mol}^{-1}$$

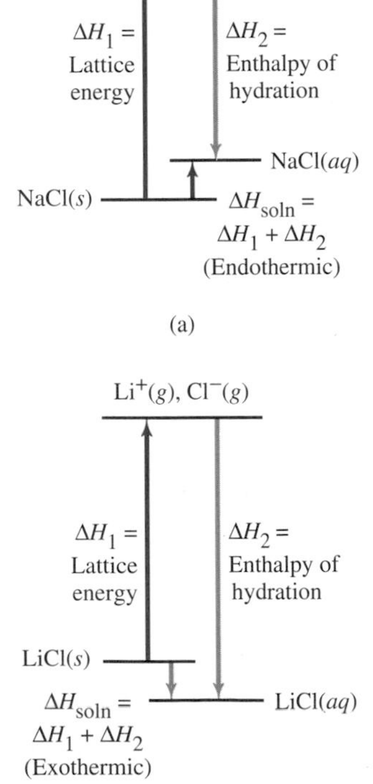

FIGURE 11.30 Enthalpy of Solution of an Ionic Solid in Water (Not drawn to scale) (a) Sodium chloride: The energy needed to form gaseous ions from the solid—the lattice energy—is greater than the energy released by the hydration of the ions; the enthalpy of solution is positive (endothermic). (b) Lithium chloride: The lattice energy is smaller than the energy released in hydrating the ions, so the enthalpy of solution is negative (exothermic).

The overall enthalpy change for the process, the enthalpy of solution, ΔH_{soln}, is

$$\Delta H_{soln} = \Delta H_1 + \Delta H_2 = (786 - 783)\ \text{kJ}\cdot\text{mol}^{-1} = 3\ \text{kJ}\cdot\text{mol}^{-1}$$

We see that sodium chloride has a small endothermic heat of solution. The corresponding small decrease in the entropy of the surroundings is more than compensated by the entropy increase when the ions are mixed with water. Thus, there is an overall entropy increase, and sodium chloride is soluble in water. Some salts, such as ammonium nitrate, have a larger endothermic enthalpy of solution in water than does sodium chloride; when we dissolve them in water, there is a considerable decrease in temperature. This property of ammonium nitrate is utilized in portable cold packs (Figure 11.31) typically used for relieving sore muscles. However, many salts, such as lithium chloride (Figure 11.30b) have an exothermic enthalpy of solution, and the solution becomes warmer when lithium chloride is dissolved in water. Because the lithium ion is smaller than the sodium ion, the negatively charged oxygen atom of a polar water molecule can get closer to the positive charge of the lithium ion than to the positive charge of the sodium ion and is attracted more strongly (Appendix P). In other words, the lithium ion is more strongly hydrated than the sodium ion and so has a larger (more exothermic) enthalpy of hydration. Hence the enthalpy of solution of lithium chloride is negative (exothermic), whereas that of sodium chloride is positive (endothermic).

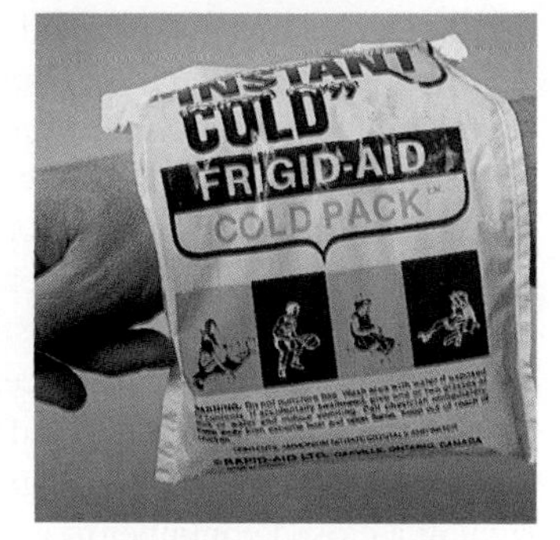

FIGURE 11.31 A Cold Pack When the pouch of water inside this cold pack is broken, the solid ammonium nitrate dissolves in the water endothermically, and the pack cools to below 0°C.

SUMMARY

Most solids form crystals with regular repeating arrays of atoms; this arrangement gives them flat faces and characteristic interfacial angles. Solids may be classified as molecular solids, with individual covalent molecules, or as network solids, which consist of a repeating arrangement of atoms or ions and not of individual molecules. There are three types of network solids that have covalent, ionic, and metallic bonding, respectively.

Crystal structures are described in terms of a regular array of points called a lattice. The smallest part of such an array that, when repeated through space, reproduces the entire lattice is called a unit cell. The three unit cells of the three types of cubic lattices are simple cubic, with lattice points at each corner of a cube; body-centered cubic, with an additional point at the center of the cube; and face-centered cubic, with additional points at the center of each face. These structures are determined by X-ray crystallography.

Most metal crystals have one of three structures: hexagonal close-packed or cubic close-packed (face-centered cubic), or body-centered cubic, which is not quite close packed. The radius of a metal atom—the metallic radius—is equal to half the distance between adjacent metal atoms in the crystal. From the distances between the atoms in a metal and its measured density, we can calculate a value for Avogadro's number.

The structures of covalent network crystals are determined mainly by the geometry of the bonds formed by their atoms. Diamond, C(*s*), and silica, $SiO_2(s)$, have AX_4 tetrahedral atoms and a structure based on the face-centered cubic lattice.

The structures of covalent molecular crystals are determined by the way the molecules can pack most efficiently, which depends mainly on their shape and polarity. The covalent radius of an atom is equal to half the distance between two covalently bound atoms of the same kind in a molecular crystal or a covalent network crystal.

The structures of ionic crystals are determined by the relative sizes of their ions. Many ionic crystals MX(*s*) have one of three different cubic structures: the sodium chloride (NaCl) structure, the cesium chloride (CsCl) structure, or the zinc sulfide (ZnS) structure. Ionic radii can be obtained from the electron density distribution in an ionic crystal, as determined by X-ray crystallography.

Covalent, ionic, and metallic radii vary in the same way through the periodic table: They increase down any group and decrease from left to right across any period.

Amorphous solids do not form crystals. They result from rapidly cooling molten solids of high melting points so that the constituent atoms or molecules have no time to adopt the regular arrangement of atoms in a crystal.

In liquids, the molecules have a less regular arrangement than they do in a crystalline solid. Although the molecules are not in fixed positions in a liquid and are continually moving, they retain some of the structure characteristic of the corresponding solid.

A phase change is any change of state among the solid, liquid, and gaseous forms. Melting (fusion) is endothermic, and the molar enthalpy of fusion, ΔH_{fus}, and the melting point of a substance are measures of the strength of intermolecular forces. At any given temperature some molecules have sufficient energy to escape (evaporate) from a liquid as gas, giving rise *at equilibrium* in a closed container to a vapor pressure. The molar enthalpy of vaporization, ΔH_v, the boiling point, and the vapor pressure of a substance are good indicators of the strength of intermolecular forces in the liquid. The normal boiling point of a liquid is the temperature at which the vapor pressure is 1 atm. In melting and evaporation, ΔS_{surr} of the surroundings decreases because melting and boiling are endothermic processes, but ΔS_{system} is large enough that in each case the overall ΔS is positive. Hence both processes are spontaneous.

Intermolecular forces are due to dipole–dipole forces between polar molecules and to London (dispersion) forces that arise from the induced fluctuating dipoles in neighboring molecules. The magnitude of London forces depends on the polarizabilities of the molecules. A polyatomic molecule with polar bonds is nonpolar if the molecule is sufficiently symmetrical that the bond dipoles cancel each other to give a zero overall dipole moment. Conversely, a polyatomic molecule is polar if the bond dipoles do not cancel each other.

The hydrogen bond is an unusually strong dipole–dipole attraction. It is formed by the interaction of the positively charged H atom in an N—H, O—H, or F—H bond with a small, localized lone pair of an electronegative atom of another molecule. Hydrogen bonding accounts for anomalous properties such as the unusually high freezing and boiling points of water and ammonia, the structure of ice, the lower density of ice compared to that of water at 0°C, and the maximum density of water at 3.98°C.

Gases, which have negligible intermolecular forces, always mix spontaneously, because mixing causes an increase in entropy. Liquids may be miscible (mixable) or immiscible (unmixable). A useful rule is that ''like dissolves like.'' Solids and liquids with polar molecules tend to be soluble in polar liquids, whereas liquids with nonpolar molecules tend to be soluble in nonpolar liquids. The enthalpy of solution of an ionic solid in water may be positive (endothermic) or negative (exothermic), depending on the relative magnitudes of the lattice energy—the enthalpy change for the breaking up of a lattice to give gaseous ions—and the enthalpy of hydration—the enthalpy change for dissolving gaseous ions.

IMPORTANT TERMS

amorphous solid (page 399)
body-centered cubic (page 392)
condensation (page 402)
covalent network solid (page 390)
covalent radius (page 396)
dipole (page 407)
dipole–dipole forces (page 408)
dipole moment, μ (page 407)
dispersion (London) forces (page 409)
enthalpy of hydration (page 416)
enthalpy of solution (page 416)
evaporation (page 402)
face-centered cubic (page 392)
hydrogen bond (page 413)
immiscible (page 415)
intermolecular forces (page 407)
ionic radius (page 396)
ionic solid (page 390)
lattice (page 391)
lattice energy (page 416)
London (dispersion) forces (page 409)
metallic radius (page 395)
metallic solid (page 391)
miscible (page 415)
molar enthalpy of fusion, ΔH_{fus} (page 401)
molar enthalpy of vaporization, ΔH_v (page 403)
molecular solid (page 390)
network solid (page 390)
nonpolar molecule (page 408)
normal boiling point (page 405)
phase change (page 401)
polarizability (page 409)
polar molecule (page 408)
unit cell (page 391)
vapor pressure (page 404)
X-ray crystallography (page 388)

REVIEW QUESTIONS

1. How does an amorphous solid differ from a crystalline solid?

2. Why is iodine, $I_2(s)$, described as a molecular solid; diamond, $C(s)$, as a covalent network solid; and sodium chloride, $NaCl(s)$, as an ionic network solid?

3. What is **(a)** a lattice; **(b)** a unit cell?

4. Describe the unit cell of the body-centered cubic lattice.

5. Describe the unit cell of the face-centered cubic lattice.

6. What problem arises in determining ionic radii from the X-ray structure of ionic crystals?

7. What lattice is adopted by the Na^+ ions in $NaCl(s)$?

8. Explain what is meant by the covalent radius of an atom.

9. What are the coordination numbers of the Cs^+ and Cl^- ions in the $CsCl(s)$ structure?

10. What phase changes are associated with each of the following?

(a) fusion **(b)** freezing **(c)** evaporation **(d)** condensation

11. Define **(a)** molar enthalpy of fusion and **(b)** molar enthalpy of vaporization.

12. Why does a liquid exert a vapor pressure?

13. Explain the terms **(a)** "polar bond," **(b)** "dipole," and **(c)** "dipole moment."

14. Explain why there are intermolecular attractions between nonpolar molecules.

15. How do the intermolecular forces between molecules depend on the molecules' polarizabilities?

16. What is it about the nature of the H atom that enables it to participate in hydrogen bonding?

17. For a substance H_nX, what are the requirements for hydrogen-bond formation between its molecules?

18. For liquids, explain the terms "miscible" and "immiscible."

19. What are **(a)** lattice energy and **(b)** enthalpy of hydration?

20. Why are ionic network substances such as $Al_2O_3(s)$ and $MgO(s)$ insoluble in water, whereas others, such as $AlCl_3(s)$ and $MgCl_2(s)$, are soluble?

PROBLEMS

Solids and Liquids

1. Which of the following are molecular solids, and which are network solids?

(a) C **(b)** S_8 **(c)** CO_2 **(d)** P_4O_6
(e) NaCl **(f)** MgO **(g)** Al

2. Give the main reasons for the differences in the melting points of each of the following pairs of substances.

(a) CO_2, SiO_2 **(b)** MgF_2, F_2 **(c)** NaF, ClF
(d) red phosphorus, white phosphorus

3. In terms of their structures, explain why each of the following pairs of substances have very different properties.

(a) silicon and aluminum **(b)** oxygen and sulfur
(c) diamond and graphite

4. Which of the following substances form network solids, and which form molecular solids?

(a) carbon monoxide **(b)** silicon carbide **(c)** chlorine
(d) magnesium **(e)** magnesium chloride

5. **(a)** What is an amorphous solid?

(b) Explain why an amorphous solid does not have a sharp melting point.

6. Sketch each of the following, and deduce how many atoms each contains.

(a) the body-centered cubic unit cell of barium

(b) the face-centered cubic unit cell of solid neon

7. The structure of aluminum is based on a cubic lattice. At 25°C, the edge length of the unit cell is 405 pm, and the density is 2.70 $g \cdot cm^{-3}$.

(a) How many Al atoms are there in the unit cell?

(b) On what type of cubic lattice must the structure of aluminum be based?

8. At 25°C, the density of platinum is 21.5 $g \cdot cm^{-3}$, and the structure is cubic close-packed with a unit cell edge length of 392 pm. Calculate the atomic mass of platinum.

9. At 24 K, solid neon has a cubic lattice with a unit cell edge length of 450 pm and a density of 1.45 $g \cdot cm^{-3}$ (for solid neon).

(a) How many Ne atoms are in the unit cell?

(b) What type of cubic lattice does crystalline neon have?

(c) What is the radius of the Ne atom?

10. Solid krypton has a cubic close-packed lattice with a unit cell edge length of 559 pm.

(a) Sketch the unit cell, showing the positions of the atoms.

(b) How many Kr atoms are there in the unit cell?

(c) What is the density of solid krypton?

(d) What is the radius of the Kr atom?

11. In crystalline sodium chloride at 25°C, the distance between the centers of neighboring Na^+ and Cl^- ions is 281 pm, and the density is 2.165 $g \cdot cm^{-3}$. Calculate a value for Avogadro's number.

12. Copper has a density of $8.930\ g \cdot cm^{-3}$. The edge length of the unit cell is 361.5 pm. Calculate a value for Avogadro's number.

13. Copper(I) chloride, $CuCl(s)$ has the zinc sulfide structure and a density of $3.41\ g \cdot cm^{-3}$.

(a) What is the edge length of the unit cell?

(b) What is the shortest distance between the centers of a Cu^+ ion and a Cl^- ion?

(c) The radius of the Cl^- ion is 181 pm. What is the radius of the Cu^+ ion?

14. Calcium fluoride, $CaF_2(s)$, of density $3.180\ g \cdot cm^{-3}$, has a cubic lattice. The edge length of the unit cell is 546.3 pm. How many formula units of CaF_2 are there per unit cell?

15. Which unit cell has the higher mass, that of cesium chloride or that of sodium chloride? Explain.

16. Cesium bromide, $CsBr(s)$, crystallizes in the cesium chloride structure. If the closest distance between the centers of oppositely charged ions is 371 pm, what is the density of cesium bromide?

17. What are the coordination numbers of **(a)** Zn^{2+} and **(b)** S^{2+} in zinc sulfide?

(c) How is the zinc sulfide structure related to the diamond structure?

18. Calculate the percentage of the total volume of the unit cell that is occupied by spheres in **(a)** a face-centered cubic structure and **(b)** a body-centered cubic structure. (Volume of a sphere $= \frac{4}{3}\pi r^3$.)

(c) Which structure has the more-closely-packed arrangement of spheres?

19. At about 1000°C, iron undergoes a transition from the body-centered cubic structure to the face-centered cubic structure. The edge length of the unit cell increases from 286 pm to 363 pm. Compare the densities of these two forms of iron, and explain the difference.

20. Barium has the body-centered structure with a unit cell edge length of 502 pm. What is the density of barium?

21. Silver has the copper (face-centered cubic) structure and a density $10.5\ g \cdot cm^{-3}$. How many silver atoms are contained in a cube of silver with an edge length of 1.00 mm?

22. Select the larger ion from each of the following pairs of ions, and justify your choice.

(a) Na^+, F^- **(b)** Na^+, K^+ **(c)** F^-, Cl^-
(d) Na^+, Mg^{2+} **(e)** S^{2-}, Cl^-

23. Repeat Problem 22 for each of the following pairs of ions.

(a) F^-, I^- **(b)** Mg^{2+}, Ca^{2+} **(c)** O^{2-}, F^-
(d) K^+, Cl^- **(e)** Al^{3+}, O^{2-}

24. In terms of their melting points and solubilities in water, describe the properties of each of the following types of solids.

(a) molecular **(b)** ionic **(c)** metallic
(d) covalent network

25.* Mercury(II) sulfide has the zinc sulfide structure. The shortest distance between the center of an Hg^{2+} ion and the center of an S^{2-} ion is 253 pm. Calculate the density of mercury(II) sulfide.

Phase Changes; Intermolecular Forces

26. Explain **(a)** evaporation, **(b)** boiling point, **(c)** molar enthalpy of condensation, and **(d)** vapor pressure.

27. Explain **(a)** phase change, **(b)** melting, and **(c)** freezing.

28. **(a)** Why does the vapor pressure of a liquid increase with increasing temperature?

(b) At 20°C, the vapor pressure of benzene is 75.0 mm Hg, and that of toluene is 50.0 mm Hg. Which of these hydrocarbons would you expect to have the higher normal boiling point?

(c) The normal boiling point of diethyl ether, $(C_2H_5)_2O$, methanol, CH_3OH, and propanone (acetone), $(CH_3)_2CO$, are 34.5, 64.5, and 56.1°C, respectively. Which of these liquids will have the highest vapor pressure at 25°C, and which will have the lowest?

29. Before the invention of the refrigerator, butter and milk were stored in the summer in porous clay pots standing in water. How does a clay pot standing in water keep the contents cool on a hot day?

30. Explain what effect (if any) each of the following has on the vapor pressure of a liquid.

(a) the surface area of the liquid

(b) the volume of the container

(c) temperature **(d)** intermolecular forces

(e) the volume of the liquid

31. Why is the vapor pressure of water at 25°C less than the vapor pressure of gasoline at the same temperature?

32. Would you expect each of the following to form a molecular solid or a network solid? What would you expect to be the dominant intermolecular force in each case? Explain your choices.

(a) nitrogen **(b)** hydrogen sulfide **(c)** iron
(d) calcium oxide **(e)** ethane, C_2H_6 **(f)** silica, SiO_2
(g) potassium hydroxide **(h)** sulfuric acid

33. What types of interactions (bonds and/or intermolecular forces) must be overcome to melt each of the following solids?

(a) BaO **(b)** diamond **(c)** I_2
(d) P_4O_{10} **(e)** copper **(f)** graphite

34. Both NaCl and MgO are ionic solids, yet $MgO(s)$ melts at a temperature about 2000 K higher than $NaCl(s)$. Explain why.

35. Show by suitable diagrams that molecules with AX_2, AX_3, AX_4, AX_5, or AX_6 geometries have zero dipole moments, whereas molecules with AX_2E, AX_2E_2, AX_3E, or AX_5E geometries have a dipole moment.

36. Which molecule in each of the following pairs has a dipole moment?

(a) $BeCl_2$, OCl_2 **(b)** PF_3, BF_3 **(c)** PF_5, ClF_5

37. Repeat Problem 36 for each of the following pairs.

(a) F_2, ClF **(b)** CO_2, OCS **(c)** CCl_4, $FCCl_3$

38. Which substance of each of the following pairs would you expect to have the higher boiling point? Justify your choice.

(a) ClF, BrF **(b)** BrCl, Cl_2 **(c)** KBr, BrCl **(d)** Na, Br_2

39. Account for the increase in boiling point from benzene,

$C_6H_6(l)$, 80°C, to chlorobenzene, $C_6H_5Cl(l)$, 132°C, to hexachlorobenzene, $C_6Cl_6(l)$, 310°C.

40. What types of intermolecular forces predominate in each of the following?

(a) $I_2(s)$ **(b)** $CaO(s)$ **(c)** $CO_2(g)$
(d) $CHCl_3(l)$ **(e)** $HF(l)$

41. Which of the following properties would you expect to depend on the strength of intermolecular forces? Justify your choice.

(a) boiling point **(b)** enthalpy of vaporization
(c) molar mass **(d)** solubility **(e)** viscosity

42. What is the strongest intermolecular attraction (or bond) that must be broken when each of the following substances melts?

(a) benzene **(b)** ethanol **(c)** ethane
(d) barium oxide **(e)** chlorine **(f)** hydrogen chloride

43. Arrange the following substances in order of the expected increase in their boiling points, and account for that order in terms of intermolecular forces.

(a) chlorine **(b)** sodium chloride **(c)** hydrogen chloride
(d) neon **(e)** oxygen **(f)** water
(g) argon **(h)** hydrogen fluoride

44. Nonpolar tetrachloromethane boils at 76.5°C, whereas polar trichloromethane (chloroform) boils at 61.7°C. Explain why the boiling point of polar chloroform is lower than that of nonpolar tetrachloromethane.

45.* Suppose that 5.00 L of air is saturated with water vapor at 25°C and then completely dried by bubbling through sulfuric acid. If this process increases the mass of the sulfuric acid by 0.115 g, what is the vapor pressure of water at 25°C?

Water and the Hydrogen Bond

46. What are the requirements for substances to be able to form hydrogen bonds?

47. Which of the following substances are associated by hydrogen bonding in the liquid state? Explain your answers.

(a) ammonia **(b)** sodium chloride
(c) hydrogen fluoride **(d)** hydrogen
(e) methane **(f)** lithium hydride
(g) methanol **(h)** acetic acid, CH_3CO_2H

48. Explain why **(a)** water does not form hydrogen bonds with either hexane or tetrachloromethane; **(b)** hydrogen fluoride, HF, has a much higher boiling point than HCl, but a much lower boiling point than H_2O.

49. Water has a greater density at 0°C than does ice, whereas liquid bromine is less dense than solid bromine at its melting point. Account for these differences.

50.* A given volume of acetic acid at 150°C and 1 atm pressure diffuses through a porous plug in 9.69 min. Repetition of this experiment with the same volume of oxygen under the same conditions gave a diffusion time of 5.00 min.

(a) What is the molar mass of acetic acid under these conditions?

(b) In terms of hydrogen bonding, account for the fact that this molar mass is not the same as that given by the molecular formula CH_3CO_2H.

Solutions

51. Make qualitative predictions about each of the following solubilities, and explain your answers.

(a) $HCl(g)$ in water and in pentane, $C_5H_{12}(l)$

(b) water in liquid HF and in gasoline

(c) chloroform (trichloromethane), $CHCl_3(l)$, in water and in carbon tetrachloride (tetrachloromethane), $CCl_4(l)$

(d) naphthalene, $C_{10}H_8(s)$, in water and in benzene, $C_6H_6(l)$

(e) $N_2(g)$ and hydrogen cyanide, $HCN(g)$, in water

(f) benzene, $C_6H_6(l)$, in toluene (methylbenzene), $C_7H_8(l)$, and in water

52. Which substance of each of the following pairs is likely to be the more soluble in water? Explain each choice and, for the less soluble of each pair, suggest a better solvent than water.

(a) hydrogen peroxide(l) or benzene (l)

(b) ethane or ethanediol, $HOCH_2CH_2OH(l)$

(c) sugar(s) or hexane, $C_6H_{14}(l)$

(d) chloroform, $CHCl_3(l)$, or magnesium chloride(s)

(e) iodine(s) or hydrogen iodide(g)

(f) sodium chloride(s) or tetrachloromethane(l)

(g) methanol, $CH_3OH(l)$, or ethane, $C_2H_6(g)$

General Problems

53. In terms of the radii of the ions involved, explain why the structure of sodium chloride is different than that of cesium chloride.

54. In terms of the kinetic molecular theory, explain **(a)** why a liquid evaporates below its boiling point; **(b)** why ice disappears without melting on a very cold, sunny day.

55. Explain why **(a)** water has a boiling point some 200°C higher than is expected by comparison with the boiling points of H_2S, H_2Se, and H_2Te; **(b)** sulfuric acid, $(HO)_2SO_2$, is a liquid, whereas F_2SO_2 is a gas; **(c)** water has its maximum density at 3.98°C; **(d)** methanol, $CH_3OH(l)$, is miscible with water, whereas hexane $C_6H_{14}(l)$, is immiscible; and **(e)** sodium oxide is readily soluble in water, whereas magnesium oxide is insoluble.

56. Explain the reasons for the differences between the boiling points and molar enthalpies of vaporization, respectively, of each of the following pairs of substances.

(a) CH_4, bp −164°C, ΔH_v 10.4 kJ · mol^{-1}; CCl_4, bp 76.7°C, ΔH_v 30.0 kJ · mol^{-1}

(b) H_2O, bp 100°C, ΔH_v 40.7 kJ · mol^{-1}; C_2H_5OH bp 78.5°C, ΔH_v 38.6 kJ · mol^{-1}

(c) benzene, C_6H_6, bp 80.2°C, ΔH_v 30.8 kJ · mol^{-1}; phenol, C_6H_5OH, bp 182°C, ΔH_v 57.4 kJ · mol^{-1}

(d) aluminum, bp 2450°C, ΔH_v 316 kJ · mol^{-1}; magnesium, bp 1110°C, ΔH_v 139 kJ · mol^{-1}

57. Explain the differences in the boiling points of each of the following pairs of substances.

	Substance	*BP (°C)*	*Substance*	*BP (°C)*
(a)	$(CH_3)_2O$	35	C_2H_5OH	79
(b)	HF	20	HCl	−85
(c)	CCl_4	76	LiCl	1360

58. Define each of the following, and clearly explain the difference between each.

(a) polarity, polarizability

(b) London forces, dipole–dipole forces

(c) polar covalent bonds, hydrogen bonds

59. **(a)** Explain why real gases deviate from ideal behavior especially at high pressures and low temperatures. To what factors can this be ascribed?

(b) From among each of the following pairs, which gas would you expect to show the greater deviation from ideal gas behavior, and why?

(i) bromine and fluorine

(ii) carbon monoxide and nitrogen

(iii) hydrogen chloride, HCl, and hydrogen fluoride, HF

60.* Elemental boron melts at 2300°C and is almost as hard as diamond. It is a poor electrical conductor as both a solid and a liquid. Would you expect **(a)** boron to be a molecular solid, an ionic solid, or a covalent network solid; **(b)** solid boron to be soluble in water?

61.* Tin tetrachloride, $SnCl_4$, boils at 114°C and melts at −33°C, whereas tin dichloride, $SnCl_2$, is a solid with a melting point of 246°C. In terms of intermolecular forces and structure, suggest a reason for these differences.

Chemical Equilibrium: Quantitative Aspects

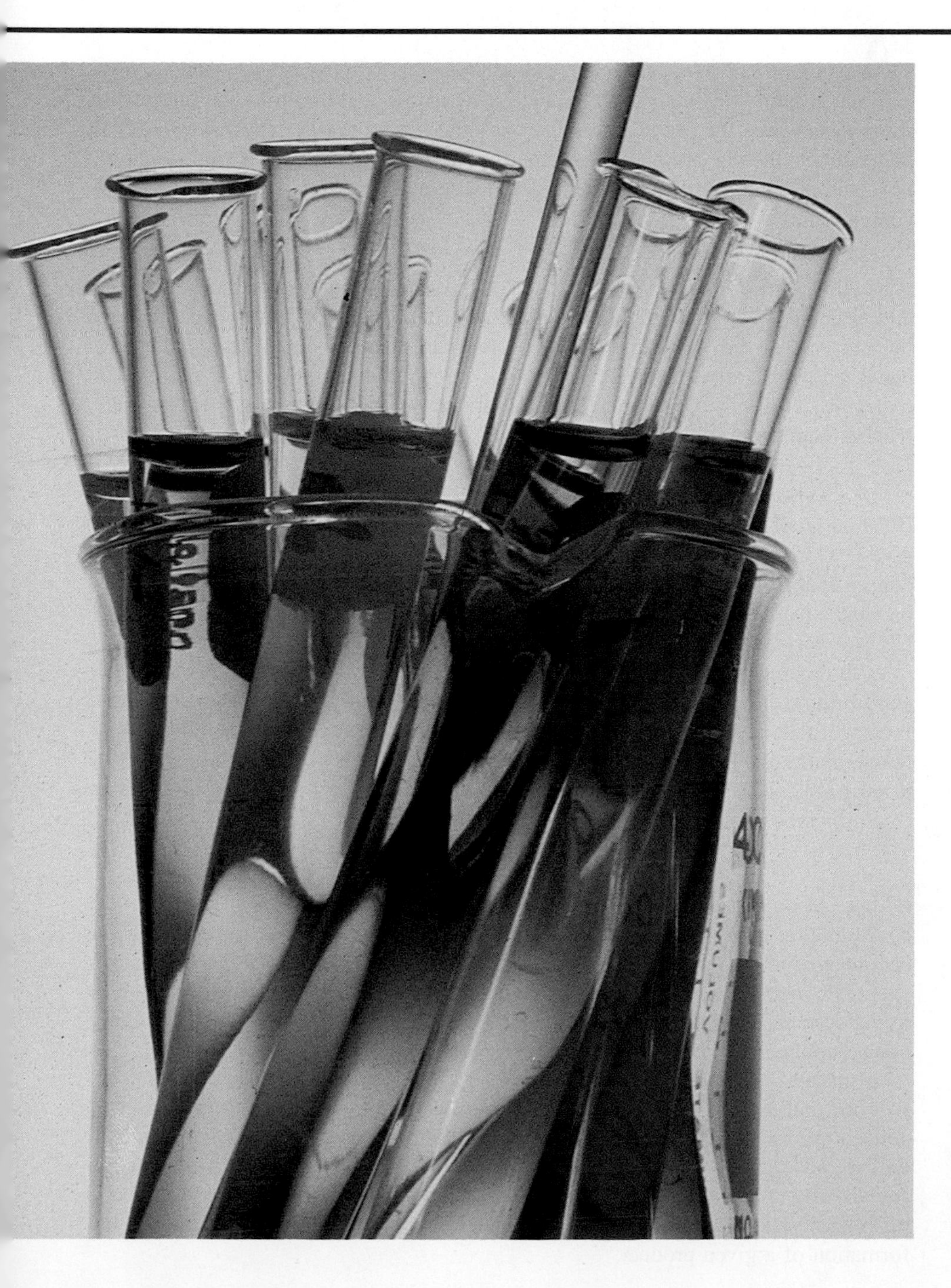

12.1 The Equilibrium Constant
12.2 Acid–Base Equilibria in Aqueous Solution
12.3 Gas-Phase Equilibria
12.4 Heterogeneous Equilibria
12.5 Gibbs Free Energy and the Equilibrium Constant

If left undisturbed, all chemical reactions eventually reach a state of equilibrium, in which the rates of the forward and reverse reactions are equal. Equilibria of acid-base reactions are important in many industrial processes and play a vital role in maintaining the environment and our body chemistry. Acid-base equilibria can be studied by using indicators, substances whose color changes with the acidity (pH) of a solution. An indicator has been added to the solutions in these test tubes. Each solution has a different color and therefore a different acidity.

As we first saw in Chapter 4, all chemical reactions carried out in a closed system eventually reach a state of dynamic equilibrium. The rate of the reaction between the products to give back the reactants (the reverse reaction) is then equal to the rate at which the reactants give the products (the forward reaction). For example, the reaction of a weak acid (such as HF) with water does not go to completion:

$$HF(aq) + H_2O(l) \rightleftharpoons H_3O^+(aq) + F^-(aq)$$

When this reaction reaches equilibrium, H_3O^+ ions, F^- ions, and HF molecules are all present in the solution. This does not mean that the reaction has stopped but only that the rate of the forward reaction and the rate of the reverse reaction are now equal, so there are no further changes in the concentrations of any of the ions or molecules. In contrast, when the reaction of a strong acid (such as HCl) with water reaches equilibrium, the reaction has virtually gone to completion:

$$HCl(aq) + H_2O(l) \rightarrow H_3O^+(aq) + Cl^-(aq)$$

Essentially all the HCl molecules have donated their protons to water molecules to give H_3O^+ and Cl^- ions, and so few HCl molecules remain in solution that their concentration is negligible. In this case the rate of the reverse reaction does not become equal to that of the forward reaction until the concentrations of H_3O^+ and Cl^- are very high and the concentration of HCl is very low.

It is important to remember that although all reactions *eventually* reach equilibrium, they may be so slow that in practice equilibrium is never attained. Hydrocarbons react with oxygen to give carbon dioxide and water, but at ordinary temperatures and in the absence of a catalyst, no reaction is observed. It is important to remember also that not all reactions are carried out in a closed system. When calcium carbonate is heated, carbon dioxide is produced:

$$CaCO_3(s) \rightarrow CaO(s) + CO_2(g)$$

If this carbon dioxide is allowed to escape into the atmosphere, the equilibrium concentration of CO_2 is never attained. Thus, the rate of the reverse reaction never becomes equal to the rate of the forward reaction, and all the calcium carbonate eventually decomposes. We will see in Chapter 17 that most reactions occurring in living systems similarly never reach equilibrium, because reactants are continually being supplied and products are continually being removed.

In this chapter we will be concerned with reactions that reach equilibrium. We will see that a quantity called the equilibrium constant enables us to calculate the concentrations of any of the reactants and products present at equilibrium. And we will see how these concentrations change as the conditions under which the reaction is carried out change. We can then choose those conditions that will give a maximum yield of product or, when necessary, those conditions that minimize the formation of a given product.

12.1 THE EQUILIBRIUM CONSTANT

Experiments have shown that for a reaction *at equilibrium*, the concentrations of the reactants and products are related to each other by an expression called the **equilibrium constant expression**. In this chapter, concentrations are expressed in moles per liter ($mol \cdot L^{-1}$) and are denoted by square brackets. Thus [X] signifies the concentration of X in moles per liter. We can write the following general equation for any reaction:

$$aA + bB + cC + \cdots \rightleftharpoons pP + qQ + rR + \cdots$$

where A, B, C, and so on are the reactants and P, Q, R, and so on are the products. The coefficients $a, b, c, \ldots, p, q, r$ represent the *number of moles* of each substance involved in the balanced equation for the reaction. For this general reaction, we express the concentration of A as [A], the concentration of B as [B], and so on. Then the equilibrium constant expression is

$$K = \left(\frac{[P]^p[Q]^q[R]^r \cdots}{[A]^a[B]^b[C]^c \cdots}\right)_{eq}$$

In this expression [A], [B], [C], . . . , [P], [Q], [R], . . . are the concentrations of the reactants and the products *at equilibrium*, and K is a constant called the **equilibrium constant**, which has a constant characteristic value at a given temperature. The subscript "eq" to the right-hand side of the expression reminds us that the concentrations are equilibrium concentrations.

We shall see in Chapter 15, where we discuss rates of reactions, that we can derive the equilibrium constant expression for a one-step reaction by setting an expression for the rate of the forward reaction equal to that for the rate of the reverse reaction.

For the reaction of HF with water,

$$HF(aq) + H_2O(l) \rightleftharpoons H_3O^+(aq) + F^-(aq)$$

$a = b = p = q = 1$, so the equilibrium constant expression is

$$K = \left(\frac{[H_3O^+][F^-]}{[H_2O][HF]}\right)_{eq}$$

This expression is valid only for dilute solutions, that is, for low concentrations of the reactants and products. In a more advanced treatment of equilibria, this expression can be modified so as to apply to more concentrated solutions. But here we restrict ourselves to dilute solutions.

12.2 ACID-BASE EQUILIBRIA IN AQUEOUS SOLUTION

We saw in earlier chapters that acid–base reactions are a very important class of reactions. An understanding of acid–base equilibria is essential not only in chemistry, but also in biochemistry, biology, geology, medicine, and environmental science. Acid–base reactions in aqueous solution are usually very fast, so equilibrium is rapidly attained.

The Acid Ionization Constant

We saw in Section 12.1 that the equilibrium constant for the reaction of the weak acid HF with water is

$$K = \left(\frac{[H_3O^+][F^-]}{[H_2O][HF]}\right)_{eq}$$

In *dilute* solutions of acids and bases in water, there is an enormous excess of water compared with the solute molecules or ions. The concentration of water in all dilute solutions is essentially the same as it is in pure water (55.5 *M*). We can therefore treat $[H_2O]$ as constant and combine it with the equilibrium constant K to give a new constant K_a:

▌The molar mass of water is 18.02 g · mol^{-1}. Therefore, 1000 g of water contains (1000 g)/(18.02 g · mol^{-1}) = 55.5 mol H_2O, and the concentration of water molecules in pure water is 55.5 *M*.

$$K_a = K[H_2O] = \left(\frac{[H_3O^+][F^-]}{[HF]}\right)_{eq}$$

K_a is called the **acid ionization constant**. The value of the acid ionization constant for any acid can be found by measuring the concentration of H_3O^+ in a solution of the acid of known concentration. We will discuss these measurements later, but we can see from the previous equation that the more completely an acid reacts with water, or as we usually say, the more completely it ionizes, the larger the ionization constant. The acid ionization constant for HF is $K_a(HF) = 3.5 \times 10^{-4}$ mol · L^{-1}. Like the values of other equilibrium constants, K_a values depend on temperature. They are normally quoted for 25°C.

Calculations and discussions of acid–base equilibria are greatly simplified if we express acid ionization constants in terms of their logarithms to the base 10 (Appendix M), thereby avoiding the need to use negative exponents. So we define the quantity

▌We denote the logarithm of x to the base 10 as $\log x$ and the logarithm of x to the base e (the natural logarithm) as $\ln x$.

$$pK_a = -\log K_a$$

For HF, $pK_a = -\log (3.5 \times 10^{-4}) = -\log 10^{-4} - \log 3.5 = 4 - 0.55 = 3.45$. The values of K_a and pK_a for some common acids are given in Table 12.1.

The strong acids at the top of Table 12.1, such as HCl, are almost completely converted to H_3O^+ ions. A *completely* ionized acid would have an ionization constant of infinity. In practice, no acid is completely ionized, but for a strong acid, the amount of remaining un-ionized acid is much too small to measure. So we can say only that a strong acid has a very large ionization constant ($K_a \gg 1$). Weak acids have ionization constants that are in most cases very much smaller than 1.0 ($K_a \ll 1$). Phosphoric acid, H_3PO_4 ($K_a = 7.5 \times 10^{-3}$, $pK_a = 2.12$), is more extensively ionized and gives a higher H_3O^+ concentration than does hydrocyanic acid, HCN ($K_a = 4.9 \times 10^{-10}$, $pK_a = 9.31$). H_3PO_4 is a stronger acid than HCN, although both are weak. For both acids the position of the equilibrium lies to the left, but it is considerably farther to the left for HCN than for H_3PO_4.

Acetic acid, also called ethanoic acid, belongs to the family of carboxylic acids that we discuss in Chapter 14. Vinegar is an approximately 5% (~0.1-*M*) solution of acetic acid in water.

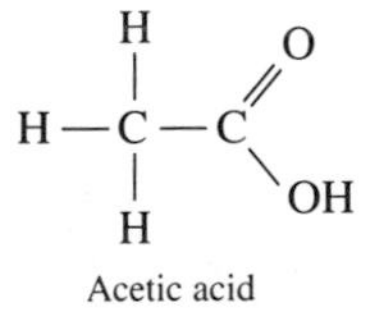

Acetic acid

We can calculate the H_3O^+ concentration in any solution of acid in water from the value of the ionization constant. Let us calculate the concentration of H_3O^+ in a 0.20-*M* solution of acetic acid, CH_3CO_2H.

The *first step* in solving any equilibrium problem is to write the equation for the equilibrium reaction. In this case the reaction is

$$CH_3CO_2H(aq) + H_2O(l) \rightleftharpoons H_3O^+(aq) + CH_3CO_2^-(aq)$$

The *second step* is to write the expression for the equilibrium constant and look up its value. From Table 12.1, K_a for acetic acid is 1.8×10^{-5} mol · L^{-1}. Thus, we can write

TABLE 12.1 Ionization Constants for Some Acids in Water at 25°C

Acid	Proton Transfer Reaction	K_a(mol · L^{-1})	pK_a(= −log K_a)
	Acid *Conjugate base*		
Perchloric acid	$HClO_4 + H_2O \rightarrow H_3O^+ + ClO_4^-$	Very large	—
Hydrochloric acid	$HCl + H_2O \rightarrow H_3O^+ + Cl^-$	Very large	—
Sulfuric acid	$H_2SO_4 + H_2O \rightarrow H_3O^+ + HSO_4^-$	Very large	—
Nitric acid	$HNO_3 + H_2O \rightarrow H_3O^+ + NO_3^-$	Very large	—
Sulfurous acid*	$SO_2 + 2H_2O \rightleftharpoons H_3O^+ + HSO_3^-$	1.2×10^{-2}	1.92
Hydrogen sulfate ion	$HSO_4^- + H_2O \rightleftharpoons H_3O^+ + SO_4^{2-}$	1.2×10^{-2}	1.92
Phosphoric acid	$H_3PO_4 + H_2O \rightleftharpoons H_3O^+ + H_2PO_4^-$	7.5×10^{-3}	2.12
Nitrous acid	$HNO_2 + H_2O \rightleftharpoons H_3O^+ + NO_2^-$	4.5×10^{-4}	3.35
Hydrofluoric acid	$HF + H_2O \rightleftharpoons H_3O^+ + F^-$	3.5×10^{-4}	3.45
Acetic acid	$CH_3CO_2H + H_2O \rightleftharpoons H_3O^+ + CH_3CO_2^-$	1.8×10^{-5}	4.74
Hydrated aluminum ion	$Al(H_2O)_6^{3+} + H_2O \rightleftharpoons H_3O^+ + Al(H_2O)_5OH^{2+}$	7.2×10^{-6}	5.14
Carbonic acid†	$CO_2 + 2H_2O \rightleftharpoons H_3O^+ + HCO_3^-$	4.3×10^{-7}	6.37
Hydrogen sulfide	$H_2S + H_2O \rightleftharpoons H_3O^+ + HS^-$	9.1×10^{-8}	7.04
Dihydrogenphosphate ion	$H_2PO_4^- + H_2O \rightleftharpoons H_3O^+ + HPO_4^{2-}$	6.2×10^{-8}	7.21
Hypochlorous acid	$HOCl + H_2O \rightleftharpoons H_3O^+ + OCl^-$	3.1×10^{-8}	7.51
Ammonium ion	$NH_4^+ + H_2O \rightleftharpoons H_3O^+ + NH_3$	5.6×10^{-10}	9.25
Hydrocyanic acid	$HCN + H_2O \rightleftharpoons H_3O^+ + CN^-$	4.9×10^{-10}	9.31
Hydrogenphosphate ion	$HPO_4^{2-} + H_2O \rightleftharpoons H_3O^+ + PO_4^{3-}$	2.1×10^{-13}	12.68
Hydrogensulfide ion	$HS^- + H_2O \rightleftharpoons H_3O^+ + S^{2-}$	1.3×10^{-13}	12.88
	Conjugate acid *Base*		

* There is no evidence for the formation of the undissociated acid H_2SO_3.

† The equilibrium given here is the sum of the two equilibria $H_2O + CO_2 \rightleftharpoons H_2CO_3$ and $H_2CO_3 + H_2O \rightleftharpoons H_3O^+ + HCO_3^-$, because the amount of H_2CO_3 formed is small and not accurately known.

$$K_a = \left(\frac{[H_3O^+][CH_3CO_2^-]}{[CH_3CO_2H]}\right)_{eq} = 1.8 \times 10^{-5}\ \text{mol} \cdot \text{L}^{-1}$$

The *third step* is to write expressions for the concentrations of each molecule or ion present at equilibrium. The *initial* concentration of acetic acid, before it transfers a proton to water, is 0.20 mol · L^{-1}. If we let x mol · L^{-1} be the concentration of H_3O^+ when equilibrium is reached, then the concentration of $CH_3CO_2^-$ is also x mol · L^{-1}, and the concentration of un-ionized CH_3CO_2H is $(0.20 - x)$ mol · L^{-1}.

	$CH_3CO_2H + H_2O \rightleftharpoons$	$H_3O^+ +$	$CH_3CO_2^-$	
Initial concentrations	0.20	0	0	mol · L^{-1}
Equilibrium concentrations	$0.20 - x$	x	x	mol · L^{-1}

The *fourth step* is to substitute the equilibrium concentrations into the expression for K_a and solve for x, the H_3O^+ concentration:

$$K_a = \left(\frac{[H_3O^+][CH_3CO_2^-]}{[CH_3CO_2H]}\right)_{eq} = \frac{(x\ \text{mol} \cdot \text{L}^{-1})(x\ \text{mol} \cdot \text{L}^{-1})}{(0.20 - x)\ \text{mol} \cdot \text{L}^{-1}}$$

$$= \frac{x^2}{0.20 - x}\ \text{mol} \cdot \text{L}^{-1} = 1.8 \times 10^{-5}\ \text{mol} \cdot \text{L}^{-1}$$

This expression can be rearranged into a quadratic equation, so it can be solved by the formula given in Appendix M. However, a simpler and much quicker approximate method can be used. Because the value of K_a is very small, we know that the position of the equilibrium is far to the left; only a very small amount of acetic acid is ionized. Therefore, x, the H_3O^+ concentration, will be very small compared with the concentration of CH_3CO_2H. We will assume then that x is much smaller than 0.20 and can be neglected with respect to 0.20. So we can write

$$0.20 - x \approx 0.20$$

Using this approximation, we have

$$\frac{x^2}{0.20} \text{ mol} \cdot \text{L}^{-1} = 1.8 \times 10^{-5} \text{ mol} \cdot \text{L}^{-1}$$

$$x^2 = 0.20 \times 1.8 \times 10^{-5} = 0.36 \times 10^{-5} = 3.6 \times 10^{-6}$$

Taking the square root of both sides, we have $x = 1.9 \times 10^{-3}$, and

$$[H_3O^+] = 1.9 \times 10^{-3} \text{ mol} \cdot \text{L}^{-1}$$

||| Always check that the assumption of ignoring x in comparison with the initial acid concentration is valid.

We see that x, or $[H_3O^+]$, is indeed much smaller than 0.20, so our approximation is justified (Appendix M). In general, if $[H_3O^+]$ is 5% or less of the initial acid concentration, the approximation gives a sufficiently accurate value of $[H_3O^+]$. In the present case $[H_3O^+]$ is $(0.0019 \text{ mol} \cdot \text{L}^{-1})/(0.20 \text{ mol} \cdot \text{L}^{-1}) \times 100\% = 1\%$ of the initial acid concentration. In other words, the concentration of acetic acid that is ionized is only 1% of the initial concentration of the acid, meaning that only 1% of the acid is ionized. The **percent ionization** of an acid HA is defined as follows:

$$\text{percent ionization} = \frac{[H_3O^+]}{[HA]_{\text{initial}}} \times 100\%$$

EXAMPLE 12.1 Calculating the Extent of Ionization of a Weak Acid

What is the H_3O^+ concentration in a 0.20-M solution of a hydrocyanic acid, HCN? What is the percent ionization of the acid?

Solution: *First*, we write the equation for the equilibrium:

$$HCN(aq) + H_2O(l) \rightleftharpoons H_3O^+(aq) + CN^-(aq)$$

Second, we write the expression for the equilibrium constant and find the value of K_a from Table 12.1:

$$K_a = \left(\frac{[H_3O^+][CN^-]}{[HCN]}\right)_{eq} = 4.9 \times 10^{-10} \text{ mol} \cdot \text{L}^{-1}$$

Third, we let $x \text{ mol} \cdot \text{L}^{-1} = [H_3O^+]$ at equilibrium. Then we have

	$HCN(aq) + H_2O(l) \rightleftharpoons$	$H_3O^+(aq) +$	$CN^-(aq)$	
Initial concentrations	0.20	0	0	$\text{mol} \cdot \text{L}^{-1}$
Equilibrium concentrations	$0.20 - x$	x	x	$\text{mol} \cdot \text{L}^{-1}$

Substituting into the expression for K_a, we have

$$K_a = \frac{x^2}{0.20 - x} \text{ mol} \cdot \text{L}^{-1} = 4.9 \times 10^{-10} \text{ mol} \cdot \text{L}^{-1}$$

If we assume that $0.20 - x \approx 0.20$,

$$\frac{x^2}{0.20} = 4.9 \times 10^{-10}$$

$$x^2 = 0.98 \times 10^{-10} = 98 \times 10^{-12}$$

Taking the square root of each side, we obtain

$$x = 9.9 \times 10^{-6}$$

Therefore,

$$[H_3O^+] = 9.9 \times 10^{-6}\ mol \cdot L^{-1}$$

We see that our assumption that $x \ll 0.20$ is certainly justified.

Because the concentration of HCN that is ionized is equal to the concentration of $[H_3O^+]$ that is formed, $9.9 \times 10^{-6}\ mol \cdot L^{-1}$, the percent ionization of the acid is

$$\frac{9.9 \times 10^{-6}\ mol \cdot L^{-1}}{0.20\ mol \cdot L^{-1}} \times 100\% = 0.005\%$$

At a concentration of 0.20 *M*, HCN is ionized to only a very small extent, namely, 0.005%. Thus, it is an extremely weak acid. Only 5 molecules in 100,000 are ionized; the rest remain as un-ionized HCN molecules.

Exercise 12.1 What are the H_3O^+ concentration and the percent ionization of HF in a 0.50-*M* solution?

The Base Ionization Constant

Ammonia is a weak base, as are many organic molecules derived from ammonia, such as methylamine, CH_3NH_2, ethylamine, $C_2H_5NH_2$, and aniline, $C_6H_5NH_2$ (Chapter 14). Anions of weak acids are another important type of weak base. For example, fluoride ion, F^-, acetate ion, $CH_3CO_2^-$, and the anions of all other weak acids are weak bases.

A weak base such as ammonia is incompletely protonated by water:

$$NH_3 + H_2O \rightleftharpoons NH_4^+ + OH^-$$

Because all the reactions in this section occur in aqueous solution, we have in most cases omitted the state designations (*aq*) and (*l*).

By analogy with the definition of an acid ionization constant, we may define the **base ionization constant**, K_b, of ammonia as

$$K_b = \left(\frac{[NH_4^+][OH^-]}{[NH_3]}\right)_{eq}, \quad \text{and} \quad pK_b = -\log K_b$$

K_b values for some weak bases are given in Table 12.2.

Strong bases are essentially completely converted to OH^-, and they have very large ionization constants ($K_b \gg 1$). Weak bases have ionization constants that are usually very much less than 1.0 ($K_b \ll 1$). We can see from Table 12.2 that phosphate ion, PO_4^{3-}, is a stronger base than ammonia, but both are weak. The position of equilibrium lies well to the left in both cases, but farther to the left for ammonia than for phosphate ion.

We can find the OH^- concentration in an aqueous solution of a weak base by a method similar to the method we used to find the H_3O^+ concentration in a solution of a weak acid, as Example 12.2 shows.

EXAMPLE 12.2 Calculating the Extent of Ionization of a Weak Base

What is the concentration of OH^- in a 0.10-*M* aqueous solution of ammonia, and what is its percent ionization?

TABLE 12.2 Ionization Constants for Some Bases in Water at 25°C

Base	Proton Transfer Reaction	K_b (mol · L^{-1})	pK_b(= −log K_b)
	Base ⇌ *Conjugate acid*		
Oxide ion	$O^{2-} + H_2O \rightleftharpoons OH^- + OH^-$	Very large	—
Sulfide ion	$S^{2-} + H_2O \rightleftharpoons OH^- + HS^-$	7.7×10^{-2}	1.11
Phosphate ion	$PO_4^{3-} + H_2O \rightleftharpoons OH^- + HPO_4^{2-}$	4.8×10^{-2}	1.32
Ethylamine	$C_2H_5NH_2 + H_2O \rightleftharpoons OH^- + C_2H_5NH_3^+$	4.7×10^{-4}	3.33
Methylamine	$CH_3NH_2 + H_2O \rightleftharpoons OH^- + CH_3NH_3^+$	3.9×10^{-4}	3.41
Carbonate ion	$CO_3^{2-} + H_2O \rightleftharpoons OH^- + HCO_3^-$	2.1×10^{-4}	3.68
Cyanide ion	$CN^- + H_2O \rightleftharpoons OH^- + HCN$	2.0×10^{-5}	4.70
Ammonia	$NH_3 + H_2O \rightleftharpoons OH^- + NH_4^+$	1.8×10^{-5}	4.74
Hydrogenphosphate ion	$HPO_4^{2-} + H_2O \rightleftharpoons OH^- + H_2PO_4^-$	1.6×10^{-7}	6.79
Hydrogencarbonate ion	$HCO_3^- + H_2O \rightleftharpoons OH^- + H_2CO_3$	2.5×10^{-8}	7.60
Aniline	$C_6H_5NH_2 + H_2O \rightleftharpoons OH^- + C_6H_5NH_3^+$	4.3×10^{-10}	9.36
Dihydrogenphosphate ion	$H_2PO_4^- + H_2O \rightleftharpoons OH^- + H_3PO_4$	1.3×10^{-12}	11.88
	Conjugate base ⇌ *Acid*		

Solution: The *first step* is to write the equation for the equilibrium reaction:

$$NH_3(aq) + H_2O(l) \rightleftharpoons NH_4^+(aq) + OH^-(aq)$$

The *second step* is to write the expression for the equilibrium constant and look up its value. From Table 12.1 we see that K_b for NH_3 is 1.8×10^{-5} mol · L^{-1}. Therefore,

$$K_b = \left(\frac{[NH_4^+][OH^-]}{[NH_3]}\right)_{eq} = 1.8 \times 10^{-5} \text{ mol} \cdot \text{L}^{-1}$$

The *third step* is to write expressions for the concentration of each molecule or ion in solution. We let the equilibrium concentration of OH^- be x mol · L^{-1}. Thus,

	NH_3 + H_2O ⇌	NH_4^+ +	OH^-
Initial concentrations	0.10	0	0 mol · L^{-1}
Equilibrium concentrations	$0.10 - x$	x	x mol · L^{-1}

The *fourth step* is to substitute these values into the expression for the equilibrium constant and then solve for x. We have

$$K_b = \left(\frac{[NH_4^+][OH^-]}{[NH_3]}\right)_{eq} = \frac{x^2}{0.10 - x} \text{ mol} \cdot \text{L}^{-1} = 1.8 \times 10^{-5} \text{ mol} \cdot \text{L}^{-1}$$

If we assume that $0.10 - x \approx 0.10$, the equation bcomes

$$\frac{x^2}{0.10} = 1.8 \times 10^{-5}$$

$$x^2 = 1.8 \times 10^{-6}$$

$$x = 1.3 \times 10^{-3}$$

Hence,

$$[OH^-] = 1.3 \times 10^{-3} \text{ mol} \cdot \text{L}^{-1}$$

Thus, $[OH^-]$ is 1.3% of the initial concentration of NH_3, 0.10 mol · L^{-1}, and the assumption that we made in solving the equation was justified. The *percent ionization* of the base is

$$\frac{[OH^-]}{[\text{base}]_{\text{initial}}} \times 100\% = \frac{1.3 \times 10^{-3}}{0.10} \times 100\% = 1.3\%$$

Exercise 12.2 What is the OH^- concentration in a 0.10-*M* solution of methylamine, CH_3NH_2? What is the percent ionization of the methylamine?

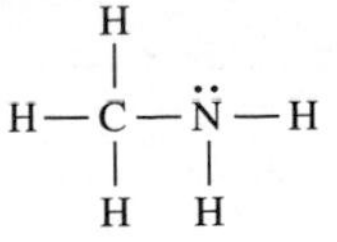

Methylamine

The Autoionization of Water

We saw in Chapter 4 that water can behave as a very weak acid and also as a very weak base. Very small concentrations of H_3O^+ and OH^- therefore form in water by the reaction

$$H_2O(l) + H_2O(l) \rightleftharpoons H_3O^+(aq) + OH^-(aq)$$

in which one water molecule behaves as an acid and the other as a base. This reaction is called the **autoionization**, or **autoprotolysis**, of water. The equilibrium constant for autoionization is

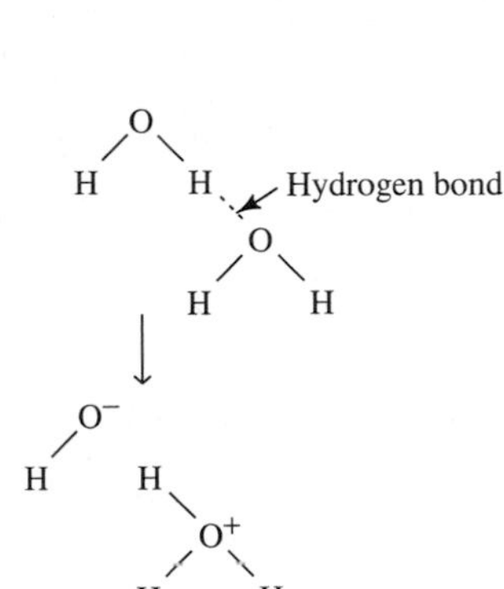

$$K = \left(\frac{[H_3O^+][OH^-]}{[H_2O]^2}\right)_{eq}$$

Because the concentration of water is very nearly constant in any dilute solution, we can multiply K by $[H_2O]^2$ to give a new constant, K_w, called the **autoionization constant of water**:

$$K_w = K[H_2O]^2 = ([H_3O^+][OH^-])_{eq}$$

Measurement of the electrical conductivity of carefully purified water has shown that, at 25°C, $[H_3O^+] = [OH^-] = 1.00 \times 10^{-7}\ mol \cdot L^{-1}$. Thus,

$$K_w = (1.00 \times 10^{-7}\ mol \cdot L^{-1})(1.00 \times 10^{-7}\ mol \cdot L^{-1})$$
$$= 1.00 \times 10^{-14}\ mol^2 \cdot L^{-2} \text{ at } 25°C$$

It is often convenient to use the quantity of $pK_w = -\log K_w$. At 25°C $pK_w = -\log(1.00 \times 10^{-14}) = 14.00$.

Like all equilibrium constants, K_w varies with temperature:
$K_w = 0.29 \times 10^{-14}$ at 10°C
$K_w = 1.00 \times 10^{-14}$ at 25°C
$K_w = 5.46 \times 10^{-14}$ at 50°C

This equilibrium constant applies to pure water and to *any* aqueous solution at 25°C. Therefore, because the product $[H_3O^+][OH^-]$ is always constant at a given temperature, if the concentration of OH^- ions is increased by adding base, the concentration of H_3O^+ ions will decrease and thereby satisfy the equilibrium constant expression. If the concentration of H_3O^+ ions is increased by adding acid, the concentration of OH^- ions will decrease correspondingly.

The pH Scale

It is very convenient to express the hydrogen ion concentration of a solution on a logarithmic scale, namely, the **pH scale**:

$$\mathbf{pH = -\log [H_3O^+]}$$

Because the pH scale is logarithmic, a pH decrease of 1.0 corresponds to a tenfold increase in $[H_3O^+]$.

In a 0.10-*M* solution of a strong acid, such as HCl, $[H_3O^+] = 1.0 \times 10^{-1}\ mol \cdot L^{-1}$, and

$$pH = -\log(1.0 \times 10^{-1}) = -(-1.00) = 1.00$$

In pure water, however, $[H_3O^+] = 1.0 \times 10^{-7}\ mol \cdot L^{-1}$, and

$$pH = -\log [H_3O^+] = -\log(1.0 \times 10^{-7}) = -(-7.00) = 7.00$$

TABLE 12.3 H_3O^+ and OH^- Concentrations and the pH Scale

pH	$[H_3O^+]$	$[OH^-]$	pH Values of Some Common Substances	
0	1	10^{-14}	1-*M* HCl solution	
1	10^{-1}	10^{-13}	Stomach acid	
2	10^{-2}	10^{-12}	Lemon juice Orange juice	Acidic pH < 7
3	10^{-3}	10^{-11}	Wine	
4	10^{-4}	10^{-10}	Soda water Tomato juice	
5	10^{-5}	10^{-9}	Rainwater	
6	10^{-6}	10^{-8}		
7	10^{-7}	10^{-7}	Pure water Milk	Neutral
8	10^{-8}	10^{-6}	Blood Sea water Baking soda solution	
9	10^{-9}	10^{-5}	Borax solution	
10	10^{-10}	10^{-4}	Toilet soap Milk of magnesia	
11	10^{-11}	10^{-3}		Basic pH > 7
12	10^{-12}	10^{-2}	Household ammonia	
13	10^{-13}	10^{-1}		
14	10^{-14}	1	1-*M* NaOH solution	

Pure water—or any solution for which $[H_3O^+] = [OH^-] = 1.0 \times 10^{-7}$ and pH = 7.00—is a *neutral solution.* A solution for which $[H_3O^+] > 10^{-7}$ and pH < 7.0 is an *acidic solution* (Table 12.3). A solution that has $[OH^-] > 10^{-7}$ and therefore $[H_3O^+] < 10^{-7}$ and pH > 7.0 is a basic solution. Table 12.3 gives the pH values of some common acidic and basic solutions.

For the purposes of some calculations it is convenient to define pOH as

$$\mathbf{pOH = -log\,[OH^-]}$$

A basic solution has pOH < 7.

The equilibrium constant expression for the autoionization of water is

$$[H_3O^+][OH^-] = 1.00 \times 10^{-14}\ \text{mol}^2 \cdot \text{L}^{-2}$$

Taking negative logarithms, we have

$$\begin{aligned} -\log\,[H_3O^+] - \log\,[OH^-] &= -\log\,(1.00 \times 10^{-14}) \\ \text{pH} + \text{pOH} &= 14.00 \end{aligned}$$

It is usual to express both the acidity and the basicity of a solution in terms of its pH. If we know the $[OH^-]$ concentration of a basic solution, we can easily find its pOH and hence its pH by using the preceding expression.

EXAMPLE 12.3 pH Calculations: Acid Solutions

What is the pH of a 0.20-*M* solution of acetic acid?

Solution: Acetic acid is a weak acid. We saw just before Example 12.1 that for a 0.20-*M* solution of acetic acid, $[H_3O^+]$ is $1.9 \times 10^{-3}\ mol \cdot L^{-1}$. So

$$pH = -\log(1.9 \times 10^{-3}) = -(-2.72) = 2.72$$

Exercise 12.3 What is the pH of a 0.10-*M* solution of hypochlorous acid, HOCl?

EXAMPLE 12.4 pH Calculations: Basic Solutions

Calculate the pH of the following aqueous solutions.

(a) 0.0100-*M* NaOH **(b)** 0.134-*M* NaOH

Solution: Because NaOH is a strong base, it is fully ionized into Na^+ and OH^-

(a)

$[OH^-] = 1.00 \times 10^{-2}\ mol \cdot L^{-1}$ *so* $pOH = -\log(1.00 \times 10^{-2}) = -(-2.00) = 2.00$

Therefore $pH = 14.00 - 2.00 = 12.00$.

(b) $[OH^-] = 0.134\ mol \cdot L^{-1}$ *or* $pOH = -\log(0.134) = -(-0.87) = 0.87$

$pH = 14.00 - 0.87 = 13.13$

EXAMPLE 12.5 pH Calculations: Basic Solutions

Aniline, $C_6H_5NH_2$, is a much weaker base than ammonia and has an ionization constant $K_b = 4.3 \times 10^{-10}\ mol \cdot L^{-1}$. What is the pH of a 0.010-*M* solution of aniline in water?

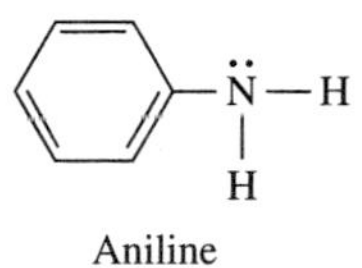

Aniline

Solution: We first write the equation for the equilibrium:

$$C_6H_5NH_2 + H_2O \rightleftharpoons C_6H_5NH_3^+ + OH^-$$

Next, we write the expression for the equilibrium constant:

$$K_b = \left(\frac{[C_6H_5NH_3^+][OH^-]}{[C_6H_5NH_2]}\right)_{eq} = 4.3 \times 10^{-10}\ mol \cdot L^{-1}$$

We then let the equilibrium concentration of OH^- be $x\ mol \cdot L^{-1}$ and write

	$C_6H_5NH_2 + H_2O \rightleftharpoons$	$C_6H_5NH_3^+$	$+ OH^-$
Initial concentrations	0.010	0	$0\ mol \cdot L^{-1}$
Equilibrium concentrations	$0.010 - x$	x	$x\ mol \cdot L^{-1}$

We can now substitute these values into the equilibrium constant expression, so we have

$$\frac{x^2}{0.010 - x}\ mol \cdot L^{-1} = 4.3 \times 10^{-10}\ mol \cdot L^{-1}$$

Because the value of the equilibrium constant is very small, we can reasonably assume that x is very small compared with 0.010. Thus, we have

$$\frac{x^2}{0.010} = 4.3 \times 10^{-10}$$

$$x^2 = 4.3 \times 10^{-12} \quad so \quad x = 2.1 \times 10^{-6}$$

$$[OH^-] = 2.1 \times 10^{-6}\ mol \cdot L^{-1}$$

We see that x is indeed very small compared with 0.010, and therefore our assumption that $0.010 - x \approx 0.010$ was justified. So

$$pOH = -\log(2.1 \times 10^{-6}) = -(-5.68) = 5.68$$

$$pH = 14.00 - 5.68 = 8.32$$

Exercise 12.4 What is the pH of a 0.05-*M* solution of methylamine, CH_3NH_2?

Conjugate Acid-Base Pairs

An acid and a base related by the equation

$$\text{base} + H^+ \rightleftharpoons \text{acid}$$

are called a **conjugate acid–base pair**. For example, acetic acid, CH_3CO_2H and the acetate ion, $CH_3CO_2^-$, constitute a conjugate acid–base pair:

$$CH_3CO_2^- + H^+ \rightleftharpoons CH_3CO_2H$$

An acid and its conjugate base differ only by a proton. When an acid donates a proton, what remains is the acid's conjugate base. When a base accepts a proton, its conjugate acid is formed.

$CH_3CO_2^-$ is the conjugate base of the acid CH_3CO_2H, and CH_3CO_2H is the conjugate acid of the base $CH_3CO_2^-$. Similarly, ammonia, NH_3, and the ammonium ion, NH_4^+, are a conjugate acid–base pair, because

$$NH_3 + H^+ \rightleftharpoons NH_4^+$$

For the ionization of a weak acid, HA, we have

$$HA + H_2O \rightleftharpoons H_3O^+ + A^- \qquad K_a(HA) = \left(\frac{[H_3O^+][A^-]}{[HA]}\right)_{eq} \tag{1}$$

and for its conjugate base A^-,

$$A^- + H_2O \rightleftharpoons HA + OH^- \qquad K_b(A^-) = \left(\frac{[HA][OH^-]}{[A^-]}\right)_{eq} \tag{2}$$

It is important to note that equation (2) is *not* the reverse of equation (1); it is the equation for the reaction of A^- with H_2O and not with H_3O^+. For convenience we will omit the designation $(\ )_{eq}$, but we must remember that these expressions for K_a and K_b are valid only when the brackets [ ] refer to an *equilibrium* concentration.

If we now multiply $K_a(HA)$ and $K_b(A^-)$, we have

$K_a(HA)$ means the acid ionization constant of the acid HA. Do not confuse it with $K_a[HA]$, which means the acid ionization constant K_a *times* the concentration of HA. Similarly, $K_b(A^-)$ means the base ionization constant of the base A^-.

$$K_a(HA)K_b(A^-) = \frac{[H_3O^+][A^-]}{[HA]} \times \frac{[HA][OH^-]}{[A^-]} = [H_3O^+][OH^-]$$

Hence,

$$K_a(HA)K_b(A^-) = [H_3O^+][OH^-] = K_w = 1.00 \times 10^{-14}\ \text{mol}^2 \cdot \text{L}^{-2} \text{ at } 25°\text{C}$$

In general, for any acid–conjugate base pair,

$$K_a(\text{acid})\ K_b(\text{conjugate base}) = K_w = 1.00 \times 10^{-14}\ \text{mol}^2 \cdot \text{L}^{-2}$$

Taking negative logarithms of both sides gives

$$pK_a(\text{acid}) + pK_b(\text{conjugate base}) = pK_w = 14.00$$

The larger K_a is (that is, the stronger the acid–see Table 12.1), the smaller K_b is (that is, the weaker the conjugate base), their product always being equal to K_w. For the strong acid, HCl, for which $K_a \gg 1$, $K_b(Cl^-) \ll 10^{-14}$. In other words, Cl^-, or the anion of any strong acid, is a weaker base than water and therefore does not behave as a base in water (Table 12.4).

It is not necessary to list values of both K_a and K_b for conjugate acid–base pairs, because one can always be obtained from the other. Many reference books

TABLE 12.4 Acid-Base Properties of Some Common Ions

	Cations	Anions
Acidic	H_3O^+, NH_4^+, $Al(H_2O)_6^{3+}$, $Fe(H_2O)_6^{3+}$	HSO_4^-, $H_2PO_4^-$
Neutral	Mg^{2+}, Ca^{2+}, Sr^{2+}, Ba^{2+}, Li^+, Na^+, K^+, Rb^+, Cs^+, Ag^+	NO_3^-, ClO_4^-, Cl^-, Br^-, I^-
Basic	None	PO_4^{3-}, CO_3^{2-}, SO_3^{2-}, F^-, CN^-, OH^-, S^{2-}, $CH_3CO_2^-$, HCO_3^-, NO_2^-, HS^-, HPO_4^{2-}, SO_4^{2-} (very weak, almost neutral)

give only K_a values, but for convenience we have listed both K_a and K_b values for some common acids and bases in Tables 12.1 and 12.2, respectively.

EXAMPLE 12.6 Calculating K_b from K_a for Conjugate Acid–Base Pairs

What is the base ionization constant for the fluoride ion, F^-?

Solution: From Table 12.1, $K_s = 3.5 \times 10^{-4}\ \text{mol} \cdot \text{L}^{-1}$.

$$K_a(\text{HF})K_b(\text{F}^-) = K_w$$

$$K_b(\text{F}^-) = \frac{1.0 \times 10^{-14}\ \text{mol}^2 \cdot \text{L}^{-2}}{3.5 \times 10^{-4}\ \text{mol} \cdot \text{L}^{-1}} = 2.9 \times 10^{-11}\ \text{mol} \cdot \text{L}^{-1}$$

Exercise 12.5 Given that K_b for the carbonate ion, CO_3^{2-}, has a value of $2.1 \times 10^{-4}\ \text{mol} \cdot \text{L}^{-1}$, calculate the value of K_a for its conjugate acid, HCO_3^-.

Acid-Base Properties of Anions, Cations, and Salts

When an acid reacts with a base to give a salt, the acid is often said to *neutralize* the base. So we might think that the solution of a salt would always be neutral, with a pH of 7. Although many salts do give neutral solutions in water, many do not, because many cations and anions behave as acids or bases. The acid–base properties of some common ions are summarized in Table 12.4.

Anions

1. *Conjugate bases of strong acids*, such as Cl^-, do not behave as bases in water. They give neutral solutions.
2. *Conjugate bases of weak acids*, such as CN^- or CO_3^{2-}, are weak bases. They give basic solutions in water. For example,

$$CN^- + H_2O \rightleftharpoons HCN + OH^-$$

3. *Conjugate bases of polyprotic acids* (acids that have two or more ionizable H atoms) may behave as either acids or bases. For example, HSO_4^- and $H_2PO_4^-$ are acids,

HSO_4^- does not act as a base in water, because its conjugate acid H_2SO_4 is a strong acid.

$$HSO_4^- + H_2O \rightleftharpoons SO_4^{2-} + H_3O^+$$
$$H_2PO_4^- + H_2O \rightleftharpoons HPO_4^{2-} + H_3O^+$$

but HPO_4^{2-} and HCO_3^- are bases,

$$HPO_4^{2-} + H_2O \rightleftharpoons H_2PO_4^- + OH^-$$
$$HCO_3^- + H_2O \rightleftharpoons H_2CO_3 + OH^-$$

How do we know whether an anion containing one or more H atoms will behave as an acid or as a base in solution in water? We simply compare its K_a and K_b values (or its pK_a and pK_b values). If $K_a > K_b$ ($pK_a < pK_b$), it behaves as an acid. If $K_b > K_a$ ($pK_b < pK_a$), it behaves as a base. For example, we see from Table 12.1 that $pK_a(H_2PO_4^-) = 7.21$ and from Table 12.2 that $pK_b(H_2PO_4^-) = 11.88$, so $pK_a(H_2PO_4^-) < pK_b(H_2PO_4^-)$. Therefore $H_2PO_4^-$ behaves as an acid and not as a base in aqueous solution.

Exercise 12.6 By comparing the K_a and K_b values for HPO_4^{2-} from Tables 12.1 and 12.2, decide whether HPO_4^{2-} behaves as a base or as an acid in aqueous solution.

Cations

1. *Hydrated metal ions* may behave as acids. For example,

$$Al(H_2O)_6^{3+} + H_2O \rightleftharpoons H_3O^+ + Al(OH)(H_2O)_5^{2+}$$

The charge on the metal ion attracts into the valence shell of the metal atom some of the electron density of an unshared pair of electrons on the oxygen atom. A small positive charge is thereby produced on the oxygen atom. This in turn attracts electrons more strongly from the hydrogen atoms, so they acquire a greater positive charge than in the free water molecule and hence become more acidic. The only common hydrated metal ions that do *not* behave as acids are Li^+, Na^+, K^+, Rb^+, Cs^+, Mg^{2+}, Ca^{2+}, Sr^{2+}, Ba^{2+}, and Ag^+. They give neutral solutions in water. All other hydrated metal ions give acidic solutions in water.

2. *Conjugate acids of weak bases* are weak acids. For example,

$$NH_4^+ + H_2O \rightleftharpoons H_3O^+ + NH_3$$

Most other acidic cations of this type are the conjugate acids of organic bases related to ammonia, such as the conjugate acid of methylamine, $CH_3NH_3^+$.

Salts We can now predict whether a salt will give an acidic, basic, or neutral solution in water by considering the acid–base properties of both its anion and its cation.

Neutral salts contain a neutral cation and a neutral anion. They include salts of Li^+, Na^+, K^+, Rb^+, Cs^+, Mg^{2+}, Ca^{2+}, Sr^{2+}, Ba^{2+}, and Ag^+ with anions of strong acids, such as Cl^- and NO_3^-; for example, KCl, $BaCl_2$, and $AgNO_3$.

Acidic salts contain an acidic cation and a neutral anion or a neutral cation and an acidic anion. They include the following:

- Salts of metal cations, except Li^+, Na^+, K^+, Rb^+, Cs^+, Mg^{2+}, Sr^{2+}, Ba^{2+}, and Ag^+ with anions of strong acids; for example, $AlCl_3$ and $Fe_2(SO_4)_3$.
- Ammonium salts of strong acids; for example, NH_4Cl.
- Some salts of polyprotic acids for example, $NaHSO_4$.

Basic salts contain a neutral cation and a basic anion. They include salts of Li^+, Na^+, K^+, Rb^+, Cs^+, Mg^{2+}, Ca^{2+}, Sr^{2+}, Ba^{2+}, and Ag^+ with anions of weak acids, such as CN^-, F^-, and CO_3^{2-}; for example, NaCN, KF, and Na_2CO_3.

For a salt of an acidic cation, such as NH_4^+, and a basic anion, such as CN^-, we cannot predict whether the solution will be acidic, basic, or neutral unless we know K_a for the cation and K_b for the anion. If $K_a(\text{cation}) > K_b(\text{anion})$, the solution will be acidic, and if $K_b(\text{anion}) > K_a(\text{cation})$, the solution will be basic.

EXAMPLE 12.7 Acid–Base Properties of Salts

Predict whether the following salts give acidic, basic, or neutral solutions when dissolved in water.

(a) NaBr **(b)** K_2CO_3 **(c)** $AlCl_3$ **(d)** NH_4ClO_4 **(e)** $(NH_4)_2S$

Solution:

(a) NaBr	Neutral cation, neutral anion	Therefore the solution is neutral.
(b) K_2CO_3	Neutral cation, basic anion	Therefore the solution is basic.
(c) $AlCl_3$	Acidic cation, neutral anion	Therefore the solution is acidic.
(d) NH_4ClO_4	Acidic cation, neutral anion	Therefore the solution is acidic.
(e) $(NH_4)_2S$	Acidic cation, basic anion	We cannot make a prediction without information on $K_a(NH_4^+)$ and $K_b(S^{2-})$.

Exercise 12.7 Predict whether the pH of an aqueous solution of each of the following salts is less than, greater than, or equal to 7.

(a) $FeCl_3$ **(b)** NH_4NO_3 **(c)** NaH_2PO_4 **(d)** K_2HPO_4

If we know the appropriate K_a or K_b value, we can calculate the pH of a solution of a salt, as Example 12.8 shows.

EXAMPLE 12.8 pH of a Salt Solution

What is the pH of a 0.10-*M* solution of sodium cyanide?

Solution: The *first step* is to recognize whether the cation or the anion reacts with water and then write the equation for that reaction. We saw previously that sodium, like the other alkali metal cations, is one of the cations that is not sufficiently strongly hydrated to give an acidic solution in water—it gives a neutral solution. But cyanide ion is the conjugate base of the weak acid HCN, and so it is a weak base. It therefore reacts with water according to the equation

$$CN^- + H_2O \rightleftharpoons HCN + OH^-$$

The *second step* is to write the equilibrium constant expression and look up the value of K_b in Table 12.2 (or calculate it from K_a, Table 12.1):

$$K_b = \frac{[HCN][OH^-]}{[CN^-]} = 2.0 \times 10^{-5}\ \text{mol} \cdot \text{L}^{-1}$$

The *third step* is to write an expression for the concentration of each molecule or ion in solution. Because NaCN is a salt and is fully ionized in aqueous solution, the initial concentrations of both Na^+ and CN^- are 0.10-*M*. If we let $[OH^-] = x\ mol \cdot L^{-1}$, we have

$$CN^- + H_2O \rightleftharpoons HCN + OH^-$$

Initial concentrations	0.10	0	0	$mol \cdot L^{-1}$
Equilibrium concentrations	$0.10 - x$	x	x	$mol \cdot L^{-1}$

Hence,

$$\frac{x^2}{0.10 - x}\ mol \cdot L^{-1} = 2.0 \times 10^{-5}\ mol \cdot L^{-1}$$

Because K_b is small, we assume that $x \ll 0.10$, and hence $0.10 - x \approx 0.10$. Thus, we have

$$\frac{x^2}{0.10} = 2.0 \times 10^{-5}$$

$$x^2 = 2.0 \times 10^{-6}$$

$$x = 1.4 \times 10^{-3}$$

$$[OH^-] = 1.4 \times 10^{-3}\ mol \cdot L^{-1}$$

Because x is less than 5% of the initial concentration of CN^-, the assumption that $0.10 - x \approx 0.10$ was justified. Thus,

$$pOH = -\log(1.4 \times 10^{-3}) = -(-2.85) = 2.85$$

$$pH = 14.00 - 2.85 = 11.15$$

Exercise 12.8 What is the pH of a 0.05-*M* solution of NH_4Cl?

Buffer Solutions

Buffer solutions. The packets contain a solid mixture of an acid and its conjugate base that, when added to a suitable amount of water, produces a buffer solution with the pH stated on the packet.

In many industrial processes and in many reactions that occur in living organisms, the pH must be held constant. Many reactions in organisms are catalyzed by enzymes, protein molecules that have well-defined shapes with cavities that can accommodate reactant molecules (Chapter 17). Proteins can be both acids and bases, changing shape as they add or lose protons. Thus, an enzyme is an effective catalyst only at the pH at which it has the necessary shape. Living organisms have developed their own methods for maintaining a constant pH. The normal pH of blood is maintained close to 7.4 (Box 12.1). But kidney failure or a disease such as emphysema can cause a significant deviation from this value and can lead to serious consequences, even death.

A solution that has the ability to maintain a very nearly constant pH even when moderate amounts of H_3O^+ or OH^- are added is called a **buffer solution**.

A buffer solution contains approximately equal amounts of either a weak acid and a salt of the weak acid or a weak base and a salt of the weak base—in other words, a conjugate acid–base pair.

For example, a buffer solution prepared from equal amounts of acetic acid and sodium acetate contains equal concentrations of CH_3CO_2H and $CH_3CO_2^-$. If H_3O^+ is added to the solution, it combines almost completely with $CH_3CO_2^-$ to form CH_3CO_2H:

$$CH_3CO_2^- + H_3O^+ \rightarrow CH_3CO_2H + H_2O \qquad (3)$$

BOX 12.1 Body Fluids: pH and Buffers

Fluids make up a significant portion of the human body. A person who weighs 70 kg contains about 50 L of body fluids, including blood, gastric fluids, saliva, and urine. Each of these body fluids is normally maintained within a narrow pH range by buffers. The rates of most reactions occurring in the body are controlled by catalysts called enzymes. Enzymes are proteins whose shape and ability to function as catalysts are strongly dependent on pH (Chapter 17). For the rates of reactions in the body to be closely controlled, the pH of body fluids must also be closely controlled.

The principal function of blood is to transport substances throughout the body (Figure A). The normal pH range of blood is 7.35 to 7.45. The main clinical effect of too high an H^+ concentration (too low a pH)—a condition called *acidosis*—is the depression of the central nervous system, which eventually results in a comatose state. Too low an H^+ concentration (too high a pH)—a condition called *alkalosis*—causes overexcitability, muscle spasms, and convulsions. Death results rapidly if acidosis and alkalosis are not treated and the blood pH falls below 6.8 or above 8.0.

The pH of blood is maintained within a very narrow range by buffers. There are three major buffer systems in blood: H_2CO_3/HCO_3^-, $H_2PO_4^-/HPO_4^{2-}$, and protein buffers. Proteins are discussed in Chapter 17. Polymers of amino acids, they have several ionizable hydrogen atoms and can therefore behave as buffers, just like any acid and its conjugate base.

Acidosis can result from metabolic or respiratory malfunctions. One of the most common causes of metabolic acidosis is diabetes mellitus, a disease in which large amounts of acidic substances such as hydroxybutyric acid are produced. In severe diarrhea, substantial quantities of HCO_3^- are lost from the intestines, increasing the H_2CO_3/HCO_3^- ratio and thereby producing acidosis. Various diseases that affect the lungs, such as pneumonia and emphysema, decrease the ability of the lungs to eliminate CO_2. When the CO_2 concentration builds up, the equilibrium concentration of H_2CO_3 increases, decreasing the pH and also producing acidosis.

A common cause of metabolic alkalosis is severe vomiting, which results in the loss of the very acidic gastric juices. Another is extremely rapid breathing—hyperventilation—as a result of drugs, fright, or hysteria, for example. Hyperventilation causes excessive amounts of CO_2 to be released by the lungs, thereby decreasing the equilibrium concentration of H_2CO_3 and producing alkalosis.

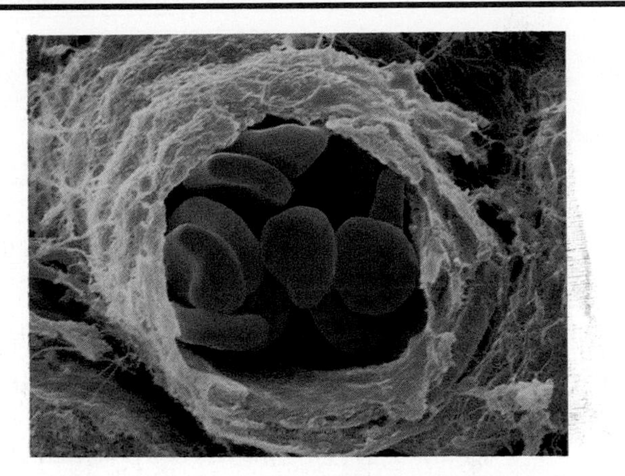

Figure A *False-color scanning electron micrograph of red blood cells passing through a small branch of an artery (about 1740× magnification)*

The stomach is lined with cells that secrete hydrochloric acid, which is needed for digestion. Stomach acids have a concentration of about 0.1 mol · L^{-1} and therefore have a pH of about 1. If stomach cells secrete too much acid, as they sometimes do when we are under stress or we eat foods that contain or produce too much acid, indigestion results. If the stomach acids move into the esophagus, which lacks a protective lining, we get heartburn. The digestion of food begins in the mouth, where the salivary glands secrete approximately 1 to 2 L of saliva per day. Saliva moistens and lubricates food in preparation for its trip to the stomach. Saliva contains enzymes that catalyze the breakdown of carbohydrates and the HCO_3^-/H_2CO_3 buffer, which removes acids from food and acids produced by bacteria in the mouth. The removal of acid by saliva helps prevent tooth decay.

Acids produced in the course of metabolism are removed from the blood in the kidneys. About 1.2 L of blood per minute circulates through the kidneys, and the metabolic acids are then excreted in the urine. Urine is normally acidic, with a pH in the range of 5.5 to 6.5. The kidneys therefore regulate the pH of blood and are the body's most powerful acid–base regulatory mechanism.

Therefore, there is almost no change in the H_3O^+ concentration. Why does this reaction go almost to completion? It is the reverse of the ionization of acetic acid. We normally think of starting with acetic acid, and so we write the equation as

$$CH_3CO_2H + H_2O \rightleftharpoons CH_3CO_2^- + H_3O^+ \qquad (4)$$

At equilibrium there are only very small concentrations of $CH_3CO_2^-$ and H_3O^+,

For any reaction

$A + B \rightleftharpoons C + D, K_f = \frac{[C][D]}{[A][B]}$

If we write the reaction in reverse,

$C + D \rightleftharpoons A + B, K_r = \frac{[A][B]}{[C][D]}$

In general, $K_r = 1/K_f$.

and most of the acetic acid remains undissociated. However, if we start with $CH_3CO_2^-$ and H_3O^+, we must reach the same equilibrium with a large concentration of CH_3CO_2H and very small concentrations of $CH_3CO_2^-$ and H_3O^+. In other words, the position of the equilibrium for reaction (3) is far to the right, so this reaction goes essentially to completion, as is shown by its large equilibrium constant:

$$K(3) = \left(\frac{[CH_3CO_2H]}{[H_3O^+][CH_3CO_2^-]}\right)_{eq} = \frac{1}{K_a(CH_3CO_2H)}$$

$$= \frac{1}{1.8 \times 10^{-5} \text{ mol} \cdot \text{L}^{-1}} = 5.6 \times 10^4 \text{ L} \cdot \text{mol}^{-1}$$

If OH^- is added to an acetic acid–acetate buffer solution, it combines almost completely with acetic acid and converts it to acetate ion:

$$CH_3CO_2H + OH^- \rightarrow CH_3CO_2^- + H_2O \quad (5)$$

So again there is almost no change in the OH^- (or H_3O^+) concentration or therefore in the pH. This reaction goes essentially to completion because it is the reverse of the reaction of acetate ion with water, that is, the reaction of the acetate ion behaving as a base. The equilibrium constant for this reaction,

$$K(5) = \left(\frac{[CH_3CO_2^-]}{[CH_3CO_2H][OH^-]}\right)_{eq} = \frac{1}{K_b(CH_3CO_2^-)} = 1.8 \times 10^9 \text{ L} \cdot \text{mol}^{-1}$$

again shows that the position of the equilibrium lies far to the right.

To make up a buffer solution to maintain a given pH, we need to know how to calculate the pH of a given buffer solution. For the equilibrium between a weak acid and its conjugate base, we can write the equation

$$HA + H_2O \rightleftharpoons A^- + H_3O^+$$

$$K_a = \left(\frac{[H_3O^+][A^-]}{[HA]}\right)_{eq}$$

This equation can be rearranged to

$$[H_3O^+] = \frac{K_a[HA]}{[A^-]}$$

Taking negative logarithms of both sides, we obtain

$$-\log [H_3O^+] = -\log K_a - \log \frac{[HA]}{[A^-]}$$

$$pH = pK_a - \log \frac{[HA]}{[A^-]} = pK_a + \log\left(\frac{[A^-]}{[HA]}\right)_{eq}$$

We have seen that the equilibrium concentrations of HA and A^- do not differ significantly from the concentrations of the weak acid and its conjugate base used to make up the buffer solution. Hence, to a very good approximation we may write

$$pH = pK_a + \log\left(\frac{[A^-]}{[HA]}\right)_{initial} \quad \textit{or} \quad pH = pK_a + \log \frac{[base]}{[acid]}$$

Here [acid] and [base] are the concentrations of acid and conjugate base or conjugate acid and base used to make up the buffer solution. This equation is called the **Henderson–Hasselbalch equation**.

The ratio [base]/[acid] has no units. Hence, we can often conveniently replace it by the ratio (moles of base)/(moles of acid) and write

$$\text{pH} = \text{p}K_a + \log\left(\frac{\text{mol B}}{\text{mol BH}^+}\right) \quad \textit{or} \quad \text{pH} = \text{p}K_a + \log\left(\frac{\text{mol A}^-}{\text{mol HA}}\right)$$

Here the ratio (moles of base)/(moles of acid) refers to the amounts of weak acid and conjugate base or weak base and conjugate acid used to make up the buffer solution. The pH of the solution can then be calculated, as illustrated in Example 12.9.

EXAMPLE 12.9 pH of a Buffer Solution

What is the pH of a buffer solution prepared from 25.00 mL of 0.100-*M* $NH_3(aq)$ and 15.00 mL of 0.200-*M* $NH_4Cl(aq)$?

Solution: The equilibrium in this solution is between the conjugate acid–base pair NH_4^+ and NH_3:

$$\underset{\text{Base}}{NH_3} + H^+ \rightleftharpoons \underset{\text{Acid}}{NH_4^+}$$

The Henderson–Hasselbalch equation then gives

$$\text{pH} = \text{p}K_a(NH_4^+) + \log\frac{[NH_3]}{[NH_4^+]} \quad \textit{or} \quad \text{pH} = \text{p}K_a(NH_4^+) + \log\frac{\text{mol } NH_3}{\text{mol } NH_4^+}$$

$$\text{moles } NH_3 = (0.025\text{ L})(0.100\text{ mol}\cdot\text{L}^{-1}) = 0.0025\text{ mol } NH_3$$

$$\text{moles } NH_4^+ = (0.015\text{ L})(0.200\text{ mol}\cdot\text{L}^{-1}) = 0.0030\text{ mol } NH_4^+$$

From Table 12.1, $\text{p}K_a(NH_4^+) = 9.25$. Hence,

$$\text{pH} = 9.25 + \log\frac{0.0025}{0.0030} = 9.25 - 0.08 = 9.17$$

You can work with moles of NH_3 and NH_4^+ rather than concentrations of NH_3 and NH_4^+ here because the ratio of moles equals the ratio of concentrations.

Exercise 12.9 Calculate the pH of each of the following buffer solutions.

(a) A solution made by dissolving 0.500 g of sodium acetate in 50.00 mL of 0.120-*M* acetic acid.

(b) A solution obtained by mixing 15.00 mL of 0.150-*M* $HCl(aq)$ with 25.00 mL of 0.180-*M* $NH_3(aq)$. You should first calculate the moles of $NH_4Cl(aq)$ and the moles of $NH_3(aq)$ in the solution.

Le Châtelier's Principle

Buffer solutions illustrate a useful qualitative rule called **Le Châtelier's principle** that was first proposed by the French chemist Henri Le Châtelier (1850–1936) in 1884. It can be stated as follows:

> **If any of the conditions affecting a system in dynamic equilibrium is changed such that the system is no longer at equilibrium, when equilibrium is again established, the original conditions are as far as possible restored.**

An equivalent briefer statement is

> **A dynamic equilibrium tends to oppose any change in the conditions.**

Le Châtelier's principle provides no explanation of the effect of changing conditions on an equilibrium; it tells us merely the qualitative result.

Among the conditions affecting an equilibrium that can be changed are

1. The concentrations of the reactants and products;
2. The temperature;
3. The pressure in gas-phase reactions.

Buffer solutions provide a good example of the effect of changing the concentration of one of the molecules or ions participating in an equilibrium in solution. If we add acid to a buffer solution and thereby increase the H_3O^+ concentration, reaction occurs, and a new equilibrium is established in which the concentration of H_3O^+ is only very slightly greater than the original H_3O^+ concentration. In general, if we increase the concentration of any of the molecules or ions at equilibrium, reaction occurs so as to give a new equilibrium in which the concentration that has been increased is reduced toward its original value. Conversely, if we decrease the concentration of any molecule or ion at equilibrium, reaction occurs to give a new equilibrium in which this concentration is increased toward its original value. If we add base to a buffer solution, thereby reducing the H_3O^+ concentration, a new equilibrium is established in which the H_3O^+ concentration is only very slightly less than the original concentration. Demonstration 12.1 shows the effect of concentration changes on the equilibrium between the complex ions $Co(H_2O)_6^{2+}$ and $CoCl_4^{2-}$.

DEMONSTRATION 12.1 Le Châtelier's Principle

The dish contains a concentrated aqueous solution of blue $CoCl_4^{2-}$ ions. They are in equilibrium with pink $Co(H_2O)_6^{2+}$ ions.

$$\underset{\text{Blue}}{CoCl_4^{2-}} + 6H_2O \rightleftharpoons \underset{\text{Pink}}{Co(H_2O)_6^{2+}} + 4Cl^-$$

As water is added, the equilibrium shifts to the right, and the blue solution becomes pink.

Sufficient water has been added to turn the solution completely pink.

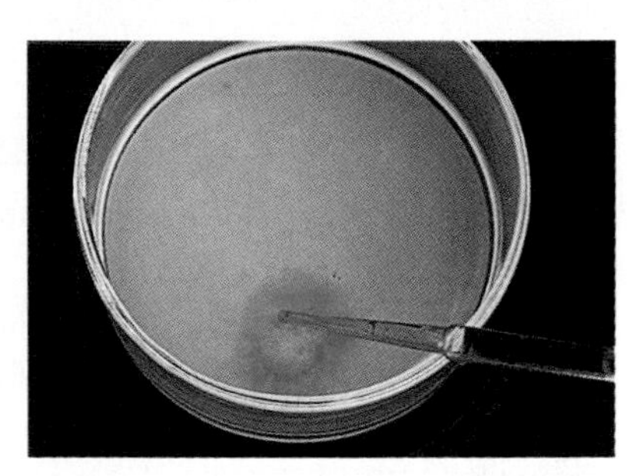

When a concentrated solution of chloride ion is added, the equilibrium shifts back to the left, and the solution again becomes blue.

The formation of pink $Co(H_2O)_6^{2+}$ is an exothermic reaction. Thus, when the solution is cooled, the equilibrium shifts to the right, and the solution turns pink. When the solution is heated, the equilibrium shifts to the left, and the solution turns blue.

We consider the effects of temperature and pressure on an equilibrium in Section 12.3.

The Reaction Quotient, *Q*

Starting with any system that is not at equilibrium, we can find which way a reaction will proceed to reach equilibrium by comparing the value of the **reaction quotient**, *Q*, with that of the equilibrium constant, *K*. The expression for the reaction quotient is the same as that for the equilibrium constant except that the concentrations are any concentrations other than the equilibrium concentrations—for example, the concentrations at the start of the reaction. For instance, when we add acid to a buffer solution containing ammonia and an ammonium salt,

$$NH_4^+ + H_2O \rightleftharpoons NH_3 + H_3O^+$$

$[H_3O^+]_{initial}$ is greater than $[H_3O^+]_{eq}$, and

$$\left\{Q = \left(\frac{[H_3O^+][NH_3]}{[NH_4^+]}\right)\right\} > \left\{K = \left(\frac{[H_3O^+][NH_3]}{[NH_4^+]}\right)_{eq}\right\}$$

For the system to reach equilibrium, $[H_3O^+]$ and $[NH_3]$ must decrease and $[NH_4^+]$ increase until $Q = K$. In other words, the reverse reaction occurs, and the position of the equilibrium shifts to the left. In general,

- If $Q > K$, reaction will occur such that the concentrations of the products decrease and the concentrations of the reactants increase until equilibrium is reached.
- If $Q < K$, reaction will occur such that the concentrations of the products increase and the concentrations of the reactants decrease until equilibrium is reached.
- If $Q = K$, the reactants and products are at equilibrium, and there will be no change in the concentrations of the reactants or products.

The Measurement of pH

The rate at which many reactions in aqueous solution reach equilibrium and the position of the equilibrium depend on the hydrogen (hydronium) ion concentration, that is, on the pH of the solution. To understand such reactions and to control them, it is important to be able to measure the pH of a solution.

The concept of pH was invented in 1909 by the Danish chemist Søren Sørensen while he was working on methods by which to determine and control the acidity of beer during brewing at the Carlsberg Brewery in Copenhagen.

Taste The taste sensors on the human tongue are very sensitive to the hydrogen ion concentration, and the sour taste of acids was one of the earliest properties of acids to be recognized. We can detect a definite sour, or acidic, taste in a solution with a pH between 4 and 5. Soda water, a solution of CO_2 in water, has a pH of about 4. Most fruit juices and soft drinks contain weak acids, such as citric acid and have a pH in the range 2 to 3. Lemon juice, with a pH of 2.8, has a sharper, distinctly more acidic taste than orange juice, with a pH of about 3.6. Basic solutions with pH greater than 7.0 have a bitter taste. (Concentrated acid and base solutions are very corrosive and should not be tasted.)

CH_2CO_2H
HO—C—CO_2H
CH_2CO_2H

Citric acid

Indicators The *approximate* pH of a solution can be easily and conveniently determined by the use of indicators.

An *indicator* is a weak acid that has a conjugate base of a different color.

DEMONSTRATION 12.2 A Natural Indicator

The compound that gives the color to a red rose dissolves in methanol to give a red solution, leaving the rose in the beaker on the left a pale pink color.

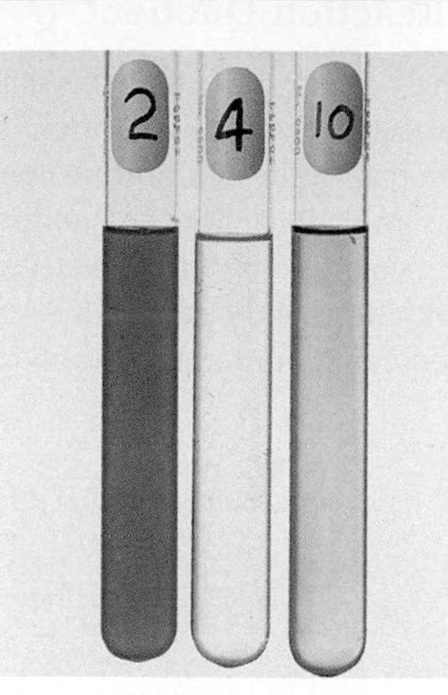

The rose extract can be used as an indicator. A small amount of the extract has been added to each of these tubes. The tube on the left is an aqueous solution at pH = 2, the middle tube is at pH = 4, and the right-hand tube is at pH = 10.

Litmus, which is extracted from certain lichens, is an indicator that has been used for centuries. In basic solutions, in which litmus is in its conjugate base form, it is blue. In acidic solutions, in which litmus is in its acid form, it is red. We saw in Demonstration 4.7 that grape juice, tea, and red cabbage also behave as indicators. Demonstration 12.2 shows that the substance that gives a red rose its color is an indicator. Two common indicators used in the laboratory are phenolphthalein and methyl red (Figure 12.1).

The pH at which the color change occurs depends on the indicator. It is determined by the position of the equilibrium between the acid form of the indicator, denoted by HIn, and its conjugate base, denoted by In^-:

$$HIn + H_2O \rightleftharpoons In^- + H_3O^+$$

According to Le Châtelier's principle, this equilibrium is shifted to the left if the H_3O^+ concentration is increased, for example, by adding acid to the solution. If the indicator is phenolphthalein, it is then very largely in its colorless form, HIn. But if the solution is made basic, the H_3O^+ concentration is much reduced, the equilibrium shifts to the right, and the indicator is converted almost entirely to the In^- form, which in the case of phenolphthalein is pink. Thus, we can distinguish between an acidic solution and a basic solution by adding a small amount of phenolphthalein. The solution will be pink if it is basic but colorless if it is acidic.

The exact pH at which the color change of an indicator occurs can be found from the acid dissociation constant of the indicator:

$$K_a(HIn) = \frac{[In^-][H_3O^+]}{[HIn]}$$

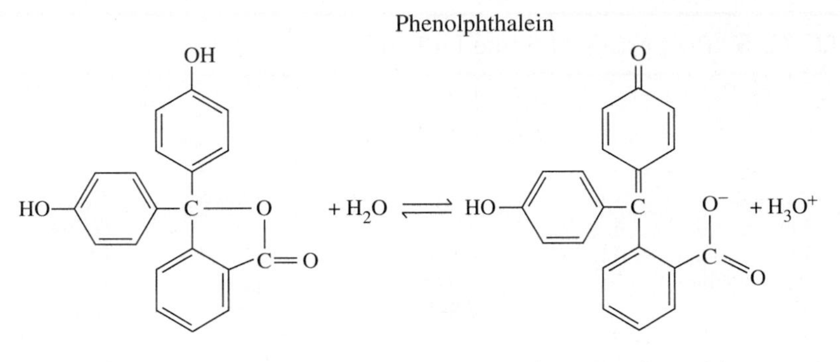

Acid form: colorless

Conjugate base form: pink

Methyl red

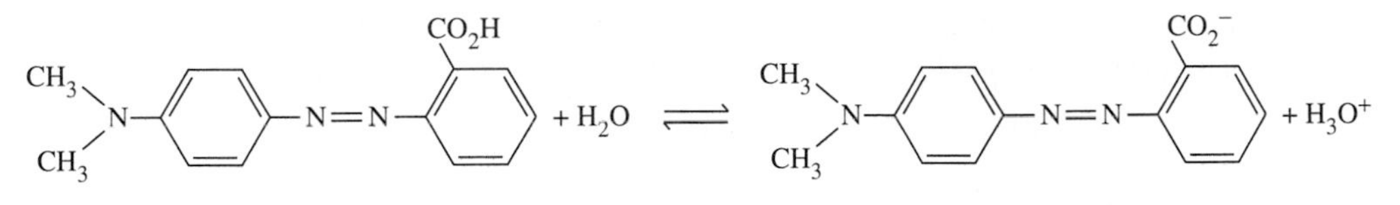

Acid form: red

Conjugate base form: yellow

FIGURE 12.1 The Indicators Phenolphthalein and Methyl Red The acid and conjugate-base forms of an indicator have different colors, so the indicator changes color as the position of the equilibrium shifts with changing pH.

Rearranging this equation, we have

$$[H_3O^+] = K_a \frac{[HIn]}{[In^-]}$$

Taking negative logarithms of both sides gives

$$-\log[H_3O^+] = -\log K_a - \log \frac{[HIn]}{[In^-]}$$

$$pH = pK_a - \log \frac{[HIn]}{[In^-]} = pK_a + \log \frac{[In^-]}{[HIn]}$$

$$pH = pK_a(HIn) + \log \frac{[In^-]}{[HIn]}$$

is a special case of the Henderson–Hasselbalch equation,

$$pH = pK_a + \log \frac{[base]}{[acid]}$$

The indicator will be half in its acid form and half in its base form—in other words, in the middle of its color change—when

$$[HIn] = [In^-] \qquad or \qquad \frac{[In^-]}{[HIn]} = 1$$

The pH will then be

$$pH = pK_a + \log 1 = pK_a + 0 = pK_a$$

Hence the pH at which an indicator changes color depends on its pK_a. For phenolphthalein $K_a = 3 \times 10^{-10}$ mol · L^{-1}, so the pH at which it is half in its base form and half in its acid form is

$$pH = pK_a = -\log(3 \times 10^{-10}) = 9.5$$

Thus, we expect phenolphthalein to change color near pH 9.5. In fact, the color change occurs over a range of pH, and this range is partly determined by the sensi-

TABLE 12.5 Properties of Some Indicators

	pK_a	Effective pH Range	Color	
			Acid Form	Base Form
Methyl violet	1.6	0.0–3.0	Yellow	Violet
Methyl orange	4.2	3.3–4.6	Red	Yellow
Methyl red	5.2	4.2–6.2	Red	Yellow
Bromothymol blue	7.1	6.0–7.8	Yellow	Blue
Thymol blue	8.2	7.9–9.4	Yellow	Blue
Phenolphthalein	9.5	8.3–10.0	Colorless	Red
Alizarine yellow	11.0	10.1–12.1	Yellow	Red

tivity of the eye to the color change. In general, when $[In^-]/[HIn]$ or $[HIn]/[In^-]$ reaches a value of approximately 10, the eye cannot detect any further color change. Therefore, the visible color change occurs approximately over the range

$$pH = pK_a(HIn) \pm \log 10 \qquad or \qquad pH = pK_a(HIn) \pm 1$$

This change is half complete when $pH = pK_a(HIn)$. Methyl red, which has a pK_a of 5.2, is clearly red in a solution of pH 4.2 and clearly yellow in a solution of pH 6.2. As the acidity increases over this range, the indicator changes color from yellow through orange to red.

Table 12.5 lists several indicators together with their useful pH ranges. The colors of some of these indicators are shown in Figure 12.2. Notice that in general indicators do not change color at pH 7. If we compare the color of a solution of unknown pH containing a suitable indicator with the colors of that indicator in a number of solutions of known pH, we can make a fairly accurate determination of the pH. The smallest amount of indicator that will give a clearly observable color should be used. Adding too much indicator disturbs the acid–base equilibrium in the solution and changes the pH.

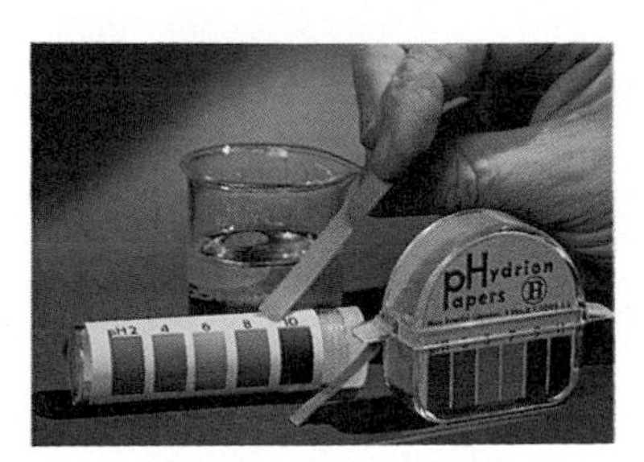

The color produced when a strip of pH paper is dipped into a solution gives an approximate value for the pH of the solution.

Universal indicator is a mixture of several indicators that has several color changes over a wide pH range. With universal indicator, we can find the approximate pH of any solution within this range. So-called *pH paper* is impregnated with universal indicator. When a strip of this paper is immersed in a solution, the pH can be judged from the resulting color.

The pH Meter For accurate pH measurements, *a pH meter* is used (Figure 12.3). The principles on which the pH meter operates are described in Chapter 13. Demonstration 12.3 shows the measurement of the pH of some familiar solutions.

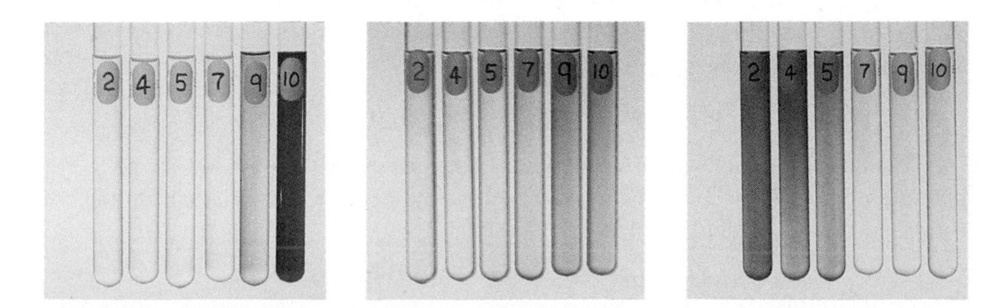

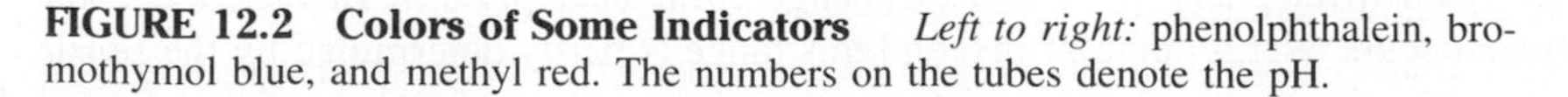
FIGURE 12.2 Colors of Some Indicators *Left to right:* phenolphthalein, bromothymol blue, and methyl red. The numbers on the tubes denote the pH.

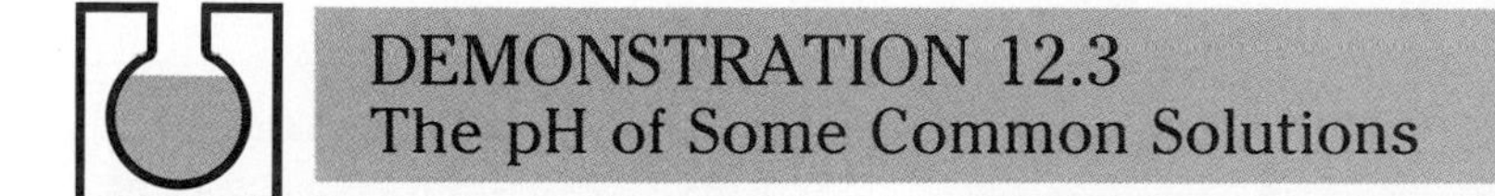

DEMONSTRATION 12.3
The pH of Some Common Solutions

The color of the universal indicator shows that solutions of lime juice in water, soda water, and vinegar are all acidic.

The color of the universal indicator shows that Drāno, household ammonia, and Milk of Magnesia are all basic (alkaline).

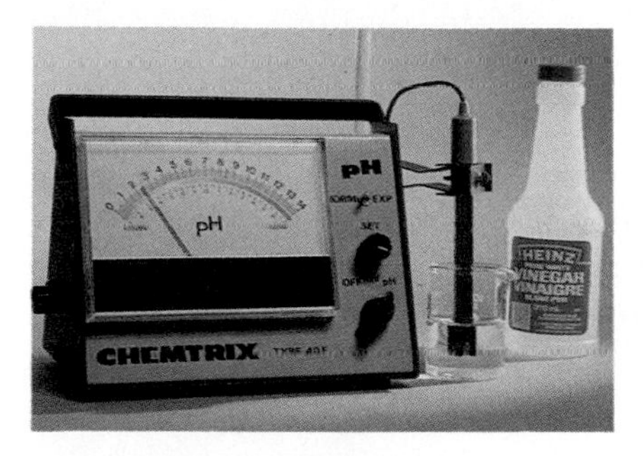

We can measure pH more accurately with a pH meter. Vinegar has a pH of 2.3.

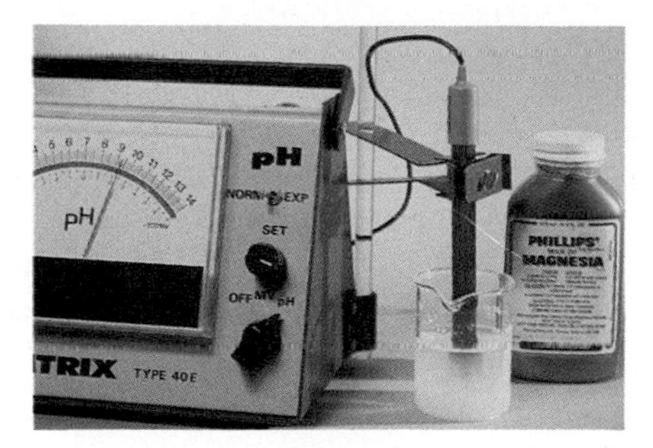

A suspension of Milk of Magnesia in water has a pH of 9.4.

Titrations

Indicators are commonly used in acid–base titrations. As we described in Chapter 5, the *equivalence point* in a titration is the point at which just enough base has been added to react completely with all the acid, or vice versa. At the equivalence point, we have a solution of the salt formed by the reaction. The pH at this point depends on the nature of the salt. We have seen that there are three possibilities:

Salt Formed at Equivalence Point	*Solution*	*pH*
Strong-acid/Strong-base	Neutral	7
Strong-acid/Weak-base	Acidic	< 7
Weak-acid/Strong-base	Basic	> 7

We can calculate the pH of a solution of the salt formed in a titration by the method of Example 12.8. To detect the equivalence point in an acid–base titration, we choose an indicator that changes color near this pH. The point at which the indicator changes color is called the *end point* of the titration. We want the experimentally observed end point to be as close as possible to the equivalence point. Fortunately, we need not be very precise in this choice, because the pH

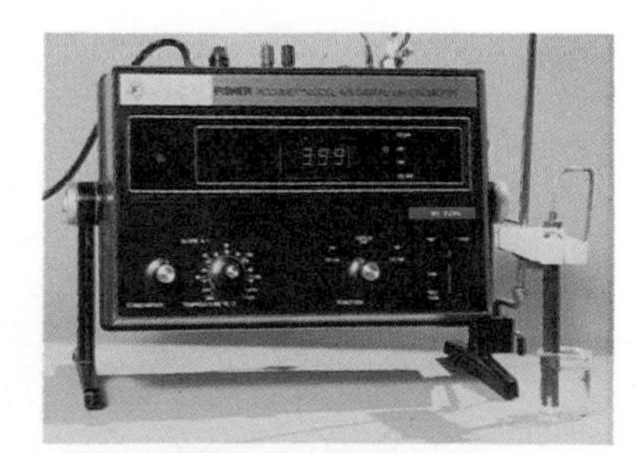

FIGURE 12.3 A pH Meter
A pH meter employs two electrodes to measure the potential difference between the solution to be tested and a standard solution of known pH. This potential difference is related to the pH of the solution being tested. The pH is read directly from the display panel of the instrument.

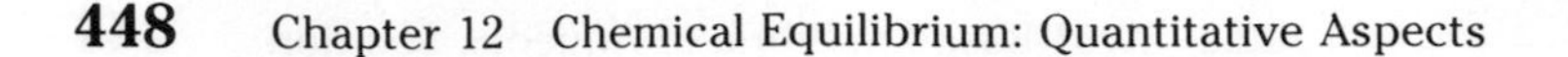

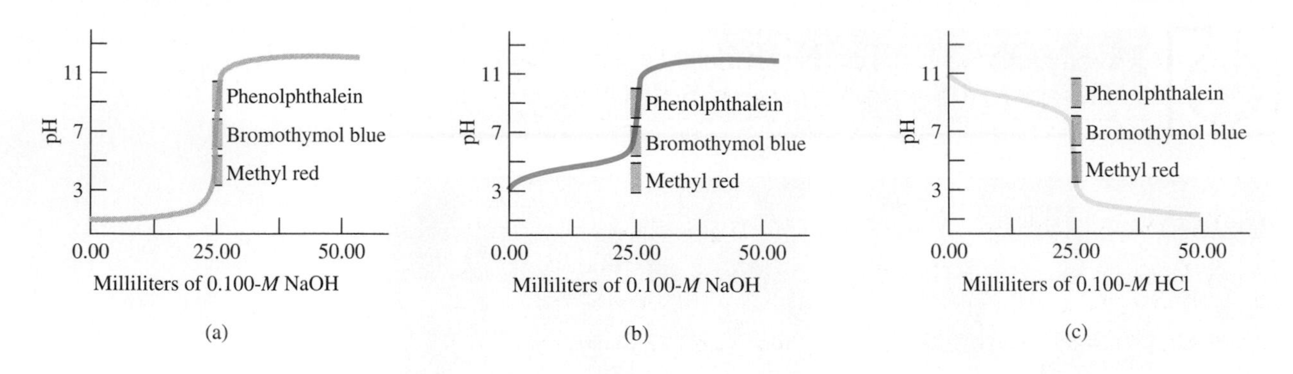

FIGURE 12.4 Change in pH in Acid-Base Titrations The range over which the indicators change color is shown in each case. (a) Strong-acid/strong-base titration: 25.00 mL of 0.100-*M* HCl with 0.100-*M* NaOH. Any of the three indicators can be used. (b) Weak-acid/strong-base titration: 25.00 mL of 0.100-*M* CH_3CO_2H with 0.100-*M* NaOH. Phenolphthalein is the best choice of the three indicators. (c) Strong-acid/weak-base titration: 25.00 mL of 0.100-*M* NH_3 with 0.100-*M* HCl. Methyl red is the best choice of the three indicators.

changes very rapidly near the end point (Figure 12.4). There is a wide choice of indicators for a strong-acid/strong-base titration. For most weak-acid/strong-base titrations, which have pH > 7 at the equivalence point, phenolphthalein ($pK_a = 9.5$) is a suitable choice, and for a weak-base/strong-acid titration, methyl red ($pK_a = 5.2$) can be used (Table 12.5).

Determination of K_a and K_b

One important application of pH measurements is the determination of the ionization constants of acids and bases. By measuring the pH of a solution of an acid or base of known concentration, the value of K_a or K_b can be calculated, as shown in Examples 12.10 and 12.11.

A monoprotic acid has only one ionizable hydrogen atom, whereas a diprotic acid has two ionizable hydrogen atoms, and a triprotic acid has three.

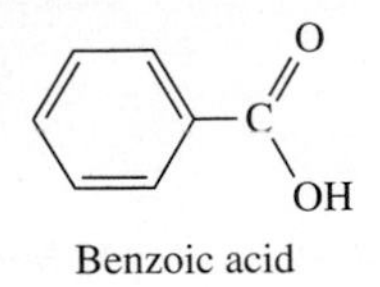

Benzoic acid

Use the *equilibrium* concentrations, not the *initial* concentrations, in evaluating equilibrium constants.

EXAMPLE 12.10 Determination of K_a from pH

A 0.125-*M* solution of benzoic acid, $C_6H_5CO_2H$, a monoprotic acid, has a pH of 2.56. What is the value of K_a for benzoic acid?

Solution: We will represent the formula of benzoic acid as HBz. The concentration of H_3O^+ in the solution can be obtained from the pH:

$$[H_3O^+] = 10^{-pH} = 10^{-2.56} = 2.8 \times 10^{-3} \text{ mol} \cdot L^{-1}$$

Now we can find the concentrations of each molecule or ion in the solution:

	HBz	+	H_2O	$\rightleftharpoons$	H_3O^+	+	Bz^-	
Initial concentrations	0.125				0		0	$mol \cdot L^{-1}$
Equilibrium concentrations	$0.125 - (2.8 \times 10^{-3})$				2.8×10^{-3}		2.8×10^{-3}	$mol \cdot L^{-1}$

From these equilibrium concentrations, we can calculate K_a:

$$K_a = \left(\frac{[H_3O^+][Bz^-]}{[HBz]}\right)_{eq} = \frac{(2.8 \times 10^{-3})(2.8 \times 10^{-3})}{0.122} = 6.4 \times 10^{-5} \text{ mol} \cdot L^{-1}$$

Exercise 12.10 Lactic acid is a monoprotic acid that accumulates in the muscles during strenuous exercise and can lead to cramps. The pH of a 0.10-*M* solution is found to be 2.43. What is the K_a of lactic acid?

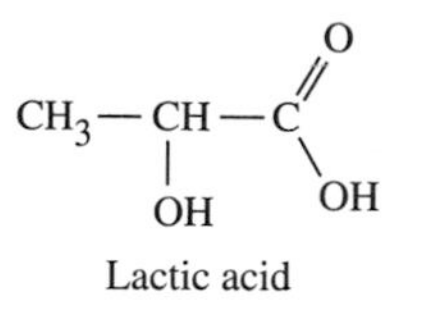

Lactic acid

EXAMPLE 12.11 Determination of K_b from pH

A 0.20-*M* solution of trimethylamine, $(CH_3)_3N$, has a pH of 11.51. What is the value of K_b for trimethylamine?

Solution: We can find the OH^- concentration from the pH as follows:

$$\text{pOH} = 14.00 - \text{pH} = 2.49 \qquad so \qquad [OH^-] = 3.2 \times 10^{-3}\ \text{mol} \cdot \text{L}^{-1}$$

Now we can find the concentrations of each molecule or ion in the solution:

	$(CH_3)_3N$ + H_2O ⇌	$(CH_3)_3NH^+$ +	OH^-	
Initial concentrations	0.20	0	0	$\text{mol} \cdot \text{L}^{-1}$
Equilibrium concentrations	$0.20 - (3.2 \times 10^{-3})$	3.2×10^{-3}	3.2×10^{-3}	$\text{mol} \cdot \text{L}^{-1}$

Hence,

$$K_b = \left(\frac{[(CH_3)_3NH^+][OH^-]}{[(CH_3)_3N]}\right)_{eq} = \frac{(3.2 \times 10^{-3})(3.2 \times 10^{-3})}{0.20}\ \text{mol} \cdot \text{L}^{-1}$$

$$= 5.1 \times 10^{-5}\ \text{mol} \cdot \text{L}^{-1}$$

Exercise 12.11 Dimethylamine, $(CH_3)_2NH$, is a weak base used in the manufacture of detergents. The pH of a 1.00-*M* aqueous solution of dimethylamine is 12.36. What is K_b for dimethylamine?

CH_3—N̈—H
|
CH_3

Dimethylamine

12.3 GAS-PHASE EQUILIBRIA

As an example of equilibrium in the gas phase, we will consider the *Haber process*, the industrial synthesis of ammonia from nitrogen and hydrogen, which we described in Chapter 2:

$$N_2(g) + 3H_2(g) \rightleftharpoons 2NH_3(g) \qquad (6)$$

The equilibrium constant expression is

$$K_c = \frac{[NH_3]^2}{[N_2][H_2]^3} = 6.0 \times 10^{-2}\ \text{mol}^{-2} \cdot \text{L}^2 \qquad \text{at } 500°\text{C}$$

Here we use K_c to designate the equilibrium constant to emphasize that the equilibrium constant expression is given in terms of concentrations. However, for reactions in the gas phase, it is often convenient to write the equilibrium constant expression in terms of partial pressures rather than concentrations. The partial pressure of a gas in a mixture is proportional to its concentration, as we can show by using the ideal gas law. As we saw in Chapter 2, if we have a mixture of n_A moles of A, n_B moles of B, . . . in a volume V, then the partial pressure of A is

$$p_A V = n_A RT \qquad or \qquad p_A = \frac{n_A}{V} RT$$

Now n_A/V is the molar concentration of A, that is, [A]. Hence, $p_A = [A]RT$. Similarly, we may write $p_B = [B]RT$, and so on, for every gas in the mixture.

At a constant temperature, RT is constant, so $p_A \propto [A]$. Similarly, for every gas in the mixture, the partial pressure is proportional to the concentration. The equilibrium constant expressed in terms of partial pressures is denoted by K_p. For the synthesis of ammonia by reaction (6), we have

$$K_p = \left(\frac{(p_{NH_3})^2}{(p_{N_2})(p_{H_2})^3}\right)_{eq}$$

EXAMPLE 12.12 Equilibrium Constant Expressions

Write the equilibrium constant expression for each of the following reactions in terms of concentrations and in terms of partial pressures.

(a) $2SO_2(g) + O_2(g) \rightleftharpoons 2SO_3(g)$

(b) $2H_2(g) + O_2(g) \rightleftharpoons 2H_2O(g)$

(c) $CH_4(g) + H_2O(g) \rightleftharpoons CO(g) + 3H_2(g)$

Solution:

(a) $K_c = \left(\frac{[SO_3]^2}{[SO_2]^2[O_2]}\right)_{eq}$ **(b)** $K_c = \left(\frac{[H_2O]^2}{[H_2]^2[O_2]}\right)_{eq}$ **(c)** $K_c = \left(\frac{[CO][H_2]^3}{[CH_4][H_2O]}\right)_{eq}$

$K_p = \left(\frac{(p_{SO_3})^2}{(p_{SO_2})^2 p_{O_2}}\right)_{eq}$ $K_p = \left(\frac{(p_{H_2O})^2}{(p_{H_2})^2 p_{O_2}}\right)_{eq}$ $K_p = \left(\frac{p_{CO}(p_{H_2})^3}{p_{CH_4} p_{H_2O}}\right)_{eq}$

For gas-phase equilibria, do not incorporate H_2O into the equilibrium constants K_c and K_p as you did for the solution equilibrium constants K_a and K_b.

Exercise 12.12 Write equilibrium constant expressions in terms of concentrations and in terms of partial pressures for both of the following reactions.

(a) the reaction of ethane with steam at high temperature,

$$C_2H_6(g) + 2H_2O(g) \rightleftharpoons 2CO(g) + 5H_2(g)$$

(b) the decomposition of NO_2 at high temperature,

$$2NO_2(g) \rightleftharpoons 2NO(g) + O_2(g)$$

In any industrial process it is important to choose conditions that give a maximum yield of product. We can qualitatively predict the effects of changing conditions by using Le Châtelier's principle (Section 12.2). For the ammonia synthesis, two important conditions that can be changed are the pressure and the temperature at which the reaction is carried out.

Pressure Changes

If we increase the total pressure of an equilibrium mixture of H_2, N_2, and NH_3 (by decreasing the volume of the reaction vessel), Le Châtelier's principle tells us that a reaction will occur to give a new equilibrium mixture that as far as possible restores the original pressure. Thus, we need to know the direction in which the reaction must proceed to decrease the total pressure. Because the total pressure is proportional to the total number of molecules in the mixture, reaction will occur to reduce the total number of molecules (Figure 12.5). Reaction occurs to the right, because four molecules of the reactants ($N_2 + 3H_2$) give two molecules of NH_3, so the

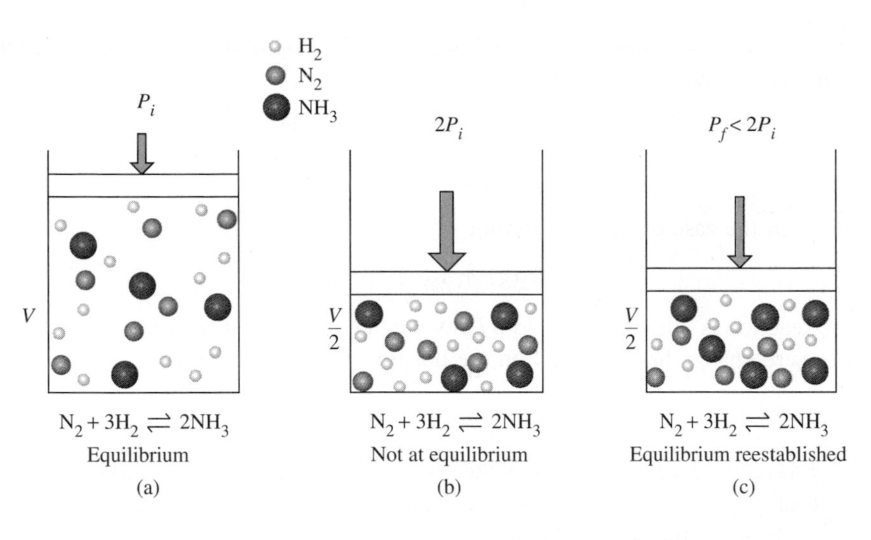

FIGURE 12.5 Le Châtelier's Principle: Effect of Pressure (a) The reaction

$$N_2(g) + 3H_2(g) \rightleftharpoons 2NH_3(g)$$

is at equilibrium. (b) Halving the volume of the reaction vessel doubles the concentrations and partial pressures of each gas and doubles the total pressure. The reaction is no longer at equilibrium. (c) So as to reduce the pressure back toward its original value, the equilibrium shifts to the right, reducing the total number of molecules and therefore decreasing the total pressure. When equilibrium is reestablished, the final pressure P_f is greater than the initial pressure P_i but less than $2P_i$.

concentration of NH_3 at equilibrium is increased. We say that the position of the equilibrium shifts to the right. The synthesis of ammonia is therefore carried out at as high a pressure as possible. In practice, a pressure in the range 200 to 400 atm is used (Figure 12.6).

In the decomposition of ethane to ethene and hydrogen,

$$C_2H_6(g) \rightarrow C_2H_4(g) + H_2(g)$$

the number of molecules increases. An increase in pressure would therefore shift the position of the equilibrium to the left, and the yield of ethene would be decreased. This reaction is therefore carried out at atmospheric pressure. At lower pressures the relative amount of C_2H_4 would increase, but the actual yield would decrease.

We can also find the effect of increasing the total pressure by comparing the value of the reaction quotient, Q, with the equilibrium constant K_p. For example, if in the synthesis of ammonia the pressure acting on the system is doubled, each of the partial pressures will double. So if p_{NH_3}, p_{N_2}, and p_{H_2} are the equilibrium partial pressures, then $Q = (2p_{NH_3})^2/(2p_{N_2})(2p_{H_2})^3 = (K_p/4) < K_p$. For Q to equal K_p, Q will have to increase, so the partial pressure of NH_3 will have to increase, and the partial pressures of H_2 and N_2 will have to decrease. In other words, the position of the equilibrium shifts to the right, and the equilibrium yield of ammonia increases (Figure 12.6).

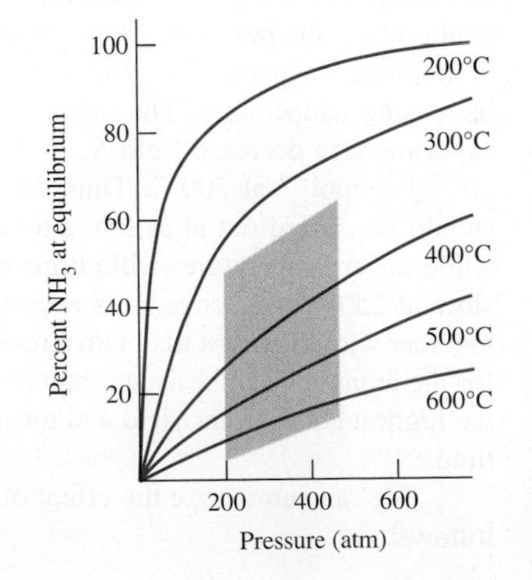

FIGURE 12.6 Effects of Pressure and Temperature on the Equilibrium $N_2 + 3H_2 \rightleftharpoons 2NH_3$ Increasing pressure increases the percentage of NH_3 at equilibrium; increasing temperature decreases the percentage of NH_3 at equilibrium. The shaded area indicates the range of conditions normally used in the Haber process.

We can summarize the effect of *pressure increase* on any gas phase equilibrium as follows:

$$\text{Reactants} \underset{\text{fewer reactant molecules}}{\overset{\text{fewer product molecules}}{\rightleftharpoons}} \text{Products}$$

If, as in the case of the equilibrium

$$H_2(g) + I_2(g) \rightleftharpoons 2HI(g)$$

the reaction is not accompanied by any change in the total number of molecules, a pressure change has no effect on the position of the equilibrium.

Exercise 12.13 Predict how the position of the equilibrium shifts when the total pressure is increased by decreasing the volume of the reacting system in each of the following reactions.

(a) $PCl_5(g) \rightleftharpoons PCl_3(g) + Cl_2(g)$

(b) $N_2(g) + O_2(g) \rightleftharpoons 2NO(g)$

(c) $2SO_2(g) + O_2(g) \rightleftharpoons 2SO_3(g)$

Temperature Changes

An equilibrium constant depends only on temperature; at any given temperature, it is independent of the values of the equilibrium concentrations and hence also of the partial pressures in a gas phase and of the total pressure: Changing the total pressure changes the partial pressures and thus the position of the equilibrium but cannot change the equilibrium constant.

The position of the equilibrium of a reaction depends on the temperature, because the equilibrium constant varies with temperature. If we measure the equilibrium constant at different temperatures, we can then predict how the concentrations or pressures of the products will depend on the temperature. However, it is often convenient to know qualitatively if there will be more or less product at equilibrium when we raise or lower the temperature. We can use Le Châtelier's principle to make such a qualitative prediction. The synthesis of ammonia is an exothermic reaction:

$$N_2(g) + 3H_2(g) \rightleftharpoons 2NH_3(g) \qquad \Delta H^\circ = -92.38 \text{ kJ}$$

If the temperature is increased, Le Châtelier's principle predicts that the reaction will occur in the endothermic direction so as to absorb heat and thereby lower the temperature; the position of equilibrium shifts to the left, and the equilibrium yield of ammonia decreases. This must mean that the equilibrium constant decreases with increasing temperature. The value of the equilibrium constant has been found by experiment to decrease from $K_c = 5.7 \times 10^5 \text{ L}^2 \cdot \text{mol}^{-2}$ at 25°C to $K_c = 6.0 \times 10^{-2} \text{ L}^2 \cdot \text{mol}^{-2}$ at 500°C. Thus, for a maximum yield of ammonia, the reaction should be carried out at as low a temperature as possible (Figure 12.6). However, while lower temperatures will ultimately result in a higher yield, the reaction is very slow at 25°C. In practice, it is necessary to use a temperature of 400°C to 500°C together with a catalyst to obtain a reasonable reaction rate. The optimum conditions for the ammonia synthesis are therefore a compromise between conditions that give the highest equilibrium yield and those needed to reach equilibrium in a reasonable time.

We can summarize the effect of a *temperature increase* on an equilibrium as follows:

Exothermic reactions	Endothermic reactions
K decreases	*K* increases
Position of equilibrium shifts to left	Position of equilibrium shifts to right
Equilibrium concentration of products decreases	Equilibrium concentration of products increases

The effects of a temperature decrease are just the reverse of the effects of a temperature increase.

EXAMPLE 12.13 Using Le Châtelier's Principle

What conditions of pressure and temperature would give a high equilibrium yield of products in the following reactions?

(a) $2SO_2(g) + O_2(g) \rightleftharpoons 2SO_3(g) \qquad \Delta H^\circ = -198 \text{ kJ}$

(b) $H_2(g) + CO_2(g) \rightleftharpoons H_2O(g) + CO(g) \qquad \Delta H^\circ = 41 \text{ kJ}$

Solution: **(a)** This reaction is accompanied by a decrease in the number of gaseous molecules. Hence, a high equilibrium concentration of the product, SO_3, would be favored by a high pressure. The reaction is exothermic, so the equilibrium yield of SO_3 would increase if we carried out the reaction at a low temperature.

(b) This reaction is not accompanied by any changes in the number of molecules, so it is not affected by a change in pressure. The reaction is endothermic, so the yield of H_2O and CO would be increased at a high temperature.

Exercise 12.14 For each of the following reactions, predict which conditions of pressure and temperature would favor a high yield of product.

(a) $N_2(g) + O_2(g) \rightleftharpoons 2NO(g) \qquad \Delta H^\circ = 173 \text{ kJ}$

(b) $CO(g) + 3H_2(g) \rightleftharpoons H_2O(g) + CH_4(g) \qquad \Delta H^\circ = -206 \text{ kJ}$

12.4 HETEROGENEOUS EQUILIBRIA

Equilibria in which all the substances involved are in the same phase are called **homogeneous equilibria**. However, many equilibria involve solids and gases, solids and liquids, or liquids and gases. These equilibria are called **heterogeneous equilibria**.

An example is the decomposition of calcium carbonate, which when heated gives calcium oxide and carbon dioxide:

$$CaCO_3(s) \rightleftharpoons CaO(s) + CO_2(g)$$

The equilibrium constant for this reaction is

$$K'_c = \left(\frac{[CaO][CO_2]}{[CaCO_3]}\right)_{eq}$$

But $CaCO_3$ and CaO are pure solids. What do we mean by the concentration of a pure solid? The number of molecules or ions in a given volume of a pure solid is fixed and cannot vary. Thus, the ''concentration'' of a pure solid, the number of moles per liter, is constant and independent of the amount of pure solid present. So

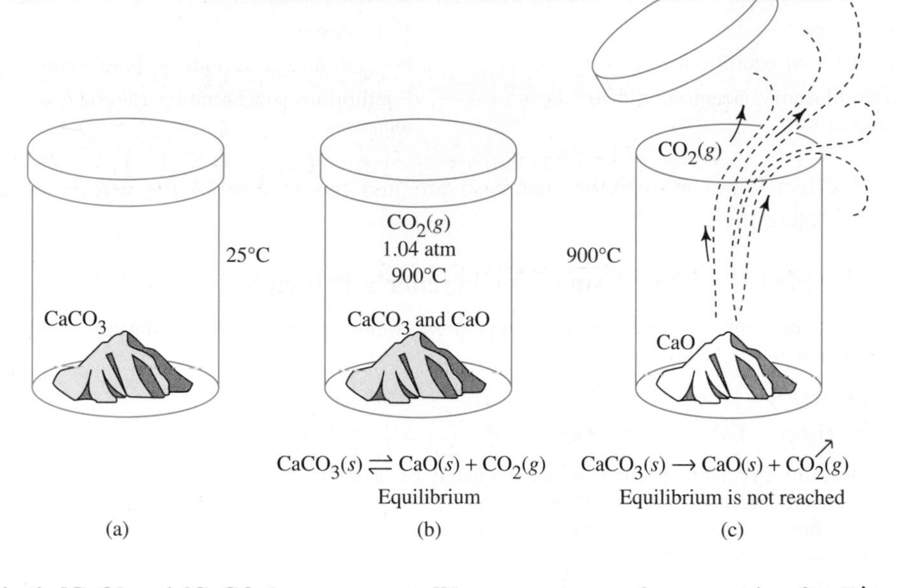

FIGURE 12.7 Heterogeneous Equilibrium (a) At 25°C there is no observable tendency for solid $CaCO_3$ to decompose. (b) When solid $CaCO_3$ is heated in a closed vessel at 900°C, it decomposes into solid CaO and gaseous CO_2. When the pressure of CO_2 reaches 1.04 atm, equilibrium is established. (c) When $CaCO_3$ is heated in an open vessel, the CO_2 diffuses away. The equilibrium pressure of CO_2 is never reached, and $CaCO_3$ is completely converted to CaO.

both [CaO] and [$CaCO_3$] are constant. We can rearrange the expression for K'_c to give

$$\frac{K'_c[CaCO_3]}{[CaO]} = [CO_2]_{eq}$$

Because [$CaCO_3$] and [CaO] are constant, $K'_c[CaCO_3]/[CaO]$ is a constant, which we will write as K_c. Thus,

$$K_c = [CO_2]_{eq}$$

or in terms of partial pressures,

$$K_p = (p_{CO_2})_{eq}$$

At 900°C, $K_p = 1.04$ atm. Thus, if $CaCO_3$ is heated in a closed vessel, it will decompose to CaO and CO_2 until the pressure of CO_2 reaches 1.04 atm. The system will then have reached equilibrium, and the amounts of $CaCO_3$, CaO, and CO_2 will remain constant. However, if $CaCO_3$ is heated in a vessel open to the atmosphere, the CO_2 diffuses away, and it never attains a partial pressure of 1.04 atm. So the reaction is driven to completion, and all the $CaCO_3$ decomposes to CaO (Figure 12.7).

The same considerations apply to pure liquids, so we can state that

The pure solids and liquids taking part in heterogeneous equilibria are not included in the equilibrium constant expression.

Exercise 12.15 Write the expressions for the equilibrium constants for the following reactions in terms of concentrations and in terms of partial pressures.

(a) $C(s) + CO_2(g) \rightleftharpoons 2CO(g)$

(b) $FeO(s) + CO(g) \rightleftharpoons Fe(s) + CO_2(g)$

(c) $PCl_5(s) \rightleftharpoons PCl_3(l) + Cl_2(g)$

EXAMPLE 12.14 Heterogeneous Equilibria

Carbon dioxide is reduced to carbon monoxide by heating with carbon at a high temperature:

$$C(s) + CO_2(g) \xrightleftharpoons{\text{heat}} 2CO(g) \qquad K_p = 1.90 \text{ atm}$$

In a particular experiment the total pressure at equilibrium was found to be 2.00 atm. What were the partial pressures of CO and CO_2?

Solution: If we let the partial pressure of CO be x atm, then because the total pressure is 2.00 atm, the partial pressure of CO_2 is $(2.00 - x)$ atm:

$$K = \frac{(p_{CO})^2}{p_{CO_2}} = \frac{x^2}{2.00 - x} \text{ atm} = 1.90 \text{ atm}$$

Rearranging gives

$$x^2 + 1.90x - 3.80 = 0$$

Solving this quadratic equation (see Appendix M) gives

$$x = 1.22 \qquad or \qquad x = -3.12$$

We can ignore the negative value of x, as it has no physical significance, so

$$p_{CO} = x \text{ atm} = 1.22 \text{ atm}$$

and

$$p_{CO_2} = (2.00 - x) \text{ atm} = 0.78 \text{ atm}$$

To check that the calculated partial pressures are correct, use them to evaluate K_p.

Exercise 12.16 The equilibrium constant for the reaction

$$NH_4Cl(s) \rightleftharpoons NH_3(g) + HCl(g)$$

is 1.04×10^{-2} atm^2 at 275°C. What will be the partial pressures of NH_3 and HCl in equilibrium with $NH_4Cl(s)$ in a closed vessel at 275°C?

12.5 GIBBS FREE ENERGY AND THE EQUILIBRIUM CONSTANT

We saw in Section 12.3 that by comparing the equilibrium constant for a reaction and the reaction quotient, Q, for any given initial concentration of reactants and products, we can predict whether a reaction will reach equilibrium by proceeding to the left or to the right. For the gas-phase reaction

$$aA + bB + \ldots \rightarrow pP + qQ + \ldots,$$

$$Q = \frac{(p_P)^p(p_Q)^{q\cdots}}{(p_A)^a(p_B)^{b\cdots}}$$

where p_A, p_B, . . . are the partial pressures of A, B, . . . and in general are *not* the equilibrium partial pressures. Then

- If $Q < K_p$ or $Q/K_p < 1$, the reaction proceeds from left to right; in other words, the reaction is *spontaneous* as written.
- If $Q > K_p$ or $Q/K_p > 1$, the reaction proceeds from right to left; in other words, it is *not spontaneous* in the direction written but is spontaneous in the reverse direction.
- If $Q = K_p$ or $Q/K_p = 1$, the reaction is at *equilibrium*.

In Chapter 9 we saw that the free energy change for a reaction also tells us whether that reaction will proceed spontaneously to the left or to the right. There must therefore be a relationship between Q/K_p and ΔG. Thermodynamics shows that this relationship is

$$\Delta G = RT \ln \frac{Q}{K_p}$$

or

$$\Delta G = RT \ln Q - RT \ln K_p$$

If all the reactants and products are in their standard states (that is, 1 atm pressure) $\Delta G = \Delta G^\circ$ and $Q = 1$, or $\ln Q = 0$, so

$$\boldsymbol{\Delta G^\circ = -RT \ln K_p}$$

This equation shows that the equilibrium constant is determined by the standard Gibbs free energy change for the reaction. If ΔG° is negative, $K_p > 1$, and the reaction is spontaneous under standard conditions. If ΔG° is positive, $K_p < 1$, and the reaction is not spontaneous under standard conditions but is spontaneous in the reverse direction. If $\Delta G^\circ = 0$, $K_p = 1$, and the reaction is at equilibrium under standard conditions.

As an example, let us calculate the equilibrium constant for the formation of $NO(g)$ from the elements at 25°C. The equation for the reaction is

$$N_2(g) + O_2(g) \rightarrow 2NO(g)$$

First we find ΔG° by using the standard Gibbs free energies of formation given in Table 9.4:

$$\begin{aligned}\Delta G^\circ &= 2\,\Delta G_f^\circ(NO, g) - [\Delta G_f^\circ(N_2, g) + \Delta G_f^\circ(O_2, g)] \\ &= (2\text{ mol})(86.6\text{ kJ}\cdot\text{mol}^{-1}) - (0 + 0) = 173.2\text{ kJ}\end{aligned}$$

Then we can find K_p from the expression

$$\ln K_p = \frac{-\Delta G^\circ}{RT} = -\frac{173.2\text{ kJ}\cdot\text{mol}^{-1}}{(0.00831\text{ kJ}\cdot\text{mol}^{-1}\cdot\text{K}^{-1})(298\text{ K})} = -69.9$$

$$K_p = 4.4 \times 10^{-31}$$

Remember that ΔG° and R must have the same energy units (here, kilojoules). Note that $\Delta G^\circ = 173.2$ kJ for the reaction as written or 173.2 kJ for 1 mole of reaction. To generalize, for this reaction $\Delta G^\circ = 173.2$ kJ per mole of reaction—that is, $\Delta G^\circ = 173.2\text{ kJ}\cdot\text{mol}^{-1}$.

This very small equilibrium constant shows that the reaction has very little tendency to proceed at room temperature. At standard conditions, because all the gas pressures are 1 atm, $Q = 1$, so $Q \gg K$, or $Q/K \gg 1$. Thus, the reaction is not spontaneous in the direction written but proceeds spontaneously in the reverse direction, although at 25°C it is exceedingly slow.

EXAMPLE 12.15 Calculating Equilibrium Constants from ΔG° Values

The standard free energy change at 25°C for the reaction

$$N_2(g) + 3H_2(g) \rightleftharpoons 2NH_3(g)$$

is $\Delta G^\circ = -32.8$ kJ. What is the equilibrium constant, K_p, for this reaction?

Solution:

$$\Delta G^\circ = -RT \ln K_p$$

Therefore,

$$\ln K_p = \frac{-\Delta G^\circ}{RT} = \frac{32.8\ \text{kJ} \cdot \text{mol}^{-1}}{(0.00831\ \text{kJ} \cdot \text{mol}^{-1} \cdot \text{K}^{-1})(298\ \text{K})} = 13.2$$

$$K_p = 5.7 \times 10^5\ \text{atm}^{-2}$$

This large value of the equilibrium constant tells us that the position of the equilibrium is far to the right; that is, at equilibrium there will be a high pressure (or concentration) of NH_3 and low pressures (or concentrations) of N_2 and H_2. We can reach essentially the same conclusion from the negative ΔG° value, which tells us that at standard conditions—all the gases at 1 atm pressure—the reaction is spontaneous, or proceeds to the right until equilibrium is reached.

The natural log of a quantity, such as $\ln K_p$, has no units. However, when we look at the balanced equation for the synthesis of NH_3, we realize that K_p must have units of atm^{-2}.

Now we can see one reason for the importance of the Gibbs free energy. From a table of standard free energies of formation, we can calculate the equilibrium constant for many thousands of reactions. In considering possible reactions by which a substance might be synthesized, we look for a reaction that has a large negative ΔG° and therefore a large equilibrium constant and a large concentration of products when equilibrium is reached.

If we cannot find a reaction with a sufficiently large negative ΔG° at 25°C, we can see if the ΔG° at some other temperature is more favorable. Or we can see if some nonstandard conditions will give a more favorable value of ΔG, as illustrated in Example 12.16.

EXAMPLE 12.16 Calculating ΔG for a Reaction at Nonstandard Conditions

Calculate ΔG at 298 K for the reaction

$$N_2(g) + 3H_2(g) \rightleftharpoons 2NH_3(g)$$

if the reaction mixture consists of 10 atm N_2, 10 atm H_2, and 1 atm NH_3.

Solution:

$$Q = \frac{(p_{NH_3})^2}{(p_{N_2})(p_{H_2})^3} = \frac{(1\ \text{atm})^2}{10\ \text{atm} \times (10\ \text{atm})^3} = \frac{1}{10^4}\ \text{atm}^{-2}$$

and

$$K_p = 5.7 \times 10^5\ \text{atm}^{-2}\ \text{(see Example 12.15)}$$

$$\Delta G = -RT \ln \frac{Q}{K_p} = (8.31\ \text{J} \cdot \text{mol}^{-1} \cdot \text{K}^{-1})(298\ \text{K})\left(\ln \frac{1}{5.7 \times 10^9}\right)$$

$$= (8.31\ \text{J} \cdot \text{mol}^{-1} \cdot \text{K}^{-1})(298\ \text{K})(-22) = -56\ \text{kJ} \cdot \text{mol}^{-1}$$

Increasing the pressure of both N_2 and H_2 from 1 to 10 atm changes ΔG from −32.8 to −55.7 kJ. Thus, the reaction has a still greater tendency to proceed to the right. In other words, the position of the equilibrium shifts to the right, as we could have predicted from Le Châtelier's principle.

Finally, we need to remember that a reaction that has a large negative ΔG° and a large equilibrium constant may nevertheless be very slow and therefore not very practical or very economical. We discuss the factors influencing the rates of reactions in Chapter 15.

SUMMARY

For any homogeneous reaction (all reactants and products in the same phase),

$$aA + bB + cC + \ldots \rightleftharpoons pP + qQ + rR + \ldots$$

the concentrations of the reactants and products at equilibrium are related by the equilibrium constant expression

$$K = \left(\frac{[P]^p[Q]^q[R]^r \ldots}{[A]^a[B]^b[C]^c \ldots}\right)_{eq}$$

where K is the equilibrium constant. K varies only with temperature.

For a weak acid,

$$HA(aq) + H_2O(l) \rightleftharpoons H_3O^+(aq) + A^-(aq)$$

and the acid ionization constant is

$$K_a = \left(\frac{[H_3O^+][A^-]}{[HA]}\right)$$

For a weak base,

$$B(aq) + H_2O(l) \rightleftharpoons BH^+(aq) + OH^-(aq)$$

and the base ionization constant is

$$K_b = \left(\frac{[BH^+][OH^-]}{[B]}\right)$$

Strong acids HA have $K_a \gg 1$, and most weak acids have $K_a \ll 1$. Strong bases B have $K_b \gg 1$, and most weak bases have $K_b \ll 1$. K_a and K_b values are also expressed in logarithmic form:

$$pK_a = -\log K_a \quad \textit{and} \quad pK_b = -\log K_b$$

Water behaves both as a very weak acid and as a very weak base and undergoes autoionization (autoprotolysis),

$$2H_2O(l) \rightleftharpoons H_3O^+(aq) + OH^-(aq)$$

The autoionization constant of water, K_w, is given by

$$K_w = ([H_3O^+][OH^-])_{eq}$$
$$= 1.00 \times 10^{-14}\ mol^2 \cdot L^{-2} \text{ at } 25°C$$

or

$$pK_w = 14.00 \text{ at } 25°C$$

Thus, at 25°C in pure water

$$[H_3O^+] = [OH^-] = 1.00 \times 10^{-7}\ mol \cdot L^{-1}$$

Acidic solutions have $[H_3O^+] > 10^{-7}\ mol \cdot L^{-1}$, and basic solutions have $[OH^-] > 10^{-7}\ mol \cdot L^{-1}$.

H_3O^+ concentrations are given by the pH scale as

$$pH = -\log [H_3O^+]$$

and OH^- concentrations as

$$pOH = -\log [OH^-]$$

In water and in any aqueous solution, pH + pOH = 14.00 at 25°C. Acidic solutions have pH < 7, and basic solutions have pH > 7 (pOH < 7).

A conjugate acid–base pair is an acid and a base related by the equation

$$\text{base} + H^+ \rightleftharpoons \text{acid}$$

For all conjugate acid–base pairs,

$$K_a(\text{acid})\ K_b(\text{conjugate base}) = K_w$$

or

$$pK_a(\text{acid}) + pK_b(\text{conjugate base}) = pK_w = 14.00 \text{ at } 25°C$$

The stronger the acid, the weaker its conjugate base, and vice versa. Conjugate bases of strong acids have no basic properties in water, and conjugate acids of strong bases have no acidic properties in water.

The acid–base properties of salt solutions depend on the acid or base behavior of their constituent anions and cations.

A solution that maintains a very nearly constant pH when moderate amounts of H_3O^+ or OH^- are added to it is a buffer solution. It contains either a weak acid and conjugate base *or* a weak base and its conjugate acid. Its pH is given by the Henderson–Hasselbalch equation:

$$pH = pK_a + \log \frac{[\text{base}]}{[\text{acid}]}$$

Values of pH are accurately measured by using a pH meter. Approximate pH values are obtained by using an indicator, which is a weak acid HIn with a conjugate base In^- of a different color. The approximate pH range over which the color change occurs is given by $pH = pK_a(HIn) \pm 1$.

For gas-phase equilibria, equilibrium constant expressions are usually written in terms of partial pressures. The equilibrium constant for such reactions is designated as K_p.

Le Châtelier's principle states that if any of the conditions affecting a system at equilibrium is changed such that the system is no longer at equilibrium, when equilibrium is again established, the original conditions are as far as possible restored. It can be used to make qualitative predictions about the effects of concentration, pressure, and temperature on the equilibrium position of a reaction.

The reaction quotient, Q, has the same form as the equilibrium constant expression, K, except that in general the concentrations of the reactants and products are *not* the values at equilibrium. When $Q > K$, reaction occurs to increase the concentrations of the reactants. When $Q < K$, reaction occurs to increase the concentrations of the products. When $Q = K$, there is no change—the system is at equilibrium.

In homogeneous equilibria, all the reactants and products are in the same phase. In heterogeneous equilibria, all the reactants and products are not in the same phase, and any pure solid (or liquid) has a constant concentration, independent of its amount. Pure solids (or liquids) do not appear in the equilibrium constant expression for a heterogeneous equilibrium.

The Gibbs free energy change, ΔG, for a reaction is related to its equilibrium constant, K_p, and its reaction quotient, Q, by the equation

$$\Delta G = RT \ln \frac{Q}{K_p}$$

For all the reactants and products in their standard states,

$$\Delta G^\circ = -RT \ln K_p.$$

IMPORTANT TERMS

acid ionization constant, K_a (page 426)
autoionization (autoprotolysis) (page 431)
autoionization constant of water, K_w (page 431)
base ionization constant, K_b (page 429)
buffer solution (page 438)
conjugate acid–base pair (page 434)
equilibrium constant (page 425)
equilibrium constant expression (page 425)
Henderson–Hasselbalch equation (page 440)
heterogeneous equilibrium (page 453)
homogeneous equilibrium (page 453)
indicator (page 443)
Le Châtelier's principle (page 441)
pH (page 431)
pH scale (page 431)
reaction quotient, Q (page 443)

REVIEW QUESTIONS

1. When a reaction has reached equilibrium, what can be said about the rate of the forward reaction and the rate of the backward reaction?

2. Why does a reaction never reach equilibrium if one or more of the products are removed as they form?

3. Write the expression for the ionization constant of an acid HA in water.

4. **(a)** How are pK_a and pK_b values defined?
(b) Of HF ($pK_a = 3.45$) and HCN ($pK_a = 9.31$), which is the weaker acid?
(c) Of NH_3 ($pK_b = 4.74$) and methylamine ($pK_b = 3.41$), which is the stronger base?

5. What are the strongest acid species and the strongest basic species that can exist in appreciable amounts in aqueous solutions?

6. **(a)** Write the expression for the autoionization constant of water, K_w. **(b)** What are the values of K_w and pK_w at 25°C?

7. Write the expressions for pH, pOH, and the relationship between them.

8. What are the pH values for **(a)** pure water, **(b)** 1.00-*M* $HCl(aq)$, and **(c)** 1.00-*M* $NaOH(aq)$?

9. For NH_3, H_2O, and HSO_4^-, what are **(a)** their conjugate acids and **(b)** their conjugate bases?

10. Why is an aqueous solution of NH_4Cl acidic, whereas a solution of sodium acetate, CH_3CO_2Na, is basic?

11. Are the following solutions expected to be acidic, neutral, or basic?
(a) $NaBr(aq)$ **(b)** $NaF(aq)$ **(c)** $AlCl_3(aq)$

12. What is a buffer solution?

13. **(a)** What is an acid–base indicator?
(b) Over what approximate pH range does phenolphthalein ($pK_a = 9.5$) change color?

14. Write the expressions for K_c and K_p for the synthesis of ammonia from $N_2(g)$ and $H_2(g)$ (the Haber process).

15. State Le Châtelier's principle.

16. **(a)** What is the reaction quotient expression Q for the reaction of Review Question 14?
(b) When does Q equal the equilibrium constant?

17. For the reaction

$$2SO_2(g) + O_2(g) \rightleftharpoons 2SO_3(g), \quad \Delta H^\circ = -199.8 \text{ kJ}$$

will the equilibrium concentration of $SO_3(g)$ increase or decrease upon increasing **(a)** the concentration of SO_3, **(b)** the total pressure, and **(c)** the temperature?

18. What is meant by the term ''heterogeneous equilibrium''?

19. For the reaction $CaCO_3(s) \rightleftharpoons CaO(s) + CO_2(g)$, write the expressions for K_p and K_c.

20. If $Q/K_p < 1$ for a reaction, is ΔG positive or negative?

PROBLEMS

Acid-Base Equilibria in Aqueous Solution; The pH Scale

1. What are the molar concentrations of the ions in each of the following aqueous solutions?
(a) 10^{-5}-*M* HNO_3 (b) 0.0023-*M* HCl
(c) 0.113-*M* $HClO_4$ (d) 0.034-*M* HBr
(e) 10^{-3}-*M* NaOH (f) 0.145-*M* $Ba(OH)_2$

2. Repeat Problem 1 for the following aqueous solutions.
(a) 0.1234-*M* HI (b) 10^{-5}-*M* $Ca(OH)_2$
(c) 0.204-*M* LiOH (d) 0.324-*M* HBr
(e) 0.0023-*M* $Sr(OH)_2$ (f) 10^{-4}-*M* Na_2O

3. What are the pH values of the solutions in Problem 1?

4. What are the pH values of the solutions in Problem 2?

5. What are **(a)** the H_3O^+ ion concentration, **(b)** the percent ionization, and **(c)** the pH of a 0.010-*M* solution of hydrocyanic acid, HCN(*aq*)?

6. Repeat Problem 5 for a 0.150-*M* solution of hydrofluoric acid, HF(*aq*).

7. Of a 0.0010-*M* solution of HNO_3(*aq*) and a 0.200-*M* solution of acetic acid, which has the lower pH?

8. What are **(a)** the OH^- ion concentration, **(b)** the percent ionization, and **(c)** the pH of a 0.040-*M* solution of aqueous ammonia?

9. Repeat Problem 8 for a 0.080-*M* solution of aniline, $C_6H_5NH_2$.

10. The pH of a 0.100-*M* solution of hypochlorous acid, HOCl(*aq*), is 4.2. What are the K_a and the pK_a of hypochlorous acid?

11. Your muscles may ache after strenuous exercise because lactic acid, $pK_a = 3.08$, forms faster than it is metabolized to CO_2 and water. What is the pH of the fluid in muscle when the lactic acid concentration reaches 1.0×10^{-3} mol · L^{-1}?

12. Equal volumes of hydrochloric acid solutions of pH 2.00 and pH 3.00 are mixed. What is the pH of the resulting solution?

13. Citric acid has a pK_a of 3.10. What is the pH of a 0.10-*M* solution of citric acid?

Acid-Base Properties of Anions, Cations, and Salts

14. Write a balanced equation for the reaction, if any, of each of the following with water, and decide whether the resulting solution is acidic, neutral, or basic.
(a) carbon dioxide (b) sulfur trioxide (c) calcium oxide
(d) sodium oxide (e) sodium hydrogensulfate

15. Give the formulas and names of the conjugate acids of each of the following bases.
(a) NH_3 (b) F^- (c) OH^- (d) H_2O (e) H^-

16. Give the formulas and names of the conjugate bases of each of the following acids.
(a) HF (b) HNO_3 (c) $HClO_4$ (d) H_2O (e) H_3O^+

17. What are the conjugate acids of each of the following?
(a) O^{2-} (b) PO_4^{3-} (c) CO_3^{2-} (d) S^{2-} (e) SO_4^{2-}

18. What are the conjugate bases of each of the following?
(a) H_3O^+ (b) NH_4^+ (c) HCl (d) H_3PO_4 (e) H_2SO_4

19. For each of the following acids, name the conjugate base, give its formula, and calculate its pK_b at 25°C.
(a) H_2CO_3 (b) H_2S (c) HNO_2 (d) H_3PO_4

20. What is the pH of a 0.050-*M* aqueous solution of KF?

21. Arrange 0.10-*M* aqueous solutions of each of the following in order of increasing pH. (No calculations are needed.)
(a) NaCN (b) KCl (c) KOH (d) HBr
(e) NH_3 (f) NH_4Cl (g) K_2O

22. **(a)** What are the equilibrium concentrations of NH_4^+, NH_3, OH^-, and H_3O^+ in a 0.020-*M* solution of NH_4Cl(*aq*)? **(b)** What is the pH of the solution in (a)?

23. What is the pH of each of the following aqueous solutions?
(a) 0.010-*M* acetic acid (b) 0.10-*M* hydrofluoric acid
(c) 0.0030-*M* ammonia (d) 0.10-*M* sodium acetate
(e) 0.20-*M* ammonium chloride

24. Suggest a suitable acid–base reaction by which each of the following salts could be prepared. Would you expect a 0.10-*M* solution of each salt to be acidic, neutral, or basic? Explain your answers.
(a) ammonium nitrate (b) ammonium chloride
(c) calcium sulfate (d) potassium acetate
(e) aluminum chloride (f) sodium iodide

25. A 30.0-g sample of phosphoric acid, H_3PO_4, is dissolved in water to give 500 mL of solution.
(a) What volume of 0.200-*M* NaOH(*aq*) is needed to react completely with the H_3PO_4 to form Na_3PO_4(*aq*)?
(b) Will the resulting solution be acidic, neutral, or basic?

26. Sodium carbonate is often added to swimming pools to change the pH. Does this raise or lower the pH of the pool? Explain.

27. What is the pH of a 0.100-*M* solution of **(a)** $AlCl_3$(*aq*) and **(b)** sodium hydrogensulfate, $NaHSO_4$(*aq*)?

Buffer Solutions

28. What are the pH values of **(a)** a buffer solution containing 0.30-*M* ammonia and 0.25-*M* ammonium chloride; **(b)** a buffer solution made from equal volumes of 0.10-*M* HCN(*aq*) and 0.10-*M* NaCN(*aq*)?

29. What ratio of masses of ammonia and ammonium chloride is needed to give a buffer solution with a pH of 9.0?

30. What are the pH values of solutions prepared by mixing **(a)** 25.0 mL of 0.100-*M* NaOH(*aq*) and 50.0 mL of 0.100-*M* acetic acid; **(b)** 15.0 mL of 0.0100-*M* NH_3(*aq*) and 25.0 mL of 0.0100-*M* NH_4Cl(*aq*)?

31. A phosphate buffer is frequently used to maintain a constant pH in biological experiments. What is the pH of the phosphate buffer prepared by dissolving 3.40 g of potassium dihydrogenphosphate and 3.55 g of disodium hydrogenphosphate in sufficient water to give 500 mL of solution?

32.* What is the change in pH when 1.00 mL of 1.00-*M* $NaOH(aq)$ is added to 100 mL of a solution containing 0.18-*M* $NH_3(aq)$ and 0.10-*M* $NH_4Cl(aq)$?

33.* **(a)** Use the Henderson–Hasselbalch equation to calculate the mole ratio of acetic acid to sodium acetate required to give a buffer solution of pH 4.50.

(b) How many grams of sodium acetate must be added to 1.00 L of 0.200-*M* $CH_3CO_2H(aq)$ to give a solution of pH 4.50?

(c) What effect does diluting the solution in (b) to 2.00 L have on its pH?

34.* **(a)** What is the pH of a buffer solution prepared by mixing 25.0 mL of a 0.020-*M* solution of aniline, $C_6H_5NH_2$, with 10.0 mL of a 0.030-*M* solution of anilinium chloride, $C_6H_5NH_3{}^+Cl^-$?

(b) What is the pH of the solution in **(a)** after 1.00 mL of 0.040-*M* $HNO_3(aq)$ is added?

(c) What will the pH be if 2.00 mL of 0.030-*M* $KOH(aq)$ is added to the solution in (a) rather than nitric acid?

The Measurement of pH

35. **(a)** In using an indicator, why is it important to add the smallest amount possible to the solution being investigated?

(b) For methyl orange indicator ($pK_a = 4.2$), at what pH values is $[In^-]/[HIn] = 5{:}1$, $1{:}1$, and $1{:}5$?

36. **(a)** Estimate the pH of a colorless aqueous solution that turns yellow when methyl red is added to it and yellow when bromothymol blue is added to it.

(b) What color would you expect a solution containing methyl red to be when the pH is 5.0?

37. From Table 12.5, select a suitable indicator for detecting the equivalence point (when moles of added base equals initial moles of acid) in each of the following titrations of 0.10-*M* acid with 0.10-*M* base.

(a) $HCl(aq)$ with $NaOH(aq)$

(b) $HF(aq)$ with $KOH(aq)$

(c) $NH_3(aq)$ with $HCl(aq)$

(d) $CH_3NH_2(aq)$ with $HCl(aq)$

38. When each of the following indicators is added to a 0.10-*M* solution of a weak acid, the colors are as follows: methyl violet—violet; methyl orange—yellow; methyl red—orange; bromothymol blue—yellow. What is the approximate K_a of the acid?

39. What color would each of the following indicators give in a 0.10-*M* solution of aqueous potassium fluoride?

(a) thymol blue **(b)** phenolphthalein

(c) bromothymol blue

40. Repeat Problem 39 for a 0.10-*M* solution of aqueous ammonium chloride.

(a) methyl red **(b)** methyl orange **(c)** bromothymol blue

41. Blood plasma is obtained by centrifuging samples of blood to remove the red blood cells. What colors would be given by adding to blood plasma small amounts of **(a)** bromothymol blue and **(b)** thymol blue?

Gas-Phase Equilibria

42. Write the equilibrium constant expressions K_c and K_p for each of the following reactions.

(a) $2SO_2(g) + O_2(g) \rightleftharpoons 2SO_3(g)$

(b) $SO_2(g) + \frac{1}{2}O_2(g) \rightleftharpoons SO_3(g)$

(c) $P_4(g) + 5O_2(g) \rightleftharpoons P_4O_{10}(g)$

(d) $PCl_5(g) \rightleftharpoons PCl_3(g) + Cl_2(g)$

43. Repeat Problem 42 for the following reactions.

(a) $CH_4(g) + H_2O(g) \rightleftharpoons 3H_2(g) + CO(g)$

(b) $2NH_3(g) \rightleftharpoons 3H_2(g) + N_2(g)$

(c) $2CO(g) + O_2(g) \rightleftharpoons 2CO_2(g)$

(d) $C_3H_8(g) \rightleftharpoons CH_4(g) + C_2H_4(g)$

44. At 425°C, $K_c = 300\ L^2 \cdot mol^{-2}$ for the reaction in which methanol, $CH_3OH(g)$, is synthesized from hydrogen and carbon monoxide:

$$2H_2(g) + CO(g) \rightleftharpoons CH_3OH(g)$$

(a) When the concentrations of H_2, CO, and CH_3OH are each $0.10\ mol \cdot L^{-1}$, is the system at equilibrium at 425°C?

(b) If the reaction is not at equilibrium, will the concentration of CH_3OH be greater than or less than $0.10\ mol \cdot L^{-1}$ when equilibrium is established?

45. For the reaction $H_2(g) + CO_2(g) \rightleftharpoons H_2O(g) + CO(g)$ at 600 K, the following concentrations, in moles per liter, were found at equilibrium: $[H_2] = 0.600$, $[CO_2] = 0.459$, $[H_2O] = 0.500$, $[CO] = 0.425$. Calculate the values of **(a)** K_c and **(b)** K_p.

46. For the reaction in Problem 45, at the same temperature, what would be the equilibrium concentrations if initially 1.00 mol of $H_2(g)$ and 1.00 mol of $CO_2(g)$ were placed in a sealed 5.00-L vessel?

47. At 1000 K, iodine molecules dissociate into iodine atoms, with $K_c = 3.76 \times 10^{-5}\ mol \cdot L^{-1}$ for the equilibrium

$$I_2(g) \rightleftharpoons 2I(g)$$

At this temperature, what are **(a)** the equilibrium concentrations of $I_2(g)$ and $I(g)$ after 1.00 mol of I_2 has initially been introduced into a 2.00-L flask and **(b)** the percent ionization of I_2?

48. The equilibrium constant at 490°C for the reaction

$$2HI(g) \rightleftharpoons H_2(g) + I_2(g)$$

is $K_c = 0.022$. What are the equilibrium concentrations of HI, H_2, and I_2 when an initial amount of 2.00 mol of $HI(g)$ is placed in a 4.3-L flask at 490°C?

49. The equilibrium constant K_c at 698 K is 54.4 for the reaction

$$H_2(g) + I_2(g) \rightleftharpoons 2HI(g)$$

A reaction mixture contains $0.10\ mol \cdot L^{-1}$ of both H_2 and I_2 and $1.00\ mol \cdot L^{-1}$ of HI.

(a) Has the reaction reached equilibrium?

(b) If not, calculate the concentrations of H_2, I_2, and HI at equilibrium.

(c) What would be the concentrations at equilibrium if we started with just 1.2 mol · L^{-1} of HI alone?

50. For the reaction

$$N_2(g) + 3H_2(g) \rightleftharpoons 2NH_3(g)$$

the equilibrium constant K_c is 6.0×10^{-2} L^2 · mol^{-2} at 500°C. Initially the concentrations of N_2 and H_2 each 1.0 mol · L^{-1}, and that of NH_3 was 0.10 mol · L^{-1}. In answering the following questions, do not explicitly solve for the equilibrium concentrations.

(a) Is the initial system at equilibrium?

(b) A student calculated that, for equilibrium to be reached, 0.010 mol of the N_2 would have to be converted to NH_3. Was this student correct? If not, was the student's estimate too large or too small?

51. A 5.00-L flask containing 1 mol of HI(g) is heated to 800°C. What is the percent dissociation of the HI, given that the value of the equilibrium constant K_c is 6.34×10^{-4} at 800°C for the reaction $2HI(g) \rightleftharpoons H_2(g) + I_2(g)$?

52. The equilibrium constant for a reaction increases as the temperature is increased.

(a) Is the forward reaction exothermic or endothermic?

(b) What can we say about the enthalpy change for the reverse reaction?

53. The equilibrium constant $K_p = 3.2 \times 10^2$ atm$^{-\frac{1}{2}}$ at 425°C for the reaction

$$SO_2(g) + \tfrac{1}{2}O_2(g) \rightleftharpoons SO_3(g)$$

At 525°C the value of K_p decreases to 33 atm$^{-\frac{1}{2}}$. Is the reaction as written exothermic or endothermic?

54. For each of the following reactions, calculate the standard reaction enthalpy for the forward reaction from the data in Table 9.1. In each case, how do changing temperature and changing pressure affect the position of equilibrium? What combination of temperature and pressure would maximize the yield of the product(s)?

(a) $2SO_2(g) + O_2(g) \rightleftharpoons 2SO_3(g)$

(b) $N_2(g) + 3H_2(g) \rightleftharpoons 2NH_3(g)$

(c) $CO(g) + H_2O(g) \rightleftharpoons CO_2(g) + H_2(g)$

55. For the equilibrium

$$C_2H_4(g) + H_2(g) \rightleftharpoons C_2H_6(g), \quad \Delta H° = -137 \text{ kJ}$$

how would each of the following changes affect the equilibrium concentration of $C_2H_6(g)$?

(a) doubling the volume of the reaction vessel

(b) increasing the temperature at constant volume

(c) adding more H_2 to the reaction vessel

(d) increasing the pressure by adding helium to the reaction vessel at constant volume

Heterogeneous Equilibria

56. Write the expressions for the equilibrium constants K_c and K_p for each of the following reactions.

(a) $H_2O(g) \rightleftharpoons H_2(g) + \tfrac{1}{2}O_2(g)$

(b) $H_2O(l) \rightleftharpoons H_2(g) + \tfrac{1}{2}O_2(g)$

(c) $2H_2O(l) \rightleftharpoons 2H_2(g) + O_2(g)$

(d) $H_2(g) + O_2(g) \rightleftharpoons H_2O_2(l)$

(e) $2HgO(s) \rightleftharpoons 2Hg(l) + O_2(g)$

57. Repeat Problem 56 for the following reactions.

(a) $C(s) + O_2(g) \rightleftharpoons CO_2(g)$

(b) $MgCO_3(s) \rightleftharpoons MgO(s) + CO_2(g)$

(c) $2NaHCO_3(s) \rightleftharpoons Na_2CO_3(s) + CO_2(g) + H_2O(g)$

(d) $ZnO(s) + CO(g) \rightleftharpoons Zn(s) + CO_2(g)$

(e) $3Fe(s) + 4H_2O(g) \rightleftharpoons Fe_3O_4(s) + 4H_2(g)$

58. In which direction would the equilibrium

$$C(s) + 2H_2(g) \rightleftharpoons CH_4(g), \quad \Delta H° = -75 \text{ kJ}$$

shift in response to each of the following changes?

(a) increasing the temperature

(b) increasing the volume of the reaction vessel

(c) increasing the partial pressure of hydrogen

(d) adding more carbon

59. At 25°C, $K_p = 9.1$ atm^{-2} for the reaction

$$NH_3(g) + H_2S(g) \rightleftharpoons NH_4HS(s)$$

(a) Calculate K_c at 25°C.

(b) How would K_c change if the above equation were multiplied throughout by 2?

(c) If 1.00 mol of solid ammonium hydrogensulfide, $NH_4HS(s)$, is placed in an evacuated 1.00-L flask at 25°C, what will be the total pressure of gases at equilibrium?

(d) Does your answer to (c) depend on which balanced equation you use? Explain.

Gibbs Free Energy and the Equilibrium Constant

60. **(a)** From the data in Table 9.4, calculate the standard reaction Gibbs free energy for the reaction

$$2SO_2(g) + O_2(g) \rightarrow 2SO_3(g)$$

(b) What is the value of K_p at 298 K?

(c) Is this reaction spontaneous at 298 K?

61. For the reaction

$$C(s, \text{graphite}) + CO_2(g) \rightarrow 2CO(g)$$

what is the value of K_p at 700°C? (Assume that $\Delta H°$ and $\Delta S°$ are independent of temperature, and use data from Tables 9.1 and 9.3.)

62. Consider the reaction

$$CH_4(g) + 2O_2(g) \rightarrow CO_2(g) + 2H_2O(g)$$

(a) Using data from Table 9.4, decide if this reaction is spontaneous at 298 K.

(b) What is the value of K_p at 298 K?

(c) Explain the fact that a mixture of methane and oxygen shows a negligible extent of reaction at room temperature, even after a very long time.

63.* From data in Table 9.4, calculate the standard reaction Gibbs free energy associated with the combustion of liquid methanol:

$$2CH_3OH(l) + 3O_2(g) \rightarrow 2CO_2(g) + 4H_2O(l)$$

(a) Is this reaction spontaneous under standard conditions?

(b) What is the value of K_p at 298 K?

(c) Does the value of K_p favor the formation of reactants or of products?

(d) What effect would an increase in pressure have on the spontaneity of the reaction?

(e) What effect would an increase in temperature have on the spontaneity of the reaction?

64.* For the reaction

$$2C(s, \text{graphite}) + H_2(g) \rightarrow C_2H_2(g)$$

the standard reaction Gibbs free energy is 209 kJ.

(a) Is this reaction a practical way to synthesize ethyne (acetylene), $C_2H_2(g)$, at room temperature?

(b) Would the reaction be expected to be spontaneous at high temperature?

(c) Assuming that $\Delta H°$ and $\Delta S°$ are independent of temperature, calculate the equilibrium constant K_p at 1200 K.

General Problems

65. **(a)** Use bond energy data from Table 9.2 to estimate $\Delta H°$ for the gas–phase reaction

$$CO(g) + H_2(g) \rightarrow H_2CO(g)$$

(b) Decide whether high or low temperatures, and high or low pressures would favor a high yield of methanal, $H_2CO(g)$.

66. From the data that follow, **(a)** calculate the pH of pure water at the temperatures given and **(b)** deduce whether the autoionization of water is exothermic or endothermic.

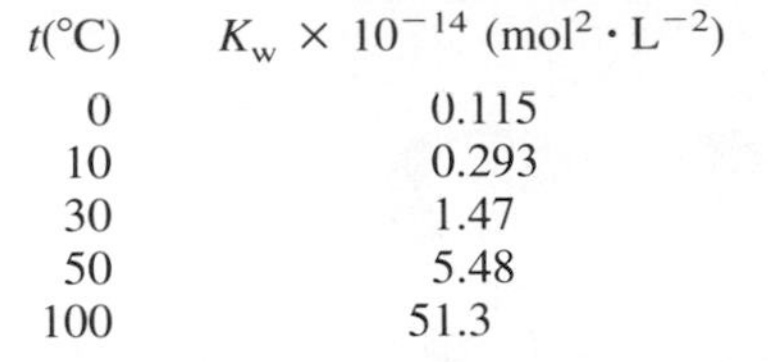

t(°C)	$K_w \times 10^{-14}$ ($mol^2 \cdot L^{-2}$)
0	0.115
10	0.293
30	1.47
50	5.48
100	51.3

67. Sulfur dioxide and oxygen in the mole ratio 2:1 were allowed to reach equilibrium in the presence of a catalyst at a pressure of 5.00 atm. At equilibrium, 33% of the SO_2 was converted to $SO_3(g)$. For the reaction

$$2SO_2(g) + O_2(g) \rightleftharpoons 2SO_3(g)$$

what is the value of K_p?

68. Which of the following are buffer solutions? Explain.

(a) 25 mL 0.10-*M* $HNO_3(aq)$ and 25 mL 0.10-*M* $NaNO_3(aq)$

(b) 25 mL 0.10-*M* $HNO_2(aq)$ and 25 mL 0.10-*M* $NaNO_2(aq)$

(c) 25 mL 0.10-*M* acetic acid and 25 mL 0.15-*M* KOH(aq)

(d) 25 mL 0.10-*M* acetic acid and 25 mL 0.05-*M* KOH(aq)

69. Explain each of the following terms and the differences between them.

(a) the "strength" of an acid and its "concentration"

(b) the "equivalence point" and the "end point" in a titration

70.* From Table 12.5, select suitable indicators for detecting the equivalence point in each of the following titrations.

(a) 0.10-*M* NaOH(aq) with 0.10-*M* HCl(aq)

(b) 0.10-*M* acetic acid with 0.10-*M* NaOH(aq)

(c) 0.10-*M* $NH_3(aq)$ with 0.10-*M* HCl(aq)

71.* Repeat Problem 70 for the following titrations.

(a) 0.10-*M* KOH(aq) with 0.05-*M* $HNO_3(aq)$

(b) 0.20-*M* HF(aq) with 0.10-*M* NaOH(aq)

(c) 0.10-*M* $CH_3NH_2(aq)$ with 0.10-*M* HCl(aq)

Chapter 13 Electrochemistry

13.1 Electrochemical Cells
13.2 Applications of Electrochemical Cells
13.3 Electrolysis

Electrochemistry is the study of the use of chemical reactions to produce electricity, as in a battery, and of the use of electricity to drive chemical reactions, as in the production of aluminum. The reaction between silver ions and copper is an oxidation-reduction (redox) reaction that can supply electricity. These crystals of silver are being deposited on a copper coil in a solution of silver nitrate. The copper ions formed in the reaction have colored the solution blue.

Oxidation–reduction reactions, of which we have seen many examples in earlier chapters, involve a transfer of electrons. Because a flow of electrons constitutes an electric current, we can, at least in principle, use any redox reaction to produce an electric current. Flashlights, portable radios, pocket calculators, portable computers, watches, automobiles, and many other devices rely on batteries that use redox reactions to produce an electric current. Conversely, we can use an electric current to carry out redox reactions that do not proceed spontaneously. The industrial production of many metals, such as sodium and aluminum, depends on the use of an electric current to reduce a positive metal ion to the corresponding metal.

In this chapter we first discuss how an electric current can be produced by a redox reaction in a device called an electrochemical cell, and we will see that an understanding of electrochemical cells enables us to treat oxidation–reduction reactions quantitatively. We will also see that electrochemical cells are important to understanding the corrosion of metals. In the second part of the chapter we consider some examples of the use of electrical energy to carry out nonspontaneous redox reactions, some of which are important industrial processes.

13.1 ELECTROCHEMICAL CELLS

Cells and Cell Reactions

When we place a piece of zinc in a solution of copper(II) sulfate, the zinc soon becomes coated with a red-brown deposit of copper, and the blue color of the solution, due to the hydrated copper ion, fades. If enough zinc is present, the solution eventually becomes colorless (Demonstration 13.1). These color changes result from the reaction

$$Zn(s) + Cu^{2+}(aq) \rightarrow Zn^{2+}(aq) + Cu(s)$$

in which zinc is oxidized to $Zn^{2+}(aq)$ while $Cu^{2+}(aq)$ is reduced to copper. The sulfate ions take no part in this reaction, in which electrons are transferred from zinc to copper. When the reaction is carried out in this way, the transfer of electrons occurs on the surface of the zinc and cannot be directly observed. But if the oxidation and reduction reactions are separated from each other, the electron transfer can occur through an external circuit. We then have an **electrochemical cell** that is producing an electric current.

An electrochemical cell is a device that uses a spontaneous chemical reaction to produce an electric current.

DEMONSTRATION 13.1 The Reaction of Zinc with an Aqueous Solution of Copper Sulfate

A strip of zinc and an aqueous solution of copper sulfate

When the zinc strip is placed in the copper sulfate solution, it rapidly becomes coated with copper.

When the zinc strip is removed from the solution, the characteristic red-brown color of the copper deposit is clearly seen.

Figure 13.1 shows how an electric current can be produced by using the reaction in which zinc reduces Cu^{2+} ions. A piece of zinc, called the zinc **electrode** (a solid metal rod connected to the external circuit), is immersed in a solution of zinc(II) sulfate, and a piece of copper, the copper electrode, is immersed in a solution of copper(II) sulfate. The two solutions are kept apart by a porous barrier that permits ions to move from one solution to the other but prevents the two solutions from mixing. A conducting metal wire connects the two electrodes. At the surface of the zinc electrode, zinc atoms are oxidized to Zn^{2+} ions, which pass into

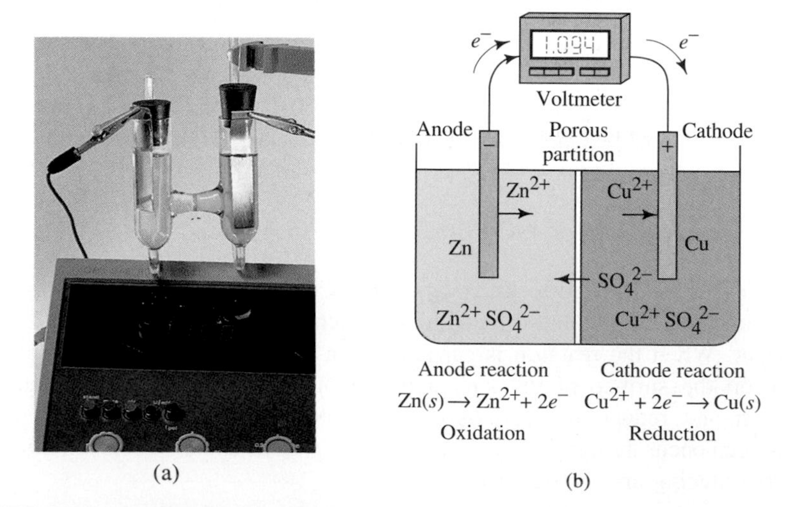

FIGURE 13.1 An Electrochemical Cell This cell is based on the overall reaction $Zn(s) + Cu^{2+}(aq) \rightarrow Zn^{2+}(aq) + Cu(s)$ (a) The experimental arrangement. The voltage of the cell is 1.100 V. (b) Schematic diagram showing the movement of the ions and the reactions that take place at the electrodes.

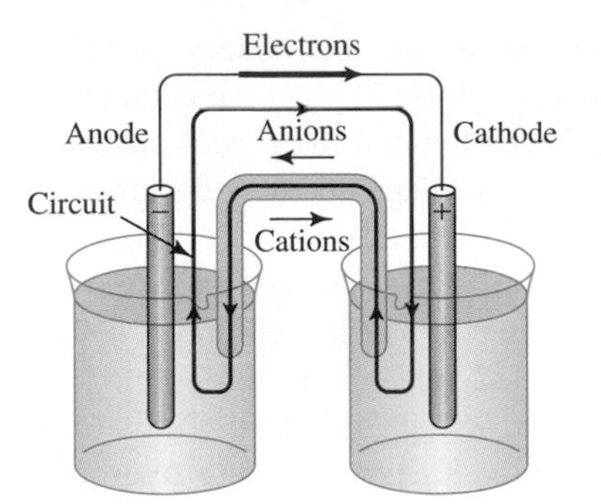

FIGURE 13.2 An Electrochemical Cell The cell consists of two half-cells connected by a salt bridge. The salt bridge contains a conducting solution, such as a concentrated KCl solution. Electrons are produced by an oxidation reaction at the anode. They travel through the external circuit and cause reduction at the cathode. Anions move from the cathode half-cell toward the anode half-cell, and cations move in the opposite direction and thereby complete the circuit inside the cell. The salt bridge allows movement of ions through the cell but prevents the solution in the two half-cells from mixing.

solution. The electrons lost by the zinc atoms leave the electrode and flow through the wire to the copper electrode. Here they reduce Cu^{2+} ions, so a deposit of copper forms on the surface of the electrode.

The electrode at which reduction occurs is called the *cathode*.
The electrode at which oxidation occurs is called the *anode*.

In the copper–zinc cell the electrode reactions are as follows:

Anode	$Zn(s) \rightarrow Zn^{2+}(aq) + 2e^-$	(Oxidation)
Cathode	$Cu^{2+}(aq) + 2e^- \rightarrow Cu(s)$	(Reduction)
Overall reaction	$Zn(s) + Cu^{2+}(aq) \rightarrow Zn^{2+}(aq) + Cu(s)$	

The two compartments of the cell are called *half cells*. The zinc electrode dipping into a zinc sulfate solution is one half-cell, and the copper electrode dipping into a copper sulfate solution is the other half-cell.

Any electrochemical cell consists of two half-cells; oxidation takes place in one, and reduction takes place in the other (Figure 13.2). Neither half-cell reaction can take place by itself, because each must be accompanied by another half-cell reaction that uses up or supplies the necessary electrons. These electrons constitute the electric current that flows through the external circuit. Not only must the circuit be complete outside the cell, but it must be complete inside as well; that is, ions must be able to move from one electrode to the other. This movement is made possible by separating the half-cells with a porous partition (of clay pottery, for example) as in Figure 13.1, or by a salt bridge, as in Figure 13.2. A **salt bridge** is an inverted U-tube containing a concentrated salt solution (such as potassium chloride) that is conveniently converted to a jelly by the addition of gelatin.

An electrochemical cell is formed whenever two different metals are inserted into a conducting medium. Here strips of copper and zinc inserted into a lemon produce a voltage of 0.9 V.

Cell Potentials

The electric current produced by an electrochemical cell is the result of electrons being pushed out of the anode and around the external circuit to the cathode, where they are used up. When a current flows between two points, we say that there is a *potential difference* between the two points. This potential difference is measured in volts (V) and is often called the *voltage* of the cell. When we speak of a 6-V battery, we mean a battery that has a potential difference of 6 V between its terminals. The potential difference is a measure of the work that the battery can do by pushing charge around an external circuit. If 1 J of work is done by moving a charge of 1 coulomb (C) through a potential difference, the potential difference is 1 V:

$$1\ V = 1\ J \cdot C^{-1}$$

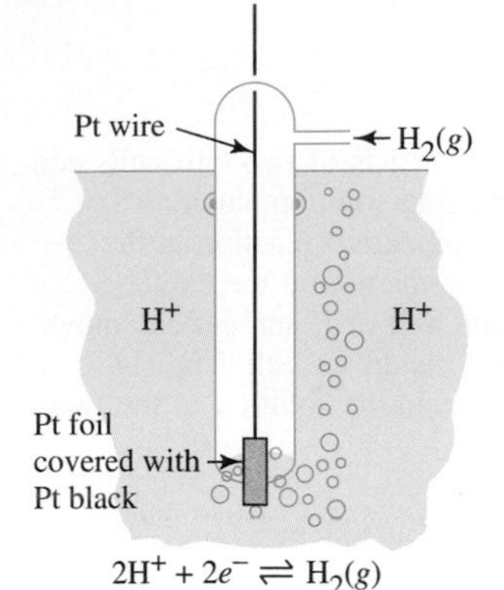

FIGURE 13.3 A Hydrogen Electrode A hydrogen electrode consists of a piece of platinum foil covered with a layer of powdered platinum (platinum black), attached to a Pt wire and surrounded by an atmosphere of hydrogen.

The potential or voltage between the electrodes of an electrochemical cell is called the **cell potential**, E_{cell}. For a given cell reaction, the cell potential depends on the concentrations of the ions in the cell, the temperature, and the partial pressures of any gases that might be involved in the cell reactions. When all the concentrations are 1 mol · L^{-1}, all partial pressures of gases are 1 atm, and the temperature is 25°C, the cell potential is called the **standard cell potential**, E°_{cell}. The cell potential can be measured by connecting a high-resistance voltmeter across the cell. The standard cell potential for the zinc–copper cell (Figure 13.1) is found to be 1.10 V.

In principle, any spontaneous redox reaction can be used as the basis of an electrochemical cell. For example, zinc reduces dilute aqueous acids to hydrogen:

$$Zn(s) + 2H^+(aq) \rightarrow Zn^{2+}(aq) + H_2(g)$$

To construct an electrochemical cell based on this reaction we would need a zinc electrode and a *hydrogen electrode*. To set up an electrode involving hydrogen, we use a piece of platinum foil coated with a fine powder of platinum (platinum black). The platinum serves as a catalyst for the reaction

$$2H^+(aq) + 2e^- \rightarrow H_2(g)$$

but is otherwise inert. This electrode is immersed in a solution of $H^+(aq)$—for example, a dilute sulfuric acid solution. Hydrogen ions combine with electrons from the metal to give hydrogen gas, which forms bubbles on the surface of the platinum. Thus, a hydrogen electrode consists of bubbles of hydrogen gas on the surface of platinum (see Figure 13.3).

A cell based on the reaction between zinc and dilute acid is shown in Figure 13.4. It is composed of a half-cell consisting of a hydrogen electrode immersed in a solution of sulfuric acid and a half-cell consisting of a zinc electrode in a solution of zinc sulfate. Zinc dissolves to give Zn^{2+} ions, and the electrons released flow around the wire of the external circuit and enter the platinum electrode, where they combine with hydrogen ions at its surface to give H_2. The standard cell potential is 0.76 V.

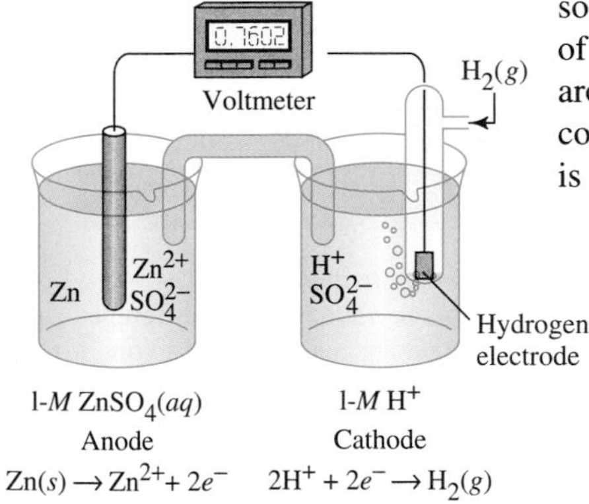

FIGURE 13.4 Electrochemical Cell Using a Hydrogen Electrode
This cell is based on the reaction

$$Zn(s) + 2H^+(aq) \rightarrow Zn^{2+}(aq) + H_2(g)$$

Anode	$Zn(s) \rightarrow Zn^{2+}(aq) + 2e^-$	(Oxidation)
Cathode	$2H^+(aq) + 2e^- \rightarrow H_2(g)$	(Reduction)
Overall reaction	$Zn(s) + 2H^+(aq) \rightarrow Zn^{2+}(aq) + H_2(g)$	$E^\circ = 0.76$ V

Rather than represent a cell by a detailed drawing, as in Figure 13.1, 13.3, and 13.4, we can use a shorthand representation called a **cell diagram**. For example, the cell in Figure 13.1 would be represented as

$$Zn(s)|ZnSO_4(aq)||CuSO_4(aq)|Cu(s)$$

and the cell shown in Figure 13.4 would be represented as

$$Zn(s)|ZnSO_4(aq)||H_2SO_4(aq)|H_2(g)|Pt(s)$$

In these diagrams the single vertical lines separate the electrodes from the solution with which they are in contact and the different components of each electrode. The double vertical lines indicate a salt bridge or porous barrier between the two solutions. In such a cell diagram it is conventional to place the anode—the electrode at which oxidation occurs—on the left and the cathode—the electrode at which reduction occurs—on the right.

Standard Reduction Potentials

Just as we can think of the overall cell reaction as the sum of two half-reactions, we can think of the cell potential as the sum of two *half-cell potentials:* E_{ox}, due to the oxidation half-reaction, and E_{red}, due to the reduction half-reaction:

$$E_{cell} = E_{ox} + E_{red}$$

Under standard conditions,

$$E^\circ_{cell} = E^\circ_{ox} + E^\circ_{red}$$

It would be useful to have values for half-cell potentials, but no potential can be measured unless two half-cells are connected to give a cell in which an overall reaction can occur. Therefore, we cannot measure experimentally the potential associated with any individual half-reaction; we can measure only the *sum* of two half-cell potentials.

However, we can obtain relative values of standard half-cell potentials by making an arbitrary assumption about the potential of one particular half-cell. The half-cell chosen as the standard with which all other half-cell potentials are compared is the **hydrogen half-cell**, also known as the **hydrogen electrode**. This half-cell is arbitrarily assigned a standard potential of exactly 0 V:

$$E^\circ_{2H^+ + 2e^- \rightarrow H_2} = 0 \text{ V}$$

The measured standard cell potential for the zinc–hydrogen cell is 0.76 V. Because

$$E^\circ_{cell} = E^\circ_{ox} + E^\circ_{red}$$

we see that

$$0.76 \text{ V} = E^\circ_{Zn \rightarrow Zn^{2+} + 2e^-} + E^\circ_{2H^+ + 2e^- \rightarrow H_2}$$

and we can then say that

$$0.76 \text{ V} = E^\circ_{Zn \rightarrow Zn^{2+} + 2e^-} + 0$$

or

$$E^\circ_{Zn \rightarrow Zn^{2+} + 2e^-} = 0.76 \text{ V}$$

Thus, we see that the standard potential for the half-cell reaction

$$Zn(s) \rightarrow Zn^{2+} + 2e^-$$

is $E^\circ_{ox} = 0.76$ V.

Now we can use this value to find the half-cell potential for the copper electrode by combining this electrode with the zinc electrode. We have already seen that the standard potential of the zinc–copper cell is 1.10 V. Therefore,

$$\begin{aligned} 1.10 \text{ V} &= E^\circ_{ox} + E^\circ_{red} = E^\circ_{Zn \rightarrow Zn^{2+} + 2e^-} + E^\circ_{Cu^{2+} + 2e^- \rightarrow Cu} \\ &= 0.76 \text{ V} + E^\circ_{Cu^{2+} + 2e^- \rightarrow Cu} \\ E^\circ_{Cu^{2+} + 2e^- \rightarrow Cu} &= 0.34 \text{ V} \end{aligned}$$

Thus, for the reaction

$$Cu^{2+} + 2e^- \rightarrow Cu(s)$$

we have

$$E^\circ_{red} = 0.34 \text{ V}$$

Because all the reactions in the cells we consider are in aqueous solutions, we have omitted the designation (*aq*) in the rest of this section. For half-cell reactions written as subscripts, we have for simplicity omitted phase designations.

We could alternatively have obtained the standard potential for the copper half-cell by combining it with a hydrogen electrode. However, we know that copper does not dissolve in dilute aqueous acids. The equilibrium

$$Cu(s) + 2H^+ \rightleftharpoons Cu^{2+} + H_2(g)$$

lies far to the left. In other words, the reverse reaction occurs when we combine a hydrogen half-cell with a copper half-cell; that is,

$$Cu^{2+} + H_2(g) \rightarrow Cu(s) + 2H^+$$

The experimentally measured potential of this cell is 0.34 V. Thus,

$$0.34 \text{ V} = E^\circ_{ox} + E^\circ_{red} = E^\circ_{H_2 \rightarrow 2H^+ + 2e^-} + E^\circ_{Cu^{2+} + 2e^- \rightarrow Cu}$$

$$= 0 + E^\circ_{Cu^{2+} + 2e^- \rightarrow Cu}$$

So for the reaction

$$Cu^{2+} + 2e^- \rightarrow Cu(s)$$

we have

$$E^\circ_{red} = 0.34 \text{ V}$$

This is the same value we obtained previously.

By combining any half-cell with a hydrogen electrode, we can determine its standard half-cell potential. Some of the potentials obtained in this way are for oxidation half-reactions; others are for reduction half-reactions. It is convenient when tabulating values to list them all in the same way. By convention, we list the potentials for reduction reactions; they are called **standard reduction potentials**, E°_{red}. We convert an oxidation reaction into a reduction reaction by reversing it, at the same time changing the sign of its potential: $E^\circ_{red} = -E^\circ_{ox}$.

EXAMPLE 13.1 Standard Reduction Potentials

Iron dissolves in dilute acid to give Fe^{2+}:

$$Fe(s) + 2H^+ \rightarrow Fe^{2+} + H_2(g)$$

The potential of a cell constructed from a standard hydrogen electrode and a half-cell consisting of an iron electrode in a 1-*M* solution of $FeSO_4$ is 0.44 V. What is the standard reduction potential of the Fe half-cell?

Solution:

$$E^\circ_{cell} = E^\circ_{ox} + E^\circ_{red}$$

$$0.44 \text{ V} = E^\circ_{Fe \rightarrow Fe^{2+} + 2e^-} + 0$$

Therefore,

$$Fe(s) \rightarrow Fe^{2+} + 2e^- \qquad E^\circ_{ox} = 0.44 \text{ V}$$

For the reverse reaction,

$$Fe^{2+} + 2e^- \rightarrow Fe(s) \qquad E^\circ_{red} = -0.44 \text{ V}$$

The standard potential for a reduction half-reaction is E°_{red}. The standard potential for the same half-reaction written as an oxidation is E°_{ox}. And

$$E^\circ_{red} = -E^\circ_{ox}$$

Exercise 13.1 The standard cell potential for the cell

$$Zn(s)|ZnSO_4(aq)||AgNO_3(aq)|Ag(s)$$

is 1.56 V. What is the standard reduction potential for the half-reaction $Ag^+ + e^- \rightarrow Ag(s)$?

Table 13.1 gives the value of the standard reduction potentials for a number of half-reactions. Recall that these are the half-cell potentials when the partial pressures of any gases are 1 atm and the concentrations of ions are 1 *M*. By convention, the most negative reduction potentials are listed at the top of the table, the most positive at the bottom.

The significance of a negative standard reduction potential is that, relative to H_2, the substances on the right side tend to give up electrons, and the half-reaction tends to proceed from right to left. The larger the negative value, the greater the tendency of the reaction to proceed from right to left. Thus, the reaction $Li^+ + e^- \rightarrow Li(s)$ has a very strong tendency to proceed in the direction opposite that in which it is written. A positive standard reduction potential indicates that, relative to H^+, the substances on the left tend to acquire electrons, and the half-

TABLE 13.1 Standard Reduction Potentials

Increasing oxidizing strength ↓	Reaction* (Oxidizing Agents → Reducing Agents)	Increasing reducing strength ↑	E°_{red} (V)
Very weak	$Li^+ + e^- \rightarrow Li(s)$	Very strong	−3.05
	$K^+ + e^- \rightarrow K(s)$		−2.93
	$Ca^{2+} + 2e^- \rightarrow Ca(s)$		−2.87
	$Na^+ + e^- \rightarrow Na(s)$		−2.71
	$Mg^{2+} + 2e^- \rightarrow Mg(s)$		−2.36
	$Al^{3+} + 3e^- \rightarrow Al(s)$		−1.66
	$2H_2O + 2e^- \rightarrow H_2(g) + 2OH^-$		−0.83
	$Zn^{2+} + 2e^- \rightarrow Zn(s)$		−0.76
	$Cr^{3+} + 3e^- \rightarrow Cr(s)$		−0.74
	$Fe^{2+} + 2e^- \rightarrow Fe(s)$		−0.44
	$Cr^{3+} + e^- \rightarrow Cr^{2+}$		−0.41
	$Ni^{2+} + 2e^- \rightarrow Ni(s)$		−0.25
	$Sn^{2+} + 2e^- \rightarrow Sn(s)$		−0.16
	$Pb^{2+} + 2e^- \rightarrow Pb(s)$		−0.13
	$2H^+ + 2e^- \rightarrow H_2(g)$		0
	$S(s) + 2H^+ + 2e^- \rightarrow H_2S$		+0.14
	$Cu^{2+} + e^- \rightarrow Cu^+$		+0.15
	$AgCl(s) + e^- \rightarrow Ag(s) + Cl^-$		+0.22
	$Cu^{2+} + 2e^- \rightarrow Cu(s)$		+0.34
	$Cu^+ + e^- \rightarrow Cu(s)$		+0.52
	$I_2(s) + 2e^- \rightarrow 2I^-$		+0.54
	$O_2(g) + 2H^+ + 2e^- \rightarrow H_2O_2$		+0.68
	$Fe^{3+} + e^- \rightarrow Fe^{2+}$		+0.77
	$Ag^+ + e^- \rightarrow Ag(s)$		+0.80
	$NO_3^- + 2H^+ + e^- \rightarrow NO_2(g) + H_2O$		+0.80
	$NO_3^- + 4H^+ + 3e^- \rightarrow NO(g) + 2H_2O$		+0.97
	$Br_2(l) + 2e^- \rightarrow 2Br^-$		+1.09
	$O_2(g) + 4H^+ + 4e^- \rightarrow 2H_2O$		+1.23
	$Cr_2O_7^{2-} + 14H^+ + 6e^- \rightarrow 2Cr^{3+} + 7H_2O$		+1.33
	$Cl_2(g) + 2e^- \rightarrow 2Cl^-$		+1.36
	$MnO_4^- + 8H^+ + 5e^- \rightarrow Mn^{2+} + 4H_2O$		+1.49
	$Au^{3+} + 3e^- \rightarrow Au(s)$		+1.50
	$H_2O_2 + 2H^+ + 2e^- \rightarrow 2H_2O$		+1.78
	$Co^{3+} + e^- \rightarrow Co^{2+}$		+1.81
Very strong	$F_2(g) + 2e^- \rightarrow 2F^-$	Very weak	+2.87

* The designation (*aq*) has been omitted after each ion and molecule in solution.

reaction tends to proceed from left to right. The larger the positive value, the greater the tendency of a reaction to proceed to the right, so the reaction $F_2 + 2e^- \rightarrow 2F^-$ (E°_{red} = +2.87 V) tends very strongly to proceed in the direction in which it is written. The substances on the right in Table 13.1 are reducing agents. The strongest reducing agents, such as Li and Mg, are at the top, and the weakest, such as Cl^- and F^-, are at the bottom. The substances on the left are oxidizing agents. The weakest oxidizing agents, such as Li^+ and Mg^{2+}, are at the top; the strongest, such as NO_3^-, MnO_4^-, O_2, and F_2, are at the bottom.

Table 13.1 includes half-reactions other than the reduction of a metal ion to the corresponding metal. For example, it includes the reduction of Cl_2 to Cl^- and Fe^{3+} to Fe^{2+}. Half-cells for these reactions consist of a solution of the ions (and gases) taking part in the reaction and a platinum electrode that supplies or removes the electrons. We can use the standard reduction potentials listed in Table 13.1 to predict the relative oxidizing and reducing strengths of any of the substances in the table. For example, because zinc has a more negative reduction potential than does iron and is above iron in the table, zinc is the stronger reducing agent. Because Br_2 has a larger positive reduction potential than does Fe^{3+} and is below iron in the table, Br_2 is the stronger oxidizing agent.

EXAMPLE 13.2 Relative Strengths of Oxidizing Agents

Place Cu^{2+}, MnO_4^-, Br_2, and Zn^{2+} in order of their strengths as oxidizing agents.

Solution: From Table 13.1, the reduction potentials are

$$Cu^{2+} + 2e^- \rightarrow Cu \qquad +0.34\ V$$

$$MnO_4^- + 8H^+ + 5e^- \rightarrow Mn^{2+} + 4H_2O \qquad +1.49\ V$$

$$Br_2 + 2e^- \rightarrow 2Br^- \qquad +1.09\ V$$

$$Zn^{2+} + 2e^- \rightarrow Zn \qquad -0.76\ V$$

The substance with the highest positive reduction potential is the strongest oxidizing agent. Therefore the order of oxidizing strength is $MnO_4^- > Br_2 > Cu^{2+} > Zn^{2+}$.

III The stronger an oxidizing agent is, the more easily it is reduced. Thus, the closer to the bottom of the left column of Table 13.1 a substance appears, the better an oxidizing agent it is.

Exercise 13.2 Place Zn^{2+}, Co^{3+}, H_2O, and NO_3^- in order of their strengths as oxidizing agents in aqueous solution.

Exercise 13.3 Place H_2S, Al(*s*), H_2O_2, and Fe^{2+} in order of their strengths as reducing agents in aqueous solution.

Direction of Oxidation-Reduction Reactions

When we combine any two half-reactions from Table 13.1 to obtain an overall reaction, the reaction that is higher in the table proceeds from right to left, and the reaction that is lower in the table proceeds from left to right. For example, if we combine the two half-reactions

$$Zn^{2+} + 2e^- \rightarrow Zn(s) \qquad and \qquad Fe^{2+} + 2e^- \rightarrow Fe(s)$$

we must reverse the first reaction so that it proceeds from right to left

$$Zn(s) \rightarrow Zn^{2+} + 2e^-$$

before we can combine it with

$$Fe^{2+} + 2e^- \rightarrow Fe(s)$$

to give the overall reaction

$$Zn(s) + Fe^{2+} \rightarrow Zn^{2+} + Fe(s)$$

Zinc, near the top of the column of reducing agents in Table 13.1, is a stronger reducing agent than Fe, so it reduces Fe^{2+} to Fe.

We can see similarly that because the half-reaction

$$Cu^{2+} + 2e^- \rightarrow Cu(s)$$

is above the half-reaction

$$Ag^+ + e^- \rightarrow Ag(s)$$

in Table 13.1, the overall reaction is

$$Cu(s) + 2Ag^+ \rightarrow Cu^{2+} + 2Ag(s)$$

From Table 13.1 we can also see that Fe(*s*) will reduce Cu^{2+} to Cu(*s*) and that Zn(*s*) will reduce Pb^{2+} to Pb(*s*) and Sn^{2+} to Sn(*s*) (Demonstration 13.2).

Table 13.1 puts the *activity series* of metals that we discussed in Chapter 10 on a quantitative basis. The most reactive metals, such as Na and Mg, are at the top of the table and have the largest negative reduction potentials. These are the most easily oxidized metals (the strongest reducing agents). The least reactive metals, such as Cu and Ag, are at the bottom of the table and have the largest positive reduction potentials. These metals are the most difficult to oxidize (the weakest reducing agents). All the metals with a negative reduction potential react with dilute aqueous acids and thereby reduce H^+ to hydrogen, whereas none of the metals with a positive reduction potential react with aqueous acids to give hydrogen.

DEMONSTRATION 13.2 Metal Displacement Reactions

An iron nail dipped in copper sulfate solution becomes coated with a reddish-brown layer of copper.

A coil of copper wire suspended in silver nitrate solution becomes covered with "whiskers" of silver, and the solution slowly turns blue as copper ions form.

A strip of zinc in lead nitrate solution becomes coated with a spongy layer of lead.

A strip of zinc in an acidic solution of tin(II) chloride, $SnCl_2$, becomes coated with fine crystals of tin.

Calculation of Cell Potentials

We can calculate the voltage of a cell made up of any two half-cells by adding the half-cell potentials. Consider a cell made by combining the two half-cells

$$Zn^{2+} + 2e^- \rightarrow Zn(s) \qquad \textit{and} \qquad Fe^{2+} + 2e^- \rightarrow Fe(s)$$

We have seen that the first reaction will proceed in the reverse direction and that the overall reaction in the cell will be

$$Zn(s) + Fe^{2+} \rightarrow Zn^{2+} + Fe(s)$$

The cell diagram is

$$Zn(s)|Zn^{2+}(aq)||Fe^{2+}(aq)|Fe(s)$$

The half-reactions and the corresponding potentials are

Anode	$Zn(s) \rightarrow Zn^{2+} + 2e^-$	$E^\circ_{ox} = -E^\circ_{red} = +0.76$ V
Cathode	$Fe^{2+} + 2e^- \rightarrow Fe(s)$	$E^\circ_{red} = -0.44$ V

The potential for the oxidation of zinc is found by changing the sign of the potential for the reduction reaction given in Table 13.1. Adding the half-cell reactions and potentials gives, for the complete cell,

$$Zn(s) + Fe^{2+} \rightarrow Zn^{2+} + Fe(s) \qquad E^\circ_{cell} = +0.32 \text{ V}$$

Now consider the cell

$$Cu(s)|Cu^{2+}(aq)||Ag^{+}(aq)|Ag(s)$$

obtained by combining the two half-cells

$$Cu^{2+} + 2e^- \rightarrow Cu(s) \qquad \textit{and} \qquad Ag^+ + e^- \rightarrow Ag(s)$$

The copper half-cell is higher in the table, and therefore the copper half-cell reaction must be reversed and written as an oxidation before we can combine it with the silver half-cell reaction:

Anode	$Cu(s) \rightarrow Cu^{2+} + 2e^-$	$E^\circ_{ox} = -E^\circ_{red} = -0.34$ V
Cathode	$2[Ag^+ + e^- \rightarrow Ag(s)]$	$E^\circ_{red} = +0.80$ V
Overall	$Cu(s) + 2Ag^+ \rightarrow Cu^{2+} + 2Ag(s)$	$E^\circ_{cell} = +0.46$ V

Notice that when we multiply the equation for a half-reaction by an appropriate coefficient to obtain an equation for the overall reaction, we do *not* multiply the half-cell potential by this coefficient. The half-cell potential is always the value given in the table and is independent of any coefficient by which we must multiply the corresponding equation to obtain a balanced equation for the overall reaction. The standard potential does not depend on the amounts of the reactants and the products but only on their concentrations, which are 1 $mol \cdot L^{-1}$. It therefore does *not* depend on *how many* electrons are transferred. A potential difference may be regarded as a difference in level between which the electrons flow; it does not depend on how many electrons flow between the levels.

Let us calculate the cell potential for the cell

$$Ag(s)|Ag^{+}(aq)||Cu^{2+}(aq)|Cu(s)$$

that is, for the reaction

$$2Ag(s) + Cu^{2+} \rightarrow 2Ag^+ + Cu(s)$$

which is the reverse of the reaction that we just considered. The half-cell reactions and their appropriate potentials are as follows:

Anode	$2[Ag(s) \rightarrow Ag^+ + e^-]$	$E^\circ_{ox} = -E^\circ_{red} = -0.80$ V
Cathode	$Cu^{2+} + 2e^- \rightarrow Cu(s)$	$E^\circ_{red} = +0.34$ V
Overall	$2Ag(s) + Cu^{2+} \rightarrow 2Ag^+ + Cu(s)$	$E^\circ_{cell} = -0.46$ V

The overall cell potential has a negative value, whereas it has a positive value for the reverse reaction. The negative value indicates that the reaction does not proceed in the direction written but rather in the reverse direction, that is,

$$2Ag^+ + Cu(s) \rightarrow Cu^{2+} + 2Ag(s)$$

as we have seen.

- The cell potential for a redox reaction that proceeds spontaneously from left to right as written is positive.
- The cell potential for a redox reaction that does not proceed spontaneously from left to right as written but proceeds in the reverse direction is negative.

So for redox reactions we have another criterion for judging whether a reaction will proceed spontaneously or not: the sign of the cell potential.

EXAMPLE 13.3 Calculating Standard Cell Potentials

What will be the spontaneous reaction when the following half-reactions are combined? What is the value of E°_{cell}?

(a) $Fe^{3+} + e^- \rightarrow Fe^{2+}$

(b) $MnO_4^- + 8H^+ + 5e^- \rightarrow Mn^{2+} + 4H_2O$

Solution: Reaction (a) is higher in Table 13.1, so that reaction proceeds to the left, driving reaction (b) to the right. We therefore reverse reaction (a) and change the sign of its potential. After multiplying by the appropriate coefficient, we add it to reaction (b). We then add the half-cell potentials to obtain E°_{cell}.

$5[Fe^{2+} \rightarrow Fe^{3+} + e^-]$	$E^\circ_{ox} = -0.77$ V
$MnO_4^- + 8H^+ + 5e^- \rightarrow Mn^{2+} + 4H_2O$	$E^\circ_{red} = +1.49$ V
$5Fe^{2+} + MnO_4^- + 8H^+ \rightarrow 5Fe^{3+} + Mn^{2+} + 4H_2O$	$E^\circ_{cell} = +0.72$ V

The overall reaction is the reduction of MnO_4^- to Mn^{2+} and the oxidation of Fe^{2+} to Fe^{3+}. The cell voltage has a positive value, confirming that we combined the two half-cell reactions correctly.

Exercise 13.4 What will be the spontaneous reaction, and what is the value of E°_{cell}, when the following half-reactions are combined?

$Cr_2O_7^{2-} + 14H^+ + 6e^- \rightarrow 2Cr^{3+} + 7H_2O$

$Cu^{2+} + 2e^- \rightarrow Cu(s)$

Gibbs Free Energy and Cell Potential

We have seen that the calculated cell potential for a redox reaction tells us whether the reaction will proceed spontaneously or not. In Chapter 9 we saw that a spontaneous reaction has a negative Gibbs free energy change and that a reaction with a positive Gibbs free energy change does not proceed spontaneously. So there must be a relationship between the free energy change for a reaction and its cell potential. Thermodynamics shows that this relationship is

$$\Delta G = -nFE_{\text{cell}}$$

For standard conditions (1-M solutions, P = 1 atm for gases), a spontaneous reaction always has a positive E°_{cell}.

If all the substances are in their standard states,

$$\Delta G^\circ = -nFE^\circ_{\text{cell}}$$

Here n is the number of moles of electrons transferred in the reaction, and F is the **Faraday constant**, named after Michael Faraday (Box 13.1). The Faraday constant is the electric charge on 1 mol of electrons. The charge on one electron is $1.602\,19 \times 10^{-19}$ C. Therefore, the charge on 1 mol of electrons is

$$(6.022\,05 \times 10^{23}\ e \cdot \text{mol}^{-1})(1.602\,19 \times 10^{-19}\ \text{C} \cdot e^{-1}) = 96{,}485\ \text{C} \cdot \text{mol}^{-1}$$

BOX 13.1 Michael Faraday

Michael Faraday was one of 10 children of a blacksmith in London. He had only a basic elementary education before, at the age of 14, he was apprenticed to a bookbinder. Faraday's employer, who was uncharacteristically lenient for the times, allowed the boy to read the books in his shop, and Faraday educated himself. In 1812 he was given tickets to attend lectures given by Humphrey Davy, the director of the Royal Institution. Faraday wrote up careful, complete notes of the lectures and bound them in a book. Friends persuaded him to send the notes to Davy in support of his application for a position as an assistant to Davy. Faraday obtained the position, beginning not only his prolific scientific career but also a very fruitful collaboration with Davy. Soon after he was appointed, Faraday left with Davy on a grand tour of Europe, which did much to broaden Faraday's scientific education and gave him the opportunity to meet many famous scientists.

When they were in Florence, Italy, Davy and Faraday were able to use a very large lens belonging to the Duke of Tuscany to prove conclusively that diamond consists only of carbon—an idea that was difficult for many scientists at that time to accept. Davy and Faraday used the lens to focus the sun's rays on a diamond enclosed in a bulb of pure oxygen. After about an hour the diamond began to burn. "The diamond," Faraday wrote in his journal, "glowed brilliantly with a scarlet light and when placed in the dark continued to burn for about four minutes." They burned the diamond completely and showed that the bulb then contained nothing but carbon dioxide and excess oxygen.

In 1825 Faraday replaced Davy as director of the Royal Institution, and his reputation soon began to rival that of Davy. Faraday was the first to liquefy several gases, including carbon dioxide, hydrogen sulfide, hydrogen bromide, and chlorine; he discovered benzene and determined its composition; and he discovered the quantitative laws of electrolysis. The Faraday constant, F, was named in his honor. Faraday made even greater contributions to physics. He found that an electric current could be induced in a wire by a moving magnet. He provided both the experimental basis and basic ideas for the theory of electromagnetism, which was later developed by James Clerk Maxwell. Albert Einstein ranked Faraday with Isaac Newton, Galileo, and Maxwell as the greatest physicists of all time.

Michael Faraday (1791–1867)

Faraday was a member of a religious sect so strict that for a time he was denied membership of the church because he had accepted an invitation to have lunch with Queen Victoria on a Sunday! In accordance with his religious beliefs, he tried to live a simple life, accepting rather reluctantly the many honors that came to him. His beliefs enabled him to resolve without uncertainty a moral problem that still faces scientists. During the Crimean War, in the 1850s, the British government asked him to head an investigation of the possibility of preparing large quantities of poison gas for use on the battlefield. Faraday refused to have anything to do with the project, and nothing came of the idea at that time.

So

$$F = 96{,}485 \text{ C} \cdot \text{mol}^{-1}$$

The value of 96,500 $C \cdot mol^{-1}$, obtained by rounding 96,485 off to 3 significant figures, is accurate enough for most purposes.

Recall from Chapter 12 that

$$\Delta G = -RT \ln \frac{K}{Q} \qquad \textit{and} \qquad \Delta G^\circ = -RT \ln K$$

Here Q is the reaction quotient (Chapter 12), an expression that has the same form as the equilibrium constant expression but in which the concentrations of the reactants and products are *not* the equilibrium concentrations. At standard conditions $\Delta G = \Delta G^\circ$ and all the concentrations are 1 M, so $Q = 1$. We can now summarize the various criteria that we can use to decide if a redox reaction occurs spontaneously or not:

- A reaction will proceed spontaneously from left to right as written if $\quad \Delta G < 0, \quad Q < K, \quad E_{cell} > 0$
- A reaction will not proceed spontaneously but will proceed in the reverse direction from right to left if $\quad \Delta G > 0, \quad Q > K, \quad E_{cell} < 0$
- A reaction has reached equilibrium if $\quad \Delta G = 0, \quad Q = K, \quad E_{cell} = 0$

Effect of Concentration on Cell Voltage: The Nernst Equation

The values of the standard reduction potentials given in Table 13.1 are for solutions in which all the dissolved molecules or ions have concentrations of 1 M and the partial pressures of any gases involved are 1 atm. In practice, the concentrations of the reactants and products of cell reactions are generally not 1 M, and the pressures of gases are not necessarily 1 atm. Even if we start with 1-M solutions, the concentrations of the reactants decrease, and the concentrations of the products increase as the reaction proceeds toward equilibrium. The cell potential decreases as the reaction proceeds. It becomes zero at equilibrium, when there is no further tendency for the cell reaction to occur. After a flashlight battery has been used for some time, its voltage decreases to a very low value; we say that the battery is dead.

The quantitative relationship between the cell potential and the concentrations of the reactants and products is given by the **Nernst equation**, as first shown by the German chemist Walther Nernst (1864–1941) in 1889. We saw previously that

$$\Delta G = -nFE_{cell} = -RT \ln \frac{K}{Q}$$

so

$$E_{cell} = \frac{RT}{nF} \ln \frac{K}{Q}$$

Because at standard conditions all concentrations and gas pressures are unity, $Q = 1$. Therefore,

$$E^\circ_{cell} = \frac{RT}{nF} \ln K$$

or

$$E_{cell} = E^\circ_{cell} - \frac{RT}{nF} \ln Q \qquad \textit{Nernst equation}$$

where $E°$ is the standard potential, R is the gas constant, T is the temperature, n is the number of electrons transferred in the reaction, F is the Faraday constant, and Q is the reaction quotient. Substituting $R = 8.314\ \text{J} \cdot \text{K}^{-1} \cdot \text{mol}^{-1}$, $T = 25°\text{C} = 298\ \text{K}$, and $F = 96{,}500\ \text{C} \cdot \text{mol}^{-1} = 96{,}500\ \text{J} \cdot \text{V}^{-1} \cdot \text{mol}^{-1}$ gives the expression

$$E_{\text{cell}} = E°_{\text{cell}} - \frac{0.0257}{n} \ln Q$$

If you prefer to use logarithms to the base 10, this equation becomes

$$E_{\text{cell}} = E°_{\text{cell}} - \frac{0.0592}{n} \log Q$$

From here on, for convenience we omit the subscript "cell."

Using this equation, we can calculate the potential of a cell if we know the concentrations of the reactants and the standard cell potential, $E°$. Consider, as an example, a cell based on the reaction

$$\text{Zn}(s) + \text{Cu}^{2+} \rightarrow \text{Zn}^{2+} + \text{Cu}(s)$$

for which $E° = 1.10$ V. In this reaction 2 mol of electrons are transferred per mole of Cu^{2+} that is reduced, so $n = 2$. Hence, the Nernst equation becomes

$$E = 1.10\ \text{V} - \frac{0.0257\ \text{V}}{2} \ln \frac{[\text{Zn}^{2+}]}{[\text{Cu}^{2+}]}$$

Recall from Chapter 12 that solids have a constant "concentration," so they do not appear in the expression for Q.

Consider a case in which the cell started to operate with $[Zn^{2+}] = [Cu^{2+}] = 1.00\ M$, and after a certain time the concentrations were $[Cu^{2+}] = 0.10\ M$ and $[Zn^{2+}] = 1.90\ M$. Then

$$E = 1.10\ \text{V} - \frac{0.0257}{2} \ln \frac{1.9}{0.1}\ \text{V} = (1.10 - 0.038)\ \text{V} = 1.06\ \text{V}$$

We see that the cell voltage has decreased but only by a rather small amount.

Let us calculate the voltage for a later time when the concentrations have reached the values $[Cu^{2+}] = 0.001$ and $[Zn^{2+}] = 1.999$:

$$E = 1.10\ \text{V} - \frac{0.0257}{2} \ln \frac{1.999}{0.001}\ \text{V} = (1.10 - 0.10)\ \text{V} = 1.00\ \text{V}$$

The cell voltage continually decreases as the cell reaction proceeds.

Equilibrium Constants from the Nernst Equation The potential of a cell becomes zero when the cell reaction has attained equilibrium. The reaction quotient, Q, is then equal to the equilibrium constant, K, and the Nernst equation may be written in the form

$$E = E° - \frac{RT}{nF} \ln K = 0$$

This can be rearranged to

$$\ln K = \frac{nFE°}{RT} = \frac{n\ (96{,}500\ \text{J} \cdot \text{V}^{-1} \cdot \text{mol}^{-1})E°\ \text{V}}{(8.314\ \text{J} \cdot \text{K}^{-1} \cdot \text{mol}^{-1})(298\text{K})} = \frac{nE°}{0.0257}$$

or

$$\log K = \frac{nE°}{0.0592} \quad \text{at } 25°\text{C}$$

This equation enables us to calculate the equilibrium constant for any oxidation–reduction reaction in aqueous solution from the corresponding standard cell potential, which in turn can often be obtained from the tabulated standard reduction potentials of the two half-cell reactions (Table 13.1).

EXAMPLE 13.4 Equilibrium Constant and the Nernst Equation

What is the equilibrium constant for the following reaction at 25°C?

$$Cu(s) + Br_2 \rightarrow Cu^{2+} + 2Br^-$$

Solution: First we write the two half-cell reactions and calculate the standard cell potential:

$$\begin{array}{rlr} Br_2 + 2e^- \rightarrow 2Br^- & & E^\circ_{red} = 1.09\text{ V} \\ Cu(s) \rightarrow Cu^{2+} + 2e^- & & E^\circ_{ox} = -E^\circ_{red} = -0.34\text{ V} \\ \hline Br_2 + Cu(s) \rightarrow Cu^{2+} + 2Br^- & & E^\circ_{cell} = 0.75\text{ V} \end{array}$$

For this reaction $n = 2$, so we have

$$\log K = \frac{nE^\circ}{0.0592} = \frac{2(0.75)}{0.0592} = 25$$

$$K = 1 \times 10^{25}$$

and the reaction goes to completion.

III The two half-reactions must involve the same number of electrons so that no electrons appear in the balanced overall cell reaction.

Exercise 13.5 What is the equilibrium constant for the following reaction in aqueous solution at 25°C?

$$5Fe^{2+} + MnO_4^- + 8H^+ \rightarrow 5Fe^{3+} + Mn^{2+} + 4H_2O$$

Example 13.4 and Exercise 13.5 illustrate an important use of the Nernst equation, namely, to obtain values of equilibrium constants for reactions that go essentially to completion—that is, for reactions that have very large (or very small) equilibrium constants—and that are difficult to determine in any other way.

Concentration Cells The Nernst equation shows that a cell will have a potential even if the two half-cell reactions are the same, provided that the solutions have a different concentration. A cell of this type is called a **concentration cell**. Its standard potential is zero, because the concentrations in both half-cells are then 1 *M*. If the cell is

$$Cu(s)|Cu^{2+}(aq\ dil)||Cu^{2+}\ (aq\ conc)|Cu(s)$$

the "reaction" in the cell is the transfer of Cu^{2+} ions from the more concentrated to the more dilute solution,

$$Cu^{2+}\ (conc) \rightarrow Cu^{2+}\ (dil)$$

as a result of the natural tendency of the two concentrations to become equal. Two moles of electrons are transferred per mole of Cu^{2+} transferred, so the cell potential is

$$E = 0\text{ V} - \frac{0.0257}{2} \ln \frac{[Cu^{2+}\ dil]}{[Cu^{2+}\ conc]}\text{ V}$$

The label "(*aq dil*)" signifies a dilute aqueous solution, and "(*aq conc*)" a concentrated aqueous solution.

For example, if one half-cell has $[Cu^{2+}] = 1.0\ M$ and the other half-cell has $[Cu^{2+}] = 0.010\ M$,

$$E = 0\text{ V} - \frac{0.0257}{2}\ln\frac{0.010}{1.0}\text{ V} = 0.059\text{ V} = 59\text{ mV}$$

An electric eel

The principle of the concentration cell is very important in living systems, where biological membranes separate regions of different ion concentrations and behave like salt bridges. For example, the transmission of an electrical impulse along a nerve cell depends on a potential difference resulting from different concentrations of potassium and sodium ions inside and outside the cell. Some fishes and eels have a special organ that behaves as an electrochemical cell that generates an electric potential used as a defense mechanism. The electric eel found in the Amazon River has a bank of such cells that can generate a potential difference of as much as 350 V between its head and its tail.

pH Meter An important application of the Nernst equation is the measurement of hydrogen ion concentration or pH. In principle, this could be done with a cell in which one of the electrodes is a hydrogen electrode. But a hydrogen electrode is not very convenient to use, so in practice a *glass electrode* is used (Figure 13.5). This electrode consists of a thin glass bulb that contains a dilute aqueous acid solution. When this electrode is dipped into a solution in which the H^+ concentration is different from that in the bulb, a potential is set up across the glass, as in a concentration cell:

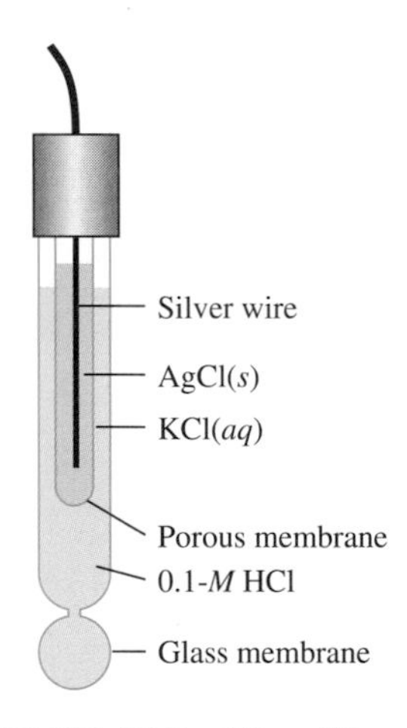

FIGURE 13.5 Glass Electrode The potential of this electrode depends on the concentration of hydrogen ion in any solution in which it is immersed.

$$E_{\text{glass electrode}} = 0 - \frac{0.0592}{1}\log\frac{[H^+]_{\text{inside}}}{[H^+]_{\text{outside}}}$$

For the solution being measured, $\text{pH} = -\log[H^+]_{\text{outside}}$. Because the hydrogen ion concentration inside the glass electrode is constant, we have

$$E_{\text{glass electrode}} = 0.0592\text{ pH} + \text{constant}$$

In a pH meter (Figure 12.3), the voltage of a cell in which one electrode is the glass electrode is measured relative to the constant potential of the other (standard) electrode. The measured cell voltage is proportional to the pH. The pH meter is calibrated by measuring the voltages obtained from a set of buffer solutions of known pH.

13.2 APPLICATIONS OF ELECTROCHEMICAL CELLS

Batteries

A **battery** stores energy in the form of oxidizing and reducing agents and then releases the energy as electricity when it is needed. The term ''battery'' was originally used to indicate a collection of cells connected together, but now the term is commonly used to mean a single cell. In principle, any oxidation–reduction reaction can be used as the basis of an electrochemical cell, but there are many limitations to the use of most reactions as the basis of a practical battery. To be useful, a battery should be reasonably light and compact and easily transported. It should have a reasonably long life, both when it is being used and when it is not. Also, for many applications, the voltage of the cell must stay constant during use.

There are two main types of batteries: (1) **primary cells**, in which the reaction reaches equilibrium and the battery is then dead and cannot be re-used; and

(2) **secondary cells**, which can be recharged by reversing the reaction by passing a current through them so that they can be used again and again.

Primary Cells The most familiar battery is the type that is widely used in portable radios and flashlights. It is called a *dry cell* or a *Leclanché cell*, after its inventor, George Leclanché (1839–1882) (Figure 13.6). The anode consists of a zinc can; the cathode is a graphite rod surrounded by powdered MnO_2, which is a conductor. The space between the electrodes is filled with a moist paste of NH_4Cl and $ZnCl_2$. Strictly speaking, the cell is not dry; it contains the moist paste in place of an aqueous solution. The electrode reactions are complex but may be written approximately as

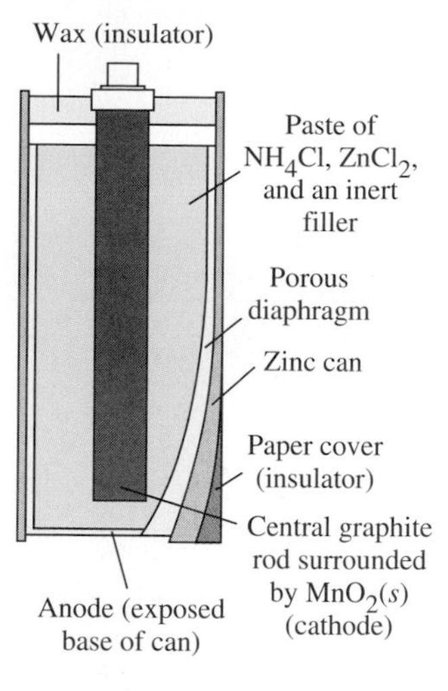

FIGURE 13.6 The Dry Cell The cell consists of a zinc anode (the battery container) and a cathode that is a graphite rod surrounded by powdered solid MnO_2. The conducting solution is a moist paste of NH_4Cl and $ZnCl_2$. This cell has a voltage of 1.5 V.

Anode $\quad Zn(s) \rightarrow Zn^{2+} + 2e^-$

Cathode $\quad MnO_2(s) + NH_4^+ + e^- \rightarrow MnO(OH)(s) + NH_3$

In the cathode reaction, manganese is reduced from the +IV oxidation state to the +III oxidation state. Ammonia is not liberated as a gas but combines with Zn^{2+} to form the complex ion $Zn(NH_3)_4^{2+}$. The dry cell has a finite life, even when not used, because the acidic NH_4Cl corrodes the zinc can.

An improved version of the dry cell, known as an *alkaline battery*, has a zinc rod anode and a cathode of MnO_2. The electrolyte is KOH in the form of a thick gel. The battery is enclosed in a steel container. The alkaline battery has a longer life than the dry cell but is more expensive. Both cells have a potential of 1.5 V.

Secondary Cells One of the most common and useful secondary cells is the **lead–acid cell**, or lead storage battery (Figure 13.7). The anode is a grid of an alloy of lead. The cathode is a lead grid packed with lead(IV) oxide, PbO_2. The electrolyte is a dilute sulfuric acid solution. At the anode, lead is oxidized to Pb^{2+}, and insoluble $PbSO_4$ forms. The $PbSO_4$ adheres to the plates, which is the main reason the cell is rechargeable. At the cathode, PbO_2 is reduced to Pb^{2+}, and $PbSO_4$ forms. The reactions are as follows:

Anode $\quad Pb(s) + SO_4^{2-} \rightarrow PbSO_4(s) + 2e^-$

Cathode $\quad PbO_2(s) + SO_4^{2-} + 4H^+ + 2e^- \rightarrow PbSO_4(s) + 2H_2O$

Overall $\quad Pb(s) + PbO_2(s) + 2H_2SO_4 \rightarrow 2PbSO_4(s) + 2H_2O$

Thus, lead sulfate is formed at each electrode, and sulfuric acid is used up.

An important advantage of a secondary cell is that it can be recharged. A

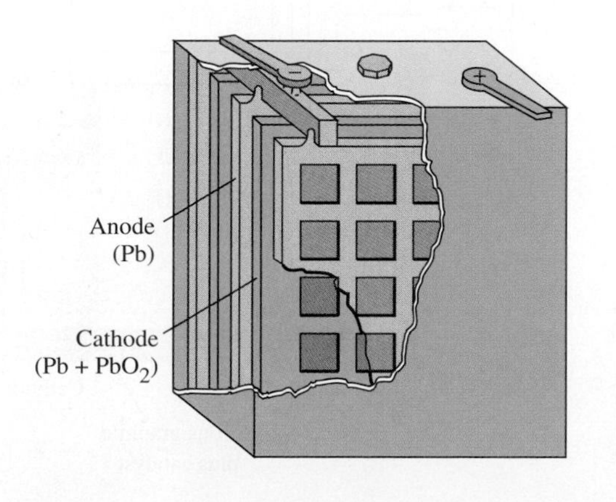

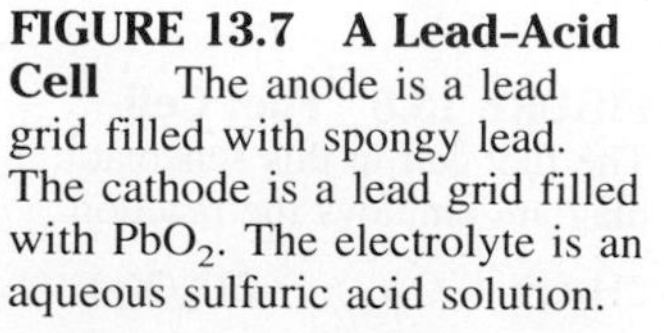
FIGURE 13.7 A Lead-Acid Cell The anode is a lead grid filled with spongy lead. The cathode is a lead grid filled with PbO_2. The electrolyte is an aqueous sulfuric acid solution.

voltage opposite in sign to and larger than the cell voltage is used to send a current through the cell and reverse the electrode reactions. During recharging, the $PbSO_4$ adhering to one electrode is oxidized to PbO_2, and at the other electrode, $PbSO_4$ is reduced to Pb and sulfuric acid is re-formed. The overall reaction during recharging is

$$2PbSO_4(s) + 2H_2O \rightarrow Pb(s) + PbO_2(s) + 2H_2SO_4$$

A single lead cell has a potential of 2.0 V. Normally, three or six such cells are connected in series (in a row) to give a 6-V or a 12-V battery, respectively. The lead–acid cell has proved to be a very practical source of electric power, particularly for applications in which it is used many times but can be easily recharged between. It is used, for example, to start the engine of an automobile and then is recharged by the generator as soon as the engine is running.

The rechargeable **nickel–cadmium cell**, or *NiCad battery*, is widely used in portable electronic equipment. The anode reaction is

$$Cd(s) + 2OH^- \rightarrow Cd(OH)_2(s) + 2e^-$$

The cathode reaction is

$$Ni(OH)_3(s) + e^- \rightarrow Ni(OH)_2(s) + OH^-$$

Charging reverses these reactions. Because no gases are evolved either upon charging or discharging and because these cells are much lighter than a lead–acid cell, they are ideal for use in portable equipment. However, they cannot deliver currents as large as can a lead–acid cell.

Fuel Cells The oxidizing and reducing agents in a secondary cell can be regenerated by recharging, but batteries can also be made in which the reactants are fed continuously to the electrodes. Such batteries are called **fuel cells**. One of the most successful fuel cells uses the reaction of hydrogen with oxygen to form water (Figure 13.8). The electrodes are hollow tubes made of porous compressed graphite impregnated with a catalyst. The electrolyte is a concentrated aqueous KOH solution. At the anode, hydrogen is oxidized to water. At the cathode, oxygen is reduced to the −II oxidation state in the hydroxide ion:

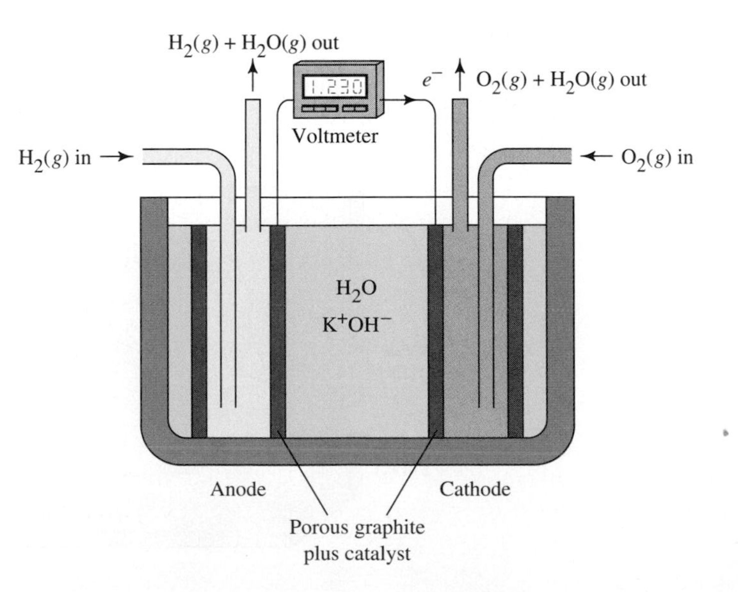

FIGURE 13.8 Fuel Cell
The fuel cell in this schematic diagram employs the reaction

$$2H_2(g) + O_2(g) \rightarrow 2H_2O(g)$$

Anode	$2[H_2(g) + 2OH^- \rightarrow 2H_2O(l) + 2e^-]$
Cathode	$O_2(g) + 2H_2O(l) + 4e^- \rightarrow 4OH^-$
Overall	$2H_2(g) + O_2(g) \rightarrow 2H_2O(l)$

Such a cell runs continuously as long as the reactants are supplied. Because fuel cells convert the energy of a fuel directly to electricity, they are potentially more efficient than conventional methods of generating electricity on a large scale by burning hydrocarbon fuels or by using nuclear reactors. At present fuel cells are too expensive for largescale use, but their importance is growing. They have been used very successfully in some special applications, for example, in spacecraft.

Corrosion

Corrosion, an inconvenient, costly, and sometimes unavoidable property of metals, is an electrochemical process that we can understand with the help of the electrochemical series. The rusting of iron, the tarnishing of silver, and the development of a patina, a green coating on copper and brass, are all familiar examples of corrosion. It causes enormous damage to bridges, ships, and cars. The damage and the efforts made to prevent it cost billions of dollars each year.

For the half-reaction

$$O_2(g) + 4H^+ + 4e^- \rightarrow 2H_2O \qquad E^\circ_{red} = 1.23 \text{ V}$$

so moist air will oxidize any metal of lower reduction potential—any metal that is higher in Table 13.1. It will, for example, oxidize Fe to Fe^{2+} or Fe^{3+}.

The formation of an oxidized layer on the surface of a metal is not always harmful. Aluminum forms a tough, dense, transparent layer of Al_2O_3, which protects the metal underneath from further oxidation. Iron, however, forms an oxide coating called **rust** that is relatively porous and easily cracks and flakes as it thickens. As a result, oxygen and moisture can continue to reach the metal, which continues to oxidize until it is completely destroyed.

Rusting Rust is hydrated iron(III) oxide, $Fe_2O_3 \cdot xH_2O$. It forms only in the presence of oxygen and water. The reactions involved are complex and not completely understood, but a simplified description of the process is shown in Figure 13.9. This figure depicts a drop of water on the surface of iron, in which a small electrochemical cell is set up. The iron on one part of the metal surfacc in the interior of the drop behaves as an anode, and iron is oxidized to Fe^{2+}

$$Fe(s) \rightarrow Fe^{2+} + 2e^- \qquad E^\circ_{ox} = 0.44 \text{ V}$$

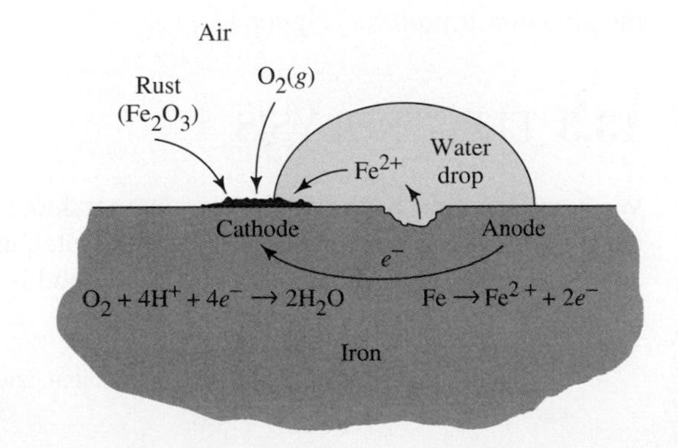

FIGURE 13.9 The Rusting of Iron Iron in contact with water forms the anode, where iron is oxidized to Fe^{2+}. Iron in contact with air forms the cathode, where oxygen is reduced to water. Thus, a cell is set up, and iron continues to go into solution as Fe^{2+}. Where the Fe^{2+} solution is in contact with air, the Fe^{2+} is oxidized to insoluble Fe_2O_3 (rust).

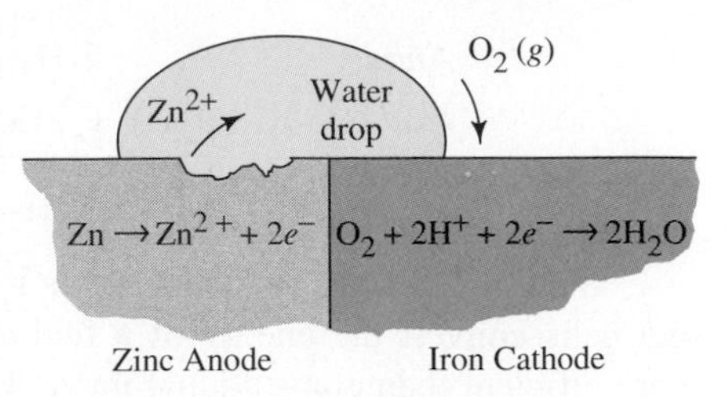

FIGURE 13.10 Galvanized Iron Iron coated with zinc is protected from corrosion even if some of the zinc wears off or is scratched off to expose the iron. An electrochemical cell is then set up in which zinc is the anode and iron is the cathode, so it is the zinc that is oxidized, not the iron.

The electrons produced in this half-reaction flow through the metal to a part of the surface closer to the edge of the drop, where oxygen is available. Here the iron acts as a cathode, and oxygen is reduced to water:

$$O_2(g) + 4H^+ + 4e^- \rightarrow 2H_2O \qquad E^\circ_{red} = 1.23 \text{ V}$$

The circuit is completed by the movement of ions through the water drop. This explains why rusting is particularly rapid in salt water, which has a high concentration of ions and can therefore carry a large current. Salt spread on roads in the winter similarly speeds up the rusting of cars. The overall reaction is the sum of the anode and cathode reactions:

$$2Fe(s) + O_2(g) + 4H^+ \rightarrow 2Fe^{2+} + 2H_2O \qquad E^\circ_{cell} = 1.67 \text{ V}$$

The Fe^{2+} is then further oxidized by atmospheric oxygen to rust:

$$4Fe^{2+} + O_2(g) + 4H_2O \rightarrow 2Fe_2O_3(s) + 8H^+$$

Preventing Corrosion An obvious method of preventing the corrosion of iron is to coat the surface with paint or with another metal. Zinc and chromium are often used for this purpose. Chromium does not corrode because its surface, like that of aluminum, becomes coated with a hard coherent layer of its oxide. Coating iron by dipping it into molten zinc is called **galvanizing**. The surface of the zinc becomes coated with an unreactive film of $Zn(OH)_2 \cdot ZnCO_3$. Because zinc is above iron in the table of reduction potentials (Table 13.1), it is more easily oxidized. So if the zinc layer is cracked or broken, it continues to protect the iron, because it becomes the anode in a cell in which iron is the cathode. Hence, the zinc is oxidized rather than the iron (Figure 13.10).

It is not practical to galvanize large structures such as ships, bridges, and pipelines, but rusting can be minimized by a method called **cathodic protection**. A large block of an active metal, such as zinc or magnesium, is connected to the structure to be protected. It becomes the anode in a cell and supplies the electrons to reduce oxygen. The active metal therefore corrodes, but it is cheaper to replace than the structure it protects (Figure 13.11).

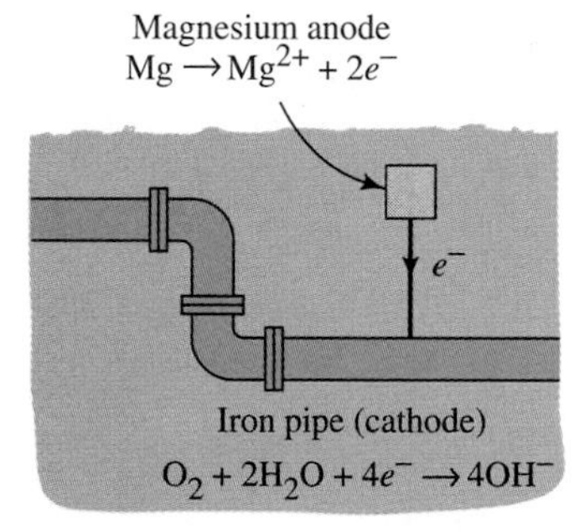

FIGURE 13.11 Cathodic Protection When a block of magnesium or zinc is attached to an iron pipe or tank buried underground, the magnesium forms the anode in an electrolytic cell, and the iron becomes the cathode. Thus, the magnesium or zinc is slowly oxidized and goes into solution as Mg^{2+} or Zn^{2+}. Oxygen is reduced at the iron cathode, leaving the iron uncorroded.

13.3 ELECTROLYSIS

We mentioned in Chapter 10 that certain metal ions, such as Na^+, Mg^{2+}, and Al^{3+}, are difficult to reduce to the corresponding metal, using even the strongest reducing agents. In many cases the most practical method is to use electrons directly, that is, to use a process called electrolysis.

Electrolysis **is a process in which electrical energy is used to produce chemical change.**

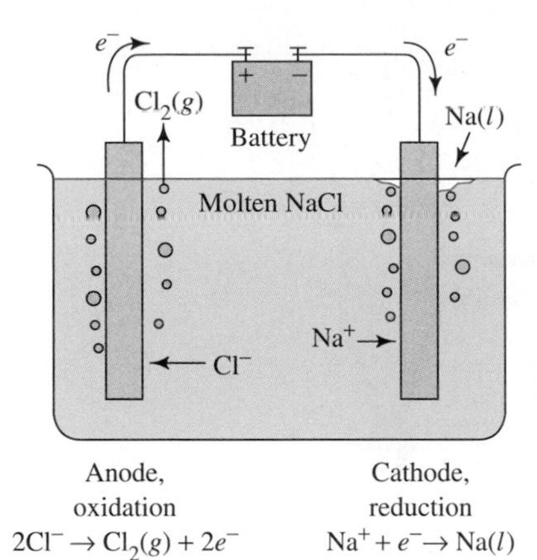

FIGURE 13.12 Electrolysis of Molten Sodium Chloride The electric current outside the electrolytic cell is carried by the electrons, which are pushed around the circuit by the battery. Inside the molten electrolyte, the current is carried by the movement of the positive and negative ions toward the electrodes. At the anode, Cl^- ions give up electrons to the electrode, to give Cl atoms, which combine to form Cl_2 molecules. At the cathode, Na^+ ions accept electrons from the electrode, to give Na atoms, which form liquid sodium metal.

Electrolysis of Molten Sodium Chloride

Sodium and chlorine combine in a spontaneous, strongly exothermic reaction to give sodium chloride (Demonstration 4.4):

$$2Na(s) + Cl_2(g) \rightarrow 2NaCl(s) \qquad \Delta H^\circ = -822 \text{ kJ}, \quad \Delta G^\circ = -768 \text{ kJ}$$

The position of the equilibrium lies far to the right, and the reverse reaction is not spontaneous. However, the reverse reaction can be made to take place by passing an electric current through molten sodium chloride, that is, by carrying out an electrolysis (Figure 13.12). Two electrodes made of a solid conducting material, such as a metal or graphite, are inserted into the molten sodium chloride and connected to a battery. The battery pushes electrons into one electrode, which becomes negatively charged, and withdraws electrons from the other electrode, which becomes positively charged. The negatively charged electrode attracts positive sodium ions to its surface, where each sodium ion acquires an electron from the electrode and is reduced to a sodium atom. The sodium atoms combine to form sodium metal, which rises to the top of the molten sodium chloride. Negative chloride ions are attracted to the positive electrode, where each chloride ion gives up an electron to the electrode and is oxidized to a chlorine atom. These chlorine atoms combine and form chlorine molecules, and $Cl_2(g)$ bubbles to the surface.

As in an electrochemical cell, the electrode at which oxidation occurs in an **electrolytic cell** is called the **anode**, and the electrode at which reduction occurs is called the **cathode**. The conducting liquid or solution is called the **electrolyte**. The reactions taking place at the electrode surfaces are

In an electrolytic cell, anions are attracted to the anode and cations are attracted to the cathode.

Anode	$2Cl^- \rightarrow Cl_2(g) + 2e^-$	(Oxidation)
Cathode	$Na^+ + e^- \rightarrow Na(l)$	(Reduction)

The overall reaction that takes place in the electrolytic cell is called the *cell reaction*. The equation for the complete cell reaction is obtained by combining the equation for the anode half-reaction with the equation for the cathode half-reaction:

Cathode	$2Na^+(l) + 2e^- \rightarrow 2Na(l)$
Anode	$2Cl^-(l) \rightarrow Cl_2(g) + 2e^-$
Overall	$2Na^+(l) + 2Cl^-(l) \rightarrow Cl_2(g) + 2Na(l)$

The industrial preparation of sodium is carried out in a cell such as that shown

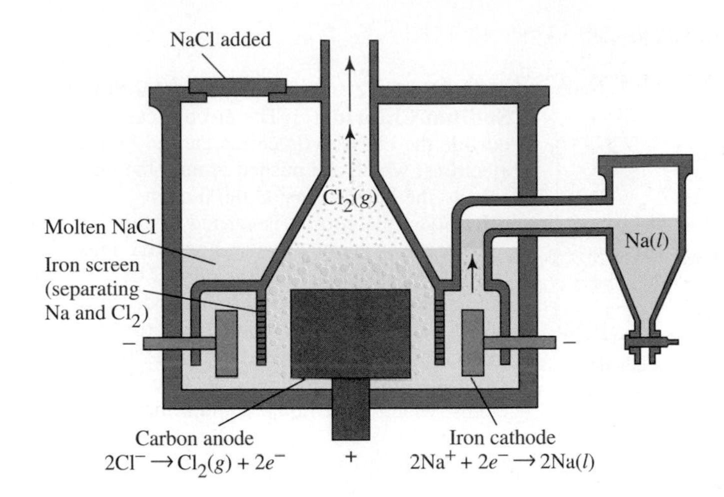

FIGURE 13.13 Electrolytic Cell for the Commercial Production of Sodium The iron cathode in the form of a ring surrounds the carbon anode. The cell is designed to prevent the sodium produced at the cathode from coming into contact with the chlorine formed at the anode to re-form sodium chloride.

in Figure 13.13. The cell is designed to prevent the sodium and chlorine produced from coming into contact and re-forming sodium chloride. This process is the most important method for making sodium and is also a source of chlorine.

For both electrochemical cells and electrolysis, reduction occurs at the cathode and oxidation at the anode.

In any electrolysis, electric current is supplied by an external source. The electrons enter and leave the electrolyte by means of suitable conducting electrodes. At the surface of each electrode, electrons are transferred from the electrode to the electrolyte or from the electrolyte to the electrode by means of chemical reactions. *Electrons are transferred to the electrolyte from the cathode in a reduction half-reaction*—for example, the reduction of Na^+ to Na. *Electrons are transferred from the electrolyte to the anode in an oxidation half-reaction*—for example, the oxidation of Cl^- to Cl_2. Charge is carried across the electrolyte not by the movement of free electrons but by the movement of positive ions toward the negative electrode, or cathode, and negative ions toward the positive electrode, or anode. This type of conduction in which the charge carriers are ions is called **ionic conduction**. In the electrodes and in the external circuit (the wires connecting the electrodes to the source of power), the charge carriers are free electrons. This is *electronic*, or *metallic, conduction* (Chapter 10). The charge carriers change from electrons to ions or vice versa at the surfaces between the electrodes and the electrolyte. It is this change that produces the chemical reactions at the electrodes—oxidation at the anode and reduction at the cathode.

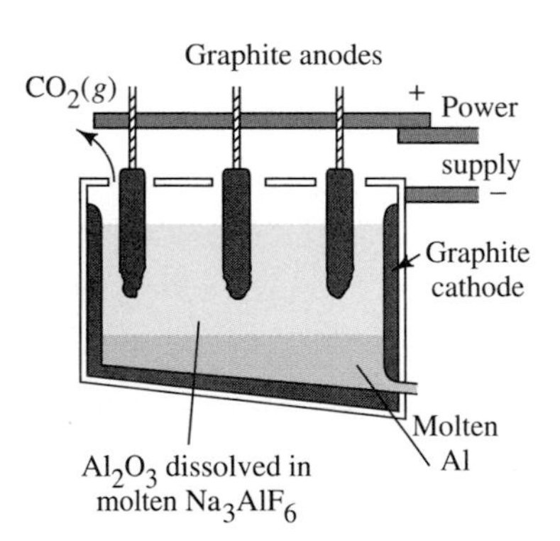

FIGURE 13.14 Electrolytic Cell for the Production of Aluminum by the Hall Process Graphite anodes are oxidized to $CO_2(g)$. As they are consumed, they are lowered farther into the molten mixture of Al_2O_3 and Na_3AlF_6. Aluminum ions are reduced to $Al(l)$ at the graphite cathode that lines the tank. Because molten aluminum is denser than the molten Al_2O_3–Na_3AlF_6 mixture, aluminum collects at the bottom of the cell and may be run off.

Preparation of Aluminum by Electrolysis

Among the metals that are prepared by reducing the corresponding positive metal ion by electrolysis, aluminum is the most important commercially. Aluminum is made by the electrolysis of its oxide Al_2O_3. This process is called the **Hall process** after its inventor, Charles Hall (Box 13.2). The oxide is obtained from the mineral bauxite, $Al_2O_3 \cdot xH_2O$, which is an impure hydrated form of the oxide. The melting point of Al_2O_3 (2050°C) is too high for it to be conveniently and economically used for electrolysis in the molten state. However, a mixture of bauxite with the aluminum mineral cryolite, Na_3AlF_6, which contains the complex ion AlF_6^{3-}, has a much lower melting point of about 1000°C. It is this mixture that is electrolyzed. Cryolite is a rather rare mineral, so some of the bauxite is converted to Na_3AlF_6 to give the desired mixture. A typical cell used for the Hall process is shown in Figure 13.14. The anodes are graphite rods dipping into the molten electrolyte. The cathode is a steel vessel, lined with graphite, that contains the molten electrolyte. Neither the

BOX 13.2 Charles Martin Hall

The process we use today for manufacturing aluminum was invented by a young American while he was still an undergraduate at Oberlin College. Inspired by a professor's remark that anyone who could invent a cheap process for producing aluminum would make a fortune, Charles Hall set out to try in 1885. At the time, aluminum cost $90 a pound and was more expensive than either silver or gold. It is said that the very rich flaunted their wealth by dining with aluminum knives and forks.

Hall worked in a woodshed, using homemade and borrowed equipment. After about a year he found that Al_2O_3 dissolves in molten cryolite to give a conducting solution, from which aluminum can be deposited by passing an electric current. He used an iron frying pan as a container for the molten cryolite–alumina mixture, which he melted over a blacksmith's forge. The electric current came from electrochemical cells that he had made from jars that his mother used to can fruit.

By an odd coincidence Paul Héroult, who was the same age as Hall, made the same discovery independently in France about the same time. As a result of the discovery of Hall and Héroult, the large-scale production of aluminum became economically feasible for the first time, and aluminum became a common metal.

Charles Martin Hall (1863–1914)

As his professor had predicted, Hall died a wealthy man. Today the company that was founded to produce aluminum by Hall's process—the Aluminum Company of America (Alcoa)—is worth billions of dollars. A life-size aluminum sculpture of Hall stands in the science building of Oberlin College.

nature of the ions present in the molten mixture nor the electrode reactions are completely understood. For simplicity we will assume that the Al_2O_3 is ionized to Al^{3+} and O^{2-}. We can represent the reaction at the cathode by the equation

$$\textit{Cathode} \qquad Al^{3+}(l) + 3e^- \rightarrow Al(l)$$

At the anode several reactions occur in which carbon dioxide, oxygen, and fluorine form. The principal reaction at this electrode may be approximately represented by the equation

$$\textit{Anode} \qquad C(s) + 2O^{2-}(l) \rightarrow CO_2(g) + 4e^-$$

The approximate overall reaction is therefore

$$\textit{Overall} \qquad 4Al^{3+}(l) + 6O^{2-}(l) + 3C(s) \rightarrow 4Al(l) + 3CO_2(g)$$

The graphite electrodes are gradually converted to carbon dioxide and must be replaced from time to time. The energy consumption is high, and the Hall process is economically feasible only when carried out near a cheap source of electric power, for example, at a site where hydroelectric power is produced.

Quantitative Aspects of Electrolysis

The principal cost in the electrolytic preparation of aluminum and other metals is the cost of electricity. Therefore, it is important to know how much electricity is needed to produce a certain amount of metal. The amount of electricity needed can be calculated from the equation for the appropriate electrode reaction. In the electrolysis of molten sodium chloride, the reactions are

$$\textit{Cathode} \quad Na^+ + e^- \rightarrow Na(l)$$

$$\textit{Anode} \quad 2Cl^- \rightarrow Cl_2(g) + 2e^-$$

The passage of one electron produces one sodium atom; the passage of 1 mol of electrons produces 1 mol of sodium. The passage of two electrons produces one molecule of Cl_2; the passage of 2 mol of electrons produces 1 mol of Cl_2. In summary,

Electrode reaction	*Product*
$Na^+ + e^- \rightarrow Na\ (l)$	1 mol electrons → 1 mol Na = 23.0 g Na
$2Cl^- \rightarrow Cl_2(g) + 2e^-$	2 mol electrons → 1 mol Cl_2 = 70.9 g Cl_2

The charge Q of n moles of electrons is

$$Q = nF$$

where F is the Faraday constant. The production of 1 mol, or 23.0 g, of sodium by the reduction of sodium ions requires 1 mol of electrons, so it requires an amount of charge

$$Q = nF = (1\ \text{mol})(96{,}500\ \text{C} \cdot \text{mol}^{-1}) = 96{,}500\ \text{C}$$

The production of 1 mol of Cl_2 requires 2 mol of electrons and therefore an amount of charge equal to 2 × 96,500 C. In summary,

Electrode reaction	*Charge*	*Product*
$Na^+ + e^- \rightarrow Na(l)$	96,500 C →	1 mol Na = 23.0 g Na
$2Cl^- \rightarrow Cl_2(g) + 2e^-$	2 × 96,500 C →	1 mol Cl_2 = 70.9 g Cl_2

In practice charge is usually determined by measuring a current flow for a given time. A charge of 1 C passes a given point when a current of 1 ampere (A) flows for 1 s:

$$1\ \text{coulomb} = 1\ \text{ampere} \times 1\ \text{second}$$

$$1\ \text{C} = 1\ \text{A} \cdot \text{s}$$

Hence if molten NaCl is electrolyzed for 1.00 h with a current of 50.0 A, the number of coulombs passed through it is

$$50.0\ \text{A} \times 3600\ \text{s} = 180{,}000\ \text{A} \cdot \text{s} = 180{,}000\ \text{C}$$

Thus, the number of moles of electrons passing through the sodium chloride is

$$n = \frac{Q}{F} = \frac{180{,}000\ \text{C}}{96{,}500\ \text{C} \cdot \text{mol}^{-1}} = 1.87\ \text{mol}$$

This 1.87 mol of electrons produces

$$1.87\ \text{mol electrons}\left(\frac{1\ \text{mol Na}}{1\ \text{mol electrons}}\right) = 1.87\ \text{mol Na}$$

The same 1.87 mol of electrons produces

$$1.87\ \text{mol electrons}\left(\frac{1\ \text{mol}\ Cl_2}{2\ \text{mol electrons}}\right) = 0.935\ \text{mol}\ Cl_2$$

III To convert from moles of electrons to moles of Cl_2, use $2Cl^- \rightarrow Cl_2 + 2e^-$. Because each mole of Cl_2 generates 2 moles of electrons, the conversion factor is ($1\ \text{mol}\ Cl_2$)/(2 mol electrons).

EXAMPLE 13.5 Calculating the Amounts of Electrolysis Products

What mass of aluminum will be produced in 1.00 h by the electrolysis of molten $AlCl_3$, using a current of 10.0 A?

Solution: First, from the current in amps and the time in seconds, we calculate the number of coulombs:

$$Q = (10.0\text{A})(1.00\text{ h})\left(\frac{3600\text{ s}}{1\text{ h}}\right)\left(\frac{1\text{ C}}{1\text{ A}\cdot\text{s}}\right) = 3.60 \times 10^4\text{C}$$

Then we find the number of moles of electrons:

$$n = \frac{Q}{F} = (3.60 \times 10^4\text{ C})\left(\frac{1\text{ mol}}{96{,}500\text{ C}}\right) = 0.373\text{ mol electrons}$$

The half-reaction for the reduction of aluminum is

$$Al^{3+} + 3e^- \rightarrow Al(l)$$

Thus, 3 mol of electrons is needed to produce 1 mol (26.98 g) of Al, so 0.373 mol of electrons will produce

$$(0.373\text{ mol electrons})\left(\frac{1\text{ mol Al}}{3\text{ mol electrons}}\right)\left(\frac{26.98\text{ g Al}}{1\text{ mol Al}}\right) = 3.36\text{ g}$$

EXAMPLE 13.6 Calculating the Amounts of Electrolysis Products

What volume of chlorine at STP is produced when a current of 20.0 A is passed through molten sodium chloride for 2.00 h?

Solution: We can calculate the number of moles of electrons in one step:

$$n = (20.0\text{ A})(2.00\text{ h})\left(\frac{3600\text{ s}}{1\text{ h}}\right)\left(\frac{1\text{ C}}{1\text{ A}\cdot\text{s}}\right)\left(\frac{1\text{ mol electrons}}{96{,}500\text{ C}}\right) = 1.49\text{ mol electrons}$$

At the anode,

$$2\,Cl^- \rightarrow Cl_2(g) + 2e^-$$

so

$$\text{mol } Cl_2 = (1.49\text{ mol electrons})\left(\frac{1\text{ mol } Cl_2}{2\text{ mol electrons}}\right) = 0.745\text{ mol}$$

We know from the ideal gas law that 1 mol of an ideal gas occupies 22.41 L at STP. Therefore,

$$\text{volume } Cl_2 \text{ at STP} = (0.745\text{ mol } Cl_2)\left(\frac{22.41\text{ L}}{1\text{ mol } Cl_2}\right) = 16.7\text{ L}$$

Exercise 13.6 What mass of magnesium and what volume of chlorine at STP will be produced by the electrolysis of molten magnesium chloride, using a current of 3.00 A for 24.0 h?

Electrolysis of Aqueous Solutions

In the electrolysis of an aqueous solution, there is the possibility that water, rather than the solute ions, may be oxidized or reduced. For example, in the electrolysis of an aqueous sodium chloride solution, hydrogen is produced at the cathode by the

An aqueous solution of potassium iodide behaves like an NaCl(aq) solution when electrolyzed. Water is reduced to $H_2(g)$ and $OH^-(aq)$ at the cathode. The bubbles of hydrogen can be seen rising from the cathode, and the OH^- ion causes phenolphthalein in the solution to turn pink. At the anode, iodide ion is oxidized to iodine, which combines with more iodide ion to form the brown I_3^- ion.

reduction of water, rather than Na^+ ions being reduced to sodium. The cathode reaction is

$$2H_2O + 2e^- \rightarrow H_2(g) + 2OH^-$$

We could have predicted this result by referring to Table 13.1, where we find the following standard reduction potentials:

$$Na^+ + e^- \rightarrow Na(s) \qquad E^\circ_{red} = -2.71 \text{ V}$$

$$2H_2O + 2e^- \rightarrow H_2(g) + 2OH^- \qquad E^\circ_{red} = -0.83 \text{ V}$$

We see that water, which is lower in Table 13.1 than Na^+, is much more easily reduced than Na^+, and so it is water rather than Na^+ that is reduced in the electrolysis of an aqueous solution of sodium chloride.

At the anode, chlorine is produced by the oxidation of Cl^-:

$$2Cl^- \rightarrow Cl_2(g) + 2e^- \qquad E^\circ_{ox} = -E^\circ_{red} = -1.36 \text{ V}$$

It may seem surprising that water is not oxidized instead, because the standard potential for the oxidation of water,

$$2H_2O \rightarrow O_2(g) + 4H^+ + 4e^- \qquad E^\circ_{ox} = -1.23 \text{ V}$$

is less negative than that for the oxidation of Cl^- (H_2O is above Cl^- in Table 13.1). That Cl^- is more easily oxidized is due to a phenomenon called **overvoltage**. For reasons too complex to discuss here, the evolution of a gas often requires a higher potential (voltage) than the standard potential. For the evolution of oxygen, the overvoltage is sufficiently high that Cl^- is preferentially oxidized. Because of overvoltage, we cannot always accurately predict the products of electrolysis of aqueous solutions from the data in Table 13.1.

Electrolytic Preparation of Chlorine and Sodium Hydroxide

The electrolysis of aqueous sodium chloride is the principal method by which chlorine and sodium hydroxide are made industrially. The chemical industry based on this process is called the **chloralkali industry**. The annual production of sodium hydroxide (caustic soda) in the United States is about 10 million metric tons, 50% of which is consumed by the chemical industry and a further 20% by pulp and paper plants. From the equation for the overall cell reaction,

$$2NaCl(aq) + 2H_2O \rightarrow Cl_2(g) + H_2(g) + 2NaOH(aq)$$

we see that hydrogen is another important product of the process, although, as we saw in Chapter 2, there are other important methods for making hydrogen. The electrolytic cell illustrated in Figure 13.15 is called a **diaphragm cell**, because the

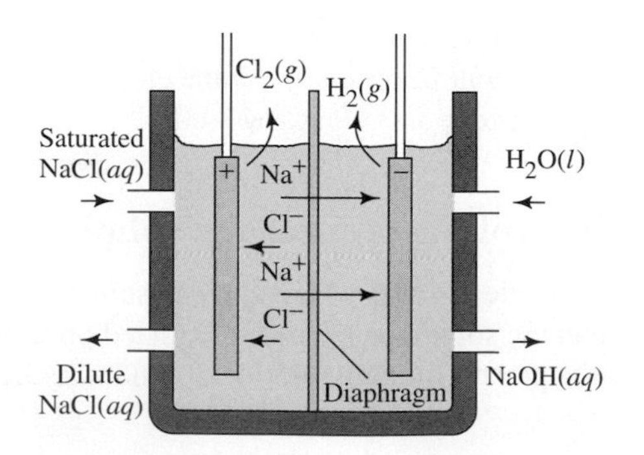

FIGURE 13.15 A Diaphragm Chloralkali Cell The cell is operated continuously. A saturated NaCl solution flows in at the top of the anode compartment, and a more dilute NaCl solution flows out at the bottom. Water enters the top of the cathode compartment, and an NaOH solution flows out at the bottom. Chlorine is liberated at the anode, and hydrogen at the cathode. The cathode and anode compartments are separated by a diaphragm (membrane) that allows Na^+ ions to pass but prevents the passage of Cl^- ions.

two electrode compartments are separated by a porous diaphragm. This diaphragm allows the passage of Na^+ ions but not Cl^- or OH^- ions, so the NaOH solution formed is not contaminated with NaCl. Formerly, the diaphragm was made of asbestos, but it is now made of fluorocarbon polymers (Chapter 20) that are highly alkali resistant.

A cell that had been widely used but is now being phased out is called a **mercury cell**; it employs a mercury cathode. It is being replaced by the diaphragm cell, because the inevitable leakage of some mercury into rivers has had serious environmental consequences.

Electrolytic Refining of Copper

In most of the examples of electrolysis that we have considered so far, the electrodes were inert. In other words, they served only to conduct the electric current to and from the solution or molten electrolyte and were not themselves oxidized or reduced. However, not all electrodes are inert. We have seen that in the manufacture of aluminum, the carbon anodes are oxidized to carbon dioxide.

Oxidation of the anode also occurs when copper is used as an anode in the electrolysis of aqueous solutions. We can see from the oxidation potentials (obtained by reversing the sign of the reduction potentials in Table 13.1) that copper is more easily *oxidized* than water:

$$Cu(s) \rightarrow Cu^{2+} + 2e^- \qquad E^\circ_{ox} = -0.34 \text{ V}$$
$$2H_2O \rightarrow O_2(g) + 4H^+ + 4e^- \qquad E^\circ_{ox} = -1.23 \text{ V}$$

Therefore, in a cell with a copper anode and a copper cathode in a copper(II) sulfate solution, copper goes into solution at the anode. Copper is deposited at the cathode, because Cu^{2+} is more easily *reduced* than water: Water is above copper in Table 13.1. The anode and cathode reactions are

$$\textit{Anode} \qquad Cu(s) \rightarrow Cu^{2+} + 2e^-$$
$$\textit{Cathode} \qquad Cu^{2+} + 2e^- \rightarrow Cu(s)$$

In the sulfate ion, sulfur is in its highest (+VI) oxidation state and hence cannot be further oxidized. Moreover, it is not as readily reduced as water, so it plays no part in the electrolysis. This cell, in which copper goes into solution at the anode and is redeposited on the cathode, is the basis for an important method for the purification (refining) of copper (Figure 13.16). Impure copper is the anode of the cell. Less easily oxidized metals, such as silver and gold, which are below copper in Table 13.1, are not oxidized and fall to the bottom of the cell as a sludge. Other metals that are more easily oxidized, such as zinc and iron, go into solution but are not redeposited, because their ions are less easily reduced than Cu^{2+}. Copper that is 99.95% pure can be obtained in this way.

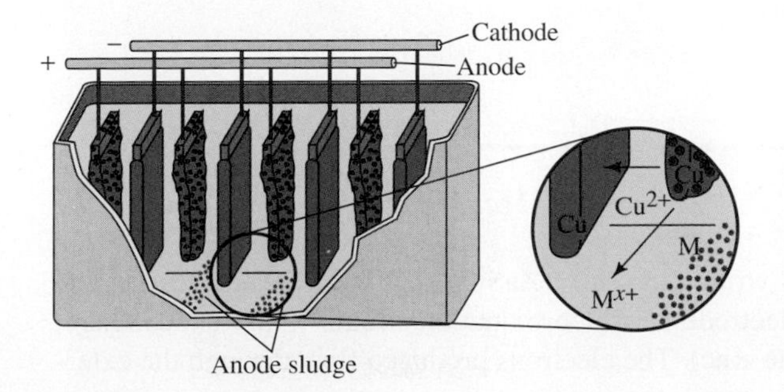

FIGURE 13.16 Electrolytic Refining of Copper Bars of impure copper are anodes. Thin sheets of pure copper are cathodes. As copper goes into solution at the anode, less easily oxidized impurities, such as Ag and Au, fall to the bottom of the cell as a sludge. More easily oxidized impurities, such as Zn and Fe, go into solution but are not deposited on the cathode. Only Cu^{2+} ions are reduced at the cathode, forming a deposit of pure copper. Refining the sludge to give metals such as silver and gold helps to pay for the process.

DEMONSTRATION 13.3 Electroplating

A current is passed between strips of copper in an aqueous solution of dichromic acid, $H_2Cr_2O_7$, obtained by adding chromium trioxide, CrO_3, to water.

After a few minutes the copper strip has become plated with a shiny layer of chromium.

Electroplating

As we have already seen for Cu^{2+}, a metal ion that is more easily reduced than water can be reduced to the corresponding metal at a cathode. This deposition of a metal is called **electroplating**. The object to be plated is the cathode, and the plating metal is the anode, of an electrolytic cell. Chromium and silver are commonly used to plate metal objects to improve their appearance and corrosion resistance (Demonstration 13.3).

Electrolysis and Gibbs Free Energy

Electrolysis is a method for carrying out a nonspontaneous reaction, that is, a reaction for which ΔG is positive. In Chapter 9 we discussed another method of doing this, namely, coupling the reaction to one with a sufficiently large negative free energy change to give an overall negative free energy change. The overall process of electrolysis, including the production of the electric current that drives it, must similarly have a negative free energy change. Because the electrolysis has a positive free energy change, the process of producing the electric current must have a sufficiently large negative free energy change to make the overall free energy change negative. This process might be the burning of fuel to drive a generator, or it might be a chemical reaction occurring in a battery. The process of electrolysis is essentially two coupled reactions in which the burning of fuel or the reaction in the battery drives the electrolysis.

SUMMARY

An electrochemical cell consists of two half-cells, each with a conducting electrode in contact with an electrolyte solution and separated by a porous barrier or salt bridge. An electrochemical cell can be represented by a cell diagram. For example, the cell diagram for the reaction

$$Zn(s) + Cu^{2+}(aq) \rightarrow Zn^{2+}(aq) + Cu(s)$$

is written as $Zn(s)|ZnSO_4(aq)||CuSO_4(aq)|Cu(s)$. The electrode at which oxidation occurs is the anode (here, the zinc). The electrons produced flow through the external circuit to the cathode, the electrode at which reduction takes place (here, the copper). The circuit is completed by cations moving to the cathode and anions moving to the anode by ionic conduction. Adding equations for the two half-reactions so that the electrons are eliminated gives the overall cell reaction. For the zinc–copper cell, the electrode reactions are

Anode	$Zn(s) \rightarrow Zn^{2+}(aq) + 2e^-$	(Oxidation)
Cathode	$Cu^{2+}(aq) + 2e^- \rightarrow Cu(s)$	(Reduction)

Under *standard conditions*, the standard cell potential, $E^\circ_{cell} = E^\circ_{ox} + E^\circ_{red}$, is the sum of the half-cell potentials of the oxidation half-reaction and of the reduction half-reaction. Half-cell potentials are measured relative to E°_{red} for the hydrogen half-cell, or hydrogen electrode. At this electrode, for which E°_{red} is defined as 0 V, the half-reaction is $2H^+(aq) + 2e^- \rightarrow H_2(g)$. All half-cell potentials are listed as standard reduction potentials, E°_{red}. For a given half-reaction, $E^\circ_{red} = -E^\circ_{ox}$. In a table of standard reduction potentials, reducing agents are on the right, with reducing strength decreasing from top to bottom; oxidizing agents are on the left, with oxidizing strength increasing from top to bottom. Only metals with negative E°_{red} values reduce $H^+(aq)$ to $H_2(g)$.

For spontaneous redox reactions, E°_{cell} has a *positive* value. Thus, when $E^\circ_{ox} + E^\circ_{red} > 0$ for a reaction, it proceeds spontaneously from left to right to achieve equilibrium. When $E^\circ_{ox} + E^\circ_{red} < 0$, it is the reverse reaction that is spontaneous.

For a cell under nonstandard conditions, $\Delta G = -nFE_{cell}$. From this equation we can derive the Nernst equation,

$$E_{cell} = E^\circ_{cell} - \frac{RT}{nF} \ln Q$$

Here R is the gas constant, T is the temperature, n is the number of electrons transferred in the reaction, F is the Faraday constant, (96,500 $C \cdot mol^{-1}$, the charge on 1 mol of electrons) and Q is the reaction quotient.

A concentration cell is a cell with two identical half-cell reactions but with solutions of different concentrations. A pH meter is an electrochemical cell in which one of the electrodes is a glass electrode. The potential of such a cell depends on the pH of the solution into which the glass electrode is dipping.

Batteries are electrochemical cells used as energy sources. In primary cells, such as the dry cell, the reaction eventually reaches equilibrium, and the battery is then dead and cannot be re-used. Secondary cells, such as the lead–acid battery, are rechargeable—the cell reaction can be reversed by passing a current through the cell. A fuel cell is a battery in which the reactants, such as $H_2(g)$ and $O_2(g)$, are fed constantly to the electrodes.

Corrosion is the deterioration of metals due to reaction with the environment—commonly oxidation to the metal oxide, as in the rusting of iron. Common methods of preventing corrosion include painting, coating with another metal, and cathodic protection.

The process of electrolysis uses electrical energy to drive oxidation–reduction reactions with positive ΔG values. Easily oxidized metals whose cations are difficult to reduce are prepared by electrolysis: For example, sodium (and chlorine) are prepared by the electrolysis of molten sodium chloride in the reaction

$$2Na^+Cl^-(l) \rightarrow 2Na(l) + Cl_2(g).$$

In an electrolytic cell, two electrodes dipping into an electrolyte (molten salt or an aqueous solution) are connected to a current source. The current is carried by the ions in the electrolyte (ionic conduction) and by electrons in the external circuit (metallic conduction). Reduction occurs at the cathode, and oxidation at the anode. The overall cell reaction is obtained by adding the anode and cathode half-reactions. For example,

Cathode	$2Na^+(l) + 2e^- \rightarrow 2Na(l)$	(Reduction)
Anode	$2Cl^-(l) \rightarrow Cl_2(g) + 2e^-$	(Oxidation)
Overall	$2Na^+(l) + 2Cl^-(l) \rightarrow 2Na(l) + Cl_2(g)$	

Aluminum is prepared in the Hall process by the electrolysis of a molten mixture of bauxite, $Al_2O_3 \cdot xH_2O$, and cryolite, Na_3AlF_6. In this process graphite anodes are oxidized to $CO_2(g)$, and a graphite-lined steel reaction vessel is the cathode.

The charge of n moles of electrons is $Q = nF$. By measuring the current flowing for a known time, we can find the number of moles of electrons that pass through an electrolytic cell and thereby determine the amounts of products produced at the electrodes from the equations for the electrode reactions.

In the electrolysis of aqueous solutions with inert electrodes, water rather than the solute ions may be oxidized or reduced, depending on the relative E°_{red} values. Overvoltage, a phenomenon in which the evolution of a gas requires a higher-than-standard potential, must also be taken into account. In the electrolytic refining of copper, the electrodes are not inert: An impure copper anode is oxidized, rather than the water or SO_4^{2-} of the $CuSO_4(aq)$ electrolyte, and $Cu^{2+}(aq)$ is reduced to $Cu(s)$ at the cathode. In electroplating, a metal ion that is more readily reduced than water is reduced at the cathode to the corresponding metal. The metal is deposited onto the cathode.

IMPORTANT TERMS

anode (page 467)
battery (page 480)
cathode (page 467)
cathodic protection (page 484)
cell diagram (page 468)
cell potential, E_{cell} (page 468)
concentration cell (page 479)
corrosion (page 483)
electrochemical cell (page 465)
electrode (page 466)
electrolysis (page 484)
electrolytic cell (page 485)
electroplating (page 492)
Faraday constant, F (page 476)
fuel cell (page 482)
half-cell potential (page 469)
Hall process (page 486)
hydrogen half-cell (electrode) (page 469)
Nernst equation (page 477)
oxidation potential (page 469)
primary cell (page 481)
reduction potential (page 469)
salt bridge (page 467)
secondary cell (page 481)
standard cell potential, E°_{cell} (page 468)
standard oxidation potential, E°_{ox} (page 470)
standard reduction potential, E°_{red} (page 470)
volt (page 467)

REVIEW QUESTIONS

1. Distinguish between the processes that occur in **(a)** an electrochemical cell and **(b)** an electrolytic cell.

2. Starting with pieces of zinc and copper and solutions of $ZnSO_4(aq)$ and $CuSO_4(aq)$, describe how an electrochemical cell is constructed in which the overall reaction is

$$Zn(s) + Cu^{2+}(aq) \rightarrow Zn^{2+}(aq) + Cu(s)$$

3. In the cell described in Review Question 2, **(a)** which metal is the anode, and which is the cathode? **(b)** What reactions occur at each electrode?

4. Draw the cell diagram for the cell described in Review Question 2.

5. Why must the two half-cells that constitute an electrochemical cell be separate but nevertheless joined by a porous barrier (salt bridge)?

6. For what conditions is a standard cell potential, E°_{cell}, defined?

7. Relative to what reduction half-reaction are values of standard reduction potentials measured?

8. **(a)** What is the relationship between the reduction potential E°_{red} of a half-reaction and the oxidation potential E°_{ox} of the reverse half-reaction?
(b) How is a cell potential, E_{cell}, found from the potentials of its two half-cells?

9. Draw the cell diagram for an electrochemical cell with the overall reaction

$$Zn(s) + H_2SO_4(aq) \rightarrow Zn^{2+}(aq) + SO_4^{2-}(aq) + H_2(g)$$

10. For standard conditions, why is $Li^+(aq) < Mg^{2+}(aq) < NO_3^-(aq) < O_2(g)$ the order of increasing strength as an oxidizing agent?

11. Under standard conditions, why are the metals Al, Zn, Cr, and Ni, but not Cu or Ag, expected to react with $HCl(aq)$?

12. What is the Faraday constant, F?

13. What can we say about the value of E_{cell} for redox reactions that **(a)** proceed spontaneously and **(b)** have attained equilibrium?

14. What is a concentration cell?

15. What is a fuel cell?

16. What is cathodic protection?

17. In electrolysis that uses inert electrodes, at which electrode does **(a)** reduction occur and **(b)** oxidation occur?

18. How is aluminum produced industrially?

19. In the electrolysis of molten sodium chloride, how much sodium is produced at the cathode if 1 mol of $Cl_2(g)$ is produced at the anode?

PROBLEMS

Electrochemical Cells

1. Using Table 13.1, explain why **(a)** metals such as magnesium, aluminum, zinc, iron, and nickel react readily with dilute sulfuric acid, whereas **(b)** copper and silver react only with hot concentrated sulfuric acid.

2. Using Table 13.1, explain why copper dissolves in dilute nitric acid but not in dilute hydrochloric acid.

3. Explain why silver, copper, lead, and iron have been known since ancient times, but other metals, such as sodium, potassium, and aluminum, were not prepared in elemental form until the nineteenth century.

4. **(a)** Place 1.00-M solutions of the metal ions Ca^{2+}, Fe^{3+}, Al^{3+}, Cu^{2+}, Ni^{2+}, and Na^+ in order of increasing oxidizing strength.
(b) State which of the ions in (a) are stable in 1.00-M acid.

5. Use Table 13.1 to predict what happens when the following metals are dipped into 1.00-*M* solutions of the electrolytes indicated.

(a) zinc and (i) $NiCl_2(aq)$; (ii) $HCl(aq)$

(b) iron and (i) $CuSO_4(aq)$; (ii) 0.5-*M* $HBr(aq)$

6. Under standard conditions, which of the following will be oxidized by dichromate ion, $Cr_2O_7^{2-}(aq)$, in acidic solution?

(a) $F^-(aq)$ **(b)** $Cl^-(aq)$ **(c)** $Br^-(aq)$

(d) $I^-(aq)$ **(e)** $Fe^{2+}(aq)$

7. Use standard half-cell reduction potentials to predict which of the following reactions will occur at standard conditions in acidic aqueous solution. For each predicted reaction, write the balanced equation.

(a) $H_2O_2(aq) + Cu^{2+}(aq) \rightarrow Cu(s) + O_2(g)$

(b) $Ag^+(aq) + Fe^{2+}(aq) \rightarrow Ag(s) + Fe^{3+}(aq)$

(c) $I^-(aq) + NO_3^-(aq) \rightarrow I_2(s) + NO(g)$

8. Predict the reaction that occurs, if any, between the following substances under standard conditions in acidic solution.

(a) $Fe^{3+}(aq)$ and $I^-(aq)$ **(b)** $Ag^+(aq)$ and $Cu(s)$

(c) $Fe^{3+}(aq)$ and $Br^-(aq)$ **(d)** $Ag(s)$ and $Fe^{3+}(aq)$

(e) $Br_2(aq)$ and $Fe^{2+}(aq)$

9. Predict which of each of the following reactions occurs under standard conditions. For those that do take place, complete and balance the equations.

(a) $Mn^{2+}(aq) + Cr_2O_7^{2-}(aq) \rightarrow MnO_4^-(aq) + Cr^{3+}(aq)$ (acidic solution)

(b) $O_2(g) + Br^-(aq) \rightarrow Br_2(aq)$ (acidic solution)

(c) $Br_2(aq) + Cl^-(aq) \rightarrow Br^-(aq) + Cl_2(aq)$

(d) $I^-(aq) + Cl_2(g) \rightarrow I_2(s) + Cl^-(aq)$

10. Predict which of the following metals should be oxidized by $O_2(g)$ at a pressure of 1 atm in 1-*M* acidic aqueous solution at room temperature. (In each case, the reduction product is water.)

(a) Ag **(b)** Cu **(c)** Ca **(d)** Zn **(e)** Al

11. According to a book about the construction of stained glass, the metallic framework holding the glass pieces can be given a bronzelike finish instead of the usual gray solder finish by wiping the solder (usually about 60% Pb by mass) with dilute copper sulfate solution.

(a) Explain how this process works.

(b) What would you expect to happen when the bronzelike finish is wiped with a dilute solution of aqueous silver nitrate?

12. Draw the electrochemical cell in which the overall reaction is

$$Zn(s) + 2Ag^+(aq) \rightarrow Zn^{2+}(aq) + 2Ag(s)$$

(a) Show the cathode, the anode, and the directions in which the ions move in the solution and the electrons move in the external circuit, and give the electrode reactions.

(b) Calculate the standard cell voltage.

13. **(a)** Illustrate the construction of the electrochemical cell

$$Zn(s)|Zn^{2+}(aq)||Ni^{2+}(aq)|Ni(s)$$

showing the direction of electron flow in the external circuit.

(b) What is the standard cell voltage, E°_{cell}?

14. For each of the following standard electrochemical cells, give the two half-cell reactions and the overall cell reaction, and calculate the standard cell voltage, E°_{cell}.

(a) $Al(s)|Al^{3+}(aq)||Cu^{2+}(aq)|Cu(s)$

(b) $Pb(s)|Pb^{2+}(aq)||Ag^+(aq)|Ag(s)$

(c) $Ag(s)|Ag^+(aq)||Cl^-(aq)|Cl_2(g)|Pt(s)$

15. Consider the reaction

$$Fe(s) + 2H^+(aq) \rightarrow Fe^{2+}(aq) + H_2(g)$$

(a) Draw an electrochemical cell in which this reaction takes place.

(b) What are the charge carriers in the wire that connects the two electrodes?

(c) At which electrode does reduction occur? Is this the anode or the cathode?

(d) Give the equation for the reaction that occurs at the cathode.

16. Draw the cell diagram for each of the standard electrochemical cells for which the overall reactions are as follows. In each case, place the anode compartment on the left and indicate the direction of electron flow in the external circuit.

(a) $Zn(s) + Br_2(aq) \rightarrow Zn^{2+}(aq) + 2Br^-(aq)$

(b) $Pb(s) + 2Ag^+(aq) \rightarrow Pb^{2+}(aq) + 2Ag(s)$

17. Write balanced equations for the reactions that occur at the anode and at the cathode of each of the following cells, and calculate the standard cell potential. For each example, draw the cell, label the anode and the cathode, and show the direction in which the electrons move in the external circuit.

(a) a lead wire dipping into 1-*M* $PbCl_2(aq)$ and a copper wire dipping into 1-*M* $CuSO_4(aq)$

(b) $Cl_2(g)$ bubbling over a platinum wire in 1-*M* $NaCl(aq)$ and a silver wire coated with $AgCl(s)$ dipping into a similar solution

(c) two platinum wires dipping into 1-*M* solutions of $HI(aq)$, with one of the wires having $H_2(g)$ bubbling over it

18. What are the standard reaction Gibbs free energies for each of the reactions in Problem 14?

19. Repeat Problem 18 for the reactions in Problem 16.

20. Separate beakers contain a piece of iron immersed in 1.00-*M* $FeSO_4(aq)$ and a piece of copper immersed in 1.00-*M* $CuSO_4(aq)$; the solutions are then connected by a salt bridge, and the metals are connected by a conducting wire.

(a) Which metal dissolves, and which metal increases in mass?

(b) What will be the initial voltage between the metals?

(c) As the reaction proceeds, which of the two solutions increases and which decreases in concentration?

(d) Does the initial voltage increase or decrease with time?

21. Consider a room-temperature electrochemical cell composed of a $Cu(s)|Cu^{2+}(aq)$ half-cell and an $Ag(s)|Ag^+(aq)$ half-cell. What is the cell voltage when **(a)** all the ions have concentrations of 1 *M*; **(b)** $[Cu^{2+}]$ is 2.0 *M* and $[Ag^+]$ is 0.05 *M*?

22. The voltage of the electrochemical cell

$$Zn(s)|Zn^{2+}(aq,\ x\ M)||Cr^{3+}(aq,\ 0.001\ M)|Ni(s)$$

is 0.00 V at 25°C. What is the $Zn^{2+}(aq)$ concentration x $mol \cdot L^{-1}$?

23. Consider the following electrochemical cell at 25°C:

$$Ag(s)|Ag^+(aq, 0.01\ M)||Br^-(aq, 0.50\ M)|Br_2(l)|Pt$$

(a) Calculate the standard cell voltage.

(b) Write the equilibrium expression for the cell reaction.

(c) Calculate the value of the equilibrium constant.

24. For the electrochemical cell

$$Pb(s)|Pb^{2+}(aq)||Cu^{2+}(aq)|Cu(s)$$

at 25°C, calculate the cell voltage for each of the following molar ion concentrations.

(a) $[Pb^{2+}]$, 1.0 M; $[Cu^{2+}]$, 1.0 M

(b) $[Pb^{2+}]$, 1.0 M; $[Cu^{2+}]$, 1.0×10^{-5} M

(c) $[Pb^{2+}]$, 1.0×10^{-3} M; $[Cu^{2+}]$, 1.0×10^{-2} M

(d) $[Pb^{2+}]$, 6.0×10^{-5} M; $[Cu^{2+}]$, 2.0×10^{-2} M

25. Repeat Problems 23 for the electrochemical cell

$$Cu(s)|Cu^{2+}(aq, 0.50\ M)||Fe^{2+}(aq, 0.10\ M), Fe^{3+}(aq, 0.01\ M)|Pt(s)$$

Applications of Electrochemical Cells

Batteries

26. **(a)** What is a battery?

(b) How does a primary cell differ from a secondary cell?

(c) Describe the electrode reactions and the overall reactions that occur in (i) a dry cell and (ii) a lead-acid battery.

27. What problems have to be overcome to invent a suitable battery to replace gasoline for powering vehicles such as automobiles?

28. A possible source of power for a cardiac pacemaker is to implant a zinc electrode and a platinum electrode directly into the body. When inserted into the oxygen-containing body fluid, these electrodes form a ''biogalvanic'' cell in which zinc is oxidized and oxygen is reduced.

(a) Write equations for the anode and cathode reactions.

(b) Estimate the standard voltage of such a cell.

(c) If a current of 40 μA is drawn from the cell, how often will a zinc electrode of mass 5.0 g have to be replaced?

29. **(a)** Write the two half-reactions for a fuel cell in which ethane, $C_2H_6(g)$, is oxidized in acid solution by $O_2(g)$ to $CO_2(g)$ and water.

(b) How many liters of ethane at STP would be needed to generate a current of 0.500 A for 6.00 h?

(c) How many liters of oxygen at STP would be consumed at the cathode during this time?

Corrosion

30. **(a)** What sort of reactions cause metals to corrode?

(b) Which of iron and aluminum is expected to corrode more easily in the presence of oxygen and water, and how do the oxides of these metals differ in their subsequent behavior?

31. Explain why iron can be protected, despite the fact that the metals used are more readily oxidized than iron, by **(a)** electrodepositing a layer of chromium on the surface; **(b)** dipping it in molten zinc.

32. Write equations for the half-reactions and overall cell reactions to show how metals such as **(a)** magnesium and **(b)** zinc protect iron by cathodic protection.

Electrolysis

33. **(a)** Explain each of the following.

(i) electrolysis (ii) electrolyte (iii) nonelectrolyte

(b) In what way does ionic conduction differ from electronic conduction?

34. Define each of the following.

(a) anode **(b)** cathode **(c)** the Faraday constant, F

35. For inert electrodes, write equations for the electrode reactions that occur during the electrolysis of **(a)** molten aluminum chloride and **(b)** a dilute aqueous solution of aluminum chloride.

(c) Explain why the products of electrolysis in (a) are different from those in (b).

36. **(a)** Draw a cell for the electrolysis of $HBr(aq)$ with platinum electrodes.

(b) Label the anode and the cathode.

(c) Write equations for the electrode reactions and for the overall cell reaction.

(d) Indicate the direction of electron flow in the external circuit and the directions in which the ions move in the solution.

37. Repeat Problem 36 for the electrolysis of $CuSO_4(aq)$ with copper electrodes.

38. How many coulombs of electricity are required for each of the following reductions?

(a) 1 mol of $Cu^{2+}(aq)$ to $Cu(s)$

(b) 1 mol of $Fe^{3+}(aq)$ to $Fe^{2+}(aq)$

(c) 1 mol of $MnO_4^-(aq)$ to $Mn^{2+}(aq)$

(d) 1 mol of $Cr_2O_7^{2-}(aq)$ to $Cr^{3+}(aq)$

39. How many coulombs of electricity are required for each of the following oxidations?

(a) 1 mol of $H_2O(l)$ to $O_2(g)$

(b) 1 mol $Cl_2(g)$ to $ClO_3^-(aq)$

(c) 1 mol $FeO(s)$ to $Fe_2O_3(s)$

(d) 1 mol $S(s)$ to $H_2SO_4(l)$

40. How many coulombs of electricity are required to produce each of the following?

(a) 50.0 mL of $O_2(g)$ at STP from $Na_2SO_4(aq)$

(b) 50.0 kg of $Al(s)$ from molten Al_2O_3

(c) 20.0 g of calcium from molten calcium chloride

(d) 5.00 g of silver from aqueous silver nitrate

41. Using a current of 2.00 A, how long will it take to produce 1.00 kg of nickel by the electrolysis of a solution of nickel(II) chloride?

42. What masses of sodium hydroxide and chlorine are produced by the electrolysis of aqueous sodium chloride for 3.00 h, using a current of 0.200 A?

43. What volumes of $H_2(g)$ and $O_2(g)$ at 27°C and a pressure 740 torr are produced by the electrolysis of $Na_2SO_4(aq)$ for 30.0 min, using a current of 4.00 A?

44. How long will it take to deposit 16.0 g of silver from $AgNO_3(aq)$ solution, using a current of 6.00 A?

45. A metal tray has dimensions of 24 cm × 12 cm. How long will it take to plate the tray with a layer of silver (density 10.54 $g \cdot cm^{-3}$) to a thickness of 0.020 mm, using a current of 7.65 A? (Neglect the amount of silver needed to coat the edges of the tray.)

46. How long would an electric current of 1.50 A have to be passed through a solution containing chromium(III) sulfate for a steel object of surface area 0.10 m^2 to be plated with a chromium (density 7.1 $g \cdot cm^{-3}$) to a thickness 0.10 mm?

47. When a current of 1.487 A was passed through aqueous hydroiodic acid for 1 h 1 min 43 s, 7.2428 g of iodine resulted at the anode. From these data, determine a value for the Faraday constant, and compare this value with the accepted value.

48. When a current of 0.500 A was passed through a solution containing an unknown $M^{2+}(aq)$ ion for exactly 2 h, 1.98 g of the metal M were deposited.

(a) What is the atomic mass of M? **(b)** What is the metal?

49. A Hall cell for the production of aluminum from alumina, $Al_2O_3(s)$, operates with a current of 1.300×10^5 A.

(a) What mass of aluminum is produced in exactly 1 min from 100 such cells?

(b) How much aluminum will be produced each day, and

(c) What mass of carbon will be consumed per day at the anodes?

50. Two electrolysis cells are connected such that the same current passes through each. In the first, $Fe^{3+}(aq)$ is reduced to $Fe(s)$; in the second, $Cu^{2+}(aq)$ is reduced to $Cu(s)$. After passage of the current for 30.00 min, 1.030 g of iron were deposited in the first cell. How many grams of copper were deposited in the second cell?

51. When an aqueous solution of sodium chloride was electrolyzed for 1.500 h at a current of 2.500 A, $Cl_2(g)$ was produced at the anode, and $H_2(g)$ and $OH^-(aq)$ at the cathode. At the end of the electrolysis, the cathode compartment contained 1.000 L of solution. What was its pH?

52. The same electric current was passed for 1.00 h through each of three electrolytic cells fitted with platinum electrodes and containing $CuSO_4(aq)$, $AgNO_3(aq)$, and $H_2SO_4(aq)$, respectively. During this time, 0.106 g of copper was deposited at the cathode in the first cell. Calculate **(a)** the average current, in milliamps; **(b)** the mass of $Ag(s)$ deposited at the cathode in the second cell; **(c)** the total volume of gas, measured at 20°C and a pressure of 750 mm Hg, liberated in the third cell.

General Problems

53. **(a)** How long would a current of 1.50 A have to be passed through a 5.00-*M* $Cr_2(SO_4)_3(aq)$ solution to coat a metal object of surface area 1.00 m^2 with a 0.0100-mm layer of chromium (density 7.21 $g \cdot cm^{-3}$)?

(b) Why do you suppose the coating on chromium-plated steel is generally very thin?

54. In the electrolysis of $KI(aq)$ with inert electrodes, a brown color is observed in the solution at one of the electrodes.

(a) To what molecule or ion is the brown color due?

(b) Is it observed at the cathode or at the anode?

(c) Write balanced equations for the reaction at each electrode and for the overall cell reaction.

55. Calculate the equilibrium constant for the reaction in which $I^-(aq)$ is oxidized by $Fe^{3+}(aq)$ to give $I_2(s)$ and $Fe^{2+}(aq)$ under standard conditions.

56.* Suppose the standard silver chloride electrode,

$$AgCl(s) + e^- \rightarrow Ag(s) + Cl^-(aq, 1\ M)$$

rather than the standard hydrogen electrode, had been assigned a potential of 0 V at 25°C.

(a) What would be the value of the standard reduction potential at 25°C for each of the following reactions?
(i) $Cl_2(g) + 2e^- \rightarrow 2Cl^-(aq)$
(ii) $2H^+(aq) + 2e^- \rightarrow H_2(g, 1\ atm)$

(b) With this new standard, calculate the standard cell potential for

$$Cl_2(g) + H_2(g) \rightarrow 2H^+(aq) + 2Cl^-(aq)$$

(c) If a different half-cell potential is assigned a value of zero, does this choice affect the standard cell potential E°_{cell} of a cell?

57.* Predict qualitatively the effect of adding each of the following on the cell voltage of the cell

$$Zn(s)|Zn^{2+}(aq, 1\ M)||Cu^{2+}(aq, 1\ M)|Cu(s)$$

(a) 2-*M* $Zn^{2+}(aq)$ to the $Zn(s)|Zn^{2+}(aq)$ half-cell

(b) $Zn(s)$ to the $Zn(s)|Zn^{2+}(aq)$ half-cell

(c) a drop or two of dilute $NaOH(aq)$ to the $Zn(s)|Zn^{2+}(aq)$ half-cell

(d) concentrated $NH_3(aq)$ to the $Cu(s)|Cu^{2+}(aq)$ half-cell

58.* An electrochemical cell has an $Ag(s)|Ag^+(aq, 0.10\ M)$ half-cell and a hydrogen electrode, with an $H_2(g)$ pressure of 1 atm, dipping into a solution of unknown pH. When the two half-cells are connected by a salt bridge, the cell voltage is 0.859 V. What is the pH of the solution?

59.* **(a)** At 25°C, what is the voltage of the concentration cell

$$Zn(s)|Zn^{2+}(aq, 1.00\ M)||Zn^{2+}(aq, 0.001\ M)|Zn(s)$$

(b) Explain why such a cell produces a current.

(c) Why is the principle of the concentration cell important in living systems?

Organic Chemistry

14.1 Functional Groups
14.2 Haloalkanes (Alkyl Halides)
14.3 Alcohols and Ethers
14.4 Thiols and Disulfides
14.5 Aldehydes and Ketones
14.6 Carboxylic Acids and Esters
14.7 Amines, Amides, and Amino Acids
14.8 Determining the Structure of Organic Compounds

Organic compounds, compounds that contain carbon, are the subject of organic chemistry. The synthesis of an organic compound from readily available substances may require many steps and numerous separations and purifications of the intermediate compounds. This apparatus is typical of the equipment used in a chemical research laboratory to carry out an organic synthesis.

Organic chemistry is the chemistry of carbon compounds. The study of all the compounds of all the other elements is called **inorganic chemistry**. Carbon stands out among the elements for the great number and diversity of its compounds. The name "organic" derives from the early belief that the carbon compounds found in living matter could be synthesized only in a living organism. Although we now know that organic compounds can be made from inorganic substances, it has proved convenient to retain the name "organic" for the vast majority of the compounds of carbon and to describe their chemistry as organic chemistry. Just a few compounds of carbon, such as carbon monoxide, carbon dioxide, and the metal carbonates that have long been known in nonliving matter, are traditionally classified as inorganic compounds.

Although some carbon compounds are still obtained from plant and animal sources, chemists can now synthesize most of the ones we know today. Even those we can obtain from natural sources are often more conveniently and more economically obtained by synthesis. Methanol, CH_3OH, was originally obtained by heating wood in the absence of air and so was called "wood alcohol." Today methanol is made industrially by heating carbon monoxide and hydrogen, two inorganic substances, at a high temperature in the presence of a catalyst:

$$CO(g) + 2H_2(g) \xrightarrow{\text{catalyst}} CH_3OH(g)$$

The most important starting materials for synthesizing organic compounds are the hydrocarbons (compounds of carbon and hydrogen only; see Chapter 8) obtained from reserves of petroleum, natural gas, and coal in the earth's crust. Substances derived from petroleum are frequently referred to as **petrochemicals**, and the industry that produces them is called the **petrochemical industry**. The petrochemical industry produces a vast number of organic substances that have become essential to modern life, such as solvents, synthetic fabrics, plastics, and medicines.

The great number and diversity of carbon compounds result from the ability of a carbon atom to form strong bonds to other carbon atoms to give an enormous variety of chains, branching chains, and rings of different shapes and sizes, as we saw in Chapter 8 when we discussed the hydrocarbons. Carbon–carbon single bonds are stronger than the single bonds between the atoms of any other element, except for the H—H bond, as we see from the bond energies in Tables 14.1 and 9.2. The atoms of other elements, such as sulfur (Chapter 7), can form chains and rings, but the resulting molecules have weaker bonds and are more reactive than molecules made of carbon chains and rings. Moreover, a sulfur atom, for example, can form only two bonds to other sulfur atoms; therefore, branching chains, and chains and rings containing multiple bonds, are not possible.

When we consider that, as well as hydrogen, many other atoms such as oxygen, nitrogen, sulfur, and the halogens can be attached to the carbon atoms of

TABLE 14.1 Bond Energy Values ($kJ \cdot mol^{-1}$)

		H—H 436		
B—B 293	C—C 348	N—N 159	O—O 138	F—F 155
	Si—Si 196	P—P 197	S—S 266	Cl—Cl 243

these chains and rings, we can understand why more than 10 million organic compounds are now known. Just the number of compounds that carbon can form—which is far greater than for any other element except perhaps hydrogen—justifies treating carbon compounds as a separate division of chemistry. Another reason is that an organic molecule can be transformed step by step to another organic molecule: In each of a series of reactions, a particular atom or group of atoms is removed from or added to the carbon framework without disturbing the remainder of the structure (Figure 14.1). The stability of the chains and rings of carbon atoms makes these types of reactions possible. The synthesis of a new compound can therefore be planned and carried out systematically in a step-by-step process, which has not yet proved possible for many compounds of other elements. Yet another reason for treating the chemistry of carbon separately is that carbon compounds are the basis of life, as we shall see in Chapter 17.

14.1 FUNCTIONAL GROUPS

Hydrocarbons are the simplest organic compounds. Many other organic compounds can be regarded as hydrocarbons in which one or more hydrogen atoms have been replaced by other atoms or groups of atoms. Because the hydrocarbon part of an organic molecule is relatively unreactive, most of the reactions of an organic compound are reactions of these other atoms or groups, which are called **functional groups**.

A functional group is an atom or group of atoms in an organic molecule that gives the molecule characteristic chemical properties.

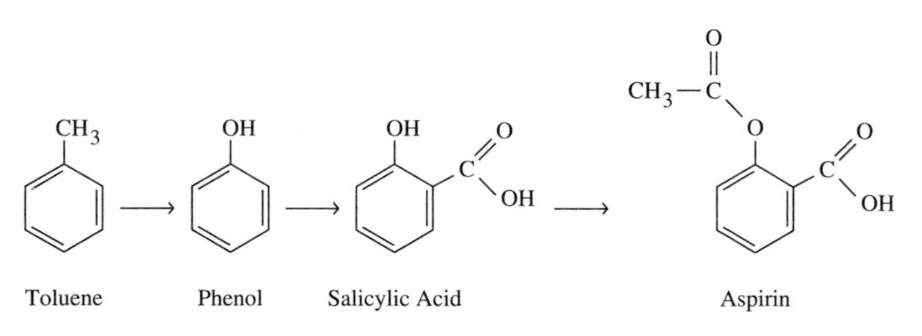

FIGURE 14.1 Synthesis of Aspirin Starting with toluene, a product of the petrochemical industry, aspirin (acetylsalicylic acid) is synthesized in these steps in which the groups attached to the benzene ring are changed but the benzene ring remains intact.

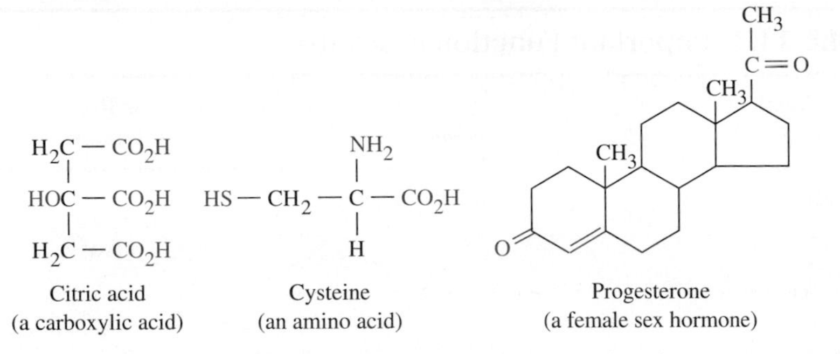

FIGURE 14.2 Functional Groups in Some Organic Molecules These molecules contain the hydroxyl, OH; carboxyl, CO_2H; thiol, SH; amino, NH_2; carbonyl, CO; and alkene, C=C, functional groups, indicated in red.

The OH and COOH groups in salicylic acid (Figure 14.1) are examples of functional groups. Figure 14.2 shows the functional groups in some other organic molecules. All molecules containing a given functional group have a similar set of characteristic properties that are largely independent of the hydrocarbon part of the molecule. Organic compounds are therefore conveniently classified on the basis of the functional groups they contain. The most important functional groups are listed in Table 14.2. As we saw in Chapter 8, carbon–carbon double and triple bonds have characteristic reactions, such as addition reactions, so they are also considered to be functional groups.

14.2 HALOALKANES (ALKYL HALIDES)

Preparation of Haloalkanes

Alkanes are rather unreactive compounds particularly at ordinary temperatures. They react with oxygen at high temperatures (combustion) and with the halogens at temperatures above about 350°C or in the presence of ultraviolet light. The resulting compounds, called **haloalkanes** or **alkyl halides**, RX (where R stands for an alkyl group), have a wide array of uses in industry and in the synthesis of other organic molecules. For example, methane and chlorine react rapidly to give a mixture of compounds in which hydrogen atoms are replaced by one or more chlorine atoms (the chloromethanes):

A group obtained from a hydrocarbon by the removal of a hydrogen atom is an *alkyl group*. CH_3 is the methyl group (an alternative name for chloromethane, CH_3Cl, is methyl chloride), and C_2H_5 is the ethyl group.

$$CH_4(g) + Cl_2(g) \rightarrow HCl(g) + CH_3Cl(g)$$ Chloromethane (methyl chloride)

$$CH_3Cl(g) + Cl_2(g) \rightarrow HCl(g) + CH_2Cl_2(l)$$ Dichloromethane (methylene chloride)

$$CH_2Cl_2(l) + Cl_2(g) \rightarrow HCl(g) + CHCl_3(l)$$ Trichloromethane (chloroform)

$$CHCl_3(l) + Cl_2(g) \rightarrow HCl(g) + CCl_4(l)$$ Tetrachloromethane (carbon tetrachloride)

The HCl is removed by dissolving it in water, and the chloromethanes are then condensed and separated by fractional distillation (Chapter 1). Their boiling points increase from 24°C for CH_3Cl to 79°C for CCl_4 as the size and polarizability of the

TABLE 14.2 Important Functional Groups

Functional Group	Name	Suffix or Prefix Used in Systematic Name
$-OH$	Alcohol (phenol)‡	-ol
$-OR^{*}$	Ether	alkoxy-
$-C(=O)H$	Aldehyde	-al
$-C(=O)R$	Ketone	-one
$-C(=O)OH$	Carboxyl	-oic acid
$-C(=O)OR$	Ester	-oate
$-X^{\dagger}$	Haloalkane	halo-
$-NH_2$	Amine	amino-
$-C(=O)NH_2$	Amide	-amide
$>C=C<$	Alkene	-ene
$-C\equiv C-$	Alkyne	-yne
$-C_6$ ring (alternating C=C and C—C bonds)	Arene	—

* R is an alkyl or an aryl group.
† X is F, Cl, Br, or I.
‡ An aromatic alcohol is called a phenol.

molecules increase and the strength of the intermolecular (London) forces increases correspondingly (Chapter 11).

The chloromethanes are good solvents for grease and for many organic compounds. Tetrachloromethane (carbon tetrachloride) was once widely used as a dry-cleaning agent. Trichloromethane (chloroform) is a volatile, sweet-tasting liquid that has been used as an inhalation anesthetic and as a solvent for cough syrups and other medicines. These uses have, however, decreased in recent years, because we now know that the chloromethanes are toxic.

Chloroethane, C_2H_5Cl, can be made by the reaction of chlorine with ethane, but it is more commonly made by the addition of HCl to ethene:

$$\underset{\text{Ethene}}{CH_2{=}CH_2(g)} + HCl(g) \rightarrow \underset{\text{Chloroethane}}{CH_3{-}CH_2Cl(g)}$$

The addition of chlorine to ethene gives 1,2-dichloroethane:

$$\underset{\text{Ethene}}{CH_2{=}CH_2(g)} + Cl_2(g) \rightarrow \underset{\text{1,2-Dichloroethane}}{CH_2Cl{-}CH_2Cl(l)}$$

This product is used to make chloroethene (commonly known as vinyl chloride) by heating to eliminate HCl:

$$\underset{\text{1,2-Dichloroethane}}{CH_2Cl{-}CH_2Cl(l)} \rightarrow \underset{\text{Chloroethene (vinyl chloride)}}{CH_2{=}CHCl(g)} + HCl(g)$$

Vinyl chloride is polymerized to give the widely used plastic poly(vinylchloride), or PVC (Chapter 20).

Bromotrifluoromethane, $CBrF_3$, is used as a fire extinguisher. The chlorofluoromethanes and ethanes, such as CF_2Cl_2, CCl_3F, and $CClF_2CClF_2$, which are commonly known as CFCs (chlorofluorocarbons) or Freons, have achieved considerable notoriety since the early 1980s (Chapter 16). They have been widely used as refrigerants, as propellants in aerosol cans, and in making foamed plastics, such as those used in insulation and mattresses. They are suitable for these purposes because they have low boiling points ($-30°C$ for CF_2Cl_2) and are inert, nonflammable, and nontoxic. However, this very lack of reactivity is the origin of a serious problem that was not originally suspected, namely, the destruction of the ozone layer in the upper atmosphere (Chapter 16). Consequently, the manufacture and use of these compounds is now being phased out.

Reactions of Haloalkanes

The halogen atoms in many haloalkanes are readily replaced by other atoms or groups of atoms. Thus, the haloalkanes are important starting materials for the preparation of many other organic compounds, by reactions called substitution reactions (Chapter 8).

A *substitution reaction* is a reaction in which an atom or group of atoms is replaced by another atom or group.

For example, when chloromethane reacts with an aqueous NaOH solution, the chlorine atom is replaced by an OH group to give methanol, CH_3OH:

$$\underset{\text{Chloromethane}}{CH_3{-}Cl} + OH^- \rightarrow \underset{\text{Methanol}}{CH_3{-}OH} + Cl^- \qquad (1)$$

H
|
H—C$^{\delta+}$—Cl$^{\delta-}$
|
H

Chloromethane

Because chlorine is more electronegative than carbon, CH_3Cl is a polar molecule in which the carbon atom has a small positive charge and the chlorine a small negative charge. These charges are much smaller than the full positive and negative charges on Na^+ and Cl^- ions, for example, but they are very important in determining the reactivity of the molecule. In reaction (1), the negatively charged oxygen atom of the hydroxide ion is attracted to the positively charged carbon atom in CH_3Cl. Because OH^- is a stronger base than the chloride ion, it donates one of its unshared electron pairs to the carbon atom and displaces the more weakly bonded chlorine atom as a chloride ion:

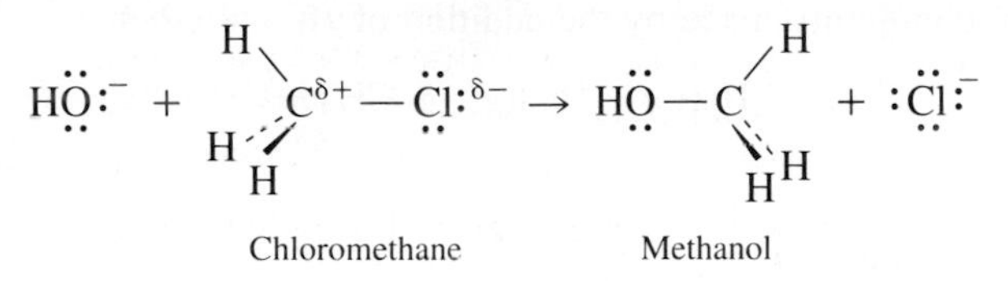

Because the hydroxide ion is attracted to a positive charge, in this case the positively charged carbon atom, it is called a **nucleophile** (''nucleus-lover'') or **nucleophilic reagent**. A nucleophile is a molecule or ion that has an atom carrying a negative charge and one or more unshared electron pairs that attacks a positively charged atom in another molecule. When the attack results in the substitution of an atom or group by the nucleophile, as in the previous example, the reaction is called a **nucleophilic substitution**. Such a reaction is generalized as

$$R—X + Y^- \rightarrow R—Y + X^-$$

where R stands for an alkyl group. By this useful type of reaction, many functional groups, such as halogen atoms, can be replaced by another functional group. For example, the chlorine in an alkyl chloride can be replaced by iodide to give alkyl iodide, by ammonia to give an alkyl ammonium salt, and by cyanide to give an alkyl cyanide:

I^-, NH_3, and CN^- are all nucleophiles.

$$R—Cl + I^- \rightarrow R—I + Cl^-$$

$$R—Cl + NH_3 \rightarrow R—NH_3^+\ Cl^-$$

$$R—Cl + CN^- \rightarrow R—CN + Cl^-$$

We will discuss some of these reactions in more detail later in this chapter.

14.3 ALCOHOLS AND ETHERS

In industry and in hardware stores, methanol is commonly known as methyl alcohol or as methyl hydrate.

Alcohols are molecules containing the OH (*hydroxyl*) group. They have the general formula ROH, where R represents an alkyl group. The simplest alcohol is **methanol**, CH_3OH:

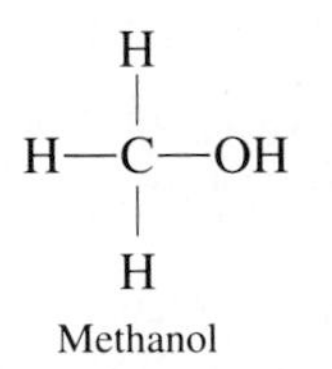

Methanol

We can imagine methanol as being derived from methane by replacing a hydrogen atom with an OH group. Methanol is a primary product of the petrochemical industry and is used in the manufacture of many other organic substances, including plastics and pharmaceuticals. It is manufactured from *synthesis gas* (Chapter 8), a 2:1 mixture of carbon monoxide and hydrogen, by heating at about 250°C under a pressure of 50 to 100 atm in the presence of a catalyst:

$$CO(g) + 2H_2(g) \xrightarrow{\text{catalyst}} CH_3OH(g) \qquad \Delta H° = -91\ kJ$$

The reaction is very slow at 25°C, so it is carried out at a higher temperature to increase the rate. However, because the reaction is exothermic (has a negative reaction enthalpy, $\Delta H°$) the position of equilibrium, which is very favorable at 25°C, shifts to the left with increasing temperature. So, just as we saw for the Haber process (Chapter 12), the reaction temperature (250°C) is a compromise chosen to give the best yield in a reasonable time.

A characteristic reaction of alkenes is addition to the double bond (Chapter 8). An important reaction of this type is the addition of water. The addition of water to ethene gives **ethanol**, CH_3CH_2OH:

$$\underset{\text{Ethene}}{CH_2{=}CH_2} + H_2O \rightarrow \underset{\text{Ethanol}}{\underset{H\qquad OH}{CH_2{-}CH_2}}$$

This reaction is extremely slow at 25°C, so a temperature of 250 to 300°C and a catalyst are used to obtain a sufficiently rapid reaction. Ethanol is commonly known as ethyl alcohol or simply alcohol.

For thousands of years ethanol in the form of beer and wine has been made by fermenting grain and sugar (Figure 14.3). **Fermentation** is a reaction, catalyzed by enzymes found in yeast, in which sugars and carbohydrates decompose into ethanol and carbon dioxide in the absence of air:

$$C_6H_{12}O_6 \xrightarrow{\text{enzyme}} 2CH_3CH_2OH + 2CO_2$$

About 13% (by volume) is the maximum alcohol concentration that can be produced by fermentation, because beyond this concentration, yeast becomes inactive. Beverages such as whiskey that have a higher alcohol content are made by distilling the solution obtained through fermentation. The intoxicating effect of ethanol is well known. Prolonged and excessive consumption of ethanol can cause permanent liver damage. Methanol and other alcohols are considerably more toxic. Small amounts of methanol can lead to blindness and even death.

FIGURE 14.3 Wine Making in Ancient Egypt The making of wine by fermentation is one of the oldest applications of chemistry.

A useful laboratory method for preparing alcohols is the reaction of halo-alkanes with hydroxide ion that we discussed previously; see equation (1) of Section 14.2:

$$R{-}Cl + OH^- \rightarrow R{-}OH + Cl^-$$

Naming Alcohols

Alcohols are named by replacing the *-e* ending in the name of the alkane from which they are derived by the ending *-ol*. For example,

method*e* → methan*ol*   ethan*e* → ethan*ol*

The position of the hydroxyl group is indicated by numbering the longest carbon chain in the direction that gives the hydroxyl group the smallest number.

Two different propanols can be derived from propane, C_3H_8: 1-propanol (not 3-propanol) and 2-propanol. In other words, there are two structural *isomers* of propanol:

Recall from Chapter 8 that molecules with the same molecular formula but different structures are called structural isomers.

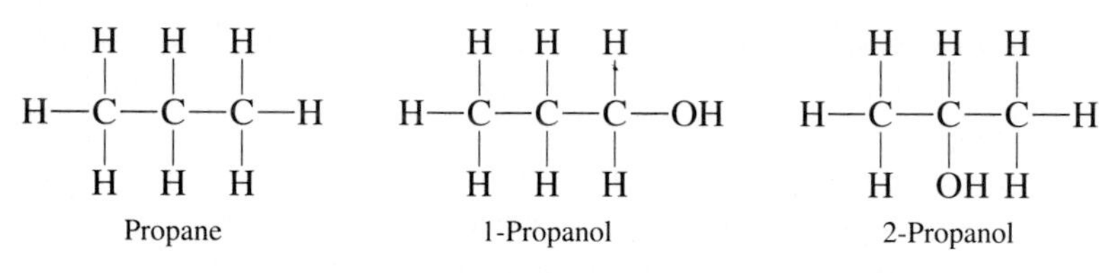

Four alcohols have the formula C_4H_9OH. They can be derived from the two isomeric alkanes C_4H_{10}:

$CH_3{-}CH_2{-}CH_2{-}CH_3$

Butane

↓

$CH_3{-}CH_2{-}CH_2{-}CH_2{-}OH$

1-Butanol

and

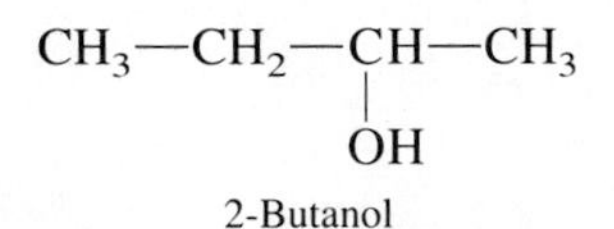

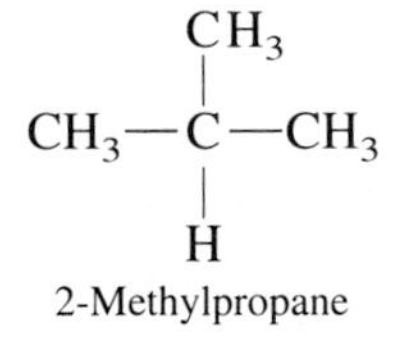

$CH_3{-}CH(CH_3){-}CH_2{-}OH$

2-Methyl-1-propanol

and

$CH_3{-}C(CH_3)(OH){-}CH_3$

2-Methyl-2-propanol

An alcohol containing a $-CH_2OH$ group is known as a *primary alcohol*. An alcohol containing a $>CHOH$ group is a *secondary alcohol*, and an alcohol containing a $\geqslant COH$ group is a *tertiary alcohol*.

EXAMPLE 14.1 Naming and Classifying Alcohols

Give the systematic name of the following alcohol, and classify it as primary, secondary, or tertiary:

$$CH_3-CH_2-\underset{\displaystyle CH_3}{\overset{\displaystyle CH_3}{\overset{|}{\underset{|}{C}}}}-OH$$

Solution: First, identify the longest carbon chain and number the carbon atoms beginning at the end nearest OH group. The longest carbon chain has four carbon atoms, so we name this alcohol as a butanol:

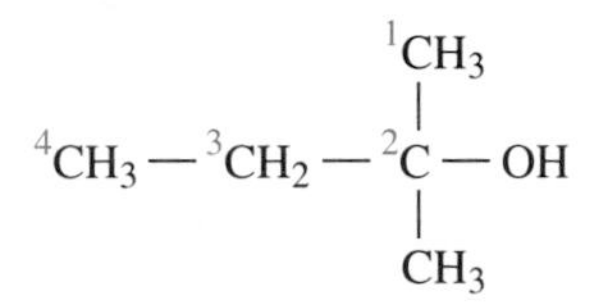

Then write the full name and indicate the position of the substituent groups:

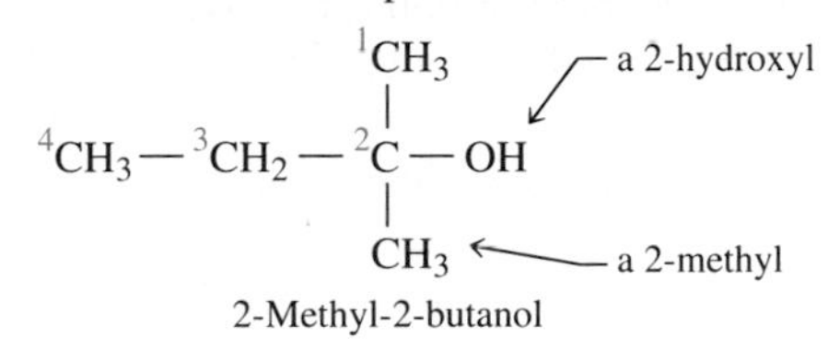

The OH group is bonded to a carbon atom that is itself bonded to three other carbon atoms, so this is a tertiary alcohol.

Exercise 14.1 Draw the structures of the following alcohols, and identify each as primary, secondary, or tertiary.

(a) 3-methyl-3-pentanol **(b)** cyclohexanol
(c) 2-methyl-4-heptanol **(d)** 3-chloro-1-propanol

To decide if an alcohol is primary, secondary, or tertiary, focus on the carbon bearing the OH group, and determine how many alkyl groups are bonded directly to that carbon.

Diols and Triols

Alcohols containing two OH groups are called **diols**; those with three OH groups are **triols**; and so on. Two important compounds of this type are *1,2-ethanediol*, also known as *ethylene glycol* or simply *glycol*, and *1,2,3-propanetriol*, also known as *glycerol* or *glycerin*:

$$\underset{\displaystyle OH}{\underset{|}{CH_2}}-\underset{\displaystyle OH}{\underset{|}{CH_2}} \qquad \underset{\displaystyle OH}{\underset{|}{CH_2}}-\underset{\displaystyle OH}{\underset{|}{CH}}-\underset{\displaystyle OH}{\underset{|}{CH_2}}$$

1,2-Ethanediol 1,2,3-Propanetriol

Ethylene glycol is the main component of automobile antifreeze. A major use is in the manufacture of polyester fibers, such as Dacron (Chapter 20). Glycerol is nontoxic and has a sweet taste, so it is used to sweeten foods. It is also used in the manufacture of plastics and nitroglycerine (Box 9.1) and in antifreeze and shock absorber fluids. As we shall see in Section 14.6, natural fats and oils are compounds derived from glycerol.

Antifreeze is ethylene glycol (1,2-ethanediol).

TABLE 14.3 Properties of Some Alcohols

Alcohol	Name	BP (°C)	Solubility (g/100 g H_2O), 25°C
CH_3OH	Methanol	65	Soluble in all proportions; miscible
CH_3CH_2OH	Ethanol	78	Soluble in all proportions; miscible
$CH_3CH_2CH_2OH$	1-Propanol	97	Soluble in all proportions; miscible
$CH_3CH_2CH_2CH_2OH$	1-Butanol	117	9.0
$CH_3CH_2CH_2CH_2CH_2OH$	1-Pentanol	138	2.7
$CH_3CH_2CH_2CH_2CH_2CH_2OH$	1-Hexanol	158	0.6

Properties of Alcohols

At 25°C and 1 atm methanol, ethanol, and other alcohols containing up to about 12 carbon atoms are liquids. The boiling point increases regularly as the length of the carbon chain increases, because the strength of the intermolecular forces increases as the number of carbon atoms increases (Table 14.3). The boiling points of alcohols are much higher than for the corresponding alkanes. For example, methane boils at −164°C, whereas methanol boils at 65°C; ethane boils at −89°C, but ethanol boils at 78°C. The high boiling points of alcohols result from hydrogen bonding. The hydrogen atom bonded to the highly electronegative oxygen atom of one molecule is hydrogen bonded to the oxygen of a neighboring molecule:

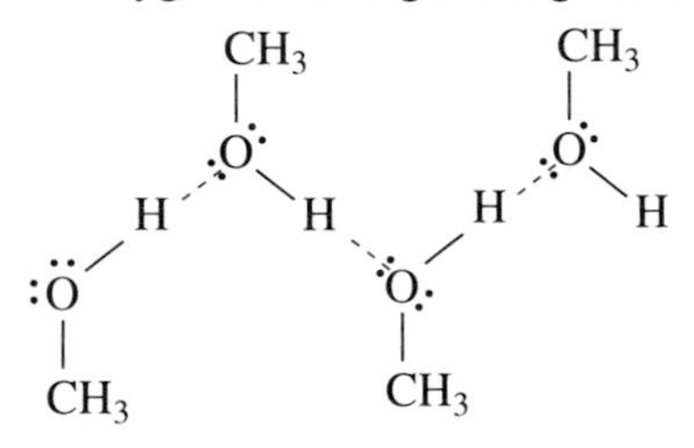

Methanol, ethanol, and the propanols are soluble in water in all proportions. But as the hydrocarbon chains become longer, the solubility of the alcohols progressively decreases.

Reactions of Alcohols

Acid-Base Reactions Like water, alcohols behave as very weak acids and bases. They are too weakly acidic to give up a proton to water, but they donate an H^+ to the OH^- ion in an aqueous hydroxide solution, forming H_2O and the *alkoxide* ion, RO^-. For example,

$$\underset{\text{Ethanol}}{C_2H_5OH} + OH^- \rightarrow \underset{\text{Ethoxide ion}}{C_2H_5O^-} + H_2O$$

The ethoxide ion can also be made by the redox reaction of sodium with ethanol. The reaction is similar to the reaction of sodium with water but is slower:

$$2C_2H_5OH + 2Na \rightarrow \underset{\text{Sodium ethoxide}}{2C_2H_5O^-Na^+} + H_2$$

Sodium reacts with ethanol but less vigorously than it does with water, producing sodium ethoxide and hydrogen.

As bases, alcohols are similar in strength to water and are protonated by strong acids to give ROH_2^+. For example,

$$CH_3-\ddot{\underset{..}{O}}-H + H-Br \rightarrow CH_3-\underset{\displaystyle H}{\underset{|}{\ddot{O}^+}}-H + Br^-$$

Dehydration Reactions Most alcohols can be dehydrated by heating them in the presence of sulfuric acid:

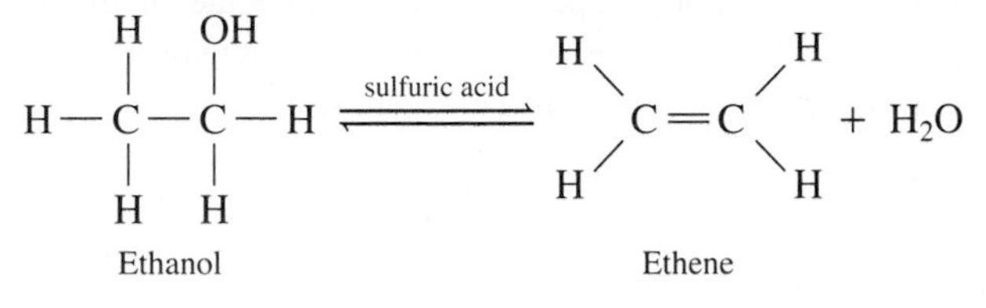

This reaction is reversible. In the reverse direction, H_2O is added to a double bond—that is, an alcohol is formed from an alkene. The direction in which the reaction proceeds is determined by the conditions. If sulfuric acid is in excess and the reaction mixture is heated, the volatile alkene is driven off. Under these conditions, water is removed by reaction with sulfuric acid to give $H_3O^+HSO_4^-$ (Chapter 7). Hence equilibrium is never attained, and most of the alcohol is converted to alkene. But in the presence of a large amount of water, the position of the equilibrium lies to the left, and most of the alkene is converted to the alcohol.

Substitution Reactions Alcohols react with hydrogen halides to form the corresponding haloalkanes (alkyl halides). For example,

$$\underset{\text{Ethanol}}{CH_3CH_2OH} + HCl \rightarrow \underset{\text{Chloroethane}}{CH_3CH_2Cl} + H_2O$$

The substitution reaction is carried out either by passing the gaseous hydrogen halide into the liquid alcohol or by using a concentrated aqueous solution of the hydrogen halide. The reaction does *not* take place with halide ion alone. This is not surprising, because the products of this reaction, if it were to occur, would be CH_3CH_2Cl and OH^-, and we have already seen that hydroxide ion reacts with haloalkanes to give the corresponding alcohol. In other words, the position of the equilibrium of the reversible reaction

$$CH_3CH_2OH + Cl^- \rightleftharpoons CH_3CH_2Cl + OH^-$$

lies far to the left.

Why then does the reaction occur with the hydrogen halide? Because alcohols are weak bases and are protonated by strong acids such as HCl, the first step in the reaction gives the protonated alcohol (the conjugate acid):

$$CH_3CH_2OH + H^+ \rightarrow CH_3CH_2-\underset{\displaystyle H}{\underset{|}{\overset{+}{O}}}-H$$

Then a water molecule is replaced by a chloride ion in a nucleophilic substitution reaction:

$$CH_3CH_2\overset{+}{O}H_2 + Cl^- \rightarrow CH_3CH_2Cl + H_2O$$

This reaction occurs readily, because a water molecule is a much weaker base than the hydroxide ion and is therefore more easily replaced by the weakly basic chloride ion. Thus, the substitution of OH by Cl is not a simple one-step reaction but has a two-step mechanism—a protonation or acid–base reaction followed by a substitution reaction.

Oxidation Reactions Alcohols can be oxidized to aldehydes and ketones (Section 14.5). Complete oxidation to carbon dioxide and water occurs when alcohols burn in air or oxygen. Alcohols represent an intermediate or incomplete stage of oxidation of a hydrocarbon, so their enthalpies of combustion, although large, are not as large as those of the corresponding alkanes:

The products of the combustion of alcohols are the same as those for the combustion of alkanes: CO_2 and H_2O.

$$CH_4(g) + 2O_2(g) \rightarrow CO_2(g) + 2H_2O(l) \qquad \Delta H^\circ = -891\ \text{kJ} \cdot \text{mol}^{-1}$$

$$CH_3OH(l) + 1\tfrac{1}{2}O_2(g) \rightarrow CO_2(g) + 2H_2O(l) \qquad \Delta H^\circ = -726\ \text{kJ} \cdot \text{mol}^{-1}$$

$$C_2H_6(g) + 3\tfrac{1}{2}O_2(g) \rightarrow 2CO_2(g) + 3H_2O(l) \qquad \Delta H^\circ = -1560\ \text{kJ} \cdot \text{mol}^{-1}$$

$$C_2H_5OH(l) + 3O_2(g) \rightarrow 2CO_2(g) + 3H_2O(l) \qquad \Delta H^\circ = -1367\ \text{kJ} \cdot \text{mol}^{-1}$$

Because we are rapidly exhausting the earth's supplies of hydrocarbon fuels, there is interest in replacing gasoline with ethanol produced by the fermentation of plant material, such as sugar cane or corn, which is a renewable source. Mixtures of ethanol and gasoline—gasohol—are now sold as automobile fuel in some parts of the United States and in other countries such as Brazil. Ethanol is used as a fuel in some camp stoves and in the home for heating and cooking food, such as cheese fondue, at the table. Unlike hydrocarbons, ethanol and other alcohols do not form explosive mixtures with air. Any fire caused by burning alcohol is easily extinguished with water.

Exercise 14.2 Write the formula for the organic product of each of the following reactions.

(a) $(CH_3)_2CHI(l) + NaOH(aq) \rightarrow$

(b) $CH_3OH(l) + K(s) \rightarrow$

(c) $(CH_3)_2CHOH(l) + HI(g) \rightarrow$

(d) (cyclohexane ring)—OH $\xrightarrow{\text{heat, } H_2SO_4}$

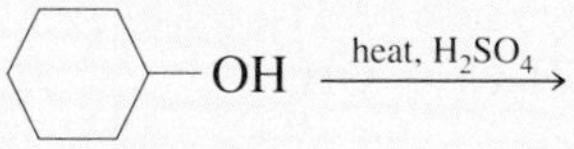

Phenols

When a hydrogen atom in an arene (aromatic hydrocarbon) is replaced by an OH group, the resulting compound is called a **phenol**. *Phenol* is also the name of the simplest compound of this type:

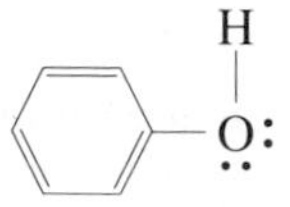

Phenol

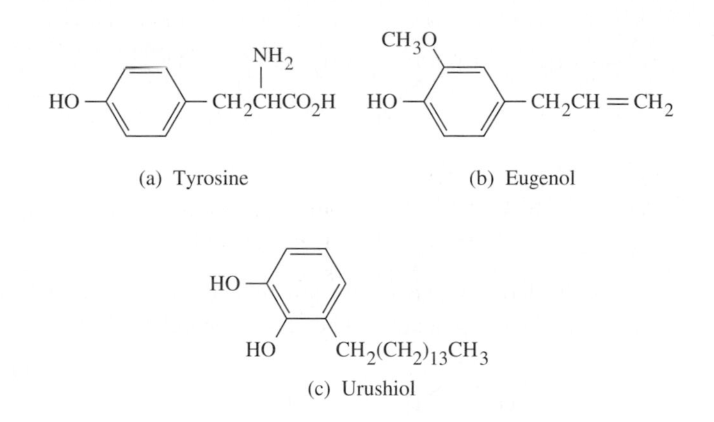

(a) Tyrosine (b) Eugenol (c) Urushiol

FIGURE 14.4 Some Biologically Important Phenols (a) Tyrosine is an amino acid, one of the building blocks of proteins. (b) Eugenol is found in cloves, bananas, and other fruits. (c) Urushiol is the skin irritant in poison ivy.

Originally called *carbolic acid*, it was first used as an antiseptic by Joseph Lister in 1867. Although phenol is very effective in killing bacteria, it has been replaced by other antiseptics because it can cause severe skin burns and is toxic. However, phenol and substituted phenols, such as *p*-methylphenol, are common disinfectants. Many biomolecules are phenols. Some examples are given in Figure 14.4.

CH_3—C_6H_4—OH

p-Methylphenol

Phenols are weak acids but are considerably stronger than alcohols. Phenol itself ($K_a = 1.3 \times 10^{-10}\ \mathrm{mol \cdot L^{-1}}$) is comparable in strength to HCO_3^- ($K_a = 5 \times 10^{-11}\ \mathrm{mol \cdot L^{-1}}$). Phenol dissolves in dilute aqueous NaOH solution to give a solution of sodium phenoxide:

A **disinfectant** is used to destroy or prevent the growth of harmful microorganisms on inanimate objects only, as it is usually dangerous to living tissue.

$$C_6H_5\text{—}OH + Na^+OH^- \rightarrow C_6H_5\text{—}O^-Na^+ + H_2O$$

Phenol Sodium phenoxide

Ethers

Ethers are a family of compounds with the general formula ROR′,

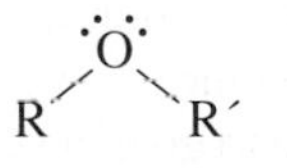

where R and R′ are alkyl or aryl groups. The common names for simple ethers are based on the two alkyl groups attached to the oxygen atom: for instance,

An aryl group is a functional group obtained from an aromatic hydrocarbon (arene) by the removal of one hydrogen atom. For example, the removal of a hydrogen atom from benzene gives the phenyl group, C_6H_5-.

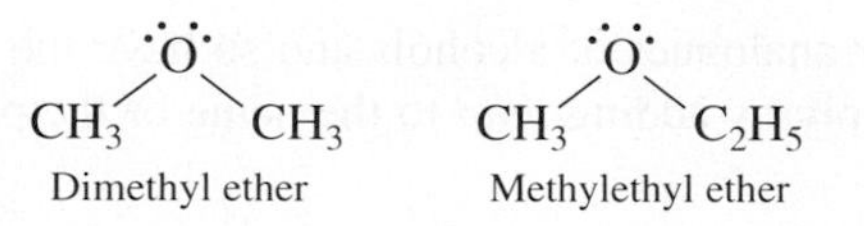

Dimethyl ether Methylethyl ether

Systematic names are based on the alkyl group with the longest carbon chain. The other alkyl group together with the oxygen atom is called an **alkoxy group**. For example, $CH_3OC_2H_5$ is methoxyethane, but its more common name is methylethyl ether.

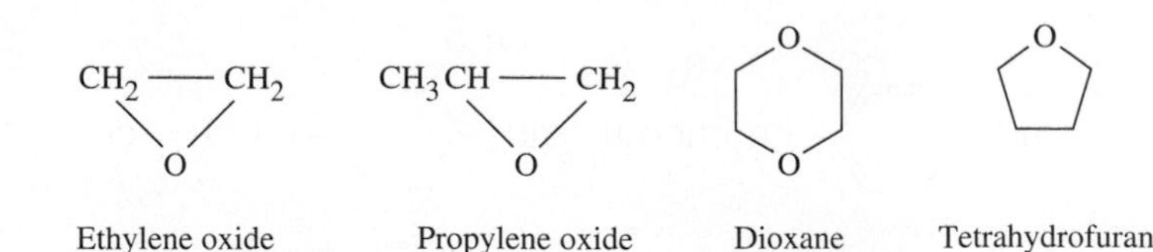

FIGURE 14.5 Some Cyclic Ethers Ethylene oxide and propylene oxide are used in the manufacture of plastics. Dioxane and tetrahydrofuran are useful solvents.

Both alcohols and ethers are related to water and have the same angular structure at the oxygen atom. They are all angular AX_2E_2 molecules—for instance,

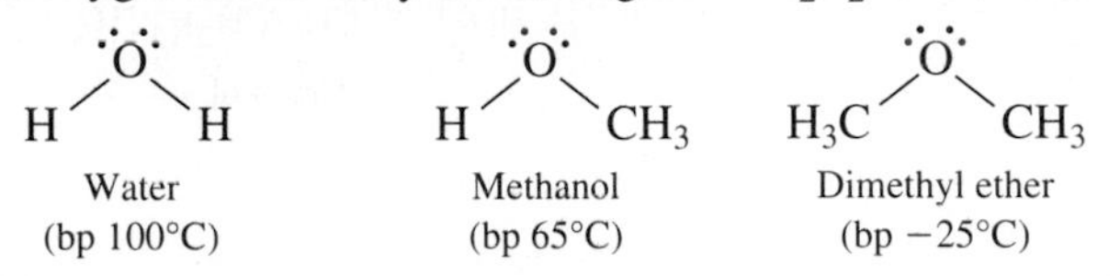

Both methanol and water are liquids, but dimethyl ether is a gas, because unlike water and methanol, dimethyl ether cannot form hydrogen bonds. For that reason, the boiling point of dimethyl ether is not very different from that of a simple alkane with a similar molecular mass ($CH_3—O—CH_3$, bp −25°C; $CH_3—CH_2—CH_3$, bp −42°C).

Diethyl ether, commonly known just as ''ether,'' is a useful solvent for many organic compounds. It was one of the earliest known anesthetics, but it is no longer used as such because the vapor forms explosive mixtures with air that are very easily ignited. Other much less flammable ethers and other compounds are now used as anesthetics (Box 14.1).

Some important ethers are cyclic, that is, they contain the C—O—C group in a ring (Figure 14.5). The cyclic ethers ethylene oxide, C_2H_4O, and propylene oxide, C_3H_6O, are important compounds in the synthesis of polymers, and dioxane and tetrahydrofuran are useful solvents.

One method of preparing ethers is the reaction of the sodium salt of an alcohol, RO^-Na^+, with a haloalkane. For example,

$$\underset{\text{Sodium ethoxide}}{C_2H_5—O^-Na^+} + \underset{\text{Bromomethane}}{CH_3—Br} \rightarrow \underset{\text{Methylethyl ether}}{C_2H_5—O—CH_3} + Na^+Br^-$$

This is a nucleophilic substitution reaction in which the ethoxide ion, $C_2H_5O^-$, is the nucleophile displacing the bromine atom as Br^-. It is analogous to the reaction of hydroxide ion with a haloalkane to give the corresponding alcohol (Section 14.2).

14.4 THIOLS AND DISULFIDES

Thiols are the sulfur analogues of alcohols and so have the general formula RSH. They are named simply by adding *thiol* to the name of the parent hydrocarbon, for example,

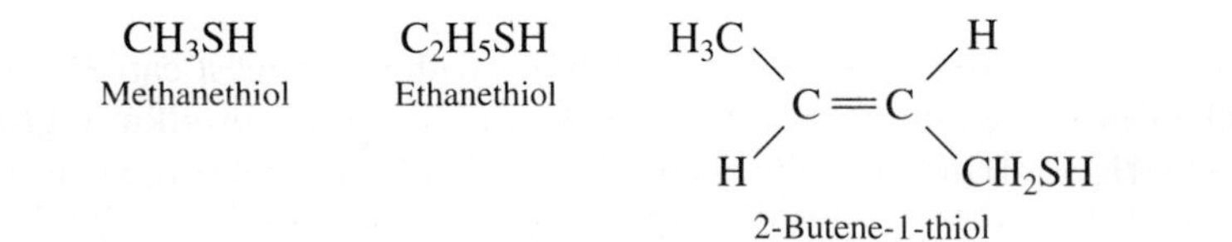

BOX 14.1 Inhalation Anesthetics

The introduction in the 1840s of diethyl ether as an inhaled anesthetic was a major advance in surgery. Before that time, operations were carried out with the patient fully conscious or intoxicated with alcohol. Other early anesthetics included dinitrogen monoxide (nitrous oxide), N_2O, which when breathed in small amounts induces a mild intoxication and therefore came to be known as laughing gas, and trichloromethane (chloroform), $CHCl_3$:

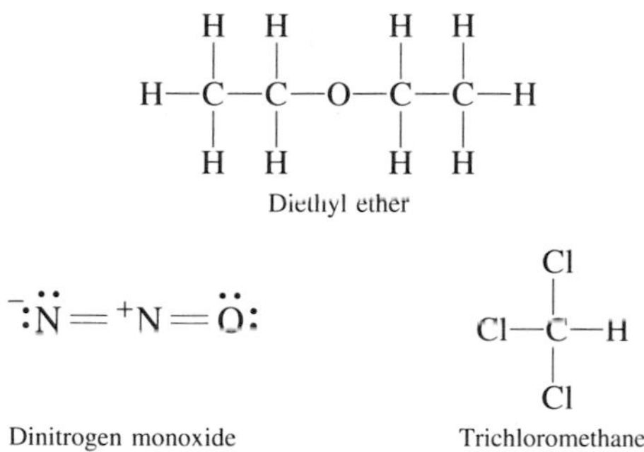

The ideal anesthetic should quickly make the patient unconscious but allow a quick return to consciousness, have few side effects, and be safe to handle. Although ether is a quick and effective anesthetic, it is far from ideal because it often induces nausea, and recovery is not quick. Moreover, it is dangerous to handle because it is a very volatile and flammable liquid and forms explosive mixtures with air that are very easily ignited. Dinitrogen monoxide is a useful anesthetic that is still used occasionally in dentistry. For most surgery, however, it has been replaced by new anesthetics introduced in the 1960s, when an intensive effort was made to synthesize better anesthetics. As a consequence, a variety of compounds are now in use (Figure A). These include halogen-substituted ethers such as enflurane and isoflurane and haloalkanes such as halothane:

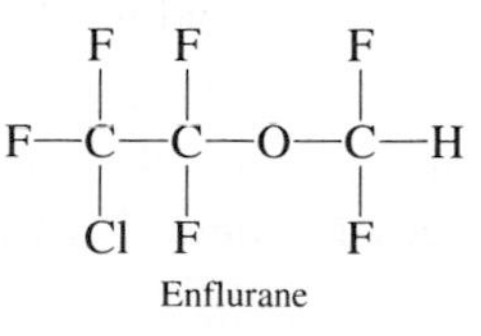

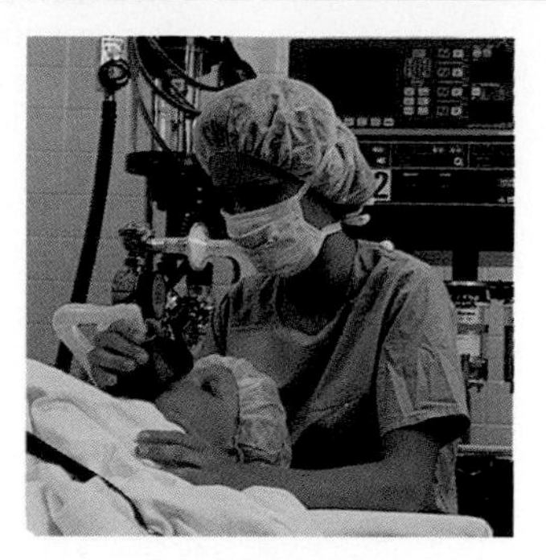

Figure A *The anesthetist is holding a mask over the mouth and nose of the child inhaling the anesthetic.*

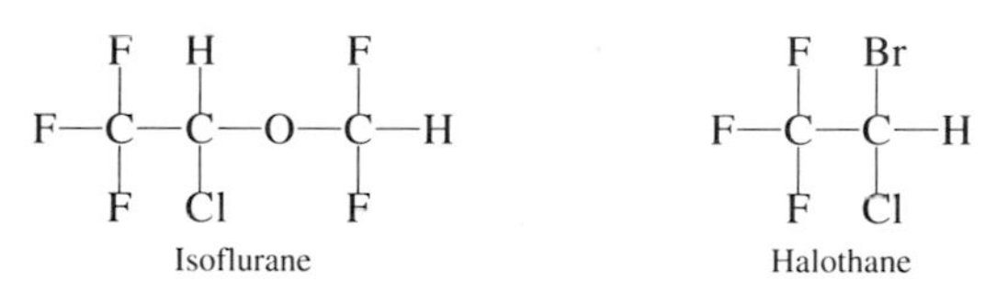

Despite their great importance and utility, surprisingly little is known about how inhaled anesthetics work. Many scientists believe that they dissolve in the fatty membranes surrounding nerve cells and that the resultant change in the membrane decreases the ability of sodium ions to pass through the membrane, thereby blocking the firing of nerve impulses. This theory is supported by the observation that the efficiency of an anesthetic correlates fairly well with its solubility in fat but does not seem to be associated with a particular group or structural feature of a molecule. Interestingly, it has been found that the noble gas xenon is a good anesthetic—ideal from the point of view of inertness and nontoxicity. But it is far too expensive for ordinary use. Because xenon atoms are large and polarizable, xenon is also soluble in nonpolar solvents such as fats.

The solubility of N_2O in fats is also exploited in cans of instant whipped cream. The cream is packaged in the can with N_2O under pressure. When the pressure is released, the cream is forced out, filled with tiny bubbles of N_2O.

The most outstanding property of the thiols is their very unpleasant odor. One substance that contributes to the smell produced by a skunk is 2-butene-1-thiol. A very small amount of methanethiol is added to natural gas, which is mainly methane and is odorless, to make it easy to detect a gas leak that might otherwise lead to a

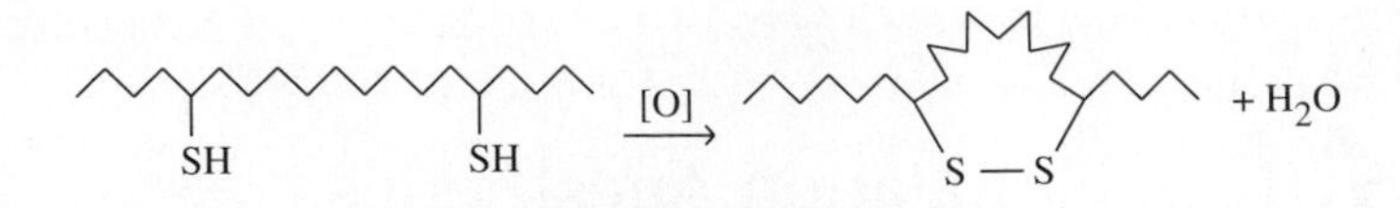

FIGURE 14.6 Oxidation of Thiols to Give Disulfide Bridges Disulfide bridges play an important role in determining the shapes of protein molecules. When hair is "permed," a mild oxidizing agent causes disulfide bonds to form between SH groups of its protein molecules, resulting in the introduction of bends and kinks.

dangerous explosion. Thiols are readily oxidized to give **disulfides**, RSSR. For example,

$$\underset{\text{Methanethiol}}{CH_3{-}S{-}H} + H{-}S{-}CH_3 \xrightarrow{[O]} \underset{\text{Dimethyldisulfide}}{CH_3{-}S{-}S{-}CH_3} + H_2O$$

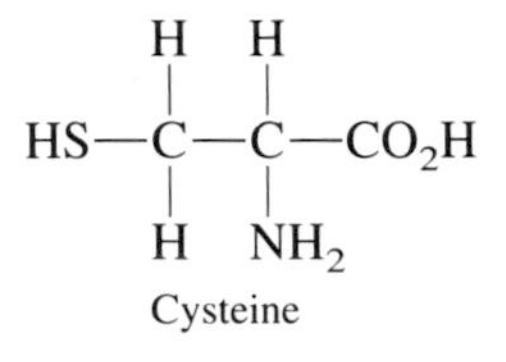

where [O] represents an oxygen atom provided by the oxidizing agent. The SH group is an important functional group in the amino acid cysteine (Section 14.7), which is a part of many proteins. The formation of an —S—S— bond by the oxidation of two cysteine groups is one way in which a large protein molecule is held in the shape needed to carry out its biological function (Chapter 17). The effect of the formation of disulfide bridges on protein shape is also seen in the process of "permanent waving" of hair (Figure 14.6).

Exercise 14.3 Identify each of the following compounds as an alcohol, a thiol, a phenol, or an ether, and name each.

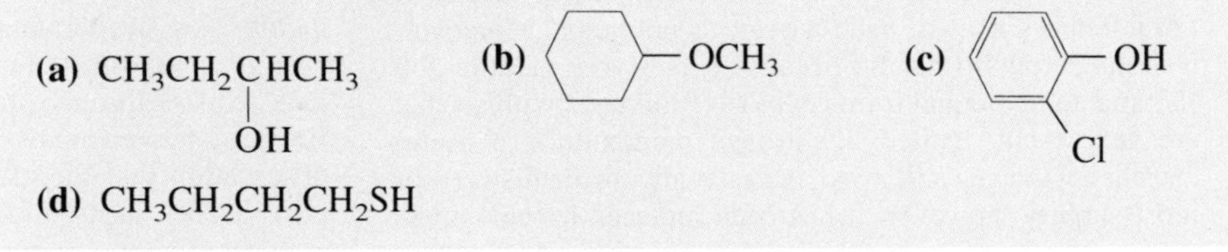

(d) $CH_3CH_2CH_2CH_2SH$

14.5 ALDEHYDES AND KETONES

Aldehydes and ketones are compounds containing the C=O, *carbonyl,* group. **Aldehydes**, RCHO, have one hydrogen atom and one alkyl group attached to the carbon atom of the carbonyl group, whereas **ketones**, RR′CO, have two alkyl groups attached to this carbon atom:

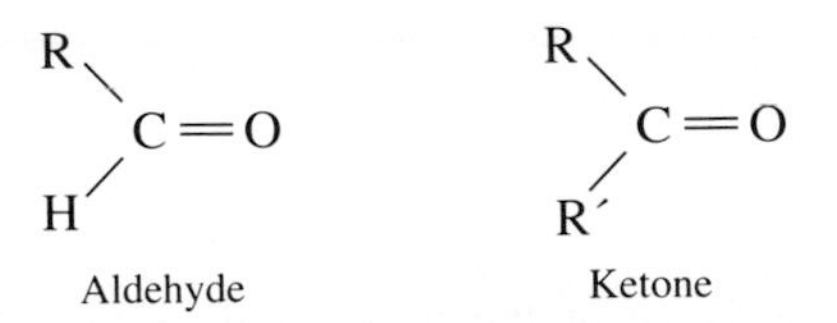

Aldehydes are named by replacing the *-e* ending of the corresponding alkane with *-al*. Ketones are named by replacing the *-e* ending of the corresponding alkane with *-one*. Thus, CH_3CHO is ethanal, and CH_3COCH_3 is propanone.

Ketones and aldehydes have a planar AX_3 (sp^2) geometry at the carbon atom of the C═O group with approximately 120° bond angles. Because oxygen is more electronegative than carbon, the carbonyl group is polar. The oxygen atom has a small negative charge, and the carbon atom a small positive charge:

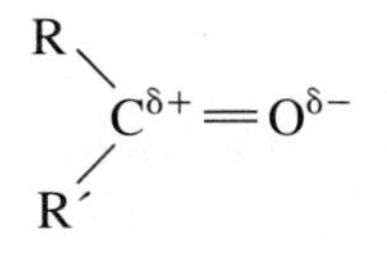

Preparation of Aldehydes and Ketones

Primary and secondary alcohols decompose at 550 to 600°C in the presence of a suitable catalyst (such as copper or silver) to form hydrogen and an aldehyde or ketone. This reaction, called a **dehydrogenation reaction**, is an industrial method for preparing aldehydes and ketones. Some examples are

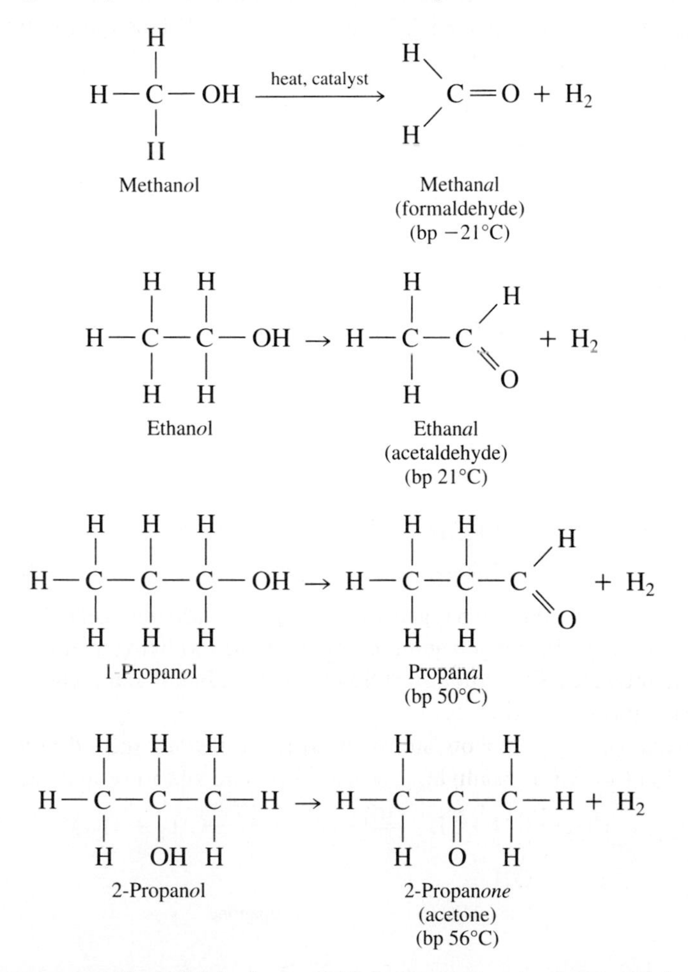

The common names formaldehyde, acetaldehyde, and acetone are widely used and should be remembered.

In the laboratory, aldehydes and ketones may be prepared by oxidizing alcohols with a strong oxidizing agent, such as $K_2Cr_2O_7$. Primary alcohols are oxidized to aldehydes. For example, ethanol is oxidized to ethanal:

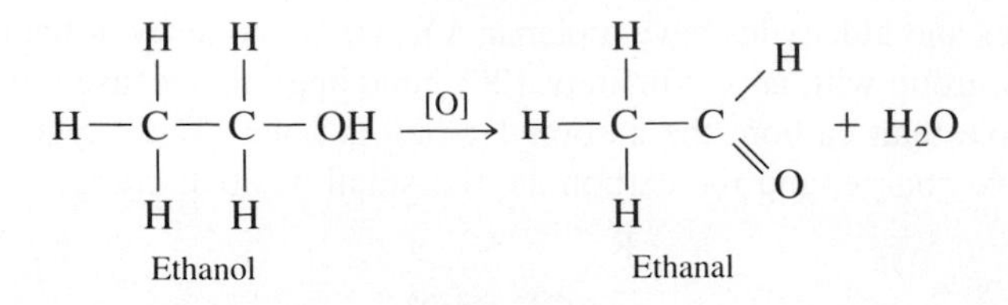

We can see that this is an oxidation because two hydrogen atoms are removed. In organic reactions, most oxidations involve the loss of hydrogen or the addition of oxygen, and reductions are the reverse processes. The concepts of oxidation number and electron loss and gain are less useful for carbon compounds than they are for inorganic compounds, because in its compounds, carbon can have many different oxidation numbers. However, as we shall see now for the oxidation of methanol by dichromate, the concept of electron loss and gain is still valid, if less convenient. By the rules for assigning oxidation numbers, the oxidation number of hydrogen is +1 and that of oxygen is −2, so the oxidation number of the carbon in methanol, CH_3OH, is −2 and that in methanal, CH_2O, is 0. The half-reaction for the oxidation of methanol is

$$CH_3OH \rightarrow CH_2O + 2H^+ + 2e^-$$

The half-reaction for the reduction of the dichromate ion is

$$Cr_2O_7^{2-} + 14H^+ + 6e^- \rightarrow 2Cr^{3+} + 7H_2O$$

So the complete equation for the oxidation of methanol to methanal by dichromate ion is

$$3CH_3OH(l) + Cr_2O_7^{2-}(aq) + 8H^+(aq) \rightarrow 3CH_2O(aq) + 2Cr^{3+}(aq) + 7H_2O(l)$$

It is clearly simpler to write

In organic reactions, we generally focus on the reactants and products rather than on the nature of the oxidizing agent.

$$CH_3OH \xrightarrow{K_2Cr_2O_7,\ H^+} CH_2O$$

or

$$\underset{\text{Methanol}}{CH_3OH} \xrightarrow{[O]} \underset{\text{Methanal}}{CH_2O} + H_2O$$

where again [O] represents an oxygen atom from the oxidizing agent. Two hydrogen atoms combine with the oxygen atom provided by the oxidizing agent and appear as water in the products. The similar reaction between ethanol and dichromate is used in the breathalyzer text (Box. 5.1)

Whereas primary alcohols are oxidized to aldehydes, secondary alcohols are oxidized to ketones. For example, 2-propanol is oxidized to propanone (acetone):

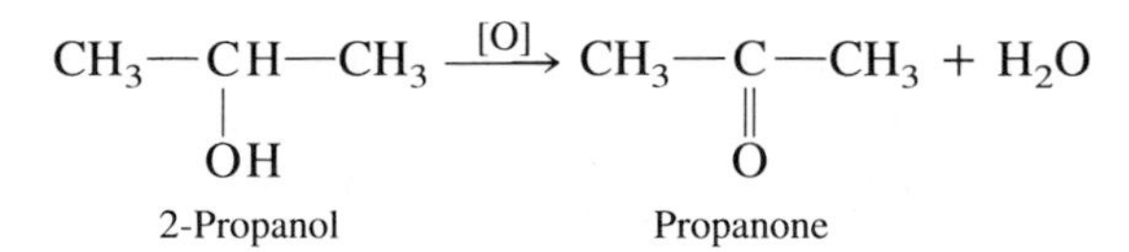

Properties of Aldehydes and Ketones

The negatively charged oxygen atom of the polar C=O group of aldehydes and ketones can form hydrogen bonds with water molecules. Therefore, the simpler

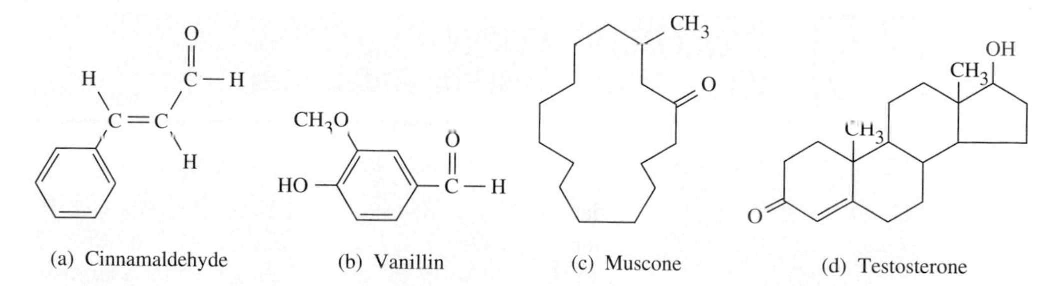

FIGURE 14.7 Some Naturally Occuring Aldehydes and Ketones (a) The aldehyde cinnamaldehyde provides a cinnamon flavor in foods and drugs and is obtained from the Chinese cinnamon plant. (b) The aldehyde vanillin is a flavoring agent obtained from the oil of the vanilla bean. (c) The ketone muscone is responsible for a musky odor in perfumes and is obtained from the scent glands of the musk deer. (d) The ketone testosterone is a male hormone.

aldehydes and ketones are soluble in water. Many aldehydes and ketones are found naturally. Some of them have characteristic odors (Figure 14.7).

The aldehyde **methanal (formaldehyde)**, HCHO, is a gas with a penetrating odor that is very soluble in water. It is normally stored and sold as a 40% aqueous solution known as *formalin*, which is used as a disinfectant and for preserving biological specimens. A major use of formaldehyde is in the manufacture of polymers, including a variety of plastics, adhesives, and insulating materials (Chapter 20). The aldehyde **ethanal (acetaldehyde)**, CH_3CHO (bp 21°C), is a sweet-smelling, flammable liquid present in ripe fruits. It is used mainly for synthesizing other organic compounds. The ketone **propanone (acetone)**, CH_3COCH_3, is a volatile liquid that is an excellent solvent for a variety of substances, both polar and nonpolar. It is completely miscible with water.

Reactions of Aldehydes and Ketones

Reduction (Hydrogenation) Aldehydes can be reduced to the corresponding primary alcohol by heating them with hydrogen in the presence of a catalyst. The hydrogen adds across the double bond, so this reaction is an addition reaction as well as a reduction. For example, ethanal (acetaldehyde) can be reduced to ethanol:

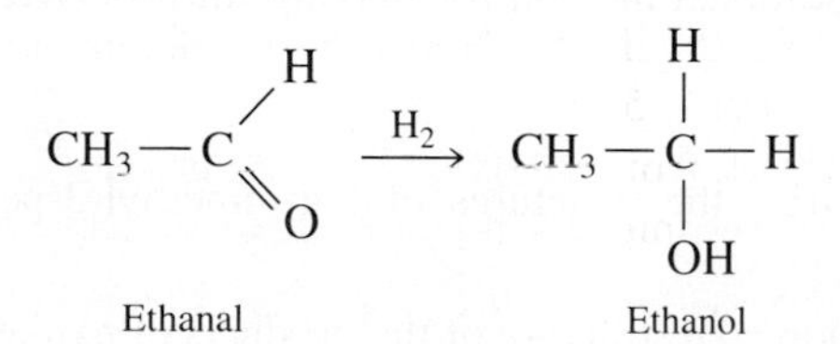

And cyclohexanone can be reduced to cyclohexanol:

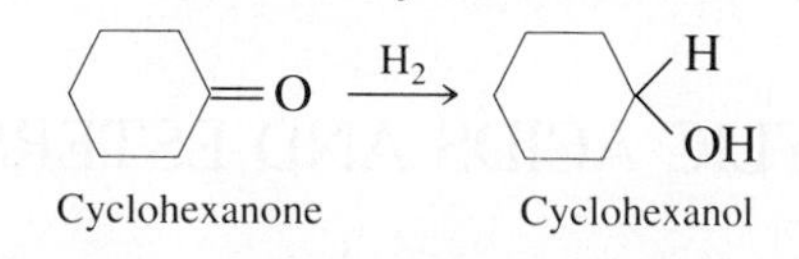

Oxidation of Aldehydes Ketones are not oxidized even by strong oxidizing agents such as $K_2Cr_2O_7$ or $KMnO_4$. But we shall see in Section 14.6 that aldehydes

DEMONSTRATION 14.1
Silver Mirror Test for Aldehydes

A clean beaker contains a solution of silver nitrate in aqueous ammonia. An aqueous solution of ethanal is added from a dropper, and the solution is stirred.

The solution rapidly darkens as ethanal is oxidized to ethanoic acid and Ag^+ is reduced to silver.

The inside of the beaker finally becomes coated with a shiny metallic silver mirror. Other aldehydes give the same result.

are easily oxidized to carboxylic acids, and many different oxidizing agents can be used. A useful test for aldehydes is to use the Ag^+ ion as an oxidizing agent:

$$\underset{\text{Propanal}}{CH_3CH_2CHO} \xrightarrow{Ag^+} \underset{\text{Propanoic acid}}{CH_3CH_2CO_2H}$$

Silver ion is reduced to silver, and under the right conditions, a silver mirror forms on the inner surface of the reaction vessel. The complete balanced equation is

$$C_2H_5CHO + 2Ag^+ + H_2O \rightarrow C_2H_5CO_2H + 2Ag + 2H^+$$

Indeed, this is the traditional method for making mirrors (Demonstration 14.1).

Exercise 14.4 Draw the structures of 2,4-dimethyl-3-pentanone and 2,2-dimethylbutanal.

Exercise 14.5 Draw the structure of the products of oxidation of each of the following. **(a)** cyclohexanol **(b)** 2-butanol **(c)** 1-butanol

14.6 CARBOXYLIC ACIDS AND ESTERS

In Chapters 4 and 12 we discussed acetic acid as an example of a weak acid in aqueous solution. It is a member of the class of compounds known as **carboxylic acids**, which have the general formula RCO_2H:

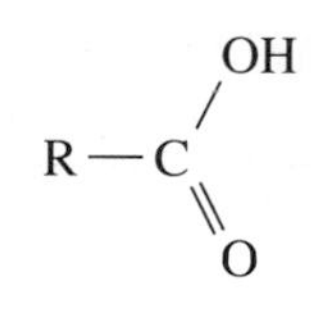

For instance, acetic acid is

$$CH_3-C(=O)-OH$$

Acetic acid

The COOH (or CO_2H) group is called the **carboxyl group**. The carboxyl group has a planar AX_3 (sp^2) geometry around the carbon atom.

Carboxylic acids are named systematically by replacing the *-e* ending in the name of the parent hydrocarbon with *-oic acid*. Thus, HCO_2H is methanoic acid, and CH_3CO_2H is ethanoic acid. If other functional groups are present, the carbon chain is numbered from the carboxyl group, as in 2-methylpropanoic acid. Because many carboxylic acids have been known for a long time, their traditional names are still widely used, particularly for the simplest and most common acids. For example, ethanoic acid is usually called acetic acid. The common and systematic IUPAC names of some carboxylic acids are given in Table 14.4.

TABLE 14.4 Carboxylic Acids

Hydrocarbon	Carboxylic Acid	Common Name of Carboxylic Acid
CH_4 (H—C(H)(H)—H) Methane	$H-C(=O)-OH$ Methanoic acid	Formic acid
CH_3-CH_3 Ethane	$CH_3-C(=O)-OH$ Ethanoic acid	Acetic Acid
$CH_3-CH_2-CH_3$ Propane	$CH_3-CH_2-C(=O)-OH$ Propanoic acid	Propionic acid
$CH_3-CH_2-CH_2-CH_3$ Butane	$CH_3-CH_2-CH_2-C(=O)-OH$ Butanoic acid	Butyric acid
$CH_3-CH(CH_3)-CH_3$ 2-Methylpropane	$CH_3-CH(CH_3)-C(=O)-OH$ 2-Methylpropanoic acid	Isobutyric acid

Preparation of Carboxylic Acids

Carboxylic acids can be prepared by the oxidation of primary alcohols and aldehydes:

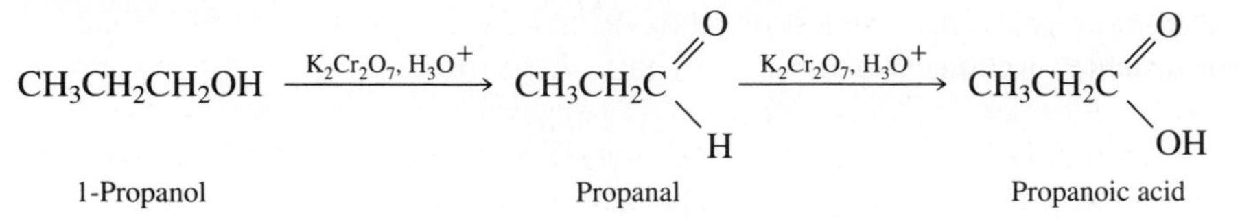

The relationships between alcohols, aldehydes, and carboxylic acids can be summarized as follows:

Some Common Carboxylic Acids

Acetic Acid An industrial method for preparing acetic acid is the oxidation of ethanal (acetaldehyde) with air at 60 to 80°C and 5 atm pressure in the presence of a catalyst:

$$2CH_3CHO + O_2 \xrightarrow{\text{catalyst}} 2CH_3CO_2H$$

Ethanal — Acetic acid (ethanoic acid)

Acetic acid can also be made by the enzyme-catalyzed oxidation of ethanol by air. This process has been used since ancient times to produce vinegar from wine and cider, but today most vinegar is made by diluting manufactured acetic acid. Vinegar is an approximately 5% solution of acetic acid in water. The structures of some other common carboxylic acids, which we describe next, are given in Figure 14.8.

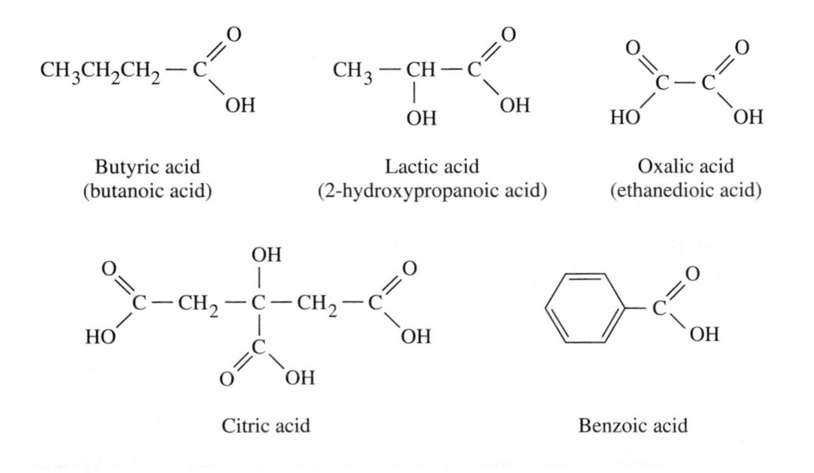

FIGURE 14.8 Some Common Carboxylic Acids Butyric acid and lactic acid are monocarboxylic acids. Oxalic acid and citric acid are dicarboxylic acids. Benzoic acid is an aromatic carboxylic acid.

Other Carboxylic Acids **Butyric acid (butanoic acid)** is obtained from butter fat. It is responsible for the odor of rancid butter. **Lactic acid (2-hydroxypropanoic acid)** is found in sour milk. It accumulates in the muscles of the human body after strenuous exercise and is responsible for the subsequent soreness. **Oxalic acid (ethanedioic acid)**, the simplest dicarboxylic acid, is found in plants of the genus *Oxalis*, which includes rhubarb and spinach. **Citric acid**, a tricarboxylic acid, is found in the juice of citrus fruits and is responsible for their tart flavor. It is produced by almost all plants and animals during metabolism (Chapter 17). **Benzoic acid**, $C_6H_5CO_2H$, is the simplest aromatic carboxylic acid.

Rhubarb leaves contain oxalic acid. Pure oxalic acid is a white crystalline solid.

Properties of Carboxylic Acids

Carboxylic acids behave as weak acids in water, as is shown by the pK_a values in Table 14.5. Carboxylic acids give up a proton to a water molecule to form a *carboxylate ion*:

$$\underset{\text{Acetic acid}}{CH_3-\overset{\overset{\displaystyle O}{\|}}{C}-OH} + H_2O \rightleftharpoons H_3O^+ + \underset{\text{Acetate ion}}{CH_3-\overset{\overset{\displaystyle O}{\|}}{C}-O^-}$$

Carboxylic acids are much stronger acids than are alcohols. Their greater acid strength is attributed to the presence of the polar C═O group. The partial positive charge on the carbon atom increases its electronegativity, so, compared with an alcohol, there is a shift of electron density away from the OH group. This electron density shift increases the positive charge on the hydrogen, thereby facilitating its donation as a proton to a water molecule (Figure 14.9).

Carboxylic acids react completely with strong bases such as sodium hydroxide to form carboxylic acid salts, $RCO_2^-M^+$. For example,

$$\underset{\text{Acetic acid}}{CH_3CO_2H} + NaOH \rightarrow \underset{\text{Sodium acetate}}{CH_3CO_2^-\ Na^+} + H_2O$$

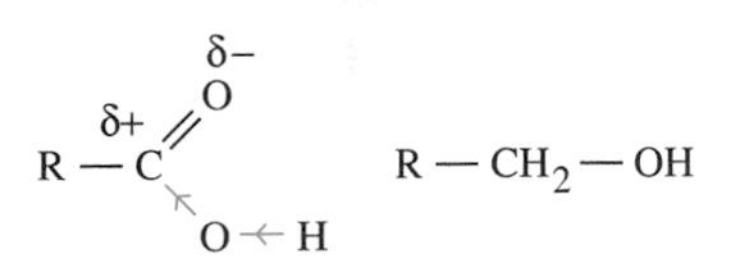

FIGURE 14.9 Acid Strength of Carboxylic Acids and Alcohols Carboxylic acids are stronger acids than alcohols. Arrows show the movement of electrons away from the OH group in a carboxylic acid as a result of the positively charged carbon atom of the carbonyl group. This increases the positive charge on the hydrogen of the OH, so it is greater than the positive charge on the hydrogen of the OH group of an alcohol. Thus, the hydrogen atom of a carboxylic acid is more readily transferred to a base as a proton than is the hydrogen atom of an alcohol.

TABLE 14.5 Melting Points, Boiling Points, and pK_a Values of Some Carboxylic Acids

Acid	Formula	MP (°C)	BP (°C)	pK_a
Methanoic (formic)	HCO_2H	8.4	110.5	3.77
Ethanoic (acetic)	CH_3CO_2H	16.6	118	4.76
Propanoic (propionic)	$CH_3CH_2CO_2H$	−22	141	4.88
Butanoic (butyric)	$CH_3CH_2CH_2CO_2H$	−5	163	4.82
Ethanedioic (oxalic)	$(CO_2H)_2$	187	—	1.46
2-Hydroxypropanoic (lactic)	$CH_3CH(OH)CO_2H$	18	—	3.87
Benzoic	$C_6H_5CO_2H$	122	249	4.17
Phthalic	$C_6H_4(CO_2H)_2$	200*	—	3.00
Salicylic	$C_6H_4(OH)(CO_2H)$	159	—	3.00

* Decomposes at this temperature

The salt of a carboxylic acid is usually much more soluble in water than is the acid itself.

Exercise 14.6 From what alcohol can pentanoic acid be prepared by oxidation? Write the structures of both compounds.

Exercise 14.7 Give the names and structures of the products of the following reactions.

(a) $CH_3CH_2CH_2CO_2H + KOH \rightarrow$

(b) 2-methylpentanoic acid + $Ba(OH)_2 \rightarrow$

Esters

Carboxylic acids react with alcohols in the presence of an acid (as a catalyst) to form compounds called **esters**. For example, acetic acid reacts with ethanol to give ethyl acetate (ethyl ethanoate) and water:

$$\underset{\text{Acetic acid}}{CH_3\overset{\overset{O}{\|}}{C}{-}OH} + \underset{\text{Ethanol}}{HO{-}C_2H_5} \rightarrow \underset{\text{Ethyl acetate}}{CH_3\overset{\overset{O}{\|}}{C}{-}O{-}C_2H_5} + H_2O$$

Butanoic acid reacts with ethanol to give *ethylbutanoate*:

$$\underset{\text{Butanoic acid}}{CH_3CH_2CH_2\overset{\overset{O}{\|}}{C}{-}OH} + \underset{\text{Ethanol}}{HO{-}C_2H_5} \rightarrow \underset{\text{Ethylbutanoate}}{CH_3CH_2CH_2\overset{\overset{O}{\|}}{C}{-}O{-}C_2H_5} + H_2O$$

To generalize, a carboxylic acid, RCO_2H, reacts with an alcohol, $R'OH$, to give the ester RCO_2R':

$$R{-}\overset{\overset{O}{\|}}{C}{-}O{-}R'$$

The systematic names for esters are obtained from the names of the acid and the alcohol from which they may be prepared. The *-oic* ending of the acid is replaced by *-oate*, and the name is preceded by that of the alkyl group in the alcohol. For example, *methylpropanoate* is prepared from methanol and propanoic acid:

$$\underset{\text{Propanoic acid}}{CH_3CH_2\overset{\overset{O}{\|}}{C}{-}OH} + \underset{\text{Methanol}}{H{-}O{-}CH_3} \rightarrow \underset{\text{Methylpropanoate}}{CH_3CH_2\overset{\overset{O}{\|}}{C}{-}O{-}CH_3} + H_2O$$

The formation of an ester from a carboxylic acid and an alcohol is a reversible reaction. If we start with 1 mol each of acetic acid and ethanol, the equilibrium mixture has the following composition at 25°C:

$$\underset{\substack{\text{Acetic acid}\\ 0.3\text{ mol}}}{CH_3CO_2H} + \underset{\substack{\text{Ethanol}\\ 0.3\text{ mol}}}{CH_3CH_2OH} \rightleftharpoons \underset{\substack{\text{Ethyl acetate}\\ 0.7\text{ mol}}}{CH_3CO_2CH_2CH_3} + \underset{0.7\text{ mol}}{H_2O}$$

The position of the equilibrium can be shifted by changing the conditions. When there is a large excess of ethanol, the position of the equilibrium shifts to the right, and most of the acetic acid is converted to the corresponding ester. When there is a large excess of water, the position of the equilibrium shifts to the left, and most of the ester is hydrolyzed to the corresponding alcohol and carboxylic acid. Esters are more readily hydrolyzed by using $OH^-(aq)$ rather than water as the nucleophile:

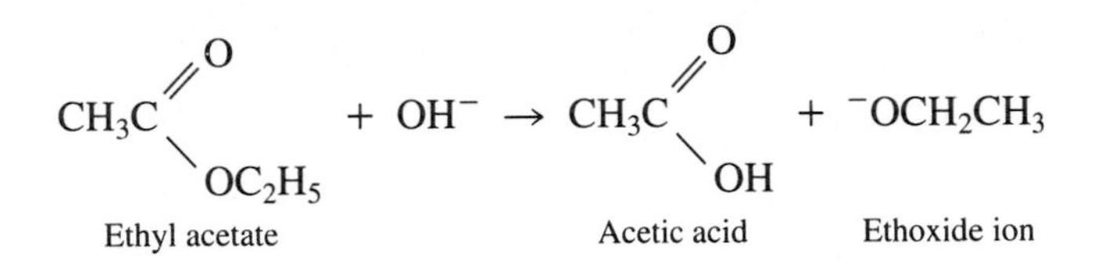

In this case the reaction goes to completion, because the carboxylic acid is removed as it forms by conversion to the corresponding carboxylate ion:

$$CH_3CO_2H + OH \rightarrow CH_3CO_2^- + H_2O$$

Acetic acid — Acetate ion

The formation and hydrolysis of an ester are further examples of nucleophilic substitution reactions at a positively charged carbonyl carbon atom. In ester formation, the alcohol is the nucleophile that attacks the carbon of the carbonyl group and replaces the OH group with an OR group. In hydrolysis, the water molecule or the OH^- ion is the nucleophile replacing the OR group with an OH group.

The alcohol ROH is polarized $RO^{\delta-}H^{\delta+}$; its oxygen attacks the positive carbonyl carbon atom of the acid $RC^{\delta+}O^{\delta-}OH$.

Esters are responsible primarily for the taste and fragrance of many fruits and flowers (Table 14.6). Synthetic esters are widely used in perfumes and for flavoring foods. Certain polymers such as Dacron are also esters, called polyesters (Chapter 20).

Salicylic acid has a CO_2H and an OH group, so it can form esters in two ways. It reacts as an alcohol with acetic acid to give the well-known *analgesic* (pain reliever) **aspirin (acetylsalicylic acid)**. It also reacts with methanol to give the methyl ester **methylsalicylate**, commonly known as *oil of wintergreen* and used, for example, in liniments for relieving muscle soreness.

The pleasant taste and odor of many fruits is due to the esters they contain.

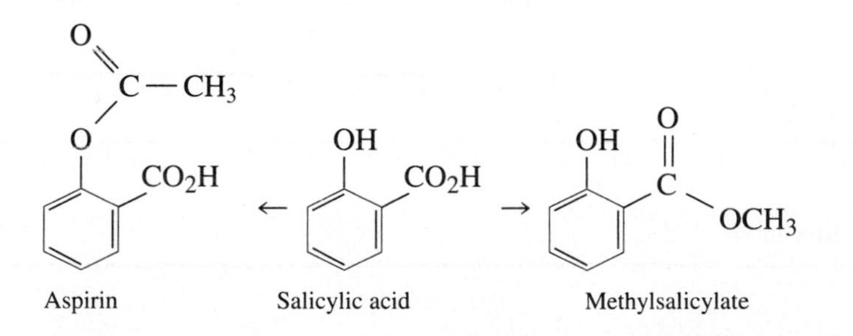

Fats are esters of glycerol and some long-chain carboxylic acids called **fatty acids**. Fatty acids consist of a long hydrocarbon chain that may contain one or more double bonds with a carboxyl group at one end. Carboxylic acids with no double bonds are called **saturated fatty acids**, and those with double bonds are called **unsaturated fatty acids**. Some typical fatty acids are listed in Table 14.7. Animal fats (such as lard and butter) and vegetable oils (such as corn oil, soybean oil,

TABLE 14.6 Odors of Some Esters

Formula	Name	Odor
$CH_3C(=O)OCH_2CH_2CH_2CH_2CH_3$	Pentylethanoate (amyl acetate)	Banana
$CH_3CH_2CH_2C(=O)OCH_2CH_3$	Ethylbutanoate (ethyl butyrate)	Pineapple
$CH_3C(=O)OCH_2(CH_2)_6CH_3$	Octylethanoate	Orange
$CH_3CH_2CH_2C(=O)OCH_2CH_2CH_2CH_2CH_3$	Pentylbutanoate (amyl butyrate)	Apricot
$C_6H_4(NH_2)C(=O)OCH_3$ (benzene ring with $COCH_3$ (C=O) and NH_2 substituents)	Methyl anthranilate	Grape
$C_6H_4(OH)C(=O)OCH_3$ (benzene ring with $COCH_3$ (C=O) and OH substituents)	Methyl salicylate	Oil of wintergreen

linseed oil, olive oil, and peanut oil) are mixtures of triglycerides. A **triglyceride** is an ester of glycerol (1,2,3-propanetriol) with three fatty acids that, in general, are not all the same. The approximate composition of some common fats and oils is given in Table 14.8.

The melting points of the fatty acids in Table 14.7 show that the greater the number of double bonds, the lower the melting point. The same trend holds for the triglycerides. The more unsaturated a triglyceride, the lower its melting point.

TABLE 14.7 Some Common Fatty Acids

Name	Number of Carbons	Number of Double Bonds	Structure	MP (°C)
Saturated				
Lauric	12	0	$CH_3(CH_2)_{10}COOH$	44
Myristic	14	0	$CH_3(CH_2)_{12}COOH$	58
Palmitic	16	0	$CH_3(CH_2)_{14}COOH$	63
Stearic	18	0	$CH_3(CH_2)_{16}COOH$	70
Unsaturated				
Oleic	18	1	$CH_3(CH_2)_7CH{=}CH(CH_2)_7COOH$ (*cis*)	4
Linoleic	18	2	$CH_3(CH_2)_4CH{=}CHCH_2CH{=}CH(CH_2)_7COOH$ (all *cis*)	−5
Linolenic	18	3	$CH_3CH_2CH{=}CHCH_2CH{=}CHCH_2CH{=}CH(CH_2)_7COOH$ (all *cis*)	−11
Arachidonic	20	4	$CH_3(CH_2)_4(CH{=}CHCH_2)_4CH_2CH_2COOH$ (all *cis*)	−50

TABLE 14.8 Approximate Composition of Some Common Fats and Oils

	Saturated Fatty Acids (%)*				Unsaturated Fatty Acids (%)*	
Source	C_{12} Lauric	C_{14} Myristic	C_{16} Palmitic	C_{18} Stearic	C_{18} Oleic	C_{18} Linoleic
Animal Fat						
Lard	—	1	25	15	50	6
Butter	2	10	25	10	25	5
Human fat	1	3	25	8	46	10
Whale blubber	—	8	12	3	35	10
Vegetable Oil						
Corn	—	1	8	4	46	42
Olive	—	1	5	5	83	7
Peanut	—	—	7	5	60	20
Soybean	—	—	7	4	34	53

* Where totals are less than 100%, small quantities of several other acids are present.

Vegetable oils have a higher proportion of double bonds than animal fats and so are usually liquids at ordinary temperatures, whereas animal fats are solids.

Why do double bonds affect the melting point? In the hydrocarbon chains in saturated acids, there is free rotation about each carbon–carbon bond that allows the chains to adopt a conformation with an approximately linear shape. These linear molecules can easily pack together in a regular arrangement to form a crystal (Figure 14.10). But there is no free rotation about a double bond, so there is a kink in the chain at this point. Because such nonlinear molecules cannot pack together so closely when they form a crystal, the intermolecular forces are weaker, and the crystal has a correspondingly lower melting point.

Triglycerides, like all esters, can be hydrolyzed to the corresponding alcohol and carboxylic acid. The mixture of sodium salts of carboxylic acids formed by hydrolysis with NaOH is known as **soap** after is has been purified (Box 14.2).

Esters of inorganic acids are formed in just the same way as those of carboxylic acids—namely, by the elimination of a molecule of water between the alcohol and the acid. Esters of phosphoric acid, diphosphoric acid, and triphosphoric acid (Chapter 7) are essential to metabolism in all living organisms. Just as phosphoric acid can form three different salts, such as NaH_2PO_4, Na_2HPO_4, and Na_3PO_4, it can form three different esters with a given alcohol:

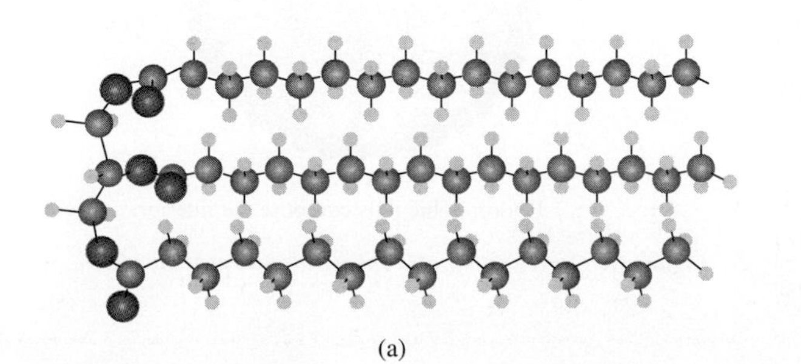

FIGURE 14.10 Saturated and Unsaturated Fatty Acids Molecular models of (a) a saturated and (b) an unsaturated triacylglycerol. The double bond (red arrow) puts a kink in the chain and prevents the molecules from packing closely in a regular arrangement and crystallizing easily. Fats are solids because of their high proportion of saturated hydrocarbon chains, and oils are liquids because of their high proportion of unsaturated hydrocarbon chains.

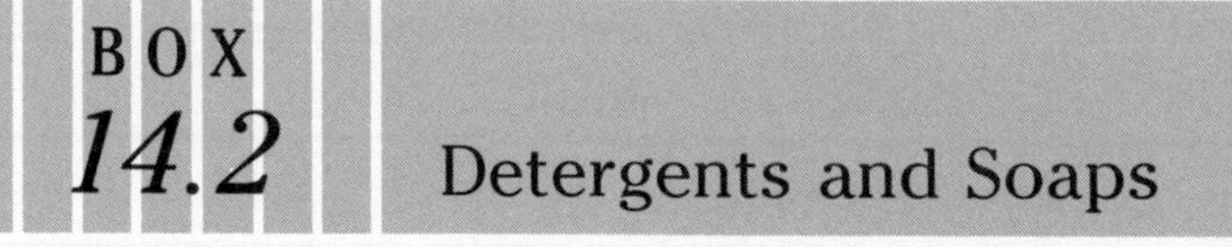

BOX 14.2 Detergents and Soaps

Detergents are substances that improve the cleaning properties of water. Soap was the first detergent ever produced, and it is still one of the commonest. It is a mixture of the sodium (and sometimes potassium) salts of long-chain carboxylic acids. A typical example is sodium octadecanoate (sodium stearate), $CH_3(CH_2)_{16}COO^-Na^+$ (Figure A). The nonpolar hydrocarbon portion of the carboxylate ion in soap is not attracted to polar water molecules. The negative end of the molecule is, however, attracted to the positively charged hydrogen atoms of the water molecule. This difference between one end of the molecule and the other gives soap its special properties.

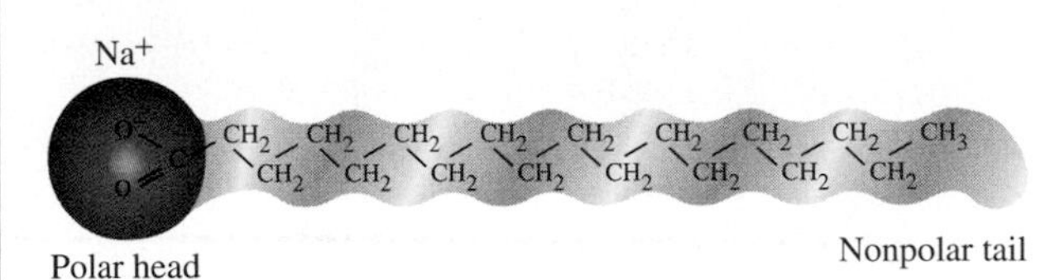

Figure A *A typical soap molecule, sodium stearate*

In pioneer days, families saved wood ashes from the stove and animal fats from the kitchen and from time to time boiled them together in a large iron pot to make soap. These pioneers were actually carrying out the same chemical reaction that is used on a large scale by modern soap manufacturers. The ashes of wood contain potassium carbonate, which reacts with water to give a basic solution. Animal fats are mixtures of esters of glycerol called triglycerides. These triglycerides are hydrolyzed by basic solutions containing hydroxide ions to give glycerol and a mixture of sodium salts of carboxylic acids—soap. Modern soap makers hydrolyze a variety of glycerides with sodium hydroxide to give glycerol and soap:

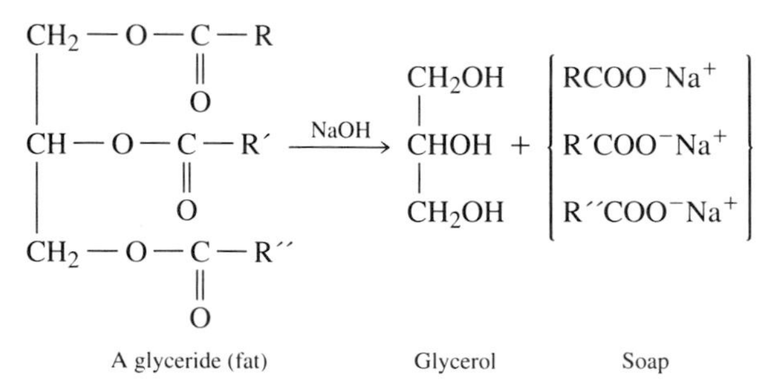

The Ca^{2+} and/or Mg^{2+} ions in hard water give a precipitate of insoluble calcium and magnesium salts with soap. This precipitate is left as a ''ring'' in a bathtub or ''scum'' on clothes. For instance,

$$2CH_3(CH_2)_{16}COO^-(aq) + Ca^{2+}(aq) \rightarrow (CH_3(CH_2)_{16}COO)_2Ca(s)$$

Such reactions explain the problems associated with the use of soap in regions where the water is hard.

Because animal fats are relatively expensive, molecules similar to soap, made from the byproducts of oil refining, have been developed for use as detergents. These soapless detergents are gradually replacing soap. An example of a synthetic detergent used domestically is sodium lauryl benzenesulfonate, with the formula

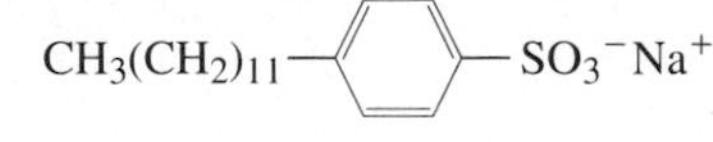

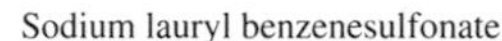

Other synthetic detergents have a positive group, or cation, to provide attraction for water. These *cationic detergents* include alkyltrimethylammonium bromide, $[RN(CH_3)_3]^+Br^-$. The R group has from 12 to 20 carbons. Synthetic detergents have the advantage of not forming a soap scum in hard water, because the calcium and magnesium salts are soluble in water.

When detergents dissolve in water, they form droplets called *micelles* in which all the hydrocarbon ends (called *hydrophobic,* or ''water-hating,'' ends) come together at the center. The CO_2^- groups (which are *hydrophilic,* or ''water loving'') point to the surface of the micelle to make contact with the water (Figure B). Micelles formed by soap are thus aggregates of fatty acid anions that have water-incompatible tails in the interior and anionic heads pointing outward to interact with polar water molecules. Soapy water is not a true solution but rather a suspension of relatively large micelles in water. Hence, soapy water looks cloudy.

Soaps and detergents work by enabling grease and oil to mix with water and therefore to be easily removed. Dirt on clothing and human skin usually consists of a mixture of natural fats with a variety of insoluble solids. Fats are not soluble in water and prevent the insoluble solids from being

Hydrophilic heads form the micelle surface

Hydrophobic tails compose the interior

Figure B *Cross section of a spherical detergent micelle*

washed away by the water. When a detergent is present, micelles form around the fat-containing dirt. The dirt then mixes with the water and is removed when the detergent solution is rinsed away, leaving the clothes or skin clean (Figure C).

Various special soaps can be made by controlling the choice of fatty acid used, because the salts of different fatty acids have different properties. As the length of the hydrocarbon chain increases, the soap feels harder to the touch. Thus, sodium stearate is harder than sodium laurate, $CH_3(CH_2)_{11}COO^-Na^+$. The longer the alkyl chain, however, the less soluble the soap is in water. Sodium stearate has a suitable firmness but is not very soluble in cold water. Salts with 12 or 14 carbon atoms in the chain are considerably more soluble.

Modern detergents have additives that improve their cleaning power. These are called *builders* and are added to promote the breakup of dirt, to keep the pH in the range in which micelles easily form, and to help prevent dirt from returning to the surface being cleansed. In the past many builders contained phosphate. Phosphate-containing detergents are now being phased out, because when waste water containing phosphate enters rivers and lakes, it encourages the growth of algae. This growth may be so rapid that it depletes the water of other nutrients needed by the growing algae, causing much of the algae to die. The decaying algae then use up the oxygen in the water and may deplete the oxygen concentration to the point where fish can no longer survive.

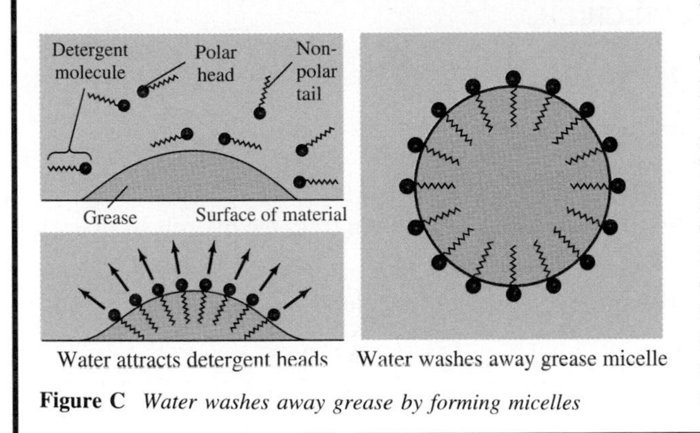

Figure C *Water washes away grease by forming micelles*

$$O{=}P(OCH_3)(OH)(OH) \qquad O{=}P(OCH_3)(OCH_3)(OH) \qquad O{=}P(OCH_3)(OCH_3)(OCH_3)$$

The monoesters and diesters have acidic OH groups. In neutral or alkaline solutions, including body fluids, they are present mainly as the corresponding anion:

$$O{=}P(OCH_3)(O^-)(O^-) \qquad O{=}P(OCH_3)(OCH_3)(O^-)$$

Among the many phosphate esters that play important roles in living organisms are ADP, adenosine diphosphate—an ester of adenosine (an alcohol) and diphosphoric acid—and ATP, an ester of adenosine and triphosphoric acid. The hydrolysis of ATP is an exothermic reaction that releases energy on demand for use in other reactions in the human body (Chapter 17):

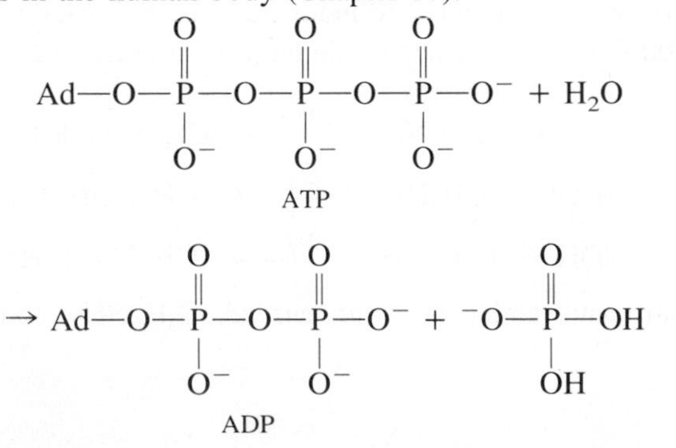

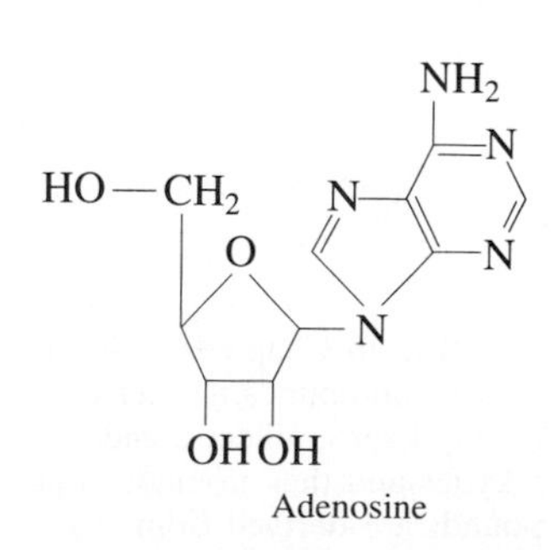

Adenosine

Here Ad stands for adenosine.

Recall that *o*- stands for ortho-, *m*- for meta-, and *p*- for para.

Exercise 14.8 Oil of wintergreen can be made by the reaction of *o*-hydroxybenzoic acid with methanol. What are its structure and its systematic name?

Exercise 14.9 What carboxylic acid and what alcohol are needed to make each of the following esters?

(a) C_6H_5—O—C(=O)$CH_2CH(CH_3)_2$

(b) $CH_3CH_2CH_2C$(=O)—O—$CH(CH_3)_2$

Exercise 14.10 What products would you obtain from the hydrolysis of each of the following esters?

(a) ethylformate **(b)** propyl *p*-bromobenzoate

14.7 AMINES, AMIDES, AND AMINO ACIDS

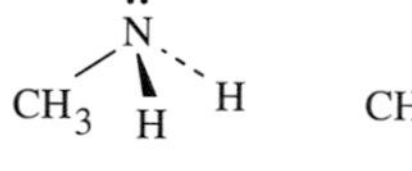

Methylamine (primary amine)

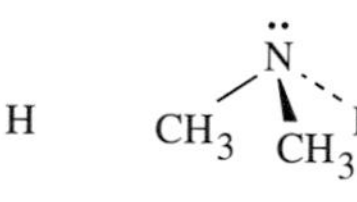

Dimethylamine (secondary amine)

Trimethylamine (tertiary amine)

FIGURE 14.11 Primary, Secondary, and Tertiary Amines A primary amine has one alkyl group attached to nitrogen. A secondary amine has two alkyl groups, and a tertiary amine has three alkyl groups.

Amines

In the same way that alcohols and ethers may be regarded as derivatives of water, analogous compounds called **amines** are derivatives of ammonia. We can distinguish primary (RNH_2), secondary (RR′NH), and tertiary (RR′R″N) amines depending on the number of alkyl groups (one, two, or three) attached to the nitrogen atom. For instance,

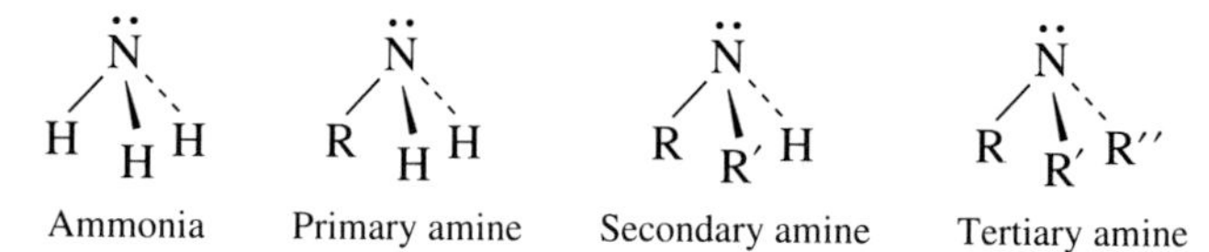

Examples are given in Figure 14.11. All the amines have the same pyramidal AX_3E geometry around the nitrogen as the ammonia molecule.

The *methylamines* are obtained industrially by the reaction of methanol with ammonia at 400°C in the presence of aluminum oxide as a catalyst:

$$CH_3OH + NH_3 \xrightarrow{Al_2O_3} CH_3NH_2 + H_2O$$

$$CH_3OH + CH_3NH_2 \xrightarrow{Al_2O_3} (CH_3)_2NH + H_2O$$

$$CH_3OH + (CH_3)_2NH \xrightarrow{Al_2O_3} (CH_3)_3N + H_2O$$

Recall from Chapter 8 that aliphatic compounds are derived from alkanes, alkenes, and alkynes and that aromatic compounds are derived from benzene and other aromatic hydrocarbons.

The simplest aromatic amine is *aminobenzene*, $C_6H_5NH_2$, commonly known as *aniline*:

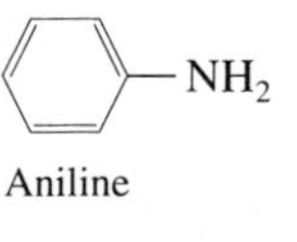

Aniline

TABLE 14.9 Properties of Ammonia and Some Amines

Compound	Formula	MP (°C)	BP (°C)	pK_b
Ammonia	NH_3	−77.7	−33.4	4.75
Methylamine	CH_3NH_2	−92.5	−6.5	3.43
Dimethylamine	$(CH_3)_2NH$	−96	7.4	3.27
Trimethylamine	$(CH_3)_3N$	−124	3.5	4.19
Aniline	$C_6H_5NH_2$	−6	184	9.37

Many amines have strong odors. Some smell like ammonia, others like fish, and others like rotting meat. Indeed, two amines produced by the decomposition of proteins in rotting flesh have the descriptive names of putrescine and cadaverine:

$$\underset{\text{Putrescine}}{H_2NCH_2CH_2CH_2CH_2NH_2} \qquad \underset{\text{Cadaverine}}{H_2NCH_2CH_2CH_2CH_2CH_2NH_2}$$

The simple aliphatic amines are generally soluble in water. Like ammonia, they are weak bases (Table 14.9):

$$\underset{\text{Methylamine}}{CH_3\text{—}NH_2} + H_2O \rightleftharpoons \underset{\text{Methylammonium ion}}{CH_3NH_3^+} + OH^-$$

They react with acids such as HCl to give salts:

$$\underset{\text{Aniline}}{C_6H_5NH_2} + HCl \rightarrow \underset{\text{Anilinium ion}}{C_6H_5NH_3^+} + Cl^-$$

In many cyclic amines, a nitrogen atom replaces a CH group in a benzene ring or a CH_2 group in a cyclic alkane (Figure 14.12). Molecules that have atoms other than carbon in a ring are called **heterocyclic molecules**. The cyclic ethers we described in Section 14.3 and the molecules in Figure 14.12 are heterocyclic molecules. DNA (Chapter 17) is a polymer of molecules called *nucleotides*, each of which contains one of the four heterocyclic amines in Figure 14.13 bonded to the phosphate ester of a sugar molecule.

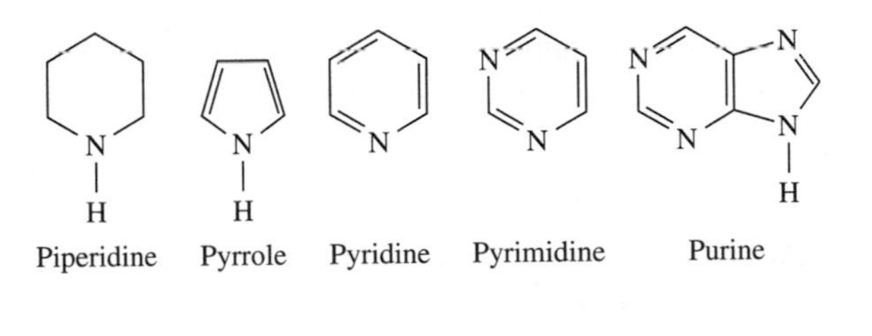

FIGURE 14.12 Some Cyclic Amines Molecules such as these that have one or more kinds of atoms in a ring other than carbon are called heterocyclic molecules.

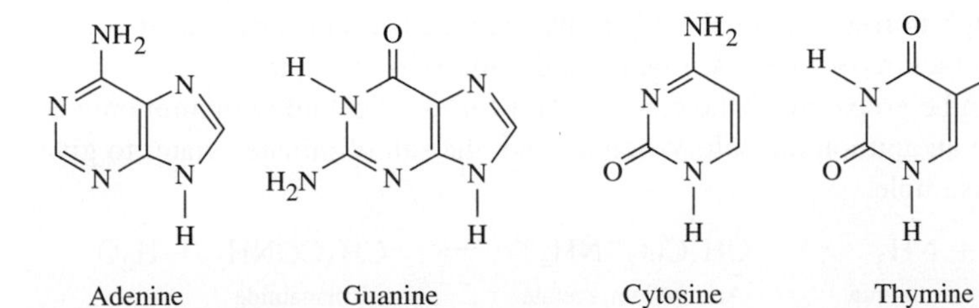

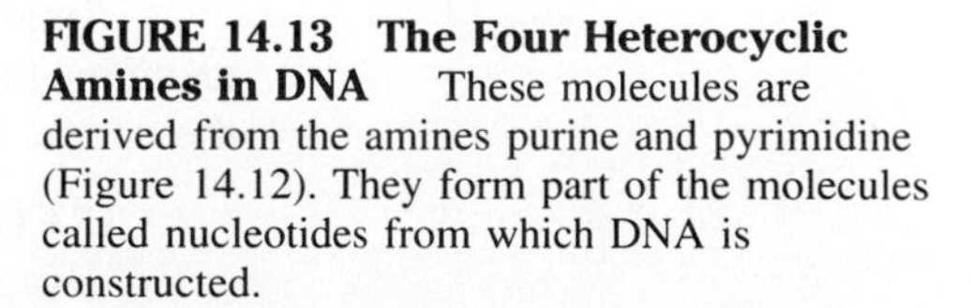
FIGURE 14.13 The Four Heterocyclic Amines in DNA These molecules are derived from the amines purine and pyrimidine (Figure 14.12). They form part of the molecules called nucleotides from which DNA is constructed.

Exercise 14.11 Write an equation for the acid–base equilibrium of dimethylamine with water.

Exercise 14.12 Complete the following equations.

(a) $C_6H_5{-}NH_2(l) + HCl(aq) \rightarrow$

(b) $CH_3CH_2NH_2(aq) + CH_3CO_2H(aq) \rightarrow$

(c) $CH_3NH_3{}^+Cl^-(aq) + NaOH(aq) \rightarrow$

Amides

Replacement of the OH group of a carboxylic acid by an NH_2 group gives an **amide**, $RCONH_2$:

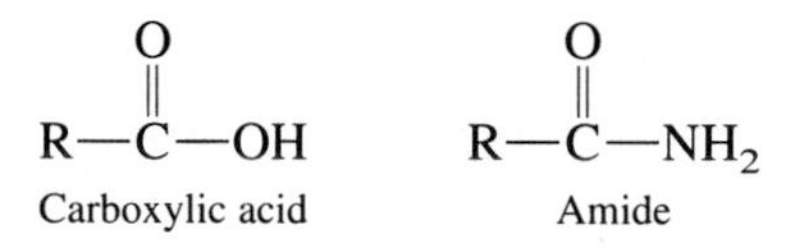

The simplest members of the amide family of compounds are

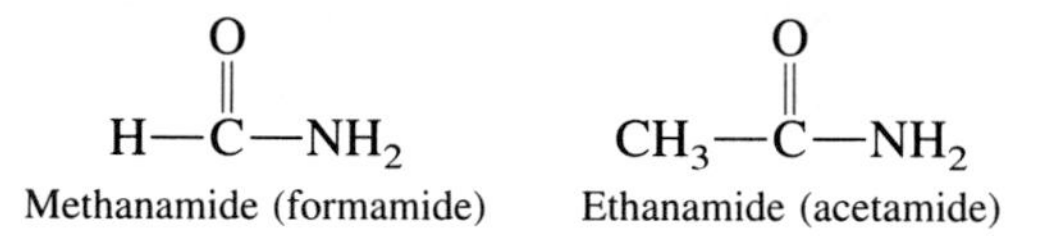

The **amide group**

$$-\overset{\overset{\displaystyle O}{\|}}{C}-\overset{\overset{\displaystyle H}{|}}{N}-$$

is the key functional group in the structure of proteins and some important synthetic polymers (Chapters 17 and 20).

The amide **urea**, $(H_2N)_2CO$, is a major animal waste product that is found in urine. It is made industrially by heating ammonia with carbon dioxide under pressure at 175 to 200°C:

$$\underset{\text{Ammonia}}{2NH_3} + CO_2 \xrightarrow{\text{heat}} \underset{\text{Urea}}{(H_2N)_2C{=}O} + H_2O$$

Because of its high nitrogen content (46%), urea is an excellent fertilizer. It is also used in the manufacture of plastics, foams, and adhesives (Chapter 20).

Amides can be prepared by the reaction of a carboxylic acid with ammonia or an amine to form an ammonium salt. When heated, the salt eliminates water to give the amide. For example,

$$\underset{\text{Acetic acid}}{CH_3COOH} + \underset{\text{Ammonia}}{NH_3} \rightarrow \underset{\text{Ammonium acetate}}{CH_3CO_2{}^-NH_4{}^+} \rightarrow \underset{\text{Ethanamide}}{CH_3CONH_2} + H_2O$$

Amino Acids

Amino acids are molecules with two functional groups, the carboxyl group and the **amino** (NH_2) group. The α-amino acids, $RCH(NH_2)CO_2H$, in which the amino group is on the same carbon atom as the carboxyl group,

$$\begin{array}{c} NH_2 \\ | \\ R{-}CH{-}CO_2H \end{array}$$

are the molecules from which proteins are constructed. Twenty different amino acids are found in proteins (Chapter 17). The two simplest are glycine and alanine:

$$\underset{\text{Glycine}}{\begin{array}{c} NH_2 \\ | \\ H{-}CH{-}CO_2H \end{array}} \qquad \underset{\text{Alanine}}{\begin{array}{c} NH_2 \\ | \\ CH_3{-}CH{-}CO_2H \end{array}}$$

Chirality and Amino Acids

The reflection of an object in a mirror gives a mirror image. In many cases a mirror image can be superimposed on the original object; coffee mugs are such objects (Figure 14.14a). We just need to turn the mirror-image mug around, and it can be merged exactly into the original mug. Other mirror-image objects cannot be superimposed on each other but are related to each other as the left hand is to the right (Figure 14.14b). There is no way that you can turn your left hand so that it could, in

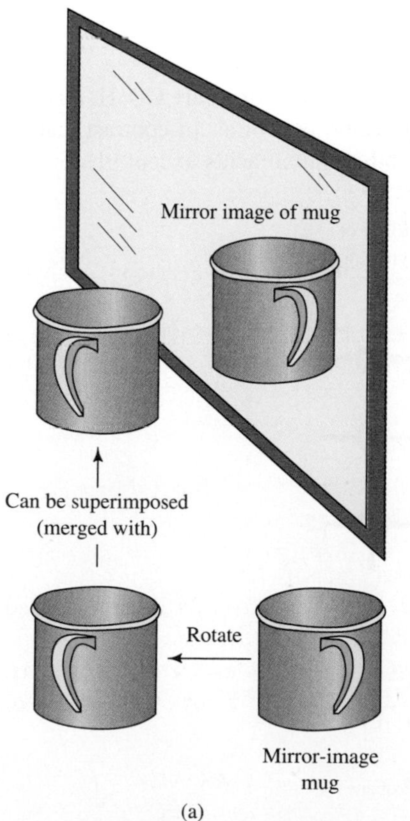

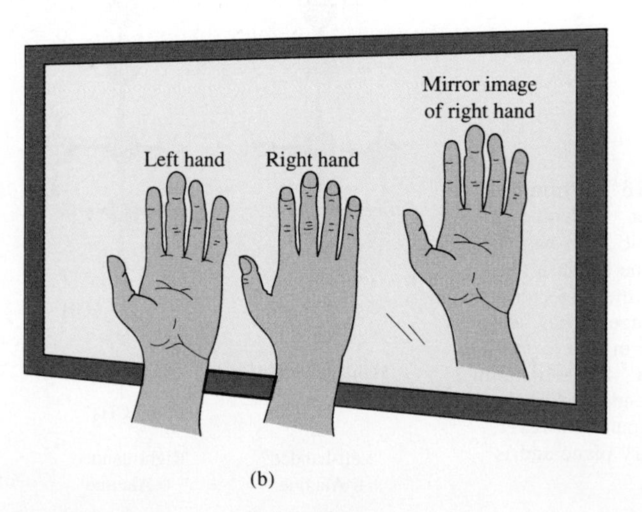

FIGURE 14.14 Mirror Images and Chirality (a) The mirror image of a coffee mug can be superimposed on itself after it is rotated. A coffee mug is an achiral object. (b) Your right hand has a mirror image that cannot be superimposed on your right hand. A hand is a chiral object. The mirror image of your right hand is identical with your left hand, so your right and left hands are mirror images of each other; both are chiral objects.

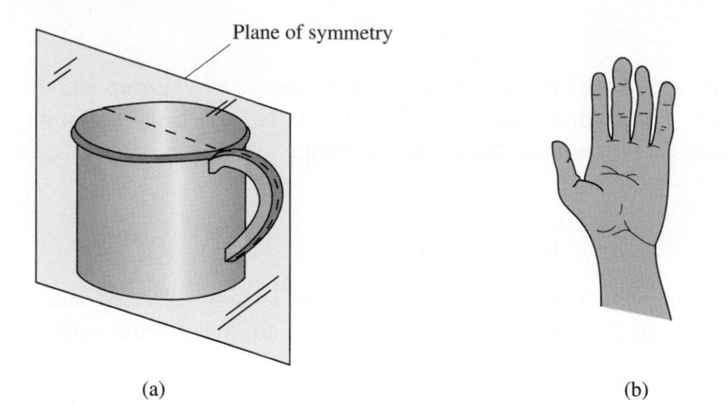

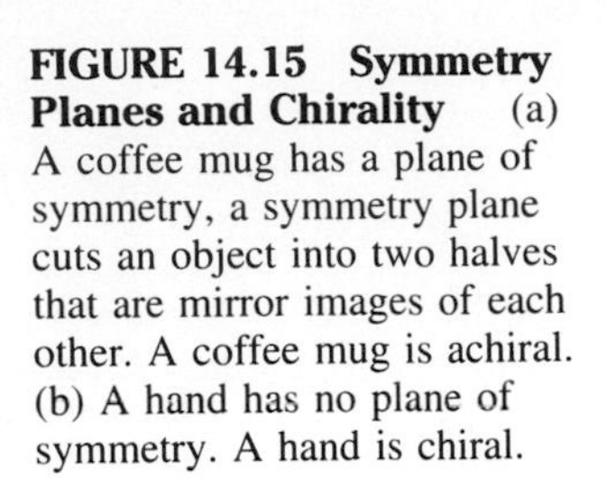

FIGURE 14.15 Symmetry Planes and Chirality (a) A coffee mug has a plane of symmetry, a symmetry plane cuts an object into two halves that are mirror images of each other. A coffee mug is achiral. (b) A hand has no plane of symmetry. A hand is chiral.

principle, be merged exactly into your right hand. Such nonsuperimposable mirror images are said to be **chiral** (from the Greek *cheir,* for ''hand''). Objects that have a superimposable mirror image are **achiral**. An achiral object is easily recognized because it has a plane of symmetry—half the object is the reflection of the other half (Figure 14.15a). A chiral object has no plane of symmetry (Figure 14.15b).

Alanine is a chiral molecule and exists in left- and right-handed forms designated as L-alanine and D-alanine (Figure 14.16). No matter how we turn the D-alanine molecule, it cannot be superimposed on or exactly merged into the L-alanine molecule. For any molecule containing one or more carbon atoms, it is simple to determine whether or not the molecule is chiral.

Whenever one of the carbon atoms in a molecule is bonded to four different atoms or groups, the molecule is chiral.

In glycine (Figure 14.17a) the groups attached to carbon atom 2 are CO_2H, NH_2, H, and H, two of which are identical, so the molecule is achiral. In contrast, carbon atom 2 of alanine (Figure 14.17b) and of all other amino acids except glycine has

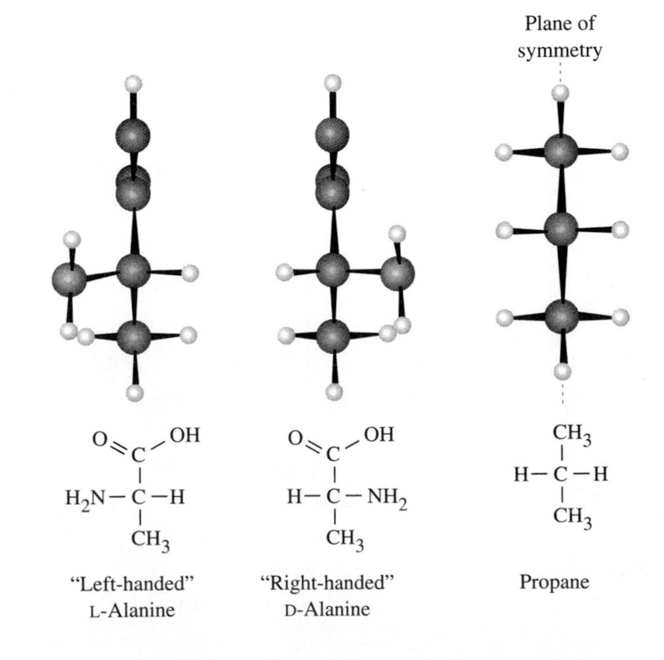

FIGURE 14.16 Symmetry in Molecules Alanine (2-aminopropanoic acid) has no symmetry plane because the two halves of the molecule are not mirror images. Thus, alanine can exist in two forms—a ''right-handed'' form, D-alanine, and a ''left-handed'' form, L-alanine. Propane, however, has a symmetry plane and is achiral.

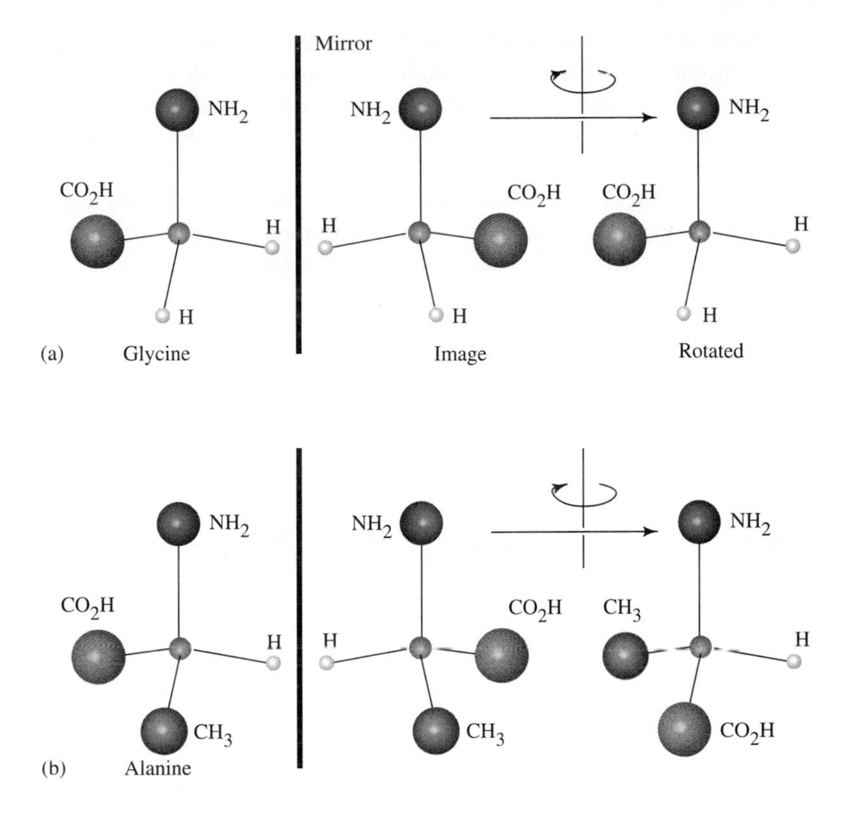

FIGURE 14.17 Chirality and Amino Acids
(a) When the mirror image of glycine is rotated, it can be superimposed on the original molecule. Glycine is achiral. Glycine has a plane of symmetry—can you find it?
(b) When the mirror image of alanine is rotated, it does not superimpose on the original molecule. Only two of the four groups can be superimposed no matter how the mirror-image molecule is rotated. Alanine is chiral—it has no plane of symmetry.

four different groups attached to it (CO_2H, NH_2, CH_3, and H for alanine). These amino acids are chiral.

The left- and right-handed forms of a chiral molecule are examples of isomers. They are called **enantiomers** or **optical isomers**, because they have the property of rotating the plane of polarized light (Figure 14.18). The D-isomer rotates the plane

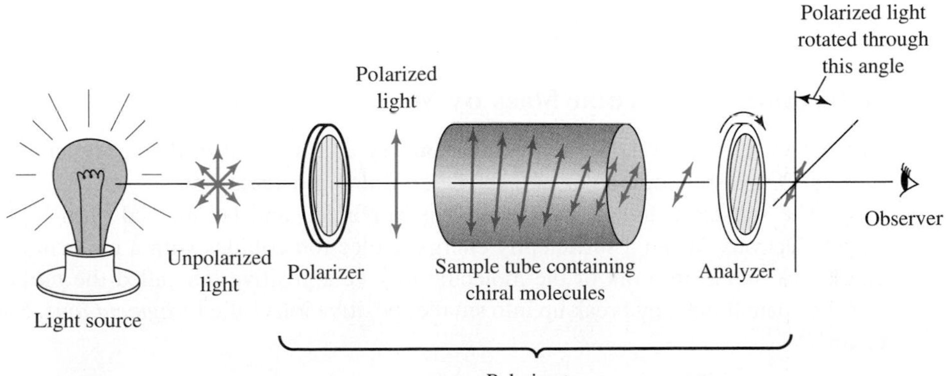

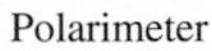

FIGURE 14.18 Polarized Light and the Polarimeter Light as we usually see it—unpolarized light—consists of electromagnetic waves oscillating in planes at all angles around the direction of travel. Light is polarized by passing it through a polarizer, such as a sheet of Polaroid film, which allows only the light oscillating in a single plane to pass through. When polarized light is passed through a solution containing chiral molecules, the angle of the plane of polarization changes. The angle of rotation is measured by finding the angle through which it is necessary to rotate another polarizer to let the light through.

of polarized light in one direction, and the L-isomer rotates it by the same amount in the opposite direction. The rotation is measured with an instrument called a *polarimeter*. Laboratory synthesis of a chiral compound almost always leads to a mixture of equal amounts of the two enantiomers called a **racemic mixture**. A racemic mixture gives an overall zero rotation of the plane of polarized light. However, a remarkable feature of nature that has not been fully explained is that living organisms synthesize only one enantiomer of a given chiral molecule. For example, all proteins are made from the L-enantiomers of α-amino acids.

Exercise 14.13

(a) Which of the amino acids $CH_3CH_2CH(NH_2)CO_2H$ and $(CH_3)_2C(NH_2)CO_2H$ are chiral?

(b) Is lactic acid chiral?

14.8 DETERMINING THE STRUCTURE OF ORGANIC COMPOUNDS

Whenever a compound is synthesized in the laboratory or isolated from a natural source, its identity must be checked by determining its composition and some characteristic properties. If it is a compound that has not been made before, its composition and structure must be determined.

We saw in Chapter 5 how we can determine the percentage of carbon and hydrogen in an organic compound by burning it to carbon dioxide and water. Methods are also available for determining the percentages of other elements such as nitrogen and sulfur. This information gives us the empirical formula of the compound.

To find the molecular formula, we must determine the molecular mass. We can obtain the molecular mass of a volatile compound by measuring the volume of a known mass—that is, the density—of the compound in the gaseous state at a known temperature and pressure, as described in Chapter 2. A more modern and usually convenient method for determining molecular mass is *mass spectrometry*.

Because

$$PV = nRT = \frac{mRT}{M},\ M = \frac{mRT}{PV}$$

Thus, knowing the mass m, pressure P, and temperature T, by measuring the volume V, we can calculate the molecular mass M.

Determining Molecular Mass by Mass Spectrometry

In Chapter 1, we saw how **mass spectrometry** can be used to determine atomic masses. It is also extremely useful for determining molecular masses.

The substance under investigation is vaporized and bombarded with high-speed electrons. When a sufficiently energetic electron collides with a molecule, it can knock an electron out of the molecule to give a positive ion called the *parent ion*. The parent ion may break up into smaller positive ions called *fragment ions*. For example,

$$CH_3{-}C(CH_3)_2{-}CH_3 + e^- \rightarrow 2e^- + C_5H_{12}^+ \rightarrow \begin{cases} C_4H_9^+ & \text{Mass 57} \\ C_3H_5^+ & \text{Mass 41} \\ C_2H_5^+ & \text{Mass 29} \\ C_2H_3^+ & \text{Mass 27} \end{cases}$$

2,2-Dimethylpropane; Parent ion Mass 72; Fragment ions

The positive ions are accelerated by an electric field and pass into a mass spectrome-

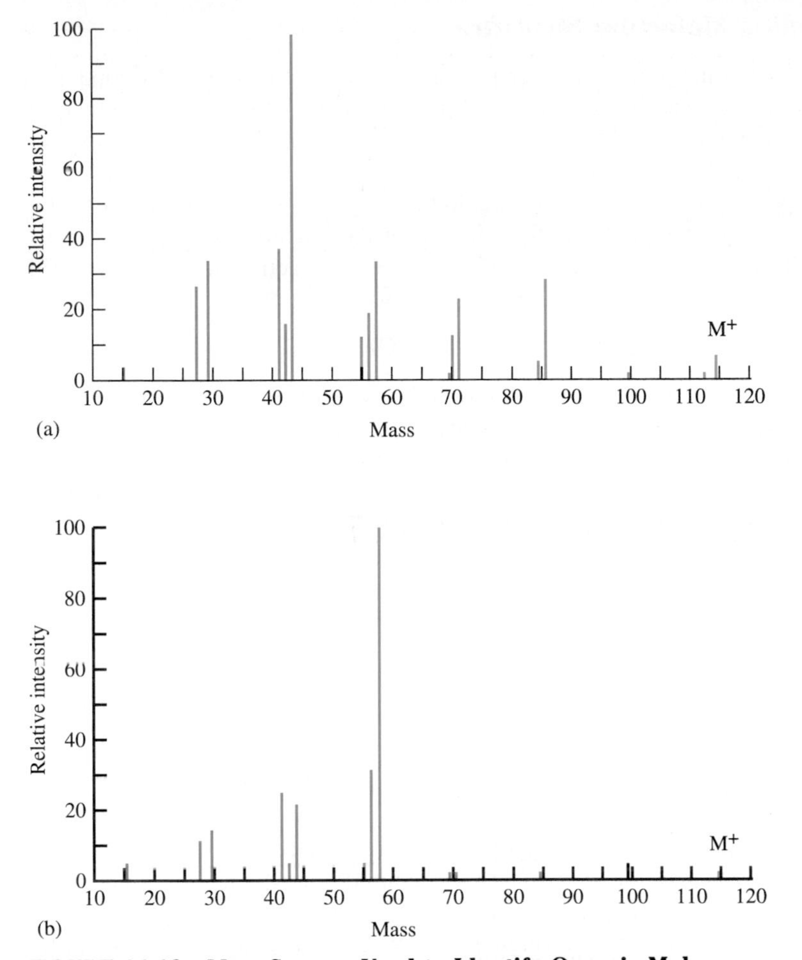

FIGURE 14.19 Mass Spectra Used to Identify Organic Molecules Relative number of ions of a given mass versus mass. The spectra of the two isomeric alkanes (a) octane and (b) 2,2,4-trimethylpentane are very different and can be used to distinguish these molecules. The relative abundance of the parent ion, M^+, is very low because it is unstable and decomposes into many fragments.

ter, where their masses are determined from the amount by which they are deflected in a magnetic field, as described in Box 1.3. The results are recorded as a plot of the relative number of ions of each mass versus the mass that is called the **mass spectrum** of the substance (Figure 14.19).

The line in the spectrum that corresponds to the greatest mass gives the mass of the parent ion and therefore the mass of the molecule from which it is derived. The masses of the fragment ions and their relative abundances are different for each substance. In other words, each substance has its own characteristic mass spectrum. This enables us to identify a substance by comparing its mass spectrum with previously recorded spectra. We can also distinguish between substances that otherwise have very similar properties, such as two isomeric alkanes. Figure 14.19 shows that the mass spectrum of octane $CH_3(CH_2)_6CH_3$ is quite different from that of its isomer 2,2,4-trimethylpentane, because the parent ions fragment in different ways.

Although we cannot always predict which fragment ions are produced from a parent ion, the fragment pattern (mass spectrum) of a given parent ion is unique.

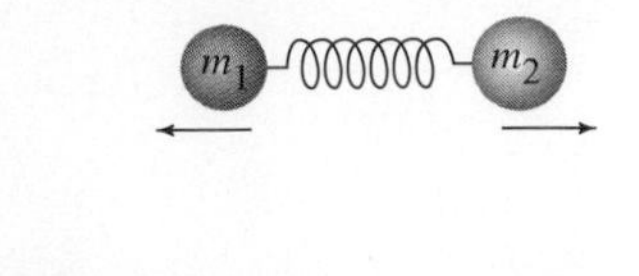

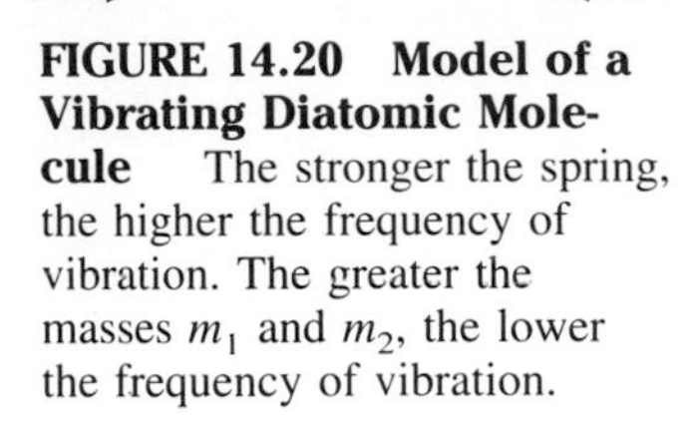

FIGURE 14.20 Model of a Vibrating Diatomic Molecule The stronger the spring, the higher the frequency of vibration. The greater the masses m_1 and m_2, the lower the frequency of vibration.

Determining Molecular Structure

If a substance has not previously been prepared, we cannot identify it by comparing its mass spectrum with known spectra. But the masses of the fragment ions often enable us to identify parts of the molecule and give us valuable information about its overall structure. More information can be obtained by determining which functional groups are present in the molecule. In the early days of organic chemistry this was done by using the characteristic reactions of the functional groups. For example, an OH group can be detected by means of its reaction with sodium: This reaction produces hydrogen according to the equation

$$2ROH + 2Na \rightarrow 2Na^+OR^- + H_2$$

For large, complicated molecules with many functional groups, this method was slow and sometimes uncertain. Today the functional groups present in a molecule and much other information about a molecule's structure can be determined more rapidly and with much greater certainty by instrumental techniques such as infrared and nuclear magnetic resonance (NMR) spectroscopy.

Infrared Spectroscopy

In addition to the translational kinetic energy they possess by virtue of their motion, molecules also have vibrational energy. The two atoms in a diatomic molecule vibrate in and out just like two balls connected by a spring (Figure 14.20). The natural frequency with which this model of a molecule vibrates depends on the stiffness of the spring (the strength of the bond) and the masses of the two balls (atoms). The energy needed to cause a diatomic molecule to vibrate with its characteristic frequency ν is $E = h\nu$. This amount of energy can be provided by a quantum of radiation of frequency ν that is usually in the infrared region, between approximately 10^{12} and 10^{14} Hz (Figure 6.4). When you stand in the sun and feel yourself getting warm, the molecules of your body are absorbing infrared radiation and are therefore vibrating more vigorously.

Only molecules such as HCl that have a dipole moment absorb infrared radiation, so diatomic molecules such as O_2 and N_2 do not have an infrared spectrum. However, their vibrational frequencies can be obtained from other types of spectra.

Each diatomic molecule has its own characteristic *vibrational frequency*. The stronger the bond, the higher the vibrational frequency; the heavier the atoms, the lower the frequency. For example, the vibrational frequencies of HF, HCl, and HBr decrease with increasing mass from 12.4×10^{13} to 8.6×10^{13} to 7.9×10^{13} Hz. Vibrational frequencies (ν) in infrared spectroscopy are usually quoted in terms of **wave numbers**, $\tilde{\nu}$, which are obtained by dividing the frequency by the speed of light. For example,

III Because 1 Hz = 1 s^{-1}, the speed of light c in cm · s^{-1} also has units of cm Hz. Then

$$\nu(\text{Hz}) = c(\text{cm Hz}) \times \tilde{\nu}\ (\text{cm}^{-1})$$

$$\tilde{\nu}(\text{HF}) = \frac{12.4 \times 10^{13}\ \text{Hz}}{3.00 \times 10^{10}\ \text{cm} \cdot \text{s}^{-1}} = 4140\ \text{cm}^{-1}$$

Similarly,

$$\tilde{\nu}(\text{HCl}) = 2886\ \text{cm}^{-1} \quad \text{and} \quad \tilde{\nu}(\text{HBr}) = 2650\ \text{cm}^{-1}$$

The vibrations of a polyatomic molecule are more complex than those of a diatomic molecule. But to a first approximation, each bond or each functional group of a polyatomic molecule may be considered to have its own characteristic vibrational frequency or frequencies. Hydrogen atoms are very light, and so they vibrate at high frequencies. A C—H bond, for example, has a characteristic vibrational frequency, called a C—H stretch, of approximately 9.0×10^{13} Hz ($\tilde{\nu}$ = 3000 cm^{-1}). Chlorine is a heavier atom, so the C—Cl stretch has a lower frequency of 2.1×10^{13} Hz ($\tilde{\nu}$ = 700 cm^{-1}).

TABLE 14.10 Characteristic Infrared Wave Numbers for Stretching Vibrations

Group	$\tilde{\nu}$ (cm^{-1})
O—H	3400–3700
N—H	3100–3500
C—H	2800–3200
C≡N	2200–2400
C≡C	2100–2300
C=O	1650–1800
C=C	1600–1700
C—O	1000–1250
C—Cl	600–800

A double bond is stronger and stiffer that a single bond and vibrates at a correspondingly higher frequency. For example, a C—C single bond absorbs radiation at about 1000 cm^{-1}, a C=C double bond near 1650 cm^{-1}, and a C≡C triple bond at about 2200 cm^{-1}. These differences in vibrational frequency provide a means for distinguishing among molecules by the technique of **infrared spectroscopy**. Table 14.10 gives the stretching frequencies of several groups, and Figure 14.21 shows some typical infrared spectra.

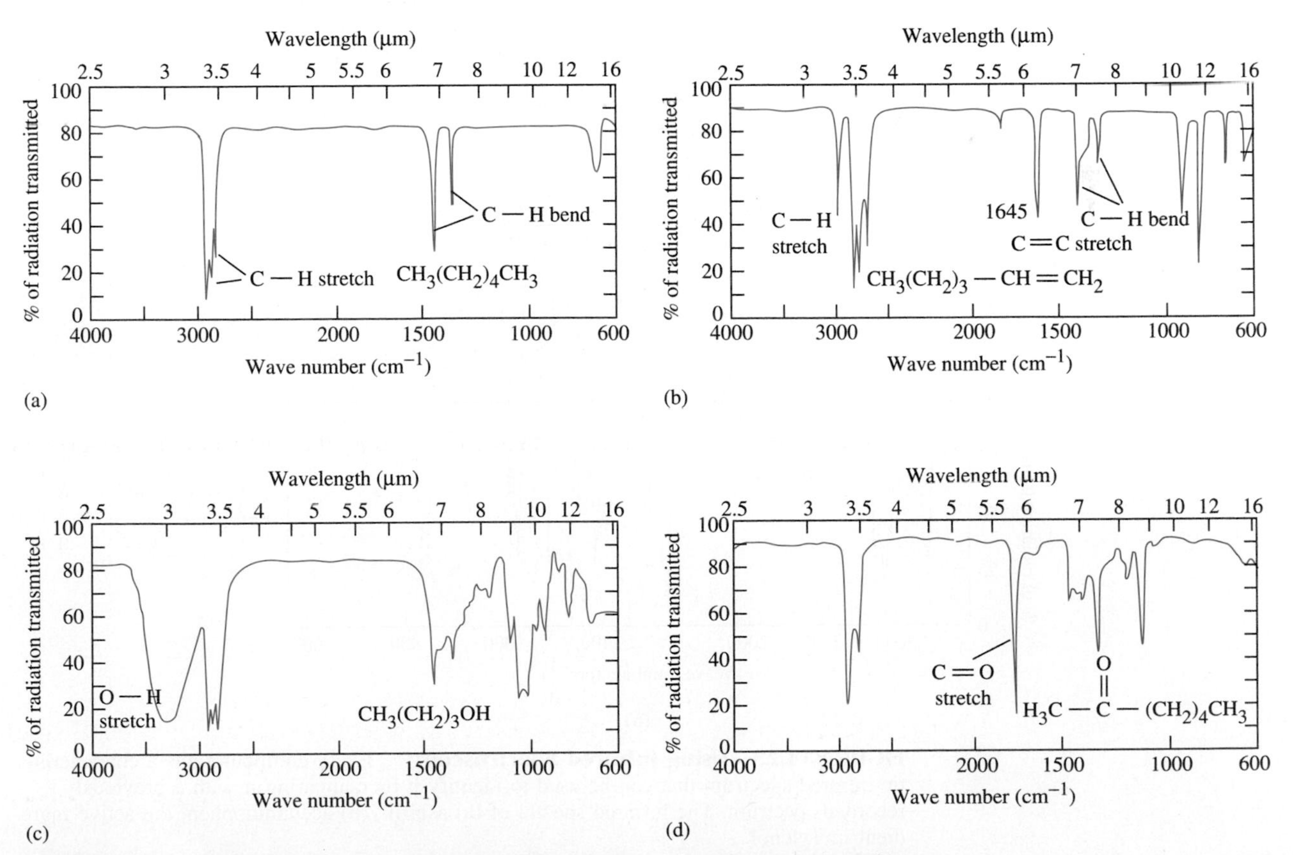

FIGURE 14.21 Infrared Spectra of Organic Molecules These spectra show the characteristic frequencies of the C—H, C=C, O—H, and C=O groups for (a) an alkane, (b) an alkene, (c) an alcohol, and (d) a ketone.

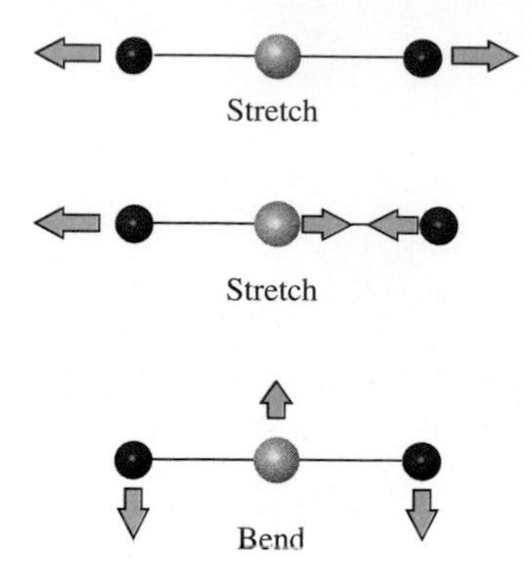

FIGURE 14.22 Vibrations of a Triatomic Molecule Two stretching and a bending vibration of a linear triatomic molecule.

In addition to stretching motions, bending motions of bonds that increase and decrease bond angles (Figure 14.22) also contribute to the infrared spectrum. Because a molecule may have several different types of bonds and several different types of vibrational motion, some parts of the spectrum, particularly between 800 cm^{-1} and 1400 cm^{-1}, may be quite complex and difficult to analyze completely in terms of specific functional groups. Nevertheless, this region of the spectrum, called the *fingerprint region*, may be very useful for identifying a previously known compound. Each molecule has a characteristic infrared spectrum in this region, from which it can often be identified by comparison with previously recorded spectra (Figure 14.23).

Nuclear Magnetic Resonance Spectroscopy

Nuclear magnetic resonance (NMR) spectroscopy is a very useful technique for studying the structures of molecules in the liquid state and in solution. Many nuclei have magnetic properties that we can think of as arising from the spin of the nucleus, just as the magnetic properties of an electron are due to its spin (Chapter 6). It is possible to observe the NMR spectra given by many different nuclei, including ^{1}H, ^{13}C, ^{19}F, and ^{31}P. We will consider only those arising from ^{1}H nuclei (protons). These spectra are particularly useful for organic compounds because they generally contain hydrogen.

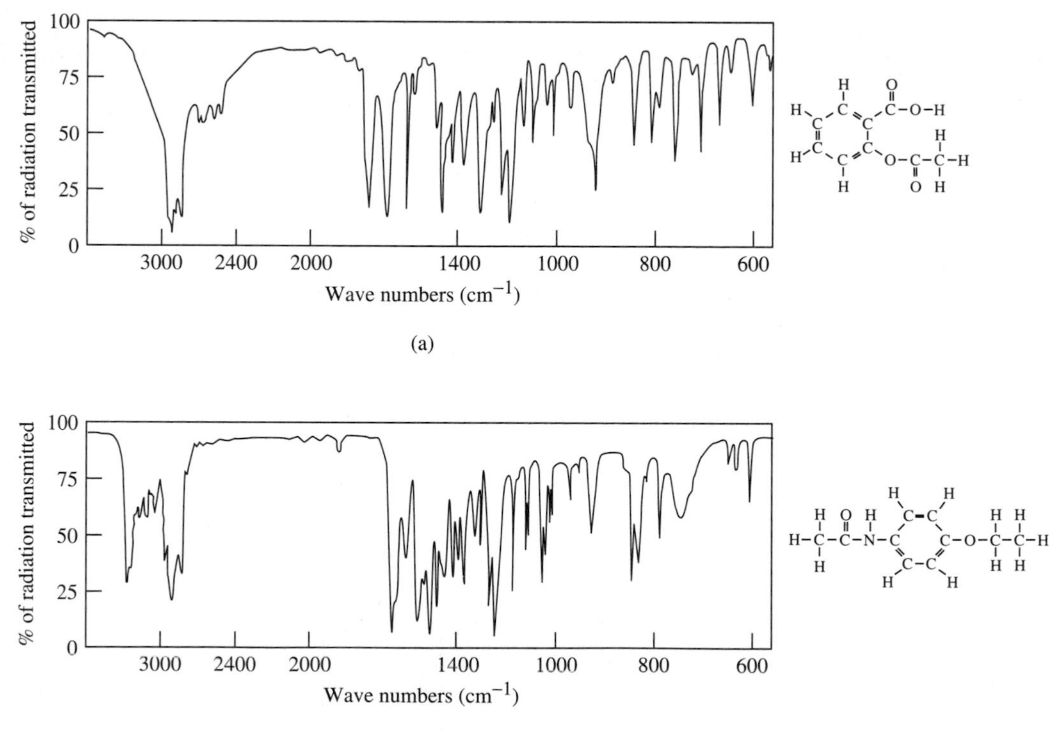

FIGURE 14.23 Using Infrared Spectroscopy Every compound has a characteristic infrared spectrum that can be used to identify it by comparing it with a previously recorded spectrum. The infrared spectra of (a) aspirin; (b) acetaminophen, the active ingredient in Tylenol.

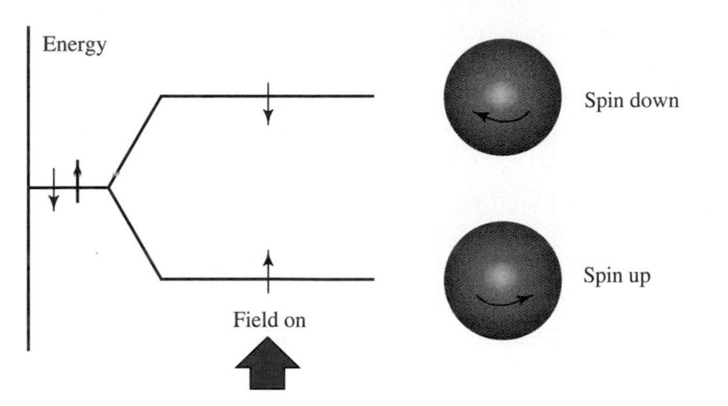

FIGURE 14.24 Nuclear Spin The two possible orientations of the spin of a proton have different energies in a magnetic field.

A spinning nucleus, like an electron, behaves like a tiny bar magnet. In the presence of an external magnetic field, this nuclear magnet can point in either of two directions: *with* the external field or *opposed to* the external field (Figure 14.24; Figure 6.19). The latter orientation has a slightly higher energy than the former. As a result, energy is required to change the spin state of the nucleus, that is, to cause the transition of the nucleus from its lower energy state to its higher energy state. The required energy is in the radio-frequency region of the spectrum (10^7–10^8 Hz; see Figure 6.4).

In an NMR spectrometer (Figure 14.25), the sample to be studied is placed in the field of a powerful electromagnet. Radio-frequency radiation is supplied to the sample, and its frequency may be steadily changed. When the frequency matches the energy separation of the two nuclear spin states, the radiation is strongly absorbed. More commonly, the radio frequency is kept constant, and the separation of the energy levels is varied by varying the magnetic field until this separation matches the radio frequency and the radiation is strongly absorbed. This radiation is detected by a receiver, which produces a spectrum like that shown in Figure 14.26.

FIGURE 14.25 An NMR Spectrometer The large vessel contains a superconducting magnet.

The frequency of an NMR transition depends on the *local* magnetic field the nucleus experiences. This field is the sum of the external field from the magnet and the magnetic field that arises from the electrons surrounding the nucleus. The electron cloud surrounding the nucleus is slightly different for hydrogen nuclei in different, nonequivalent positions in a molecule. Hydrogen nuclei in different

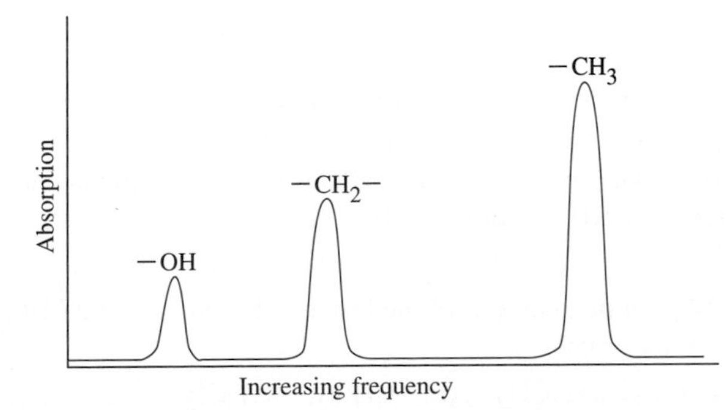

FIGURE 14.26 Low-Resolution NMR Spectrum of Ethanol The spectrum has three peaks, arising from the CH_3, CH_2, and OH hydrogen atoms, respectively. The peaks have relative areas 3:2:1.

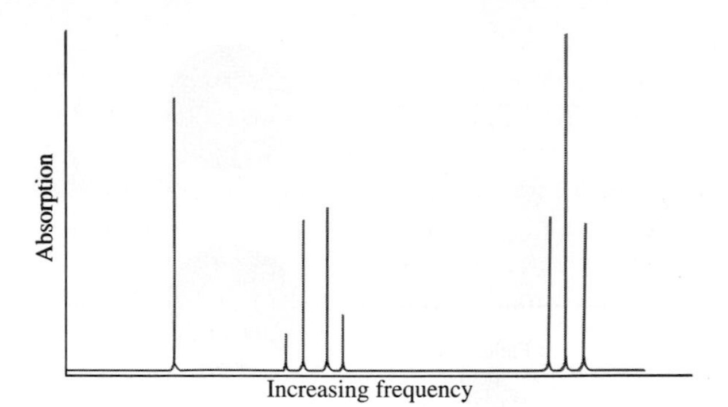

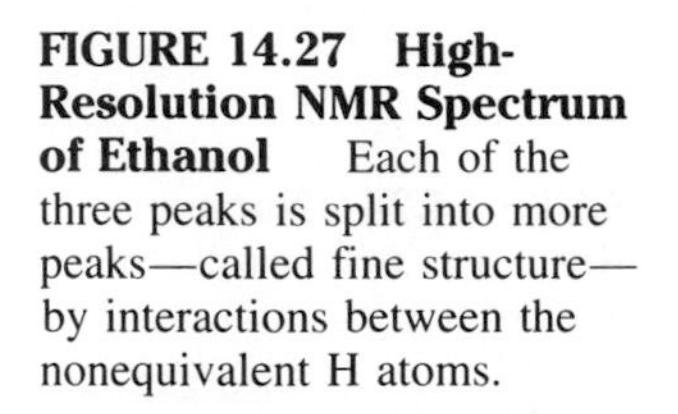

FIGURE 14.27 High-Resolution NMR Spectrum of Ethanol Each of the three peaks is split into more peaks—called fine structure—by interactions between the nonequivalent H atoms.

positions are therefore in different local fields and thus absorb radio-frequency radiation of slightly different energies or frequencies. For example, ethanol has three sets of nonequivalent hydrogen atoms—those in the CH_3 group, those in the CH_2 group, and that in the OH group—so it absorbs radiation of three different frequencies, giving three peaks in the NMR spectrum (Figure 14.26). These peaks are shifted by different amounts from the peak of a standard reference sample. These shifts in the absorption frequency are called **chemical shifts**. A key feature of NMR spectra is that the intensity of each peak in a spectrum is proportional to the number of hydrogen atoms contributing to the particular peak. Thus, in the spectrum of ethanol the three peaks have the intensity ratio 3:2:1.

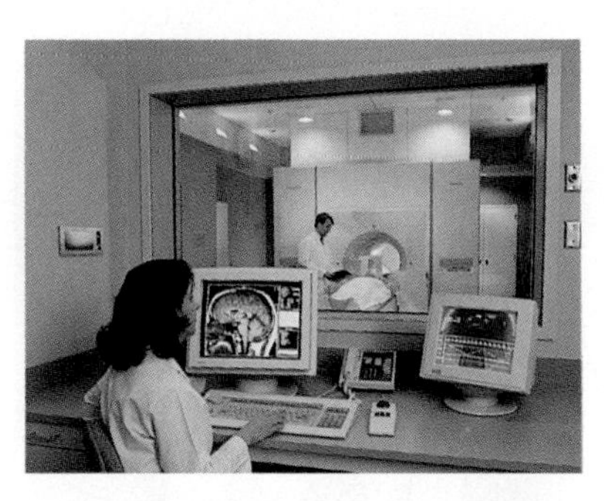

FIGURE 14.28 Magnetic Resonance Imaging (MRI) *Background:* A patient is placed inside a large electromagnet. *Foreground:* A physician examines the image on a computer screen.

The number of peaks in an NMR spectrum and their relative intensities very often enable us to determine the structure of an organic molecule. Most peaks in the spectra obtained by modern instruments, called high-resolution spectra, are split into two or more closely spaced lines due to interactions between the nonequivalent hydrogen nuclei (Figure 14.27). The peaks are said to have *fine structure*, which gives more structural information.

In addition to providing information that typically enables us to find the structure of a molecule, an NMR spectrum, like an infrared spectrum, is a useful fingerprint of a molecule. So an NMR spectrum can be used for identification by comparing it with previously recorded spectra.

A proton NMR spectrum gives us information on the arrangement of the hydrogen atoms in a molecule that typically enables us to write a structural formula. But an NMR spectrum tells us nothing about the lengths of bonds and the angles between them. Most detailed information of this type comes from X-ray crystallography.

In naming magnetic resonance imaging, the medical profession dropped the word "nuclear" to avoid the public's fear and concern that this technique might involve radioactivity or nuclear reactions and that therefore it might be hazardous.

In recent years the technique of nuclear magnetic resonance has been developed to such an extent that signals from the hydrogen nuclei in living matter, such as the human body, can be recorded, combined, and displayed on a computer screen as an image of a particular part of the body. This useful technique is called **magnetic resonance imaging** (MRI) (Figure 14.28).

Exercise 14.14 Draw diagrams of the low-resolution proton NMR spectra of the following molecules.

(a) C_2H_6 **(b)** $CH_3CH_2CH_3$ **(c)** $(CH_3)_2CHCHCl_2$

Exercise 14.15 How could NMR and infrared spectroscopy be used to show whether a compound with the molecular formula C_4H_6 is 1,3-butadiene or 2-butyne?

X-Ray Crystallography

The method that gives the most complete and detailed structural information is X-ray crystallography (Chapter 10). No preliminary information, such as composition and molecular mass, is required. This method not only shows how all the atoms are arranged in a molecule but gives accurate information on all the bond distances and bond angles as well. However, to use this method it is essential to have suitable crystals. This criterion is not always easy to meet for organic compounds, particularly those that are liquids or gases at room temperature. Moreover, the method requires relatively expensive equipment and is more time consuming than other methods, although the increasing power and speed of modern computers is drastically reducing the time required for an X-ray structure determination. The more complex the molecule, the more time is needed to work out the structure. But for a relatively simple molecule, only a few days or even a few hours may be needed.

Gas Chromatography

Before the composition and structure of a compound can be determined, the compound must be obtained in a pure state. A useful technique of doing this for organic compounds, which are typically gases or volatile liquids or solids, is **gas chromatography**. In this technique a long tube is packed with a powdered inert solid such as aluminum oxide. The surface of the solid is coated with a liquid that does not react with the substances to be separated (Figure 14.29). A stream of an inert gas such as He, Ar, or N_2 is passed continuously down the tube. The mixture of substances to be separated is injected into the inert gas stream and is vaporized by heating. The vapor is swept down the tube by the inert gas. The more-soluble components in the liquid lag behind, and the less-soluble components move more rapidly down the tube. Eventually, all the components separate from each other and emerge one by one from the end of the tube, where they can be identified and collected separately.

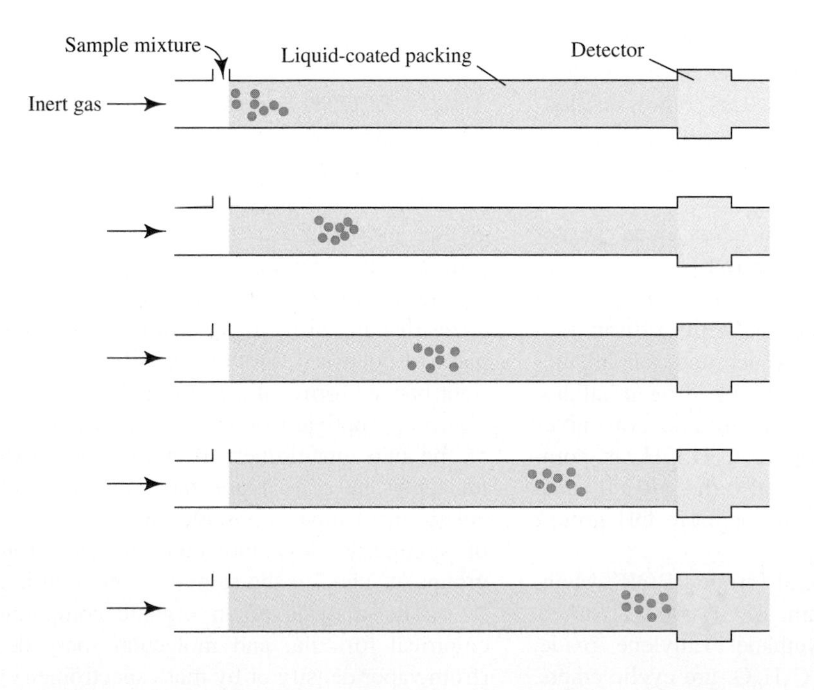

FIGURE 14.29 Schematic Diagram of a Gas Chromatography Apparatus The dark circles represent molecules of an organic compound that move more slowly than the molecules of another compound, represented by the light circles. The two (or more) components of the mixture separate completely.

SUMMARY

A functional group is an atom or group of atoms in an organic molecule that gives the molecule characteristic chemical properties. Organic molecules are classified into families by the functional groups they contain. The important families of organic compounds we have discussed are haloalkanes, RX; alcohols and phenols, ROH; ethers, ROR′; thiols, RSH; aldehydes, RCHO; ketones, RR′CO; carboxylic acids, RCO_2H; esters, RCO_2R'; amines, RNH_2; amides, $RCONH_2$; and α-amino acids, $RCH(NH_2)CO_2H$. Here R and R′ are alkyl or aryl groups. An important organic reaction type is nucleophilic substitution ($RX + Y^- \rightarrow RY + X^-$), in which the positively charged carbon atom of a polar $^{\delta+}C—X^{\delta-}$ bond attracts an unshared pair of electrons of a nucleophile Y, and Y replaces X.

The photochemical or high-temperature reaction of halogens with hydrocarbons gives haloalkanes (alkyl halides). Haloalkanes also result from addition of HX or X_2 to a C═C or C≡C bond.

Alcohols are molecules that contain the OH group. Their names have the ending *-ol*. Methan*ol*, CH_3OH, is manufactured from synthesis gas. Ethan*ol*, C_2H_5OH, results from the addition of H_2O to the C═C double bond of ethene or from the fermentation of sugars and carbohydrates. A general method of preparing alcohols is by the reaction of OH^- with an alkyl halide. Alcohols with more than two C atoms have isomers (such as 1-propanol and 2-propanol) with the same molecular formula but different structures. Primary, secondary, and tertiary alcohols contain a $-CH_2OH$, >CHOH, and a ⋛COH group, respectively. 1,2-ethanediol (ethylene glycol, or glycol), $HOCH_2CH_2OH(l)$, is a diol, an alcohol with two OH groups; 1,2,3-propanetriol (glycerol, or glycerin), $HOCH_2CH(OH)CH_2OH(l)$, is a triol, with three OH groups.

Alcohols are weak acids and bases, even weaker acids than water. They donate a proton to $OH^-(aq)$ to give alkoxide ions, RO^-, and are protonated by strong acids, giving ROH_2^+. Dehydration with sulfuric acid eliminates water to give an alkene, and haloalkanes result from substitution reactions with hydrogen halides. The complete oxidation of alcohols (combustion) gives CO_2 and water. Hydroxybenzene, C_6H_5OH, is commonly known as phenol, which is also the general name for aromatic compounds with one or more OH groups substituted on an aromatic ring.

The ethers have the general formula ROR′. Methylethyl ether is the common name of $CH_3OC_2H_5$, but its systematic name is methoxyethane. Ethylene oxide, C_2H_4O, and propylene oxide, C_3H_6O, are cyclic ethers. The reaction of a sodium alkoxide (RO^-Na^+) with a haloalkane (R′Cl) gives an ether, ROR′, in a nucleophilic substitution reaction.

Thiols, RSH, are the sulfur-containing analogues of alcohols. They are readily oxidized to disulfides, RSSR. Disulfide bonds, —S—S—, are important in some biomolecules.

Aldehydes, RCHO, with one alkyl group, result from oxidizing primary alcohols with a strong oxidizing agent, and ketones, RR′CO, with two alkyl groups, result from oxidizing secondary alcohols. The reverse reactions occur upon reduction with H_2. Aldehydes and ketones contain the polar carbonyl group, $^{\delta+}C{=}O^{\delta-}$. Aldehydes (but not ketones) are oxidized to carboxylic acids, RCO_2H.

Carboxylic acids are weak acids, although stronger acids than alcohols. They result from the oxidation of primary alcohols or aldehydes. The simplest aromatic carboxylic acid is benzoic acid, $C_6H_5CO_2H$. With strong bases carboxylic acids form carboxylate salts $RCO_2^-M^+$.

Carboxylic acids react with alcohols to form esters with the elimination of H_2O. Fats (triglycerides) are esters of glycerol and some long-chain carboxylic acids called fatty acids. The important biomolecules adenosine diphosphate, ADP, and adenosine triphosphate, ATP, are esters of diphosphoric acid and triphosphoric acid, respectively.

The reaction of an alcohol, ROH, with ammonia, NH_3, gives primary, secondary, and tertiary amines, RNH_2, RR′NH, and RR′R″N. Amines are weak bases. Amides, $RCONH_2$, result from the elimination of water from the ammonium salts of carboxylic acids upon heating. Amino acids have both amino (NH_2) and carboxyl (COOH) functional groups. In α-amino acids, such as glycine and alanine, both groups are attached to the same carbon atom. Alanine is chiral—has a nonsuperimposable mirror image—and exists in left- and right-handed forms that are mirror images of each other and rotate the plane of polarized light in opposite directions. Left- and right-handed forms of a chiral molecule are called enantiomers (or optical isomers). A mixture of equal amounts of the two enantiomers is a racemic mixture. Chiral molecules have no plane of symmetry. Achiral molecules, with a superimposable mirror image, have a plane of symmetry. Any molecule that has four different groups attached to the same carbon atom is chiral.

The analysis of an organic compound gives its empirical formula, and molecular mass determination (from vapor density or by mass spectrometry) then gives its molecular formula. The structural formula can often be inferred by identifying its functional groups chemically or from infrared or nuclear magnetic resonance (NMR) spectroscopy. Infrared spectroscopy gives vibrational frequencies typical of particular functional groups. NMR spectroscopy distinguishes between atoms (such as hydrogen atoms) whose nuclei have spin when such atoms are in structurally nonequivalent positions in a molecule. Bond lengths and bond angles in molecules are determined by X-ray crystallography.

IMPORTANT TERMS

achiral (page 532)
alcohol (page 504)
aldehyde (page 514)
amide (page 530)
amine (page 528)
amino acid (page 531)
carbonyl group (page 514)
carboxyl group (page 519)
carboxylic acid (page 518)
chemical shift (page 540)
chiral (page 532)
dehydrogenation reaction (page 515)
diol (page 507)
disulfide (page 514)
enantiomer (optical isomer) (page 533)
ester (page 522)
ether (page 511)
fatty acid (page 523)
functional group (page 500)
haloalkane (alkyl halide) (page 501)
heterocyclic molecule (page 529)
infrared spectroscopy (page 536)
inorganic chemistry (page 499)
ketone (page 514)
mass spectrometry (page 534)
NMR spectroscopy (page 538)
nucleophile (page 504)
nucleophilic substitution (page 504)
organic chemistry (page 499)
phenol (page 510)
racemic mixture (page 534)
substitution reaction (page 503)
thiol (page 512)
triglyceride (page 524)
triol (page 507)
wave number (page 536)

REVIEW QUESTIONS

1. Give two reasons why the compounds of carbon are so numerous.

2. Explain what is meant by the term ''functional group,'' and give five examples of functional groups.

3. **(a)** What is a nucleophile?

(b) Give an example of a nucleophilic substitution reaction.

4. Write equations for two reactions by which ethanol can be prepared.

5. **(a)** Draw the structures and name the four alcohols with molecular formula C_4H_9OH.

(b) Classify each as a primary, secondary, or tertiary alcohol.

6. Briefly explain why ethane (bp −89°C), chloroethane (bp 12°C), and ethanol (bp 78°C) have such different boiling points.

7. Starting with ethanol, sodium, and ethyl bromide, write equations for the two-step preparation of diethyl ether.

8. What are the systematic names and structural formulas of formaldehyde, acetaldehyde, and acetone?

9. What are the products of oxidation by $Cr_2O_7^{2-}$ in acidic aqueous solution of **(a)** 1-propanol and **(b)** 2-propanol?

10. Write structural formulas for 1-butanol, butanal, butanone, and butanoic acid.

11. Write the balanced equation for the reaction of ethanol and propanoic acid, and name the product.

12. Write the balanced equation for the reaction of methylethanoate with NaOH (*aq*).

13. Give one example each of **(a)** a saturated fatty acid and **(b)** an unsaturated fatty acid.

14. Write the structural formulas of the esters that can be formed from phosphoric acid and ethanol and from diphosphoric acid and ethanol.

15. Give the balanced equations and name the products of all the amines that can be formed by the high-temperature reaction of ammonia with methanol.

16. What is meant by the term ''chiral molecule''?

17. Name two methods that can be used to measure the molar mass of a volatile substance.

18. How can we distinguish between structural isomers such as butane and 2-methyl propane by mass spectrometry?

19. What structural information is obtained from the stretching frequencies observed in the infrared spectrum of a molecule?

20. **(a)** Why does the NMR spectrum of ethanol show three main peaks with different chemical shifts?

(b) What are their relative intensities?

PROBLEMS

Functional Groups

1. Write the structural formulas of and name each of the following noncyclic compounds containing three carbon atoms.

(a) a primary alcohol (b) a secondary alcohol
(c) an aldehyde (d) a ketone
(e) a carboxylic acid (f) an ether
(g) an amide (h) a primary amine

2. Write the formulas for the following functional groups.

(a) chloro- (b) iodo- (c) hydroxyl
(d) carbonyl (e) carboxyl (f) ether
(g) secondary amino- (h) amido-

3. Name the functional groups in each of the following.

(a) monosodium glutamate (a flavor enhancer)

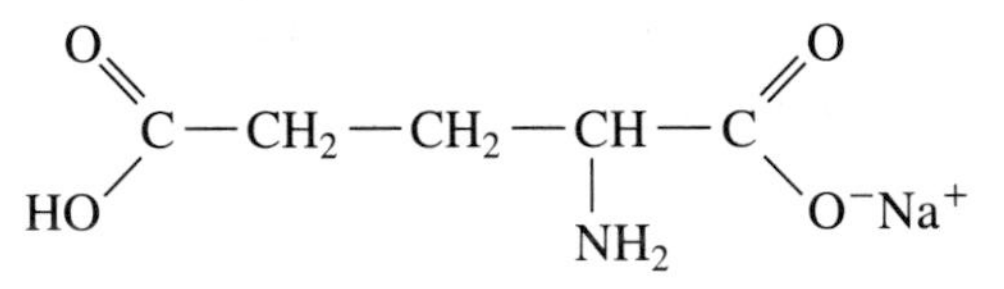

(b) vanillin (used as vanilla flavoring)

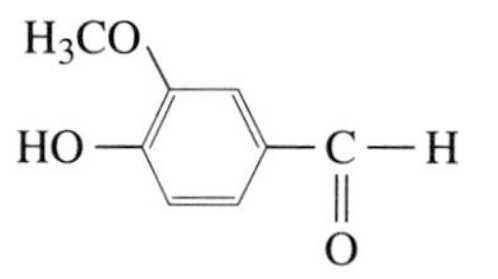

4. Identify and name the functional group in each of the following.

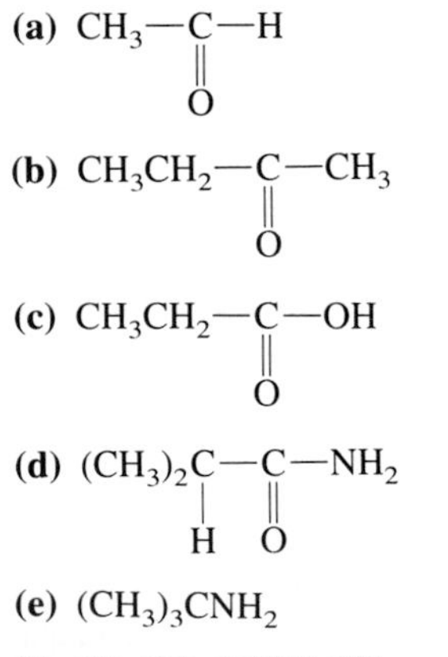

(a) $CH_3—C(=O)—H$

(b) $CH_3CH_2—C(=O)—CH_3$

(c) $CH_3CH_2—C(=O)—OH$

(d) $(CH_3)_2C(H)—C(=O)—NH_2$

(e) $(CH_3)_3CNH_2$

(f) $CH_3CH_2CH(OH)CH_2CH_3$

5. Name each of the compounds in Problem 4.

6. Name the compounds with the following molecular formulas.

(a) CH_3OH (b) CH_3CH_2OH
(c) HCO_2H (d) $CH_3CH_2CH(OH)CH_3$
(e) CH_3CH_2CHO (f) $HOCH_2CH_2OH$
(g) CH_3COCH_3 (h) $(CH_3)_2CHCHO$
(i) $ClCH_2CH(Br)CH_3$

7. Identify the functional groups in each of the following.

(a) the thyroid gland hormone thyroxine

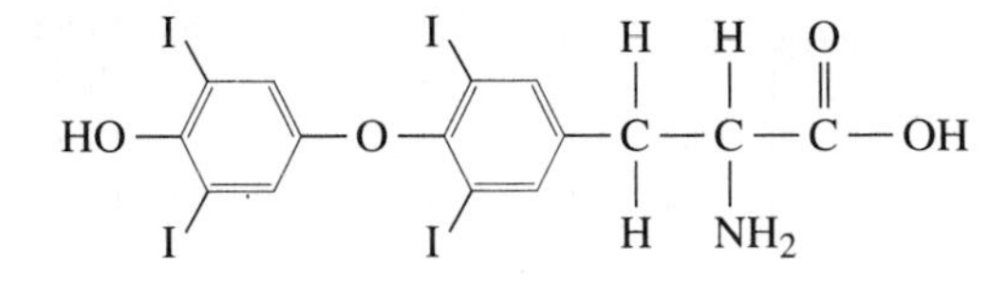

(b) ascorbic acid (vitamin C)

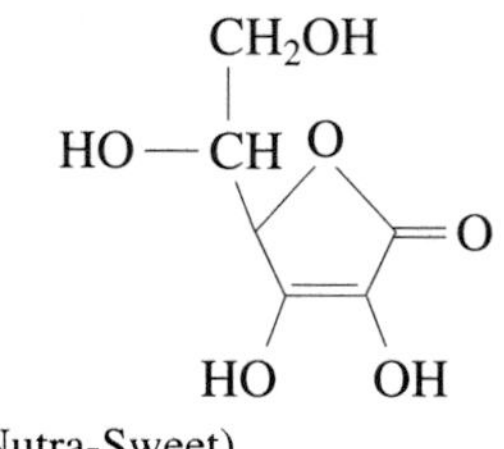

(c) aspartame (Nutra-Sweet)

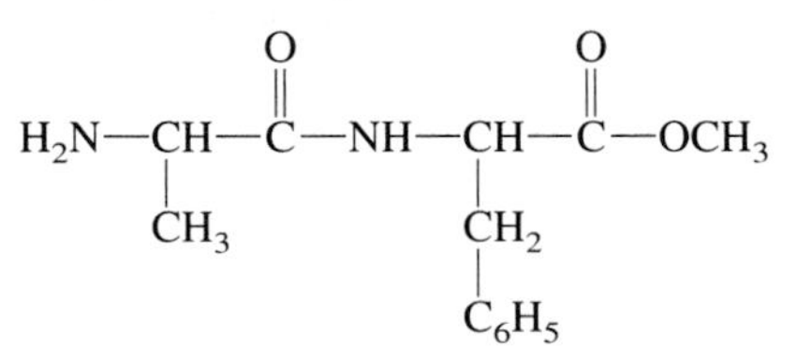

Haloalkanes (Alkyl Halides)

8. Arrange the following in order of expected increase in boiling point, and explain your choice.

(a) bromoethane, chloromethane, iodomethane

(b) tetrachloromethane, trichloromethane, dichloromethane, chloromethane, methane

9. What reaction or reactions and other reactants would you use to synthesize each of the following from ethyl bromide?

(a) ethyl iodide (b) ethyl acetate
(c) diethyl ether (d) acetic acid

Alcohols and Ethers

10. Write structural formulas for each of the following.

(a) 2-methyl-2-butanol (b) 5-methyl-2-hexanol
(c) 2-chloro-1-propanol (d) 3-chlorocyclohexanol
(e) 6-methyl-1-octanol (f) 1,3-propanediol

11. Give the systematic names of

(a)

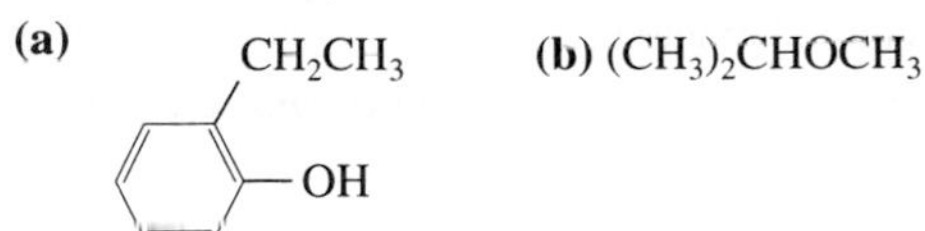

(b) $(CH_3)_2CHOCH_3$

(c) $(CH_3CH_2CH_2)_2O$

(d) $HS—CH_2—CH(NH_2)—C(=O)—OH$

(e) $(CH_3)_2C(H)—SH$

12. Write the structural formulas and give the systematic names of primary, secondary, and tertiary butyl alcohol.

13. Classify each of the following alcohols as a primary, secondary, or tertiary alcohol, and give the systematic name.

(a) $(CH_3)_2C(OH)CH_2CH_3$ (b) $CH_3CH(OH)CH_2CH_3$
(c) $(CH_3)_3CCH_2OH$ (d) $(CH_3)_2CH(OH)$
(e) $(CH_3CH_2)_2CH(OH)$

14. Write balanced equations for the reaction of ethanol with each of the following, and name the products.

(a) OH^- (*aq*) (b) sodium
(c) H_2SO_4(*conc*) (d) CH_3CO_2H

15. What alkene(s) might be formed by the dehydration of each of the following?

(a) $CH_3CH_2CH(OH)CH(CH_3)CH_3$ (b) $(CH_3CH_2)_3COH$

(c) $C_6H_5—CH(OH)CH_2CH_3$

16. Write the overall balanced equation for the production of methanol from methane and water via synthesis gas.

17. What reactions occur when phenol is dissolved in (a) $HCl(aq)$ and (b) $NaOH(aq)$?

18. Complete and balance each of the following equations, and name the organic product(s).

(a) $C_2H_5OH(g) \rightarrow$ (using a Cu catalyst at 600°C)
(b) $CH_3(CH_2)_2OH(l) + NaOH(s) \rightarrow$
(c) $CH_3OH(l) + Na(s) \rightarrow$
(d) $C_2H_5OH(l) + H_2SO_4(conc) \rightarrow$
(e) $CH_3CH{=}CH_2(g) + H_2O(g) \rightarrow$ (using a catalyst at high T and P)

19. (a) Why is the boiling point of ethanol (78°C) less than that of water (100°C)?

(b) Why are ethanol and water soluble in each other in all proportions (miscible)?

20. (a) Explain why the structural isomers dimethyl ether (bp −23°C) and ethanol (bp 78.3°C) differ significantly in their boiling points.

(b) Describe a simple chemical test that could distinguish ethanol from dimethyl ether.

21. Arrange each of the following groups of compounds in order of expected increase in boiling point, and explain your choice.

(a) 1-butanol, pentane, methoxyethane
(b) 1-butanol, 1-butanethiol, diethyldisulfide

22. Name the product of heating 2-butanol with (a) sulfuric acid; (b) hydrobromic acid; (c) oxalic acid.

23. The scent of roses is due to geraniol,

$$CH_3C(CH_3){=}CHCH_2CH_2C(CH_3){=}CHCH_2OH$$

Oxidation of the alcohol group gives the aldehyde *citral*, one of the compounds responsible for lemon scent.

(a) Write the structure of citral.
(b) Give the systematic names of geraniol and citral.

24.* Suggest ways you could prepare each of the following.

(a) 2-propanol from 1-propanol
(b) ethylene glycol from ethanol
(c) oxalic acid, $(CO_2H)_2$, from ethylene glycol

Thiols and Disulfides

25. (a) What is the most noticeable property of thiols?
(b) How are thiols related structurally to alcohols?
(c) Why is ethanethiol only slightly soluble in water, whereas ethanol is very soluble?

26. Explain why 1-propanol (bp 97°C), ethanethiol (bp 37°C), and chloroethane (bp 13°C), all of comparable molar mass, have such different boiling points.

27. The α-amino acid cysteine

$$HS—CH_2—CH(NH_2)—C(=O)—OH$$

Cysteine

forms a disulfide when oxidized. What is the structure of this disulfide?

Aldehydes and Ketones

28. Draw structural formulas for each of the following aldehydes and ketones.

(a) butanal (b) 2-pentanone (c) 3-methyl-2-butanone
(d) 3,3-dimethylhexanal (e) acetone (f) formaldehyde

29. Repeat Problem 28 for the following aldehydes and ketones.

(a) 4,4-dimethylpentanal (b) 3-heptanone
(c) cyclohexanone (d) 2-methyl-3-hexanone
(e) benzaldehyde

30. The following aldehydes and ketones may be prepared by the oxidation of a suitable alcohol. In each case name the alcohol, and draw its structural formula.

(a) ethanal **(b)** propanone **(c)** 2-methylpropanal

(d) 2-pentanone **(e)** cyclopentanone

31. Draw structures for the products of oxidation with $KMnO_4(aq)$ in acidic aqueous solution of each of the following.

(a) butanol **(b)** propanal **(c)** 2-methyl-1-butanol

(d) 3-methyl-2-butanol **(e)** 3,3-dimethylbutanal

32. How are each of the following made industrially?

(a) formaldehyde **(b)** acetaldehyde

(c) acetone **(d)** 2-butanone

33. What ketones or aldehydes might be reduced to give the alcohols **(a)** 2-propanol; **(b)** 2,2-dimethyl-1-hexanol; **(c)** 2-methyl-1-pentanol; **(d)** 2-butanol; **(e)** cyclohexanol?

34.* A compound was found by analysis and molar mass determination to have the molecular formula $C_5H_{12}O$. Oxidation converts it to a compound with the molecular formula $C_5H_{10}O$, which gives the characteristic reactions of a ketone. Suggest two or more structural formulas for the original compound.

Carboxylic Acids and Esters

35. Draw and name the four different carboxylic acids with the molecular formula $C_5H_{10}O_2$.

36. Draw structural formulas for each of the following.

(a) 3,4-dimethylpentanoic acid

(b) 2,2-dichlorobutanoic acid

(c) 3-hydroxyhexanoic acid

(d) 3,3-dimethyl-4-phenylpentanoic acid

37. Name the carboxylic acids formed by the oxidation of the following alcohols and aldehydes.

(a) 3-ethyl-1-hexanol **(b)** 4,4-dimethylpentanal

(c) 4-methylbenzaldehyde **(d)** 2,3,3-trimethylbutanol

(e) heptanal

38. Draw the structural formulas of the three different esters with the molecular formula $C_5H_{10}O_2$, and name them.

39. Write the balanced equation for the reaction by which the ester methyl methanoate (methyl formate) can be obtained from the appropriate carboxylic acid and alcohol.

40. Name the ester that could be obtained from each of the following reactions, and draw its structure.

(a) methanoic acid and methanol

(b) butanoic acid and ethanol

(c) ethanoic acid and butanol

(d) propanoic acid and 2-propanol

(e) 1 mol of phosphoric acid and 1 mol of ethanol

41. Name and draw the structures of the esters formed by the reaction of each of the following pairs of compounds.

(a) ethanoic acid and 2-propanol

(b) ethanoic acid and 1-butanol

(c) benzoic acid, $C_6H_5CO_2H$, and methanol

(d) methanoic acid and ethanol

42. Write equations to describe how ethyl ethanoate could be made from ethene as the starting material.

Amines, Amides, and Amino Acids

43. Draw the structures for **(a)** methylamine; **(b)** diethylamine; **(c)** tripropylamine; **(d)** anilinium chloride.

44. Draw structures for and name **(a)** primary, **(b)** secondary, and **(c)** tertiary amines with the molecular formula C_3H_9N.

45. Complete the following equations.

(a) $CH_3CH_2NH_2 + HBr \rightarrow$

(b) $C_6H_5NH_3{}^+Cl^- + NaOH \rightarrow$

(c) $(CH_3)_3N(aq) + H_2O(l) \rightleftharpoons$

(d) $(CH_3)_2NH + H_3O^+ \rightarrow$

(e) [pyridine ring] N + HCl →

46. Draw the structure of **(a)** PABA, para-aminobenzoic acid, used in sunscreens; **(b)** mescalin, 3,4,5-trimethoxyphenylethylamine, a powerful hallucinogen derived from the peyote cactus; and **(c)** L-dopa, 2-amino-3-(3,4-dihydroxyphenyl) propanoic acid, used medicinally for its potent activity against Parkinson's disease.

47. **(a)** Draw the structural formulas of three different amides with the molecular formula C_4H_9NO, and name them.

(b) From what carboxylic acids and amines could each of the amides in part (a) be prepared?

48. **(a)** Arrange the following in the order of increasing basicity: ammonia (pK_b 4.74), methylamine (pK_b 3.41), dimethylamine (pK_b 3.27), trimethylamine (pK_b 4.19), and aniline (pK_b 9.39).

(b) What is the pH of a 0.100-*M* solution of (i) the strongest and (ii) the weakest of these bases?

49. **(a)** What functional groups are found in all amino acids?

(b) What is the significance of the "α" in "α-amino acid"?

50. **(a)** What do the terms "chiral" and "achiral" mean?

(b) Use the terms "chiral" or "achiral" to describe a shoe, a house key, a baseball, a nail, and a screw.

51. Categorize as chiral or achiral **(a)** 2-butanol, **(b)** 2-chloro-1-propanol, **(c)** 2-bromo-2-chlorobutane, **(d)** 2-aminoethanoic acid, and **(e)** 2-aminopropanoic acid.

Determining the Structure of Organic Compounds

52. Name the isomers with the following molecular formulas, and suggest how they might be distinguished from one another by NMR spectroscopy.

(a) $C_2H_4Br_2$ **(b)** C_2H_6O **(c)** C_3H_8O

53. Suggest structures for each of the following.

(a) a compound of molecular formula C_4H_8O with three peaks in its NMR spectrum of relative intensity 3:3:2 and a strong stretching frequency in its infrared spectrum at 1720 cm^{-1}

(b) a compound of molecular formula $C_4H_7ClO_2$ with four peaks in its NMR spectrum of relative intensity 3:2:1:1 and an infrared spectrum with a broad absorption band in the range 2500 to 3000 cm^{-1} and an intense band at 1715 cm^{-1}

54. The peak of the parent ion in the mass spectrum of an aromatic hydrocarbon corresponds to a mass of 102 u, and its proton NMR spectrum has two peaks of intensity 1:5. What is the hydrocarbon's structure?

55. The mass peaks of the parent ion of a chloroalkane correspond to masses of 78 u and 80 u. There are two important isotopes of chlorine, ^{35}Cl and ^{37}Cl.

(a) What are the possible structures of the compound?

(b) How could they be distinguished by proton NMR spectroscopy?

56.* Suggest a likely structure for each of the compounds with the following molecular formulas and numbers of peaks in their proton NMR spectra, with the indicated relative intensities.

(a) C_2H_6O, a single peak

(b) C_2H_6O, three peaks of intensity 3:2:1

(c) $C_4H_{10}O$, three peaks of intensity 6:3:1

(d) $C_5H_{10}O_2$, two peaks of intensity 3:2

57.* Repeat Problem 56 for the following derivatives of benzene.

(a) C_9H_{12} with three peaks of intensity 5:1:6

(b) C_9H_{10} with three peaks of intensity 2:2:1

(c) $C_8H_{11}N$ with four peaks of intensity 5:2:3:1

General Problems

58. In terms of the AX_nE_m nomenclature of the VSEPR model, classify each of the following molecules and ions in terms of the geometry around the starred atom. Draw a diagram to show this geometry, and classify the starred atom as *sp*, sp^2, or sp^3.

(a) $^*CH_3OH$ **(b)** $CH_3{}^*OCH_3$ **(c)** H^*CO_2H

(d) $H^*CO_2^-$ **(e)** $CH_3{}^*CO_2H$ **(f)** $(CH_3)_2{}^*CO$

(g) $CH_3{}^*NH_2$ **(h)** $(CH_3)_4{}^*N^+$ **(i)** $(CH_3)_3{}^*N$

59. Classify each of the following as an acid or a base, as both, or as having neither acidic nor basic properties, and explain your choice.

(a) ethanol **(b)** ethylamine

(c) pentane **(d)** bromoethane

(e) phenol **(f)** ethylacetate

(g) methyl dihydrogenphosphate

(h) aminobenzene (aniline) **(i)** ethanamide

60. Give reaction schemes (one or more steps) for the preparation of each of the following. For each, name the starting organic reactant, and give the reaction conditions.

(a) ethyl bromide from an alkane

(b) ethyl bromide from an alkene

(c) ethanal from an alcohol

(d) diethyl ether from an alcohol

(d) propanoic acid from an alcohol

61. Repeat Problem 60 for the following.

(a) ethyl propanoate from an alcohol

(b) 3-methyl-1-butene from an alcohol

(c) methyl formate from an alkyl halide

(d) ethylmethyl disulfide from two thiols

(e) acetone from an alkyl halide

62. In terms of suitable balanced equations, describe the manufacture of each of the following.

(a) synthesis gas

(b) methanol from synthesis gas

(c) methanal from methanol

(d) methylamine from methanol

63. Complete and balance each of the following equations.

(a) $CH_3NH_2(g) + H_2O(l) \rightarrow$

(b) $CH_3NH_2(aq) + HCl(aq) \rightarrow$

(c) $CH_3CH_2CO_2H(l) + (CH_3)_2CHOH(l) \xrightarrow{\text{heat}}$

(d) $CH_3OH(l) + NH_3(g) \xrightarrow{\text{heat}}$

(e) $CH_3CO_2H(l) + C_2H_5NH_2(l) \xrightarrow{\text{heat}}$

(f) $CH_3CH_2CHO(l) + Cr_2O_7^{2-}(aq) + H_3O^+(aq) \xrightarrow{\text{heat}}$

64. Each of the compounds with the following structural formulas may be synthesized from an alkene, or an alkyne, and another reactant. In each case give the name and structure of the alkene or alkyne, and name the other reactant.

(a) $CH_3CH(OH)CH_3$ **(b)** $CH_3CBr_2CBr_2CH_3$

(c) $CH_3CH{=}CHCH_3$ **(d)** $CH_3C(Br){=}CH_2$

65. How would you prepare each of the following?

(a) 1,2-dichloroethane from ethene

(b) propanoic acid from 1-propanol

(c) ethyl ethanoate from ethanal

66.* A mass of 1.00 g of a primary alcohol was completely oxidized to a carboxylic acid, which required 83.3 mL of 0.200-*M* NaOH(*aq*) to reach the equivalence point in a titration.

(a) Calculate the molar mass of the alcohol, deduce its molecular formula, write its structural formula, and give its systematic name.

(b) Name and give the structure of another alcohol that is isomeric with the alcohol in (a), and discuss the possible products of its oxidation with acidic $K_2Cr_2O_7(aq)$.

67.* Ethylene was oxidized directly in air by means of a silver catalyst to give a compound containing 40.0% C, 6.69% H, and 53.3% O by mass, with a boiling point of 14°C. A 0.9656-g sample of the compound filled a 1.00-L flask at 100°C and a pressure of 0.985 atm.

(a) Determine the empirical formula and the molecular formula of the compound.

(b) Draw possible Lewis structures.

(c) Name the compound.

68.* A mass of 256 mg of a compound containing carbon, hydrogen, and oxygen was burned in excess oxygen to give 512 mg CO_2 and 209 mg H_2O. At 100°C, 156 mg of the compound occupied a volume of 93.3 mL at a pressure of 882 torr.

(a) Suggest a structure for the compound.

(b) Identify its functional group.

(c) Suggest tests that would confirm the presence of this functional group.

69.* An organic compound A contains 52.1% C, 13.1% H, and 34.8% O by mass. At 100°C and 1.00 atm, 0.230 g of A occupies a volume of 153 mL.

(a) Find the empirical and molecular formulas of A, and draw the possible structural formulas.

(b) If the compound reacts with sodium to give hydrogen and a sodium salt, write the balanced equation for the reaction, and name A and its salt.

(c) What volume of hydrogen at 25°C and a pressure of 730 torr would result from the reaction of 0.250 g of A with excess sodium?

70.* Describe chemical reactions by which you could distinguish **(a)** ethyl bromide from cyclohexane; **(b)** dipropyl ether from 1-propanol; **(c)** butanoic acid from 2-butanone; **(d)** 2-butanone from butanal; **(e)** propanoic acid from propylamine.

71.* **(a)** Of three isomers with the molecular formula C_4H_6, two have strong infrared frequencies close to 2200 cm^{-1}, and the third has a strong infrared frequency at 1630 cm^{-1}. Suggest their likely structures, and name them.

(b) Given a sample known to be one of the isomers in (a), could it be unambiguously identified from a combination of data from infrared and proton NMR spectroscopic data?

The Rates and Mechanisms of Reactions

15.1 Reaction Rate
15.2 Effect of Reactant Concentration on Reaction Rate
15.3 Activation Energy and the Effect of Temperature on Reaction Rate
15.4 Catalysis
15.5 Chain Reactions

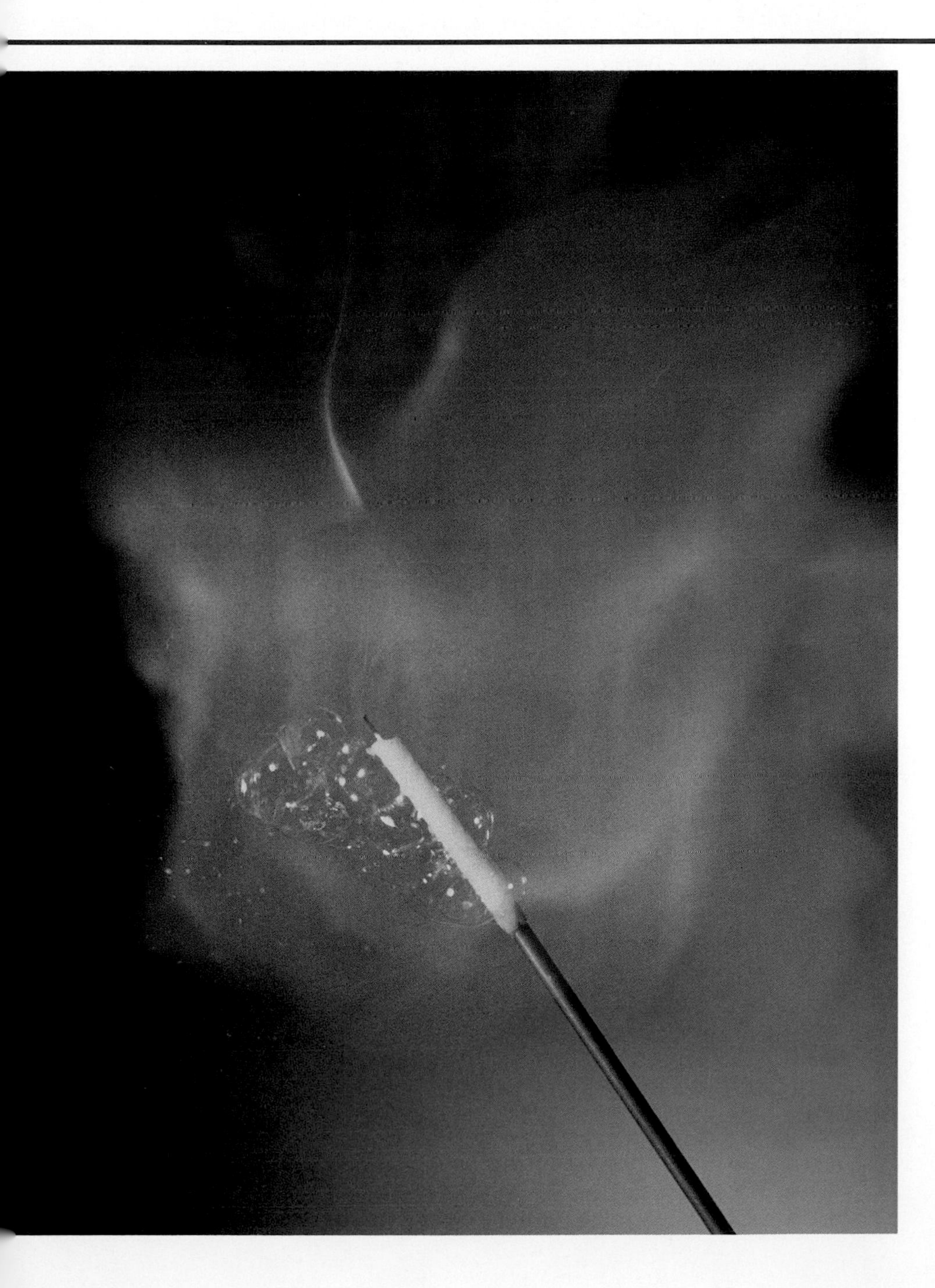

Some reactions, such as rusting and the radioactive decay of uranium-238 to thorium-234, are very slow, whereas other reactions, such as explosions and the combustion of rocket fuel, are very fast. When a hydrogen-filled soap bubble is touched with a flame, the hydrogen in the bubble combines with the oxygen in the air in a rapid reaction. In this reaction, a large amount of energy is liberated as heat, light, and sound.

The rates of reactions vary widely, from those that occur in a fraction of a second to those that reach equilibrium in days or weeks. The explosive reaction between oxygen and a hydrocarbon in the cylinders of an automobile, the explosive decomposition of TNT (trinitrotoluene), and the reaction between a strong acid and a strong base, which occurs as fast as acid is added to base, are all very fast reactions. The souring of milk and the fermentation of sugar to ethanol are much slower and take days at ordinary temperatures. The rusting of iron is still slower.

The study of the rates of chemical reactions is called **chemical kinetics** or **reaction kinetics**. An understanding of reaction rates is important in the chemical industry and in biology, medicine, geology, and environmental science. For example, we need to know the rates and mechanisms of the many reactions occurring in the atmosphere if we are to solve problems caused by atmospheric pollution (Chapter 16). The rates of most of the reactions that go on in living organisms are closely controlled by catalysts called enzymes. We will consider such reactions in Chapter 17.

Sometimes we would like to slow down reactions such as the rusting of iron and the deterioration of foods. More often we would like to speed up reactions so that we can obtain a useful amount of a desired product in a reasonable time. To be able to control the rates of reactions, we need to understand the factors that influence these rates. In Chapter 4 we briefly considered the effects of the concentrations of the reactants, the temperature, and catalysts on reaction rate, and thereafter we discussed many examples qualitatively. In this chapter we consider the effects of these factors on reaction rate quantitatively and in more detail.

15.1 REACTION RATE

The rate (or speed) of a chemical reaction is defined in the same way that we define the speed of an automobile. The speed is equal to the change in the automobile's position (distance traveled) divided by the time taken for the change (time between initial and final positions):

$$\text{speed of automobile} = \frac{\text{change in position}}{\text{time interval of change}}$$

The rate of a chemical reaction is defined as the change in the concentration of a reactant (or product) in a given time interval.

Consider the reaction between an alkyl halide (haloalkane), such as ethyl chloride (chloroethane), C_2H_5Cl, and the hydroxide ion to give an alcohol and a chloride ion (Chapter 14):

$$R{-}Cl + OH^- \rightarrow ROH + Cl^-$$

If the concentration of ROH at time t_1 is $[ROH]_1$ and it has increased to $[ROH]_2$ by time t_2, then the concentration of ROH has changed by

$$\Delta[ROH] = [ROH]_2 - [ROH]_1$$

in the time interval $\Delta t = t_2 - t_1$. The rate of the reaction is then

$$\frac{\text{change in concentration of ROH}}{\text{time interval of change}} = \frac{\Delta[ROH]}{\Delta t}$$

Recall that square brackets denote concentrations. Thus, [A] means the concentration of A.

The concentrations of the products *increase* during a reaction, so the change in the concentration of a product, Δ[product], is positive. The concentrations of reactants *decrease*, so Δ[reactant] is negative. Because rate is defined as a positive quantity, we can write the following for the hydrolysis of an alkyl halide:

$$\text{rate of reaction} = \frac{-\Delta[RCl]}{\Delta t} \quad or \quad \frac{-\Delta[OH^-]}{\Delta t} \quad or \quad \frac{\Delta[ROH]}{\Delta t} \quad or \quad \frac{\Delta[Cl^-]}{\Delta t}$$

The changes in the concentrations of the products, $\Delta[ROH]$ and $\Delta[Cl^-]$, are equal because the coefficients in the balanced chemical equation for ROH and Cl^- are the same.

In general, for a balanced equation $aA + bB + \cdots \rightarrow pP + qQ + \cdots$,

$$-\frac{1}{a}\frac{\Delta[A]}{\Delta t} = -\frac{1}{b}\frac{\Delta[B]}{\Delta t} = \cdots = \frac{1}{p}\frac{\Delta[P]}{\Delta t} = \frac{1}{q}\frac{\Delta[Q]}{\Delta t} = \cdots$$

Speed has units of distance divided by time and may be expressed, for example, in miles per hour ($mi \cdot h^{-1}$), kilometers per hour ($km \cdot h^{-1}$), or meters per second ($m \cdot s^{-1}$). Reaction rate has units of concentration divided by time. We express concentrations in moles per liter ($mol \cdot L^{-1}$), but time may be measured in any convenient unit—seconds, minutes, hours, or even years. Therefore, the units of reaction rate may be $mol \cdot L^{-1} \cdot s^{-1}$, $mol \cdot L^{-1} \cdot min^{-1}$, $mol \cdot L^{-1} \cdot h^{-1}$, and so on.

We can measure the rate of a reaction by following the change in concentration of any of the reactants or products. For the reaction of an alkyl halide with hydroxide ion, for example, it would be convenient to measure the change in concentration of the OH^- ion by titrating a sample of the solution with an acid or by measuring the change in the pH of the solution.

As a reaction proceeds and the concentrations of the reactants decrease, the rate decreases correspondingly. Table 15.1 and Figure 15.1 give some data for the

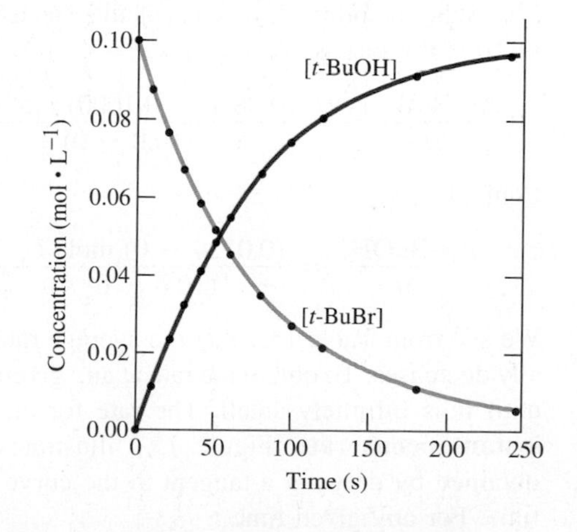

FIGURE 15.1 Concentration Changes during the Reaction of *t*-Butyl Bromide with Hydroxide Ion The reaction was carried out with excess water at 55°C in 20% aqueous ethanol.

TABLE 15.1 Average Rates of Hydrolysis of *t*-Butyl Bromide*

Time (s)	[*t*-BuBr] ($mol \cdot L^{-1}$)	[*t*-BuOH] ($mol \cdot L^{-1}$)	Rate $\frac{-\Delta[t\text{-BuBr}]}{-\Delta t}$ ($mol \cdot L^{-1} \cdot s^{-1}$)
0	0.1000	0	
			1.24×10^{-3}
10.0	0.0876	0.0124	
			1.08×10^{-3}
20.0	0.0768	0.0232	
			0.96×10^{-3}
30.0	0.0672	0.0328	
			0.82×10^{-3}
40.0	0.0590	0.0410	
			0.73×10^{-3}
50.0	0.0517	0.0483	
			0.64×10^{-3}
60.0	0.0453	0.0547	
			0.53×10^{-3}
80.0	0.0348	0.0652	
			0.41×10^{-3}
100.0	0.0267	0.0733	
			0.31×10^{-3}
120.0	0.0205	0.0795	
			0.19×10^{-3}
180.0	0.0093	0.0907	
			0.09×10^{-3}
240.0	0.0042	0.0958	

* The reaction of *t*-butyl bromide was carried out with excess water at 55°C in 20% aqueous ethanol.

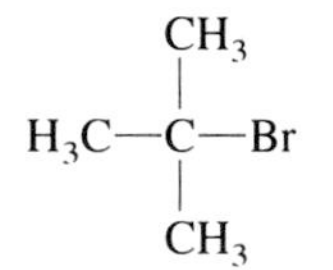

t-Butyl bromide
(2-Bromo-2-methylpropane)

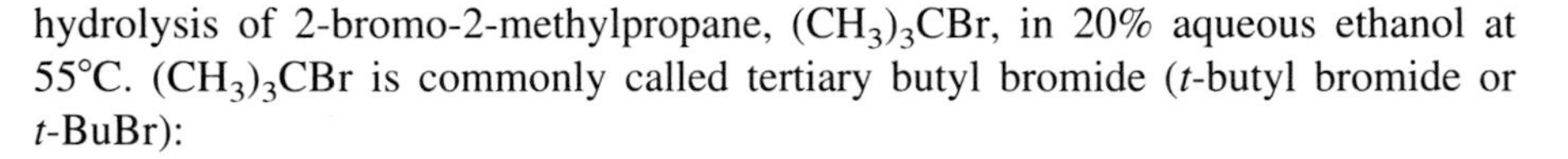

hydrolysis of 2-bromo-2-methylpropane, $(CH_3)_3CBr$, in 20% aqueous ethanol at 55°C. $(CH_3)_3CBr$ is commonly called tertiary butyl bromide (*t*-butyl bromide or *t*-BuBr):

$$t\text{-BuBr} + H_2O \rightarrow t\text{-BuOH} + HBr$$

For each time interval, we can calculate an **average rate**. For the initial time interval of 10 s, the rate is

$$-\frac{\Delta[t\text{-BuBr}]}{\Delta t} = -\frac{(0.0876 - 0.1000)\ mol \cdot L^{-1}}{(10.0 - 0)\ s} = 1.24 \times 10^{-3}\ mol \cdot L^{-1} \cdot s^{-1}$$

Equivalently,

$$\frac{\Delta[t\text{-BuOH}]}{\Delta t} = \frac{(0.0124 - 0)\ mol \cdot L^{-1}}{(10.0 - 0)\ s} = 1.24 \times 10^{-3}\ mol \cdot L^{-1} \cdot s^{-1}$$

We see from Table 15.1 that the average rate for successive time intervals continually decreases. To obtain the rate at any given time, we must reduce the time interval until it is infinitely small. The rate for an infinitely small interval is called the **instantaneous rate**. Figure 15.2 illustrates how the instantaneous rate may be obtained by drawing a tangent to the curve of the concentration of *t*-BuBr versus time. For any given time *t*,

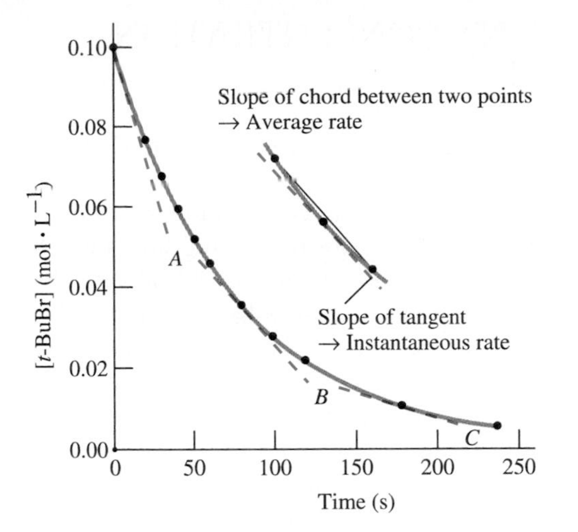

FIGURE 15.2 Rate of Hydrolysis of *t*-Butyl Bromide The slope of the tangent to the curve at any time t gives the instantaneous rate at this time. Tangent A gives the rate at $t = 0$, the initial rate. Tangent B gives the rate at 80 s, and tangent C gives the rate at 180 s. The inset shows an enlargement of a small part of the curve. The slope of a chord between two points gives the average rate over that interval. As the time interval is decreased to zero, the chord becomes a tangent to the curve, and the slope of the tangent gives the instantaneous rate at this point.

$$\text{rate at time } t = \text{limiting value of } -\frac{\Delta[t\text{-BuBr}]}{\Delta t} \quad \text{as } \Delta t \rightarrow 0$$

In the language of calculus,

$$\text{rate at time } t = \text{slope of tangent to curve at time } t = -\frac{d[t\text{-BuBr}]}{dt}$$

Rates obtained by this method from Figure 15.2 are given in Table 15.2.

Exercise 15.1 For the reaction $N_2(g) + 3H_2(g) \rightarrow 2NH_3(g)$, if the rate of formation of $NH_3(g)$ in a given time interval was $0.0010 \text{ mol} \cdot L^{-1} \cdot \text{min}^{-1}$, what was the rate of disappearance of $N_2(g)$ and $H_2(g)$, respectively, in the same time interval?

TABLE 15.2 Instantaneous Rates of Hydrolysis of *t*-Butyl Bromide

Time (s)	Rate ($\text{mol} \cdot L^{-1} \cdot s^{-1}$) = slope of tangent
0	1.32×10^{-3}*†
20	1.01×10^{-3}
40	0.78×10^{-3}
60	0.60×10^{-3}
80	0.46×10^{-3}*
100	0.35×10^{-3}
120	0.27×10^{-3}
180	0.12×10^{-3}*
240	0.06×10^{-3}

Rate of hydrolysis of *t*-butyl bromide = slope of tangent to the plot of [*t*-BuBr] versus time (see Table 15.1 and Figure 15.2).

* Tangents shown in Figure 15.2

† Initial rate

15.2 EFFECT OF REACTANT CONCENTRATION ON REACTION RATE

We expect the rate of a reaction to increase with increased concentration of each of the reactants, because the greater the concentrations of the reactants, the greater the rate of collisions between their molecules. Because the concentrations of *all* the reactants decrease during a reaction, a useful procedure for finding how the rate depends on the concentration of any *one* reactant is to measure the **initial rate**, that is, the rate before any significant change in the concentrations has occurred. As we have just seen, we can find the initial rate from the slope of the tangent to the plot of reactant or product concentration against time at $t = 0$ (Figure 15.2). We then change the initial concentration of just one reactant and measure the initial rate again. The same procedure is followed for each reactant.

The data in Table 15.3 were obtained by this initial rate method for the hydrolysis of 1-bromobutane with hydroxide ion in aqueous ethanol at 60°C:

$$CH_3CH_2CH_2CH_2Br + OH^- \rightarrow CH_3CH_2CH_2CH_2OH + Br^-$$

We see that if we double the concentration of 1-bromobutane (Experiment B), the rate doubles. If we halve the concentration of OH^- (Experiment C), the rate decreases by half, so the rate is also proportional to the OH^- concentration. Doubling the concentration of both reactants increases the rate four times (Experiment D), confirming that the rate is proportional to the concentrations of both OH^- and 1-bromobutane. So for this reaction we can write

$$\text{rate} = k[CH_3CH_2CH_2CH_2Br][OH^-]$$

where k is a constant called the **rate constant**. The value of k at a given temperature is a characteristic of a given reaction. It is independent of the concentrations of the reactants but depends on the temperature. For the reaction between 1-bromobutane and OH^-,

$$k = 1.0 \times 10^{-3}\ L \cdot mol^{-1} \cdot s^{-1} \quad \text{at } 60°C$$

In general, if rate = $k[A][B]$,

$$k = \frac{\text{rate}}{[A][B]}$$

so k has units of

$$\frac{mol \cdot L^{-1} \cdot s^{-1}}{(mol \cdot L^{-1})^2} = L \cdot mol^{-1} \cdot s^{-1}$$

The n in n-BuBr stands for *normal* and designates a straight chain with no branching.

We will find it convenient in the rest of this chapter to use the alkyl halide nomenclature for haloalkanes. Hence we will use the alternative name n-butylbromide and the abbreviation n-BuBr for 1-bromobutane. We rewrite its rate law as

$$\text{rate} = k[n\text{-BuBr}][OH^-]$$

TABLE 15.3 Rate of Reaction of 1-Bromobutane with Hydroxide Ion

Experiment	Initial Concentrations $(mol \cdot L^{-1})$ $[CH_3CH_2CH_2CH_2Br]$	$[OH^-]$	Initial Rate $(mol \cdot L^{-1} \cdot s^{-1})$
A	0.10	0.10	1.0×10^{-5}
B	0.20	0.10	2.0×10^{-5}
C	0.20	0.05	1.0×10^{-5}
D	0.20	0.20	4.0×10^{-5}

The reaction was carried out at 60°C in aqueous ethanol.

Rate Laws and Reaction Mechanisms

An expression such as

$$\text{rate} = k[n\text{-BuBr}][\text{OH}^-]$$

that relates the rate of a reaction to the concentrations of the reactants, is called a **rate law**.

> **A rate law is an equation expressing the rate of a reaction in terms of the concentrations of the substances taking part in the reaction.**

In general, the rate law for any reaction is

$$\text{rate} = k[\text{A}]^x[\text{B}]^y[\text{C}]^z \cdots$$

where A, B, C, . . . are reactants and x is the **order** with respect to A, y the order with respect to B, z the order with respect to C, and so on. The sum $x + y + z + \cdots$ is the overall order of the reaction. For the reaction between n-BuBr and hydroxide ion, the rate is directly proportional to *two* concentrations ($x = y = 1$ and $x + y = 2$); such a rate law is called a **second-order rate law**. The rate is said to be **first order** with respect to n-BuBr ($x = 1$), first order with respect to OH^- ($y = 1$), and second order overall ($x + y = 2$). In general,

> **The overall order of a reaction is the sum of the powers to which the individual concentrations are raised in the rate law.**

The rate law

$$\text{rate} = k[n\text{-BuBr}][\text{OH}^-]$$

is what we expect if the reaction between n-BuBr and OH^- occurs by collisions between n-BuBr molecules and hydroxide ions, as is suggested by the equation for the reaction. Increasing the concentration of either reactant will correspondingly increase the number of collisions that reactant molecules make in a given time with the molecules of the other reactant and will correspondingly increase the rate of the reaction.

However, the observed rate law does not always follow from the equation for the reaction. We have seen that the reaction between OH^- and n-BuBr follows the second-order rate law

$$\text{rate} = k[\text{OH}^-][n\text{-BuBr}]$$

But the reaction between t-BuBr and OH^-,

$$t\text{-BuBr} + \text{OH}^- \rightarrow t\text{-BuOH} + \text{Br}^-$$

follows the first-order rate law

$$\text{rate} = k[t\text{-BuBr}]$$

The rate is independent of the OH^- concentration, and so this rate law does not correspond to the equation for the reaction. This result appears rather mysterious at first sight, but it gives us some important information. It tells us that the reaction cannot be occurring by collisions between t-BuBr molecules and OH^- ions, because in that case the rate law would be

$$\text{rate} = k[t\text{-BuBr}][\text{OH}^-]$$

Instead the reaction takes place in two steps. The first step is the ionization of a $(CH_3)_3CBr$ molecule to give a positive ion called a *carbocation*, in this case the trimethylcarbocation $(CH_3)_3C^+$:

$$(CH_3)_3C\text{—}Br \rightarrow (CH_3)_3C^+ + Br^-$$

The trimethylcarbocation is a very reactive ion that is attacked by a hydroxide ion as fast as it forms and, in the second step, is converted to the corresponding alcohol:

$$(CH_3)_3C^+ + OH^- \rightarrow (CH_3)_3COH$$

The (two) steps by which this reaction takes place constitute what is called the **mechanism** of the reaction. The rate of an overall reaction cannot be greater than the rate of the slowest step, so the rate and the rate law for the overall reaction are determined by the slowest step. In this case, the first step of the reaction is much slower than the second step, so the first step is called the **rate-determining step** or the **rate-limiting step**. The first step involves only the ionization of the *t*-butyl bromide molecule, so we expect its rate law to be

$$\text{rate} = k[t\text{-BuBr}]$$

Because this rate law is also the observed rate law, we can say that this two-step mechanism is consistent with the observed rate law.

Many reactions take place in two or more steps, each of which is called an *elementary reaction*. Consider the equation for the oxidation of iodide ion by Fe^{3+} in aqueous solution (Chapter 10):

$$2Fe^{3+}(aq) + 2I^-(aq) \rightarrow 2Fe^{2+}(aq) + I_2(s) \qquad (1)$$

It is highly improbable that this reaction occurs in a single step in which all four reactants collide at the same time. The mechanism of this reaction is thought to involve the following three steps:

$$\begin{aligned} Fe^{3+} + I^- &\rightarrow FeI^{2+} \\ FeI^{2+} + I^- &\rightarrow Fe^{2+} + I_2^- \\ Fe^{3+} + I_2^- &\rightarrow Fe^{2+} + I_2 \end{aligned} \qquad (2)$$

The sum of the equations for these three steps in the reaction gives the overall equation (1). In general,

A reaction mechanism is a series of steps or elementary reactions that is proposed to account for the rate law of the overall reaction.

As the preceding examples have shown, each of the steps or elementary reactions that make up the overall reaction usually involves only one or two molecules. If only one molecule or ion is involved, as in the first step of the reaction of *t*-butyl bromide with hydroxide ion, this step of the reaction is called a **unimolecular reaction**. If two molecules or ions are involved in a single step of a reaction, as in the hydrolysis of *n*-butylbromide, it is called a **bimolecular reaction**. Occasionally a single reaction step might involve three molecules or ions, in which case it is called a **termolecular reaction**.

The *molecularity* of a single step in a reaction (an elementary reaction) is the number of reactant molecules taking part in it.

Most reactions take place in a series of two or more steps, in each of which, one, two, or occasionally three molecules or ions take part. *Any reaction in which there are four or more molecules or ions in the balanced equation*, such as the oxidation of I^- by Fe^{3+}, *must take place in at least two steps, because the simultaneous collision of four or more molecules is highly improbable.* We can often propose a plausible mechanism for a reaction on the basis of the observed rate law for the reaction, and there are other ways to obtain evidence about reaction mechanisms. But we can never be absolutely certain that the proposed mechanism is the correct one. New evidence may disprove a particular mechanism and lead to the proposal of a new mechanism.

Many reactions in living systems, such as the oxidation of carbohydrates to CO_2 and water, occur in a lengthy series of steps (Chapter 17). The elucidation of the mechanism of such reactions presents an exciting challenge to chemists and biochemists.

Reaction Intermediates

In most multi-step reactions, some molecules or ions are formed and used up again during the reaction. Hence they do not appear in the overall equation for the reaction or in the rate law. Examples are $(CH_3)_3C^+$ in the hydrolysis of *t*-BuBr and I_2^- in the oxidation of iodide ion by Fe^{3+}. These molecules or ions, called **reaction intermediates**, are so reactive that they are present only in very small concentrations. Because they are so reactive, it is often impossible, or at least very difficult, to isolate them. Thus, they are not familiar to us as stable compounds. Nevertheless, they play an important role in reactions. We will meet other examples of reactive reaction intermediates in Chapter 16, including free atoms such as $\cdot Cl$ and $\cdot O$ and odd-electron molecules such as $\cdot CH_3$.

We denote the single unpaired electron by a single dot, as in $Cl\cdot$ and $\cdot CH_3$.

Exercise 15.2 What are the rate laws and the units of the rate constants for the following reactions?

(a) the isomerization of cyclopropane to propene at 500°C,

$$C_3H_6(g) \rightarrow CH_3CH{=}CH_2(g)$$

which is first order in cyclopropane

(b) the reaction of hydrogen with nitrogen monoxide at 800°C,

$$2H_2(g) + 2NO(g) \rightarrow 2H_2O(g) + N_2(g)$$

which is first order in H_2 and second order in NO

Exercise 15.3 A knowledge of the rate of the reaction between oxygen, $O_2(g)$, in air and nitrogen monoxide, $NO(g)$, from automobile exhausts,

$$2NO(g) + O_2(g) \rightarrow 2NO_2(g)$$

is important in air pollution studies. Several experiments gave the following data:

Experiment	Initial Concentration ($mol \cdot L^{-1}$) [NO]	$[O_2]$	Initial Rate ($mol \cdot L^{-1} \cdot s^{-1}$)
1	0.001	0.001	7.0×10^{-6}
2	0.001	0.002	1.4×10^{-5}
3	0.001	0.003	2.1×10^{-5}
4	0.002	0.003	8.4×10^{-5}

What are **(a)** the rate law for the reaction and **(b)** the value of the rate constant?

Exercise 15.4 If the first step of the three-step mechanism for the oxidation of I^- by Fe^{3+} in aqueous solution, equation (2), is the rate-determining step, what is the rate law for the reaction?

Integrated Form of the First-Order Rate Law

We have seen that the rate law for a reaction can be ascertained from the method of initial rates, in which reaction orders are determined by varying the initial concentration of one reactant at a time. This method uses data obtained only at the very beginning of a reaction, but it is preferable and more reliable to use data obtained throughout the reaction. We can do this by using equations called **integrated rate laws**. For a first-order rate law,

$$\text{rate} = k_1[\text{A}] \qquad or \qquad -\frac{d[\text{A}]}{dt} = k_1[\text{A}]$$

where [A] is the concentration of a reactant A, and k_1 is the rate constant. Using calculus (Appendix M), we can integrate this equation to give

$$\ln[\text{A}]_t = -k_1 t + \ln[\text{A}]_0 \tag{3}$$

where $[\text{A}]_t$ is the concentration of A at time t and $[\text{A}]_0$ is the initial concentration of A. This equation is called the **integrated form of the first-order rate law**. It has the form of an equation for a straight line (Appendix M),

$$y = mx + b$$

Thus, for a first-order reaction, a plot of $\ln[\text{A}]_t$ versus t is a straight line with a slope $-k_1$. Figure 15.3 shows a plot of ln[t-BuBr] versus t for the hydrolysis of t-BuBr,

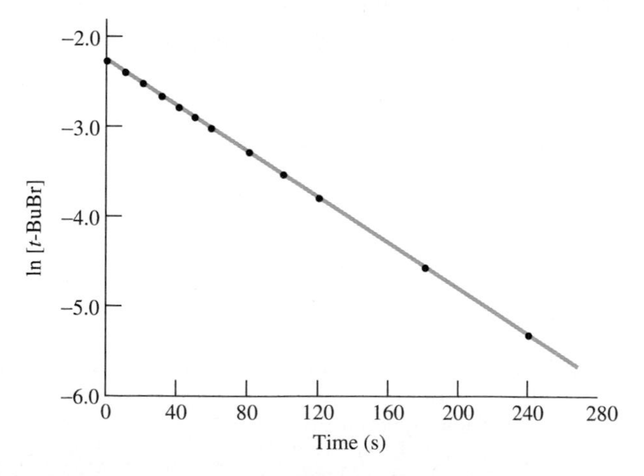

FIGURE 15.3 Determination of the Order of the Reaction of *t*-Butyl Bromide with Water The reaction was carried out with excess water at 55°C. The plot of ln[t-BuBr] versus time is a straight line. Hence the reaction is first order with respect to t-BuBr and has the rate law rate = k_1[t-BuBr]. The slope of the line gives the value of the rate constant $k_1 = 1.32 \times 10^{-2}\ \text{s}^{-1}$ at 55°C.

Time (s)	0	20	40	60	100	180
[t-BuBr] (mol · L⁻¹)	0.1000	0.0768	0.0590	0.0453	0.0267	0.0093
ln[t-BuBr]	−2.30	−2.57	−2.83	−3.09	−3.62	−4.68

using the data in Table 15.1. That the plot is a straight line confirms that this reaction follows the first-order rate law

$$\text{rate} = k_1[t\text{-BuBr}]$$

The slope of this line, $-k_1$, is $-1.32 \times 10^{-2}\ s^{-1}$, which gives $k_1 = 1.32 \times 10^{-2}\ s^{-1}$ for the value of the rate constant at 55°C.

Half-Life of a First-Order Reaction

The half-life of a first-order reaction is a very convenient and useful property.

The *half-life* $t_{\frac{1}{2}}$ of a reaction is the time required for the concentration of a reactant to decrease to half its initial value.

We can use the integrated form of the first-order rate law, equation (3), to derive an expression for the half-life of a first-order reaction. We can rearrange this equation as

$$\ln[A]_0 - \ln[A]_t = \ln\frac{[A]_0}{[A]_t} = k_1 t$$

At $t = t_{\frac{1}{2}}$, $[A] = \frac{1}{2}[A]_0$, so we have

$$\ln\frac{[A]_0}{\frac{1}{2}[A]_0} = \ln 2 = k_1 t_{\frac{1}{2}}$$

$$t_{\frac{1}{2}} = \frac{\ln 2}{k_1} = \frac{0.693}{k_1}$$

For a first-order reaction, $t_{\frac{1}{2}}$ is independent of the concentration of A. Whatever its initial concentration, the concentration of A will drop to half its initial value in the same time inverval. If $t_{\frac{1}{2}} = 10$ min, [A] will decrease to half its initial value in 10 minutes; then it will decrease again by a factor of 2 in the next 10 minutes. After 20 minutes [A] will be one-quarter its initial value, and so on (Figure 15.4).

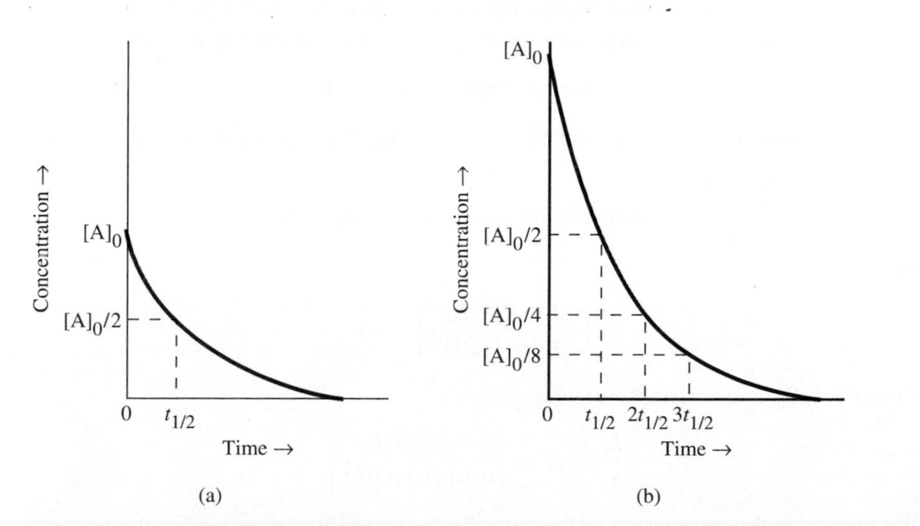

FIGURE 15.4 Half-Life of a First-Order Reaction The half-life is independent of the initial concentration. The initial concentration of the reactant $[A]_0$ is less in (a) than it is in (b), but the time $t_{\frac{1}{2}}$ needed for the initial concentration to decrease to half its initial value is the same. The time needed for the initial concentration to decrease to one-quarter its initial value is $2t_{\frac{1}{2}}$ and to one-eighth its initial value is $3t_{\frac{1}{2}}$.

EXAMPLE 15.1 Half-Life of a First-Order Reaction

The hydrolysis of sucrose to glucose and fructose is catalyzed by the enzyme sucrase. It is first order in the concentration of sucrose. If the half-life is 80 min at 20°C, what will be the concentration of a solution of sucrose after 160 min and 320 min in a solution whose initial concentration was 0.0400 *M*?

Solution: For this first-order reaction, $t_{\frac{1}{2}}$ is 80 min, and the concentration is halved every 80 min. Because 160 min is two half-lives, after 160 min the initial concentration is reduced to $\frac{1}{2} \times \frac{1}{2} = \frac{1}{4}$ of the original value, and the concentration of sucrose is $\frac{1}{4}(0.0400) = 0.0100\ M$. Similarly, after 320 min, or four half-lives, the initial concentration is reduced to $(\frac{1}{2})^4 = \frac{1}{16}$ of its original value, and the concentration of sucrose is $\frac{1}{16}(0.0400) = 0.0025\ M$.

After n half-lives, the amount of the original concentration $[A]_0$ that remains is $(\frac{1}{2})^n\ [A]_0$.

Exercise 15.5 The rate constant for the first-order isomerization of cyclopropane to propene at 500°C is $6.7 \times 10^{-4}\ s^{-1}$. Find the time needed for the concentration of cyclopropane to drop to **(a)** half and **(b)** one-quarter its initial value.

Equilibrium Constant and Reaction Mechanism

When a reaction reaches equilibrium, the rate of the forward reaction equals the rate of the reverse reaction. We have seen that the reaction of *n*-BuBr with hydroxide ion to give *n*-BuOH (1-butanol) is a single-step bimolecular reaction with a second-order rate law:

$$n\text{-BuBr} + OH^- \rightleftharpoons n\text{-BuOH} + Br^-$$

So we can write

$$\text{rate of forward reaction} = k_f[n\text{-BuBr}][OH^-]$$

The reaction is reversible, but the position of the equilibrium lies well to the right in a basic solution. However, as we saw in Chapter 14, if the concentration of OH^- is decreased to a very low value by making the solution acidic, then *n*-BuOH can be converted to *n*-BuBr. In other words, the position of the equilibrium then lies to the left. Because this is a one-step reaction, the rate law for the reverse reaction must be

$$\text{rate of reverse reaction} = k_r[n\text{-BuOH}][Br^-]$$

At equilibrium the rate of the forward reaction must equal the rate of the reverse reaction:

$$k_f[n\text{-BuBr}][OH^-] = k_r[n\text{-BuOH}][Br^-]$$

so

$$\frac{[n\text{-BuOH}][Br^-]}{[n\text{-BuBr}][OH^-]} = \frac{k_f}{k_r}$$

Because k_f/k_r is a constant, we may write

$$\frac{k_f}{k_r} = K = \frac{[n\text{-BuOH}][Br^-]}{[n\text{-BuBr}][OH^-]}$$

We recognize this expression as the equilibrium constant expression for the reaction (Chapter 12). Thus, the equilibrium constant expression for a single-step reaction is a consequence of the rate laws for the forward and reverse reactions and of the equality of the rates of the forward and reverse reactions at equilibrium.

Even if a reaction does not take place in the single step suggested by the balanced equation, it is always correct to write the equilibrium constant in terms of this equation, as we did in Chapter 12. For example, we have seen that the reaction of *t*-butylbromide with hydroxide ion,

$$t\text{-BuBr} + OH^- \rightleftharpoons t\text{-BuOH} + Br^-$$

takes place in the following two steps:

$$t\text{-BuBr} \rightleftharpoons t\text{-Bu}^+ + Br^- \qquad (4)$$

$$t\text{-Bu}^+ + OH^- \rightleftharpoons t\text{-BuOH} \qquad (5)$$

When the overall reaction reaches equilibrium, both of these reactions must also be at equilibrium. We can write the following equilibrium constants for these two steps:

$$K_4 = \frac{[t\text{-Bu}^+][Br^-]}{[t\text{-BuBr}]} \quad \textit{and} \quad K_5 = \frac{[t\text{-BuOH}]}{[t\text{-Bu}^+][OH^-]}$$

Multiplying these constants gives

$$K_4K_5 = K = \frac{[t\text{-BuOH}][Br^-]}{[t\text{-BuBr}][OH^-]}$$

which is the equilibrium constant for the overall reaction.

We see that deriving the equilibrium constant from the individual steps in the reaction gives the same equilibrium constant as is obtained from the balanced equation for the overall reaction.

The equilibrium constant expression for any reaction is independent of the mechanism of the reaction.

We can always write the correct expression for the equilibrium constant for a reaction from the balanced equation, whether or not we know the detailed mechanism of the reaction.

15.3 ACTIVATION ENERGY AND THE EFFECT OF TEMPERATURE ON REACTION RATE

The rates of almost all chemical reactions increase with increasing temperature. We saw many examples of this fact in earlier chapters, and we see examples all around us every day. When we cook, we increase the rates of the desired reactions in food by raising the temperature. We store milk and other foods in the refrigerator or freezer to decrease the rate of reactions that cause them to deteriorate. The effect of temperature on reaction rates is quite marked. For many reactions, the rate approximately doubles for every 10-K rise in temperature.

A bimolecular reaction can occur only when the reacting molecules collide. Thus, the rate of a biomolecular reaction depends on the rate at which collisions occur and on what fraction of these collisions are effective in leading to reaction. A quantitative calculation based on the kinetic theory of gases shows that in a gas at 25°C and 1 atm pressure, each molecule collides with another within about 10^{-10} s. If each collision led to a reaction, the half-life of any gas-phase reaction at approximately 1 atm would be about 10^{-9} to 10^{-10} s. We know that a great many reactions are much slower than this, so it is clear that not every collision leads to reaction. Some of the colliding molecules just bounce off each other unchanged.

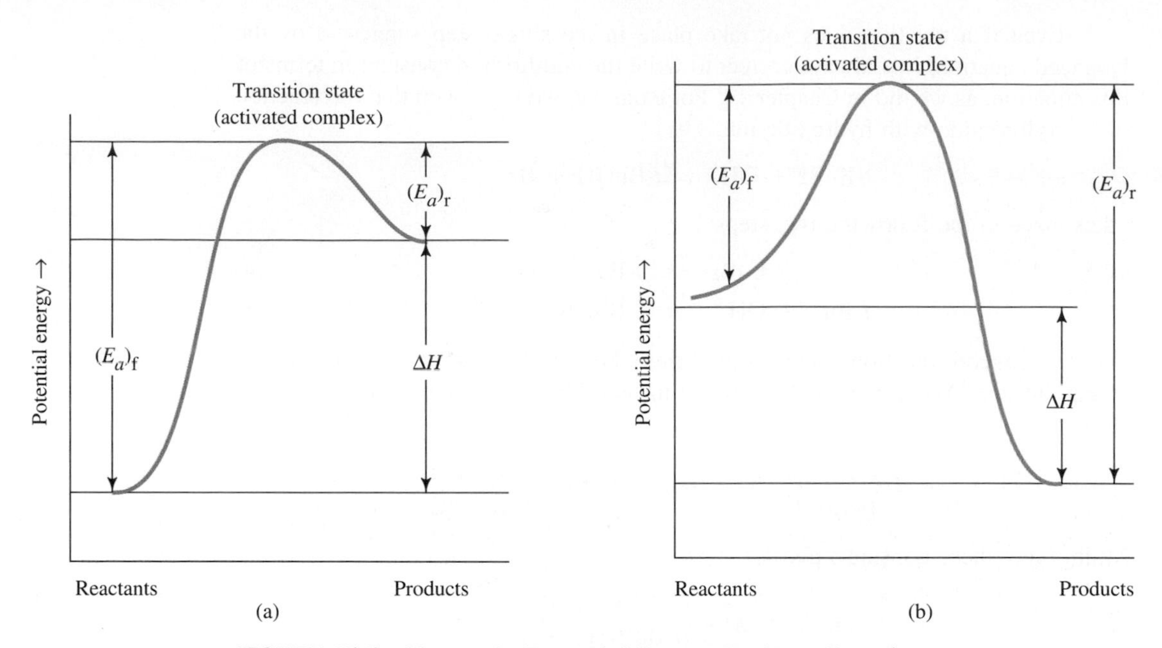

FIGURE 15.5 Change in Potential Energy During a Reaction (a) An endothermic reaction; (b) an exothermic reaction. In both, the reaction enthalpy ΔH is equal to the difference between the activation energies of the forward and reverse reactions. For an endothermic reaction, $(E_a)_f > (E_a)_r$. For an exothermic reaction, $(E_a)_f < (E_a)_r$.

Activation Energy

The most important reason that a reaction does not occur on every collision is that the molecules must collide with sufficient force. That is, the energy of the collision must be great enough that bonds in the colliding molecules are broken or weakened enough to allow new bonds to form.

> **The minimum energy of collision that is needed for reaction to occur is called the *activation energy*, E_a.**

The activation energy can be thought of as an energy barrier that the reacting molecules must surmount in order to react (Figure 15.5). The state of the reacting system at which its energy is a maximum is called the **transition state**, or **activated complex**, of the reaction. The activation energy E_a is the difference in energy between the reactants and the transition state. Figure 15.6 shows the progress of the hydrolysis of methyl bromide with hydroxide ion. In the transition state, the hydroxide ion has begun to form a bond with the carbon atom, and the carbon–bromine bond has lengthened and has begun to break. At the same time, the C—H bonds have been pushed into a planar geometry with 120° bond angles instead of the

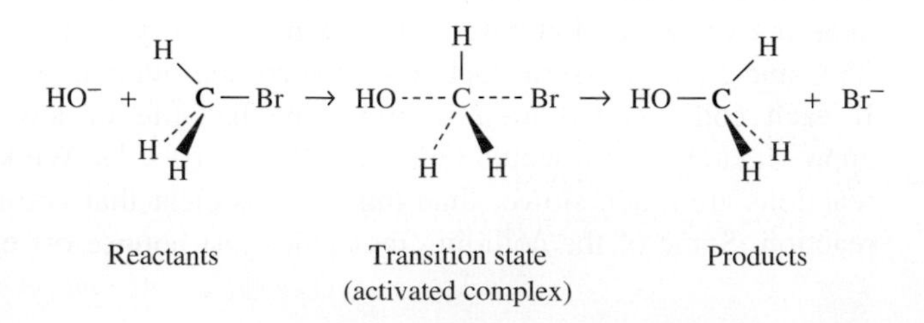

FIGURE 15.6 The Reaction of Hydroxide Ion with Methyl Bromide The hydroxide ion begins to form a bond with the carbon atom while the C—Br bond becomes longer and weaker and the three C—H bonds move into the same plane. At some point in this process, the energy of the system reaches a maximum, and the system is said to be in its transition state.

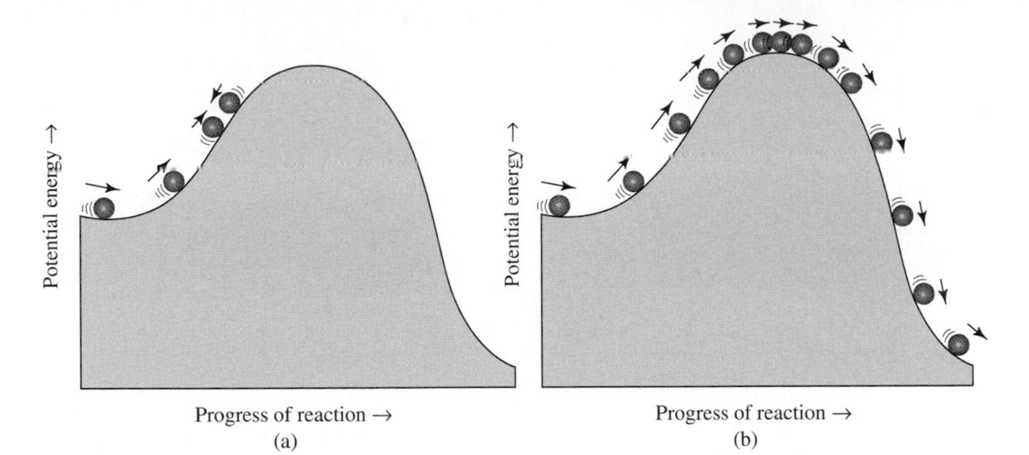

FIGURE 15.7 Analogy for the Activation Energy of a Chemical Reaction Energy is required to move a ball to the top of a barrier, just as energy is required for chemical reactants to form a transition state, or activated complex. (a) Here the ball has insufficient energy to climb the potential energy barrier and rolls back down the same side of the barrier. No reaction has occurred. (b) Here the ball has sufficient energy to reach the top of the potential energy barrier and may then roll down the other side. A reaction has occurred.

usual 109° angles. This transition state has a higher energy than either the reactants or the products.

As an analogy, imagine a small ball that moves on the surface shown in Figure 15.7. The movement of the ball from left to right represents the progress of the reaction, and the barrier corresponds to the activation energy. If the ball initially has a low speed, all its kinetic energy is converted to potential energy before it reaches the top of the barrier and rolls back again. This situation corresponds to a collision in which the kinetic energy of the reacting molecules is too small for reaction to occur; the molecules simply bounce apart unchanged. The ball must have sufficient initial speed and kinetic energy to enable it to reach the top of the barrier, which corresponds to the formation of the transition state. It may then roll down the other side, and the potential energy gained upon climbing to the top of the energy barrier is reconverted into kinetic energy. In a chemical reaction, the energy released as the products are formed is converted to the kinetic energy of the product molecules as they move apart.

Although energy is always needed to reach the transition state, the overall reaction may be exothermic or endothermic. The activation energy is lower when the reaction proceeds in the exothermic direction and higher when the reaction proceeds in the endothermic direction (Figure 15.5). The reaction enthalpy ΔH is equal to the difference in the activation energies for the forward and reverse reactions:

$$\Delta H = (E_a)_f - (E_a)_r$$

The Collision Model

To understand the effect of temperature on reaction rate, we must consider how the number of collisions with an energy equal to or higher than the activation energy increases with temperature. We will use as an example the bimolecular reaction between nitrogen monoxide and ozone,

$$NO + O_3 \rightarrow NO_2 + O_2$$

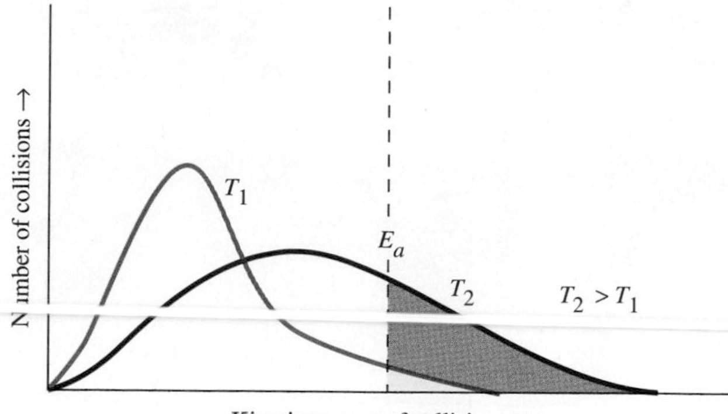

FIGURE 15.8 Collision Energy Distributions Distribution of collision energies at temperatures T_1 and T_2. The shaded areas show that the fraction f of molecules with kinetic energies of collision higher than the activation energy, E_a, increases very rapidly with an increase in temperature.

which is one of the reactions responsible for the destruction of ozone in the stratosphere (Chapter 16). We saw in Chapter 3 that the kinetic energies of the molecules of a gas have the type of distribution shown in Figure 2.15. Figure 15.8 shows that the collision energies have a similar distribution. With increasing temperature the average kinetic energy increases, the distribution of energies broadens, and more molecules have higher energies. In particular, with increasing temperature there is a very large increase in the number of molecules with a minimum energy E_a. Hence, there is a very large increase in reaction rate with increasing temperature.

In the 1860s the Scottish physicist James Clerk Maxwell and the Austrian physicist Ludwig Boltzmann (1844–1906) worked out the mathematics of the kinetic theory of gases and calculated the curves for the distribution of collision energies, such as those in Figure 15.8. These curves are often called the *Maxwell–Boltzmann distribution curves*. They showed that the fraction f of the total number of collisions in which the energy exceeds a certain minimum energy E_a is given by the exponential function (Appendix M)

$$\ln f = \frac{-E_a}{RT}, \qquad f = e^{-E_a/RT} \tag{6}$$

where R is the gas constant ($8.314\ \mathrm{J \cdot K^{-1} \cdot mol^{-1}}$) and T is the temperature in kelvins. The fraction f increases exponentially with temperature.

Let us see now how we can derive an expression for the variation of the rate constant k with temperature for a bimolecular reaction, such as that between NO and O_3. The number of collisions per second at a given temperature—the rate of collisions—is proportional to the concentrations of NO and O_3,

$$\text{rate of collisions} = Z[\mathrm{NO}][\mathrm{O_3}]$$

where Z is a constant for any bimolecular reaction at a given temperature and pressure—the collision rate for concentrations of the reactants of $1\ \mathrm{mol \cdot L^{-1}}$. But only a small fraction f of the collisions leads to reaction, so

$$\text{rate of reaction} = f \times (\text{rate of collisions}) = fZ[\mathrm{NO}][\mathrm{O_3}]$$

Because this is a one-step bimolecular reaction, it is governed by the second-order rate law

$$\text{rate of reaction} = k[\mathrm{NO}][\mathrm{O_3}]$$

where k is the rate constant. Thus,

$$k = fZ$$

We saw in equation (6) that the fraction f of collisions with a minimum energy E_a varies with temperature according to the expression

$$f = e^{-E_a/RT}$$

Substituting, we have

$$k = Ze^{-E_a/RT} \qquad or \qquad \ln k = \ln Z - \frac{E_a}{RT} \tag{7}$$

These equations show how the rate constant k varies with temperature. Z is the rate of collisions for concentrations of 1 $mol \cdot L^{-1}$, and E_a is the activation energy.

The Arrhenius Equation

An equation of the same form as equation (7), known as the **Arrhenius equation**, was first proposed by the Swedish chemist Svante Arrhenius (1859–1927) in 1887 as a result of his experimental studies of the variation of temperature with the rates of reactions in solution. He wrote the equation as

$$k = Ae^{-E_a/RT} \qquad or \qquad \ln k = \ln A - \frac{E_a}{R}\left(\frac{1}{T}\right) \tag{8}$$

where A is a constant called the *Arrhenius constant*, obtained from the experimental results. According to this equation, a plot of $\ln k$ versus $1/T$ should be a straight line with a slope of $-E_a/R$, from which we can obtain a value of the activation energy, E_a. Table 15.4 shows the results of the experimental determination of the rate constant k at different temperatures for the reaction between ethyl bromide and hydroxide ions in water:

$$C_2H_5Br + OH^- \rightarrow C_2H_5OH + Br^-$$

Figure 15.9 shows that the plot of $\ln k$ versus $1/T$ is indeed a straight line, with a slope of -1.07×10^4 K. Hence, the activation energy for this reaction is

$$E_a = -R(\text{slope}) = -(8.31\ J \cdot mol^{-1} \cdot K^{-1})(-1.07 \times 10^4\ K) = 89\ kJ \cdot mol^{-1}$$

Thus, in collisions between C_2H_5Br and OH^- ions, these reactants must have at least this much energy for reaction to occur. As the temperature increases, more of the molecules have this energy, and so the rate of the reaction increases accordingly. Reactions with high activation energies proceed at reasonable speeds only at high temperatures, and their rate increases rapidly with increasing temperature. In contrast, reactions with a low activation energy are fast at low temperatures, and their rate increases more slowly with increasing temperature.

TABLE 15.4 Rate Constants for the Reaction of Ethyl Bromide with Hydroxide Ion

t (°C)	k ($L \cdot mol^{-1} \cdot s^{-1}$)	T (K)	$1/T$ (K^{-1})	$\ln k$
25	8.8×10^{-5}	298	3.36×10^{-3}	−9.34
30	1.6×10^{-4}	303	3.30×10^{-3}	−8.74
35	2.8×10^{-4}	308	3.25×10^{-3}	−8.18
40	5.0×10^{-4}	313	3.19×10^{-3}	−7.60
45	8.5×10^{-3}	318	3.14×10^{-3}	−7.07
50	1.4×10^{-3}	323	3.10×10^{-3}	−6.57

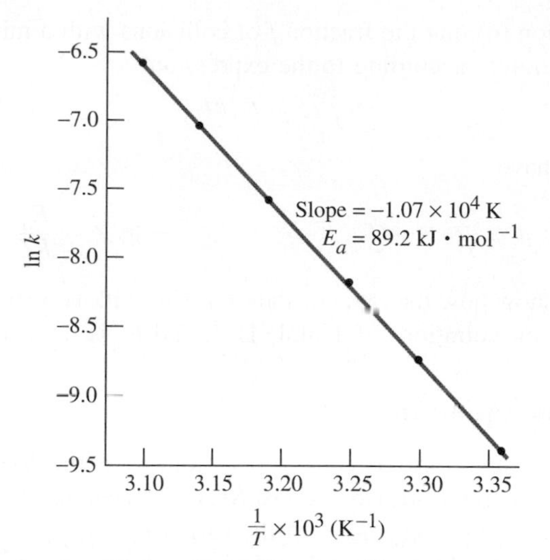

FIGURE 15.9 Arrhenius Plot Arrhenius plot of ln k, for rate constants k at different temperatures versus $1/T$, for the reaction between ethyl bromide and hydroxide ions in water (data from Table 15.4). The slope of the line is -1.07×10^4 K and equals $-E_a/R$, giving a value of 89.2 kJ · mol^{-1} for the activation energy, E_a.

If we know the activation energy of a reaction and the rate constant k_1 at some temperature T_1, we can use the Arrhenius equation to calculate the rate constant k_2 at some other temperature T_2. At the two temperatures,

$$\ln k_1 = \ln A - \frac{E_a}{RT_1}$$

$$\ln k_2 = \ln A - \frac{E_a}{RT_2}$$

Subtracting the first equation from the second gives

$$\ln k_2 - \ln k_1 = \left(-\frac{E_a}{RT_2}\right) - \left(-\frac{E_a}{RT_1}\right)$$

which can be rearranged to

$$\ln\frac{k_2}{k_1} = \frac{E_a}{R}\left(\frac{1}{T_1} - \frac{1}{T_2}\right) \quad or \quad \ln\frac{k_2}{k_1} = \frac{E_a}{R}\left(\frac{T_2 - T_1}{T_1T_2}\right) \qquad (9)$$

It is a property of logarithms that

$$\ln x - \ln y = \ln\frac{x}{y}$$

(Appendix M).

EXAMPLE 15.2 Calculating the Rate Constant at One Temperature from Its Value at Another Temperature

The hydrogenation of ethene

$$C_2H_4(g) + H_2(g) \rightarrow C_2H_6(g)$$

has an activation energy of 181 kJ · mol^{-1}. The rate constant at 700 K is 1.30×10^{-3} L · mol^{-1} · s^{-1}. What is the rate constant at 750 K?

Solution: The second temperature is higher than the first, so we expect the reaction to be faster and therefore the rate constant to be higher. Because E_a = 181 kJ · mol^{-1}, equation (9) becomes

$$\ln\frac{k_2}{k_1} = \frac{(181 \text{ kJ} \cdot \text{mol}^{-1})(10^3 \text{ J/1 kJ})}{8.31 \text{ J} \cdot \text{K}^{-1} \cdot \text{mol}^{-1}}\left(\frac{1}{700 \text{ K}} - \frac{1}{750 \text{ K}}\right) = 2.07$$

R and E_a must be expressed in the same energy units. With R = 8.314 J · mol^{-1} · K^{-1}, we must use the factor (10^3 J)/(1 kJ) to convert the units of E_a to J · mol^{-1}.

$$\ln \frac{k_2}{k_1} = 2.07, \qquad \text{so } \frac{k_2}{k_1} = e^{2.07} = 7.96$$

Because $k_1 = 1.30 \times 10^{-3}\ \text{L} \cdot \text{mol}^{-1} \cdot \text{s}^{-1}$, $k_2 = 1.03 \times 10^{-2}\ \text{L} \cdot \text{mol}^{-1} \cdot \text{s}^{-1}$. Thus, k_2 is indeed larger than k_1. A temperature increase of 50 K increases the reaction rate by a factor of about 8.

Exercise 15.6 For the hydrolysis of methyl bromide by OH^- ion, the activation energy is 83.7 $\text{kJ} \cdot \text{mol}^{-1}$ and the rate constant is 3.44×10^{-2} $\text{L} \cdot \text{mol}^{-1} \cdot \text{s}^{-1}$ at 55°C. What is the rate constant at 25°C?

Exercise 15.7 For the hydrolysis of ethyl chloride by OH^- ion, the activation energy is 96 $\text{kJ} \cdot \text{mol}^{-1}$. By what factor will the rate increase from 25°C to 35°C?

We can also use equation (9) to calculate the activation energy from the rate constants at two temperatures, as shown in Example 15.3. This method is in effect the same as determining the slope of a line such as that in Figure 15.9 but is generally not as accurate.

EXAMPLE 15.3

As we saw in Chapter 14, ethers can be made by the substitution of the halogen in an alkyl halide by an alkoxide ion. For example, the reaction between methyl iodide and sodium ethoxide gives methylethyl ether:

$$\underset{\text{Methyliodide}}{CH_3I} + \underset{\text{Sodium ethoxide}}{C_2H_5ONa} \rightarrow \underset{\text{Methylethyl ether}}{CH_3OC_2H_5} + NaI$$

For this reaction in solution in ethanol at 12°C, $k_1 = 2.45 \times 10^{-4}\ \text{L} \cdot \text{mol}^{-1} \cdot \text{s}^{-1}$, and at 30°C, $k_2 = 2.08 \times 10^{-3}\ \text{L} \cdot \text{mol}^{-1} \cdot \text{s}^{-1}$. What is the activation energy of this reaction?

Solution: Converting the temperatures to the Kelvin scale, we have 12°C = 285 K, and 30°C = 303 K. Substituting into equation (9) then gives

$$\ln \frac{k_2}{k_1} = \frac{E_a}{R}\left(\frac{T_2 - T_1}{T_1 T_2}\right)$$

$$\ln \left(\frac{2.08 \times 10^{-3}}{2.45 \times 10^{-4}}\right) = \frac{E_a}{R}\left(\frac{303\ K - 285\ K}{(285\ K)(303\ K)}\right)$$

$$= \ln 8.490 = 2.139 = \frac{E_a}{R}\,\frac{18.0\ \text{K}}{(285\ \text{K})(303\ \text{K})}$$

$$E_a = 2.139(8.31\ \text{J} \cdot \text{K}^{-1} \cdot \text{mol}^{-1})\,\frac{(285\ \text{K})(303\ \text{K})}{18.0\ \text{K}}$$

$$= (8.53 \times 10^4\ \text{J} \cdot \text{mol}^{-1})\left(\frac{1\ \text{kJ}}{10^3\ \text{J}}\right) = 85.3\ \text{kJ} \cdot \text{mol}^{-1}$$

Exercise 15.8 Find the activation energy for the reaction between ethyl bromide and hydroxide ions, using the rates at 25°C and 35°C given in Table 15.4. Compare this value with that obtained from Figure 15.9.

Our simple collision model of a bimolecular reaction led to equation (7),

$$\ln k = \ln Z - \frac{E_a}{RT}$$

Comparing this with the Arrhenius equation, equation (8), leads us to conclude that the Arrhenius constant, A, should equal the constant Z. However, this is rarely found to be the case, and A is often considerably smaller than Z. There are several reasons for the difference between A and Z. The details are complex, but some of the reasons are as follows:

1. Even when molecules collide with sufficient energy, they may not react because they do not collide with the appropriate orientation to enable them to exchange atoms.
2. In solution, the presence of the solvent molecules may strongly affect the rate of collision between the reactant molecules. In the gas phase, intermolecular attractions may affect the collision rate.
3. In our simple treatment of the collision theory, we have assumed that the collision rate, Z, is independent of the temperature. In fact, Z increases with increasing temperature, although this contribution to the increased reaction rate is small compared with the increase in rate due to the increase in the fraction of molecules with sufficient energy to react.

Unimolecular Reactions

If a bimolecular reaction is a consequence of collisions between the reacting molecules, how does a unimolecular reaction, such as the isomerization of cyclopropane to propene,

$$\underset{\text{Cyclopropane}}{\overset{\quad CH_2}{CH_2-CH_2}} \rightarrow \underset{\text{Propene}}{CH_2=CH-CH_3}$$

take place? The activation energy necessary for this and other unimolecular reactions is not provided by the collision between two molecules. In many cases it comes from the absorption of radiation, as in a photochemical reaction. In other cases it comes from the redistribution of energy, such as vibrational energy, that the molecule already possesses as a result of previous collisons. The vibrational energy of a polyatomic molecule such as cyclopropane is distributed among many stretching and bending vibrations. During the course of the complex vibrational motions of the molecule, much of this energy may become concentrated in just one vibration, such as the stretching of a particular bond. If sufficient energy becomes concentrated in one of the C—C bonds of the cyclopropane molecule in this way, the bond may break, and the molecule rearranges to propene. This is then a unimolecular reaction with a first-order rate law.

15.4 CATALYSIS

Although raising the temperature increases the rate of a reaction, this is not a practical approach in all cases. For example, raising the temperature of an exothermic reaction decreases the equilibrium yield of the product (Chapter 12). In other cases high temperatures cause the products to decompose or undergo other undesirable reactions. In such situations, catalysts are often essential to speed up reaction rates. **Catalysis** is the use of catalysts to increase reaction rates. The rates of virtually all reactions in living organisms are controlled by catalysts called enzymes

(Chapter 17). In previous chapters we saw many other examples of the use of catalysts to increase reaction rates. We are now in a position to look at how a catalyst works.

> **A *catalyst* is a substance that increases the rate of a reaction but is not used up in the reaction; its final concentration is equal to its initial concentration.**

A catalyst does not affect the position of the equilibrium of a reaction. It increases the rate of both the forward and reverse reactions by providing an alternative mechanism for the reaction with a lower activation energy. A catalyst clearly must take part in a reaction, or it could not affect the reaction rate. But it is regenerated during the reaction and so does not appear in the overall equation for the reaction. It may be regarded as both a reactant and a product. In Demonstration 1.1 we saw that the decomposition of hydrogen peroxide is catalyzed by solid manganese dioxide and also by an enzyme in blood called catalase. Both the manganese dioxide and the catalase remain unchanged after the reaction. Demonstration 15.1 shows that water catalyzes the reaction between sulfur dioxide and hydrogen sulfide to give sulfur:

Catalase is needed in blood because hydrogen peroxide, a dangerous byproduct of some reactions in the body, must be rapidly destroyed.

$$SO_2(g) + 2H_2S(g) \rightarrow 3S(s) + 2H_2O(l)$$

Water is a product of this reaction as well as a catalyst. This is an example of an **autocatalytic reaction**—a reaction that, once it gets going, catalyzes itself.

There are two different types of catalysts: homogeneous and heterogeneous catalysts. A **homogeneous catalyst** is in the same phase as the reactants; usually both are in solution or in the gas phase. A **heterogeneous catalyst** is in a different phase from the reactants. Typically the catalyst is a solid, and the reactants are in

DEMONSTRATION 15.1 Catalysis

The upper jar contains SO_2 gas. The lower jar contains H_2S gas. The two gases are separated by a glass plate. The plate is removed, but no reaction is observed even after 10 min.

A few drops of water are then added, and the pair of jars is inverted a few times to mix the two gases quickly.

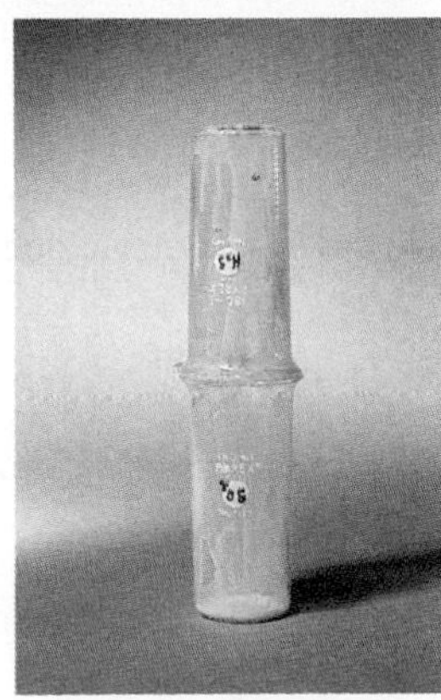

Water catalyzes the reaction between H_2S and SO_2, so the sides of the jar are rapidly coated with yellow sulfur. In this reaction water is a product of the reaction as well as a catalyst. ●

solution or in the gas phase. In Demonstration 1.1 solid manganese dioxide behaves as a heterogeneous catalyst, whereas the enzyme catalase is in solution and behaves as a homogeneous catalyst.

Homogeneous Catalysis

The rate of decomposition of an aqueous solution of hydrogen peroxide can be increased by iodide ion as well as by manganese dioxide or the enzyme catalase. Because iodide ion is soluble, it behaves as a homogeneous catalyst. The catalyzed reaction is believed to occur by a two-step mechanism. First, hydrogen peroxide is reduced to water by iodide ion, which is oxidized to hypoiodite ion, IO^-:

$$H_2O_2(aq) + I^-(aq) \rightarrow IO^-(aq) + H_2O(l) \qquad \text{(Slow)} \qquad (10)$$

Then hypoiodite ion oxidizes hydrogen peroxide to oxygen and is reduced back to iodide ion:

$$H_2O_2(aq) + IO^-(aq) \rightarrow O_2(g) + H_2O(l) + I^-(aq) \qquad \text{(Fast)}$$

The equation for the overall reaction is the sum of the preceding two equations:

$$2H_2O_2(aq) \rightarrow 2H_2O(l) + O_2(g)$$

The iodide ion takes part in the first step but is regenerated in the second. It is not used up in the reaction, and so it does not appear in the overall equation. In other words, it behaves as a catalyst. The catalyzed reaction has an activation energy of 57 $kJ \cdot mol^{-1}$ and is therefore much faster than the uncatalyzed reaction, which has an activation energy of 76 $kJ \cdot mol^{-1}$ (Figure 15.10).

Heterogeneous Catalysis

We have seen several examples of heterogeneous catalysis, such as the oxidation of SO_2 to SO_3 in the contact process for making H_2SO_4 (Chapter 7) and the Haber process for making ammonia (Chapter 2), in which a solid is used to catalyze the reaction between two gases. In these and many similar cases, the catalyst is a solid, and the reactants are gases. The reaction occurs on the surface of the solid catalyst, so this type of catalyst is often referred to as a **surface catalyst**.

The most common types of heterogeneous catalysts for gas-phase reactions are metals, metal oxides, and zeolites (aluminosilicates). Metals such as Fe, Ni, Pd, and Pt are particularly useful for hydrogenation and dehydrogenation reactions of hydrocarbons. Oxides such as ZnO, MnO_2, V_2O_5, and Al_2O_3 are used for oxidation–reduction and dehydration reactions. Zeolites (Chapter 19) are used for isomerization, polymerization, and other reactions of hydrocarbons (Chapter 20).

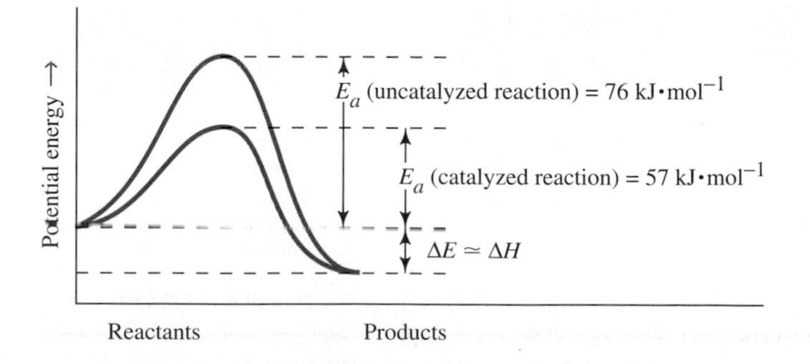

FIGURE 15.10 Activation Energy of the Iodide-Ion-Catalyzed Decomposition of Hydrogen Peroxide A catalyst provides an alternative mechanism for a reaction that has a lower activation energy than does the uncatalyzed reaction.

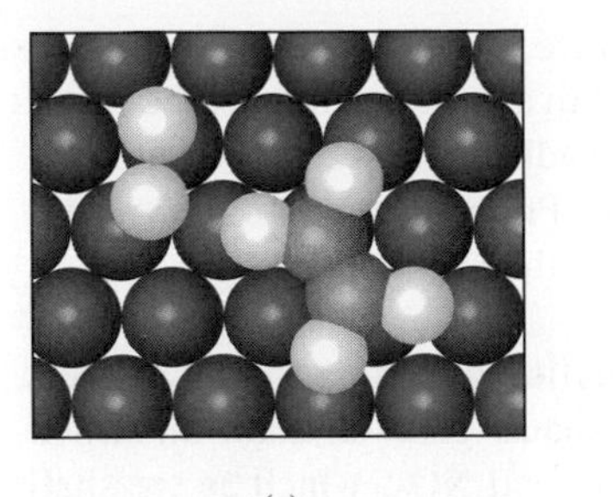

(a)

(b)

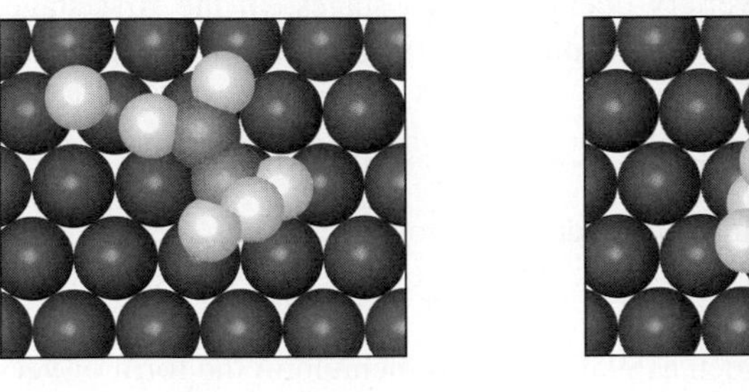

(c)

(d)

FIGURE 15.11 Heterogeneous (Surface) Catalysis The reaction between ethene, C_2H_4, and hydrogen, H_2, on a metal surface. (a) Hydrogen and ethene molecules are adsorbed onto a metal surface. (b) As they are adsorbed, the hydrogen molecules are dissociated into hydrogen atoms. (c) A hydrogen atom moves across the surface, collides with an ethene molecule, and forms a C_2H_5 molecule, which remains attached to the surface. (d) Another hydrogen atom moves across the surface, collides with the C_2H_5 molecule, and combines with it to form an ethane molecule, C_2H_6, which then leaves the surface.

Although heterogeneous catalysis is of enormous practical importance, the detailed mechanism by which many heterogeneous catalysts work is not completely understood. We do know that the reactant molecules become attached to the surface of the catalyst, which brings them closer together and weakens some of their bonds so that they react more readily with other molecules. The activation energy for the reaction on the surface of a solid is therefore lower than for the same reaction in the gas phase.

An important example of heterogeneous catalysis is the use of the surface of platinum or nickel as a catalyst for the addition of hydrogen to a carbon–carbon double bond, as in the hydrogenation of ethene to ethane (Figure 15.11). In the absence of a catalyst, this reaction is too slow to observe. But at a pressure of several hundred atmospheres and in the presence of powdered platinum or nickel, it occurs rapidly at room temperature:

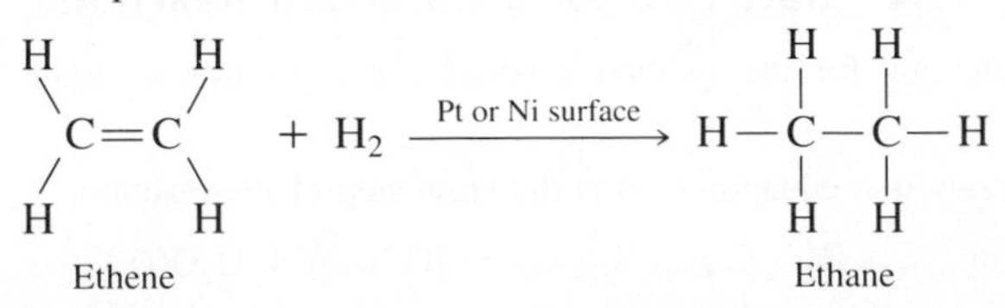

The C_2H_4 and H_2 molecules become attached to the metal surface; they are said to be adsorbed. In this process the H_2 molecules are dissociated into hydrogen atoms,

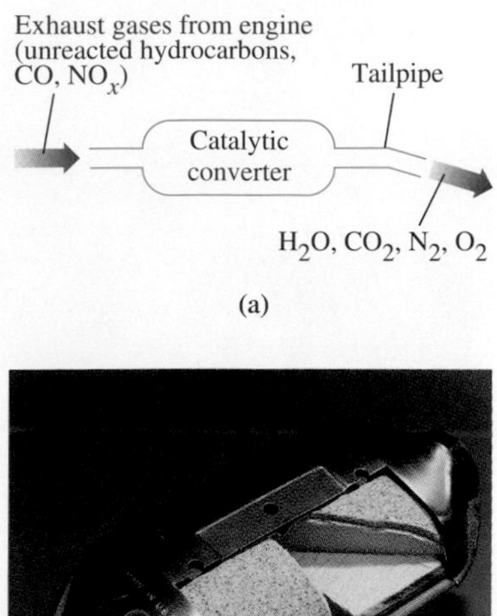

(a)

(b)

FIGURE 15.12 An Automobile Catalytic Converter (a) In a catalytic converter, CO is oxidized to CO_2; unburned hydrocarbons are oxidized to CO_2 and H_2O; and nitrogen oxides, NO, and NO_2 are decomposed to N_2 and O_2. The catalyst is a porous solid through which the exhaust gases pass. (b) Cutaway of a catalytic converter to reveal the cover, insulator, and catalyst.

which become attached to platinum atoms on the surface. These hydrogen atoms can move across the surface rather easily from one platinum atom to another. When they encounter an ethene molecule, they combine with it readily. The successive addition of two hydrogen atoms gives the final product, ethane. Probably due to its nonplanar shape, ethane is adsorbed less readily on the Pt surface than is ethene and so escapes from the surface.

Heterogeneous catalysts are used in the catalytic converters in the exhaust system of most automobiles (Figure 15.12). The exhaust gases contain unburned hydrocarbons and CO and the oxides of nitrogen NO and NO_2, which as we shall see in Chapter 16 are undesirable pollutants in the atmosphere. The function of the catalytic converter is to remove these gases from the exhaust by converting them to less harmful products. Most converters operate in two stages. In the first stage, hydrogen is produced by the catalyzed reaction between water vapor and unburned hydrocarbons present in the exhaust gases:

$$H_2O(g) + CH_4(g) \xrightarrow{\text{catalyst}} CO(g) + 3H_2(g)$$

The hydrogen then reduces the nitrogen oxides to nitrogen:

$$2H_2(g) + 2NO(g) \xrightarrow{\text{catalyst}} N_2(g) + 2H_2O(g)$$

The catalyst is a transition metal such as platinum or ruthenium in the form of very fine particles coating a porous metal oxide. In the second stage, carbon monoxide and unburned hydrocarbons are oxidized to carbon dioxide and water with oxygen from the air in the presence of a catalyst and at a temperature sufficiently low that very little NO is formed:

$$C_8H_{18}(g) \xrightarrow{O_2,\ \text{catalyst}} CO_2(g) + H_2O(g)$$

$$CO(g) \xrightarrow{O_2,\ \text{catalyst}} CO_2(g)$$

Enzymes

Enzymes are large protein molecules that catalyze reactions in living organisms. It has been estimated that the human body contains more than 100,000 different enzymes, each of which is specific to only a very few or even just one reaction. Enzymes are amazingly efficient and may increase reaction rates by a factor as great as 10^{10}, so reactions occur rapidly at body temperature. For example, an aqueous solution of sugar in a beaker at 37°C does not oxidize at a significant rate. Yet in the body at this temperature, sugar is rapidly oxidized to CO_2 and water:

$$C_{12}H_{22}O_{11} + 12O_2 \rightarrow 12CO_2 + 11H_2O$$

This reaction occurs in a long series of steps, each of which is catalyzed by a specific enzyme catalyst. An enzyme functions by what is known as a "lock-and-key" mechanism, which we will consider in more detail in Chapter 17.

EXAMPLE 15.4 Rate Law for a Catalyzed Reaction

Deduce the rate law for the iodide-catalyzed decomposition of hydrogen peroxide in aqueous solution.

Solution: We saw that equation (10) is the slow step of the reaction:

$$H_2O_2(aq) + I^-(aq) \rightarrow IO^-(aq) + H_2O(l)$$

So the rate law is predicted to be

$$\text{rate} = k[H_2O_2][I^-]$$

The reaction is second order overall: first order in H_2O_2 and first order in I^-. The catalyst appears in the rate law but not in the overall equation for the reaction. Because this rate law is also the observed rate law for the reaction, we can say that the observed rate law is consistent with the proposed mechanism.

Exercise 15.9 Dinitrogen monoxide, $N_2O(g)$, decomposes at 600°C

$$2N_2O(g) \rightarrow 2N_2(g) + O_2(g)$$

according to the following mechanism:

$$N_2O \rightarrow N_2 + O \qquad \text{(Slow)}$$

$$O + N_2O \rightarrow N_2 + O_2 \qquad \text{(Fast)}$$

The reaction is catalyzed by a trace of $Cl_2(g)$, and the catalyzed reaction follows the mechanism

$$Cl_2 \rightleftharpoons 2Cl \qquad \text{(Fast equilibrium)}$$

$$N_2O + Cl \rightarrow N_2 + ClO \qquad \text{(Slow)}$$

$$2ClO \rightarrow Cl_2 + O_2 \qquad \text{(Fast)}$$

(a) Derive the overall equation for the catalyzed reaction, and thus confirm that Cl_2 behaves as a catalyst for this reaction.
(b) Derive the rate laws for the uncatalyzed and the catalyzed reactions.

Exercise 15.10 Given that the activation energy is 76 $kJ \cdot mol^{-1}$ for the uncatalyzed, and 57 $kJ \cdot mol^{-1}$ for the catalyzed decomposition of $H_2O_2(aq)$ by $I^-(aq)$, by what factor does the Arrhenius equation predict that the rate of the catalyzed reaction will be greater than that of the uncatalyzed reaction at 25°C?

15.5 CHAIN REACTIONS

We saw in Chapter 4 that the reaction between hydrogen and chlorine at room temperature is extremely slow: No measurable amount of hydrogen chloride forms, even over a long period. However, if the mixture of gases is exposed to a bright light, a very rapid reaction occurs, which we observe as an explosion. This reaction belongs to an important class of reactions called chain reactions.

> **A *chain reaction* is a multi-step reaction with an initial step that produces a reactive intermediate, chain-propagation steps in which the product is formed and the reactive intermediate is regenerated, and one or more chain-termination steps, in which the reactive intermediate is removed.**

In the reaction of H_2 with Cl_2 absorption of light of maximum wavelength λ = 493 nm causes the dissociation of chlorine molecules into chlorine atoms (Chapter 6):

1. $$Cl_2 \rightarrow 2Cl\cdot$$

This first step is called **chain initiation**. Like almost all free atoms, chlorine atoms are very reactive. A Cl atom reacts rapidly with an H_2 molecule, producing an HCl molecule and an H atom:

2. $$Cl\cdot + H_2 \rightarrow HCl + H\cdot$$

The very reactive H atom then attacks a Cl_2 molecule, forming an HCl molecule and regenerating a Cl atom:

3. $$H\cdot + Cl_2 \rightarrow HCl + Cl\cdot$$

The Cl atom can attack another H_2 molecule as in step 2, producing another H atom, which can react with a Cl_2 molecule, as in step 3. The reactions in steps 2 and 3 can repeat themselves many times, giving a chain of successive reactions called **chain-propagation steps**. In this reaction the H and Cl atoms are reactive intermediates—they both have a single unpaired electron ($H\cdot$, $:\ddot{\underset{..}{Cl}}\cdot$). The overall reaction

$$H_2 + Cl_2 \rightarrow 2HCl$$

is the sum of the reactions in steps 2 and 3. These two reactions involve two reactive intermediates, the atoms H and Cl, that are continually regenerated. So both reactions continue indefinitely until the Cl and H atoms are removed by combining with another Cl or H atom in what are called **chain-termination** reactions. The mechanism of the H_2–Cl_2 reaction can therefore be summarized by the following sequence, which is typical of chain reactions:

$$Cl_2 \rightarrow 2Cl\cdot \quad \textit{Chain initiation}$$

$$\left.\begin{array}{r} Cl\cdot + H_2 \rightarrow HCl + H\cdot \\ H\cdot + Cl_2 \rightarrow HCl + Cl\cdot \end{array}\right\} \textit{Chain propagation}$$

$$\left.\begin{array}{r} 2Cl\cdot \rightarrow Cl_2 \\ 2H\cdot \rightarrow H_2 \\ H\cdot + Cl\cdot \rightarrow HCl \end{array}\right\} \textit{Chain termination}$$

The formation of a single Cl atom by the photolysis of a Cl_2 molecule may cause the rapid formation of many thousands of HCl molecules before it is removed from the reacting system by the chain-termination reactions (Figure 15.13).

The chlorination of methane (Chapter 14) provides us with another example of a chain reaction:

$$Cl_2 + CH_4 \rightarrow CH_3Cl + HCl$$

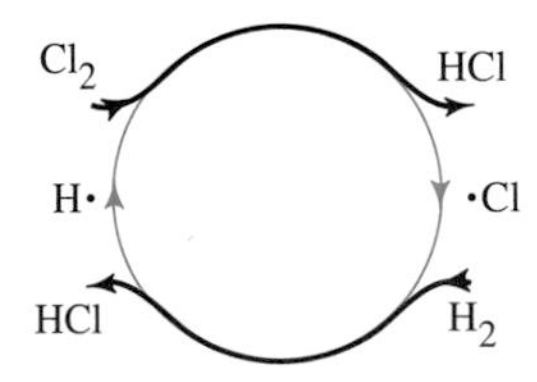

FIGURE 15.13 Hydrogen-Chlorine Chain Reaction
A Cl atom reacts with an H_2 molecule to form an HCl molecule and an H atom. The H atom reacts with a Cl_2 molecule to form another HCl molecule and another Cl atom, which reacts with another H_2 molecule, and so on. One Cl atom can cause the formation of many HCl molecules.

The reaction is carried out at high enough temperatures (350 to 750°C) to generate a small number of chlorine atoms:

$$Cl_2 \rightarrow 2Cl\cdot \quad \textit{Chain initiation}$$

These very reactive chlorine atoms remove a hydrogen atom from methane to form HCl and $\cdot CH_3$. The molecule $\cdot CH_3$ has a single unpaired electron—an odd electron—in the valence shell of the carbon atom. Molecules with an odd number of electrons are called *free radicals*. Like free atoms with one or more unpaired electrons, free radicals are generally very reactive. The $\cdot CH_3$ free radical attacks a chlorine molecule to form CH_3Cl and another reactive chlorine atom, so a chain is set up:

$$\left.\begin{array}{r} Cl\cdot + CH_4 \rightarrow H\text{—}Cl + \cdot CH_3 \\ \cdot CH_3 + Cl_2 \rightarrow CH_3Cl + Cl\cdot \end{array}\right\} \textit{Chain propagation}$$

In this chain, many hundreds of thousands of molecules react until the chain is finally terminated by one of the following reactions:

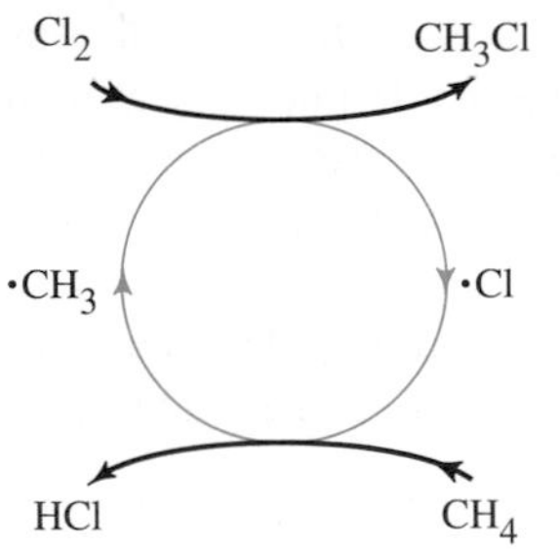

FIGURE 15.14 The Methane-Chlorine Chain Reaction A chlorine atom produced by the dissociation of a chlorine molecule reacts with a methane molecule to give a methyl radical and an HCl molecule. The methyl radical reacts with a chlorine molecule to give a chloromethane molecule, CH_3Cl, and another chlorine atom, and so on. Thus, one chlorine atom can cause the formation of many CH_3Cl molecules. The methyl radical and the chlorine atom are the free radical intermediates in this reaction.

$$\left.\begin{array}{r} Cl\cdot + Cl\cdot \rightarrow Cl_2 \\ \cdot CH_3 + \cdot CH_3 \rightarrow C_2H_6 \\ \cdot CH_3 + Cl\cdot \rightarrow CH_3Cl \end{array}\right\} \quad \textit{Chain termination}$$

This chain reaction can be summarized by the diagram in Figure 15.14.

SUMMARY

A reaction rate is the change in the concentration of any reactant or product in a given time interval Δt, given by $-\Delta[\text{reactant}]/\Delta t$ or $+\Delta[\text{product}]/\Delta t$. For a balanced equation $a\text{A} + b\text{B} + \cdots \rightarrow p\text{P} + q\text{Q} + \cdots$,

$$-\frac{1}{a}\frac{\Delta[\text{A}]}{\Delta t} = -\frac{1}{b}\frac{\Delta[\text{B}]}{\Delta t} = -\cdots = \frac{1}{p}\frac{\Delta[\text{P}]}{\Delta t} = \frac{1}{q}\frac{\Delta[\text{Q}]}{\Delta t} = \cdots$$

For an infinitely small time interval dt, the instantaneous reaction rate at time t is the slope of the tangent to the graph of concentration versus time at time t. Rate laws are equations that relate the rate of a reaction to the concentrations of the reactants. Established experimentally, they have the form rate $= k[\text{A}]^x[\text{B}]^y[\text{C}]^z\cdots$, where k is the rate constant at a given temperature and x, y, z, . . . are the reaction orders with respect to the reactants A, B, C, ... The sum $x + y + z + \cdots$ is the overall reaction order.

The mechanism of a reaction is the series of steps (elementary reactions) by which reactants are converted to products. Any proposed mechanism must be consistent with the experimentally observed rate law. The slowest step of a reaction, called the rate-determining or rate-limiting step, determines the overall rate and the form of the rate equation.

The molecularity of a reaction step is the number of participating reactant molecules. It is unimolecular if there is one reactant molecule, bimolecular if there are two molecules, and termolecular if there are three. Bimolecular reactions result from molecular collisions, whereas unimolecular reactions involve the breaking of chemical bonds by light or as a result of the accumulation of vibrational energy due to previous collisions. Because the simultaneous collision of four or more molecules is highly improbable, reactions with four or more reactants in the balanced equation involve at least two steps. Reaction intermediates (such as t-Bu^+) are reactive species that do not appear in the rate law but are formed in one reaction step and used up in a later step of a reaction mechanism.

The rate law for a reaction can be found by using the initial rate method or by fitting rate data to an integrated rate law, an equation that relates the concentrations of reactants at a given time to the time t. For a first-order reaction, the integration of $-d[\text{A}]/dt = k_1[\text{A}]$ gives the equation of a straight line, $\ln[\text{A}]_t = -k_1t + \ln[\text{A}]_0$: for this rate law, a plot of $\ln[\text{A}]_t$ versus t is a straight line with slope $-k_1$. From this equation, the half-life $t_{\frac{1}{2}}$ of a first-order reaction is related to the rate constant k_1 by the equation $t_{\frac{1}{2}} = 0.693/k_1$, and $t_{\frac{1}{2}}$ is independent of the initial concentration of the reactant.

For a reaction at equilibrium,

$$\text{rate forward} = \text{rate reverse}$$

The equilibrium constant expression for a reaction, K, can always be derived from the balanced equation for the reaction. It is independent of the reaction mechanism.

The reaction rate of a bimolecular reaction depends on the rate of collisions between the reacting molecules A and B—collision rate $= Z[\text{A}][\text{B}]$, where Z is a constant for any bimolecular reaction at a given temperature and pressure—and on the energies of collision. For reaction to occur, the collision energy must be greater than a minimum amount, the activation energy, E_a. E_a is the energy barrier that reacting molecules must surmount to reach the transition state, or activated complex, of a reaction and is the difference in energies between the reactants and the transition state. The fraction f of collisions with energy greater than E_a is given by the exponential function

$$\ln f = \frac{-E_a}{RT}, \qquad f = e^{-E_a/RT}$$

where R, the gas constant, equals $8.314\ \text{J} \cdot \text{K}^{-1} \cdot \text{mol}^{-1}$ and T is the temperature in kelvins. This means that for a bimolecular reaction,

$$k = fZ = Ze^{-E_a/RT}$$

The equation $k = Ae^{-E_a/RT}$, or

$$\ln k = \ln A - \frac{E_a}{R}\left(\frac{1}{T}\right)$$

is the Arrhenius equation, where A is the Arrhenius constant. A is not identical to the constant Z for reasons that include the need for molecules to collide with appropriate orientations for them to exchange atoms, the effect on the collision rate of interactions with solvent molecules in solution or intermolecular attractions in the gas phase, and the relatively slow increase of Z with increasing temperature.

The activation energy of a reaction is calculated from values of the rate constant at different temperatures. The Arrhenius equation shows that a plot of $\ln k$ versus $1/T$ gives a straight line with slope $-E_a/R$. Reactions with large E_a values have a reasonable rate only at high temperatures, and their rates increase rapidly with increasing temperature. Reactions with small E_a values are fast at low temperatures, and their rates increase more slowly with increasing temperature. The rate constants k_1 and k_2 at two temperatures T_1 and T_2 are related by the equation

$$\ln \frac{k_2}{k_1} = \frac{E_a}{R}\left(\frac{1}{T_1} - \frac{1}{T_2}\right)$$

Catalysis is the use of catalysts to increase reaction rates. A catalyst speeds up the rate at which a reaction reaches equilibrium by providing an alternative mechanism with a lower activation energy. The catalyst is regenerated during the reaction, does not appear in the overall equation for the reaction, and may be regarded as both a reactant and a product. Homogeneous catalysts are in the same phase as, and heterogeneous catalysts are in a different phase than, the reactants. Enzymes are large protein molecules that catalyze reactions in living organisms.

A chain reaction is a multi-step reaction with a chain-initiation step that produces a reactive intermediate, followed by chain propagation steps in which the product is formed and the reactive intermediate is regenerated, and then one or more chain-termination steps in which the reactive intermediate is removed.

IMPORTANT TERMS

activated complex (page 562)
activation energy (page 562)
Arrhenius equation (page 565)
bimolecular reaction (page 556)
catalysis (page 568)
catalyst (page 569)
chain reaction (page 573)
enzyme (page 572)
first-order rate law (page 555)
half-life (page 559)
heterogeneous catalyst (page 569)
homogeneous catalyst (page 569)
initial rate (page 554)
instantaneous rate (page 552)
integrated first-order rate law (page 558)
molecularity (page 556)
rate constant (page 554)
rate-determining (rate-limiting) step (page 556)
rate law (page 555)
reaction intermediate (page 557)
reaction mechanism (page 556)
reaction order (page 555)
reaction rate (page 550)
second-order rate law (page 555)
transition state (page 562)
unimolecular reaction (page 568)

REVIEW QUESTIONS

1. How is the rate of the reaction

$$C_2H_6(g) + 2H_2O(g) \rightarrow 2CO(g) + 5H_2(g)$$

expressed in terms of the changes in the concentrations of each reactant and product in a given time Δt, and how are these rates related?

2. How is the instantaneous rate of a reaction at a given time t defined?

3. How is the instantaneous rate of a reaction at time t obtained from a plot of the concentration of any reactant (or product) versus time?

4. For the reaction

$$2H_2(g) + 2NO(g) \rightarrow 2H_2O(g) + N_2(g)$$

if the rate law is rate $= k[H_2][NO]^2$, what are **(a)** the reaction orders with respect to H_2 and NO, **(b)** the overall reaction order, and **(c)** the units of the rate constant k?

5. Explain the initial rate method for determining the order of a reaction with respect to each reactant.

6. If the rate of a reaction increases by a factor of 9 when the concentration of a reactant [A] is tripled, what is the order of the reaction with respect to A?

7. What is meant by reaction mechanism?

8. What is meant by the rate-determining step of a reaction mechanism?

9. What is meant by **(a)** a unimolecular reaction; **(b)** a bimolecular reaction?

10. What is meant by the half-life of a reaction?

11. How does the equilibrium constant for a reaction depend on the mechanism of the reaction?

12. How does an increase in temperature affect most reaction rates?

13. **(a)** What is a transition state (activated complex)? **(b)** What is the activation energy, E_a, of a reaction?

14. Explain why the fraction of bimolecular collisions with energy greater than E_a increases very rapidly with increasing temperature.

15. How can the activation energy of a reaction be found from values of the rate constant at different temperatures?

16. What is a catalyst?

17. For a catalyzed reaction, does a catalyst appear in **(a)** the balanced equation; **(b)** the rate law?

18. What are **(a)** a homogeneous catalyst and **(b)** a heterogeneous catalyst?

19. What is a chain reaction?

20. Explain chain initiation, chain propagation, and chain termination.

PROBLEMS

Reaction Rate

1. For the reaction

$$C_2H_6(g) + 2H_2O(g) \rightarrow 2CO(g) + 5H_2(g)$$

the initial rate of formation of $CO(g)$ in a particular experiment was $1.00\ mL \cdot s^{-1}$. What was the initial rate expressed as **(a)** the formation of $H_2(g)$ and **(b)** the disappearance of $C_2H_6(g)$?

2. For each of the following reactions, how is the reaction rate expressed in terms of each reactant and each product, and how are these rates related?

(a) $C_3H_8(g) + 3H_2O(g) \rightarrow 3CO(g) + 7H_2(g)$

(b) $2C_4H_{10}(g) + 13O_2(g) \rightarrow 8CO_2(g) + 10H_2O(g)$

3. For each of the following reactions, express the rate in terms of the change in concentration of the reactant with time, and relate this to the rate of formation of each product.

(a) $2HI(g) \rightarrow H_2(g) + I_2(g)$

(b) $2N_2O(g) \rightarrow 2N_2(g) + O_2(g)$

(c) $2N_2O_5(g) \rightarrow 4NO_2(g) + O_2(g)$

4. For the reaction

$$4NH_3(g) + 3O_2(g) \rightarrow 2N_2(g) + 6H_2O(g)$$

it was found that at a given temperature and time, the rate of formation of $N_2(g)$ was $0.27\ mol \cdot L^{-1} \cdot s^{-1}$. At what rate was **(a)** water being formed; **(b)** $NH_3(g)$ being consumed; **(c)** $O_2(g)$ being consumed?

5. For each of the following reactions, express the rate in terms of the change in concentration of oxygen, and relate this to the rate of change of the concentrations of each of the other reactants and products.

(a) $H_2(g) + O_2(g) \rightarrow H_2O_2(g)$

(b) $2H_2(g) + 2O_2(g) \rightarrow 2H_2O_2(g)$

(c) $2NO(g) + O_2(g) \rightarrow 2NO_2(g)$

(d) $4PH_3(g) + 8O_2(g) \rightarrow P_4O_{10}(s) + 6H_2O(g)$

6. The reaction

$$CH_3OH(aq) + HCl(aq) \rightarrow CH_3Cl(aq) + H_2O(l)$$

was followed by measuring the change in hydronium ion concentration with time to give the results in the following table. Calculate the average reaction rate for each time interval.

Time (min)	*$[H_3O^+]$ $(mol \cdot L^{-1})$*
0	1.85
80	1.66
159	1.53
314	1.31
628	1.02

Effect of Reactant Concentration on Reaction Rate

7. **(a)** How is the rate of a reaction defined?

(b) What is the rate law for the hydrolysis of an alkyl bromide, RBr, which is first order in RBr and first order in OH^- ion?

(c) Express the rate constant for the reaction in (b) in terms of the reaction rate and the concentrations of RBr and OH^-.

(d) What are the units of this rate constant?

8. **(a)** What is the rate law for a reaction with the balanced equation

$$aA + bB + cC \rightarrow \text{products}$$

when the reaction is found experimentally to be first-order in A, first-order in B, and second-order in C?

(b) What are the units of this rate constant?

9. The rate constant for a first-order reaction at a certain temperature is $3.7 \times 10^{-2}\ s^{-1}$. For an initial concentration of $0.040\ mol \cdot L^{-1}$, what is the initial rate **(a)** in moles per liter per second and **(b)** in moles per liter per hour?

10. The rate law for the formation of octafluorocyclobutane from tetrafluoroethene according to the equation

$$2C_2F_4(g) \rightarrow C_4F_8(g)$$

is rate $= k[C_2F_4]^2$, with $k = 4.48 \times 10^{-2}\ L \cdot mol^{-1} \cdot s^{-1}$ at 450 K. What is the reaction rate when **(a)** $[C_2F_4] = 0.010\ mol \cdot L^{-1}$ and **(b)** $[C_2F_4] = 0.0050\ mol \cdot L^{-1}$?

11. For the reaction

$$2NO(g) + H_2(g) \rightarrow N_2O(g) + H_2O(g)$$

the following initial rate data were obtained at a given temperature. What are **(a)** the rate law for the reaction and **(b)** the value of the rate constant?

Expt.	*[NO] ($mol \cdot L^{-1}$)*	*[H_2] ($mol \cdot L^{-1}$)*	*Initial Rate ($mol \cdot L^{-1} \cdot min^{-1}$)*
1	0.150	0.800	0.500
2	0.075	0.800	0.125
3	0.150	0.400	0.250

12. A method for preparing nitrogen in the laboratory is by the reaction

$$NH_4^+(aq) + NO_2^-(aq) \rightarrow N_2(g) + 2H_2O(l)$$

The following initial rate data were obtained at 25°C:

Expt.	*[NO_2^-] ($mol \cdot L^{-1}$)*	*[NH_4^+] ($mol \cdot L^{-1}$)*	*Rate ($mol \cdot L^{-1} \cdot s^{-1}$)*
1	0.0100	0.200	5.40×10^{-7}
2	0.0400	0.200	2.15×10^{-6}
3	0.200	0.0200	1.08×10^{-6}
4	0.200	0.0600	3.24×10^{-6}

What are **(a)** the rate equation for the reaction and **(b)** the value of the rate constant at 25°C?

13. One of the major irritants in smog is formaldehyde (methanal), $CH_2O(g)$, formed by the reaction between ethene and ozone, $O_3(g)$, in the atmosphere:

$$2C_2H_4(g) + 2O_3(g) \rightarrow 4CH_2O(g) + O_2(g)$$

From the following initial rate data, deduce the rate equation for this reaction:

Expt.	*[O_3] ($mol \cdot L^{-1}$)*	*[C_2H_4] ($mol \cdot L^{-1}$)*	*Initial Rate ($mol \cdot L^{-1} \cdot s^{-1}$)*
1	0.5×10^{-7}	1.0×10^{-8}	1.0×10^{-12}
2	1.5×10^{-7}	1.0×10^{-8}	3.0×10^{-12}
3	1.0×10^{-7}	2.0×10^{-8}	4.0×10^{-12}

14. For the isomerization of cyclopropane to propene,

$$C_3H_6(g) \rightarrow CH_3CH{=}CH_2(g)$$

the following rate data were obtained at 500°C:

t (min)	*[C_3H_6] ($mol \cdot L^{-1}$)*
0	1.500×10^{-3}
5	1.218×10^{-3}
10	0.988×10^{-3}
15	0.802×10^{-3}
20	0.649×10^{-3}

(a) Using the integrated first-order rate equation, confirm that the rate law is rate $= k[C_3H_6]$ by plotting $\ln[C_3H_6]$ versus time t.

(b) Calculate the value of the rate constant at 500°C from the slope of the graph plotted in (a).

(c) Confirm the value of the rate constant in (b) from the value of the time of half-reaction.

15. The decomposition of gaseous hydrogen peroxide to $O_2(g)$ and $H_2O(g)$ is a first-order reaction. Experimentally, at a given temperature, the initial concentration of H_2O_2 decreased to half in 17.0 min.

(a) What is the rate constant of the reaction?

(b) What fraction of the initial H_2O_2 would remain after (i) 51.0 min and (ii) 10 half-lives?

16. In an experiment like that in Problem 15 but at a higher temperature, one-fourth the initial H_2O_2 remained after 8.0 min.

(a) What is the half-life of H_2O_2 under these conditions?

(b) What is the value of the rate constant?

(c) How long will it take for the concentration of H_2O_2 to decrease to about 3% of its initial value?

17. The rate of decomposition of H_2O_2 in aqueous solution at a particular temperature was measured by withdrawing samples of the reaction mixture after given times of reaction. These were immediately titrated with acidified $KMnO_4(aq)$ to determine the concentration of H_2O_2 remaining, which gave the following results:

t (min)	0	10	20
mL $KMnO_4$(aq)	22.8	13.8	8.3

(The balanced equation for the reaction between $KMnO_4$ and H_2O_2 is not required.) Without plotting the data, **(a)** show that the reaction is first order in H_2O_2.

(b) Find the value of the rate constant.

(c) Calculate the half-life of the reaction.

18. At 45°C N_2O_5 decomposes in solution in tetrachloromethane according to the equation

$$2N_2O_5 \rightarrow 4NO_2 + O_2$$

In a particular experiment, the following rate data were obtained:

$[N_2O_5]$ $(mol \cdot L^{-1})$	Rate $(mol \cdot L^{-1} \cdot min^{-1})$
2.00	1.26×10^{-3}
1.80	1.13×10^{-3}
1.51	0.94×10^{-3}
0.92	0.57×10^{-3}

(a) Determine the rate law for the reaction.

(b) Calculate the value of the rate constant.

(c) How long will it take for $[N_2O_5]$ to decrease to 0.50 $mol \cdot L^{-1}$?

19. Annual production of the insecticide DDT amounted to about 7.5×10^7 kg in the 1960s. In 1972, DDT was banned for general use in the United States. At ordinary temperatures the half-life of DDT in soil is about 10 yr. How long will it take for 1000 kg of DDT sprayed on the ground in 1965 to decrease to 1.0 g?

20. The thermal decomposition of phosphine to phosphorus and hydrogen,

$$4PH_3(g) \rightarrow P_4(g) + 6H_2(g)$$

is a first-order reaction with a half-life of 35.0 s at 680°C. Calculate **(a)** the rate constant and **(b)** the time required for exactly 90% of the phosphine to decompose at 680°C.

21. Explain the terms **(a)** "multi-step reaction mechanism" and **(b)** "reaction intermediate."

22. For the reaction between chlorine and carbon monoxide to form the poisonous war gas phosgene,

$$Cl_2(g) + CO(g) \rightarrow COCl_2(g)$$

the following mechanism has been proposed:

$Cl_2 \rightleftharpoons 2Cl$	(Fast equilibrium)	(1)
$Cl + CO \rightleftharpoons COCl$	(Fast equilibrium)	(2)
$COCl + Cl_2 \rightarrow COCl_2 + Cl$	(Slow)	(3)

(a) In terms of the reactants CO and Cl_2, what is the expected rate law?

(b) What name is given to molecules such as COCl and Cl?

23. A proposed mechanism for the first-order decomposition of $N_2O_5(g)$ consists of the steps

$N_2O_5 \rightarrow NO_2 + NO_3$	(Slow)	(1)
$NO_3 \rightarrow NO + O_2$	(Fast)	(2)
$NO + N_2O_5 \rightarrow NO_2 + N_2O_4$	(Fast)	(3)
$N_2O_4 \rightleftharpoons 2NO_2$	(Fast equilibrium)	(4)

(a) What is the molecularity of each step of this reaction?

(b) Is the proposed mechanism consistent with an observed first-order rate law?

24.* The conversion of ozone to molecular oxygen in the upper atmosphere,

$$2O_3(g) \rightarrow 3O_2(g)$$

is thought to occur via the mechanism

$O_3 \rightleftharpoons O_2 + O$	(Fast equilibrium)
$O + O_3 \rightarrow 2O_2$	(Slow)

(a) What rate law is consistent with this mechanism?

(b) Explain why the reaction rate decreases as the concentration of O_2 increases—in other words, why O_2 occurs in the rate law with a negative order.

25.* For the reaction

$$2NO(g) + O_2(g) \rightarrow 2NO_2(g)$$

the rate law is

$$\text{rate} = k[NO]^2[O_2]$$

(a) Explain why this reaction is unlikely to occur by a one-step termolecular process.

(b) Devise two multi-step mechanisms for this reaction that are consistent with the observed rate law and do not involve the simultaneous collision of three molecules.

26.* In acidic solution bromate ion, $BrO_3^-(aq)$, slowly oxidizes bromide ion, $Br^-(aq)$, to bromine.

(a) Write the balanced ionic equation for this reaction.

(b) Find the rate law from the following initial rate data:

Expt.	$[BrO_3^-]$ $(mol \cdot L^{-1})$	$[Br^-]$ $(mol \cdot L^{-1})$	$[H_3O^+]$ $(mol \cdot L^{-1})$	Relative Rate
1	0.05	0.25	0.30	1
2	0.05	0.25	0.60	4
3	0.10	0.25	0.60	8
4	0.05	0.25	0.60	2
5	0.05	0.50	0.30	2

Activation Energy and the Effect of Temperature on Reaction Rate

27. Explain how the rate of a reaction is affected by **(a)** the frequency and kinetic energy of collisions; **(b)** the orientation of the molecules during a collision.

28. Rate constants at several different temperatures for the reaction

$$2HI(g) \rightarrow H_2(g) + I_2(g)$$

are given in the following table:

T (°C)	k $(L \cdot mol^{-1} \cdot s^{-1})$
302	1.18×10^{-6}
356	3.33×10^{-5}
374	8.96×10^{-5}
410	5.53×10^{-4}
427	1.21×10^{-3}

(a) Plot ln k versus $1/T$.

(b) Obtain the value for the activation energy, E_a, of the reaction from the slope of the plot.

(c) Calculate the value of the rate constant at 400°C.

29. The activation energy for the hydrogenation of ethene is 180 $kJ \cdot mol^{-1}$. Calculate the temperature at which the rate of the reaction will be 10 times the rate at 200°C.

30. The activation energy of a reaction is 100 kJ · mol^{-1}. To what temperature must the reaction mixture be raised for its rate constant to have exactly twice the value it has at 27°C?

31. Nitrogen dioxide decomposes to nitrogen monoxide at high temperature:

$$2NO_2(g) \rightarrow 2NO(g) + O_2(g)$$

Find the activation energy for this reaction from the following rate constant data:

T (K)	k (L · mol^{-1} · s^{-1})
650	3.16
730	28.2
800	1.58×10^2
900	1.12×10^3
1000	5.01×10^3

32. For the reaction

$$2N_2O(g) \rightarrow 2N_2(g) + O_2(g)$$

the rate constant is 1.1×10^{-3} L · mol^{-1} · s^{-1} at 565°C and 3.8×10^{-3} L · mol^{-1} · s^{-1} at 728°C. What are the values of **(a)** the activation energy and **(b)** the rate constant at 780°C?

33. For the reaction

$$CO(g) + NO_2(g) \rightarrow CO_2(g) + NO(g)$$

the rate constant at 425°C has the value 1.3 L · mol^{-1} · s^{-1}, and at 525°C the value is 23 L · mol^{-1} · s $^{-1}$. Find **(a)** the activation energy of the reaction and **(b)** the rate constant at 298°C.

34. For the first-order reaction

$$C_2H_5Br(g) \rightarrow C_2H_4(g) + HBr(g)$$

the activation energy is 226 kJ · mol^{-1}, and the value of the rate constant at 650 K is 2.0×10^{-3} s^{-1}.

(a) What is the value of the rate constant at 600 K?

(b) At this temperature, how long will it take the concentration of C_2H_5Br to decrease from 1.00×10^{-2} mol · L^{-1} to 0.25×10^{-2} mol · L^{-1}?

35. What is the activation energy for the first-order reaction

$$C_2H_5Cl(g) \rightarrow C_2H_4(g) + HCl(g)$$

if the value of the rate constant is 3.5×10^{-8} s^{-1} at 600 K and 1.6×10^{-6} s^{-1} at 650 K?

Catalysis

36. State the effect that a catalyst has on each of the following.

(a) the rate of a reaction

(b) the activation energy of a reaction

(c) the enthalpy change of a reaction

(d) the temperature at which a reaction has a given rate

(e) the equilibrium position of a reaction

37. For a catalyzed reaction, does the concentration of a homogeneous catalyst appear in **(a)** the balanced equation; **(b)** the rate law? Explain why.

38. For a reaction at a given temperature, does a catalyst affect **(a)** the value of the rate constant; **(b)** the equilibrium position?

39. By what factor does a platinum catalyst increase the rate of the reaction

$$2HI(g) \rightarrow H_2(g) + I_2(g)$$

if it reduces the activation energy from 183 kJ · mol^{-1} to 58 kJ · mol^{-1} at 300°C?

40. A platinum catalyst reduces the activation energy for the hydrogenation of ethene from 180 kJ · mol^{-1} to 80 kJ · mol^{-1}. At what temperature will the rate of the catalyzed reaction be the same as that of the uncatalyzed reaction at 1000°C?

41. The uncatalyzed decomposition of hydrogen peroxide,

$$2H_2O_2(aq) \rightarrow 2H_2O(l) + O_2(aq)$$

has an activation energy of 75 kJ · mol^{-1}, and a platinum catalyst reduces that activation energy to 50 kJ · mol^{-1}. At 25°C, by what factor is the catalyzed reaction faster than the uncatalyzed reaction?

42. Urea, $(NH_2)_2CO$, is converted to ammonium ion and hydrogencarbonate ion in acidic aqueous solution:

$$(NH_2)_2CO(aq) + H_3O^+(aq) + H_2O(l) \rightarrow 2NH_4^+(aq) + HCO_3^-(aq)$$

The rate of this reaction catalyzed by the enzyme urease at 25°C is 10^{13} times faster than the uncatalyzed reaction. To what change in the activation energy of the reaction does this change in rate correspond?

43.* The reaction of acetone (propanone) with iodine in aqueous solution occurs according to the equation

$$CH_3COCH_3(aq) + I_2(aq) \rightarrow CH_3COCH_2I(aq) + H^+(aq) + I^-(aq)$$

In the presence of a base B, the rate law is

$$\text{rate} = k[CH_3COCH_3][B]$$

(a) Suggest a three-step mechanism that is consistent with this rate law.

(b) What is the function of the base B?

Chain Reactions

44. **(a)** Describe each step in the following mechanism for the reaction of bromine with methane as chain initiation, chain propagation, or chain termination.

$$Br_2 \rightarrow 2Br\cdot \quad (1)$$

$$Br\cdot + CH_4 \rightarrow \cdot CH_3 + HBr \quad (2)$$

$$\cdot CH_3 + Br_2 \rightarrow CH_3Br + Br\cdot \quad (3)$$

$$2Br\cdot \rightarrow Br_2 \quad (4)$$

(b) What is the maximum wavelength of light that could initiate this reaction?

45.* The following three-step mechanism has been proposed for the gas-phase reaction in which tetrachloromethane is formed from chlorine and trichloromethane (chloroform):

$$CHCl_3(g) + Cl_2(g) \rightarrow CCl_4(g) + HCl(g)$$

$$Cl_2 \rightleftharpoons 2Cl\cdot \quad \text{(Fast equilibrium)} \quad (1)$$

$$Cl\cdot + CHCl_3 \rightarrow \cdot CCl_3 + HCl \quad (2)$$

$$Cl\cdot + \cdot CCl_3 \rightarrow CCl_4 \quad (3)$$

The observed rate law is

$$\text{rate} = k[CHCl_3][Cl_2]^{\frac{1}{2}}$$

Is the proposed mechanism consistent with the observed rate law when step 2 is much slower than step 3 or when step 3 is much slower than step 2?

General Problems

46. **(a)** What is a single-step reaction?

(b) Why is a single-step bimolecular reaction necessarily second order?

(c) If a multi-step reaction is found to be second order, what can be said about the rate-determining step?

47. In a reaction of ozone with ammonia at 500 K according to the equation

$$5O_3(g) + 6NH_3(g) \rightarrow 6NO(g) + 9H_2O(g)$$

the rate of increase in the partial pressure of $NO(g)$ in a given time interval was 1095 (mm Hg) $\cdot$ s^{-1}. For the same interval, what was **(a)** the rate of disappearance of ozone in moles per liter per second and **(b)** the rate of disappearance of ammonia?

48. For the reaction

$$C_2H_4(g) + H_2(g) \rightarrow C_2H_6(g) \qquad \Delta H° = -137 \text{ kJ}$$

explain why low temperature, high pressure, and a catalyst are beneficial to obtaining a high yield of ethane.

49. **(a)** In a gas-phase bimolecular reaction, why is it that not all collisions between reactant molecules lead to the formation of product molecules?

(b) For a single-step reaction, how can we distinguish experimentally between a unimolecular mechanism and a bimolecular mechanism?

50. Define the terms **(a)** "reaction order"; **(b)** "rate constant"; and **(c)** "molecularity."

51. The reaction of an alkyl halide with hydroxide ion,

$$RX + OH^- \rightarrow ROH + X^-$$

occurs by one of two mechanisms, depending on the nature of the alkyl group R. Describe **(a)** the two possible mechanisms and **(b)** how the actual mechanism could be established for a particular alkyl halide.

52. Common table sugar (sucrose) is hydrolyzed to two other sugars, glucose and fructose. In an experiment at 25°C, the following rate data were obtained:

t (h)	0.00	0.50	1.00	1.50	2.00	3.00	4.00	5.00
10^2 [sucrose] (mol $\cdot$ L^{-1})	10.00	9.00	8.10	7.29	6.57	5.32	4.31	3.50

(a) Show that the reaction is first order in sucrose.

(b) What two methods could be used to obtain a value for the rate constant?

(c) If the experiment was repeated at 37°C (body temperature), how could a value for the activation energy be obtained?

(d) Why is it important that the reaction rate is vastly increased in human metabolism by an enzyme catalyst?

53. At high temperatures chloroethane (ethyl chloride) decomposes in a first-order reaction to ethene and hydrogen chloride gas. In an experiment at 800 K, an initial concentration of chloroethane of 0.0200 mol $\cdot$ L^{-1} decreased to 0.0033 mol $\cdot$ L^{-1} in 340 s. At 800 K, what are **(a)** the balanced equation for the reaction, **(b)** the value of the rate constant, and **(c)** the half-life of the reaction?

54.* In acidic aqueous solution, hydrogen peroxide oxidizes bromide ions to bromine. Experimentally, when the initial concentration of H_2O_2 was doubled and $[Br^-]$ and the pH were held constant, the rate doubled. The rate also doubled when $[Br^-]$ was doubled and $[H_2O_2]$ and the pH were held constant. When the pH alone was changed from 1.00 to 0.400, the rate increased four-fold.

(a) Write the balanced equation for the reaction.

(b) Find the rate law.

(c) If under given conditions the rate of disappearance of Br^- is 7.2×10^{-3} mol $\cdot$ L^{-1} $\cdot$ s^{-1}, what are the rate of disappearance of H_2O_2 and the rate of appearance of Br_2?

(d) What is the effect on the rate constant of increasing the pH?

(e) If the initial solution is diluted with water so that its volume is doubled, what would be the effect on the initial reaction rate?

Chemistry of the Environment

16.1 Oxides and Oxoacids of Nitrogen
16.2 Ozone and Peroxides
16.3 Pollution of the Stratosphere
16.4 Photochemical Smog
16.5 Acid Rain
16.6 The Greenhouse Effect

The study of the chemistry of the environment is essential if our planet is to continue to support humans and all other forms of life. This magnificent bird, a male resplendent quetzal, is native to Mexico, Panama, and Costa Rica. It belongs to one of the ever-growing numbers of endangered species whose habitats have become threatened by the activities of humans, such as the widespread destruction of rain forests and the pollution of the atmosphere.

To the best of our knowledge, life exists at only one location in the universe: the biosphere of our planet earth. The **biosphere** is that part of the earth that supports life. A very thin layer on the planet's surface, it is visible in its entirety from an orbiting spacecraft (Figure 16.1). Life as we know it is possible only in an environment that satisfies some very specific requirements. It must not be too hot or too cold; there must be an atmosphere with the right amounts of oxygen and of carbon dioxide; and there must be accessible sources of all the other essential elements, such as phosphorus, iron, magnesium and sodium. But the supply of these elements is limited, so there must also be ways of continually regenerating them. On earth this regeneration is accomplished by many self-sustaining natural cycles.

In the biological carbon cycle, animals breathe oxygen and exhale carbon dioxide; plants use carbon dioxide for photosynthesis and produce oxygen as a byproduct. Other cycles maintain the availability of other essential elements, such as nitrogen and phosphorus. In the hydrological cycle, water evaporates from rivers, lakes and oceans and returns as rain on the land, ensuring a continuous supply of life-sustaining fresh water. During the last several hundred million years, these cycles have maintained the conditions required to support life.

Since the Industrial Revolution, humans have been producing increasing amounts of waste materials of all kinds. Animals have always returned their waste to the environment, and this waste has always been a part of natural recycling processes. Until quite recently, humans assumed that the environment could also absorb all our waste, but we are becoming increasingly aware that it will not continue to do so indefinitely. In some cases we are simply increasing the amounts of substances generated naturally. For example, volcanoes produce about 10^8 metric tons of sulfur dioxide per year, but some 8×10^7 metric tons of SO_2 is now

FIGURE 16.1 The Earth from Space The biosphere is that part of the earth that we can observe directly from an orbiting spacecraft. It consists of the lower atmosphere (the troposphere), the hydrosphere, the surface of the earth's crust, and all the organic matter contained in these other parts.

produced by the burning of coal and smelting of sulfide ores to obtain metals. So the concentration of SO_2 in the troposphere has been approximately doubled worldwide and may be many times greater near industrial sites. Natural control mechanisms are overburdened, and the problems associated with the acid rain that the SO_2 produces are a consequence. In other cases, substances not found in nature are being allowed to escape into the biosphere. Natural control mechanisms for these substances may not exist, so even small amounts can have serious consequences. The chlorofluorocarbons (CFCs; Chapter 14) that are damaging the atmosphere's ozone layer are an example.

Substances such as CFCs that were not present before human activities began, and substances such as SO_2, that are now present in much greater amounts are called **pollutants**. The aim of environmental science is to understand how the natural environment is regulated and how it is affected by pollutants and to seek ways to minimize these negative effects. Environmental problems are inevitably complex and involve many disciplines: chemistry, physics, biology, ecology, geology, and climatology are all relevant to environmental science. Chemistry plays a central role, because almost all the fundamental processes that affect the environment are chemical reactions.

Understanding the fundamental science behind an environmental problem cannot guarantee its solution. But not understanding the fundamentals will almost certainly prevent us from achieving satisfactory solutions. A knowledge and understanding of the basic chemistry underlying environmental concerns are needed by everyone from politicians to environmental activists to economists to business executives. Only with this knowledge can intelligent decisions be made.

In this chapter we discuss the chemistry associated with just a few of the environmental problems of current concern. The earth's atmosphere and its lakes, rivers, and oceans play an essential role in the natural recycling processes, and so environmental chemistry is particularly concerned with the atmosphere and with water. Most of our discussion will focus on the atmosphere. We gave a brief introduction to the atmosphere in Chapter 2, where we concentrated on its two major components, oxygen and nitrogen. Some of the minor components of the atmosphere, such as the oxides of nitrogen and ozone, are very important to an understanding of environmental problems. So we first discuss some of the chemistry of the oxides of nitrogen and of ozone and peroxides.

16.1 OXIDES AND OXOACIDS OF NITROGEN

We saw in Chapter 7 that phosphorus forms the oxides P_4O_6 and P_4O_{10} (empirical formulas P_2O_3 and P_2O_5). Nitrogen, the first element in Group V, forms the two analogous oxides N_2O_3 and N_2O_5, but these are not the only oxides of nitrogen nor even the most stable or the most important. Nitrogen in its oxides exhibits every oxidation state from I to V. The formulas and physical properties of these oxides are listed in Table 16.1.

Nitrogen Monoxide

Nitrogen monoxide, NO, often called nitric oxide, is a colorless reactive gas. A small amount of nitrogen monoxide is present in the atmosphere as a result of the reaction between nitrogen and oxygen:

$$N_2(g) + O_2(g) \rightleftharpoons 2NO(g) \qquad K(25°C) = 10^{-30} \qquad \Delta H° = 180 \text{ kJ}$$

Because $K = [NO]^2/[N_2][O_2]$, the concentration units cancel and K is dimensionless.

TABLE 16.1 Properties of the Oxides of Nitrogen

Oxidation State	Formula	Name	Physical State at 25°C	MP (°C)	BP (°C)
+V	N_2O_5	Dinitrogen pentaoxide	White solid	30	47*
+IV	N_2O_4	Dinitrogen tetraoxide	White solid }	−11.2	21.2
+IV	NO_2	Nitrogen dioxide	Brown gas }		
+III	N_2O_3	Dinitrogen trioxide	Deep blue liquid	−102	—*
+II	NO	Nitrogen monoxide (nitric oxide)	Colorless gas	−90.8	−88.5
+I	N_2O	Dinitrogen monoxide (nitrous oxide)	Colorless gas	−163.7	−151.8

* Decomposes

The position of the equilibrium in this reaction lies far to the left at 25°C, so there is virtually no NO present at equilibrium. However, because the reaction is endothermic, the equilibrium constant increases with increasing temperature, and the position of the equilibrium shifts to the right. For example, at 2400 K the equilibrium constant is 2.5×10^{-2}. Although still small, it is large enough to give an appreciable equilibrium concentration of NO. At the high temperatures produced by the lightning in a thunderstorm or in a forest fire, some NO forms and is liberated into the much cooler surrounding atmosphere. The rate of the reverse reaction (and the rate of the forward reaction) is very slow at this much lower temperature, so it takes the NO formed a very long time to decompose into oxygen and nitrogen. Therefore, there is always a very small concentration of NO in the atmosphere.

There are about 100 cloud-to-ground lightning strokes per second over the surface of the globe. Each stroke is estimated to produce approximately 1000 mol of NO, leading to a total annual production of more than 10^{10} metric tons of NO.

Nitrogen monoxide can be conveniently prepared in the laboratory by reducing *dilute* nitric acid with copper:

$$3Cu(s) + 8HNO_3(aq) \rightarrow 3Cu(NO_3)_2(aq) + 2NO(g) + 4H_2O(l)$$

Nitrogen monoxide reacts with oxygen to form nitrogen dioxide, NO_2, as we see in Demonstration 16.1:

$$2NO(g) + O_2(g) \rightarrow 2NO_2(g)$$

Although this reaction is quite fast for the relatively high concentration of NO in Demonstration 16.1, it is very slow at the normally low concentration of NO in the atmosphere. Nitrogen monoxide is even more readily oxidized by ozone:

$$NO(g) + O_3(g) \rightarrow NO_2(g) + O_2(g)$$

This reaction, as we shall see, is one of the reactions that leads to the destruction of ozone in the stratosphere.

Nitrogen Dioxide and Dinitrogen Tetraoxide

Nitrogen dioxide is a red-brown gas that is mainly responsible for the color of photochemical smog. Two molecules of NO_2 combine to form colorless dinitrogen tetraoxide, N_2O_4:

$$2NO_2(g) \rightleftharpoons N_2O_4(g) \qquad K(25°C) = 6.9 \text{ atm}^{-1} \qquad \Delta H° = -57 \text{ kJ}$$

DEMONSTRATION 16.1 Nitrogen Monoxide, NO

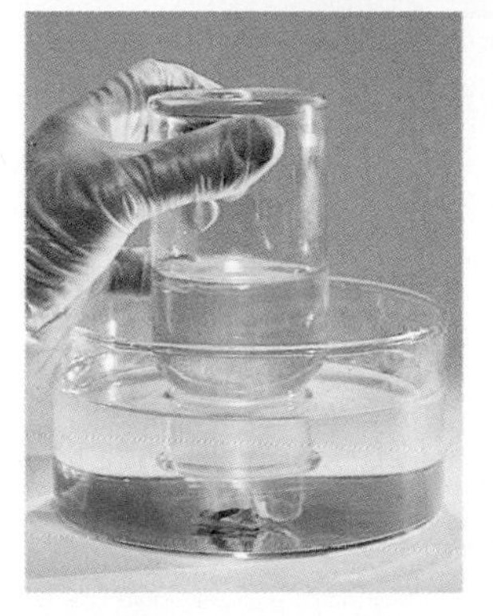

Nitrogen monoxide can be prepared by reacting copper with 6-*M* nitric acid. In this experiment a jar containing 6-*M* nitric acid has been inverted over pieces of copper in a trough of water. Colorless NO is evolved and collects in the jar, because it is only slightly soluble in water. The Cu^{2+} ion formed turns the nitric acid blue.

A jar of colorless NO

When the stopper is removed, the NO reacts with oxygen of the atmosphere to form brown NO_2.

The jar of NO_2 is then inverted over water containing a little methyl red indicator. The water rises rapidly in the jar as the NO_2 dissolves to form a solution of HNO_2 and HNO_3. This solution changes the color of the indicator from yellow to red.

In the liquid and gaseous states, nitrogen(IV) oxide is an equilibrium mixture of NO_2 and N_2O_4 molecules. Because the reaction is exothermic, more N_2O_4 forms with decreasing temperature, as predicted by Le Châtelier's principle (Chapter 12). At 21°C the gas condenses to a deep red-brown liquid. As the temperature decreases, the liquid becomes less intensely colored. At −11°C it freezes to a white solid consisting entirely of N_2O_4 molecules (Demonstration 16.2). At the very low concentrations of NO_2 in the atmosphere, the position of the equilibrium is far to the left, and there is essentially no N_2O_4.

Nitrogen dioxide is conveniently made in the laboratory by reducing *concentrated* nitric acid with copper (Figure 10.16)

$$Cu(s) + 4HNO_3(aq) \rightarrow Cu(NO_3)_2(aq) + 2NO_2(g) + 2H_2O(l)$$

or by heating some metal nitrates, for example

$$2Pb(NO_3)_2(s) \rightarrow 2PbO(s) + 4NO_2(g) + O_2(g)$$

The Structures of NO and NO_2: Free Radicals

Both nitrogen monoxide and nitrogen dioxide have an odd number of valence electrons—11 and 17, respectively. But the vast majority of stable molecules have an *even* number of valence electrons arranged in pairs—shared (bonding) pairs and unshared (nonbonding) pairs.

Because NO and NO_2 have odd numbers of valence electrons, we cannot draw Lewis structures for them in which all the electrons are paired and all the atoms have

DEMONSTRATION 16.2 Nitrogen Dioxide and Dinitrogen Tetraoxide

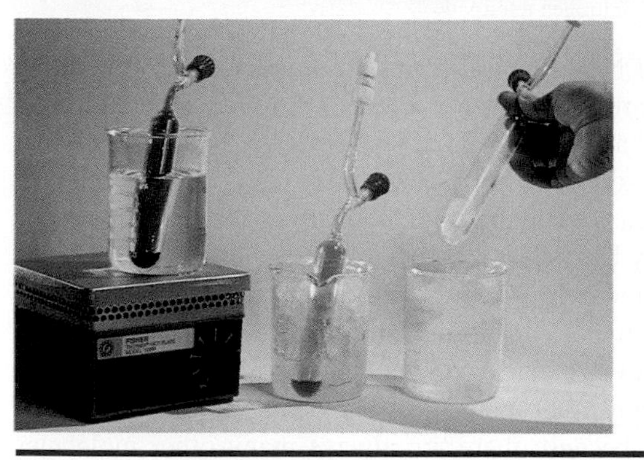

The equilibrium $2NO_2 \rightleftharpoons N_2O_4$ shifts to the right with decreasing temperature. A large amount of dark brown NO_2 vapor can be seen in the tube on the left, which is in hot water. The vapor in the tube in the middle, which is in ice, is much paler, because the equilibrium has been shifted to the right, and there is therefore less NO_2 vapor in the tube. The yellow-brown liquid at the bottom of the tube consists of N_2O_4 with a little NO_2. The tube on the right has been kept in a freezing mixture at a temperature of approximately −15°C. It contains white solid N_2O_4.

an octet of electrons, as is usually the case. We can draw two Lewis (resonance) structures for NO in which either the nitrogen atom or the oxygen atom has only seven electrons in its valence shell:

$$:\dot{N}=\ddot{O}: \leftrightarrow {}^{-}:\ddot{N}=\dot{O}:^{+}$$

Nitrogen monoxide

Alternatively, we can think of the odd electron as being shared between the oxygen and the nitrogen so that both atoms attain an octet of electrons:

$$:\dot{N}\overset{\cdot}{=}\dot{O}:$$

Nitrogen monoxide

Sharing a single electron corresponds to half a bond, so NO has a bond order of 2.5. This bond order is consistent with the bond length, which is intermediate between the bond lengths of N_2 and O_2:

	N_2	NO	O_2
Bond length	109 pm	116 pm	121 pm
Bond order	3	2.5	2

Recall from Chapter 8 that as bond order increases, bond length decreases.

The NO_2 molecule is angular with a bond length of 119 pm and a bond angle of 134°. The following two resonance structures can be written for the NO_2 molecule, in which the odd electron is on the nitrogen atom:

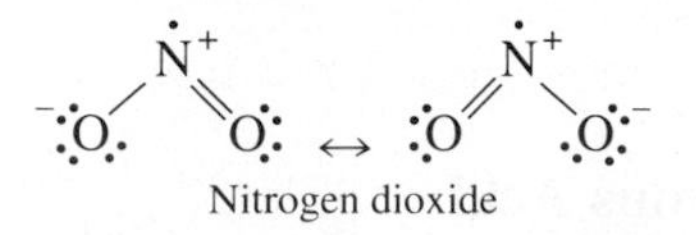

Nitrogen dioxide

Because a single electron has a smaller domain than does an unshared pair, the bond angle is considerably larger than the ideal angle of 120° for an AX_2E molecule. The odd electrons on two NO_2 molecules pair to form the bond between the nitrogen atoms in the N_2O_4 molecule, which has the following structure:

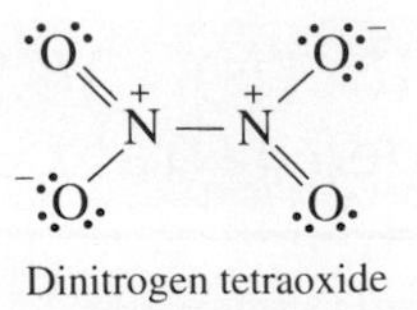

Dinitrogen tetraoxide

Molecules such as NO and NO_2 that have an odd number of electrons and therefore one unpaired electron are called **free radicals**. Nitrogen monoxide and nitrogen dioxide are unusual in that they are much less reactive than more typical free radicals, such as the methyl radical, $\cdot CH_3$. The structure $:N\dot{=}\dot{O}:$, in which a single electron is shared between two atoms and both atoms have an octet of electrons, helps us understand why NO is less reactive than many other free radicals, such as the $\cdot CH_3$ radical. In the methyl radical the odd electron is confined to the carbon atom, which has only seven electrons in its valence shell. Hence the methyl radical has a much greater tendency to acquire another electron than does NO. Most free radicals react rapidly with many other molecules. They exist for only very brief times under ordinary conditions, because there are always many other molecules present with which they can react. So most free radicals are not substances that we can find in a bottle or in a gas cylinder in the chemistry laboratory. Nevertheless, as we shall see, free radicals play a very important role in the chemistry of the troposphere. There they take part in many reactions, often as catalysts, and so are continually regenerated.

```
    H
    |
  •C—H
    |
    H
```
The methyl radical

Other Oxides of Nitrogen

Three other oxides of nitrogen are known: dinitrogen monoxide, N_2O (often called nitrous oxide); dinitrogen trioxide, N_2O_3; and dinitrogen pentaoxide, N_2O_5 (Table 16.1). Dinitrogen monoxide is a product of the bacterial decay of organic matter and nitrogen compounds in the soil and is present in very small concentrations in the atmosphere. In the laboratory it can be made by heating ammonium nitrate:

$$NH_4NO_3(s) \rightarrow N_2O(g) + 2H_2O(g)$$

Dinitrogen monoxide has an even number of electrons and can be described by the following two resonance structures:

$$:\ddot{O}=\overset{+}{N}=\ddot{N}:^- \longleftrightarrow \,^-\!:\ddot{\underset{\cdot\cdot}{O}}-\overset{+}{N}\equiv N:$$

Dinitrogen monoxide

Dinitrogen monoxide has the useful property of being rather soluble in fats, a property exploited in making whipped cream. Cream is packaged with N_2O under pressure to increase its solubility. When the pressure is released, some of the N_2O escapes to form tiny bubbles, which whip the cream.

It is isoelectronic with CO_2 and has a similar AX_2 linear structure. In small amounts it is a mild intoxicant, and because of this property it has been called laughing gas. In larger amounts it acts as an anesthetic (Box 14.1).

Exercise 16.1 At low temperatures, NO and NO_2 combine to form the deep-blue liquid oxide N_2O_3. Draw a Lewis structure for N_2O_3.

Nitric Acid and Nitrous Acid

Nitric acid, an important constituent of acid rain, is found in the environment wherever nitrogen oxides are present. It is produced industrially from ammonia by the **Ostwald process**. In this process, ammonia and oxygen are passed over a platinum–rhodium wire gauze heated to 900°C. The wire gauze acts as a catalyst for the reaction

$$4NH_3(g) + 5O_2(g) \rightarrow 4NO(g) + 6H_2O(g) \qquad \Delta H^\circ = -906 \text{ kJ}$$

Because this reaction is strongly exothermic, it supplies the heat necessary to keep the catalyst at the operating temperature. NO then reacts with the oxygen in air to give NO_2:

$$2NO(g) + O_2(g) \rightarrow 2NO_2(g)$$

Finally, in the industrial process the nitrogen dioxide is passed into water to give nitric acid and NO:

$$3NO_2(g) + H_2O(l) \rightarrow 2HNO_3(aq) + NO(g)$$

The NO is reoxidized to NO_2 by air and used to make more nitric acid. Demonstration 16.3 shows how the catalysis of the oxidation of ammonia by a copper wire can be observed in the laboratory.

Pure nitric acid boils at 84.1°C and freezes at −41.6°C. It is a strong acid in dilute aqueous solution:

$$HNO_3(aq) + H_2O(l) \rightarrow H_3O^+(aq) + NO_3^-(aq)$$

The salts of nitric acid are called **nitrates**, and they contain the nitrate ion, NO_3^-. All the common metal nitrates are soluble in water. Much of the nitric acid produced by industry is combined with ammonia to give ammonium nitrate, which is widely used as a fertilizer:

$$NH_3(g) + HNO_3(l) \rightarrow NH_4NO_3(s)$$

The nitric acid molecule has a planar AX_3 geometry around the nitrogen atom and can be represented by two equivalent resonance structures. The nitrate ion, NO_3^-, is isoelectronic with the CO_3^{2-} ion. It has a planar triangular AX_3 geometry and can be represented by three equivalent resonance structures:

DEMONSTRATION 16.3 Preparation of NO by the Catalytic Oxidation of NH_3

When a coil of copper wire is heated in a flame and then suspended over a warm, concentrated, aqueous solution of ammonia, the copper coil continues to glow brightly for a long time as the ammonia is oxidized on the metal surface in a strongly exothermic reaction. The heat produced in the reaction is usually sufficient to melt the copper wire, so molten copper drops into the ammonia solution. The solution turns blue due to the formation of the $Cu(NH_3)_4^{2+}$ ion.

When you draw resonance structures, move only electron *pairs*, which are thereby converted from bonding pairs to nonbonding pairs or from nonbonding pairs to bonding pairs.

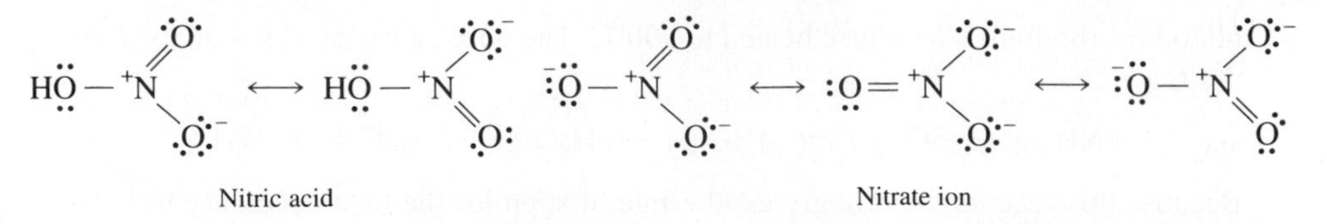

Nitric acid contains nitrogen in the highest +V oxidation state and is a good oxidizing agent, with the nitrogen generally being reduced to NO or NO_2. We saw in Chapter 10 that copper, which is not oxidized by acids such as HCl(*aq*), is oxidized by nitric acid. In the reaction with HCl(*aq*), the oxidizing agent is $H_3O^+(aq)$, but HNO_3 is a stronger oxidizing agent. Indeed, HNO_3 is a sufficiently strong oxidizing agent to dissolve all the metals except for a very few, such as gold and platinum, by oxidizing them to their positive ions (Table 13.1).

A dilute aqueous solution of **nitrous acid**, HNO_2, is obtained when a mixture of NO and NO_2 is passed into water:

$$NO(g) + NO_2(g) + H_2O(l) \rightarrow 2HNO_2(aq)$$

The salts of nitrous acid are the *nitrites*. The nitrite ion is an angular AX_2E molecule for which two resonance structures can be written:

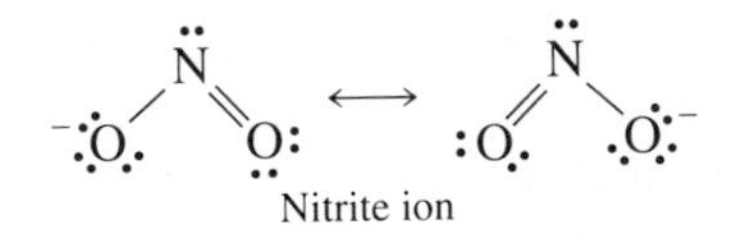
Nitrite ion

Because they are readily oxidized to nitric acid and nitrate ion, nitrous acid and the nitrites are good reducing agents:

$$NO_2^- + H_2O \rightarrow NO_3^- + 2H^+ + 2e^-$$

Sodium nitrite and sodium nitrate have long been used in the preservation of meat. Meat darkens when it is stored, because the hemoglobin in the blood is oxidized. Sodium nitrite and sodium nitrate are reduced to NO by compounds in the meat. The NO combines with the hemoglobin and retards its oxidation, thus maintaining a fresh red appearance to the meat. Nitrates and nitrites also prevent the growth of the bacterium that causes botulism, a dangerous and sometimes fatal form of food poisoning. However, concern has arisen about the possibility of a reaction between nitrite ion and certain organic compounds in the meat or in the digestive system to form nitrosamines, which are cancer-causing compounds.

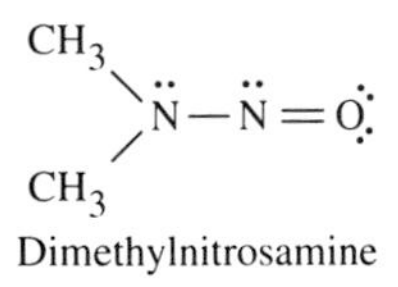
Dimethylnitrosamine

Exercise 16.2 Write equations for the half-reaction in which nitrate ion is reduced to **(a)** $NO_2(g)$, **(b)** NO(*g*), and **(c)** NH_4^+ in acidic aqueous solution.

16.2 OZONE AND PEROXIDES

Ozone

Ozone (trioxygen) is an allotrope of oxygen consisting of O_3 molecules. It is a pale blue gas that condenses to a deep blue liquid at −112°C. When an electric discharge is passed through oxygen, a small percentage of the oxygen is converted to ozone.

$$3O_2(g) \rightarrow 2O_3(g)$$

The characteristic odor of ozone is sometimes noticed during electric storms and near electric motors and laser printers.

Ozone is an angular AX_2E molecule that is isoelectronic with the nitrite ion. It may be represented by the following two resonance structures:

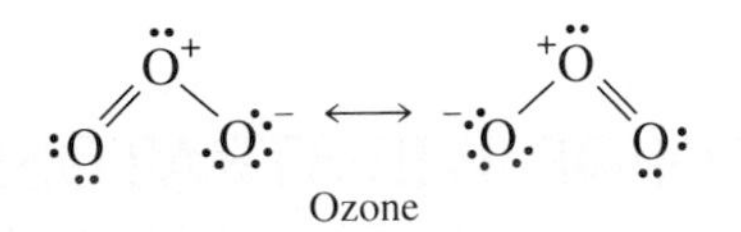

Ozone

The high voltage used in a computer laser printer produces an electric discharge that may cause ozone to form. This ozone is destroyed by an "ozone filter," but when the filter is used up, excess ozone is released and can be detected by its odor. A laser printer should therefore be well ventilated.

Ozone is an extremely powerful oxidizing agent. The only common oxidizing agent that is stronger is fluorine, F_2. Ozone can be used to replace chlorine for destroying bacteria in water by oxidation. Unlike chlorine, it leaves no taste but is generally less convenient and more expensive for small-scale use.

Peroxides and the Hydroxyl and Hydroperoxyl Radicals

When sodium is heated in a limited supply of air, it forms the expected Na_2O. With excess oxygen it forms instead the pale yellow peroxide Na_2O_2, which contains the **peroxide ion**, O_2^{2-}:

$$^{-}:\ddot{\underset{\cdot\cdot}{O}}-\ddot{\underset{\cdot\cdot}{O}}:^{-}$$

Peroxide ion

Barium behaves in the same way to give barium peroxide BaO_2. Barium peroxide, BaO_2, reacts with aqueous sulfuric acid to give a precipitate of insoluble barium sulfate and an aqueous solution of **hydrogen peroxide**, H_2O_2:

$$BaO_2(s) + H_2SO_4(aq) \rightarrow BaSO_4(s) + H_2O_2(aq)$$

Hydrogen peroxide in aqueous solution has many uses. Its use for bleaching hair is well known, but it is also important as a bleaching agent for textiles and for wood pulp and waste paper in paper making. Its bleaching action is due to its strong oxidizing properties. The oxygen in hydrogen peroxide has the unusual oxidation number -1 (rather than the usual -2). It is reduced to water, in which it has the oxidation number -2, and in so doing it oxidizes colored compounds to colorless compounds. Aqueous solutions of hydrogen peroxide decompose very slowly at room temperature:

$$2H_2O_2(aq) \rightarrow 2H_2O(l) + O_2(g)$$

This decomposition is catalyzed by many different substances, such as Fe(II) salts, manganese dioxide, and the enzyme catalase in blood (Demonstration 1.1).

Two free radicals that play important roles in the chemistry of the atmosphere are related to hydrogen peroxide. They are the **hydroxyl radical**, HO·, which we can think of as half a hydrogen peroxide molecule, and the **hydroperoxyl radical**, HOO·, which we can think of as a hydrogen peroxide molecule that has lost a hydrogen atom.

Exercise 16.3 Draw Lewis structures for **(a)** hydrogen peroxide, **(b)** the hydroxyl radical, **(c)** the hydroperoxyl radical, and **(d)** the ethyl radical. Assign oxidation numbers to the oxygen and carbon atoms in these molecules.

Exercise 16.4 What is the bond order for the bonds in the ozone molecule? Show that the observed bond length of 128 pm is consistent with this bond order.

Exercise 16.5 Write equations for the half reactions in which **(a)** ozone is reduced to oxygen and **(b)** hydrogen peroxide is reduced to water in acidic aqueous solution.

16.3 POLLUTION OF THE STRATOSPHERE

Solar Radiation and Atmospheric Photochemistry

All the energy needed to sustain life on earth is provided by the sun in the form of electromagnetic radiation. However, not all this radiation is beneficial to life; some of it is harmful and even lethal. The atmosphere plays a vital role in protecting life by absorbing the harmful short-wavelength radiation while letting through the longer-wavelength radiation on which life depends. Figure 16.2 shows as a function of wavelength the intensity of the radiation reaching the outer atmosphere and of the radiation that penetrates to the earth's surface. We saw in Chapter 2 that the temperature of the atmosphere varies in a complex way with increasing altitude (Figure 2.1). In the lowest layer, the **troposphere**, the temperature decreases with increasing distance from the warm surface of the earth. But about 15 km above the surface, the temperature begins to increase again, marking the bottom of a layer called the **stratosphere**. Radiation from the sun is absorbed by molecules in the stratosphere and converted to heat, which results in the observed temperature increase.

When the quinine molecules in tonic water are exposed to ultraviolet radiation, they absorb high-energy photons and fluoresce, emitting lower-energy photons in the visible region of the spectrum.

When a molecule absorbs visible or ultraviolet light, it gains energy and is excited to a higher-energy state—a state in which the electrons have a different arrangement than in the ground state. We discussed in detail the analogous process for free atoms in Chapter 6. The excited molecule can lose this excess electron energy in several different ways: (1) It can be converted to vibrational energy in the molecule, and this energy can then be converted in collisions with other molecules to translational energy—that is, into heat; (2) it can emit a photon in a process called *fluorescence*; (3) it can take part in reactions with other molecules that the molecule in its ground state does not undergo; (4) a bond can be broken if the energy of the photon is equal to or greater than the dissociation energy of the bond, which would

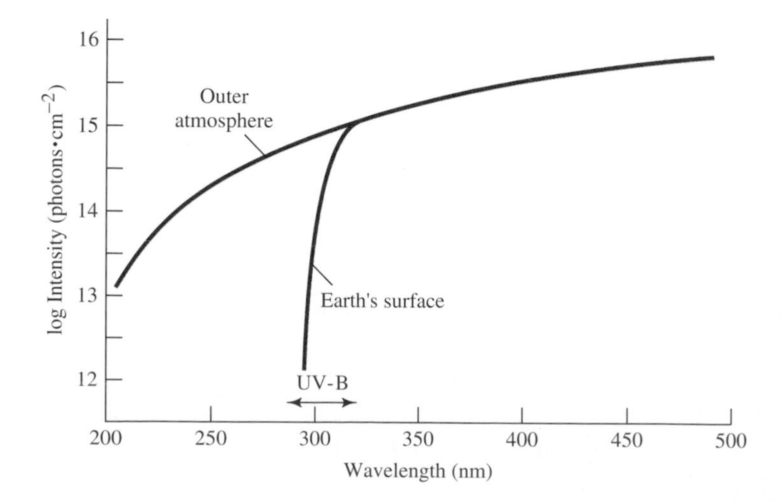

FIGURE 16.2 The Spectrum of Sunlight Most of the radiation that reaches earth from the sun is visible light (400 to 750 nm), but there is some ultraviolet ($\lambda < 400$ nm) and some infrared ($\lambda > 750$ nm). The difference between the upper and lower curves represents radiation absorbed by the atmosphere.

cause the molecule to dissociate into two fragments in a process called **photodissociation** or **photolysis**. Whether or not photodissociation can occur depends on whether or not the molecule absorbs radiation in the appropriate frequency range, which in turn depends on the energies of the excited states of the molecule. If the light absorbed is too low in frequency (too long in wavelength), then it may produce an electronically excited molecule, but photodissociation will not occur.

Electronically excited molecules and highly reactive atoms and free radicals produced by photodissociation are responsible for the very large number of reactions that occur in the atmosphere. Although our knowledge and understanding of these reactions has increased greatly since the 1970s, a great deal of further research is needed if we are to understand fully the effect of human activities on the atmosphere and the actions we must take to protect it.

As we saw in Chapter 6, the higher the frequency of light, the shorter the wavelength and the higher the energy. We saw that the minimum frequency of light needed to break a bond with a bond energy *BE* is given by $h\nu = BE/N_A$ and that the corresponding maximum wavelength is $\lambda = hcN_A/BE$. Table 16.2 gives the maximum wavelengths of radiation needed to break a variety of chemical bonds. These wavelengths are only approximate values because, as we saw in Chapter 9, the bond energy of a given bond varies somewhat from molecule to molecule. The molecules that are primarily involved in the absorption of ultraviolet radiation in the stratos-

TABLE 16.2 Bond Energies and the Interaction of Radiation with Molecules in the Atmosphere

Bond	Bond Energy ($kJ \cdot mol^{-1}$)	Wavelength* (nm)	Type of Radiation	Interaction with the Atmosphere	
N≡N	941	127			
C≡C	812	148			
C=O	707	169	Ultraviolet (UV)	Absorbed in the mesosphere	
C=C	619	194			
				205 nm Absorbed by $O_2 \rightarrow 2O$	in the stratosphere (205–320 nm)
O=O	498	242		240 nm	
H—H	436	275	UV-C	Absorbed by $O_3 \rightarrow O_2 + O$	
C—H	413	289		290 nm	
			UV-B	Partially absorbed by O_3 320 nm	
C—C	348	345		Penetrates to the	
C—O	335	358	UV-A	troposphere and	
C—Cl	326	367		earth's surface 400 nm	
C—N	293	410			
Cl—Cl	242	492	Visible		
Br—Br	190	632		750 nm	
I—I	149	803	Infrared		

* Maximum wavelength of radiation that can break the bond

phere are oxygen and ozone. The photodissociation of oxygen molecules into oxygen atoms absorbs all the radiation up to 240 nm:

$$O_2 \xrightarrow{\lambda < 240\text{ nm}} 2O$$

Ozone molecules absorb radiation up to a wavelength of 320 nm and dissociate into an oxygen molecule and an oxygen atom:

$$O_3 \xrightarrow{\lambda < 320\text{ nm}} O_2 + O$$

Ozone also absorbs, although less strongly, radiation up to a wavelength of 350 nm, which is not energetic enough to dissociate the molecule but raises it to an excited state:

Electronically excited atoms and molecules are usually denoted by an asterisk. Thus O* is an excited oxygen atom, and O_2* is an excited oxygen molecule.

$$O_3 \xrightarrow{\lambda < 350\text{ nm}} O_3^*$$

The excited ozone molecule eventually loses this absorbed energy by transferring it to other molecules in collisions:

$$O_3^* + M \rightarrow O_3 + M(\text{increased kinetic energy})$$

Here M represents any molecule with which O_3* does not react. Thus, oxygen and ozone absorb essentially all the radiation up to 320 nm, and ozone absorbs most of the radiation in the range 320 to 350 nm.

From Table 16.2 we can see that C—H, C—C, C—O, and C—N bonds can be broken by the ultraviolet (UV) radiation that reaches the earth, so organic and biomolecules are susceptible to decomposition by ultraviolet light. The most dangerous radiation reaching the earth's surface is that in the 290 to 320 nm range, called UV-B by biological scientists. UV-B is harmful to many species of plants, animals, and microorganisms. It causes sunburn and can cause skin cancer in humans. And it is believed to be a cause of eye cataracts and reduced efficiency of the body's immune system. Any decrease in the amount of ozone in the atmosphere that would allow more UV-B to reach the earth's surface could lead to an increase in the incidence of such illnesses as skin cancer and, if continued long enough, could eventually threaten life itself. Next we shall discuss the evidence that the ozone concentration has been decreasing in recent years and the reasons for the decrease.

Exercise 16.6 **(a)** Derive the relationship

$$\lambda_{max} = \frac{1.20 \times 10^5}{BE}$$

between the bond energy, *BE* (the energy needed to break one mole of bonds, in kilojoules per mole), and the maximum wavelength of radiation, λ_{max} (in nanometers), that will break the bond.

(b) Use this relationship to show that the bonds in CO_2 cannot be broken in the troposphere but may be broken in the stratosphere.

Exercise 16.7 **(a)** Calculate the maximum wavelength of radiation that should dissociate a chlorine atom from a CF_2Cl_2 molecule if the energy needed to dissociate the Cl atom is 318 $kJ \cdot mol^{-1}$?

(b) Why is CF_2Cl_2 not dissociated in the troposphere?

The Ozone Layer

We have seen how ozone plays a vital role in screening out harmful ultraviolet radiation. Although ozone is continuously destroyed as it absorbs ultraviolet radiation because there is a virtually constant concentration of ozone in the stratosphere, it must also continuously form. Oxygen atoms produced by the photochemical decomposition of oxygen molecules react with other oxygen molecules to give ozone:

$$O_2 \xrightarrow{\lambda < 240\text{ nm}} 2O$$

$$O_2 + O + M \rightarrow O_3 + M(\text{increased kinetic energy}) \qquad \Delta H^\circ = -105\text{ kJ}\cdot\text{mol}^{-1} \qquad (1)$$

Because reaction (1) is exothermic, a third molecule M is needed to carry away some of the energy produced. Otherwise this energy would cause the ozone molecule to decompose back into an O_2 molecule and an O atom. M can be any other unreactive molecule. Its kinetic energy is increased in the collision, thereby increasing the temperature of the surroundings. The heat given off in this and similar reactions causes the temperature of the stratosphere to increase with increasing altitude (Figure 2.1).

As we saw, ozone is decomposed when it absorbs ultraviolet radiation with a wavelength of < 320 nm:

$$O_3 \xrightarrow{\lambda < 320\text{ nm}} O_2 + O \qquad (2)$$

Ozone is also destroyed by reaction with oxygen atoms:

$$O + O_3 \rightarrow 2O_2 \qquad \Delta H^\circ = -390\text{ kJ} \qquad (3)$$

This is another of the reactions that contributes to the heating of the stratosphere. The rate of ozone loss by reactions (2) and (3) is just balanced by the rate of formation by reaction (1). Thus, there is a constant steady-state concentration—*not* an equilibrium concentration—of ozone (Figure 16.3). If equilibrium were attained,

B attains a constant *equilibrium* concentration when it forms A as fast as it is formed from A:

$$A \underset{2}{\overset{1}{\rightleftharpoons}} B \qquad \text{rate 2 = rate 1}$$

B attains a constant *steady-state* concentration when it reacts to give C as fast as it is formed from A:

$$A \xrightarrow{1} B \xrightarrow{2} C \qquad \text{rate 2 = rate 1}$$

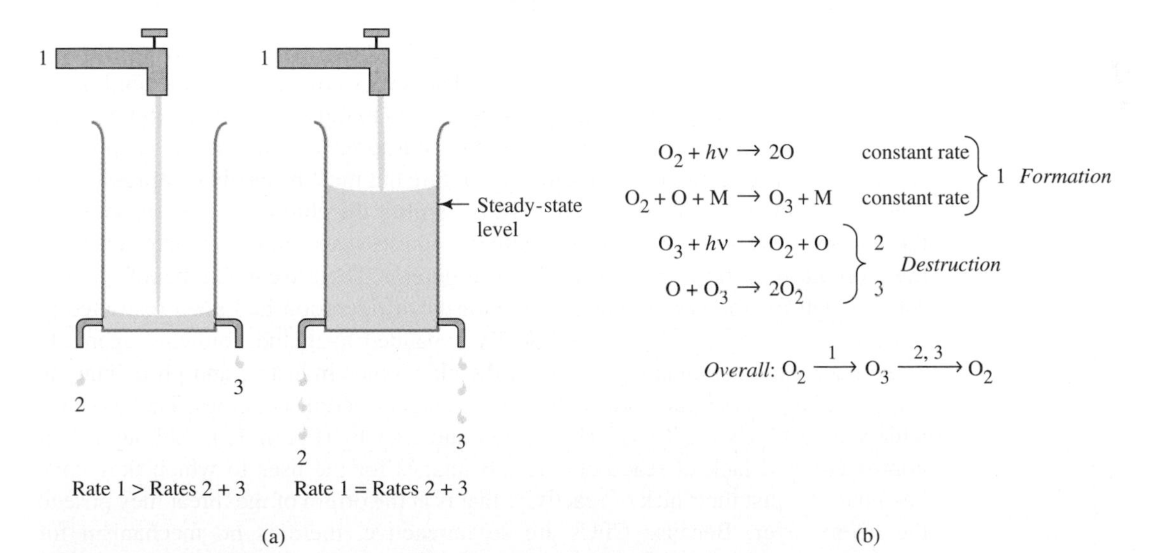

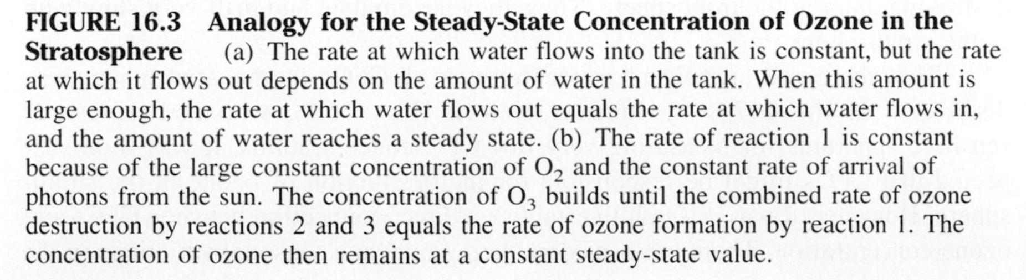

FIGURE 16.3 Analogy for the Steady-State Concentration of Ozone in the Stratosphere (a) The rate at which water flows into the tank is constant, but the rate at which it flows out depends on the amount of water in the tank. When this amount is large enough, the rate at which water flows out equals the rate at which water flows in, and the amount of water reaches a steady state. (b) The rate of reaction 1 is constant because of the large constant concentration of O_2 and the constant rate of arrival of photons from the sun. The concentration of O_3 builds until the combined rate of ozone destruction by reactions 2 and 3 equals the rate of ozone formation by reaction 1. The concentration of ozone then remains at a constant steady-state value.

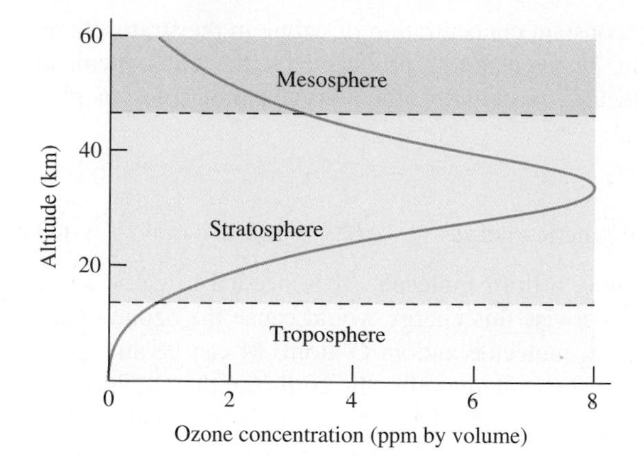

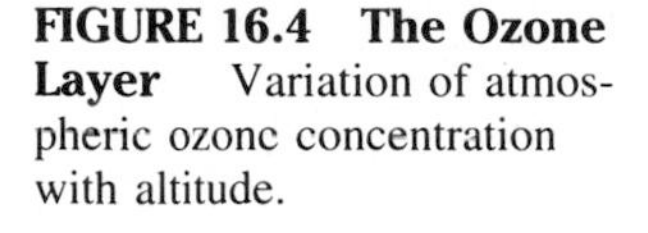
FIGURE 16.4 The Ozone Layer Variation of atmospheric ozone concentration with altitude.

the concentration of ozone in the stratosphere would be negligibly small. The greater the height above the earth, the more ultraviolet radiation is available to form ozone, but because the pressure is lower, fewer oxygen molecules are present to react. As a result, the concentration of ozone molecules reaches a maximum of 20 to 30 ppm at a height of 25 to 35 km. It is this band of increased ozone concentration that is called the **ozone layer** (Figure 16.4). Although the actual concentration of ozone is very low, it is this ozone layer that protects life on earth by absorbing harmful ultraviolet radiation.

If all the ozone in the stratosphere were brought to the surface of the earth, a layer of pure ozone only 3 mm thick would form.

A current concern is that atmospheric pollutants may be decreasing the amount of ozone in the ozone layer. One type of pollutant that poses a major threat to the ozone layer is the chlorofluorocarbons.

Chlorofluorocarbons and the Depletion of the Ozone Layer

Chlorofluoroalkanes such as CCl_2F_2 and CCl_3F are commonly known as **chlorofluorocarbons (CFCs)** or Freons. These substances were first manufactured in the 1930s for use as the operating fluid in refrigerators to replace the toxic and odorous SO_2 and NH_3 that had been used for this purpose. The operating fluid for a refrigerator must be gaseous at room temperature but must be easily compressible to a liquid, as is the case for SO_2 and NH_3. Several of the chlorofluoromethanes have these properties and in addition are nontoxic, odorless, and nonflammable. They are therefore ideal as the working fluid in a refrigerator. Their use made the refrigerator safe enough for the home, whereas previously refrigeration had been restricted to industry. The uses of CFCs subsequently expanded to include blowing agents to make plastic foams, cleaning agents in the electronics industry, and propellants in aerosol spray cans of hair sprays, deodorants, paints, and insecticides. The two most widely used CFCs are CCl_3F (Freon 11) and CCl_2F_2 (Freon 12). Although their nontoxicity and lack of reactivity are advantages for the uses to which they have been put, it is just their lack of reactivity that is at the origin of the threat they pose to the ozone layer. Because CFCs are so unreactive, there is no mechanism for destroying them in the troposphere. Thus, they accumulate and drift very slowly up to the stratosphere.

That there are significant concentrations of CFCs in the atmosphere was discovered in the 1970s by James Lovelock (Box 16.1), who had developed sensitive apparatus for measuring very low CFC concentrations. It was soon suspected that CFCs might be responsible for the destruction of ozone in the stratosphere. However, it was difficult to establish a direct connection between CFCs and ozone concentration. There are considerable regional and seasonal variations in the

BOX 16.1 James Lovelock and the Gaia Hypothesis

The relationship between life and the environment is a topic of considerable current interest. One of the most famous models to describe this relationship is the so-called Gaia hypothesis suggested by James Lovelock (Figure A). Lovelock is an independent scientist specializing in the design of very sensitive analytical instruments. He produced instruments used to measure the very small concentrations of CFCs in the atmosphere. It was these measurements that led to the realization that the stability of CFCs poses a threat to the ozone layer.

Lovelock was a consultant to NASA on the project to search for evidence of life on Mars. He asked the question, How could an alien observing the earth from a distance deduce that the planet supports life? He realized that the unique composition of the earth's atmosphere provided the necessary information. The earth's atmosphere is far from chemical equilibrium, whereas the atmospheres of the other planets have reached equilibrium. Our atmosphere contains too much oxygen and not nearly enough carbon dioxide to be in equilibrium. Such a composition could result only from the activities of living organisms, because all inanimate processes eventually lead to equilibrium. That the atmospheres of all other planets have reached equilibrium is, according to Lovelock, sufficient evidence that no life is present there. He believed that no evidence of life would be detected by the instruments that were landed on Mars, and that turned out to be the case.

Rather than being in equilibrium, the earth's atmosphere is in a stable steady state that provides the conditions needed to sustain life as we know it. There is the right amount of oxygen to support the respiration of animals and the right amount of CO_2 for plants to photosynthesize. The biosphere is not too hot and not too cold to support life. Fresh water is renewed by the hydrologic cycle, and nitrogen needed for proteins is continuously provided by the nitrogen cycle.

Figure A James Lovelock (1919–) *James Lovelock is a specialist in sensitive analytical techniques, including gas chromatography (Chapter 14).*

The conditions for life are controlled by life itself. Lovelock suggested that "The earth's living matter, air, oceans, and land surface form a complex system which can be seen as a single living organism and which has the capacity to keep our planet a fit place to live." This statement is the Gaia hypothesis, named after the ancient Greek goddess of the earth. The biosphere and the atmosphere appear to have existed in their present steady states during the last 500 million years. Only within the last century have humans developed large-scale technology that could disturb the balance by exposing life to ultraviolet radiation, interfere with the temperature maintenance of the biosphere, and introduce toxic substances into the air and water. These are the problems facing environmental scientists—and all people—today.

ozone concentration in the stratosphere and variations at times of increased sunspot activity. However, it became clear by 1987 that an unexpected decrease of as much as 50% in the ozone concentration occurs over Antarctica each spring. This severe thinning of the ozone layer has been called the *ozone hole*. This description is an exaggeration, because the ozone is depleted rather than removed entirely (Figure 16.5).

When CFCs reach the stratosphere, they are photodissociated by ultraviolet radiation ($\lambda \leq 250$ nm). This reaction breaks a C—Cl bond to produce a chlorine atom and a CF_2Cl radical:

$$CF_2Cl_2 + h\nu \rightarrow \cdot CF_2Cl + Cl\cdot \qquad (4)$$

This reaction initiates a chain reaction, because the chlorine atoms react with ozone molecules, producing an oxygen molecule and chlorine monoxide, which is a reactive free radical:

$$Cl\cdot + O_3 \rightarrow ClO\cdot + O_2 \qquad (5)$$

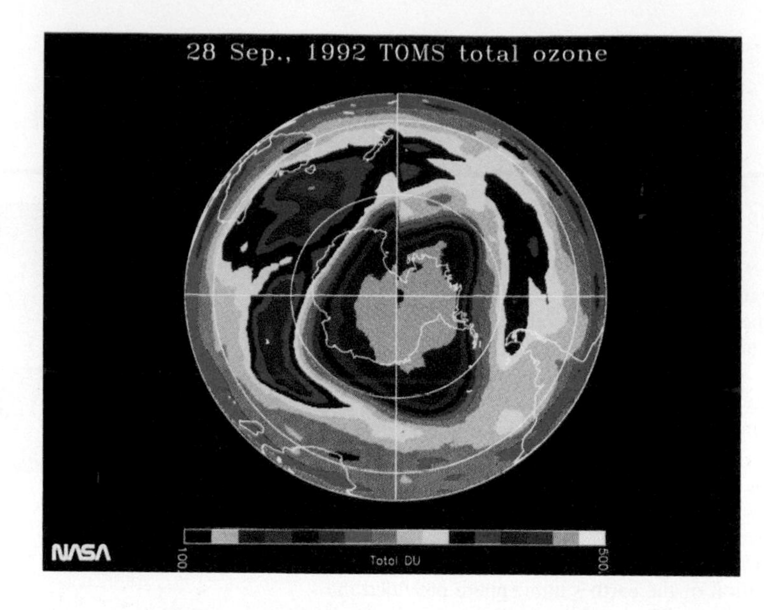

FIGURE 16.5 The Ozone "Hole" Map of total ozone in the Southern Hemisphere on September 28, 1992, illustrating the Antarctic ozone "hole." This oval feature (light pink area) covers most of the Antarctic.

The ClO· then reacts with an oxygen atom from the photodissociation of O_2, producing an oxygen molecule and regenerating the chlorine atom:

$$\text{ClO}\cdot + :\dot{\text{O}}\cdot \rightarrow \text{Cl}\cdot + \text{O}_2 \tag{6}$$

In this way, a chain reaction is set up, the result of which is that oxygen atoms and ozone molecules are converted to oxygen molecules:

$$\text{O} + \text{O}_3 \rightarrow 2\text{O}_2$$

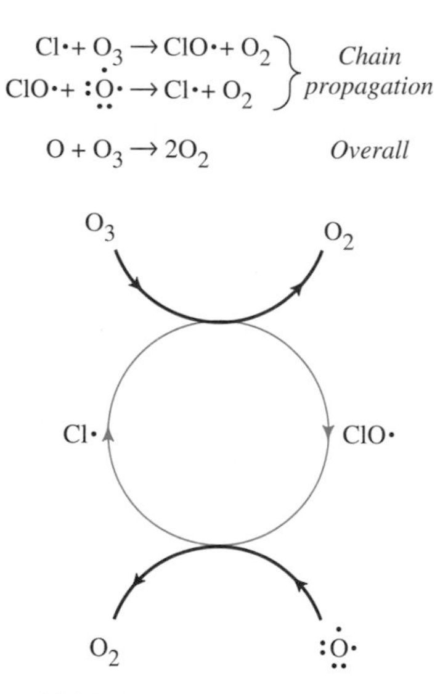

FIGURE 16.6 Cl-Atom-Catalyzed Decomposition of Ozone In this reaction chlorine atoms react with O_3 molecules to form O_2 molecules and ClO· radicals. The ClO· radicals react with O atoms to form O_2 molecules and re-form Cl atoms. These are the two chain-propagation steps of a chain reaction. The result is that O_3 molecules and O atoms produce O_2 molecules in a reaction in which Cl atoms behave as a catalyst.

In this way, a single chlorine atom can destroy many ozone molecules, probably as many as 100,000 (Figure 16.6). This chain reaction provides a mechanism for the destruction of ozone, in addition to the natural mechanism described by reactions (2) and (3). The overall rate of ozone loss is therefore increased, and the steady-state concentration of ozone is correspondingly decreased.

That chlorine from CFCs is responsible for the depletion of ozone was unambiguously established by observations made first in 1987 during stratospheric flights through the Antarctic ozone hole. During these flights, it was found that the higher the concentration of ClO, the lower the ozone concentration. Although the details are not completely understood, the mechanism of the large-scale ozone depletion in the stratosphere above the Antarctic is now fairly clear. The very cold conditions of the dark Antarctic winter lead to the formation of clouds of ice crystals called *polar stratospheric clouds*. Several chlorine-containing molecules (such as HCl and chlorine nitrate, $ClONO_2$) that arise from the chlorine atoms produced by the photodissociation of CFCs condense on the ice crystals and remain trapped there during the dark polar winter. With the arrival of the sun in the spring, these molecules are photodissociated to give Cl atoms, which react with ozone or oxygen atoms to give ClO radicals in locally high concentrations. These radicals then destroy a large amount of ozone as described by reactions (5) and (6). The creation of the Antarctic ozone hole is seasonal—as summer progresses, the ozone concentration increases again. Nevertheless, it now seems clear that ozone is being depleted on a global scale, although to a much smaller extent than in the Antarctic, and that the Cl atoms from CFCs are largely responsible for the depletion. However, it has not yet been clearly established that there has been a corresponding increase in the amount of ultraviolet radiation reaching the earth's surface.

$$\left.\begin{array}{l} NO + O_3 \rightarrow NO_2 + O_2 \\ NO_2 + O \rightarrow NO + O_2 \end{array}\right\} \textit{Chain propagation}$$

$$O_3 + O \rightarrow 2O_2 \qquad \textit{Overall}$$

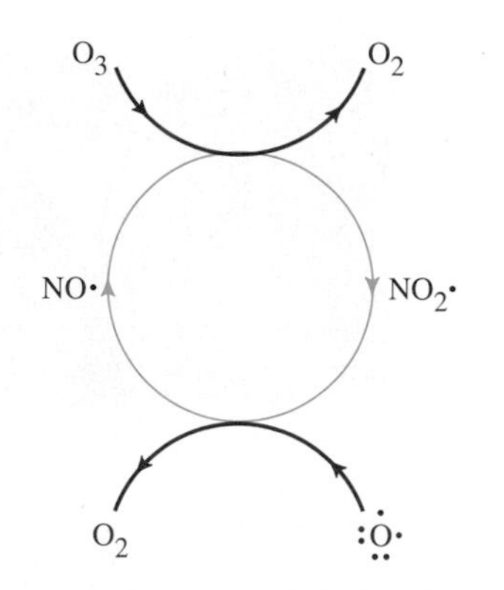

FIGURE 16.7 NO-Catalyzed Decomposition of Ozone In this chain reaction, NO is first used up and then re-formed so that it behaves as a catalyst, just as Cl atoms behave as a catalyst for the same reaction.

Ozone Depletion by Nitrogen Oxides

Another catalyst for the decomposition of ozone is the homogeneous catalyst NO, which reacts with ozone to form O_2 and NO_2:

$$NO + O_3 \rightarrow NO_2 + O_2$$

The NO_2 thus formed reacts with an oxygen atom to form O_2 and regenerate NO:

$$NO_2 + O \rightarrow NO + O_2$$

This reaction leads to a chain reaction (Figure 16.7) that removes ozone molecules and O atoms in the overall reaction

$$O + O_3 \rightarrow 2O_2$$

Nitrogen monoxide forms in the stratosphere from dinitrogen monoxide, N_2O, which forms initially on the earth's surface by the oxidation of ammonia from decaying organisms, fertilizers, and animal and human wastes. The dinitrogen monoxide is released into the troposphere and, because it is relatively unreactive, slowly diffuses into the stratosphere. There it reacts with oxygen atoms to form nitrogen monoxide, NO:

$$N_2O + O \rightarrow 2NO$$

The concentration of N_2O in the atmosphere has been found to be slowly increasing, presumably as a consequence of the increased use of nitrogen fertilizers and the growing world population. So the destruction of ozone by NO in the stratosphere is probably another factor contributing to the thinning of the ozone layer.

Solutions to the Problem of Ozone Depletion

One obvious solution to the ozone problem is to prohibit the use of CFCs. In an unusually rapid political acceptance of scientific predictions, the use of CFCs as aerosol propellants was banned in the United States and Canada in 1978. And in 1987 an international treaty called the Montreal Protocol on Substances that Deplete the Ozone Layer was signed in which a large number of countries agreed to cut back the production of CFCs. Following this agreement, many governments in Europe and North America agreed in 1989 to phase out the production of CFCs entirely by 1995. A challenge for chemists is to find suitable replacement substances.

Protecting the ozone layer from depletion by nitrogen oxides is another challenging problem. The probable need for increased amounts of fertilizers to produce enough food for an ever-increasing world population will make it difficult to decrease the amount of N_2O that enters the atmosphere.

16.4 PHOTOCHEMICAL SMOG

In Section 16.3, we discussed reactions occurring in the stratosphere that are a cause for environmental concern. Here and in Section 16.5, we consider the effects of pollutants in the troposphere, the part of the atmosphere in which we live and breathe.

(a) (b)

FIGURE 16.8 Photochemical Smog A view of Los Angeles (a) on a clear day and (b) on a smoggy day. The brown color of the smog is due to nitrogen dioxide, NO_2.

Smog is derived from the words "*sm*oke" and "f*og*."

Humans have been polluting the atmosphere for centuries. In the nineteenth century and in the first half of the twentieth century, sulfur-containing coal, burned in an open hearth, was the principal means of heating in cities in Britain. The cold, damp weather of a British winter typically produces mist and fog, which, when combined with copious smoke and sulfur dioxide, SO_2, from coal fires, forms a suffocating smog. In 1952 an estimated 4000 Londoners were killed by a particularly severe smog. After this environmental disaster, the burning of coal was banned in London and other cities, and smog is no longer a major problem in London. But similar conditions have occurred much more recently in Eastern Europe and China.

Another kind of smog—namely, the **photochemical smog** formed by the action of sunlight on substances in automobile exhaust—is especially common in cities with a dry, sunny climate and a high density of automobiles. Los Angeles and Mexico City are the most notorious examples, but photochemical smog now occurs in an increasing number of cities around the world. It is characterized by a yellow-brown haze and the presence of substances that irritate the respiratory tract and the eyes (Figure 16.8). The main products of the combustion of gasoline are carbon dioxide and water, but there are also a number of minor products—particularly oxides of nitrogen, produced by the combination of N_2 and O_2 at the high temperatures in an internal combustion engine. These oxides of nitrogen undergo photochemical reactions that produce ozone and other toxic substances. The toxicity of ozone is, of course, not a problem in the stratosphere, as life does not exist there.

Ozone and Other Constituents of Photochemical Smog

Nitrogen dioxide is decomposed by visible light of wavelengths shorter than about 400 nm to give nitrogen monoxide and oxygen atoms:

$$NO_2 + h\nu \xrightarrow{(\lambda < 400\text{ nm})} NO + O$$

The oxygen atoms produced in this way combine with oxygen molecules to give ozone, exactly as they do in the stratosphere:

$$O + O_2 \rightarrow O_3$$

However, we saw in Section 16.3 that nitrogen monoxide *destroys* ozone to re-form nitrogen dioxide,

$$NO + O_3 \rightarrow NO_2 + O_2$$

so we might expect that there would be no overall production of ozone. Unfortunately, the other components of photochemical smog, carbon monoxide and hydrocarbons from the incomplete combustion of gasoline, catalyze the conversion of nitrogen monoxide to nitrogen dioxide. Thus, the rate of formation of O_3 is greater than its rate of removal by NO, and the ozone concentration builds up. In addition, oxygen atoms react with water molecules to form hydroxyl radicals:

$$:\dot{\underset{\cdot\cdot}{O}}\cdot + H_2O \rightarrow 2\cdot OH$$

Hydroxyl radicals attack hydrocarbons to give alkyl radicals:

$$CH_4 + \cdot OH \rightarrow \cdot CH_3 + H_2O$$

Then, in a lengthy series of reactions, aldehydes and peroxyacetylnitrate (PAN) are formed.

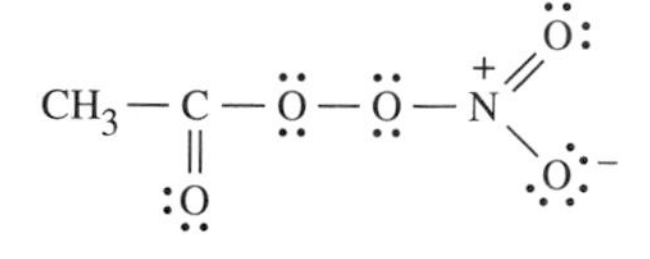

Peroxyacetylnitrate

Toxic Effects of Photochemical Smog

Overall, it is the combination of nitrogen oxides, carbon monoxide, unburned hydrocarbons, and sunlight that leads to the production of photochemical smog. Table 16.3 gives the active ingredients in a typical photochemical smog. The most noxious are the very strong oxidizing agents ozone and PAN, both of which attack the respiratory system and cause severe eye irritation. Although their effects are very unpleasant, they probably cause no permanent harm to most individuals, but they can be life-threatening to people with respiratory ailments. These compounds are more toxic to plants, because they can severely reduce the rate of photosynthesis. For example, ozone adds to a double bond to form an *ozonide*, which in the presence of water decomposes to give aldehydes and ketones:

Because it is the total amount of nitrogen oxides (mainly NO and NO_2) in the atmosphere that is important and the relative amounts of NO and NO_2 vary widely with time and location, atmospheric concentrations of nitrogen oxides are typically reported as $[NO_x]$, where NO_x stands for all oxides of nitrogen.

$$RHC{=}CR'R'' \xrightarrow{O_3} \underset{\text{Ozonide}}{RHC{-}CR'R''\ (\text{O–O–O ring})} \xrightarrow{H_2O} \underset{\text{Aldehyde}}{RHC{=}O} + \underset{\text{Ketone}}{O{=}CRR'}$$

Cell membranes are particularly susceptible to this type of attack, because they are partly composed of esters of unsaturated carboxylic acids (Chapter 17). Although ozone in the stratosphere (often called ''upper-atmosphere ozone'') is essential to

TABLE 16.3 Components of Photochemical Smog

Gas	Major Source	Concentration (ppm)
Carbon monoxide, CO	Automobiles	200–2000
Hydrocarbons	Automobiles, incomplete combustion	20–50
NO_x gases	Automobiles, coal burning	5–35
Ozone, O_3	Photochemical reactions	2–20
Peroxyacetylnitrate (PAN)	Photochemical reactions	1–4

life, ozone in the troposphere (often called "ground-level ozone"), particularly in the air we breathe, is a most undesirable pollutant.

Solutions to the Problem of Photochemical Smog

The most obvious solution to the problem of photochemical smog is to eliminate or at least reduce the emissions of unburned hydrocarbons, carbon monoxide, and nitrogen oxides from the exhaust gases of automobiles. The state of California took the lead in instituting emissions controls. Many other jurisdictions have since done the same, and regulations are becoming increasingly strict. These regulations are being met by improving the design of engines to give a cleaner exhaust and by the installation of catalytic converters (Chapter 15), which remove NO_x, CO, and hydrocarbons from the exhaust.

A possible longer-term solution is the development of battery-powered or hydrogen-powered vehicles. However, much development is needed before these vehicles become a practical solution.

16.5 ACID RAIN

Unpolluted rainwater is slightly acidic, with a pH of about 5.6, because it is a dilute solution of carbonic acid, H_2CO_3, in equilibrium with dissolved carbon dioxide:

$$CO_2(aq) + H_2O(l) \rightleftharpoons H_2CO_3(aq)$$

$$H_2CO_3(aq) + H_2O(l) \rightleftharpoons H_3O^+(aq) + HCO_3^-(aq)$$

Rain falling in industrialized regions is more acidic. It is this rain with a pH < 5.6 that we call **acid rain**. In parts of eastern North America, for example, the average pH of rainwater is about 4.2, representing a factor of 25 increase in acidity (Figure 16.9). Individual rainfalls with a pH as low as 2.1, approximately 5000 times more acidic than normal rain and as acidic as vinegar, have been recorded.

For pH = 5.6, $[H_3O^+] = 2.5 \times 10^{-6}\ mol \cdot L^{-1}$. For pH = 4.2, $[H_3O^+] = 6.3 \times 10^{-5}\ mol \cdot L^{-1}$. The $[H_3O^+]$ ratio is

$$\frac{6.3 \times 10^{-5}}{2.5 \times 10^{-6}} = 25$$

Sources of Acid Rain

The main causes of acid rain are oxides of sulfur and oxides of nitrogen present in the atmosphere in higher than normal amounts in industrialized regions. The iron and steel industry and coal-burning power plants use enormous quantities of coal. Coal typically contains 2 to 3% sulfur. When coal is burned, the sulfur is emitted to the atmosphere as sulfur dioxide, SO_2. Sulfur dioxide is also produced in vast quantities by the smelting of metal sulfide ores, as in the production of copper (Chapter 10):

$$Cu_2S(l) + O_2(g) \rightarrow 2Cu(s) + SO_2(g)$$

Sulfur dioxide is oxidized in the atmosphere to sulfur trioxide, which then reacts with water droplets in clouds to form a solution of sulfuric acid:

$$SO_2(g) \xrightarrow{\text{oxidation}} SO_3(g)$$

$$SO_3(g) + H_2O(l) \rightarrow H_2SO_4(aq)$$

We saw in Chapter 7 that the direct reaction between SO_2 and oxygen is very slow. However, it appears that the rate of oxidation of SO_2 in the atmosphere is increased in several different ways—for example, by reaction with hydroxyl radicals, with

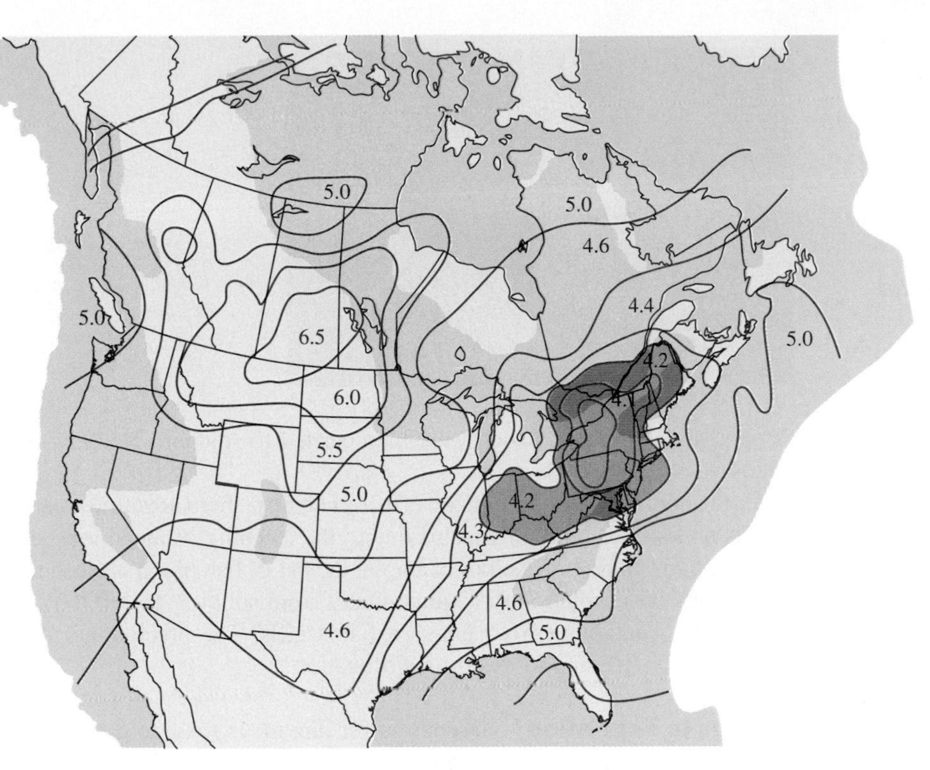

FIGURE 16.9 Average pH of Rain in North America Rain in parts of the northeastern United States and southern Canada has a pH of 4.2 or less. This poses a severe threat of acid-rain damage, particularly in areas where the underlying rock is granite (shown in pale pink). [After *Environmental Chemistry*, N. Bunce, Wertz Pub. Co., 1991.]

hydrogen peroxide dissolved in raindrops, and with O_2 on the surface of suspended metal oxide particles, which act as catalysts, just as V_2O_5 does in the industrial contact process.

Another major source of acid rain is the oxides of nitrogen emitted in the exhaust gases of automobiles, which are converted to nitric acid by hydroxyl radicals and by other oxidizing agents in the atmosphere:

$$\cdot NO_2 + \cdot OH \rightarrow HNO_3$$

Effects of Acid Rain

The detrimental effects of acid rain are threefold: the acidification of lakes and rivers, with the subsequent loss of fish and other aquatic life; damage to trees and other plants; and the corrosion of stonework and glass in buildings.

Acidification of Lakes and Rivers The damaging effects of acid rain on lakes and rivers depend very much on the type of underlying rock. A basic rock such as limestone neutralizes the acid, whereas aluminosilicate rocks (Chapter 19) such as granite do not. Thus, lakes in the northeastern United States, eastern Canada, and Scandinavia, which are underlain by granite, are the most susceptible. The normal pH of lakes in these regions is in the range 6.5 to 7.0, but today many of these lakes have a pH of 5.0—a pH at which fish cannot survive. The role of acid rain in killing fish was first realized in Norway in the 1970s. It immediately became a political issue: The acids were traced to power stations in Britain, built with tall chimneys to

FIGURE 16.10 Effects of Acid Rain on Trees This photo was taken in 1990 on Mt. Mitchell, North Carolina.

disperse the pollutants, which were then carried to southern Norway by prevailing winds.

In addition to the direct toxic effect of acidity, there are almost certain indirect effects on fish and other organisms due to the leaching of metal ions such as Al^{3+} from the surrounding rocks. For example, the pH of fish blood is close to 7.4, and in this slightly basic solution Al^{3+} precipitates from solution as $Al(OH)_3$. This occurs when water containing Al^{3+} enters a fish's gills. The precipitated $Al(OH)_3$ then clogs the gills, leading to death by suffocation.

Damage to Vegetation Gaseous sulfur dioxide is toxic to plants. At concentrations of 0.1 to 1 ppm, it inhibits growth and causes observable damage to leaves and flowers. Leaves are also damaged by acid rain of pH 3.5 or less, and excessive acidification of soil is harmful to most plants. In the Black Forest of Germany, 30 to 50% of the trees are dead or dying, and although the proof is not yet conclusive, the link to acid rain seems very likely. This problem first appeared in the 1980s and particularly affects coniferous trees (Figure 16.10). It has been suggested that magnesium, an essential component in chlorophyll, is removed from pine needles by the combined effects of ozone and acids. In Canada and the northeastern United States, there has been similar but less drastic damage to maple trees. Another harmful effect of acid rain may be that it leaches essential metal ions such as Ca^{2+} and Mg^{2+} from soil as soluble salts. Whatever the mechanism, it is very probable that acid rain is a factor leading to the death of trees.

The Taj Mahal in India is built entirely of marble and is therefore prone to acid-rain damage.

Destruction of Building Materials For centuries limestone and marble have been used as building materials. Even unpolluted rain dissolves limestone and marble by the same process that forms caves (Box 10.1):

$$CaCO_3(s) + H_2CO_3(aq) \rightleftharpoons Ca^{2+}(aq) + 2HCO_3^-(aq)$$

The position of this equilibrium lies far to the left, so the equilibrium concentration of Ca^{2+} is only a few ppm, and $CaCO_3$ dissolves exceedingly slowly. However, the hydrogen ion concentration in acid rain is much higher, and the position of the equilibrium

$$CaCO_3(s) + H^+(aq) \rightleftharpoons Ca^{2+}(aq) + HCO_3^-(aq)$$

is much farther to the right. So the equilibrium concentration of Ca^{2+} is considerably greater and the rate of dissolution of $CaCO_3$ correspondingly faster. Ancient monuments that have withstood centuries or even millennia of natural weathering are now deteriorating rapidly (Figure 16.11). It has been claimed that ancient buildings in Athens have deteriorated more since 1970 than in the previous 2000 or more years.

FIGURE 16.11 Effects of Acid Rain on Limestone These gargoyles decorating the Notre Dame Cathedral in Paris have been badly damaged by acid rain.

Solutions to the Problem of Acid Rain

Because we already discussed possible ways for decreasing NO_x emissions from automobiles in Chapter 15, we will concentrate here on SO_2. Possible ways to minimize the harmful effects of SO_2 include

1. Minimizing the production of SO_2;
2. Diluting the SO_2 so that it is no longer harmful;
3. Removing the SO_2 before the gases are released to the atmosphere by converting it to a harmless substance.

Emissions from coal-burning power stations can be minimized by decreasing the amount of sulfur in the coal used. Sulfur in the form of metal sulfides can be removed by grinding the coal to a fine powder and then using a flotation method to separate the metal sulfides, which are much denser than coal. However, this does not remove the sulfur that occurs as organic compounds in the coal. Recent research has suggested that it may be possible to use genetically engineered microorganisms to remove as H_2S the sulfur in the organic sulfur compounds in coal without breaking down the carbon skeleton of the coal.

Dilution is the cheapest method. It can be achieved by emitting the SO_2 from a very tall stack so that it is considerably diluted before it reaches ground level (Figure 16.12). This alleviates the local problem but does nothing to reduce the total amount of SO_2 added to the atmosphere.

FIGURE 16.12 "Superstack" at a Nickel Smelter near Sudbury, Ontario Constructed in 1972, this superstack towers above the town of Copper Cliff, Ontario. Its function is to disperse the SO_2 from the smelting operation over a wide area, thereby diluting it and reducing its damaging effects. Before this stack was built, the surrounding landscape was so devoid of vegetation that the area was used as a training ground for astronauts planning to land on the moon. Nevertheless, the SO_2 emitted by this stack spreads SO_2 over a wide area of Canada and the United States.

Because SO_2 is an acidic oxide, it can be removed from the gases produced by the combustion of coal by reacting it with a base. Bases such as NaOH and KOH are too expensive for large-scale use, so industrial processes almost invariably use calcium oxide, CaO, or calcium hydroxide, $Ca(OH)_2$, obtained from limestone, $CaCO_3$. A slurry of water and $Ca(OH)_2$ is sprayed down a tower in which the hot gases containing SO_2 are rising. The SO_2 reacts with the $Ca(OH)_2$ to form calcium sulfite, $CaSO_3(s)$:

$$Ca(OH)_2(aq) + SO_2(g) \rightarrow CaSO_3(s) + H_2O(l)$$

Limestone is cheap, but $CaSO_3$, with no practical uses, poses a disposal problem.

In the gases emitted during the smelting of metal sulfide ores, the concentration of SO_2 is quite high. It is economically feasible to separate at least some of the SO_2 by liquefaction. Liquid SO_2 is sold as such or is converted to sulfuric acid.

These and other methods are all expensive to use on the large scale, and expensive solutions are not attractive to industries or to politicians. Nevertheless, the governments of Canada and of many countries in the European Economic Community have agreed to reduce acidic emissions by 50%. But it remains a challenging problem for chemists to find better and more economical solutions to the acid-rain problem.

16.6 THE GREENHOUSE EFFECT

Potentially the most serious of all the changes occurring in the atmosphere is the so-called **greenhouse effect**, which is due to the presence of carbon dioxide and certain other gases in the atmosphere. The carbon dioxide in the atmosphere controls the earth's surface temperature. We saw in Chapter 14 that the vibrational energy of some molecules is increased by the absorption of infrared radiation. Carbon dioxide and water absorb infrared radiation in this way. They absorb some of the infrared radiation emitted by the earth and prevent it from returning to space, thus keeping the temperatures of the troposphere and the earth's surface higher than they otherwise would be. It has been calculated that if there were no CO_2 and H_2O in the atmosphere, the average surface temperature of the earth would be $-30°C$ rather than $+15°C$. The very high surface temperature of Venus is thought to be a result of the greenhouse effect exerted by its dense, carbon-dioxide–containing atmosphere (Chapter 19).

The warming of the troposphere by the absorption of infrared radiation is referred to as the "greenhouse effect" because the temperature in a greenhouse is similarly maintained higher than that outside. The glass covering a greenhouse lets in the visible light from the sun but absorbs the infrared radiation emitted by the warm plants and soil. The glass becomes warm and emits infrared radiation back into the greenhouse, raising the temperature inside the greenhouse (Figure 16.13). The analogy is not perfect, however, because the glass also physically prevents the warmer air in the greenhouse from mixing with the colder air outside.

The greenhouse effect is not a new phenomenon; it has been going on as long as the "greenhouse gases" (such as CO_2 and H_2O) have been a part of our atmosphere. The present concern about the greenhouse effect arises from the clear evidence that, as we saw in Chapter 2 (Figure 2.2), the concentration of carbon dioxide in the troposphere is increasing. Therefore, it is very probable that the average surface temperature of the earth is increasing correspondingly. It is almost certain that industrial activities are responsible for the increase in the carbon dioxide

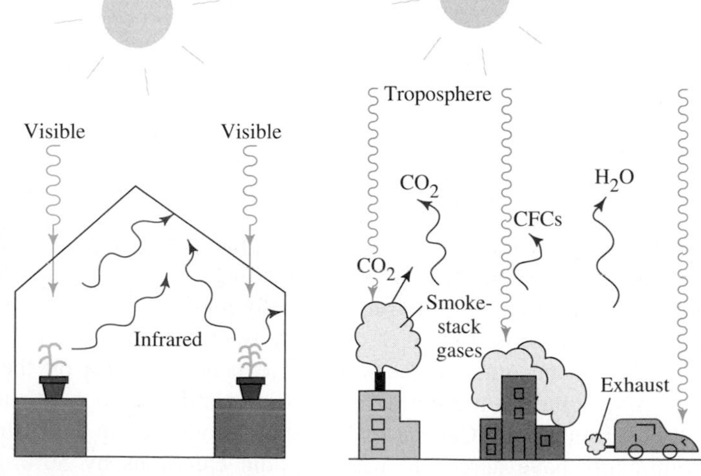

FIGURE 16.13 The Greenhouse Effect
(a) Visible light from the sun passes through the glass. Energy from this light is absorbed by the objects in the greenhouse, and their temperature increases. Warm objects emit infrared radiation that is absorbed by the glass. Some of it is reemitted into the interior of the greenhouse. In this way, some of the infrared radiation is trapped inside, and the greenhouse remains warmer than the exterior. (b) Gases such as CO_2 and H_2O that absorb infrared radiation and are present in the troposphere act like the glass of a greenhouse: They absorb infrared radiation emitted by the surface of the earth. These gases prevent the radiation from being lost into space, thus keeping the temperature of the troposphere and the earth's surface warmer than it would be in their absence. An increased CO_2 concentration resulting from the combustion of fossil fuels is therefore expected to raise the earth's average surface temperature.

concentration. The concentration of CO_2 before the Industrial Revolution can be estimated from measurements of air samples trapped at various depths in glaciers. Until about 1850, it had been constant at 270 to 280 ppm for many thousands of years; the amount of carbon dioxide added to the atmosphere by respiration, decay, and natural forest fires was just balanced by the amount removed by photosynthesis. Since then, however, human activities have tipped the balance, so the amount of carbon dioxide in the atmosphere is now steadily increasing. In 1958 it had risen to 316 ppm, and by 1990 it had reached 350 ppm (Figure 2.2). It is estimated that, if our present rate of industrial activity continues, the CO_2 concentration will be double its preindustrial value by late in the twenty-first century.

The carbon dioxide concentration is increasing for two main reasons. First, we are adding carbon dioxide to the atmosphere by our use of fossil fuels—coal, oil, natural gas, and gasoline. Second, we are reducing the amount of carbon dioxide removed from the atmosphere by photosynthesis by destroying vast areas of vegetation, particularly tropical rain forests.

Carbon dioxide and water are not the only greenhouse gases. Methane, the CFCs, N_2O, O_3, and CO also absorb infrared radiation. Although they are present in the atmosphere at very low concentrations, they absorb infrared radiation relatively strongly and in regions of the spectrum not absorbed by CO_2 and H_2O. So these gases, especially methane, are also contributing to the greenhouse effect. Until about 1850 the concentration of methane was relatively constant at about 0.7 ppm. It has now reached a concentration of about 1.6 ppm and is increasing by 1 to 2% annually. As its old name ''marsh gas'' implies, wetlands are an important source of methane, which is produced by the anaerobic decay of plant material. The anaerobic fermentation of plant material in the digestive system of cattle is also a major source. Increased agriculture, including the wetland cultivation of rice, and an increase in the numbers of cattle are probably responsible for the rise in the methane concentration.

''Anaerobic'' means ''without oxygen.'' The earliest bacteria evolved when oxygen was absent from the earth's atmosphere. Today such bacteria are found only in environments that lack oxygen—deep in soil or in the digestive systems of animals. Oxygen is a poison to these bacteria.

The effect an increasing concentration of CO_2 and other greenhouse gases will have on the surface temperature of the earth is difficult to predict, but most recent estimates indicate that a doubling of the CO_2 concentration alone would lead to an average temperature increase of 3 to 5°C. At first glance this might seem a rather small increase compared with the considerably larger seasonal variations. But the difference between the average global temperature today and that of the last ice age is on the order of magnitude of 3 to 5°C, so it is clear that such a temperature increase could cause considerable climatic changes. Such changes would include the melting of the polar icecaps, which would in turn cause severe permanent flooding in low-lying coastal regions and thereby destroy many major cities. However, there is not yet any conclusive evidence that the average temperature of the earth has begun to increase despite the increase in the CO_2 concentration.

Probably the only effective method of counteracting greenhouse warming would be to stop, or at least drastically reduce, the use of fossil fuels. Eventually our fossil fuel reserves will be exhausted, and we will be forced to develop alternative energy sources. But until then, we may have to adjust to climatic changes and move people and agricultural activities to new locations.

Exercise 16.8 Molecules absorb infrared radiation only if their dipole moment changes during at least one of their vibrations. Explain why CO, CO_2, and H_2O absorb infrared radiation but, O_2 and N_2 do not.

Postscript

In this chapter we have discussed several environmental problems of current concern. It is important to realize that a given problem can rarely be considered in isolation. Acid rain, photochemical smog, the thinning of the ozone layer, and greenhouse warming are all interrelated; this is true of most problems involving ecology and the biosphere. The mechanisms controlling the operation of the biosphere are largely chemical in nature. To understand the malfunctions of the system, which appear as environmental problems, we therefore need to understand the chemistry and related science behind the problem. Only then can wise political and economic decisions be made concerning the appropriate actions to take.

SUMMARY

Nitrogen monoxide (nitric oxide), $NO(g)$, forms readily at high temperatures in lightning discharges, forest fires, and engines that burn hydrocarbons. At normal temperatures it decomposes very slowly to N_2 and O_2, so it is present in small concentrations in the atmosphere. $NO(g)$ reacts with $O_2(g)$ to give nitrogen dioxide, $NO_2(g)$, a red-brown gas. The exothermic reaction by which $NO_2(g)$ forms dinitrogen tetraoxide, $N_2O_4(g)$, is favored as temperature decreases. $NO(g)$ and $NO_2(g)$ may be prepared in the laboratory by reducing with $Cu(s)$ dilute and concentrated $HNO_3(aq)$, respectively. NO and NO_2 are unusually unreactive free radicals. Dinitrogen monoxide (nitrous oxide), $N_2O(g)$, isoelectronic with CO_2, results from the bacterial decay of nitrogen compounds in soil and from heating $NH_4NO_3(s)$. The nitric acid, $HNO_3(aq)$, formed from nitrogen oxides in the atmosphere is an important constituent of acid rain. Industrially it is manufactured from $NH_3(g)$ by the Ostwald process. Pure nitric acid is a liquid and a strong acid in aqueous solution. Nitrate ion, NO_3^-, is isoelectronic with CO_3^{2-}. Both HNO_3 and NO_3^- are strong oxidizing agents and are usually reduced to NO or NO_2. Nitrous acid, $HNO_2(aq)$, is a weak acid. HNO_2 and the nitrite ion, NO_2^-, are good reducing agents; they are readily oxidized to NO_3^-.

Ozone, $O_3(g)$, is a pale blue gas and a powerful oxidizing agent. It results from subjecting $O_2(g)$ to an electric discharge. $Na_2O_2(s)$ and $BaO_2(s)$, from reaction of the corresponding metals with a limited amount of air, contain the $^-O—O^-$ peroxide ion and react with aqueous acid to give hydrogen peroxide, $H_2O_2(aq)$. Hydroxyl, HO·, and hydroperoxyl, HOO·, radicals are important in atmospheric chemistry.

When molecules in the atmosphere absorb solar radiation, chemical bonds can be broken or the molecules may be excited to higher-energy states. The maximum wavelength that will break a bond with a bond energy *BE* is given by $\lambda = (1.20 \times 10^5)/BE$ (with λ in nm and *BE* in $kJ \cdot mol^{-1}$). The dissociation of O_2 to O atoms and of O_3 to O_2 and O atoms absorbs almost all solar radiation with wavelengths up to 320 nm before it reaches the surface of the earth. Radiation in the range 320 to 350 nm, which raises O_3 molecules to an excited state, O_3^*, is substantially reduced in intensity. C—H, C—C, C—O, and C—N single bonds can all be broken by ultraviolet radiation that reaches the earth.

$O_3(g)$ forms in the stratosphere when O atoms (from the photodissociation, or photochemical dissociation, of O_2) and O_2 molecules collide with a third molecule M that carries away some of the energy liberated. The formation of ozone leads to an increase in the temperature of the stratosphere with increased altitude. O_3 is decomposed by photodissociation and by reaction with other O atoms. The rates of its formation and decomposition just balance, giving a constant steady-state $O_3(g)$ concentration, which is maximum at an altitude of 25 to 35 km and constitutes the ozone layer. Chemically inert CFCs (Freons), chlorofluorocarbons, are ideally suited for use as refrigerants, but they escape into the troposphere and eventually diffuse into the stratosphere, where their photodissociation produces Cl atoms. Cl atoms catalyze the decomposition of ozone in a chain reaction that decreases its steady-state concentration. In a similar chain reaction, $NO(g)$ also catalyzes the depletion of ozone. $N_2O(g)$ formed on the earth's surface is rather unreactive and drifts up to the stratosphere, where it reacts with O atoms to form NO.

UV-B radiation in the 290 to 320 nm range is particularly damaging to plants and animals. An increase in the intensity of this radiation due to ozone depletion could have serious consequences.

Photochemical smog is a major concern in the troposphere. At the high temperature in a gasoline engine, $N_2(g)$ and $O_2(g)$ combine to form oxides of nitrogen, NO_x. Photodissociation of NO_2 by sunlight gives NO and O atoms, which combine with $O_2(g)$ in a reaction catalyzed by $CO(g)$ and hydrocarbons from the incomplete combustion of gasoline, to give ozone. This catalyzed reaction is faster than the reaction of NO and O_3 to regenerate NO_2 and O_2, so the concentration of ozone builds up near the earth's surface. A byproduct is peroxyacetylnitrate, PAN. Ozone, PAN, and other toxic compounds are irritating to people and severely damage plants. Emission controls and catalytic converters reduce the problem.

$SO_2(g)$ is directly toxic to plants. Oxides of sulfur and nitrogen decrease the pH of rainwater below its natural value of 5.6. This acid rain damages plants, stonework, and glass and increases the acidity of lakes and rivers, thereby destroying aquatic life. $SO_2(g)$ in the atmosphere is catalytically oxidized to $SO_3(g)$ by $O_2(g)$. $SO_3(g)$ then reacts with water droplets in clouds to give $H_2SO_4(aq)$. $NO(g)$ is oxidized to $NO_2(g)$, which also further reacts with water droplets to form $HNO_3(aq)$. Lakes in regions with limestone rocks are relatively immune, but those in regions with granite (aluminosilicate) rocks can have a pH as low as 4 or 5, which directly kills plant life. Acid rain also increases the solubility of some metal salts. Consequences that are understood are the killing of fish in lakes—due to the precipitation of $Al^{3+}(aq)$ as $Al(OH)_3(s)$ in their gills at pH 7.4—and the depletion of magnesium in the chlorophyll of trees—due to the leaching of Ca^{2+} and Mg^{2+} ions from soil.

The greenhouse effect is due to increases in the concentrations of greenhouse gases, particularly CO_2, methane, CFCs, N_2O, O_3, and CO, in the troposphere due to human activity. These gases trap infrared radiation from the earth's surface and prevent it from radiating into space, thus raising atmospheric temperature. Two main causes for the increase in $CO_2(g)$ are recognized: the increased use of fossil fuels and the decreased efficiency of photosynthetic conversion of CO_2 to oxygen by vegetation, due principally to deforestation. An increase in methane concentrations is attributed mainly to an increase in anaerobic processes involving plants, such as the decay of plant material in wetlands and digestive fermentation by cattle.

IMPORTANT TERMS

acid rain (page 602)
biosphere (page 583)
chlorofluorocarbon (CFC) (page 596)
free radical (page 588)
greenhouse effect (page 606)
Ostwald process (page 588)
ozone (page 590)
ozone layer (page 596)
photochemical smog (page 600)
photodissociation (photolysis) (page 593)
pollutant (page 584)
stratosphere (page 592)
troposphere (page 592)

REVIEW QUESTIONS

1. By what reactions of nitric acid can **(a)** $NO(g)$ and **(b)** $NO_2(g)$ be made in the laboratory?

2. Why are molecules such as NO and NO_2 known as free radicals?

3. Why do low temperatures favor the formation of $N_2O_4(g)$ from $NO_2(g)$?

4. **(a)** How can $N_2O(g)$ be prepared?
(b) What is its Lewis structure?

5. What are the three steps by which nitric acid is produced from ammonia in the Ostwald process?

6. Draw Lewis structures for the peroxide ion, O_2^{2-}, and hydrogen peroxide, H_2O_2.

7. What is photodissociation?

8. N_2 has a stronger bond than O_2. Will it take a photon of longer or shorter maximum wavelength to break the bond in N_2 than to break the bond in O_2?

9. What are the troposphere and the stratosphere?

10. Why does $O_2(g)$ dissociate into O atoms in the stratosphere but not in the troposphere?

11. How does ozone form in the stratosphere?

12. Could the bonds in ozone (*BE* 414 kJ · mol^{-1}) be broken by sunlight at the earth's surface?

13. Why is the destruction of ozone initiated by chlorine atoms described as a chain reaction?

14. Why do CFCs, which are inert under ordinary conditions, cause problems in the stratosphere?

15. What factors are responsible for the formation of smog in cities such as Los Angeles?

16. What ingredients of automobile exhaust are catalytic converters designed to remove?

17. What is meant by the term "acid rain"?

18. How do **(a)** sulfuric acid and **(b)** nitric acid form in the troposphere?

19. What kind of stone used in buildings and monuments is most affected by acid rain, and why?

20. Why does acid rain affect the pH of lakes in different regions to different extents?

21. Why are some atmospheric gases called greenhouse gases?

PROBLEMS

Oxides and Oxoacids of Nitrogen

1. Write the formula of each of the following compounds.

(a) nitric acid **(b)** nitrous acid

(c) potassium nitrate **(d)** sodium nitrite

(e) nitric oxide **(f)** nitrous oxide

(g) nitrogen dioxide **(h)** dinitrogen tetraoxide

2. Name the compounds with each of the following formulas.

(a) $NaNO_3(s)$ **(b)** $KNO_2(s)$ **(c)** $N_2O_4(s)$ **(d)** $NO_2(g)$

(e) $NO(g)$ **(f)** $N_2O(g)$ **(g)** $HNO_2(aq)$ **(h)** $HNO_3(l)$

3. What are the oxidation numbers of the nitrogen atoms in each compound in Problem 2?

4. Draw Lewis structures, including resonance structures where appropriate, for each of the following. Find the bond orders, and hence arrange the molecules in order of expected decreasing NO bond length:

(a) N_2O **(b)** NO **(c)** NO_2^- **(d)** NO_3^- **(e)** NO^+

5. Write balanced equations for the reaction of copper with **(a)** dilute nitric acid and **(b)** concentrated nitric acid.

6. Using the VSEPR model, categorize each of the following (where nitrogen is the central atom) in terms of the AE_nX_m nomenclature, and predict the approximate geometry of each.

(a) NO_2^- **(b)** NO_2 **(c)** NO_2^+

7. The following reactions are methods by which $N_2(g)$ may be prepared in the laboratory. In each case, write the balanced equation, and indicate which substances are reduced and which are oxidized.

(a) the reaction of ammonia with hot copper(II) oxide to give nitrogen, water, and copper metal

(b) the reaction of nitrogen monoxide with ammonia, using a red-hot copper catalyst, to give nitrogen and water

(c) the decomposition of $(NH_4)_2Cr_2O_7(s)$, ammonium dichromate, upon heating to give chromium(III) oxide, nitrogen, and water

8. Write a balanced equation for each of the following reactions. State whether each is an acid–base or an oxidation–reduction reaction. For the acid–base reactions, indicate which reactant is the acid and which is the base; for the redox reactions, state which element is oxidized and which is reduced.

(a) the photochemical decomposition of nitric acid to $NO_2(g)$, $O_2(g)$, and $H_2O(l)$

(b) the decomposition upon heating of lead(II) nitrate to give $PbO(s)$, $NO_2(g)$, and $O_2(g)$

(c) the reaction of nitrous acid with water

9. How may a sample of each of the following be prepared in the laboratory?

(a) dinitrogen monoxide, $N_2O(g)$

(b) nitrogen dioxide, $NO_2(g)$

(c) dinitrogen tetraoxide, $N_2O_4(s)$

10. A useful method for the quantitative determination of nitrate ion in aqueous solution is to determine the amount of nitrogen monoxide gas formed by the reaction

$$2NO_3^-(aq) + 3H_2SO_4(aq) + 3Hg(l) \rightarrow 3HgSO_4(s) + 2H_2O(l) + 2OH^-(aq) + 2NO(g)$$

If 3.26 L of NO(g) at STP was obtained from 2.00 L of a solution containing nitrate ion, what was the concentration of nitrate ion in the solution?

11. **(a)** Calculate the standard enthalpy change for the reaction

$$4NH_3(g) + 5O_2(g) \rightarrow 4NO(g) + 6H_2O(g)$$

(Use data from Table 9.1 to calculate $\Delta H°$.)

(b) Under what conditions of temperature and pressure would the equilibrium concentration of NO(g) be maximized?

12. For the room-temperature reaction between nitrogen monoxide and oxygen,

$$2NO(g) + O_2(g) \rightarrow 2NO_2(g)$$

the following initial rate data were obtained.

[NO] ($mol \cdot L^{-1}$)	[O_2] ($mol \cdot L^{-1}$)	Initial Rate ($mol \cdot L^{-1} \cdot s^{-1}$)
0.010	0.020	0.014
0.010	0.010	0.007
0.020	0.040	0.114
0.040	0.020	0.227

Determine **(a)** the order of the reaction with respect to NO and O_2 and **(b)** the value of the rate constant.

13. Nitrogen dioxide decomposes to nitrogen monoxide and oxygen at high temperature. Find the activation energy for this reaction from the following rate constant data.

T (K)	*k* ($L \cdot mol^{-1} \cdot s^{-1}$)
650	3.16
730	28.2
800	1.58×10^2
900	1.12×10^3
1000	5.01×10^3

Ozone and Peroxides

14. Write the formulas of each of the following compounds.

(a) ozone (b) sodium oxide
(c) sodium peroxide (d) barium oxide
(e) barium peroxide (f) hydrogen peroxide

15. Write balanced equations for each of the following oxidation reactions of ozone.

(a) iodide ion to iodine in basic aqueous solution

(b) Fe^{2+} (aq) to $Fe^{3+}(aq)$ in acidic solution

16. (a) Draw Lewis resonance structures for the ozone molecule, O_3, and calculate the bond order.

(b) Why is ozone not given the following Lewis structure?

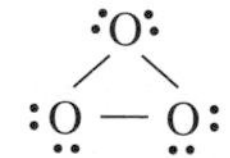

17. (a) The standard enthalpy of formation of ozone, $O_3(g)$, is 142 $kJ \cdot mol^{-1}$, and the dissociation energy of $O_2(g)$ is 498 $kJ \cdot mol^{-1}$. What is the average bond energy of the two bonds in the O_3 molecule?

(b) Why does the average bond energy calculated in (a) not correspond to the energy of the photons of maximum wavelength 320 nm required to dissociate $O_3(g)$ to $O_2(g)$ and $O(g)$?

18. (a) Describe how an aqueous solution of hydrogen peroxide can be prepared in the laboratory from $BaO_2(s)$.

(b) Explain why $H_2O_2(aq)$ sometimes behaves as an oxidizing agent and sometimes as a reducing agent.

(c) Write balanced equations for each of the following reactions of $H_2O_2(aq)$ in acidic solution.

(i) oxidation of sulfur dioxide to sulfate ion

(ii) reduction of ozone to water

(iii) oxidation of iodide ion to iodine

(iv) oxidation of nitrite ion to nitrate ion

19. The standard reduction potential for the half-reaction

$$O_3(g) + 2H^+(aq) + 2e- \rightarrow O_2(g) + H_2O(l)$$

is 2.07 V. On the basis of standard reduction potentials from Table 13.1, decide which of the oxidations (a) H_2O to H_2O_2, (b) Mn^{2+} to MnO_4^-, (c) Cr^{3+} to $Cr_2O_7^{2-}$, (d) Co^{2+} to Co^{3+}, (e) NO (g, 1 atm) to NO_3^- could be achieved under standard conditions in acidic solution by using (i) ozone and (ii) oxygen.

20. The decomposition of gaseous hydrogen peroxide to $O_2(g)$ and $H_2O(g)$ is a first-order reaction. At 25°C, the initial concentration of H_2O_2 was found experimentally to decrease to half in 17.0 min.

(a) What is the rate constant at that temperature?

(b) If the activation energy is 75 $kJ \cdot mol^{-1}$, what is the half-life at 50°C?

Pollution of the Stratosphere

21. (a) Why is the spectrum of solar radiation different at the earth's surface than it is outside the atmosphere?

(b) Account for the fact that, of the major constituents of the atmosphere, nitrogen and argon remain unchanged in the stratosphere, whereas oxygen decomposes although it is stable in the troposphere.

22. (a) By what reactions does nitric oxide, NO(g), form naturally, and by what reactions is it removed (b) in the stratosphere and (c) in the troposphere?

23. Use standard enthalpies of formation from Table 9.1 to find the reaction enthalpies for both steps in the conversion of $O_3(g)$ to $O_2(g)$ catalyzed by NO(g).

(a) $NO(g) + O_3(g) \rightarrow NO_2(g) + O_2(g)$

(b) $NO_2(g) + O_3(g) \rightarrow 2O_2(g) + NO(g)$

24. (a) Using Table 9.1, calculate $\Delta H°$ for the reaction

$$O_3(g) \rightarrow O_2(g) + O(g)$$

(b) Estimate whether 300-nm radiation is capable of bringing about this reaction.

(c) What is the maximum wavelength radiation capable of bringing about this reaction?

25. For the reaction

$$O(g) + O_3(g) \rightarrow 2O_2(g)$$

the rate constant is 5.56×10^6 $L \cdot mol^{-1} \cdot s^{-1}$ at 300 K and 1.38×10^5 $L \cdot mol^{-1} \cdot s^{-1}$ at 200 K.

(a) What is the activation energy for this reaction?

(b) What is the rate constant for the reaction at −55°C, the temperature of the stratosphere?

26. For the reaction

$$2NO(g) + O_2(g) \rightarrow 2NO_2(g)$$

the rate law is

$$\text{rate} = k[NO]^2[O_2]$$

(a) Explain why this reaction is unlikely to occur by an elementary termolecular mechanism.

(b) Devise a mechanism for this reaction that is consistent with the observed rate law and does not involve the simultaneous collision of three molecules.

27. (a) Give the major reasons why the maximum ozone concentration is found at an altitude of about 30 km.

(b) Describe the principal reactions that are most probably responsible for the destruction of the ozone layer.

28. (a) Discuss the reasons why CFCs were used so extensively in the past and why their use is now being strongly discouraged.

(b) What properties should possible replacements for CFCs have?

29. (a) Estimate $\Delta H_f°$(ClO, g) from the following bond energy data (in $kJ \cdot mol^{-1}$): Cl_2 243; O_2 498; ClO 205.

(b) How is the ClO· radical formed in the atmosphere, and why is it important in atmospheric chemistry?

(c) Use your value of $\Delta H_f°$(ClO, g) to estimate the standard reaction enthalpies for the reactions

$$O_3(g) + Cl(g) \xrightarrow{1} ClO(g) + O_2(g)$$
$$O(g) + ClO(g) \xrightarrow{2} O_2(g) + Cl(g)$$

involved in the catalytic decomposition of ozone by CFCs.

30. Dinitrogen monoxide, $N_2O(g)$, can undergo the photochemical decomposition

$$N_2O(g) \rightarrow N_2(g) + O^*(g)$$

where $O^*(g)$ is an excited state of $O(g)$.

(a) Given $\Delta H_f^\circ(N_2O, g) = 82\ kJ \cdot mol^{-1}$, $\Delta H_f^\circ(O, g) = 249\ kJ \cdot mol^{-1}$, and that the excitation energy of O* is 188 $kJ \cdot mol^{-1}$, what is the maximum wavelength of radiation that will bring about this reaction?

(b) Why does this reaction occur in the stratosphere but not in the troposphere?

31.* The conversion of ozone to molecular oxygen in the upper atmosphere,

$$2O_3(g) \rightarrow 3O_2(g)$$

is thought to follow the mechanism

$$O_3 \rightleftharpoons O_2 + O \text{ (Fast equilibrium)}$$
$$O + O_3 \rightarrow 2O_2 \text{ (Slow)}$$

(a) What rate law is consistent with this mechanism?

(b) Explain why the reaction rate decreases as the concentration of O_2 increases. In other words, why does $[O_2]$ occur in the rate law with a negative order?

32.* **(a)** Calculate K_p from ΔG at $-55°C$ for the reaction

$$\tfrac{3}{2}O_2(g) \rightleftharpoons O_3(g)$$

from Table 9.1 and $S°(O_3, g) = 238.8\ J \cdot K^{-1} \cdot mol^{-1}$, $S°(O_2, g) = 205.0\ J \cdot K^{-1} \cdot mol^{-1}$, and $S°(O, g) = 160.9\ J \cdot K^{-1} \cdot mol^{-1}$. Assume that ΔH_f° and $S°$ do not vary with temperature.

(b) If the pressure of $O_2(g)$ is 2.0×10^{-3} atm in the stratosphere, what is the equilibrium pressure of $O_3(g)$?

(c) The steady-state concentration of $O_3(g)$ is 3.0×10^{15} molecules per liter. Show by calculation whether the system O_3/O_2 is at equilibrium, and comment on the significance of your result.

33.* The mechanism for the reaction of Cl atoms with ozone in the stratosphere (at $-55°C$) involves the steps

$$O_3(g) + Cl(g) \xrightarrow{1} ClO(g) + O_2(g)$$
$$O(g) + ClO(g) \xrightarrow{2} O_2(g) + Cl(g)$$

(a) Why is the overall reaction described as a chain reaction, and why are the Cl atoms described as a catalyst?

(b) At $-55°C$ the rate constant for step (1) is $5.2 \times 10^9\ L \cdot mol^{-1} \cdot s^{-1}$; that for step (2) is $2.5 \times 10^{10}\ L \cdot mol^{-1} \cdot s^{-1}$. The steady-state concentrations are $[O_3] = 5.3 \times 10^{-9}\ mol \cdot L^{-1}$, $[O] = 8.3 \times 10^{-14}\ mol \cdot L^{-1}$, $[Cl] = 1.7 \times 10^{-16}\ mol \cdot L^{-1}$, and $[ClO] = 1.1 \times 10^{-13}\ mol \cdot L^{-1}$. What is the overall rate of the reaction?

(c) For the direct reaction

$$O_3(g) + O(g) \rightarrow 2O_2(g)$$

the rate constant is $3.5 \times 10^6\ L \cdot mol^{-1} \cdot s^{-1}$. What fraction of the ozone is destroyed under these conditions by the reaction with Cl atoms?

Photochemical Smog

34. **(a)** What is photochemical smog?

(b) What are the principal sources of oxides of nitrogen, NO_x, in the troposphere?

(c) Why does the concentration of ozone continue to increase in conditions in which photochemical smog is formed?

35. **(a)** What reactions of exhaust gases from automobile engines are catalyzed by catalytic converters?

(b) Why would replacement of hydrocarbon-burning automobiles by electrically powered vehicles be beneficial but not necessarily a perfect environmental solution?

36. **(a)** Explain why the concentrations of oxides of nitrogen are reported as NO_x rather than separately as NO and NO_2.

(b) Close to an urban highway, the concentration of NO_x is 50 $\mu g \cdot m^{-3}$. Express this as atm, ppm, and $mol \cdot L^{-1}$.

(c) Suppose that a city is circular and 20 km across. What is the number of moles of NO_x in the atmosphere to an altitude of 1.0 km if the average NO_x concentration is uniformly 0.04 ppm? Assume that the average molar mass of NO_x is 38 $g \cdot mol^{-1}$.

37. The rate constant for the reaction

$$O_3(g) + NO(g) \rightarrow O_2(g) + NO_2(g)$$

is $1.07 \times 10^7\ L \cdot mol^{-1} \cdot s^{-1}$ at 25°C and $1.25 \times 10^7\ L \cdot mol^{-1} \cdot s^{-1}$ at 35°C.

(a) What is the activation energy?

(b) What is the rate constant at 20°C?

(c) Calculate the initial rate of the reaction when the initial concentrations of $O_3(g)$ and $NO(g)$ are 1.0 and 5.4 ppm, respectively.

38. An oil-fired power station consumes 10^6 L of oil daily. The oil has the average composition $C_{15}H_{32}(l)$ and a density of $0.80\ g \cdot cm^{-3}$, and the gas emitted from the stacks contains 70 ppm of $NO(g)$.

(a) How much $NO(g)$ is emitted per day?

(b) Assuming that the stack gases become uniformly mixed to an altitude of 1 km above a circular city 20 km across, what will be the increase in the NO_x concentration caused by the $NO(g)$?

39. One of the processes in the catalytic decomposition of $O_3(g)$ is the chain reaction

$$NO(g) + O_3(g) \rightarrow NO_2(g) + O_2(g)$$
$$NO_2(g) + O(g) \rightarrow NO(g) + O_2(g)$$

Calculate $\Delta H°$ for each step in the reaction, using Table 9.1.

Acid Rain

40. **(a)** Why does rain normally have a pH of 5.6?

(b) Why does $Al^{3+}(aq)$ remain in solution in water at pH 5 but precipitate as $Al(OH)_3(s)$ at pH 7.4?

(c) What is the relation between the answer to (b) and the death of fish in acidified lakes?

41. Assuming that the acidity of a lake is due entirely to sulfuric acid from acid rain, approximately how many kilograms of calcium carbonate (limestone) is required to reduce the acidity of 10^6 L of lake water to pH 5.6 if its initial pH is 4.2?

42. Calculate the pH of rainwater if SO_2 is the only acidic gas present and its concentration is 0.120 ppm.

43. A power plant burns 10^7 kg per day of coal containing 2.35% sulfur by mass. The stack gases contain $SO_2(g)$ and 150 ppm of $NO_x(g)$.

(a) Taking the average molar mass of NO_x as 38 g · mol^{-1}, calculate the total number of moles of acid emissions from the plant per day.

(b) How do the ratio and the absolute amounts of these emissions change if the plant switches to clean coal with a sulfur content of only 0.30% by mass?

44. A nickel ore has the partial composition Ni 1.4%, Cu 1.3%, Fe 7.2%, and S 9.1% by mass. A plant processes 3.5×10^7 kg of ore per day; 17% of the sulfur is converted to H_2SO_4, and 30% of the sulfur is released to the atmosphere as $SO_2(g)$. Calculate the following.

(a) the volume of SO_2 in liters released into the atmosphere per day

(b) the mass in kilograms of H_2SO_4 per day

(c) the mass of SO_2 emitted per kilogram of nickel produced

45.* At 450°C, $K_p = 24\ \text{atm}^{-\frac{1}{2}}$ for the reaction

$$SO_2(g) + \tfrac{1}{2}O_2(g) \rightleftharpoons SO_3(g)$$

At 450°C and initial pressures of 2.0 atm of $SO_2(g)$ and 20 atm of air passed over a catalyst, 97% of the SO_2 is converted to SO_3. Did the reaction reach equilibrium?

The Greenhouse Effect

46. Why is the temperature inside a greenhouse higher than that outside? What is the greenhouse effect in the atmosphere?

47. **(a)** What is a greenhouse gas?

(b) Which gases are thought to be the main culprits responsible for global warming?

(c) Why has the greenhouse effect not been found to be a problem until quite recently?

48. **(a)** What human activities are leading to an ever-increasing concentration of greenhouse gases in the atmosphere?

(b) What steps could be taken to alleviate the problem?

(c) Why is any solution, even when the science involved is reasonably well understood, proving difficult to implement?

General Problems

49. A hot coil of platinum wire is inserted into a flask containing a mixture of ammonia and oxygen. Brown fumes are formed at the surface of the wire. Write balanced equations for the reactions taking place.

50. In terms of the bond orders in the following molecules and ions, account for the observed trends in the observed bond lengths.

(a) O_2, 121 pm **(b)** O_3, 128 pm **(c)** O_2^{2-}, 149 pm

51. Upon heating, ammonium nitrite decomposes to nitrogen and water. What maximum volume of nitrogen at 60°C and 750 torr is formed when 10.0 g of ammonium nitrite is heated?

52. For the reaction

$$2N_2O(g) \rightarrow 2N_2(g) + O_2(g)$$

the rate constant is 1.1×10^{-3} L · mol^{-1} · s^{-1} at 565°C and 3.8×10^{-3} L · mol^{-1} · s^{-1} at 728°C. What is **(a)** the activation energy of the reactions; **(b)** the value of the rate constant at 25°C?

53. For the reaction

$$CO(g) + NO_2(g) \rightarrow CO_2(g) + NO(g)$$

the rate constant at 425°C has the value 1.3 L · mol^{-1} · s^{-1}; at 525°C the value is 23 L · mol^{-1} · s^{-1}. Calculate **(a)** the activation energy of the reaction and **(b)** the expected value of the rate constant at 1000°C.

54. Balance each of the following equations, and in each case classify NO_2 as an oxidizing agent or a reducing agent.

(a) $NO_2(g) + I^-(aq) + H_2O(l) \rightarrow NO(g) + I_3^-(aq) + OH^-(aq)$

(b) $MnO_4^-(aq) + NO_2(g) + H_2O(l) \rightarrow Mn^{2+}(aq) + H_3O^+(aq) + NO_3^-(aq)$

(c) $NO_2(g) + H_2O(l) \rightarrow HNO_2(aq) + HNO_3(aq)$

55. Write balanced equations for the half-reactions for the reduction of $NO_3^-(aq)$ in acidic solution to each of the following.

(a) $NO_2(g)$ **(b)** $NO(g)$ **(c)** $N_2(g)$ **(d)** $NH_4^+(aq)$

56. List known or possible natural sources of each of the following gases found in the atmosphere.

(a) CO **(b)** CO_2 **(c)** SO_2
(d) NO **(e)** N_2O **(f)** CH_4

57.* A gas X that is an oxide of nitrogen supports the combustion of a variety of substances. X may be prepared by gently heating ammonium nitrate. When 0.1020 g of white phosphorus was burned completely in X, another gas Y and 0.2337 g of an oxide of phosphorus were obtained. Starting with a given volume of X, the volume of Y produced was identical at a given temperature and pressure, but Y was found not to support combustion. The relative densities of X and Y were in the ratio 1.571:1.000.

(a) Identify X and Y.

(b) Draw their Lewis structures.

(c) Write balanced equations for the reactions described.

58.* A possible way in which ClO(g) may behave as a catalyst in the decomposition of $O_3(g)$ is by the chain reaction

$$ClO(g) + O_3(g) \rightarrow ClO_2(g) + O_2(g)$$
$$ClO_2(g) + O(g) \rightarrow ClO(g) + O_2(g)$$

(a) Calculate the overall enthalpy change for each step in the reaction, using Table 9.1, ΔH_f° (ClO, g) = 101 kJ · mol^{-1}, and ΔH_f°(ClO_2, g) = 102 kJ · mol^{-1}.

(b) Is this a possible mechanism for the decomposition of ozone in the stratosphere?

CHAPTER 17 Biochemistry: The Chemistry of Life

17.1 Proteins
17.2 Enzymes
17.3 Carbohydrates
17.4 Energy in Biochemical Reactions: ATP and ADP
17.5 DNA and RNA
17.6 Lipids
17.7 Nutrition

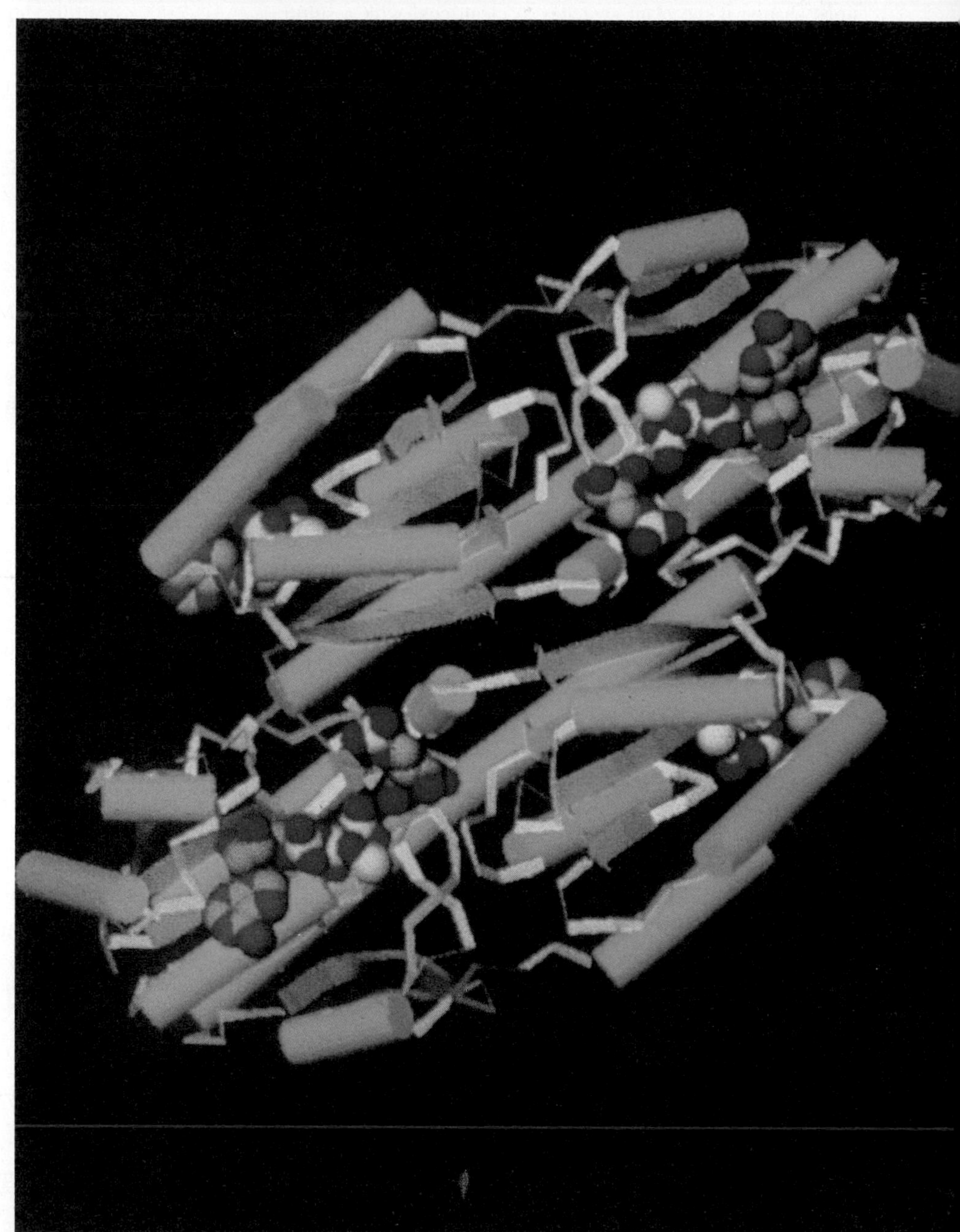

Life as we know it could not exist if biochemical reactions were not controlled by protein catalysts called enzymes. This computer-generated model represents the enzyme phosphofructokinase in action—binding to fructose-6-phosphate, the substrate, or reactant molecule on which the enzyme acts. This reaction is one of the steps in which the sugar fructose (that is, food) is converted to carbon dioxide and water and releases the energy all organisms need to survive and grow.

Period \ Group	I	II											III	IV	V	VI	VII	VIII
1	H																	He
2	Li	Be											B	C	N	O	F	Ne
3	Na	Mg											Al	Si	P	S	Cl	Ar
4	K	Ca	Sc	Ti	V	Cr	Mn	Fe	Co	Ni	Cu	Zn	Ga	Ge	As	Se	Br	Kr
5	Rb	Sr	Y	Zr	Nb	Mo	Tc	Ru	Rh	Pd	Ag	Cd	In	Sn	Sb	Te	I	Xe
6	Cs	Ba	La	Hf	Ta	W	Re	Os	Ir	Pt	Au	Hg	Tl	Pb	Bi	Po	At	Rn
7	Fr	Ra	Ac	104	105	106	107	108	109									

Metals   Nonmetals   Semimetals

Biochemistry is the study of the molecules and chemical reactions on which life is based. It is a branch of chemistry that is now in a very active and exciting stage of development. Recent research has shown that the same molecular structures and chemical reactions underlie all the diverse expressions of life, from bacteria to human beings. Biochemists can now tackle such fundamental questions as, How does a fertilized egg give rise to cells as different as those in muscle tissue and in the brain, and What is the mechanism of memory? Because of this increased understanding of life processes, biochemistry is having a profound influence on medicine. We now know the molecular mechanisms for many diseases, enabling us to devise better treatments and to synthesize more effective drugs.

Molecules and the reactions between them are the basis of biochemistry. At any instant many thousands of chemical reactions are occurring in your body. In principle, a chemist could set up reactions that would accomplish the same chemical objectives as each of the reactions occurring in the human body. But the result would not be life. The chemical reactions in a living organism are not independent of each other but are associated with each other so as to give increasingly complex structures. Molecules form membranes and cell nuclei, which are successively organized into the cells, tissues, organs, and organ systems of living organisms. All organisms, including plants and animals, in turn interact, forming complex ecological systems. The whole biosphere is dynamic and is maintained by a continuous input of energy. It is this extensive organization that differentiates the chemistry of life from the chemistry discussed in other chapters of this book.

Two of the characteristic features of life are reproduction and metabolism. All living organisms have the power of *reproduction*—the ability to produce offspring similar to themselves. The sum of all reactions occurring in a living organism is called the *metabolism* of the organism. Organisms need a continuous supply of energy and raw materials. A plant absorbs energy in the form of light and raw materials such CO_2 and water and uses them in chemical reactions to make the substances it needs to grow and sustain itself. Animals get their energy and the substances they need from the plants and other animals they eat.

The molecules of which living organisms are composed are often called **biomolecules**. Biomolecules are organic molecules with the same functional groups as those that we studied in Chapter 14; the structure and reactivity of these molecules depends on the principles we discussed in that chapter. Biomolecules differ from most other organic molecules only in that the majority of them are very large and are typically polymers, made by joining many small molecules together. Therefore, they often have special properties not possessed by the small organic molecules we have considered so far. Although living organisms are incredibly diverse, very similar molecules and very similar reactions control and maintain the diverse forms of life. Evolution has modified, but not fundamentally changed, the reactions on which life is based.

A major characteristic of the chemistry of life is that it operates far from equilibrium. Energy and reactants (food) are continually supplied, and products are continually removed. The substances of which an organism is composed are intermediates in the overall process. They have steady-state concentrations (not equilibrium concentrations) in which their rate of formation equals their rate of destruction. We discussed a similar situation in Chapter 16 in regard to ozone in the stratosphere. While energy from the sun is supplied, a steady-state concentration of ozone is maintained. If the sun were turned off, the ozone concentration would revert to a very low equilibrium value. Similarly, if an organism's food supply is cut off, it dies, and its components eventually reach equilibrium with their environment.

In this chapter we describe (1) the structure and properties of some typical biomolecules, such as proteins and carbohydrates, and the role that these molecules play in reproduction and metabolism (particularly in humans and other mammals); (2) how the information required for an organism to reproduce is passed on from one generation to the next; (3) how reaction rates are controlled by the catalysts called enzymes; and (4) nutrition, the study of the food that supplies the initial reactants for the reactions from which other biomolecules are made.

17.1 PROTEINS

Proteins play crucial roles in virtually all biological processes. They are present in all cells. Skin, hair, and muscle are proteins. So are the enzymes needed to catalyze reactions. The hemoglobin that transports oxygen in blood to the sites where it is needed and hormones (such as insulin) that regulate body processes are also proteins.

All proteins in all species, from bacteria to humans, are constructed from the same set of 20 α-**amino acids** (Table 17.1). (Recall from Chapter 14 that the "α" indicates that the amino group is attached to the same carbon atom to which the carboxyl group is attached.) Life has made use of this same set of amino acids for at least 2 billion years. The α-amino acids have the same basic structure, differing only in the nature of the side chain R. But, as we shall see, the properties of these R groups have a profound effect on the shapes and properties of proteins.

Two amino acid molecules can undergo a condensation reaction in which water is eliminated and an amide is formed:

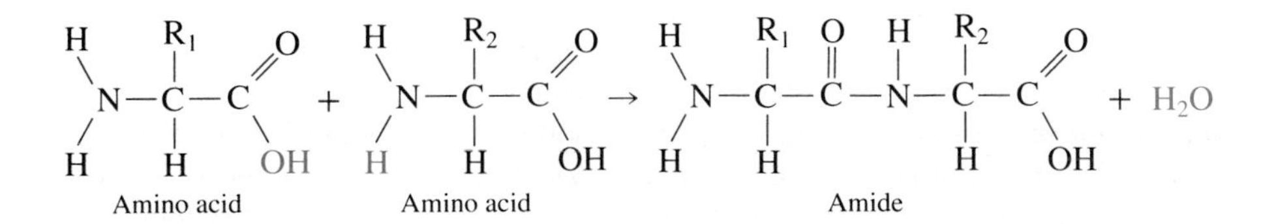

Because this amide still has a carboxylic acid group and an amino group, it can undergo further condensation reactions at each end and thereby build up a polymer:

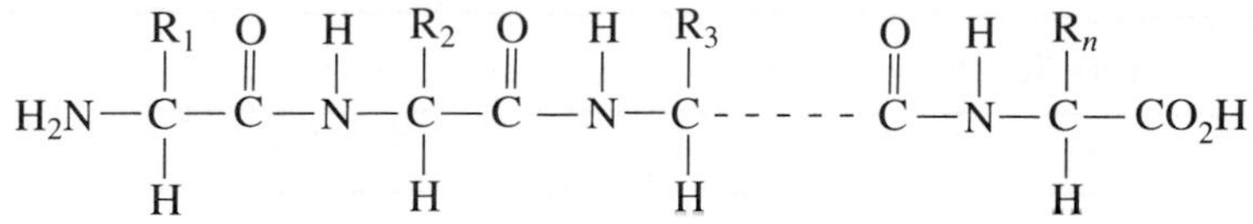

$$H_2N-\underset{\displaystyle H}{\overset{\displaystyle R_1}{C}}-\overset{\displaystyle O}{\overset{\|}{C}}-\overset{\displaystyle H}{N}-\underset{\displaystyle H}{\overset{\displaystyle R_2}{C}}-\overset{\displaystyle O}{\overset{\|}{C}}-\overset{\displaystyle H}{N}-\underset{\displaystyle H}{\overset{\displaystyle R_3}{C}}\text{-----}\overset{\displaystyle O}{\overset{\|}{C}}-\overset{\displaystyle H}{N}-\underset{\displaystyle H}{\overset{\displaystyle R_n}{C}}-CO_2H$$

The bond between the C=O group and the N—H group in an amide is called an *amide bond* or, more usually in biochemistry, a **peptide bond**, and the polymers are called **polypeptides. Proteins** are large polypeptides formed by the condensation of 50 or more amino acids. The remarkable range of protein functions results from the enormous variety of structures that can be built up from the 20 α-amino acids in Table 17.1. Although billions of different structures are possible, nature makes use of only a few thousand of these possibilities: The proteins in many different organisms are very similar. The sequence of amino acids in the proteins of different organisms is very similar if they have a common ancestor. The study of the evolution of animals and plants by the determination of the sequence of amino acids in their proteins is a rapidly developing field of research.

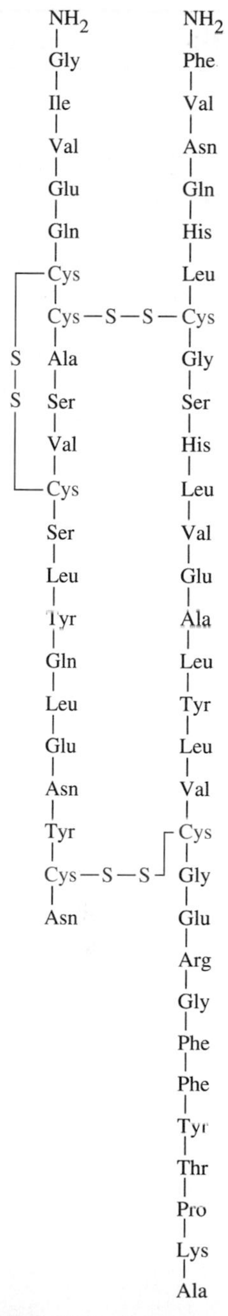

Primary Structure

The **primary structure** of a protein is the order of the amino acids from which the protein is formed. It is expressed by using a three-letter abbreviation for each of the amino acids (Table 17.1). The primary structure of a protein was first determined by the British biochemist Frederick Sanger (1918–). Between 1945 and 1952 he elucidated the structure of the protein *insulin* (Figure 17.1), which helps the body use sugar. Insulin contains 51 amino acids in two chains held together by *disulfide bridges*, or —S—S— disulfide linkages, derived from the sulfur-containing amino acid cysteine (Table 17.1). The determination of the structure of insulin was a considerable achievement at that time. Today determining the primary structure of a protein is a routine procedure.

Exercise 17.1 Draw the structure of the polypeptide formed by the condensation of four glycine molecules.

Exercise 17.2 Draw the primary structure of a segment of a polypeptide containing the sequence —Gly—His—Val—Glu—.

Secondary Structure

The polypeptide chains in a protein are more compact than is implied by Figure 17.1. They are folded such that nearby segments of a chain orient into a regular pattern called the **secondary structure**. There are two common types of secondary structure—the α-helix and the β-pleated sheet. In the **α-helix structure**, the polypeptide chain is coiled into a helix and is held in this shape by hydrogen bonds between C=O and N—H groups. The side chains of the amino acids then protrude from the outside of the helix (Figure 17.2). In the **β-pleated sheet**, part of the polypeptide chain is stretched out and folded back upon itself such that several parts of the chain lie side by side, connected by hydrogen bonds between C=O and N—H groups (Figure 17.3). Both of these structures were first proposed in 1951 by Linus Pauling (Box 3.2) and Robert Corey, who measured the bond distances and bond angles in many amino acids and small peptides by X-ray crystallography.

NH_2—Gly—Ile—Val—Glu—Gln—Cys—Cys—Ala—Ser—Val—Cys—Ser—Leu—Tyr—Gln—Leu—Glu—Asn—Tyr—Cys—Asn

NH_2—Phe—Val—Asn—Gln—His—Leu—Cys—Gly—Ser—His—Leu—Val—Glu—Ala—Leu—Tyr—Leu—Val—Cys—Gly—Glu—Arg—Gly—Phe—Phe—Tyr—Thr—Pro—Lys—Ala

(Cys—S—S—Cys bridges between the chains; S—S bridge within the first chain)

FIGURE 17.1 Primary Structure of Insulin
Insulin is a protein that contains two polypeptide chains held together by disulfide bridges between cysteines. Insulin is one of the principal hormones that regulates glucose metabolism.

TABLE 17.1 Structures of the 20 α– Amino Acids

Name	Abbreviation	Structure
Basic Amino Acids		
Arginine	Arg	$H_2NC(=NH)—CH_2CH_2CH_2—C(NH_2)(H)—COOH$
Histidine	His	(imidazole ring: N, N–H)$—CH_2—C(NH_2)(H)—COOH$
Lysine	Lys	$H_2NCH_2CH_2CH_2CH_2—C(NH_2)(H)—COOH$
Neutral Amino Acids–Nonpolar Side Chains		
Alanine	Ala	$CH_3—C(NH_2)(H)—COOH$
Glycine	Gly	$H—C(NH_2)(H)—COOH$
Isoleucine	Ile	$CH_3CH_2CH(CH_3)—C(NH_2)(H)—COOH$
Leucine	Leu	$CH_3CH(CH_3)CH_2—C(NH_2)(H)—COOH$
Methionine	Met	$CH_3SCH_2CH_2—C(NH_2)(H)—COOH$
Phenylalanine	Phe	(benzene ring)$—CH_2—C(NH_2)(H)—COOH$
Proline	Pro	ring: H_2C, $C H_2$, NH, $C H_2$; $—CH—COOH$

TABLE 17.1 Structures of the 20 α– Amino Acids *(continued)*

Name	Abbreviation	Structure
Valine	Val	$CH_3CH(CH_3)-C(NH_2)(H)-COOH$
Neutral Amino Acids–Polar Side Chains		
Asparagine	Asn	$H_2NC(=O)CH_2-C(NH_2)(H)-COOH$
Cysteine	Cys	$HSCH_2-C(NH_2)(H)-COOH$
Glutamine	Gln	$H_2NC(=O)CH_2CH_2-C(NH_2)(H)-COOH$
Serine	Ser	$HOCH_2-C(NH_2)(H)-COOH$
Threonine	Thr	$CH_3CH(OH)-C(NH_2)(H)-COOH$
Tryptophan	Trp	(indole ring, N–H)$-CH_2-C(NH_2)(H)-COOH$
Tyrosine	Tyr	$HO-C_6H_4-CH_2-C(NH_2)(H)-COOH$
Acidic Amino Acids		
Aspartic acid	Asp	$HOC(=O)CH_2-C(NH_2)(H)-COOH$
Glutamic acid	Glu	$HOC(=O)CH_2CH_2-C(NH_2)(H)-COOH$

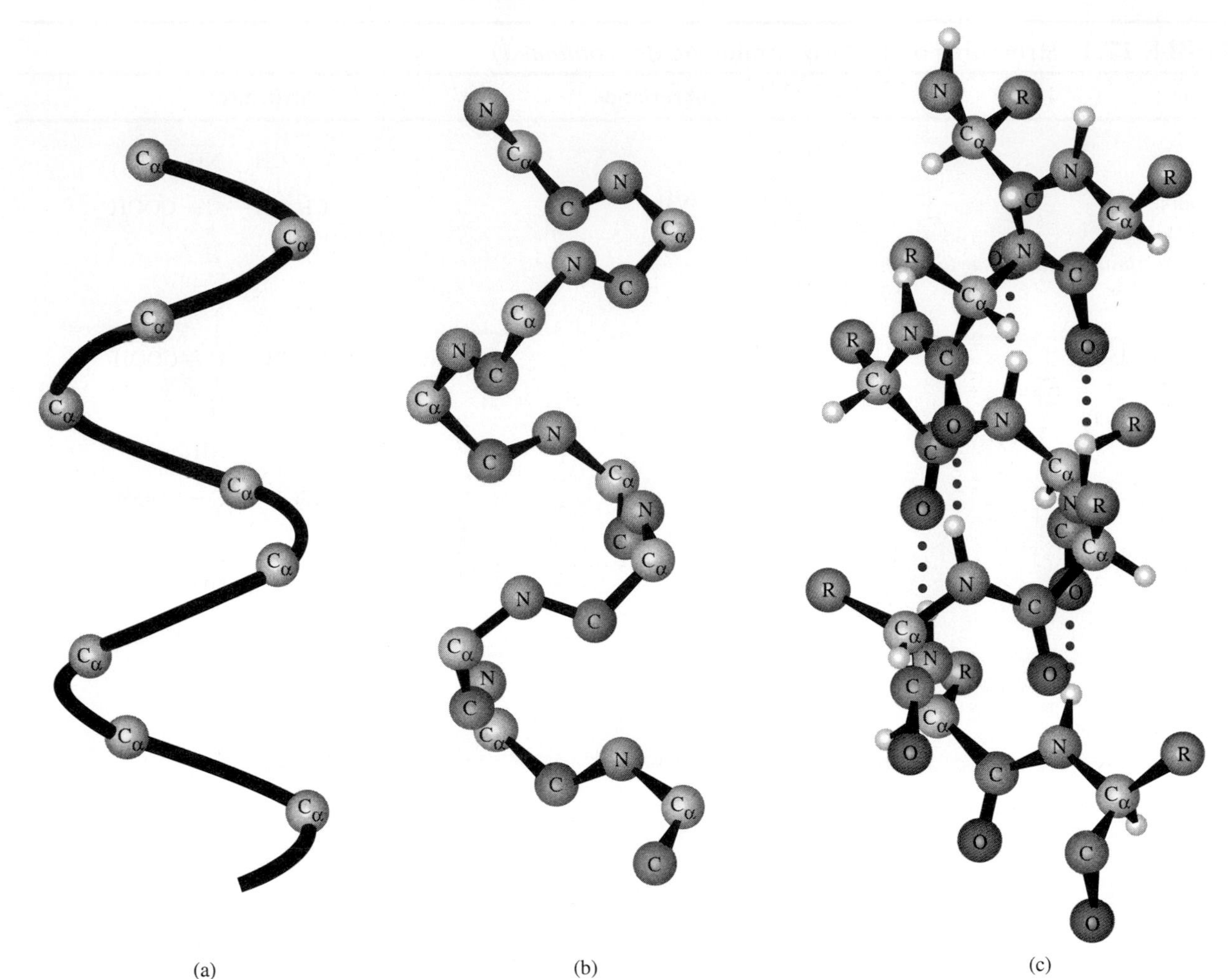

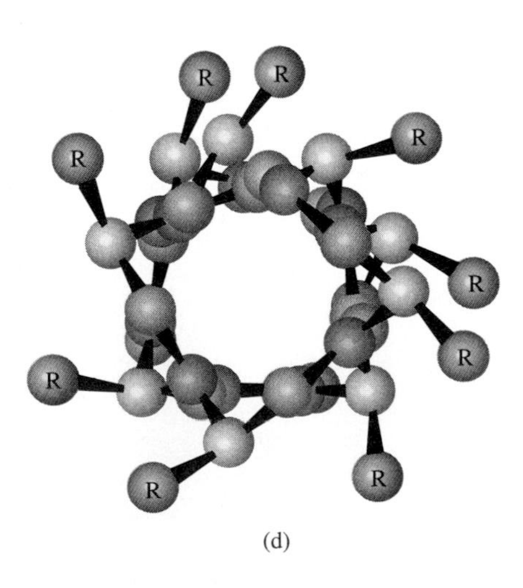

FIGURE 17.2 The α-Helix Secondary Structure of Proteins Model of an α-helix: (a) Only the α-carbon atoms are shown on a helical chain; (b) only the backbone nitrogen (N), α-carbon (C_α), and carbonyl carbon (C) atoms are shown; (c) the entire helix, stabilized by hydrogen bonds (denoted by red dots) between N—H and C=O groups; (d) cross-sectional view—the side chains (shown in green) are on the outside of the helix.

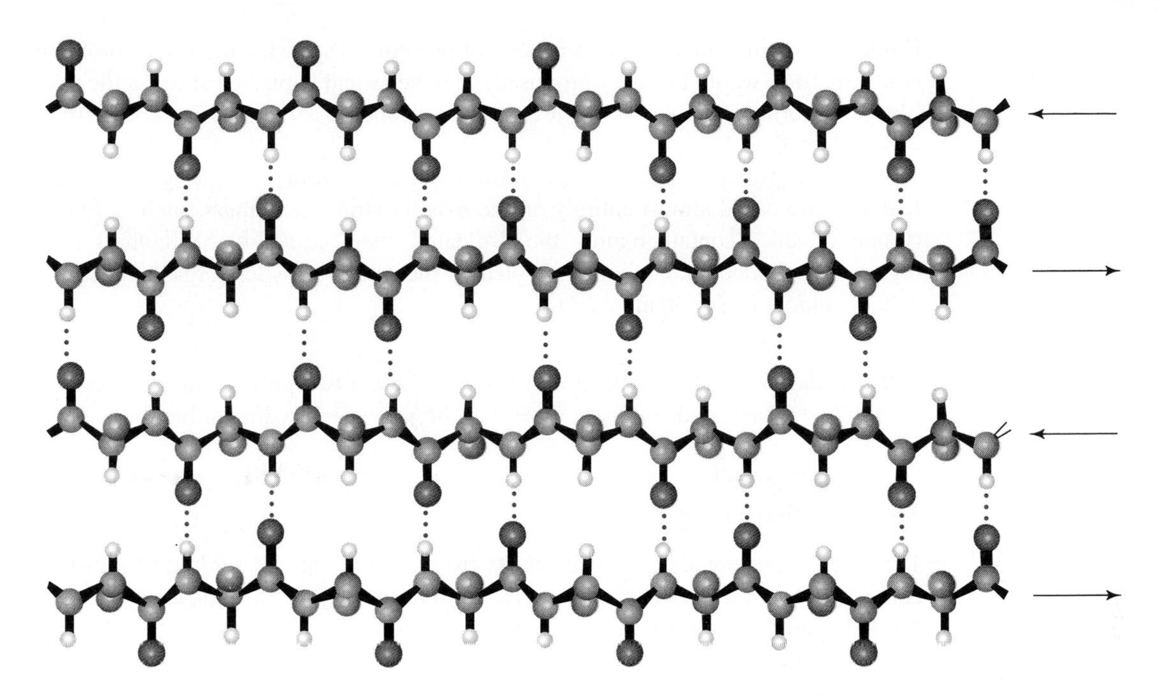

(a)

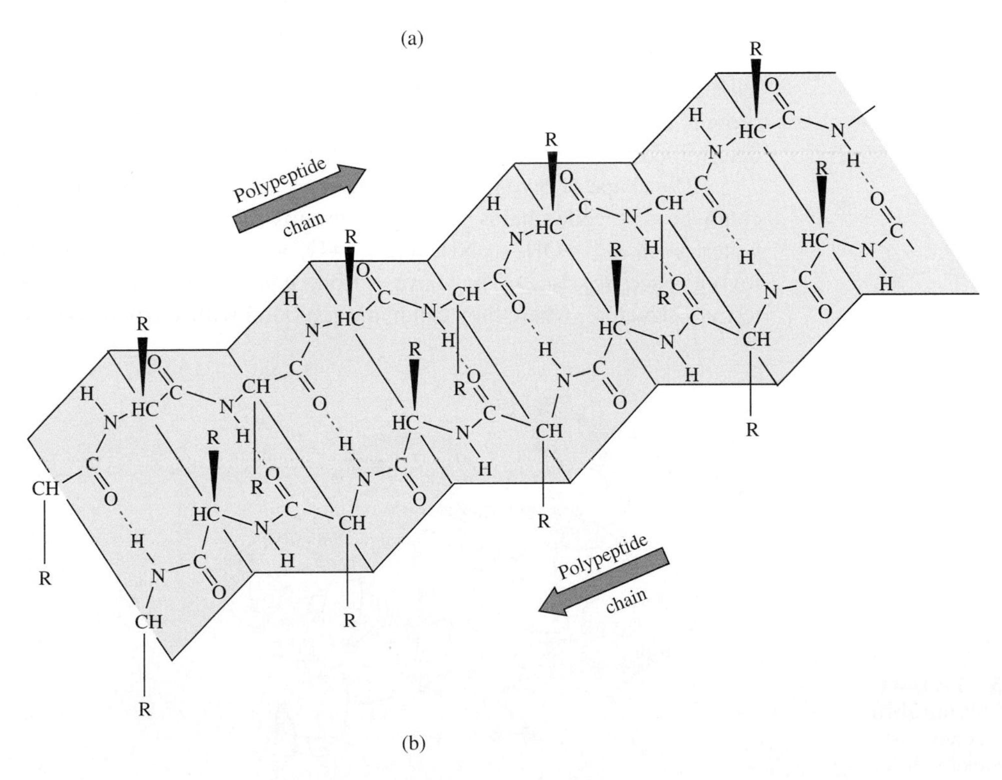

(b)

FIGURE 17.3 The β-Pleated Sheet Secondary Structure of Proteins Two models of the β-pleated sheet structure, in which a protein chain is extended and folded back upon itself. Different segments of the chain run side by side in opposite directions and are held together by N—H---O=C hydrogen bonds. (a) Four chains held together side by side with the side-chain groups (shown in green) projecting above and below the sheet. (b) A version of two chains that emphasizes the pleated (corrugated) nature of the structure.

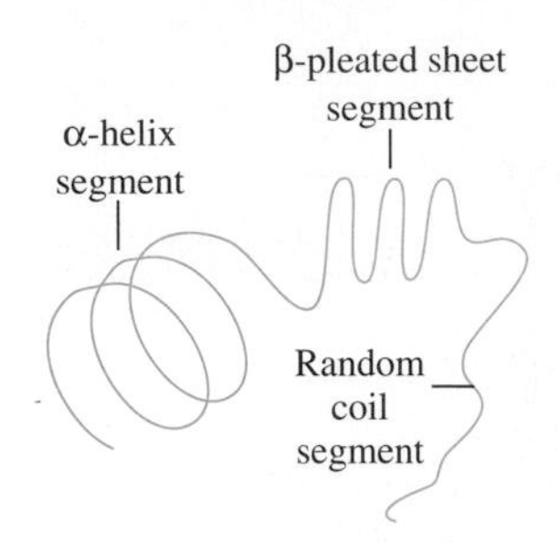

FIGURE 17.4 Types of Secondary Structure in Proteins Some proteins have more than one secondary structure. The protein in this figure has three different segments—α-helix, β-pleated sheet, and random coil.

From this information they made models of proteins to see what their most probable conformations would be. They proposed the α-helix and β-pleated sheet as the most likely conformations. As we shall see, these proposed structures were confirmed experimentally a few years later.

Some proteins, such as *α-keratin* (found in wool, hair, fingernails, and feathers) are based almost entirely on the α-helix structure. Others, such as *fibroin* (found in silk), contain mainly the β-pleated sheet structure. Still others have regions of α-helices and regions of β-pleated sheets joined by less organized regions called *random coils* (Figure 17.4).

> **Exercise 17.3** Draw a diagram to show how the two polypeptide segments —Gly—Ala— and —Gly—Ser— can be joined by hydrogen bonds.

Tertiary Structure

The three-dimensional shape that results from the coiling and folding of a protein chain is called the **tertiary structure**. **Myoglobin**, a protein with a single chain formed from 153 amino acids, provides a good example of tertiary structure (Figure 17.5). Myoglobin is located in muscle, where it stores oxygen until it is needed. Sea mammals rely on the oxygen stored in myoglobin to sustain them on long dives. It was the first protein whose complete structure was determined by X-ray crystallography, in 1956 (Box 17.1). This structure provided the first evidence that the α-helix proposed by Pauling and Corey is indeed a favored conformation for a protein chain. Since then the structures of more than 2000 proteins have been determined by X-ray crystallography (Box 11.1).

The shape of myoglobin is determined primarily by the properties of the side-chain groups. Side chains with polar groups that can form hydrogen bonds with water, such as —OH, —NH, and C=O, are described as **hydrophilic** (''water loving''; see Box 14.2). They have a strong tendency to remain on the outside of the tertiary structure, where they can hydrogen bond with surrounding water molecules.

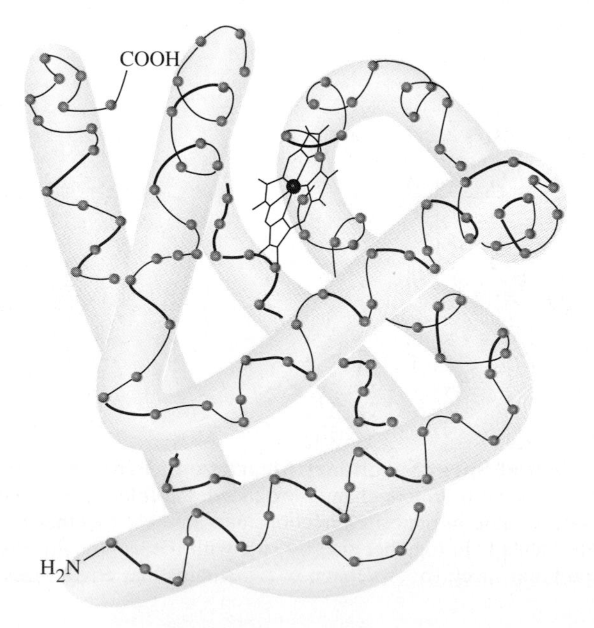

FIGURE 17.5 Tertiary Structure of Myoglobin The tertiary structure consists of straight, rodlike α-helix segments (secondary structure) separated by random-coil segments. The red complex embedded in the protein is a molecule of heme, an iron-containing group to which O_2 binds.

BOX 17.1 Molecular Biology

Until the middle of the twentieth century, the living cell was largely a chemical black box. We knew what went in and what came out, but the actual chemical processes occurring in the cell were for the most part unknown. It was not until the structures of two key types of substances, proteins and nucleic acids, were determined that biological results could be related to chemical causes. The resulting science is known as **molecular biology**. It bridges the frontier between chemistry and biology.

Molecular biology was born in Cambridge, England, in the 1950s when the structures of both nucleic acids and proteins were discovered. The double-helix structure of DNA was suggested by Watson and Crick in a famous paper published in the journal *Nature* in 1953. James Watson was born in Chicago in 1928 and Francis Crick in Northampton, England, in 1916 (Figure A). Together in Cambridge they built a model of DNA composed of two spirals to explain how DNA could store genetic information, replicate, and pass the information to subsequent generations of cells (Figure B). The chemical evidence for the structure was provided by X-ray crystallographic studies carried out at the University of London by Maurice Wilkins and (independently) by Rosalind Franklin. Wilkins shared the Nobel Prize in medicine and physiology with Watson and Crick in 1962. Rosalind Franklin had died in 1958 and was therefore denied the possibility of Nobel recognition. The determination of the structure of DNA is regarded as one of the most significant breakthroughs in science in recent times. Watson wrote a popular and highly successful account of the discovery in his book *The Double Helix* (1969).

At the time Watson and Crick were developing their model of DNA, John Kendrew, born in 1917, was working in an adjacent laboratory in Cambridge on protein structures. He used X-ray crystallography to obtain the three-dimensional structure of myglobin, the first protein structure to be determined, in 1957 (Figure C). A little later, Max Perutz, also working in Cambridge, was able to determine the structure of the larger and more complicated protein hemoglobin. Once protein structures had been determined, chemists could sensibly discuss how molecular oxygen is bound to substances such as myglobin and hemoglobin, and the chemical mechanisms of enzymatic catalysis. Kendrew and Perutz were awarded the Nobel Prize in chemistry in 1962. After these initial discoveries, molecular biology became one of the most rapidly growing sciences. Molecular biologists can now cut a gene from one strand of DNA and splice it into a different DNA molecule. In this way the genetic properties of a plant such as corn can be modified, for example, to increase its disease resistance. Such applications of molecular biology have given rise to the new industry of biotechnology.

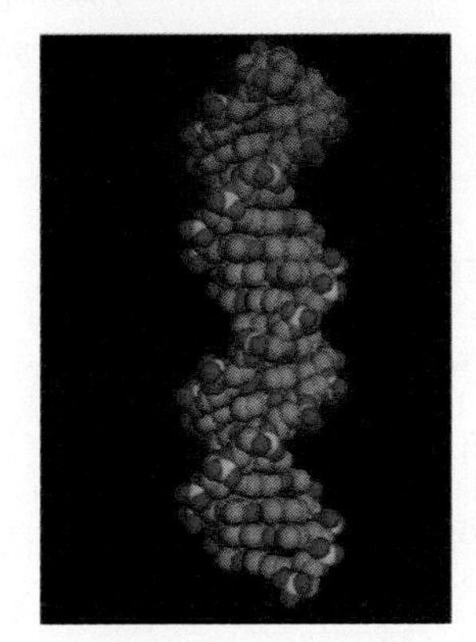

Figure B *Computer-generated model of the DNA double helix.*

Figure A *Watson (left) and Crick examining their model of DNA. Contrast this model, built in 1953, with the model "constructed" on a computer (Figure B).*

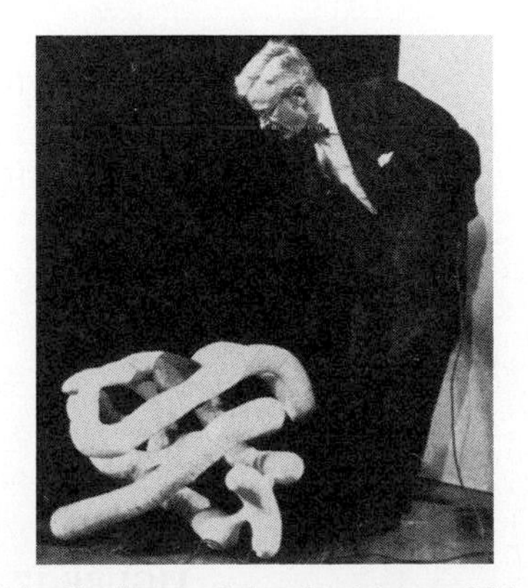

Figure C *John Kendrew and the plasticine model he constructed to represent the myoglobin molecule. The heme group is shown (half-hidden) at the top.*

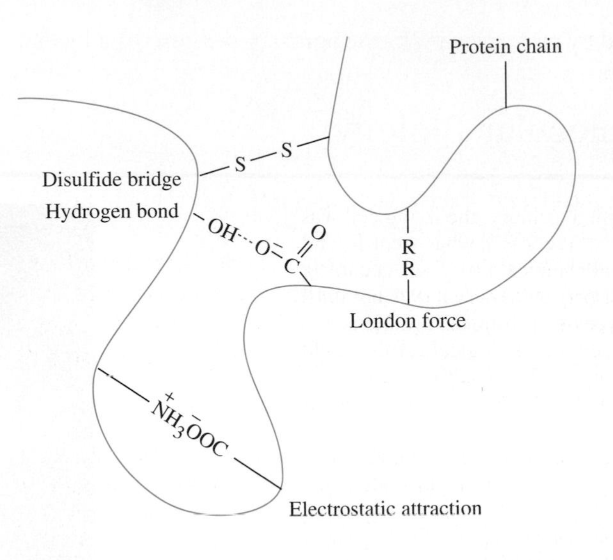

FIGURE 17.6 The Various Interactions Maintaining the Tertiary Structure of Proteins The four principal interactions within protein molecules are electrostatic attraction, disulfide bridges, London (dispersion) forces, and hydrogen bonds.

Nonpolar side chains that cannot hydrogen bond to water, such as alkyl and aryl groups, are described as **hydrophobic** (''water hating''). They tend to disrupt the hydrogen bonding between water molecules and are repelled to the inside of the tertiary structure, where they attract each other by London (dispersion) forces (Chapter 11). In addition, the tertiary structure is maintained by the electrostatic attraction between charged atoms in polar groups, by hydrogen bonds, and by —S—S— disulfide bridges (Figure 17.6).

Like many other proteins, myoglobin has a nonpolypeptide part. In myoglobin this is a **heme** group. Heme contains a central iron atom coordinated to a planar arrangement of four nitrogen atoms. A fifth coordination position on the iron atom is occupied by a nitrogen of the amino acid histidine in the polypeptide chain (Figure 17.7). The sixth position that completes the octahedral coordination around the iron

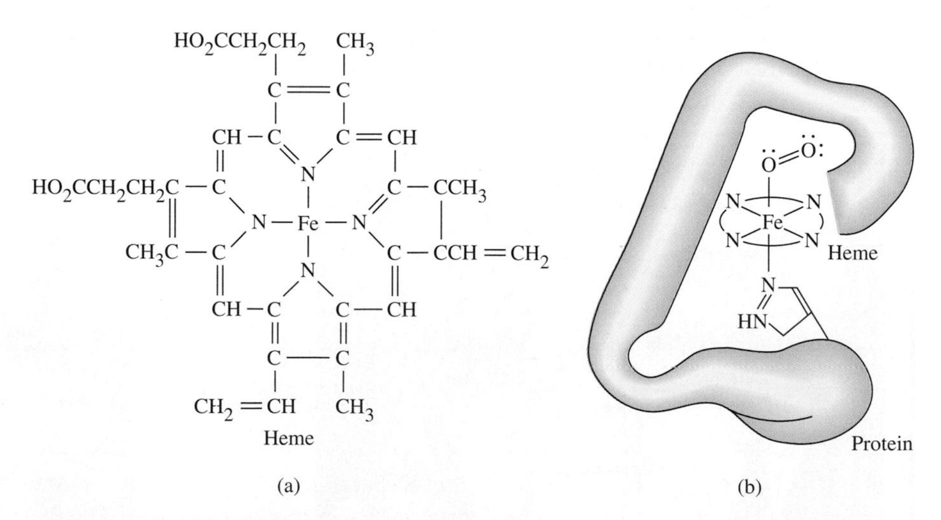

FIGURE 17.7 The Heme Group in Myoglobin (a) Heme is a planar molecule in which a central iron atom is coordinated to four nitrogen atoms in a square planar arrangement. (b) The heme group is bound to the polypeptide chain in myoglobin by a nitrogen atom of histidine. This nitrogen atom, the four nitrogen atoms of heme, and an oxygen molecule are coordinated to the iron atom with an octahedral geometry.

is either empty or is occupied by an oxygen molecule bound to the iron atom by one of its unshared electron pairs.

Quaternary Structure

Hemoglobin is the protein that carries oxygen around the body in the red blood cells of animals. It consists of four polypeptide chains, each with a tertiary structure very similar to that of myoglobin and each with a heme group that can combine with an oxygen molecule. The four polypeptides are held, primarily by London forces, in a tetrahedral arrangement and form an almost spherical molecule overall. When two or more polypeptides combine in a single functional unit, the shape of the resulting protein is referred to as its **quaternary structure**.

Many proteins lose their secondary, tertiary, and quarternary structure rather easily, for example, upon gentle heating or upon changing the pH. Such a protein is then said to be **denatured** and can no longer fulfill its normal function. But it regains its original shape and biological function if returned to its normal environment. Thus, we see that the shape and properties of a protein are fully determined by its primary structure, that is, by the sequence of its amino acids. The secondary and tertiary structures represent the most stable geometry for a particular sequence of amino acids and are adopted automatically when the protein is in its natural environment.

Globular and Fibrous Proteins

Myoglobin and hemoglobin are **globular proteins**. Globular proteins have compact, nearly spherical shapes with hydrophilic groups on the outside and hydrophobic groups folded in toward the center. Most globular proteins are soluble in water and are mobile within cells. Almost all enzymes are globular proteins. In contrast, **fibrous proteins** consist of polypeptide chains arranged side by side in long filaments. Because these proteins form very strong structures and are insoluble in water, nature uses them for structural materials. The fibrous protein collagen (Figure 17.8), the most abundant protein in mammals, constitutes about one-quarter of their total weight. It is the major component of skin, bone, tendon, cartilage, and blood vessels. Collagen consists of three helical protein chains made from glycine, proline, and hydroxyproline that spiral around each other and are held together by hydrogen bonds to give a stiff, ropelike structure.

Glycine
Hydroxyproline
Proline
(a)
(b)
(c)

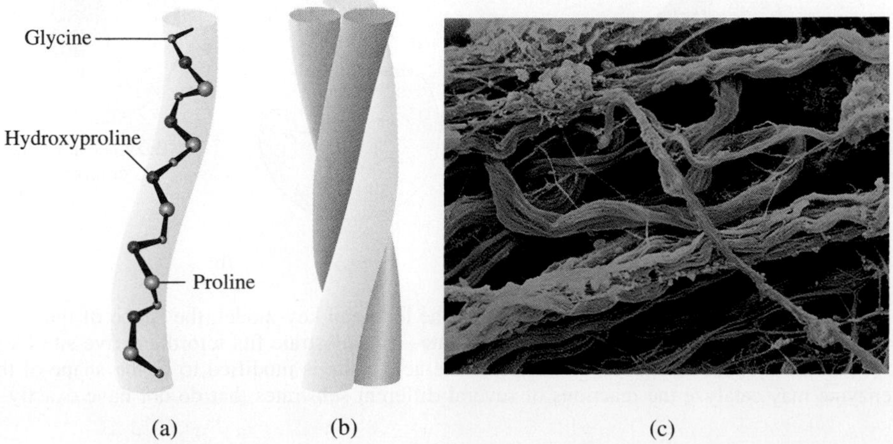

FIGURE 17.8 The Fibrous Protein Collagen (a) Each chain of the collagen triple helix has a helical conformation. (b) The three chains are wrapped around each other and held together by hydrogen bonds to form a stiff, ropelike structure. (c) False-color scanning electron micrograph of bundles of collagen and elastic fibers (magnified 4100×).

17.2 ENZYMES

Enzymes are catalysts for biochemical reactions. They provide a means for controlling the rates of reactions in an organism. In the laboratory or in industry we can control the rates of reactions by adjusting conditions such as the temperature, solvent, and pH. In an organism, however, these conditions can change only within very narrow limits. In the human body the temperature must be close to 37°C (about 98°F), the pH must be close to 7.4, and water must be the solvent.

Enzymes differ in two ways from inorganic catalysts, such as acids, metals, and metal oxides. First, enzymes are very large molecules that, with a very few exceptions, are proteins. Second, enzymes are very specific in their action. An inorganic catalyst such as an acid will generally catalyze many reactions—for example, the formation of an ester from *any* carboxylic acid and *any* alcohol (Chapter 14). In contrast, most enzymes catalyze only a single reaction or a few very similar reactions. Third, enzymes have a much greater catalytic power than most inorganic catalysts and often accelerate reactions by a factor of a million or more.

Most enzymes are globular proteins on the surface of which is a cleft or crevice of a very specific shape, called the **active site**. The **substrate** (one of the reactants) is held in the active site by London forces, dipole forces, and very often hydrogen bonds. To fit into the active site, the shape of the substrate must match that of the active site. The shape of the active site is like a lock into which only a specific key, the substrate, will fit. This model of enzyme activity is called the **lock-and-key model** (Figure 17.9a). However, not all enzymes are so specific; some are thought to be flexible, such that the shape of the active site can change somewhat to accommodate the substrate. This model is called the **induced-fit model** (Figure 17.9b). According to both models, the formation of the **enzyme–substrate complex** has two important consequences:

1. The substrate molecule is held in exactly the right orientation for reaction with a second substrate, such as a water molecule, or for the addition or removal of H^+ ions for acid–base reactions or electrons for redox reactions.
2. The substrate molecule is distorted in such a way that one or more of its bonds are weakened, so it reacts more readily. In other words, the activation energy for the reaction is decreased (Chapter 15).

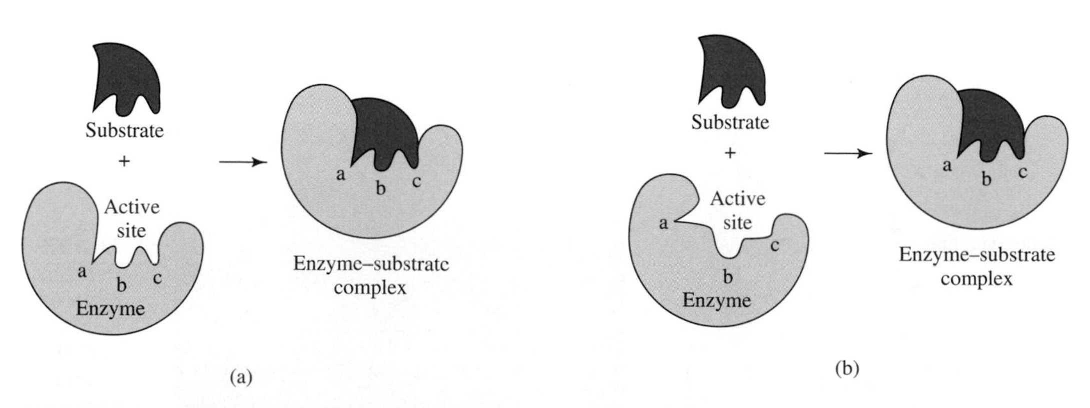

FIGURE 17.9 Models for Enzyme Catalysis (a) In the lock-and-key model, the shape of the active site on the enzyme is complementary to that of substrate—the substrate fits into the active site like a key in a lock. (b) In the induced-fit model, the shape of the active site is modified to fit the shape of the substrate. This enzyme may catalyze the reactions of several different substrates that do not have exactly the same shape.

Exercise 17.4 An uncatalyzed reaction has an activation energy of 100 $kJ \cdot mol^{-1}$. If an enzyme increases the rate of this reaction by a factor of 1 million at 37°C, to what value does the enzyme reduce the activation energy?

17.3 CARBOHYDRATES

Carbohydrates make up most of the organic matter on earth and have many different roles in all forms of life. They are used for energy generation, energy storage, structural purposes, and as components of ATP, nucleic acids, and many related compounds. Their name arises from the early name for glucose, $C_6H_{12}O_6$, which was thought to be a hydrate of carbon $C_6(H_2O)_6$. Although this was soon found to be incorrect, the name "carbohydrate" has been retained to describe glucose and related compounds. Carbohydrates have the general formula $C_xH_{2y}O_y$, or $C_x(H_2O)_y$.

Monosaccharides, Disaccharides, and Polysaccharides

The structure of carbohydrates are based on five- or six-membered rings containing four or five carbon atoms and an oxygen atom with one or more hydroxyl group substituents on the ring. Depending on the number of rings in the structure, they are called **monosaccharides**, **disaccharides**, or **polysaccharides**. The monosaccharides α-*glucose* and β-*glucose* have a six-membered ring and differ only in the orientation of an OH group relative to the ring (Figure 17.10a). *Fructose*, *ribose*, and *deoxyribose* have five-membered rings (Figure 17.10b).

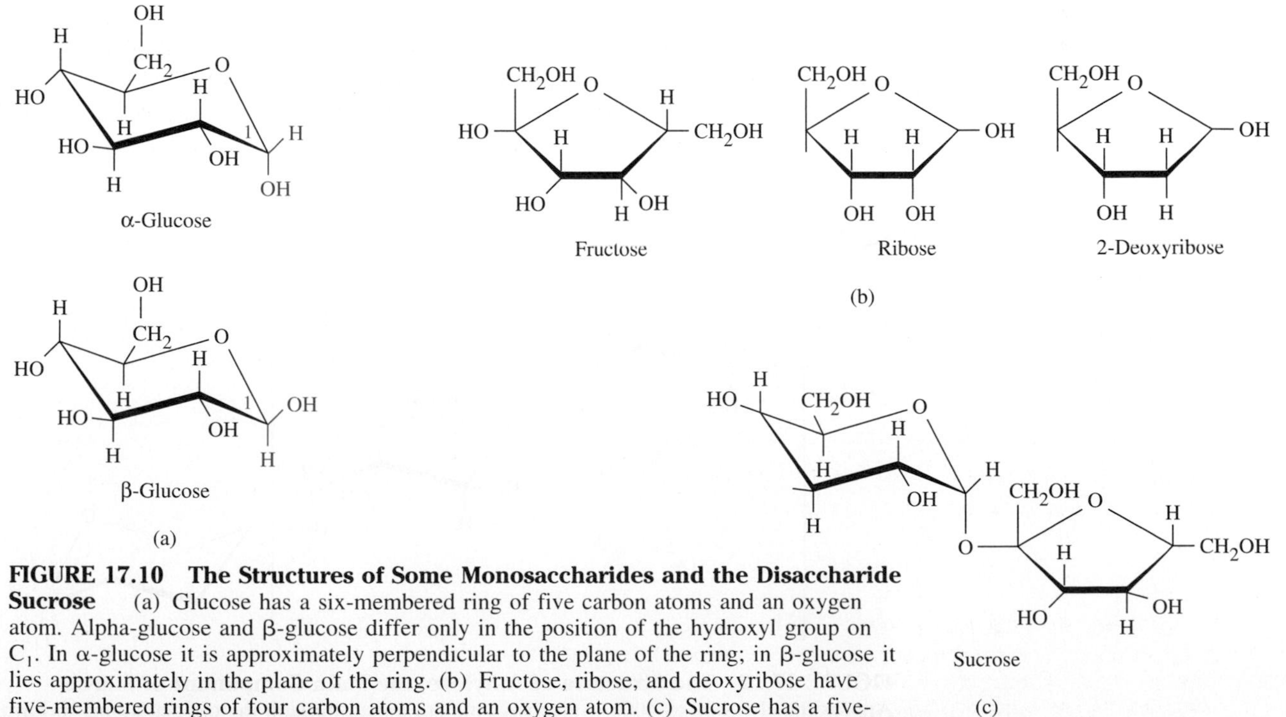

FIGURE 17.10 The Structures of Some Monosaccharides and the Disaccharide Sucrose (a) Glucose has a six-membered ring of five carbon atoms and an oxygen atom. Alpha-glucose and β-glucose differ only in the position of the hydroxyl group on C_1. In α-glucose it is approximately perpendicular to the plane of the ring; in β-glucose it lies approximately in the plane of the ring. (b) Fructose, ribose, and deoxyribose have five-membered rings of four carbon atoms and an oxygen atom. (c) Sucrose has a five-membered ring and a six-membered ring.

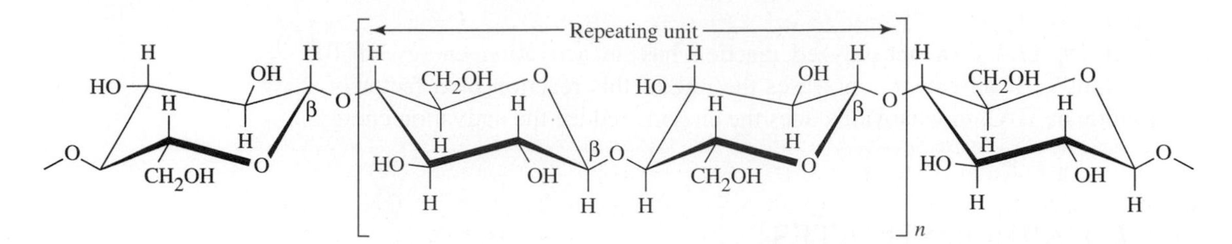

FIGURE 17.11 The Structure of Cellulose On average about 3000 β-glucose units are linked through oxygen atoms to form a cellulose molecule.

Two monosaccharides can be linked in a condensation reaction in which a molecule of water is eliminated from two OH groups, leading to a C—O—C bridge between the two rings. Joining together glucose and fructose in this manner gives the disaccharide *sucrose* (common table sugar) (Figure 17.10c).

Polysaccharides, such as starch and cellulose, are formed by the elimination of water between many monosaccharide molecules. **Cellulose** is a continuous-chain (unbranched) polymer of β-glucose containing on average about 3000 glucose units (Figure 17.11). The chains in cellulose fit together to form tough fibers. Cellulose is the major structural component of trees and other plants and accounts for more than half of all living matter. Humans do not possess the enzymes needed to speed up the hydrolysis of cellulose to glucose, and so we cannot digest cellulose. But many ruminant animals, such as cattle and sheep, feed on plant material, because bacteria in their intestines have the enzymes needed to break down cellulose.

Starch consists mainly of *amylose*, a continuous-chain polymer of α-glucose (Figure 17.12), mixed with some branched polymers of α-glucose. Starch is much more readily hydrolyzed than cellulose and is rapidly broken down to glucose in the human digestive tract with the aid of enzymes. It serves as an energy storage compound in plants.

Glycogen is a branched-chain polymer of α-glucose. It is the main energy-storage compound in humans and other animals.

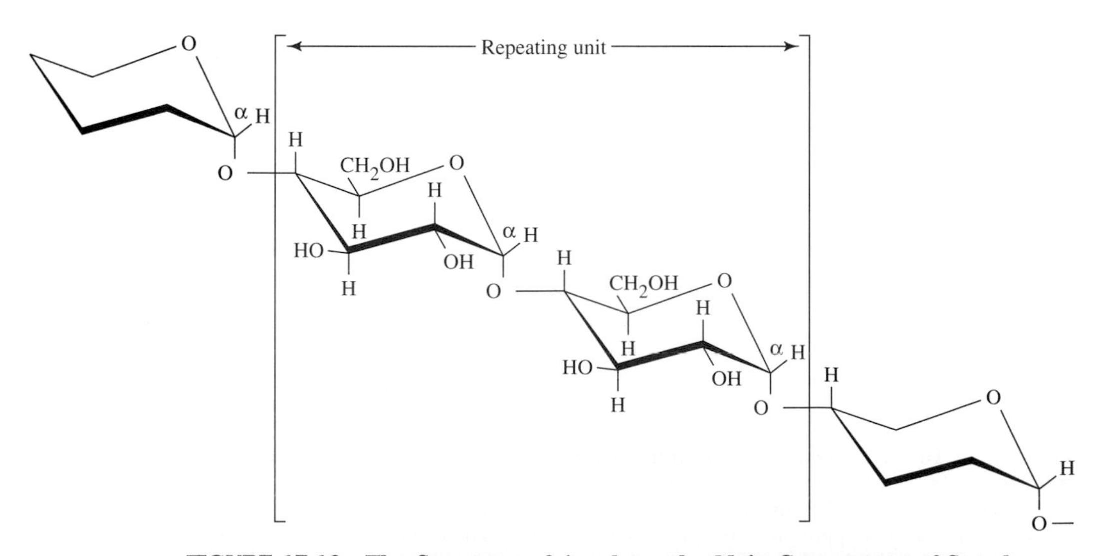

FIGURE 17.12 The Structure of Amylose, the Main Component of Starch Approximately 1000 to 4000 α-glucose units are linked through the oxygen atoms to form amylose.

17.4 ENERGY IN BIOCHEMICAL REACTIONS: ATP AND ADP

Carbohydrate Metabolism

All living organisms require a constant input of energy for three major purposes: to do the mechanical work needed to move or grow, to do the work needed for the internal movement of molecules and ions, and to synthesize biomolecules. Animals obtain this energy by the oxidation of foodstuffs, whereas plants obtain it by trapping light energy.

In the process of digestion in animals, carbohydrates such as starch, glycogen, and sucrose are hydrolyzed to monosaccharides such as glucose that pass into the bloodstream. The principal source of energy for animals is the oxidation of glucose, a strongly exothermic reaction with a large negative free energy change:

$$C_6H_{12}O_6 + 6O_2 \rightarrow 6CO_2 + 6H_2O \qquad \Delta H^\circ = -2816 \text{ kJ}, \quad \Delta G^\circ = -2867 \text{ kJ}$$

As we saw in Chapter 9, the maximum work that can be performed by a reaction is $-\Delta G^\circ$ (2867 kJ of work can be performed by the oxidation of 1 mol of glucose). Clearly, the oxidation of glucose cannot occur in the body as it does when it burns in air. Even if this *were* possible, it would produce a large amount of heat and almost no energy for doing work. In the body this reaction occurs in a long series of steps, each of which releases only a small amount of energy. In each step, most of the energy is not released as heat but is used to do work—for example, to drive other nonspontaneous reactions. The total energy evolved in this long, complicated series of reactions is exactly the same as would be obtained by burning glucose in air (Chapter 9). Much of this energy is used to synthesize **adenosine triphosphate, ATP**, which transfers energy to where it is required.

Adenosine Triphosphate, ATP

The structure of ATP is shown in Figure 17.13. It is the triphosphoric acid ester of ribose to which the heterocyclic amine adenine is attached. This very important type

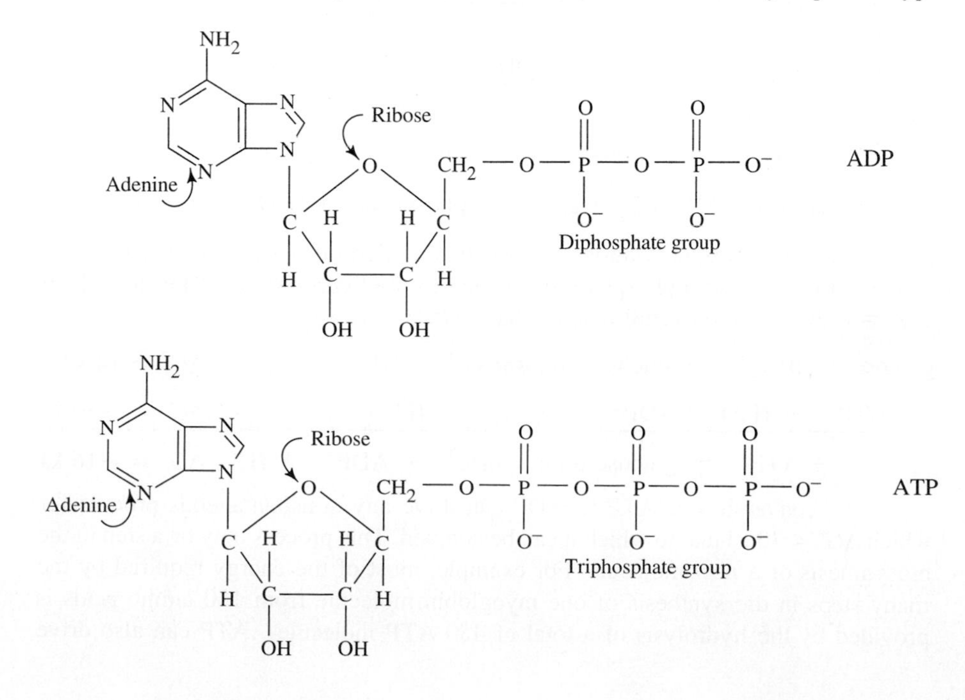

FIGURE 17.13 The Structures of ADP and ATP
The molecules adenosine diphosphate (ADP) and adenosine triphosphate (ATP) are made up of three parts: the heterocyclic amine (nitrogenous base) adenine, the sugar ribose, and a diphosphate or a triphosphate group.

of molecule is called a *nucleotide*; we will meet it again in our discussion of DNA and RNA in Section 17.5. The function of ATP is to store energy temporarily and to transport it to where it is required—for example, for biomolecule synthesis in which nonspontaneous steps are driven by coupling with the hydrolysis of ATP to **adenosine diphosphate**, ADP:

$$\mathrm{Ad{-}O{-}\overset{\overset{\displaystyle O}{\|}}{\underset{\underset{\displaystyle O^-}{|}}{P}}{-}O{-}\overset{\overset{\displaystyle O}{\|}}{\underset{\underset{\displaystyle O^-}{|}}{P}}{-}O{-}\overset{\overset{\displaystyle O}{\|}}{\underset{\underset{\displaystyle O^-}{|}}{P}}{-}O^- + H_2O}$$

$$\rightarrow \mathrm{Ad{-}O{-}\overset{\overset{\displaystyle O}{\|}}{\underset{\underset{\displaystyle O^-}{|}}{P}}{-}O{-}\overset{\overset{\displaystyle O}{\|}}{\underset{\underset{\displaystyle O^-}{|}}{P}}{-}O^- + HO{-}\overset{\overset{\displaystyle O}{\|}}{\underset{\underset{\displaystyle O^-}{|}}{P}}{-}O^- + H^+}$$

$$ATP^{4-} + H_2O \rightarrow ADP^{3-} + HPO_4^{2-} + H^+ \quad \Delta G^\circ = -30 \text{ kJ} \qquad (1)$$

ΔG must be *negative* for a spontaneous reaction. Here both the ΔH and $-T\Delta S$ contributions are negative, and ΔG is negative, so the breakdown of ATP is spontaneous.

ATP reacts spontaneously with many biomolecules, particularly substances containing hydroxyl groups such as carbohydrates, by the transfer of a phosphate group. Such *phosphorylated* molecules (molecules to which phosphate has been added) are often more reactive than the molecule from which they are derived and subsequently participate in a reaction in which the phosphate group is removed and a new product formed. The overall reaction results in the hydrolysis of ATP to ADP and the synthesis of a new biomolecule. For example, the first step in the metabolism of glucose is the formation of glucose 6-phosphate. This is a nonspontaneous reaction for which ΔG° is 14 kJ:

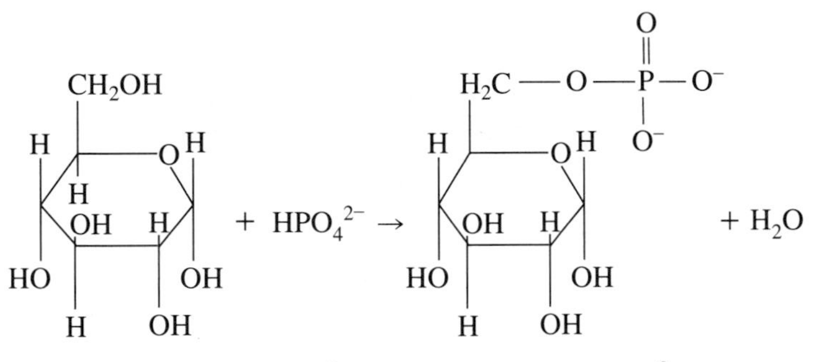

$$\text{glucose} + HPO_4^{2-} \rightarrow \text{glucose 6–phosphate}^{2-} + H_2O \quad \Delta G^\circ = 14 \text{ kJ}$$

However, if the phosphorylation of glucose is carried out by ATP—that is, if the formation of glucose 6-phosphate is coupled to the hydrolysis of ATP, for which $\Delta G^\circ = -30$ kJ—the overall reaction has $\Delta G^\circ = -16$ kJ:

$$\begin{array}{ll} \text{glucose} + HPO_4^{2-} \rightarrow \text{glucose 6-phosphate}^{2-} + H_2O & \Delta G^\circ = 14 \text{ kJ} \\ ATP^{4-} + H_2O \rightarrow ADP^{3-} + HPO_4^{2-} + H^+ & \Delta G^\circ = -30 \text{ kJ} \\ \hline \text{glucose} + ATP^{4-} \rightarrow \text{glucose 6-phosphate}^{2-} + ADP^{3-} + H^+ & \Delta G^\circ = -16 \text{ kJ} \end{array}$$

The conversion of ATP to ADP can drive any nonspontaneous process for which $\Delta G^\circ < 30$ kJ and to which it can be coupled. This process may be a step in the biosynthesis of a new molecule. For example, most of the energy required by the many steps in the synthesis of one myoglobin molecule from 150 amino acids is provided by the hydrolysis of a total of 450 ATP molecules. ATP can also drive many other processes, such as the change in the conformation of a protein that occurs when muscles contract or a change in the relative concentrations of ions inside and outside a cell. ATP is the principal means by which energy is transferred from the molecules in which it is stored to processes in which it is needed.

After it has been formed, ADP must be converted back to ATP. This reaction—the reverse of reaction (1) has $\Delta G^\circ = 30$ kJ. To drive the reverse reaction, the same principle of coupling is used. For example, in a later step of the oxidation of glucose, a phosphate group is lost from phosphoenolpyruvate to give pyruvate in a reaction that has $\Delta G^\circ = -62$ kJ:

$$H_2O + CH_2{=}C(O{-}PO_3^{2-})(COO^-) \rightarrow CH_3{-}C(=O){-}COO^- + HPO_4^{2-}$$

Phosphoenolpyruvate   Pyruvate

This reaction is used to drive the synthesis of ATP from ADP:

$$\text{phosphoenolpyruvate}^{3-} + H_2O \rightarrow \text{pyruvate}^- + HPO_4^{2-} \quad \Delta G^\circ = -62\text{ kJ}$$

$$ADP^{3-} + HPO_4^{2-} + H^+ \rightarrow ATP^{4-} + H_2O \quad \Delta G^\circ = 30\text{ kJ}$$

$$\text{phosphoenolpyruvate}^{3-} + ADP^{3-} + H^+ \rightarrow \text{pyruvate}^- + ATP^{4-} \quad \Delta G^\circ = -32\text{ kJ}$$

Overall, a glucose molecule is converted in a series of many steps to two pyruvate ions. Although ATP is used up in some of the steps, more is produced in other steps. A total of two ATP molecules are synthesized:

$$\text{glucose} + 2ADP^{3-} + 2HPO_4^- \rightarrow 2CH_3C(=O){-}COO^- + 2ATP^{4-} + 2H^+$$

$$\text{glucose} + 2ADP^{3-} + 2HPO_4^- \rightarrow 2(\text{pyruvate}^-) + 2ATP^{4-} + 2H^+ \quad \Delta G^\circ = -73\text{ kJ}$$

In the complete oxidation of one molecule of glucose to CO_2 and H_2O, 38 molecules of ATP are synthesized. These ATP molecules are then ready to transfer their stored energy to where it is required.

17.5 DNA AND RNA

We said earlier that a characteristic feature of living organisms is their ability to reproduce. How does this happen? How does a seed ''know'' what kind of plant to become? A seed or a human embryo must contain all the information necessary to enable it to grow into a particular kind of plant or into a human being.

Early studies led to the conclusion that the necessary information is carried by structures called *chromosomes* in the nucleus of a cell. Each portion of a chromosome that controls a specific inheritable trait, such as yellow seeds or blue eyes, is called a *gene*. But for a long time no one knew how this genetic information is carried in a chromosome.

We now know that a chromosome contains an immensely long, threadlike macromolecule (polymer) called **deoxyribonucleic acid (DNA)**. In humans a DNA molecule can be as long as 12 cm when stretched out, but it has a width of only a few atoms. A gene corresponds to a particular short segment of this molecule. Human DNA has been estimated to contain about 100,000 genes.

Nucleic Acids

A **nucleic acid** such as DNA is a polymer of **nucleotides**—a polynucleotide. Nucleotides are molecules made up of three parts: a heterocyclic amine (Chapter 14), usually called a *nitrogenous base* in biochemistry; a sugar (monosaccharide); and a phosphate group. There are four different nitrogenous bases in DNA—**adenine (A)**, **guanine (G)**, **cytosine (C)**, and **thymine (T)** (Figure 17.14)—and the sugar is deoxyribose (Figure 17.10). The sugar molecules are linked by the phosphate groups to form the long chain that is the backbone of the nucleic acid molecule, and one of the four nitrogenous bases is attached to each sugar molecule (Figure 17.15). In a human DNA molecule there are several billion nucleotides. Even simpler DNA molecules, such as those in bacteria, may contain several million nucleotides.

ATP is also a nucleotide, and its close and readily apparent relationship to DNA cannot be accidental. ATP and DNA presumably had a common ancestor, but their functions must have separated in the early stages of the development of life. Whereas the triphosphate group is the important reactive part of the ATP molecule, in DNA the phosphate group is merely part of the backbone of the molecule; the important part of the DNA molecule is the heterocyclic amine (nitrogenous base).

Both proteins and nucleic acids are long-chain polymers. A protein has a polyamide backbone with different side chains attached to it at regular intervals, whereas a nucleic acid has a sugar–phosphate backbone with one of the four nitrogenous bases attached at regular intervals (Figure 17.16).

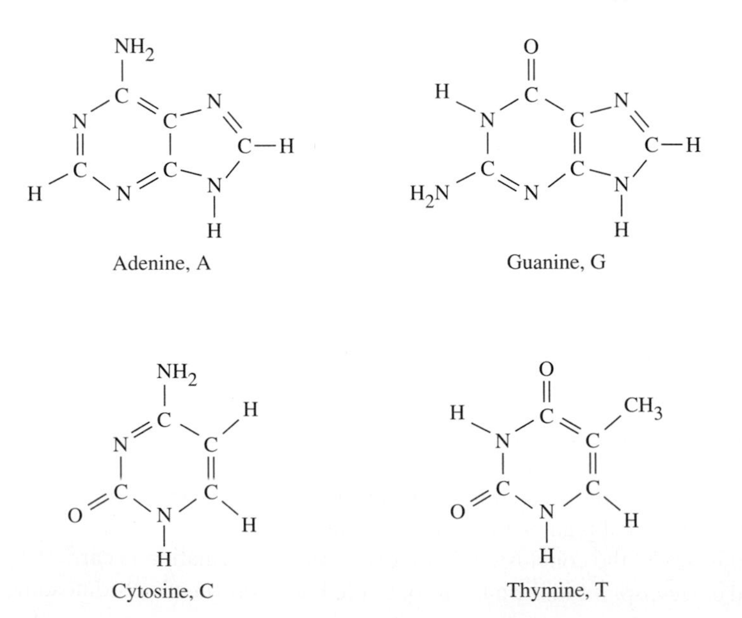

FIGURE 17.14 The Nitrogenous Bases in the Polynucleotide DNA
These bases are heterocyclic amines.

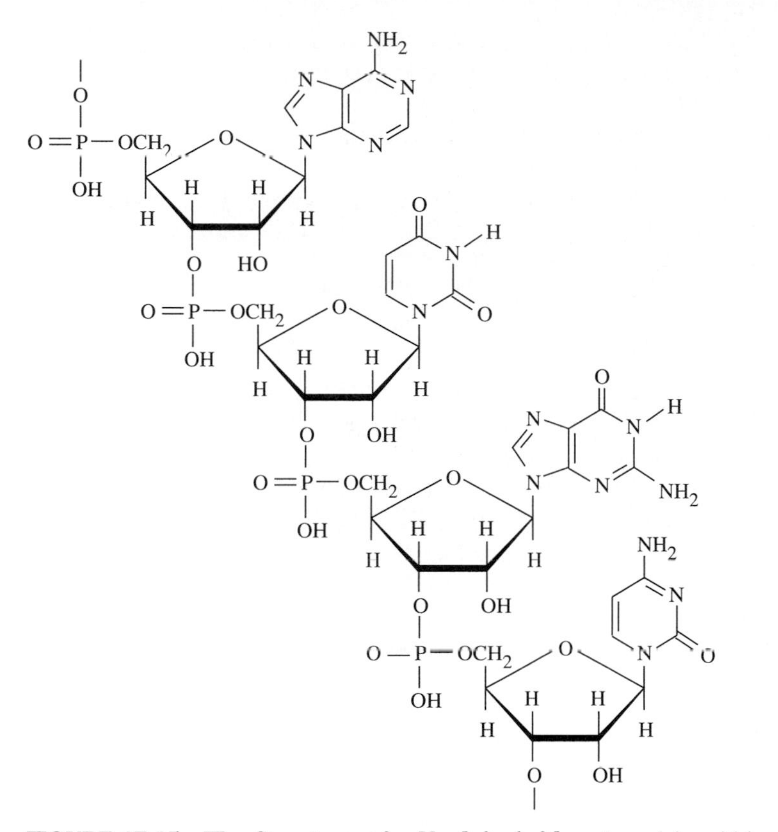

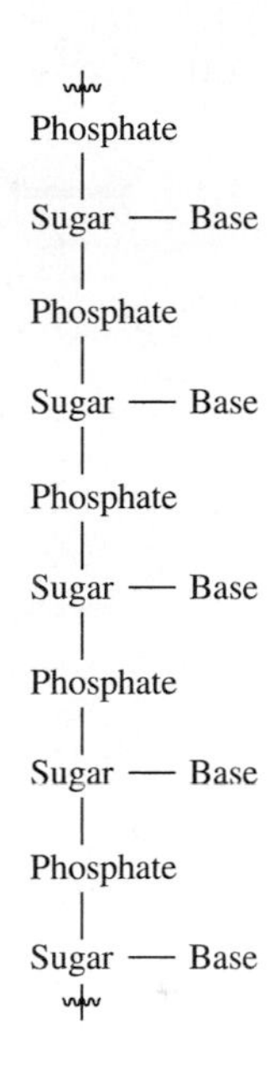

FIGURE 17.15 The Structure of a Nucleic Acid A nucleic acid is a polymer with a backbone consisting of alternating phosphate groups and sugar molecules and a heterocyclic amine (nitrogenous base) attached to each sugar molecule. (The squiggle indicates that the chain continues.)

The structure of DNA was elucidated by James Watson (1928–) and Francis Crick (1916–) in 1953 (Box 17.1). They showed that DNA is made of two polynucleotide strands wound around each other in the form of a double helix (Figure 17.17). The sequence of bases runs in one direction in one strand and in the opposite direction in the other strand. Watson and Crick deduced this structure from

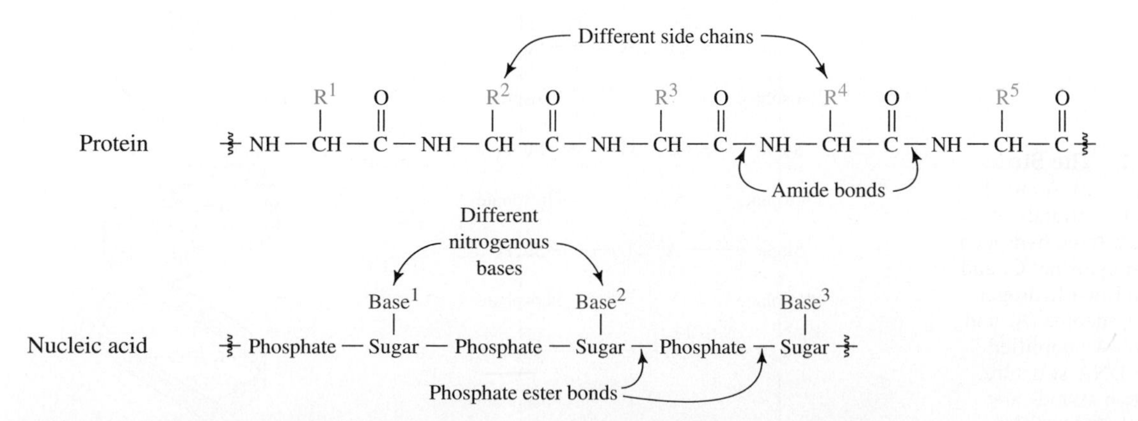

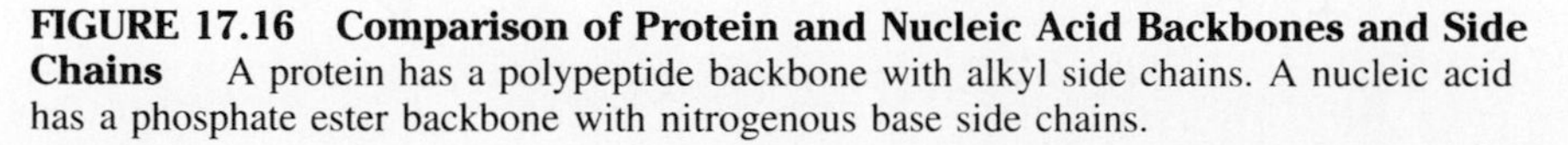

FIGURE 17.16 Comparison of Protein and Nucleic Acid Backbones and Side Chains A protein has a polypeptide backbone with alkyl side chains. A nucleic acid has a phosphate ester backbone with nitrogenous base side chains.

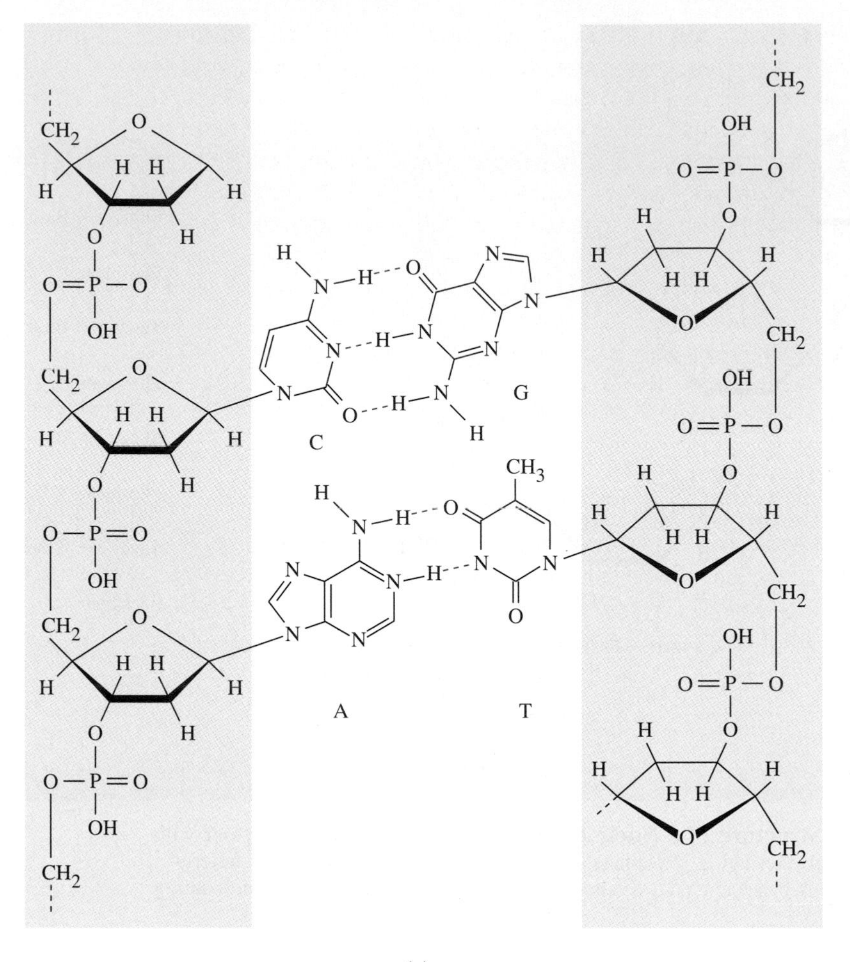

(a)

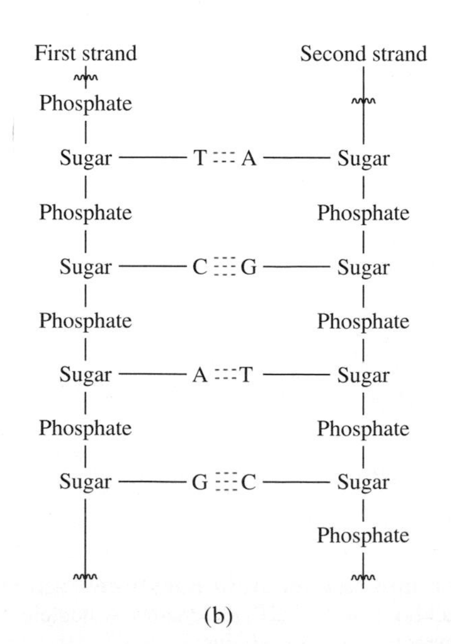

(b)

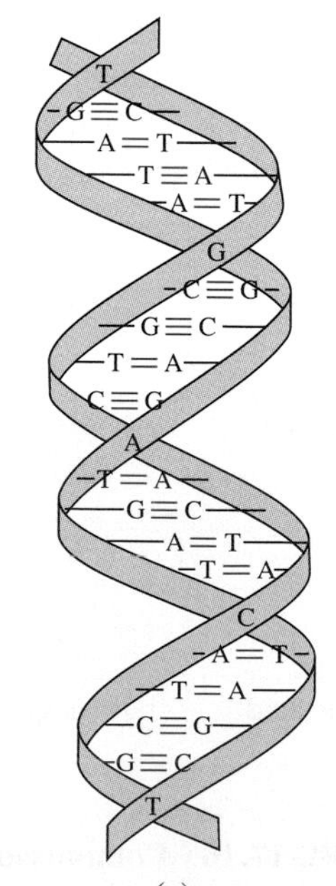

(c)

FIGURE 17.17 The Structure of DNA (a) A small portion of the two strands of DNA. There are three hydrogen bonds between cytosine, C, and guanine, G, and two hydrogen bonds between adenine, A, and thymine, T. (b) A simplified version of the DNA structure. (c) How the two strands are held together in the form of a double helix by hydrogen bonds.

X-ray diffraction photographs and from the earlier observations that the amount of adenine in DNA equals the amount of thymine and that the amount of guanine equals that of cytosine (which suggested that these pairs of bases occur together in the structure). Hydrogen bonds between these base pairs hold the two polynucleotides together. The geometry of these bases is such that guanine can hydrogen bond only to cytosine, and adenine can hydrogen bond only to thymine. This geometry causes one polynucleotide to spiral around the other, forming a double helix.

III The H atom covalently bonded to an N atom of one base has a $\delta+$ charge and is attracted to a lone pair on an N atom or an O atom of another base.

The deduction of the structure of DNA by Watson and Crick was one of the most significant accomplishments in the history of biology, because it led immediately to an understanding of genetics in molecular terms. It became clear how each DNA molecule can give rise to two exact copies of itself, that is, how it can *replicate* itself, and how the genetic information is contained in the DNA molecule.

Replication and Protein Synthesis

The process of **replication** begins when the relatively weak hydrogen bonds that hold the two strands of the DNA molecule together are broken and the two strands uncoil. As they uncoil, new nucleotides are added to each separate strand, forming a new DNA molecule (Figure 17.18). Because cytosine can hydrogen bond only to guanine and adenine can hydrogen bond only to thymine, this new DNA molecule exactly complements the first. In other words, each half of the DNA molecule serves as a template for the synthesis of the other half. Two new identical daughter molecules are formed, each of which contains one old chain and one newly synthesized chain.

The sequence of bases in a DNA molecule constitutes the **genetic code**. The sequence of bases in each segment of the DNA molecule provides the information necessary for the synthesis of a particular protein. By replication the DNA molecule is able to repeatedly duplicate itself. Each new molecule carries with it all the

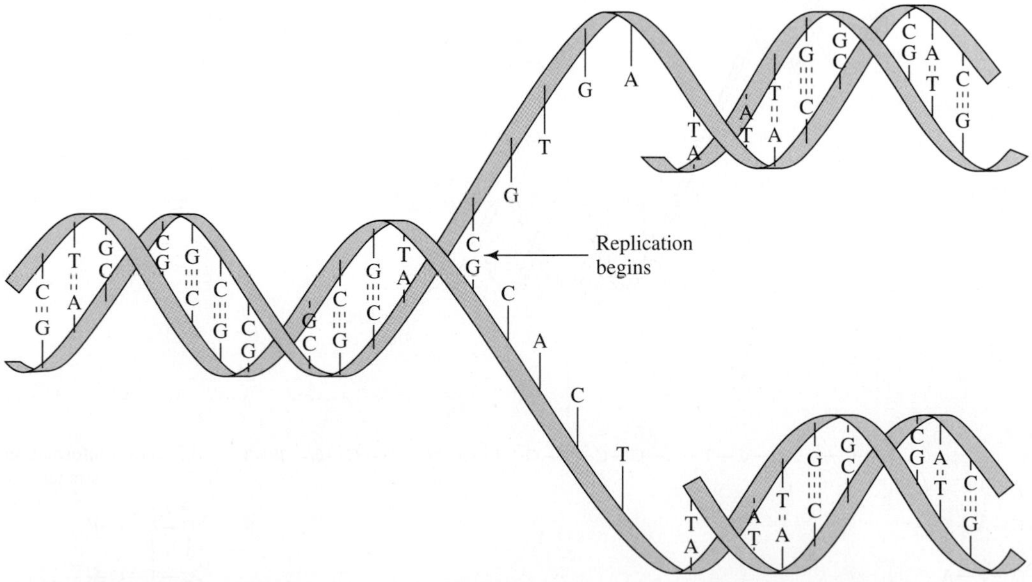

FIGURE 17.18 DNA Replication A schematic representation of DNA replication. The original DNA double helix unzips at a point called the replication fork. Complementary new nucleotides line up on each strand such that identical new DNA molecules form.

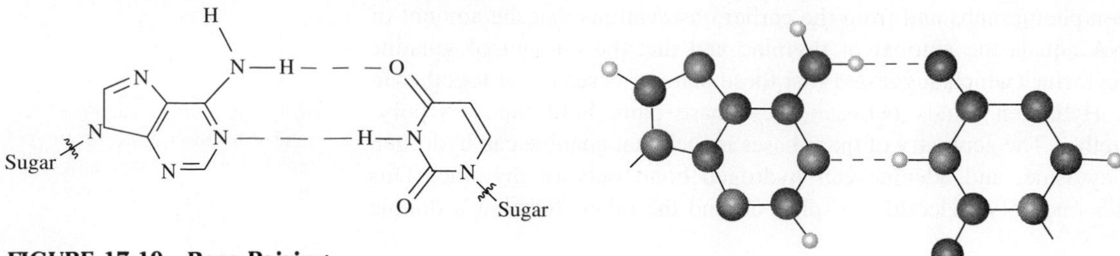

FIGURE 17.19 Base Pairing in RNA Hydrogen bonds between adenine and uracil in RNA.

information needed to control the synthesis of the proteins that enable an organism to grow and function.

The process of **protein synthesis**, the conversion of the information contained in DNA into proteins, begins with the synthesis of **RNA, ribonucleic acid**. RNA is similar to DNA except that it is a single-strand polynucleotide with the sugar ribose, rather than a double-strand polynucleotide with the sugar deoxyribose, and the base thymine is replaced by uracil (Figure 17.19). Uracil pairs with adenine, just as thymine does. RNA molecules are much shorter than DNA, containing no more than a few thousand nucleotides. RNA molecules have several different functions. One function is to carry the information in DNA to the site where the proteins are synthesized; these RNA molecules are called **messenger RNA (mRNA)**. They obtain the information from DNA by the process of **transcription**, which is similar to DNA replication. In transcription, a portion of the DNA double helix first unwinds (Figure 17.20). The unwound portion of one of these strands, called the **template strand**, is copied by the construction of an mRNA molecule complementary to the template strand. This mRNA molecule contains all the information in the unwound portion of the *other* DNA strand, called the **informational strand**; it is an

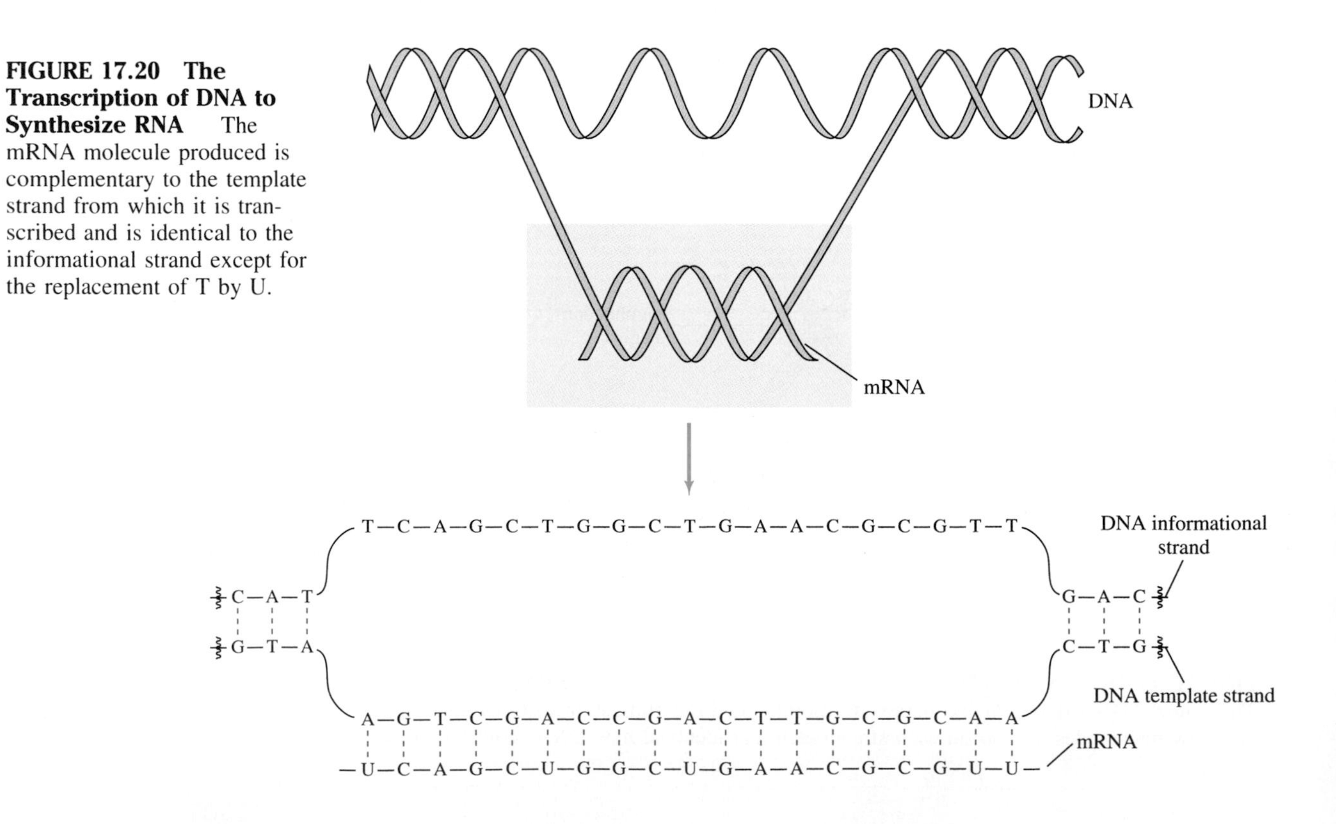

FIGURE 17.20 The Transcription of DNA to Synthesize RNA The mRNA molecule produced is complementary to the template strand from which it is transcribed and is identical to the informational strand except for the replacement of T by U.

exact copy of this portion of the informational strand except that U replaces T. The mRNA molecule then migrates to the site of protein synthesis. There the message contained in the mRNA is read, and the corresponding protein is synthesized in a process involving other RNA molecules (the details of which we cannot go into here).

The Genetic Code

How is the genetic information stored in the DNA molecule? The identity and sequence of the bases in the DNA molecule specifies the amino acids needed for the synthesis of a particular protein and the order in which the amino acids must be joined together. This information is written in the language of the **genetic code**, which uses only four "letters"—A, U, G, and C (representing the bases adenine, uracil, guanine, and cytosine). All the "words" in this language are three-letter words called **codons**. Each successive set of three letters, or triplet, is the code for inserting a particular amino acid into the protein at that point. A sequence of bases such as

$$\underbrace{\text{CGC}}_{A_1}\ \underbrace{\text{CGC}}_{A_1}\ \underbrace{\text{UGC}}_{A_2}\ \underbrace{\text{AAA}}_{A_3}\ \underbrace{\text{CGU}}_{A_4}\ \underbrace{\text{UGC}}_{A_2}\ \ldots \text{ is read as}$$

where A_1, A_2, . . . represent different amino acids. For example, CGU is the code for arginine, and AAA is the code for lysine; the sequence AAACGU corresponds to an Arg–Lys segment of a polypeptide. There are 64 possible three-base combinations. Of these, 61 code for specific amino acids (most of which are specified by more than one codon), and 3 code for chain termination—that is, they indicate where the chain stops (Table 17.2). It is difficult to imagine the magnitude of the replication process. There are several billion base pairs of human DNA. Yet the base sequence is faithfully copied during replication, and an error occurs only once in about 10 to 100 billion bases.

TABLE 17.2 Codon Assignments of Base Triplets

First Base	Second Base	Third Base U	Third Base C	Third Base A	Third Base G
U	U	Phe	Phe	Leu	Leu
	C	Ser	Ser	Ser	Ser
	A	Tyr	Tyr	Stop	Stop
	G	Cys	Cys	Stop	Trp
C	U	Leu	Leu	Leu	Leu
	C	Pro	Pro	Pro	Pro
	A	His	His	Gln	Gln
	G	Arg	Arg	Arg	Arg
A	U	Ile	Ile	Ile	Met
	C	Thr	Thr	Thr	Thr
	A	Asn	Asn	Lys	Lys
	G	Ser	Ser	Arg	Arg
G	U	Val	Val	Val	Val
	C	Ala	Ala	Ala	Ala
	A	Asp	Asp	Glu	Glu
	G	Gly	Gly	Gly	Gly

When an error does occur in the replication of DNA, the deficient DNA may give rise to a disease—called a genetic disease—that is passed on from one generation to another. Many such genetic diseases have now been identified. We are gradually learning how to detect them and in some cases how to treat them with appropriate drugs (Box 17.2).

Exercise 17.5 **(a)** For which polypeptide segment does the base sequence GCGUUUGGA code? **(b)** Draw the structure of this segment.

Exercise 17.6 How many possible genetic codes are there for **(a)** the —Cys—Phe— and **(b)** —Trp—Lys—Glu—Met— segments of polypeptides?

We have seen that the functions of nucleic acids such as DNA, of proteins such as enzymes, and of many other biomolecules depend in a very critical way on their shapes. Our ability to understand the shapes of biomolecules and how they affect the molecules' properties has been very much enhanced in recent years by the development of computer molecular modeling programs (Box 17.3). Such programs enable us to examine molecular shapes quickly and easily and to avoid the tedium of building conventional large ball-and-stick or similar models.

BOX 17.2 Chemistry in Medicine

As we achieve a better understanding of the chemistry behind biological processes, the possibility of correcting defects that can lead to certain types of disease increases. Gertrude B. Elion (Figure A) spent her career investigating the role of one class of substances, the purines, in human metabolism and their use in the treatment of disease. In 1988 she was awarded the Nobel Prize in medicine for her achievements. Purines, such as adenine (Figure B), are heterocyclic bases that contain nitrogen. They are components of many biomolecules such as nucleic acids and ATP. Gout is one of the diseases that can be treated with purines. Recorded cases of gout date back to Roman times, but it was particularly prevalent among the wealthy classes in the sixteenth to eighteenth centuries. It was discovered in 1847 that gout is associated with high blood concentrations of uric acid (Figure C). This substance is not very soluble and tends to crystallize in the joints, particularly the big toe, leading to great pain with any movement of the joint. Purines are both the cause of and the remedy for gout.

Purines are a valuable part of a normal diet. They are obtained by the decomposition of nucleic acids contained in foods such as red meat, fish, and red wines. They can be used to make more nucleic acids and ATP, but, if our diet is too "rich," excess purines must be eliminated. They are converted first to uric acid and then to urea, which is

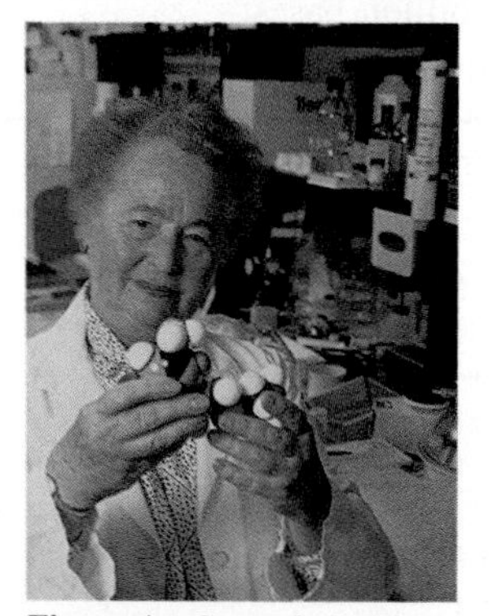

Figure A *Gertrude B. Elion (1918–)*

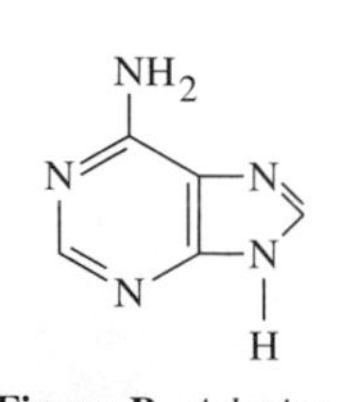

Figure B *Adenine*

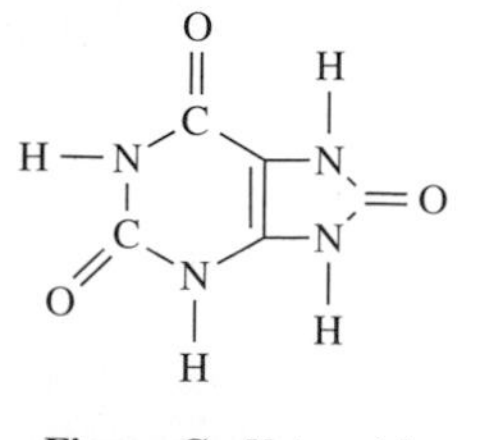

Figure C *Uric acid*

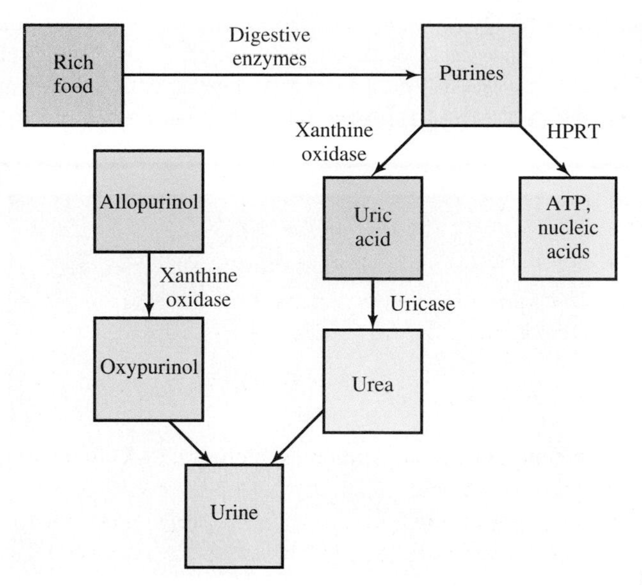

Figure D *The digestion of purines*

excreted in the urine (Figure D). Gout results from a malfunction of the enzymes used in the processing of purines.

Uric acid is transported in the blood, and its concentration can be measured in blood serum, the fluid that remains after the red and white blood cells have been removed. All "normal" men are on the brink of gout: The average concentrations of uric acid are 2 to 10 mg per 100 ml of blood serum, and precipitation can occur at concentrations greater than 7 mg per 100 ml. Gout sufferers are usually in the range 6 to 13 mg per 100 ml. The amount of uric acid in the blood is the result of a balance between the rate of production (which depends on diet), the rate of destruction (catalyzed by the enzyme uricase), and the conversion rate of purines to useful biomolecules (catalyzed by the enzyme HPRT). The enzyme HPRT is derived from a gene on the X chromosome, which may be deficient. Because men have only one X chromosome, whereas women have two, the deficiency is more likely to affect men than women.

Elion discovered that the rate of production of uric acid from a purine, such as adenine, can be controlled by the drug allopurinol (Figure E). The structure of allopurinol is like that of adenine, but the NH_2 substituent of adenine is replaced by OH. Allopurinol competes with adenine as a substrate for the enzyme xanthine oxidase, which catalyzes the first step in the conversion of adenine to uric acid.

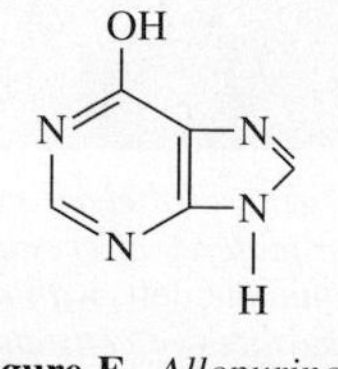

Figure E *Allopurinol*

Furthermore, the oxidation product of allopurinol, oxypurinol (Figure F), binds very tightly to xanthine oxidase and greatly restricts its ability to aid in the formation of uric acid. Oxypurinol can be directly excreted through the kidneys without further processing and therefore does not pose the same threat as uric acid. Hence, if it is combined with dietary moderation, allopurinol is very effective in controlling gout.

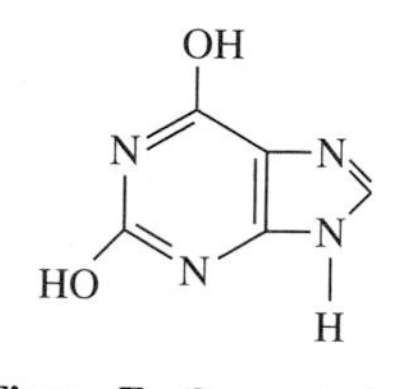

Figure F *Oxypurinol*

Allopurinol is only one of the vast number of substances available for treating disease. A very early discovery of such a drug was that of penicillin by Alexander Fleming in 1929. The original drug was found as a product of the growth of molds on food exposed to the air at room temperature. It proved effective as an antibacterial agent (antibiotic), particularly in the prevention of infection in wounds. There is a whole family of penicillin drugs with different R groups (Figure G). Today they are largely made by laboratory chemistry.

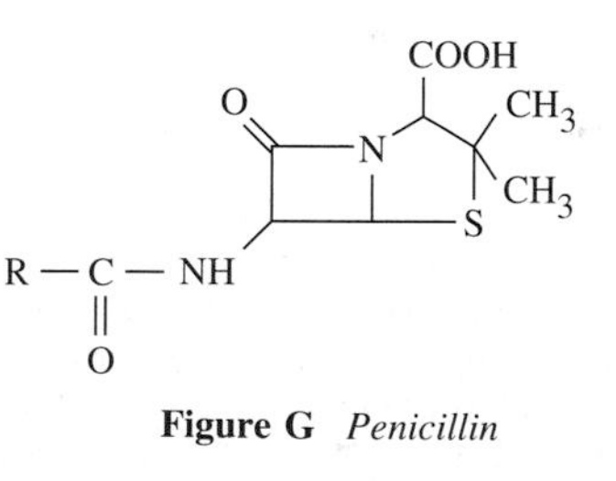

Figure G *Penicillin*

A more recent objective has been the development of drugs effective against cancer. Success in this area has been elusive. The major problem is that most substances that will destroy a cancerous cell will also destroy a normal cell. They therefore have side effects of varying degrees of seriousness and must be used sparingly, often in combination with radiation treatment. One such anti-cancer drug is the simple inorganic compound *cis*-platin (Figure H).

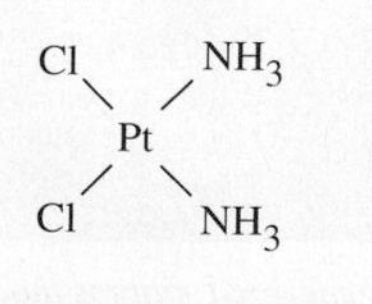

Figure H cis-*Platin*

BOX 17.3 Computer Modeling of Biomolecules

In Box 17.1 we saw Francis Crick and James Watson with their model of DNA and John Kendrew with his model of myoglobin. Such models are crucial to our understanding of the three-dimensional shape of a molecule, because two-dimensional representations are hard to draw and to interpret. We saw earlier that simple ball-and-stick and space-filling models help us explain the properties of even simple molecules, and models are essential for understanding the properties of large biomolecules such as enzymes. But building models of large molecules is difficult and time consuming.

Fortunately, the computer has come to the aid of molecular modelers. During the 1980s molecular modeling programs were developed that can be used even on personal computers. If the computer is supplied with data such as the coordinates of each atom in a molecule, the combination of fast data-processing with high-quality computer graphics results in a model of the molecule displayed on screen in a matter of seconds. We can rotate the computer model about any axis to visualize its three-dimensional shape accurately, just as if we were handling a true three-dimensional model.

The coordinates of each atom in a molecule are available if the molecule's structure has been determined by X-ray crystallography. Alternatively, if just the structural formula is known, the computer can use stored data on typical bond lengths and angles to construct a model of the molecule. For simple molecules, programs are available to calculate molecular structure directly from the properties of the component atoms.

Different types of model displays are useful for different purposes. For proteins it is often the secondary structure—the way the protein chain is folded—that is of interest. A fairly simple program can produce a ribbon diagram showing the folding of the chain but not the individual atoms (Figure A).

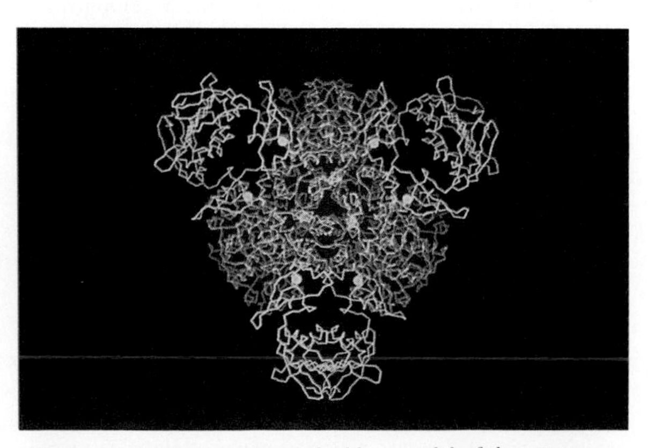

Figure A *Computer-generated ribbon model of the enzyme ATCase*

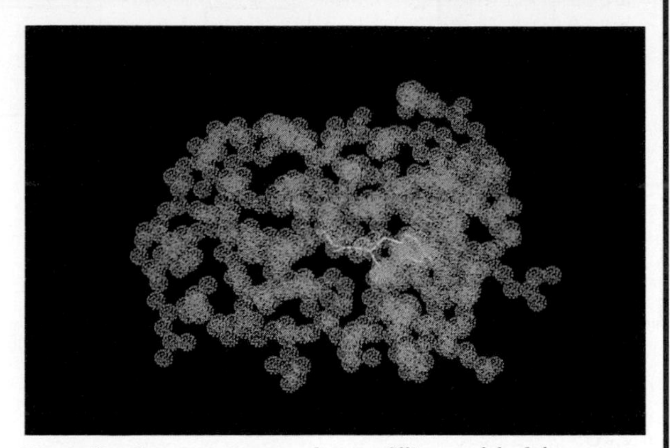

Figure B *Computer-generated space-filling model of the enzyme carboxypeptidase A bound to the substrate glycyl-tyrosine*

A space-filling model of an enzyme gives important information about the shape of the enzyme's active site—the cavity that accommodates the substrate molecule. With a computer-generated space-filling model, we can study in detail the operation of the enzyme (Figure B). In this way new drug molecules can be designed, for example. As we saw in Box 17.2, allopurinol and its oxidation product, oxypurinol, operate by blocking the active site of xanthine oxidase, the enzyme that catalyzes the production of uric acid. Other potential drug molecules can be tested by seeing how well they fit the active site of xanthine oxidase and thereby block it. Computer graphics are assisting in the design of new drugs to treat such diverse disorders as

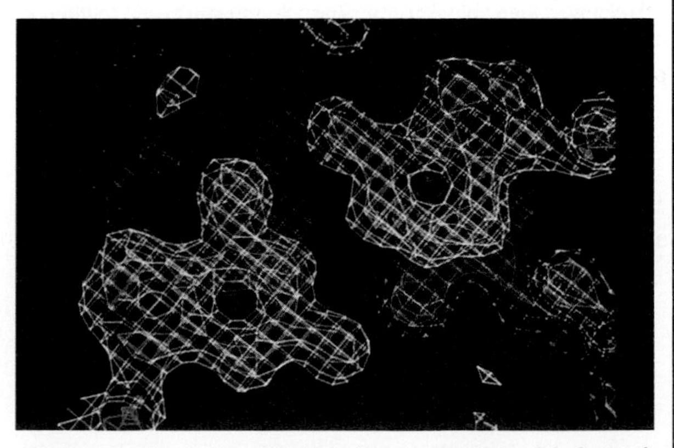

Figure C *Computer-generated electron density model obtained from X-ray diffraction patterns of crystals (here, an unusual incorrect pairing of guanine,* left, *with adenine,* right*). The colored lines represent surfaces of constant electron density, increasing from green to yellow to red.*

hypertension, emphysema, glaucoma, various forms of cancer, and acquired immune deficiency syndrome (AIDS).

Another type of display (Figure C) shows how the electron density varies throughout a molecule, which gives information on the reactive sites in the molecule. Sites that are deficient in electron density are potential sites for attack by electron-donating molecules—molecules such as the water molecule or the hydroxide ion—with unshared pairs of electrons. Sites that have excess electron density are potential sites for attack by molecules that are deficient in electron density, such as the ethyl cation $C_2H_5^+$ or atoms with a partial positive charge, such as the carbon atom in a carbonyl group, $C^{\delta+}{=}O^{\delta-}$.

In recent years the speed and capacity of computer hardware have been rising very rapidly. This increasing power is especially valuable in applications involving animation, as enthusiasts of computer games are well aware. With animation programs, a chemist can simulate the molecular motions that occur as a molecule vibrates and can follow a chemical reaction from reactants to products. Virtual reality simulators that immerse the viewer in a molecular world through which he or she can move will soon be available. The day may be approaching when a computer simulation of the reactions involved in various possible routes for the synthesis of a desired compound will enable the probable best route to be chosen before laboratory experiments are carried out, thereby saving much time and expense. But ultimately any synthesis would have to be tested before being carried out on a large scale, because nature often surprises us: New reactions do not always proceed exactly as existing theory predicts.

17.6 LIPIDS

Lipids are a class of biomolecules that are soluble in nonpolar organic solvents but insoluble in water. They have a variety of biological roles; for example, they serve as fuel, store energy, act as thermal and electrical insulators, and are the major components of membranes, the boundaries of cells and some cellular components. We will consider only two of the many different types of lipids—*triacylglycerols* and *glycerophospholipids*.

Triacylglycerols

Triacylglycerols are the long-chain carboxylic acid esters of glycerol. In Chapter 14 we briefly discussed these compounds, which are commonly known as *fats* and *oils*. Figure 17.21 shows the structure of a typical triacylglycerol in which the carboxylic acids, the so-called fatty acids are the saturated carboxylic acids palmitic acid, $CH_3(CH_2)_{14}CO_2H$, and stearic acid, $CH_3(CH_2)_{16}CO_2H$, and the unsaturated acid oleic acid, $CH_3(CH_2)_7CH{=}CH(CH_2)_7CO_2H$. The glycerol esters of shorter, more unsaturated carboxylic acids are oils, which have lower melting points; the glycerol esters of longer, more saturated carboxylic acids are fats, which have higher melting points.

Triacylglycerols are commonly called triglycerides.

Triacylglycerols are highly concentrated stores of metabolic energy.

The complete oxidation of fats yields about 40 $kJ \cdot g^{-1}$ of energy, in contrast with about 17 $kJ \cdot g^{-1}$ for carbohydrates and proteins. Fats have this high-energy yield

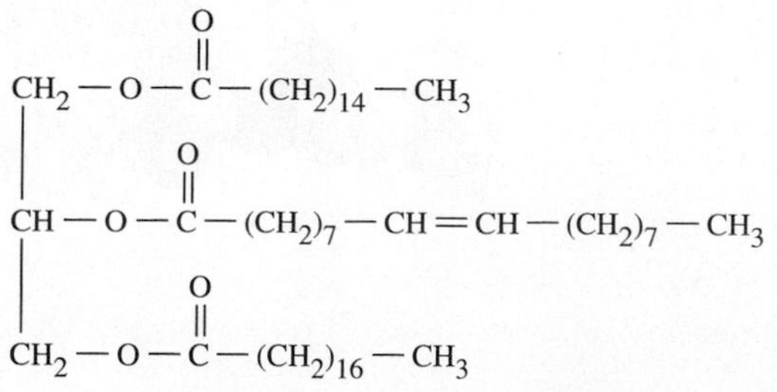

$$CH_2-O-\overset{O}{\overset{\|}{C}}-(CH_2)_{14}-CH_3$$
$$CH-O-\overset{O}{\overset{\|}{C}}-(CH_2)_7-CH{=}CH-(CH_2)_7-CH_3$$
$$CH_2-O-\overset{O}{\overset{\|}{C}}-(CH_2)_{16}-CH_3$$

FIGURE 17.21 A Triacylglycerol This triacylglycerol is the ester of glycerol with the long-chain carboxylic acids palmitic acid, oleic acid, and stearic acid.

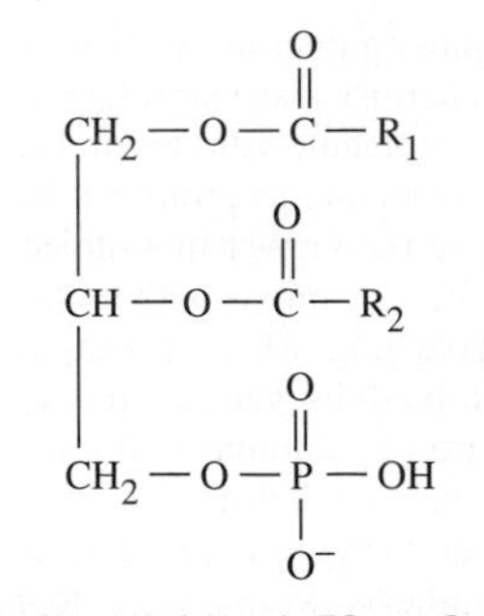

FIGURE 17.22 A Phosphatidic Acid A phosphatidic acid is an ester of glycerol with two long-chain carboxylic acids and phosphoric acid. R_1 and R_2 are alkyl or alkenyl (a hydrocarbon group containing a double bond) chains.

because they are much more *reduced* (they contain a smaller proportion of oxygen) than carbohydrates and proteins. Moreover, because fats are nonpolar, they are stored in a nearly *anhydrous* form, whereas proteins and carbohydrates are much more polar and are therefore more hydrated. Consequently, a gram of nearly anhydrous fat stores about six times as much energy as does a gram of hydrated carbohydrate. In mammals triacylglycerols are synthesized from glucose and stored in the adipose cells (fat cells), which provide thermal insulation as well as an energy reserve. The triacylglycerols are broken down first to glycerol and the fatty acids and ultimately to carbon dioxide and water, thereby providing energy for the synthesis of ATP.

Glycerophospholipids and Membranes

Glycerophospholipids resemble triacylglycerols in that they are derivatives of phosphatidic acid (Figure 17.22), which is an ester of glycerol with two carboxylic acids and phosphoric acid. One of the three OH groups in phosphoric acid is used in the formation of phosphatidic acid; in a glycerophospholipid, a second OH group forms an ester with a small alcohol ROH, such as choline, $(CH_3)_3N^+CH_2CH_2OH$ (Figure 17.23). At the pH of body fluids, the third OH group is ionized. Hence

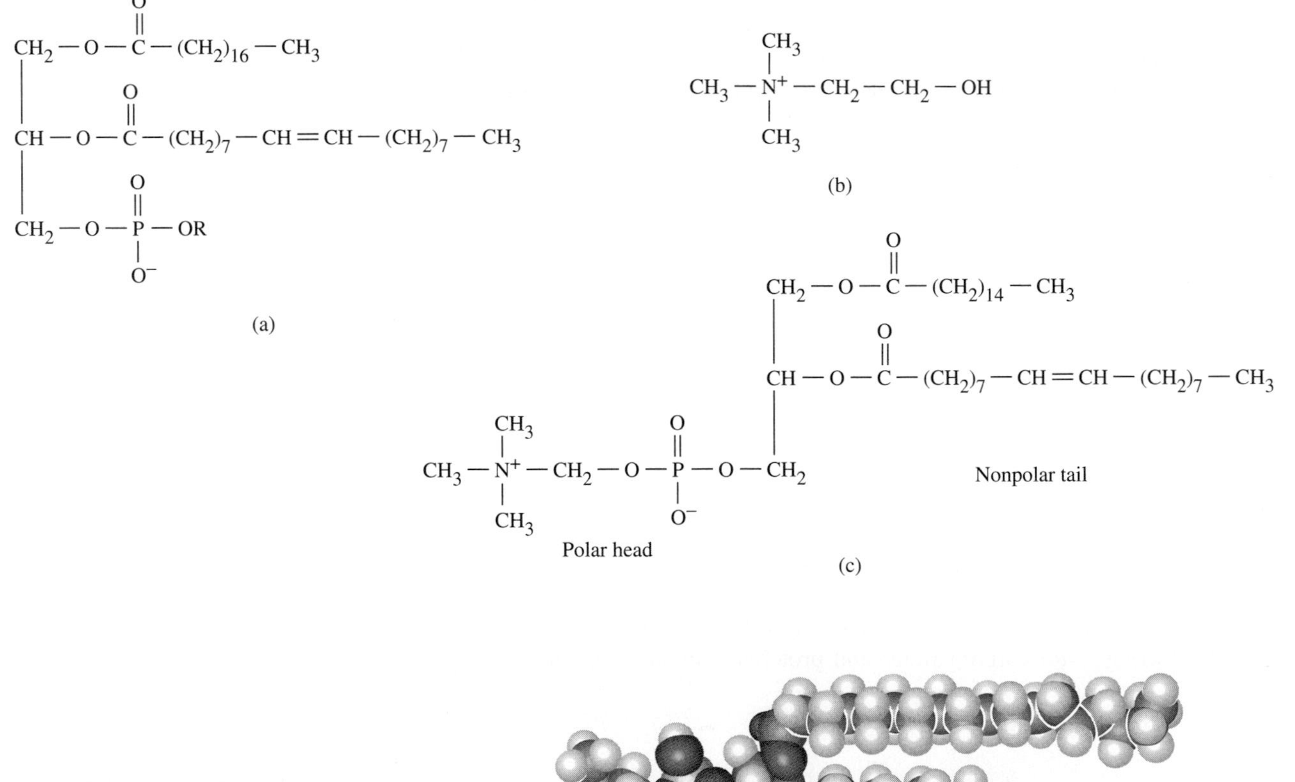

FIGURE 17.23 Esters of Phosphatidic Acid (a) An ester of phosphatidic acid; (b) choline; (c) a phosphatidic ester of choline; (d) a space-filling model of a phosphatidic ester of choline.

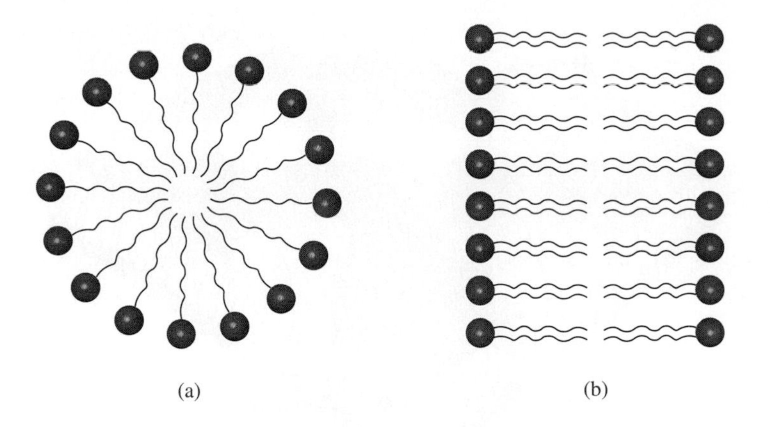

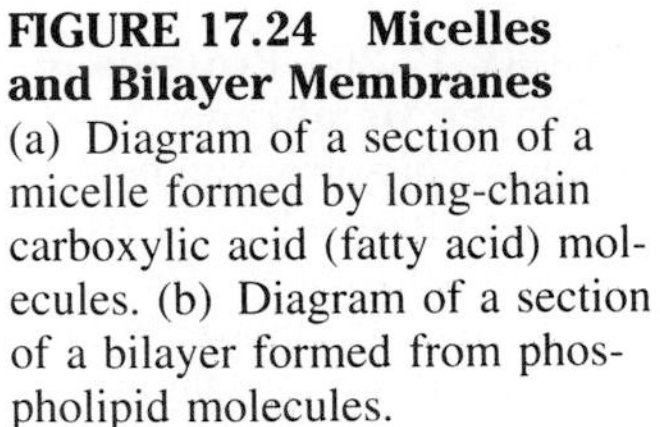
FIGURE 17.24 Micelles and Bilayer Membranes (a) Diagram of a section of a micelle formed by long-chain carboxylic acid (fatty acid) molecules. (b) Diagram of a section of a bilayer formed from phospholipid molecules.

glycerophospholipids have a polar hydrophilic end—the phosphoric acid group—and a nonpolar hydrophobic end—the hydrocarbon chain of the carboxylic acid. These molecules therefore share similar properties with the anions of long-chain carboxylic acids that we discussed in Chapter 14. There we saw that the carboxyl group (the hydrophilic group) of long-chain carboxylic acids (fatty acids) of soap interacts strongly with water molecules, whereas the hydrocarbon chains (the hydrophobic groups) avoid water molecules. Thus, the anions of long-chain fatty acids form micelles in water (Figure 17.24a) with the hydrophilic carboxyl groups on the outside and the hydrophobic hydrocarbon chains in the interior. But phospholipids do not form micelles, because the two hydrocarbon chains are too bulky to be crowded into the interior of a micelle. Instead, in an aqueous solution, they spontaneously form a double layer, or **bilayer** (Figure 17.24b), with the phosphoric acid groups on the outside and the hydrocarbon chains on the inside. Such a bilayer curves around on itself to form a closed structure that separates an aqueous solution on the inside from the aqueous solution on the outside (Figure 17.25).

The *membranes* that surround all living cells are bilayers of glycerophospholipids and other related lipids.

The membranes of cells also are embedded with many other molecules, particularly protein molecules (Figure 17.26). These lipid bilayers can be thought of as a two-dimensional liquid in which the lipid and protein molecules can move easily in the plane of the bilayer but cannot move across the bilayer. Most molecules and ions cannot pass through a cell's membrane, which is said to be *semi-permeable*. An important function of the proteins in the membrane is to facilitate the movement of particular ions and molecules across the membrane when required.

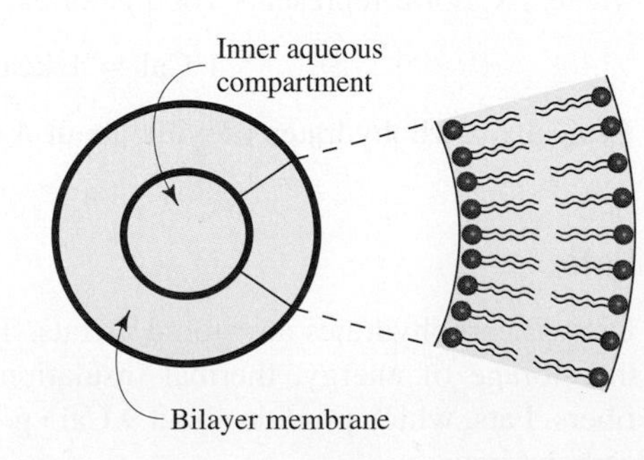

FIGURE 17.25 A Lipid Bilayer Membrane The lipid bilayer membrane closes on itself, forming the walls of a cell.

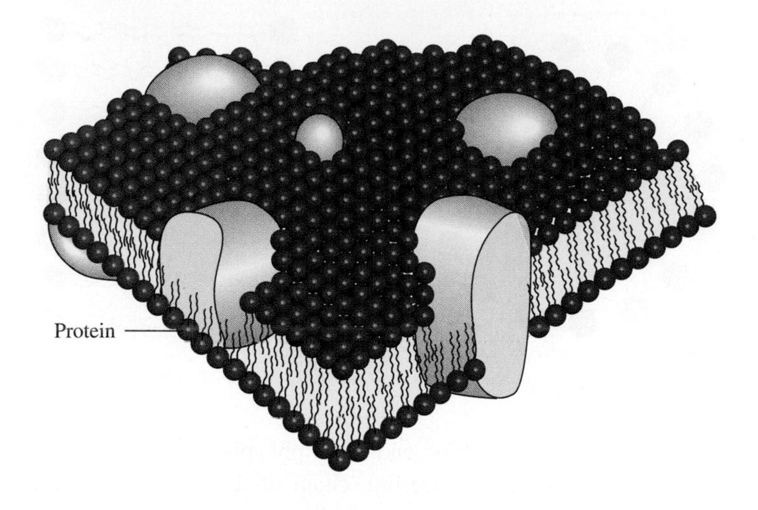

FIGURE 17.26 Proteins in a Bilayer Membrane Proteins "dissolve" in the bilayer to form a kind of two-dimensional solution. They can move within the bilayer but in general cannot move through it. The proteins facilitate and control the movement of ions and molecules through the membrane.

17.7 NUTRITION

We need to eat to stay alive, and we need to eat the right foods to remain healthy. The study of what we need to eat, and of what we are better off not eating, is called **nutrition**. Five classes of substances are usually identified as being the necessary components of a healthy diet: carbohydrates, fats, proteins, vitamins, and "minerals" (in this context, simple inorganic substances containing elements such as calcium and iron).

Carbohydrates

Carbohydrates are our primary source of energy. Starch is hydrolyzed to glucose in stages by enzymes in saliva and then in the stomach and the small intestine (Box 12.1). Finally, in further steps glucose is oxidized to CO_2 and H_2O. Humans cannot digest cellulose, the main constituent of plant tissues, but it serves the useful purpose of providing the "bulk" necessary for the efficient elimination of solid waste.

The SI unit of energy is the joule. Another, older unit for energy is the calorie, defined as the amount of heat required to raise the temperature of 1 g of water by 1°C:

$$1 \text{ cal} = 4.184 \text{ J}$$

The energy obtained from food is, for historical reasons, usually given in Calories, where 1 Calorie represents 1000 calories, or 1 kcal:

$$1 \text{ Cal} = 1 \text{ kcal} = 4.184 \text{ kJ}$$

Typically, carbohydrates provide about 4 $\text{Cal} \cdot \text{g}^{-1}$.

Fats

Unused carbohydrates are stored as fats. The primary uses of fats in mammals are the storage of energy, thermal insulation, and the electrical insulation of nerve fibers. Fats, which provide about 9 $\text{Cal} \cdot \text{g}^{-1}$, are better suppliers of energy than are carbohydrates.

TABLE 17.3 Estimated Amino Acid Requirements

Amino Acid	Daily Requirement (mg per kg body weight) Infant	Adult
Arginine	?	?
Histidine	33	10
Isoleucine	83	12
Leucine	135	16
Lysine	99	12
Methionine*	49	10
Phenylalanine†	141	16
Threonine	68	8
Tryptophan	21	3
Valine	92	14

* Also synthesized from cysteine in diet
† Also synthesized from tyrosine in diet

Proteins

Proteins are necessary in our diet to supply the amino acids needed to make many of the molecules that sustain life. Of the 20 amino acids required, only 10 can be synthesized by the human body. The remaining 10 amino acids, called the ''essential amino acids,'' must be supplied in the diet (Table 17.3). It is estimated that as many as 1 billion people in the world suffer from various ailments caused by a diet deficient in the essential amino acids.

Vitamins

Vitamins are organic molecules needed in small amounts in the diets of many animals because they play essential roles in the reactions by which foods are used. Vitamins serve nearly the same purposes in all forms of life, but through evolution primates have lost the ability to synthesize most of them. Not all the functions of vitamins are clearly understood, and there may be vitamins that have not yet been identified. Some of the more common vitamins are described in Box 17.4.

Minerals

In addition to the organic substances described above, we also need small amounts of many inorganic substances in our food. Inorganic substances supply the small amounts of elements other than C, H, O, and N that are needed for life processes. In nutritional science these elements are called **minerals**. The minerals needed in greatest amounts, the *major minerals*, are calcium, phosphorus, sulfur, sodium, potassium, chlorine, and magnesium. They are listed in Table 17.4 together with some of the *trace minerals*—minerals required in only very small, or trace, amounts. Calcium and phosphorus are needed for bones and teeth, a major component of which is calcium phosphate hydroxide (hydroxyapatite), $Ca_5(PO_4)_3OH$ (Chapter 7). As we have seen, phosphorus in the form of phosphate is also an essential component of nucleotides and nucleic acids, such as DNA, and other biomolecules, such as ATP. Sulfur is required for the synthesis of the sulfur-containing amino acid cysteine (Table 17.1) and some other sulfur-containing biomolecules. Sodium chloride and potassium chloride are the major electrolytes in body fluids. The Na^+ and K^+ ions play an essential role in the transmission of nerve

The use of the term ''mineral'' in nutrition (for example, in the nutritional information on a box of cereal) differs from that in chemistry and geology, where the term means a naturally occurring pure crystalline inorganic solid.

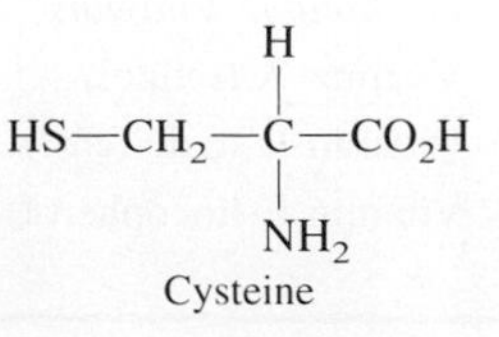

Cysteine

BOX 17.4 Vitamins

You are probably aware that a lack of certain vitamins in your diet may cause illness. Here we take a closer look at a few of the common vitamins—the structures of their molecules and their functions in the body. Vitamins can be broadly classified as water-soluble or fat-soluble; some examples are given in Table A. Lack of vitamins in the diet leads to certain diseases. For example, the absence of *vitamin C* (ascorbic acid) (Figure A) in the diet causes scurvy. This disease, historically associated with sailors, caused numerous deaths during the long voyages of the seafaring explorers in the sixteenth to eighteenth centuries. The symptoms of scurvy were vividly described by Jacques Cartier in 1536 when his men were afflicted with it as they explored the Saint Lawrence River:

> Some did lose all their strength and could not stand on their feet. . . . Others also had all their skin spotted with spots of blood of a purple color: then did it ascend up to their ankles, knees, thighs, shoulders, arms and necks. Their mouths became stinking, their gums so rotten, that all the flesh did fall off, even to the roots of the teeth, which did also almost all fall out.

In 1753 the Scottish physician James Lind recognized that scurvy can be prevented by eating fresh vegetables and fruits, which contain ascorbic acid, but they were usually not available to sailors on long voyages. Lind recommended that lemon juice be included in the diet of sailors. Forty years later his advice was adopted by the British Navy.

When collagen, the fibrous protein that is an essential component of skin and blood vessels, is synthesized in the absence of vitamin C, it does not contain enough OH groups to provide the hydrogen bonding that allows it to form strong fibers. This abnormal collagen gives rise to the symptoms of scurvy. Only 20 mg per day of vitamin C will prevent scurvy, but 60 mg per day is recommended for a healthy diet. It has been claimed, particularly by Linus Pauling (Box 3.2), that much larger doses, 1 to 10 *grams* per day, prevent colds and cancer, but the evidence for these claims is not strong and is certainly controversial. However, moderately large doses may not be harmful because, being water soluble, excess vitamin C is readily excreted.

Vitamin A (retinol) (Figure B) is a fat-soluble vitamin. Vitamin A is converted to retinal, the light-absorbing molecule in the retina of the eye. A deficiency of this vitamin results in poor vision, particularly in weak light, often called night blindness. Vitamin A is also required for growth in young mammals. Carrots and other yellow-orange vegetables are a good source of vitamin A, because they

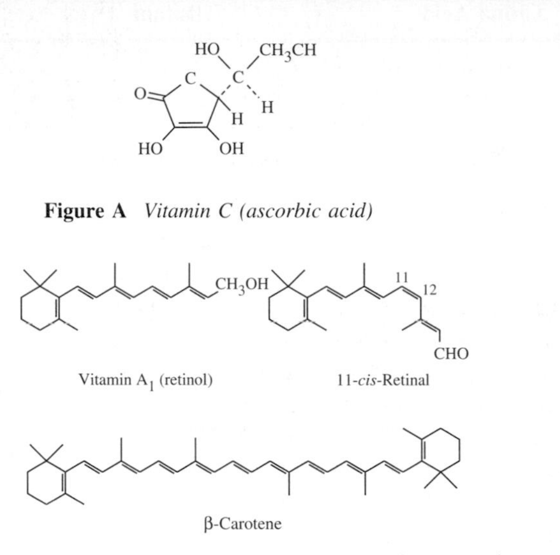

Figure A *Vitamin C (ascorbic acid)*

Figure B *Vitamin A (retinol) and some related compounds*

TABLE A Vitamins

Vitamin	Deficiency Disease and Symptoms
Water-Soluble Vitamins	
Vitamin C (ascorbic acid)	Scurvy: bleeding gums, bruises
Vitamin B_1 (thiamine)	Beriberi: fatigue, depression
Vitamin B_6 (pyridoxine)	Anemia, irritability, retarded growth
Vitamin H (biotin)	Dermatitis, anorexia, depression
Fat-Soluble Vitamins	
Vitamin A (retinol)	Night blindness
Vitamin D (calciferol)	Rickets, osteomalacia
Vitamin E (tocopherol)	Hemolysis of red blood cells

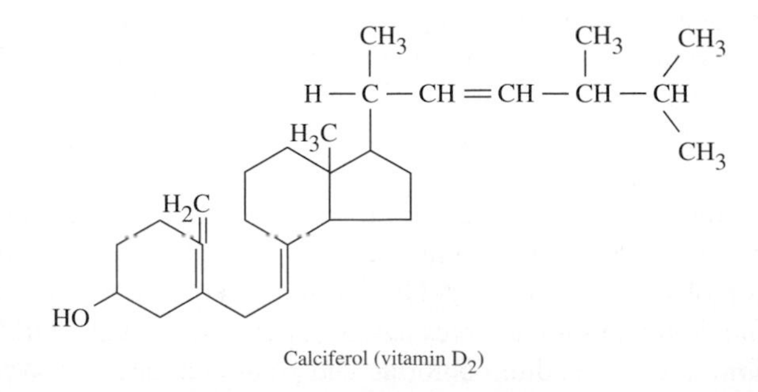

Figure C *Vitamin D (calciferol)*

contain β-carotene (Figure B), which is readily converted to vitamin A.

Vitamin D (calciferol) (Figure C) is a fat-soluble vitamin that plays an essential role in the metabolism of calcium and phosphorus. Most naturally occurring foods have a low vitamin D content. It is synthesized in the body from cholesterol, but only in the presence of ultraviolet light. A deficiency of vitamin D in childhood produces the disease rickets, in which the bones become soft—so soft in acute cases that the weight of the body deforms them (Figure D). The disease was common in the nineteenth century in Britain and other places with a cloudy climate and short winter days and therefore a low incidence of ultraviolet light. In adults the lack of vitamin D also causes bones to become soft and weak, a condition found among Bedouin women who are clothed head to toe and wear veils so that only their eyes are exposed to sunlight. Fish-liver oil is one of the few food sources rich in Vitamin D, and for many years cod-liver oil was used as a food supplement. Today in most developed countries, the necessary amount is obtained from milk fortified with vitamin D to a level of 400 μg per quart.

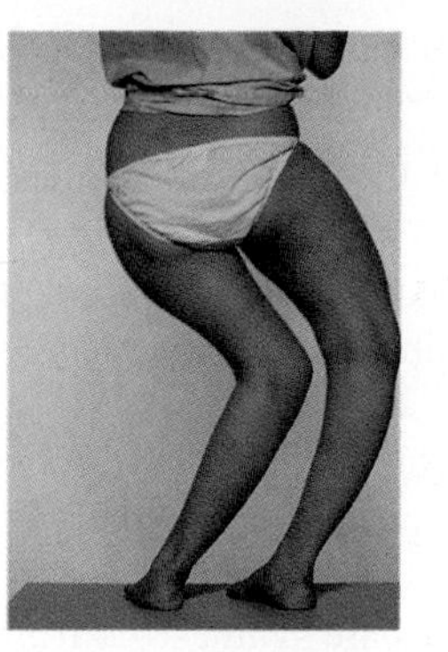

Figure D *A child with rickets*

impulses (Chapter 13), in the control of the passage of molecules through membranes, and in the maintenance of blood pressure. Magnesium is an essential mineral because the enzymes involved in the reactions of ATP are effective only in the presence of Mg^{2+}.

Among the trace elements, iron is an essential component of the hemoglobin in red blood cells, iodine is required by the thyroid gland, and zinc is a constituent of some 90 enzymes. In general, all these elements are present in sufficient quantities in a normal diet so that no supplements are necessary.

TABLE 17.4 Elements Essential to Human Life

Element	Symbol	Function
Carbon	C	Needed as structural and functional components of living organisms
Hydrogen	H	
Oxygen	O	
Nitrogen	N	
Boron	B	Aids in the use of Ca, P, and Mg
Calcium	Ca	Needed for growth of teeth and bones
Chlorine	Cl	Needed for maintaining salt balance in body fluids
Chromium	Cr	Aids in carbohydrate metabolism
Cobalt	Co	Component of vitamin B_{12}
Copper	Cu	Needed to maintain blood chemistry
Fluorine	F	Aids in the development of teeth and bones
Iodine	I	Needed for thyroid function
Iron	Fe	Needed for oxygen-carrying ability of blood
Magnesium	Mg	Needed for bones, teeth, and muscle and nerve action
Manganese	Mn	Needed for carbohydrate metabolism and bone formation
Molybdenum	Mo	Component of enzymes needed in metabolism
Nickel	Ni	Aids in the use of Fe and Cu
Phosphorus	P	Needed for growth of bones and teeth; component of DNA and RNA
Potassium	K	Component of body fluids; needed for nerve action
Selenium	Se	Aids in vitamin E action and fat metabolism
Silicon	Si	Helps form connective tissue and bone
Sodium	Na	Component of body fluids; needed for nerve and muscle action
Sulfur	S	Component of proteins; needed for blood clotting
Zinc	Zn	Needed for growth, healing, and overall health

SUMMARY

The molecules of life are biomolecules, many of which are polymers. Life processes involve steady-state concentrations of molecules that are formed and removed at the same rate; allowing them to achieve equilibrium is synonymous with death.

Proteins (polypeptides) are polymers formed by the elimination of water between CO_2H and NH_2 groups of α-amino acids, which produces peptide bonds. In living cells proteins are synthesized from the 20 α-amino acids found in all living matter. The primary structure of a protein is the sequence of amino acids from which the polypeptide is formed. The secondary structure describes the way in which the polypeptide chains are folded or coiled and held by hydrogen bonds between C═O and N—H groups. There are two common types of secondary structure: the α-helix and the β-pleated sheet. The three-dimensional shape resulting from the coiling and folding of a polypeptide chain is called the tertiary structure. A globular protein has a nearly spherical shape; myoglobin is an example. Because the polar (hydrophilic) groups are on the outside and nonpolar (hydrophobic) groups are on the inside, globular proteins are soluble in water. In contrast, a fibrous protein has a filamentous shape; α-keratin, which occurs in hair and muscle, is an example. The quaternary structure of a protein describes the arrangement of two or more polypeptides that act as a unit. Hemoglobin consists of four polypeptides, each with a structure like that of myoglobin held together in a single functional unit. Nonpolypeptide constituents, such as the oxygen-carrying heme group in myglobin (in muscles) and hemoglobin (in red blood cells) have important biological functions.

An enzyme (most often a globular protein) usually catalyzes only one specific reaction. Enzymes have a cleft or crevice in their surface called an active site. A substrate (one of the reactants) fits into the active site and forms an enzyme–substrate complex, according to the lock-and-key model and the induced-fit model. The substrate molecule is held in exactly the right position for reaction. It may be distorted so that some of its bonds are weakened, thus decreasing the activation energy for the reaction.

Carbohydrates, $C_xH_{2y}O_y$, have five- or six-membered rings containing an oxygen atom and have one or more hydroxyl-group substituents. Monosaccharides (one ring) polymerize by the elimination of water to give disaccharides (two rings) or polysaccharides (many rings). Monosaccharides include the isomers α- and β-glucose, with six-membered rings, and fructose, ribose, and deoxyribose, with five-membered rings. Sucrose is a disaccharide. Polysaccharides include cellulose, found in the tough fibers of plants, and starch, the energy-storage compound in plants. Animals obtain energy by the hydrolysis of polysaccharides to monosaccharides, such as glucose. Monosaccharides pass into the bloodstream and are oxidized to CO_2 and water in a long series of reactions, many involving the synthesis of ATP. ATP is a nucleotide, the triphosphoric acid ester of ribose bonded to the heterocyclic amine adenine. ATP^{4-} is synthesized from ADP^{3-} and HPO_4^{2-}, and its hydrolysis back to ADP^{3-} releases free energy for synthesizing new molecules and for other processes, such as muscle contraction.

DNA (deoxyribonucleic acid) is a two-strand polynucleotide. Each strand results from condensing deoxyribose and phosphate groups of adjacent nucleotides to form a backbone. Nitrogenous bases (heterocyclic amines) attached to the deoxyribose are adenine (A) and thymine (T), in equal amounts, and cytosine (C) and guanine (G), also in equal amounts. The two strands form a double helix held together by A--T or C--G hydrogen bonds.

Replication occurs when, with the help of an enzyme, the two strands of DNA uncoil by breaking their weak A--T and C--G hydrogen bonds. Nucleotides with complementary bases then add to each strand to form new A--T and C--G hydrogen bonds. This gives two new DNA molecules identical to the original; each DNA strand acts as a template for the synthesis of a new complementary strand. Thus, a DNA molecule can repeatedly duplicate itself and carry with it all the information in the original. RNA (ribonucleic acid) is similar to DNA, but the sugar is ribose instead of deoxyribose and the nitrogenous base T is replaced by uracil U, which forms A--U hydrogen bonds. Messenger RNA (mRNA), a short polymer, is formed from the strands of DNA (informational strands) as templates by transcription, a process similar to replication. The mRNA carries the information stored in the DNA—the genetic code—to the sites of protein synthesis. The bases in the mRNA and their sequence provide the information needed to assemble the correct amino acids in the correct sequence to give a particular protein. Codons are triplet series of the bases A, U, C, and G that provide the information to arrange amino acids in the proper sequence.

Lipids are biomolecules that are soluble in nonpolar organic solvents but insoluble in water. Two types of lipids are triacylglycerols (fats), which store energy, and glycerophospholipids, which form the bilayer membranes of cells.

The five necessary components of a healthy diet are carbohydrates, fats, proteins, vitamins, and minerals.

IMPORTANT TERMS

active site (page 626)
adenosine diphosphate (ADP) (page 630)
adenosine triphosphate (ATP) (page 629)
α-amino acid (page 616)
α-helix (page 617)
β-pleated sheet (page 617)
bilayer (page 643)
carbohydrate (page 627)
codon (page 637)
deoxyribonucleic acid (DNA) (page 632)
disaccharide (page 627)
enzyme (page 626)
enzyme–substrate complex (page 635)
fats (page 641)
genetic code (page 635)
glycerophospholipid (page 642)
hemoglobin (page 625)
hydrophilic (page 622)
hydrophobic (page 624)
induced-fit model (page 626)
lipid (page 641)
lock-and-key model (page 626)
membrane (page 643)
messenger RNA (mRNA) (page 636)
mineral (page 645)
monosaccharide (page 627)
myoglobin (page 622)
nucleic acid (page 632)
nucleotide (page 632)
peptide bond (page 617)
polypeptide (page 617)
polysaccharide (page 627)
primary structure (page 617)
protein (page 616)
quaternary structure (page 625)
replication (page 635)
ribonucleic acid (RNA) (page 626)
secondary structure (page 617)
starch (page 628)
substrate (page 626)
tertiary structure (page 622)
transcription (page 636)
triacylglycerol (page 641)
vitamin (page 645)

REVIEW QUESTIONS

1. Why is a living system described as being "far from equilibrium"?

2. **(a)** Draw the structure of an α-amino acid with a side chain R.

(b) Show how two amino acids form a peptide bond.

3. What is a protein?

4. What groups are involved in hydrogen bonding in the α-helix and β-pleated sheet structures of proteins?

5. **(a)** What is the geometry of the five N atoms that surround the Fe atom in hemoglobin?

(b) What position does an oxygen molecule occupy when it combines with hemoglobin?

6. How does a globular protein differ from a fibrous protein? Give one example of each.

7. What is an enzyme?

8. What is meant by the active site of an enzyme?

9. How do the monosaccharides α-glucose and β-glucose differ structurally?

10. How do the polysaccharides starch, cellulose, and glycogen differ structurally and chemically?

11. **(a)** What are the structures of ADP and ATP?

(b) How are they interconverted?

12. What are the three important structural components of a nucleotide?

13. **(a)** What is a nucleic acid?

(b) What are DNA and mRNA?

14. How does DNA differ from RNA in terms of **(a)** the sugar components and **(b)** the four nitrogenous base components?

15. How are the two strands of the double helix of DNA held together?

16. How is DNA replicated?

17. How does mRNA function in protein synthesis?

18. What are **(a)** the genetic code and **(b)** a codon?

19. What are the five classes of substances essential to a healthy diet?

20. What are the essential amino acids?

PROBLEMS

Proteins

1. Draw the structures of each of the following α-amino acids. Which are dicarboxylic acids?

(a) alanine (b) glycine
(c) asparagine (d) glutamic acid

2. What α-amino acid does each of the following three-letter codes represent?

(a) GAU (b) GCU (c) GGA
(d) AAC (e) AGC (f) UAU

3. An amino acid isolated from animal tissue was believed to be glycine (aminoethanoic acid), $NH_2CH_2CO_2H$. When 0.500 g of the amino acid was completely converted to ammonia, which was then absorbed into 50.0 mL of 0.0500-*M* $HCl(aq)$, the excess $HCl(aq)$ required 30.57 mL of 0.0600-*M* $NaOH(aq)$ for neutralization.

(a) How many moles of ammonia were neutralized by the $NaOH(aq)$?

(b) How many grams of nitrogen were in the 0.500-g sample of the amino acid?

(c) Is the mass percentage of nitrogen in the sample that expected for glycine?

4. (a) Give an example of a reaction in which a peptide link forms.

(b) Why is the reaction in (a) described as a condensation reaction?

(c) Why are polypeptides described as condensation polymers?

5. Draw the structures of (a) a dipeptide formed from the α-amino acids histidine and cysteine and (b) a tripeptide formed from the α-amino acids alanine, glycine, and histidine. (c) Express these peptides in terms of their three-letter codes.

6. How many tripeptides are possible from the condensation of each of the following numbers of different amino acids?

(a) three (b) four (c) six

7. (a) Consider the segment —Ala—Gly—Asp—Glu— of a polypeptide. Draw its structure, and show how two such segments can be linked by hydrogen bonds.

(b) Draw the structure for the segments —Val—Cys—Gly— and —Tyr—Cys—Asn— of two polypeptide chains, and show how they can be connected by a disulfide bridge.

8. What are meant by the (a) primary, (b) secondary, and (c) tertiary structures of a protein?

9. What are the two common types of secondary structure in proteins, and what is responsible for each?

10. (a) How does a globular protein differ from a fibrous protein? Give one example of each.

(b) What functions does each type of protein serve in a living organism?

11. (a) Explain the meaning of the terms "hydrophobic" and "hydrophilic."

(b) Give two examples each of hydrophobic and hydrophilic side chains found in polypeptides.

(c) How are hydrophobic and hydrophilic interactions revealed in the tertiary structure of a globular protein?

12. For the heme group in hemoglobin and myoglobin, describe (a) its structure and (b) its biological function in blood and in muscles.

Enzymes

13. (a) Explain the principal differences between how the rates of chemical reactions inside and outside the body are regulated.

(b) What conditions in the body require that enzymes be very efficient?

14. Describe the lock-and-key and induced-fit models of enzyme action.

15. (a) The enzyme in blood that catalyzes the decomposition of hydrogen peroxide in aqueous solution is catalase, which speeds up the reaction at 25°C by a factor of 5×10^6. If this is due entirely to a change in the activation energy, which is 23 $kJ \cdot mol^{-1}$ for the catalyzed reaction, what is the activation energy for the uncatalyzed reaction?

(b) To what value would the temperature have to be raised to increase the rate of the uncatalyzed reaction to that of the catalyzed reaction at 25°C?

Carbohydrates

16. (a) What general formula gives the composition of all carbohydrates?

(b) Draw a five-membered and a six-membered ring found in a carbohydrate.

(c) Draw structural formulas to show the formation of sucrose from glucose and fructose.

17. (a) Sugars are carbohydrates. Why was the name "carbohydrate" originally given to such compounds, and how appropriate is it?

(b) Give one example each of a monosaccharide, a disaccharide, and a polysaccharide.

(c) Both glucose and fructose have the molecular formula $C_6H_{12}O_6$. How do they differ structurally?

18. What is the principal difference between starch and cellulose (a) structurally and (b) biochemically?

19. Amylose (starch) is a polysaccharide of β-glucose and has an average molar mass of about $3.0 \times 10^5\ g \cdot mol^{-1}$.

(a) Approximately how many glucose units does an amylose molecule contain?

(b) What is the structure of starch?

Energy in Biochemical Reactions: ATP and ADP

20. (a) Using Table 9.1, calculate $\Delta H°$ for the photosynthesis reaction

$$6CO_2(g) + 6H_2O(l) \rightarrow C_6H_{12}O_6(s) + 6O_2(g)$$

(a) per mole and **(b)** per gram of glucose.

(b) If leaves of plants absorb 2.0×10^{-2} kJ · cm^{-2} of appropriate light energy per day, what leaf area is required to synthesize 1 g of glucose per day?

21. **(a)** Draw the structures of mono-, di-, and triphosphoric acids and the monoester that each forms with ethanol.

(b) Why are ATP and ADP described as esters of di- and triphosphoric acids, respectively?

22. Explain how ATP functions in transferring energy in living systems.

23. **(a)** To what class of compounds does ATP belong?

(b) How does ATP differ from DNA?

24. **(a)** Draw the structures of ADP and ATP.

(b) Write equations showing how they are interconverted.

25. **(a)** How is the maximum work that can be performed by a reaction measured?

(b) What is the maximum amount of work that can be obtained from the oxidation of glucose?

$$C_6H_{12}O_6(s) + 6O_2(g) \rightarrow 6CO_2(g) + 6H_2O(l)$$

(c) How is this reaction achieved in metabolism?

DNA and RNA

26. **(a)** Describe the structural components of a nucleotide.

(b) Draw the structure of the nucleotide formed from the base cytosine, the sugar α-glucose, and phosphoric acid.

(c) How are nucleic acids formed from nucleotides?

27. DNA and mRNA have three common nitrogenous bases, and both contain one additional base.

(a) Name these bases; **(b)** indicate how they are usually represented; and **(c)** draw their structures.

28. What special characteristics of adenine, thymine, guanine, and cytosine make them important as constituent groups of DNA?

29. **(a)** Why does DNA form a double helix?

(b) Are the base pairs in DNA found inside the helix or outside the helix, or do they constitute part of the backbone of the structure?

30. Explain why two strands of DNA are linked by interactions between two specific pairs of nitrogenous bases rather than between all six possible combinations of these four bases.

31. **(a)** How does the structure of an mRNA molecule differ from that of a DNA molecule?

(b) What is a template strand?

(c) Describe the replication and transcription processes involving DNA and RNA.

32. **(a)** What is the genetic code?

(b) How is information stored and passed on to successive generations of organisms?

33. If a section of a single strand of DNA contains the base sequence ACTCGC, what is the corresponding base sequence in **(a)** the complementary DNA strand and **(b)** the mRNA strand formed by transcription?

Lipids

34. **(a)** Draw the structure of a typical triglyceride.

(b) What is the function of triglycerides in the body?

35. **(a)** What is a glycerophospholipid?

(b) Explain why it forms a bilayer in water.

Nutrition

36. **(a)** What five classes of substances constitute the essential components of a healthy diet?

(b) What class of substances is the primary source of energy for humans, and in what units is this energy usually expressed?

(c) In what form is energy stored in the body?

37. **(a)** Why are proteins essential components of the diet?

(b) How are the essential amino acids obtained for growth?

38. Give some of the important functions in the body of each of the following vitamins. **(a)** A **(b)** C **(c)** D

General Problems

39. Photosynthesis in plants requires light of maximum wavelength 700 nm, the wavelength of red light.

(a) To what energy does this correspond?

(b) Is this energy sufficient to break directly the O—H bond in water, $BE(OH) = 463$ kJ · mol^{-1}, to produce oxygen?

40. **(a)** Why is fluoride ion, F^-, added to the water supplies of most communities?

(b) Why is iodine in the form of potassium or sodium iodide added to table salt?

41. **(a)** Ammonia is poisonous, so most mammals eliminate nitrogen, a result of the oxidation of proteins, in the form of urea. With what substance must ammonia react to form urea?

(b) Fish usually eliminate nitrogen as ammonia. Suggest a reason for this difference between fish and mammals.

42. Air is bubbled through a solution of iron(II) sulfate and through a solution of hemoglobin. What observations would you expect to make in the two cases? Account for the different behavior.

Nuclear Chemistry

18.1 Radioactivity
18.2 Radioactive Decay Rates
18.3 Artificial Radioisotopes
18.4 Nuclear Energy

Nuclear chemistry deals with radioactive decay and other nuclear reactions. Despite its bad reputation, radioactivity has numerous applications in everyday life—from radiation therapy to glow-in-the-dark clock faces to radiocarbon dating. The sterilization of medical instruments by irradiation is another important application. Each tube in the rack shown here, stored under water, contains the radioactive isotope cobalt-60. Radiation from the cobalt-60 produces the blue glow in the surrounding water and is being used to sterilize syringes.

We noted in Chapter 1 that some nuclei are unstable and change into other nuclei by emitting particles such as α particles (helium nuclei) or electrons. The spontaneous disintegration of a nucleus is called **radioactivity.** An unstable nucleus that decomposes spontaneously is said to be **radioactive** and to undergo **radioactive decay.** Many stable, nonradioactive nuclei can be changed into other nuclei by bombarding them with particles such as α particles and neutrons. Any process in which a nucleus is changed into another nucleus, either spontaneously or when bombarded with other particles, is called a **nuclear reaction.** The study of such changes is usually considered a part of physics. Nevertheless, chemistry is very much involved in the study of nuclear reactions and their use for generating power. Moreover, radioactivity has important applications in chemistry and in many other fields, such as medicine, biology, geology, and archeology. The use and applications of chemistry in these areas constitute the subject of **nuclear chemistry.**

We start by discussing the stability of nuclei and the distintegration of unstable nuclei that is known as radioactivity. Then we examine the production and uses of radioactive isotopes that do not occur in nature. Finally, we consider the enormous quantities of energy released in nuclear reactions and touch upon some of the social, ethical, economical, and political questions that nuclear energy raises.

18.1 RADIOACTIVITY

Nuclear Structure and Stability

The nucleus of an isotope is often called a **nuclide.** Each nuclide has a particular mass number, *A*, and atomic number, *Z*, and is represented by the symbol $^{A}_{Z}E$. For instance, $^{12}_{6}C$ (carbon-12) is a nuclide of carbon, and $^{235}_{92}U$ (uranium-235) is a nuclide of uranium. Strictly speaking, the atomic number *Z* is redundant, because the symbol E implies the atomic number *Z* (Chapter 1). All the elements with atomic numbers of 83 or less—with the exceptions of technetium ($Z = 41$) and promethium ($Z = 61$)—have at least one stable isotope. But the nuclei of all the isotopes of the elements with atomic numbers greater than 83 are unstable and so are radioactive. A radioactive isotope is called a **radioisotope.**

A nuclide may be represented in one of three ways. For example, the carbon nuclide with mass number 12 is denoted as $^{12}_{6}C$, as ^{12}C, or as carbon-12.

$^{209}_{83}Bi$ is the heaviest stable (nonradioactive) isotope.

What makes some nuclei stable and others unstable? To answer this question, we must first consider why *any* nucleus with more than one proton should hold together. The distances between the particles in a nucleus are very small—less than 10^{-15} m—so the electrostatic repulsion between the protons is extremely strong. However, there are very strong *attractive* forces between all the nuclear particles (both the protons and the neutrons) called **nuclear forces.** The nature of these forces is the subject of continued active investigation by nuclear physicists. As the number of protons increases, the total electrostatic repulsion between them increases very

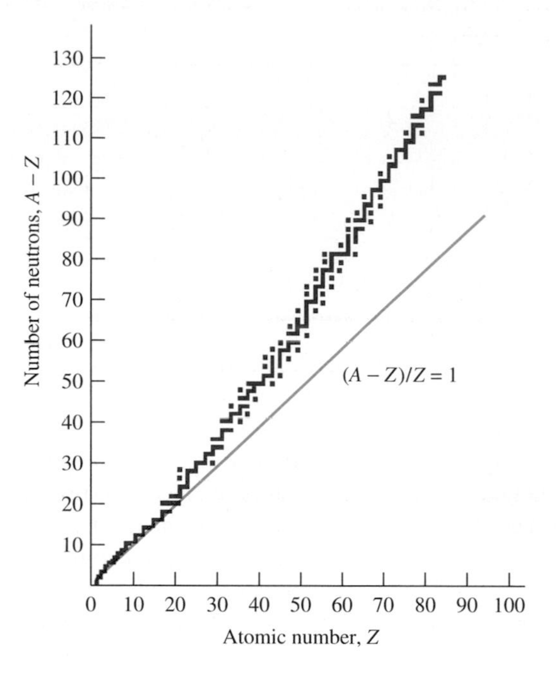

FIGURE 18.1 The Band of Stability All stable nuclei lie in a narrow band in this plot of the number of neutrons $(A-Z)$ versus the number of protons Z. Light nuclei have the same number of neutrons as protons, $A-Z = Z$, or nearly so. Heavier nuclei contain more neutrons than protons.

rapidly, with the result that an increasing proportion of neutrons is needed to provide sufficiently strong attractive forces to keep the nucleus stable. As the plot in Figure 18.1 of the number of neutrons $(A - Z)$ versus the number of protons (Z) for stable nuclei shows, all stable nuclei fall within a narrow region called the **band of stability.** For stable nuclei with low atomic numbers, the number of protons is equal to the number of neutrons. As Z increases, however, the number of neutrons exceeds the number of protons, and the neutron–proton ratio increases from 1.0 to close to 1.5 for the heaviest stable nuclei. However, the total electrostatic repulsive forces are so large for nuclei with $Z > 83$ that all these nuclei are unstable.

Radioactive Decay Processes

We mentioned in Chapter 1 that when an unstable nucleus decomposes or disintegrates, it gives off a particle such as a helium nucleus (α particle) or an electron (β particle) and changes into another, more stable nucleus, called a **daughter nucleus.** We say that the unstable parent nucleus decays to the daughter nucleus. There are several other important types of radioactivity, including positron emission, electron capture, and the emission of gamma radiation (Table 18.1).

α-Particle Emission Emission of an α particle leads to a decrease in the atomic mass of a nucleus by four units and a decrease in the atomic number by two units. For example,

$$^{238}_{92}U \rightarrow {}^{234}_{90}Th + {}^{4}_{2}He$$

β-Particle Emission There are no electrons in nuclei; the emission of a β

TABLE 18.1 Radioactive Decay Processes

Particle Emitted		Change in Mass Number	Change in Atomic Number	Example
Helium nucleus (α particle)	$^{4}_{2}He$	Decreases by 4	Decreases by 2	$^{238}_{92}U \rightarrow ^{234}_{90}Th + ^{4}_{2}He$
Electron (β particle)	$^{0}_{-1}e$	No change	Increases by 1	$^{14}_{6}C \rightarrow ^{14}_{7}N + ^{0}_{-1}e$
Positron	$^{0}_{1}e$	No change	Decreases by 1	$^{64}_{29}Cu \rightarrow ^{64}_{28}Ni + ^{0}_{1}e$
Electron capture		No change	Decreases by 1	$^{195}_{79}Au + ^{0}_{-1}e \rightarrow ^{195}_{78}Pt$
γ-ray photon	γ	No change	No change	$^{87}_{38}Sr^{*} \rightarrow ^{87}_{38}Sr + \gamma$

* The asterisk denotes a nucleus in an excited state.

particle from a nucleus results from the transformation of a neutron into a proton and an electron:

$$^{1}_{0}n \rightarrow ^{1}_{1}H + ^{0}_{-1}e$$

The emission of an electron leaves the mass of the nucleus unchanged but increases its atomic number by 1. For example,

$$^{14}_{6}C \rightarrow ^{14}_{7}N + ^{0}_{-1}e$$

Positron Emission A **positron** is a particle with the same mass as an electron but with a positive charge. The symbol for a positron in an equation for a nuclear reaction is $^{0}_{1}e$. Two examples of positron emission are

$$^{39}_{19}K \rightarrow ^{39}_{18}Ar + ^{0}_{1}e$$

$$^{11}_{6}C \rightarrow ^{11}_{5}B + ^{0}_{1}e$$

The emission of a positron decreases the nuclear charge by one and leaves the mass number unchanged. There are no positrons in nuclei. The emission of a positron can be considered to result from the conversion of a proton to a neutron:

$$^{1}_{1}H \rightarrow ^{1}_{0}n + ^{0}_{1}e$$

Positrons exist only for a very short time. Within about 10^{-9} s, they combine with an electron and are converted to high-energy radiation called gamma (γ) radiation, which has a shorter wavelength and higher frequency than X-rays.

Electron Capture Some nuclei undergo radioactive transformation without emitting particles. Instead, one of the inner electrons of an atom, such as a $1s$ electron, enters the nucleus. This process is called **electron capture.** Rubidium-81 is transformed in this way to krypton-81:

$$^{81}_{37}Rb + ^{0}_{-1}e \rightarrow ^{81}_{36}Kr$$

Electron capture has the same result as positron emission—it decreases the nuclear charge by one and leaves the mass number unchanged.

When a nucleus captures an inner shell electron, the atom is left in an electronically excited state. Another electron quickly drops from the valence shell to fill the vacant orbital, and energy is emitted in the form of an X-ray.

γ-Ray Emission The new nucleus formed in a radioactive decay process may be in an excited state—in other words, its constituent neutrons and protons may not have their most stable arrangement. Such a nucleus returns to its ground state by

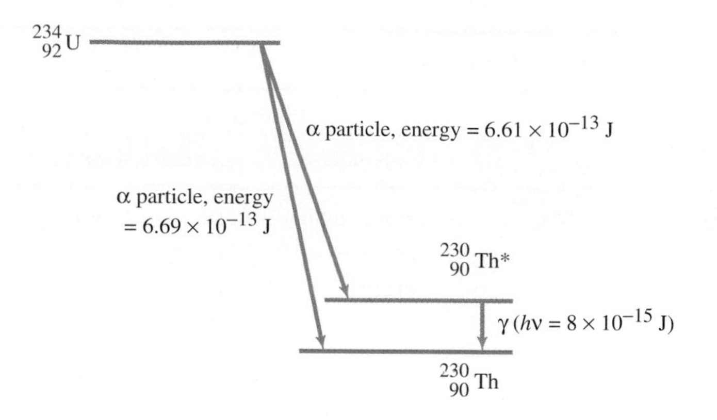

FIGURE 18.2 γ-Ray Emission In the decay of ^{234}U, α particles of two different energies are produced. The emission of α particles of energy 6.61×10^{-13} J leads to the formation of ^{230}Th in an excited state. This excited state decays to the ground state by the emission of a γ-ray photon of energy 8×10^{-15} J.

emitting γ radiation. For example, in the decay of $^{234}_{92}U$ to $^{230}_{90}Th$ by α-particle emission, 77% of the $^{230}_{90}Th$ nuclei are produced in the ground state by the emission of an α particle with an energy of 6.69×10^{-13} J, but 23% are produced in an excited state by the emission of an α particle with an energy of only 6.61×10^{-13} J. This excited Th atom then emits γ radiation of energy 8×10^{-15} J and returns to the ground state (Figure 18.2). In some cases an excited state produced in this manner can be relatively long-lived; such nuclei are said to be **metastable**. The mass number of a metastable nucleus is given a label "m." An example is $^{99m}_{43}Tc$—it takes 6 h for half the atoms to decay to the ground state $^{99}_{43}Tc$ by γ-ray emission. Metastable isotopes such as $^{99m}_{43}Tc$ are very useful in medicine, as we will discuss later in the chapter.

Decay Processes and Stability A nucleus with a composition outside the band of stability decomposes to give a nucleus with a more favorable neutron-to-proton ratio. Thus, a nuclide that lies above the band of stability must decrease its neutron-to-proton ratio to become more stable. The nuclide ^{14}C, which has more neutrons than the stable ^{12}C nuclide, decays by the emission of an electron, because this process converts a neutron into a proton, changing the neutron-to-proton ratio from 8:6 to 7:7:

$$^{14}_{6}C \rightarrow ^{14}_{7}N + ^{0}_{-1}e$$

Conversely, a nucleus located below the band of stability must increase its neutron-to-proton ratio to achieve stability. This can be accomplished by positron emission or by electron capture, both of which convert a proton into a neutron. An example of positron emission is the decay of ^{11}C:

$$^{11}_{6}C \rightarrow ^{11}_{5}B + ^{0}_{1}e$$

An example of electron capture is the decay of ^{7}Be:

$$^{7}_{4}Be + ^{0}_{-1}e \rightarrow ^{7}_{3}Li$$

Heavy nuclei frequently attain greater stability by emitting an α particle, thereby losing 2 protons and 2 neutrons, which decreases the atomic number by 2 and the mass number by 4. For example, $^{238}_{92}U$ decays to $^{234}_{90}Th$, which is also radioactive and decays to $^{234}_{91}Pa$. This isotope is also unstable and decays to $^{234}_{92}U$. The 14 successive nuclear reactions shown in Figure 18.3 finally give the stable lead isotope $^{206}_{82}Pb$. Such a series of nuclear reactions that eventually leads to a stable nucleus is called a **radioactive decay series.**

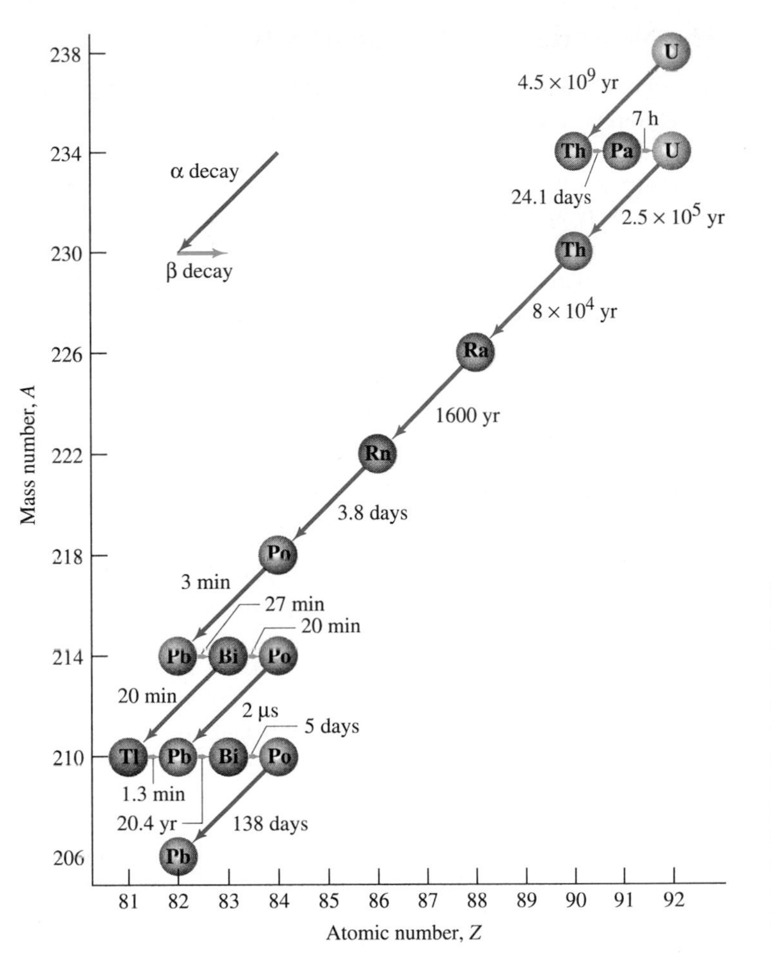

FIGURE 18.3 The Uranium-238 Decay Series
The nuclei of the ^{238}U radioactive decay series are represented on a graph of mass number versus atomic number. The emission of an α particle reduces the mass number by four and the atomic number by two. The emission of a β particle raises the atomic number by one but has no effect on the mass number.

EXAMPLE 18.1 Writing Equations for Radioactive Decay Reactions

Complete the following equations.

(a) $^{32}_{15}P \rightarrow ? + {}^{0}_{-1}e$ **(b)** $^{43}_{19}K \rightarrow {}^{43}_{20}Ca + ?$ **(c)** $^{210}_{84}Po \rightarrow {}^{206}_{82}Pb + ?$ **(d)** $^{17}_{9}F \rightarrow ? + {}^{0}_{1}e$

Solution: **(a)** The emission of an electron does not change the mass number but increases Z by 1, so the nucleus formed is $^{32}_{16}S$.

(b) The mass number remains unchanged, but the atomic number increases by 1; the particle emitted must be an electron, $^{0}_{-1}e$.

(c) The mass number decreases by 4 and the atomic number decreases by 2, so the particle emitted must be an α particle, $^{4}_{2}He$.

(d) The emission of a positron means that the mass number does not change, but the charge decreases by 1. Therefore, the nucleus formed is $^{17}_{8}O$.

The sums of the mass numbers on the reactant and product sides are equal in nuclear reactions, as are the sums of the atomic numbers.

Exercise 18.1 Complete the following equations.

(a) $^{22}_{11}Na \rightarrow {}^{0}_{1}e + ?$

(b) $^{27}_{12}Mg \rightarrow {}^{0}_{-1}e + ?$

(c) $^{36}_{17}Cl \rightarrow {}^{36}_{18}Ar + ?$

(d) $^{239}_{94}Pu \rightarrow {}^{235}_{92}U + ?$

The Measurement of Radioactivity

A variety of methods can be used to detect the emissions from radioactive materials. Photographic film and plates are sensitive not only to light but also to the α particles, electrons, and γ rays emitted by radioactive substances. The greater the exposure to radioactive emissions, the greater the blackening of the negative. Persons who work with radioactive substances wear a film badge to record the extent of their exposure to radiation.

An important instrument for detecting and accurately measuring radioactivity is the **Geiger counter** (Figure 18.4). A Geiger counter consists of a metal tube filled with a gas such as argon. One end of the tube has a thin mica window that allows fast-moving electrons, α particles, and γ rays to pass through. A potential difference (voltage) of about 1000 V is maintained between the tube wall and a central wire. Any high-energy electron, α particle, or γ-ray photon that enters the tube through the window knocks electrons out of the atoms in its path. The electrons and ions thus formed are accelerated to high speeds by the voltage, and they in turn ionize other atoms, producing more electrons and ions, which in turn ionize more atoms, and so on. Thus, a single high-energy electron, α particle, or γ-ray photon entering the tube produces an avalanche of ions and electrons that gives a brief pulse of electric current in the external circuit. This current pulse is amplified and recorded or is made audible as a click. The number of clicks in a given time is a direct measure of the number of particles entering the Geiger counter. Because a Geiger counter can record individual particles or photons, it is an extremely sensitive device.

The Biological Effects of Radiation

The discovery of radioactivity at the end of the nineteenth century marked an important milestone in the development of modern science (Box 18.1). Marie Curie, one of the pioneers in the field, died from radiation sickness after decades of exposure to radioactive materials. The hazards of radioactivity were not understood at that time, but they are now well recognized and have become an important and controversial social issue. To understand the dangers of the radiation from radioactive materials, we need some appreciation of how these dangers depend on the type, energy, and penetrating power of the radiation as well as on the length of exposure.

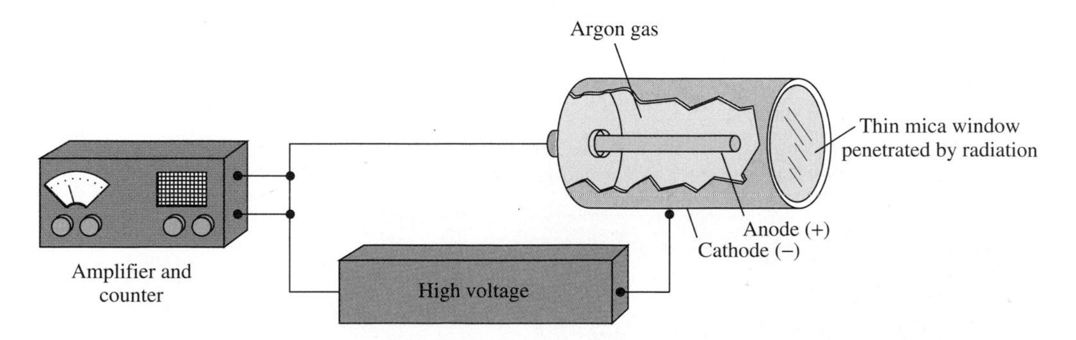

FIGURE 18.4 Schematic Representation of a Geiger Counter Radiation enters the counter through the mica window and ionizes the argon. This ionization causes a current to flow between the container walls (the cathode) and the central wire (the anode).

γ Radiation Gamma radiation is the most penetrating and the most damaging type of radiation. A γ-ray photon can penetrate deep into the body, causing damage by knocking electrons out of molecules to form highly reactive radicals and ions. Typically it knocks an electron out of a water molecule, thereby producing the free radical $H_2O\cdot^+$ ion:

$$H_2O + \gamma \rightarrow H_2O\cdot^+ + e^-$$

H_2O^+ then reacts with another water molecule to give a hydroxyl radical:

$$H_2O\cdot^+ + H_2O \rightarrow H_3O^+ + HO\cdot$$

Electrons produced in the initial reaction also give reactive free radicals by reactions such as

$$e^- + H_2O \rightarrow H\cdot + OH^-$$

$$e^- + O_2 \rightarrow \cdot O_2^-$$

$H_2O\cdot^+$, $\cdot OH$, and $\cdot O_2^-$ are free radicals with an odd number of electrons:

$$H{-}\ddot{\underset{|}{\underset{H}{O}}}\cdot^{+} \quad H{-}\ddot{\underset{\cdot\cdot}{O}}\cdot \quad \cdot\ddot{\underset{\cdot\cdot}{O}}{-}\ddot{\underset{\cdot\cdot}{O}}\colon^{-}$$

These free radicals attack virtually any organic molecule, including proteins and nucleic acids, usually by removing a hydrogen atom. We saw in Chapter 16 that the $\cdot OH$ radical reacts readily with hydrocarbons that have been liberated into the atmosphere:

$$RH + \cdot OH \rightarrow R\cdot + H_2O$$

Similar reactions occur with the hydroxyl radicals formed by γ rays and can cause substantial damage in the body. The ability of protein molecules and DNA damaged in this way to carry out their normal functions is impaired, a situation that can result in radiation sickness and cancer. But, if properly directed to kill cancerous cells, γ radiation can also be used to treat cancer.

β Radiation The fast electrons of β radiation can penetrate only a few millimeters of flesh before they are stopped by collisions with nuclei and other electrons. They produce essentially the same effects as those of γ rays but are less damaging because of their poorer penetrating power and lower energy.

α Radiation The least penetrating radiation is α radiation. Because α particles have high mass and charge, when they collide with a molecule, they break it into fragments. They are soon slowed down by these collisions and by capturing two electrons are converted to inert helium atoms. For this reason, α particles present little hazard if they are absorbed by the skin, which is continually dying and being lost. But α-emitting particles are a serious hazard if inhaled or ingested.

Activity of Radioactive Sources The intensity of the radiation from a radioactive element depends on the element's *activity*, the number of nuclear disintegrations that occur per second. In honor of Marie Curie, the fundamental unit of activity, the number of disintegrations per second of 1 g of radium, or 3.7×10^{10} disintegrations per second, is called a **curie** and is given the symbol Ci:

1 Ci = 3.7×10^{10} disintegrations per second

A more convenient unit is the millicurie (mCi), one-thousandth of a curie.

As we have seen, not all radiation carries the same amount of energy, and the biological effects of radiation depend on the amount of energy absorbed. Therefore, knowing the number of curies to which a person has been exposed does not give

BOX 18.1 The Curie Family and Radioactivity

Although the French physicist Antoine-Henri Becquerel (1852–1908) discovered radioactivity, the credit must go to Marie Sklodowska Curie (Figure A) for most of our knowledge and understanding of this phenomenon. Born in Russian-ruled Poland in 1867, Marie Sklodowska decided to study science at an early age, but at that time in Poland she was unable to get any education beyond high school. After working for several years, she eventually saved enough money to go to Paris and enter the Sorbonne as a student of chemistry and physics in 1891. There she met the physicist Pierre Curie, whom she married in 1895.

In 1896 Becquerel accidentally found that a sample of potassium uranyl sulfate, KUO_2SO_4, left in a drawer with a photographic plate caused the plate to fog, just as if it had been exposed to light. Other uranium compounds were soon found to produce the same result. It appeared that uranium emits some type of radiation, such as X-rays, that could affect a photographic plate in the same way as visible light. Becquerel's discovery inspired Marie Curie to begin working on the new phenomenon, which she named radioactivity. Curie, Becquerel, and Ernest Rutherford all studied the effect of electric fields on the radiation emitted by uranium. They showed that it consisted of both positive and negative particles, which Rutherford called α and β particles, and a radiation that was not deflected by an electric field, which he called γ radiation. It was later shown that α and β particles are helium nuclei and electrons, respectively, and that γ radiation is electromagnetic radiation of shorter wavelength than X-rays.

Curie found that some samples of the uranium ore pitchblende were much more radioactive than could be accounted for by their uranium content. Because none of the other elements known to be in the ore were radioactive, she concluded that a then-unknown element or elements must be present in very small amounts and must therefore be much more radioactive than uranium. Pierre Curie, recognizing the importance of his wife's work, abandoned his own research and joined her as an assistant. They isolated a small amount of a new element hundreds of times as radioactive as uranium that they called "polonium" in honor of Marie's native country. But polonium did not account for all the radioactivity of the ore, and they continued to try to isolate what had to be a still more radioactive element. After a monumental effort involving thousands of recrystallizations over a period of four years, they finally obtained 1 g of the new and highly radioactive element radium from 10 tons of pitchblende they had transported from a mine in Czechoslovakia. In 1903 Marie and Pierre Curie shared the Nobel Prize in physics with Becquerel for their work on radioactivity.

In 1906 Pierre was killed in a street accident. Marie became the first woman to teach at the Sorbonne when she took over his lecture course there. Even so, she had to contend all her life with prejudice against women in science. When she was nominated for membership in the prestigious all-male Academie Française, she lost by one vote because of her gender. In 1911 she was awarded the Nobel Prize in chemistry for her discovery of the elements polonium and radium. She and her work became widely known, particularly after it was found that the radiation from radium could be used to treat cancer. She spent her last years as director of the Paris Radium Institute. She died of leukemia in 1934 as a result of overexposure to radiation from radio-

Figure A *Marie Curie (1867–1934) in her laboratory at the Sorbonne*

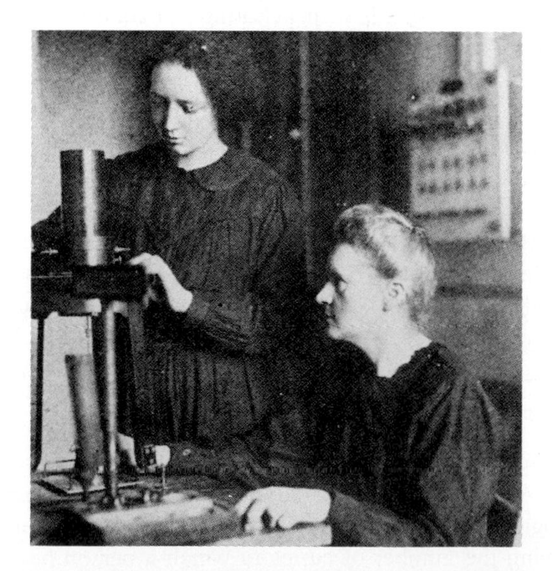

Figure B *Marie and Irène Curie working in their laboratory*

active substances, the dangers of which were not realized during the early years of her work.

The Curies' daughter Irène had been born in 1897. She acted as her mother's assistant (Figure B). After Marie Curie's death, Irène continued her mother's work, in collaboration with her husband Frédéric Joliot, who changed his name to Joliot-Curie. Like Rutherford they studied the effects of bombarding the nuclei of atoms with α particles, which, as Rutherford had shown, could produce new nuclei. Upon bombarding aluminum, they produced phosphorus that was radioactive and emitted positrons:

$$^{27}_{13}\text{Al} + {}^{4}_{2}\text{He} \rightarrow {}^{30}_{15}\text{P} + {}^{1}_{0}n$$

$$^{30}_{15}\text{P} \rightarrow {}^{30}_{14}\text{Si} + {}^{0}_{1}e$$

The Joliot-Curies had produced the first artificial radioactive isotope. For the first time it was realized that radioactivity is not confined to the heaviest elements but that any element can be radioactive if the proper isotope is prepared. Since that time hundreds of other radioactive isotopes have been prepared. For this work, Irène and Frédéric Joliot-Curie were awarded the Nobel Prize in chemistry in 1935. In 1939 they were the first to show that more neutrons are produced when uranium is bombarded with neutrons, which suggested the possibility of a chain reaction. This discovery led to the construction of the first atomic reactor in 1942 and of the atomic bomb, first used in 1945. Despite her scientific renown, Irène's application for membership in the American Chemical Society was rejected in 1954 because of her association with Communist organizations.

adequate information for assessing possible damage. A common unit used to measure the *dose* of radiation is the **rad** (**r**adiation **a**bsorbed **d**ose):

1 rad is the amount of radiation that results in the absorption of 10^{-2} J of energy per kilogram of body tissue.

This may seem like a minute amount of energy, but we must remember that it is highly localized and that each incoming particle very probably has sufficient energy to break a chemical bond.

The extent of radiation damage to living tissue depends not only on the energy but also on the type of radiation. The radiation dose expressed in rads is therefore multiplied by an approximate and somewhat arbitrary factor Q, called RBE (relative biological effectiveness), to give the *dose equivalent*, which is measured in a unit called a **rem** (**r**oentgen **e**quivalent for **m**an):

$$\text{dose equivalent in rem} = Q \times \text{absorbed dose in rad}$$

A unit of millirems (mrem) is often more convenient.

The damage caused by various dose equivalents of radiation is summarized in Table 18.2. We each typically receive a radiation dose of about 200 mrem per year. About 30% of this comes from cosmic rays (radiation that reaches the earth from outer space), and 20% comes from the radioactive ^{40}K in our bodies. The remaining 50% is made up of variable amounts from medical and dental X-rays, from radioactivity ingested in food or breathed in as $^{14}CO_2$, and from radon, which is an α emitter that seeps up from the ground (where it is formed by the decay of uranium minerals). Emissions from nuclear facilities contribute only about 0.1% of the total, although locally they may constitute a significantly larger contribution.

Potassium ions in the body occur mainly in the fluid inside cells. They are involved for example in the transmission of electrical impulses in nerves. A β emitter with a half-life of 1.3×10^9 yr, ^{40}K has a natural abundance of 0.012%.

TABLE 18.2 Health Hazards of Radiation

Dose Equivalent (rem)	Effect
0–25	None observable
25–50	Decrease of white blood cell count
100–200	Nausea; marked decrease in white blood cell count
500	50% likelihood of death within 30 days

18.2 RADIOACTIVE DECAY RATES

Nuclear Half-Life

The instant at which any given radioactive nucleus will decay is inherently unpredictable, but each nucleus of the same kind has the same probability of decaying in a certain time interval. As a result, the rate of decay of a sample of a radioactive nuclide—that is, the number of nuclei that disintegrate in a given time—is proportional to the number of nuclei, N, in the sample:

$$\text{decay rate} \propto N$$

or

$$\text{decay rate} = kN$$

This is the equation for a first-order rate law (Chapter 15), where k is a first-order rate constant that we call the **decay constant**. If ΔN is the number of nuclei disintegrating in a time interval Δt, then

$$\text{decay rate} = -\frac{\Delta N}{\Delta t} = kN$$

This expression for the decay rate is another form of the expression for a first-order rate law.

The integrated form of this equation is

$$\ln N = -kt + \ln N_0 \qquad \textit{or} \qquad \ln \frac{N_0}{N} = kt$$

where N_0 is the initial number of nuclei at time $t = 0$ and N is the number that remain at time t.

Figure 18.5 is a plot of the decay of ^{222}Rn and the buildup of the daughter nuclide ^{218}Po. Some radioactive nuclei decay much faster than others, but for each isotope a specific time is required for half a given sample to disintegrate. This time is the **half-life**, $t_{\frac{1}{2}}$. We can find the relationship between $t_{\frac{1}{2}}$ and the decay constant k, just as we did in Chapter 15, by setting $N = \frac{1}{2}N_0$:

III The units of k times the units of $t_{\frac{1}{2}}$ must cancel each other, because ln 2 is dimensionless.

$$\ln \frac{N_0}{N} = \ln 2 = 0.693 = kt_{\frac{1}{2}}$$

or

$$t_{\frac{1}{2}} = \frac{0.693}{k}$$

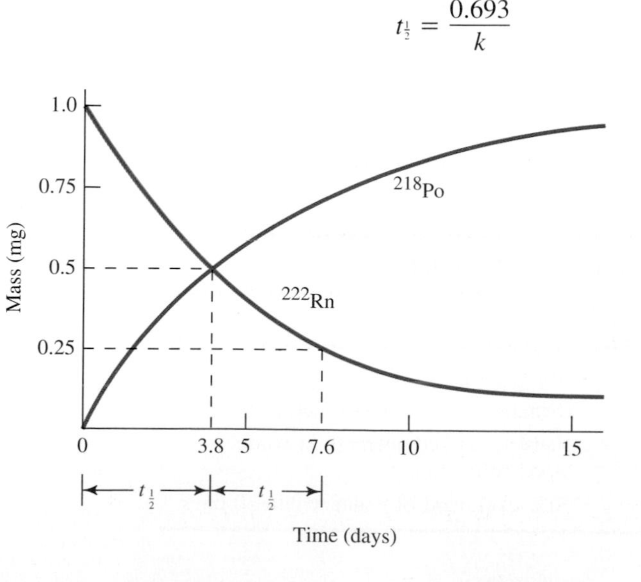

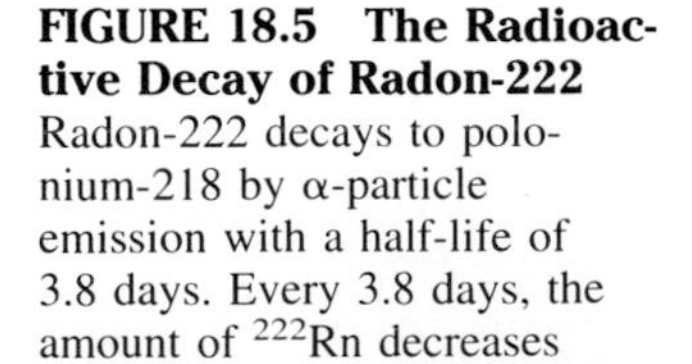

FIGURE 18.5 The Radioactive Decay of Radon-222 Radon-222 decays to polonium-218 by α-particle emission with a half-life of 3.8 days. Every 3.8 days, the amount of ^{222}Rn decreases by 50%.

TABLE 18.3 Half-Lives of Some Radioisotopes

Isotope	Half-Life	Mode of Decay
$^{214}_{84}Po$	164 s	α
$^{25}_{11}Na$	1.0 min	β
$^{131}_{53}I$	8.0 days	β
$^{222}_{86}Rn$	3.8 days	α
$^{32}_{15}P$	14.3 days	β
$^{60}_{27}Co$	5.3 yr	β
$^{90}_{38}Sr$	28.8 yr	β
$^{14}_{6}C$	5730 yr	β
$^{230}_{94}Pu$	2.4×10^4 yr	α
$^{40}_{19}K$	1.3×10^9 yr	β
$^{238}_{92}U$	4.5×10^9 yr	α
$^{232}_{90}Th$	1.4×10^{10} yr	α

From Figure 18.5 we see that the half-life of ^{222}Rn is 3.8 days. After one half-life (3.8 days), 50% of the original ^{222}Rn remains; after two half-lives (7.6 days), only 25% is left, and so on. If we know the decay constant, we can find the half-life, and vice versa. Half-lives may vary from fractions of a second to millions of years (Table 18.3).

The energies involved in the binding of nuclear particles are much greater than the energies of chemical bonds or the energies of thermal motion (except at temperatures of millions of degrees in the interiors of stars). The decay constant and the half-life of any nuclide are therefore independent of temperature and of the nature and physical state of the substance containing the nuclide.

The number of nuclei in a sample of a radioactive isotope is proportional to its mass. If m_0 is the initial mass of a sample of a radioactive nuclide and m is the mass of that nuclide remaining after a time t, then

$$\ln \frac{N_0}{N} = \ln \frac{m_0}{m} = kt$$

We can use this equation to calculate the amount of a radioactive isotope that remains after any given time, as Example 18.2 shows.

EXAMPLE 18.2 Radioactive Decay Rates

Cobalt-60 is used extensively in radiation therapy for cancer victims. If a hospital laboratory has a sample of 25.0 mg of $^{60}_{27}Co$, how much will remain 18 months (1.50 yr) later? The half-life of $^{60}_{27}Co$ is 5.26 yr.

Solution: From the half-life we can calculate the decay constant:

$$k = \frac{0.693}{t_{\frac{1}{2}}} = \frac{0.693}{5.26\ \text{yr}} = 0.132\ \text{yr}^{-1}$$

Then substituting $k = 0.132\ \text{yr}^{-1}$ and $m_0 = 25.0$ mg in the equation

$$\ln \frac{m_0}{m} = kt$$

we obtain

$$\ln \frac{25.0\ \text{mg}}{m} = 0.132\ \text{yr}^{-1} \times 1.50\ \text{yr} = 0.198$$

So

Amount remaining = initial amount − amount decayed

$$\frac{25.0\ \text{mg}}{m} = 1.22 \qquad \text{and} \qquad m = 20.5\ \text{mg}$$

Thus, 20.5 mg, or 82.0% of the initial amount, still remains.

Exercise 18.2 The radioactive noble gas $^{222}_{86}Rn$ formed by the decay of uranium seeps out of the ground. It may collect in tightly closed basements and pose a health hazard, as it is believed to cause lung cancer. How much $^{222}_{86}Rn$ will remain after 12.0 days in a sample initially containing 30 μg of $^{222}_{86}Rn$? The half-life of radon-222 is 3.8 days.

EXAMPLE 18.3 Calculating the Half-Life of a Radioisotope

The radioisotope $^{131}_{53}I$ is used in medical studies and tests on the thyroid gland. A sample that originally contained 1.00 μg of $^{131}_{53}I$ contained 0.32 μg of the isotope after 13.3 days. What is the half-life of $^{131}_{53}I$?

Solution: Because $k = 0.693/t_{\frac{1}{2}}$,

$$\ln \frac{m_0}{m} = kt = \frac{0.693t}{t_{\frac{1}{2}}}$$

and therefore

$$t_{\frac{1}{2}} = \frac{0.693t}{\ln(m_0/m)}$$

Substituting $t = 13.3$ days, $m_0 = 1.00$ μg, $m = 0.32$ μg gives $t_{\frac{1}{2}} = 8.1$ days.

Exercise 18.3 A sample that originally contained 0.30 mg of ^{60}Co was found to contain only 0.25 mg of ^{60}Co 1.40 yr later. What is the half-life of ^{60}Co?

Using Radioisotopes for Dating

Because the half-life of a radioisotope is a constant that is independent of temperature and all other conditions, the decay of a radioisotope can be used to determine the time at which certain events occurred in the past. We will consider two examples: the dating of rocks and the dating of archeological specimens.

The Dating of Rocks All the radioactive elements that were present when the earth first formed have been decaying steadily ever since. Potassium-40, which constitutes 0.012% of naturally occurring potassium, is a useful isotope for dating ancient rocks because potassium is widely distributed in rocks and ^{40}K has a half-life of 1300 million yr. Potassium-40 decays to ^{40}Ar according to the equation

$$^{40}_{19}K \rightarrow {}^{40}_{20}Ar + {}^{0}_{-1}e$$

Early in its history the earth was largely molten, so any gaseous argon formed from the decay of ^{40}K at that time would have been lost. Once the earth's rocks solidified, the ^{40}Ar would have been trapped in the solid rock. When we analyze a rock sample for potassium and argon, the amount of argon tells us how much potassium has decayed. The total amount of argon and potassium tells us how much potassium was present in the rock when it solidified. Knowing the half-life of ^{40}K, we can determine the time when the rock solidified—by the procedure known as *potassium/argon dating* (see Example 18.4).

EXAMPLE 18.4 Potassium/Argon Dating

Mass spectrometric analysis showed that a sample of rock from the moon contained 100 μg of ^{40}K and 456 μg of Ar. The half-life of ^{40}K is 1.3×10^9 yr. When did the moon rock solidify?

Solution: Because both ^{40}K and ^{40}Ar have atomic mass 40,

$$\text{amount of } ^{40}\text{K present initially} = \text{amount remaining today} + \text{amount converted to } ^{40}\text{Ar}$$
$$= 100\ \mu g + 456\ \mu g = 556\ \mu g$$

So the ratio of the number of K atoms at the time of formation of the rock (N_0) to the number of K atoms now (N) is

$$\frac{N_0}{N} = \frac{100 + 456}{100} = 5.56$$

The decay constant is

$$k = \frac{0.693}{t_{\frac{1}{2}}} = \frac{0.693}{1.3 \times 10^9\ \text{yr}} = 5.3 \times 10^{-10}\ \text{yr}^{-1}$$

Rearranging the rate law $\ln(N_0/N) = kt$ gives

$$t = \frac{\ln(N_0/N)}{k} = \frac{\ln 5.56}{5.3 \times 10^{-10}\ \text{yr}^{-1}} = 3.2 \times 10^9\ \text{yr}$$

We conclude that this rock sample solidified 3.2 billion yr ago.

Not all kinds of rocks are suitable for potassium/argon dating, because argon gradually diffuses out of permeable rocks. Igneous rocks (rocks that cooled from a molten state) give the most reliable results. However, even igneous rocks may have lost some of their argon if at any time they were reheated until they were molten. Dates obtained by this method are therefore considered minimum dates: If some argon has been lost, the rock is older than the calculated value.

Other isotopes useful for radioactive dating of rocks are rubidium-87, which decays to strontium-87 and has a half-life of 3.21 billion yr; uranium-238, which decays through a series of short-lived intermediate isotopes to lead-206 and has a half-life of 4.51 billion yr; and uranium-235, which decays to lead-207 and has a half-life of 713 million yr.

Exercise 18.4 A sample of rock contains 13.2 μg of uranium-238 and 3.42 μg of lead-206. If the half-life of uranium-238 is 4.51×10^9 yr, what is the age of the rock? In other words, when did the rock solidify?

The rocks in meteorites have been determined to be about 4.6 billion yr old. Meteorites are believed to be the remnants of the material from which the earth and the other planets formed. Thus, we can say that the age of the earth is also about 4.6 billion yr. This date is consistent with ages of 3.5 to 4.0 billion yr found by potassium/argon dating for the oldest rocks on earth, which reflect the longer time a molten body the size of the earth required to cool to solid rock.

Radiocarbon Dating It has been only a very short time on the geological time scale since *Homo sapiens* appeared on earth, so the dating of archeological samples must be based on a radioactive isotope with a much shorter half-life than that of ^{40}K or ^{235}U. Carbon-14, with a half-life of 5730 yr, is very suitable. Carbon is present in all formerly living material, such as bones, cloth, charcoal from wood fires, or wooden objects made by ancient peoples. Any ^{14}C that was originally present on earth has long since disappeared, but ^{14}C is continually being formed by cosmic rays that reach the earth from outer space. Cosmic rays are composed of subatomic particles of very high energies. In the outer atmosphere, cosmic rays break up any atom with which they collide, giving high-energy neutrons, protons, and electrons, which in turn react with other atoms in the atmosphere. Nitrogen-14 atoms are relatively abundant in the upper atmosphere, and the reaction of a neutron with a ^{14}N atom gives ^{14}C and a proton:

$$^{14}N + {}^{1}n \rightarrow {}^{14}C + {}^{1}H$$

Because ^{14}C decays back to ^{14}N by the emission of an electron,

$$^{14}_{6}C \rightarrow {}^{14}_{7}N + {}^{0}_{-1}e$$

there is a steady-state concentration of ^{14}C in the atmosphere, much of which reacts with oxygen to form carbon dioxide. Carbon dioxide is incorporated by photosynthesis into plants, which are in turn eaten by animals, so all the carbon in living organisms contains a small constant concentration of ^{14}C. When an organism dies, incorporation of radiocarbon ceases, but radioactive decay continues. Thus, the living cells of a tree contain the same amount of ^{14}C as does the atmosphere, but dead wood contains less—some of its ^{14}C has been lost by radioactive decay. The extent of the loss measures the length of time since death. Careful measurements have shown that the current steady-state concentration of ^{14}C corresponds to 15.3 disintegrations per gram of carbon per minute. Assuming that the steady-state concentration of ^{14}C was the same when the tree was alive, we can determine how long ago the tree died, as shown in Example 18.5.

EXAMPLE 18.5 Carbon-14 Dating

A sample of charcoal from one of the earliest Polynesian settlements in Hawaii was found in 1990 to have a disintegration rate of 13.6 disintegrations per minute per gram. The steady-state rate of disintegration of ^{14}C is 15.3 disintegrations per gram per minute, and the half-life of ^{14}C is 5730 yr. What is the age of the charcoal?

Solution: The rate of disintegration is proportional to the number of ^{14}C nuclei remaining in 1 g of the sample. Therefore,

$$\frac{N_0}{N} = \frac{15.3}{13.6} = 1.13$$

From Example 18.4 we know that

$$t = \frac{t_{\frac{1}{2}}}{0.693} \ln \frac{N_0}{N} = \frac{5730 \text{ yr}}{0.693} \ln 1.13 = 974 \text{ yr}$$

This result suggests that the Polynesians had already arrived in Hawaii by about the year A.D. 1020.

Exercise 18.5 A wooden bowl found in an archeological excavation has a ^{14}C disintegration rate of 11.2 per gram per minute. What is the bowl's age?

The calculation of dates that we have just described assumes that the current atmospheric concentration of ^{14}C has been constant during the last 10,000 yr or so. In fact, small variations in this concentration have occurred because of changes in the amount of cosmic radiation reaching the earth. Box 18.2 describes an ingenious method for correcting for these changes.

BOX 18.2 Calibrating the ^{14}C Clock

Our description of ^{14}C dating rests on the assumption that the steady-state concentration of ^{14}C has remained constant over the past 20,000 years or so. However, there are various reasons why we would *not* expect the steady-state concentration of ^{14}C to have remained constant—most notably, fluctuations in the earth's magnetic field, which diverts some of the incoming cosmic rays that produce ^{14}C. To obtain accurate results from ^{14}C dating, we must therefore calibrate the ^{14}C clock. Fortunately this has proved to be possible by using the tree rings of bristlecone pines, which grow in the southwestern United States (Figure A).

Trees grow rapidly in the spring, producing large cells and light-colored wood, and slowly in the summer and fall, producing a darker band of wood. Each year's growth therefore produces a *tree ring*. Counting the rings after a tree has been felled determines its age. The rings differ somewhat from year to year. A warm, wet spring produces a broad ring, and a cool, dry spring a much narrower ring. The climate in the southwestern United States is uniform over a wide area, so the ring patterns for all trees are similar. Bristlecone pines are very long-lived trees; counting tree rings has revealed that some of these pines have lived as long as 4900 years. The characteristic ring patterns allow the old rings of contemporary trees to be matched to the younger rings of trees long dead, so the time span covered can be extended. In this way, a characteristic ring pattern that stretches back to 5500 B.C. has been identified. If the ^{14}C content of each ring is determined, and if we determine the age of each ring by counting rings, we can correct dates obtained by ^{14}C dating. In other words, we have a method for calibrating the ^{14}C clock. This calibration method gives ages for artifacts from ancient civilizations that agree well with the dates established by historical records (Figure B).

Figure A *(1) Bristlecone pine, the oldest living tree. (2) A researcher examining the rings in a cross section of a bristlecone pine.*

Figure B *A researcher in a museum in Israel working on the Dead Sea Scrolls, ancient Hebrew religious writings determined by ^{14}C dating to date from A.D. 30 ± 200 yr*

18.3 ARTIFICIAL RADIOISOTOPES

Substantial amounts of very-long-lived radioisotopes, such as ^{40}K and ^{238}U that were present when the earth formed, still remain today. Most of the other, shorter-lived radioactive nuclei that were present at the formation of the earth have long since disappeared. However, a large number of radioactive nuclei as well as stable nuclei have been produced by nuclear reactions in recent times.

Synthesis of Radioisotopes

Radioisotopes are most commonly made by nuclear reactions, using the neutrons available from a nuclear reactor or the very energetic (high-speed) protons or α particles from a particle accelerator. Ernest Rutherford (Box 1.2) in 1917 was the first to carry out a nuclear reaction in the laboratory. When he bombarded nitrogen with a stream of α particles obtained from a radioactive element, he produced $^{17}_{8}O$ and protons by the reaction

$$^{14}_{7}N + ^{4}_{2}He \rightarrow ^{17}_{8}O + ^{1}_{1}H$$

This was the first observed transmutation of one element into another. The product of the reaction carried out by Rutherford is the stable isotope $^{17}_{8}O$, but many radioisotopes can be formed by similar reactions (Box 18.1). For example, $^{27}_{13}Al$ can be transformed into radioactive phosphorus-30 by bombardment with α particles:

$$^{27}_{13}Al + ^{4}_{2}He \rightarrow ^{30}_{15}P + ^{1}_{0}n$$

Because of the strong electrostatic repulsion between the positively charged α particles and positively charged nuclei, it is usually necessary to use α particles that have been accelerated to very high energies (and hence high speeds) in a particle accelerator to bring the nuclei close enough together for a nuclear reaction to occur.

Because they are neutral, neutrons are not repelled by nuclei, so they need not be moving at high speeds to get close enough to the nucleus to react. Neutrons are conveniently obtained from a nuclear reactor, in which many neutrons are produced in fission reactions. The first step in the preparation of cobalt-60, which is used in some forms of radiation therapy, is the reaction of neutrons with iron-58 to form iron-59:

$$^{58}_{26}Fe + ^{1}_{0}n \rightarrow ^{59}_{26}Fe$$

Iron-59 decays to cobalt-59,

$$^{59}_{26}Fe \rightarrow ^{59}_{27}Co + ^{0}_{-1}e$$

which absorbs another neutron to form cobalt-60:

$$^{59}_{27}Co + ^{1}_{0}n \rightarrow ^{60}_{27}Co$$

Neutron bombardment can also be used to make technetium, a radioactive element not found in nature, from molybdenum:

$$^{98}_{42}Mo + ^{1}_{0}n \rightarrow ^{99}_{42}Mo$$

$$^{99}_{42}Mo \rightarrow ^{99}_{43}Tc + ^{0}_{-1}e$$

Uses of Radioisotopes

Many artificially produced radioisotopes have important applications in chemistry, medicine, agricultural research, oil exploration, and many other fields.

Radiolabeling in Chemistry In chemistry, one of the important applications of radioisotopes is the study of reaction mechanisms—that is, the determination of the successive steps by which many reactions take place (Chapter 15). If a very small percentage of the atoms in a compound is exchanged for a radioactive isotope of the element, the element is said to be *labeled*, or radiolabeled. Because a radioactive isotope of an element has the same chemical properties as the stable isotopes of the element, the radioactivity of the labeled element can be used to follow the movement of the element through a complex series of reactions.

An early and important application of the radioisotope labeling technique was the elucidation by Melvin Calvin (Figure 18.6) of the mechanism of photosynthesis

FIGURE 18.6 Melvin Calvin (1911–) Melvin Calvin, the son of Russian immigrants, graduated from the Michigan College of Mining and Technology in 1931 and earned his Ph.D. at the University of Minnesota in 1935. He became director of the Lawrence Radiation Laboratory in Berkeley, California, in 1946. He was awarded the Nobel Prize in chemistry in 1961 for elucidating the mechanism of photosynthesis.

in plants. The overall reaction for photosynthesis is represented by the equation

$$6CO_2 + 6H_2O \rightarrow C_6H_{12}O_6 + 6O_2$$

Calvin and his co-workers exposed growing plants to ^{14}C-labeled CO_2. Then at various times later, they extracted compounds from the plants and determined which of these compounds contained ^{14}C. In this way, by much painstaking work between 1949 and 1957, they were able to work out the complicated series of steps by which carbon dioxide is converted to carbohydrates. One related question that they answered was, Do the oxygen atoms in the evolved oxygen come from water, from carbon dioxide, or both? For this study they used as a label the stable isotope ^{18}O, which can be detected and measured in a mass spectrometer, for example. When they used water labeled with $H_2{}^{18}O$, they found ^{18}O in the oxygen given off by the plant. But when they used carbon dioxide containing $C(^{18}O)_2$, they found none of the ^{18}O in the evolved oxygen. Thus, they determined that all the oxygen produced in photosynthesis comes from water.

Radiolabeling in Medicine Because the radiation emitted by a radioisotope is easily detected, it is easy to follow the movement of a radioisotope through the body and to obtain a photographic or computer image of any organ that the radioisotope reaches. Only very small amounts of the radioisotope are needed; choosing a radioisotope with a short half-life further reduces the possibility of harmful effects. For example, sodium-24 (a β emitter with a half-life of 14.8 h) injected into the blood in the form of a sodium chloride solution can be used to follow the flow of the blood through the body and to indicate blockages or constrictions in the circulatory system. Iodine-131 (a β emitter with a half-life of 8 days) is used to monitor the functioning of the thyroid gland. The radioactivity of the thyroid gland can be measured after the patient has drunk a solution of sodium iodide containing $Na(^{131}I)$ to see if the iodine is absorbed by the thyroid gland at the normal rate (Figure 18.7). Technetium-99 (a γ emitter with a half-life of 6 h) is similarly used for brain imaging. It can be administered as sodium pertechnate, $NaTcO_4$, which is carried in the blood to the brain and provides an image of the brain. It is concentrated in the fastest-growing cells and therefore in tumors and allows such malignancies to be located.

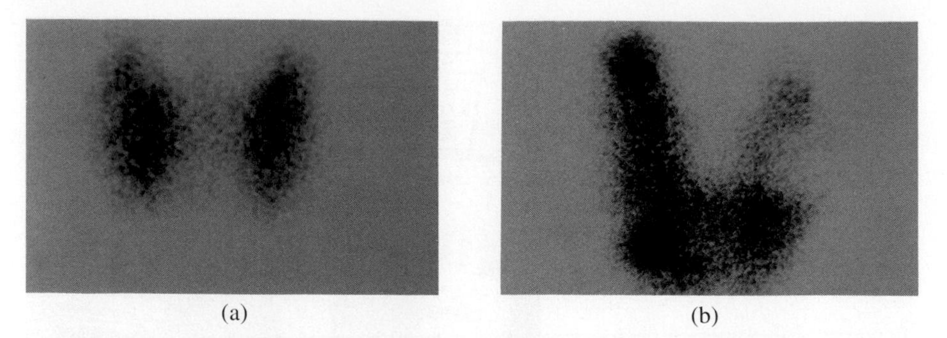

FIGURE 18.7 Images of the Thyroid Gland Produced by Iodine-131 (a) A normal thyroid. (b) An enlarged thyroid resulting from an iodine-deficient diet. These images were produced by using radioactive iodine-131, which is a β emitter with a half-life of 8 days.

Radiation Therapy Another important medical application of radioisotopes is radiation therapy, the use of radiation for destroying cancerous tissue. The most widely used isotope for this purpose is ^{60}Co, which decays by electron emission to ^{60}Ni in an excited state. The excited ^{60m}Ni nucleus then emits γ radiation when it falls to its ground state:

$$^{60}_{27}\mathrm{Co} \rightarrow {}^{60m}_{28}\mathrm{Ni} + {}^{0}_{-1}e$$

$$^{60m}_{28}\mathrm{Ni} \rightarrow {}^{60}_{28}\mathrm{Ni} + \gamma$$

As we have seen, high-energy radiation, particularly γ rays, generates large numbers of free radicals, which can rapidly destroy living tissue. Great care has to be taken to localize the radiation so as to destroy only the tumor and leave surrounding healthy tissue unaffected. Iodine concentrates in the thyroid gland, so radioactive iodine can be used not only for monitoring the activity of the thyroid gland, but also for destroying thyroid tumors.

Transuranium Elements

The heaviest naturally occurring element is uranium, with atomic number 92. But 17 heavier elements, with atomic numbers up to 109, have been prepared in nuclear reactors or in particle accelerators (Table 18.4). An example is the reaction of ^{238}U with α particles to produce plutonium:

$$^{238}_{92}\mathrm{U} + {}^{4}_{2}\mathrm{He} \rightarrow {}^{239}_{94}\mathrm{Pu} + 3\,{}^{1}_{0}n$$

Most of these heavier nuclei are intensely radioactive; except for neptunium and plutonium, they have short half-lives. They are very difficult to handle because of their strong radioactivity, so most of these so-called **transuranium elements** have been obtained only in very small quantities, and their chemistry has not been extensively studied. After neptunium and plutonium, the naming of which followed the planetary theme of uranium, the elements up to atomic number 103 (except for americium) have been named after distinguished nuclear scientists. There has, however, been controversy over appropriate names for the remaining transuranium elements, and as a temporary measure names based on their atomic numbers have been proposed (Table 18.4).

The names and symbols of the elements 104 to 109 are composed from the following roots, which represent the digits of the respective atomic numbers: 1, un; 2, bi; 3, tri; 4, quad; 5, pent; 6, hex; 7, sept; 8, oct; 9, enn; 0, nil. For example, element 108 is named unniloctium and has the symbol Uno.

TABLE 18.4 The Transuranium Elements

Atomic Number	Name	Symbol	Preparation
93	Neptunium	Np	$^{238}_{92}U + ^{1}_{0}n \rightarrow ^{239}_{93}Np + ^{0}_{-1}e$
94	Plutonium	Pu	$^{239}_{93}Np \rightarrow ^{239}_{94}Pu + ^{0}_{-1}e$
95	Americium	Am	$^{239}_{94}Pu + ^{1}_{0}n \rightarrow ^{240}_{95}Am + ^{0}_{-1}e$
96	Curium	Cm	$^{239}_{94}Pu + ^{4}_{2}He \rightarrow ^{242}_{96}Cm + ^{1}_{0}n$
97	Berkelium	Bk	$^{241}_{95}Am + ^{4}_{2}He \rightarrow ^{243}_{97}Bk + 2^{1}_{0}n$
98	Californium	Cf	$^{242}_{96}Cm + ^{4}_{2}He \rightarrow ^{245}_{98}Cf + ^{1}_{0}n$
99	Einsteinium	Es	$^{238}_{92}U + 15^{1}_{0}n \rightarrow ^{253}_{99}Es + 7^{0}_{-1}e$
100	Fermium	Fm	$^{238}_{92}U + 17^{1}_{0}n \rightarrow ^{255}_{100}Fm + 8^{0}_{-1}e$
101	Mendelevium	Md	$^{253}_{99}Es + ^{4}_{2}He \rightarrow ^{256}_{101}Md + ^{1}_{0}n$
102	Nobelium	No	$^{246}_{96}Cm + ^{12}_{6}C \rightarrow ^{254}_{102}No + 4^{1}_{0}n$
103	Lawrencium	Lr	$^{252}_{98}Cf + ^{10}_{5}B \rightarrow ^{257}_{103}Lr + 5^{1}_{0}n$
104	Unnilquadium	Unq	$^{249}_{98}Cf + ^{12}_{6}C \rightarrow ^{257}_{104}Unq + 4^{1}_{0}n$
105	Unnilpentium	Unp	$^{249}_{98}Cf + ^{15}_{7}N \rightarrow ^{260}_{105}Unp + 4^{1}_{0}n$
106	Unnilhexium	Unh	$^{249}_{98}Cf + ^{18}_{8}O \rightarrow ^{263}_{106}Unh + 4^{1}_{0}n$
107	Unnilseptium	Uns	$^{209}_{83}Bi + ^{54}_{24}Cr \rightarrow ^{262}_{107}Uns + ^{1}_{0}n$
108	Unniloctium	Uno	$^{208}_{82}Pb + ^{58}_{26}Fe \rightarrow ^{265}_{108}Uno + ^{1}_{0}n$
109	Unnilennium	Une	$^{209}_{83}Bi + ^{58}_{26}Fe \rightarrow ^{266}_{109}Une + ^{1}_{0}n$

18.4 NUCLEAR ENERGY

One of the most important scientific discoveries of the twentieth century was that a great deal of energy is released in certain types of nuclear reactions. The use of nuclear energy in both war and peace has raised many difficult questions. In this section we discuss the physics and chemistry of nuclear energy and its applications.

Mass Defect and Binding Energy

We saw in Chapter 1 that the mass of a ^{4}He nucleus is not exactly equal to the total mass of its constituents. Two protons and two neutrons have a combined mass of 4.031 88 u, but the mass of a helium nucleus is only 4.001 50 u. The difference of 0.030 38 u = 5.045×10^{-29} kg, called the **mass defect**, appears as energy when a helium nucleus forms from its constituent particles. The amount of energy involved can be calculated from Einstein's equation

$$\Delta E = (\Delta m)c^2$$

where Δm is the mass defect:

$$\begin{aligned}\Delta E &= (5.045 \times 10^{-29}\ \text{kg})(2.998 \times 10^{8}\ \text{m} \cdot \text{s}^{-1})^2 \\ &= 4.534 \times 10^{-12}\ \text{kg} \cdot \text{m}^2 \cdot \text{s}^{-2} = 4.534 \times 10^{-12}\ \text{J} \quad (1\ \text{J} = 1\ \text{kg} \cdot \text{m}^2 \cdot \text{s}^{-2})\end{aligned}$$

Both ΔE and Δm are negative here, but by convention a negative ΔE is called a positive nuclear binding energy.

The energy released in the formation of a nucleus from its constituent particles is called the **nuclear binding energy**.

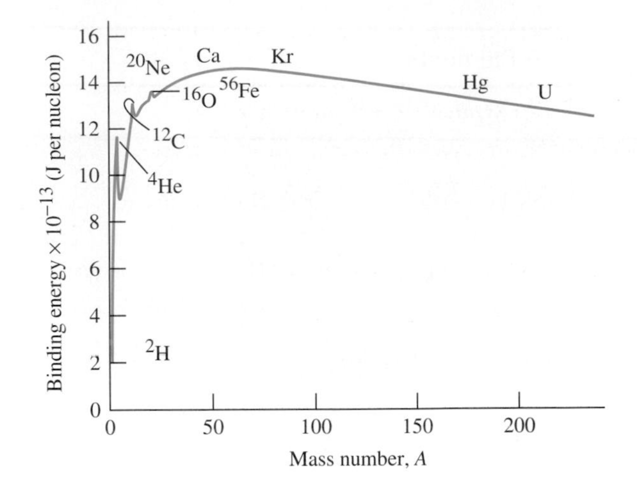

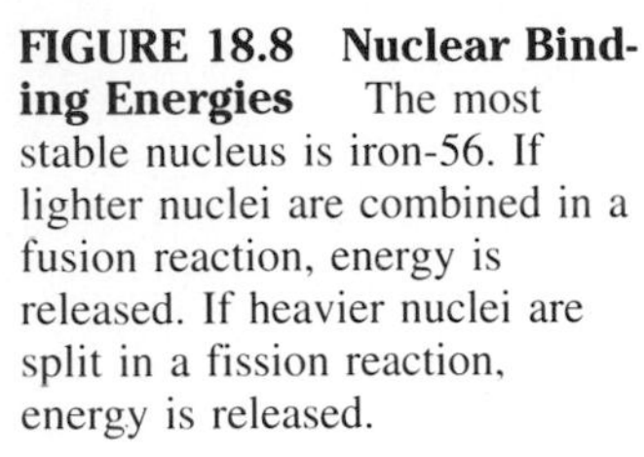

FIGURE 18.8 Nuclear Binding Energies The most stable nucleus is iron-56. If lighter nuclei are combined in a fusion reaction, energy is released. If heavier nuclei are split in a fission reaction, energy is released.

The nuclear binding energy of helium is 4.534×10^{-12} J. We can more easily appreciate the enormous quantity of energy produced in the formation of a helium nucleus by finding the energy released in the formation of 1 mol of He atoms (N_A atoms of He):

$$E = (4.534 \times 10^{-12}\ \text{J})(6.022 \times 10^{23}\ \text{mol}^{-1})$$
$$= 2.730 \times 10^{12}\ \text{J} \cdot \text{mol}^{-1} = 2.730 \times 10^{9}\ \text{kJ} \cdot \text{mol}^{-1}$$

This energy release corresponds to a mass loss of

$$(5.045 \times 10^{-29}\ \text{kg})(6.022 \times 10^{23}\ \text{mol}^{-1}) = 3.038 \times 10^{-5}\ \text{kg} \cdot \text{mol}^{-1}$$
$$= 3.038 \times 10^{-2}\ \text{g} \cdot \text{mol}^{-1}$$

This amount is more than a million times the energy released in a typical chemical reaction, where the corresponding mass loss is completely negligible. Burning 1 mol of hydrogen to form water, for example, liberates only 285.8 kJ of energy (Chapter 9), and the corresponding mass loss is $2.57 \times 10^{-9}\ \text{g} \cdot \text{mol}^{-1}$, which is much too small to be detected.

To compare the binding energies of different nuclei, it is convenient to quote values of the binding energy per nucleon:

$$\text{binding energy per nucleon} = \frac{\text{binding energy}}{\text{number of nucleons}}$$

Figure 18.8 is a plot of binding energy per nucleon versus mass number. This plot has a maximum at ^{56}Fe, which is therefore the most stable nucleus. Energy can be obtained either by synthesizing elements lighter than Fe from smaller nuclei, a process known as **fusion**, or by decomposing nuclei heavier than Fe into smaller nuclei, a process known as **fission**. In either case the mass of the products is less than that of the reacting nuclei. The energy produced in this way is called **nuclear energy**. The energy of the sun is produced primarily by the fusion of hydrogen nuclei to give helium nuclei. The energy produced by nuclear power stations comes from fission reactions, which we discuss next.

EXAMPLE 18.6 Calculating the Binding Energy of Nuclei

The mass of an $^{56}_{26}$Fe atom is 55.934 93 u. What are its binding energy and binding energy per nucleon?

Solution: The masses of the proton and the neutron are given in Table 1.3;

$$\begin{aligned}\text{mass of nucleus} &= \text{mass of atom} - \text{mass of electrons}\\ &= 55.934\,93\ \text{u} - 0.014\,263\ \text{u} = 55.920\,67\ \text{u}\end{aligned}$$

The $^{56}_{26}Fe$ nucleus has 26 protons and 30 neutrons, which gives a mass defect of

$$(26 \times 1.007\,28) + (30 \times 1.008\,66) - 55.920\,67 = 0.5284\ \text{u} = 8.774 \times 10^{-28}\ \text{kg}$$

Hence,

$$\begin{aligned}\Delta E &= (\Delta m)c^2 = (8.774 \times 10^{-28}\ \text{kg})(2.998 \times 10^8\ \text{m}\cdot\text{s}^{-1})^2\\ &= 7.886 \times 10^{-11}\ \text{kg}\cdot\text{m}^2\cdot\text{s}^{-2} = 7.886 \times 10^{-11}\ \text{J}\end{aligned}$$

This is the binding energy of $^{56}_{26}Fe$. This isotope has 56 nucleons (26 protons + 30 neutrons), so the binding energy per nucleon is

$$\frac{7.886 \times 10^{-11}}{56} = 1.408 \times 10^{-12}\ \text{J}\cdot\text{nucleon}^{-1}$$

Recall that for $^{56}_{26}Fe$, $Z = 26$ and $A = 56$ indicate there are 26 protons and $A - Z = 56 - 26 = 30$ neutrons, respectively, in the nucleus.

The mass number A equals the total number of nucleons.

Exercise 18.6 The mass of a ^{16}O nucleus is 15.990 53 u. What are its binding energy and binding energy per nucleon?

Tables of isotopic masses give the mass of an atom. To obtain the mass of a nucleus, you must subtract the total mass of the electrons.

Nuclear Fission

Some heavy nuclei, such as those of uranium, can be caused to undergo fission by bombarding them with neutrons. For example, a uranium-235 nucleus that reacts with a neutron breaks up into a variety of lighter nuclei and more neutrons (Figure 18.9). Two of these fission reactions are

$$^{235}_{92}U + ^{1}_{0}n \rightarrow \begin{cases} ^{137}_{52}Te + ^{97}_{40}Zr + 2\,^{1}_{0}n \\ ^{142}_{56}Ba + ^{91}_{36}Kr + 3\,^{1}_{0}n \end{cases}$$

The fission of a uranium-235 nucleus gives an average of 2.4 neutrons. Because each fission produces additional neutrons, the reaction once initiated continues as a chain reaction (Chapter 15) that may be self-sustaining. Some of the neutrons produced in such a reaction inevitably escape to the surroundings. If enough escape so that less than one neutron per fission is available to cause another fission, the reaction will slow down and eventually stop. However, if more than one neutron per fission is available to cause further fissions, the reaction then becomes a *branching chain reaction* (Figure 18.10), and its speed continually increases. Indeed, if more than one neutron per fission is available to cause further fissions, the rate of the reaction increases so rapidly that an explosion may result. The number of neutrons that escape without causing fission depends on the surface-to-volume ratio of the uranium—the smaller the volume of the uranium, the more neutrons escape. So there is a critical mass of uranium, or of any fissionable element, below which so many neutrons escape that a branching chain reaction cannot occur. However, once

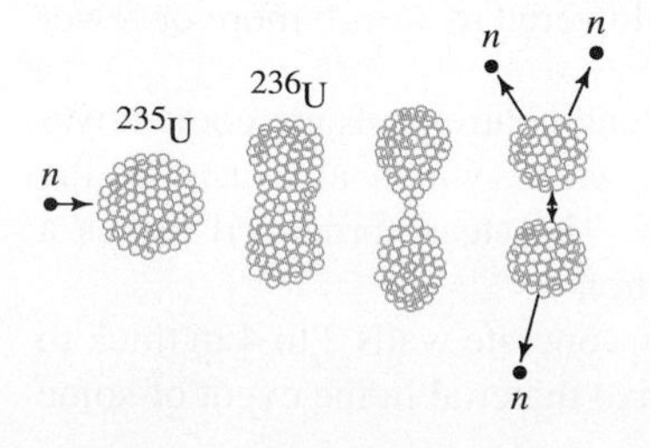

FIGURE 18.9 The Neutron-Induced Fission of Uranium-235 The absorption of a neutron produces a uranium-236 nucleus in an excited state. The oscillations of this nucleus in its excited state cause it to split apart in much the same way as a vibrating drop of liquid might split in two. Simultaneously, several neutrons are emitted.

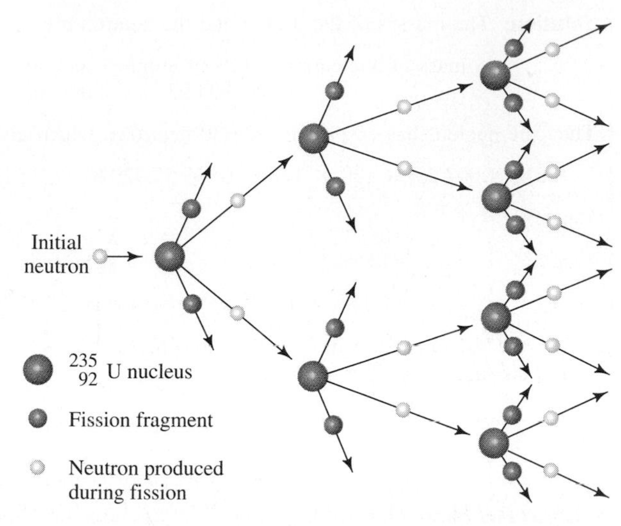

FIGURE 18.10 Fission of Uranium-235 A self-sustaining branching chain reaction. Each fission produces two or more neutrons. These two neutrons cause the fission of two uranium-235 nuclei, which produces four neutrons. The neutrons then cause the fission of four uranium-235 nuclei, and so on. Unless some of the neutrons are lost from the sample or are removed in other reactions, the number of fissions increases very rapidly, resulting in an explosive release of energy.

this mass is exceeded, an explosion inevitably results. This happens, for example, in an atomic bomb in which two ''subcritical'' masses are brought together by means of a conventional explosion to form a ''supercritical'' mass. In a nuclear reactor, however, the rate of the reaction is carefully controlled by limiting the supply of neutrons.

Nuclear Reactors

Uranium is the only naturally occurring element that has an isotope—^{235}U—capable of undergoing fission, so most reactors use uranium as a fuel. Uranium is usually used in the form of pellets of the oxide UO_2 placed in a zirconium alloy tube to give a **fuel element**, several of which are packed together to form the **reactor core** (Figure 18.11). Naturally occurring uranium is a mixture of isotopes. One of these isotopes, ^{238}U, strongly absorbs high-energy neutrons and is converted in several steps to plutonium. If this is allowed to happen, too few neutrons are available to sustain the fission of ^{235}U. However, if these fast-moving neutrons are slowed by collisions with other atoms, they are no longer absorbed by ^{238}U but are still effective in decomposing ^{235}U. For this reason the fuel elements are surrounded by a **moderator**, a substance such as water, heavy water, or graphite that slows down neutrons. A good moderator should have light atoms to optimize the transfer of kinetic energy but should not itself react with neutrons. The best moderator is heavy water, D_2O, which does not react at all with neutrons. ''Light'' water, H_2O, or graphite are also commonly used, but they do react with some of the neutrons; it is then necessary to enrich the fuel in ^{235}U to provide more neutrons. The supply of neutrons is regulated by **control rods** made from materials that readily absorb neutrons, such as steel containing boron or cadmium. The rods are inserted into the reactor between the fuel elements and are raised or lowered to absorb more or fewer neutrons, as required.

In commercial nuclear reactors, the fuel rods and control rods are cooled by a system of circulating water under pressure. This water, which also acts as the moderator, then passes through a steam generator. The steam produced drives a turbine, as in a conventional power generating station.

The core of a nuclear reactor is surrounded by concrete walls 2 to 4 m thick to absorb radiation and prevent the escape of radioactive material in the event of some

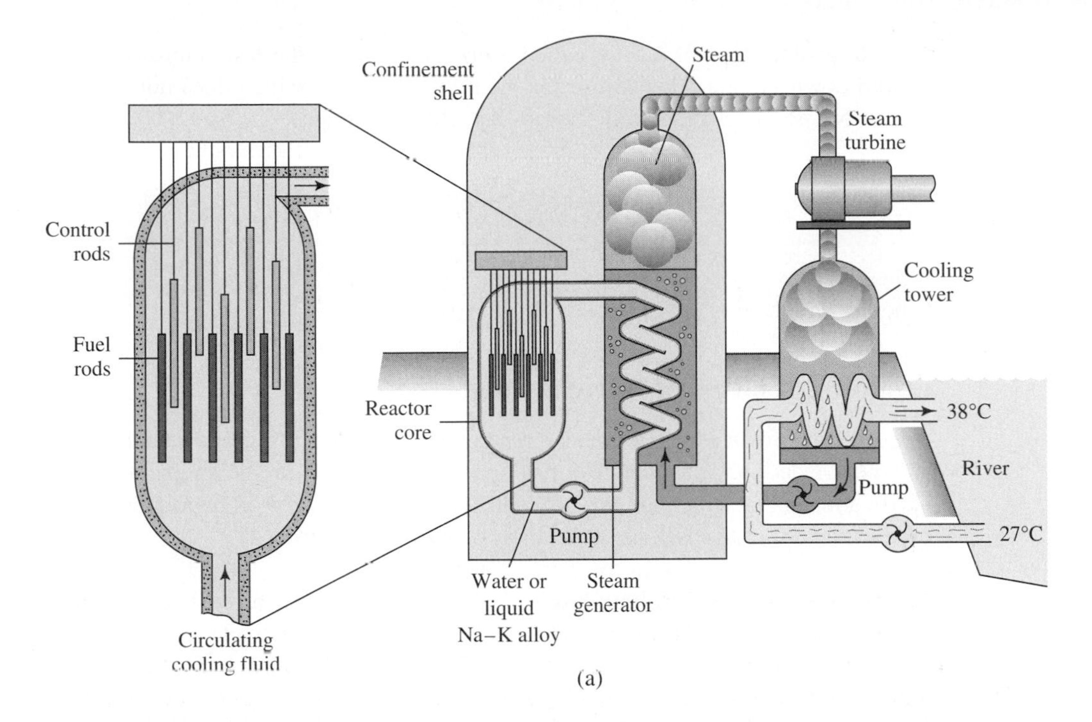

(a)

(b)

FIGURE 18.11 A Nuclear Reactor (a) The heat produced in the core is transferred to water that is heated to its boiling point, forming steam. The steam is used to drive a turbine, which produces electricity. The steam from the turbine is condensed to water in a cooling tower, and the resulting water is pumped back into the steam generator. (b) The nuclear reactor and processing plant at Sellafield, England.

malfunction. In addition, the reactor is housed in a building made of 1-m thick steel-reinforced concrete designed to withstand the force of a chemical explosion or an earthquake.

In another type of reactor called a *breeder reactor*, the fast neutrons are not slowed down but are allowed to react with uranium-238, which does not undergo fission but produces plutonium-239 in the following stages:

$$^{238}_{92}U + ^{1}_{0}n \rightarrow ^{239}_{92}U$$

$$^{239}_{92}U \rightarrow ^{239}_{93}Np + ^{0}_{-1}e \qquad t_{\frac{1}{2}} = 23.4 \text{ min}$$

$$^{239}_{93}Np \rightarrow ^{239}_{94}Pu + ^{0}_{-1}e \qquad t_{\frac{1}{2}} = 2.35 \text{ days}$$

When ^{239}Pu absorbs a proton, it undergoes fission like ^{235}U.

Chemical Aspects of Nuclear Power

The energy produced by a nuclear reactor comes from a nuclear rather than a chemical reaction. But chemical reactions are involved in the preparation and purification of the uranium fuel as well as in the recovery of important fission products and the disposal and utilization of waste products.

Preparation of the Nuclear Fuel Uranium is a member of the actinide series of elements, which, like the transition metals, have several oxidation states. The most stable oxidation states of uranium are IV and VI. Uranium occurs mainly as the mineral pitchblende, which is impure uranium(IV) oxide, UO_2. Pure UO_2 is obtained by oxidizing the UO_2 in pitchblende to the uranium(VI) compound uranyl nitrate, $UO_2(NO_3)_2$, by reaction with nitric acid. The uranyl nitrate is purified by recrystallization and then heated to give uranium(VI) oxide, UO_3:

$$2UO_2(NO_3)_2(s) \rightarrow 2UO_3(s) + 4NO_2(g) + O_2(g)$$

UO_3 is then reduced back to pure UO_2 by reaction with hydrogen.

Exercise 18.7 When nitric acid oxidizes UO_2, it is reduced to $NO(g)$. Write a balanced equation for this reaction.

Isotopic Enrichment Natural uranium contains only 0.71% of the fissionable isotope ^{235}U. If natural uranium is to be the fuel in a reactor, very efficient use must be made of the few neutrons available. To do that, the most efficient known moderator, heavy water, is required, but heavy water is expensive. The alternative is to enrich the uranium in ^{235}U so that ordinary water can be used as a moderator. To enrich the uranium isotopically, the UO_2 is converted to uranium hexafluoride, $UF_6(g)$, and the isotopes are separated by diffusion (Chapter 4). Uranium hexafluoride is made from UO_2 in two steps. First the $UO_2(s)$ is converted to $UF_4(s)$ by reaction with $HF(aq)$:

$$UO_2(s) + 4HF(aq) \rightarrow UF_4(s) + 2H_2O(l)$$

This is an acid–base reaction between a basic oxide and an acid, just like the reaction of $CaO(s)$ with $HCl(aq)$: In all such reactions, the strong base O^{2-} reacts with hydrogen ions to give water. Then the uranium(IV) fluoride, $UF_4(s)$, is oxidized to uranium(VI) fluoride, $UF_6(g)$, by reaction with fluorine:

$$UF_4(s) + F_2(g) \rightarrow UF_6(g)$$

After enrichment by diffusion, the UF_6 is converted back to UO_2 by reduction with hydrogen:

$$UF_6(g) + H_2(g) + 2H_2O(l) \rightarrow UO_2(s) + 6HF(aq)$$

The separation of ^{235}U by diffusion depends on the very slight difference in the diffusion rates of the two isotopes:

$$\frac{\text{rate of diffusion of } ^{235}UF_6}{\text{rate of diffusion of } ^{238}UF_6} = \frac{\sqrt{\text{molar mass } ^{238}UF_6}}{\sqrt{\text{molar mass } ^{235}UF_6}} = \frac{\sqrt{352 \text{ g} \cdot \text{mol}^{-1}}}{\sqrt{349 \text{ g} \cdot \text{mol}^{-1}}} = 1.004$$

To achieve an appreciable enrichment of ^{235}U, it is necessary to repeat the diffusion through porous barriers thousands of times—a technically difficult process that requires a large amount of energy. Scientists and engineers continue to look for better enrichment procedures.

Nuclear Fusion

We see from Figure 18.8 that a large amount of energy can be obtained by fusing hydrogen nuclei to give heavier nuclei. The principal difficulty is that to overcome the electrostatic repulsion between them, the protons must collide with very high kinetic energies—energies equivalent to temperatures of several million degrees. Such temperatures are obtained in a hydrogen bomb by exploding a fission bomb. To use a fusion reaction for the production of power, these high temperatures must be achieved in another way. One method is to pass a very large electric current through $H_2(g)$ to transform the gas into a **plasma**, a state of matter in which all the molecules are decomposed to nuclei and free electrons moving at very high speeds. Rather than using ^1H, it is advantageous to use the heavy isotopes deuterium, ^2H, or tritium, ^3H, because the attraction resulting from the additional neutrons helps overcome the repulsion between protons. For example, in a plasma deuterium undergoes the reactions

$$^2_1H + ^2_1H \rightarrow ^3_2H + ^1_0n$$

$$^2_1H + ^2_1H \rightarrow ^3_1H + ^1_1H$$

$$^2_1H + ^3_1H \rightarrow ^4_2He + ^1_0n$$

$$^2_1H + ^3_2He \rightarrow ^4_2He + ^1_1H$$

The overall reaction is the fusion of 6 deuterium nuclei to form 2 helium nuclei, 2 protons, and 2 neutrons:

$$6^2_1H \rightarrow 2^4_2He + 2^1_1H + 2^1_0n$$

Exercise 18.8 Find the amount of energy released when 1 mol of deuterium undergoes fusion according to the reactions just listed.

Another serious practical difficulty is to contain the high-temperature plasma, which would vaporize any ordinary container. There have been two approaches to solving this problem. The original idea, which had some early success, was to contain the plasma with very strong magnetic fields, forming what has been called a "magnetic bottle" to isolate the electrically charged particles from the surroundings. A more recent approach has been to use powerful lasers to pump an enormous amount of energy into a tiny volume of material in a very short period of time. Again some success has been claimed, but the consensus of scientists is that practical power from fusion is still at least several decades in the future.

Nuclear Power: Risks and Benefits

The development of nuclear reactors for research and for power generation has brought great benefits but has raised social, ethical, economic, and political problems that have not yet been answered. The introduction of new technologies has often been opposed on the grounds that the benefits are not worth the risks, but nuclear power has generated more continuous opposition than most new technologies.

What are the benefits and risks of nuclear reactors and nuclear power? Nuclear power does not pollute the atmosphere with carbon dioxide or the oxides of sulfur and nitrogen that cause acid rain. It does not use up our rapidly diminishing stocks of fossil fuels. Moreover, nuclear reactors are used to produce radioisotopes, which as we have seen, have many important uses.

But how safe are nuclear reactors, and are they an efficient way to generate power? A major problem is the disposal of the spent fuel elements, which must be replaced regularly. After a few years in use, they contain too many reaction products that absorb neutrons and insufficient ^{235}U to sustain a fission reaction. Of the many radioactive isotopes produced, some are useful and may be recovered, some are short-lived and can be allowed to decay in a controlled environment, and some are long-lived and must be disposed of. Plutonium is removed because, as we have seen, it can also be used as a reactor fuel. The remaining radioactive waste is stored until the short-lived radioisotopes have decayed so that their radioactivity is at a safe level. The preferred but still controversial disposal method for long-lived radioactive isotopes is to seal the fuel in a container and bury it in a geologically stable site.

An even more controversial issue is the safety of nuclear reactors. Nuclear reactors for power generation are built with many safety devices designed to shut down the reactor if the operating conditions become abnormal. Despite such devices, accidents at Three Mile Island in the United States in 1979 and at Chernobyl in the Ukraine in 1986 have caused extensive public concern. In both cases the safety devices were neutralized by human errors. The Chernobyl accident was by far the more serious, leading to about 30 immediate deaths and a large but unknown number of serious cases of radiation sickness and subsequent deaths. The explosion at Chernobyl was chemical in origin; it was not a nuclear explosion like that produced by a nuclear bomb. Overheating of the reactor raised the temperature of steam sufficiently that it reacted with the zirconium casing of the fuel elements and with other materials used in the construction of the reactor, producing hydrogen. We saw in Chapter 10 (Figure 10.7) that steam reacts with the more reactive metal magnesium. When hydrogen at high temperatures mixes with air, the very rapid formation of water causes a powerful explosion. The explosion at Chernobyl was strong enough to destroy the reactor container and release large amounts of radioactive material into the atmosphere.

Advocates of nuclear power point out that the death toll from nuclear accidents is far lower than has resulted from breakdowns of hydroelectric dams or from coal-mining accidents. Opponents stress the potential for catastrophic explosions and widespread radioactive contamination; as a result there is now considerable public opposition to the development of new nuclear power plants. Do the benefits outweigh the risks? You will have to help make this decision. The better informed you are, the more you will be able to contribute to the discussion.

SUMMARY

Nuclear chemistry studies radioactive atoms, unstable atoms that spontaneously disintegrate, and nuclear reactions, reactions in which nuclei are changed into other nuclei. Most atoms of atomic number $Z \leq 83$ have one or more stable isotopes (nuclides), in which attractive nuclear forces between nucleons outweigh strong interproton repulsions. Stable nuclei are in a band of stability with neutron-to-proton ratios increasing from 1.0 to 1.5 with increasing mass, but for $Z > 83$ no increase in the neutron-to-proton ratio gives a stable nucleus. The four radioactive processes are α-particle emission; β-particle emission; positron emission, and electron capture; we can summarize these processes as

$$\alpha\text{-particle emission: } {}^{A}_{Z}R \rightarrow {}^{A-4}_{Z-2}\text{P} + {}^{4}_{2}\text{He}$$

$$\beta\text{-particle emission: } {}^{A}_{Z}\text{R} \rightarrow {}_{Z+1}{}^{A}\text{S} + {}_{-1}^{0}e$$

$$\text{positron emission: } {}^{A}_{Z}\text{R} \rightarrow {}_{Z-1}{}^{A}\text{Q} + {}^{0}_{1}e$$

$$\text{electron capture: } {}^{A}_{Z}\text{R} + {}_{-1}^{0}e \rightarrow {}_{Z-1}{}^{A}\text{Q}$$

Product nuclei are sometimes formed in an excited (metastable) state and emit high-energy γ rays to return to their ground states. A radioactive decay series is a series of steps in which a radioactive nucleus decays to another radioactive nucleus, which also decays, and so on, until a stable nucleus finally results. Methods of measuring radioactivity include the exposure of photographic films or plates and using a Geiger counter.

Biological effects of radioactivity depend on the activity of the nuclide, the type and energy of the radiation, its penetrating power ($\gamma > \beta > \alpha$), and the exposure time. The activity of a radioactive nuclide is measured in curies (1 Ci = 3.7×10^{10} disintegrations per second), or millicuries (mCi).

Radioactive decay is a first-order process with rate $= kN$, where N is the number of radioactive nuclei. Thus, $kt_{\frac{1}{2}} = 0.693$, where $t_{\frac{1}{2}}$ is the half-life and k is the rate constant, which is independent of temperature. For a radioactive nuclide we can also write

$$\ln \frac{N_0}{N} = \ln \frac{m_0}{m} = kt$$

where m_0 is the initial mass of a sample of a radioactive nuclide and m is the mass that remains after time t.

Radioactive nuclides, often called radioisotopes, have many applications. ^{40}K (potassium/argon dating) and ^{238}U and ^{235}U are used in determining the age of rocks, and ^{14}C is used to date archeological remains. Radioisotopes are used also in the study of reaction mechanisms by using labeled compounds. In medicine, ^{131}I allows us to study the thyroid gland; ^{24}Na, the flow of blood in the body; and ^{99}Tc, the brain. ^{60}Co and ^{131}I are utilized in cancer treatment.

Artificial radioisotopes result from nuclear reactions produced by neutrons from a nuclear reactor or energetic protons or α particles from a particle accelerator. Transuranium elements are highly radioactive artificial radioisotopes with atomic numbers $Z = 93$ to 109.

A mass defect Δm is the mass difference that results when protons and neutrons combine to form a nucleus. It is given by $\Delta E = (\Delta m)c^2$, where ΔE is the nuclear binding energy and c is the speed of light. The binding energy per nucleon compares the relative stability of different nuclei and is a maximum for ^{56}Fe. Enormous amounts of energy come from the conversion of mass to energy when the mass of the reactant nuclides is greater than the mass of the products of a nuclear reaction. This occurs by fusion, in which elements lighter than Fe are synthesized from smaller nuclei, or fission, in which heavy nuclei decompose into smaller nuclei.

A neutron reacts with a ^{235}U nucleus to give several lighter nuclei and an average of 2.4 neutrons. Some of these neutrons escape, but one available neutron per fission leads to a continually self-sustaining branching chain reaction; more than one leads to a rapidly increasing rate and ultimately an explosion.

In a nuclear reactor fuel elements consist of $UO_2(s)$ pellets packed into zirconium alloy tubes. These tubes form the reactor core and are surrounded by a moderator (water, heavy water, or graphite) to slow down the high-energy neutrons from ^{235}U, which would otherwise react with ^{238}U to give ^{239}Pu. Movable control rods, which absorb neutrons, control the rate of the reaction of neutrons with ^{235}U. The heat produced in the reactor core is carried away and used to produce electricity.

Pure $UO_2(s)$ is obtained from the mineral pitchblende (impure UO_2) by the series of (unbalanced) reactions pitchblende $\rightarrow UO_2(NO_3)_2 \rightarrow UO_3 \rightarrow UO_2$. This UO_2 can be used directly as a fuel if the moderator is very efficient. Otherwise it is converted to $UF_6(g)$, which is enriched in ^{235}U by diffusion and then reconverted to $UO_2(s)$.

To achieve nuclear fusion, nuclei must be forced together with great energy to overcome their enormous electrostatic repulsion. Fusion requires temperatures of several million degrees and presents the problem of "containing" the reaction if it is to be a practical source of power.

IMPORTANT TERMS

alpha(α)-particle emission (page 654)
artificial radioisotope (page 667)
band of stability (page 654)
beta(β)-particle emission (page 654)
binding energy (page 671)
electron capture (page 655)
fission (page 672, 673)
fuel element (page 674)
fusion (page 672, 677)
gamma(γ)-ray emission (page 655)
Geiger counter (page 658)
half-life, $t_{\frac{1}{2}}$ (page 662)
mass defect (page 671)
moderator (page 674)
nuclear energy (page 671)
nuclear power (page 678)
nuclear reaction (page 653)
nuclear reactor (page 674)
nuclide (page 653)
plasma (page 677)
positron emission (page 655)
radioactive decay (page 654)
radioactive decay constant (page 662)
radioactive decay series (page 656)
radioactivity (page 653)
radiocarbon dating (page 665)
radioisotope (page 653)
transuranium element (page 670)

REVIEW QUESTIONS

1. What is radioactivity?

2. **(a)** Which elements have at least one stable isotope? **(b)** Why are some isotopes unstable?

3. What are the mass numbers and electric charges of **(a)** an α particle; **(b)** a β particle; **(c)** a positron?

4. How do the atomic number Z and nucleon number A of a nuclide ${}^{A}_{Z}X$ change in **(a)** α-particle emission; **(b)** β-particle emission; **(c)** positron emission; **(d)** electron capture?

5. What are nuclear forces?

6. When an unstable nuclide decays, how is a more favorable neutron-to-proton ratio achieved for nuclei **(a)** above and **(b)** below the band of stability?

7. How is the amount of radioactivity measured with a Geiger counter?

8. What is a metastable nucleus? Give an example.

9. **(a)** What process results in γ-ray emission?
(b) Why is γ radiation the most damaging type of radiation?

10. What rate law governs the decay of radioactive nuclei?

11. What is the relationship between $t_{\frac{1}{2}}$, the half-life of a radioisotope, and k_1, the rate constant for its decay?

12. In determining the age of rocks by potassium/argon radioactive dating, why must the amounts of both ^{40}K and ^{40}Ar in a sample be measured?

13. **(a)** How is the ^{14}C isotope formed in the atmosphere? **(b)** Why is it useful for archeological dating?

14. What is a transuranium element? Name two of these elements.

15. What are **(a)** nuclear fission and **(b)** nuclear fusion?

16. **(a)** What is the binding energy of a nucleus?
(b) Which nuclide has the largest binding energy per nucleon?

17. Explain how both nuclear fission and nuclear fusion can lead to the release of enormous amounts of energy.

18. How is an explosive chain reaction that results from the fission of ^{235}U prevented in a nuclear reactor?

19. How is natural UO_2 enriched in ^{235}U?

PROBLEMS

Radioactivity

1. Give the number of protons, neutrons, and nucleons in each of the following nuclei.
(a) ${}^{11}_{5}B$ **(b)** ${}^{13}_{6}C$ **(c)** ${}^{94}_{40}Zr$ **(d)** ${}^{137}_{56}Ba$ **(e)** ^{57}Fe

2. Repeat Problem 1 for the following nuclei.
(a) ${}^{2}_{1}H$ **(b)** ${}^{109}_{47}Ag$ **(c)** ${}^{90}_{38}Sr$ **(d)** ${}^{197}_{79}Au$ **(e)** ^{128}I

3. What is the product nucleus when a ^{80}Br nucleus decays by **(a)** β emission; **(b)** positron emission; **(c)** electron capture?

4. When the ^{233}Th nucleus undergoes two successive β-particle emissions, which isotope of uranium is formed?

5. The ^{207}Po isotope of polonium can decay in three ways: **(a)** by electron capture, **(b)** by positron emission, and **(c)** by α-particle emission. Write a balanced equation for each reaction.

6. Identify the nuclides formed in each of the following.
(a) β decay of tritium, ${}^{3}_{1}H$ **(b)** β decay of carbon-14
(c) α decay of radium-226 **(d)** α decay of gold-179
(e) positron decay of bromine-80
(f) electron capture of iridium-188

7. Complete each of the following equations.

(a) $^{32}_{15}P \rightarrow ^{32}_{16}S + ?$

(b) $^{15}_{8}O \rightarrow ^{15}_{7}N + ?$

(c) $^{52}_{26}Fe \rightarrow ^{52}_{25}Mn + ?$

(d) $^{218}_{87}Fr \rightarrow ? + ^{4}_{2}He$

(e) $^{50}_{26}Fe \rightarrow ? + ^{0}_{-1}e$

(f) $^{122}_{53}I \rightarrow ^{122}_{54}Xe + ?$

8. A container in which radium-226 was sealed was later found to contain a highly radioactive gas. The radioactive gas decays by emitting α particles. Write equations for the nuclear reactions that have taken place.

9. Thorium-231 decays to lead-207 by emitting the following particles in successive steps. Write the symbol for the isotope formed in each step.

(a) β (b) α (c) α (d) β (e) α

(f) α (g) α (h) β (i) β (j) α

10. The nucleus of nitrogen-18 lies above the band of stability. Suggest a nuclear reaction that will give an isotope of greater stability, and write the equation for this reaction.

11. For each of the following unstable nuclides, suggest what type of single-step radioactive decay each will undergo, and name the product nuclide.

(a) ^{10}Be (b) ^{12}N (c) ^{14}C (d) ^{87}Br

12. Repeat Problem 11 for each of the following unstable nuclides.

(a) ^{3}H (b) ^{24}Na (c) ^{27}Mg (d) ^{140}Xe

13. Express the activity of the radioactive sources with each of the following disintegration rates in curies, Ci, and in millicuries, mCi.

(a) 3.70×10^{10} disintegrations s^{-1}

(b) 3.50×10^{6} disintegrations s^{-1}

(c) 7.00×10^{12} disintegrations s^{-1}

(d) 4.80×10^{16} disintegrations s^{-1}

14. Measurement of the activity of various radioactive samples with a Geiger counter gave the following results. Express each in curies.

(a) 3 clicks · s^{-1} (b) 154 clicks in 10 s

(c) 1440 clicks in 1 min (d) 1.4×10^{4} clicks in 1.0 h

15. Calculate the dose of radiation (in rad) and the dose equivalent (in rem) when a 65.0-kg man absorbs 1×10^{-2} J of energy as the result of exposure to **(a)** β radiation and **(b)** α radiation. Assume that the relative biological effectiveness (RBE) is 1 for β and for γ radiation and 20 for α radiation.

16. The onset of radiation sickness in humans is apparent at a dose equivalent of about 100 rem. When should the symptoms become apparent for each of the following exposures?

(a) α radiation with a dose rate of 0.50 rad per day

(b) β radiation with a dose rate of 2.5 mrad per day
(See Problem 15 for other data.)

Radioactive Decay Rates

17. A container holds 1.00 Ci of a radioactive isotope that decays to stable products. How many curies of radiation are left after five half-lives?

18. Germanium-66 decays by positron emission with a half-life of 9.40 h.

(a) Write the balanced equation for this nuclear reaction.

(b) How much ^{66}Ge remains from a 50.0-mg sample after 47.0 h?

19. A sample of a radioactive nuclide is recorded to have 668 disintegrations per minute. After 60 min, the number of disintegrations diminished to 25 per minute. What is the half-life of the radioactive substance?

20. **(a)** If the half-life of ^{238}U is 4.5×10^{9} yr, what is the rate constant for its disintegration?

(b) If 40.0% of the ^{238}U originally in a rock has decayed, how old is the rock?

21. A cobalt source in a cancer hospital contains a mass of 500 g of ^{60}Co (half-life 5.26 yr). What mass of ^{60}Co will remain after **(a)** 1.00 yr; **(b)** 5.00 yr; **(c)** 10.00 yr?

22. The half-life of plutonium-239 is 2.40×10^{4} yr. What fraction of this plutonium in the nuclear wastes generated today will remain by A.D. 4000?

23. Two of the radioactive isotopes found in the fallout from nuclear explosions or accidents are ^{131}I, with a half-life of 8.0 days, and ^{90}Sr, with a half-life of 19.9 yr. How long will it take each isotope to decay to **(a)** 10% and **(b)** 1% of its initial concentration? **(c)** Which isotope has the most serious long-term effects?

24. Cesium-137, with a half-life of 30.2 yr, is produced from fission reactions in nuclear reactors. As a result of the Chernobyl accident in the Ukraine, ^{137}Cs was spread over large areas of Europe. How long will it take for the activity to decrease to 1.0% of its value immediately after the accident?

25. A sample known to contain 3.40 mg of radioactive ^{32}P originally was found to contain only 2.09 mg of ^{32}P after 10.0 days. What is the half-life of ^{32}P?

26. One of the principal sources of radioactivity inside your body is the disintegration of ^{40}K, one of the naturally occurring isotopes of potassium. What is the half-life of ^{40}K, given that only 7.0% remains of the ^{40}K present when the earth formed 4.5 billion yr ago?

27. A sample containing ^{42}K, a radioactive isotope with a half-life of 12.4 h, has an initial activity of 1.10×10^{9} disintegrations per minute.

(a) Calculate the activity after 30.0 h.

(b) How long will it be before the initial activity drops to 1.00×10^{5} disintegrations per minute?

28. An ancient wooden casket has radioactivity from the ^{14}C it contains corresponding to 9.6 disintegrations per gram of carbon per minute. What is its age?

29. A sample of charcoal from the Lascaux cave in France has a count rate of 2.4 disintegrations per gram per minute. Assuming that the fire that produced the charcoal was lit by the artists of the renowned cave paintings, when did these artists live?

Artificial Radioisotopes

30. Write equations for the nuclear reactions involved in preparation of **(a)** cobalt-60 from iron-58 and **(b)** technetium-99 from molybdenum-98.

31. Complete each of the following equations.

(a) $^{238}_{92}U + ^{1}_{0}n \rightarrow ^{141}_{56}Ba + ^{92}_{36}Kr + ?$

(b) $^{238}_{92}U + ^{1}_{0}n \rightarrow ^{103}_{42}Mo + ^{131}_{50}Sn + ?$

32. Fill in the missing symbols in each of the following equations for nuclear reactions.

(a) $^{35}_{17}Cl + ? \rightarrow ^{32}_{16}S + ^{4}_{2}He$

(b) $^{15}_{7}N + ? \rightarrow ^{12}_{6}C + ^{4}_{2}He$

(c) $^{12}_{6}C + ^{12}_{6}C \rightarrow ? + ^{1}_{1}H$

(d) $? + ^{4}_{2}He \rightarrow ^{7}_{4}Be + \gamma$

33. If the nuclei $^{209}_{83}Bi$ and $^{58}_{26}Fe$ were fused, what isotope would be produced?

34. Complete each of the following nuclear reactions, and name the transuranium element formed.

(a) $^{238}_{92}U + ^{4}_{2}He \rightarrow ^{239}_{94}Pu + ?$

(b) $^{239}_{94}Pu + ? \rightarrow ^{242}_{96}Cm + ^{1}_{0}n$

(c) $^{238}_{92}U + ^{1}_{0}n \rightarrow ^{239}_{93}Np + ?$

(d) $^{239}_{94}Pu + ^{1}_{0}n \rightarrow ? + ^{0}_{-1}e$

35. The transuranium radioactive isotope of curium ^{240}Cm is synthesized by bombarding the isotope ^{232}Th of thorium with the nuclei of carbon atoms. Write a balanced equation for this reaction.

36. Explain how you might use the radioactive ^{59}Fe isotope (a β emitter with a half-life of 46 days) to determine the extent to which a particular iron compound in the diet of rabbits is used to form hemoglobin?

37. The following equation depicts the reaction that occurs when sparingly soluble lead(II) iodide dissolves in water:

$$PbI_2(s) \rightleftharpoons Pb^{2+}(aq) + 2I^-(aq)$$

How would you use a radioactive iodine isotope to show that when the reaction has achieved equilibrium, it is still proceeding in both the forward and reverse directions?

38. For a system at equilibrium, no change with time is observed in the concentrations of any of the reactants, nor in any of the products. Explain how you could use the radioactive isotope ^{131}I ($t_{\frac{1}{2}}$ = 8.0 days) to show that the equilibrium

$$H_2(g) + I_2(g) \rightleftharpoons 2HI(g)$$

is a dynamic process—in other words, that the forward and reverse reactions occur at the same rate.

Nuclear Energy

39. (a) Explain the terms "mass defect" and "nuclear binding energy."

(b) What is the relationship between them?

40. The mass of chlorine-35 is 34.9689 u. Calculate its (a) binding energy and (b) binding energy per nucleon.

41. (a) Is the reaction $^{7}_{3}Li + ^{1}_{1}H \rightarrow 2^{4}_{2}He$ exothermic or endothermic?

(b) How much energy is absorbed or liberated in this reaction?

42. Every second, the sun radiates 3.9×10^{23} J of energy into space. By how much does the mass of the sun decrease each year?

43. The isotope $^{56}_{26}Fe$ has the highest binding energy of all atomic nuclei. Would you expect this isotope to be a good nuclear fuel? Explain your answer.

44. (a) Calculate the binding energy per nucleon for ^{14}N and for ^{15}N, which have atomic masses 14.003 07 u and 15.000 11 u, respectively.

(b) What do you conclude about the relative stability of these two isotopes?

(c) Which isotope has the greater natural abundance?

45. An oxygen-16 atom has a mass of 15.994 91 u.

(a) Is its mass greater than, equal to, or less than that of its constituent protons, neutrons, and electrons?

(b) Calculate the binding energy of an ^{16}O nucleus.

46. Determine which of the ^{12}C, ^{13}C, and ^{14}C isotopes of carbon is the most stable.

47. Calculate the binding energies per nucleon of (a) a ^{238}U nucleus and (b) a ^{56}Fe nucleus.

(c) Determine which is the more stable.

General Problems

48. Uranium-235 undergoes radioactive decay by α emission to give A; A decays by β emission to give B; B decays by α emission to give C; and C decays by β emission to give D. Write balanced equations for these four steps in the decay scheme for ^{235}U.

49. Explain (a) branching chain reactions and (b) critical mass.

50. Explain the functions in a nuclear reactor of (a) a moderator and (b) the control rods.

51. For a controlled fusion process, a possible nuclear reaction is the conversion of deuterium ^{2}H to the helium isotope ^{3}He:

$$^{2}_{1}H + ^{2}_{1}H \rightarrow ^{3}_{2}He + ^{0}_{1}n$$

Compare the energy released per mole of deuterium consumed in the nuclear reaction with that obtained by burning deuterium in oxygen to give heavy water, $^{2}H_2O$. Assume that the enthalpy of combustion of deuterium is the same as that of natural $H_2(g)$.

52. ^{226}Ra, with a half-life of 1.6×10^3 yr, emits α particles, which readily pick up electrons to form He(g). For a particular sample, the initial rate of emission of α particles is 9.27×10^{12} s^{-1}, and the amount of He(g) collected in 3 months is 2.97 mL at 25°C and 1 atm. Calculate a value for Avogadro's number.

53. The following series of reactions is the *Bethe chain*, responsible for the energy production of some stars:

$^{12}C + ^{1}H \rightarrow ^{13}N$

$^{13}N \rightarrow ^{0}_{1}e + X$

$X + ^{1}H \rightarrow Y$

$Y + ^{1}H \rightarrow Z$

$Z \rightarrow ^{0}_{1}e + ^{15}N$

$^{15}N + ^{1}H \rightarrow ^{12}C + ^{4}He$

(a) Identify the nuclides X, Y, and Z in these equations. (b) Give the equation for the overall reaction.

54. Either complete or give the balanced equation for each of the following nuclear reactions.

(a) ${}^{1}_{1}H + {}^{35}_{17}Cl \rightarrow {}^{4}_{2}? + ?$

(b) β emission from ${}^{60}_{27}Co$

(c) α emission from ${}^{238}_{90}Th$

(d) ${}^{32}_{16}S + {}^{1}_{0}n \rightarrow {}^{1}_{1}H + ?$

(e) ${}^{1}_{1}H + {}^{2}_{1}H \rightarrow ? + \gamma$

(f) ${}^{238}_{92}U + {}^{12}_{6}C \rightarrow ? + 4{}^{1}_{0}n$

55. Calculate (i) the total binding energy and (ii) the binding energy per nucleon for each of the following nuclides.

(a) ^{20}Ne, mass 19.992 44 u

(b) ^{64}Zn, mass 63.929 14 u

(c) ^{61}Ni, mass 60.930 06 u

(d) ^{226}Ra, mass 226.0254 u

56. **(a)** What is the band of stability for isotopes?

(b) What relationship exists between the number of protons and the number of neutrons for stable nuclei of low atomic numbers?

(c) What is the neutron-to-proton ratio for the heaviest stable nuclei?

(d) Why are nuclei with $Z > 83$ unstable?

57. Why can the concentration of ^{14}C be taken as constant per gram of carbon in living organisms?

58.* Technetium is in the same group of transition metals as manganese.

(a) Predict the highest oxidation state of technetium.

(b) Write the formula and the name of the oxoacid and the fluoride with technetium in this oxidation state.

59.* Explain how the ^{18}O isotope could be used to determine if the oxygen in the ethanol formed upon the hydrolysis of ethyl acetate (ethanoate) comes from the water or from the ester.

CHAPTER 19 Cosmochemistry and Geochemistry

19.1 The Origin of Atoms and Molecules

19.2 The Formation and Composition of the Solar System

19.3 The Structure and Composition of the Earth

19.4 The Biosphere

Humans have long been fascinated by the universe around us and by the earth beneath us; these are the realms of cosmochemistry and geochemistry, respectively. Solar eclipses were a mystery to ancient peoples, who feared that the sun might not return. To scientists today, solar eclipses provide an opportunity to observe the outer regions of our nearest star. This view of a solar eclipse shows streams of gas, consisting mainly of hydrogen atoms, that are ejected from the sun's surface.

Metals / Nonmetals / Semimetals

Period \ Group	I	II											III	IV	V	VI	VII	VIII
1	H																	He
2	Li	Be											B	C	N	O	F	Ne
3	Na	Mg											Al	Si	P	S	Cl	Ar
4	K	Ca	Sc	Ti	V	Cr	Mn	Fe	Co	Ni	Cu	Zn	Ga	Ge	As	Se	Br	Kr
5	Rb	Sr	Y	Zr	Nb	Mo	Tc	Ru	Rh	Pd	Ag	Cd	In	Sn	Sb	Te	I	Xe
6	Cs	Ba	La	Hf	Ta	W	Re	Os	Ir	Pt	Au	Hg	Tl	Pb	Bi	Po	At	Rn
7	Fr	Ra	Ac	104	105	106	107	108	109									

The biosphere that we inhabit, the entire earth, and indeed the solar system, are made of about 100 elements—100 different kinds of atoms. Where did these atoms come from? What atoms do we find in the sun and other stars and in interstellar space? When and where did the first molecules appear? What substances comprise the surface and interior of the earth, and how does their composition differ from that of other planets? How did molecules on the surface of the earth begin to replicate and organize themselves into living matter?

In this chapter we explore the chemical history of our world from its beginnings to the present to see what answers—some still speculative—we can give to these fascinating and fundamental questions. Two important branches of chemistry deal with such questions: **cosmochemistry**, the chemistry of the universe outside the earth, and **geochemistry**, the chemistry associated with the formation and structure of the earth and the changes occurring in it. We begin with the synthesis of the lightest atomic nuclei from which the stars were formed and then proceed to the synthesis of the heavier elements in the stars. We see that molecules first formed in interstellar space and often occur in enormous clouds of gaseous molecules and fine solid particles, one of which gave birth to our solar system. We examine the composition of the planets, particularly that of our own planet. Finally, in considering the origin of life, we return to biochemistry, which we discussed in Chapter 17.

19.1 THE ORIGIN OF ATOMS AND MOLECULES

Most astronomers believe that about 15 billion yr ago, the universe exploded from a pointlike region in the event called the "big bang." Initially there was only energy, but as the universe expanded and its temperature decreased, particles appeared. According to current theories, about 2 h after the big bang, the matter in the universe was mostly electrons, protons, and some helium nuclei. It is from this matter that the stars condensed and **nucleosynthesis**, the process of element building, began. Under the action of gravity, hydrogen and helium began to form aggregates of matter that increased in temperature and density as they grew in size and as gravitational energy was converted to kinetic energy. When the mass of an aggregate grows to not less than one-tenth the mass of our sun, its temperature becomes high enough that fast-moving protons collide with sufficient energy to initiate nuclear fusion reactions, and a star is born. As we shall see, each atom (other than H or He) now on earth formed in a star or during the explosion of a star that scattered these atoms through an enormous volume of space. The sun and many other stars called second-generation stars formed from these atoms.

Nucleosynthesis in the Stars

Nuclear reactions require a much higher activation energy than do chemical reactions and so require a much higher temperature, but they liberate a very large amount of energy that is sufficient to maintain the temperature needed for fusion. Thus, as we saw in Chapter 18, nuclear fusion reactions are self-sustaining as long as there is a supply of fuel.

III This reaction takes place in more than one step, but the overall reaction is equivalent to $4\,^{1}_{1}H \rightarrow \,^{4}_{2}He + 2\,^{0}_{1}e$.

In the context of nuclear reactions, ''burning'' does not mean ''combustion in oxygen'' but simply a reaction that produces energy (including heat).

The initial fusion reaction in a new star is the formation of a helium nucleus from four protons at a temperature of 10^7 to 10^8 K and a density of about 100 $g \cdot cm^{-3}$. When a star has burned (used up) most of its hydrogen in this way, the rate of helium formation and of energy generation diminishes. As a result, the nuclei do not have enough energy to overcome the attraction between them. The core of the star collapses and shrinks in volume, which leads to a corresponding increase in density—to around 10^5 $g \cdot cm^{-3}$—and an increase in temperature—to about 10^8 K. Fusion of He nuclei then becomes possible. Two He nuclei combine to give the unstable ^{8}Be isotope, which rapidly combines with a third ^{4}He to give ^{12}C:

$$^{4}_{2}He + \,^{4}_{2}He \rightarrow \,^{8}_{4}Be$$

$$^{8}_{4}Be + \,^{4}_{2}He \rightarrow \,^{12}_{6}C$$

The stable ^{12}C isotope then reacts with other helium nuclei to give ^{16}O, ^{20}Ne, and ^{24}Mg. A variety of nuclear reactions involving protons give the remainder of the lighter elements. Eventually the He supply is used up. No more energy is produced, so the star shrinks again, raising the temperature to 10^9 K. At this point the ^{12}C and ^{16}O react to give still heavier nuclei.

The ultimate fate of a star depends on its mass. Here we are concerned only with stars with masses 8 to 50 times that of our sun, which evolve to produce in their cores the nuclei of all the elements up to ^{56}Fe. We saw in Chapter 18 that the maximum binding energy per nucleon is achieved at ^{56}Fe (Figure 18.8); the synthesis of heavier nuclei is endothermic. Once this stage has been reached, no more exothermic reactions can occur, and the production of energy ceases. The nuclei then have insufficient energy to overcome the gravitational attraction between them. During the rapid core collapse that follows, electrons combine with protons to give neutrons. When the neutrons are closely packed together, the contraction comes to an abrupt halt. The result is a rebound shock wave that gives rise to an enormous explosion visible on earth as a **supernova** (Figure 19.1). The enormous amount of energy made available in a supernova is sufficient to synthesize all the heavier elements. The explosion distributes over a vast volume of space all the elements in the star and those produced during the explosion. This process is a major source of the elements, especially the heavier elements, from which second-generation stars have formed and continue to form. We discuss the formation of our sun, a second-generation star, in Section 19.2.

Whereas the stars formed early in the history of the universe were composed only of hydrogen and helium, second-generation stars contain most of the elements. The estimated abundances of the elements in the sun and in the solar system as a whole are shown in Table 19.1 and in Figure 19.2. The abundances of the elements vary enormously, decreasing rapidly from the lightest to the heaviest. There are about a thousand times more helium atoms than silicon atoms, but there are a million silicon atoms for every iodine atom in the solar system.

FIGURE 19.1 A Supernova This portion of an exploding star, the Cygnus Loop supernova remnant, is about 2600 light-years from our galaxy and marks the edge of an expanding blast that occurred some 15,000 years ago.

It is interesting that among the most abundant elements are all those with even atomic numbers up to 20, with the sole exception of boron. All these elements can be made by the fusion of helium nuclei. Most of the elements with odd atomic numbers—namely, 3, 5, 15, 17, and 19—are missing from this list. These elements are made in subsequent reactions from the even-numbered nuclei such as ^{16}O and ^{12}C.

The addition of a proton, ^1_1H, increases the atomic number Z by 1, thus converting a nucleus with Z = (even number) to a new nucleus with Z = (odd number).

TABLE 19.1 Relative Abundances of the Most Common Elements of the Universe

Element	Z	Relative Abundance*
Hydrogen	1	1.0×10^6
Helium	2	6.3×10^4
Oxygen	8	600
Carbon	6	340
Silicon	14	50
Neon	10	40
Magnesium	12	40
Iron	26	30
Aluminum	13	3
Calcium	20	3
Sodium	11	2
Nickel	28	2
Sulfur	16	1
Argon	18	1

* Atoms per 10^6 atoms of hydrogen

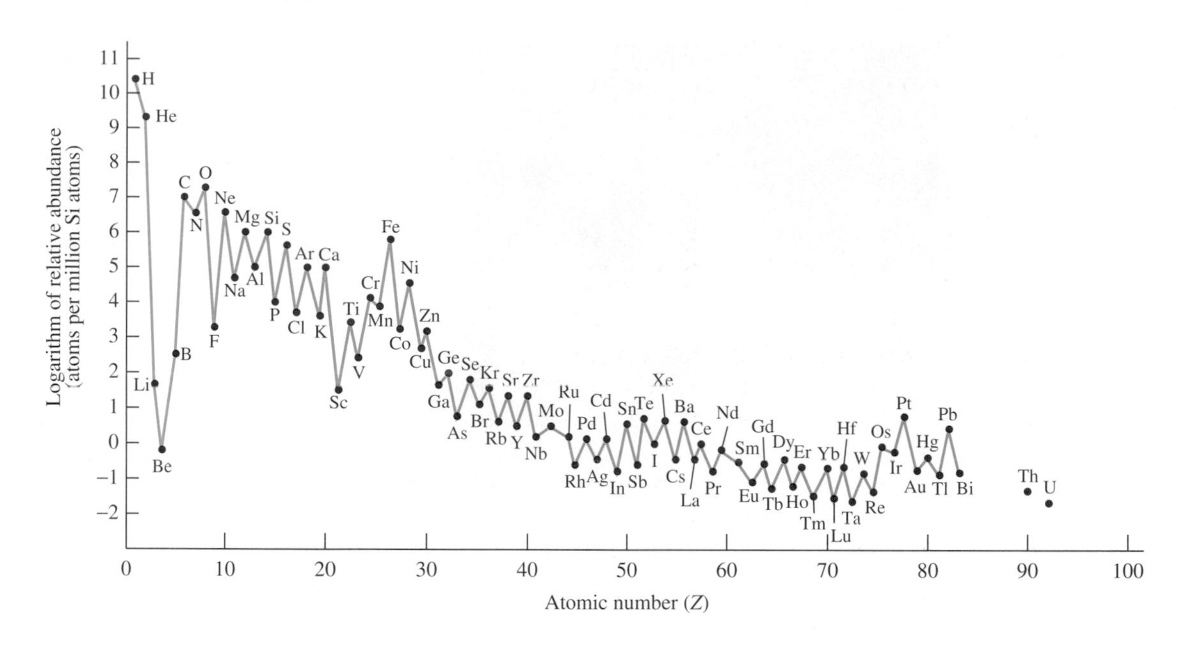

FIGURE 19.2 Estimated Relative Abundances of Elements in the Universe The abundance scale is logarithmic—hydrogen and helium are several orders of magnitude more abundant than any of the other elements.

Interstellar Molecules

Where and when did atoms, once formed, first combine together into molecules? There are no molecules in the interior of stars, where exceedingly high temperatures provide sufficient energy to break all chemical bonds. However, in the outer layers of some cool stars, simple molecules such as H_2, MgH, TiO, CaH, CH, and CN have been observed. But many more molecules form in interstellar space, where temperatures are much lower.

It has been estimated that 10% of the mass of the universe, equivalent to 10^{10} stars, is present in the form of interstellar matter. Some of this matter has very low density (1 to 100 atoms in every 1000 cm^3), but in some places it has accumulated into interstellar clouds, which have densities in the range 10^2 to 10^6 atoms per cubic centimeter. These densities are still much lower than the density of a gas such as H_2 at atmospheric pressure, which has about 10^{19} atoms per cubic centimeter. These interstellar clouds contain many kinds of molecules as well as tiny solid particles, usually called ''dust'' (Figure 19.3).

FIGURE 19.3 The Trifid Nebula in the Constellation Sagittarius The pink area, a glowing cloud of hydrogen ionized and lit up by a bright central star, is crossed by dark lines of interstellar dust, which absorb and scatter the emitted light. The blue area is a cloud of cooler hydrogen gas that reflects the light of the bright central star.

All our information about atoms and molecules in interstellar space arrives in the form of electromagnetic radiation. We saw in Chapter 6 that each kind of atom has its own characteristic spectrum that can be used to identify it; helium was detected in the sun by this method before it was found in the atmosphere. Subsequently, many different atoms such as Na, K, Ca, and Fe that are much less abundant than H and He have been detected by using powerful telescopes (Figure 19.2).

Molecules also absorb and emit electromagnetic radiation. Several small molecules and molecular ions, such as CN, CH, and CH^+, have been identified from their spectra in the visible region (Figure 19.4). But much of the visible light emitted by molecules in the interstellar clouds is absorbed by the dust and other molecules in the cloud, so most of this light does not reach the earth. However, molecules also emit infrared and longer-wavelength microwave (radio frequency) radiation (Figure

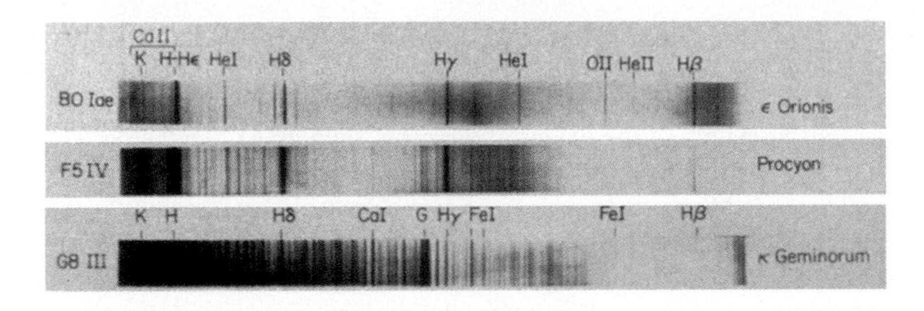

FIGURE 19.4 Spectra of Stars The dark lines on the light background result from the absorption of light by atoms in the outer layer of the star. Most of the lines are due to hydrogen and helium, but the atoms of some of the other, more abundant elements can also be detected as well as some simple molecules, such as CN.

6.4), which is not absorbed by the interstellar cloud. As we saw in Chapter 14, the infrared spectrum of a molecule is not associated with changes in its electronic energy levels but with changes in its vibrational energies; the microwave spectrum is associated with changes in the rotational energies of a molecule. Many molecules have a characteristic microwave spectrum that can be detected by radiotelescopes. More than 100 molecules, including many organic molecules, have been identified in interstellar space in this way (Table 19.2). Some of these, such as

TABLE 19.2 Molecules Observed in Interstellar Space

Two atoms			
H_2	CH^+	CH	OH
C_2	CN	CO	CO^+
NO	CS	SiO	SO
NS	SiS		
Three atoms			
H_2O	C_2H	HCN	HNC
HCO^+	N_2H^+	H_2S	HCS^+
OCS	SO_2	NaOH	
Four atoms			
NH_3	C_2H_2	H_2CO	HNCO
C_3N	H_2CS	HNCS	
Five atoms			
CH_4	CH_2NH	CH_2CO	NH_2CN
HCOOH	C_4H	HC_3N	
Six atoms			
CH_3OH	CH_3CN	NH_2CHO	CH_3SH
Seven atoms			
CH_3NH_2	CH_3C_2H	CH_3CHO	CH_2CHCN
HC_5N			
Eight atoms			
$HCOOCH_3$			
Nine atoms			
C_2H_5OH	$(CH_3)_2O$	C_2H_5CN	HC_7N
Eleven atoms			
HC_9N			
Thirteen atoms			
$HC_{11}N$			

H(C≡C)$_n$C≡N (n = 1 to 5) are unusual molecules that have not yet been found or synthesized on earth. Not all molecules emit microwave radiation, so it is probable that there are many molecules that have not been detected. The dust particles appear to consist mainly of the oxides of elements such as Al, Ca, Ti, Ni, Fe, Mg, and Si.

The reactions by which all these molecules form are by no means fully understood. Many reactive ions and free radicals are formed from some of the molecules in interstellar clouds by the interaction of the ultraviolet light and cosmic rays that bombard the clouds. Subsequent reactions of these ions and radicals lead to the formation of many other interstellar molecules. Many of the reactions probably occur on the surface of the oxide particles, which act as heterogeneous catalysts.

Exercise 19.1 **(a)** Identify the free radicals among the two- and three-atom molecules in Table 19.2.

(b) Why are so many free radicals found in space, whereas they are rare on the earth?

19.2 THE FORMATION AND COMPOSITION OF THE SOLAR SYSTEM

Most astronomers think that 5 to 6 billion years ago, a giant interstellar cloud of gases and dust was spinning in our part of the galaxy. The cloud is thought to have consisted chiefly of H and He with lesser amounts of O, N, Si, Ca, K, Al, Na, Mg, C, S, and Fe as elements or compounds such as oxides and still smaller amounts of heavier elements. Once such an interstellar cloud reaches a certain mass, it collapses gravitationally if it is perturbed (for example, by the blast from a supernova). Gravitational energy is converted to kinetic energy of the atoms and molecules, and the temperature of the contracting mass increases, causing molecules to break up into atoms. As the temperature of the contracting cloud increases further, atoms are stripped of their electrons, forming a plasma. In a **plasma**, a state of matter that exists only at very high temperatures, all the electrons have been stripped from the atoms, so the plasma consists of a mixture of electrons and nuclei moving at very high speeds. At a critical temperature of several million degrees, the fusion of hydrogen nuclei begins, and a star is born at the center of the cloud. This is how we think the sun formed. The spinning motion of the interstellar cloud—called the *solar nebula*—caused it to adopt a flat, disklike shape with the sun at its center (Figure 19.5). Eddies in the outer, cooler parts of the spinning nebula caused some

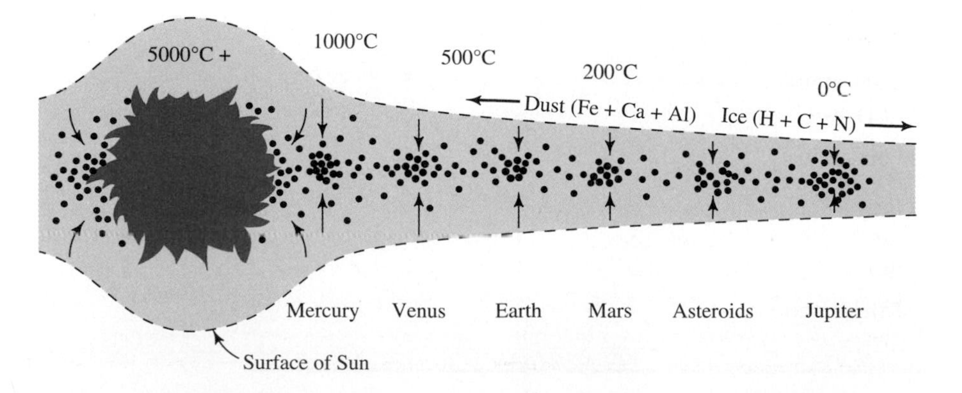

FIGURE 19.5 Formation of the Planets in the Solar Nebula Cross section of the spinning disk-shaped nebular cloud, showing the formation of the inner planets (not drawn to scale)

particles to come close together and to coalesce under their gravitational attraction to form small chunks of matter called *planetismals*. These planetismals continued to grow, coalescing with each other and sweeping up much of the remaining interstellar matter. In this way they formed the planets, which contain only about 1% of the original mass of the nebula; the remaining 99% formed the sun.

The Planets

Because of their proximity to the sun, the inner planets (Mercury, Mars, Venus, and Earth) condensed at higher temperatures than did the outer planets Saturn, Jupiter, Neptune, and Uranus. The inner planets therefore contain a higher proportion of substances with high boiling points, particularly metals such as iron and their oxides and silicates. Thus, oxygen, silicon, magnesium, and iron are the four most abundant elements in the inner planets (Table 19.3). In contrast, Jupiter, Saturn, Neptune, and Uranus are composed largely of hydrogen, helium, methane, and ammonia. It appears that these light, low-boiling-point gases were swept away from the inner planets by the *solar wind*—a stream of high-energy particles emitted by the sun. The differences in the composition of the planets are reflected in their average densities, calculated from astronomical data on the revolutions of their satellites. The average densities of Earth, Mars, and Jupiter are 5.5, 3.9, and 1.3 $g \cdot cm^{-1}$, respectively (Table 19.3). As the inner planets cooled, the heavy metals (such as iron and nickel) sank toward the center of the planet, forming a heavy metallic core. The less dense silicates were left as an outer layer, just as the silicate slag floats on top of molten iron in a blast furnace.

Space exploration has added greatly to our knowledge of the other planets. Instruments such as spectrometers have been landed on the surface of Mars and

TABLE 19.3 The Solar System

	Distance from Sun ($\times 10^6$ km)	Mass Relative to Earth	Atmospheric Pressure at Surface (atm)	Density ($g \cdot cm^{-3}$)	Composition of Atmosphere	Average Surface Temperature (K)
Sun	—	340,000	—	1.4	H, He	6000
Mercury	58	0.06	—	5.6	none	—
Venus	108	0.81	90	5.1	96.5% CO_2 3.5% N_2	732
Earth	150	1.00	1.0	5.5	78.1% N_2 20.9% O_2 0.9% Ar	288
Mars	228	0.11	0.006	3.9	95.3% CO_2 2.7% N_2 1.6% Ar	223
Jupiter	778	317	no true surface	1.3	H, He	unknown
Saturn	1430	95	unknown	0.7	H, He	unknown
Uranus	2870	14	unknown	1.6	H, He	unknown
Neptune	4500	16	unknown	2.3	H, He	unknown
Pluto	5900	0.1	unknown	unknown	unknown	unknown

Venus, and spacecraft carrying a variety of analytical instruments have flown by all the planets or are in orbit around them. In this way we have obtained much new information, particularly about planetary atmospheres. It has been said that in the short time space missions have been gathering data, more has been learned about the planets than in the rest of human history! We know that Mercury, like our moon, has no atmosphere. The compositions of the atmospheres of the earth, and the other planets are compared in Table 19.3. The atmosphere of Mars is very thin and consists mainly of carbon dioxide, with small amounts of nitrogen, oxygen, and water vapor. In contrast, the atmosphere of Venus is 90 times denser than the earth's and is 97% carbon dioxide. This large concentration of CO_2 has led to an extreme greenhouse effect (Chapter 16), which gives Venus an average surface temperature of approximately 730°C. Venus is covered by thick clouds, but they are not the familiar clouds of water droplets that we have on earth. They consist of droplets of concentrated sulfuric acid!

Comets, Asteroids, and Meteorites

Halley's comet is named after Edmund Halley (1656–1742), the seventeenth-century astronomer and colleague of Newton who first showed that the comet returns every 76 years.

Comets and asteroids are small bodies in the solar system that failed to accrete into larger planets. Comets have highly elliptical orbits and are observed only when they come close to the sun. The most celebrated is Halley's comet, which is visible from the earth periodically every 76 years. The composition of Halley's comet was probed directly by spacecraft containing mass spectrometers and other instruments when the comet returned in 1986. These observations confirmed an earlier model according to which a comet is like a huge, dirty snowball composed of a solid "ice" of volatile compounds such as H_2O, NH_3, CH_4, and CO_2 and solid dust particles made of oxides such as those of silicon, aluminum, and iron. In addition, a wide variety of organic compounds including alkanes, alkenes, alkynes, aldehydes, and aromatic compounds were detected in the gases given off by Halley's comet as it warmed up upon nearing the sun. Presumably these molecules were present in the material from which the solar system, including the planets, was formed.

Asteroids are thought to originate in the asteroid belt, the region of many small planetoids that orbit the sun between the orbits of Mars and Jupiter, and to have a composition close to the original composition of the inner planets. Asteroids that enter the earth's atmosphere are called meteorites. They are observable as "shooting stars," because they are heated to very high temperatures by friction with the earth's atmosphere. Many are completely vaporized, but those that have survived to hit the earth's surface have been analyzed in detail. Some are made predominantly of iron, whereas others are composed mainly of silicates and metal oxides. The rare but interesting meteorites called carbonaceous chondrites contain significant amounts of carbon as well as organic compounds, including hydrocarbons, carboxylic acids, amino acids, and even the nitrogenous bases that are important constituents of DNA. Some of these organic molecules may have formed by reactions catalyzed by the silicates and metal oxides of the meteorite. Hydrocarbons and water are formed when carbon monoxide and hydrogen are heated in the presence of a metal oxide or silicate catalyst:

$$CO + 3H_2 \xrightarrow{\text{catalyst}} CH_4 + H_2O$$

$$2CO + 5H_2 \xrightarrow{\text{catalyst}} C_2H_6 + 2H_2O$$

When ammonia is added to the reaction mixture, amines, amino acids, and other nitrogen-containing organic compounds of the type found in meteorites are obtained.

Exercise 19.2 How do you explain that the density of the sun is 1.4 times greater than that of water, when the sun is composed largely of hydrogen and helium?

19.3 THE STRUCTURE AND COMPOSITION OF THE EARTH

The Structure of the Earth

The earth is believed to have a central **core** composed mostly of iron alloyed with some nickel (Figure 19.6). The inner part of the core is believed to be solid and the outer part molten. The evidence for the composition of the core is based on evidence such as density, analysis of meteorites, and seismic studies—that is, the analysis of the propagation of shock waves through the earth. These studies indicate a high-density core that can be made only of a metal such as iron. As we saw in Chapter 18, iron is relatively abundant in the universe because it has the most stable nucleus; it has a maximum binding energy per nucleon. This conclusion is confirmed by the existence of the earth's magnetic field, which is believed to arise from electric currents circulating in the molten iron. The layer of silicates outside the core is called the **mantle**. The mantle is covered by a very thin **crust**, something like the skin on an apple, that contains only 0.5% of the earth's mass. The crust is about 5 to 10 km thick under the oceans and about 30 to 50 km thick beneath the continents. The crust and the upper part of the mantle are relatively rigid. Together they constitute the *lithosphere*, which may be thought of as floating on the underlying part of the mantle called the *asthenosphere*, in which the rocks are near their melting point. Under the enormous pressure exerted by the layers above, the rocks of the asthenosphere behave as if they are plastic. They slowly deform or creep, so on a geological time scale, the asthenosphere behaves like a very viscous liquid. The lithosphere is broken up into plates that float on the asthenosphere. These plates, driven by convection currents in the molten rock of the mantle, move around the earth's surface. This model is called the *plate tectonics* theory.

The rocks of which the earth's crust and mantle are composed are classified by geologists into three groups according to their origins. **Igneous rocks**, such as basalt and granite, solidify from molten rock (magma). The weathering (chemical

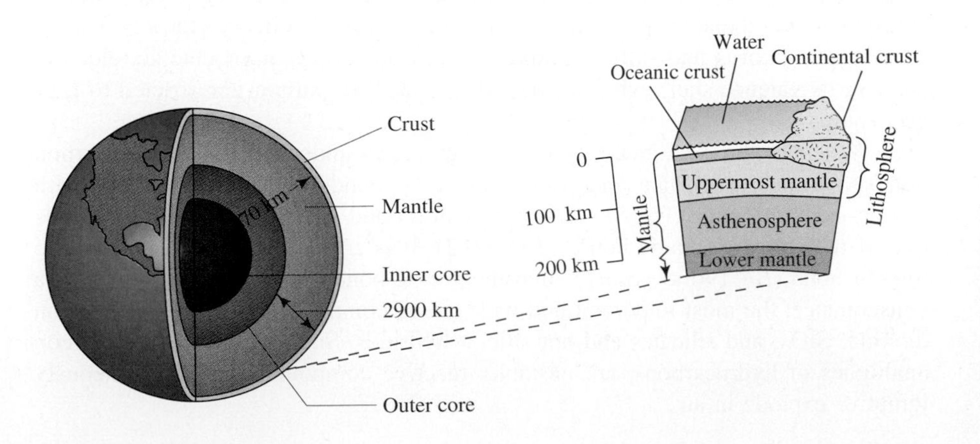

FIGURE 19.6 Structure of the Earth Cross section through the earth. The expanded section shows the relationship between the oceanic crust, the continental crust, the lithosphere, the asthenosphere, and the mantle.

and physical breakdown) of rocks produces small particles that collect in layers on the earth's surface and are called *sediments*. When compacted or cemented together, they form **sedimentary rocks**, such as sandstone and clay. Sedimentary rocks also form by the precipitation of material from solution—for example, salt is deposited when saline lakes evaporate, forming rock salt. The consolidation of the remains of dead organisms gives another form of sedimentary rock. For instance, the calcite shells of microscopic organisms deposited on the ocean floor may be consolidated into the calcium carbonate rock known as limestone. **Metamorphic rocks** form by the high-temperature and/or high-pressure modification of existing rocks. Examples are marble, a form of calcium carbonate that is produced from limestone, and slate, formed from the sedimentary rocks shale or mudstone.

Below the ocean, the crust is mostly basalt, with a density of 2.8 $g \cdot cm^{-3}$. On the continents it is mostly granite, with a density of 2.7 $g \cdot cm^{-3}$. Both basalt and granite are igneous rocks composed of mixtures of aluminosilicates. Basalt is richer in Fe, Ti, Mg, and Ca and poorer in Na, K, and silica than is granite. The density of the mantle, which consists mainly of a mixture of iron and magnesium silicates, is about 3.3 $g \cdot cm^{-3}$.

Exercise 19.3 Suggest why basalt has a higher density than granite.

Exercise 19.4 **(a)** If the earth (density 5.52 $g \cdot cm^{-3}$) were composed entirely of iron (density 7.5 $g \cdot cm^{-3}$) and silicate rock (density 3.2 $g \cdot cm^{-3}$), what would be the relative amounts of each?

(b) If the earth's core were all iron and the mantle all silicate, what would be the thickness of the mantle, relative to the radius of the iron core?

Silicon and Its Compounds

The 10 most abundant elements in the earth's crust comprise more than 99% of the total (Table 19.4). These elements are present almost exclusively in the form of silicates. Because most of the mantle and crust are made of silicate rocks, we look next at the composition and structure of silicates.

Silicon and Carbon Compared Both silicon and carbon are in Group IV of the periodic table. Their properties are nevertheless very different, because the silicon atom is larger and can accommodate more electrons in its valence shell than can carbon, and because silicon has a lower electronegativity than carbon. The covalent radius of carbon is 77 pm, and that of silicon is 117 pm; the electronegativity of carbon is 2.5, and that of silicon is 1.7. The valence shell of silicon is the $n = 3$ shell, so, like phosphorus and sulfur (Chapter 7), silicon can accommodate six electron pairs in its valence shell, whereas the valence shell of carbon is restricted to four electron pairs.

Whereas the vast majority of carbon compounds contain carbon–carbon bonds, the most important and widespread compounds of silicon are those with silicon–oxygen bonds. The Si—O bond, with a bond energy of 464 $kJ \cdot mol^{-1}$, is one of the strongest single bonds (Table 9.2). It is considerably stronger than the Si—Si bond (*BE* 196 $kJ \cdot mol^{-1}$) and the C—C bond (*BE* 348 $kJ \cdot mol^{-1}$). As a consequence, the most important and widespread compounds of silicon are silicon dioxide, SiO_2, and silicates and not silicon hydrides. Silicon hydrides, the silicon analogues of hydrocarbons, are unstable, reactive compounds that spontaneously ignite or explode in air.

TABLE 19.4 Abundance of Elements in the Earth and the Sun

Element	Z	Earth (percent by mass) Crust	Mantle	Sun*
Oxygen	8	46.9	43.9	600
Silicon	14	26.9	20.8	50
Aluminum	13	8.1	1.8	3
Iron	26	5.1	6.2	30
Calcium	20	5.0	1.8	3
Sodium	11	2.1		2
Potassium	19	2.6		
Magnesium	12	2.3	24.6	40
Hydrogen	1			1.0×10^6
Helium	2			6.3×10^4
Carbon	6			340
Neon	10			40
Nickel	28			2
Sulfur	16			1
Argon	18			1
Total		99.0	99.1	

*Atoms per 10^6 atoms of hydrogen

Silicon Dioxide (Silica) Carbon dioxide is a gas consisting of O═C═O molecules. In contrast, silicon dioxide (silica) is found in several different forms, each of which is a three-dimensional network solid. In all forms of silica, each silicon atom has a tetrahedral AX_4 geometry and is surrounded by a tetrahedral arrangement of oxygen atoms. The SiO_4 tetrahedra are joined to one another through the oxygen atoms, each of which is bonded to two silicon atoms. Figure 19.7 shows the structure of β-cristobalite, one of the simpler forms of silica. In quartz, a more common form of silica, SiO_4 tetrahedra are linked in a more complicated fashion.

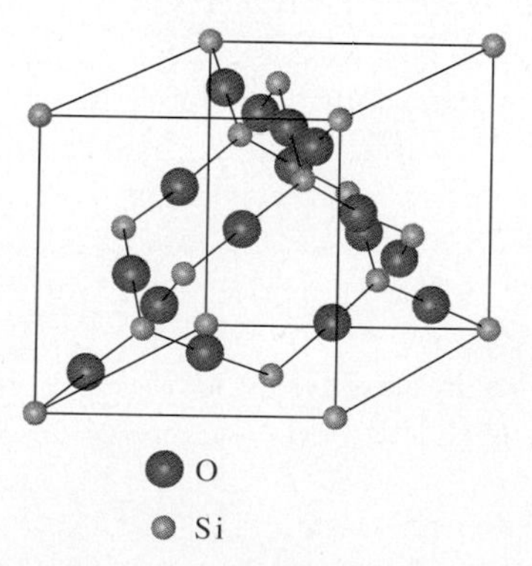

FIGURE 19.7 The Structure of β-cristobalite The structure of β-cristobalite, a form of silicon dioxide (silica), is based on a face-centered cubic lattice. Each silicon atom is surrounded by a tetrahedral arrangement of four oxygen atoms, and each oxygen atom is bonded to two silicon atoms. In quartz, the most common form of silica, the same SiO_4 tetrahedra are arranged in a more complicated way.

FIGURE 19.8 Quartz Crystals These large, well-shaped crystals formed by very slow crystallization.

The important difference in structure between CO_2 and SiO_2 reflects the tendency for the Period 2 elements carbon, nitrogen, and oxygen to form multiple bonds much more readily than do the corresponding Period 3 elements (Chapter 8). Because the electronegativity of silicon is lower than that of oxygen, silicon–oxygen bonds are quite polar. The structure of silicon dioxide may be regarded as intermediate between a purely covalent three-dimensional network, such as diamond, and an ionic crystal consisting of Si^{4+} and O^{2-} ions.

Silicon dioxide is most commonly encountered as sand, fine particles of quartz that are typically colored brown due to the presence of iron oxide. Quartz is sometimes found as large, colorless, transparent crystals (Figure 19.8). Some quartz crystals are beautifully colored by traces of impurities and are semiprecious gemstones. Amethyst, for example, is quartz that is colored violet by traces of Fe(III) (Figure 19.9). Onyx, jasper, carnelian, and flint are colored forms of noncrystalline silicon dioxide. Quartz is also a constituent of many rocks. Granite, for example, is a mixture of quartz and the silicates mica and feldspar, which we shall soon discuss.

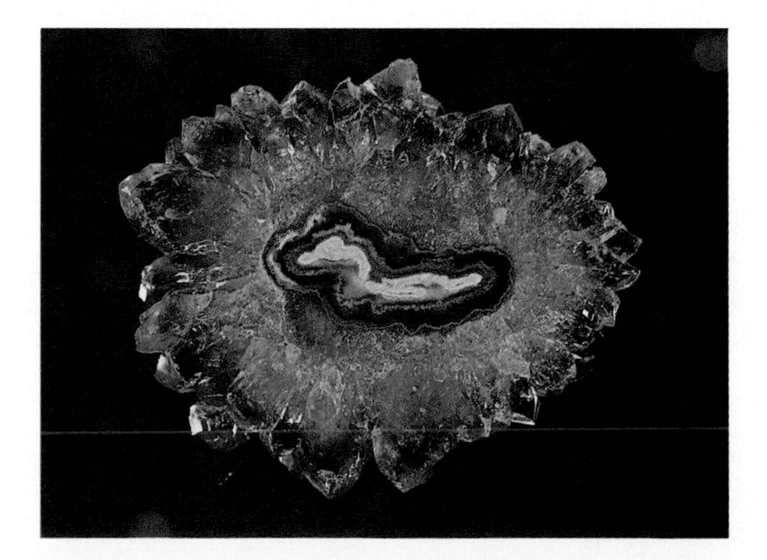

FIGURE 19.9 Amethyst Crystals Amethyst is quartz colored purple by impurities—a trace of Fe^{3+}.

Silicic Acid Silica is only very slightly soluble in water, giving very dilute solutions of the weak acid *silicic acid*, $Si(OH)_4$. It forms silicate salts by reaction with bases such as NaOH:

$$SiO_2(s) + 4NaOH(s) \xrightarrow{\text{heat}} Na_4SiO_4(s) + 2H_2O(g)$$

However, as we shall see, a great many silicates contain anions that are more complex than the simple SiO_4^{4-} anion. More concentrated solutions of silicic acid can be made by the reaction of silicon tetrachloride with water:

$$SiCl_4(l) + 4H_2O(l) \rightarrow Si(OH)_4(aq) + 4HCl(aq)$$

However, the product of this reaction, $Si(OH)_4$, rapidly condenses to give a gelatinous solid consisting of a mixture of polymeric silicic acid such as trisilicic acid:

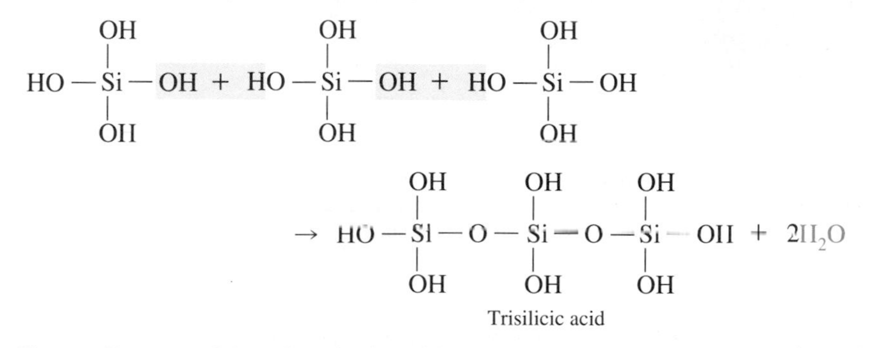

Trisilicic acid

In Chapter 7 we noted that phosphoric acid has a similar tendency to form polymeric acids by condensation, and in Chapter 17 we discussed the importance of adenosine diphosphate, ADP, and adenosine triphosphate, ATP, in living systems.

Silicic acid is a very weak acid ($K_a = 1 \times 10^{-10}$ mol · L^{-1}). It is the first member of the series of Period 3 oxoacids that increase in strength in the series $Si(OH)_4$, $PO(OH)_3$, $SO_2(OH)_2$, and $ClO_3(OH)$ (Chapter 7).

Silicates The metal salts of silicic acid are **silicates**. More than 1000 different silicates occur naturally, their number and complexity resulting from the many different ways SiO_4 tetrahedra can be linked. The simplest silicates contain the tetrahedral SiO_4^{4-} ion. *Olivine* (Figure 19.10), the principal component of the earth's mantle, consists of tetrahedral SiO_4^{4-} ions and either Fe^{2+} or Mg^{2+} or a mixture of both. Thus, we write the formula of olivine as $(Fe,Mg)_2SiO_4$. Most other silicates are the salts of polymeric silicic acids. The condensation of two molecules of silicic acid, $Si(OH)_4$, gives $H_6Si_2O_7$. The anion derived from this acid, $Si_2O_7^{6-}$, consists of two tetrahedra sharing a corner:

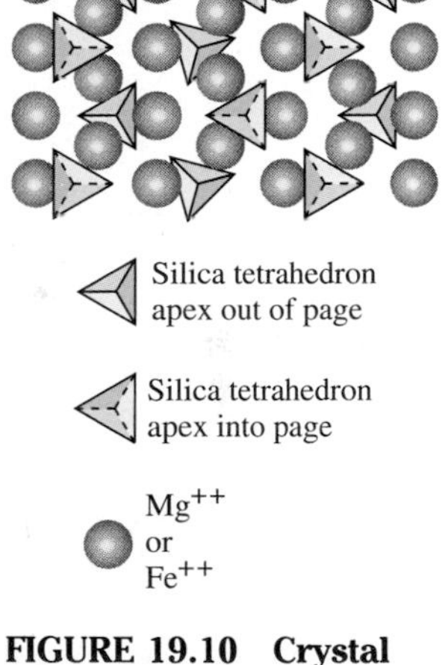

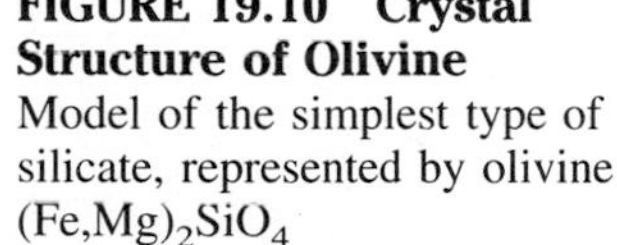

FIGURE 19.10 Crystal Structure of Olivine
Model of the simplest type of silicate, represented by olivine $(Fe,Mg)_2SiO_4$

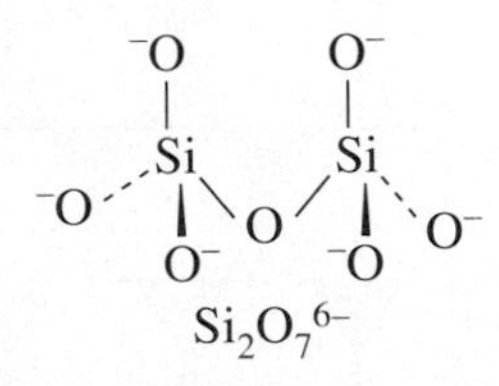

$Si_2O_7^{6-}$

Continued condensation reactions of this type can give infinitely long chains of SiO_4 tetrahedra. Each silicon atom in such a chain is bonded to four oxygen atoms, two of which are not shared and two of which are shared with other silicon atoms (Figure 19.11). Thus, there is a total of three oxygen atoms—$(2 \times 1) + (2 \times \frac{1}{2})$—per silicon atom. The empirical formula for the chain is SiO_3^{2-}, because each of the

Each oxygen that is not shared with another silicon counts as 1; each oxygen that is shared with one other silicon has a contribution of $\frac{1}{2}$.

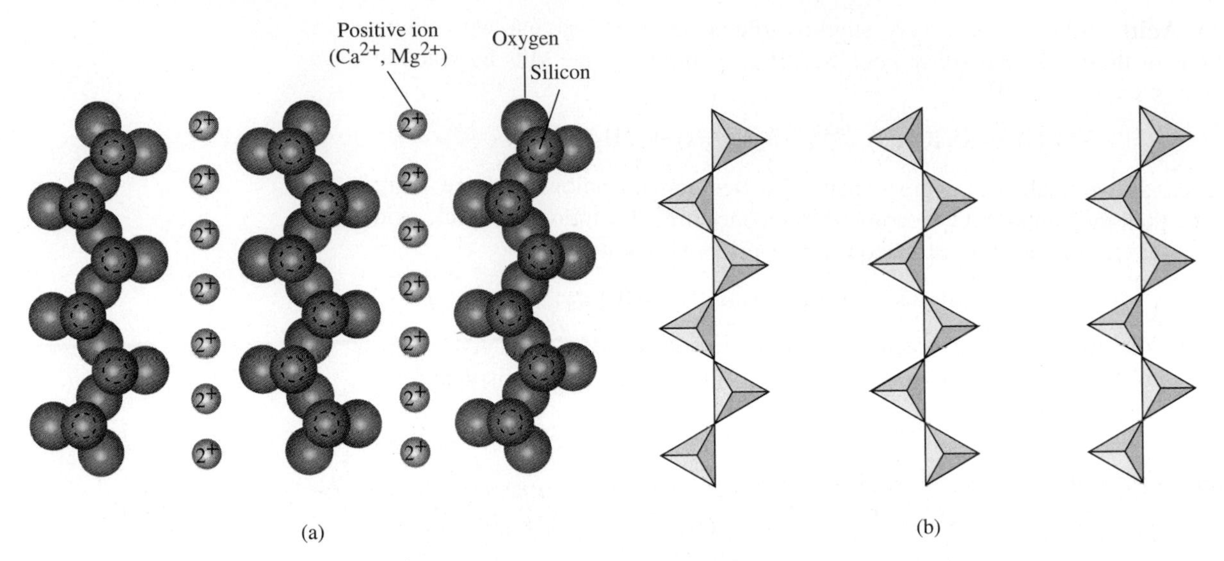

FIGURE 19.11 Structure of a Single-Chain Silicate (a) Model of a single-chain silicate such as diopside, $CaMg(SiO_3)_2$. (b) The same chain represented by linked tetrahedra; positive ions between the chains are not shown.

unshared oxygen atoms carries a negative charge. This empirical formula is like that of the carbonate ion, CO_3^{2-}, but the structure it represents, an *infinite-chain* anion, is quite different from the structure of the carbonate ion.

Silicates containing this infinite-chain anion, called **pyroxenes**, are another component of the earth's mantle (Figure 19.12a). An example is *diopside*, $CaMg(SiO_3)_2$, which is composed of Ca^{2+} and Mg^{2+} ions and the infinite-chain anion $(SiO_3^{2-})_n$. In the silicates called **amphiboles**, two $(SiO_3^{2-})_n$ chains are joined side by side by sharing oxygen atoms on alternate tetrahedra, giving a *double chain* with the empirical formula $Si_4O_{11}^{6-}$ (Figure 19.12b). Amphiboles are important constituents of the earth's crust.

If a large number of $(SiO_3^{2-})_n$ chains are joined side by side, a *sheet* of SiO_4 tetrahedra results. Clays and micas are such sheet silicates, in which each tetrahedron shares *three* corners with neighboring tetrahedra (Figure 19.12c). Each silicon atom is bonded to three other silicon atoms through an oxygen atom and to

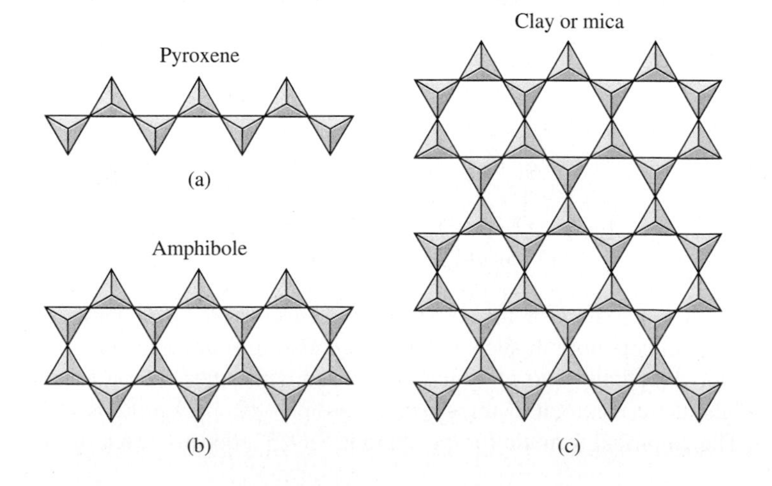

FIGURE 19.12 Polymeric Silicate Anions
(a) Pyroxenes have single chains of linked SiO_4 tetrahedra. **(b)** Amphiboles have double chains of linked SiO_4 tetrahedra. **(c)** Clays and micas have sheets of linked SiO_4 tetrahedra.

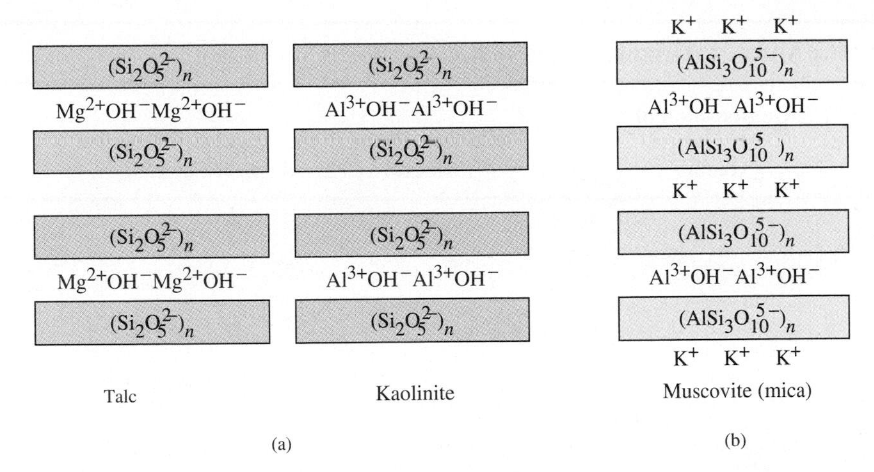

FIGURE 19.13 Structures of Some Sheet Silicates and Aluminosilicates (a) Edge view of the layer structures of talc, $Mg_3(Si_2O_5)_2(OH)_2$, and kaolinite, $Al_2(Si_2O_5)(OH)_4$. Each layer consists of a sandwich of two $(Si_2O_5^{2-})_n$ sheet anions held together by metal ions and hydroxide ions. Because each layer has a zero overall charge, there are only weak London (dispersion) forces holding the layers together. These silicates are soft because the layers slide over each other easily. (b) In muscovite, $KAl_2(AlSi_3O_{10})(OH)_2$, each layer consists of a sandwich of two $(AlSi_3O_{10}^{5-})_n$ anions held by Al^{3+} and OH^- anions. In this case, however, the layers have a negative charge and are held together strongly by potassium ions. Muscovite, a form of mica, is a relatively hard but flaky mineral.

one unshared oxygen atom bearing a negative charge. Because each silicon atom is associated with $(1 \times 1) + (3 \times \frac{1}{2}) = 2\frac{1}{2}$ oxygen atoms, one of them negatively charged, the empirical formula of this sheet anion is $Si_2O_5^{2-}$. Figure 19.13a shows the structures of **talc**, $Mg_3(Si_2O_5)_2(OH)_2$, which is used in making talcum powder, and *kaolinite*, $Al_2(Si_2O_5)(OH)_4$, a type of clay that is particularly valuable for pottery making. Talc and kaolinite consist of layers, each composed of two $(Si_2O_5^{2-})_n$ sheet anions, hydroxide ions, and sufficient cations to neutralize the total charge of the silicate and hydroxide anions. These layers slide over each other easily, giving the minerals a characteristic soft, soapy feel. *Soapstone*, an easily carved rock, is talc in a compact form.

The layered structure of a clay such as kaolinite can be penetrated by water molecules, which act as a lubricant between the layers, enabling them to slip easily over one another. Thus, wet clay is pliable and slippery. When clay objects are fired in a kiln to 1100°C or more, the water molecules are driven out, and the layers lock into a rigid structure. The resulting strong, hard substance is called a **ceramic**.

Soapstone carvings created by an Innuit inhabitant of Arctic Canada

Aluminosilicates Silicates in which some of the silicon atoms are replaced by aluminum atoms are called **aluminosilicates**. Table 19.5 summarizes the structures of the common silicates and aluminosilicates. Aluminum has one less electron than silicon. When an Al atom replaces a Si atom in an SiO_4 tetrahedron, it must have four electrons in its valence shell, and so it must be present as Al^-. This means that one extra positive charge must be provided by the metal ions in the structure to neutralize the charge of each Al^-.

Aluminosilicates in which some of the Si atoms in silicon dioxide are replaced by Al and that also contain Na, K, or Ca are called **feldspars**. Feldspars are the most abundant of all the silicates and make up about 60% of the earth's crust. They are, for instance, an essential constituent of granite. An example of a feldspar is

TABLE 19.5 The Major Silicate and Aluminosilicate Minerals

Type	Structure	Composition of Tetrahedral Groups	Si:O Ratio	Mineral	Typical Formula
Mantle					
Separate tetrahedron	Plan view	$(SiO_4)^{4-}$	1:4	Olivine	$(Mg,Fe)_2SiO_4$
Chain	used below	$(SiO_3)^{2-}$	1:3	Pyroxene	$MgSiO_3$
Double chain		$(Si_4O_{11})^{6-}$	1:2.75	Amphiboles	$Ca_2(Mg, Fe)_5(Si_8O_{22})(OH, F)_2$
Crust					
Sheet		$(Si_2O_5)^{2-}$	1:2.5	Clay minerals	
				—Kaolinite	$Al_2[Si_2O_5](OH)_4$
		$(AlSi_3O_{10})^{5-}$		Micas	$K(Mg,Al,Fe)_{2-3}(AlSi_3O_{10})(OH)_8$
3-D framework	(SiO_4) tetrahedron sharing all vertices	SiO_2	1:2	Quartz	SiO_2
		$(AlSi_3O_8)^-$		Feldspars	$(Na,K,Ca)(AlSi_3O_8)$

A sample of the mica muscovite

orthoclase, $KAlSi_3O_8$. This aluminosilicate has a structure like that of silicon dioxide except that one-quarter of the Si atoms have been replaced by Al^-. Hence it is a *three-dimensional network* anion, $(AlSi_3O_8{}^-)_n$, rather than the neutral three-dimensional network structure SiO_2.

Muscovite, $KAl_2(AlSi_3O_{10})(OH)_2$, has a structure like the amphibole double-chain anion but with one-quarter of the Si atoms replaced by Al (Figure 19.13b). If we double the empirical formula $Si_2O_5{}^{2-}$ of the amphibole anion, we obtain $Si_4O_{10}{}^{4-}$. Replacing one of the Si atoms by Al^- gives the empirical formula $AlSi_3O_{10}{}^{5-}$. Each layer of the muscovite structure consists of two of these infinite-sheet $(AlSi_3O_{10}{}^{5-})_n$ anions held together by OH^- and Al^{3+} ions. Each such double layer has an overall negative charge, and layers of positive potassium ions are interspersed between the double silicate layers. Because of the electrostatic attraction between the sheet anions and the potassium cations, the layers do not slide over one another as readily as do the layers of talc, so muscovite is not as soft as talc. However, muscovite, which is a common form of the sheet silicates known as *micas*, cleaves readily into thin, transparent sheets that are used for windows in high-temperature furnaces. "Metallic" paints, such as those often used for automobiles, contain small flecks of mica.

Zeolites, such as $Na_2Al_2Si_3O_{10} \cdot 2H_2O$, are aluminosilicates with three-dimensional network structures containing tunnels and systems of interconnected cavities (Figure 19.14). There is currently a great interest in zeolites, because they have been found to be very useful as catalysts, as we shall discuss in Chapter 20.

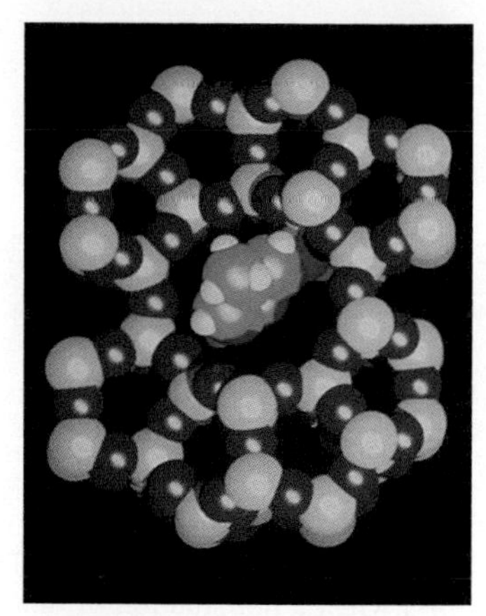

FIGURE 19.14 A Zeolite Computer-generated model of the pores in the zeolite ZSM-5, showing a trapped *o*-xylene molecule.

EXAMPLE 19.1 Empirical Formula of an Aluminosilicate

Anorthite is a feldspar containing Ca^{2+} ions and an anion formed by replacing half the silicon atoms in SiO_2 by aluminum. What is the empirical formula of anorthite?

Solution: If we double the empirical formula SiO_2 and then replace one of the silicon atoms with Al^-, we obtain the empirical formula $AlSiO_4^-$ for the infinite, three-dimensional aluminosilicate anion. For electrical neutrality, two of these empirical formula units are required for each Ca^{2+}. Thus, the empirical formula of anorthite is $Ca(AlSiO_4)_2$.

Exercise 19.5 What is the empirical formula of the mica mineral phlogopite, in which one-quarter of the silicon atoms of talc are replaced by aluminum and the additional negative charge is balanced by K^+ ions?

For any empirical formula, (charge of cation) × (number of cations) = (charge of anion) × (number of anions).

Minerals and Other Geologic Resources

Most of the rocks that form the earth's crust are heterogeneous mixtures. Each of the crystalline substances in a rock is called a **mineral**. Minerals of commercial value are called **ores**. Ores include the metals copper, silver, and gold, which occur in their elemental forms; metal oxides such as bauxite, Al_2O_3, hematite, Fe_2O_3, and cuprite, Cu_2O; and metal sulfides such as pyrrhotite, iron(II) sulfide, FeS, and pentlandite, iron nickel sulfide, (Fe,Ni)S.

Although some metal oxides, such as those of iron and chromium, probably crystallize directly from magma, most ore deposits form in the earth's crust by **hydrothermal processes**, processes that occur in solution at high temperatures. Water from the ocean percolates deep into the crust, where it is subjected to high temperatures and pressures. Many minerals are soluble under these conditions; they can then be transported to different locations and reprecipitated. Silica is extracted from silicate rocks as $Si(OH)_4$ and then, under cooler conditions, precipitates as quartz, SiO_2. Rocks such as granite typically contain veins of crystalline quartz deposited from solutions that have percolated through cracks in the rock. Gold may dissolve as the complex ions $AuCl_2^-$ or $Au(SH)_2^-$ before being deposited as the element. Metal sulfides such as copper sulfide, CuS, in the magma are dissolved by water at high temperatures and pressures and then reprecipitate at the surface, as has been observed at hot springs located on the ocean floor in regions of diverging plate boundaries. When the hot metal sulfide solutions meet the cold sea water, metal sulfides precipitate in a mound around the spring. This process has been filmed at the mid-ocean ridge in the Pacific Ocean, where hot springs spew clouds of fine-grained sulfides that look like black smoke (Figure 19.15). Very large deposits of sulfide ores may form in this way. They may subsequently be covered with sediments, which, through geologic changes, may eventually end up on dry land.

FIGURE 19.15 Black Smoker This submarine hot spring, or "black smoker," is located on the Galapagos Trench in the Pacific Ocean. The "smoke" is a fine precipitate of metal sulfides that form when a hot solution of minerals meets the cold water. The surrounding mounds are metal sulfide deposits.

Petroleum and coal are useful geologic resources commonly called *fossil fuels*. *Petroleum* and *natural gas* appear to originate from marine organisms. After they die microscopic organisms such as single-celled algae settle to the sea floor and accumulate in mud. The organic matter may partly decompose, using up the dissolved oxygen in the sediment. As soon as the oxygen is gone, decomposition stops, and the remaining organic matter is preserved. Continued sedimentation buries the organic matter and subjects it to higher temperatures and pressures, which cause physical and chemical changes that, over a period of millions of years, lead to the formation of liquid and gaseous hydrocarbons.

Coal is formed from dead vegetable matter that has not completely decayed. Swamps and bogs are the most likely environments in which dead plant material is converted to coal in the same way that microscopic marine organisms are converted to oil.

The Atmosphere

In Chapter 2 we described the present composition of the atmosphere. But where did the earth's atmosphere originate? And why is it different from that of all the other planets?

The large outer planets, such as Jupiter, have retained their original atmospheres, which consist almost entirely of hydrogen and helium with smaller amounts of methane and ammonia. But the gravitational attraction of the inner planets was not sufficient for them to hold on to these gases, particularly when, early in their history, their temperatures were much higher than they are now. Streams of intense radiation from the sun—the solar wind—produced at a time when the sun was much more active, is thought to have blown away any remaining gases from the atmospheres of the inner planets. That the earth lost not only hydrogen and helium but almost all of its original atmosphere is shown, for example, by the fact that neon is about as abundant as nitrogen in the solar system but is 10^{-5} times less abundant in the earth's atmosphere. Moreover, the argon in the sun's atmosphere is almost entirely argon-36 and argon-38, and presumably the earth's original atmosphere also had argon with this isotopic composition. But today the argon in the earth's atmosphere is mainly argon-40 arising from the radioactive decay of potassium-40, indicating that almost all the original argon-36 and argon-38 have been lost.

It is now generally believed that the gases of the present atmosphere came largely from *outgassing*; that is, they were formed by the decomposition of substances in the earth's interior and liberated by volcanic activity. The gases released were probably mainly CO_2, H_2O, and N_2 from the decomposition of carbonates, hydrates, and nitrides, with smaller amounts of many other gases, such as SO_2, SO_3, and HCl. Volcanoes expel these gases into the atmosphere, along with helium and argon formed in the mantle by radioactive decay. In the past, reactive gases such as SO_2 and HCl reacted with surface rocks; the water vapor condensed to form the oceans; and the carbon dioxide formed carbonates, leaving an atmosphere dominated by N_2.

The major differences between the atmosphere of earth and the atmospheres of the other planets are that our atmosphere contains oxygen, O_2, and that its composition is far from chemical equilibrium. Knowing the amounts of nitrogen, oxygen, carbon and other elements in the earth's crust, we can calculate what the composition of the atmosphere would be if all the possible reactions had reached equilibrium. This equilibrium atmosphere has very little oxygen because it has been converted to carbon dioxide and other oxides; it is very similar to the composition of the atmosphere of Venus, which is 98% CO_2 (Table 19.3). If this were the actual composition of our atmosphere, the resulting greenhouse effect (Chapter 16) would lead to a surface temperature of about 290°C. This temperature is somewhat lower than the surface temperature on Venus because of the greater distance of the earth from the sun but is far too hot to allow life as we know it.

In Chapter 2 we mentioned the reason for the divergence of the earth's atmosphere from chemical equilibrium: The presence of life has modified our environment. Photosynthesis, which leads to the conversion of water and carbon dioxide to oxygen and carbohydrates, has resulted in the high concentration of oxygen in the atmosphere. The continuous input of radiation from the sun maintains the composition of the biosphere far from equilibrium. If life were not present or if there were no continuous supply of energy, the composition of the atmosphere would rapidly (on a geological time scale) revert to the equilibrium composition with very little oxygen. The high O_2 concentration in our atmosphere appears to date back some 600 million years. Prior to that time there had been a gradual increase in the oxygen concentration, starting when the first photosynthetic bacteria appeared on earth about 3.4 billion years ago. It was only at the beginning of the Cambrian Period, 600 million years ago, that the O_2 concentration became high enough to support the respiration required for fungal, plant, and animal life, and a great evolutionary proliferation occurred. Figure 19.16 shows how the composition of the atmosphere is thought to have changed since the earth formed.

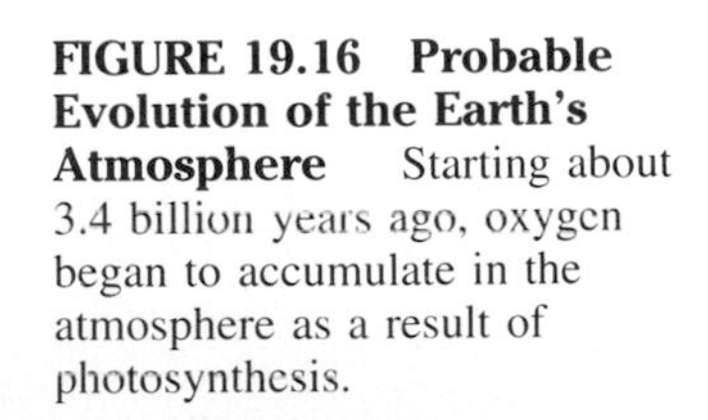

FIGURE 19.16 Probable Evolution of the Earth's Atmosphere Starting about 3.4 billion years ago, oxygen began to accumulate in the atmosphere as a result of photosynthesis.

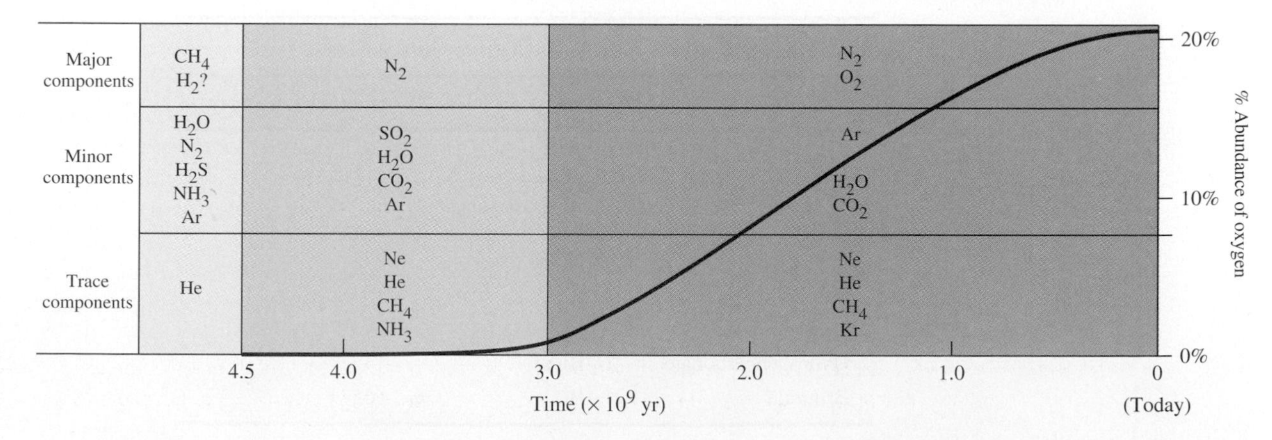

Exercise 19.6 **(a)** Explain why the concentrations of the noble gases are about 10 times lower in the atmosphere than in the universe as a whole. **(b)** Explain how and why the isotopic composition of argon on the earth differs from that in the universe as a whole.

Exercise 19.7 Briefly summarize the role of solar energy in the development of our present atmosphere from the primeval one.

The Hydrosphere

The earth is unique among the planets in possessing a **hydrosphere**—collectively, all the water on the planet's surface. The temperature of the surface of the earth early in its history was such that water vapor liberated by outgassing condensed to liquid water. None of the other planets have a surface temperature in the range required to maintain liquid water, although there is some evidence that Mars may at one time have had liquid water on its surface. Any water now on or near the surface of Mars is frozen.

The oceans cover about 71% of the earth's surface. They contain enormous amounts of many elements, such as the halogens in the form of the halide ions, F^-, Cl^-, Br^-, and I^-, and the alkali and alkaline earth metals in the form of cations. The current composition of the oceans is given in Table 19.6. This composition, like that of the atmosphere, has changed with time. Before oxygen was abundant in the atmosphere, the oceans had relatively high concentrations of soluble Fe^{2+} salts. When the concentration of oxygen in the atmosphere increased, these Fe^{2+} salts were oxidized to insoluble $Fe(OH)_3$. This hydroxide was deposited on the ocean floor and eventually became our present deposits of iron ore, such as hematite, Fe_2O_3.

The rich reddish color of the "Needles" in Canyonlands National Park in Utah is due to large amounts of iron(III) oxide, Fe_2O_3, in the rock.

Although the composition of the oceans has changed in the past and may still be changing very slowly on a geological time scale, in the shorter term the composition is essentially constant. This constant composition is maintained by a balance between the addition of molecules and ions by the weathering of rocks and from hydrothermal vents along the mid-ocean ridges and the removal of molecules and ions by precipitation and by marine life. Metal ions such as Na^+, Mg^{2+}, and Ca^{2+} from the weathering of rocks are washed into rivers and transported to the oceans. Soluble alkali metal halides removed by the evaporation of seas that become cut off from the ocean give salt deposits. Although originally formed on the surface, they are typically buried by sediments and are thus found underground. Many

TABLE 19.6 The Composition of Sea Water

Ion	Formula	Concentration (ppm)
Chloride	Cl^-	18,980
Sodium	Na^+	10,561
Sulfate	SO_4^{2-}	2,649
Magnesium	Mg^{2+}	1,272
Calcium	Ca^{2+}	400
Potassium	K^+	380
Hydrogencarbonate	HCO_3^-	142
Bromide	Br^-	65

marine organisms build skeletons or hard shells from calcium carbonate. The dead organisms eventually sink to the ocean floor, where they are compacted into the limestone deposits that form an important fraction of sedimentary rocks. The amount of chloride and bromide in the oceans appears to be too great to be accounted for by the weathering of rocks, which contain rather little of these elements. Geologists think the chlorine and bromine in the oceans probably came from volcanic activity, particularly in the early stages of the formation of the atmosphere. As we saw in Chapter 4, $HCl(g)$ is very soluble in water, so much of the HCl emitted by volcanoes would have ended up in the oceans.

19.4 THE BIOSPHERE

The Formation of the Biosphere

A major difference between earth and all the other planets of the solar system is that the earth has a very thin layer on its surface, called the **biosphere**. The biosphere consists of the hydrosphere, the troposphere, the surface layers of the crust, and all the organic matter in these parts of the earth's surface. The biosphere includes many molecules not found on other planets. These are the molecules of life, the biomolecules that we discussed in Chapter 17. We would like the answers to the following questions: (1) How did the molecules necessary for the construction of living organisms form, and (2) how did living organisms form from these molecules? Not surprisingly, we have more complete answers to the first question than to the second.

Life very probably originated in the oceans and was certainly confined there until quite recently on a geological time scale. Thus, it is not surprising to find that with few exceptions, the elements of life, those that are most abundant in living organisms (O, C, H, N, Ca, P, K, S, Cl, Na, and Mg) are also those that are most abundant in the oceans (Table 19.6).

We saw in Chapter 17 that the most important biomolecules are the amino acids and the proteins derived from them; the nitrogenous bases, such as adenine; and sugars. In 1953 Stanley Miller and Harold Urey tried to determine how these molecules might first have formed by subjecting a mixture of methane, ammonia, hydrogen, and water vapor to repeated electric discharges (Figure 19.17). The

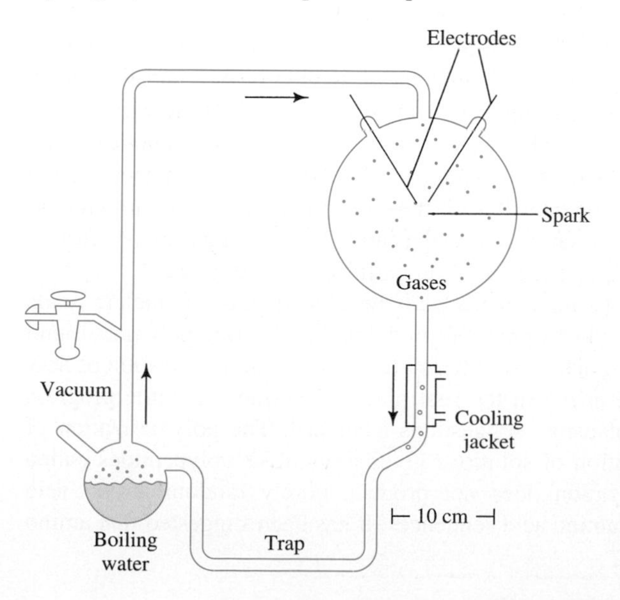

FIGURE 19.17 Apparatus Used to Study the Formation of Amino Acids on the Primordial Earth In the Miller-Urey experiment, electric discharges were passed through a mixture of carbon dioxide, methane, ammonia, and steam. Small concentrations of several amino acids were found in the trap after the experiment had run for one week.

mixture of gases was thought to resemble the composition of the earth's atmosphere before life began, and the electric discharges mimicked the lightning that was probably more common at that time than at present.

Many of the amino acids required to make proteins were formed as well as low concentrations of the nitrogenous bases needed to make nucleic acids. Later studies showed that the initial products were hydrogen cyanide, HCN, and aldehydes such as formaldehyde (methanal), CH_2O, and acetaldehyde (ethanal), CH_3CHO. They were then converted to amino acids in reactions such as

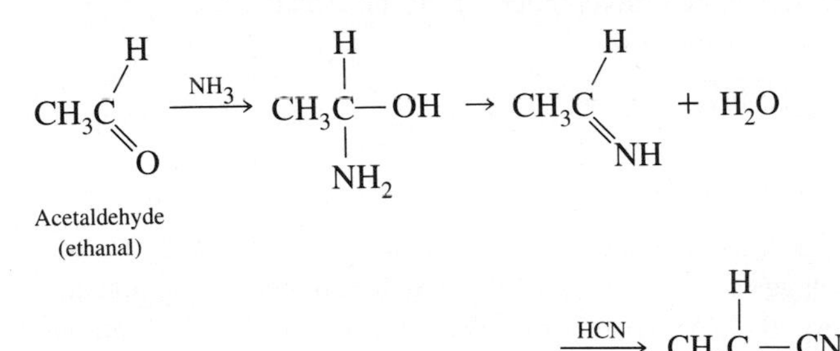

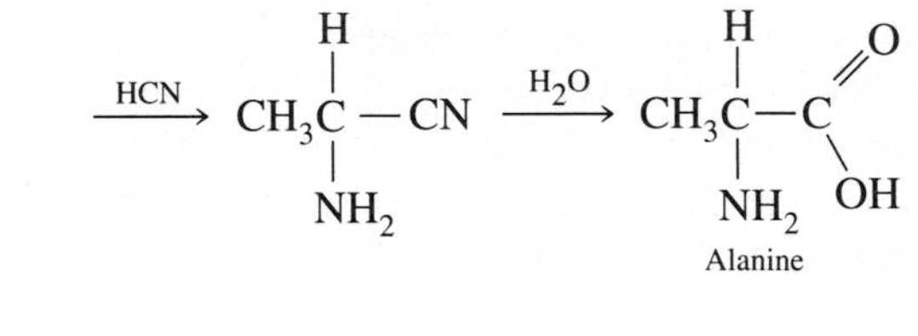

Exercise 19.8 **(a)** Write complete balanced equations for each step in the conversion of acetaldehyde to alanine.

(b) Describe each reaction step in (a) as an elimination reaction, an addition reaction, or some other type of reaction.

The nitrogenous base adenine has the formula $C_5N_5H_5$ and can therefore be regarded as a polymer of HCN. It has been shown that adenine can be formed in a concentrated aqueous HCN solution by the series of steps shown in Figure 19.18. A sugar such as glucose, $C_6H_{12}O_6$, can be regarded as a polymer of formaldehyde, CH_2O. It is believed that sugars such as glucose were originally formed by the polymerization of formaldehyde dissolved in the oceans.

Since Miller and Urey carried out their experiments, our knowledge of the molecules present in the interstellar clouds has increased considerably. We now know that all the simple molecules needed for the formation of biologically important molecules are present in these interstellar clouds (Table 19.2). So these molecules may well have been present when the earth first formed, and it is not essential to postulate that they formed in the earth's atmosphere. These simple molecules, and even amino acids and the nitrogenous bases, have been found in small amounts in meteorites, which formed at the same time as the earth. This finding provides further evidence that these molecules were present on earth in its early days.

How did a mixture of amino acids and the constituents of nucleic acids (sugars, nitrogenous bases, and phosphate) form the self-replicating polymers found in living organisms? This is much more difficult to answer than the question of how these molecules might have arisen in the first place. Nevertheless, some progress has been made and some plausible suggestions advanced. The polymerization of amino acids by the evaporation of solutions gives proteinlike polypeptides called *proteinoids*. This polymerization does not produce purely random amino acid polymers but favors certain amino acid sequences. It has been suggested that amino

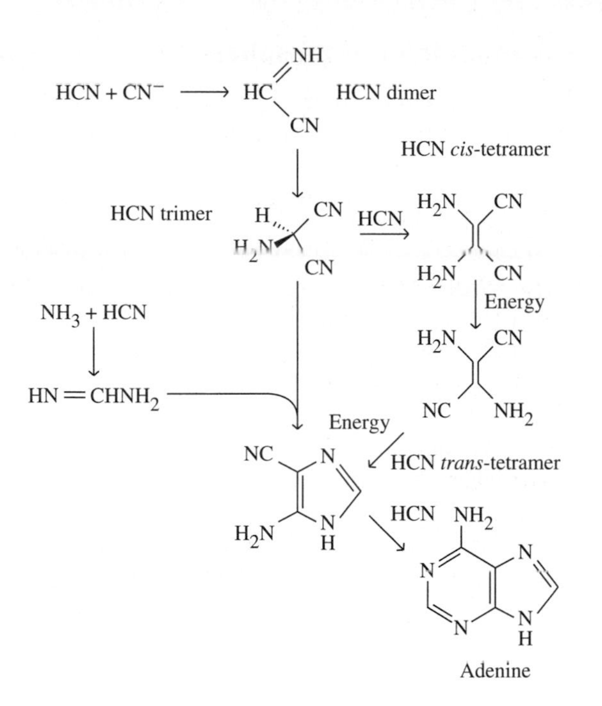

FIGURE 19.18 Formation of Adenine from HCN A proposed reaction scheme to show how the nitrogenous base adenine, a component of DNA, might have formed from the inorganic compound hydrogen cyanide.

acids could have been polymerized on the surfaces of clays acting as specific catalysts (Chapter 20) in the same way that an enzyme catalyzes only certain specific reactions. These clay-catalyzed polymerizations gave the particular amino acid sequences that eventually led to the emergence of life. How life arose is a very active area of research that as yet has produced no definite answers.

Whatever the mechanism that led to the first appearance of life, it had occurred within a billion years of the formation of the earth. The earliest remains of living material have been dated to around 3.6 billion years ago. At that time, the sun was providing 25% less radiation than at present, but the atmosphere contained no oxygen and significantly more CO_2 (200 to 1000 times as much). The enhanced greenhouse effect resulted in a surface temperature 13°C higher than at present despite the reduced energy of sunlight. The lack of oxygen also resulted in a lack of ozone, so much more ultraviolet radiation reached the earth's surface and was available to drive photochemical reactions. The earliest forms of life probably existed deep in the ocean, where they were protected from ultraviolet radiation. Perhaps these organisms relied on the heat energy supplied by volcanic vents for a metabolism based on the oxidation of hydrogen sulfide.

The first major change came with the development of photosynthesis about 3.4 billion years ago. Sunlight was then used directly to provide energy to sustain life. A side effect of photosynthesis was the conversion of CO_2 to O_2. Initially much of the oxygen was used up in oxidation reactions, such as the oxidation of Fe^{2+} salts in the ocean, thus precipitating insoluble $Fe(OH)_3$ and forming iron ore deposits. But slowly more oxygen accumulated until it had reached its present steady-state level by about 600 million years ago. The decrease in the concentration of CO_2 reduced the greenhouse effect. The oxygen provided a protective ozone layer that allowed life to emerge from the sea and, more importantly, led to the development of respiration. Because respiration is required to provide enough energy to fuel the movement of animals, the development of respiration represented a major evolutionary step.

Maintaining the Biosphere

The present composition of the biosphere is favorable for the maintenance of life as we know it. It would not remain this way unless there were mechanisms to replenish substances essential to life. These processes take the form of various cycles that take the waste products of metabolism and process them to give substances needed by organisms.

The Water Cycle The *water cycle* consists of the evaporation of sea water by solar heat and its subsequent condensation as rain or snow. Water containing waste products is eventually returned to the sea by way of streams and rivers. In this way the supply of fresh water necessary for land-based life is maintained.

The Nitrogen Cycle Nitrogen, N_2, is abundant and readily available in the atmosphere. Nitrogen is needed to make proteins and nucleic acids. But as we have seen, N_2 is a rather unreactive molecule with a very strong triple bond, so most organisms cannot utilize it directly. However, as we mentioned in Chapter 2, certain microorganisms in the root nodules of plants of the pea family are able to fix nitrogen—that is, they can oxidize nitrogen to nitrates, which can be used by plants to make proteins that are in turn used by animals. When an animal dies, the nitrogen-containing molecules are decomposed by other microorganisms to give mainly N_2, N_2O, and NO, which enter the atmosphere. The *nitrogen cycle* is now complete. Significant amounts of nitrogen in the atmosphere are also converted by lightning to NO, which is then converted to nitric acid in the atmosphere and returned to the earth's surface in rain. Indeed, if there were no life on earth, the action of lightning would eventually remove most of the nitrogen from the air, and it would ultimately be washed into the ocean and accumulate there as nitrate ion. Not only would the high concentration of nitrate ion be toxic to many forms of life, but the increased salinity of the ocean would prevent the formation of cell membranes.

The Sulfur Cycle When an organism dies, the sulfur-containing compounds in it are oxidized to sulfate ion, which is lost to the sea in river runoff. Until recently it was not clear how this sulfur was replenished—that is, how the *sulfur cycle* works. In 1971 James Lovelock (Box 16.1) was the first to suggest that sulfur is returned to the atmosphere from the ocean as dimethylsulfide, $(CH_3)_2S$. By the 1980s it was clearly established that certain microscopic marine organisms produce large quantities of dimethyl sulfide. Being a gas, dimethyl sulfide escapes to the atmosphere, where it is oxidized to sulfur dioxide, sulfuric acid, sulfites, and sulfates, which return to the land in rain.

The Carbon Cycles There are two carbon cycles—a *biological carbon cycle* and a *geological carbon cycle*. In the biological carbon cycle, photosynthesis converts CO_2 to organic compounds, and respiration oxidizes the organic compounds to CO_2. The geological carbon cycle (Figure 19.19) starts when carbon dioxide dissolved in rain water reacts with silicates containing calcium to produce soluble calcium hydrogencarbonate and silicic acid,

$$3H_2O(l) + 2CO_2(g) + CaSiO_3(s) \rightarrow Ca^{2+}(aq) + 2HCO_3^-(aq) + H_4SiO_4(aq)$$

which are carried to the oceans. There marine organisms form shells of calcium carbonate from the calcium hydrogencarbonate:

$$Ca^{2+}(aq) + 2HCO_3^-(aq) \rightarrow CaCO_3(s) + H_2O(l) + CO_2(g)$$

The liberated carbon dioxide returns to the biosphere. After the organisms die, the

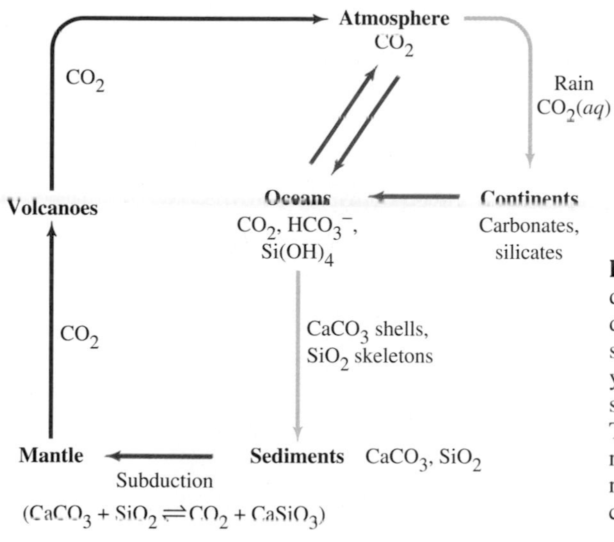

FIGURE 19.19 The Geological Carbon Cycle Carbon dioxide in the ocean is taken up by marine organisms that produce $CaCO_3$ shells or siliceous skeletons and are deposited as sediments on the ocean floor after they die. Over millions of years, oceanic crust undergoes *subduction*—it is carried (with the sediments) atop moving plates of lithosphere down to the mantle. There the sediments are heated, and the $CaCO_3$ is decomposed by reaction with silica to CO_2 and $CaSiO_3$. The CO_2 is eventually released to the atmosphere at volcanoes, and the cycle is complete.

calcium carbonate is deposited on the ocean floor and is eventually converted to limestone and chalk. Other marine organisms, including a type of algae called diatoms, make siliceous skeletons from silicic acid that is formed by the weathering of silicate rocks and is washed into the oceans. These skeletons are deposited on the ocean floor after the organisms die. Over time, the chalk, limestone, and siliceous deposits are carried down to the mantle by the sinking lithosphere in regions called subduction zones, where the overlying oceanic crust is thrust under the continental crust. There they react under high-temperature and high-pressure conditions to release carbon dioxide,

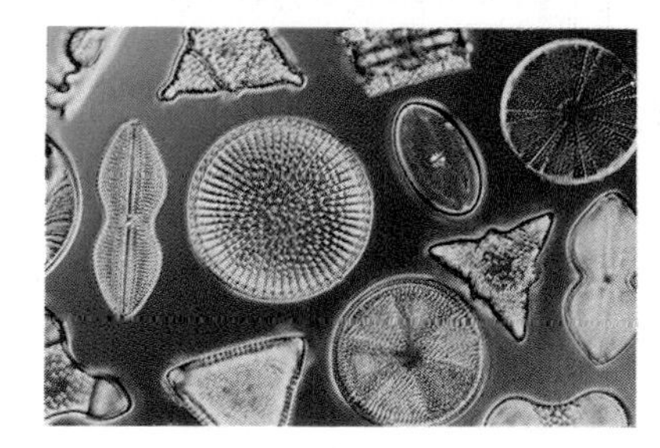

The silica skeletons of a variety of diatoms

$$SiO_2(l) + CaCO_3(s) \rightarrow CaSiO_3(l) + CO_2(g)$$

which is eventually returned to the atmosphere as volcanic gas. This is the same reaction that occurs in the formation of silicate slag in a blast furnace in the manufacture of iron and steel (Chapter 10).

The biological and geological carbon cycles are not independent of each other. Much of the remains of dead organisms are converted to coal and oil. In the absence of human activities, these fossil fuels would eventually be carried down to the magma by subduction, converted to carbon dioxide, and returned to the atmosphere. Burning of these fossil fuels returns carbon to the atmosphere more directly as carbon dioxide. Forest fires also convert much plant material directly to carbon dioxide.

The composition and conditions of the biosphere that are essential for maintaining life in its present form are held within very close limits by a complex set of interdependent chemical reactions. The biosphere appears to be a self-regulating chemical system in which life plays an essential role in the control and regulation of the conditions it needs to survive (Box 16.1).

Exercise 19.9 The rate of lightning flashes in the (entire) atmosphere is estimated at about 100 s^{-1}, and each flash produces about 10^{27} molecules of NO. What is the approximate amount of nitrogen, in metric tons, that is fixed annually by lightning?

BOX 19.1 Chemical Analysis and the Extinction of the Dinosaurs

The transition between the Cretaceous and Tertiary periods of geological time, called the K–T boundary, occurred about 65 million years ago. It seems to be marked by the extinction of the dinosaurs and of many other animal and plant species—a *mass extinction*. Dinosaurs were the dominant land animals for about 160 million years, up to and during the Cretaceous period. The only mammals at that time appear to have been small, rodentlike creatures. But by the Tertiary period, many new mammal species had evolved, and the dinosaurs had completely disappeared.

Several hypotheses have been proposed to account for the apparent mass extinction at the K–T boundary. Geologist Walter Alvarez (1940–) and his physicist father Luis Alvarez (1911–) proposed that the mass extinction was caused by the impact of an asteroid (a large meteorite). They based this suggestion on the chemical analysis of a thin layer of clay that marks the K–T boundary (Figure A). They found about 30 times more of the rare element iridium in this clay layer than is normal for the earth's crustal rocks. This same high concentration has since been found in the same clay layer at several locations separated by thousands of miles. Although iridium is very rare in the earth's crust (probably because it has been concentrated together with iron and other metals in the earth's core), it is relatively abundant in meteorites and in the universe as a whole. The Alvarezes concluded that the iridium had an extraterrestrial origin. They proposed that the impact of an asteroid about 10 km in diameter could account for the high iridium content at the K–T boundary. Such a collision could have generated a gigantic dust cloud (containing iridium) that spread around the earth, blocked out sunlight, and decreased the temperature of the earth long enough to kill many plants and animals. The dinosaurs might have perished because their food supply was wiped out or because they were unable to withstand the sudden climatic change. Some marine organisms with calcium carbonate shells went extinct at about the same time; the color change in Figure A indicates that the deposition of the shells from which the Cretaceous limestone formed had ceased abruptly. The limestone was replaced by the clay layer and then by the sandstone of the Tertiary period.

One argument against the Alvarez hypothesis has been that the asteroid would have produced a crater about 150 km in diameter, but no crater of that size and the correct age was known. However, evidence of such a crater was found in 1992 in the Yucatan Peninsula of Mexico.

Figure A *Iridium-rich clay layer at the Cretaceous–Tertiary boundary (marked by a coin)*

The Alvarez theory has received wide publicity and has generated considerable controversy. Some paleontologists have suggested that the iridium layer resulted from unusually intense volcanic activity that could have brought deeply buried iridium to the surface and released enough gas and dust to darken the skies and cool the earth for more than six months, which could have caused a mass extinction. Some eruptions from volcanoes in Hawaii do indeed have a high iridium content. Because there appear to have been similar, although some less dramatic, extinctions on average every 20 or 30 million years, other researchers have proposed that each extinction was due not to a collision with a single large asteroid but to collisions with many smaller meteorites: The earth passes through a dense meteorite cloud every 20 to 30 million years.

Whatever the correct explanation for the extinction of dinosaurs, the evidence provided by the analysis of iridium tells us that something very unusual was happening at that time in the earth's crust. This evidence is another example of the importance of chemistry in other fields. Although the abundance of iridium in the earth's crust is very low, chemists have developed very sensitive techniques such as neutron activation analysis by which all the elements can be analyzed (Box 5.1).

Postscript: Chemistry, The Central Science

In our discussions of the origins of atoms and molecules, of their distribution in interstellar space, in the stars, in the planets, on earth, and in the biosphere, and of the origins of life, we have seen how important an understanding of atoms and molecules is to the other sciences. Whether we are dealing with cosmology, astronomy, geology, biology, biochemistry, or ecology, the answers to many questions require a knowledge and understanding of atoms, molecules, and their reactions. For example, a possible answer to the question, What was responsible for the extinction of the dinosaurs 65 million years ago? has been provided by analytical chemistry (Box 19.1). This chapter illustrates particularly well that chemistry is the central science. Chemistry is everywhere. No matter what aspect of the material world we are concerned with, atoms, molecules, and their reactions are involved.

SUMMARY

Cosmochemistry is the chemistry of the universe. According to the big-bang model, the early universe consisted of a mixture of hydrogen and helium nuclei and electrons. Under the action of gravity, aggregates of increasing mass and temperature formed in which self-sustaining nuclear reactions could occur between protons to give ^{4}He nuclei, and a star was born. Then, when most of the hydrogen had been consumed, gravitational forces caused the star to collapse, increasing the density and temperature. Under these conditions, further nuclear reactions could take place to give heavier nuclei up to ^{56}Fe. The nucleosynthesis of heavier nuclei is an endothermic process, so heavier elements could have formed only when a star eventually collapsed to form an extremely dense, high-energy neutron star and exploded as a supernova, scattering its nuclei into space. The stage was then set for the formation of second-generation stars containing most of the elements.

Most of the matter in the universe is found in stars, but about 10% is found at very low densities in interstellar space, where the temperature is suitable for the formation of molecules. Some of this matter has accumulated into interstellar clouds containing gaseous molecules and dust (mainly metal oxides). Interstellar atoms and molecules can be characterized by analyzing the electromagnetic radiation that arrives from space. Atoms are identified from their atomic spectra, and molecules from their visible (electronic), infrared (vibrational), or microwave (rotational) spectra.

The solar system is thought to have been born when a giant cloud of spinning interstellar gas and dust (the solar nebula) condensed. The further collapse of this cloud under the action of gravity eventually led to the formation of the sun and the planets. The inner planets (Mercury, Venus, Earth, and Mars) condensed at higher temperatures and contain a higher proportion of high-boiling-point elements and their compounds than did the outer planets Jupiter, Saturn, Uranus, and Neptune, which are composed mainly of H_2, He, CH_4, and NH_3.

Comets and asteroids are smaller bodies in the solar system. Comets consist of a solid "ice" of H_2O, NH_3, CH_4, and CO_2 that contains particles of metal oxides and SiO_2. Meteorites are asteroids that enter the earth's atmosphere. They are composed mainly of silicates or iron, and some contain organic compounds.

The earth has a partly solid, partly molten Fe/Ni core that gives rise to its magnetic field, and an outer silicate mantle covered by a thin, rigid outer layer called the crust. Rocks are classified as igneous (such as the aluminosilicates basalt and granite), which solidify from the molten state; sedimentary (such as sandstone, clay, and limestone), which form by sediment deposition or chemical precipitation; or metamorphic (such as marble and slate), which form by the modification of existing rocks.

Geochemistry is the chemistry of the earth. The chemistry of silicon is of key importance in geochemistry. Silicon forms strong polar SiO bonds, but Si—Si bonds are weaker than C—C bonds, the strength of which is an important aspect of carbon chemistry. The chemistry of silicon differs from that of carbon (although both are Group IV elements) because of differences in their bond strengths, electronegativities, the sizes of their valence shells, and the greater ability of carbon to form multiple bonds. CO_2 is a gas containing covalent linear molecules, whereas the various forms of silica (silicon dioxide), SiO_2, such as β-cristobalite and quartz, are covalent network solids based on the tetrahedral SiO_4 group. Silica is slightly soluble in water to give weak silicic acid, $Si(OH)_4$, which reacts with bases to give silicates, such as Na_4SiO_4. Olivine, the main constituent of the mantle, is $(Fe,Mg)_2SiO_4$. The many different ways in which SiO_4 tetrahedra can be linked give a large number of different silicate minerals. Pyroxenes contain long-chain anions with the formula $(SiO_3^{2-})_n$, such as diopside, $CaMg(SiO_3)_2$. Amphiboles contain double chains that share O atoms on alternate SiO_4 tetrahedra, giving a silicate anion with the empirical formula

$Si_4O_{11}^{6-}$. Sheets of SiO_4 tetrahedra sharing three corners have the empirical formula $Si_2O_5^{2-}$ and are found in talc, $Mg_3(Si_2O_5)_2(OH)_2$, and the clay kaolinite, $Al_2(Si_2O_5)(OH)_4$. In aluminosilicates, such as feldspars and micas, Si atoms of silicate anions are partly replaced by Al^- ions. A typical feldspar is orthoclase, $K(AlSi_3O_8)$.

Rocks are usually heterogeneous mixtures. Minerals are pure crystalline substances. Minerals of commercial value are called ores. Hydrothermal processes form ore deposits by dissolving minerals at high temperatures and pressures and reprecipitating them. After millions of years, the decay of dead organisms in marine sediments and of plant material in terrestrial swamps gives oil or natural gas and coal, respectively.

The massive outer planets have retained their original atmospheres, largely H_2 and He, but these light elements have been lost from the smaller inner planets and replaced by the products of outgassing from the earth's interior. The atmosphere of the earth, unlike that of the other planets, contains $O_2(g)$ as a result of photosynthesis. Energy from the sun maintains nonequilibrium conditions. Liquid water in the earth's hydrosphere also makes it a unique planet.

Molecules essential to life probably first formed on earth by reactions between simple molecules such as HCN and CH_2O. Some of them, such as amino acids, may also have formed in interplanetary space. How these simple molecules first formed the self-replicating molecules of life is still a largely unanswered question.

The biosphere has a constant composition that is maintained by natural cycles, including the water, nitrogen, sulfur, and carbon cycles. The elements necessary for life are recycled from dead organisms by converting them to a form in which they can be reused to form new organisms.

IMPORTANT TERMS

aluminosilicate (page 699)
amphibole (page 698)
biosphere (page 705)
core (page 693)
cosmochemistry (page 685)
crust (page 693)
feldspar (page 699)
geochemistry (page 685)
hydrosphere (page 704)
hydrothermal process (page 701)
igneous rock (page 693)
mantle (page 693)
metamorphic rock (page 694)
mica (page 701)
mineral (page 701)
nucleosynthesis (page 685)
ore (page 701)
pyroxene (page 698)
sedimentary rock (page 694)
silica (page 695)
silicate (page 697)
supernova (page 686)

REVIEW QUESTIONS

1. What is nucleosynthesis?

2. How are the temperatures needed to achieve nucleosynthesis reached in a star?

3. Among the first 20 elements, elements with odd atomic numbers are much less abundant than those with even atomic numbers. Why?

4. Why have only a few different molecules been observed in stars, whereas more than 100 have been detected in interstellar space?

5. What is an interstellar cloud?

6. What spectroscopic measurements give information about **(a)** atoms and **(b)** molecules in space?

7. What are the principal differences between the compositions of the inner and outer planets?

8. Name the principal layers of the earth, and describe their compositions.

9. What are the three principal types of rocks?

10. Why are silicon dioxide and carbon dioxide so different structurally and chemically?

11. Write **(a)** the formula of silicic acid and **(b)** the equation for its condensation to give disilicic acid.

12. Explain the terms **(a)** ''chain silicate'' and **(b)** ''sheet silicate.''

13. How are aluminosilicates related to silicates?

14. What are **(a)** a mineral and **(b)** an ore?

15. What is outgassing?

16. What accounts for the great change in the earth's atmosphere since the earth formed?

17. Briefly describe the Miller–Urey experiment.

18. (a) What are meteorites?

(b) Briefly describe their composition.

19. Explain (a) the two carbon cycles; (b) the sulfur cycle.

PROBLEMS

The Origin of Atoms and Molecules

Nucleosynthesis in the Stars

1. (a) Why does nuclear fusion occur only at very high temperatures?

(b) How much energy evolves in the multi-step process in which four ^{1}H atoms combine to give a ^{4}He nucleus?

2. Describe the formation of (a) a "hydrogen" star and (b) the steps and reactions involved in its evolution to a "helium" star, a "neutron" star, and eventually a "supernova."

3. (a) Why does a "hydrogen" star collapse when most of its hydrogen has been converted to helium?

(b) Why does this collapse initiate the synthesis of heavier nuclei?

(c) Why are no nuclei heavier than ^{56}Fe formed by this process?

(d) How does a second-generation star differ from a first-generation star?

4. Why has it proved very difficult for scientists to achieve controlled nuclear fusion power as an alternative energy source on earth?

5. Is more energy released in making ^{4}He from ^{1}H or in making ^{16}O from ^{12}C plus ^{4}He? Calculate the energy released in the latter nuclear reaction.

Interstellar Molecules

6. The molecule H_2O is far more abundant on earth than is the radical OH. The opposite is probably true in parts of interstellar space. Suggest a reason for this difference.

7. (a) In the laboratory, a vacuum pump that reduces the pressure inside an apparatus to 1.00×10^{-4} mm Hg is regarded as fairly efficient. To how many molecules per liter does this pressure correspond at 25°C?

(b) What is the range of gas pressures in interstellar clouds containing from 10^5 to 10^9 H_2 molecules per liter at −270°C?

The Formation and Composition of the Solar System

8. (a) Describe how the solar system is thought to have formed.

(b) Account for the markedly different compositions of the inner planets and the outer planets.

9. How is the Earth's atmosphere different from that of (a) Venus or (b) Mars?

(c) Explain these differences.

10. Using Table 19.3, calculate the partial pressures of (a) CO_2 and N_2 in the atmospheres of Venus and Mars and (b) N_2, O_2, and Ar on Earth.

(c) How many molecules per liter do each of the partial pressures in (a) and (b) represent?

11. (a) What is a comet?

(b) What is believed to be the chemical composition of comets?

(c) What direct evidence do we have of the composition of comets?

The Structure and Composition of the Earth

12. (a) Name a sedimentary rock, and give its chemical composition.

(b) Name an igneous rock, and give its chemical composition.

13. The average density of the earth is 5.5 g · cm^{-3}, whereas surface rocks have a density of 2.8 g · cm^{-3}. Explain this difference in terms of the structure of the earth.

14. Describe the origins of (a) igneous, (b) sedimentary, and (c) metamorphic rocks.

Silicon and Its Compounds

15. (a) Write the ground-state electronic configuration of silicon, and deduce the expected valence of silicon.

(b) Explain why silicon exhibits a different valence in most of its compounds.

16. What are the principal reasons the chemistry of silicon differs in many ways from that of carbon?

17. (a) Give an example of a compound of silicon whose structure and properties are very different from those of the corresponding compound of carbon, and explain this difference.

(b) Give an example of a compound of silicon whose structure is similar to that of the corresponding carbon compound but that has at least one very different chemical property, and explain this difference.

18. Write a formula and describe the structure of each of the following compounds.

(a) silica (b) silicon tetrachloride (c) carborundum

(d) silane, the silicon analogue of methane

19. Silicon reacts with sulfur at elevated temperatures.

(a) If 0.0932 g of silicon reacts with sulfur to give 0.3060 g of silicon sulfide, determine the empirical formula of silicon sulfide.

(b) Suggest a probable structure for silicon sulfide.

(c) Would you expect silicon sulfide to be a gas, a liquid, or a solid?

20. Describe the structure of clay, and explain what happens when clay is baked to make a ceramic object such as a pot.

21. **(a)** What is the basic structural unit of silicates?

(b) Explain with examples how these basic units can be joined to give chains and sheets.

(c) What is the empirical formula of an infinite silicate chain and of an infinite silicate sheet?

22. **(a)** Use Table 9.2 and the ΔH_f° values 34.3, 80.3, and 450 $kJ \cdot mol^{-1}$ for $SiH_4(g)$, $Si_2H_6(g)$, and $Si(g)$, respectively, to calculate and compare the average bond energies of the C—H and Si—H bonds in $CH_4(g)$ and $SiH_4(g)$, respectively.

(b) Using these values, calculate and compare the C—C and Si—Si bond energies in $C_2H_6(g)$ and $Si_2H_6(g)$, respectively.

23. **(a)** How is the fundamental structural unit on which the structures of all silicates are based modified in the aluminosilicates?

(b) Draw the structures of three different silicate anions and one aluminosilicate anion.

24. **(a)** How is silicic acid prepared from $SiCl_4$?

(b) How does silicic acid differ from carbonic acid?

(c) Why cannot carbonic acid be similarly prepared by an analogous reaction of CCl_4?

25. Predict the structure of the silicate anion (chain, ring, double chain, sheet, or network) in each of the following silicates.

(a) gillespite, $BaFeSi_4O_{10}$ **(b)** dentitoite, $BaTiSi_3O_9$

(c) chrysotile, $Mg_3Si_2O_5(OH)_4$

(d) rhodonite, $CaMn_4Si_5O_{15}$

(e) vermiculite, $Mg_3Si_4O_{10}(OH)_4 \cdot xH_2O$

26. Predict the structure of the aluminosilicate anion (chain, ring, sheet, or network) in each of the following aluminosilicates.

(a) anorthite, $Ca(AlSiO_4)_2$

(b) muscovite, $KAl_2(AlSi_3O_{10})(OH)_2$

(c) amesite, $Mg_2Al(AlSiO_5)(OH)_4$

(d) thomsonite, $NaCa_2(AlSiO_4)_5 \cdot 6H_2O$

27. Describe the structure of the silicate or aluminosilicate anion in each of the following minerals.

(a) diopside, $CaMgSi_2O_6$ **(b)** orthoclase, $KAlSi_3O_8$

(c) hardystonite, $Ca_2ZnSi_2O_7$

(d) dentitoite, $BaTiSi_3O_9$

Minerals and Other Geologic Resources

28. What distinguishes a rock, a mineral, and an ore? Give an example of each.

29. Some iron ores are sulfides, such as pyrite, $FeS_2(s)$, and some are oxides, such as hematite, $Fe_2O_3(s)$. Describe how and where these ores probably originated.

30. Explain how **(a)** oil and natural gas and **(b)** coal form.

The Atmosphere

31. Explain the statement "the earth's atmosphere is not at chemical equilibrium."

General Problems

32. **(a)** What small, monomeric, organic molecules were needed before life could begin?

(b) What is the evidence that these molecules were present on earth prior to the emergence of life?

33. Name each of the oxoacids with the following molecular formulas. Place them in order of increasing strength, and explain this order.

(a) $Si(OH)_4$ **(b)** $PO(OH)_3$

(c) $SO_2(OH)_2$ **(d)** $ClO_3(OH)$

34. Germanium is the element below silicon in Group IV. What properties would you predict for **(a)** the oxide GeO_2 and **(b)** the hydroxide $Ge(OH)_4$?

35. **(a)** Explain why silicon tetrafluoride reacts with fluoride ion to form the complex ions SiF_5^- and SiF_6^{2-} but carbon tetrafluoride does not react with fluoride ion.

(b) Predict the structures of SiF_5^- and SiF_6^{2-}.

36. Write balanced equations to show how a mixture of formaldehyde, ammonia, HCN, and water can form an amino acid, and name this acid.

37. Carbon dioxide can react with silicates to give carbonates, and silicon dioxide can react with carbonates to give carbon dioxide.

(a) Under what conditions do these two types of reactions occur?

(b) How are these reactions important in maintaining the atmospheric composition of the earth?

Polymers and Materials Science

20.1 Polymer Chemistry
20.2 Materials Science

We have come to rely on synthetic polymers, such as the fastener Velcro, in many aspects of our lives. As this photomicrograph shows, one side of Velcro is covered with many tiny hooks, and the other side is covered with even more tiny loops, both made of a polymer such as nylon or a polyester. The hooks and loops become entangled and steadfastly hold the two sides together. This ingenious fastening device was designed to copy nature's own method of dispersing seed burrs.

Early civilizations used the materials they found in their surroundings to make their homes, clothes, containers for food and drink, weapons, farming implements, and ornaments. These materials included wood, stone, plant fibers, animals furs and skin, clay, and naturally occurring metals such as gold and silver. But gradually, largely through accidental discoveries, people learned how to transform naturally occurring materials into other, more useful materials. They learned, for example, to reduce metal ores with charcoal to make metals, to heat limestone to make lime, to treat animal skins to make leather, and to treat animal fats with wood ashes to make soap. In other words, they learned to make use of chemistry, although their knowledge was entirely empirical—they had no understanding of the atomic or molecular compositions of these materials. This empirical chemical knowledge has had an enormous impact on human society, but it is only since the establishment of chemistry as a science, beginning early in the nineteenth century, that chemists have been able to transform natural materials in a planned and rational way. Still more importantly, chemists have been able to synthesize an enormous number of entirely new substances.

In this chapter we discuss developments in two relatively new areas of chemisry that have transformed our lives in recent years: *polymer chemistry* and *materials science*. Both are concerned with new types of solids whose properties differ from those of the simple network and molecular solids we have considered previously. These novel and occasionally unforeseen properties have a vast number of applications that play an increasingly important role in modern life. Although polymers are new solid materials that may be considered to comprise part of materials science, they form a largely separate area of research and development. This is mainly because polymers are mostly organic compounds, whereas the substances considered in materials science are usually inorganic compounds.

20.1 POLYMER CHEMISTRY

Until about 1930, chemists did not seriously consider the existence of molecules with molecular masses of 10,000 u or more. However, after the German chemist Hermann Staudinger (1881–1965) and others showed that molecules of this size called **polymers**, can be made simply by joining many small molecules together, rapid progress was made in the synthesis and understanding of such molecules. **Polymer chemistry** is now one of the major branches of chemistry, and the polymer industry employs about half the organic chemists in the United States. Polymers, often called *plastics*, have a wide range of useful properties and are used for making an enormous variety of articles, such as clothing, carpets, water pipes, buckets, sheathing for electrical cables, skis, tennis rackets, canoes, boats, and even car bodies. We may be said to be living in the Plastics Age. Chemists often call large molecules **macromolecules**, and **macromolecular chemistry** is an alternative name for polymer chemistry.

Many polymers are products of the petrochemical industry. In each case the starting materials are hydrocarbons that are either polymerized directly or are converted to other compounds that are then polymerized. There are two principal types of synthetic polymers: *addition polymers* and *condensation polymers*.

Addition Polymers

Addition polymers are polymers formed by joining a large number of simple molecules called *monomers*. The empirical formula of the resulting polymer is the same as that of the monomer. The monomers are alkenes and their derivatives—in particular, ethene, $H_2C{=}CH_2$, and derivatives of ethene in which one or more of the hydrogen atoms have been replaced by other atoms or groups. The names of addition polymers are derived by adding the prefix *poly-* to the name of the monomer, as in *polyethylene*. Typically, the common name of the monomer (in this case, ethylene) is used rather than the systematic name (ethene). A few of the most important addition polymers formed from derivatives of ethene are listed in Table 20.1. Four of these—polyethylene, polypropylene, polyvinylchloride, and polystyrene—are discussed next.

Polyethylene *Polyethylene*, C_nH_{2n}, is a simple and well-known addition polymer formed by the addition of ethene molecules to give a long hydrocarbon chain (Figure 20.1). Most commonly the reaction is initiated by the addition of a small amount of an organic peroxide, R—O—O—R, such as benzoyl peroxide:

A collection of articles made of polyethylene

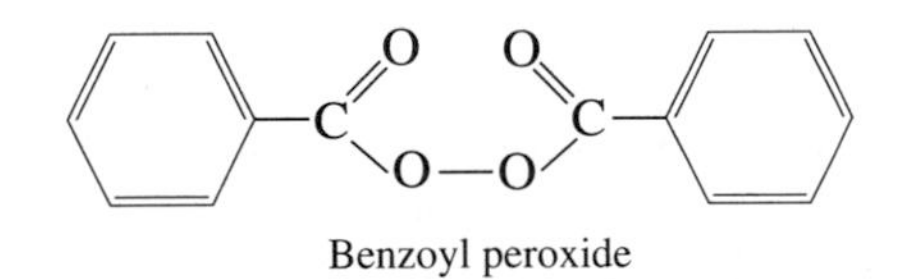

Benzoyl peroxide

Because the O—O single bond is rather weak (BE 138 $kJ \cdot mol^{-1}$), it breaks upon heating to give two free radicals:

$$R{-}O{-}O{-}R \rightarrow R{-}O\cdot + \cdot O{-}R$$

The R—O· free radical adds to an ethene molecule to give another free radical:

$$RO\cdot + H_2C{=}CH_2 \rightarrow RO{-}CH_2{-}CH_2\cdot$$

The $RO{-}CH_2{-}CH_2\cdot$ free radical then adds to another ethene molecule and so on in a series of propagation steps that give free radicals with ever-increasing chain lengths:

$$RO{-}CH_2{-}CH_2\cdot + H_2C{=}CH_2 \rightarrow RO{-}CH_2{-}CH_2{-}CH_2{-}CH_2\cdot$$

$$RO{-}CH_2{-}CH_2{-}CH_2{-}CH_2\cdot + H_2C{=}CH_2$$
$$\rightarrow RO{-}CH_2{-}CH_2{-}CH_2{-}CH_2{-}CH_2{-}CH_2\cdot$$

When one free radical containing n repeating units combines with another containing m units, the product is $RO(CH_2CH_2)_{n+m}OR$.

TABLE 20.1 Some Addition Polymers Produced from Substituted Ethenes

Monomer	Polymer	Typical Uses
$CH_2{=}CH_2$ Ethene	$\text{-}[CH_2{-}CH_2]_n\text{-}$ Polyethylene	Containers, pipes, bags, toys, wire insulation
$CH_2{=}CHCH_3$ Propene	$[CH_2{-}CH(CH_3)]_n$ Polypropylene	Fibers for carpets, artificial turf, rope, fishing nets
$CH_2{=}CHCl$ Chloroethene (vinyl chloride)	$[CH_2{-}CH(Cl)]_n$ Polyvinylchloride (PVC)	Garden hoses, floor tiles, plumbing, records, laboratory tubing
$CH_2{=}CHCN$ Propenenitrile (acrylonitrile)	$[CH_2{-}CH(CN)]_n$ Polyacrylonitrile (Orlon, Acrilan)	Fibers for cloth, carpets, upholstery
$CH_2{=}CH{-}C_6H_5$ Styrene	$[CH_2{-}CH(C_6H_5)]_n$ Polystyrene	Styrofoam, hot-drink cups, insulation
$CF_2{=}CF_2$ Tetrafluoroethylene	$\text{-}[CF_2{-}CF_2]_n\text{-}$ Teflon	Nonstick coating for kitchen utensils
$CH_2{=}C(CH_3){-}CO_2CH_3$ Methylmethacrylate	$[CH_2{-}C(CH_3)(CO_2CH_3)]_n$ Polymethylmethacrylate	Plexiglas, Lucite, headlamp lenses, sunglasses, aircraft windows

Eventually, the chain is terminated either by the combination of two $RO(CH_2CH_2)_n\cdot$ radicals to give $RO(CH_2CH_2)_nOR$ or by the extraction of a hydrogen atom from the next-to-last carbon atom in a chain by a radical $RO\cdot$ or $RO(CH_2CH_2)_n\cdot$ to give a polymer with a terminal double bond:

$$-CH_2-CH_2-CH_2-CH_2-CH_2-CH_2\cdot \xrightarrow{RO\cdot} -CH_2-CH_2-CH_2-CH_2-CH{=}CH_2 + ROH$$

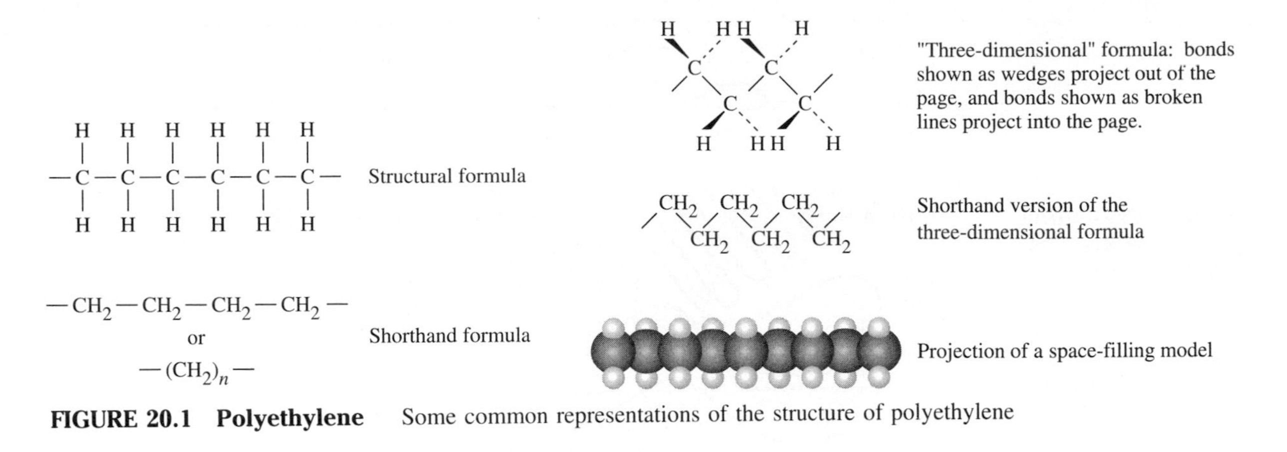

FIGURE 20.1 Polyethylene Some common representations of the structure of polyethylene

The steps we have just described produce a polymer with a single continuous chain of carbon atoms. However, a free radical can extract a hydrogen atom not only from the carbon atom one from the end of a polymer chain, as in a termination reaction, but also from a carbon atom in the middle of a growing polymer chain:

$$RO{-}CH_2{-}CH_2{-}CH_2{-}CH_2{-}CH_2{-}CH_2{-}\cdots CH_2^{\cdot}$$

$$\big\downarrow RO\cdot$$

$$RO{-}CH_2{-}CH_2{-}\dot{C}H{-}CH_2{-}CH_2{-}CH_2{-}\cdots CH_2\cdot + ROH$$

The chain now has two growing points, and a *branched-chain polymer* results. If this process occurs several times during the growth of the polymer, the final product may have many branches (Figure 20.2). Separate chains may be linked to give a *cross-linked polymer.*

The propagation steps in the polymerization reaction are usually rapid compared with the termination steps, so many propagation steps occur before the chain is terminated, giving a long-chain polymer with a high molecular mass. Because chain termination is a random process, the resulting polymer is a mixture of molecules of different chain lengths with the same empirical formula. Thus, although both a synthetic polymer and a nonpolymeric substance have a constant composition and corresponding empirical formula, a synthetic polymer differs from nonpolymeric substances in that it has molecules of many different sizes and masses. Even the statement that synthetic polymers have a constant composition is not *completely* accurate. Although we say that polyethylene has the empirical formula C_nH_{2n}, each chain has end groups such as OR or $CH{=}CH_2$ that do not have this composition. However, if the chains have thousands or tens of thousands of atoms, the effect of two end groups per chain on the overall composition is negligible.

Because all the molecules of a nonpolymeric substance have the same size and mass, every sample of such a substance has the same properties. But the properties of a polymer can vary, depending on the lengths of the chains and the degree of branching. If we can control the chain length and the degree of branching and cross linking, we can control the properties of the resulting polymer. In this way we can tailor the properties of a polymer such as polyethylene to suit our needs.

If ethene is polymerized by using a free-radical initiator such as benzoyl peroxide, at approximately 1500 atm and 200°C a form of polyethylene called *low-density polyethylene* is obtained. This form of polyethylene has a density of 0.91 to 0.94 $g \cdot cm^{-3}$ and an average molecular mass of 5×10^4 u to 3×10^5 u. Because

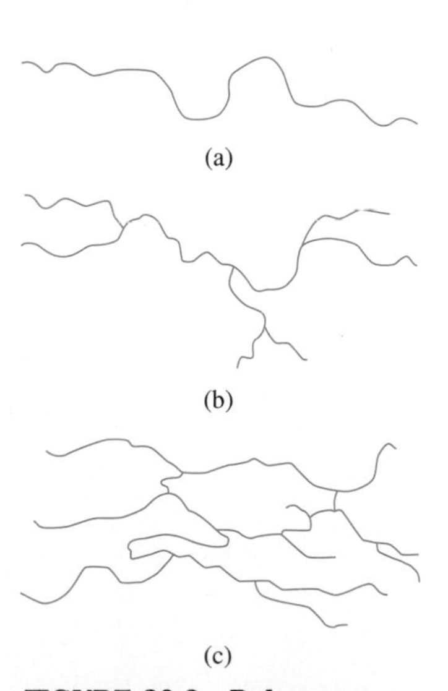

FIGURE 20.2 Polymer Types (a) A continuous-chain (linear) polymer; (b) a branched-chain polymer; (c) a cross-linked (network) polymer

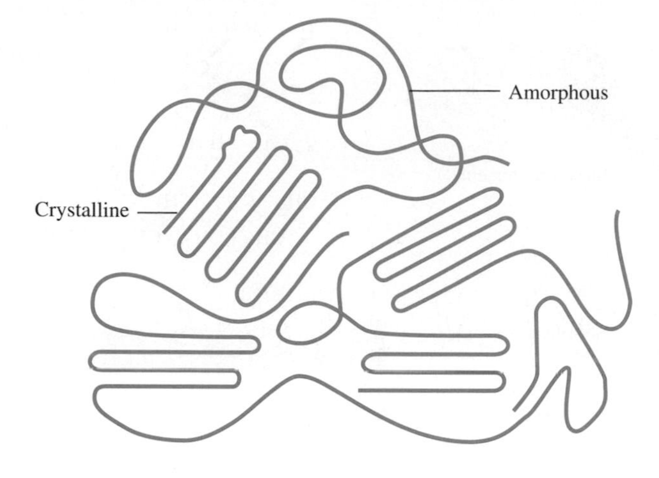

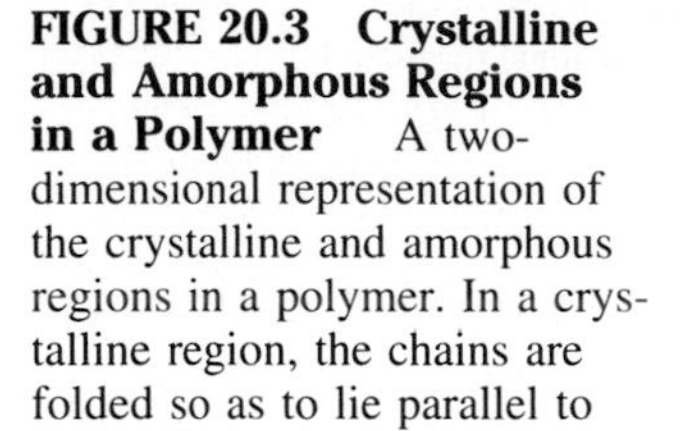
FIGURE 20.3 Crystalline and Amorphous Regions in a Polymer A two-dimensional representation of the crystalline and amorphous regions in a polymer. In a crystalline region, the chains are folded so as to lie parallel to each other.

polyethylene is a mixture of molecules of many different sizes, it does not have a sharply defined melting point but softens over a range of temperatures around 110°C. Low-density polyethylene is translucent in thin sheets, flexible, and cheap. It is the material commonly used for food wrappers, shopping bags, and garbage bags.

Ethene can also be polymerized at much lower pressures and temperatures by using transition metal compounds such as MoO_2, Cr_2O_3, or $TiCl_4$ as catalysts. This reaction proceeds by a mechanism different from that of a polymerization initiated by an organic peroxide. The product is *high-density polyethylene*, which has a density of $0.96\ g \cdot cm^{-3}$, an average molecular mass up to 3×10^6 u, and a softening point of about 130°C. This form of polyethylene is much harder and less flexible than low-density polyethylene. High-density polyethylene is used for commercial packaging and in the manufacture of containers for liquids such as soft drinks, bleach, antifreeze, and engine oil.

The differences in the properties of low- and high-density polyethylene arise from their structural differences. Low-density polyethylene is much more branched than the high-density form, which contains a higher proportion of continuous-chain polymers. Continuous-chain polymers are able to pack together more closely than branched-chain polymers and therefore attain a higher density. When molecules pack closely, they tend to do so in a regular way and to form a crystalline solid, as we saw in Chapter 10. But it is rare to find a polymer that is completely crystalline; rather, only certain regions are crystalline, and the rest is amorphous (Figure 20.3). Generally, less branching and more crystallinity produces a stronger, harder polymer. High-density polyethylene is about 90% crystalline.

Polypropylene The polymerization of propene, $CH_3CH{=}CH_2$, gives *polypropylene*:

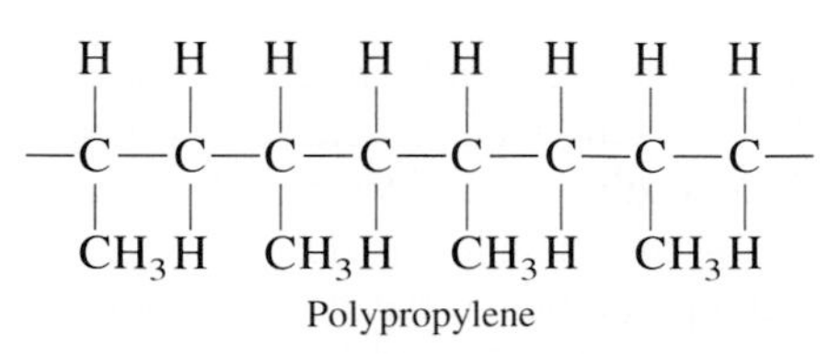

Polypropylene

The free-radical polymerization of propene at high temperatures and pressures gives a viscous liquid or soft solid without useful commercial properties. In this form of polypropylene, the methyl groups have a random arrangement along the chain

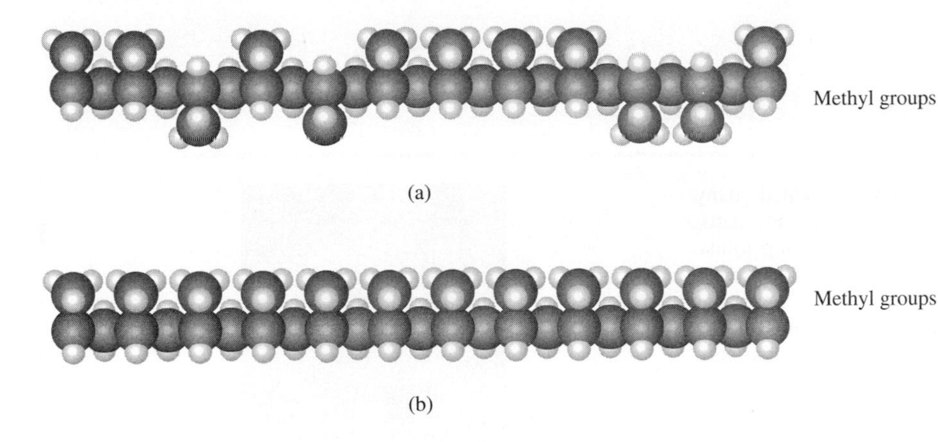

FIGURE 20.4 Polypropylene (a) Polypropylene obtained by free-radical polymerization is a viscous liquid or soft solid. The methyl groups are arranged irregularly. (b) Polypropylene obtained by transition-metal-catalyzed polymerization is a hard solid with a high crystallinity. The methyl groups are arranged regularly.

(Figure 20.4a). Polypropylene made with transition-metal catalysts under much milder temperature and pressure conditions gives a hard solid product with high crystallinity. The long chains are aligned parallel to each other, and the methyl groups are arranged in a regular pattern (Figure 20.4b). Because of its high crystallinity, polypropylene is even stronger and has a higher melting point than high-density polyethylene. Because polypropylene can be sterilized at 140°C, it is widely used for hospital equipment. Other common uses are for packaging and indoor-outdoor carpeting.

Polyvinylchloride (PVC) Chloroethene, $H_2C{=}CHCl$, which is commonly known by its nonsystematic name, vinyl chloride, is made by the reactions

$$\underset{\text{Ethene}}{H_2C{=}CH_2} + Cl_2 \rightarrow \underset{\text{1,2-Dichloroethane}}{ClH_2C{-}CH_2Cl} \xrightarrow{\text{heat}} \underset{\text{Chloroethene}}{H_2C{=}CHCl} + HCl$$

Vinyl chloride is readily polymerized by free-radical initiators to give *polyvinylchloride*, PVC, $-\!\!(CH_2{-}CHCl)\!\!-_n$:

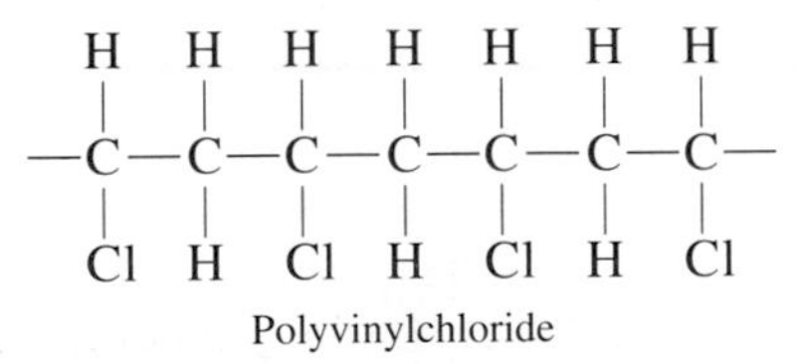

Polyvinylchloride

PVC is one of the cheapest and commonest plastics. This colorless polymer is strong, though somewhat brittle, and relatively resistant to corrosive substances, fire, and weathering, and it is easy to color. Familiar uses are for water pipes, floor tiles, insulation of electrical wire, toys, credit cards, and compact discs.

Polystyrene Phenylethene, $C_6H_5CH{=}CH_2$, commonly called styrene, is made from benzene and ethene by the following two-step process:

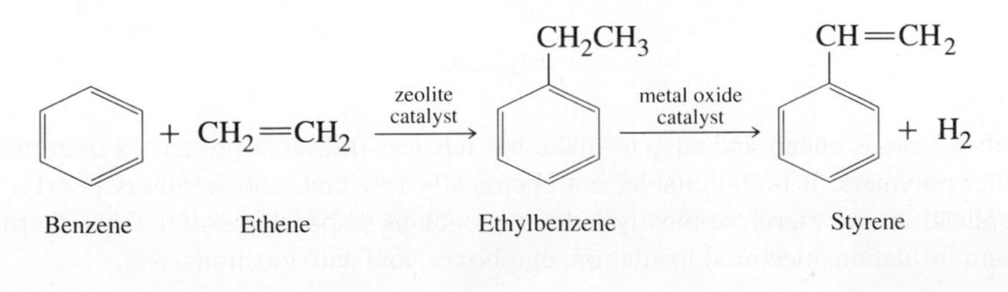

BOX 20.1 Polymers in Medicine

Medical science has in the recent past developed many procedures for replacing and improving human body parts. We have pacemakers, artificial hearts, artificial hip joints, and contact lenses. All these applications require special materials that must be "biocompatible" with the particular part of the body where they are to be located. They must not be rejected by the body and must be resistant to the substances that will surround them. They must not, for example, trigger blood clotting or immune-system responses. Polymers play an important role in this area of medicine.

Artificial hip joints are constructed of metal alloys (Figure A). To carry out a hip replacement, the top of the thigh bone is removed and replaced by a metal prosthesis (an artificial body part) that is cemented in place. The prosthesis must be seated into the hip joint with a plastic cup, which was originally constructed of Teflon. The cup is now usually made from a special high-density polyethylene developed for medical use, as it provokes no adverse reaction from biological materials. To secure the prosthesis, the cement, a mixture of the polymer polymethylmethacrylate, its monomer (methylmethacrylate), and a small amount of initiator, is made at the time of the operation and is allowed to polymerize partly in place. When polymerization is complete, the new joint is firmly anchored (Figure B).

Polymethylmethacrylate is also utilized in contact lenses and in temporary crowns in dentistry. Other polymers with medical applications include polyvinylchloride (PVC), found in dialysis tubing and catheters; poly(hydroxyethyl methacrylate), used in soft contact lenses and burn dressings; and poly(glycolic acid), used for biodegradable sutures.

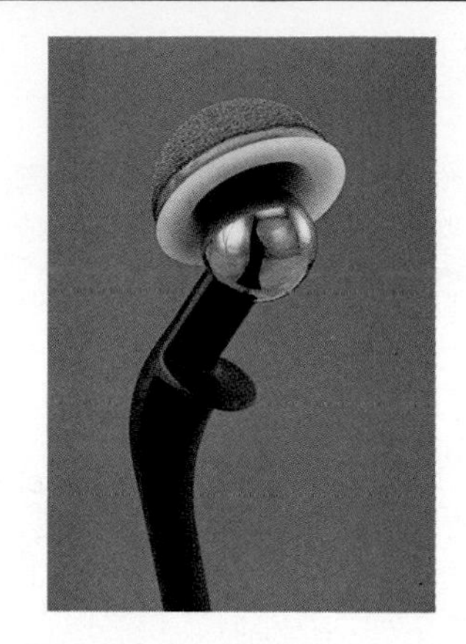

Figure A *Artificial hip joint made of a titanium–steel alloy ball and a high-density polyethylene socket*

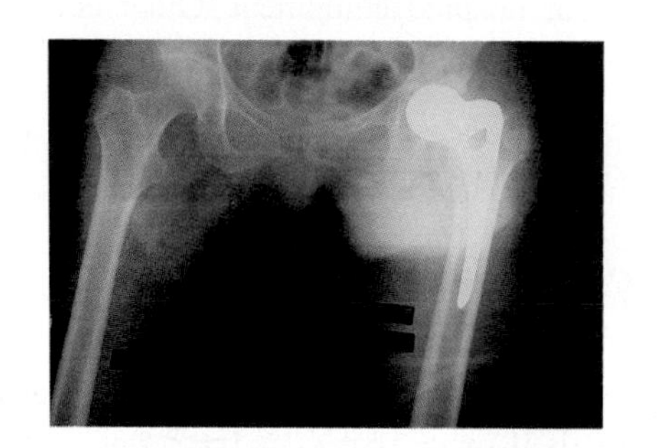

Figure B *X-ray photograph of artificial hip joint in place*

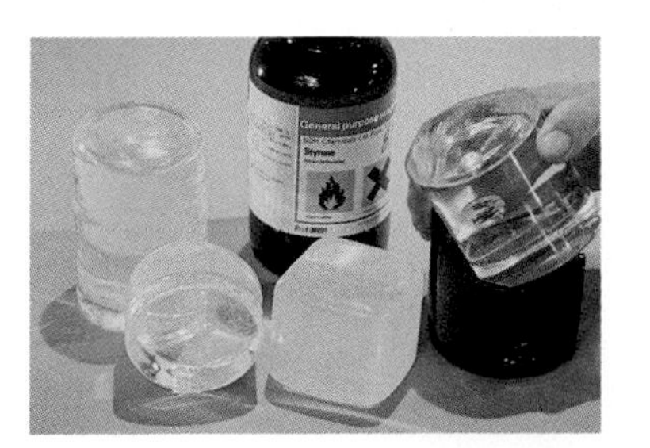

These blocks of solid transparent polystyrene formed by spontaneous polymerization in bottles of styrene that had been kept for a long time.

Styrene is polymerized by free radicals to give *polystyrene*, $(—CH_2—CH(C_6H_5)—)_n$:

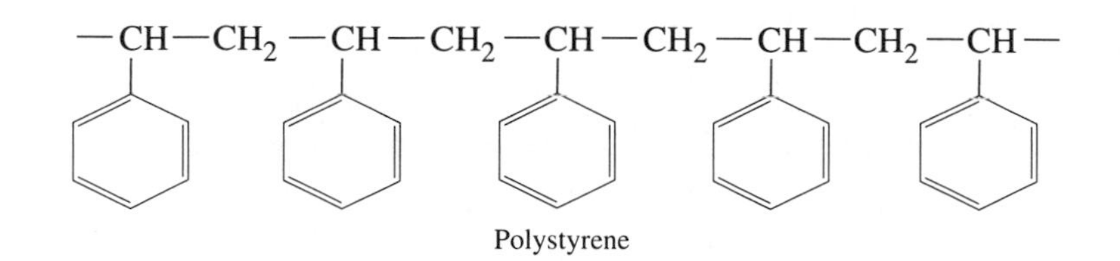

Polystyrene

Polystyrene is cheap and easy to make but has less desirable properties than many other polymers: It is flammable, not chemically resistant, and weathers poorly. Its applications are therefore mostly indoors—such as disposable coffee cups, thermal foam insulation, electrical insulation, egg boxes, and auto cushions.

Other Addition Polymers *Polymethylmethacrylate* (Table 20.1) is an addition polymer with some important medical uses. Box 20.1 discusses the medical applications of polymethylmethacrylate and other polymers. The discovery and uses of the addition polymer *polytetrafluoroethylene*, C_nF_{2n}, better known as Teflon, are described in Box 20.2.

Exercise 20.1 What is the average molecular mass of polystyrene if the average chain consists of 2000 monomers?

Exercise 20.2 Poly-4-methyl-1-pentene is a hard, transparent polymer used in the manufacture of laboratory ware such as flasks and beakers. Draw the structures of the monomer and the polymer.

Rubber and Other Elastomers *Natural rubber* is obtained from the sap of the rubber tree, which grows wild in South America and is cultivated in other parts of the world. The milky-white sap, called *latex*, is collected and then is allowed to dry and harden in the air. The crude form of rubber obtained in this way is a polymer of 2-methyl-1,3-butadiene (isoprene), $H_2C{=}C(CH_3){-}CH{=}CH_2$:

BOX 20.2 Polytetrafluoroethylene

The discovery of this useful polymer, commonly known by its tradename "Teflon," was accidental. Tetrafluoroethene is a gas and is therefore transported in metal cylinders. In 1938 Dr. Roy Plunkett, a DuPont chemist, had occasion to use the gas but found the cylinder apparently empty. Weighing the cylinder confirmed that it should be full; he then sawed the cylinder in half and revealed a solid, white polymer inside. He then found that tetrafluoroethene can be polymerized rather easily to polytetrafluoroethylene, just as ethene can be polymerized by using a free-radical initiator such as an organic peroxide.

Polytetrafluoroethylene is an expensive material, mostly because of the fluorine used in its manufacture. But its properties, particularly its high thermal stability and its extreme resistance to chemical corrosion, have ensured its widespread use in small-scale applications (Figure A). The polymer was developed during World War II, and its first application was in the manufacture of gaskets for the isotope-separation plant built to separate ^{235}U, using the highly corrosive gas UF_6. Its best-known postwar application is as the material that forms the "nonstick" coating for frying pans and other kitchen utensils.

The starting material for the manufacture of tetrafluoroethene is methane. Reaction of methane with chlorine gives chloroform, $CHCl_3$ (Chapter 14):

$$CH_4 + 3Cl_2 \rightarrow CHCl_3 + 3HCl$$

Figure A *A cable with a layer of Teflon insulation.*

Reaction of $CHCl_3$ with hydrogen fluoride, HF, in the presence of a catalyst exchanges two of the chlorine atoms for fluorine atoms to give $CHClF_2$:

$$CHCl_3 + 2HF \xrightarrow{\text{catalyst}} CHClF_2 + 2HCl$$

Upon heating, $CHClF_2$ loses hydrogen chloride to give tetrafluoroethene:

$$2CHClF_2 \xrightarrow{\text{heat}} F_2C{=}CF_2 + 2HCl$$

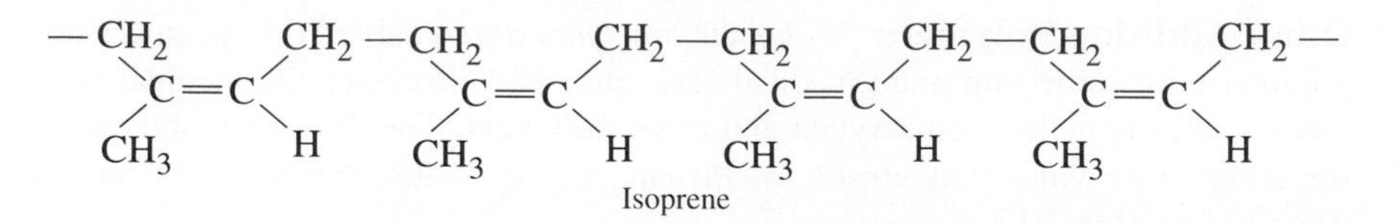

A long sharp-pointed needle can be pushed slowly through an inflated rubber balloon without causing the balloon to burst. The long, flexible rubber molecules move around the hole and seal it.

Rubber was known to the native peoples of South and Central America; the ancient Mayans used rubber balls in a game for which some of the courts are still intact. Christopher Columbus was the first to introduce this material to Europe, in 1496. But natural rubber is not particularly useful, because it becomes soft and sticky in warm weather and brittle in cold weather.

In 1893, soon after the first automobiles appeared, the American inventor Charles Goodyear discovered the process of **vulcanization**. Heating natural rubber with sulfur transforms it into a material that remains hard and maintains its elasticity and flexibility over a wide range of temperatures. This improved form of rubber quickly came to be used for automobile tires. Vulcanization results in the formation of —S—S— cross-links between neighboring polymer molecules, similar to the disulfide bridges found in many proteins (Chapter 17). In natural rubber the long-chain polymer molecules are tangled like spaghetti (Figure 20.5a). When rubber is stretched, the chains straighten out to some extent and slip past each other. When the tension on the rubber is released, the chains tend to coil again, and the rubber resumes its original shape. The coiled, more disordered chains have a higher entropy than do straighter, more ordered chains. When rubber is cross-linked with disulfide bridges, as is vulcanized rubber, the extent to which the chains can be straightened is limited, and adjacent chains cannot slip past each other (Figure 20.5b). They also have a greater tendency to resume their original shape when the tension is released. Thus, vulcanized rubber is stronger, harder, less sticky, and more elastic than natural rubber. Any material such as rubber that resumes its original shape when stretched or compressed and then released is called an **elastomer**.

Natural rubber comes mainly from large plantations in Malaysia and other parts of the Far East. During World War II, this source was cut off. But because rubber was needed for the tires of military vehicles, there was an intensive effort to develop synthetic rubbers. The polymerization of butadiene gives an elastomer, but a more useful material is obtained by polymerizing a mixture of 1,3-butadiene and styrene to give styrenebutadiene rubber (SBR). The process in which a mixture of two (or sometimes more) monomers is polymerized is called copolymerization, and the product is called a **copolymer**.

Another important synthetic rubber, **neoprene**, is obtained by polymerizing 2-chloro-1,3-butadiene (chloroprene), which is made from butadiene:

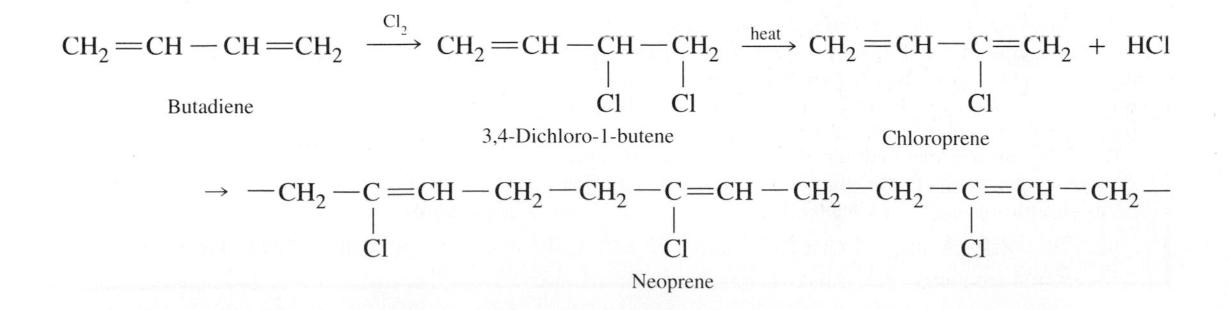

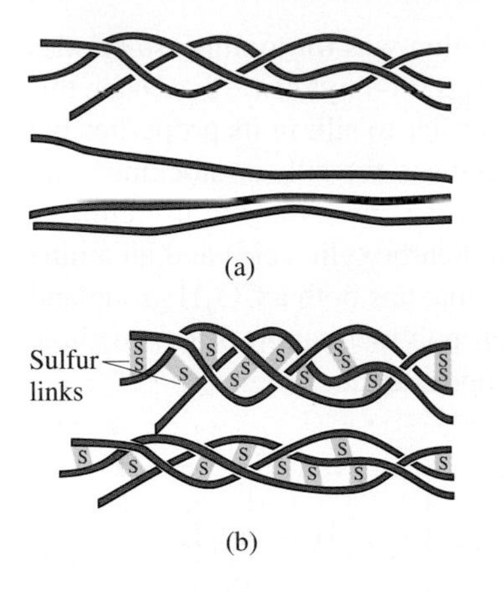

FIGURE 20.5 Natural and Vulcanized Rubber (a) When tension is applied to natural rubber, the chains straighten out and slip past each other. (b) In vulcanized rubber, the chains can straighten out but cannot slip past each other, because they are held together by disulfide bridges. Vulcanized rubber therefore stretches less and resumes its original shape better than natural rubber.

Neoprene is resistant to gasoline and oil. It is used in the manufacture of gas-pump hoses, in parts for automobile engines, and where chemical resistance is important.

Exercise 20.3 Draw the structure of styrenebutadiene (SBR) rubber. Assume that the two monomers alternate in the polymer.

Condensation Polymers

Condensation polymers are polymers formed from monomers by the loss of a small molecule, most often water, between each pair of monomers that are joined. The two most important types of condensation polymers are polyamides and polyesters.

Polyamides **Polyamides** are condensation polymers made either from an amino acid or from a diamine and a dicarboxylic acid. Proteins are polyamide polymers of amino acids that play crucial roles in nearly all biological processes (Chapter 17). The monomers in a polyamide are joined by the amide linkage —CO—NH— (called a peptide linkage in biochemistry). The fibers produced by the silkworm, which have been used for centuries to make fine fabrics, consist of proteins, so it was natural for chemists to attempt to create similar polyamides. The first to succeed was Wallace Carothers (Figure 20.6), the DuPont chemist who first synthesized the

FIGURE 20.6 Wallace H. Carothers (1896-1937) Wallace Carothers, born in Burlington, Iowa, in 1896, was responsible for the development of nylon and of neoprene, the first synthetic rubber. He joined the DuPont Company in 1928 as head of the organic chemistry division and was the first industrial chemist to be elected to the National Academy of Science. But he became depressed by a family death and committed suicide by drinking cyanide at the age of 41. Although his career was brief, Carothers made enormous contributions to our understanding of the structure and properties of macromolecules.

polyamide called **nylon**. The first consumer use of nylon was for toothbrush bristles in 1938, but it was the use of the polymer for stockings, or ''nylons,'' that made this condensation polymer an overwhelming success. Similar to silk in its properties but far less expensive, nylon became the ideal replacement for silk in stockings and other clothing. To make a polyamide such as nylon, it is necessary to use either a carboxylic acid with two carboxylic acid groups (a dicarboxylic acid) and an amine that has two NH_2 groups (a diamine), or a molecule that has both a CO_2H group and an NH_2 group. Nylon is made by the condensation polymerization of hexanedioic acid, commonly called adipic acid, and 1,6-diaminohexane:

This is a Lewis acid–base reaction in which the nonbonding electron pair on N attacks the positively polarized C of the CO_2H group.

$$n\left[HO-\overset{\overset{\displaystyle O}{\|}}{C}(CH_2)_4\overset{\overset{\displaystyle O}{\|}}{C}-OH\right] + n\left[H-\overset{\overset{\displaystyle H}{|}}{N}(CH_2)_6\overset{\overset{\displaystyle H}{|}}{N}-H\right]$$

Hexanedioic acid 1,6-Diaminohexane

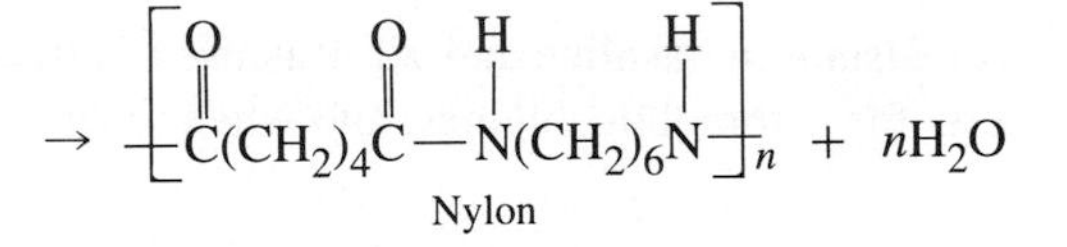

Nylon

This polymer is more accurately called *nylon 66*, because it is made from a dicarboxylic acid and a diamine that contain six carbon atoms each. Figure 20.7 shows how adipic acid and 1,6-diaminohexane are synthesized.

The name nylon is used to describe the whole family of aliphatic polyamides such as nylon 66. A second common type of nylon is *nylon 6*, which can be made by the condensation of 6-aminohexanoic acid with itself:

$$2n\left[H-\overset{\overset{\displaystyle H}{|}}{N}(CH_2)_5\overset{\overset{\displaystyle O}{\|}}{C}-OH\right] \rightarrow \left[\overset{\overset{\displaystyle H}{|}}{N}(CH_2)_5\overset{\overset{\displaystyle O}{\|}}{C}-\overset{\overset{\displaystyle H}{|}}{N}(CH_2)_5\overset{\overset{\displaystyle O}{\|}}{C}\right]_n + 2nH_2O$$

6-Aminohexanoic acid Nylon 6

In practice it is made from the cyclic amide *caprolactam*,

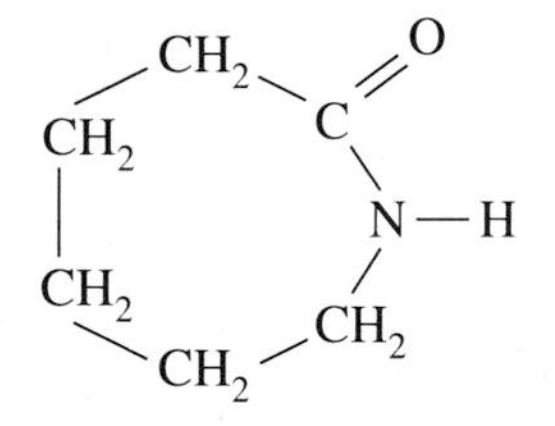

Caprolactam

because this is more easily synthesized. Caprolactam is readily polymerized to give nylon 6. The preparation of yet another nylon, nylon 610, is shown in Demonstration 20.1.

The nylons have molecular masses in the range 1.2×10^3 to 1.5×10^3 u. They are used mainly in the form of fibers, which are formed by forcing the nylon through small holes at temperatures above its softening point (265°C for nylon 66 and 215°C for nylon 6). After the fibers cool, they are stretched to four times their original length. This stretching aligns the long polyamide molecules parallel to each other, and adjacent chains become linked by hydrogen bonds between CO and NH groups (Figure 20.8). This hydrogen-bonded interaction, which considerably increases the strength of the fibers, is the same as that responsible for the secondary structures of proteins. Nylon is one of the strongest and hardest-wearing of all

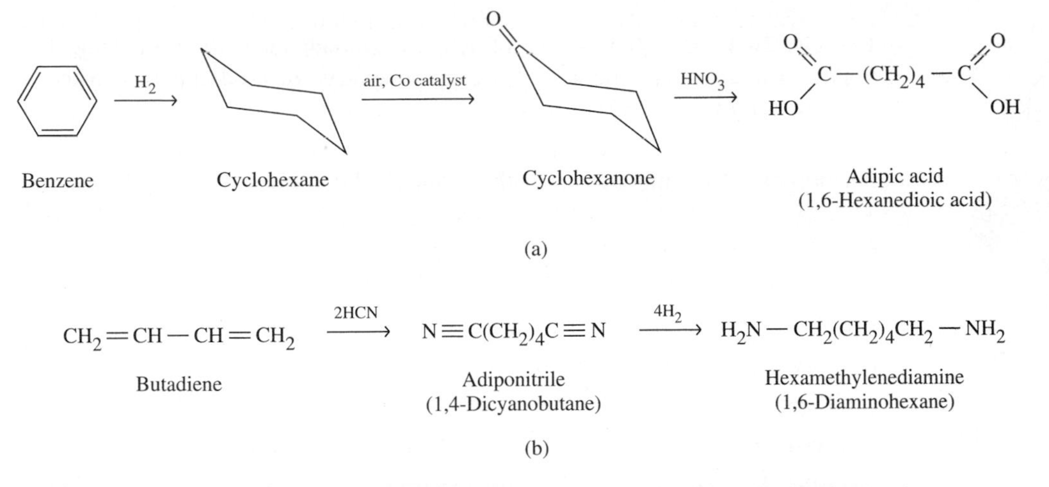

FIGURE 20.7 Synthesis of the Starting Materials for Nylon 66 (a) Synthesis of adipic acid (1,6-hexanedioic acid) from benzene; (b) synthesis of 1,6-diaminohexane from butadiene

synthetic fibers. It is used, among other things, for clothes, tire cords, carpets, and ropes.

Polyamides based on aromatic amines and carboxylic acids are stronger and harder than aliphatic polyamides such as nylon. An example is *Kevlar*:

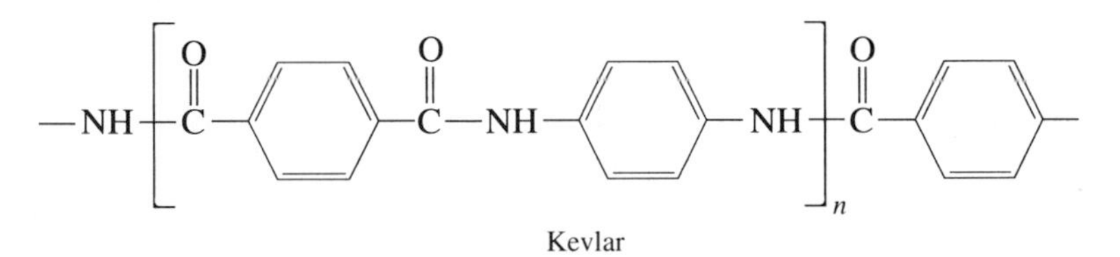

The aromatic groups make the molecule much more rigid and rodlike than a nylon molecule. It has a tensile strength much greater than that of steel but is much lighter. Most Kevlar is used in tire cords, but one speciality use is for bulletproof vests, which contain up to 18 layers of cloth woven from Kevlar fibers.

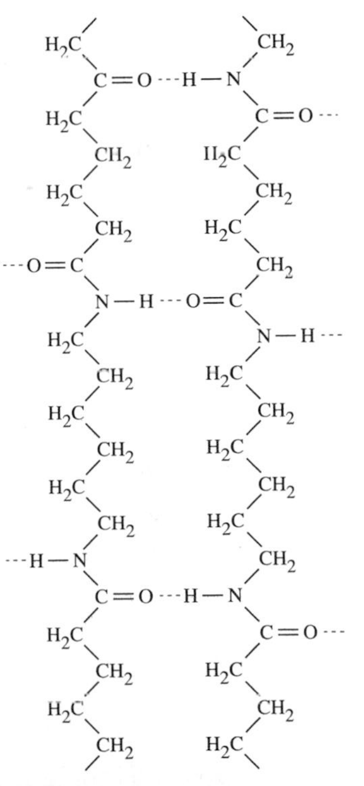

FIGURE 20.8 Hydrogen Bonding in Nylon When a nylon fiber is stretched, the polymer chains straighten and line up with each other, allowing hydrogen bonds to form between adjacent chains. Hydrogen bonding considerably increases the strength of the fiber.

DEMONSTRATION 20.1
Synthesis of Nylon 610

When a solution of 1,6-diaminohexane, $H_2N(CH_2)_6NH_2$, in aqueous sodium hydroxide is poured gently onto a solution of decanedioyl chloride, $COCl(CH_2)_8COCl$, a white film of nylon 610 forms between the two layers. The film can be grasped with tweezers and pulled up as a nylon string, which can be wound on a glass rod as shown.

Bulletproof vest made of Kevlar

Exercise 20.4 If 1.30×10^6 kg of nylon 66 is made annually in the United States, how many kilograms of 1,6-diaminohexane are used annually in the manufacture of nylon 66?

Exercise 20.5 The nylon shown in Demonstration 20.1 is made by the reaction of 1,6-diaminohexane with decanedioyl chloride,

$$\mathrm{Cl{-}\underset{\underset{O}{\|}}{C}{-}(CH_2)_8{-}\underset{\underset{O}{\|}}{C}{-}Cl}$$

(a) Write the equation for the reaction.

(b) Draw the structure of the polymer.

(c) Explain why it is called nylon 610.

Exercise 20.6 What monomers could be used to make Kevlar?

Exercise 20.7 Classify each of the reactions in Figure 20.7 as an addition reaction, an oxidation reaction, or some other type of reaction.

Polyesters **Polyesters** are condensation polymers that, as their name implies, are polymeric esters, made from a dicarboxylic acid and a diol (Chapter 14). The most important commercial polyester is *Dacron*, made from ethylene glycol (ethanediol) and either terephthalic (1,4-benzenedicarboxylic) acid or its methyl ester. Figure 20.9 shows the structure of Dacron as well as the syntheses of ethylene glycol (Chapter 14) and terephthalic acid.

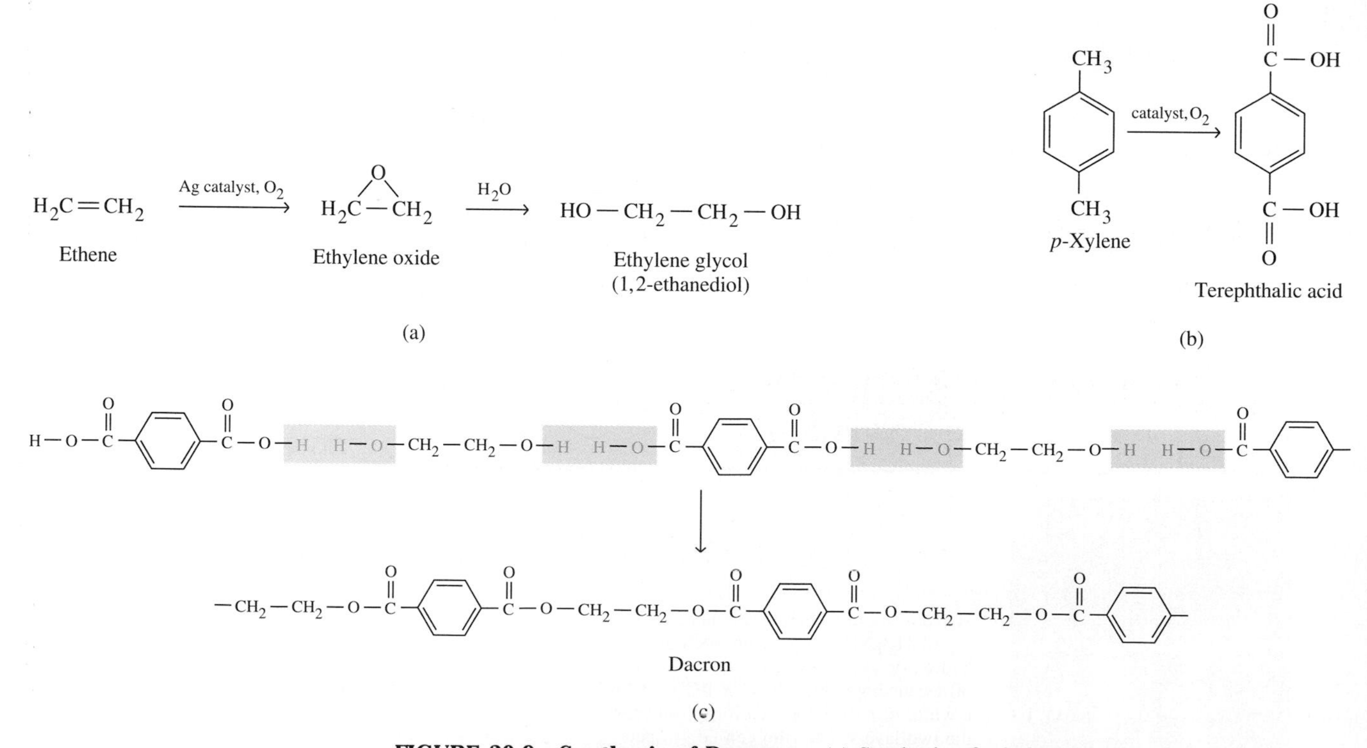

FIGURE 20.9 Synthesis of Dacron (a) Synthesis of ethylene glycol (1,2-ethanediol) from ethene; (b) synthesis of terephthalic acid from *p*-xylene; (c) synthesis of Dacron from ethylene glycol and terephthalic acid

Unlike polyamides, polyesters cannot cross-link through hydrogen bonds. However, all the useful polyesters contain aromatic rings, which add rigidity to the polymer chains. The commercial polyester fibers are made in the same way as those of nylon and have a similar range of uses, but they are less expensive to manufacture. Polyester fabrics are considerably longer-wearing than cotton fabrics, but polyester cloth has poor draping properties and a tendency to build up static electricity. Polyesters are often blended with cotton to produce a combination of desirable properties.

Phenol-Formaldehyde The first fully synthetic polymer was made from phenol and formaldehyde in 1909. Its inventor, the Belgian chemist Leo Baekeland (1863–1944), called the product Bakelite. Although this plastic material was made long before the other polymers that we have discussed, its structure was not understood then. Little progress was therefore made in making similar materials until the work of Staudinger and Carothers revealed the nature of polymers, which could then be synthesized in a planned and rational way.

The synthesis of Bakelite, or phenol–formaldehyde, is shown in Demonstration 20.2. This polymer is formed by a condensation reaction in which water is eliminated from the oxygen atom of formaldehyde (methanal); $H_2C{=}O$, and hydrogen atoms from two phenol, C_6H_5OH, molecules (Figure 20.10). Because each formaldehyde molecule bonds to two different phenol molecules and each phenol molecule bonds to three different formaldehyde molecules, the phenol–formaldehyde polymer has an infinite three-dimensional network structure.

Phenol–formaldehyde is a hard, strong material. It is heat resistant and a poor conductor of heat and electricity. Because of its network structure, it cannot be melted and reshaped—it decomposes when heated to a sufficiently high temperature. Materials of this kind are called **thermosetting polymers**. In contrast, polyamides, polyesters, and polyethylene and related polymers, all of which consist of long-chain molecules, can be softened or melted and reformed into new shapes. Such materials are called **thermoplastic polymers**.

DEMONSTRATION 20.2 The Formation of Phenol–Formaldehyde Polymer

Concentrated HCl is poured into a solution containing aqueous HCHO (formaldehyde) and phenol.

As the mixture is stirred, it warms up. It turns pink when the polymer begins to form.

After about 1 min of stirring, a solid phenol–formaldehyde condensation polymer forms, which is removed on the stirring rod.

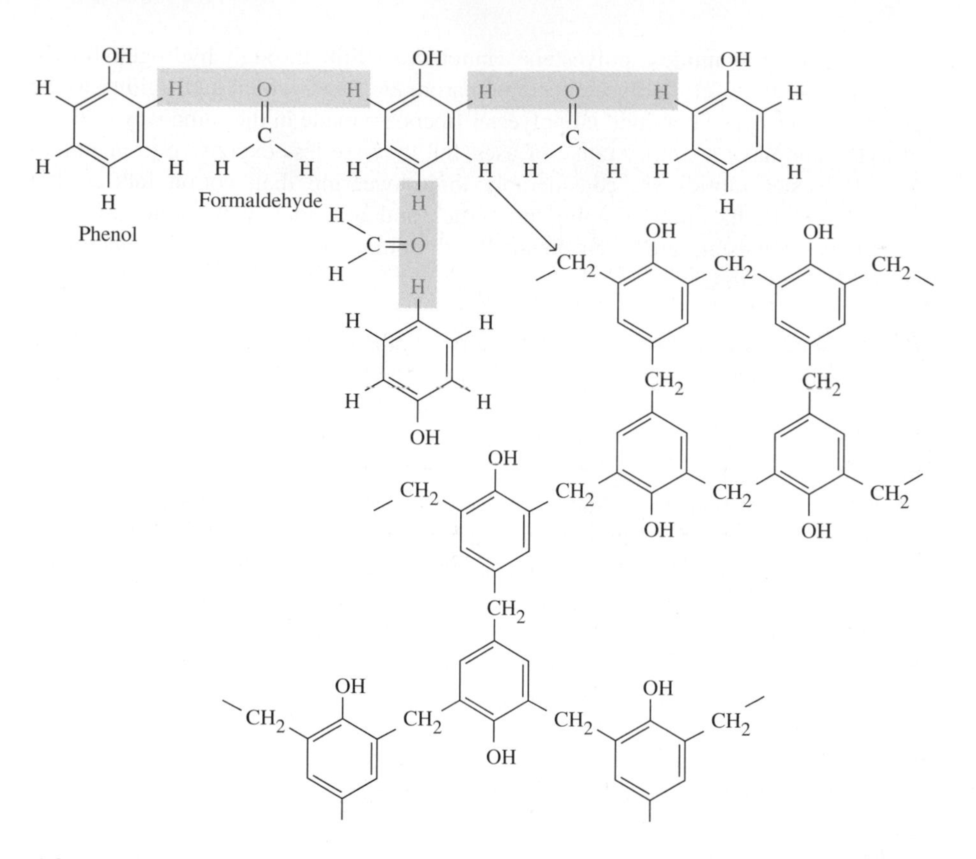

FIGURE 20.10 Synthesis of Phenol-Formaldehyde (Bakelite) Phenol and formaldehyde form a three-dimensional network polymer by the elimination of water, thus linking each phenol molecule to three others by CH_2 bridges.

20.2 MATERIALS SCIENCE

Materials science is the study of the chemistry and physics of solids, particularly solids with unusual physical properties of the kind that have led to modern technological innovations. In the past chemists studying solids have been interested primarily in the composition, structures, and reactions of solids, whereas physicists have concentrated on physical properties such as electrical, magnetic, and optical properties. For example, all metals are conductors, whereas most covalent network, molecular, and ionic network solids are insulators. Physicists could study the electrical properties of solids without paying much attention to their chemical compositions. However, the electrical properties of solids are much more varied and interesting than this simple classification into metallic conductors versus insulators would lead us to believe. To understand the electrical behaviors of solids, it is essential to take into account their chemical compositions and structures. This necessary merging of chemistry and physics has led to the development of the scientific frontier known as materials science.

We will consider just three examples of the many types of materials that are currently being studied and developed in this rapidly expanding field: zeolites, semiconductors, and superconductors. The first two depend on the chemistry of silicon, an element that is as important in the inorganic world as carbon is in the organic world.

Zeolites

In the preceding chapters we saw many illustrations of the great importance of catalysts. Many industrial processes depend on catalysts to speed up reaction rates, and chemists are constantly searching for new and better catalysts. Much attention is now being given to aluminosilicates (Chapter 19) of a type called *zeolites*, which are among the most useful and most versatile catalysts known.

The name ''zeolite'' means ''stone that boils'' and comes from the Greek words *zein* (''to boil'') and *lithos* (''stone''). It was first used to describe a type of naturally occurring mineral from which water ''boiled off'' upon heating. **Zeolites** are porous aluminosilicates that typically contain water that can be removed by heating. We saw in Chapter 19 that aluminosilicates consist of infinite linear, sheet, or network anions built of SiO_4 and AlO_4 tetrahedra (Figure 20.11a) with metal cations interspersed in the crystal lattice. The basic building block of many zeolites, which have an infinite network structure, is a cage of 12 Si atoms, 12 Al atoms, and 24 O atoms built from 12 SiO_4 tetrahedra and 12 AlO_4 tetrahedra that share corners (Figure 20.11b). This cage is called the *sodalite cage*, as it is the building block of the anion present in the mineral sodalite, $Na_4(Al_3Si_3O_{12})Cl$. In a zeolite, sodalite cages and similar cages are stacked in a variety of different ways to give three-dimensional network anions containing large cavities and tunnels, which create a very open structure (Figures 20.11c and 20.11d). The cavities and tunnels contain metal cations such as K^+ and Na^+ and commonly water molecules. If water molecules are removed from the tunnels by heating, the resulting material strongly absorbs water or other small molecules. In addition to the naturally occurring zeolites, chemists have been able to synthesize more than 70 different zeolites with varying sizes of cavities and tunnels.

Zeolites as Ion-Exchange Materials The useful properties of zeolites arise from the ease with which ions and molecules in the channels and cavities of these aluminosilicates can be replaced by other ions and molecules. One of the first applications of zeolites was as **ion-exchange materials**, which are used to replace undesirable ions in solution by other ions. For example, hard water containing Ca^{2+} or Mg^{2+} ions (Box 10.1) can be ''softened'' by passing it through a sodium zeolite. Because of their greater charge, Ca^{2+} and Mg^{2+} ions are held in the zeolite more

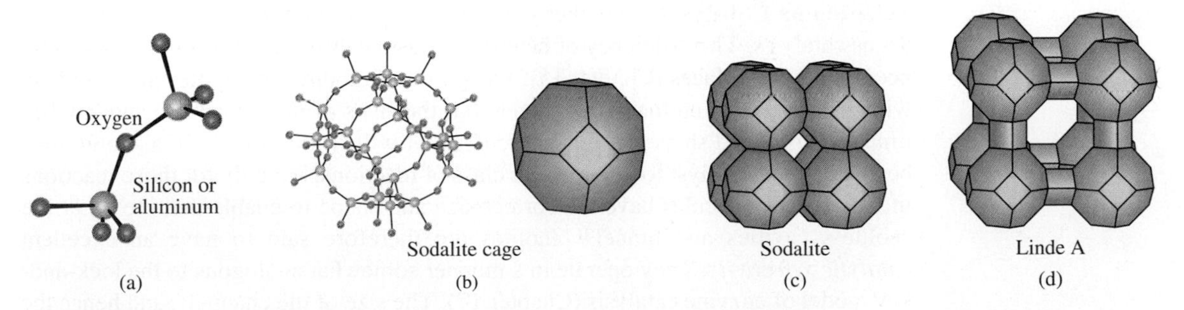

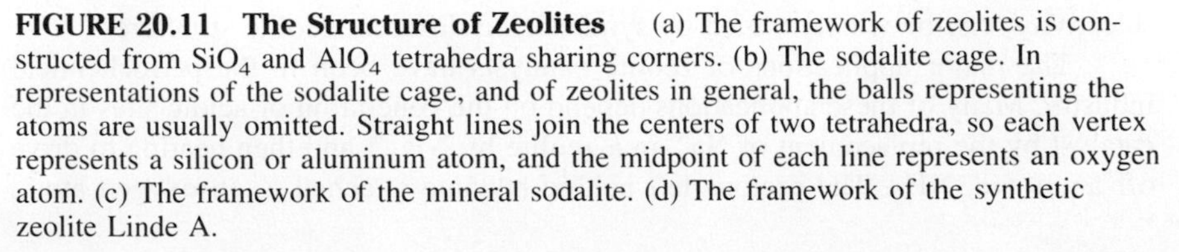

FIGURE 20.11 The Structure of Zeolites (a) The framework of zeolites is constructed from SiO_4 and AlO_4 tetrahedra sharing corners. (b) The sodalite cage. In representations of the sodalite cage, and of zeolites in general, the balls representing the atoms are usually omitted. Straight lines join the centers of two tetrahedra, so each vertex represents a silicon or aluminum atom, and the midpoint of each line represents an oxygen atom. (c) The framework of the mineral sodalite. (d) The framework of the synthetic zeolite Linde A.

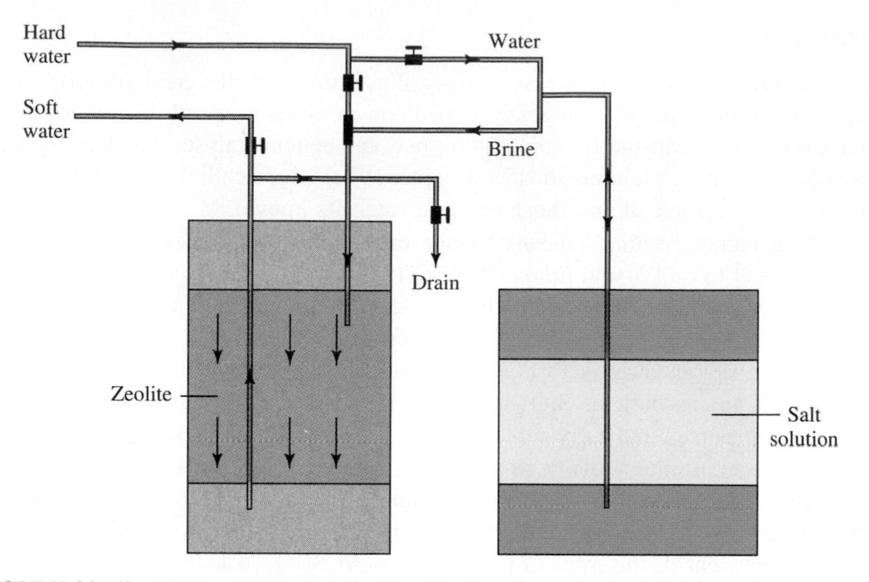

FIGURE 20.12 Water Softening by Ion Exchange In the left-hand tank, hard water flows through the zeolite, which removes calcium ions from the water and replaces them with sodium ions. A concentrated salt solution from the right-hand tank is periodically sent through the zeolite to restore its activity by replacing the calcium ions with sodium ions.

strongly than are Na^+ ions, so they replace the Na^+ ions, which go into solution in the water (Figure 20.12). When most of the Na^+ ions have been replaced, the zeolite is no longer effective as a water softener, but it can be regenerated by passing a concentrated aqueous NaCl solution through it. A similar application is the removal of radioactive $^{137}Cs^+$ and $^{90}Sr^{2+}$ ions from aqueous solutions obtained by treating nuclear wastes.

Exercise 20.8 Explain why a zeolite that has been used for softening water can be reused after a concentrated NaCl solution has been passed through it.

Zeolites as Catalysts Another important property of zeolites is their ability to act as catalysts. The efficiency of heterogeneous catalysts depends on reactions that occur on their surfaces (Chapter 15). The porous structures of zeolites provide them with an enormous internal surface area, but the sizes of their channels and cavities limit the sizes and shapes of the molecules that will fit into them. So a zeolite may be a very good catalyst for a particular class of reactions, but only for those reactions in which the molecules have the correct size and shape to enable them to enter the zeolite's cavities and tunnels. Zeolites are therefore said to have an excellent *catalytic selectivity*. They operate in a manner somewhat analogous to the lock-and-key model of enzyme catalysis (Chapter 17). The size of the channels, and hence the selectivity, can be widely varied by synthesizing different zeolite structures.

The major applications of zeolite catalysts have been in the petrochemical industry. Many of these applications depend on the generation of acidic sites in the catalyst by the replacement of Na^+ in a zeolite by NH_4^+ and then heating to drive off ammonia, NH_3. The protons left behind become attached to an oxygen atom,

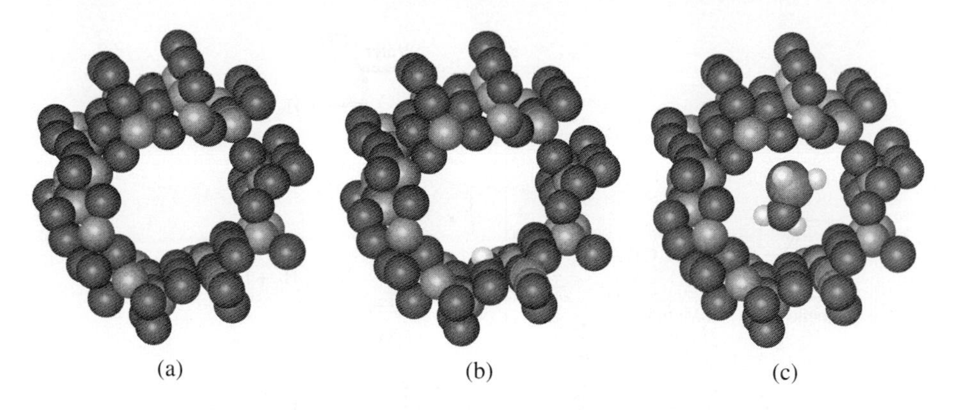

FIGURE 20.13 An Acidic Site in a Zeolite (a) A computer image of a single cavity in the zeolite ZSM–5, which has a diameter of 550 pm; silicon atoms are yellow, and oxygen atoms are red. (b) When a silicon atom is replaced by an aluminum atom (blue), a hydrogen ion (white) attaches to an oxygen atom to preserve electrical neutrality. This proton gives the cavity its ability to act as an acid catalyst. (c) When this proton adds to a molecule such as methanol, the molecule becomes a reactive intermediate.

forming an Al—O bond in the aluminosilicate framework (Figure 20.13), and are strongly acidic. A typical reaction that uses a zeolite catalyst is the synthesis of ethylbenzene from benzene and ethene (Figure 20.14). The acid catalyst protonates ethene to form the ethyl cation $CH_3CH_2^+$, which then reacts with benzene in an aromatic substitution reaction (Chapters 8 and 14). A proton is thereby displaced and returns to the catalyst.

An example of the selectivity of zeolite-catalyzed reactions is the reaction of methanol with toluene to give *o*-, *m*-, and *p*-xylene and water (Figure 20.15). The desirable product is *p*-xylene (1,4-dimethylbenzene), because it can be oxidized to terephthalic acid, used to make polyesters (Section 20.1). The symmetrical shape of *p*-xylene allows it to pass through the channels of the particular zeolite catalyst used and to escape. The *o*- and *m*-isomers are too large to pass through the channels and so are held in the cavities of the zeolite. The loss of *p*-xylene from the catalyst leads to continuous formation of this isomer, because its concentration never builds up to the equilibrium concentration. It is thus the major product of the catalyzed reaction.

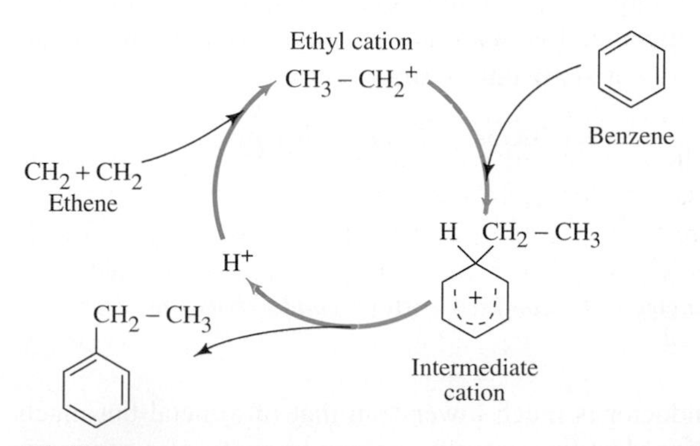

FIGURE 20.14 Synthesis of Ethylbenzene in a Zeolite Ethene is protonated by the acidic zeolite catalyst to give the ethyl cation $CH_3CH_2^+$. The ethyl cation reacts with benzene in an aromatic substitution reaction, forming an intermediate cation. This cation loses a proton back to the catalyst to become ethylbenzene.

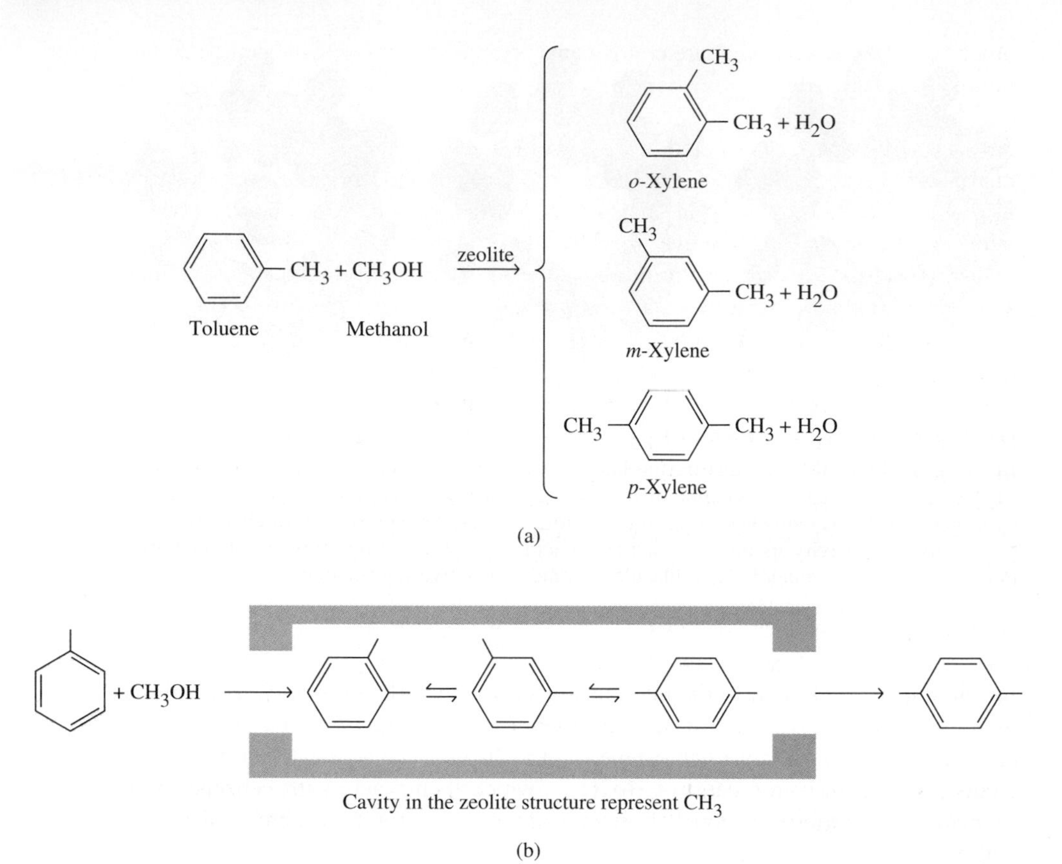

FIGURE 20.15 Selective Catalysts in the Synthesis of *p*-Xylene (a) Toluene reacts with methanol in a zeolite catalyst to give an equilibrium mixture of *o*-, *m*-, and *p*-xylene. (b) Only *p*-xylene has the right shape to enable it to escape the zeolite cavity, so *p*-xylene is continually formed to maintain the equilibrium and is the major product of the reaction. The CH_3 groups are represented by the bonds outside the rings.

Semiconductors

As we saw in Chapter 10, metals are good conductors of electricity, whereas most solid nonmetals are insulators. As we proceed down a group in the periodic table, metallic properties become more pronounced. For example, in Group IV, carbon (as the allotrope diamond) is an insulator. At the bottom of that group, tin and lead are metallic conductors, whereas silicon and germanium, the elements in between, are neither insulators nor conductors but **semiconductors**.

> **A *semiconductor* is an electronic conductor with a *conductivity* that *increases* as the temperature is raised.**

In contrast,

> **A metallic conductor is an electronic conductor with a *conductivity* that *decreases* as the temperature is raised.**

The conductivity of a semiconductor is much lower than that of a metal but much greater than that of an insulator. Most importantly, the conductivity of a semiconductor can be varied over many orders of magnitude. As we will see, it is this property that makes semiconductors so useful in the construction of transistors and other electronic devices that are components of modern computers, calculators, and radios, for example.

Semiconductivity We noted in Chapter 19 that C—C bonds are much stronger than Si—Si bonds, and Ge—Ge bonds are still weaker (Table 9.2). All the electrons in diamond are held tightly in strong bonds—there are no free electrons in the lattice. If electrons cannot move, there can be no conductivity, so diamond is an insulator. But the bonding electrons are held less strongly in silicon and germanium, and even at room temperature a few of them have enough energy to escape from an atom and move around the lattice, as in a metal. When an electron is lost from an atom such as silicon, it leaves a vacancy—a hole—in the valence shell of the atom. An atom that has lost an electron has a positive charge, so we speak of the hole as a **positive hole**. This positive hole can move through the lattice because an electron moving into the hole from a neighboring atom creates a new positive hole (Figure 20.16). Both the free electrons and the positive holes contribute to the conductivity of a semiconductor. With increasing temperature, more electrons have enough energy to escape from the atoms, and the conductivity increases.

***n*- and *p*-type Semiconductors** We can increase the conductivity of a pure semiconductor, such as silicon, by adding minute quantities of certain other elements, a process called **doping**. Suppose that we mix into molten silicon a small amount of a Group V element, such as phosphorus, and then let the mixture crystallize. When a phosphorus atom replaces a silicon atom in the silicon lattice, it needs only four of its five electrons to form four bonds to the adjacent silicon atoms. One electron is left free to roam around the lattice and contribute to the conductivity (Figure 20.17a). The addition of a very small amount of phosphorus—as little as 10^{-8} g in 1 g Si—dramatically increases the conductivity of silicon. The silicon is said to be doped with phosphorus. The phosphorus atom that has lost an electron has a positive charge. But, because it now has eight electrons in its valence shell, the phosphorus atom has no vacancy into which an electron from a neighboring atom can move. Hence this positive charge does not behave as a positive hole that can move from place to place. The conductivity of silicon doped with phosphorus is due *only* to the (negative) electrons, so silicon is called an ***n*-type semiconductor**.

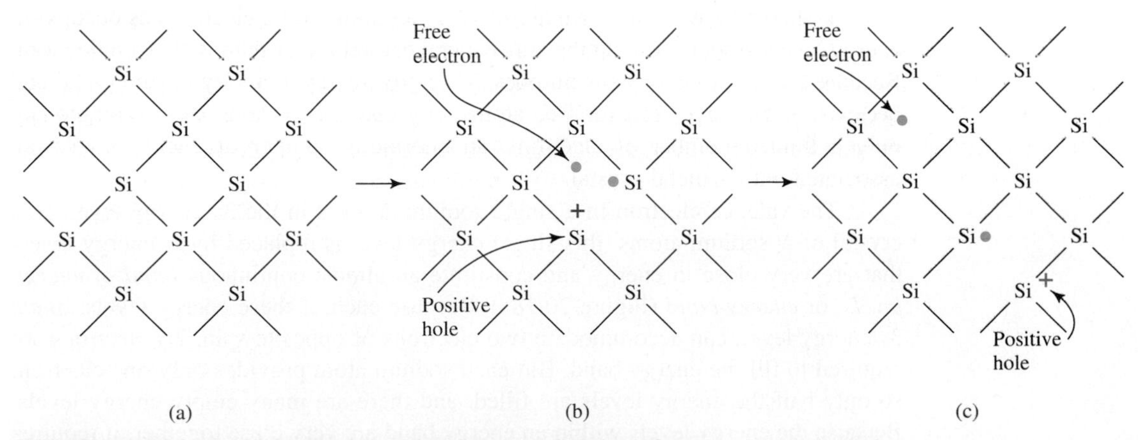

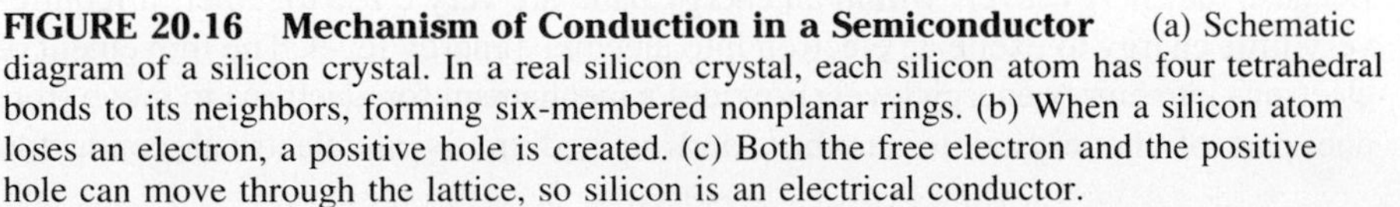
FIGURE 20.16 Mechanism of Conduction in a Semiconductor (a) Schematic diagram of a silicon crystal. In a real silicon crystal, each silicon atom has four tetrahedral bonds to its neighbors, forming six-membered nonplanar rings. (b) When a silicon atom loses an electron, a positive hole is created. (c) Both the free electron and the positive hole can move through the lattice, so silicon is an electrical conductor.

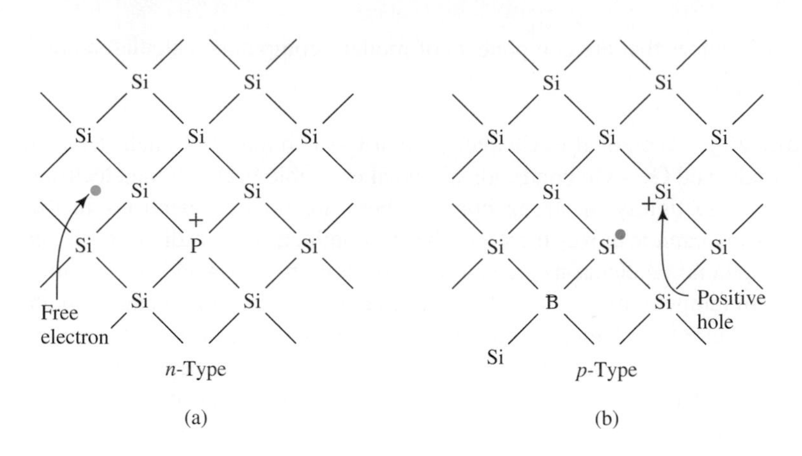

FIGURE 20.17 Doped Semiconductors (a) n-Type semiconductors are doped with a Group V element, providing free electrons that increase the conductivity. (b) p-Type semiconductors are doped with a Group III element, creating positive holes that increase the conductivity.

If instead we add to silicon a small amount of a Group III element, such as boron, a silicon atom is replaced by a boron atom in the lattice (Figure 20.17b). Boron has only three electrons in its valence shell, so there is a hole in the valence shell of the boron atom into which an electron from a neighboring silicon can move. This leaves a positive hole on a silicon atom that can move through the lattice from silicon atom to silicon atom, as in Figure 20.17. Silicon doped with boron behaves as a *positive* hole conductor and is called a ***p*-type semiconductor**. The extensive use of semiconductor devices in modern electronic equipment depends on our ability to control accurately the type and magnitude of the conductivity and to change it from one region of a device to another by changing the amounts and nature of the impurities.

Energy Levels in Solids Thus far we have discussed the conductivity of semiconductors by extending our model of a three-dimensional covalent network solid such as diamond. Semiconductors have properties intermediate between those of a covalent solid and a metal; to expand our understanding of semiconductors, it is useful to look at the relationship between semiconductors and metals. We will do so in terms of the energy levels occupied by the electrons. In Chapter 6 we discussed the energy levels in atoms in some detail. The charge cloud of an atom in a metal overlaps that of all its neighbors, so the valence electrons are not confined to one atom or shared between just two atoms. We can think of the electrons as occupying energy levels associated with the entire metal crystal rather than with a single atom. Because there is an enormous number of electrons in even a very small crystal and because, as we have seen for free atoms, any one energy level can accommodate only a limited number of electrons, an enormous number of energy levels are associated with a metal crystal.

The valence electron in a single sodium atom is in the $3s$ energy level. In a crystal of N sodium atoms, this single energy level is replaced by N energy levels that are very close in energy and constitute an almost continuous *band of energy levels*, or *energy band* (Figure 20.18a). Because each of these energy levels, like a $3s$ energy level, can accommodate two electrons of opposite spin, $2N$ electrons are required to fill the energy band. But each sodium atom provides only one electron, so only half the energy levels are filled, and there are many empty energy levels. Because the energy levels within an energy band are very close together, it requires very little energy to excite an electron into an empty energy level. The movement of electrons into empty energy levels provides a mechanism for electrons to move from one part of the crystal to another, and so sodium is an electrical conductor.

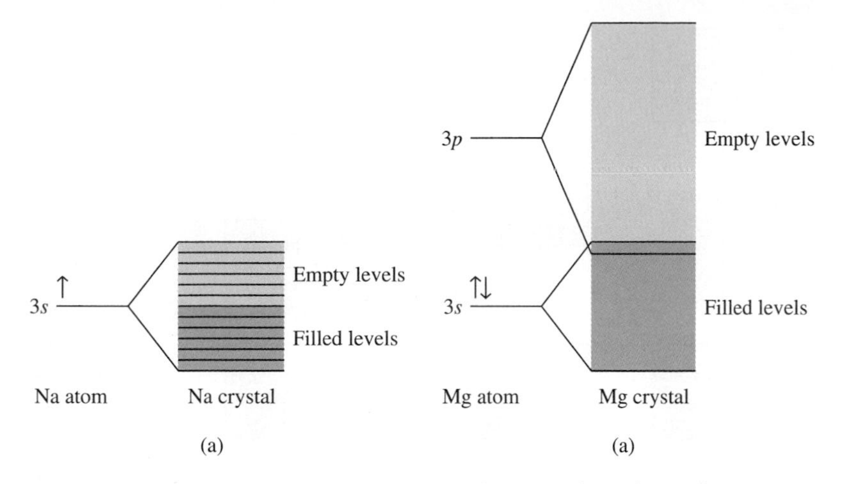

FIGURE 20.18 Energy Levels in Sodium and Magnesium Crystals (a) The $3s$ energy level of a single sodium atom is replaced by an almost continuous band of energy levels in the crystal. Only half of these energy levels are filled, so sodium is a metallic conductor. (b) In magnesium the energy band formed from the $3s$ energy level of a single atom is filled but overlaps with the empty band formed from the $3p$ orbitals. The result is a continuous energy band that is only partly filled, so magnesium is a metallic conductor.

Magnesium has two electrons in its valence shell, so the magnesium electrons fill the $3s$ band. But the band formed from the $3p$ atomic energy levels is empty and overlaps the $3s$ band. There are many empty levels into which the electrons can move. Consequently, magnesium is also a conductor (Figure 20.18b).

The situation is different in diamond. The $2s$ and $2p$ carbon atomic energy levels form two bands in a diamond crystal—a lower, filled band and an upper, empty band (Figure 20.19a). There is a large gap in energy between the lower, filled band and the upper, empty band. None of the electrons in the lower band have enough energy to move into the upper band. The lower, filled band is called a **valence band**, because it contains the electrons responsible for forming the bonds—in diamond, the carbon–carbon bonds. The electrons in the valence band are held strongly in pairs between the carbon atoms and so do not give diamond any conductivity. The upper, empty band (for reasons we shall see shortly) is called the **conduction band**. In diamond, because this band is empty, it also does not contribute to the conductivity.

A semiconductor, such as silicon, also has an energy gap between the valence and conduction bands, but the gap is small enough that, at room temperature, a few electrons have sufficient thermal energy to be excited to the conduction band (Figure 20.19b). There they can move freely from one energy level to another, so silicon is a conductor, although a considerably poorer conductor than is a metal. Moreover, when an electron moves to the conduction band, it leaves a positive hole in the valence band. The hole can move from one energy level to another, also contributing to the conductivity.

The higher the temperature, the greater the average kinetic energy of electrons, and therefore the greater the probability of excitation to the conduction band. Thus, the conductivity of a semiconductor increases with increasing temperature.

A useful analogy for understanding the difference between insulators and semiconductors is illustrated in Figure 20.20. We can liken a crystal such as diamond to a filled parking garage, with no spaces left for cars. No cars (electrons)

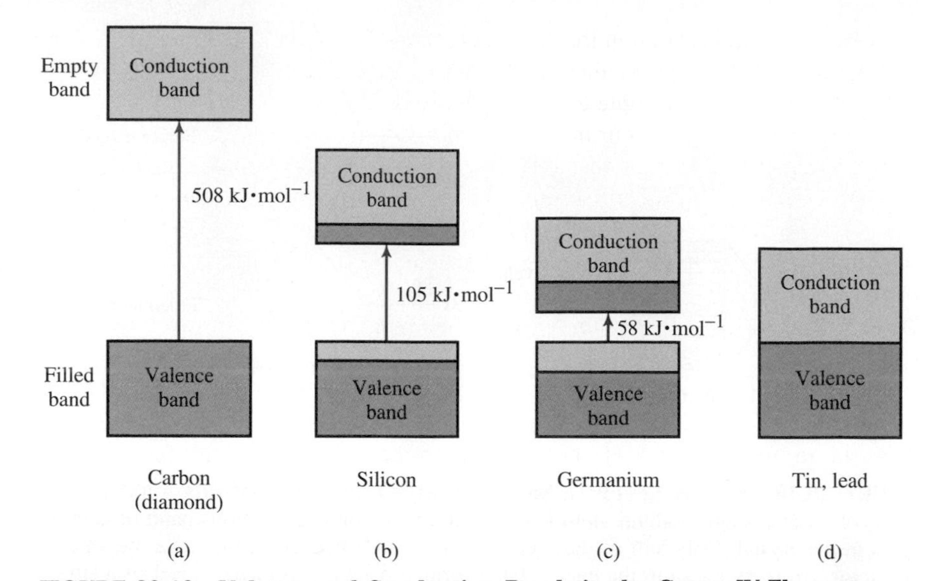

FIGURE 20.19 Valence and Conduction Bands in the Group IV Elements (a) In carbon (diamond) the valence band is full and the conduction band is empty, so carbon is an insulator. (b) In silicon the valence band and the conduction band are close enough that a few electrons have sufficient energy to enter the conduction band. The valence band is left with a few positive holes (empty energy levels), so silicon is a semiconductor. (c) Germanium is a semiconductor with a smaller energy gap than silicon, so it has more electrons in the conduction band and more positive holes (empty energy levels) in the valence band. (d) In tin and lead there is no energy gap between the valence band and the conduction band (they form one continuous partly filled band), so tin and lead are metallic conductors.

can move, because there is nowhere for them to go (Figure 20.20a). However, if a car can move up to an empty floor in the garage above the filled floor, then it can move freely. The position of the vacant space it leaves on the lower floor can shift if a car moves into it, creating another vacancy (hole) elsewhere (Figure 20.20b). In a

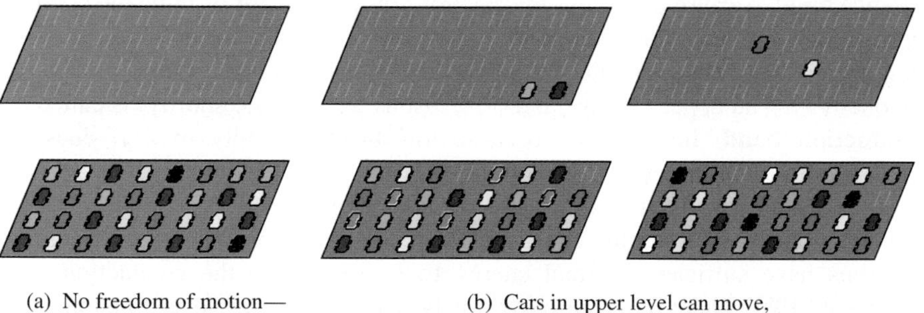

FIGURE 20.20 The Parking-Garage Analogy for Conduction in a Semiconductor (a) A garage with a filled lower floor (so there can be no movement of cars) and an empty but inaccessible upper floor is analogous to an insulator with a filled valence band, in which movement of electrons is not possible, and an empty conduction band. (b) A garage with a lower floor with two empty spaces (holes) created by the movement of two cars to the upper floor allows cars in both floors to move. This situation is analogous to a semiconductor, in which a few electrons have enough energy to reach the conduction band, leaving positive holes in the valence band. Both the movement of electrons in the upper, conduction band and the movement of holes in the lower, valence band contribute to conductivity.

semiconductor the electron in the conduction band is like the car on the upper floor, and the positive hole is like the empty parking spot in the floor below—both can move and therefore contribute to the conductivity. Pursuing the analogy further, we could say that the empty floor in the insulator parking garage is too high up and the ramp too steep for cars to reach the upper level. In the semiconductor parking garage, the upper floor is not so high and is accessible to a few cars.

The magnitude of the energy gap between the valence band and the conduction band is what determines whether a substance is an insulator, a semiconductor, or a conductor. Carbon (diamond) has a large energy gap and is an insulator (Figure 20.19a). Silicon and germanium have smaller energy gaps and are semiconductors (Figures 20.19b and 20.19c). Tin and lead have a zero energy gap, so a negligible amount of energy is needed to excite electrons into the conduction band, and these two elements, like sodium and magnesium, are metallic conductors (Figure 20.19d). Thus, in an insulator there are no electrons in the conduction band. In a semiconductor there are only a few electrons in the conduction band; the number of electrons in the conduction band—and hence the conductivity—increases with increasing temperature. In a metallic conductor there are a large number of electrons in the conduction band, even at low temperatures, and this number is not significantly affected by increasing temperature. In contrast, as we saw in Chapter 10, an increase in temperature leads to a *decrease* in the conductivity of a metal because the vibrational motions of the atoms increase in magnitude and increasingly impede the motion of the electrons.

In a pure semiconductor there are always equal numbers of positive holes and negative electrons (Figure 20.21a). In silicon doped with phosphorus (or arsenic), the extra electron from each doping atom enters the conduction band, and the conductivity is due to (negative) electrons in the conduction band. The positive charges in equal numbers in the valence band cannot move. Thus, phosphorus-doped silicon is an *n*-type semiconductor (Figure 20.21b). In silicon doped with boron (or aluminum), one electron from each boron atom is missing from the valance band. In other words, a positive hole is created for each boron atom added, and boron-doped silicon is a *p*-type semiconductor (Figure 20.21c).

Ultrapure Silicon and Zone Refining The conductivity of doped semiconductors depends strongly on the percentage of the doping atoms. Even a very small percentage can have a very large effect. Silicon cannot be doped in a controlled and reproducible way unless it is initially ultrapure. The discovery of methods for making ultrapure solids has been crucial to the development of transistors and other electronic devices based on semiconductors, such as the silicon chip.

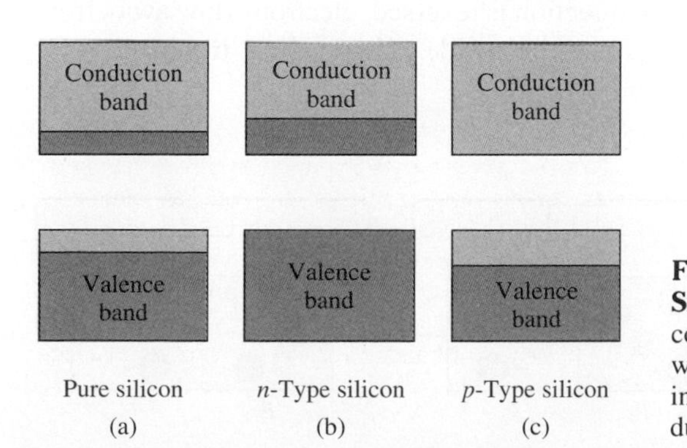

FIGURE 20.21 Semiconduction in Pure Silicon and in Doped Silicon (a) In pure silicon, conduction is due both to the electrons in the conduction band and to the holes in the valence band. (b) In silicon doped with phosphorus (an *n*-type semiconductor), conduction is due to electrons in the conduction band. (c) In silicon doped with boron (a *p*-type semiconductor), conduction is due to positive holes in the valence band.

Silicon is made from silica, SiO_2, by heating it with coke to a temperature of about 3000°C in an electric furnace:

Carbon is the reducing agent in this redox reaction and is oxidized to CO.

$$SiO_2(s) + 2C(s) \rightarrow Si(l) + 2CO(g)$$

Carbon monoxide escapes from the furnace and burns to give carbon dioxide, while molten silicon (mp 1414°C) runs out the bottom of the furnace and solidifies. Silicon produced in this way is pure enough for many purposes, such as alloying with metals, but not for use in electronic devices. To purify it further, it is first converted to silicon tetrachloride, $SiCl_4$, which is a liquid (bp 57.6°C) that can be purified by distillation:

$$Si(s) + 2Cl_2(g) \rightarrow SiCl_4(l)$$

The distilled silicon tetrachloride is then reduced to silicon by using $H_2(g)$ or magnesium:

$$SiCl_4(g) + 2H_2(g) \rightarrow Si(s) + 4HCl(g)$$

or

$$SiCl_4(g) + 2Mg(s) \rightarrow Si(s) + 2MgCl_2(s)$$

If magnesium is used, the soluble magnesium chloride can be washed away with water. Silicon obtained in this way is further purified by **zone refining** (Figure 20.22). In this process, a short segment of a rod of silicon is heated until it melts. Any impurities are more soluble in the molten silicon than in the solid and therefore concentrate in the liquid. The rod is slowly moved through the heater so that the molten zone traverses the length of the rod, removing impurities as it moves. When the impure molten zone reaches the end of the rod, it is allowed to solidify and is cut off. This technique is widely applicable to the manufacture of very pure solids.

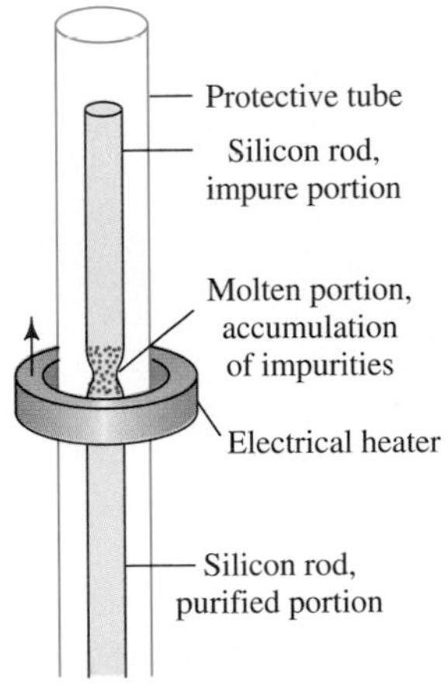

FIGURE 20.22 Purification of Silicon by Zone Refining A rod of impure silicon is drawn very slowly through the electrical heater. The molten zone, in which the impurities dissolve, is thus moved to one end of the rod, where it can be cut off.

***p-n* Junctions, Transistors, and Integrated Circuits** Many important electrical circuit elements can be made from combinations of *p*-type and *n*-type semiconductors. A simple union of one *p*-type and one *n*-type semiconductor creates a ***p-n* junction**, an interface that conducts electric current in one direction only. Figure 20.23 illustrates what happens when a voltage is applied to a *p-n* junction by means of a battery. If the negative electrode of the battery is connected to the *n*-side of the junction (Figure 20.23a), electrons flow into the *n*-side from the battery and back into the positive electrode from the *p*-side of the junction. As electrons flow out of the *p*-side, holes move in the opposite direction—that is, toward the interface. Here the positive holes meet electrons flowing into the *n*-side of the interface and are neutralized. As long as the voltage is applied, a current will continue to flow through the junction. If, however, the voltage direction is reversed, electrons flow away from the *n*-side, and holes move away from the *p*-side. There is no mechanism for

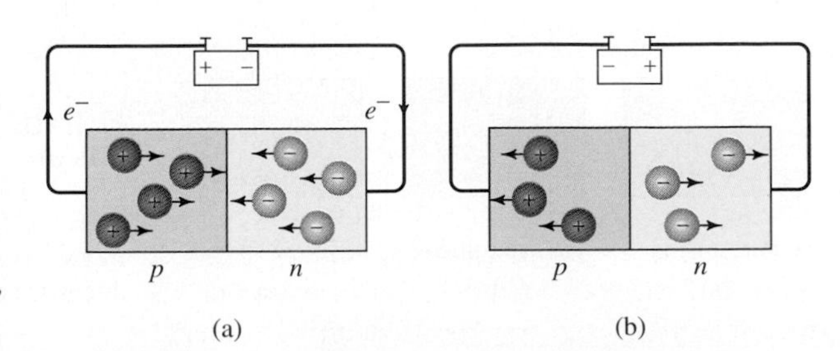

FIGURE 20.23 A *p-n* Junction of Semiconductors Current is carried in a *p*-type semiconductor by the motion of holes and in an *n*-type semiconductor by the motion of electrons. (a) With the voltage directed as shown, a large current can flow. (b) With the voltage applied in the opposite direction, little current flows.

creating new holes or new electrons at the junction, so the current ceases (Figure 20.23b). In other words, a *p-n* junction carries a current in only one direction.

We can also see from Figure 20.21 that electrons in the conduction band on the *n*-side of the junction are at a higher energy level than are the electrons on the *p*-side. Hence electrons flow from the *n*-side to the *p*-side (high energy to low energy) and not from the *p*-side to the *n*-side (low energy to high energy).

A *p-n* junction behaves as a *diode*, a circuit element that carries current in only one direction. Diodes were originally vacuum tubes, but today they are semiconductor *p-n* junctions. **Transistors**, which are circuit elements constructed from three thin layers of doped silicon to give *p-n-p* or *n-p-n junctions*, can function as oscillators and amplifiers. These semiconductor devices can be made very small and can be produced very economically and efficiently, so they have allowed for the miniaturization of electronic devices. They have completely replaced the much bulkier and more expensive vacuum tubes that were once used for the same purposes.

FIGURE 20.24 Integrated Circuit on a Silicon Chip You can see how small a silicon chip is by comparing it to a typical aspirin tablet.

The central processing unit (CPU) of a computer is based largely on an array of semiconductor transistors. Today's personal computers contain as many as 10 million such transistors. The amazingly rapid development of computers since the 1960s has been made possible by the development of the **integrated circuit**. Integrated circuits are constructed on the surface of a small piece of silicon called a **silicon chip**, which typically has dimensions of only a few millimeters (Figure 20.24). Successive layers of insulating silicon dioxide, a conducting metal, and silicon doped to give *n*- and *p*-type semiconductors are deposited on the surface of the chip. This process is accomplished by such operations as baking the chip in oxygen, evaporating a metal at very low pressures, doping a layer, subjecting it to radiation and etching away part of the layer with an acid—chemical and physical processes carried out on an ever smaller scale. During these operations a template called a *mask* is applied to the surface so that only certain areas are affected. In this way circuits containing thousands or even millions of circuit elements such as resistors, capacitors, and transistors are constructed. With the continued miniaturization of the process, the number of circuit elements that can be constructed on a single chip has dramatically increased. In 1960 the average silicon chip contained about 10 transistors; in 1975, it was more than 10,000, and in 1994 more than a million. The limits of this miniaturization process have not yet been reached. The 10 million or more transistors in a personal computer is an amazing number of manufactured items for one person to own, and yet they cost less than the keyboard and less, for example, than 1 million staples.

Materials scientists have discovered a number of semiconducting materials that are alternatives to silicon. One of the most promising is gallium arsenide, GaAs. Gallium is in Group III of the periodic table, and arsenic is in Group V, so a one-to-one combination is isoelectronic with silicon in Group IV. An advantage of gallium arsenide is that electrons travel about five times faster in it than they do in silicon, which increases the speed of computing correspondingly. However, gallium arsenide is expensive and difficult to make, so silicon is still the preferred material for most applications.

Electrical Properties of Metal Oxides

We saw in Chapter 4 that metal oxides are typically nonconducting ionic solids. However, there are many exceptions to this generalization, particularly among the oxides of the transition metals. These oxides may be insulators, semiconductors, metallic conductors, or (as we shall soon learn) even superconductors.

Some transition metal oxides can behave as metallic conductors due to the presence of *d* electrons in the valence shell of the metal ion. When these electrons are not held tightly, they may escape from the metal ion and move around the lattice just as the 3*s* electrons do in sodium or magnesium. The electron configurations of some of the Period 4 transition metals and their ions are as follows:

Ti[Ar]	$3d^24s^2$	Ti^{2+}[Ar]	$3d^2$	Ti^{4+}[Ar]	
V[Ar]	$3d^34s^2$	V^{2+}[Ar]	$3d^3$		
Fe[Ar]	$3d^64s^2$	Fe^{2+}[Ar]	$3d^6$	Fe^{3+}[Ar]	$3d^5$

The oxides TiO and VO have the ionic sodium chloride structure, but the 3*d* electrons are not held very strongly and are available to form a delocalized charge cloud (as in a metal). Hence these oxides have a high metallic conductivity. With increasing atomic number and core charge, however, the 3*d* electrons are held more tightly; they do not delocalize to form metallic bonds. Iron(II) oxide, FeO, is not therefore a metallic conductor. However, it is a *p*-type semiconductor. It owes its semiconducting properties to the fact that it is a *nonstoichiometric compound.*

A *nonstoichiometric compound* is a substance that has a variable composition but retains the same basic structure.

The composition of FeO can vary from $Fe_{0.89}O$ to $Fe_{0.96}O$, but it always has the sodium chloride structure. This compound does not have the *stoichiometric* composition FeO (that is, Fe_1O) because some of the Fe^{2+} ions are replaced in the lattice by Fe^{3+} ions and, for every two Fe^{3+} ions in the lattice, one oxide ion is missing, leaving a hole in the lattice. So there is fewer than one oxide ion per Fe ion. Each site of an Fe^{3+} ion can be regarded as a positive hole. When an electron from a neighboring Fe^{2+} moves to the Fe^{3+} ion, the positive hole moves—so FeO is a *p*-type semiconductor.

Nonstoichiometric compounds are relatively common among the oxides (and sulfides) of the transition metals. Because transition metals generally have more than one common oxidation state, it is frequently possible for a small number of metal ions to be replaced by an ion of higher oxidation state, thereby leading to a corresponding deficiency of oxide (or sulfide) ions.

Superconductivity

In 1911, the Dutch physicist H. Kamerlingh Onnes (1853–1926) discovered that if mercury is cooled to 4 K with liquid helium, it becomes a perfect conductor. At this temperature and below, mercury has no electrical resistance and is described as a superconductor.

A *superconductor* is an electronic conductor with zero resistance.

Once an electric current is started in a superconducting circuit, it will flow forever without external help. Above the *critical temperature* of 4 K, mercury behaves as an ordinary metal. Several other metals and alloys were found to behave as superconductors, some with higher critical temperatures. For many years a niobium–tin alloy held the record for the highest critical temperature, 23 K.

Not until 1957 did physicists develop a satisfactory explanation of superconductivity in metals. According to this theory, the maximum observed critical temperature of 23 K was close to the theoretical maximum. It was therefore a considerable surprise when, in 1986, the Swiss scientists J. Georg Bednorz and K. Alex Müller reported that a mixed oxide of barium, lanthanum, and copper was superconducting up to 30 K. Other investigators quickly confirmed this finding and started a search for materials with even higher critical temperatures. The mixed oxide $YBa_2Cu_3O_7$ was found to have a critical temperature of 93 K. It contains Y^{3+}, Ba^{2+}, and O^{2-} ions (Figure 20.25). For the ionic charges to balance, the formula of this compound would have to be $YBa_2Cu_3O_6$ if all the copper were present as Cu^{2+} ions. It appears therefore that one-third of the copper is present as Cu^{3+} and that the formula is best written as $YBa_2Cu(II)_2Cu(III)O_7$. The movement of an electron from a Cu^{2+} ion to a Cu^{3+} ion provides a mechanism for electronic conduction.

The significance of the critical temperature of 93 K for $YBa_2Cu_3O_6$ is that this temperature is above the boiling point of liquid nitrogen, 77 K, so liquid nitrogen can be used to cool the mixed oxide below its superconducting critical temperature. Liquid nitrogen is much cheaper and much easier to handle than liquid helium, which is needed to reach temperatures below 77 K. If liquid nitrogen can be used as a coolant, the possible applications of the material are vastly increased. Consequently, superconductors with even higher critical temperatures have been sought, and researchers hope someday to find a material that is superconducting at room temperature. The compound $Tl_2Ca_2Ba_2Cu_3O_{10}$ was reported in 1992 to have a critical temperature of 125 K.

All the high-temperature superconductors that have been prepared so far are mixed copper oxides; it appears that the layers of copper and oxide ions in all these structures are essential for superconductivity. However, as yet there is no generally accepted explanation of the superconducting properties of these oxides, so the search for higher-temperature superconductors is difficult and largely empirical. Once a good theory has been developed, we should be able to predict what types of compounds and what types of structures will have high superconducting critical temperatures. Materials science is still in its early stages of development. There are many exciting problems to be solved and many useful applications to be developed by the future generation of scientists—the present generation of students.

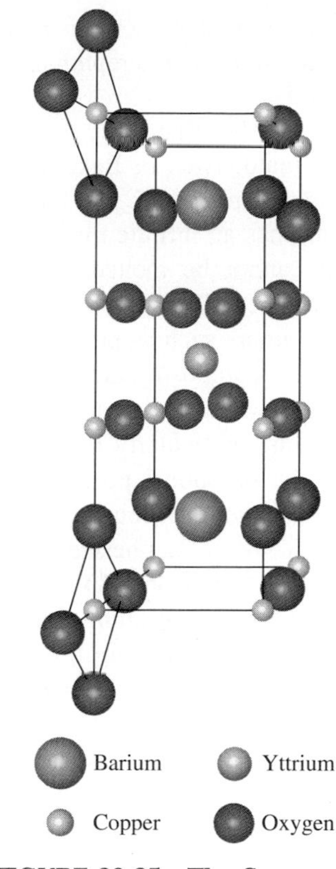

FIGURE 20.25 The Superconducting Oxide $YBa_2Cu_3O_7$ A few oxygen atoms that lie outside the unit cell are shown to illustrate the arrangement of oxygen atoms about each copper atom.

SUMMARY

Addition reactions in which many alkene (monomer) molecules are joined give addition polymers. The addition polymerization of ethene, $H_2C{=}CH_2$, can be initiated by R—O· free radicals obtained by heating organic peroxides R—O—O—R. At high pressure and 200°C, free-radical polymerization gives highly branched, soft, low-density polyethylene of low crystallinity; polymerization at lower pressures and temperatures with a transition-metal catalyst gives hard, high-density polyethylene with less branching and high crystallinity. Polypropylene, polyvinylchloride (PVC), and polystyrene result from the free-radical polymerization of propylene, $H_2C{=}CH{-}CH_3$, vinyl chloride, $H_2C{=}CHCl$, and styrene, $C_6H_5CH{=}CH_2$, respectively. Natural rubber, polyisoprene, is converted to a more useful material by vulcanization—heating with sulfur to form —S—S— cross-links between polymer molecules. Synthetic rubbers include polybutadiene, neoprene, and the copolymer styrenebutadiene rubber (SBR). A copolymer is the product of the polymerization of a mixture of two or more monomers. Materials such as rubber that return to their original shape after being stretched or compressed are called elastomers.

Condensation polymers are formed by the elimination of a small molecule, often a water molecule, between pairs of monomer molecules. Proteins are polyamides, polymers with amide, —CO—NH—, linkages. Nylon 66 is a polyamide resulting from the condensation of hexanedioic (adipic) acid and 1,6-diaminohexane. Aromatic polyamides, such as Kevlar, are stronger and harder than nylons. Polyesters, polymeric esters, result from the reaction between a dicarboxylic acid and a diol.

The polyester Dacron is made from ethylene glycol (ethane diol) and terephthalic (1,4-benzene dicarboxylic) acid.

The first fully synthetic polymer was phenol–formaldehyde, or Bakelite. It is made by the condensation of phenol, C_6H_5OH, and formaldehyde, $H_2C{=}O$, and has an infinite three-dimensional network structure. It cannot be melted, as it decomposes upon heating. Polymers of this type are called thermosetting polymers. Polymers such as polyethylene that can be melted without decomposing are called thermoplastic polymers.

Materials science is the study of new types of solids, particularly those with unusual electrical, magnetic, or optical properties. Zeolites are porous aluminosilicates first found in volcanic rocks, but many others have been synthesized. They have infinite network anions built from SiO_4 and AlO_4 tetrahedra with cavities and tunnels containing the cations and, typically, water molecules. They are used as ion-exchange materials to "soften" water by exchanging the Ca^{2+} and Mg^{2+} ions in hard water for Na^+ ion. Because of their porous structures and consequent enormous internal surface areas, zeolites are effective heterogeneous catalysts. The molecules that can enter a zeolite are limited in size and shape by the sizes of the channels and cavities in the zeolite. Zeolites are therefore highly selective catalysts.

A semiconductor is an electronic conductor with a conductivity that increases with temperature, in contrast to a metal, an electronic conductor with a conductivity that decreases with increasing temperature. The conductivity of a semiconductor is due to a small number of electrons that have sufficient energy to escape from bonds and move freely in the lattice and to the positive holes that are thereby created and also move through the lattice. Silicon is a semiconductor. When a minute amount of a Group V element such as P is added to Si (that is, when Si is doped with that element), electrons in excess of those required to form four bonds to adjacent Si atoms are free to move in the lattice, giving an *n*-type semiconductor. In contrast, a *p*-type semiconductor is created by doping Si with a Group III element such as B, creating excess positive holes.

In a solid crystal, the energy levels of a single atom are replaced by bands of closely spaced energy levels. In a metal the highest occupied energy band is only partly filled with electrons, which can therefore move easily. Metals have a high conductivity. In contrast, in an insulator (a nonconductor), such as diamond, the highest occupied band, the valence band, is full and the next-highest band, the conduction band, which is separated from the valence band by a large gap, is empty. Semiconductors have a relatively small energy gap between their valence and conduction bands, so at ordinary temperatures some valence-band electrons have sufficient energy to enter the conduction band, leaving a corresponding number of positive holes in the valence band. Both electrons and holes contribute to the conductivity. With increasing temperature, the number of electrons entering the conduction band increases, so the conductivity increases. Doping either gives a material additional electrons that enter the conduction band or depletes the valence band of electrons, creating positive holes and hence greatly increasing the conductivity.

Silicon is prepared from silica, SiO_2, by reduction with coke and is further purified by conversion to $SiCl_4$, which can be purified by distillation and then reduced back to Si by reduction with H_2 or Mg. An ultrapure material is prepared by zone refining, a process in which a rod of the impure material is passed slowly through a furnace so that a molten region travels along the rod, dissolving and accumulating impurities.

The combination of a *p*-type and an *n*-type semiconductor gives a *p-n* junction, which conducts an electric current in one direction only. Transistors are three-layered circuit elements (commonly *n-p-n* or *p-n-p* junctions) that behave as oscillators and amplifiers. Thousands of transistors and other circuit elements can be created on a single silicon chip and combined to form an integrated circuit.

Some metal oxides, particularly those of the transition metals (such as TiO and VO), whose ions have *d* electrons, are metallic conductors. Other transition metal oxides, such as FeO, are nonstoichiometric compounds: They have a fixed structure but a variable composition. Some nonstoichiometric oxides are semiconductors.

Some metals exhibit superconductivity below a certain very low critical temperature. A superconductor is an electronic conductor with zero resistance. Recently some mixed metal oxides have been discovered that are superconductors with higher critical temperatures than those of metals.

IMPORTANT TERMS

addition polymer (page 717)
condensation polymer (page 725)
conduction band (page 737)
copolymer (page 724)
doping (page 735)
elastomer (page 724)
ion exchange (page 731)
macromolecule (page 717)
nonstoichiometric compound (page 742)
n-type semiconductor (page 735)
p-n junction (page 740)
polyamide (page 725)
polyester (page 728)
positive hole (page 735)
p-type semiconductor (page 736)
semiconductor (page 734)
superconductor (page 742)
thermoplastic polymer (page 729)
thermosetting polymer (page 729)
transistor (page 741)
valence band (page 737)
vulcanization (page 724)
zeolite (page 731)
zone refining (page 740)

REVIEW QUESTIONS

1. How does an addition polymer differ from a condensation polymer?

2. Why is the polymerization of ethene to give polyethylene described as a chain reaction?

3. How are **(a)** chain termination and **(b)** chain branching achieved in the free-radical polymerization of ethene?

4. **(a)** Describe the main differences in physical properties between high-density and low-density polyethylene.

(b) Relate these differences to their structures.

5. **(a)** Explain the term "elastomer."

(b) What is vulcanization?

6. Explain and give an example of each of the terms **(a)** "copolymer," **(b)** "thermosetting polymer," and **(c)** "thermoplastic polymer."

7. **(a)** What is nylon 66?

(b) Why does it form strong, tough fibers?

8. Why is a polyester described as a condensation polymer?

9. What is a zeolite?

10. How do zeolites behave as **(a)** ion-exchange materials; **(b)** selective catalysts?

11. **(a)** What is a semiconductor?

(b) Why does its electrical conductivity increase as temperature increases?

12. What is meant by the terms **(a)** "valence band" and **(b)** "conduction band"?

13. Why is diamond an insulator, whereas both silicon and germanium are semiconductors?

14. Why is silicon doped with phosphorus an *n*-type semiconductor, whereas silicon doped with boron is a *p*-type semiconductor?

15. What is zone refining?

16. Why does a *p-n* junction conduct an electric current in only one direction?

17. What is a silicon chip?

18. What is a nonstoichiometric compound?

19. Explain why TiO is a metallic conductor.

20. What is a superconductor?

PROBLEMS

Polymer Chemistry

Addition Polymers

1. Define **(a)** "addition polymer" and **(b)** "condensation polymer," and give one example of both types of polymers.

2. What structural feature must an organic molecule have if it is to undergo addition polymerization?

3. What is meant by the terms **(a)** "linear-chain polymer," **(b)** "branched-chain polymer," and **(c)** "cross-linked polymer chain"?

4. By means of the appropriate equations, describe the mechanism of the polymerization of three molecules of ethene in the presence of benzoyl peroxide.

5. Draw a diagram to show the mechanism of chain branching in the free-radical polymerization of styrene.

6. Name the addition polymers that result from the polymerization of each of the following monomers, and draw the repeating unit in the structure of each polymer.

(a) chloroethene (vinyl chloride)

(b) phenylethene (styrene)

(c) tetrafluoroethene

(d) polymethylmethacrylate

7. Draw the repeating unit in the structures of the polymers formed from each of the following monomers.

(a) propene **(b)** acrylonitrile, $H_2C{=}CHC{\equiv}N$

(c) 1,3-butadiene

8. 2-Methyl-1,3-butadiene forms a hard natural polymer with bonds in the *trans* position. What is its structure?

9. The synthetic fiber Orlon has the polymeric chain structure

$$\left[-CH_2-\underset{\underset{C\equiv N}{|}}{CH}-CH_2-\underset{\underset{C\equiv N}{|}}{CH}-CH_2-\underset{\underset{C\equiv N}{|}}{CH}- \right]_n$$

From what monomer is this polymer synthesized?

10. **(a)** How does natural rubber differ from the synthetic rubber neoprene?

(b) Why are rubbers described as elastomers?

(c) How does vulcanization change the structure and properties of rubber?

11. Isobutylene (2-methylpropene) and isoprene (2-methylbutadiene) copolymerize to give butyl rubber. Draw the structure of the repeating unit; assume that the two monomers alternate.

12. The monomers used in the manufacture of Saran Wrap are vinyl chloride and 1,1-dichloroethene.

(a) What type of polymer is Saran Wrap?

(b) Draw the structure of the polymer, assuming that the monomers alternate.

Condensation Polymers

13. Draw the repeating unit in the structures of the polymers formed from the following monomers, and explain why the polymers are described as condensation polymers.

(a) 1,6-diaminohexane and 1,6-hexanedioic acid

(b) 1,2-ethanediol and terephthalic acid

14. Draw the repeating unit of condensation polymers made from each of the following pairs of monomers, and name the alcohol that is eliminated.

(a) 1,2-ethanediol and dimethylpropanedioate

(b) 1,3-propanediol and diethyl-1,4-butanedioate

15. Nylon stockings dissolve readily in concentrated acids, such as hydrochloric acid or sulfuric acid. Suggest a possible explanation.

16. What intermolecular forces are present between polymer chains in **(a)** an addition polymer, such as polyethylene; **(b)** a condensation polymer, such as Dacron; **(c)** a polyamide?

17. Kodel is a polyester made from terephthalic acid (1,4-benzenedicarboxylic acid) and 1,4-di(hydroxymethyl)cyclohexane. What is the structural unit of Kodel?

18. From what monomers is each of the following formed?

(a) Teflon **(b)** PVC **(c)** nylon 6 **(d)** Dacron

Materials Science

Zeolites

19. **(a)** What are the basic structural units in a zeolite?

(b) How are they joined?

(c) What is the distinguishing structural feature of a zeolite?

20. **(a)** How do aluminosilicates differ from silicates?

(b) How is sodalite, $Na_8(Al_6Si_6O_{24})Cl_2(s)$, related to silica?

(c) Why are anhydrous zeolites excellent drying agents?

21. **(a)** Which will be held more tightly in the channels and cavities of zeolites, Ca^{2+} or Na^+ ions?

(b) Explain how a zeolite can function as a water softener.

22. Why would you expect the catalytic properties of a zeolite to be greater than those of either silica, SiO_2, or aluminum oxide, Al_2O_3?

Semiconductors

23. Explain why the electrical conductivity of a metal decreases with increasing temperature, whereas that of a semiconductor increases.

24. **(a)** Why does diamond have a negligible electrical conductivity, whereas that of silicon is much greater?

(b) Why does the electrical conductivity of silicon depend strongly on its purity?

25. **(a)** How is silicon prepared from silica, $SiO_2(s)$?

(b) How is ultrapure silicon prepared?

26. **(a)** Explain the difference between the valence band and the conduction band.

(b) How does the energy gap between the valence band and the conduction band determine whether a solid is an electrical conductor, a semiconductor, or an insulator?

27. Which of the following are *n*-type conductors, and which are *p*-type conductors?

(a) Si doped with P **(b)** Si doped with B

(c) Si doped with As **(d)** Ge doped with As

(e) Ge doped with Ga

28. How does electron conduction occur in **(a)** a *p*-type conductor; **(b)** an *n*-type conductor? **(c)** Why does a *p-n* junction allow electrons to flow in one direction only?

General Problems

29. Styrene (phenylethene) may be copolymerized with maleic acid (2-butenedioic acid). What is the structure of the product, in which the two monomers provide the alternating monomer units of a copolymer?

30. The polyester formed from dimethylterephthalate and ethylene glycol contains no branched chains, whereas the same is not true of polyethylene. Why?

31. How, and from what monomers, would you synthesize the following polymers?

(a) Tedlar (polyvinylfluoride)

(b) Kel-F (polychlorotrifluoroethylene)

32. The addition of methanol to ethyne gives vinylmethyl ether, which can be polymerized to polyvinylmethylether (an adhesive).

(a) Write equations for the reactions described.

(b) Classify the final product as an addition polymer or a condensation polymer.

33. Why do polyamides such as nylon have a much greater mechanical strength than do polyalkanes such as polyethylene?

34. **(a)** What factors influence the crystallinity of a polymer?

(b) How does a "crystalline" polymer differ from an ordinary crystalline substance?

35. The energy differences between the valence bands and the conduction bands of carbon (diamond), silicon, and germanium are 502, 105, and 59 $kJ \cdot mol^{-1}$, respectively. In each case, calculate the maximum wavelength of light needed to bring about this excitation.

36. What would be the advantages and disadvantages of using superconducting material for long-distance electrical transmission lines?

37. Why are some metal oxides metallic conductors?

38. Explain why some transition metal compounds are non-stoichiometric.

Appendix C Chemistry: Important Units and Tables

1 SI UNITS

In ordinary life many different units are encountered. For example, in various parts of the world, gasoline is sold in liters, imperial gallons, or U.S. gallons; distances are measured in inches, feet, yards, and miles or centimeters, meters, and kilometers; and masses are given in ounces, pounds, and tons, metric tons, or grams and kilograms. In science the use of different units for the same quantity would be very confusing. After years of discussion, there is now almost universal agreement to use only the system of units based on the **metric system** invented in France after the French Revolution at the end of the eighteenth century. Since 1960, most scientists have used a form of the metric system called the **International System of Units** (abbreviated **SI**, from the French Système International d'Unités). The basic SI units for the quantities encountered in this book are given in Table C.1.

TABLE C.1 SI Base Units

Quantity Measured	Name	Symbol
Length	meter	m
Mass	kilogram	kg
Time	second	s
Electric current	ampere	A
Temperature	kelvin	K
Amount of substance	mole	mol

These base units are defined as follows.

- The **meter** is the distance light travels in a vacuum during a time of 1/299,792,458 second.
- The **kilogram** is the mass of a specific platinum–iridium alloy cylinder kept at the International Bureau of Weights and Measures at Sèvres, France.
- The **second** is the duration of 9,192,631,770 periods (regular oscillations) of the radiation corresponding to the transition between two specified energy levels of the cesium-133 atom.
- The **ampere** is the constant current that, if maintained in two straight, parallel conductors of infinite length and placed 1 meter apart in a vacuum, would

produce between these conductors a force equal to 2×10^{-7} newtons per meter of length.

- The **kelvin** unit of temperature is the fraction 1/273.16 of the temperature of the triple point of water.
- The **mole** is the amount of substance of a system that contains as many elementary entities as there are atoms in 0.012 kilograms of carbon-12. The entities must be specified and may be atoms, molecules, ions, electrons, or other particles.

2 DERIVED UNITS

There are important SI **derived units** (for example, units of volume, force, and energy) that can be expressed in terms of the above basic SI units (Table C.2).

TABLE C.2 SI Derived Units

Quantity Measured	Derived Unit	Symbol	SI Units
Volume	liter	L	$10^{-3}\ m^3$
Force	newton	N	$kg \cdot m \cdot s^{-2} = J \cdot m^{-1}$
Power	watt	W	$kg \cdot m^2 \cdot s^{-3} = J \cdot s^{-1}$
Pressure	pascal	Pa	$kg \cdot m^{-1} \cdot s^{-2} = N \cdot m^{-2} = J \cdot m^{-3}$
Energy, work, heat	joule	J	$kg \cdot m^2 \cdot s^{-2}$
Frequency	hertz	Hz	s^{-1}
Electric charge	coulomb	C	$A \cdot s$
Electric potential difference	volt	V	$kg \cdot m^2 \cdot s^{-3} \cdot A^{-1} = J \cdot A^{-1} \cdot s^{-1}$

Because the basic units are not always appropriate for very small or very large quantities, a series of prefixes have been defined in SI to allow us to reduce or enlarge any unit to a convenient size. Those used in this book are shown in Table C.3.

TABLE C.3 SI Prefixes

Prefix	Symbol	Multiple or Decimal Fraction	Exponential Factor
mega-	M	1,000,000	10^6
kilo-	k	1,000	10^3
—	—	1	10^0
deci-	d	0.1	10^{-1}
centi-	c	0.01	10^{-2}
milli-	m	0.001	10^{-3}
micro-	μ	0.000 001	10^{-6}
nano-	n	0.000 000 001	10^{-9}
pico-	p	0.000 000 000 001	10^{-12}

Thus, for example, 3.423 Mg = 3.423×10^6 g; 5.6 km = 5.6×10^3 m; 3.567 cm = 3.567×10^{-2} m; 5.67 μg = 5.67×10^{-6} g, and 2.5 pm = 2.5×10^{-12} m.

3 CONVERSION OF UNITS: UNIT FACTORS

Sometimes we need to convert from SI units to other units. A very convenient way of doing these conversions, as well as many other types of calculations, is the **unit factor method**. To use this method we write the units for every quantity used in a calculation and carry the units through the calculation, treating them as algebraic quantities. For example, suppose we wish to convert 3.00×10^2 s to minutes. Minutes are related to seconds by the basic relationship

$$1 \text{ min} = 60 \text{ s}$$

We can therefore write the following equalities:

$$\frac{60 \text{ s}}{1 \text{ min}} = 1 \quad \textit{or} \quad \frac{1 \text{ min}}{60 \text{ s}} = 1$$

These equalities are called **unit conversion factors**, or simply **unit factors**, because the result of multiplying a quantity of time by either factor is to multiply by 1. This act changes the units but not the absolute magnitude of a quantity. Thus, we can write

$$3.00 \times 10^2 \text{ s} = (3.00 \times 10^2 \cancel{\text{s}})\left(\frac{1 \text{ min}}{60 \cancel{\text{s}}}\right) = 5.00 \text{ min}$$

We see that the units cancel, giving the answer in the required units of minutes.

It is convenient to use unit conversion factors wherever appropriate, as illustrated in the following examples.

EXAMPLE C.1 Converting Units

Using the ideal gas equation, $PV = nRT$, calculate the volume of 0.250 g of oxygen gas at a temperature of 25.0°C and a pressure of 752 mm Hg. ($R = 0.0821 \text{ atm} \cdot \text{L} \cdot \text{mol}^{-1} \cdot \text{K}^{-1}$, and the molar mass of O_2 is $32.00 \text{ g} \cdot \text{mol}^{-1}$.)

Solution: When we use $PV = nRT$ with R in units of $\text{atm} \cdot \text{L} \cdot \text{mol}^{-1} \cdot \text{K}^{-1}$, the calculated volume (V) will be in liters if the value for the pressure (P) is given in atmospheres, the quantity of gas (n) is in moles, and the temperature (T) is in kelvins. However, in this example P is given in millimeters Hg, the quantity of gas is given in grams, and the temperature is given in degrees Celsius. We must first convert these quantities to the appropriate units:

$$1 \text{ mol } O_2 = 32.00 \text{ g}; \; n = (0.250 \cancel{\text{g } O_2})\left(\frac{1 \text{ mol } O_2}{32.00 \cancel{\text{g } O_2}}\right) = 7.813 \times 10^{-3} \text{ mol } O_2$$

$$1 \text{ atm} = 760 \text{ mm Hg}; \; P = (752 \cancel{\text{mm Hg}})\left(\frac{1 \text{ atm}}{760 \cancel{\text{mm Hg}}}\right) = 0.9895 \text{ atm}$$

and

$$T = 25.0°\text{C} = (25.0 + 273.1) \text{ K} = 298.1 \text{ K}$$

Thus,

$$PV = nRT \qquad or \qquad V = \frac{nRT}{P}$$

$$V = \frac{(7.813 \times 10^{-3}\ \cancel{mol})(0.0821\ \cancel{atm} \cdot L \cdot \cancel{mol^{-1}} \cdot \cancel{K^{-1}})(298.1\ \cancel{K})}{0.9895\ \cancel{atm}} = 0.193\ L$$

Deciding the appropriate form of the unit factor is obvious provided we keep track of the units.

Normally such calculations are done in one step, using the conversions as appropriate. For this example,

$$V = \frac{nRT}{P}$$

$$= \frac{(0.250\ \cancel{g\ O_2})\left(\frac{1\ \cancel{mol\ O_2}}{32.00\ \cancel{g\ O_2}}\right)(0.0821\ \cancel{atm} \cdot L \cdot \cancel{mol^{-1}} \cdot \cancel{K^{-1}})(25.0 + 273.1)\ \cancel{K}}{(752\ \cancel{mm\ Hg})\left(\frac{1\ \cancel{atm}}{760\ \cancel{mm\ Hg}}\right)} = 0.193\ L$$

4 TEMPERATURE CONVERSIONS

$$0\ K = -273.15°C = -459.67°F$$
$$K = °C + 273.15$$
$$°C = \tfrac{5}{9}(°F - 32)$$
$$°F = \tfrac{9}{5}(°C) + 32$$

5 ATOMIC AND MOLAR MASSES OF THE ELEMENTS

Element	Symbol	Atomic Number	Atomic Mass* (u) Molar Mass ($g \cdot mol^{-1}$)	Element	Symbol	Atomic Number	Atomic Mass* (u) Molar Mass ($g \cdot mol^{-1}$)
Actinium	Ac	89	227.0278†	Cadmium	Cd	48	112.41
Aluminum	Al	13	26.981 54	Calcium	Ca	20	40.078
Americium	Am	95	(243)**	Californium	Cf	98	(249)**
Antimony	Sb	51	121.75	Carbon	C	6	12.011
Argon	Ar	18	39.948	Cerium	Ce	58	140.12
Arsenic	As	33	74.9216	Cesium	Cs	55	132.9054
Astatine	At	85	(210)**	Chlorine	Cl	17	35.453
Barium	Ba	56	137.33	Chromium	Cr	24	51.996
Berkelium	Bk	97	(247)**	Cobalt	Co	27	58.9332
Beryllium	Be	4	9.012 18	Copper	Cu	29	63.546
Bismuth	Bi	83	208.9804	Curium	Cm	96	(247)**
Boron	B	5	10.81	Dysprosium	Dy	66	162.50
Bromine	Br	35	79.904	Einsteinium	Es	99	(252)**

Element	Symbol	Atomic Number	Atomic Mass* (u) Molar Mass ($g \cdot mol^{-1}$)	Element	Symbol	Atomic Number	Atomic Mass* (u) Molar Mass ($g \cdot mol^{-1}$)
Erbium	Er	68	167.26	Platinum	Pt	78	195.08
Europium	Eu	63	151.96	Plutonium	Pu	94	(244)**
Fermium	Fm	100	(257)**	Polonium	Po	84	208.980
Fluorine	F	9	18.998 403	Potassium	K	19	39.0983
Francium	Fr	87	(223)**	Praseodymium	Pr	59	140.9077
Gadolinium	Gd	64	157.25	Promethium	Pm	61	144.9127
Gallium	Ga	31	69.72	Protactinium	Pa	91	231.0359†
Germanium	Ge	32	72.61	Radium	Ra	88	226.0254†
Gold	Au	79	196.9665	Radon	Rn	86	(222)**
Hafnium	Hf	72	178.49	Rhenium	Re	75	186.207
Helium	He	2	4.002 60	Rhodium	Rh	45	102.9055
Holmium	Ho	67	164.9304	Rubidium	Rb	37	85.4678
Hydrogen	H	1	1.007 94	Ruthenium	Ru	44	101.07
Indium	In	49	114.82	Samarium	Sm	62	150.36
Iodine	I	53	126.9045	Scandium	Sc	21	44.9559
Iridium	Ir	77	192.22	Selenium	Se	34	78.96
Iron	Fe	26	55.847	Silicon	Si	14	28.0855
Krypton	Kr	36	83.80	Silver	Ag	47	107.8682
Lanthanum	La	57	138.9055	Sodium	Na	11	22.989 77
Lawrencium	Lr	103	(260)**	Strontium	Sr	38	87.62
Lead	Pb	82	207.2	Sulfur	S	16	32.07
Lithium	Li	3	6.941	Tantalum	Ta	73	180.9479
Lutetium	Lu	71	174.967	Technetium	Tc	43	98.91
Magnesium	Mg	12	24.305	Tellurium	Te	52	127.60
Manganese	Mn	25	54.9380	Terbium	Tb	65	158.9254
Mendelevium	Md	101	(258)**	Thallium	Tl	81	204.37
Mercury	Hg	80	200.59	Thorium	Th	90	232.0381*
Molybdenum	Mo	42	95.94	Thulium	Tm	69	168.9342
Neodymium	Nd	60	144.24	Tin	Sn	50	118.69
Neon	Ne	10	20.180	Titanium	Ti	22	47.88
Neptunium	Np	93	237.0482†	Tungsten	W	74	183.85
Nickel	Ni	28	58.69	Uranium	U	92	238.0289
Niobium	Nb	41	92.9064	Vanadium	V	23	50.9415
Nitrogen	N	7	14.0067	Xenon	Xe	54	131.29
Nobelium	No	102	(259)**	Ytterbium	Yb	70	173.04
Osmium	Os	76	190.2	Yttrium	Y	39	88.9059
Oxygen	O	8	15.9994	Zinc	Zn	30	65.39
Palladium	Pd	46	106.42	Zirconium	Zr	40	91.22
Phosphorus	P	15	30.973 76				

* The atomic masses of many elements are not invariant but depend on the origin and treatment of the material. The values given here apply to elements as they exist naturally on earth and to certain artificial elements.

† For these radioactive elements, the mass given is that for the longest-lived isotope.

** Atomic masses for these radioactive elements cannot be quoted precisely without knowledge of the origin of the element. The value given is the atomic mass number of the isotope of that element of longest known half-life.

6 ELECTRON CONFIGURATIONS OF THE ELEMENTS

Atomic Number	Element	Electron Configuration	Atomic Number	Element	Electron Configuration
1	H	$1s^1$	53	I	$[Kr]4d^{10}5s^25p^5$
2	He	$1s^2$	54	Xe	$[Kr]4d^{10}5s^25p^6$
3	Li	$[He]2s^1$	55	Cs	$[Xe]6s^1$
4	Be	$[He]2s^2$	56	Ba	$[Xe]6s^2$
5	B	$[He]2s^22p^1$	57	La	$[Xe]5d^16s^2$
6	C	$[He]2s^22p^2$	58	Ce	$[Xe]4f^15d^16s^2$
7	N	$[He]2s^22p^3$	59	Pr	$[Xe]4f^3\ 6s^2$
8	O	$[He]2s^22p^4$	60	Nd	$[Xe]4f^4\ 6s^2$
9	F	$[He]2s^22p^5$	61	Pm	$[Xe]4f^5\ 6s^2$
10	Ne	$[He]2s^22p^6$	62	Sm	$[Xe]4f^6\ 6s^2$
11	Na	$[Ne]3s^1$	63	Eu	$[Xe]4f^7\ 6s^2$
12	Mg	$[Ne]3s^2$	64	Gd	$[Xe]4f^75d^16s^2$
13	Al	$[Ne]3s^23p^1$	65	Tb	$[Xe]4f^9\ 6s^2$
14	Si	$[Ne]3s^23p^2$	66	Dy	$[Xe]4f^{10}\ 6s^2$
15	P	$[Ne]3s^23p^3$	67	Ho	$[Xe]4f^{11}\ 6s^2$
16	S	$[Ne]3s^23p^4$	68	Er	$[Xe]4f^{12}\ 6s^2$
17	Cl	$[Ne]3s^23p^5$	69	Tm	$[Xe]4f^{13}\ 6s^2$
18	Ar	$[Ne]3s^23p^6$	70	Yb	$[Xe]4f^{14}\ 6s^2$
19	K	$[Ar]4s^1$	71	Lu	$[Xe]4f^{14}5d^16s^2$
20	Ca	$[Ar]4s^2$	72	Hf	$[Xe]4f^{14}5d^26s^2$
21	Sc	$[Ar]3d^14s^2$	73	Ta	$[Xe]4f^{14}5d^36s^2$
22	Ti	$[Ar]3d^24s^2$	74	W	$[Xe]4f^{14}5d^46s^2$
23	V	$[Ar]3d^34s^2$	75	Re	$[Xe]4f^{14}5d^56s^2$
24	Cr	$[Ar]3d^54s^1$	76	Os	$[Xe]4f^{14}5d^66s^2$
25	Mn	$[Ar]3d^54s^2$	77	Ir	$[Xe]4f^{14}5d^76s^2$
26	Fe	$[Ar]3d^64s^2$	78	Pt	$[Xe]4f^{14}5d^96s^1$
27	Co	$[Ar]3d^74s^2$	79	Au	$[Xe]4f^{14}5d^{10}6s^1$
28	Ni	$[Ar]3d^84s^2$	80	Hg	$[Xe]4f^{14}5d^{10}6s^2$
29	Cu	$[Ar]3d^{10}4s^1$	81	Tl	$[Xe]4f^{14}5d^{10}6s^26p^1$
30	Zn	$[Ar]3d^{10}4s^2$	82	Pb	$[Xe]4f^{14}5d^{10}6s^26p^2$
31	Ga	$[Ar]3d^{10}4s^24p^1$	83	Bi	$[Xe]4f^{14}5d^{10}6s^26p^3$
32	Ge	$[Ar]3d^{10}4s^24p^2$	84	Po	$[Xe]4f^{14}5d^{10}6s^26p^4$
33	As	$[Ar]3d^{10}4s^24p^3$	85	At	$[Xe]4f^{14}5d^{10}6s^26p^5$
34	Se	$[Ar]3d^{10}4s^24p^4$	86	Rn	$[Xe]4f^{14}5d^{10}6s^26p^6$
35	Br	$[Ar]3d^{10}4s^24p^5$	87	Fr	$[Rn]7s^1$
36	Kr	$[Ar]3d^{10}4s^24p^6$	88	Ra	$[Rn]7s^2$
37	Rb	$[Kr]5s^1$	89	Ac	$[Rn]6d^17s^2$
38	Sr	$[Kr]5s^2$	90	Th	$[Rn]6d^27s^2$
39	Y	$[Kr]4d^15s^2$	91	Pa	$[Rn]5f^26d^17s^2$
40	Zr	$[Kr]4d^25s^2$	92	U	$[Rn]5f^36d^17s^2$
41	Nb	$[Kr]4d^45s^1$	93	Np	$[Rn]5f^46d^17s^2$
42	Mo	$[Kr]4d^55s^1$	94	Pu	$[Rn]5f^67s^2$
43	Tc	$[Kr]4d^55s^2$	95	Am	$[Rn]5f^7\ 7s^2$
44	Ru	$[Kr]4d^75s^1$	96	Cm	$[Rn]5f^76d^17s^2$
45	Rh	$[Kr]4d^85s^1$	97	Bk	$[Rn]5f^9\ 7s^2$
46	Pd	$[Kr]4d^{10}$	98	Cf	$[Rn]5f^{10}\ 7s^2$
47	Ag	$[Kr]4d^{10}5s^1$	99	Es	$[Rn]5f^{11}\ 7s^2$
48	Cd	$[Kr]4d^{10}5s^2$	100	Fm	$[Rn]5f^{12}\ 7s^2$
49	In	$[Kr]4d^{10}5s^25p^1$	101	Md	$[Rn]5f^{13}\ 7s^2$
50	Sn	$[Kr]4d^{10}5s^25p^2$	102	No	$[Rn]5f^{14}\ 7s^2$
51	Sb	$[Kr]4d^{10}5s^25p^3$	103	Lr	$[Rn]5f^{14}6d^17s^2$
52	Te	$[Kr]4d^{10}5d^25p^4$			

7 THERMODYNAMIC DATA

Elements and Inorganic Compounds at 25°C

Compound	ΔH_f° kJ · mol^{-1}	S° J · K^{-1} · mol^{-1}	ΔG_f° kJ · mol^{-1}	Compound	ΔH_f° kJ · mol^{-1}	S° J · K^{-1} · mol^{-1}	ΔG_f° kJ · mol^{-1}
$Ag(s)$	0	+42.6	0	$Na(s)$	0	+51.3	0
$AgCl(s)$	−127.1	+96.2	−109.8	$NaF(s)$	−569.0	+51.3	−546.3
$AlCl_3(s)$	−704.2	+110.7	−628.8	$NaCl(s)$	−411.1	+72.4	−384.3
$Al_2O_3(s)$	−1676	+50.9	−1582	$NaBr(s)$	−361.1	+87.2	−349.1
$B_5H_9(s)$	+73.2	+276	+175	$NaI(s)$	−287.8	+98.5	−282.4
$B_2O_3(s)$	−1273.5	+54.0	−1194.4	$NaOH(s)$	−425.6	+64.5	−379.5
$Br_2(l)$	0	+152.2	0	$Na_2O_2(s)$	−511.7	+104	−451.0
$BrF_3(g)$	−255.6	+292.4	−229.5	$NH_3(g)$	−46.2	+192.7	−16.4
$CaO(s)$	−635.1	+38.1	−603.5	$N_2H_4(l)$	+50.6	+121.2	+149.2
$CaCO_3(s)$(calcite)	−1206.9	+92.9	−1128.8	$NH_4ClO_4(s)$	−295	+186	−89
$Cl_2(g)$	0	+223.0	0	$NO(g)$	+90.3	+210.6	+86.6
$Cu(s)$	0	+33.2	0	$NO_2(g)$	+33.2	+240.0	+51.3
$F_2(g)$	0	+202.7	0	$N_2O_4(l)$	−19.5	+219.7	+97.5
$Fe(s)$	0	+27.3	0	$N_2O_4(g)$	+9.2	+304.2	+97.8
$Fe_2O_3(s)$(hematite)	−824.2	+87.4	−742.2	$HNO_3(l)$	−174.1	+155.6	−80.8
$H(g)$	+218.0	+114.6	+203.3	$NOCl(g)$	+51.7	+261.6	+66.1
$H_2(g)$	0	+130.6	0	$O_2(g)$	0	+205.0	0
$HCl(g)$	−92.3	+186.8	−95.3	$O_3(g)$	+142.7	+238.8	+163.2
$HF(g)$	−271.1	+173.8	−273.2	$P(s)$(white)	0	+41.1	0
$HI(g)$	+26.4	+206.5	+1.6	$P_4O_{10}(s)$	−3010	+231	−2724
$HBr(g)$	−36.4	+198.6	−53.5	$PCl_3(g)$	−287.0	+311.7	−267.8
$HCN(g)$	+135.1	+201.7	+124.7	$PCl_5(g)$	−374.9	+364.5	−305.0
$H_2O(g)$	−241.8	+188.7	−228.6	$PbO_2(s)$	−277.4	+68.6	−217.4
$H_2O(l)$	−285.8	+70.0	−237.2	$S(s)$(orthorhombic)	0	+32.0	0
$H_2O_2(l)$	−187.8	+109.6	−120.4	$H_2S(g)$	−20.6	+205.6	−33.4
$Hg(l)$	0	+75.9	0	$SiO_2(s)$(quartz)	−910.7	+41.5	−856.3
$I_2(s)$	0	+116.1	0	$SiCl_4(l)$	−687.0	+239.7	−619.9
$I_2(g)$	+62.4	+260.6	+19.4	$SO_2(g)$	−296.8	+248.1	−300.2
$MgO(s)$	−601.5	+27.0	−569.2	$SO_3(g)$	−395.7	+256.6	−371.1
$MnO_2(s)$	−520.0	+53.1	−465.2	$H_2SO_4(l)$	−814.0	+156.9	−690.1
$N_2(g)$	0	+191.5	0	$ZnO(s)$	−350.5	+43.6	−320.5
$N_2O_4(g)$	+9.3	+304.2	+97.8				

Carbon and Carbon Compounds at 25°C

Compound	ΔH_f° kJ · mol^{-1}	S° J · K^{-1} · mol^{-1}	ΔG_f° kJ · mol^{-1}
$C(g)$	+716.7	+158.0	+671.3
C(graphite)	0	+5.8	0
C(diamond)	+1.9	+2.4	+2.9
$CO(g)$	−110.5	+197.6	−137.2
$CO_2(g)$	−393.5	+213.7	−394.4
$CH_4(g)$	−74.5	+186.1	−50.8
$C_2H_2(g)$	+226.8	+200.8	+209.2
$C_2H_4(g)$	+52.3	+219.4	+68.1
$C_2H_6(g)$	−84.7	+229.5	−32.9
$C_3H_6(g)$(cyclopropane)	+53.3	+237	+104
$C_3H_8(g)$	−103.8	+269.9	−23.4
$C_4H_8(g)$(cyclobutane)	+28.4	+265	+100
$C_4H_{10}(g)$(*n*-butane)	−126.1	+310.1	−17.2
$C_5H_{10}(g)$(cyclopentane)	−78.4	+293	+39
$C_5H_{12}(g)$(*n*-pentane)	−146.4	+348.9	−8.4
$C_5H_{12}(l)$(*n*-pentane)	−173.2	+263.3	−9.6
$C_6H_6(l)$(benzene)	+49.0	+172.2	+124.7
$C_6H_{12}(g)$(cyclohexane)	−123.3	+298	+32
n-$C_6H_{14}(l)$	−198.6	+295.9	−4.4
n-$C_7H_{16}(l)$	−224.0	+328.5	+1.0
n-$C_8H_{18}(l)$	−250.0	+361.2	+6.4
$CH_2O(g)$	−108.7	+218.7	−113
$CH_3OH(l)$	−238.7	+126.8	−166.4
$C_2H_5OH(l)$	−277.7	+160.7	−174.9
$CH_3CO_2H(l)$	−484.5	+159.8	−390.0
$C_6H_{12}O_6(s)$(glucose)	−1268	+182.4	−919.2
$C_{12}H_{22}O_{11}(s)$(sucrose)	−2226.1	+360	−1545
$CH_2Cl_2(l)$	−124.1	+177.8	−67.3
$CHCl_3(l)$	−135.1	+201.7	−73.7
$CCl_4(l)$	−135.4	+216.4	−65.3

Appendix P
Physics Review

1 THE LAWS OF MOTION, FORCES, AND FIELDS

Newton's First Law of Motion

An object at rest will remain at rest, and an object in motion will continue to move at a constant velocity (constant speed in a straight line) unless it is acted upon by a net force.

Newton's Second Law of Motion

The acceleration, a, of an object is directly proportional to the resultant force, F, acting on it and inversely proportional to its mass, m:

$$F = ma$$

The SI unit of force is the **newton** (N); 1 N is defined as the force that, when acting on a mass of 1 kg, produces an acceleration of $1\ \text{m} \cdot \text{s}^{-2}$:

$$1\ \text{N} = 1\ \text{kg} \cdot \text{m} \cdot \text{s}^{-2}$$

Newton's Third Law of Motion

When one body exerts a force on another, the second exerts on the first a force that is equal in magnitude but opposite in direction.

Forces and Fields

In everyday life we are familiar with forces such as the force exerted on a football when it is kicked. In general,

> **A force is any influence that can produce a change in the velocity of an object.**

The only known *fundamental forces* in nature—forces from which all others ultimately derive—are (1) gravitational force, (2) electrical force, and (3) nuclear force.

> **The *gravitational force* acts between all objects and is proportional to their masses.**

The *electrical force* acts between charged objects and is proportional to their charges. Magnetic forces arise from charges in motion, but these forces are not fundamentally different from the electrical force.

The *nuclear force* acts between subatomic particles and holds together the particles (protons and neutrons) of which a nucleus is composed. (Physicists distinguish between the *strong* nuclear force and the *weak* nuclear force, but this difference is of no importance in chemistry.)

These fundamental forces act through space. Because it is difficult to visualize a force acting through empty space, Michael Faraday introduced the concept of a force *field*. When a mass m_1 is placed at a point P near a mass m_2, we say that m_1 interacts with m_2 by virtue of the gravitational field that exists at P. Gravitational, electrical, and nuclear forces are called *field forces*. A field is simply a model devised to provide a framework for understanding how forces are transmitted from one object to another through space.

We are more familiar with forces that include physical contact, called *contact forces*, such as the force exerted on a tennis ball when it is struck with a racquet. But this is not a fundamental type of force. When two macroscopic objects collide, the contact force between them is a consequence of the electrical force between the electrons and nuclei of which the objects are composed.

In chemistry only the electrical force is important. The gravitational force is so much weaker than the electrical force that, when we are considering forces between very small objects such as atoms and molecules, the gravitational force can be ignored. But the gravitational forces between objects of "normal" size and between very large objects such as the planets and the sun are of great importance. Nuclear forces are important only over distances much smaller than the distances between atoms in molecules—in particular, over the extremely small distances between the particles in a nucleus.

Pressure

Pressure is the force per unit area. Consider the force exerted by the molecules of a gas colliding with the walls of the container. The force per unit area of the walls is the pressure exerted by the gas on the walls, which equals the pressure exerted by the walls on the gas. The SI unit of pressure is newtons per square meter ($N \cdot m^{-2}$), which is called the **pascal** (Pa);

$$1 \text{ Pa} = 1 \text{ N} \cdot \text{m}^{-2}$$

2 MOMENTUM

The **momentum**, p, of a body is the product of the body's mass, m, and its velocity, v:*

$$p = mv$$

* Strictly speaking, velocity (and hence momentum) is a vector quantity—a quantity whose value includes a direction as well as a magnitude. For example, "5 $m \cdot s^{-1}$" is a speed, whereas "5 $m \cdot s^{-1}$ to the left" is a velocity. But this distinction is not important for our purposes.

The Law of Conservation of Momentum When no external force acts on a system, the total momentum of the system remains constant in magnitude and direction. When two bodies A and B collide,

$$m_A(v_A)_1 + m_B(v_B)_1 = m_A(v_A)_2 + m_B(v_B)_2$$

where $(v_X)_1$ is the velocity of body X before the collision and $(v_X)_2$ is the velocity of body X after the collision. The kinetic energy is also conserved in a collision:

$$\tfrac{1}{2}m_A(v_A)_1{}^2 + \tfrac{1}{2}m_B(v_B)_1{}^2 = \tfrac{1}{2}m_A(v_A)_2{}^2 + \tfrac{1}{2}m_B(v_B)_2{}^2$$

From these two laws it can be shown that when A and B collide, if A has a higher velocity than B, then A is slowed down while B is speeded up. Also, the greater the mass of an object, the less its velocity is changed. In the special case when A has an essentially infinite mass and is at rest (such as when A is the wall of a container of gas and B is a gas molecule), B bounces off the wall with its speed unchanged, and A remains at rest.

3 WORK AND ENERGY

Work

When a constant force F acting on an object moves the object, the **work**, w, done on the object is defined as the product of the force times the distance, d, through which the object is displaced:

$$w = F \times d$$

The SI unit of work is the **newton · meter** (N · m), which is also called the **joule** (J):

$$1 \text{ J} = 1 \text{ kg} \cdot \text{m}^2 \cdot \text{s}^{-2}$$

Energy

Energy is the ability to do work. The SI unit of energy is the joule (J). The amount of energy a body has depends on how much work it can do. For example, a moving object can do work when it collides with another object. A moving object is said to possess **kinetic energy**, *KE*. The kinetic energy of a particle of mass m and speed v is defined as

$$KE = \tfrac{1}{2}mv^2$$

Kinetic energy has the same units as work ($\text{kg} \cdot \text{m}^2 \cdot \text{s}^{-2} = \text{N} \cdot \text{m} = \text{J}$). The work done by a force F in displacing a particle equals the change in kinetic energy of the particle.

Potential energy, *PE*, is the energy possessed by an object because of its position. The term ''potential'' implies that the object has the capability, or potential, of gaining kinetic energy or doing work. There are several types of potential energy. An object has *gravitational potential energy* because of its position with respect to the surface of the earth. A compressed spring has *elastic potential energy*; if the spring is released, its potential energy is converted to kinetic energy as it expands, and it can then do work. An electric charge in the presence of a second charge has *electrical potential energy*. If it is not held in position, one charge will move toward or away from the other, and their potential energy will be converted to kinetic energy.

The Law of Conservation of Energy

If the kinetic energy of an object increases (or decreases) by some amount, then the potential energy of the object must decrease (or increase) by the same amount. The total energy, $KE + PE$, is constant. We say that total energy is conserved—it may be converted from kinetic energy to potential energy or vice versa, but the total remains constant.

4 ELECTRICITY AND MAGNETISM

We saw that electromagnetic forces—one of the fundamental forces in nature—are the forces between charged particles. The forces between charged particles at rest are called *electrostatic forces*. In contrast, the forces that arise from charges in motion are called *magnetic forces*. In chemistry we are concerned primarily, although not exclusively, with electrostatic forces.

Coulomb's Law There are two types of charge, positive and negative. Like charges repel each other, and unlike charges attract each other. The SI unit of charge is the **coulomb** (C). The basic law of force between two stationary charged particles is Coulomb's law,

$$F = k\frac{Q_1Q_2}{r^2}$$

where F is the force between two electric charges Q_1 and Q_2 separated by a distance r, and k is a constant called the Coulomb constant.

A flow of charged particles—for example, electrons moving in a metal wire—constitutes an *electric current*. The SI unit of current is the **ampere** (A) (see Appendix C). When the current in a wire is 1 A, the amount of charge that flows past a given point in 1 s is 1 C. The smallest unit of charge known in nature is the charge of an electron or a proton. The electron has a charge of $-1.602\ 19 \times 10^{-19}$ C, and the proton has a charge of $+1.602\ 19 \times 10^{-19}$ C. A charge of 1.602 19 C is denoted by the symbol e:

$$e = 1.602\ 19 \text{ C}$$

A total of 6.241×10^{18} electrons has a charge of 1 C.

Charges on particles such as the electron and the proton are usually expressed as multiples of e and not in coulombs. An electron is said to have a charge of -1 (that is, a charge of $-1e$, or -1.602×10^{-19} C). Similarly, a proton is said to have a charge of $+1$ (that is, a charge of $+1e$, or $+1.602 \times 10^{-19}$ C).

Electric Potential The **electric potential** at an arbitrary point is the work done in moving unit positive charge (in SI, a charge of 1 C) from infinity (zero potential) to that point. Electric potential is a measure of the energy per unit charge, and the SI unit of potential is joules per coulomb, defined to be equal to a unit called the **volt** (V):

$$1 \text{ V} = 1 \text{ J} \cdot \text{C}^{-1}$$

Instead of speaking about the electric potential difference between two points, we often refer to the *voltage* difference between them.

Electromagnetic Radiation **Electromagnetic radiation** is radiation generated by accelerating electric charges. It can be described in terms of waves that consist of an oscillating electric field and an oscillating magnetic field that are at right angles to each other and to the direction of wave propagation. Electromagnetic waves propagate through space at a speed of $3.00 \times 10^8\ \mathrm{m \cdot s^{-1}}$. Light is an electromagnetic wave, and $3.00 \times 10^8\ \mathrm{m \cdot s^{-1}}$ is often referred to as the speed of light.

Electromagnetic radiation can instead be described as consisting of *photons*, particles with energy $E = h\nu$. Here ν is the frequency of the radiation and $h = 6.63 \times 10^{-34}\ \mathrm{J \cdot s}$ is the Planck constant.

Appendix M
Mathematics Review

1 SCIENTIFIC (EXPONENTIAL) NOTATION

Although we commonly write both large and small numbers in decimal form, such as 5,343,147.51 or 0.000 003 71, it is customary in scientific work to use **scientific notation**, a method that expresses any quantity as a number between 1 and 10 that is either multipled by 10 or divided by 10 an appropriate number of times. For example,

$$5{,}343{,}147.51 = 5.343\,147\,51 \times 10 \times 10 \times 10 \times 10 \times 10 \times 10$$

$$0.000\,003\,71 = \frac{3.71}{10 \times 10 \times 10 \times 10 \times 10 \times 10}$$

In exponential form these are written as

$$5.343\,147\,51 \times 10^6 \qquad \textit{and} \qquad 3.71 \times 10^{-6}$$

To write any decimal quantity in scientific notation,

1. First move the decimal point so that it comes immediately after the first number between 1 and 9, and count the number of places you have moved the decimal point.
2. Then express the number as $n \times 10^x$, where x is the number of places you moved the decimal point. The exponent x is a positive integer if you moved the decimal point to the left, and a negative integer if you moved the decimal point to the right.

EXAMPLE M.1 Scientific Notation

(a) The number 5,343,147.51 becomes 5.343 147 51 by moving the decimal point six places to the left. Therefore, $x = +6$, and

$$5{,}343{,}147.51 = 5.343\,147\,51 \times 10^6$$

(b) The number 0.000 003 71 becomes 3.71 by moving the decimal point six places to the right. Thus, $x = -6$, and

$$0.000\,003\,71 = 3.71 \times 10^{-6}$$

Exercise M.1 Express each of the following numbers in scientific notation.

(a) 5,371.1 **(b)** 63,000 **(c)** 0.0233 **(d)** 0.000 57 **(e)** 64.000 57
(f) 0.000 000 000 0061

Answer: **(a)** 5.3711×10^3 **(b)** 6.3000×10^4 **(c)** 2.33×10^{-2}
(d) 5.7×10^{-4} **(e)** $6.400\,57 \times 10^1$ **(f)** 6.1×10^{-12}

2 SIGNIFICANT FIGURES

Two types of numbers are used in chemistry: *exact numbers* and *inexact numbers*. Exact numbers are numbers whose values are precisely known or exactly fixed by definition. For example, we may refer to 12 beakers, 10 atoms in a molecule, 60 seconds in 1 minute, or 12 atomic units for the mass of one ^{12}C atom; these numbers are exact. Inexact numbers are numbers associated with quantities measured experimentally—the accuracy with which an experimental quantity can be given is limited by the precision of the instruments used to measure that quantity. This accuracy is indicated by the number of **significant figures** quoted. For example, a pressure given as 752.5 mm Hg contains four significant figures and is interpreted as having an uncertainty only in the last digit on the right. The mass of a sample quoted as 5.4321 g (to five significant figures) is known reliably to ±0.0001 g.

For numbers expressed in exponential form, the number of significant figures is given by the number of digits before the exponential term. Hence 2.998×10^8 $\text{m} \cdot \text{s}^{-1}$ has four significant figures, and $1.600\,56 \times 10^{-27}$ kg has six.

Exercise M.2 How many significant figures are there in each of the following?

(a) 5.32 g **(b)** 1.51 $\text{g} \cdot \text{cm}^{-3}$ **(c)** 3.005×10^{-2} mol **(d)** 4.5 cm

Answer: **(a)** 3 **(b)** 3 **(c)** 4 **(d)** 2

In calculations that involve several different quantities, the answer can be given with reliability only to the number of significant figures associated with the least precise quantity involved in the calculation. For example, suppose we want to determine the volume of 25.4 g of lead, which has a density of 11.34 $\text{g} \cdot \text{cm}^{-3}$. Although the density is given to four significant figures, the mass is known only to three significant figures, and the volume can be given only to three significant figures:

$$\text{volume of lead} = \frac{25.4 \text{ g}}{11.34 \text{ g} \cdot \text{cm}^{-3}} = 2.239\,85 \ldots \text{cm}^3$$

Some calculators would give this answer with even more digits to the right of the decimal. But this number must be rounded off to three significant figures, like the least precise quantity—in this case, the mass. Hence the answer is 2.24 cm^3, which has three significant figures.

EXAMPLE M.2 Significant Figures

How many moles of sulfur (molar mass 32.07 g · mol^{-1}) are there in 2.5432 g of sulfur?

Solution: The mass of sulfur is known to five significant figures, but the molar mass is given only to four significant figures. So the number of moles of sulfur can be given to no more than four significant figures:

$$\text{moles S} = \frac{2.5432 \text{ g}}{32.07 \text{ g} \cdot \text{mol}^{-1}} = 7.930 \times 10^{-2} \text{ mol (four significant figures)}$$

Rounding Off

When the final calculated quantity contains more figures than are significant, it should be **rounded off** as follows:

1. If the figures following the last significant figure are 499 . . . or less, discard them and leave the last number unchanged. For example, to three significant figures 3.6246 is 3.62, and to four significant figures, 187.4499 is 187.4.
2. If the figures following the last significant figure are 500 . . . or greater, discard them and increase the last number by one. For example, to three significant figures, 7.635 becomes 7.64, and to five significant figures, 2.87257×10^{-3} becomes 2.8726×10^{-3}.

To avoid error, when a calculated quantity is the result of a sequence of separate calculations, the answer carried forward from one step to the next should contain one more significant figure than will appear in the final answer.

Addition and Subtraction

When numbers are added or subtracted, the number of decimal places in the result should equal the smallest number of *decimal places*—not necessarily significant figures—of all terms in the sum. For example, when we compute 123 + 5.24, the answer is 128 and not 128.52, because 123 has no decimal. In the sum 1.0001 + 0.0004 = 1.0005, the result has five significant figures because both terms have four decimal places even though the term 0.0004 has only one significant figure. Likewise, if we perform the subtraction 1.002 − 0.998 = 0.004, the result correctly has three decimal places according to the rule but only one significant figure.

The Significance of Zeros in a Number

The presence of zeros in a number may be misinterpreted. Suppose that the mass of an object is measured to be 1500 g. This value is ambiguous because it is not known whether the last two zeros are being used to locate the decimal point or whether they represent significant figures in the measurement. To remove this ambiguity we use scientific notation to indicate the number of significant figures. We would express a mass of 1500 g as 1.5×10^3 g if it is known to two significant figures and as 1.50×10^3 g if it is known to three significant figures. A number such as 0.00015 would be expressed in scientific notation as 1.5×10^{-4} g to two significant figures and as 1.50×10^{-4} to three significant figures. The three zeros following the decimal point in the number 0.00015 are not counted as significant figures; they merely locate the decimal point. The zero in a number such as 30,567 would, however, be counted as a significant figure.

Multiplication, Division, Powers, and Roots

With numbers expressed in scientific notation, it can be convenient to handle the powers of 10 separately, using the rules for exponents:

Multiplication $10^x \times 10^y = 10^{x+y}$

Division $\dfrac{10^x}{10^y} = 10^{x-y}$

Powers $(10^x)^y = 10^{xy}$

Roots $\sqrt[y]{10^x} = (10^x)^{\frac{1}{y}} = 10^{\frac{x}{y}}$

EXAMPLE M.3 Working with Exponents

Evaluate each of the following.

(a) $57{,}000 \times 0.0032$ **(b)** $\dfrac{57{,}000}{0.0032}$ **(c)** $(3000)^4$ **(d)** $\sqrt[3]{0.027}$

Solution: **(a)** $(57{,}000)(0.0032) = (5.7 \times 10^4)(3.2 \times 10^{-3}) = 18 \times 10^1 = 1.8 \times 10^2$

(b) $\dfrac{57{,}000}{0.0032} = \dfrac{5.7 \times 10^4}{3.2 \times 10^{-3}} = 1.8 \times 10^{(4-(-3))} = 1.8 \times 10^7$

(c) $(3000)^4 = (3.000 \times 10^3)^4 = (3.000)^4 \times (10^3)^4 = 81 \times 10^{12} = 8.1 \times 10^{13}$

(d) $\sqrt[3]{0.027} = (0.027)^{\frac{1}{3}} = (2.7 \times 10^{-2})^{\frac{1}{3}} = (2.7)^{\frac{1}{3}} \times (10^{-3})^{\frac{1}{3}} = 3.0 \times 10^{-1}$

Exercise M.3 Evaluate each of the following, and give each answer in scientific notation.

(a) $\dfrac{(534{,}600)(43.41)}{0.003\,52}$ **(b)** $\sqrt{9.54} \times 0.000\,0062$ **(c)** $(36.54 \times 63.54)^{-5}$

Answer: **(a)** 6.59×10^9 **(b)** 1.9×10^{-5} **(c)** 1.482×10^{-17}

Here the answers are rounded off to the appropriate number of significant figures.

Order-of-Magnitude Estimation of an Answer

Because you can easily make errors even when you use a calculator, it is prudent to estimate your answers to calculations to see if the order of magnitude is appropriate. For example, round off numbers to the nearest integers to get a rough answer:

$$\frac{(2.11 \times 10^6)(4.14 \times 10^{-5})}{8.34 \times 10^{-2}} \simeq \frac{(2 \times 10^6)(4 \times 10^{-5})}{8 \times 10^{-2}} = \frac{8 \times 10^1}{8 \times 10^{-2}} = 1 \times 10^3$$

It is then a simple matter to check that your real answer (1.05×10^3) is of the correct magnitude—that is, correct within a factor of 10. Make it a habit to check the order of magnitude of answers to all calculations in this way.

3 LOGARITHMS

Suppose that a number x is expressed as a power y of some number a:

$$x = a^y$$

The number a is called the *base* number, and the number y is an *exponent.* The

logarithm of x with respect to the base a is equal to the power y to which base a must be raised to give x:

$$y = \log_a x$$

Conversely, the **antilogarithm** of y is the number x.

The two bases used most often are base 10, called the *common* logarithm base, and the ''natural'' number $e = 2.718 \ldots$, called the *natural* logarithm base. With common logarithms the abbreviations ''log'' and ''antilog'' are used, and

$$y = \log_{10} x \qquad or \qquad y = \log x$$

For instance, $\log 52 = 1.716$, so $\text{antilog } 1.716 = 10^{1.716} = 52$.

If your calculator has a LOG key, you can easily compute common logarithms with it.

Exercise M.4 Evaluate **(a)** log 38.43 and **(b)** $\log(8.4 \times 10^{-2})$.

Answer: **(a)** 1.5847 **(b)** -1.076

To calculate the antilogarithm of a common logarithm, use the 10^x key or first the INV key and then the LOG key.

Exercise M.5 What is the value of x if **(a)** $\log x = -1.076$ and **(b)** $\log x = -3.843$?

Answer: **(a)** 8.40×10^{-2} **(b)** 1.44×10^{-4}

With natural logarithms the abbreviations ''ln'' and ''antiln'' are used, and

$$y = \ln_e x \qquad or \qquad y = \ln x$$

For instance, $\ln 52 = 3.951$, so $\text{antiln } 3.951 = e^{3.951} = 52$. The LN key on your calculator makes easy work of determining natural logarithms.

Exercise M.6 Evaluate **(a)** ln(38.43) and **(b)** $\ln(8.4 \times 10^{-2})$.

Answer: **(a)** 3.6488 **(b)** -2.48

You can compute antilogarithms of natural logarithms by using the e^x key or by first pressing the INV key and then LN, thus ''inverting'' the LN operation.

Exercise M.7 Given that $\ln x = -3.58$, find the value of x.

Answer: 2.79×10^{-2}

You can convert between base 10 and base e with the expression

$$\ln x = 2.303 \log x$$

Some other useful relationships of logarithms are

If $x = 10^y$, then $\log x = y$.

If $x = e^y$, then $\ln x = y$.

If $x = a^b$, then $\log x = b \log a$ and $\ln x = b \ln a$.

If $x = ab$, then $\log x = \log a + \log b$ and $\ln x = \ln a + \ln b$.

If $x = \dfrac{a}{b}$, then $\log x = \log a - \log b$ and $\ln x = \ln a - \ln b$.

4 LINEAR EQUATIONS AND STRAIGHT-LINE GRAPHS

A linear equation has the general form

$$y = mx + b$$

where m and b are constants, y is the *dependent variable*, and x is the *independent variable*. This equation is said to be *linear* because the graph of y versus x is a straight line (Figure M.1). The constant b, called the *y-intercept*, is the value of y at which the straight line intercepts the y-axis. The constant m is called the *slope* of the straight line. If any two points on the straight line are specified by the coordinates (x_1, y_1) and (x_2, y_2), respectively, as in Figure M.1, then the slope of the straight line can be expressed as

$$\text{slope} = \frac{y_2 - y_1}{x_2 - x_1} = \frac{\Delta y}{\Delta x} = m$$

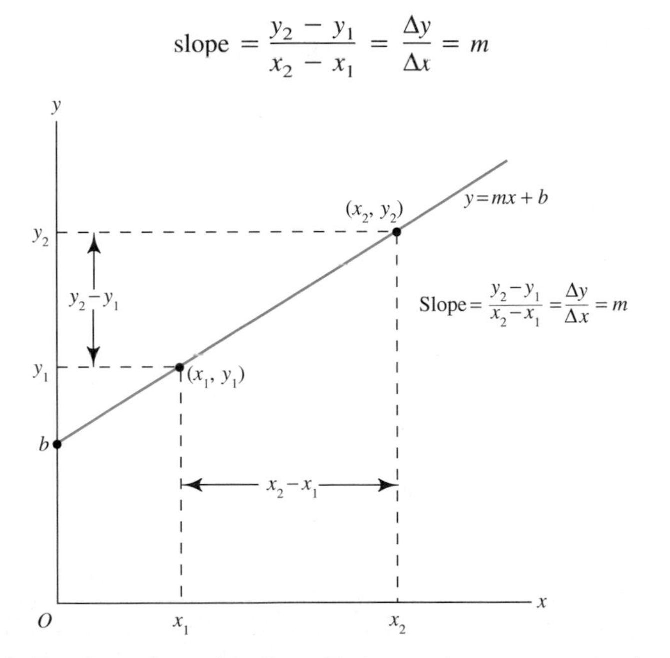

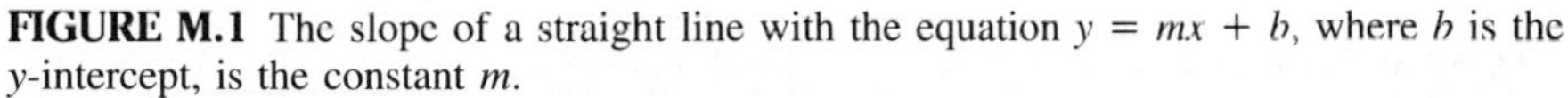

FIGURE M.1 The slope of a straight line with the equation $y = mx + b$, where b is the y-intercept, is the constant m.

Exercise M.8 Draw the graphs of the following straight lines, and find their slopes.

(a) $y = 5x + 4$ **(b)** $y = -2x + 3$ **(c)** $y = -3x - 5$

Answer: **(a)** 5 **(b)** -2 **(c)** -3

5 QUADRATIC EQUATIONS

In working equilibrium problems we often have to solve equations of the form

$$\frac{x^2}{M - x} = K \qquad or \qquad \frac{x(N + x)}{M - x} = K \tag{1}$$

We solve such equations in Chapter 12 by assuming that $x \ll M$ and $x \ll N$ so that

$$M - x \approx M \qquad \textit{and} \qquad \frac{N + x}{M - x} \approx \frac{N}{M}$$

With these assumptions, equations (1) reduce to

$$x = \sqrt{MK} \qquad \textit{or} \qquad x = \frac{MK}{N}$$

In most situations in this book this approximation method gives a sufficiently accurate value for x provided that $x < (5\%)M$ and $x < (5\%)N$. When x is not much smaller than M or N, this method cannot be used. In the latter case we can multiply both sides of either equation by $M - x$ and rearrange the terms to get the standard form of a **quadratic equation**, an equation of the form

$$ax^2 + bx + c = 0$$

where a, b, and c are constants. The general solutions to this equation are

$$x = \frac{-b \pm \sqrt{b^2 - 4ac}}{2a}$$

EXAMPLE M.4 Quadratic Equations

Solve

$$\frac{x^2}{10^{-2} - x} = 5.0 \times 10^{-4}$$

Solution: To get this equation into the standard form, we first multiply both sides of the equation by $10^{-2} - x$ and get

$$x^2 = (5.0 \times 10^{-6}) - (5.0 \times 10^{-4})x$$

Next we rearrange all the terms on the left-hand side:

$$x^2 + (5.0 \times 10^{-4})x - (5.0 \times 10^{-6}) = 0$$

According to the standard form of the quadratic equation, $a = 1$, $b = 5.0 \times 10^{-4}$, and $c = -5.0 \times 10^{-6}$. Finally, we substitute these values into the equation:

$$x = \frac{-b \pm \sqrt{b^2 - 4ac}}{2a} = \frac{-(5.0 \times 10^{-4}) \pm \sqrt{(5.0 \times 10^{-4})^2 - 4\,(1)(-5.0 \times 10^{-6})}}{2\,(1)}$$

$$= \frac{-(5.0 \times 10^{-4}) \pm \sqrt{(25.0 \times 10^{-8}) + (20.0 \times 10^{-6})}}{2} = \frac{-(5.0 \times 10^{-4}) \pm \sqrt{2.025 \times 10^{-5}}}{2}$$

$$= \frac{-(5.0 \times 10^{-4}) \pm 4.5 \times 10^{-3}}{2} = +2.0 \times 10^{-3} \textit{ or } -2.5 \times 10^{3}$$

If x were the value of a concentration, for instance, the negative answer would have no physical significance and could be ignored.

Exercise M.9 Solve the equation $\dfrac{x^2}{(2 \times 10^{-4}) - x} = 2 \times 10^{-6}$.
Answer: $+1.90 \times 10^{-5}$ *or* 2.10×10^{-5}

Answers to Exercises

CHAPTER 1

1.1 (a) H **(b)** He **(c)** Li **(d)** Be **(e)** B **(f)** C **(g)** N **(h)** O **(i)** F **(j)** Ne

1.2 (a) hydrogen **(b)** helium **(c)** boron **(d)** beryllium **(e)** carbon **(f)** calcium **(g)** nitrogen **(h)** sodium **(i)** fluorine **(j)** iron **(k)** potassium **(l)** krypton **1.3 (a)** hydrogen, E **(b)** nitrogen, E **(c)** oxygen, E **(d)** nitrogen monoxide, C **(e)** sulfur, E **(f)** chlorine, E **(g)** water, C **(h)** ammonia, C **(i)** methane, C **(j)** carbon dioxide, C

1.4 (a) HO **(b)** CH_4 **(c)** CH_3 **(d)** S **(e)** CO_2 **(f)** N **(g)** $C_{12}H_{22}O_{11}$ **(h)** CH_2

1.5 (a) P **(b)** P **(c)** P **(d)** C **(e)** P **(f)** P **(g)** P **(h)** C **(i)** C **1.6** 99.97% **1.7 (a)** S **(b)** H **(c)** H **(d)** S **(e)** H **(f)** S **(g)** S

1.8 (a) $2CO(g) + O_2(g) \rightarrow 2CO_2(g)$

(b) $CH_4(g) + H_2O(g) \rightarrow CO(g) + 3H_2(g)$

1.9 (a) +15 **(b)** 31 times **(c)** 15 **1.10** 10.81 u

1.11 5.55×10^{-2} mol H_2O molecules, 3.34×10^{22} molecules H_2O **1.12 (a)** 16.04 g · mol^{-1} **(b)** 34.02 g · mol^{-1} **(c)** 256.6 g · mol^{-1} **(d)** 342.3 g · mol^{-1}

1.13 60.09 g · mol^{-1}; 58.44 g · mol^{-1}

CHAPTER 2

2.1 (a) zinc monoxide **(b)** disodium oxide **(c)** sulfur trioxide **(d)** calcium monoxide **(e)** dichlorine monoxide **(f)** dialuminum trioxide **(g)** tetraphosphorus hexaoxide **2.2 (a)** $CH_4(g) + H_2O(g) \rightarrow CO(g) + 3H_2(g)$; CH_4 is oxidized, H_2O is reduced. **(b)** $2Fe_2O_3(s) + 3C(s) \rightarrow 3CO_2(g) + 4Fe(s)$, C is oxidized; Fe_2O_3 is reduced. **2.3** 0.980 atm, 1.03 atm, 0.987 atm, 298.1 K, 373.1 K, 123 K **2.4** 317.8 mL

2.5 24.5 L **2.6** 0.0326 mol H_2O, 1.97×10^{22} molecules H_2O **2.7** 0.0801 mol CO_2; 44.0 g · mol^{-1}

2.8 C_6H_6 **2.9** 0.588 g · L^{-1} **2.10** 58.1 g · mol^{-1}; C_4H_{10} **2.11** 0.16 L **2.12** $H_2(g)$, 490.8 mm Hg; $O_2(g)$, 245.4 mm Hg **2.13 (a)** 0.7097 **(b)** 0.2682 **(c)** 0.2510 **(d)** 0.2140 **(e)** 0.1775 **(f)** 0.1123

2.14 $\approx$ 63 g · mol^{-1}; SO_2

CHAPTER 3

3.1 Metals: (a), (b), (f), (i); nonmetals: (c), (d), (e), (g), (h)

3.2 (a) carbon; Group IV, Period 2 **(b)** sodium; Group I, Period 3 **(c)** nitrogen; Group V, Period 2 **(d)** chlorine, Group VII, Period 3 **(e)** magnesium; Group II, Period 3 **(f)** silicon, Group IV, Period 3 **(g)** aluminum; Group III, Period 3 **(h)** oxygen, Group VI, Period 2 **(i)** fluorine; Group VII, Period 2 **(j)** calcium; Group II, Period 4 **3.3 (a)** CaH_2 **(b)** K_2O **(c)** MgS **(d)** Na_3N **(e)** Al_2S_3 **3.4** Physical State, solid; molecular formula, At_2; empirical formulas: HAt, At_2O, NaAt, $CaAt_2$ **3.5 (a)** 2, 4 **(b)** 2, 6 **(c)** 2, 7 **(d)** 2, 8, 2 **(e)** 2, 8, 8, 1 **(f)** 2, 8, 5 **(g)** 2, 8, 6 **(h)** 2, 8, 7 **3.6 (a)** VII; 7 **(b)** V; 5 **(c)** VII; 7 **(d)** III; 3 **(e)** I; 1 **(f)** VI; 6 **(g)** II; 2 **(h)** V; 5 **(i)** IV; 4 **(j)** III; 3 **3.7 (a)** +7 **(b)** +5 **(c)** +7 **(d)** +3 **(e)** +1 **(f)** +6 **(g)** +2 **(h)** +5 **(i)** +4 **(j)** +3

3.8 (a) H· **(b)** :F̤̈· **(c)** ·Ċ̣· **(d)** ·Al̇· **(e)** :Ṇ̇· **(f)** :N̤̈e: **(g)** ·Ca· **(h)** :Ö̤·

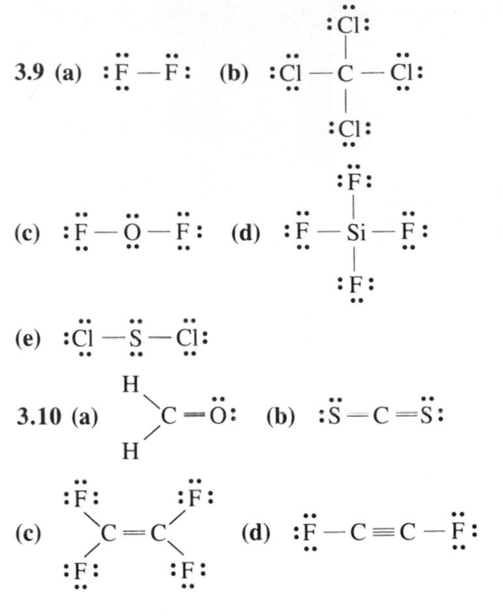

(e) H—C≡N:

3.11 (a) nonpolar covalent **(b)** polar covalent **(c)** polar covalent **(d)** ionic **(e)** nonpolar covalent **(f)** ionic **(g)** polar covalent **(h)** polar covalent

3.12 (a) AX_4, tetrahedron **(b)** AX_3E, triangular pyramid **(c)** AX_3, triangular planar **(d)** AX_2E_2, angular **(e)** AXE_3, linear **3.13 (a)** AX_2, linear **(b)** AX_3, triangular planar **(c)** AX_2, linear **(d)** AX_2, linear **(e)** AX_2E, angular

CHAPTER 4

4.1 (a) D **(b)** S **(c)** S **(d)** D

4.2 $2Na(s) + Cl_2(g) \rightarrow 2NaCl(s)$; $2Fe(s) + 3Cl_2(g) \rightarrow 2FeCl_3(s)$; $2Sb(s) + 3Cl_2(g) \rightarrow 2SbCl_3(s)$; $Mg(s) + Cl_2(g) \rightarrow MgCl_2(s)$ **4.3 (a)** $Sr(s) + Br_2(l) \rightarrow Sr^{2+}[Br^-]_2(s)$; solid **(b)** $2As(s) + 3Cl_2(g) \rightarrow 2AsCl_3(g$ or $l)$; liquid or gas **(c)** $2Rb(s) + F_2(g) \rightarrow 2Rb^+F^-(s)$; solid **(d)** $Br_2(l) + Cl_2(g) \rightarrow 2BrCl(g$ or $l)$; liquid or gas

4.4 I^- (iodide ion) **4.5** Br_2 (bromine)

4.6 (a) Cu is the reducing agent, which is oxidized; O_2 is the oxidizing agent, which is reduced. **(b)** Cs is the reducing agent, which is oxidized; I_2 is the oxidizing agent, which is reduced. **(c)** Zn is the reducing agent, which is oxidized; S is the oxidizing agent, which is reduced. **(d)** Zn is the reducing agent, which is oxidized; Cu^{2+} is the oxidizing agent, which is reduced.

4.7 (a) $LiOH(s) + HI(aq) \rightarrow LiI(aq) + H_2O(l)$ **(b)** $Ca(OH)_2(s) + 2HBr(aq) \rightarrow CaBr_2(aq) + 2H_2O(l)$ **(c)** $Mg(OH)_2(s) + 2HNO_3(aq) \rightarrow Mg(NO_3)_2(aq) + 2H_2O(l)$ **4.8 (a)** insoluble **(b)** insoluble **(c)** soluble **(d)** insoluble **(e)** soluble

4.9 (a) $MgF_2(s)$ forms. **(b)** No precipitate forms. **(c)** $FeS(s)$ forms. **4.10 (a)** OR **(b)** OR **(c)** AB **(d)** OR **(e)** AB **(f)** AB **(g)** AB and P

CHAPTER 5

5.1 0.6922 g $CaCl_2(s)$ **5.2 (a)** Al **(b)** 62.0 g **(c)** 11 g (Fe_2O_3) **5.3 (a)** 0.390 g **(b)** 64.1%

5.4 molar mass: 97.99 $g \cdot mol^{-1}$; mass percentage composition: H, 3.086; P, 31.61; O, 65.31 **5.5** C_6H_6O

5.6 $C_6H_{12}O_6$ **5.7** 15.00 L **5.8** 2.10 L

5.9 $BaCl_2(aq)$, 0.240 *M*; $Ba^{2+}(aq)$, 0.240 *M*; $Cl^-(aq)$, 0.480 *M*

5.10 (a) 8.50 g **(b)** 51.5 g **(c)** 20.8 g

5.11 0.0772 *M* **5.12** 0.0694 *M*

CHAPTER 6

6.1 7.50×10^{14} Hz (violet) to 4.00×10^{14} Hz (red)

6.2 240 nm (ultraviolet) **6.3 (a)** $1.26 \times 10^6\ J \cdot mol^{-1}$ **(b)** 3.16×10^{15} Hz, 9.50 nm **(c)** ultraviolet

6.4 (a) $(Z = 3)\ 1s^22s^1$ **(b)** $(Z = 6)\ 1s^22s^22p^2$ **(c)** $(Z = 10)\ 1s^22s^22p^6$ **(d)** $(Z = 13)\ 1s^22s^22p^63s^23p^1$ **(e)** $(Z = 15)\ 1s^22s^22p^63s^23p^3$ **(f)** $(Z = 17)\ 1s^22s^22p^63s^23p^5$ **(g)** $(Z = 20)\ 1s^22s^22p^63s^23p^64s^2$

6.5

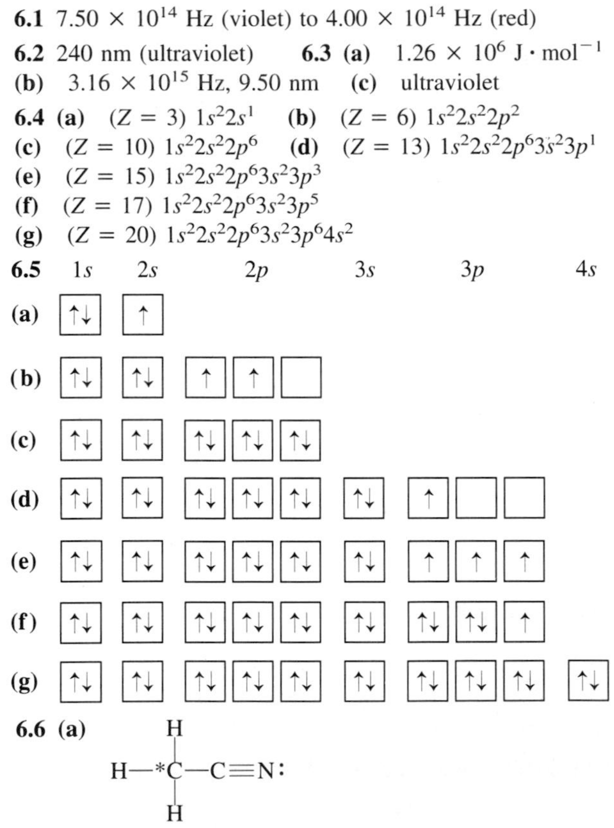

(b) <HCH = 109.5°, <*CCN = 180°

(c) The *C—H bonds are σ bonds formed by overlap of sp^3 hybrid atomic orbitals on *C and H 1*s* atomic orbitals. The *C—C bond is a σ bond formed from overlap of a sp^3 hybrid atomic orbital on *C and an *sp* hybrid atomic orbital on C. The C≡N bond consists of one σ bond formed by overlap between *sp* hybrid orbitals on C and N and two π bonds formed by overlap of mutually perpendicular pairs of *p* orbitals on C and N, respectively.

CHAPTER 7

7.1 $2CsOH(s) + H_2SO_4(aq) \rightarrow 2Cs^+(aq) + SO_4^{2-}(aq) + 2H_2O(l)$
$2NH_3(g) + H_2SO_4(aq) \rightarrow 2NH_4^+(aq) + SO_4^{2-}(aq)$
$Ba(OH)_2(aq) + H_2SO_4(aq) \rightarrow BaSO_4(s) + 2H_2O(l)$
$MgO(s) + H_2SO_4(aq) \rightarrow Mg^{2+}(aq) + SO_4^{2-}(aq) + H_2O(l)$

7.2 (a) Na, +1; O, −2 **(b)** Al, +3; O, −2 **(c)** Ba, +2; H, −1 **(d)** Ca, +2; S, −2 **(e)** S, +6; F, −1 **7.3 (a)** S, +6; O, −2 **(b)** S_2^{2-}, S −1 **(c)** Cl, +7; O, −2 **(d)** Al, +3; S, +6; O, −2

7.4

$$2(H_2S(aq) \rightarrow S(s) + 2e^- + 2H^+)$$
$$O_2(g) + 4H^+ \rightarrow 2H_2O(l)$$
$$\overline{2H_2S(g) + O_2(g) \rightarrow 2S(s) + 2H_2O(l)}$$

$$H_2S(g) \rightarrow S(s) + 2e^- + 2H^+$$
$$Br_2(aq) + 2e^- \rightarrow 2Br^-$$
$$\overline{H_2S(g) + Br_2(aq) \rightarrow S(s) + 2HBr(aq)}$$

7.5 (a) $PO(OH)_3$, weak **(b)** ClO_2OH, strong **(c)** ClOH, weak **(d)** $Si(OH)_4$, weak **(e)** $B(OH)_3$ weak

7.6 $SO_3^{2-}(aq) + HOCl(aq) \rightarrow SO_4^{2-}(aq) + HCl(aq)$

7.7 (a) N: +1; H: 0 **(b)** O: −1; H: 0

7.8 (a) 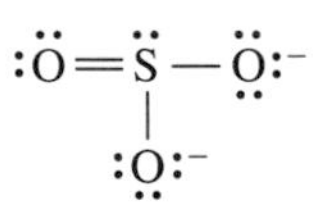 $:\ddot{O}=\ddot{S}-\ddot{O}:^-$ with $:\ddot{O}:^-$ bonded below S **(b)** 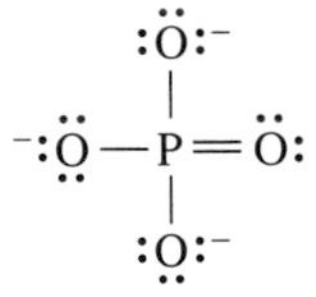 $^-:\ddot{O}-P=\ddot{O}:$ with $:\ddot{O}:^-$ bonded above and below P

(c) 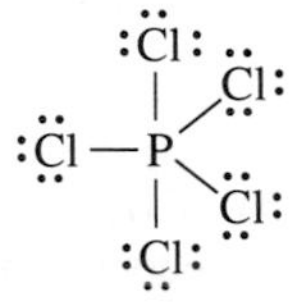 P bonded to five Cl atoms, each $:\ddot{Cl}:$

CHAPTER 8

8.1 (a) $MgCO_3(s) \xrightarrow{\text{heat}} MgO(s) + CO_2(g)$
Neither redox nor acid–base (decomposition)
(b) $PbO(s) + CO(g) \xrightarrow{\text{heat}} Pb(s) + CO_2(g)$ *Redox*
(c) $MgO(s) + 3C(s) \rightarrow MgC_2(s) + CO(g)$ *Redox*
(d) $MgC_2(s) + 2H_2O(l) \rightarrow Mg(OH)_2(s) + HC{\equiv}CH(g)$ *Acid–base*

8.2 weak bases: (a), (b), (d), and (g); strong bases: (e) and (f); no basic properties: (c) **8.3** nitrogen, $:N{\equiv}N:$; carbon monoxide, $^-:C{\equiv}O:^+$; and carbide ion, $^-:C{\equiv}C:^-$

8.4 2,6-dimethyloctane

8.5 (a) 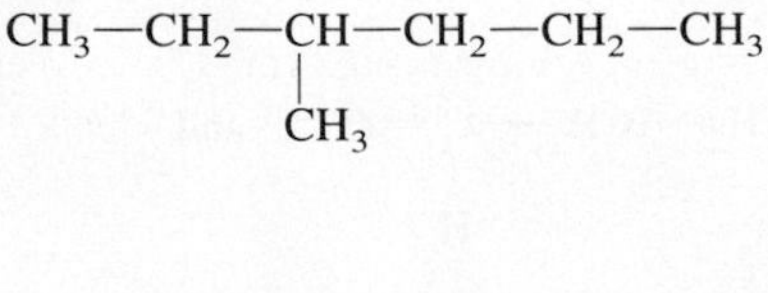 $CH_3-CH_2-CH-CH_2-CH_2-CH_3$ with CH_3 on the third carbon

(b) 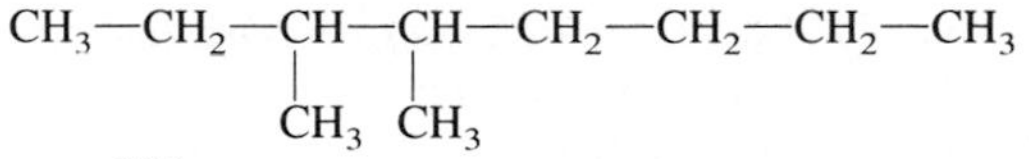 $CH_3-CH_2-CH-CH-CH_2-CH_2-CH_2-CH_3$ with CH_3 on the third and fourth carbons

(c) 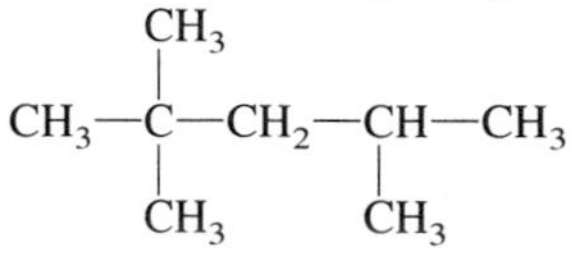 $CH_3-C-CH_2-CH-CH_3$ with two CH_3 on the second carbon and CH_3 on the fourth carbon

8.6 (a) 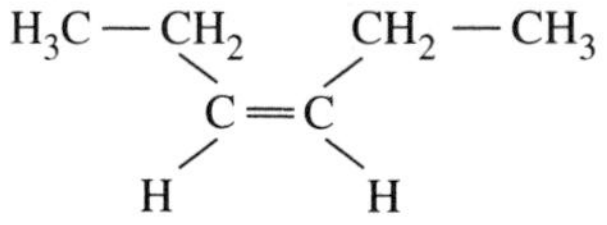 H_3C-CH_2 and CH_2-CH_3 on the same side of C=C, H and H on the other side

(b) 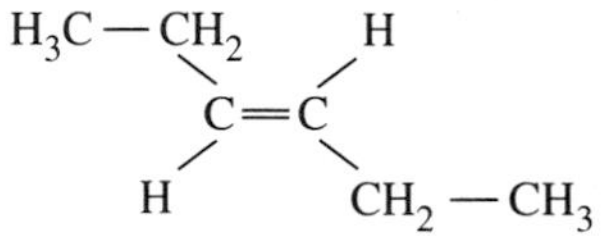 H_3C-CH_2 and H on one carbon, H and CH_2-CH_3 on the other (trans) of C=C

(c) $H_3C-C{\equiv}C-CH_3$

8.7 (a) 1,2-dichloropropane **(b)** *cis-* and trans-1,2-dichloropropene and 2,2,4,4-tetrachloropropane

8.8 Six resonance structures of the sulfate ion, each with S bonded to four O atoms, two by S=O double bonds and two by S—O single bonds carrying a negative charge (:Ö:⁻), connected by ↔

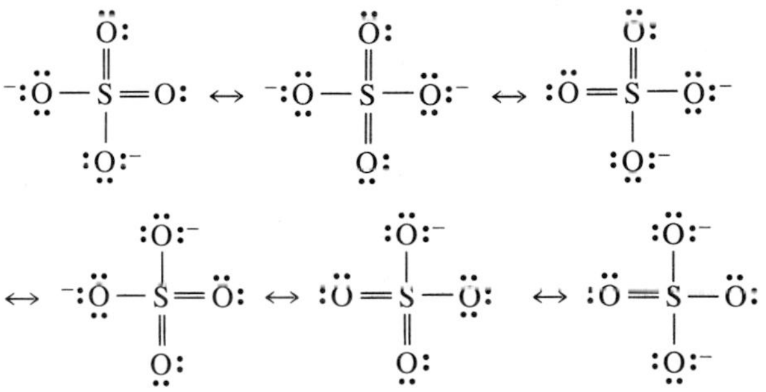

$1\frac{1}{2}; -\frac{1}{2}$

CHAPTER 9

9.1 $\Delta H° = -1.1 \times 10^2$ kJ · mol^{-1}—almost twice the −54 kJ · mol^{-1} obtained for the neutralization of the strong monoprotic acid HCl(*aq*), because $H_2SO_4(aq)$ is a strong diprotic acid. **9.2** -2.7×10^3 kJ · mol^{-1} **9.3** −750 kJ

9.4 −550.8 kJ; 22.4 kJ **9.5** 589 kJ · mol^{-1}

9.6 −1399 kJ · mol^{-1}; from standard enthalpies of formation, the value is −1428 kJ · mol^{-1}. **9.7** +15.5 kJ · mol^{-1}

9.8 (a) negative **(b)** negative **(c)** negative

9.9 (a) +130.9 kJ, not spontaneous **(b)** −702.4 kJ, spontaneous

CHAPTER 10

10.1 (1) Burn Mg(*s*) in oxygen (air): $2Mg(s) + O_2(g) \rightarrow 2MgO(s)$ *Redox* (2) Dissolve MgO(*s*) in HCl(*aq*): $MgO(s) + 2HCl(aq) \rightarrow MgCl_2(aq) + H_2O(l)$ *Acid–base* (3) Add a solution of a base (such as

NaOH(*aq*)) to $MgCl_2(aq)$: $MgCl_2(aq) + 2OH^-(aq) \rightarrow Mg(OH)_2(s) + 2Cl^-(aq)$ *Precipitation*

10.2 (a) $2KOH(aq) + H_2SO_4(aq) \rightarrow K_2SO_4(aq) + 2H_2O(l)$
(b) $2NaOH(aq) + H_3PO_4(aq) \rightarrow Na_2HPO_4(aq) + 2H_2O(l)$
(c) $Ba(OH)_2(aq) + 2HCl(aq) \rightarrow BaCl_2(aq) + 2H_2O(l)$

10.3 (a) $CO_2(g) + MgO(s) \rightarrow MgCO_3(s)$
(b) $SO_3(g) + CaO(s) \rightarrow CaSO_4(s)$ **(c)** $SO_2(g) + Na_2O(s) \rightarrow Na_2SO_3(s)$ **10.4 (a)** $2Al(s) + 6HBr(aq) \rightarrow 2AlBr_3(aq) + 3H_2(g)$ *Redox* **(b)** $2Al(OH)_3(s) \xrightarrow{\text{heat}} Al_2O_3(s) + 3H_2O(g)$ *Decomposition* **(c)** $2Al(OH)_3(s) + 3H_2SO_4(aq) \rightarrow Al_2(SO_4)_3(aq) + 6H_2O(l)$ *Acid–base*

10.5 (a) $Fe(s) + H_2SO_4(aq) \rightarrow FeSO_4(aq) + H_2(g)$
(b) $2Fe(s) + 6H_2SO_4(aq) \rightarrow Fe_2(SO_4)_3(aq) + 3SO_2(g) + 6H_2O(l)$ **10.6** $2Cu(s) + O_2(g) \rightarrow 2CuO(s)$ *Redox* $CuO(s) + H_2SO_4(aq) \rightarrow CuSO_4(aq) + H_2O(l)$ *Acid–base* $CuSO_4(aq) + H_2S(aq) \rightarrow CuS(s) + H_2SO_4(aq)$ *Precipitation* $CuS(s) + 2HCl(aq) \rightarrow CuCl_2(aq) + H_2S(g)$ *Acid–base*

10.7
$$Cr_2O_7^{2-} + 14H^+ + 6e^- \rightarrow 2Cr^{3+} + 7H_2O$$
$$Fe^{2+} \rightarrow Fe^{3+} + e^-$$
$$Cr_2O_7^{2-}(aq) + 14H^+(aq) + 6Fe^{3+}(aq) \rightarrow 2Cr^{3+}(aq) + 6Fe^{3+}(aq) + 7H_2O(l)$$

10.8 (a) $2FeCl_2(s) + Cl_2(g) \rightarrow 2FeCl_3(s)$ *Redox*
(b) $FeCl_2(aq) + H_2S(aq) \rightarrow FeS(s) + 2HCl(aq)$ *Precipitation* **(c)** $CuSO_4(aq) + 2NaOH(aq) \rightarrow Cu(OH)_2(s) + Na_2SO_4(aq)$ *Precipitation*
(d) $Fe(s) + 2HClO_4(aq) \rightarrow Fe(clO_4)_2(aq) + H_2(g)$ *Redox* **(e)** $Fe^{3+}(aq) + 6CN^-(aq) \rightarrow Fe(CN)_6^{3-}(aq)$ *Lewis acid–base*

CHAPTER 11

11.1 328 pm; 142 pm **11.2** 6.022×10^{23}
11.3 4 Na^+ and 4 Cl^- **11.4** 1 Cs^+ and 1 Cl^-
11.5 (a) 400 mm Hg **(b)** 200 mm Hg **11.6** (c), (d), and (e) **11.7** $CH_4 < C_2H_6 < C_2H_5Cl < C_2H_5Br$
11.8 (a), (c), (e) and (f)

CHAPTER 12

12.1 1.32×10^{-2} mol · L^{-1}; 2.6%
12.2 6.2×10^{-3} mol · L^{-1}; 6.2% **12.3** 4.25
12.4 11.65 **12.5** 4.8×10^{-11} mol · L^{-1} **12.6** base
12.7 (a) < 7 **(b)** < 7 **(c)** < 7 **(d)** > 7
12.8 5.28 **12.9** (a) 4.75 **(b)** 9.26 **12.10** 1.4×10^{-4} mol · L^{-1} **12.11** 5.4×10^{-4} mol · L^{-1}

12.12 (a) $K_c = \left(\dfrac{[CO]^2[H_2]^5}{[C_2H_6][H_2O]^2}\right)_{eq}$; $K_p = \left(\dfrac{(p_{CO})^2(p_{H_2})^5}{p_{C_2H_6}(p_{H_2O})^2}\right)_{eq}$

(b) $K_c = \left(\dfrac{[NO]^2[O_2]}{[NO_2]^2}\right)_{eq}$; $K_p = \left(\dfrac{(p_{NO})^2 p_{O_2}}{(p_{NO_2})}\right)_{eq}$

12.13 (a) in favor of more PCl_5 **(b)** no change **(c)** in favor of more SO_3

12.14 (a) high temperature (pressure indeterminate) **(b)** low temperature and high pressure.

12.15 (a) $K_c = \left(\dfrac{[CO]^2}{[CO_2]}\right)_{eq}$; $K_p = \left(\dfrac{(p_{CO})^2}{p_{CO_2}}\right)_{eq}$

(b) $K_c = \left(\dfrac{[CO_2]}{[CO]}\right)_{eq}$; $K_p = \left(\dfrac{p_{CO_2}}{p_{CO}}\right)_{eq}$

(c) $K_c = ([Cl_2]_{eq}$; $K_p = (p_{Cl_2})_{eq}$

12.16 $p_{NH_3} = p_{HCl} = 0.102$ atm

CHAPTER 13

13.1 +0.80 V **13.2** $NO_3^- > Co^{3+} > Zn^{2+} > H_2O$
13.3 $Al(s) > H_2S > H_2O_2 > Fe^{2+}$
13.4 $Cr_2O_7^{2-} + 14H^+ + 3Cu^{2+} \rightarrow 2Cr^{3+} + 3Cu(s) + 7H_2O$; 0.99 V **13.5** 1×10^{61} **13.6** 54.0 g Mg, 49.9 L $Cl_2(g)$

CHAPTER 14

14.1 (a) **(b)**

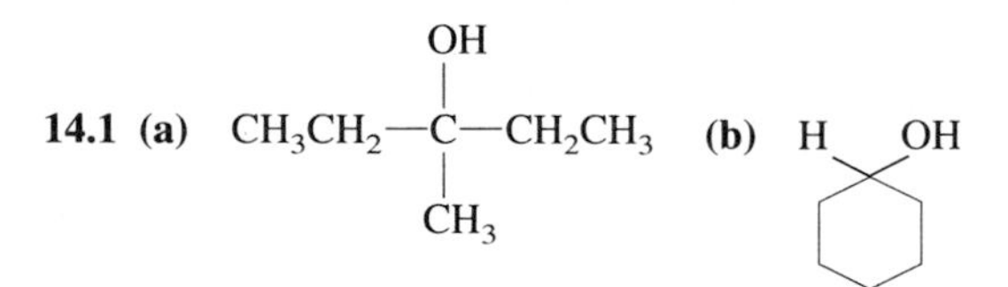

(c) $CH_3CHCH_2CHCH_2CH_2CH_3$ (with CH_3 on C-2 and OH on C-4)
secondary

(d) Cl—$CH_2CH_2CH_2$—OH
primary

14.2 (a) $(CH_3)_2C(H)OH$, 2-propanol **(b)** $CH_3O^-K^+$, potassium methoxide **(c)** $(CH_3)_2C(H)I$, 2-iodopropane

(d) , cyclohexene

14.3 (a) alcohol, 2-butanol **(b)** ether, methyl cyclohexylether, or cyclohexylmethoxide **(c)** phenol, 2-chlorophenol **(d)** thiol, 1-butanethiol

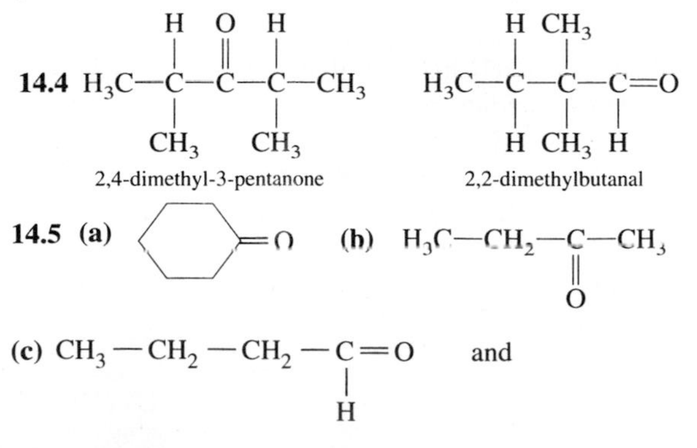

(c) $CH_3—CH_2—CH_2—C{=}O$ (with H on the carbonyl carbon) and

$$CH_3-CH_2-CH_2-\underset{\displaystyle OH}{\underset{|}{C}}=O$$

14.6 1-pentanol; $CH_3CH_2CH_2CH_2CH_2OH \rightarrow$
1-Pentanol

$$\underset{\text{Pentanal}}{CH_3CH_2CH_2CH_2-\underset{\displaystyle H}{\underset{|}{C}}=O} \rightarrow \underset{\text{Pentanoic acid}}{CH_3CH_2CH_2CH_2-\underset{\displaystyle OH}{\underset{|}{C}}=O}$$

14.7 **(a)** potassium butanoate, $CH_3CH_2CH_2CO_2^-\ K^+$ **(b)** barium-2-methylpentanoate, $[(CH_3)_2CHCH_2CH_2CO_2^-]_2\ Ba^{2+}$

14.8 (o-HO-C_6H_4)$-\underset{\displaystyle O}{\underset{\|}{C}}-OCH_3$ Methyl-*o*-hydroxybenzoate

14.9 **(a)** 2-methylbutanoic acid and phenol **(b)** butanoic acid and 2-propanol **14.10** **(a)** ethanol and methanoic (formic) acid **(b)** *p*-bromobenzoic acid and 1-propanol

14.11 $(CH_3)_2NH(aq) + H_2O(l) \rightleftharpoons (CH_3)_2NH_2^+(aq) + OH^-(aq)$ **14.12** **(a)** $C_6H_5NH_2(l) + HCl(aq) \rightarrow C_6H_5NH_3^+(aq) + Cl^-(aq)$ **(b)** $CH_3CH_2NH_2(aq) + CH_3CO_2H(aq) \rightarrow CH_3CH_2NH_3^+(aq) + CH_3CO_2^-(aq)$ **(c)** $CH_3NH_3^+\ Cl^-(aq) + NaOH(aq) \rightarrow CH_3NH_2(aq) + NaCl(aq) + H_2O(l)$ **14.13** **(a)** $CH_3CH_2CH(NH_2)CO_2H$ **(b)** yes **14.14** **(a)** a single peak **(b)** two peaks with relative intensities 6:2 **(c)** three peaks with relative intensities 6:1:1 **14.15** NMR spectroscopy: 1,3-butadiene, two peaks of relative intensity 2:1; 2-butyne, a single peak Infrared spectoscopy: 1,3-butadiene, a C=C frequency close to 1650 cm^{-1}; 2-butyne, a C≡C stretching frequency close to 2200 cm^{-1}

CHAPTER 15

15.1 $-\dfrac{\Delta[N_2]}{\Delta t} = 0.0010\ mol \cdot L^{-1}$; $-\dfrac{\Delta[H_2]}{\Delta t} = 0.0005\ mol \cdot L^{-1}$ **15.2** **(a)** rate $= k[C_3H_6]$ **(b)** rate $= k[H_2][NO]^2$ **15.3** **(a)** rate $= k[NO]^2[O_2]$ **(b)** $k = 7 \times 10^3\ L^2 \cdot mol^{-2} \cdot s^{-1}$ **15.4** rate $= k[Fe^{3+}][I^-]$ **15.5** **(a)** 17.2 min **(b)** 34.5 min

15.6 $k = 1.56 \times 10^{-3}\ L \cdot mol^{-1} \cdot s^{-1}$ **15.7** 3.5

15.8 89 kJ **15.9** **(a)** Cl_2 does not appear in the balanced equation for the reaction: $2N_2O(g) \rightarrow N_2(g) + O_2(g)$ **(b)** uncatalyzed: rate $= k[N_2O]$; catalyzed: rate $= k[N_2O][Cl_2]^{\frac{1}{2}}$ **15.10** 2.1×10^3

CHAPTER 16

16.1 $:\ddot{O}=\ddot{N}-\underset{\displaystyle \underset{\cdot\cdot}{O}:}{\underset{\|}{N^+}}-\ddot{\underset{\cdot\cdot}{O}}:^-$

16.2 **(a)** $NO_3^-(aq) + e^- + 2H^+(aq) \rightarrow NO_2(g) + H_2O(l)$ **(b)** $NO_3^-(aq) + 3e^- + 4H^+(aq) \rightarrow NO(g) + 2H_2O(l)$ **(c)** $NO_3^-(aq) + 8e^- + 10H^+(aq) \rightarrow NH_4^+(aq) + 3H_2O(l)$

16.3 **(a)** $H-\ddot{\underset{\cdot\cdot}{O}}-\ddot{\underset{\cdot\cdot}{O}}-H$; O:−1 **(b)** $H-\ddot{\underset{\cdot\cdot}{O}}\cdot$; O, −1

(c) $H-\ddot{\underset{\cdot\cdot}{O}}-\ddot{\underset{\cdot\cdot}{O}}\cdot$; O, $-\frac{1}{2}$

(d) $H-\underset{\displaystyle H}{\overset{\displaystyle H}{C}}-\underset{\displaystyle H}{\overset{\displaystyle H}{C}}\cdot$; C, $-2\frac{1}{2}$

16.4 $1\frac{1}{2}$; intermediate in length between a double bond and a single bond **16.5** **(a)** $O_3(g) + 2e^- + 2H^+(aq) \rightarrow O_2(g) + H_2O(l)$ **(b)** $H_2O_2(aq) + 2e^- + 2H^+(aq) \rightarrow 2H_2O(l)$

16.6 **(a)** $BE = E = \dfrac{N_A hc}{\lambda} = \dfrac{1.196 \times 10^5\ kJ \cdot mol^{-1}}{\lambda_{max}}$

so $\lambda_{max} = \dfrac{1.20 \times 10^5\ nm \cdot kJ}{BE\ kJ}$

(b) For C=O bonds (BE 707 $kJ \cdot mol^{-1}$),

$\lambda_{max} = \dfrac{1.20 \times 10^5\ nm \cdot kJ}{707\ kJ} = 170\ nm$

Radiation with $\lambda \leq 170$ dissociates C=O bonds; bonds in CO_2 will be broken in the stratosphere, where solar radiation of this wavelength penetrates, but not in the troposphere, where no radiation of this wavelength penetrates (Figure 16.2).

16.7 **(a)** 377 nm **(b)** Although 377-nm light should break a C—Cl bond, CF_2Cl_2 must absorb 250-nm light to give an excited state that could then dissociate. **16.8** CO is polar, and its dipole moment changes when it vibrates; CO_2 (AX_2 linear) is nonpolar but can undergo vibrations (such as bending motions) in which there is a changing dipole moment; H_2O (angular AX_2E_2) has a dipole moment that changes, for example, in vibrational motion, when the O—H bonds stretch. N_2 and O_2 are nonpolar diatomic molecules; there is no change in dipole moment when they vibrate.

CHAPTER 17

17.1

$$H_2N-CH_2-\overset{\displaystyle O}{\overset{\|}{C}}-\overset{\displaystyle H}{\overset{|}{N}}-CH_2-\overset{\displaystyle O}{\overset{\|}{C}}-\overset{\displaystyle H}{\overset{|}{N}}-CH_2-\overset{\displaystyle O}{\overset{\|}{C}}-\overset{\displaystyle H}{\overset{|}{N}}-CH_2-CO_2H$$

17.2

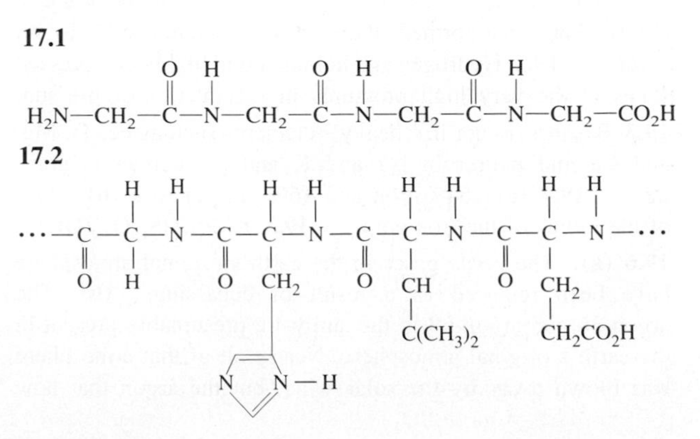

17.3

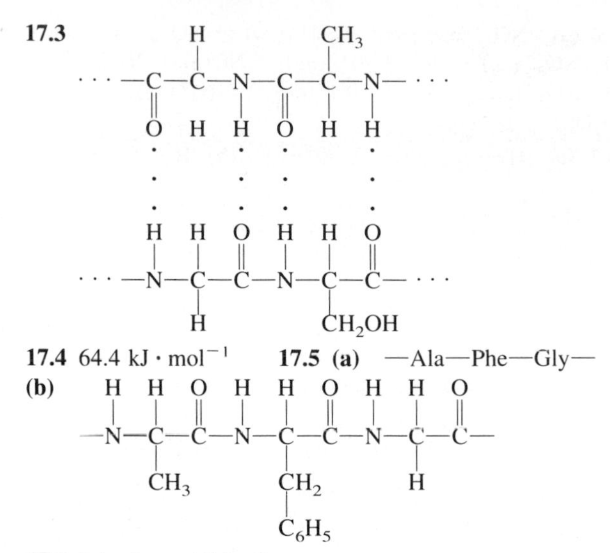

17.4 64.4 kJ · mol^{-1} **17.5 (a)** —Ala—Phe—Gly— **(b)**

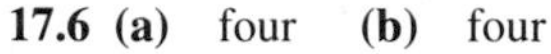

17.6 (a) four **(b)** four

CHAPTER 18

18.1 (a) $^{22}_{11}Na \rightarrow ^{0}_{1}e + ^{22}_{10}Ne$ **(b)** $^{27}_{12}Mg \rightarrow ^{0}_{-1}e + ^{27}_{13}Al$ **(c)** $^{36}_{17}Cl \rightarrow ^{36}_{18}Ar + ^{0}_{-1}e$ **(d)** $^{239}_{94}Pu \rightarrow ^{235}_{92}U + ^{4}_{2}He$ **18.2** 3.5 μg **18.3** 5.33 yr **18.4** 1.72×10^9 yr **18.5** 2.58×10^3 yr **18.6** 1.841×10^{-11} J, 1.115×10^{-12} J · nucleon^{-1} **18.7** $3UO_2 + 2HNO_3 + 6H^+ \rightarrow 3UO_2^{2+} + 2NO + 4H_2O$ **18.8** 1.391×10^8 kJ · mol^{-1}

CHAPTER 19

19.1 (a) Two atoms: CH, OH, CN, CO^+, NO, NS; three atoms: C_2H **(b)** Free radicals are formed by breaking a covalent bond in a molecule to give two radicals, both with an unpaired electron, or by ionization of molecules with paired electrons. Such processes readily occur at the very high temperatures at which interstellar dust clouds formation and reactions between free radicals are infrequent because their partial pressures are low. Free radicals are formed on earth largely by photolysis, and the solar radiation that reaches the earth is in general insufficiently energetic to break all but the weakest bonds. And once formed, their rates of reaction are fast on earth. **19.2** Hydrogen and helium are in highly compressed states at the very high pressures in the interior of the sun. **19.3** Basalt is richer in ''heavy'' elements such as Fe, Ti, Mg, and Ca and poorer in Na and K and silica than is granite. **19.4 (a)** 54% iron and 46% silicate rock **(b)** 20% of the radius of the iron core **19.5** $KMg_3AlSi_3O_{10}(OH)_2$ **19.6 (a)** The noble gases in the earth's original atmosphere have been replaced as a result of degassing. **(b)** The argon-36 and argon-38 in the sun were presumably present in the earth's original atmosphere. Nearly all of that atmosphere was blown away by the solar wind, but the argon that now constitutes about 1% of the atmosphere is mainly argon-40 formed over time from the radioactive decay of potassium-40 in rocks: $^{40}_{19}K \rightarrow ^{40}_{18}Ar + ^{0}_{+1}e$ **19.7** The earth's primeval atmosphere contained little oxygen. This very reactive element would have entirely combined with other elements to give oxides in the surface rocks. The unique feature of the present atmosphere is its high steady-state oxygen content, maintained by living organisms. In respiration, animals use oxygen and generate carbon dioxide; carbon dioxide is used by plants in photosynthesis, which utilizes solar energy to form carbohydrates and regenerates oxygen as a byproduct.

19.8 (a)

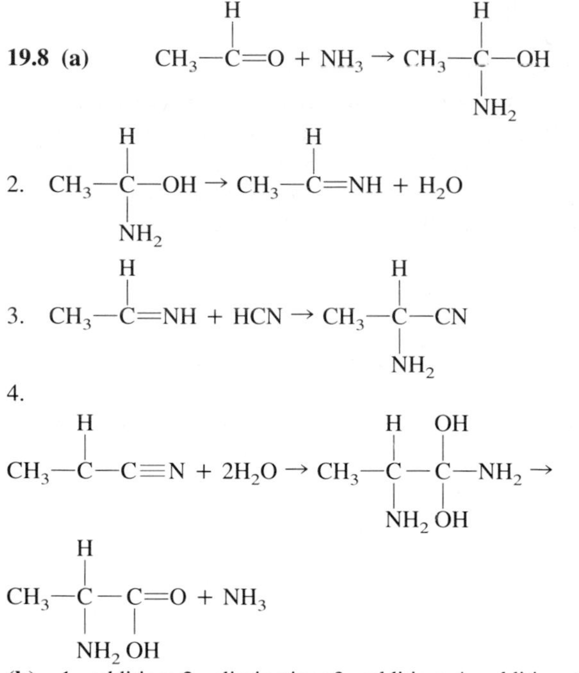

(b) 1. addition; 2. elimination; 3. addition; 4. addition, then elimination **19.9** 1.57×10^8 metric tons

CHAPTER 20

20.1 2.082×10^5 u

20.2 monomer:

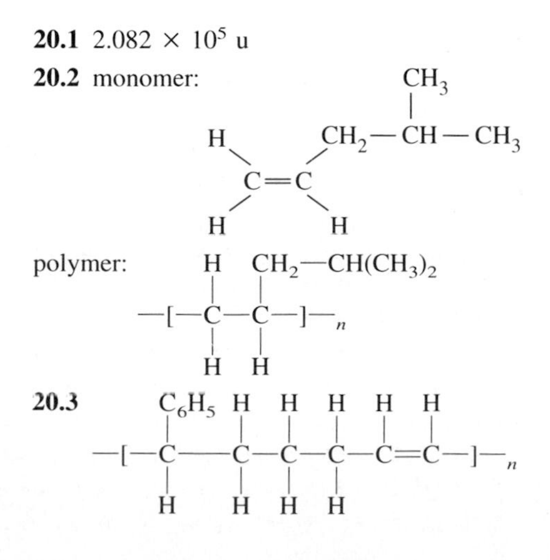

polymer:

20.3

20.4 5.94×10^5 kg **20.5** (a), (b)

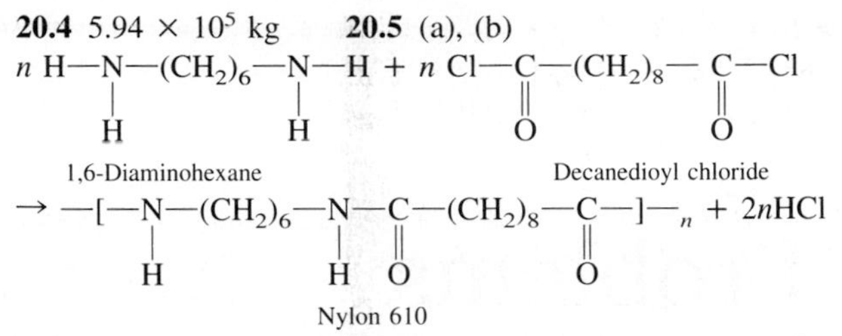

(c) The name nylon 610 informs us that the diamine forming the polymer (1,6-diaminohexane) has a chain of six carbon atoms and that the dicarboxylic acid derivative decanedioyl chloride has a chain of ten carbon atoms.

20.6 1,4-benzenediamine and 1,4-benzenedicarboxylic (terephthalic) acid **20.7 (a)** Benzene is hydrogenated to cyclohexane, which is oxidized to the ketone cyclohexanone, which is oxidized to 1,6-hexanedioic acid. **(b)** Butadiene reacts with HCN in an addition reaction to give 1,4-dicyanobutane, which is hydrogenated to give 1,6-diaminohexane.

Answers to Selected Problems

CHAPTER 1

1. (a) Na **(b)** K **(c)** Hg **(d)** Au **(e)** Sn **(f)** Pb **(g)** Fe **3. (a)** hydrogen **(b)** helium **(c)** neon **(d)** fluorine **(e)** magnesium **(f)** aluminum **(g)** phosphorus **(h)** sulfur **(i)** potassium **(j)** sodium **4. (a)** S_8 **(b)** H_2SO_4 **(c)** $H_2SO_4(l)$ **(d)** $H_2SO_4(aq)$ **(e)** $SiC(s)$ **5. (a)** P_4, P, $P_4(s)$ **(b)** P_n, P, $P(s)$ **(c)** C_3H_6, CH_2, $C_3H_6(g)$ **(d)** $C_2F_2H_4$, CFH_2, $C_2F_2H_4(g)$

7. (a) H_2, $H_2(g)$ **(b)** O_2, $O_2(g)$ **(c)** N_2, $N_2(g)$ **(d)** H_2O, $H_2O(l)$ **(e)** H_2O_2, $H_2O_2(l)$ **(f)** CH_4, $CH_4(g)$ **(g)** CO_2, $CO_2(g)$ **(h)** MgO, MgO(*s*) **(i)** SiO_2, $SiO_2(s)$

10. (a) A mixture is a blend of different substances. ''Pure'' describes a single substance; ''impure'' describes a substance contaminated with other substances. **(b)** Substances that occur naturally or are made by chemists contain impurities, even after careful purification, and are mixtures. **(c)** The sample contains impurities to the extent of 0.02 g in a 100-g sample. **11.** Elements: (b), (g); compounds: (a), (d), (f), (j); mixtures: (c), (e), (h), (i) **12.** Pure substances: elements (a), (g), (j); compounds (d), (e), (h); mixtures: homogenous (b), (c), (f), (i); heterogeneous (k), (l)

14. (a) The separation of liquids from solids by passage through a porous filter. **(b)** The separation of liquids by evaporation and condensation, utilizing their differences in boiling point. **(c)** A single-phase mixture with uniform properties and constant composition throughout; a solution. **(d)** A homogeneous, usually liquid, mixture. **(e)** A mixture with a nonuniform composition and different properties in different parts. **16. (a)** Distilling off the water **(b)** distillation or the mechanical separation of the two layers **(c)** attracting the iron to a magnet **(d)** dissolving the sugar in water, followed by filtration **(e)** paper chromatography **17. (a)** Reactants are the substances that react with each other; products are the new substances that form. **(b)** All the products and their formulas must be known—in this case, whether $CO(g)$ or $CO_2(g)$ is the (dominant) product.

19. (a) $2SO_2 + 2H_2O + O_2 \rightarrow 2H_2SO_4$ **(b)** $2CH_3OH + 3O_2 \rightarrow 2CO_2 + 4H_2O$ **(c)** $2H_2O_2 \rightarrow 2H_2O + O_2$ **(d)** balanced

21. (a) $2S + 3O_2 \rightarrow 2SO_3$ **(b)** $2C_2H_2 + 5O_2 \rightarrow 4CO_2 + 2H_2O$ **(c)** $2Cu_2S + 3O_2 \rightarrow 2Cu_2O + 2SO_2$ **(d)** $Na_2SO_4 + 4H_2 \rightarrow Na_2S + 4H_2O$ **(e)** $Na_2CO_3 + Ca(OH)_2 \rightarrow 2NaOH + CaCO_3$

23. (a) H, $Z = 1$ **(b)** O, $Z = 8$ **(c)** F, $Z = 9$ **(d)** Ne, $Z = 10$ **(e)** Mg, $Z = 12$ **(f)** P, $Z = 15$ **(g)** Cl, $Z = 17$ **(h)** Ca, $Z = 20$ **(i)** Zn, $Z = 30$

25. (a) B, 5 **(b)** N, 7 **(c)** H, 1 **(d)** Ne, 10 **(e)** Cl, 17 **(f)** O, 8 **(g)** S, 16 **(h)** K, 19 **(i)** Fe, 26 **27. (a)** ${}^{40}_{19}K$ **(b)** ${}^{30}_{14}Si$ **(c)** ${}^{40}_{18}Ar$ **(d)** ${}^{15}_{7}N$ **(e)** ${}^{32}_{16}S$ **(f)** ${}^{23}_{11}Na$ **(g)** ${}^{27}_{13}Al$

29. (a) ${}^{2}_{1}H$: 1, 1, 1 **(b)** ${}^{19}_{9}F$: 9, 10, 9 **(c)** ${}^{40}_{20}Ca$: 20, 20, 20 **(d)** ${}^{112}_{48}Cd$: 48, 64, 48 **(e)** ${}^{117}_{50}Sn$: 50, 67, 50 **(f)** ${}^{235}_{92}U$: 92, 143, 92

30.

Atomic symbol	${}^{9}Be$	${}^{15}N$	${}^{18}O$	${}^{12}C$	${}^{23}Na$
Mass number (*A*)	9	15	18	12	23
Atomic number (*Z*)	4	7	8	6	11
Number of protons (*Z*)	4	7	8	6	11
Number of electrons (*Z*)	4	7	8	6	11
Number of neutrons ($A - Z$)	5	8	10	6	12

32. ${}^{1}H{}^{1}H{}^{16}O$; ${}^{1}H{}^{2}H{}^{16}O$; ${}^{2}H{}^{2}H{}^{16}O$; ${}^{1}H{}^{1}H{}^{17}O$; ${}^{1}H{}^{2}H{}^{17}O$; ${}^{2}H{}^{2}H{}^{17}O$; ${}^{1}H{}^{1}H{}^{18}O$; ${}^{1}H{}^{2}H{}^{18}O$; ${}^{2}H{}^{2}H{}^{18}O$

33. (a) Mg-24 isotope, mass number 24, with 12 protons, 12 neutrons, and 12 electrons **(b)** 24.31 u

35. 0.70% by mass **37.* (a)** ${}^{79}Br$ and ${}^{81}Br$ **(b)** ${}^{79}Br{}^{35}Cl$, ${}^{79}Br{}^{37}Cl$, ${}^{81}Br{}^{35}Cl$, ${}^{81}Br{}^{37}Cl$

39. (a) N_A = number of ${}^{12}C$ atoms in 12 lb of ${}^{12}C$ atoms **(b)** $N_A = 2.732 \times 10^{26}$ entities $\cdot$ mol^{-1}

41. (a) 19.00 u **(b)** 19.00 g $\cdot$ mol^{-1} **(c)** ${}^{19}_{9}F$

42. (a) 18.02 g $\cdot$ mol^{-1} **(b)** 34.02 g $\cdot$ mol^{-1}

(c) NaCl, 58.44 $g \cdot mol^{-1}$ **(d)** $MgBr_2$, 184.1 $g \cdot mol^{-1}$
(e) CO, 28.01 $g \cdot mol^{-1}$ **(f)** CO_2, 44.01 $g \cdot mol^{-1}$
(g) CH_4, 16.04 $g \cdot mol^{-1}$ **(h)** C_2H_6, 30.08 $g \cdot mol^{-1}$
(i) NH_3, 17.03 $g \cdot mol^{-1}$ **(j)** HCl, 35.46 $g \cdot mol^{-1}$

45. (a) 294.3 $g \cdot mol^{-1}$ **(b)** 2.11×10^{-2} mol **(c)** 72.1 g **(d)** 8.76×10^{18} molecules

47. (a) 4.4×10^{-4} mol **(b)** 192.2 g

49. (a) 180.2 u **(b)** 2.77×10^{-3} mol, 1.67×10^{21} molecules **51. (a)** 0.4 mol cells **(b)** 1.7×10^{27} H_2O molecules **52.** 7.68×10^{23} CO molecules **53. (a)** Nuclear neutron-proton attractions must be strong enough to overcome the large repulsive forces between the closely packed, positively charged nuclear protons. **(b)** When electrons, protons, and neutrons combine to form a nucleus, part of the mass is converted to energy to bind the nucleons together, given by $E = \Delta mc^2$, where Δm is the mass lost and c is the speed of light. **55.* (a)** Loss of an α particle (helium nucleus, ^4_2He) decreases the mass number by 4; because 2 protons constitute part of this mass, the atomic number Z also decreases by 2. **(b)** Loss of a β particle (an electron, $^{0}_{-1}e$), of negligible mass, is the result of conversion of a nuclear neutron to a proton; the mass remains unchanged, but the atomic number, Z, increases by 1.

57. $N_2(g) + 3H_2(g) \rightarrow 2NH_3(g)$; $2NH_3(g) + H_2SO_4(aq) \rightarrow (NH_4)_2SO_4(aq)$ [$\rightarrow (NH_4)_2SO_4(s)$]

59. (a) $^{28}_{14}Si$ **(b)** $^{29}_{14}Si$ and $^{30}_{14}Si$ **(c)** 28.09 u

60. (a) Colorless, liquid, hexagonal crystal structure, density 1.105 $g \cdot cm^{-3}$, mp 3.8°C, bp 101.4°C **(b)** Electrolysis: $2D_2O(l) \rightarrow 2D_2(g) + O_2(g)$; reaction with SO_3: $D_2O(l) + SO_3(g) \rightarrow D_2SO_4(l)$ **61. (a)** N_A atomic mass units (N_A u) equals exactly 1 g; multiplication of any number of atomic mass units by N_A gives the same number of grams. **(b)** 3.342×10^{22} H_2O molecules **(c)** 2.98×10^7 pm^3

63.* (a) Dissolve the $KNO_3(s)$ in water and filter the solution; dry the mixture of S(s) and C(s), and then dissolve the S(s) in a solvent such as carbon disulfide and filter the solution. **(b)** $4KNO_3(s) + S(s) + C(s) \rightarrow 4KNO_2(s) + SO_2(g) + CO_2(g)$

CHAPTER 2

2. (a) N_2, O_2, and Ar **(b)** Any two of oxygen, carbon dioxide, water vapor, methane, hydrogen, dinitrogen monoxide. **(c)** As part of the carbon-cycle involving photosynthesis.

5. (a) $S(s) + O_2(g) \rightarrow SO_2(g)$ **(b)** $2Mg(s) + O_2(g) \rightarrow 2MgO(s)$ **(c)** $CH_4(g) + 2O_2(g) \rightarrow CO_2(g) + 2H_2O(g)$ **(d)** $2H_2(g) + O_2(g) \rightarrow 2H_2O(g)$

7. (a) $2Fe(s) + 3H_2O(g) \rightarrow Fe_2O_3(s) + 3H_2(g)$
(b) $Mg(s) + H_2O(g) \rightarrow MgO(s) + H_2(g)$
(c) $CH_4(g) + H_2O(g) \rightarrow CO(g) + 3H_2(g)$ (synthesis gas)

8. (a) $N_2(g) + 3H_2(g) \rightarrow 2NH_3(g)$, ammonia
(b) $N_2(g) + O_2(g) \rightarrow 2NO(g)$, nitrogen monoxide

10. Electrolysis; any reaction in which water is reduced to $H_2(g)$—for example, $H_2O(g) + C(s) \rightarrow H_2(g) + CO(g)$; $H_2O(g) + CH_4(g) \rightarrow 3H_2(g) + CO(g)$; $3H_2O(g) + 2Fe(s) \rightarrow 3H_2(g) + Fe_2O_3(s)$ **12. (a)** $2C(s) + O_2(g) \rightarrow 2CO(g)$ **(b)** $2Ca(s) + O_2(g) \rightarrow 2CaO(s)$
(c) $C_3H_8(g) + 5O_2(g) \rightarrow 3CO_2(g) + 4H_2O(g)$
(d) $C_2H_6O(l) + 3O_2(g) \rightarrow 2CO_2(g) + 3H_2O(g)$

14. (a) (s) **(b)** (g) **(c)** (g) **(d)** (s) **(e)** (s) **(f)** (g) **(g)** (l) **17.** 64 L **19.** 2.5 atm

21. 0.28 atm **23.** 10 K (−263°C) **25. (a)** Both have the same number of molecules; **(b)** $O_2(g)$

27. (a) C_3H_6 **(b)** $2C_3H_6(g) + 9O_2(g) \rightarrow 6CO_2(g) + 6H_2O(g)$ **30. (a)** Halving the volume doubles the density of molecules, and the number of collisions with the wall of the vessel doubles; hence the pressure doubles. **(b)** Increasing the temperature increases the average kinetic energy of the molecules and hence the force of collisions, so the pressure increases. **31.** 0.163 mol $CH_4(g)$

33. 1.50 atm **35. (a)** 2.59 atm **(b)** 2.24 atm

37. 0.0859 g HCl(g) **39.** 102 g He(g)

41. 5.03 $g \cdot L^{-1}$ **42.** Ar(g) **44. (a)** 28.1 $g \cdot mol^{-1}$ **(b)** C_2H_4 **46.** $N_2(g)$, 5.9×10^2 mm Hg; $O_2(g)$, 1.6×10^2 mm Hg; Ar(g), 8 mm Hg **48. (a)** He, 5.47 atm; H_2, 10.9 atm **(b)** 16.4 atm **50.** 1.4×10^{21} O_2 molecules **52.** $CO_2(g)$, 4.7 torr; $N_2(g)$, 0.2 torr; Ar(g), 0.1 torr

53. (a) 1.79×10^3 $m \cdot s^{-1}$ **(b)** 302 $m \cdot s^{-1}$ **(c)** 481 $m \cdot s^{-1}$ **(d)** 600 $m \cdot s^{-1}$ **(e)** 384 $m \cdot s^{-1}$

55. 1.414:1 **57. (a)** 0.996 $mi \cdot s^{-1}$ **(b)** 0.267 $mi \cdot s^{-1}$ **(c)** 0.250 $mi \cdot s^{-1}$ **58.** $F_2(g)$

60. (a) 4.10×10^{10} pm^3, **(b)** 4.19×10^6 pm^3, **(c)** ratio = 9.78×10^3 (~10,000) **61.** 1.63×10^3

63. 23.5 mm Hg **64. (a)** $(NH_4)_2Cr_2O_7(s) \rightarrow Cr_2O_3(s) + N_2(g) + 4H_2O(g)$ **(b)** 1.38 atm **(c)** $N_2(g)$, 0.275 atm; $H_2O(g)$, 1.10 atm

CHAPTER 3

2. (a) helium, Group VIII, Period 1, nonmetal
(b) phosphorus, Group V, Period 3, nonmetal
(c) potassium, Group I, Period 4, metal
(d) calcium, Group II, Period 4, metal
(e) sulfur, Group VI, Period 3, nonmetal
(f) bromine, Group VII, Period 4, nonmetal
(g) aluminum, Group III, Period 3, metal
(h) fluorine, Group VII, Period 2, nonmetal

4. A metal; titanium **5. (a)** main group, VI, nonmetal **(b)** main group, V, nonmetal **(c)** transition element (metal) **(d)** main group, VIII, nonmetal **(e)** transition element (metal) **(f)** main group, III, metal **(g)** main group, IV, metal **7. (a)** metal, Group I, 1 electron, valence 1 **(b)** metal, Group II, 2 electrons, valence 2 **(c)** nonmetal, Group VI, 6 electrons, valence 2 **(d)** nonmetal, Group V, 5 electrons, valence 3 **(e)** nonmetal, Group VII, 7 electrons, valence 1 **(f)** nonmetal, Group VIII, 8 electrons, valence 0 **(g)** nonmetal, Group V, 5 electrons, valence 3 **(h)** nonmetal, Group VI, 6 electrons, valence 2 **(i)** nonmetal, Group VII, 7 electrons, valence 1 **(j)** metal, Group II, 2 electrons, valence 2 **9.** BCl_3, $AlCl_3$, $GaCl_3$, $InCl_3$, $TlCl_3$ **11. (a)** nonmetal, Group IV, CH_4 **(b)** metal, Group II, CaH_2 **(c)** nonmetal, Group VIII, none **(d)** nonmetal, Group III, BH_3 **(e)** nonmetal, Group VII, HCl **(f)** metal, Group I, LiH **(g)** nonmetal, Group VI, H_2O **(h)** nonmetal, Group VII, HF **(i)** nonmetal, Group V, PH_3 **(j)** metal, Group II, MgH_2

12. (a) $Mg(s) + Br_2(l) \rightarrow MgBr_2(s)$
(b) $Ca(s) + O_2(g) \rightarrow CaO(s)$
(c) $2Na(s) + I_2(s) \rightarrow 2NaI(s)$
(d) $3Mg(s) + N_2(g) \rightarrow Mg_3N_2(s)$

14. Na and K (Group I); Mg and Ca (Group II); C and Si (Group IV); Cl and F (Group VII) **16. (a)** The outer shell. **(b)** All have the same number of valence electrons. **(c)** Period 2: Li: 2, 1; Be: 2, 2; B: 2, 3; C: 2, 4; N: 2, 5; O: 2, 6; F: 2, 7; Ne: 2, 8
Period 3: Na: 2, 8, 1; Mg: 2, 8, 2; Al: 2, 8, 3; Si: 2, 8, 4; P: 2, 8, 5; S: 2, 8, 6; Cl: 2, 8, 7; Ar: 2, 8, 8 **18. (a)** 2, 1; +1 **(b)** 2, 8, 2; +2 **(c)** 2, 8, 6; +6 **(d)** 2, 8, 5; +5

20. (a) (i) 2, 5 (ii) 2, 8 (iii) 2, 2 (iv) 2 (v) 2, 6 (vi) 2, 8 **(b)** N^{3-} and O^{2-} **23. (a)** +4 **(b)** +2 **(c)** +10 **(d)** +4 **(e)** +6 **(f)** +6 **(g)** +6 **(h)** +7 **25. (a)** K· **(b)** ·Ca· **(c)** ·Ḃ· **(d)** ·Sn· **(e)** :Sb· **(f)** :Te· **(g)** :Br· **(h)** :Xe: **(i)** :As· **(j)** ·Ge·

26. (a) 1, 3, 1, 2, 2, 1, 2 (Li^+, Al^{3+}, Na^+, Ca^{2+}, Mg^{2+}, Rb^+, Sr^{2+}, respectively) **(b)** 2, 1, 1, 3, 3, 2, 1, 1 (O^{2-}, H^-, Cl^-, N^{3-}, P^{3-}, S^{2-}, F^-, I^-, respectively) **28. (a)** $(NH_4)_2S$ **(b)** Fe_2O_3 **(c)** Cu_2O **(d)** $AlCl_3$ **29. (a)** Li_2S **(b)** BaO **(c)** $MgBr_2$ **(d)** NaH **(e)** AlI_3

31. (a) BaI_2, $Ba(s) + I_2(s) \rightarrow BaI_2(s)$
(b) $AlCl_3$, $2Al(s) + 3Cl_2(g) \rightarrow 2AlCl_3(s)$
(c) CaO, $2Ca(s) + O_2(g) \rightarrow 2CaO(s)$
(d) Na_2S, $2Na(s) + S(s) \rightarrow Na_2S(s)$ **(e)** Al_2O_3, $4Al(s) + 6O_2(g) \rightarrow 2Al_2O_3(s)$ **32.** (a) and (d), MgH_2; (a) and (e), AlH_3; (a) and (f), CaH_2; (b) and (d), MgO; (b) and (e), Al_2O_3; (b) and (f), CaO; (c) and (d), MgF_2; (c) and (e), AlF_3; (c) and (f), CaF_2

34. (a) H—H (b) H—C̈l: (c) H—P̈—H (with H bonded below P)
(d) Si bonded to four :F̈: atoms (e) :F̈—Ö—F̈: (f) :C̈l—C̈l:

35.* (a) H—C=Ö: (with H bonded below C) (b) :P≡P: (c) ⁻:C≡N:
(d) H—N̈—N̈—H (with H bonded below each N) (e) H—Ö—Ö—H

37.* (b), (c) **39. (a)** F **(b)** F **(c)** S **(d)** C **(e)** O **(f)** Br **(g)** P **40. (a)** covalent **(b)** polar covalent **(c)** ionic **(d)** polar covalent **(e)** ionic **43. (a)** magnesium chloride, ionic
(b) sulfur dichloride, covalent, δ− :C̈l—S̈(2δ+)—C̈l: δ−
(c) phosphorus trichloride, covalent, δ− :C̈l—P̈(3δ+)—C̈l: δ−, with :C̈l: δ− bonded below P
(d) hydrogen fluoride, covalent, δ+ H—F̈: δ−
(e) oxygen dichloride, covalent, δ+ :C̈l—Ö(2δ−)—C̈l: δ+
(f) carbon disulfide, covalent, δ+ :S̈=C(2δ−)=S̈: δ+
(g) lithium hydride, ionic
(h) nitrogen trifluoride, covalent, δ− :F̈—N̈(3δ+) with :F̈: δ− above and :F̈: δ− below N

45. When Na^+Cl^- forms, each ion attracts ions of opposite charge by electrostatic forces and surrounds itself with as many of these ions as possible—six for both Na^+ and Cl^- in octahedral arrangements. Thus, at room temperature NaCl is a solid consisting of the same very large numbers of Na^+ and Cl^- ions strongly attracted to each other by ionic bonds.

47. For diagrams, see Figures 3.17 through 3.21. **(a)** triangular pyramidal **(b)** angular **(c)** linear **(d)** triangular planar **(e)** linear
49. (a) AX_2E_2 **(b)** AX_3E **(c)** AX_3 **(d)** AX_4
52. (a) tetrahedral **(b)** linear **(c)** triangular planar **(d)** linear **(e)** planar (triangular at each C atom)
54.* 109.47° **56.*** InO gives atomic mass 76.6 u and locates In between As and Se in Period 4, which is not possible. In_2O_3 gives atomic mass 115 u and locates In in Period 5 and Group III, between Cd and Sn, which is reasonable.

CHAPTER 4

1. The rate of a reaction may be increased by (1) raising the temperature, which increases the rate of collisions between the reactant molecules and the energy of the collisions, (2) increasing the concentrations of the reactants, which increases the rate of collisions; (3) using a catalyst.

3. (a) F < Cl < Br < I < At **(b)** All occur as covalent diatomic, X_2, molecules: F_2 as a greenish-yellow gas; Cl_2 as a green gas; Br_2 as a red-brown liquid, and I_2 as a dark purple solid. **(c)** Melting points and boiling points increase with the increasing strength of intermolecular forces between nonpolar X—X molecules, which in turn increase with the size of the halogen atoms in descending Group VII (as successive filled shells of electrons are added). Electronegativity increases in the same order; all the Group VII elements have a core charge of +7, but the valence electrons become increasingly distant from the nucleus and are thus less strongly attracted to it as the atoms increase in size. **6.** Ionic: (a), (b), (e), and (h); polar covalent: (c), (d), (f), (g), and (i).

(c) :F̤—Ö—F̤: **(d)** :C̤l—P̈(—C̤l:)—C̤l: **(f)** CCl_4: C bonded to four :C̤l: **(i)** :C̤l—F̤:

8. (a) $Ba(s) + Cl_2(g) \rightarrow BaCl_2(s)$; barium chloride
(b) $2Al(s) + 3Br_2(g) \rightarrow 2AlBr_3(l)$; aluminum bromide
(c) $2K(s) + I_2(s) \rightarrow 2KI(s)$; potassium iodide
(d) $2P(s) + 3Cl_2(g) \rightarrow 2PCl_3(l)$; phosphorus trichloride
(e) $2P(s) + 3I_2(g) \rightarrow 2PI_3(s)$; phosphorus triiodide
11. (a) $2K(s) + Br_2(l) \rightarrow 2KBr(s)$
(b)

$$2K \rightarrow 2K^+ + 2e^- \quad \textit{Oxidation}$$
$$Br_2 + 2e^- \rightarrow 2Br^- \quad \textit{Reduction}$$
$$2K + Br_2 \rightarrow 2K^+Br^- \quad \textit{Overall}$$

(c) Upon oxidation, a substance loses electrons. Upon reduction, a substance gains electrons. An oxidizing agent accepts electrons from the reactant that is oxidized, and a reducing agent donates electrons to the reactant that is reduced. **13. (a)** By the electrolysis of aqueous (or molten) sodium chloride, NaCl: $2Cl^-(aq) \rightarrow Cl_2(g) + 2e^-$
(b) By the electrolysis of molten sodium fluoride, NaF: $2F^-(l) \rightarrow F_2(g) + 2e^-$ **15. (a)** NR (no reaction)
(b) $2NaCl(s) + F_2(g) \rightarrow Cl_2(g) + 2NaF(s)$
RA OA
(RA: reducing agent; OA: oxidizing agent) **(c)** NR
(d) $CaBr_2(s) + F_2(g) \rightarrow Br_2(l) + CaF_2(s)$
RA OA
(e) NR **(f)** $MgI_2(aq) + Cl_2(g) \rightarrow I_2(s) + MgCl_2(aq)$*
RA OA
*$I_2(s)$ dissolves due to the reaction $I_2(s) + I^-(aq) \rightarrow I_3^-(aq)$.
17. (a) $Cl_2(g) + 3KI(aq) \rightarrow 2KCl(aq) + KI_3(aq)$
(b) NR **(c)** $Br_2(l) + 3NaI(aq) \rightarrow 2NaBr(aq) + NaI_3(aq)$ **(d)** $2F_2(g) + 2H_2O(l) \rightarrow 4HF(aq) + O_2(g)$
19. (a) $2Rb(s) + I_2(s) \rightarrow 2RbI(s)$
RA, SO OA, SR
(SO: substance oxidized; SR: substance reduced)
(b) $4Al(s) + 3O_2(g) \rightarrow 2(Al^{3+})_2(O^{2-})_3(s)$
RA, SO OA, SR
(c) $Zn(s) + S(s) \rightarrow Zn^{2+}S^{2-}(s)$
RA, SO OA, SR
(d) $Mg(s) + 2HCl(aq) \rightarrow MgCl_2(aq) + H_2(g)$
RA, SO OA, SR

21. (a) In terms of proton transfer, an acid is a proton donor and a base is a proton acceptor. In the reaction $HBr(g) + :NH_3(g) \rightarrow NH_4^+Br^-(s)$, $HBr(g)$ is a proton donor and thus an acid, whereas $NH_3(g)$ is a proton acceptor and thus a base, utilizing its lone pair of electrons to form another N—H bond and hence the NH_4^+ ion. **(b)** In the reaction of a metal with an aqueous acid, the metal atoms lose electrons to give metal ions, and the hydronium ions, $H_3O^+(aq)$, accept electrons to give neutral H_3O molecules, which decompose to water and hydrogen gas:

$$Mg(s) \rightarrow Mg^{2+}(aq) + 2e^-$$
$$2H_3O^+ + 2e^- \rightarrow [2H_3O(aq)] \rightarrow 2H_2O(l) + H_2(g)$$
$$Mg(s) + 2H_3O^+(aq) \rightarrow Mg^{2+}(aq) + 2H_2O(l) + H_2(g)$$

Thus, this is an electron transfer (oxidation–reduction) reaction rather than a proton transfer (acid–base) reaction.
23. (a) $Na_2O(s) + H_2O(l) \rightarrow 2Na^+(aq) + 2OH^-(aq)$ (strong base)
(b) $KOH(s) \rightarrow K^+(aq) + OH^-(aq)$ (strong base)
(c) $NH_3(aq) + H_2O(l) \rightleftarrows NH_4^+(aq) + OH^-(aq)$ (weak base)
(d) $LiH(s) + H_2O(l) \rightarrow Li^+(aq) + H_2(g) + OH^-(aq)$ (strong base)
25. (a) $Ca(OH)_2(s) + H_2SO_4(l) \rightarrow CaSO_4(s) + 2H_2O(l)$
$CaO(s) + H_2SO_4(l) \rightarrow CaSO_4(s) + H_2O(l)$
(b) $LiOH(s) + HF(aq) \rightarrow LiF(aq) + H_2O(l)$
$Li_2O(s) + 2HF(aq) \rightarrow 2LiF(aq) + H_2O(l)$
(c) $NH_3(aq) + HNO_3(aq) \rightarrow NH_4^+(aq) + NO_3^-(aq)$
(d) $Mg(OH)_2(s) + 2HClO_4(aq) \rightarrow Mg(ClO_4)_2(aq) + H_2O(l)$
$MgO(s) + 2HClO_4(aq) \rightarrow Mg(ClO_4)_2(aq) + H_2O(l)$
26. (a) (i) strong acid (ii) weak acid (iii) strong acid (iv) no basic or acidic properties (v) weak acid **(b)** (i) negligible basicity (ii) weak base (iii) negligible basicity (iv) weak base (v) strong base **28. (a)** $H_3O^+(aq) + OH^-(aq) \rightarrow 2H_2O(l)$; all acids give H_3O^+ and all bases give OH^- in aqueous solution. **(b)** $HF(aq)$ is only partially ionized:

$$HF(aq) + H_2O(l) \rightleftharpoons H_3O^+(aq) + F^-(aq)$$

But as $Na^+OH^-(aq)$ is added, the $OH^-(aq)$ removes $H_3O^+(aq)$ as water, allowing more $HF(aq)$ to form F^- and H_3O^+, which in turn reacts. Eventually all the $HF(aq)$ is converted to $F^-(aq)$, leaving only $Na^+(aq)$ and $F^-(aq)$ in solution.

30. (a) H—N̤⁻—H **(b)** H—F̤⁺—H

(c) PH_4^+: H—P⁺—H with H above and below **(d)** BF_4^-: :F̤—B⁻—F̤: with :F̤: above and below

32. (a) $Ag^+(aq) + Br^-(aq) \rightarrow AgBr(s)$ **(b)** In solutions where precipitation reactions occur, ions such as $NO_3^-(aq)$ and $Na^+(aq)$ in the solution in (a) that take no part in the reaction are described as spectator ions.
33. Insoluble: (a), (b), (f), (h); soluble: (c), (d), (e), (g)
35. (a) $FeCl_3(aq) + 3NaOH(aq) \rightarrow Fe(OH)_3(s) + 3NaCl(aq)$; $Fe^{3+}(aq) + 3OH^-(aq) \rightarrow Fe(OH)_3(s)$

(b) no reaction **(c)** $Pb(NO_3)_2(aq) + H_2SO_4(aq) \rightarrow PbSO_4(s) + 2HNO_3(aq)$; $Pb^{2+}(aq) + SO_4^{2-}(aq) \rightarrow PbSO_4(s)$
(d) $2AgNO_3(aq) + Na_2S(aq) \rightarrow Ag_2S(s) + 2NaNO_3(aq)$; $2Ag^+(aq) + S^{2-}(aq) \rightarrow Ag_2S(s)$
(e) $AgNO_3(aq) + HI(aq) \rightarrow AgI(s) + HNO_3(aq)$; $Ag^+(aq) + I^-(aq) \rightarrow AgI(s)$ **(f)** $Pb(NO_3)_2(aq) + 2HCl(aq) \rightarrow PbCl_2(s) + 2HNO_3(aq)$; $Pb^{2+}(aq) + 2Cl^-(aq) \rightarrow PbCl_2(s)$ **37. (a)** $2AgNO_3(aq) + BaCl_2(aq) \rightarrow Ba(NO_3)_2(aq) + 2AgCl(s)$; precipitation **(b)** $2NH_3(aq) + H_2SO_4(aq) \rightarrow (NH_4)_2SO_4(aq)$; acid–base
(c) $Zn(s) + 2HCl(aq) \rightarrow ZnCl_2(aq) + H_2(g)$; oxidation–reduction **(d)** $NaHCO_3(aq) + HCl(aq) \rightarrow NaCl(aq) + CO_2(g) + H_2O(l)$; acid–base

39. (a) $Cl_2(aq) + 2I^-(aq) \rightarrow 2Cl^-(aq) + I_2(s)$;
OA RA
oxidation–reduction
(b) $HCl(aq) + H_2O(l) \rightarrow Cl^-(aq) + H_3O^+(aq)$; acid–base
acid base
(c) $Zn(s) + 2HCl(aq) \rightarrow ZnCl_2(aq) + H_2(g)$;
RA OA
oxidation–reduction
(d) $HCO_3^-(aq) + H_3O^+(aq) \rightarrow CO_2(aq) + 2H_2O(l)$;
base acid
acid–base

42. (a) solid **(b)** the weakest **(c)** (i) $2Na(s) + At_2(s) \rightarrow 2Na^+At^-(s)$ (ii) $Ca(s) + At_2(s) \rightarrow Ca^{2+}(At^-)_2(s)$ (iii) $2P(s) + 3At_2(s) \rightarrow 2PAt_3(s)$
(iv) $H_2(g) + At_2(s) \rightarrow 2HAt(g)$ (v) $Br_2(l) + At_2(s) \rightarrow 2AtBr(s \text{ or } l)$ **(d)** strong acid; $HAt(g) + H_2O(l) \rightarrow H_3O^+(aq) + At^-(aq)$ **(e)** polar covalent
(f) ionic, K^+At^-; solid

43. (a) (i) CH_4 (ii) NH_3 (iii) H_2O (iv) HF
(b) (i) $CH_4(g)$, neither acid nor base (insoluble)
(ii) $NH_3(g)$, $NH_3(g) + H_2O(l) \rightleftharpoons NH_4^+(aq) + OH^-(aq)$; weak base (iii) $H_2O(l)$, $H_2O(l) + H_2O(l) \rightleftharpoons H_3O^+(aq) + OH^-(aq)$; weak acid and weak base (iv) HF(g), $HF(aq) + H_2O(l) \rightleftharpoons H_3O^+(aq) + F^-(aq)$; weak acid

45.* Magnesium bromide, $MgBr_2(s)$ **47.*** 2.24×10^{19} Na^+ ions, 2.24×10^{19} Cl^- ions

CHAPTER 5

1. (a) 18.02 g · mol^{-1} **(b)** 16.04 g · mol^{-1}
(c) 17.03 g · mol^{-1} **(d)** 44.01 g · mol^{-1}
(e) 58.44 g · mol^{-1} **3. (a)** 5.549 mol
(b) 0.3117 mol **(c)** 0.4313 mol
(d) 5.378×10^{-2} mol **(e)** 0.2139 mol

5. (a) $2HCl + Ba(OH)_2 \rightarrow BaCl_2 + 2H_2O$; 23.49 g $Ba(OH)_2$
(b) $Cl_2 + 2NaBr \rightarrow 2NaCl + Br_2$; 29.03 g NaBr
(c) $2NaOH + H_2SO_4 \rightarrow Na_2SO_4 + 2H_2O$; 12.26 g H_2SO_4
(d) $2HNO_3 + CaO \rightarrow Ca(NO_3)_2 + H_2O$; 4.449 g CaO

7. $NH_3(g) + HCl(g) \rightarrow NH_4Cl(s)$; 0.428 g

9. 50.0% Mg and 50.0% Zn by mass

11. 0.326 kg **13.** 2.58 L **15.** 5.6 mol
17. 4.87 g $AlCl_3$; 1.71 g Al **19.** 77.94%

21. (a) 11.19% H, 88.79% O
(b) 39.34% Na, 60.66% Cl
(c) 79.89% C, 20.11% H **(d)** 13.20% Mg, 86.80% Br
(e) 27.29% C, 72.71% O **23. (a)** 85.63% C, 14.37% H **(b)** 92.26% C, 7.74% H **(c)** 11.96% Mg, 34.87% Cl, 5.95% H, 47.22% O **(d)** 20.09% Fe, 11.53% S, 5.08% H, 63.31% O **25. (a)** 21.20%
(b) 471.7 g **27.** C_7H_5; 89.11 u **29. (a)** CH_3O
(b) $C_2H_6O_2$ **31.** C_6H_7N **33.** C_2H_6S

35. Cu_2O (red); CuO (black)

37. 116 g · mol^{-1}; $C_6H_{12}O$

38. 6.58×10^2 L $N_2(g)$; 1.97×10^3 L $H_2(g)$

40. 400 L synthesis gas; 100 L CO(g); 300 L $H_2(g)$

42. (a) 0.0100 mol **(b)** 2.50×10^{-3} mol
(c) 2.52×10^{-4} mol **(d)** 3.37×10^{-4} mol

44. 0.0380-*M* $KMnO_4(aq)$ **46.** 4.75 mL

48. (a) 189 mL 0.100-*M* $Ba(OH)_2(aq)$ is diluted with distilled water to a volume of 6.30 L **(b)** 14.9 mL 35.0% $Cr_2(SO_4)_3(aq)$ is diluted with distilled water to a volume of 750 mL. **50.** $H_3O^+(aq) + OH^-(aq) \rightarrow 2H_2O(l)$
(a) 0.333-*M* **(b)** 14.6 g **52. (a)** 21.6 mL
(b) 183 mL **54. (a)** 0.204 g **(b)** 0.127 g

56. 87.4% by mass **58.** $CuSO_4 \cdot 5H_2O$

60. 104.5 u; 104.5 u; CF_3Cl **62.** 8.64 g

CHAPTER 6

1. 4.0×10^{-7} m (400 nm) **3.** 11.0 m

5. red; 4.74×10^{14} Hz **6. (a)** 2.97×10^{-19} J · photon^{-1}
(b) 1.79×10^2 kJ · mol^{-1}

8. (a) 3.64×10^{-19} J · photon^{-1}
(b) 2.19×10^2 kJ · mol^{-1} **10. (a)** 1.24×10^{15} Hz, colorless (ultraviolet light) **(b)** 1.73×10^{-19} J

12. No (except in bright sunlight) **13. (a)** Yes
(b) 630 nm **15.** 411 nm; the visible spectrum (blue light)

17. $n = 3$ **19.** 657 nm **21.** $n = 6$

23. (a) In photoelectron spectroscopy, gaseous samples are irradiated with high-energy photons of known energy, $h\nu$, which eject photoelectrons with kinetic energies $\frac{1}{2}mv^2$ from the atoms. These kinetic energies are then measured by deflection in a magnetic field. For a large number of atoms, $h\nu = IE + \frac{1}{2}mv^2$ gives the ionization energy, *IE*, of electrons from each energy level, provided suitably energetic photons are used.
(b) For Ar ($Z = 18$), there are five energy levels, designated 1*s*, 2*s*, 2*p*, 3*s*, and 3*p* (from highest to lowest energy).
25. 0.62 MJ · mol^{-1} **27. (a)** 2 ($n = 1$); 8 ($n = 2$); 18 ($n = 3$) **(b)** one, 1*s*; two, 2*s* and 2*p*; three, 3*s*, 3*p*, and 3*d*
(c) *ns*, 2; *np*, 6; *nd*, 10; *nf*, 14 **28.** (b), (d)

30. (a) Be ($Z = 4$) has the ground-state configuration $1s^22s^2$. The configuration $1s^4$ is not possible because the Pauli exclusion principle allows any orbital to have a maximum of two electrons with opposite spins. **(b)** Nitrogen ($Z = 7$) has the ground-state configuration $1s^22s^22p^3$. Each of the three $2p$ orbitals has the same energy level, so the three electrons in the $2p$ level occupy these orbital separately with the same spin and thereby minimize their electrostatic repulsions in accordance with Hund's rule.

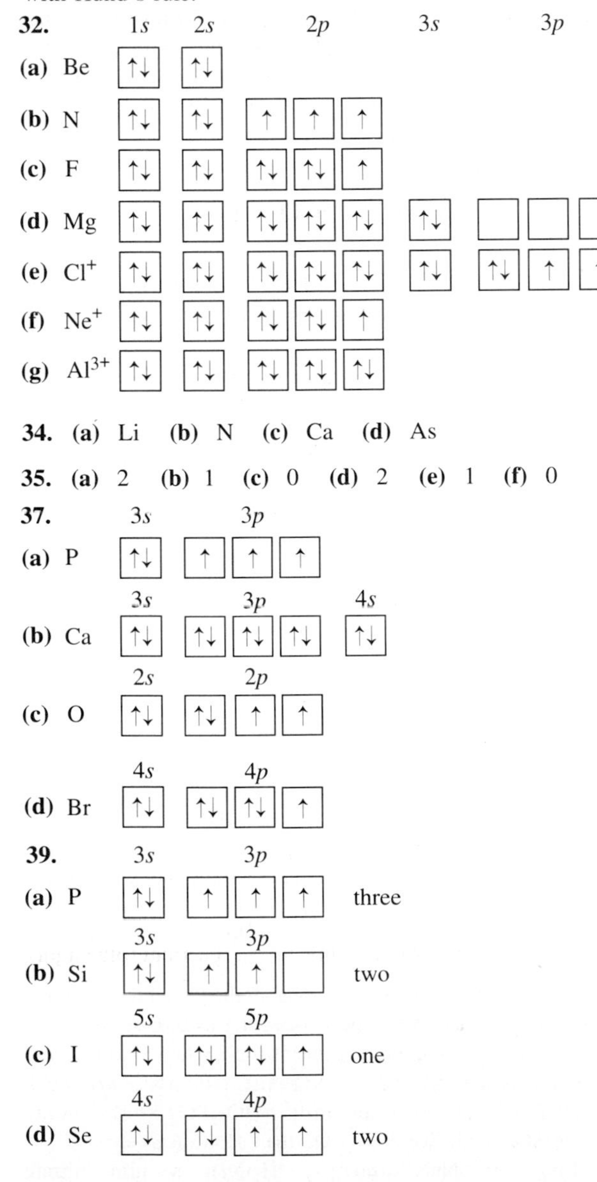

41. (a) Group V, Period 4 **(b)** eight **(c)** three **(d)** nitrogen, $[He]2s^22p^3$; phosphorus, $[Ne]3s^23p^3$

42. (a) $1s^22s^22p^1$; boron, *p*-block
(b) $1s^22s^22p^63s^1$; sodium, *s*-block
(c) $1s^22s^22p^63s^23p^64s^1$; potassium, *s*-block
(d) $1s^22s^22p^63s^23p^64s^23d^2$; titanium, *d*-block
(e) $1s^22s^22p^63s^23p^64s^23d^3$; vanadium, *d*-block
(f) $1s^22s^22p^63s^23p^64s^13d^{10}$; copper, *d*-block
(g) $1s^22s^22p^63s^23p^63d^{10}4s^24p^3$; arsenic, *p*-block
(h) $1s^22s^22p^63s^23p^64s^23d^{10}4p^6$; krypton, *p*-block

45. (a) $[Ne]3s^23p^2$ **(b)** two **(c)** +2 **(d)** $SiCl_4$ and SiH_4 form from Si in the $[Ne]3s^13p^3$ excited state (valence +4).

47. (a) The 2*s* and 2*p* atomic orbitals do not have the geometry required to form four tetrahedral bonding orbitals by overlap with hydrogen 1*s* orbitals. **(b)** In terms of a set of tetrahedral sp^3 hybrid orbitals on the carbon atom, each of which can overlap with a hydrogen 1*s* atomic orbital to form four tetrahedral bonding orbitals. **49.** C_2H_2: According to the σ–π model, each C atom forms two colinear σ bonds (one to an H atom, one to the other C atom), and there are two π bonds between the C atoms; HCN: The C atom forms two colinear σ bonds (one to the H atom, one to the N atom), and there are two π bonds between the C and the N atoms; CO_2: The C atom forms two colinear σ bonds (one to each O atom), and there is a π bond between the C atom and each O atom.

51. (a) 1: 120°; 2: 120° **(b)** 1: 109.5°; 2: 120°; 3: 109.5° **(c)** 1: 109.5°; 2: 120° **53. (a)** 5 σ bonds, 1 π bond **(b)** 7 σ bonds, 1 π bond **(c)** 9 σ bonds, 1 π bond **55. (a)** Green (λ = 549 nm)

57. (a) $1.12\ kJ \cdot mol^{-1}$ **(b)** 4.23×10^{-19} J **(c)** $254\ kJ \cdot mol^{-1}$

60. Nitrogen has electrons in three singly occupied 2*p* atomic orbitals and attracts another N atom to form three bonding molecular orbitals, each containing an electron pair (a triple bond). In each pair there is a small increase in the electron density in the internuclear region, which pulls the two atoms together by electrostatic attraction. Oxygen has electrons in two singly occupied atomic orbitals and attracts another O atom to form two bonding molecular orbitals, each containing an electron pair (a double bond). Fluorine has one electron in a singly occupied atomic orbital and combines with another F atom to form only one bonding molecular orbital (a single bond). In contrast, neon, with four lone pairs in filled atomic orbitals in its valence shell, has no unpaired electrons and cannot form bonds.

CHAPTER 7

2. (a) $H_2SO_4(aq) + H_2O(l) \rightarrow H_3O^+(aq) + HSO_4^-(aq)$
$HSO_4^-(aq) + H_2O(l) \rightleftarrows H_3O^+(aq) + SO_4^{2-}(aq)$
$NH_3(g) + H_2SO_4(aq) \rightarrow NH_4^+(aq) + HSO_4^-(aq)$
$NaOH(aq) + H_2SO_4(aq) \rightarrow Na^+(aq) + H_3O^+(aq) + 2HSO_4^-(aq)$
$CaO(s) + 2H_2SO_4(aq) \rightarrow Ca^{2+}(aq) + H_3O^+(aq) + 2HSO_4^-(aq)$
(b) $2Br^-(aq) + 5H_2SO_4(l) \rightarrow Br_2(aq) + SO_2(g) + 2H_3O^+(aq) + 4HSO_4^-(aq)$
$2I^-(aq) + 5H_2SO_4(l) \rightarrow I_2(s) + SO_2(g) + 2H_3O^+(aq) + 4HSO_4^-(aq)$
$Cu(s) + 5H_2SO_4(l) \rightarrow Cu^{2+}(aq) + SO_2(g) + 2H_3O^+(aq) + 4HSO_4^-(aq)$

$2Ag(s) + 5H_2SO_4(l) \rightarrow 2Ag^+(aq) + SO_2(g) + 2H_3O^+(aq) + 4HSO_4^-(aq)$

(c) $C_{12}H_{22}O_{11}(s) + 11H_2SO_4(l) \rightarrow 12C(s) + 11H_3O^+(aq) + 11HSO_4^-(aq)$

$CuSO_4 \cdot 5H_2O(s) \rightarrow CuSO_4(s) + 5H_2O$(absorbed by H_2SO_4)
blue white

4. (a) $Zn(s) + H_2SO_4(aq) \rightarrow Zn^{2+}(aq) + SO_4^{2-}(aq) + H_2(g)$ **(b)** $2NaI(s) + 5H_2SO_4(l) \rightarrow I_2(s) + SO_2(g) + 2Na^+ + 2H_3O^+ + 4HSO_4^-$ **(c)** $Cu(s) + 5H_2SO_4(l) \rightarrow Cu^{2+} + SO_2(g) + 2H_3O^+ + 4HSO_4^-$ **(d)** $Mg(OH)_2(s) + H_2SO_4(aq) \rightarrow Mg^{2+}(aq) + SO_4^{2-}(aq) + 2H_2O(l)$

6. Sulfate ion: $Ba^{2+}(aq) + SO_4^{2-}(aq) \rightarrow BaSO_4(s)$
Sulfite ion:

$$Br_2(aq) + SO_3^{2-}(aq) + 3H_2O(l) \rightarrow 2Br^-(aq) + SO_4^{2-}(aq) + 2H_3O^+(aq)$$

or

$$SO_3^{2-}(aq) + 2H_3O^+(aq) \rightarrow SO_2(g) + 3H_2O(l)$$

8. (a) $Cl_2(aq) + H_2S(aq) \rightarrow 2Cl^-(aq) + 2H^+(aq) + S(s)$ **(b)** 229 g **10. (a)** K_2SO_4 **(b)** $Ca(HSO_4)_2$ **(c)** $CaSO_3$ **(d)** $K_4P_2O_7$ **(e)** $Al_2(SO_4)_3$

12. (a) Sodium sulfide, ionic **(b)** magnesium sulfide, ionic **(c)** orthorhombic sulfur, covalent [S_8 ring structure]

(d) carbon disulfide, covalent :S̈=C=S̈:

(e) sulfur dioxide, covalent :Ö=S̈=Ö:

(f) sulfur trioxide, covalent :Ö=S=Ö: (with =Ö: below S)

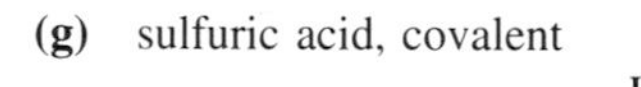

(g) sulfuric acid, covalent [$(HO)_2SO_2$ structure]

14. (a) $5H_2SO_4(conc) + Cu(s) \rightarrow Cu^{2+} + SO_2(g) + 2H_3O^+(aq) + 4HSO_4^-(aq)$ **(b)** $H_2SO_4(aq) + CuS(s) \rightarrow Cu^{2+}(aq) + SO_4^{2-}(aq) + H_2S(g)$ **(c)** $Na_2SO_3(aq) + H_2SO_4(aq) \rightarrow 2Na^+(aq) + SO_4^{2-}(aq) + H_2O(l) + SO_2(g)$ **(d)** $H_2S(aq) + Cl_2(aq) \rightarrow 2HCl(aq) + S(s)$ **(e)** $H_2S(aq) + Pb^{2+}(aq) + 2H_2O(l) \rightarrow PbS(s) + 2H_3O^+(aq)$ **(f)** $5H_2SO_4(conc) + 2NaI(s) \rightarrow I_2(s) + SO_2(g) + 2Na^+ + 2H_3O^+ + 4HSO_4^-$

16.* **(a)** SO_2Cl_2, SO_2Cl_2

(b) :C̈l—S—C̈l: (with =Ö: above and below S)

18. (a) White phosphorus consists of P_4 molecules and is a molecular solid. To melt it, energy is required only to overcome weak intermolecular forces between the P_4 molecules. Red phosphorus is an infinite two-dimensional covalent network solid. To melt it, many stronger P—P covalent bonds have to be broken. $P_4(s)$ is much more volatile than red phosphorus because P_4 molecules can easily break away from the solid—no covalent bonds must be broken. **(b)** The relatively small P_4 molecules of white phosphorus go into solution in the covalent carbon disulfide, $CS_2(l)$, solvent. Red phosphorus is insoluble in $CS_2(l)$, because many covalent P—P bonds must be broken for it to go into solution.

20. $P_4S_3(s) + 8O_2(g) \rightarrow P_4O_{10}(s) + 3SO_2(g)$; 50.7 mL

22. (a) A gas-phase acid–base reaction **(b)** $PH_3(g) + HCl(g) \rightarrow PH_4^+Cl^-(s)$ **(c)** $[PH_4]^+$ (P bonded to four H) :C̈l:$^-$ phosphonium chloride

(d) An ionic solid **24. (a)** P_4O_6 **(b)** $Ca(H_2PO_4)_2$ **(c)** Ca_3P_2 **(d)** H_3PO_3 **(e)** $Na_5P_3O_{10}$

26. (a) $P_4(s) + 3O_2(g) \rightarrow P_4O_6(s)$ (limited supply of air) **(b)** $P_4(s) + 10Cl_2(g) \rightarrow 4PCl_5(s)$ (excess $Cl_2(g)$) **(c)** $P_4(s) + 5O_2(g) \rightarrow P_4O_{10}(s)$ (excess oxygen), $P_4O_{10}(s) + 6H_2O(l) \rightarrow 4H_3PO_4(l)$ **(d)** $3Ca(s) + 2P(s) \rightarrow Ca_3P_2(s)$ and $[Ca^{2+}]_3[P^{3-}](s) + 6H_3O^+(aq) \rightarrow 3Ca^{2+}(aq) + 2PH_3(g) + 6H_2O(l)$

28.* **(a)** POF_3 **(b)** :F̈—P=Ö: (with :F̈: above and below P)

30. Potassium (Group I) and calcium (Group II) have low electronegativities compared to oxygen. Therefore the X—O bond is highly polar (ionic) as $X^{\delta+}—O^{\delta-}$. K^+OH^- and $Ca^{2+}(OH^-)_2$ dissociate fully into their cations and OH^- ions in water and thus behave as strong bases. In contrast, silicon (Group IV) has a higher electronegativity and is a nonmetal, and silicic acid, $SI(OH)_4$, is a covalent molecule. The Si—O bond is correspondingly less polar. Both the O and the Si atoms pull the bonding electrons away from the hydrogen atoms, so the O—H bond becomes polar as $Si—O^{\delta-}—H^{\delta+}$, and silicic acid is a weak acid. **32. (a)** $HOSO_3^-(aq)$, weak **(b)** $HOPO_3^{2-}(aq)$, weak **(c)** $(HO)_2PO_2^-(aq)$, weak **(d)** $HOSO_2^-(aq)$, weak **34. (a)** $HNO_3(aq) + NaOH(aq) \rightarrow NaNO_3(aq) + H_2O(l)$; sodium nitrate **(b)** $H_2SO_4(aq) + NaOH(aq) \rightarrow NaHSO_4(aq) + H_2O(l)$; sodium hydrogensulfate $H_2SO_4(aq) + 2NaOH(aq) \rightarrow Na_2SO_4(aq) + 2H_2O(l)$; sodium sulfate **(c)** $H_3PO_4(aq) + NaOH(aq) \rightarrow NaH_2PO_4(aq) + H_2O(l)$; sodium dihydrogenphosphate $H_3PO_4(aq) + 2NaOH(aq) \rightarrow Na_2HPO_4(aq) + 2H_2O(l)$; sodium hydrogenphosphate $H_3PO_4(aq) + 3NaOH(aq) \rightarrow Na_3PO_4(aq) + 3H_2O(l)$; sodium phosphate **(d)** $HClO_4(aq) + NaOH(aq) \rightarrow NaClO_4(aq) + H_2O(l)$; sodium perchlorate **36. (a)** H_2O, water; H_3O^+, hydronium ion; HSO_4^-, hydrogensulfate ion; SO_4^{2-}, sulfate ion **(b)** H_2O, water; $SO_2(aq)$, sulfur dioxide; H_3O^+, hydronium ion; H_2SO_3, sulfurous acid; HSO_3^-, hydrogensulfite ion; SO_3^{2-}, sulfite ion **(c)** H_2O, water; H_3O^+, hydronium ion; H_3PO_4, phosphoric acid; $H_2PO_4^-$, dihydrogenphosphate ion; HPO_4^{2-}, hydrogenphosphate ion; PO_4^{3-}, phosphate ion **(d)** H_2O, water; $CO_2(g)$, carbon dioxide; H_3O^+, hydronium ion; HCO_3^-, hydrogencarbonate ion; CO_3^{2-}, carbonate ion

38. $ClO_3^- + 6e^- + 6H^+ \rightarrow Cl^- + 3H_2O$
$ClO_2^- + 4e^- + 4H^+ \rightarrow Cl^- + 2H_2O$
$ClO^- + 2e^- + 2H^+ \rightarrow Cl^- + H_2O$

40.* **(a)** $Ca_3O_4Cl_4H_2$ or $(Ca^{2+})_3(Cl^-)_2(OCl^-)_2(OH^-)_2$
(b) $3Ca(OH)_2(s) + 2Cl_2(g) \rightarrow (Ca^{2+})_3(Cl^-)_2(OCl^-)_2(OH^-)_2$

42.* **(a)** :F̈—C̈l—F̈: (with :F̈: below Cl) and ClF_5 (Cl with five F and one lone pair)

AX_3E_2, three structures AX_5E, one structure (square pyramid)

(b) :F̈: above, :Cl—F̈:, :F̈: below

The three Cl–F bonds form the shape of the letter T.

44. (a) −II (H_2S, HS^-, S^{2-}); 0 (S_8, S_n), +II (SO, SCl_2); +IV (SO_2, SO_3^{2-}, HSO_3^-); +VI (SO_3, H_2SO_4, HSO_4^-, SO_4^{2-}, SF_6) **(b)** −III (PH_4^+, PH_3, P^{3-}); 0 (P_4, P_n); +III (P_4O_6, PCl_3, PF_3, H_3PO_3, $H_2PO_3^-$, HPO_3^{2-}); +V (P_4O_{10}, H_3PO_4, $H_2PO_4^-$, HPO_4^{2-}, PO_4^{3-}, PCl_5, PF_5, $POCl_3$, POF_5)

46. (a) +2 **(b)** +4 **(c)** +5 **(d)** +6 **(e)** −1 **(f)** +3 **48. (a)** N, O **(b)** H, +1; C, +4; O, −2 **(c)** N, −3; H, +1 **(d)** P, −3; H, +1 **(e)** Mn, +4; O, −2 **(f)** H, +1; N, +5; O, −2

50. (a) increase **(b)** decrease **(c)** (i) $Cu \rightarrow Cu^{2+} + 2e^-$ (ii) $S + 2e^- \rightarrow S^{2-}$ (iii) $NO_3^- + 3e^- + 4H^+ \rightarrow NO + 2H_2O$ **52. (a)** $2Br^- + H_2SO_4 + 2H^+ \rightarrow Br_2 + SO_2 + 2H_2O$ **(b)** $Br_2 + SO_2 + 2H_2O \rightarrow 2Br^- + SO_4^{2-} + 4H^+$

53. (a) :Ö=S̈=Ö: AX_2E, angular

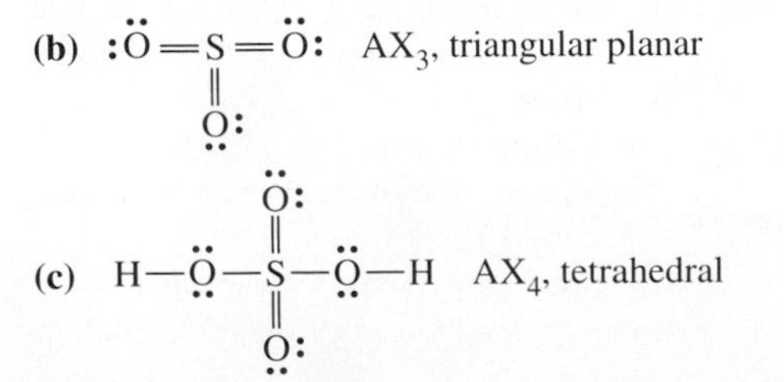

(d) :Ö=S̈—Ö:⁻ (with :Ö:⁻ below S) AX_3E, triangular pyramidal

55. (a) AX_5: square pyramidal; AX_6: octahedral
(b) AX_5: PCl_5 or PF_5; AX_6: SF_6

57.* **(a)** H_3PO_4 structure: P=Ö: bonded to three —Ö—H groups **(b)** $H_4P_2O_7$ structure: H—Ö—P(—Ö—H)(=O)—Ö—P(—Ö—H)(=O)—Ö—H

(c) :F̈—S(=Ö:)(=Ö:)—:Ö:⁻ **(d)** :F̈—P(=Ö:)(—:Ö:⁻)(—:Ö:⁻)

58. (a) $2SO_2(g) + O_2(g) \rightarrow 2SO_3(g)$
$SO_3(g) + H_2O(g/l) \rightarrow H_2SO_4(aq)$
(b) 153 kg **(c)** 116 kg **60.*** **(a)** A: S(*s*); B: $SO_2(g)$; C: $H_2SO_3(aq)$; D: FeS(*s*); E: $H_2S(g)$
(b) $S(s) + O_2(g) \rightarrow SO_2(g)$
$SO_2(g) + H_2O(l) \rightarrow H_2SO_3(aq)$
$S(s) + Fe(s) \rightarrow FeS(s)$
$FeS(s) + H_2SO_4(aq) \rightarrow FeSO_4(aq) + H_2S(g)$
$2H_2S(aq) + H_2SO_3(aq) \rightarrow 3S(s) + 3H_2O(l)$

62.* **(a)** $P_4(s)$ **(b)** tetrahedral P_4, each P with one lone pair

(c) $P_4(g) + 6Cl_2(g) \rightarrow 4PCl_3(g)$

CHAPTER 8

1. The position of the equilibrium between diamond and graphite favors graphite at ordinary temperatures and pressures, but at high temperatures and pressures hundreds of kilometers below the earth's surface, the formation of diamond is favored. Diamonds are carried to the surface by volcanic activity. The infinite three-dimensional covalent lattice structure of diamond in which each C atom is joined by four strong covalent bonds into a tetrahedron with four surrounding C atoms, makes it very hard and chemically unreactive. In contrast, graphite consists of infinite two-dimensional sheets of C atoms stacked on each other at relatively large distances, held together only by weak intermolecular forces. The layers can slide over each other relatively easily, accounting for its softness and a slippery feel.

3. All the bonds in diamond are single C—C bonds with the normal single bond length of 154 pm. In contrast, the bonding in graphite must be described by an infinite number of resonance structures in which each C atom is bonded to three others by one C=C bond and two C—C bonds, giving each bond a bond order of $1\frac{1}{3}$. Thus, all the bonds in graphite have the same length, intermediate between a C—C bond (bond order 1, length 154 pm) and a C=C bond (bond order 2, length 134 pm).

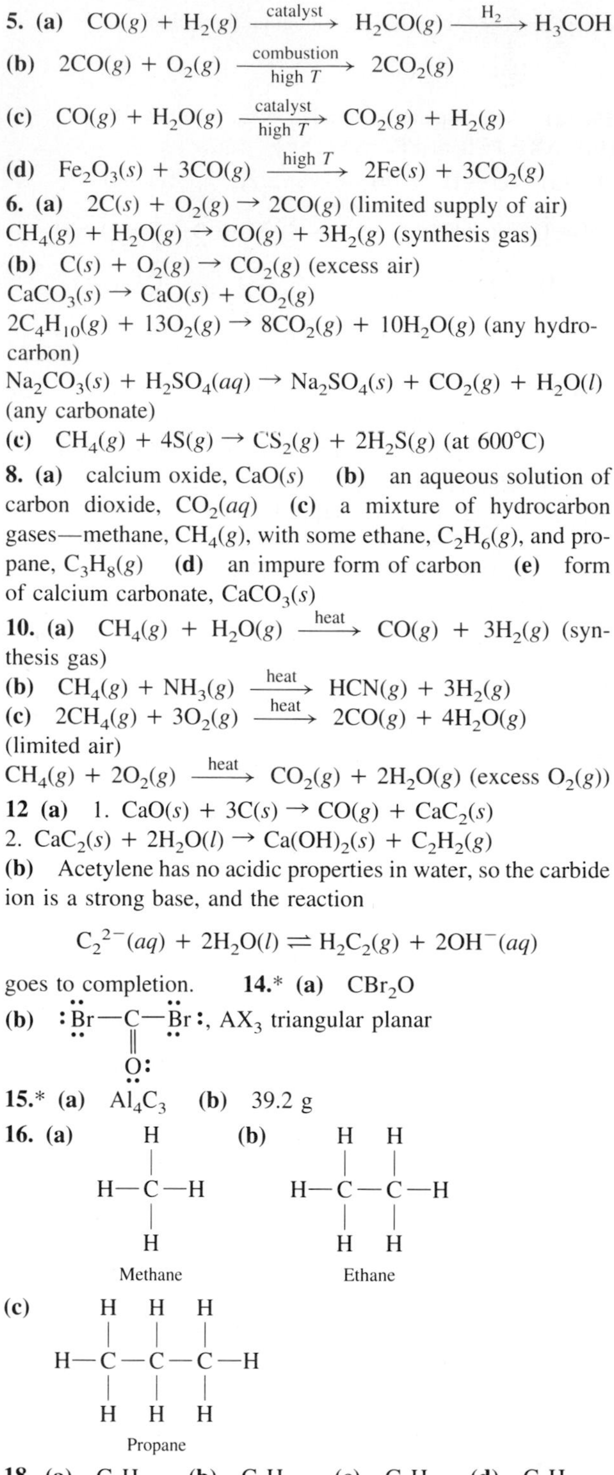

5. (a) $CO(g) + H_2(g) \xrightarrow{\text{catalyst}} H_2CO(g) \xrightarrow{H_2} H_3COH$

(b) $2CO(g) + O_2(g) \xrightarrow[\text{high } T]{\text{combustion}} 2CO_2(g)$

(c) $CO(g) + H_2O(g) \xrightarrow[\text{high } T]{\text{catalyst}} CO_2(g) + H_2(g)$

(d) $Fe_2O_3(s) + 3CO(g) \xrightarrow{\text{high } T} 2Fe(s) + 3CO_2(g)$

6. (a) $2C(s) + O_2(g) \rightarrow 2CO(g)$ (limited supply of air)
$CH_4(g) + H_2O(g) \rightarrow CO(g) + 3H_2(g)$ (synthesis gas)
(b) $C(s) + O_2(g) \rightarrow CO_2(g)$ (excess air)
$CaCO_3(s) \rightarrow CaO(s) + CO_2(g)$
$2C_4H_{10}(g) + 13O_2(g) \rightarrow 8CO_2(g) + 10H_2O(g)$ (any hydrocarbon)
$Na_2CO_3(s) + H_2SO_4(aq) \rightarrow Na_2SO_4(s) + CO_2(g) + H_2O(l)$ (any carbonate)
(c) $CH_4(g) + 4S(g) \rightarrow CS_2(g) + 2H_2S(g)$ (at 600°C)

8. (a) calcium oxide, $CaO(s)$ **(b)** an aqueous solution of carbon dioxide, $CO_2(aq)$ **(c)** a mixture of hydrocarbon gases—methane, $CH_4(g)$, with some ethane, $C_2H_6(g)$, and propane, $C_3H_8(g)$ **(d)** an impure form of carbon **(e)** form of calcium carbonate, $CaCO_3(s)$

10. (a) $CH_4(g) + H_2O(g) \xrightarrow{\text{heat}} CO(g) + 3H_2(g)$ (synthesis gas)
(b) $CH_4(g) + NH_3(g) \xrightarrow{\text{heat}} HCN(g) + 3H_2(g)$
(c) $2CH_4(g) + 3O_2(g) \xrightarrow{\text{heat}} 2CO(g) + 4H_2O(g)$ (limited air)
$CH_4(g) + 2O_2(g) \xrightarrow{\text{heat}} CO_2(g) + 2H_2O(g)$ (excess $O_2(g)$)

12 (a) 1. $CaO(s) + 3C(s) \rightarrow CO(g) + CaC_2(s)$
2. $CaC_2(s) + 2H_2O(l) \rightarrow Ca(OH)_2(s) + C_2H_2(g)$
(b) Acetylene has no acidic properties in water, so the carbide ion is a strong base, and the reaction

$$C_2^{2-}(aq) + 2H_2O(l) \rightleftharpoons H_2C_2(g) + 2OH^-(aq)$$

goes to completion. **14.* (a)** CBr_2O
(b) $:\ddot{Br}-C(=\ddot{O}:)-\ddot{Br}:$, AX_3 triangular planar

15.* (a) Al_4C_3 **(b)** 39.2 g

16. (a)
```
    H
    |
H — C — H
    |
    H
```
Methane

(b)
```
    H   H
    |   |
H — C — C — H
    |   |
    H   H
```
Ethane

(c)
```
    H   H   H
    |   |   |
H — C — C — C — H
    |   |   |
    H   H   H
```
Propane

18. (a) C_8H_{18} **(b)** C_6H_{12} **(c)** C_5H_8 **(d)** C_6H_{12}

20. (a) 3-methylpentane **(b)** 2,2-dimethylbutane **(c)** pentane **(d)** correct **(e)** 2-methylbutane

21.

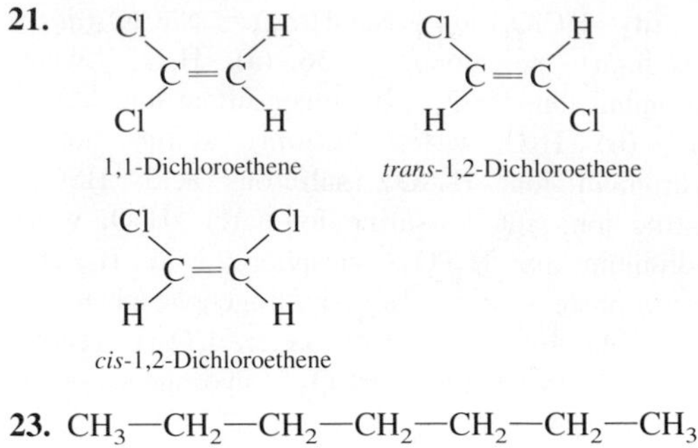

1,1-Dichloroethene   *trans*-1,2-Dichloroethene

```
Cl      Cl
  \    /
   C=C
  /    \
H       H
```
cis-1,2-Dichloroethene

23. $CH_3-CH_2-CH_2-CH_2-CH_2-CH_2-CH_3$
Heptane

$CH_3-CH(CH_3)-CH_2-CH_2-CH_2-CH_3$
2-Methylhexane

$CH_3-CH_2-CH(CH_3)-CH_2-CH_2-CH_3$
3-Methylhexane

$CH_3-CH(CH_3)-CH(CH_3)-CH_2-CH_3$
2,3-Dimethylpentane

$CH_3-CH(CH_3)-CH_2-CH(CH_3)-CH_3$
2,4-Dimethylpentane

$CH_3-C(CH_3)_2-CH_2-CH_2-CH_3$
2,2-Dimethylpentane

$CH_3-CH_2-C(CH_3)_2-CH_2-CH_3$
3,3-Dimethylpentane

$CH_3-CH_2-CH(CH_2-CH_3)-CH_2-CH_3$
3-Ethylpentane

$CH_3-C(CH_3)(CH_3)-C(CH_3)(H)-CH_3$
2,2,3-Trimethylbutane

24.

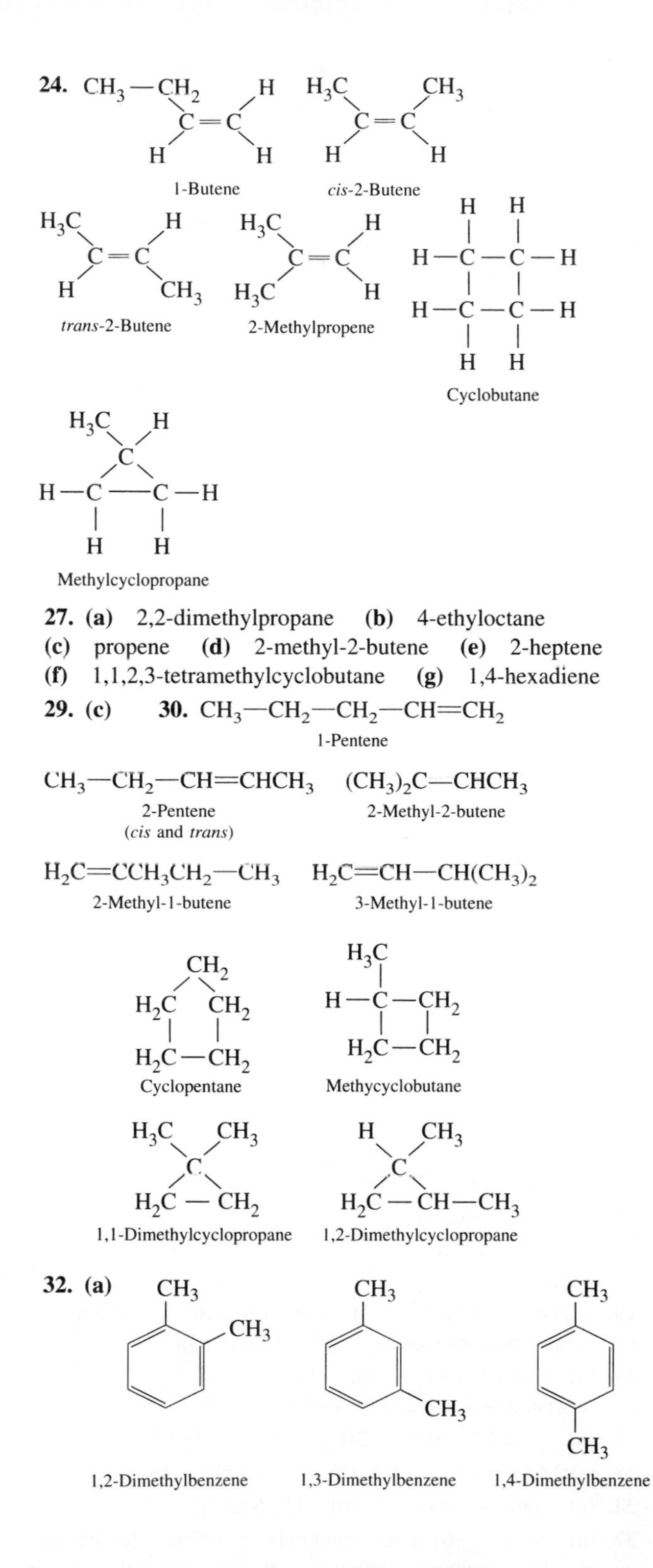

27. (a) 2,2-dimethylpropane **(b)** 4-ethyloctane **(c)** propene **(d)** 2-methyl-2-butene **(e)** 2-heptene **(f)** 1,1,2,3-tetramethylcyclobutane **(g)** 1,4-hexadiene

29. (c) **30.** $CH_3—CH_2—CH_2—CH{=}CH_2$ (1-Pentene)

$CH_3—CH_2—CH{=}CHCH_3$ (2-Pentene, *cis* and *trans*) $(CH_3)_2C—CHCH_3$ (2-Methyl-2-butene)

$H_2C{=}CCH_3CH_2—CH_3$ (2-Methyl-1-butene) $H_2C{=}CH—CH(CH_3)_2$ (3-Methyl-1-butene)

32. (a)

(b)

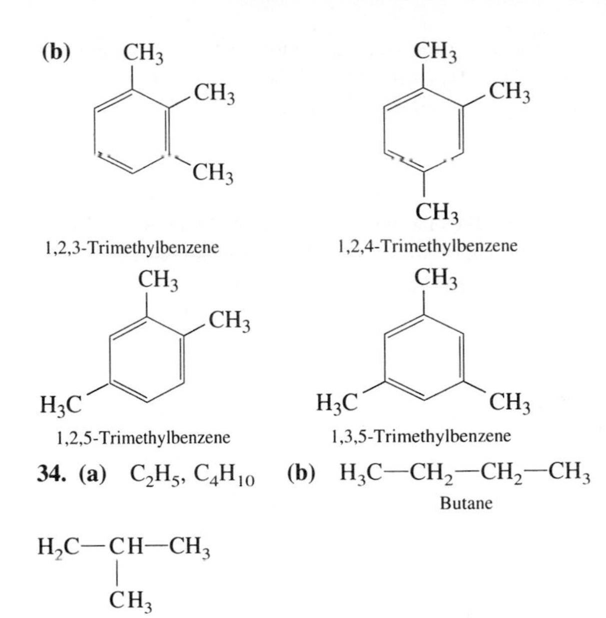

34. (a) C_2H_5, C_4H_{10} **(b)** $H_3C—CH_2—CH_2—CH_3$ (Butane)

$H_3C—CH(CH_3)—CH_3$ (2-Methylpropane)

35. (a) CH_2, C_4H_8 **(b)** See the answer to problem 24.

37. (a) $CH_4(g) + 2O_2(g) \rightarrow CO_2(g) + 2H_2O(l)$

$C_5H_{12}(g) + 8O_2(g) \rightarrow 5CO_2(g) + 6H_2O(l)$

(b) $H_3C—CH_3(g) \rightarrow H_2C{=}CH_2(g) + H_2(g)$

$H_3C—CH_2—CH_3(g) \rightarrow H_2C{=}CH_2(g) + CH_4(g)$

(c) $H_2C{=}CH_2(g) + H_2(g) \rightarrow H_3C—CH_3(g)$

$H_2C{=}CH_2(g) + HBr(g) \rightarrow H_3C—CH_2Br(l)$

$H_2C{=}CH_2(g) + H_2O(g) \rightarrow H_3C—CH_2OH(l)$

(d) $H—C{\equiv}C—H(g) + Br_2(g) \rightarrow H—C(Br){=}C(Br)—H(g)$

and $H—C(Br){=}C(Br)—H(g) + Br_2(g) \rightarrow H—CBr_2—CBr_2—H(g)$

(e) $n(CH_2{=}CH_2)(g) \rightarrow {+}CH_2—CH_2{+}_n(s)$

39. (a) Ethane, $H_3C—CH_3$, is a saturated hydrocarbon; ethyne, $HC{\equiv}CH$, is unsaturated. Thus, for instance, Br_2 will add to the triple bond in ethyne but will not react with ethane under ordinary conditions. If you bubble the gases through bromine water, $Br_2(aq)$, its red-brown color would be decolorized by ethyne but not by ethane:

$$H—C{\equiv}C—H \xrightarrow{Br_2} HBrC{=}CHBr \xrightarrow{Br_2} HBr_2C—CHBr_2$$

(b) Propane burns in air; carbon dioxide does not—the carbon is in its highest oxidation state:

$$C_3H_8(g) + 5O_2(g) \rightarrow 3CO_2(g) + 4H_2O(g)$$

Alternatively, when the gases are bubbled through an aqueous solution of $Ca(OH)_2$ (limewater), only $CO_2(g)$ reacts to give a white precipitate of calcium carbonate, $CaCO_3(s)$:

$$Ca(OH)_2(aq) + CO_2(g) \rightarrow CaCO_3(s) + H_2O(l)$$

41. (a) $C_2H_5OH + H_2SO_4 \rightarrow H_2C{=}CH_2 + H_3O^+ + HSO_4^-$ **(b)** By heating 2-methyl-2-propanol or 2-methyl-1-propanol with concentrated sulfuric acid.

43.

(a) $H_3C{-}CH{=}CH{-}H + H_2O \rightarrow H_3C{-}CH(OH){-}CH_3$

Propene Water 2-Propanol

(b) $H_3C{-}C{\equiv}C{-}CH_3 + 2Br_2 \rightarrow H_3C{-}CBr_2{-}CBr_2{-}CH_3$

2-Butyne Bromine 2,2,3,3-Tetrabromobutane

(c) $H{-}C{\equiv}C{-}CH_2{-}CH_3 + H_2 \rightarrow H_2C{=}CH{-}CH_2{-}CH_3$

1-Butyne Hydrogen 1-Butene

(d) $H_3C{-}C{\equiv}C{-}H + HBr \rightarrow H_3C{-}CBr{=}CH_2$

Propyne Hydrogen bromide 2-Bromopropene

45. (a) $:\ddot{O}{=}C{=}\ddot{O}:$ carbon dioxide **(b)** $^-:C{\equiv}O:^+$, carbon monoxide **(c)** $^-:C{\equiv}N:$, cyanide ion **(d)** $^-:C{\equiv}C:^-$, carbide ion **(e)** $H{-}C{\equiv}N:$, hydrogen cyanide

47. The Lewis structure $^-:\ddot{O}{-}C({=}\ddot{O}:){-}\ddot{O}:^-$ implies that there there is one short double bond and two longer single bonds, but the observed structure has three carbon–oxygen bonds of equal length (intermediate between the lengths of a C—O and a C=O bond). The structure is better described in terms of three equivalent resonance structures with three equivalent bonds, each of bond order $1\frac{1}{3}$.

49. (a) $(H{-}\ddot{O})_2S({=}\ddot{O}:)_2$

S=O bond order: 2
S—OH bond order: 1
AX_4 tetrahedral

(b) $:\ddot{O}{=}\ddot{S}{=}\ddot{O}:$
S=O bond order: 2
AX_2E angular

(c) $^-:\ddot{O}{-}S({=}\ddot{O}:)({-}\ddot{O}:^-){=}\ddot{O}: \leftrightarrow {}^-:\ddot{O}{-}S({=}\ddot{O}:)({=}\ddot{O}:){-}\ddot{O}:^- \leftrightarrow :\ddot{O}{=}S({=}\ddot{O}:)({-}\ddot{O}:^-){-}\ddot{O}:^-$

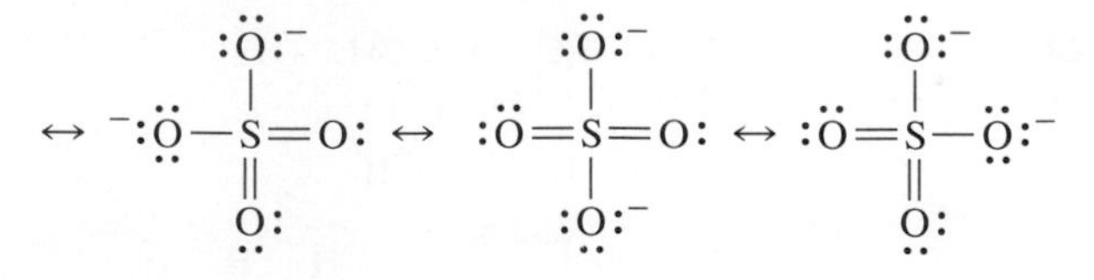

SO bond order: $1\frac{1}{2}$
AX_4 tetrahedral

(d) $:\ddot{O}{=}\ddot{S}(\ddot{O}:^-){-}\ddot{O}{-}H \leftrightarrow {}^-:\ddot{O}{-}\ddot{S}({=}\ddot{O}:){-}\ddot{O}{-}H$

S—OH bond order: 1
SO bond order: $1\frac{1}{2}$
AX_3E triangular pyramidal

(e) $:\ddot{O}{=}S({=}\ddot{O}:){=}\ddot{O}:$

S=O bond order: 2
AX_3 triangular planar

51.*

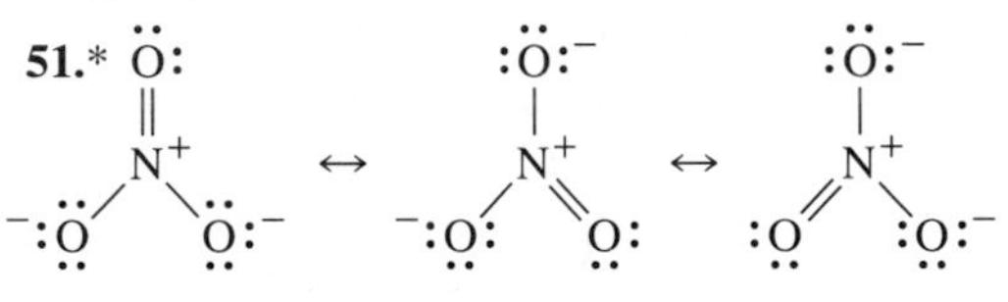

NO bond order: $1\frac{1}{3}$

52. (a) C_2H_5, C_4H_{10} **(b)** No, there are two isomers—butane and 2-methylpropane. **54.*** C_4H_8; for possible isomers, see answer to Problem 24. **56.*** ethylene; 36.4% $CH_4(g)$ and 63.6% $C_2H_4(g)$ by mass

57. (a) CN, C_2N_2 **(b)** $:N{\equiv}C{-}C{\equiv}N:$, linear **(c)** $:\ddot{O}{=}C{=}C{=}C{=}\ddot{O}:$, linear

CHAPTER 9

1. $-5450\ kJ \cdot mol^{-1}$ **3.** $0.239\ mol \cdot L^{-1}$ **5.** $+4.89$ K **6.** -512 kJ **8.** $+226$ kJ **10.** 70 kJ **12.** $H_2O(l)$: $-285.8\ kJ \cdot mol^{-1}$; $H_2O(g)$: $-241.8\ kJ \cdot mol^{-1}$; $NH_3(g)$: $-46.2\ kJ \cdot mol^{-1}$ **13.** $-224\ kj \cdot mol^{-1}$ **15.** $+227\ kJ \cdot mol^{-1}$ **17.** $-103.7\ kJ \cdot mol^{-1}$ **19.** $+26.4\ kJ \cdot mol^{-1}$ **21.** $+1.9\ kJ \cdot mol^{-1}$; graphite is more stable than diamond. **23.** $254\ kJ \cdot mol^{-1}$ **25. (a)** $463\ kJ \cdot mol^{-1}$ **(b)** $141\ kJ \cdot mol^{-1}$ **(c)** a weak bond **27.** $-84.6\ kJ \cdot mol^{-1}$ **29. (a)** -78 kJ **(b)** -218 kJ **(c)** $+119$ kJ **31.** $-92\ kJ \cdot mol^{-1}$ **32. (a)** natural gas **(b)** no **34. (a)** $2803\ kJ \cdot mol^{-1}$ **(b)** $15.55\ kJ \cdot g^{-1}$ **37.** (a) water **(b)** cards randomly shuffled **(c)** jigsaw pieces **(d)** aqueous solution **39. (a)** $\Delta S > 0$ **(b)** $\Delta S < 0$ **(c)** $\Delta S < 0$ **41. (a)** $\Delta S < 0$ **(b)** $\Delta S < 0$ **(c)** $\Delta S > 0$ **(d)** $\Delta S < 0$

42. (a) $\Delta S > 0$ **(b)** $\Delta S < 0$ **(c)** $\Delta S < 0$ **(d)** $\Delta S < 0$ **44. (a)** $2.9\ J \cdot K^{-1} \cdot mol^{-1}$ **(b)** $-138.3\ J \cdot K^{-1} \cdot mol^{-1}$ **(c)** $289.8\ J \cdot K^{-1} \cdot mol^{-1}$ **(d)** $-166.3\ J \cdot K^{-1} \cdot mol^{-1}$

46. (a) $-232.5\ J \cdot K^{-1} \cdot mol^{-1}$ **(b)** $178.6\ J \cdot K^{-1} \cdot mol^{-1}$ **(c)** $114.4\ J \cdot K^{-1} \cdot mol^{-1}$

48. (a) -1302.6 kJ, spontaneous **(b)** -457.2 kJ, spontaneous **(c)** -142.0 kJ, spontaneous **49. (a)** most often positive **(b)** always negative **(c)** most often negative **(d)** always positive **51. (a)** $-397.3\ kJ \cdot mol^{-1}$ **(b)** The reaction is spontaneous **(c)** No, the reaction must be negligibly slow under standard conditions.

54. (a) oxidizing agent: SO_2; reducing agent: H_2S **(b)** 145.6 kJ **55. (a)** 324.0 kJ **(b)** -70.4 kJ Reaction (a) is not spontaneous, but the reverse reaction is, so F_2 is a stronger oxidizing agent than Cl_2. Reaction (b) is spontaneous as written, so Cl_2 is a stronger oxidizing agent than Br_2.

56.* (a) ΔH_f° ($MgCO_3$, *s*) is ΔH° for the reaction

$$Mg(s) + C(s, \text{graphite}) + 1\tfrac{1}{2}O_2(g) \rightarrow MgCO_3(s)$$

under standard conditions. **(b)** -4270 J, $-511\ kJ \cdot mol^{-1}$ **(c)** $-1100\ kJ \cdot mol^{-1}$ **58.* (a)** $C_2H_6(g)$: $-84.6\ kJ \cdot mol^{-1}$; $C_3H_8(g)$: $-103.8\ kJ \cdot mol^{-1}$ **(b)** ΔH_f° $-123\ kJ \cdot mol^{-1}$; ΔH° (combustion) = $-2880\ kJ \cdot mol^{-1}$ **60.*** -128.0 kJ

CHAPTER 10

2. Metal production depends on the ease of reduction of metal cations. The activity series of common metals gives the order $Au^+ > Ag^+ > Cu^{2+} > Fe^{2+} > Al^{3+} > Ca^{2+} > K^+ > Na^+$ for the ease of reduction of ions of the metals in question. The earliest-known metals were formed in the earth by naturally occurring reducing agents, such as $S_8(s)$ and S^{2-} or were obtained by humans using readily available reducing agents, such as charcoal. Because these reducing agents are not strong enough to reduce alkali metal, alkaline earth, or Al^{3+} cations, these metals were not produced until much stronger reducing agents were available. The use of the strongest reducing agent—electrons—was possible only after electrical batteries were invented. **4.** Metals may be regarded as composed of metal cations, M^{n+}, and *n* electrons per atom forming a charge cloud that is delocalized over all the atoms of a metal crystal. The metallic bond is the electrostatic attraction between this charge cloud and the metal ions. **6. (a)** According to the electron-gas model, electrons, which have a very low mass, move at high speeds between the positively charged metal ions, both in the solid and liquid phases. Heat is conducted very rapidly through a metal by these fast moving electrons. When a potential difference (voltage) is applied to a metal, electrons enter the metal at one terminal to replace electrons as they are removed from the metal at the other terminal, setting up an electric current through the metal. **(b)** In metallic electrical conduction, the free movement of electrons through a metal is impeded only by collisions between the electrons and vibrating metal ions. As temperature decreases, the amplitude of vibration of the ions decreases, so there are fewer collisions of electrons with metal ions and therefore an increase in the electrical conductivity. **7. (a)** The melting points of metals are related to the strength of their metallic bonding, which depends on the number of electrons per metal atom available for bonding. Na, Mg, and Al atoms have, respectively, 1, 2, and 3 easily ionized valence electrons. Thus, the strength of the metallic bonding increases in the order Na < Mg < Al, which is the order of increasing melting point. **(b)** Iron has the valence-shell electron configuration $3d^6 4s^2$ and has available for metallic bonding more valence electrons per iron atom than aluminum has. **11. (a)** The reactivities of metals depend on the ease of cation formation, which is measured by their ionization energies. Group I metals, with 1 valence electron, easily lose this electron to form M^+ ions. Reactivity decreases down the group because the ionization energies decrease in this order, with increasing distance of the valence shell from the nucleus. Similar reasoning applies to Group II metals, which have two valence electrons and form M^{2+} ions. **(b)** It takes more energy to form a M^{2+} ion than to form an M^+ ion, so Group II metals are less reactive than Group I metals, as exemplified by the ease of their reactions with water. Group I metals react readily with water according to the equation $2M(s) + 2H_2O(l) \rightarrow 2M^+OH^-(aq) + H_2(g)$. Li and Na react slowly at room temperature, K reacts with enough energy to ignite the H_2 as it is formed, and Rb and Cs react explosively. In Group II, Mg reacts only with steam, Ca reacts slowly with water, and Sr and Ba react vigorously but not as strongly as any of the Group I metals, by the equation $M(s) + 2H_2O(l) \rightarrow M^{2+}(OH^-)_2(s) + H_2(g)$. **13. (a)** Na > Mg > Al > Cu **(b)** In terms of the ease with which they react with oxidizing agents of increasing strength to form cations, only Na is oxidized by water at room temperature; Na and Mg are oxidized by steam, $H_2O(g)$; Na, Mg, and Al are oxidized by the $H_3O^+(aq)$ of dilute acids; and all four are oxidized by concentrated sulfuric acid (Cu only upon heating).

16. (a) $2K(s) + H_2(g) \rightarrow 2KH(s)$; $Ca(s) + H_2(g) \rightarrow CaH_2(s)$ **(b)** $3Mg(s) + N_2(g) \xrightarrow{\text{heat}} Mg_3N_2$

18. (a) Any metal above H_2 in the activity series for metals (Na, K, Ca, Mg, Zn, Al, and Fe) reacts with $H_3O^+(aq)$ in dilute $HCl(aq)$; for example,

$$Zn(s) + 2H_3O^+(aq) \rightarrow Zn^{2+}(aq) + 2H_2O(l) + H_2(g)$$
$$Fe(s) + 2H_3O^+(aq) \rightarrow Fe^{2+}(aq) + 2H_2O(l) + H_2(g)$$

(b) Metals below H_2 in the activity series (Cu, Ag, and Au) do not react with $H_3O^+(aq)$, but dilute nitric acid, $HNO_3(aq)$, contains $NO_3^-(aq)$ as well as $H_3O^+(aq)$. Any metal for which $NO_3^-(aq)$ is a sufficiently strong oxidizing agent, including Cu and Ag, will reduce $NO_3^-(aq)$.

20. (a) 2.13 L **(b)** 3.81 g

22. (a) $2[Na_5(CO_3)_2(HCO_3) \cdot 2H_2O(s)] \xrightarrow{heat} 5Na_2CO_3(s) + CO_2(g) + 5H_2O(g)$ **(b)** 2.16 metric tons

26. (a) $K_2O(s) + H_2O(l) \rightarrow 2K^+(aq) + 2OH^-(aq)$; basic oxide **(b)** $SrO(s) + H_2O(l) \rightarrow Sr^{2+}(aq) + 2OH^-(aq)$; basic oxide **(c)** $SO_2(g) + H_2O(l) \rightarrow H_2SO_3(aq)$; acidic oxide
(d) $SO_3(g) + H_2O(l) \rightarrow H_2SO_4(aq)$; acidic oxide
(e) $CO_2(g) + H_2O(l) \rightarrow H_2CO_3(aq)$; acidic oxide
(f) $P_4O_6(s) + 6H_2O(l) \rightarrow 4H_3PO_3(aq)$; acidic oxide
Note: In water the acids formed by acidic oxides will be dissociated to give $H_3O^+(aq)$ and the appropriate anions.

27. (a) $Na_2O(s) + SO_2(g) \rightarrow Na_2SO_3(s)$
(b) $CaO(s) + SO_3(g) \rightarrow CaSO_4(s)$
(c) $6Na_2O(s) + P_4O_{10}(s) \rightarrow 4Na_3PO_4(s)$
(d) $Al_2O_3(s) + 3SO_3(g) \rightarrow Al_2(SO_4)_3(s)$

29. (a) $CaCO_3(s) \xrightarrow{heat} CaO(s) + CO_2(g)$
Calcium oxide, Carbon dioxide
(b) $Ca(OH)_2(s) \xrightarrow{heat} CaO(s) + H_2O(g)$
Calcium oxide, Water
(c) $Ca(HCO_3)_2(s) \xrightarrow{heat} CaCO_3(s) + CO_2(g) + H_2O(g)$
Calcium carbonate, Carbon dioxide, Water
(d) $2Al(OH)_3 \xrightarrow{heat} Al_2O_3(s) + 3H_2O(g)$
Aluminum oxide, Water

31.* Although $AlCl_3$ can be regarded as consisting of Al^{3+} and Cl^- ions in the solid, in the gaseous state both $AlCl_3$ and $AlBr_3$ form covalent molecules with polar bonds, with only three pairs of electrons in the valence shell of Al. They form dimers in which a lone electron pair of a halogen atom of one AX_3 monomer is donated to the valence shell of the Al atom of the other to complete its octet—and vice versa. Each $AlCl_3$ molecule behaves as both a Lewis acid and a Lewis base.

33.* (a) $2Al(s) + 2NaOH(aq) + 6H_2O(l) \rightarrow 2NaAl(OH)_4(aq) + 3H_2(g)$
(b) $Al_2O_3(s) + 3C(s) + 3Cl_2(g) \rightarrow 2AlCl_3(s) + 3CO(g)$
(c) $2NH_4Al(SO_4)_2 \cdot 12H_2O(s) \xrightarrow{heat} 2NH_3(g) + 4H_2SO_4(l) + Al_2O_3(s) + 9H_2O(g)$

34. (a) $Ca(s) + 2H_2O(l) \rightarrow Ca(OH)_2(s) + H_2(g)$
(b) $2Fe(s) + 3H_2O(g) \rightarrow Fe_2O_3(s) + 3H_2(g)$
(c) $2Al(s) + 3H_2SO_4(aq) \rightarrow 2Al^{3+}(aq) + 3SO_4^{2-}(aq) + 3H_2(g)$
(d) $Cu(s) + 2H_2SO_4(conc) \rightarrow Cu^{2+}(aq) + SO_4^{2-}(aq) + SO_2(g) + 2H_2O(l)$

36. (a) Al: +3; O: −2 **(b)** Al: + 3; Cl: −1
(c) Fe: +3; S: +6; O: −2
(d) Cu: +2; S: +6; O: −2; H: +1
(e) Cu: + 2; N: −3; H: + 1; S: +6; O: −2

38. (a) Lewis base

$$Cu^{2+}(aq) + 4NH_3(aq) \rightarrow Cu(NH_3)_4^{2+}(aq)$$

(b) Lewis acid

$$Cu^{2+}(aq) + 4H_2O(l) \rightarrow Cu(H_2O)_4^{2+}(aq)$$
$$Cu^{2+}(aq) + 4NH_3(aq) \rightarrow Cu(NH_3)_4^{2+}(aq)$$
$$Cu^{2+}(aq) + 4Cl^-(aq) \rightarrow CuCl_4^{2-}(aq)$$

(c) Lewis acid

$$Fe^{3+}(aq) + 6H_2O(l) \rightarrow Fe(H_2O)_6^{3+}(aq)$$
$$Fe^{3+}(aq) + 6CN^-(aq) \rightarrow Fe(CN)_6^{3-}(aq)$$

(d) Lewis base; see the final equation of part (b).
(e) Lewis acid

$$AlCl_3(s) + NaCl(s) \rightarrow Na^+AlCl_4^-(s)$$

40. (a) $Al(OH)_3(s)$ and many transition-metal hydroxides, such as $Fe(OH)_3$ and $Cu(OH)_2$
(b) $2Al(OH)_3(s) \rightarrow Al_2O_3(s) + 3H_2O(g)$
$2Fe(OH)_3(s) \rightarrow Fe_2O_3(s) + 3H_2O(g)$
$Cu(OH)_2(s) \rightarrow CuO(s) + H_2O(g)$

42. $CuSO_4 \cdot (NH_4)_2SO_4 \cdot 6H_2O$ or $(NH_4)_2Cu(SO_4)_2 \cdot 6H_2O$

43. (a) Addition of $NaOH(aq)$ to $CuSO_4(aq)$ gives a pale blue precipitate of $Cu(OH)_2(s)$. When heated strongly, $Cu(OH)_2(s)$ gives black copper(II) oxide, $CuO(s)$, which may be reduced to copper with hydrogen, carbon, or carbon monoxide.
(b) $Cu(OH)_2(s)$ or $CuO(s)$ prepared in part (a), when dissolved in hydrochloric acid, $HCl(aq)$, gives a solution from which $CuCl_2 \cdot 2H_2O(s)$ could be crystallized after concentration by evaporation. Or copper from part (a), when heated with chlorine gas, gives anhydrous $CuCl_2(s)$. **(c)** By heating $CuO(s)$ with $Cu(s)$ (both obtained in part (a))
(d) $Cu(NH_3)_4^{2+}$ ion is formed when the appropriate amount of concentrated ammonia solution is added to a solution of $CuSO_4(aq)$. $Cu(NH_3)_4SO_4 \cdot H_2O(s)$ crystallizes when the solution is concentrated by evaporation.

45. (a) Sc: + III; Ti: +IV; V: +V; Cr: + VI; Mn: +VII
(b) (i) all: +II (ii) Ni: +III; Co: +III; Cu: +III; Zn: +II

46.* (a) 1. $2Fe(s) + 3Cl_2(g) \rightarrow 2FeCl_3(s)$
2. $FeCl_3(aq) + 3OH^-(aq) \rightarrow Fe(OH)_3(s) + 3Cl^-(aq)$
3. $2Fe(OH)_3(s) \rightarrow Fe_2O_3(s) + 3H_2O(g)$
(b) $FeCl_3(s)$: iron(III) chloride; $Fe(OH)_3(s)$: iron(III) hydroxide; $Fe_2O_3(s)$: iron(III) oxide **(c)** 2.14 g $Fe_2O_3(s)$

47.* 68.0% Cu, 29.0% Zn, 3.0% Al

50. (a) 4.8×10^6 tons **(b)** 210 kg **(c)** 135 kg

52. -33 kJ mol^{-1} **54.** $x = 3$ **56.** $\Delta G = \Delta H - T\Delta S$ must be negative for the reaction to be spontaneous. ΔS is expected to be positive, because a gas and solid are produced from a solid. The reaction is endothermic, so ΔH is positive. At room temperature $T\Delta S < \Delta H$, $\Delta H - T\Delta S > 0$, and $\Delta G > 0$—so the reaction is not spontaneous. However, at a temperature of 1100°C, $T\Delta S > \Delta H$, so $\Delta H - T\Delta S < 0$, $\Delta G < 0$, and the decomposition of $CaCO_3(s)$ becomes spontaneous.

58. (a) $Cr_2O_7^{2-}(aq) + 6e^- + 14H^+(aq) \rightarrow 2Cr^{3+}(aq) + 7H_2O(l)$ **(b)** $3SO_2(g) + Cr_2O_7^{2-}(aq) + 2H^+(aq) \rightarrow 3SO_4^{2-}(aq) + 2Cr^{3+}(aq) + H_2O(l)$ **60.* (a)** 54.6% **(b)** $x = 7$ **61.*** −2675 kJ

CHAPTER 11

1. Molecular solids: (b), (c), and (d); network solids: (a) (both diamond and graphite), (e), (f), and (g)

3. (a) Silicon is a three-dimensional covalent solid, whereas aluminum is a three-dimensional metallic solid. **(b)** Oxygen consists of covalent O_2 molecules and is a gas, whereas sulfur consists of more polarizable covalent S_8 molecules and is a solid. **(c)** Diamond is a three-dimensional network solid with all C—C bonds, whereas graphite is a two-dimensional network solid with layers of C atoms held together only by weak London forces. **4.** Network solids: (b), (d), and (e); molecular solids: (a) and (c) **6. (a)** Figure 11.8, but with Ba atoms per unit cell **(b)** Figure 11.9, but with Ne atoms per unit cell. **7. (a)** 4; **(b)** face-centered cubic

9. (a) 4 **(b)** face-centered cubic **(c)** 159 pm

11. 6.08×10^{23} **13. (a)** 578 pm **(b)** 250 pm **(c)** 69 pm **15.** NaCl, because the mass of four NaCl units (empirical formulas) is greater than the mass of one CsCl unit (empirical formula). **16.** $4.51\ g \cdot cm^{-3}$

19. < 1000°C: $7.93\ g \cdot cm^{-3}$; > 1000°C: $7.76\ g \cdot cm^{-3}$; although the low-temperature form has two Fe atoms per unit cell and the high-temperature form has four, the increase in volume of the unit cell dominates. **21.** 5.86×10^{19} atoms

22. (a) $F^- > Na^+$ **(b)** $K^+ > Na^+$ **(c)** $Cl^- > F^-$ **(d)** $Na^+ > Mg^{2+}$ **(e)** $S^{2-} > Cl^-$

25.* $7.76\ g \cdot cm^{-3}$ **28. (a)** The proportion of molecules with sufficient energy to escape from a liquid increases with temperature **(b)** toluene **(c)** highest: diethyl ether; lowest: methanol **30. (a)** no effect **(b)** no effect **(c)** an increase **(d)** The stronger the intermolecular forces in the liquid, the lower the vapor pressure. **(e)** no effect

33. (a) ionic bonds **(b)** covalent bonds **(c)** London forces **(d)** London forces **(e)** metallic bonds **(f)** covalent bonds

34. In both ionic bonds have to be broken, but those in MgO are stronger than those in NaCl, because Mg^{2+} has twice the charge and is a smaller ion than Na^+, and O^{2-} has twice the charge and is a smaller ion than Cl^-. **36. (a)** OCl_2 **(b)** PF_3 **(c)** ClF_5 **38. (a)** Covalent BrF, because it is more polarizable than covalent ClF; **(b)** covalent BrCl, because it is more polarizable than covalent Cl_2 and has a dipole moment; **(c)** KBr, because it is ionic, whereas BrCl is polar covalent; **(d)** Na, because it is a metal, whereas Br_2 is a covalent liquid. **40. (a)** London forces **(b)** ionic bonds **(c)** London forces **(d)** London forces (strongest) and dipole–dipole forces **(e)** hydrogen bonds (strongest), dipole–dipole forces, and London forces

42. (a) London forces **(b)** hydrogen bonds **(c)** London forces **(d)** ionic bonds **(e)** London forces **(f)** London forces and dipole–dipole forces

44. Tetrachloromethane is a covalent liquid, whereas Trichloromethane is a polar covalent liquid, but CCl_4 is more polarizable than $CHCl_3$. **45.*** 23.7 mm Hg **47.** (a), (c), (g), and (h): all satisfy the criteria for hydrogen bond formation.

49. $Br_2(s)$ denser than $Br_2(l)$ at its melting point because its molecules are more closely packed (as is typical of most substances). Ice is less dense than liquid water at 0°C because its hydrogen bonding creates a cagelike structure that partly collapses on melting, so the H_2O molecules in ice are less closely packed than are those in liquid water.

51. (a) Polar HCl is soluble in polar water but not very soluble in nonpolar pentane. **(b)** Polar water forms hydrogen bonds with liquid HF and dissolves but is insoluble in gasoline (a mixture of nonpolar hydrocarbons. **(c)** Chloroform is insoluble in hydrogen-bonded water but dissolves in covalent carbon tetrachloride. **(d)** Naphthalene is insoluble in hydrogen-bonded water but soluble in covalent benzene. **(e)** Nonpolar N_2 is not very soluble in water, but polar HCN forms hydrogen bonds with water and is very soluble. **(f)** Covalent, nonpolar benzene is soluble in covalent, nonpolar toluene but insoluble in polar, hydrogen-bonded water.

53. The structures are different because more Cl^- ions can pack around a Cs^+ ion (eight) than around the smaller Na^+ ion (six). **55. (a)** Among the hydrides of Group VI, only water can form strong hydrogen bonds. **(b)** Of $(HO)_2SO_2$ and F_2SO_2, only sulfuric acid can form hydrogen bonds. **(c)** When ice melts, hydrogen bonds are broken, the H_2O molecules then pack more closely than in ice, and the density decreases. Concurrently, thermal vibrations increase with temperature, and the expansion of water, which decreases the density, overtakes the collapse of the hydrogen-bonded structure above 3.98°C. **(d)** Of methanol and hexane, only methanol forms hydrogen bonds with water. **(e)** Ionic bonds between Na^+ ions and O^{2-} ions in Na_2O are weaker than ionic bonds between the smaller Mg^{2+} ions and O^{2-} ions, so Na^+ and O^{2-} ions are more readily separated than Mg^{2+} and O^{2-} ions, and only $Na_2O(s)$ dissolves and reacts to give $NaOH(aq)$. **57. (a)** C_2H_5OH is a hydrogen-bonded liquid, whereas $(CH_3)_2CO$ is not. **(b)** HF forms a hydrogen-bonded liquid, whereas HCl does not. **(c)** LiCl is an ionic substance, whereas CCl_4 is a covalent liquid.

59. (a) Ideal behavior assumes zero molecular volume and zero intermolecular forces, which is not true for real gases. Both effects become significant when the molecules are relatively closely packed (at high pressures) and when they are moving relatively slowly (at low temperatures).
(b) (i) Bromine, because it is a larger molecule and has a greater polarizability than fluorine. (ii) Carbon monoxide, because although CO and N_2 are comparable in size and polarizability, only CO has polar molecules. (iii) Hydrogen fluoride, because although HF and HCl are both polar covalent molecules, HF has the greater polarity and alone can form hydrogen bonds. **60.* (a)** a covalent network solid
(b) No, because the strength of intermolecular forces with water is insufficient to break strong B—B bonds in $B(s)$ or hydrogen bonds in water.

CHAPTER 12

1. (a) 10^{-5}-*M* $H_3O^+(aq)$, 10^{-5}-*M* $NO_3^-(aq)$ **(b)** 0.0023-*M* $H_3O^+(aq)$, 0.0023-*M* $Cl^-(aq)$ **(c)** 0.113-*M* $H_3O^+(aq)$, 0.113-*M* $ClO_4^-(aq)$ **(d)** 0.034-*M* $H_3O^+(aq)$, 0.034-*M* $Br^-(aq)$ **(e)** 10^{-3}-*M* $Na^+(aq)$, 10^{-3}-*M* $OH^-(aq)$ **(f)** 0.145-*M* $Ba^{2+}(aq)$, 0.290-*M* $OH^-(aq)$ **3. (a)** 5.00 **(b)** 2.64 **(c)** 0.95 **(d)** 1.47 **(e)** 11.00 **(f)** 13.46 **5. (a)** 2.2×10^{-6} mol · L^{-1} **(b)** 0.022% **(c)** 5.66 **7.** 0.200-*M* acetic acid **9. (a)** 5.9×10^{-6} mol · L^{-1} **(b)** 0.007% **(c)** 8.77 **11.** 3.23 **13.** 2.07

15. (a) NH_4^+, ammonium ion **(b)** HF, hydrogen fluoride **(c)** H_2O, water **(d)** H_3O^+, hydronium ion **(e)** H_2, hydrogen **16. (a)** F^-, fluoride ion **(b)** NO_3^-, nitrate ion **(c)** ClO_4^-, perchlorate ion **(d)** OH^-, hydroxide ion **(e)** H_2O, water

19. (a) hydrogencarbonate ion, HCO_3^-, 2.9×10^{-8} mol · L^{-1} **(b)** hydrogensulfide ion, HS^-, 1.1×10^{-7} mol · L^{-1} **(c)** nitrite ion, NO_2^-, 2.2×10^{-11} mol · L^{-1} **(d)** dihydrogenphosphate ion, $H_2PO_4^-$, 1.3×10^{-12} mol · L^{-1} **20.** 8.08 **22.** $[NH_4^+] = 0.020$ *M*, $[NH_3] = [H_3O^+] = 3.4 \times 10^{-6}$ *M*, $[OH^-] = 2.9 \times 10^{-9}$ *M*

24. (a) $NH_3(aq) + HNO_3(aq) \rightarrow NH_4NO_3(aq)$; acidic due to the reaction $NH_4^+(aq) + H_2O(l) \rightleftharpoons NH_3(aq) + H_3O^+(aq)$. **(b)** $NH_3(aq) + HCl(aq) \rightarrow NH_4Cl(aq)$; acidic due to the reaction $NH_4^+(aq) + H_2O(l) \rightleftharpoons NH_3(aq) + H_3O^+(aq)$. **(c)** $Ca(OH)_2(s) + H_2SO_4(aq) \rightarrow CaSO_4(aq) + 2H_2O(l)$; very weakly basic due to the reaction $SO_4^{2-}(aq) + H_2O(l) \rightleftharpoons HSO_4^-(aq) + OH^-(aq)$ **(d)** $CH_3CO_2H(aq) + KOH(aq) \rightarrow CH_3CO_2K(aq) + H_2O(l)$; basic due to the reaction $CH_3CO_2^-(aq) + H_2O(l) \rightleftharpoons CH_3CO_2H(aq) + OH^-(aq)$ **(e)** $Al(OH)_3(s) + 3HCl(aq) \rightarrow AlCl_3(aq) + 3H_2O(l)$; acidic due to the reaction $Al(H_2O)_6^{3+}(aq) + H_2O(l) \rightleftharpoons Al(H_2O)_5(OH)^{2+}(aq) + H_3O^+(aq)$ **(f)** $NaOH(aq) + HI(aq) \rightarrow NaI(aq) + H_2O(l)$; neutral **25. (a)** 4.59 L 0.200-*M* NaOH(*aq*) **(b)** basic **27. (a)** 3.07 **(b)** 1.53 **28. (a)** 9.33 **(b)** 9.31 **30. (a)** 4.74 **(b)** 9.03 **32.*** From pH 9.51 to pH 9.58.

34.* (a) 4.86 **(b)** 4.77 **(c)** 5.01 **35. (a)** Not to change its pH significance. **(b)** 4.9, 4.2, and 3.7 **36. (a)** $6.0 < pH < 6.2$ **(b)** orange **38.** 4.0×10^{-10} mol · L^{-1} **40. (a)** orange **(b)** yellow **(c)** yellow

42. (a) $K_c = \dfrac{[SO_3]^2}{[SO_2]^2[O_2]}$, $K_p = \dfrac{p_{SO_4}{}^2}{p_{SO_2}{}^2 \cdot p_{O_2}}$

(b) $K_c = \dfrac{[SO_3]}{[SO_2]\,[O_2]^{1/2}}$, $K_p = \dfrac{p_{SO_3}}{p_{SO_2} \cdot p_O{}^{1/2}}$

(c) $K_c = \dfrac{[P_4O_{10}]}{[P_4]\,[O_2]^5}$, $K_p = \dfrac{p_{P_4O_{10}}}{p_{P_4} \cdot p_{O_2}{}^5}$

(d) $K_c = \dfrac{[PCl_3]\,[Cl_2]}{[PCl_5]}$, $K_p = \dfrac{p_{PCl_3} \cdot p_{Cl_2}}{p_{PCl_5}}$

44. (a) $Q < K_{eq}$, not at equilibrium **(b)** > 0.10 mol · L^{-1} **45. (a)** 0.772 **(b)** 0.772 **47. (a)** $[I_2] = 0.498$ mol · L^{-1}, $[I] = 4.34 \times 10^{-3}$ mol · L^{-1} **(b)** 0.434% **49. (a)** $Q > K_c$, not at equilibrium **(b)** $[H_2] = [I_2] = 0.13$ mol · L^{-1}, $[HI] = 0.94$ mol · L^{-1} **(c)** The same as in part (b). **51.** 4.80% **53.** exothermic **55.** $[C_2H_6]$: **(a)** decreases, **(b)** decreases, **(c)** increases, **(d)** remains unchanged.

56. (a) $K_c = \dfrac{[H_2]\,[O_2]^{1/2}}{[H_2O]}$, $K_p = \dfrac{p_{H_2} \cdot p_{O_2}{}^{1/2}}{p_{H_2O}}$

(b) $K_c = [H_2]\,[O_2]^{1/2}$, $K_p = p_{H_2} \cdot p_{O_2}{}^{1/2}$

(c) $K_c = [H_2]^2\,[O_2]$, $K_p = p_{H_2}^2 \cdot p_{O_2}$

(d) $K_c = \dfrac{1}{[H_2]\,[O_2]}$, $K_p = \dfrac{1}{p_{H_2} \cdot p_{O_2}}$

(e) $K_c = [O_2]$, $K_p = p_{O_2}$

58. (a) to the left **(b)** to the left **(c)** to the right **(d)** no change **60. (a)** −142.0 kJ **(b)** 7.7×10^{24} atm^{-1} **(c)** yes **62. (a)** spontaneous **(b)** 10^{140} atm **(c)** The reaction is negligibly slow. **64.* (a)** No, because $\Delta G° > 0$ **(b)** yes, above 3890 K **(c)** $\Delta G =$ 157.7 kJ; $K_p = 1.4 \times 10^{-7}$ **66. (a)** 7.47 (0°C), 7.27 (10°C), 6.92 (30°C), 6.63 (50°C), 6.15 (100°C) **(b)** exothermic **68. (a)** No—a solution of strong acid and a salt of the strong acid is not a buffer solution. **(b)** Yes—a solution of a weak acid and a salt of a weak acid is a buffer solution. **(c)** No—this is a solution of potassium acetate with excess KOH(*aq*). **(d)** Yes—this is a solution of a weak acid and a salt of a weak acid. **70.* (a)** bromothymol blue **(b)** thymol blue **(c)** methyl red

CHAPTER 13

1. (a) $E°_{ox}$ values for Mg, Al, Zn, Fe, and Ni are positive relative to $E°_{red} = 0$ for H_3O^+, so all are spontaneously oxidized by $H_3O^+(aq)$ from $H_2SO_4(aq)$. **(b)** $E°_{ox}$ values for Cu and Ag are negative relative to $E°_{red}$ for H_3O^+, so Cu and Ag are not oxidized by $H_3O^+(aq)$ from HCl(*aq*) but are oxidized by H_2SO_4 molecules in H_2SO_4(*conc*). **2.** $E°_{ox}$ for Cu is negative relative to $E°_{red}$ for H_3O^+, so Cu is not oxidized by $H_3O^+(aq)$ from HCl(*aq*) or $HNO_3(aq)$; it is positive relative to $E°_{red}$ for $NO_3^-(aq)$, so Cu is oxidized by $NO_3^-(aq)$ from $HNO_3(aq)$ and dissolves.

4. (a) $Ca^{2+} < Na^+ < Al^{3+} < Ni^{2+} < Cu^{2+} < Fe^{3+}$ **(b)** All of them. **6.** (c), (d), and (e)

7. (b) $Ag^+(aq) + Fe^{2+}(aq) \rightarrow Ag(s) + Fe^{3+}(aq)$, $E°_{cell} =$ +0.03 V **(c)** $6I^-(aq) + 2NO_3^-(aq) + 8H^+(aq) \rightarrow 3I_2(s) + 2NO(g) + 4H_2O(l)$, $E°_{cell} = +0.43$ V

9. (b) $O_2(g) + 4Br^-(aq) + 4H^+(aq) \rightarrow 2H_2O(l) + 2Br_2(aq)$, $E°_{cell} = +0.14$ V **(d)** $2I^-(aq) + Cl_2(g) \rightarrow I_2(s) + 2Cl^-(aq)$, $E°_{cell} = +0.82$ V **10.** All are oxidized. **13. (a)** Show two half-cells separated by a salt bridge or a porous barrier, with a $Zn(s)$ anode dipping into $Zn^{2+}(aq)$ in the anode compartment, a $Ni(s)$ cathode dipping into $Ni^{2+}(aq)$ in the cathode compartment, and the anode and cathode connected by a wire in which electrons flow from the Zn anode to the Ni cathode. **(b)** +1.01 V

15. (a) Show two half cells separated by a salt bridge or a porous barrier, with an $Fe(s)$ anode dipping into $Fe^{2+}(aq)$ in the anode compartment and a cathode, consisting of a $Pt(s)$ wire over which $H_2(g)$ bubbles, dipping into acid, $H^+(aq)$, in the cathode compartment. **(b)** electrons **(c)** the platinum cathode **(d)** $2H^+(aq) + 2e^- \rightarrow H_2(g)$

16. (a) $Zn(s) \mid Zn^{2+}(aq) \parallel Br^-(aq) \mid Br_2(aq), Pt(s)$; electrons flow in the external circuit from the Zn(s) anode to the Pt(s) cathode. **(b)** $Pb(s) \mid Pb^{2+}(aq) \parallel Ag^+(aq) \mid Ag(s)$; electrons flow in the external circuit from the Pb(s) anode to the Ag(s) cathode. **18. (a)** -1.16×10^3 kJ **(b)** -179 kJ **(c)** -108 kJ **20. (a)** Iron dissolves and the copper increases in mass. **(b)** +0.78 V **(c)** $[Fe^{2+}]$ increases and $[Cu^{2+}]$ decreases. **(d)** decrease **21. (a)** +0.46 V **(b)** 0.37 V **22.** 0.047 $mol \cdot L^{-1}$ **24. (a)** +0.47 V **(b)** +0.29 V **(c)** +0.50 V **(d)** +0.54 V

28. (a) anode: $Zn(s) \rightarrow Zn^{2+}(aq) + 2e^-$ cathode: $O_2(g) + 4H^+(aq) + 4e^- \rightarrow 2H_2O(l)$ **(b)** 1.99 V **(c)** every 12 years **29. (a)** oxidation: $C_2H_6(g) + 4H_2O(l) \rightarrow 2CO_2(g) + 14H^+(aq) + 14e^-$ reduction: $O_2(g) + 4H^+(aq) + 4e^- \rightarrow 2H_2O(l)$ **(b)** 0.18 L **(c)** 0.63 L

30. (a) oxidation–reduction **(b)** Aluminum, but subsequent reactions give the oxides $Al_2O_3(s)$, which forms a hard, protective layer, and $Fe_2O_3(s)$, which flakes easily and exposes more metal to further corrosion.

32. (a)

anode	$2[Mg(s) \rightarrow Mg^{2+}(aq) + 2e^-]$	*oxidation*
cathode	$O_2(g) + 4e^- + 4H^+(aq) \rightarrow 2H_2O(l)$	*reduction*
	$2Mg(s) + O_2(g) + 4H^+(aq) \rightarrow 2Mg^{2+}(aq) + 2H_2O(l)$	

(b)

anode	$2[Zn(s) \rightarrow Zn^{2+}(aq) + 2e^-]$	*oxidation*
cathode	$O_2(g) + 4e^- + 4H^+(aq) \rightarrow 2H_2O(l)$	*reduction*
	$2Zn(s) + O_2(g) + 4H^+(aq) \rightarrow 2Zn^{2+}(aq) + 2H_2O(l)$	

35. (a)

cathode	$2[Al^{3+} + 3e^- \rightarrow Al(l)]$
anode	$3[2Cl^- \rightarrow Cl_2(g) + 2e^-]$
overall	$2AlCl_3(l) \rightarrow 2Al(l) + 3Cl_2(g)$

(b)

cathode	$3[2H_2O(l) + 2e^- \rightarrow H_2(g) + 2OH^-]$
anode	$3[2Cl^- \rightarrow Cl_2(g) + 2e^-]$
	$2AlCl_3(l) + 6H_2O(l) \rightarrow 2Al(OH)_3(s) + 3H_2(g) + 3Cl_2(g)$

(c) In part (b), water is more easily reduced than $Al^{3+}(aq)$. **36. (a)** Show two Pt electrodes dipping into HBr(*aq*) and joined to a current source (battery), with (b) the electrode on the left labeled "anode" (+) and that on the right labeled "cathode" (−).

(c)

cathode	$2H^+ + 2e^- \rightarrow H_2(g)$	*reduction*
anode	$2Br^- \rightarrow Br_2(l) + 2e^-$	*oxidation*
overall	$2HBr(aq) \rightarrow H_2(g) + Br_2(l)$	

(d) In the external circuit, electrons flow from anode to cathode; in the solution $H^+(aq)$, ions move to the cathode, where they are reduced, and Br^-(aq) ions move to the anode, where they are oxidized. **38. (a)** 1.93×10^5 C **(b)** 9.65×10^4 C **(c)** 4.83×10^5 C **(d)** 5.79×10^5 C **40. (a)** 861 C **(b)** 5.37×10^8 C **(c)** 9.63×10^4 C **(d)** 4.47×10^3 C **42.** 0.896 g NaOH, 0.794 g Cl_2 **44.** 39.7 min **46.** 3.1 days **48. (a)** 106 $g \cdot mol^{-1}$ **(b)** palladium, Pd **50.** 1.758 g **52. (a)** 89.4 mA **(b)** 0.360 g **(c)** 62.5 mL of $O_2(g)$ and $H_2(g)$ **53. (a)** 3.10 days **(b)** Chromium compounds are quite rare and costly and a large amount of current is needed, so the process expensive. **55.** $K_{eq} = 6.0 \times 10^6$ $mol^{-2} \cdot L^{-2}$ **57.* (a)** The voltage decreases. **(b)** no change **(c)** The voltage increases. **(d)** The voltage decreases.

CHAPTER 14

1. (a) $CH_3CH_2CH_2OH$, 1-propanol
(b) CH_3CHCH_3 (OH on the central C), 2-propanol
(c) $CH_3CH_2C{=}O$ (H on C), propanal
(d) CH_3CCH_3 (C=O on the central C), propanone
(e) $CH_3CH_2{-}C{=}O$ (OH on C), propanoic acid
(f) $CH_3{-}O{-}CH_3$, methoxyethane
(g) $CH_3CH_2{-}C{=}O$ (NH_2 on C), propanamide
(h) $CH_3CH_2CH_2NH_2$, propylamine **3. (a)** carboxylic acid, $-CO_2H$; carboxylate, $-CO_2^-$; amino-, $-NH_2$ **(b)** methoxy-, $-OCH_3$; phenoxy-, $-OH$; aldehyde, $-CHO$

5. (a) ethanal (acetaldehyde) **(b)** 2-butanone **(c)** propanoic acid **(d)** 2-methylpropanamide **(e)** 2-amino-2-methylpropane (2-methyl-2-propanamine) **(f)** 3-pentanol **7. (a)** iodo-, I; hydroxyl (phenoxyl), $-OH$; ether, $-O-$; primary amino-, $-NH_2$; carboxylic acid, $-CO_2H$ **(b)** primary alcohol, $-CH_2-OH$; secondary alcohol, C—CH(OH); *in ring*: ether, $-O-$; keto-, C=O; alcohol, $-OH$; alkene, C=C **(c)** primary amino-, $-NH_2$; amido-, —C(O)N(H)—; ester, $-C(O)OCH_3$.

9. (a) $C_2H_5Br + NaI \rightarrow C_2H_5I + NaBr$
(b) $C_2H_5Br + NaOH \rightarrow C_2H_5OH + NaBr$
$C_2H_5OH + 2[O] \rightarrow CH_3C(O)OH + H_2O$
$C_2H_5OH + CH_3CO_2H \rightarrow C_2H_5O{-}(O)CH_3 + H_2O$
(c) $C_2H_5Br + NaOH \rightarrow C_2H_5OH + NaBr$
$C_2H_5OH + NaOH \rightarrow C_2H_5O^-Na^+ + H_2O$
$C_2H_5O^-Na^+ + C_2H_5Br \rightarrow (C_2H_5)_2O + NaBr$
(d) $C_2H_5Br + NaOH \rightarrow C_2H_5OH + NaBr$
$C_2H_5OH + [O] \rightarrow CH_3C(O)H + H_2O$
$CH_3C(O)H + [O] \rightarrow CH_3C(O)OH + H_2O$

11. (a) 2-ethylphenol (2-ethylhydroxybenzene) **(b)** methyl-2-propyl ether (2-methoxypropane) **(c)** dipropyl ether (propoxypropane) **(d)** 2-amino-3-thiol-propanoic acid **(e)** propane-2-thiol

13. (a) tertiary, 2-methyl-2-butanol **(b)** secondary, 2-butanol **(c)** primary, 2,2-dimethylpropanol **(d)** secondary, 2-propanol **(e)** secondary, 3-pentanol

15. (a) *cis*- or *trans*-4-methyl-2-pentene and 2-methyl-2-pentene **(b)** 3-ethyl-2-pentene **(c)** 2-methyl-3-phenyl-2-propene

17. (a) $C_6H_5OH(aq) + HCl(aq) \rightleftharpoons C_6H_5OH_2^+(aq) + Cl^-(aq)$ **(b)** $C_6H_5OH(aq) + NaOH(aq) \rightleftharpoons C_6H_5O^-(aq) + Na^+(aq) + H_2O(l)$

18. (a) $2C_2H_5OH(g) + O_2(g) \xrightarrow{Cu} 2CH_3CHO(g) + 2H_2O(g)$, ethanal **(b)** $CH_3(CH_2)_2OH(l) + NaOH(s) \rightarrow CH_3(CH_2)_2O^-Na^+(s) + H_2O(l)$, sodium propoxide **(c)** $2CH_3OH(l) + 2Na(s) \rightarrow 2CH_3O^-Na^+ + H_2(g)$, sodium methoxide **(d)** $C_2H_5OH(l) + H_2SO_4(conc) \rightarrow H_2C{=}CH_2(g) + H_3O^+ + HSO_4^-$, ethene **(e)** $CH_3CH{=}CH_2(g) + H_2O(g) \xrightarrow{catalyst} CH_3CH_2CH_2OH(g)$ or $CH_3CH(OH)CH_3(g)$, 1-propanol or 2-propanol

20. (a) Ethanol is a hydrogen-bonded liquid, but dimethyl ether is not. **(b)** When ethanol is reduced by sodium, $H_2(g)$ is evolved

$$2C_2H_5OH + 2Na \rightarrow 2C_2H_5O^-Na^+ + H_2,$$

and oxidizes acidified orange $Cr_2O_7^{2-}(aq)$ ion to green $Cr^{3+}(aq)$,

$$3CH_3CH_2OH + Cr_2O_7^{2-} + 8H^+ \rightarrow 3CH_3C(H)O + 2Cr^{3+} + 7H_2O$$

$$3CH_3CHO + Cr_2O_7^{2-} + 8H^+ \rightarrow 3CH_3CO_2H + 2Cr^{3+} + 4H_2O$$

22. (a) 1-butene and *cis*- and *trans*-2-butene **(b)** 2-bromobutane **(c)** 1-methylpropyl oxalate and di(1-methylpropyl) oxalate

24.* (a) Dehydration with $H_2SO_4(conc)$ to give 1-propene, which when reacted with $HBr(g)$ gives 1-bromopropane and 2-bromopropane; then hydrolysis of the 2-bromopropane with $NaOH(aq)$ gives 2-propanol. **(b)** Dehydration of ethanol to give ethene, which is reacted with bromine to give 1,2-dibromoethane, which upon hydrolysis with $NaOH(aq)$ gives ethylene glycol (1,2-ethanediol). **(c)** Complete oxidation of ethylene glycol (1,2-ethanediol) with acidified $Cr_2O_7^{2-}(aq)$ gives oxalic acid.

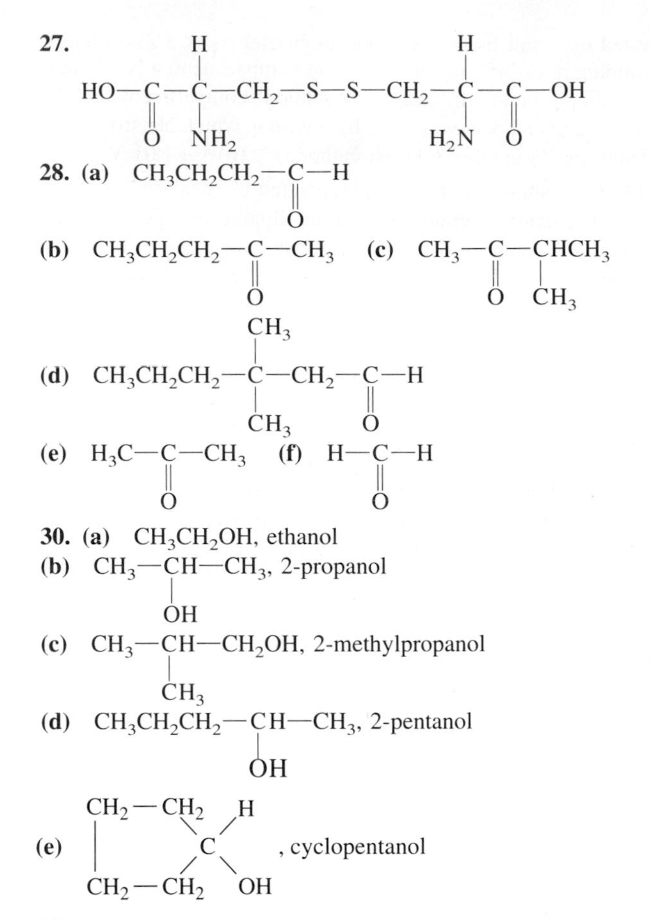

32. All these aldehydes and ketones result at 500–600°C from the dehydrogenation of the appropriate alcohol by using a copper or silver catalyst.

(a) $CH_3OH \rightarrow H_2C{=}O + H_2$ (Methanol)

(b) $C_2H_5OH \rightarrow CH_3CHO + H_2$ (Ethanol)

(c) $CH_3CH(OH)CH_3 \rightarrow (CH_3)_2C{=}O + H_2$ (2-Propanol)

(d) $CH_3CH_2CH(OH)CH_3 \rightarrow CH_3CH_2C(O)CH_3 + H_2$ (2-Butanol)

34.* $CH_3{-}CH(OH){-}CH_2CH_2CH_3$ (2-Pentanol) or $CH_3CH_2{-}CH(OH){-}CH_2CH_3$ (3-Pentanol) or $(CH_3)_2CH{-}CH(OH){-}CH_3$ (3-Methyl-2-butanol)

35. $H_3C{-}CH_2{-}CH_2{-}CH_2{-}C(OH){=}O$

Pentanoic acid

$$\begin{array}{c} CH_3 \\ | \\ H_3C—C—CH_2—C{=}O \\ |\qquad\qquad\ \ | \\ H\qquad\qquad OH \end{array}$$

3-Methylbutanoic acid

$$\begin{array}{c} CH_3 \\ | \\ H_3C—CH_2—C—C{=}O \\ \qquad\qquad |\quad\ \ | \\ \qquad\qquad H\quad OH \end{array}$$

2-Methylbutanoic acid

$$\begin{array}{c} CH_3 \\ | \\ H_3C—C—C{=}O \\ |\quad\ \ | \\ CH_3\ OH \end{array}$$

2,2-Dimethyl propanoic acid

37. (a) 3-ethylhexanoic acid **(b)** 4,4-dimethylpentanoic acid **(c)** 4-methylbenzoic acid **(d)** 2,3,3-trimethylbutanoic acid **(e)** heptanoic acid

38. Any three of

$$\begin{array}{c} H_3C—O—C—CH_2CH_2CH_3 \\ \| \\ O \end{array}$$

Methyl butanoate

$$\begin{array}{c} \qquad CH_3 \\ \qquad | \\ H_3C—O—C—C—CH_3 \\ \qquad \|\quad\ | \\ \qquad O\quad H \end{array}$$

Methyl-2-methylpropanoate

$$\begin{array}{c} CH_3CH_2CH_2—O—C—CH_3 \\ \qquad\qquad\qquad \| \\ \qquad\qquad\qquad O \end{array}$$

Propyl ethanoate

$$\begin{array}{c} \qquad\qquad CH_3 \\ \qquad\qquad | \\ H_3C—C—O—C—CH_3 \\ \|\qquad\ \ | \\ O\qquad\ H \end{array}$$

2-Propyl ethanoate

$$\begin{array}{c} H_3C—CH_2—O—C—CH_2—CH_3 \\ \| \\ O \end{array}$$

Ethyl propanoate

40. (a)

$$\begin{array}{c} H_3C—O—C—H \\ \qquad\ \| \\ \qquad\ O \end{array}$$

Methyl methanoate

(b)

$$\begin{array}{c} CH_3CH_2—O—C—CH_2CH_2CH_3 \\ \| \\ O \end{array}$$

Ethyl butanoate

(c)

$$\begin{array}{c} CH_3CH_2CH_2CH_2—O—C—CH_3 \\ \qquad\qquad\qquad\quad \| \\ \qquad\qquad\qquad\quad O \end{array}$$

Butyl ethanoate

(d)

$$\begin{array}{c} CH_3 \\ | \\ H_3C—C—O—C—CH_2CH_3 \\ |\qquad\ \| \\ H\qquad\ O \end{array}$$

2-Propyl propanoate

(e)

$$\begin{array}{c} \qquad\qquad OH \\ \qquad\qquad | \\ CH_3CH_2—O—P{=}O \\ \qquad\qquad | \\ \qquad\qquad OH \end{array}$$

Ethyl dihydrogenphosphate

42. (1) $H_2C{=}CH_2 + H_2O \rightarrow H_3C—CH_2—OH$,
(2) $C_2H_5OH + 2[O] \rightarrow CH_3CO_2H + H_2O$
(3) $C_2H_5OH + CH_3CO_2H \rightarrow C_2H_5O—C(O)—CH_3 + H_2O$

44. (a) $CH_3CH_2CH_2—NH_2$

Propylamine

$$\begin{array}{c} CH_3—CH—NH_2 \\ | \\ CH_3 \end{array}$$

2-Propylamine

(b)

$$\begin{array}{c} CH_3CH_2—N—H \\ \qquad\quad | \\ \qquad\quad CH_3 \end{array}$$

Ethylmethylamine

(c)

$$\begin{array}{c} H_3C—N—CH_3 \\ | \\ CH_3 \end{array}$$

Trimethylamine

47. (a)

$$\begin{array}{c} CH_3CH_2CH_2—C—NH_2 \\ \qquad\qquad \| \\ \qquad\qquad O \end{array}$$

Butanamide

$$\begin{array}{c} CH_3CH—C—NH_2 \\ \quad |\quad\ \ \| \\ \quad CH_3\ O \end{array}$$

2-Methylpropanamide

$$\begin{array}{c} H_3C—C—N—CH_2CH_3 \\ \quad \|\quad | \\ \quad O\quad H \end{array}$$

N-Methylpropanamide

$$\begin{array}{c} CH_3CH_2—C—N—CH_3 \\ \qquad\quad \|\quad | \\ \qquad\quad O\quad H \end{array}$$

N-Methylpropanamide

$$\begin{array}{c} H_3C—C—N—CH_3 \\ \quad \|\quad | \\ \quad O\quad CH_3 \end{array}$$

N,N-Dimethylethanamide

(b) React the appropriate amine, RR′NH, with the appropriate carboxylic acid, R″CO_2H, to give the salt R″CO_2^- RR′NH_2^+, which is then decomposed to the amide and water by heating:

$$RR'NH_2 + R''CO_2H \rightarrow R''CO_2^-\ RR'NH_2^+ \rightarrow R''C(O)NRR' + H_2O$$

butanoic acid + ammonia → butanamide
2-methylpropanoic acid + ammonia → 2-methylpropanamide
ethanoic acid + ethylamine → N-ethylethanamide
propanoic acid + methylamine → N-methylpropanamide
ethanoic acid + dimethylamine → N,N-dimethylethanamide

48. (a) aniline < ammonia < trimethylamine < methylamine < dimethylamine **(b)** (i) aniline, pH 8.80 (ii) dimethylamine, pH 11.86 **50. (a)** Lacking a plane of symmetry—chiral molecules (or objects) are nonsuperimposable mirror images of each other and are isomers; having a plane of symmetry—achiral molecules (or objects) are superimposable mirror images. **(b)** chiral: a shoe, a house key, a baseball, and a screw; achiral: a nail

51. chiral: (a), (b), (c), and (e); achiral: (d)

53. (a)

$$\begin{array}{c} H_3C—C—CH_2CH_3 \\ \| \\ O \end{array}$$

, 2-butanone

(b)

$$\begin{array}{c} CH_3CH_2—CH—C—OH \\ \qquad\qquad |\quad\ \| \\ \qquad\qquad Cl\quad O \end{array}$$

, 2-chlorobutanoic acid

$$\begin{array}{c} CH_3—CH—CH_2—C—OH \\ \quad |\qquad\qquad \| \\ \quad Cl\qquad\qquad O \end{array}$$

, 3-chlorobutanoic acid

Cl—CH_2—CH(CH_3)—C(=O)—OH, 3-chloro-2-methylpropanoic acid

55. **(a)** $CH_3CH_2CH_2Cl$ or CH_3—CH(Cl)—CH_3

1-Chloropropane 2-Chloropropane

(b) 1-Chloropropane should have three NMR peaks of relative intensities 3:2:2, whereas 2-chloropropane should have two NMR peaks of intensities 6:1.

56.* **(a)** H_3C—O—CH_3, dimethyl ether
(b) CH_3CH_2OH, ethanol **(c)** $(CH_3)_2CH$—OCH_3, 2-methoxypropane
(d) 1,1-dimethoxycyclopropane

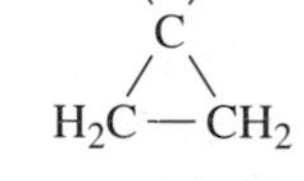

59. **(a)** weak acid and weak base **(b)** weak base
(c) neither acid nor base **(d)** neither acid nor base
(e) weak acid and weak base **(f)** weak base
(g) weak acid **(h)** weak base **(i)** weak base

60. **(a)** photochemical bromination of ethane, $C_2H_6(g)$
(b) addition of HBr(g) to ethene, $C_2H_4(g)$
(c) high-temperature catalytic dehydrogenation, or oxidation, of ethanol, $C_2H_5OH(l)$ **(d)** reaction of sodium ethoxide, $C_2H_5O^-Na^+$ (from reaction of ethanol with sodium or with strong NaOH(aq)), with ethyl bromide, $C_2H_5Br(l)$ (from reaction of ethanol with HBr(aq)) **(e)** oxidation (via propanal) of 1-propanol, $CH_3CH_2CH_2OH$

62. **(a)** $CH_4(g) + H_2O(g) \xrightarrow[\text{Ni catalyst}]{700\text{–}800°C} CO(g) + 3H_2(g)$

(b) $CO(g) + 2H_2(g) \xrightarrow[Al_2O_3(s)\text{ catalyst}]{250°C\text{ at }50\text{–}100\text{ atm}} CH_3OH(g)$

(c) $CH_3OH(g) \xrightarrow[\text{Cu catalyst}]{500\text{–}600°C} H_2C{=}O(g) + H_2(g)$

(d) $CH_3OH(g) + NH_3(g) \xrightarrow[Al_2O_3(s)\text{ catalyst}]{400°C} CH_3NH_2(g) + H_2O(g)$

64. **(a)** propene, $CH_3CH{=}CH_2$, and water **(b)** 2-butyne, $H_3CC{\equiv}CCH_3$, and bromine **(c)** 2-butyne, $H_3CC{\equiv}CCH_3$, and hydrogen **(d)** propyne, $CH_3{\equiv}CH$, and hydrogen bromide **66.*** **(a)** 60.0 g · mol^{-1}, C_3H_8O, $CH_3CH_2CH_2OH$, 1-propanol **(b)** 2-propanol, $(CH_3)_2CH(OH)$; only 2-propanone, $(CH_3)_2CO$ **68.*** **(a)** CH_3CHO, ethanal (molar mass 44.1 g · mol^{-1}) **(b)** —CHO, aldehyde **(c)** silver mirror test or reduction of orange $Cr_2O_7^{2-}(aq)$ to green $Cr^{3+}(aq)$ in acidic solution **69.*(a)** C_2H_6O, C_2H_6O; CH_3CH_2OH and $(CH_3)_2O$
(b) $2C_2H_6O + 2Na \rightarrow 2C_2H_5O^-Na^+ + H_2$, ethanol, sodium ethoxide **(c)** 69.1 mL

71.* **(a)** $H_3CC{\equiv}CCH_3$, 2-butyne; $HC{\equiv}CCH_2CH_3$, 1-butyne; cyclobutene, H_2C—CH_2 / H—C=C—H (ring) **(b)** yes

CHAPTER 15

1. **(a)** $\frac{\Delta[H_2]}{\Delta t} = \frac{5}{2}\frac{\Delta[CO]}{\Delta t} = \frac{5}{2}(1.00\text{ mL}\cdot s^{-1}) = 2.50\text{ mL}\cdot s^{-1}$

(b) $-\frac{\Delta[C_2H_6]}{\Delta t} = \frac{1}{2}\frac{\Delta[CO]}{\Delta t} = \frac{1}{2}(1.00\text{ mL}\cdot s^{-1}) = 0.500\text{ mL}\cdot s^{-1}$

3. **(a)** $-\frac{\Delta[HI]}{\Delta t} = \frac{1}{2}\frac{\Delta[H_2]}{\Delta t} = \frac{1}{2}\frac{\Delta[I_2]}{\Delta t}$

(b) $-\frac{\Delta[N_2O]}{\Delta t} = \frac{\Delta[N_2]}{\Delta t} = \frac{1}{2}\frac{\Delta[O_2]}{\Delta t}$

(c) $-\frac{\Delta[N_2O_5]}{\Delta t} = 2\frac{\Delta[NO_2]}{\Delta t} = \frac{1}{2}\frac{\Delta[O_2]}{\Delta t}$

6.

Time (min)	$[H_3O^+]$ (mol · L^{-1})	Δt (min)	$\Delta[H_3O^+]$ (mol · L^{-1})	Average Rate (mol · L^{-1} · min^{-1})
0	1.85			
80	1.66	80	−0.19	2.4×10^{-3}
159	1.53	79	−0.13	1.6×10^{-3}
314	1.31	155	−0.22	1.4×10^{-3}
628	1.02	314	−0.29	0.92×10^{-3}

9. (a) 1.5×10^{-3} mol · L^{-1} · s^{-1}
(b) 54 mol · L^{-1} · h^{-1} **11.** **(a)** rate = $k[NO]^2[H_2]$
(b) 27.8 L^2 · mol^{-2} · min^{-1} **13.** rate = $k[C_2H_4][O_3]$
15. **(a)** $k = 4.08 \times 10^{-2}$ min^{-1} **(b)** (i) $\frac{1}{8}$ (ii) $\frac{1}{1024}$
17. **(a)** Show that calculating the slope $-k_1$ for any two sets of data, from the equation $\ln V_t = \ln V_0 - kt$, gives the same value of k_1. **(b)** $k_1 = 0.050$ min^{-1} **(c)** 13.9 min
19. 200 years **22.** **(a)** rate = $k[CO][Cl_2]^{3/2}$
(b) reaction intermediates **23.** **(a)** (1), (2), and (4): unimolecular; (3): bimolecular **(b)** yes, rate = $k[N_2O_5]$
25.* **(a)** The simultaneous collision of three molecules with an appropriate spatial orientation and energy is unlikely.
(b) 1. $NO + NO \rightleftharpoons N_2O_2$ *fast equilibrium* (1)
$N_2O_2 + O_2 \rightarrow 2NO_2$ *slow* (2)
rate = $k_2[N_2O_2][O_2] = k[NO]^2[O_2]$
2. $NO + O_2 \rightleftharpoons NO_3$ *fast equilibrium* (1)
$NO_3 + NO \rightarrow 2NO_2$ *slow* (2)
rate = $k_2[NO_3][NO] = k[NO]^2[O_2]$
28. **(a)** Your plot should be a straight line.
(b) 186 kJ · mol^{-1} **(c)** 3.42×10^{-4} L · mol^{-1} · s^{-1}
30. 305 K (32°C) **33.** **(a)** 133 kJ · mol^{-1}
(b) 7.9×10^{-3} L · mol^{-1} · s^{-1}
35. 456 kJ · mol^{-1} **36.** **(a)** an increase
(b) a decrease **(c)** none **(d)** a decrease
(e) no effect **38.** **(a)** k increases
(b) K_{eq} remains the same. **40.** 565 K (292°C).
43.* **(a)**
$B\!: + CH_3COCH_3 \rightarrow BH^+ + {}^-\!:CH_2COCH_3$ *slow* (1)
$I_2 + {}^-\!:CH_2COCH_3 \rightarrow I$—$CH_2COCH_3 + I^-$ *fast* (2)
$BH^+ \rightleftharpoons B\!: + H^+$ *fast equilibrium* (3)

(b) B is a catalyst.

44. (a) (1): chain initiation; (2), (3): chain propagation; (4): chain termination **(b)** 630 nm (to break the bond in $Br_2(g)$, $BE = 190\ kJ \cdot mol^{-1}$)

45.* When step 2 is much slower than step 3.

47. (a) $2.92 \times 10^{-2}\ mol \cdot L^{-1} \cdot s^{-1}$
(b) $3.51 \times 10^{-2}\ mol \cdot L^{-1} \cdot s^{-1}$

48. Among the forward and reverse reactions, low temperature favors the exothermic, forward reaction; high pressure favors the reaction that decreases the number of molecules (the forward reaction); and a catalyst does not affect the position of the reaction but speeds up the reaction rate.

51. (a) 1. $H\ddot{O}:^- + :\ddot{X}—CH_3 \rightarrow$

$$^{\delta-}H\ddot{O}—C(H)(H)(H)—\ddot{X}:^{\delta-} \rightarrow H\ddot{O}—CH_3 + :\ddot{X}:^-$$

(a single-step bimolecular nucleophilic substitution mechanism)

2. $(CH_3)_3C—X \rightarrow (CH_3)_3C^+ + X^-$ *slow* (1)
$(CH_3)_3C^+ + OH^- \rightarrow (CH_3)_3C—OH$ *fast* (2)

(a two-step nucleophilic substitution mechanism in which the first step is unimolecular)

(b) The rate law for the first mechanism in (a) is rate = $k_1[RX]$; the rate law for the second mechanism is rate = $k_2[RX][OH^-]$. Which is followed in a particular case can be distinguished by using the respective integrated laws.

53. (a) $C_2H_5Cl(g) \rightarrow H_2C{=}CH_2 + HCl(g)$
(b) $5.3 \times 10^{-3}\ s^{-1}$ **(c)** 131 s

54.* **(a)** $2Br^- + H_2O_2 + 2H^+ \rightarrow Br_2 + 2H_2O$
(b) rate = $k[H_2O_2][Br^-][H^+]$
(c) $-\dfrac{\Delta[H_2O_2]}{\Delta t} = \dfrac{\Delta[Br_2]}{\Delta t} = 3.6 \times 10^{-3}\ mol \cdot L^{-1} \cdot s^{-1}$
(d) pH affects the rate but not the value of the rate constant.
(e) Initial rate is reduced by one-eighth.

CHAPTER 16

2. (a) sodium nitrate **(b)** potassium nitrite
(c) dinitrogentetraoxide **(d)** nitrogen dioxide
(e) nitric oxide (nitrogen monoxide) **(f)** nitrous oxide
(g) nitrous acid **(h)** nitric acid

4. (a) $^-:\ddot{N}{=}N^+{=}\ddot{O}: \leftrightarrow :N{\equiv}N^+—\ddot{O}:^-$; bond order $2\frac{1}{2}$

(b) $^-:\dot{N}{=}\ddot{O}:^+ \leftrightarrow :\dot{N}{=}\ddot{O}:$; bond order 2

(c) $:\ddot{O}{=}\ddot{N}—\ddot{O}:^- \leftrightarrow {}^-:\ddot{O}—\ddot{N}{=}\ddot{O}:$; bond order $1\frac{1}{2}$

(d) 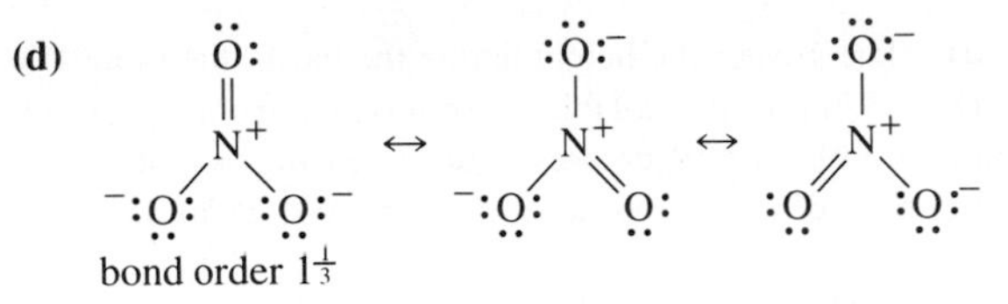
bond order $1\frac{1}{3}$

(e) $:N{\equiv}O:^+$; bond order 3
NO bond lengths: $NO_3^- > NO_2^- > NO > N_2O > NO^+$

6. (a) AX_2E, angular **(b)** AX_2E, angular
(c) AX_2, linear **8. (a)** $4HNO_3 \rightarrow 4NO_2 + O_2 + 2H_2O$; redox; N in HNO_3 is reduced, and O in HNO_3 is oxidized.
(b) $2Pb(NO_3)_2 \rightarrow 2PbO + 4NO_2 + O_2$; redox; N in NO_3^- and is reduced, and O in NO_3^- is oxidized.
(c) $3HNO_2 \rightarrow HNO_3 + 2NO + H_2O$; redox; N in HNO_2 is oxidized and reduced. **10.** $0.0727\ mol \cdot L^{-1}$

11. (a) −904.8 kJ **(b)** low temperatures and pressures

12. (a) second order in NO and first order in O_2
(b) $7.1 \times 10^3\ L^2 \cdot mol^{-2} \cdot s^{-1}$

14. (a) O_3 **(b)** Na_2O **(c)** Na_2O_2 **(d)** BaO
(e) BaO_2 **(f)** H_2O_2

16. (a) $:\ddot{O}{=}\ddot{O}^+—\ddot{O}:^- \leftrightarrow {}^-:\ddot{O}—\ddot{O}^+{=}\ddot{O}:$;
bond order $1\frac{1}{2}$

(b) The equilateral triangular structure has O—O single bonds and O—O—O bond angles of 60°. Each O atom would have AX_2E_2 angular geometry, for which the expected bond angle is close to the tetrahedral angle of 109.5°, which is inconsistent with the 60° angle of the triangular structure.

18. (a) Reaction of $BaO_2(s)$ with $H_2SO_4(aq)$ gives a solution of $H_2O_2(aq)$ and a precipitate of $BaSO_4(s)$ that can be filtered off. **(b)** H_2O_2 contains oxygen in the −I oxidation state, which can be oxidized to O_2 (oxygen ON 0) or reduced to H_2O (oxygen ON −2). **(c)** (i) $SO_2 + H_2O_2 \rightarrow SO_4^{2-} + 2H^+$ (ii) $H_2O_2 + O_3 \rightarrow 2O_2(g) + H_2O$ (iii) $3I^- + H_2O_2 + 2H^+ \rightarrow I_3^- + 2H_2O$ (iv) $NO_2^- + H_2O_2 \rightarrow NO_3^- + H_2O$

19. (i) All the oxidations are achieved by using $O_3(g)$. (ii) Only oxidation of NO(*g*), part (e), is achieved by using $O_2(g)$.

21. (a) Almost all the solar radiation with $\lambda < 240$ nm is absorbed by O_2 molecules, and ozone absorbs most of the radiation in the range from 240 to 350 nm. As a result, most of the low-wavelength UV light does not reach the earth's surface. **(b)** Ar atoms can interact with solar radiation to form only excited states. The triple bond of N_2 and the double bond of O_2 require radiation with $\lambda \leq 128$ nm and $\lambda \leq 240$ nm, respectively, for their dissociation into atoms. In the stratosphere, no radiation penetrates that is sufficiently energetic to dissociate N_2, but O_2 is dissociated into O atoms. In the troposphere, radiation with $\lambda < 240$ pm has already been absorbed, so there O_2 molecules remain undissociated. **23. (a)** −199.8 kJ
(b) −192 kJ **25. (a)** $18.4\ kJ \cdot mol^{-1}$
(b) $3.44 \times 10^5\ L \cdot mol^{-1} \cdot s^{-1}$

27. **(a)** The greater the height above the earth, the more UV light ($\lambda < 320$ nm) is available to form ozone from $O_2(g)$. But the pressure of $O_2(g)$ decreases with increasing height, so the concentration of ozone reaches a maximum at 25 to 35 km to form the ozone layer. **(b)** Reactions catalyzed by either Cl atoms or NO molecules. Cl atoms result largely from the photochemical dissociation of CFCs and catalyze the reaction $O_3 + O \rightarrow 2O_2$ by the chain mechanism $O_3 + Cl \rightarrow ClO + O_2$, followed by $ClO + O \rightarrow Cl + O_2$. NO molecules are formed largely by the reaction $N_2O + O \rightarrow NO + O_2$, where the N_2O is a product of oxidation of ammonia on earth. NO catalyzes the reaction $O_3 + O \rightarrow 2O_2$ by the two step mechanism $NO + O_3 \rightarrow NO_2 + O_2$, followed by $NO_2 + O \rightarrow NO + O_2$.
29. **(a)** 165 $kJ \cdot mol^{-1}$ **(b)** ClO is formed by the reaction $O_3 + Cl \rightarrow O_2 + ClO$ and then reacts with O atoms, $ClO + O \rightarrow Cl + O_2$, to re-form Cl atoms, thereby catalyzing the decomposition of ozone. **(c)** 1:-99 kJ; 2:-293 kJ
31.* **(a)** rate $= k[O_3]^2/[O_2]$; **(b)** The rate is inversely proportional to $[O_2]$, and since O_2 is a product in the first reaction step, an increase in $[O_2]$ shifts the position of this equilibrium to the left, thereby decreasing the concentration of O atoms. Since the rate of the rate determining step is proportional to [O], any decrease in its concentration will decrease the rate.
32.* **(a)** 1.63×10^{-38} $atm^{-\frac{1}{2}}$ **(b)** 10^{-42} atm **(c)** The system is not at equilibrium; the steady-state concentration of O_3 is much higher than the equilibrium concentration and is responsible for screening out most of the UV light from the sun.
33.* **(a)** The reaction is a chain reaction because, once Cl atoms are generated in an initiation step, the ensuing steps regenerate Cl atoms, which then continue to react. They are a catalyst because they do not appear in the balanced equation for the overall reaction.
(b) rate $= 1.5 \times 10^{-15}$ $mol \cdot L^{-1} \cdot s^{-1}$ **(c)** 76%
36. **(a)** NO readily forms NO_2 with oxygen, so it is not meaningful to report the concentrations separately.
(b) 3.5×10^{-8} atm; 3.5×10^{-2} ppm; 1.6×10^{-9} $mol \cdot L^{-1}$
(c) 1.4×10^5 mol **38.** **(a)** 8.3×10^3 mol
(b) 2.6×10^{-11} $mol \cdot L^{-1}$
40. **(a)** Due to the dissolved $CO_2(g)$ (carbonic acid).
(b) Al occurs as $Al(H_2O)_6^{3+}(aq)$ in acidic solution (pH < 5) but precipitates as $Al(OH)_3(s)$ in basic solution (pH > 7).
(c) $Al^{3+}(aq)$ from acidified lake water enters the gills of fish, where it is precipitated as $Al(OH)_3(s)$(at pH 7.4) and suffocates them. **42.** 5.7 **44.** **(a)** 2.4×10^7 $L \cdot day^{-1}$
(b) 1.82×10^5 $kg \cdot day^{-1}$ **(c)** 3.9 kg

45.* no ($Q < K_p$)
49. $4NH_3(g) + 5O_2(g) \rightarrow 4NO(g) + 6H_2O(g)$; $2NO(g) + O_2(g) \rightarrow 2NO_2(g)$ (brown fumes) **50.** The bond orders are **(a)** $O_2 = 2$ **(b)** $O_3 = 1\frac{1}{2}$, and **(c)** $O_2^{2-} = 1$; bond length increases with decreasing bond order.
52. **(a)** 53 $kJ \cdot mol^{-1}$ **(b)** 1.2×10^{-9} $L \cdot mol^{-1} \cdot s^{-1}$
54. **(a)** $3I^- + NO_2 + H_2O \rightarrow I_3^- + NO + 2OH^-$; oxidizing agent **(b)** $MnO_4^- + 5NO_2 + H_2O \rightarrow Mn^{2+} + 5NO_3^- + 2H_3O^+$; reducing agent **(c)** $2NO_2 + H_2O \rightarrow HNO_3 + HNO_2$; oxidizing and reducing agent **57.*** **(a)** $N_2O(g)$ and $N_2(g)$ **(b)** $^-\!:\ddot{N}{=}N^+{=}\ddot{O}:$ $\;:N{\equiv}N:$
(c) $NH_4NO_3(s) \rightarrow N_2O(g) + 2H_2O(g)$
$P_4(s) + 10N_2O(g) \rightarrow P_4O_{10}(s) + 10N_2(g)$

CHAPTER 17

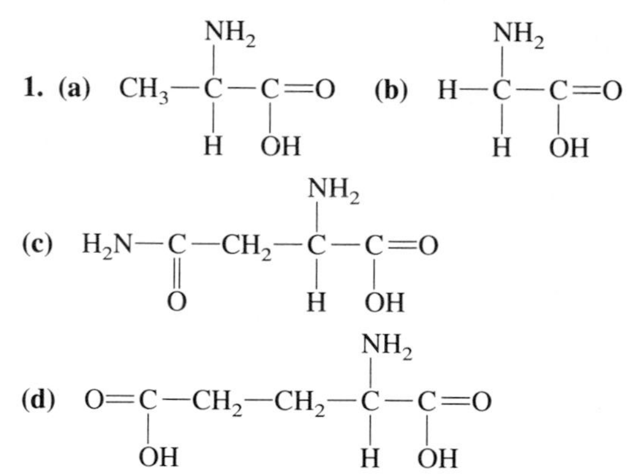

Glutamic acid is a dicarboxylic acid (two $-CO_2H$ groups).
3. **(a)** 6.7×10^{-4} mol **(b)** 9.4×10^{-3} g; **(c)** yes
5.

```
       N—CH            NH2         H
      //   \\           |          |
(a) HC       C—CH2—C—C—N—C—CH2—SH
      \     /           |  ||  |   |
        NH              H  O   H   CO2H

       N—CH            CO2H        H
      //   \\           |          |
or  HC       C—CH2—C—N—C—C—CH2—SH
      \     /           |  |  ||  |
        NH              H  H  O   NH2

          NH2                  CO2H              CH—N
           |                    |               //    \\
(b) CH3—C—C—N—CH2—C—N—C—CH2—C          CH
           |  ||  |          ||  |  |              \    /
           H  O   H          O   H  H                NH
```

(c) His—Cys; Ala—Gly—His
6. **(a)** 6 **(b)** 24 **(c)** 720

7.

(a)

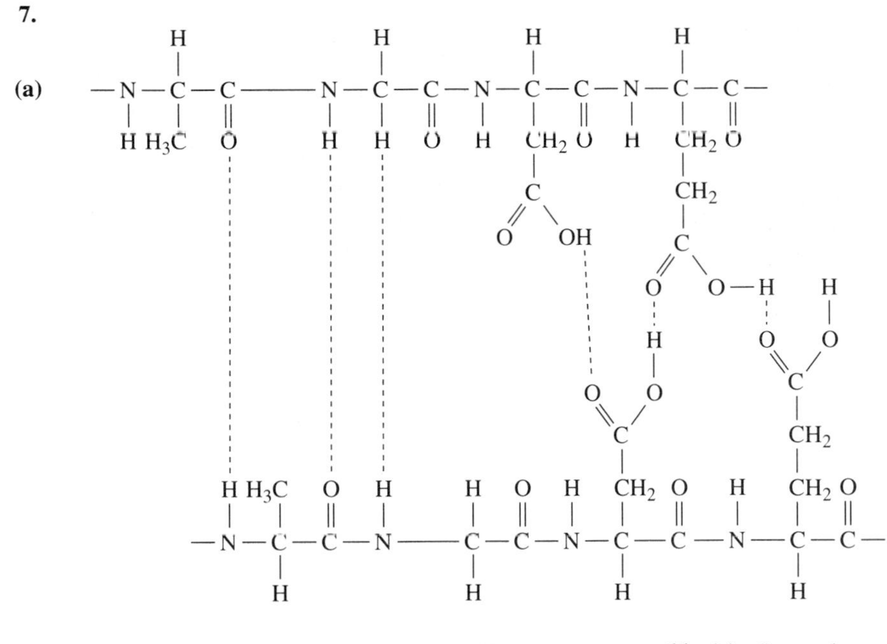

(b)

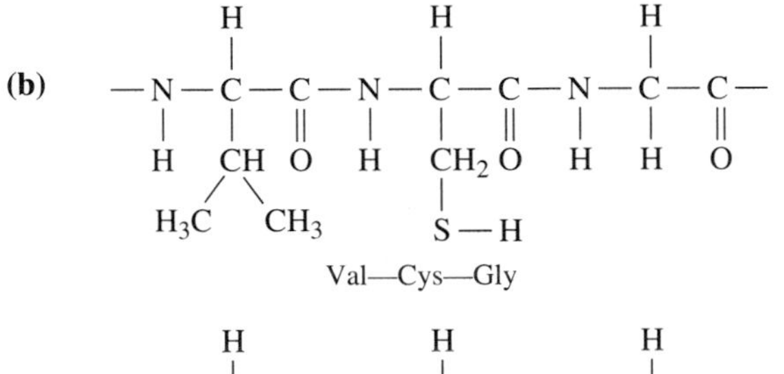

Val—Cys—Gly

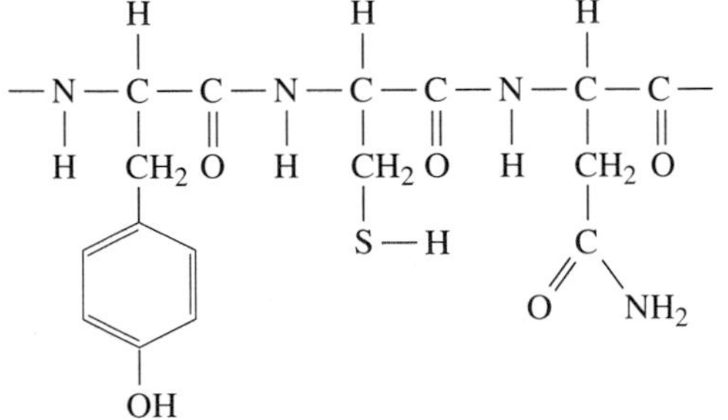

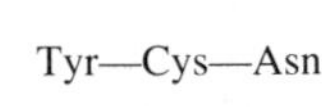

Tyr—Cys—Asn

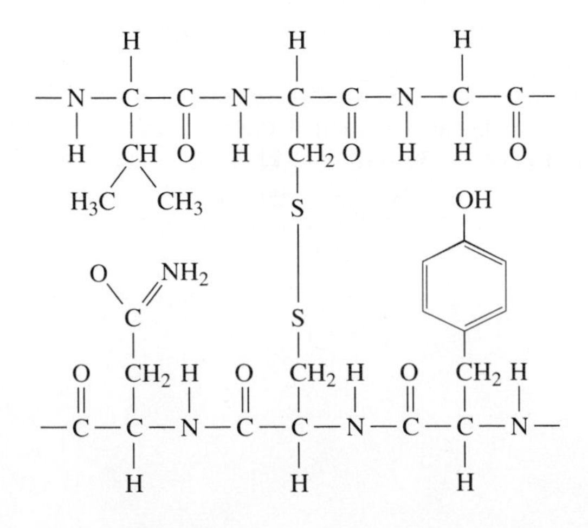

11. (a) In an interaction with water, ''hydrophobic'' describes nonpolar groups that do not form hydrogen bonds, and ''hydrophilic'' describes polar groups that form hydrogen bonds. **(b)** Hydrophobic: alkyl or aryl side chains; hydrophilic: chains containing polar —HN—, $—NH_2$, —OH, or $>$C=O groups. **(c)** Hydrophilic side chains remain on the outside of the structure, where they attract surrounding water molecules by hydrogen bonds. Hydrophobic side chains are repelled to the inside of the structure, where they attract each other by London (dispersion) forces.

13. (a) Rates of reactions outside the body are changed by large factors by changing the temperature, by using a catalyst, or by changing the solvent. Reactions in the healthy body occur at a relatively low temperature (37°C) in aqueous media and constant pH and can be regulated only by biologically stable catalysts (enzymes). **(b)** Processes in the body occur as the result of many successive reactions; an enzyme must be biologically stable, very specific, and change the rate of the uncatalyzed reaction by a specific amount.

15. 61 $kJ \cdot mol^{-1}$ **(b)** (5.2×10^{-2})°C

17. (a) Carbohydrates have the empirical formulas of ''carbon hydrates,'' $C_x(H_2O)_y$, but there is no relationship between this formula and their actual molecular structures.
(b) monosaccharides: fructose, glucose, ribose, deoxyribose; disaccharides: sucrose; polysaccharides: starch and cellulose
(c) Both are cyclic ethers with —OH functional groups, but fructose has a five-membered ring, and glucose has a six-membered ring. **19. (a)** 1.85×10^3 units **(b)** a mixture of straight and branched-chain polymers of β glucose.

20. (a) 2808 $kJ \cdot mol^{-1}$ **(b)** 15.58 $kJ \cdot g^{-1}$
(c) 7.8×10^2 cm^2

21. (a)

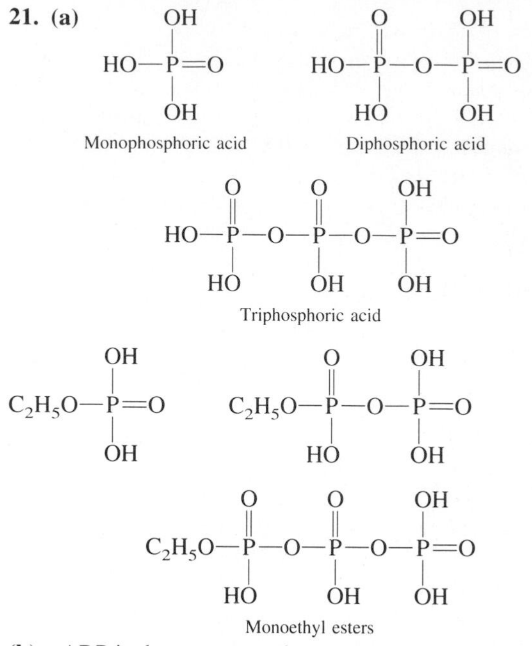

Monoethyl esters

(b) ADP is the monoester formed between diphosphoric acid and the sugar ribose to which the heterocyclic amine adenine is attached, and ATP is the monoester formed between triphosphoric acid and the sugar ribose to which adenine is attached; both are nucleotides. **23. (a)** ATP is a nucleotide formed from triphosphoric acid, ribose, and adenine units. **(b)** DNA is a polynucleotide in which the nucleotide units are formed from monophosphoric acid, deoxyribose, and a heterocyclic base; the bases are adenine, guanine, cytosine, and thymine. **25. (a)** By the Gibbs free energy change, ΔG. **(b)** 2870 $kJ \cdot mol^{-1}$ **(c)** In a sequence of many steps involving the formation of ATP, which transfers energy to where it is needed. **27. (a)** Adenine, guanine, cytosine, and thymine (in DNA), and thymine is replaced by uracil in RNA. **(b)** A, G, C, T, and U

(c)

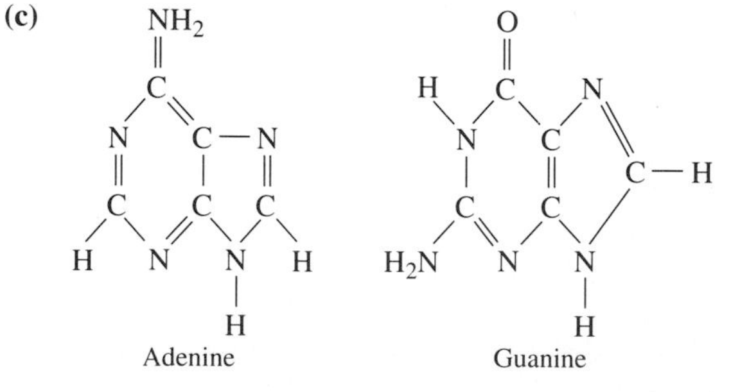

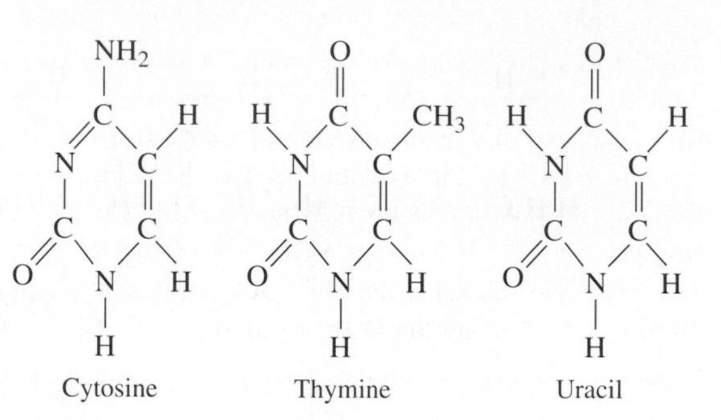

29. (a) DNA has two polynucleotide strands in which the amount of the nitrogenous base adenine, A, is the same as the base thymine, T, and the amount of the base cytosine, C, is the same as the amount of the base guanine, G. The double helix forms because A units and T units, and C units and G units, on opposite strands form A---T and C---G hydrogen bonds that give exactly the same spacing between the two polynucleotide strands when they are wound together. **(b)** The A---T and C---G base pairs are found on the inside of the double helix, because it is the hydrogen bonds formed between each pair that holds the double helix together. **30.** A double helix forms only when the hydrogen bonds between base units on opposite strands give the same constant spacing between the two polynucleotide chains, which occurs only for A---T, joined by two hydrogen bonds, and C---G, joined by three hydrogen bonds.

33. (a) TGAGCG
(b) ACUCGC

34. (a)

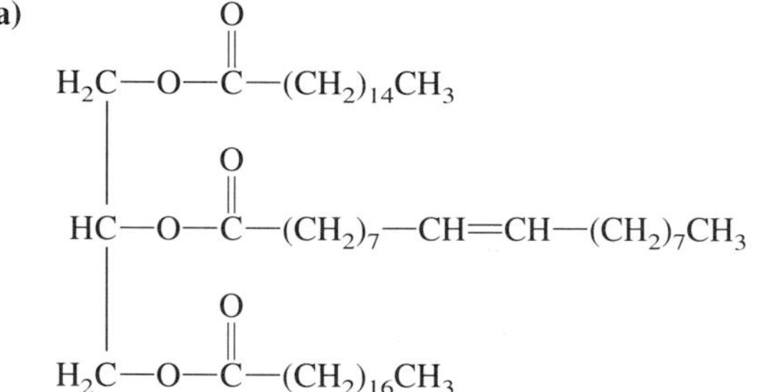

(b) Triacylglycerols (triglycerides) are highly concentrated stores of metabolic energy, which in mammals are synthesized from glucose and stored in adipose (fat) cells. They are hydrolyzed first to glycerol and their constituent fatty acids and eventually to carbon dioxide and water, thereby providing energy for the snythesis of ATP. **36. (a)** carbohydrates, fats, proteins (α-amino acids), vitamins, and minerals
(b) carbohydrates (starch); 1 Cal = 1 kcal = 4.187 kJ
(c) as fats **39. (a)** 171 $kJ \cdot mol^{-1}$ **(b)** no

41. (a) carbonic acid, $(HO)_2C{=}O$; $(HO)_2C{=}O + 2NH_3 \rightarrow 2NH_4^+\ CO_3^{2-} \rightarrow (NH_2)_2C{=}O + 2H_2O$
(b) Fish live in an aqueous environment, so it is convenient for ammonia from the oxidation of their proteins to pass into the water that surrounds them, where it is immediately diluted and swept away. Thus, it never achieves a lethal concentration.
42. $FeSO_4(aq)$ contains the $Fe^{2+}(aq)$ ion that is readily oxidized to $Fe^{3+}(aq)$ by the $O_2(g)$ from air,

$$4Fe^{2+}(aq) + O_2(g) + 2H_2O(l) \rightarrow 4Fe^{3+}(aq) + 4OH^-(aq)$$

and some of the $Fe^{3+}(aq)$ formed will be precipitated as brown $Fe(OH)_3(s)$. Iron is also present in hemoglobin as Fe(II), where it is coordinated to five N atoms, leaving the sixth position of the octahedral coordination sphere empty. When $O_2(g)$ is bubbled through a hemoglobin solution, the O_2 molecule bonds to the iron atom at this site, giving the bright red complex oxyhemoglobin, rather than oxidizing Fe(II) to Fe(III).

CHAPTER 18

1.

	$^{11}_{5}B$	$^{13}_{6}C$	$^{94}_{40}Zr$	$^{137}_{56}Ba$	^{57}Fe
Protons	5	6	40	56	26
Neutrons	6	7	54	81	31
Nucleons	11	13	94	137	57

3. (a) $^{80}_{36}Kr$ **(b)** $^{80}_{34}Se$ **(c)** $^{80}_{34}Se$
5. (a) $^{207}_{84}Po + {}^{0}_{-1}e \rightarrow {}^{207}_{83}Bi$ **(b)** $^{207}_{84}Po \rightarrow {}^{207}_{83}Bi + {}^{0}_{1}e$
(c) $^{207}_{84}Po \rightarrow {}^{203}_{82}Pb + {}^{4}_{2}He$ **7. (a)** $^{32}_{15}P \rightarrow {}^{32}_{16}S + {}^{0}_{-1}e$
(b) $^{15}_{8}O \rightarrow {}^{15}_{7}N + {}^{0}_{1}e$ **(c)** $^{52}_{26}Fe \rightarrow {}^{52}_{25}Mn + {}^{0}_{1}e$
(d) $^{218}_{87}Fr \rightarrow {}^{214}_{85}At + {}^{4}_{2}He$ **(e)** $^{50}_{26}Fe \rightarrow {}^{50}_{27}Co + {}^{0}_{-1}e$
(f) $^{122}_{53}I \rightarrow {}^{122}_{54}Xe + {}^{0}_{-1}e$
9. (a) $^{231}_{91}Pa$ **(b)** $^{227}_{89}Ac$ **(c)** $^{223}_{87}Fr$ **(d)** $^{223}_{88}Ra$
(e) $^{219}_{86}Rn$ **(f)** $^{215}_{84}Po$ **(g)** $^{211}_{82}Pb$ **(h)** $^{211}_{83}Bi$
(i) $^{211}_{84}Po$ **(j)** $^{207}_{82}Pb$
11. (a) $^{10}_{4}Be \rightarrow {}^{10}_{5}B + {}^{0}_{-1}e$; boron-5
(b) $^{12}_{7}N \rightarrow {}^{12}_{6}C + {}^{0}_{1}e$ or $^{12}_{7}N + {}^{0}_{-1}e \rightarrow {}^{12}_{6}C$; carbon-12
(c) $^{14}_{6}C \rightarrow {}^{14}_{7}N + {}^{0}_{-1}e$; nitrogen-14
(d) $^{87}_{35}Br \rightarrow {}^{87}_{36}Kr + {}^{0}_{-1}e$; krypton-87
13. (a) 1.00 Ci, 1.00×10^3 mCi
(b) 9.46×10^{-5} Ci; 9.46×10^{-2} mCi
(c) 1.89×10^2 Ci, 1.89×10^5 mCi
(d) 1.30×10^6 Ci; 1.30×10^9 mCi
15. (a) 1.54×10^{-2} rad, 1.54×10^{-2} rem
(b) 1.54×10^{-2} rad; 0.308 rem **17.** 3.13×10^{-2} Ci
19. 13 min **21. (a)** 438 g **(b)** 259 g **(c)** 134 g
23. (a) ^{131}I: 27 days; ^{90}Sr: 93.3 yr
(b) ^{131}I: 53 days; ^{90}Sr: 187 yr
(c) ^{90}Sr **25.** 14.2 days
27. (a) 2.06×10^9 disintegrations · min^{-1}
(b) 6.94 days **29.** about 13,000 B.C.
31. (a) $^{238}_{92}U + {}^{1}_{0}n \rightarrow {}^{141}_{56}Ba + {}^{92}_{36}Kr + 6\,{}_{0}^{1}n$
(b) $^{238}_{92}U + {}^{1}_{0}n \rightarrow {}^{103}_{42}Mo + {}^{131}_{50}Sn + 5\,{}_{0}^{1}n$
33. $^{267}_{109}Unn$, unnilennium-266
35. $^{232}_{90}Th + {}^{12}_{6}C \rightarrow {}^{240}_{96}Cm + 4\,{}_{0}^{1}n$
37. By using $PbI_2(s)$ enriched with the radioactive ^{131}I isotope and showing that, in the $PbI_2(s)$–$PbI_2(aq)$ system, the radioactivity of the $PbI_2(s)$ and the $PbI_2(aq)$ both eventually achieve constant values. **40. (a)** 4.778×10^{-11} J
(b) 1.365×10^{-12} J · nucleon^{-1}
42. 1.368×10^{14} kg · yr^{-1}
44. (a) ^{14}N: 1.197×10^{-12} J · nucleon^{-1};
^{15}N: 1.234×10^{-12} J
(b) ^{15}N is the more stable. **(c)** ^{14}N **46.** ^{12}C
48. $^{235}_{92}U \rightarrow {}^{231}_{90}Th + {}^{4}_{2}He$; $^{231}_{90}Th \rightarrow {}^{231}_{91}Pa + {}^{0}_{-1}e$;
$^{231}_{91}Pa \rightarrow {}^{227}_{89}Ac + {}^{4}_{2}He$; $^{227}_{89}Ac \rightarrow {}^{227}_{90}Th + {}^{0}_{-1}e$
51. 1.58×10^8 kJ · mol^{-1} (fusion) versus
2.42×10^2 kJ · mol^{-1} (combustion)
53. (a) X: C-13; Y: N-14; Z: O-15
(b) $4\,{}^{1}_{1}H \rightarrow {}^{4}_{2}He + 2\,{}^{0}_{1}e$
55. (a) (i) 2.574×10^{-11} J (ii) 1.287×10^{-12} J · nucleon^{-1}
(b) (i) 8.973×10^{-11} J (ii) 1.402×10^{-12} J · nucleon^{-1}
(c) (i) 8.581×10^{-11} J (ii) 1.409×10^{-12} J · nucleon^{-1}
(d) (i) 2.773×10^{-10} J (ii) 1.227×10^{-12} J · nucleon^{-1}
58. (a) +VII **(b)** $HTcO_4$, TcF_7

CHAPTER 19

1. (a) Nuclei have to be forced together with sufficient kinetic energy to overcome strong repulsions between their positive charges. **(b)** 2.48×10^9 kJ · mol^{-1}
3. (a) Due to decreasing temperature and increasing gravitational forces. **(b)** Collapse eventually raises the temperature sufficiently to allow further nucleosynthesis.
(c) Because nucleosyntheses of nuclei heavier than ^{56}Fe are endothermic processes. **(d)** Second-generation stars contain the entire range of nuclides. **5. (a)** Nucleosynthesis of ^4He releases more energy; 2.49×10^{11} J · mol^{-1}.
7. (a) 3.24×10^{15} molecules · L^{-1} **(b)** 3.26×10^{-17} to 3.26×10^{-13} mm Hg **10. (a)** Venus: 87 atm, 3.2 atm; Mars: 5.7×10^{-3} atm, 1.6×10^{-4} atm **(b)** Earth: 0.781 atm, 0.209 atm, 0.009 atm **13.** The surface rocks are mainly silicates and aluminosilicates, whereas the core has a high proportion of iron and nickel. **15. (a)** $[Ne]3s^23p^2$ with two unpaired $3p$ electrons and a valence of 2.
(b) Because most compounds are formed from the $[Ne]3s^13p^3$ excited state (with four unpaired electrons).
17. (a) Silica, $SiO_2(s)$, with Si—O single bonds and a three-dimensional covalent network structure related to diamond; in contrast to $CO_2(g)$, $:\ddot{O}{=}C{=}\ddot{O}:$, composed of CO_2 molecules. The difference is due to the inability of Si to form multiple bonds. **(b)** $SiCl_4(l)$, wich is readily hydrolized to $Si(OH)_4(s)$; unlike $CCl_4(l)$ which in unreactive. They differ because Si in Period 3 has an incomplete valence shell; whereas in CCl_4 the valence shell is filled. **19. (a)** SiS_2 **(b)** A structure analogous to that of $SiO_2(s)$. **(c)** solid

22. (a) C—H: 416 kJ · mol^{-1}; Si—H: 322 kJ · mol^{-1} **(b)** C—C: 330 kJ · mol^{-1}; Si—Si: 196 kJ · mol^{-1}

24. (a) $SiCl_4(l) + 4H_2O(l) \rightarrow Si(OH)_4(aq) + 4HCl(aq)$ **(b)** The structures are tetrahedral $Si(OH)_4$ and triangular planar $(HO)_2C{=}O$; both are weak acids, but $Si(OH)_4$ is weaker than H_2CO_3. **(c)** CCl_4 does not react with water because the valence of C is filled with an octet of electrons.

26. (a) network **(b)** sheet **(c)** sheet **(d)** network

27. (a) chain **(b)** network **(c)** disilicate chain **(d)** trisilicate ring **29.** Hydrothermally: $Fe_2O_3(s)$ from iron and water to give $Fe(OH)_2(s)$, which is oxidized by weathering, and $FeS_2(s)$ from hot magma reacting with water at high temperatures and pressures at suboceanic tectonic plate boundaries. **33. (a)** silicic acid < **(b)** phosphoric acid < **(c)** sulfuric acid < **(d)** perchloric acid

35. (a) Both tetrafluorides are in Group IV, but C is in Period 2 and Si is in Period 3, so C in CF_4 has a completed valence shell and is unreactive; Si in SiF_4 has an incomplete valence shell and reacts as a Lewis acid with F^- ion. **(b)** triangular bipyramidal; octahedral.

36.

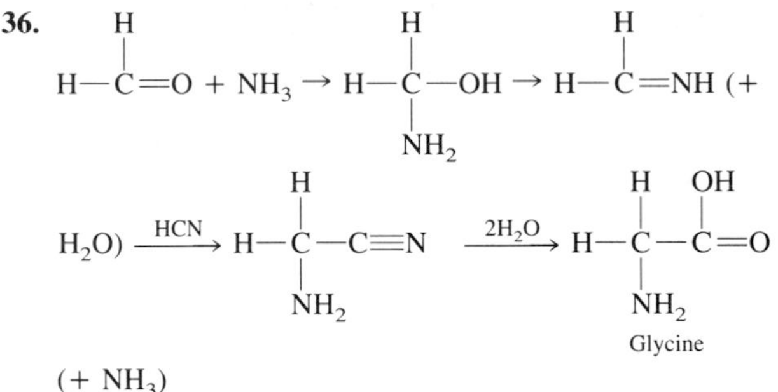

CHAPTER 20

1. (a) A mixture of macromolecules containing all the atoms of the monomers from which it is formed (such as polyethylene, $(CH_2)n$; polytetrafluoroethylene (Teflon), $(CF_2)_n$, and polyvinyl chloride, $(CH_2CHCl)_n$, PVC). **(b)** A mixture of macromolecules formed by elimination of small molecules, such as H_2O, between monomers (for example, polyamides, such as nylon and polypeptides, and polyesters, such as Dacron and Bakelite.) **3. (a)** A polymer containing macromolecules formed by head-to-tail addition of monomers to give a single continuous chain. **(b)** A polymer containing macromolecules formed not only by continuous chain growth but also by intermittent addition of monomers at various points in the chain, giving also branched chains that can take part in further branching. **(c)** The linking of polymer chains—for example, by the formation of disulfide, —S—S—, bridges between neighboring a polymer chains, which occurs in vulcanization of rubbers and in polypeptides with sulfur-containing α-amino acid units. **5.** Reaction of (I) with RO · gives a free radical (II) with the odd electron on a C atom in the middle of the chain, which grows to form a branched chain by reacting with another styrene molecule to give the free radical (III), which continues to grow by adding further styrene molecules:

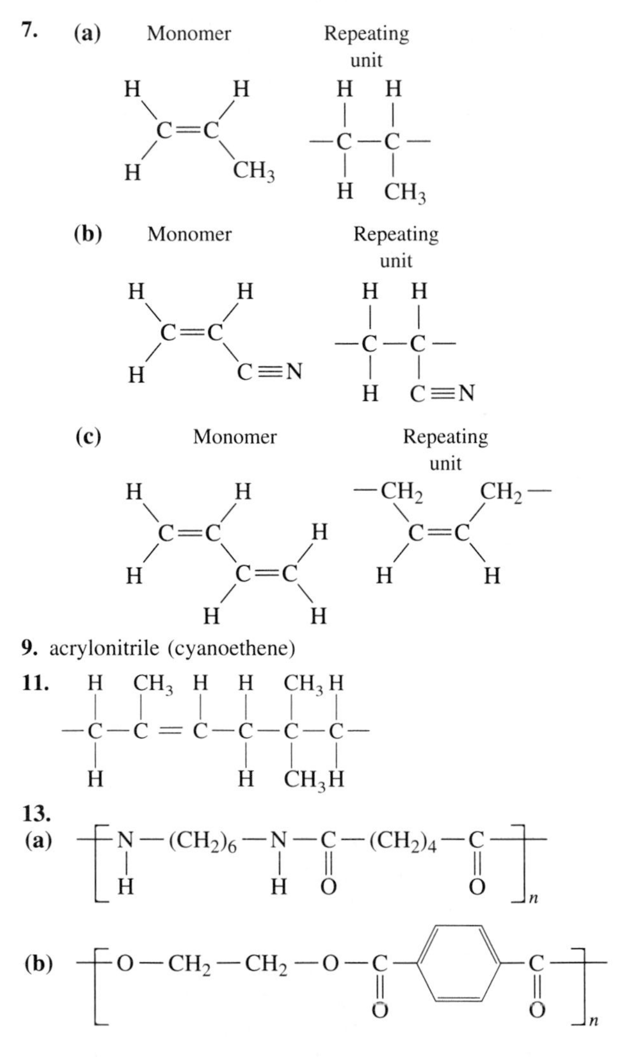

9. acrylonitrile (cyanoethene)

Both are formed by the condensation of H_2O molecules between the monomers.

14. (a) $-\mathrm{O}-\mathrm{CH_2}-\mathrm{CH_2}-\underset{\underset{\mathrm{O}}{\|}}{\mathrm{C}}-\mathrm{CH_2}-\underset{\underset{\mathrm{O}}{\|}}{\mathrm{C}}-$, methanol

(b) $-\mathrm{O}-(\mathrm{CH_2})_3-\underset{\underset{\mathrm{O}}{\|}}{\mathrm{C}}-(\mathrm{CH_2})_2-\underset{\underset{\mathrm{O}}{\|}}{\mathrm{C}}-$, ethanol

16. (a) London forces **(b)** dipole–dipole and London forces **(c)** hydrogen bonds **18. (a)** tetrafluoroethene **(b)** vinyl chloride **(c)** caprolactam **(d)** 1,2-ethanediol and terephthalic acid

20. (a) Aluminosilicates have similar structures to silicates but with a proportion of the Si atoms replaced by Al^- ions and additional cations to neutralize the extra charges. **(b)** Half the Si atoms in the silica polymer $(SiO_2)_{24}$ are replaced by Al^- ions, to give the $Al_{12}Si_{12}O_{48}{}^{12-}$ ion. **(c)** Empty channels in anhydrous zeolites are receptive to absorbing water molecules. **22.** $SiO_2(s)$ and $Al_2O_3(s)$ have compact structures, and catalytic activity is largely confined to their external surfaces. Zeolites have channels and cavities that give them an enormous ''surface'' area and, thus, a much greater catalytic activity. **23.** The electrical conductivity of metals is due to the movement of electrons (the ''electron gas''), which is impeded only by the thermal vibrations of their metal cations, which decrease with increasing temperature. Electrical conduction by semiconductors is a consequence of the number of electrons promoted to the conduction band, which increases with temperature. **25. (a)** SiO_2 is reduced at high temperatures with coke. **(b)** By conversion to $SiCl_4$, which is reduced with hydrogen or magnesium and ultrapurified by zone refining. **27.** *n*-type: (a), (c), and (d); *p*-type: (b) and (e)

29. 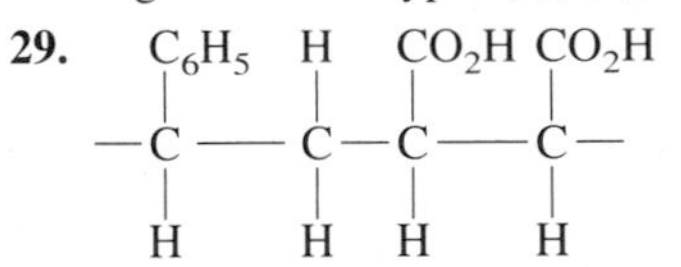

31. (a) vinyl fluoride, $CH_2 = CHF$ **(b)** chlorotrifluoroethene, $CF_2{=}CFCl$ **33.** Polyamides contain $>\!C{=}O$ and $>\!N{-}H$ groups that can form hydrogen bonds between adjacent polymer chains, whereas in polyalkanes the intermolecular forces are confined to weak London forces.

35. C, 239 nm Si, 1.14 μm Ge, 2.0 μm

37. Because in some metal oxides *d* electrons in the valence shell of the metal cations can easily enter the lattice to form a delocalized charge cloud, or the oxide is nonstoichiometric.

Glossary

Absolute zero of temperature The lowest attainable temperature, (0 K = −273.15°C); the temperature at which the kinetic energy of the molecules of a substance is zero.

Absorption spectrum The spectrum, consisting of characteristic series of discrete dark lines in a continuous background, obtained when atoms absorb light from white light. Used for analyzing for the elements present in almost any sample of a substance.

Achiral A molecule or other object that has a plane of symmetry and therefore a superimposable mirror image.

Acid A proton donor; in aqueous solution an acid transfers a proton to a water molecule forming a hydronium ion.

Acid ionization constant, K_a Also called acidity constant. The equilibrium constant for the reaction of an acid in water, $HA + H_2O \rightleftharpoons H_3O^+ + A^-$, given by the expression

$$K_a = \left(\frac{[H_3O^+][A^-]}{[HA]}\right)$$

Acid rain Rain that has a pH of less than the normal value of 5.6.

Acid–base reaction A reaction in which a proton is transferred from an acid to a base; a proton transfer reaction.

Acid–base titration A procedure for the determination of the concentration of a solution of a base by reacting it with a known volume of a solution of an acid of known concentration, or vice-versa.

Acidic oxide An oxide that reacts with water to give an acidic solution (containing $H_3O^+(aq)$ ions); an oxide that reacts with a basic oxide to give a salt.

Actinide One of 14 elements within Period 7 of the periodic table from Z = 90 to Z = 103.

Activated complex (transition state) An unstable arrangement of atoms that has the highest energy reached during the rearrangement of the reactant atoms to give the products of a reaction.

Activation energy, E_a The minimum energy of collision between molecules required for reaction to occur; the difference between the energy of the activated complex and that of the reactants.

Active site The part of an enzyme to which a substrate binds and at which reaction takes place.

Activity series The arrangement of metals (and hydrogen) in order of decreasing strength as oxidizing agents.

Addition polymer A polymer formed by joining a large number of simple molecules (monomers).

Addition reaction A reaction in which a molecule adds across a double or triple bond of another molecule.

Adenine One of the purine bases found in nucleic acid.

Adenosine diphosphate (ADP) The substance produced when adenosine triphosphate loses a phosphate group in providing the energy needed for a reaction in which it takes part.

Adenosine triphosphate (ATP) The universal energy carrier in metabolic processes. The energy liberated when ATP loses a phosphate group is used to drive another reaction to which its conversion to ADP is coupled.

Alcohol An organic compound containing an OH group.

Aldehyde An organic compound containing an CH=O group.

Aliphatic compound An alkane, alkene, or alkyne or any compound derived from these hydrocarbons by replacing hydrogen atoms with other atoms or nonaromatic groups of atoms.

Alkali metal An element of Group I of the periodic table.

Alkaline earth metal An element of Group II of the periodic table.

Alkane A hydrocarbon in which all the carbon–carbon bonds are single bonds.

Alkene A hydrocarbon that has one or more carbon–carbon double bonds.

Alkyl group A group composed of carbon and hydrogen atoms derived from an alkane by the removal of a hydrogen atom.

Alkyne A hydrocarbon that has one or more carbon–carbon triple bonds.

Allotropes Forms of an element that differ in chemical and physical properties.

Alpha (α) amino acid An amino acid in which the NH_2 group is bonded to the carbon atom to which the CO_2H group is attached.

Alpha (α) helix A common secondary protein structure in which the protein chain is in the form of a helix or coil stabilized by hydrogen bonds between NH and CO groups.

Alpha (α) particle, $^4_2$He A helium, He^{2+}, nucleus resulting from a nuclear reaction.

Alpha-(α-)particle emission The emission of α particles (helium nuclei) in a radioactive decay process.

Aluminosilicate A silicate in which a certain proportion of the silicon atoms are replaced by Al^- ions and a sufficient number of extra metal ions such as Na^+ and Mg^{2+} to balance the charge of the aluminum ions.

Amide An organic compound containing a

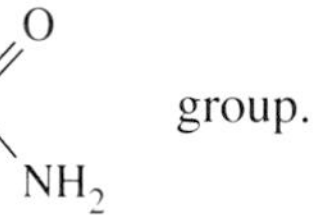

group.

Amine An organic compound containing an NH_2 group.

Amino acid An organic compound containing both an NH_2 (amine) group and a CO_2H (carboxyl) group.

Amorphous solid A noncrystalline solid in which the atoms, molecules or ions have a random disordered arrangement rather than the regular ordered arrangement characteristic of crystalline solids.

Amphiboles A family of minerals containing a double silicate chain with the formula $(Si_4O_{11}{}^{6-})_n$.

Amphoteric substance A substance that can behave both as an acid and as a base.

Anion A negatively charged atom or group of atoms.

Anode The electrode at which oxidation occurs in an electrochemical or electrolytic cell.

Arene A hydrocarbon derived from benzene by substituting one or more of its hydrogen atoms with alkyl groups or by fusing together two or more benzene rings.

Aromatic compound An arene or any compound derived from an arene by replacing hydrogen atoms with other atoms or groups of atoms.

Aromatic substitution reaction The reaction of an arene in which a hydrogen atom is replaced by another atom or group of atoms.

Artificial radioisotope A radioactive isotope that does not occur naturally but is made in a nuclear reaction.

Atmosphere The envelope of gases surrounding the earth and held to it by gravity.

Atmospheric pressure The force exerted by the atmosphere on unit area of the earth's surface.

Atom The smallest indivisible (by chemical reaction) particle of an element that can either exist alone or enter into chemical combination. An atom consists of a central positively charged nucleus surrounded by a sufficient number of negatively charged electrons to balance the charge of the nucleus.

Atomic mass The *average* atomic mass of an atom of an element, allowing for the relative fractional abundances of the various isotopes found on earth; the *relative average* atomic mass is numerically the same as the average atomic mass but has no units.

Atomic mass unit, u The unit of mass defined as $\frac{1}{12}$ the mass of a single $^{12}_{6}$C (carbon-12) atom, 1 u = 1.6605×10^{-24} g, based on assigning one carbon-12 atom a mass of exactly 12 u.

Atomic number, Z The number of protons in the nuclei of the atoms of an element.

Atomic orbital The volume of space around an atom in which an electron has a 90% probability of being found; described mathematically as a probability function defining the spatial distribution of the electron.

Atomic spectrum The range of frequencies of radiation absorbed or emitted by an atom.

Autoionization The term used to describe the self-ionization reaction of water in which one H_2O molecule behaves as an acid and another H_2O molecule as a base:

$$H_2O(l) + H_2O(l) \rightleftharpoons H_3O^+(aq) + OH^-(aq)$$

Avogadro's law The volume of a sample of gas at a given temperature and pressure is proportional to the number of moles of molecules in the sample; V/n = constant (at constant T and P).

Avogadro's number, N_A The number of carbon-12 atoms in exactly 12 g (1 mol) of carbon-12. The number of elementary entities (atoms, molecules, electrons, and so on) in 1 mol of entities; $N_A = 6.022\,05 \times 10^{23}\ \mathrm{mol}^{-1}$.

Balanced equation An equation showing all the reactants and products of a chemical reaction and having the same total number of atoms of each kind on both sides of the equation.

Band of stability The narrow band in which the nuclei of all the stable isotopes fall when the number of neutrons ($A - Z$) in a stable isotope is plotted against its number of protons, Z.

Barometer An instrument for measuring the pressure of the atmosphere.

Base A proton acceptor; in aqueous solution a base accepts a proton from a water molecule forming an hydroxide ion.

Base ionization constant, K_b The equilibrium constant for the reaction of a base in water, $B + H_2O \rightleftharpoons BH^+ + OH^-$, given by the expression:

$$K_b = \left(\frac{[BH^+][OH^-]}{[B]}\right)_{eq}$$

Basic oxide An oxide that reacts with water to give a basic solution (containing $OH^-(aq)$ ions); an oxide that reacts with an acidic oxide to give a salt.

Battery An electrochemical cell used as an energy source.

Beta (β) particle, $_{-1}^{0}e$ An electron emitted from a nucleus as a result of the transformation of a neutron into a proton (which is retained in the nucleus) and an electron.

Beta-(β-)particle emission The emission of β particles (electrons) in a radioactive decay process.

Beta (β) pleated sheet A common secondary protein structure in which segments of a protein chain fold back on themselves to form parallel strands held together by hydrogen bonds between NH and CO groups.

Bilayer The double layer of phospholipid molecules that constitute the membrane of a cell.

Bimolecular reaction A one-step reaction involving the collision of two molecules.

Binding energy (nuclear) The energy released in the formation of a nucleus from its constituent particles.

Biosphere That part of the earth that supports life, consisting of the troposphere, the hydrosphere and the upper part of the crust and the living matter that it contains.

Body-centered cubic structure A structure based on a unit cell consisting a lattice point at the center of a cube and one at each of the eight corners.

Bohr model A model of the hydrogen atom in which an electron rotates around the nucleus in one of a limited number of circular orbits.

Bond energy (*BE*) The bond energy for a diatomic molecule is the energy required to dissociate 1 mol of molecules in the gas phase into atoms in the gas phase; for a polyatomic molecule AX_n the *average bond energy* is the energy per bond required to dissociate 1 mol of molecules in the gas phase into atoms in the gas phase.

Bond order The total number of electron pairs forming a bond.

Bonding electron pair A pair of electrons shared between two atoms that are held together by the bond.

Bonding orbital An orbital localized between two atoms in a molecule; a molecular orbital occupied by a bonding pair of electrons. It is lower in energy than the atomic orbitals from which it is formed.

Boyle's law The volume of a given amount of a gas is inversely proportional to the pressure; PV = constant (at constant n and T).

Branched chain A chain of atoms (usually carbon atoms) with attached side chains.

Buffer solution A solution that has the ability to maintain a very nearly constant pH when H_3O^+ or OH^- ions are added.

Calorimetry The measurement of the heat absorbed or liberated in a chemical reaction or a physical change using an apparatus called a calorimeter.

Carbohydrate Member of a large class of naturally occurring compounds that includes sugars and starch consisting of five- and six-membered carbon rings with one or more hydroxyl group substituents.

Carbonyl group The $>C{=}O$ group.

Carboxyl group The $-C(=O)OH$ group.

Carboxylic acid An organic compound containing the $-C(=O)OH$ group.

Catalysis The use of a catalyst to increase the rate of a reaction.

Catalyst A substance that increases the rate of a reaction but is not used up in the reaction.

Cathode The electrode at which reduction occurs in an electrochemical or electrolytic cell.

Cathodic protection A method of protecting a metal from corrosion by making it the cathode of a cell in which a more easily oxidized metal acts as the anode and is preferentially oxidized.

Cation A positively charged atom or group of atoms.

Cell diagram The shorthand representation of an electrochemical cell showing the two half-cells connected by a salt bridge or porous barrier, such as, $Zn(s) | ZnSO_4(aq) || CuSO_4(aq) | Cu(s)$.

Cell potential, E_{cell} The potential difference between the electrodes of an electrochemical cell in volts.

Chain reaction A multi-step reaction in which an initiation step gives a reactive intermediate that participates in ensuing chain-propagation steps. These steps form the products and regenerate the reactive intermediate, which is eventually removed in one or more chain-termination steps.

Change of state The change of a substance from one physical state (gas, liquid, or solid) to another.

Charles' law At constant pressure, the volume of a given amount of gas is proportional to its temperature on the Kelvin scale; V/T = constant (constant n and P).

Chemical bond A chemical bond is said to exist between two atoms whenever they are held strongly together.

Chemical equation The shorthand description of a chemical reaction showing the formulas of the reactants on the left side and the formulas of the products on the right side.

Chemical properties Characteristic properties of a substance that relate to its participation in chemical reactions.

Chemical reaction A process in which one or more substances are changed into one or more different substances; a process in which atoms are rearranged to form new combinations.

Chemical shift The difference in frequency between the absorption peak of a particular hydrogen atom in a molecule and a reference peak.

Chlorofluorocarbon (CFC) A chlorofluoroalkane; also known as Freons, these unreactive substances appear to be destroying ozone in the stratosphere.

Chromatography A process for separating mixtures into their components that depends on the different tendencies of substances to be absorbed onto (stick to) a solid surface or to dissolve in nonvolatile liquid.

***Cis*-isomer** A structural (geometric) isomer of an alkene in which substituents at each end of the double bond are on the same side of the double bond.

Codon A sequence of three nucleotide bases on mRNA that specifies a given amino acid.

Combustion A self-sustaining chemical reaction between oxygen and other reactants.

Complex ion (complex) A polyatomic ion (or molecule) consisting of a central metal ion to which are bonded several other ions or molecules, called ligands.

Compound A substance with a fixed characteristic composition.

Condensation The transformation of a gas (vapor) to a liquid.

Condensation polymer A polymer formed from monomers by a condensation reaction in which molecules of small molecular mass, such as H_2O, are eliminated.

Condensation reaction The reaction of two or more molecules to form a larger molecule with the elimination of a small molecule, such as water.

Conduction band An incompletely filled band of energy levels, or an empty band into which electrons are easily excited, in which electrons can move freely and therefore contribute to metallic conductance.

Conformations The different arrangements of the atoms and the groups in a molecule (different shapes) that are possible as a consequence of rotation about single bonds.

Conjugate acid-base pair An acid and a base that are related by the equation base + $H^+ \rightleftharpoons$ acid.

Continuous chain A single (nonbranching) chain of atoms (usually carbon atoms).

Coordination compound A complex ion or complex.

Coordination number The number of ligands bound to the central metal ion in a complex ion or complex.

Copolymer A polymer produced by the polymerization of two or more types of monomers.

Core (of the earth) The innermost zone of the earth composed largely of iron.

Core (of an atom) The nucleus of an atom plus all the electrons in its filled inner shells; an atom minus its valence shell electrons.

Core charge The positive charge carried by an atomic core; equal to the atomic number Z minus the number of electrons in its filled inner shells.

Corrosion The deterioration of metals as a result of their reactions with the environment.

Cosmochemistry The chemistry of the other planets, stars and interstellar space.

Covalent bond A directional bond consisting of one or more pairs of electrons (bonding electron pairs) shared between two atoms.

Covalent molecule A molecule in which the atoms are joined by covalent bonds.

Covalent network solid A network solid in which the atoms are joined together in three dimensions by covalent bonds.

Covalent radius Half the distance between two atoms held together by a covalent bond.

Cracking reaction A reaction in which alkanes are converted to shorter chain alkanes and alkenes by heating.

Crust The outer layer of rock forming a thin skin over the earth's surface.

Cubic close-packed structure A structure in which atoms are close-packed in layers and stacked on each other in an ABCABC. . . arrangement such that the atoms in any layer lie directly above those three layers below.

Cycloalkane An alkane in which the carbon chain forms a ring.

Dalton's law of partial pressures The total pressure of a mixture of gases is the sum of the partial pressures of its components; $P_{total} = \sum p_i$.

Decomposition reaction A reaction in which a compound decomposes into simpler products; the reverse of a synthesis reaction.

Dehydrating agent A substance that removes water, or the elements of water, from other substances.

Delocalized electrons Bonding electrons that are not localized in a bond between two atoms but are spread among more than two atoms.

Density The mass of unit volume of a substance.

Deoxyribonucleic acid (DNA) The molecule that stores the genetic code.

Diffusion The process by which one gas mixes with another.

Diol An alcohol containing two —OH groups.

Dipole Two separated charges of equal magnitude and opposite sign.

Dipole moment, μ The product of the magnitude of the charges, Q, and the distance, r, between the charges of a dipole: $\mu = Qr$.

Dipole–dipole forces The intermolecular forces that result from the attraction between two molecules that each have a dipole moment.

Diprotic acid An acid with two ionizable hydrogen atoms.

Disaccharide A carbohydrate that contains two carbon rings; a carbohydrate derived from two monosaccharides.

Dispersion forces See **London forces**.

Dissociation reaction A reaction in which a substance splits into two or more simpler species.

Distillation A method of separating mixtures of liquids with different boiling points.

Disulfide An organic compound containing the —S—S— functional group.

Doping The addition of a very small amount of an impurity to very pure silicon to greatly enhance its conductivity.

Double bond A covalent bond consisting of two pairs of shared electrons.

Effusion The process by which a gas escapes from a container through a very small hole.

Elastomer A polymer that has the property of resuming its original shape after being stretched or compressed; a rubberlike material.

Electrochemical cell A device which uses a spontaneous chemical reaction to produce an electric current.

Electrode A strip of metal or other electronic conducting material by means of which electrons are transferred to or from a solution (liquid) in an electrochemical or electrolytic cell.

Electrolysis A process in which electrical energy is used to produce chemical change, in which one substance is oxidized at the anode and another is reduced at the cathode.

Electrolyte A substance that dissolves in water to give ions and a conducting solution.

Electrolytic cell An apparatus in which electrolysis occurs, in which a chemical change is carried out by using an external source of electrical energy.

Electromagnetic spectrum The complete range of electromagnetic radiation (γ-rays, X-rays, ultraviolet light, visible light, infrared radiation, and radiowaves).

Electron A elementary particle with a charge of -1.6022×10^{-19} C and a mass of 9.1096×10^{-31} kg (0.000 548 u).

Electron capture A process in which a nucleus absorbs an inner shell electron and is thereby converted to a new nucleus.

Electron configuration The shorthand representation of the occupancy of the energy levels (shells and subshells) of an atom by electrons.

Electron spin A model that accounts for the magnetic properties of electrons; a spherical electron is imagined as spinning around its own axis in either a clockwise or an anticlockwise direction.

Electron-dot symbol See Lewis symbol.

Electron-gas model The model that describes a metal as consisting of a regular arrangement of positive metal ions (atomic cores) and electrons that behave like a gas moving freely between the metal ions and forming a continuous electron cloud.

Electron-pair domain The approximate region of space occupied by a pair of electrons.

Electronegativity The relative ability of an atom in a molecule to attract the electrons of a covalent bond to itself.

Electroplating Deposition of a layer of a metal on the cathode in an electrochemical cell containing a solution of the salt of the metal to be deposited.

Element A substance composed of atoms of only one kind, all of which have the same atomic number.

Emission spectrum The spectrum of frequencies of radiation emitted by an atom or a molecule in a transition from a higher to a lower energy state.

Empirical formula The simplest formula of a substance that expresses its composition in terms of the ratios of the numbers of different kinds of atoms.

Enantiomers (optical isomers) The left- and right-handed forms of a chiral molecule.

Endothermic reaction A reaction in which heat is absorbed; a reaction for which ΔH has a positive value.

Energy level One of the allowed quantized energy states of an atom or molecule.

Enthalpy change, ΔH The heat absorbed in a reaction at constant pressure; $\Delta H = q$ (at constant pressure) where q is the heat transferred.

Enthalpy of hydration The enthalpy change for the process of dissolving gaseous ions in water.

Enthalpy of solution The enthalpy change when a solute dissolves in a solvent.

Entropy, S A function that measures the extent of disorder in a system.

Enzyme A large molecule—generally a globular protein—that acts as a catalyst for biological reactions.

Enzyme–substrate complex A complex of enzyme and a substrate in which the enzyme holds the substrate in a position to react.

Equilibrium constant, K_{eq} A ratio giving the equilibrium position of a chemical reaction at a given temperature; the product of the concentrations of the reaction products divided by the product of the concentrations of the reactants for a chemical reaction at equilibrium.

Equilibrium constant expression The expression defining the equilibrium constant, K_{eq}, for a specific reaction at a given temperature; the reaction

$$aA + bB + cC \ldots \rightleftharpoons pP + qQ + rR$$

$$K_{eq} = \left(\frac{[P]^p[Q]^q[R]^r \cdots}{[A]^a[B]^b[C]^c \cdots}\right)_{eq}$$

Equivalence point (acid–base titration) The point in an acid–base titration at which moles of added acid equals moles of initial base or vice versa.

Ester An organic compound with functional group $-C(=O)-OR$, formed by the elimination of water between an alcohol and an acid.

Ether Common name for the alkoxyalkanes, R—O—R′.

Evaporation The escape of molecules from the surface of a liquid; also called vaporization.

Excited state A state in which an atom has one or more of its electrons in energy levels other than the lowest available (ground-state) energy levels.

Exothermic reaction A reaction in which heat is liberated; a reaction for which ΔH has a negative value.

Face-centered cubic structure A structure based on a unit cell with one lattice point at each of the eight corners of a cube and one at the center of each of the six faces, giving a total of four lattice points associated entirely with one cell.

Faraday Constant, F The charge on 1 mol of electrons; $1\ F = 96{,}485\ C \cdot mol^{-1}$.

Fat A mixture of triglycerides that is solid because it contains a high proportion of saturated fatty acids.

Fatty acid A long-chain carboxylic acid that may contain one or more double bonds and a carboxylic acid group at one end.

Feldspars A common family of aluminosilicates containing the network anion $(AlSiO_4^-)_n$.

Ferromagnetic substance A substance that can be permanently magnetized.

Filtration A method whereby a solid and a liquid or solution can be separated by pouring them through a filter funnel containing porous filter paper, or a sintered glass disk, that retains the solid.

First law of thermodynamics The energy of an isolated system is constant; an alternative name for the law of conservation of energy.

First-order rate law The rate law for a reaction with a rate proportional to the concentration of only one reactant; rate $= k_1[A]$.

Fission A nuclear reaction in which a heavy nucleus splits into two lighter nuclei with the production of a large amount of energy.

Formal charge The charge on each atom of a molecule, assuming that the shared electrons are equally divided between the bonded atoms of a pair.

formal charge = (core charge) − (number of unshared electrons) − $\frac{1}{2}$ (number of shared electrons)
= (group number) − (number unshared electrons) − (number of bonds)

Free radical A molecule or ion with an odd number of electrons.

Free rotation The property of a single bond that allows the atoms and the groups attached at one end of the bond to rotate freely with respect to those at the other end.

Frequency ν The number of wave crests that pass a given point in one second; related to wavelength λ and velocity v by $\lambda v = \nu$; units: s^{-1} (Hertz, Hz).

Fuel cell An electrochemical cell in which the substances that are oxidized and reduced are fed continuously to the anode and cathode, respectively.

Functional Group An atom or group of atoms in an organic molecule that gives the molecule characteristic chemical properties.

Fusion A nuclear reaction in which two light nuclei combine to form a heavier nucleus with the production of a large amount of energy.

Gamma-(γ-)ray emission A radioactive decay process in which a nucleus in an excited state decays to the ground state by the emission of γ rays.

Gas A state of matter characterized by its fluidity, lack of a surface, easy compressibility, and the ability to completely fill any container it occupies.

Gas constant, *R* The proportionality constant in the ideal gas law, $PV = nRT$; the gas constant equals 0.082 06 $atm \cdot L \cdot mol^{-1} \cdot K^{-1}$, 8.314 $J \cdot K^{-1} \cdot mol^{-1}$, or 8.314 $kPa \cdot dm^3 \cdot mol^{-1} \cdot K^{-1}$.

Geiger counter An instrument for detecting the α particles and γ rays emitted by radioactive nuclei.

Genetic code The complete set of the 64 messenger RNA codons needed to specify amino acids (or stop signals).

Geochemistry The study of the chemical composition of the earth.

Geometic isomers Molecules with the same composition in which all the atoms are connected to each other in the same way but that differ in the positions of atoms or groups of atoms in space and cannot be converted one into another without breaking bonds.

Gibbs free energy change, ΔG A thermodynamic function that takes into account the entropy change and enthalpy change of a closed system. In a spontaneous process the Gibbs free energy decreases: $\Delta G < 0$, where $\Delta G = \Delta H - T\,\Delta S$. A reaction at equilibrium has $\Delta G = 0$.

Glycerophospholipid An ester of glycerol with two fatty acids and a monoester of phosphoric acid; a derivative of phosphatidic acid.

Graham's law The rate of diffusion or effusion of a gas is inversely proportional to the square root of its molar mass.

Greenhouse effect The warming of the troposphere by the absorption of infrared radiation by atmospheric gases such as CO_2 and H_2O.

Ground state The lowest energy state of an atom or molecule.

Group One of the vertical columns of the periodic table.

Haber process The industrial process of nitrogen fixation in which $N_2(g)$ reacts with $H_2(g)$ to give $NH_3(g)$.

Half-cell potential The potential produced by the cathode half or the anode half of an electrochemical cell.

Half-life, $t_{\frac{1}{2}}$ (of a radioactive nucleus) The time it takes for half the nuclei in a given sample to decay.

Half-life, $t_{\frac{1}{2}}$ (of a reaction) The time it takes the concentration of a reactant to decrease to half its value.

Half-reaction The oxidation half or the reduction half of an oxidation–reduction reaction.

Halide A compound of a halogen and another element.

Haloalkane (alkyl halide) An organic molecule containing one or more halogen atoms.

Halogen An element of group VII of the periodic table.

Halogenation A substitution reaction in which a hydrogen atom is replaced by a halogen atom. Most frequently encountered halogenations involve alkanes or arenes.

Heat, *q* Energy transferred as a result of a temperature difference.

Heat capacity The amount of heat required to raise the temperature of a system by 1 K.

Hemoglobin An iron-containing protein that functions as an oxygen carrier.

Henderson–Hasselbalch equation The expression used to calculate the pH of a buffer solution:

$$pH = pK_a + \log \frac{[base]}{[acid]}$$

where [acid] and [base] are the concentrations of acid and conjugate base, or base and conjugate acid, used to make up the buffer solution.

Hess's law The enthalpy change for a reaction is the sum of the enthalpy changes for the individual steps of the reaction and is independent of the path by which the reaction is carried out.

Heterocyclic molecule A molecule with one or more atoms other than carbon replacing carbon atoms in a cyclic organic molecule.

Heterogeneous catalyst A catalyst that is in a different phase from the reactants.

Heterogeneous equilibrium A state of equilibrium in which some the participating substances are in different phases.

Heterogeneous mixture A mixture of two or more substances that has a nonuniform composition. Different parts of the mixture have different physical and chemical properties.

Hexagonal close-packed structure A structure in which layers of close packed atoms are stacked one upon another in an ABABAB . . . arrangement such that alternate layers lie directly above each other.

Homogeneous catalyst A catalyst that is in the same phase as the reactants.

Homogeneous equilibrium A state of equilibrium in which all the participating substances are in the same phase, commonly all in the gas phase or all in solution.

Homogeneous mixture A mixture of two or more substances that has throughout the same uniform composition and physical and chemical properties but whose composition and properties depend on the relative amounts of the substances in the mixture.

Hund's rule The rule that states that electrons in the same energy level as far as possible occupy separate orbitals and have the same spin.

Hybrid orbital A wave pattern (orbital) produced by the mathematical combination of the wave patterns (orbitals) of two or more electrons in an atom.

Hybridization The mathematical operation in which the wave patterns of the electrons (orbitals) in an atom are combined to produce new wave patterns (hybrid orbitals).

Hydrated salt A salt with crystals containing water molecules as part of its three-dimensional network structure. A salt containing water of crystallization.

Hydride A compound of hydrogen with another element (usually a metal).

Hydrogen bond The intermolecular attraction between a hydrogen atom that is bonded to an electronegative atom, and therefore has a partial positive charge, and an unshared electron pair of another electronegative atom.

Hydrogen electrode (hydrogen half-cell) Consists of a platinum electrode covered with a fine powder of platinum around which hydrogen, $H_2(g)$ is bubbled. Its potential is defined as zero.

Hydrogenation reaction Addition of hydrogen across a double or triple bond.

Hydronium ion, H_3O^+ The ion formed by all acids in aqueous solution.

Hydrophilic Water loving; tending to dissolve in water.

Hydrophobic Water hating; tending not to dissolve in water.

Hydrosphere All the water on and near the earth's surface.

Hydrothermal process The process by which elements and compounds are dissolved from rocks by water at high temperature and pressure, transported to another location and precipitated as concentrated mineral deposits.

Hydroxide ion, OH^- The ion formed by all bases in aqueous solution.

Ideal gas A hypothetical gas that obeys the ideal gas law exactly.

Ideal gas law The law that relates the pressure, volume and temperature of an ideal gas; $PV = nRT$.

Igneous rock A rock that crystallizes from or solidifies from molten rock (magma).

Immiscible Unmixable, for liquids; two liquids that do not mix are immiscible.

Impurity A substance in a mixture that is present in a small amount with respect to the major component of the mixture; a contaminant.

Indicator A weak acid HIn that has a conjugate base In^- which has a different color from the acid HIn and whose color therefore depends on the pH of the solution.

Induced-fit model A model that pictures an enzyme as having an active site that can change its shape to accommodate the substrate.

Infrared (IR) spectroscopy The study of the spectra obtained from substances that absorb infrared radiation.

Initial rate The rate of a reaction at the start of the reaction (at time zero).

Inorganic chemistry The chemistry of all the elements except carbon.

Inorganic compound A compound that does not contain carbon; however, a few compounds containing carbon (such as CO_2) are, for historical reasons, regarded as inorganic compounds.

Integrated first-order rate law The integrated form of the rate law for a first-order reaction, $\ln[A]_t = \ln [A]_0 - k_1t$, that relates $[A]_0$, the initial concentration of a reactant to the rate constant k_1 and to $[A]_t$, the concentration of this reactant at time t.

Intermolecular forces The weak forces of attraction between molecules.

Internal energy, *E* The energy stored in a substance; the sum of the kinetic and potential energies of all the nuclei and electrons of which a substance is composed.

Internal energy change, ΔE The change in internal energy of a system in a reaction; $\Delta E = q + w$, the sum of the heat absorbed by the system and the amount of work done on the system.

Ion An atom or group of atoms carrying a charge.

Ion exchange The removal of an ion from solution and replacement by another when the solution is passed through a zeolite or other ion exchange material.

Ionic bond A bond resulting from the electrostatic attraction between oppositely charged ions.

Ionic conduction Conduction resulting from the movement of positive and negative ions—generally in solution; also called electrolytic conduction.

Ionic radius The radius of a monatomic ion.

Ionic solid (crystal) A crystalline solid composed of a regular array of positive and negative ions.

Ionization energy The energy required to remove an electron from a particular energy level of an atom.

Isoelectronic Atoms or ions that have the same electron arrangement are said to be isoelectronic.

Isolated system A system to or from which no energy or matter is transferred.

Isotopes Atoms with the same atomic number but with different masses.

Isotopic abundance The fractional or percentage abundance of each of the naturally occurring isotopes of which an element is composed.

Kelvin, K The unit of temperature on the Kelvin scale; $1 \text{ K} = 1°\text{C}$.

Kelvin temperature scale Temperature scale based on the absolute zero of temperature defined as 0 K; $T \text{ K} = t°\text{C} + 273.15$.

Ketone An organic compound containing a carbonyl, $>C{=}O$, group.

Kinetic energy The energy that a body has as a consequence of its motion. Kinetic energy $= \frac{1}{2}mv^2$, where m is the mass and v is the velocity.

Kinetic theory (of gases) A model based on the assumption that gases consist of particles of negligible volume between which there are no forces and that are in constant motion at any temperature above 0 K.

Lanthanide One of 14 elements within Period 6 of the periodic table from $Z = 58$ to $Z = 71$.

Lattice The three-dimensional array of points in space used to describe the positions of atoms in crystals.

Lattice energy The energy required to separate the ions in an ionic solid.

Law of conservation of mass The total mass of the products of a chemical reaction is equal to the total mass of the reactants. In a chemical reaction, mass is conserved because the total number of atoms remains unchanged.

Law of conservation of mass and energy In a nuclear reaction, mass and energy are conserved.

Le Châtelier's principle If any of the conditions affecting a dynamic equilibrium is changed so the system is no longer in equilibrium, when equilibrium is once again established the original conditions are as far as possible restored.

Lewis acid A substance that can accept one or more lone pairs of electrons into its valence shell; an electron-pair acceptor.

Lewis base A substance that can donate a lone pair of electrons into the valence shell of a Lewis acid; an electron-pair donor.

Lewis structure A diagram showing the locations of all the valence-shell bonding and nonbonding electron pairs in a molecule or ion.

Lewis symbol The symbol of an element surrounded by single (unpaired) and/or paired dots representing the electrons in the valence shell.

Ligand A molecule or ion (Lewis base) bonded to a metal ion.

Limiting reactant The reactant among two or more reactants whose amount determines the maximum amount of products that may be formed in a given reaction, leaving excess amounts of the other reactants unused.

Line spectrum An absorption or emission spectrum consisting of radiation of only certain specific frequencies rather than a continuous range of frequencies.

Lipid A naturally occurring molecule found in plants and animals that is insoluble in water but soluble in nonpolar organic solvents.

Liquid A state of matter characterized by its fluidity, relative incompressibility, and an ability to flow to take the shape of the part of the container it occupies.

Lock-and-key model A model for enzyme catalysis that pictures an enzyme as a large molecule with a cleft into which only specific substrate molecules can fit.

London (dispersion) forces Attractive intermolecular forces resulting from the interaction of the fluctuating instantaneous dipoles in atoms and molecules.

Lone pair A nonbonding pair of electrons.

Macromolecule A giant molecule such as a polymer.

Mantle A thick layer of largely silicate rock that separates the earth's crust from the core below.

Mass defect The difference between the observed mass of an atom and the sum of the masses of its constituent particles (protons, neutrons, and electrons).

Mass number, *A* The total number of nucleons (protons and neutrons) in the nucleus of an atom of an element.

Mass spectrometer An instrument for measuring the masses of atoms and molecules by converting them to positive ions and passing the ions through a magnetic field.

Mass spectrometry A method of analysis using the mass spectrometer to determine the masses of ions produced by bombarding the vapor of a substance with fast-moving electrons.

Mean free path The average distance traveled by a molecule between collisions.

Membrane The bilayer of glycerophospholipids with embedded proteins that encloses a cell.

Messenger RNA (mRNA) The kind of ribonucleic acid whose function is to carry genetic messages transcribed from deoxyribonucleic acid and to direct protein synthesis.

Metal A solid element that has a high electrical conductivity, reflects light, is malleable and ductile, and exhibits photoelectric and thermionic effects. Mercury is the only metal that is liquid at room temperature.

Metallic bond The type of bond that holds atoms together in a metal. It results from the electrostatic attraction between a negative charge cloud of electrons and positive metal ions.

Metallic conduction Electrical conduction due to the movement of electrons through a metal under the influence of an applied voltage.

Metallic radius The radius of an atom in a metal.

Metalloid A semimetal.

Metamorphic rock Rock that has been modified by the effects of heat and/or pressure.

Micas A family of minerals with sheet silicate structures.

Mineral (in chemistry and geology) A naturally occurring crystalline inorganic compound or element.

Mineral (in nutrition) An inorganic substance required in metabolism.

Miscible Unmixable, for liquids; two liquids that form a homogeneous mixture (solution) are miscible.

Mixture A material containing two or more substances.

Moderator A substance used in a nuclear reactor to slow down neutrons.

Molar enthalpy of fusion, ΔH_{fus} The enthalpy change for the fusion of 1 mole of a substance.

Molar enthalpy of vaporization, ΔH_v The enthalpy change for the vaporization of 1 mole of a liquid.

Molar heat capacity The amount of heat required to raise the temperature of 1 mole of a substance by 1 kelvin.

Molar mass The mass of one mole of a substance.

Molar volume The volume of 1 mole of a substance; 22.41 L for an ideal gas at standard temperature and pressure.

Molarity, M ($mol \cdot L^{-1}$) The number of moles of solute contained in 1 liter (1 dm^3) of solution.

Mole (mol) The amount of a substance that contains Avogadro's number, N_A, of entities (such as atoms, molecules, or ions).

Molecular formula The formula of a substance that gives the numbers of atoms of each different kind in one molecule of the substance.

Molecular mass The sum of the masses of all the constituent atoms in a molecule.

Molecular shape (geometry) The shape or geometry that describes the spatial arrangement of the atoms of a molecule.

Molecular solid A solid consisting of individual covalent molecules held together by intermolecular forces.

Molecularity The number of reactant molecules taking part in a single-step reaction.

Molecule A combination of atoms that has its own characteristic set of properties.

Monoprotic acid An acid with one ionizable H atom.

Monosaccharide A simple carbohydrate that contains a single carbon ring.

Multiple bond A covalent bond containing more than one pair of electrons.

Myoglobin An iron-containing protein that stores oxygen in muscle.

***n*-type semiconductor** The type of semiconductor formed by doping silicon with a Group V atom; its semiconductivity is due only to (*n*egative) electrons.

Nernst equation An expression that gives the relation between the voltage of a cell, E_{cell}, and the concentrations of the reactants and products:

$$E_{cell} = E^{\circ}{}_{cell} - \frac{RT}{nF} \ln Q$$

where $E^{\circ}{}_{cell}$ is the standard cell potential, R is the gas constant, T is the temperature, n is the number of electrons transferred in the reaction, and F is the Faraday constant.

Net ionic equation An equation expressing the actual reaction occurring between electrolytes in solution in which some of the ions are spectator ions.

Network solid A solid with an infinite three-dimensional network of atoms or ions in which each atom or ion is strongly bonded to its neighbors in a definite repeating arrangement in which no individual molecules can be detected.

Neutralization reaction A reaction between an acid and a base to give a salt and water.

Neutron A fundamental particle that has a mass of $1.674\ 95 \times 10^{-27}$ kg (1.008 66 u) and a zero charge; found in the nucleus.

Newton Unit of force; the force that accelerates a 1-kg mass by $1\ m \cdot s^{-2}$.

Nitration reaction A reaction in which a hydrogen atom of an arene is replaced by a nitro (NO_2) group.

Nitrogen fixation A process for converting $N_2(g)$ of the air into useful compounds.

Nuclear magnetic resonance (NMR) spectroscopy The study of the spectra resulting from the absorption of radio-frequency radiation by substances containing magnetic nuclei and situated in a magnetic field.

Noble gas An element of Group VIII of the periodic table.

Nonbonding pair A pair of electrons in the valence shell of an atom that is not involved in the bonding in a molecule; also called an unshared pair, or a lone pair.

Nonelectrolyte A substance that does not conduct an electric current in solution in water.

Nonmetal An element of low electrical conductivity, high ionization energy, and relatively high electronegativity; located on the right-hand side of the periodic table.

Nonpolar molecule A molecule that has no permanent dipole moment.

Normal boiling point The temperature at which the vapor pressure of a liquid is 1 atmosphere.

Nuclear charge The charge on the nucleus of an atom, equal to the number of protons in the nucleus.

Nuclear energy The energy released in nuclear reactions.

Nuclear force The very strong attractive force between nucleons that holds them together in an atomic nucleus.

Nuclear power The power generated by a nuclear reactor.

Nuclear reaction A reaction in which one or more nuclei are transformed into other nuclei.

Nuclear reactor A device for generating energy by means of nuclear reactions and converting it to a useable form.

Nucleic acid A biological polymer formed from nucleotides.

Nucleon A nuclear particle—a proton or a neutron.

Nucleophile A molecule or ion carrying a negative charge and one or more unshared electron pairs that attacks a postively charged atom in another molecule.

Nucleophilic substitution A substitution reaction in which an atom or group of atoms is replaced by a nucleophile.

Nucleosynthesis The nuclear reactions leading to the synthesis of the elements in stars.

Nucleotide A building block for nucleic acids, consisting of a five-carbon sugar bonded to a heterocyclic amine and phosphoric acid.

Nucleus The very small, positively charged particle composed of protons and neutrons at the center of an atom; accounts for almost all the mass of an atom.

Nuclide The nucleus of a isotope; a nucleus with particular values of A and Z.

Octet A valence shell containing eight electrons.

Octet rule The octet rule states that in compound formation an atom as far as possible gains or loses electrons, or shares pairs of electrons, until it has eight electrons in its valence shell.

Ore A mineral that is economically valuable to mine.

Organic chemistry The chemistry of the element carbon and its compounds.

Organic compound A compound that contains carbon (with a few exceptions, such as CO_2, for historical reasons).

Ostwald process The industrial process for the manufacture of nitric acid from ammonia.

Oxidation A half-reaction in which a reactant loses electrons; a half-reaction in which the oxidation number of an atom increases; a reaction in which an element or compound combines with oxygen (limited definition).

Oxidation number A number that indicates the extent of the oxidation of an element in its compounds; the charge an atom would have if the electrons in each of its bonds were transferred to the more electronegative atom of each bond.

Oxidation potential, E_{ox} The potential produced by a half-cell in which oxidation is occurring.

Oxidation state An element is said to be in the same oxidation state in all its compounds in which it has the same oxidation number.

Oxidation–reduction reaction An electron-transfer reaction; a reaction in which one reactant is oxidized and another reactant is reduced.

Oxide A compound of an element with oxygen.

Oxidizing agent A substance that gains electrons; a substance that is reduced.

Oxidizing strength The relative power of a substance to gain electrons (to be reduced).

Oxoacid A substance whose molecules have one or more OH groups attached to an electronegative atom, with the general formula $XO_m(OH)_n$, where $m = 0, 1, 2, 3, \ldots$ and $n = 1, 2, 3, \ldots$.

Ozone Trioxygen, O_3, an allotrope of oxygen.

Ozone layer A layer of the stratosphere at a height of about 25 to 35 km in which the ozone concentration reaches a maximum value of 20 to 30 ppm; this layer protects the earth from harmful ultraviolet radiation.

***p-n* junction** A junction between a *p*-type semiconductor and an *n*-type semiconductor.

***p*-type semiconductor** The type of semiconductor formed when silocon is doped with a Group III element; its semiconductivity is due only to *p*ositive holes.

Paramagnetism The magnetic properties of an ion or molecule that result from one or more unpaired electrons.

Partial pressure The pressure exerted by one gas in a mixture of gases; the pressure it would exert alone under the same conditions.

Pascal, Pa The SI unit of pressure; $1\ Pa = 1\ N \cdot m^{-2}$.

Pauli exclusion principle No orbital can accommodate more than two electrons, and these two electrons must have opposite spins.

Peptide bond An amide bond between amino acids in a protein.

Percent composition The composition of a substance expressed as the mass percentage of each of its constituent elements.

Percent yield The actual amount of a product obtained in a reaction expressed as a percentage of the amount that should be obtained according to the balanced equation for the reaction.

Period A horizontal row of the periodic table.

Periodic table The arrangement of the elements, in order of increasing atomic number, in eight principal vertical columns (groups) and seven horizontal rows (periods), which shows a regular repeating variation in the properties of the elements.

pH A convenient way to express the concentration of H_3O^+ in solution;

$$ph = -\log_{10}[H_3O]^+$$

Phase change The transformation of a substance from one physical state to another.

Phenol An aromatic organic compound containing an —OH group.

Photochemical smog Smog (a mixture of solid particles, liquid droplets and gases) formed by the action of sunlight on substances emitted in automobile exhaust.

Photodissociation (photolysis) The dissociation of a molecule into two fragments when a bond is broken by the energy of a photon absorbed by the molecule.

Photoelectric effect The emission of electrons from a surface in response to the absorption of electromagnetic radiation.

Photoelectron spectroscopy A method for determining the ionization energies of the electrons in an atom by measuring the kinetic energies of the electrons ejected from an atom by high energy ultraviolet or X-ray photons.

Photon A particle (quantum) of electromagnetic radiation (light) with energy $E = h\nu$, where h is Planck's constant and υ is the frequency.

Physical property A property of a substance that can be observed and measured without the substance changing into other substances.

Physical state Solid, liquid, or gas.

Planck constant, *h* The proportionality constant relating the energy of a photon to its frequency, $E = h\nu$; $h = 6.626\ 18 \times 10^{-34}$ J · s.

Plasma A state of matter that exists only at very high temperatures in which all the electrons have been stripped from atoms so that it consists of a mixture of free electrons and nuclei.

Polar molecule A molecule that has a permanent dipole moment.

Polar covalent bond A covalent bond between two atoms of different electronegativities in which the atom of higher electronegativity has a partial negative charge, and the other atom of lower electronegativity has a partial positive charge of equal magnitude; a bond in which the bond electrons are unequally shared.

Polarizability The ease with which the electrons in atoms and molecules can be displaced from their equilibrium positions.

Pollutant (atmospheric) Substance not normally present in the atmosphere that results from human activities.

Polyamide A condensation polymer with units joined by amide bonds, formed from an amino acid or from a dicarboxylic acid and a diamine.

Polyatomic ion A positive or negative ion containing more than one atom.

Polyester A condensation polymer with ester bonds between units, formed from a dicarboxylic acid and a diol.

Polyprotic acid An acid with more than one ionizable hydrogen.

Polysaccharide A carbohydrate composed of many monosaccharides bonded together.

Positive hole A positively charged silicon atom formed when silicon is doped with a Group III element.

Positron A particle with the same mass as an electron but with a positive charge.

Positron emission A radioactive decay process in which positrons are emitted.

Precipitate An insoluble compound formed in a precipitation reaction.

Precipitation reaction A reaction in which an insoluble compound is formed on mixing solutions of two soluble compounds.

Pressure Force per unit area.

Primary structure (of a protein) The sequence in which amino acids are linked together in a protein.

Primary cell A nonrechargeable battery.

Product One of the substances produced in a chemical reaction.

Protein A large biomolecule formed by the polymerization of many amino acids.

Proton A fundamental particle with a mass of $1.627\ 26 \times 10^{-27}$ kg (1.007 28 u) and a charge of $+1.6022 \times 10^{-19}$ C; found in the nucleus.

Pure substance A substance that consists entirely of one kind of molecule; a substance that contains no impurities.

Pyroxenes A family of silicate minerals that contain single chain silicate anions with the formula $(SiO_3\ )_n$.

Quantum mechanics The mechanics used to describe the behavior of very small particles, in particular the behavior of electrons in atoms and molecules.

Quaternary structure The way in which two or more protein chains are combined together to form a larger structure.

Racemic mixture A mixture of equal amounts of the two enantiomers of a chiral molecule.

Radioactive decay The process in which a radioactive nucleus emits particles (such as electrons, positrons, or α particles) and is transformed into another nucleus.

Radioactive decay constant The rate constant for the first order process of radioactive decay.

Radioactive decay series The series of radioactive decay processes by which an unstable nucleus is transformed to a stable nucleus.

Radioactivity The spontaneous disintegration of an unstable (radioactive) nucleus.

Radiocarbon dating The determination of the age of formerly living material by determining its content of carbon-14.

Radioisotope A radioactive isotope.

Rate constant, *k* The proportionality constant that appears in a rate law.

Rate law An equation of the form rate $= k[A]^x[B]^y$ that relates the rate of a reaction to the concentrations of the reactants.

Rate-determining (-limiting) step The slowest step in the series of steps that constitute the mechanism of a reaction; the step that determines the overall reaction rate.

Reactant A substance taking part in a chemical reaction.

Reaction enthalpy The enthalpy change for a reaction.

Reaction intermediate A reactive species that is formed in small concentrations in one step of a multistep reaction mechanism, is used up in a following step, and does not appear in the balanced equation for the reaction or the rate law.

Reaction mechanism A series of successive reaction steps proposed to explain how the reactants are converted to the products of a reaction.

Reaction order The number of concentration terms in the rate law for a reaction. For the rate law rate $= k[A]^x[B]^y$, the reaction order is $x + y$.

Reaction quotient, Q An expression with the same form as the equilibrium constant expression, except that the concentrations are generally not the equilibrium concentrations.

Reaction rate The speed at which a reaction proceeds; the change in the concentration of a reactant, or a product, in a given time interval divided by that time interval.

Reducing agent A substance that gives up electrons; a substance that is oxidized.

Reducing strength The relative power of a substance to lose electrons (to be oxidized).

Reduction A half-reaction in which a reactant gains electrons; a half-reaction in which the oxidation number of an atom decreases; a reaction in which a compound loses oxygen (limited definition).

Reduction potential, E_{red} The potential of any half-cell in which reduction is occurring.

Replication The process whereby a deoxyribonucleic acid molecule makes copies of itself.

Resonance structures A set of two or more Lewis structures used to represent the structure of a molecule that cannot be adequately represented by a single structure because some of the electrons are more delocalized than is indicated by any single Lewis structure.

Reversible reaction A reaction for which both reactants and products are present when it reaches equilibrium; a reaction for which the reverse reacation is not negligibly slow compared with the forward reaction.

Ribonucleic acid (RNA) Nucleic acid responsible for putting the genetic information to use in protein synthesis.

Salt An ionic compound consisting of a positive ion (cation) and a negative ion (anion); the product of the neutralization of a base with an acid.

Salt bridge A tube containing an electrolyte in gelatin, used to form a conducting path between the two halves of an electrochemical cell.

Saturated hydrocarbon An alkane; a hydrocarbon with no multiple bonds.

Second law of thermodynamics In any spontaneous change the total entropy of a system and its surroundings is positive (increases):

$$\Delta S_{total} = \Delta S_{system} + \Delta S_{surroundings} > 0.$$

Second-order rate law The rate law for a reaction in which the rate is proportional to the concentrations of two reactants; rate $= k_2[A][B]$ or rate $= k_2[A]^2$.

Secondary cell A rechargeable battery.

Secondary structure The relative orientation of segments of a protein chain to form, for example, an α helix or a β sheet.

Sedimentary rock Rock formed from compacted or cemented sediments, layers of small particles that result from the breakdown of rocks or from precipitation from solution.

Semiconductor An electronic conductor with a resistance that decreases with increasing temperature.

Semimetal (metalloid) An element with properties intermediate between those of a metal and a nonmetal; found in the periodic table on the border between the metals and the nonmetals.

Shell model The model of an atom that describes it as a nucleus surrounded by concentric spherical shells of electrons.

Side chain An alkyl group or a substituted alkyl group attached to a longer continuous hydrocarbon chain.

Silica Silicon dioxide, SiO_2.

Silicate A salt of silicic acid or polysilicic acid; silicates are the dominant rock-forming minerals on earth.

Single bond A covalent bond consisting of one shared electron pair.

Solubility The concentration of a solution containing the maximum amount of solute that will dissolve in a given amount of solvent at a specified temperature.

Solution A homogeneous mixture.

Specific heat The amount of heat required to raise the temperature of 1 gram of a substance by 1 kelvin.

Spin See **electron spin**.

Spontaneous process A process that proceeds without any external intervention; a process in which the total entropy of a system and its surroundings increases; a process in which the Gibbs free energy of the system decreases, $\Delta G < 0$.

Standard atmosphere 1 atm = 760 mm Hg = 760 torr = 101.325 kPa.

Standard cell potential, E°_{cell} The potential of a cell in volts when the concentrations of all solutes are 1 mol · L^{-1}, all the partial pressures of any gases are 1 atm, and the temperature is 25°C.

Standard enthalpy of formation, ΔH_f° The reaction enthalpy of a reaction in which 1 mole of a compound is formed from its elements in their standard states.

Standard molar entropy, S° The entropy of 1 mole of a substance in its standard state (1 atm and 25°C).

Standard oxidation potential, E_{ox} The potential under standard conditions (25°C with all ions at 1-*M* concentration and all gases at 1 atm pressure) of any half-cell in which oxidation is occurring.

Standard reaction enthalpy, ΔH° The enthalpy change of a reaction under standard conditions:

$$\Delta H^\circ = \sum \Delta H^\circ(\text{products}) - \sum \Delta H^\circ(\text{reactants})$$

Standard reaction entropy, ΔS° The standard entropy change of a reaction:

$$\Delta S^\circ = \sum S^\circ(\text{products}) - \sum S^\circ(\text{reactants})$$

Standard reaction Gibbs free energy, ΔG° The Gibbs free energy change of a reaction under standard conditions:

$$\Delta G^\circ = \sum \Delta G^\circ(\text{products}) - \sum \Delta G^\circ(\text{reactants})$$

Standard reduction potential, E°_{red} The potential under standard conditions (25°C with all ions at 1-*M* concentration and all gases at 1 atm pressure) of a half-cell in which reduction is occurring.

Standard temperature and pressure (STP) O°C (273.15 K) and 1 atm (101.33 kPa).

Starch A polymer of glucose that plays an energy storage role in plants.

Stoichiometry The quantitative aspects of chemical composition and reactions.

Stratosphere The layer of the atmosphere above the troposphere in which the temperature increases with increasing height because of the absorption of ultraviolet radiation from the sun.

Strong acid An acid that is fully ionized; a substance that quantitatively forms H_3O^+ ions is aqueous solution.

Strong base A base that is fully ionized; a substance that quantitatively forms OH^- ions in aqueous solution.

Structural formula The molecular formula of a substance written to show how the atoms are joined (bonded).

Structural isomers Molecules that have the same molecular formula but different structures.

Subshell An alternative name for each of the energy levels in a particular shell; designated by *s*, *p*, *d*, *f*, . . . in order of increasing energy.

Substance An element or a compound; a homogenous portion of matter that has a fixed, constant composition and a characteristic set of chemical and physical properties.

Substrate The reactant in an enzyme-catalyzed reaction.

Superconductor A solid that below a certain critical temperature has zero electrical resistance.

Surroundings The rest of the universe surrounding the reaction vessel and its contents.

Synthesis reaction A reaction in which two or more simpler substances combine to give a more complex substance.

System A reaction vessel and its contents.

Temperature A physical property that is proportional to the average kinetic energy of the molecules of a substance.

Tertiary structure The way in which an entire protein chain is coiled and folded into its specific three-dimensional shape.

Theoretical yield The maximum amount of a product that can be formed in a reaction according to the balanced equation for the reaction.

Thermionic effect The emission of electrons or ions from a metal as a result of heating.

Thermochemistry The quantitative study of the heat changes accompanying chemical reactions.

Thermodynamics The study of energy and its transformations.

Thermoplastic polymer A polymer that can be softened and melted by heating without decomposition so that it can be formed into a variety of different shapes including fibers and films.

Thermosetting polymer A polymer that decomposes on heating so that it cannot be softened or melted to form into other shapes.

Thiol An organic compound containing an —SH group.

Titration A procedure by which the volume of a solution needed to react completely with a known volume of another solution is determined.

Torr Alternative name for a pressure of 1 mm Hg; defined as $\frac{1}{760}$ of a standard atmosphere.

***Trans*-isomer** A structure (geometric) isomer of an alkene in which the substituents at each end of a double bond are on the opposite sides of the double bond.

Transcription The process by which the information in deoxyribonucleic acid is read and used to synthesize messenger RNA.

Transistor A circuit element constructed from thin layers of n and *p*-type silicon that can function as an oscillator or amplifier.

Transition elements Metals that fall between Groups II and III, in Periods 4, 5, 6, and 7 of the periodic table.

Transition state See **Activated Complex**.

Transuranium element An element with an atomic number greater than 92.

Triacylglycerol (triglyceride) A lipid; a triester of glycerol with three fatty acids.

Triple bond A covalent bond consisting of three shared electron pairs.

Triprotic acid An acid with three ionizable hydrogen atoms.

Troposphere The lowest layer of the earth's atmosphere, of almost uniform and constant composition, extending to 10–15 km above the surface, in which temperature decreases with increasing altitude.

Ultraviolet (UV) radiation That part of the electromagnetic spectrum with wavelengths in the range 0.4 μm to 0.04 μm (between the violet light of the visible spectrum and the X-ray region).

Unimolecular reaction A reaction step involving only one molecule.

Unit cell The smallest part of the lattice of a crystal that when repeated regularly in three dimensions gives the complete structure.

Unsaturated hydrocarbon An alkene or an alkyne.

Unshared pair A nonbonding pair of electrons; a lone pair.

Valence The combining power of an element for other elements.

Valence band A filled band of electronic levels that contains the electrons responsible for the bonding in a semiconductor.

Valence shell The outer electron shell of an atom.

Vapor An alternative word for "gas," commonly used when the liquid substance is also present.

Vapor pressure The constant pressure exerted by the vapor above a liquid when equilibrium is established between liquid and vapor.

Vitamin A small organic molecule that must be present in the diet and that is essential in small amounts for proper biological functioning.

Volt, V The SI unit of potential difference, if 1 J of work is done by moving a charge of 1 C through a potential difference, the potential difference is 1 V.

VSEPR (Valence-shell electron-pair repulsion) model A model that describes the geometries of molecules and ions in terms of the number of ligands (n) and the number of nonbonding electron pairs E (m) surrounding a central atom A in an AX_nE_m ion or molecule.

Vulcanization The process of heating rubber with sulfur to form disulfide cross links between the hydrocarbon chains to give improved elastic and other physical properties.

Wave number, $\tilde{\nu}$ The reciprocal of wavelength, in cm^{-1}; the frequency divided by the speed.

$$\tilde{\nu}\,(cm^{-1}) = \frac{\nu\,(s^{-1})}{c\,(cm \cdot s^{-1})}$$

where $c = 2.998 \times 10^{10}\ cm \cdot s^{-1}$.

Wavelength, λ The distance between successive wave crests, or points of equal displacement on successive waves; related to frequency ν and velocity v by $\lambda\nu = v$.

Weak acid An acid that is incompletely ionized; an acid that is only partially converted to H_3O^+ ions in aqueous solution.

Weak base A base that is incompletely ionized; a base that is only partially converted to OH^- ions in aqueous solution.

Work, *w* The mechanical work done on a system is a product of the force acting on the system and the distance through which it acts. In chemical systems the work done on a system is the pressure on the system times the decrease in volume of the system.

X-ray crystallography A technique by which the structures of solids are determined from the diffraction patterns produced when X-rays interact with a crystal.

Zeolite An aluminosilicate mineral that typically contains water that can be removed by heating; an aluminosilicate with a network anion with tunnels and cavities that can contain water and other molecules.

Zone refining A technique for purifying solids in which a molten zone is moved along a rod of solid material. Impurities concentrate in the molten zone and are thus removed to one end of the rod, which can then be cut off.

Index

NOTE: *D* indicates an experiment; *t* indicates a table; *f* indicates a figure; *M* indicates margin notes; and *B* indicates a box.

Absolute zero of temperature, 57
Absorption spectrum, 195*f*, 195
Acetaldehyde, 515, 517
Acetamide, 530
Acetate ions, 141
Acetic acid, 426*M*, 519
  buffer solution, effects of, 439–41
  ionization of, 141
  production of, 520
Acetone, 515, 517
Acetylene. *See* Ethyne
Acid–base equilibria, 425–48
  acid ionization constant, 426–29
  and anions, 435–36
  autoionization of water, 431
  base ionization constant, 429–30
  buffer solutions, 438–41
  and cations, 436–37
  conjugate acid–base pairs, 434–35
  pH scale, 431–33
Acid–base reactions, 122, 138–46
  alcohols, 508–9
  in aqueous solutions, 44, 138–40
  characteristics of, 139–40
  and hydronium ion, 138, 139
  Lewis acid–base reaction, 363–64
  metals, 379
  and natural indicators, 139*D*
  proton transfer in, 149
  titrations, 173*f*, 173–74, 174*f*, 447–48
Acidic oxides, 227, 243, 354, 605
Acid ionization constant, 426–29
  calculation of, 426–28
  for selected acids in water, 427*t*
Acid rain, 602–5
  and coal, 320, 602, 605
  solutions to problem, 605
  sources of, 602–3
  toxic effects of, 603–4
Acids, 138–42
  acidic solutions, 432
  adding base to, 144
  aqueous solutions of, 138–39
  Brønsted–Lowry definition, 139–40
  carboxylic acids, 141
  cations as, 94, 436–37
  diphosphoric acid, 246
  diprotic acid, 228, 229
  electrical conductivity of, 141
  hydroxides of nonmetals as, 249
  ionization of, 140, 426–29
  Lewis acids, 363–64
  oxidation–reduction reactions, 142
  pH, calculation of, 433
  polyprotic acids, 436
  properties of, 138
  reaction with copper, 372
  reaction with iron, 368–69
Acids (*cont.*)
  reaction with metals, 354–55
  strong acids, 140, 141*t*
  triprotic acids, 244
  weak acids, 140, 141*t*
Actinides, 84, 203
Activated charcoal, 268*D*
Activated complex, 563
Activation energy:
  analogy for, 563f
  and Arrhenius equation, 565–66
  and catalysts, 568–69
  and reaction rate, 562–63
Activity series, 354–55, 473
  for selected metals, 355*t*
Addition polymers, 717–24
  nature of, 717
  types of, 717–24
Addition reactions, 288
Adenine, 632, 637
Adenosine diphosphate (ADP), 630–31
Adenosine triphosphate (ATP), 87, 629–31
  conversion to ADP, 630–31
  structure of, 246*f*
Air:
  composition of, 44*t*
  See also Atmosphere
Air pollution. *See* Environmental chemistry
Alanine, 618*t*
  molecular structure, 532*f*
Albite, 87*M*
Alcoholic beverages, fermentation of, 505
Alcohols, 504–11
  acid–base reactions, 508–9
  dehydration reactions, 509
  diols, 507
  ethanol, 505
  methanol, 504–5
  naming of, 506–7
  oxidation reactions, 510
  primary/secondary/tertiary alcohols, 506
  properties of, 508*t*, 508
  substitution reactions, 509–10
  triols, 507
Aldehydes, 514–18
  naming of, 514
  natural occurrence of, 517*f*
  oxidation of, 517–18
  preparation of, 515–16
  silver mirror test for, 518*D*
  types of, 515, 517
Aliphatic compounds, 290
Alkali metals (Group I), 86, 203
  atomic spectrum, 189–91
  atoms, properties of, 350*t*
  boiling and melting points, 345*t*
  densities of, 345*t*
  electrical conductivity, 345*t*
Alkali metals (*cont.*)
  flame tests, 359–60*t*
  hydroxides, 354
  oxides, 354
  properties of, 86, 344*D*, 344–45*t*
  reactions of, 86, 350–51
  types of, 86
Alkaline earth metals (Group II), 86, 203
  atoms, properties of, 351*t*
  flame tests, 359–60*t*
  hydroxides, 354
  oxides, 354
  properties of, 86
  reactions, products of, 86
  reactions of, 351
  salts of, 355
  types of, 86
Alkanes, 274–84
  boiling/melting points of, 274*t*
  branched-chain alkane, 276
  butane, 274
  conformations, 280
  continuous-chain alkanes, 275
  cycloalkanes, 281–83
  drawing structure of, 279–80
  eclipsed conformation, 280, 281*f*
  ethane, 274
  methane, 274
  naming of, 277–79
  natural gas, 283
  petroleum, 283–84
  propane, 274
  reactions of, 281
  staggered conformation, 280, 281*f*
  structure of, 274–75*f*
Alkenes, 284–89
  ethene, 284, 286
  naming of, 286–88
  propene, 286
  reactions of, 288–89
  types of, 287*t*
Alkoxy group, 511–12
Alkyl groups, 278
Alkyl halides. *See* Haloalkanes
Alkynes, 290
  ethyne, 290
Allopurinol, 639*B*
Allotropes:
  of carbon, 265–68
  of phosphorus, 240–41
  of sulfur, 224–26
Alloy, nature of, 12
α-amino acids, 616–17
α-helix structure, of proteins, 617, 620*f*
α particles, 33
  α-particle emission, 654–55
α radiation, 659
Alternative energy sources, 319–21

Alternative energy sources (*cont.*)
  biomass, 320–21
  coal, 319–20
  hydrogen, 320
  nuclear energy, 321
  solar energy, 321
Altitude, and atmospheric pressure, 56
Alumina, 361
Aluminosilicates, 87, 699–701
  structure of, 700*t*
  types of, 699–701
  zeolites, 731–33
Aluminum, Al, 86, 360–65
  aluminosilicates, 699–701
  atoms, properties of, 360*t*
  compounds of, 361–63
  preparation by electrolysis, 486–87
  properties of, 345
  reactions of, 360–65
  uses of, 345
Aluminum bromide, production of, 363
Aluminum chloride:
  production of, 362
  reactions of, 362–63
Aluminum hydroxide, 361–62
  as amphoteric, 362
  production of, 361–62
Aluminum oxide, 361
  production of, 361
  uses of, 361
Alvarez, Luis, 710
Alvarez, Walter, 710
Amides, 530
  amide bond, 617
  amide group, 530
  preparation of, 530
  types of, 530
  urea, 530
Amines, 528–29
  cyclic amines, 529*f*
  heterocyclic amines, 529*f*
  primary/secondary/tertiary, 528
  properties of, 529*t*
  types of, 528–29
Amino acids, 531–34
  and chirality, 531–33*f*
  condensation reaction, 616
  nutritional requirements, 645
  and proteins, 616, 618–19
  structure of, 531, 618–19*t*
Aminobenzene, 528
Ammonia, 5, 143
  amine derivatives of, 528–29
  in fertilizers, 51
  and production of nitric acid, 588–89
  synthesis of, 452
Amorphous solids, 399–400
  glasses as, 399–400
Amphiboles, 698
Amphoteric, meaning of, 249, 362
Amplitude, of electromagnetic radiation, 184
Amylose, structure of, 628*f*
Anaerobic, meaning of, 607
Anesthetics, inhaled type, 513*B*
Angular molecular shape, 109
Anhydrous, meaning of, 230
Aniline, 528, 529*t*
Animal fats, 523, 525*t*
Anions, 94
  acid–base properties, 435–36
Anode, 136, 467, 485
Anthracene, 294
Antifreeze, 16, 507
Aqueous solutions, 12
  acid–base reactions in, 144
  electrolysis of, 489–90
  of hydrogen halides, 138
  precipitation reactions in, 146*D*–47
Arachidonic acid, 524*t*
Arenes, 290
  benzene, 291–94
  graphite, 265–66
  naphthalene, 294
  nature of, 294
  polycyclic arenes, 294
Arginine, 618*t*
Argon, 88
  in sun's atmosphere, 702
Aristotle, 2*B*
Aromatic compounds, 290
  benzene, 291–94
  naming of, 293–94
Aromatic properties, 292
Aromatic ring, 292
Aromatic substitution reactions, 293
Arrhenius, Svante, 565
Arrhenius constant, 565
Arrhenius equation, 565–68
  Arrhenius plot, 566*f*
Aryl group, 511*M*
Asparagine, 619*t*
Aspartic acid, 619*t*
Aspirin:
  production of, 523
  synthesis of, 500f
Astatine, 88
Asteroids, formation of, 692
Asthenosphere, 693
Atmosphere, 43–44, 702–5
  carbon dioxide concentrations, 45*f*
  dry air, composition of, 44*t*
  earth compared to other planets, 691*t*, 703
  formation of, 702–3
  greenhouse effect, 45
  layers of, 43–44, 592
  and life forms of earth, 44–45
  modifications in, 703
  noble gases, 88
  photochemical aspects, 592–94
  temperature variations in, 43*f*
Atmospheric pressure, 55
  and altitude, 56
  and boiling point, 405
  formula for, 54*f*
  measurement of, 55*f*, 55
  nature of, 55
  standard atmosphere, 35
Atomic absorption spectroscopy, 164, 195
Atomic mass, 25, 27–28
  and atomic mass unit (u), 25
  average atomic mass, 25, 27–28
  characteristics of, 25
Atomic mass (*cont.*)
  compared to atomic weight, 29*B*
  Dalton, 25*M*
  mass defect, 35
  mass equivalent, 35
  measurement with mass spectrometer, 25, 26*B*
  molecular mass, 30
  relative average atomic mass, 28, 29*B*
Atomic number, 24
  notation for, 24
Atomic orbitals, 1*s* and 2*s* orbital, 206
Atomic radius, 395*f*, 396
Atomic spectra, 189–95
  absorption spectrum, 195*f*, 195
  and analysis of elements, 194–95
  Bohr model for hydrogen atom, 192–94
  colored flames and, 190*D*
  electron configurations, 200–207
  line spectrum, 189*f*, 189–90
  origin of, 190–91
Atomic theory, elements of, 2*B*
Atomic weight, 28
  compared to atomic mass, 29*B*
Atoms, 1–2
  atomic mass, 25, 27–28
  atomic number, 24
  chemical bonds, 92–101
  close-packed arrangement, 346–47
  core charge of, 92–93
  core of, 92
  electrical nature of, 7*M*
  electronegativity, 101–4
  electrons of, 21–22
  energy levels, 191*f*, 191
  energy states of, 191
  formal charges of, 254
  formulation of model of, 22–23*B*
  Greek conception of, 2*B*
  ions, 7, 94
  isotopes, 24–25
  multi-electron atoms, 211
  neutrons of, 22
  nucleons of, 24
  nucleus of, 22
  observation of, 2–3*B*
  origin of, 685–87
  prefixes denoting number, 47*M*
  protons of, 21
  and radioactivity, 33–34
  shell model of, 90–92
  size in metals, 395–97
  stable and unstable, 33–34
  structure of, 21–22
Autocatalytic reaction, 569
Average atomic mass, 25, 27–28
  from isotopic abundance, 28
Average rate, chemical reactions, 552
Avogadro, Amedeo, 59
Avogadro's law, 59–60
  and gas reactions, 168–69
Avogadro's number, 30–31
  from unit cell dimensions, 393–94

Baekeland, Leo, 729
Bakelite, 720

Band of stability, nucleus of atom, 34, 654, 656
Barium peroxide, 591
Barium sulfate, production of, 356*f*
Barometer, 55*f*, 55
Basalt, 694
Base ionization constant, 429–30
for selected bases in water, 430*t*
Bases, 142–44
adding to acid, 144
basic solution, 432
Brønsted–Lowry definition, 139–40, 142
hydroxides of metals as, 249
ionization of, 429–30
Lewis bases, 363–64
pH, calculation of, 433
strong base, 143
weak base, 143
Basic oxides, 354
Basic oxygen furnace, 376, 377*f*
Batteries, 480–83
dry cell, 481*f*, 481
fuel cells, 482*f*, 482–83
lead–acid cell, 481*f*, 481–82
nickel–cadmium cells, 482
primary cells, 480, 481
secondary cells, 481–82
Becquerel, Antoine-Henri, 660
Bednorz, J. Georg, 743
Bent bonds:
cycloalkanes, 282
molecular shape, 110*f*, 110
Benzene, 291–94
reactions of, 293
structure of, 291–92
Benzoic acid, 521
Beryllium, 200
β particles, 33
β-particle emission, 33, 654–55
β-pleated sheet structure, of proteins, 617, 621*f*
β radiation, 659
Bimolecular reaction, 556
mechanism of, 561
Biochemical reactions, carbohydrate: metabolism, 629–31
Biochemistry, nature of study, 615
Biogas, 320
Biological carbon cycle, 708–9
Biomass, as energy source, 320–21
Biomolecules, 615
carbohydrates, 627–28
computer modeling of, 640–41*B*
lipids, 641–43
proteins, 616–25
Biosphere, 583, 705–11
carbon cycles, 708–9
formation of, 705–6
and maintenance of life, 709, 711
method of study of, 705–6
nitrogen cycle, 708
sulfur cycle, 708
water cycle, 708
Bismuth, 33, 34
Blast furnace, 375*f*, 375
Bleaching powder, 250
Blood:
Blood (*cont.*)
hemoglobin, 370, 371*B*
pH of, 439
Body-centered cubic structure, 348, 392
Bohr, Niels, 192*M*
Bohr model, spectrum of hydrogen atom, 192–94
Boiling point, 405
and atmospheric pressure, 405
normal boiling point, 405
Boltzmann, Ludwig, 564
Bond energies, 314–19
average bond energies, 315*t*
and bond strength, 316–17
calculation of, 316
estimation enthalpy from, 318–19
Bonding orbitals, 208–9
versus nonbonding orbitals, 210
Bond order:
definition of, 295
and resonance structures, 295
Bonds, chemical. *See* Chemical bonds
Boron, B, 86, 200
Boyle, Robert, 56
Boyle's law, 56
apparatus in formulation of, 56*f*
elements of, 56
pressure/volume relationship in, 56, 57*f*
Bragg, Lawrence, 388*B*
Bragg, William, 388*B*
Branched-chain alkane, 276
Breathalyzer, 165
Broglie, Louis de, 195*f*
Bromine, Br_2, 125
compounds of, 125
diffusion of bromine vapor, 70*D*
in oxidation-reduction reactions, 133–34
preparation of, 136
properties of, 124
sources of, 124
states of, 54*f*
uses of, 125
Bromotrifluoromethane, 503
Brønsted, Johannes, 139
Brønsted–Lowry acids and bases, 139–40
Buckminsterfullerene, 267*B*
Buffer solutions, 438–41
of bodily fluids, 439
components of, 438
and La Châtelier's principle, 441–42
pH, determination of, 441
Bulk sample, nature of, 5
Bulk substances:
representation of, 5
states of, 5
Burning, combustion as, 47–48
Butane, 274
mass calculations, 32
structure of, 276*f*
Butanoic acid, 519
t-butyl bromide:
average rates, 552*t*
instantaneous rates, 553*t*
rate of hydrolysis, 552*t*, 553*f*, 553*t*
Butyric acid, sources of, 521

Calcium, Ca, reaction with water, 353*D*
Calcium carbonate, 358
forms of, 358
production of, 358
Calorimeter, 307, 308*f*
coffee-cup calorimeter, 307, 308*f*, 309
flame calorimeter, 308*f*, 309
vacuum flask calorimeter, 308*f*, 309
Calorimetry, 307–9
Calvin, Melvin, 668–69*f*
Carbides, 271–72
formation of, 271–72
Carbocation, 556
Carbohydrate metabolism, 629–31
and adenosine diphosphate (ADP), 630–31
and adenosine triphosphate (ATP), 629–31
Carbohydrates, 627–28
atoms of, 230*M*
cellulose, 628
disaccharides, 627
energy obtained from, 644
formula for, 44*M*, 627
monosaccharides, 627–28
polysaccharides, 627, 628
starch, 628
Carbolic acid, 511
Carbon, C, 87
allotropes (forms) of, 265–68
buckminsterfullerene, 267*B*
carbides, 271–72
carbon black, 266
carbon dioxide, 268–69
carbon disulfide, 271
carbonic acid, 269–70
carbon monoxide, 270–71
charcoal, 266, 268
coke, 268
compared to silicon, 694
diamond, 265
graphite, 265
hydrogen cyanide, 271
multiple bonding, 272–73
and organic and inorganic compounds, 264
reaction with oxygen, 48*t*
valence of, 283*M*
Carbon black, 266
Carbon compounds. *See* Organic compounds
Carbon cycles, 708–9
biological carbon cycle, 708–9
geological carbon cycle, 708–9*f*
Carbon dioxide, 268–69
as acid in water, 269*D*
as fire extinguisher, 270
formation of, 268–69
as greenhouse gas, 607
and nonpolar molecule, 408
shape of molecule, 110
solid structure, 269*f*
structure of, 395*f*
Carbon disulfide, 271
formation of, 271
Carbon fixation, 44
Carbon-14 dating, 165, 665–66
and carbon-14 clock, 667*B*

Carbonic acid, 269–70
  formation of, 269
  Lewis structures, 270
Carbon monoxide, 270–71
  formation of, 270
  toxicity of, 270, 371*B*
Carbon steel, 376
Carboxylic acids, 141, 518–22, 519*t*, 530
  acetic acid, 520
  benzoic acid, 521
  butyric acid, 521
  citric acid, 521
  lactic acid, 521
  naming of, 519
  oxalic acid, 521
  preparation of, 520
  properties of, 521–22
  strength of, 521
Carothers, Wallace H., 725*f*
Catalysts and catalysis, 50–51, 120–21
  actions of, 51, 569*D*, 569
  and activation energy, 568–69
  autocatalytic reaction, 569
  enzymes, 121, 572, 626
  heterogeneous catalyst, 569, 570–72, 571*f*
  homogeneous catalyst, 569, 570
  surface catalyst, 570
  zeolites as, 732–33
Catalytic converters, 284
  automobile, 572*f*
  operation of, 572
  type of catalyst in, 572
Catalytic selectivity, 732–33
Cathode, 136, 467, 485
Cathodic protection, from rust, 484*f*, 484
Cations, 94
  acid–base properties, 436–37
Caustic potash, 354
Caustic soda, 354
Cell diagrams, 470
Cell potentials:
  calculation of, 474–75
  concentration cells, 479–80
  and Gibbs free energy, 475–77
  half-cell potentials, 469–72
  and Nernst equation, 477–80
  potential difference, 467–68
  standard cell potential, 468
  standard reduction potentials, 470–72
Cells:
  electrochemical, 465–83
  unit cells, 391*f*, 391, 393
Cellulose, 628
  structure of, 628*f*
Celsius temperature scale, 58
Ceramic, 699
Cesium chloide, structure of, 398–99*f*
Chain initiation, 573–74
Chain-propagation steps, 574
Chain reactions, 188, 573–75
  branching chain reaction, 673
  definition of, 573
  examples of, 574f, 575*f*
  steps in, 573–74
Chain-termination reactions, 574
Chalk, 358
Charcoal, 266, 268
  activated charcoal, 268*D*
Charge:
  core charge, 92–93
  of electron, 21, 22*t*
  formal charges, 254
  of proton, 22*t*
Charles, Jacques, 57
Charles' law, 57–58, 60, 63
  application of, 59*B*
  elements of, 57–58
  and Kelvin scale, 58
  and liquefaction of air, 58*D*
Chemical, use of term, 9
Chemical analysis, 164–65*B*
  applications of, 164–65
  instruments used, 164
  See also specific instruments
Chemical bonds, 7, 92–101, 348*f*, 348, 391
  bond energies, 314–19
  bonding orbitals, 208–9
  bonding region of electrons, 208
  covalent bonds, 98–99
  hydrogen bond, 412*f*, 413
  ionic bonds, 95
  and Lewis acid, 363–64
  metals, 348*f*, 348, 391
  multiple bonds, 100, 272–73
  octet rule, 93, 94, 97
  polar covalent bond, 104
  single bonds, 100
  triple bonds, 100
Chemical change, meaning of, 15
Chemical energy, 302
Chemical equations, 18–20
  balancing of, 18–20
  mistakes related to balancing, 19–20
  products in, 18
  reactants in, 18
Chemical equilibria. *See* Acid–base equilibria
Chemical formulas. *See* Formulas
Chemical kinetics, 550
Chemical properties, of substances, 10
Chemical reactions, 1, 18–20, 119–23
  acid–base reactions, 122, 138–46
  addition reactions, 288
  aromatic substitution reactions, 293
  bimolecular reaction, 556, 561
  and catalysts, 120–21, 568–73
  chain reactions, 573–75
  characteristics of, 18
  combustion, 47–48
  condensation reactions, 245
  conditions necessary for, 119–20
  coupled reactions, 334–35, 335*f*
  cracking reactions, 284
  decomposition reactions, 122, 123*D*
  elementary reaction, 556
  endothermic reactions, 48, 302
  equilibrium, 121–22
  exothermic reactions, 48, 302
  half-reactions, 131
  identification of reaction type, 149–50
  mechanism of reaction, 120
  oxidation-reduction reactions, 122, 131–38
Chemical reactions (*cont.*)
  photochemical reactions, 187–89
  precipitation reactions, 122, 146–49
  rate of. *See* Reaction rate
  spontaneous reactions, 321–23*D*
  study of, 302, 550
  substitution reactions, 503–4
  synthesis reactions, 122
  termolecular reaction, 556
  unimolecular reaction, 556, 568
Chemical shifts, 540
Chemistry, nature of science, 1
Chernobyl accident, 678
Chirality, 531*f*, 532*f*
  and amino acids, 531–33*f*
  and racemic mixture, 534
Chloralkali cell, 490*f*
Chloralkali industry, 490
Chlorates, 251
  formation of, 251
Chloric acid, 244*M*, 251
  formation of, 251
Chlorides, 86
  aluminum chloride, 362
  iron chloride, 368
  properties of, 129*t*
Chlorine, Cl, 88, 124–25, 250–52
  average mass of atom, 27
  chlorates, 251
  chloric acid, 251
  compounds of, 125
  and fluorides, 252
  hypochlorites, 250
  in oxidation–reduction reaction, 133–34
  oxidation states, 250*t*
  oxides, 252
  perchlorates, 252
  perchloric acid, 252
  photochemical reaction, 188
  preparation by electrolysis, 136, 490–91
  reactions of metals with, 128*D*
  sources of, 124
  uses of, 125
Chloroethane, 503
Chlorofluorocarbons, 125
  and ozone depletion, 596–99
  uses of, 503
Chlorofluoromethane, 503
Chloromethane, 501, 503
Chromate ions, 374*f*
Chromium, Cr, 346
  in electroplating, 492
Chromatography, 16–17*D*, 164
  gas chromatography, 164
  liquid-column chromatography, 17*f*, 17
  paper chromatography, 16–17*D*
Citric acid, sources of, 521
Closed system, 304
Close-packed arrangement, 346*f*, 346–47, 391–92
  body-centered cubic structure, 392
  cubic, 347*f*, 347
  example of, 347*D*
  face-centered cubic structure, 392
  hexagonal, 347*f*, 347
Coal:
  and acid rain, 320, 602, 605

Coal (*cont.*)
as energy source, 319–20
origin of, 702
Codons, 637
assignments of base triplets, 637
Coffee-cup calorimeter, 307, 308*f*, 309
Coinage metals, 346
Coke, 268
Collision model, reaction rate, 563–65
Combustion:
apparatus for, 167*f*
and enthalpy change, 309–10
of organic compounds, empirical formula determined from, 166–67
and oxygen, 47–48
examples of combustion reactions, 47–48
Comets:
formation of, 692
Halley's comet, 692*M*
Complex ions, 367
of copper, 373*f*, 373
Compounds, 4–5
characteristics of, 4–5
inorganic compounds, 264
molecular formulas of, 4–5
organic compounds, 264
Computer models, 640–41
electron density model, 640–41*B*
molecular models, 106*B*
ribbon model, 640*B*
space-filling model, 640*B*
Concentration cells, 479–80
Concentration of solution, 170
parts per billion, 173
parts per million, 172
Condensation, 402
Condensation polymers, 725–29
nature of, 725
types of, 725–29
Condensation reactions:
amino acids, 616
nature of, 245
Conduction band, 737, 738*f*
Conductivity. *See* Electrical conductivity
Conductors:
characteristics of, 739
types of, 738*f*, 739
Conformations, alkanes, 280
Conjugate acid–base pairs, 434–35
Conservation of mass and energy, law of, 35–36
Contact process, 228
Continuous-chain alkanes, 275
Control rods, of nuclear reactor, 674
Coordination compounds, 367
Coordination number, 367
Copolymers, 724
Copper, Cu, 350, 371–73
complex ions, 373*f*, 373
copper salts, 372–73
oxidation states, 371
production of, 377
properties of, 346
reaction with acids, 372
reaction with oxygen, 48*t*, 372
reaction with sulfur, 372
Copper Cu (*cont.*)
refining by electrolysis, 491
uses of, 346
Copper iodide, 373
production of, 373
Copper nitrate, 372–73
production of, 373
Copper oxide:
formation of, 48*t*
production of, 372
reduction of, 49*D*
Copper salts, types of, 372–73
Copper sulfate, 230*f*, 372–73*f*
production of, 373
reaction with zinc, 466*D*
Copper sulfide, production of, 372
Core, of atom, 92
Core, of earth, 693
Core charge, 92–93
Corey, Robert, 617
Corrosion, 483–84
prevention of, 484
rusting, 483–84
Corundum, 361
Cosmochemistry:
asteroids, formation of, 692
atmosphere, 702–5
comets, formation of, 692
interstellar molecules, 688–90
meteorites, formation of, 692
nature of study, 685
nucleosynthesis, 685–87
planets, formation of, 691–92
Coulombs (C), 21*M*
Coupled reactions, 334–35, 335*f*
Covalent bonds, 98–99
formation of, 98*f*, 98
and hydrogen atom, 208–9
and Lewis structures, 98–99
multiple bonds, 100
polar covalent bond, 104
single bonds, 100
and unshared pairs, 99
Covalent molecular crystals, 395
Covalent network crystals, 394
Covalent network solids, 390
Covalent radius, 395*f*, 396
Cracking reactions, 284
Crick, Francis, 623*B*
Cross-linked polymer, 719
Crude oil, 283
Crust of earth, 693, 694
Crystals, 387–99
covalent molecular crystals, 395
covalent network crystals, 394
determination of structure of, 387, 388–89*B*
diamond structure, 394*f*
ionic crystals, 397–99
liquid crystals, 401–2
metal crystals, 346*f*, 391
shapes of, 387*f*
two-dimensional view of, 387*f*
Cubic close-packed structure, 347*f*, 347
body-centered, 392
face-centered, 392
Curie, Irène, 661*B*
Curie, Marie, 660–61*B*
Curie, Pierre, 660*B*
Cycloalkanes, 281–83
structures of, 282*f*
Cysteine, 619*t*
Cytochromes, 371
Cytosine, 632, 637

Dacron, production of, 728*f*, 728–29
Dalton (Da), unit of mass, 25*M*
Dalton, John, 2*B*
Dalton's law of partial pressure, 68
Davisson, Clinton, 197*B*
d-block elements, 203
Decomposition reactions, 122, 123*D*
process of, 150
Dehydrating agent, sulfuric acid as, 230, 231*D*
Dehydration reactions, alcohols, 509
Dehydrogenation reaction, aldehyde production, 515
Delocalized electrons, 292
Delta, meaning of, 304*M*
Denatured protein, proteins, 625
Density, formula for, 51*M*
Deoxyribonucleic acid (DNA), 632–35
discovery of shape of, 623*B*, 633, 635
and genetic code, 635, 637–38
informational strand in, 636–37
replication of, 635–37
shape of, 104, 105*f*, 634
template strand in, 636
transcription of, 636
Desiccator, 230
Detergents. *See* Soap
Deuterium, ${}^{2}_{1}H$, 25
Diamond, 265
structure of, 265*f*
Diaphragm cell, 490*f*, 490–91
Dichloromethane, 501
Dichromate ions, 374*f*
Diethyl ether, 512, 513*B*
Diffraction:
diffraction grating, 197
waves, 197*B*
Diffusion, 69–73
of bromine vapor, 70*D*
molecular speeds, 71–72
process of, 69*f*, 69–60
rate of, 72–73
Dimethylamine, 529*t*
Dinitrogen monoxide, 588
production of, 588
properties of, 588
Dinitrogen tetraoxide, 585–86
preparation of, 585–86, 587*D*
Diols, 507
Diopside, 698
Dioxane, 512
Diphosphoric acid, 246
Dipole, 407
Dipole–dipole forces, 407–8
hydrogen bond, 413
and polyatomic molecules, 408*f*, 408
Dipole moment, 407–8
Diprotic acids, 228, 229, 238
Disaccharides, 627
structure of, 627*f*

Disinfectant, function of, 511*M*
Disorder:
  and entropy, 323–26, 325*D*
  and spontaneous processes, 323–26
Distance, measurement in chemistry, 21*M*
Distillation, 15*f*, 15
  apparatus for, 15*f*
  distillate in, 15
  fractional distillation, 16*f*, 16
Distilled water, 15
Disulfides, 514
  formation of, 514
DNA. *See* Deoxyribonucleic acid (DNA)
Domain, of electron pair, 105–6
Doping, 735
Double bonds, 100
Dry cell, batteries, 481*f*, 481
Dry ice, 269*f*
Duet rule, 97
Dynamite, discovery of, 317*B*

Earth:
  atmosphere, 43–44, 702–5
  core of, 693
  crust of, 693, 694
  elements of surface of, 50*t*
  mantle of, 693
  rocks, types of, 693–94
  structure of, 693*f*
Eclipsed conformation, 280, 281*f*
Effusion, 69–73
  process of, 69*f*, 69–73
Einstein, Albert, 35*M*, 186*B*
Elastomer, 724
Electrical conductivity:
  of acids, 141
  of halogens, 129–30
  of metal oxides, 741–42
  metals, 345*t*, 349
  superconductivity, 742–43
  and temperature, 349*f*
Electrochemical cells:
  batteries, 480–83
  calculation of cell potentials, 474–75
  cell potentials, 467–68
  construction of, 466*f*, 467*f*
  corrosion, 483–84
  and Gibbs free energy, 475–77
  Nernst equation, 477–80
  operation of, 465–67
  standard reduction potentials, 469–72
Electrodes, 136, 466–67
  anode, 467, 485
  cathode, 467, 485
  glass electrode, 480*f*, 480
  hydrogen electrode, 468*f*, 469
Electrolysis, 484–92
  aluminum, preparation of, 486–87
  of aqueous solutions, 489–90
  calculation of amounts of products, 489
  calculation of electrolysis products, 489
  chlorine, preparation of, 490–91
  copper refining, 491
  electroplating, 492
  and Gibbs free energy, 492
  ionic conduction in, 486
  of molten sodium chloride, 485–86
Electrolysis (*cont.*)
  process of, 136
  quantitative aspects of, 487–89
  of sodium chloride, 136*f*
  sodium hydroxide, preparation of, 490–91
  of water, 46
Electrolytes, 130, 485
Electromagnetic radiation, 181–89
  amplitude of, 184
  frequency of, 183
  photochemical reactions, 187–89
  photoelectric effect, 185–87
  photons, 185
  wavelength of, 183
Electromagnetic spectrum, 184*f*, 184
  light, 184
  visible spectrum, 184
Electromagnetic wave:
  composition of, 182*f*
  production of, 182
Electron capture, in radioactive process, 655
Electron configurations, 200–207
  and energy levels, 201*f*, 203
  examples of, 201, 202*t*
  ground-state configurations, 207*t*
  hybrid orbitals, 211–16
  and Pauli exclusion principle, 206
  and periodic table, 203–4*f*
  rules of, 201, 203
  shells and subshells in, 200, 201–2
  and valence, 209–10
Electron density distribution:
  and atomic orbitals, 210*f*, 211*f*
  and bonding region, 208
  in hydrogen atom, 195*f*, 208*f*, 208–9, 209*f*
  in sodium chloride, 396*f*
Electron density model, computer models, 640–41*B*
Electron dot symbols. *See* Lewis symbols
Electronegativity, 101–4
  bond type and polarity, 103–4
  definition of, 101
  and periodic table, 102
  and polar covalent bond, 104
Electron-gas model, 348
Electron pairs, 105–6
  domain of, 105–6
  models of, 107*D*
  and shape of molecules, 108–10
Electrons, 21–22
  charge on, 21, 22*t*
  delocalized electrons, 292
  electron pairs, 105–6
  energy levels, 200
  ionization energy, 198–201
  and Lewis symbols, 94
  and oxidation, 132–33
  photoelectrons, 198
  and radioactivity, 33
Electron spin, 204*f*, 204–7
  and atomic orbitals, 206
  bonding orbitals, 208–9
  experimental evidence for, 205*B*
  Hund's rule, 206–7
Electron spin (*cont.*)
  hydrogenlike orbitals, 211
  p orbitals, 211
  s orbitals, 210–11
Electron-transfer reaction, 132
  See also Oxidation–reduction reactions
Electroplating, 492*D*
  process of, 492
Elementary reactions, 556
Elements, 2–3
  atomic numbers of, 24
  of earth's surface, 50*t*
  isotopes, 25
  molecular formulas, 3*B*
  names and symbols of, 4*t*
  nature of, 2–3
  periodic table, 81–90
  relative abundances in universe, 687*t*, 688*f*
  valence of, 88–89
Elion, Gertrude B., 638*B*
Emission spectrum, 190
Empirical formulas, 6–7, 165–67
  calculation of molecular formula from, 66
  determination by combustion of organic compound, 166–67
  determination by synthesis, 165–66
  examples of, 6–7
  for ionic compounds, 96
  valence in writing of, 89
Enantiomers, 533*f*
Endothermic reactions, 48, 302, 306*f*, 307
  change in potential energy, 562*f*
  spontaneous type, 322–23
Energy:
  alternate sources. *See* Alternative energy sources
  bond energies, 314–19
  and closed system, 304
  conservation of mass and energy, 35–36
  Einstein's formula, 35*M*
  entropy, 326–30
  first law of thermodynamics, 304
  Gibbs free energy, 331–34
  heat, 303*f*
  internal energy, 304
  of isolated system, 304
  kinetic energy, 43
  lattice energy, 416
Energy levels, 191*f*, 191, 200
  and electron configurations, 201*f*, 203
  of hydrogen atom, 192*f*
  representation of, 203*f*
  in solids, 736–39
  subshells, 200, 201–2
Energy states:
  energy level as, 191*f*, 191
  excited states, 191
  ground state, 191
Engines:
  internal combustion engine, 285*B*
  and mechanical energy, 334
Enthalpy, *H*, 306–11
  and combustion reaction, 309–10
  enthalpy change, 306, 308–11
  estimation from bond energies, 318–19

Enthalpy *H* (*cont.*)
  and Hess's law, 310–11
  of hydration, 416–17
  and neutralization reaction, 308–9
  of solutions, 416–17
  standard enthalpy of formation, 312–14
  standard reaction enthalpy, 306–7, 312
Entropy, *S*, 326–30
  changes in reacting system, 326–28
  and disorder, 323–26, 325*D*
  meaning of, 326
  and phase changes, 406–7
  in reactions with gases, 328
  standard molar entropy, 327*t*, 327
  standard reaction entropy, 327–28
  and temperature, 326*f*, 326–27, 329*f*
  total entropy change, calculation of, 329–30
Environmental chemistry:
  acid rain, 602–5
  greenhouse effect, 606
  oxides/oxoacids of nitrogen, 584–90
  ozone, 591–92
  ozone depletion, 594–99
  peroxides, 592
  photochemical smog, 599–602
Enzymes, 626
  active site, 626
  compared to inorganic catalysts, 626
  enzyme–substrate complex, 626
  functions of, 572
  induced-fit model, 626
  lock-and-key model, 626
  nature of, 121
  and reaction rate, 572
Equations. *See* Chemical equations
Equilibrium:
  acid–base equilibria, 425–48
  affecting factors, 442
  chemical reactions, 121–22
  equilibrium constant, 425
  gas-phase equilibria, 449–53
  Gibbs free energy, 333, 455–57
  heterogeneous equilibria, 453–55
  homogeneous equilibria, 453
  Le Châtelier's principle, 441–42
  reaction quotient, *Q*, 443
Equilibrium constant, K_{eq}, 425
  calculation of, 456–57
  and endothermic reactions, 585
  and Gibbs free energy, 455–57
  and Nernst equation, 478–79
  and reaction rate, 560–61
  and temperature, 452*M*
Equilibrium constant expression, 425
Equivalence point, in titrations, 173, 448
Esters, 522–27
  fats, 523–25
  formation of, 522–23
  of inorganic acids, 525, 527
  odors of, 524*t*
  uses of, 523
Ethanal, 515, 517
Ethanamide, 530
Ethane, 274
  combustion of, 310*f*, 310–11
  shape of molecule, 111
Ethane (*cont.*)
  structure of, 275*f*
Ethanedioic acid, sources of, 521
Ethanoic acid. *See* Acetic acid
Ethanol, 505
  production of, 320, 505
Ethene, 284, 286
  and hybrid orbitals, 214–15*f*
  polymerization of, 719–20
  structure of, 286*f*
  uses of, 289
Ethers, 511–12
  as anesthetic, 512, 513*B*
  cyclic ethers, 512
  preparation of, 512
Ethyl alcohol. *See* Ethanol
Ethylene glycol, 16, 507
Ethylene oxide, 512
Ethyne, 272, 290
  and hybrid orbitals, 215*f*, 215
  preparation/combustion of, 290*f*
  properties of, 290
  structure of, 290*f*
Eugenol, 511*f*
Euler, Leonhard, 267*B*
Evaporation, 402–3
  process of, 403
Excited state, of atoms, 191
Exothermic reactions, 48, 302, 306*f*, 306
  change in potential energy, 562*f*
Explosives, development of, 317*B*

Face-centered cubic structure, 392
Faraday, Michael, 476*B*
Faraday constant, 476
Fats, 644–45
  animal, 523, 525*t*
  composition of, 525*t*
  uses in mammals, 644
  vegetable, 523, 525*t*
Fatty acids, 523–25
  saturated fatty acids, 523–24*t*
  structure of, 523, 524*t*, 525*f*
  types of, 524*t*, 525*t*
  unsaturated fatty acids, 523–24*t*
Feldspars, 699–700
Fermentation, alcohol production, 505
Ferromagnetism, 366, 367*f*
Fertilizers:
  ammonia in, 51, 589
  phosphates as, 244, 245*B*
  urea, 530
Fibroin, 622
Fibrous proteins, 625
Filtration, 15*f*, 15
Fire extinguisher, carbon dioxide as, 270
Fireworks, salts used, 359*B*
First law of thermodynamics, 304–5, 311
First-order rate law, 555
  integrated form of, 558
Fission, nuclear, 672, 673–76
  of uranium 235, 673*f*, 673–74*f*
Flame calorimeter, 308*f*, 309
Flame tests, metals, 359–60*t*
Fleming, Alexander, 639
Fluorescence, 592
Fluoridation of water, 165
Fluorides:
  formation of, 252
  properties of, 129*t*
Fluorine, 88, 125
  atomic structure of, 24*f*
  compounds of, 125
  in oxidation–reduction reaction, 135
  preparation of, 137
  properties of, 124
  shape of molecule, 110
  sources of, 124
  uses of, 125
Formal charges, 254
Formaldehyde, 515, 517
Formamide, 530
Formic acid, 519
Formulas:
  empirical formulas, 6–7
  molecular formulas, 3, 5
  structural formulas, 7–8
Fossil fuels, 702
  formation of, 702
  types of, 702
Fractional abundances, 27
Fractional distillation, 16*f*, 16
  of petroleum, 283–84
Fractionating column, in distillation, 16*f*
Fragment ions, 534
Fragrance, of esters, 523, 524*t*
Free energy. *See* Gibbs free energy
Free radicals, 252
  hydroperoxyl radical, 591
  hydroxyl radical, 591
  and nitrogen oxides, 586–88
  and peroxides, 591
  reactivity of, 574
Frequency:
  of electromagnetic radiation, 183
  measurement of, 183
Fuel cells, 482*f*, 482–83
Fuel element, of nuclear reactor, 674
Fuels:
  fossil fuels, 702
  hydrogen as, 52*B*
Functional groups:
  alkoxy group, 511
  aryl group, 511
  important groups, 502*t*
  organic compounds, 500–501
Fusion, nuclear, 672, 677–78
  and formation of stars, 686–87
  molar enthalpy of, 401*t*
  process of, 401, 677

Gaia hypothesis, 597*B*
Gallium, 203
Galvanizing, rust protection, 484*f*, 484
Gamma radiation, 659
Gamma-ray emission, 655–56
Gas chromatography, 164, 541
  apparatus for, 541*f*
  method in, 541
Gas constant, 60
Gas density:
  calculation of, 66–67
  calculation of molar mass from, 67
  and molar mass, 66–67

Gaseous solution, 12
Gases:
  characteristics of, 53
  expansion in vacuum, 323–24
  hydrogen, 51–52
  inert gases, 88*M*
  and intermolecular forces, 73–74
  kinetic theory of, 61, 63, 64
  molar volumes of, 60*t*, 60
  nitrogen, 50–51
  noble gases, 88
  oxygen, 46–49
  pressure effects, 56–60
  synthesis gas, 51
  temperature effects, 56–60
Gas laws, 56–60
  Avogadro's law, 59–60
  Boyle's law, 56
  Charles' law, 57–58, 60
  ideal gas law, 60
Gasohol, 321
Gas-phase equilibria, 449–53
  Haber process as example, 449–50
  and le Châtelier's principle, 450, 451*f*, 452
  and pressure changes, 450–52
  temperature changes, 452–53
Gas pressure, 55
Gas reactions, and Avogadro's law, 168–69
Geiger, Ernest, 22–23*B*
Geiger counter, 658
  operation of, 658*f*
Genetic code:
  codons, 637
  and DNA, 635, 637–38
Genetic disease, 638
Geochemistry:
  biosphere, formation of, 705–11
  earth, structure of, 693–94
  geologic resources, 702
  hydrothermal processes, 701
  nature of study, 685
  silicon, element in earth, 694–701
Geological carbon cycle, 708–9*f*
Geologic resources, 702
  fossil fuels, 702
  hydrothermal processes, 701
  minerals, 701
Gerlach, Walter, 205*B*
Germanium, 87
  as semiconductor, 738*f*, 739
Germer, Lester, 197*B*
Gibbs free energy, 331–34, 455–57
  calculation of, 332
  and cell potentials, 475–77
  coupled reactions, 334–35
  definition of, 331
  and electrolysis, 492
  energy change, 331
  and equilibrium, 333
  and equilibrium constant, 455–67
  formation at 25 degrees, 331*t*
  standard Gibbs free energy of formation, 331–32
  and work, 333–34
Gibbs, Josiah Willard, 330*f*
Glass, as amorphous solid, 399–400
Glass electrode, 480*f*, 480
Global warming, and greenhouse effect, 607
Globular proteins, 625
Glucose, in carbohydrate metabolism, 629, 631
Glutamic acid, 619*t*
Glutamine, 619*t*
Glycerol, uses of, 507
Glycerophospholipids, 642–43
Glycine, 618*t*
  molecular structure, 533*f*
Glycogen, 628
Goodyear, Charles, 724
Goudsmit, Sam, 205*B*
Gout:
  and purines, 638–39
  and urea, 638–39
Graham, Thomas, 72
Graham's law, 72
Grams, conversion to moles, 32
Granite, 14*f*, 87*M*, 696
Graphite, 129*M*, 265
  bonding in, 266*f*
  structure of, 266*f*, 391*f*
Gravimetric analysis, 157
Greenhouse effect, 45, 606
  effects of, 607
  and global warming, 607
  greenhouse gases, 606–7
  process in, 606*f*
Ground state, of atoms, 191
Group I elements. *See* Alkali metals
Group II elements. *See* Alkaline earth metals
Group III elements, 86–87
Group IV elements, 87
Group V elements, 87
Group VI elements, 87–88
Group VII elements. *See* Halogens
Group VIII elements. *See* Noble gases
Groups, in periodic table, 83–84, 86–88
Guanine, 632, 637
Gunpowder, 317B

Haber process, 505
  and gas-phase equilibria, 449–50
  and nitrogen fixation, 50
Half-cell potentials, 469–72
Half-life:
  calculation of, 663
  of first order reaction, 559*f*, 559–60
  nuclear, 662–64
  of selected isotopes, 663*t*
Half-reactions, 131
Halides, 110
  hydrogen halides, 126–27
  metal halides, 128
  nonmetal halides, 127
  phosphorus, 247
  sources of, 125
  sulfur, 239
Halite, 5*M*
Hall, Charles Martin, 487*B*
Halley, Edmund, 692
Halley's comet, 692*M*
Hall process, 486–87
Haloalkanes (alkyl halides), 501–4
  preparation of, 501, 503
  reactions of, 503–4
  types of, 501, 503
Halogenation, 293
Halogens, 88, 123–31, 704
  bromine, 125
  chlorine, 124–25
  composition of, 123
  electrical conductivity of, 129–30
  fluorine, 125
  hydrogen halides, 126–27
  industrial preparation of, 135–37
  iodine, 125
  metal halides, 128
  nonmetal halides, 127
  oxidation number of, 233
  in oxidation–reduction reactions, 133–35
  properties of, 124
  reactions with metals, 130–31, 351–52
  reactions with nonmetals, 130–31
  reactivity of, 123
Hard water, 357*B*
  softening of, 356, 357*B*
Heat:
  calorimetry, 307–9
  combustion, 47–48
  definition of, 64
  endothermic reactions, 48
  enthalpy, 306–7
  exothermic reactions, 48
  flow of, 303*f*
  nature of, 303
Heat capacity, 307
  molar and specific capacities, 307
Helium, He, 88
  in atmosphere, 45
  atomic structure of, 24*f*
Heme group, in myoglobin, 624*f*
Hemoglobin, 370, 371*B*, 625
Henderson–Hasselbalch equation, 440
Hematite, 368
Hertz (Hz), 183
Hess, Germain, 311
Hess's law, 310–11, 313*f*
Heterocyclic molecules, 529
Heterogeneous catalyst, 569, 570–72, 571*f*
Heterogeneous equilibria, 453–55, 454*f*
  example of, 453–54
Heterogeneous mixtures, 12, 14
  example of, 14
Hexagonal close-packed structure, 347*f*, 347
Histidine, 618*t*
Homogeneous catalyst, 569, 570
Homogeneous equilibria, 453
Homogeneous mixtures, 12
Hot air balloons, 59*B*
Hund, Fritz, 207
Hund's rule, 206–7
Hybridization, 212
Hybrid orbitals, 211–16
  ethene and ethyne, 214–15*f*
  sp hybrid orbitals, 213, 214
  sp^2 hybrid orbitals, 213, 214
  sp^3 hybrid orbitals, 212*f*, 212, 214*M*

Hydrated salts, 229
  types of, 230
Hydrides, 86, 87
Hydrobromic acid, 244*M*
Hydrocarbons, 264, 500–501
  alkanes, 274–84
  alkenes, 284–89
  alkynes, 290
  benzene, 291–94
  industrial use of, 264
  saturated and unsaturated, 288
  uses of, 284
Hydrochloric acid, 139, 244*M*
Hydrodermal processes, nature of, 701
Hydrogen, H, 51–52
  in atmosphere, 45
  atom, 21*f*
  Bohr model of atomic spectrum, 192–94
  characteristics of, 51
  electron density in, 195*f*, 208*f*, 208–9, 209*f*
  energy levels of, 192*f*
  and formation of water, 52
  as fuel, 52*B*, 320
  hydrides, 51
  industrial production of, 51
  in periodic table, 86
  photochemical reaction, 188
  reaction with oxygen, 48*t*
Hydrogenation, process of, 288
Hydrogen bond, 412*f*, 413
  nature of, 413
Hydrogen chloride, solubility of, 126*D*
Hydrogen-chlorine chain reaction, 574*f*
Hydrogen cyanide, 271
  formation of, 271
Hydrogen electrode, 468*f*, 469
Hydrogen halides, 126–27
  aqueous solutions of, 138
  formation of, 126
Hydrogenlike orbitals, 211
Hydrogen peroxide:
  decomposition reactions, 123*D*
  empirical formula of, 6
  properties of, 10*D*, 591
  uses of, 591
Hydrogen sulfide, 238–39
  properties of, 238*D*
Hydrogensulfite ion, 236
Hydroiodic acid, 244*M*
Hydronium ions, 138, 139
  Lewis structure of, 138*f*, 138
Hydroperoxyl radical, 591
Hydrophilic, 622
Hydrophobic, 624
Hydrosphere:
  composition of, 704*t*, 704–5
  size of, 704
Hydrothermal processes, 701
Hydroxide ion, 142
Hydroxides, 354
  aluminum hydroxide, 361–62
  iron hydroxide, 369
  production of, 354
Hydroxyl radical, 591
Hypochlorites, 250
  formation of, 250
Hypothesis, formulation of, 63*B*

Ice, structure of, 414*f*, 414
Ideal gas law, 60, 64–68
  application of, 64–69
  basis of, 73
Igneous rocks, 693–94
Immiscible solutions, 415, 416*D*
Impure substance, 11
Indicators, 444–46
  properties of, 446*t*
  types of, 138, 444–46
  universal, 446
Induced-fit model, 626
Inert gases, 88*M*
Infrared spectroscopy, 536–38
  fingerprint region of spectra, 538
  infrared spectra, examples of, 537*f*, 538*f*
Initial rate, chemical reactions, 554
Inorganic compounds:
  nature of, 264
  study of
Instantaneous rate, chemical reactions, 552–53
Insulators, 349
  characteristics of, 739
  compared to semiconductors, 737–38*f*
  type of, 738*f*, 739
Insulin, structure of, 617*f*
Integrated circuits, 741
Interference:
  constructive and destructive, 196
  interference pattern, 196*f*, 196, 197*f*
  waves, 196–97*B*
Intermolecular forces, 73–74, 390, 392, 407–10
  dipole–dipole forces, 407–8
  hydrogen bond, 413
  compared to intramolecular forces, 124*M*, 407*M*
  London (dispersion) forces, 409–10
Internal combustion engine, 285*B*
Internal energy, nature of, 304
Interstellar molecules, 688–90
  and electromagnetic radiation, 688–90
  identified molecules, 689*t*
Intramolecular forces, compared to intermolecular forces, 124*M*, 407*M*
Iodides, copper iodide, 373
Iodine, I, 125
  nutritional function of, 125
  in oxidation–reduction reaction, 135
  properties of, 124
Iodine-131 radiolabeling, 669, 670*f*
Ion-exchange materials:
  function of, 731
  zeolites as, 731
Ionic bonds, 95
  formation of, 95
Ionic compounds, 95–96
  formation of, 95
  formulas for, 96
  structure of, 96
Ionic conduction, 486
Ionic crystals:
  cesium chloride, 398
Ionic crystals (*cont.*)
  sodium chloride, 398
  zinc sulfide, 399
Ionic radius, 397–98
Ionic solids, 390–91
  solubility of, 415–16
Ionization:
  of acids, 426–29
  of acid in water, 140
  autoionization of water, 431
  of bases, 429–30
Ionization energy, 198–201
  from photoelectron spectroscopy, 199*f*
  of selected elements, 199*t*, 200*f*
Ions, 7
  complex ions, 367
  formation of, 94
  isoelectric ions, 97
  measurement of size in metals, 395–97
  positive and negative ions, 94
Iron, Fe, 366–71
  biochemistry of, 370–71
  complexes of, 370
  and hemoglobin, 371*B*
  magnetic properties, 366–67
  oxidation states, 367
  as oxidizing agent, 370
  production of, 346, 375–76
  properties of, 345–46
  reaction with acids, 368–69
  reaction with oxygen, 368
  rusting, 370, 483–84
  rust prevention, 484
Iron chloride, production of, 368
Iron hydroxide, 369
  production of, 369*D*, 369
Iron oxide:
  production of, 368
  uses of, 368
Iron sulfate, production of, 368
Iron sulfide, production of, 368
Isobutane, 276
  structure of, 276*f*
Isoelectric ions, 97
Isolated systems, 304
Isoleucine, 618*t*
Isomers, 533–34
  optical isomers, 533
Isotopes, 24–25
  fractional abundances, 27
  mass of selected isotopes, 27*t*
  nature of, 25
  of oxygen, 25*f*
  percent abundances, 27
  radioactive, 33
  in radioactive dating, 664–66

Jeweler's rouge, 368
Joliot-Curie, Frédéric, 661*B*
Joule, 185*M*

Kaolinite, 699
Kelvin, Lord, 58
Kelvin temperature scale, 58
Kendrew, John, 623*B*
Keratin, 622

Ketones, 514–18
  naming of, 514
  natural occurrence, 517*f*
  preparation of, 515–16
  reduction of, 517
Kevlar, 727, 728*f*
Kilopascals, 55
Kinetic energy:
  average kinetic energy, 64
  nature of, 43, 64
Kinetic theory of gases, 61, 63, 64
  assumptions in, 61
  average kinetic energy, formulation of, 64
  and ideal gas behavior, 61
  and temperature, 64
Krypton, Kr, 88, 203

Lactic acid, sources of, 521
Lakes, acid rain effects, 603
Lanthanides, 84, 203
Lattice, and structure of solids, 391
Lattice energy, 416
Laue, Max von, 388*B*
Lauric acid, 524*t*, 525*t*
Lavoisier, Antoine Laurent, 35*f*
Law of conservation of energy, 304
Laws, formulation of, 62*B*
Lead, Pb, 87
  as conductor, 738*f*, 739
Lead–acid cell, 481*f*, 481–82
Le Châtelier, Henri, 441
Le Châtelier's principle, 441–442
  application of, 442*D*
  and gas-phase equilibria, 450, 451*f*, 452
Leclanché, George, 481
Leclanché cell, 481
Leucine, 618*t*
Lewis, Gilbert Newton, 93*f*, 257*B*, 363
Lewis acid, 363–64
Lewis acid–base reaction, 363–64
Lewis base, 363–64
Lewis structures, 252–57
  choosing best approximation, 257*B*
  and covalent bonds, 98–99
  difference from Lewis symbols, 98
  examples of, 98–99
  method for drawing of, 252–56
  molecules with multiple bonds, 100
  of nonmetal halides, 127*f*
  of polyatomic ions, 145–46
  resonance structures, 292, 294–95
Lewis symbols, 94
  examples of, 94*t*
Ligands, 367
Light:
  atomic spectra, 189–95
  as visible spectrum, 184
  See also Electromagnetic radiation
Lightning, 4
Limestone, 358, 605
  and hard water, 357*B*
Limiting reactants, 160–62
  calculations related to, 161–62
  characteristics of, 160–61
Linear molecular shape, 107*D*, 110
Line spectrum, 189*f*, 189–90
Linoleic acid, 524*t*, 525*t*
Lipids, 641–43
  energy and oxidation of, 641–42
  glycerophospholipids, 642–43
  triacylglycerols, 641–42
Liquid-column chromatography, 17*f*, 17
Liquid crystals, 401–2
  molecules of, 402*B*
Liquids:
  boiling point, 405
  characteristics of, 53
  condensation, 402
  evaporation of, 402–3
  structure of, 400
  vapor pressure of, 404–5
Lithium, Li, 200
Lithium chloride, electrical conductivity of, 130*D*
Lithosphere, 693
Litmus, 138, 143*M*, 444
Lock-and-key model, enzymes, 626
London (dispersion) forces, 409–10
  polarizability, 409, 410*t*
Lovelock, James, 597*B*, 708
Lowry, Thomas, 139
Lysine, 618*t*

Macromolecular chemistry, 717
Macromolecules, 717
Magnesium, Mg, reaction with oxygen, 46, 48*t*
Magnetic resonance imaging, 540
Magnetism, 366–67
  ferromagnetism, 366, 367*f*
  and iron, 366–67
  paramagnetic compounds, 366
Manganese, 346
Mantle, of earth, 693
Marble, 358
Marsden, Ernest, 22–23*B*
Mass:
  Einstein's formula, 35
  law of conservation of mass, 35–36
  measurement of, 21*M*, 29*B*
  of object, determination of, 29*B*
Mass defect, 35, 671
Mass equivalent, 35
Mass percentage:
  formula for, 11*M*
  purity of substance, 11
Mass spectrometry, 25, 26*B*, 164, 534–35
  determinining molecular mass, 534–35
  development of, 26*B*
  molecular formula determined by, 168
Mass spectrum, 535
  identification of organic molecules, 535*f*
Matches, burning of, 242*D*
Materials science:
  metal oxides, electrical properties of, 741–43
  nature of study, 730
  semiconductors, 734–41
  zeolites, 731–33
Maxwell, James Clerk, 181, 564
Maxwell–Boltzmann distribution curves, 564
Mean free path, of molecules, 72
Melting, 401
  process of, 401
Melting point, metals, 348
Mendeleev, Dimitri, 81, 82*B*
Mercury, Hg:
  mercury barometer, 55*f*, 55
  reaction with oxygen, 48*t*
Mercury cell, 491
Mesosphere, characteristics of, 44
Messenger RNA (mRNA), 636–37
Metabolism, 615
Metal halides, 128
  formation of, 128
  as ionic compounds, 128
Metallic bond, 348*f*, 348, 391
Metallic conduction, 349
Metallic radius, 395*f*, 395
Metallic solids, 391
Metallurgy, 375–77
  copper production, 377
  iron and steel production, 375–77
  thermite process, 375, 376*D*
Metal oxides, 47*D*, 47–48, 351–52
  electrical properties of, 741–42
Metals:
  acid–base reactions, 379
  activity series, 354–55
  alkali metals, 86
  alkaline earth metals, 86
  bonding, 348*f*, 348
  and complex ions, 367
  crystals of, 346*f*, 391
  d-block elements, 203
  electrical conductivity, 345*t*, 349
  flame tests, 359–360*t*
  Group I, 86
  Group II, 86
  Group III, 86–87
  Group IV, 87
  Group V, 87
  Group VI, 87
  halogen reactions with, 128
  historical view, 343
  magnetic properties, 366–67
  mechanical properties, 350
  melting point, 348
  oxidation states, 365, 378
  and periodic table, 84–85, 86
  photoelectric effect, 185–87
  precipitation reactions, 379
  properties of, 84–85, 344–46, 348–50
  reactions with acids, 354–35
  reactions with halogens, 351–52
  reactions with oxygen, 47*D*, 47–48, 351–52
  reactions with water, 352–54
  reduction–oxidation reactions, 378–79
  salts, production of, 355–58
  s-block elements, 203
  structure of, 346–48
  thermionic effect, 350
  transition metals, 203, 365–75
  See also individual metals
Metal sulfides, 237, 239
Metamorphic rocks, 694
Metastable nucleus, 656

Meteorites, formation of, 692
Methanal, 515, 517
Methanamide, 530
Methane, 5, 274
combustion of, 47, 306–7
as energy source, 320
as greenhouse gas, 607
as limiting reactant, 160
polymerization of, 723*B*
shape of molecule, 108
structure of, 275*f*
Methane–chlorine chain reaction, 575*f*
Methanioc acid, 519
Methanol, 504–5
production of, 504–5
Methionine, 618*t*
Methylamines, 528, 529*t*
Methyl red indicator, 445*f*
Methylsalicylic acid, production of, 523
Micas, 700–701
Micelles, 527*B*
Microscope:
function of, 2–3*B*
scanning tunneling microscope, 2–3*B*
Minerals, 701–2
major minerals, 645, 647*t*
nutritional requirements, 645, 647
ores, 701
sources of, 701
trace minerals, 645, 647*t*
Mirror image, chirality, 531*f*, 531–33, 532*f*
Miscible solutions, 415, 416*D*
Mixtures, 12–17
examples of, 14*f*
heterogeneous mixtures, 12, 13*f*, 14
homogeneous mixtures, 12
separation of, 13*D*, 14–16
Models:
formulation of, 62*B*
of molecules, 1*f*
Moderator, of nuclear reactor, 674
Molar enthalpy of fusion, 401
values in selected elements, 401*t*
Molar enthalpy of vaporization, 403
values in selected elements, 403*t*
Molar heat capacity, 307
Molarity, of solutions, 169–71
Molar mass, 31–32, 65–67
calculation from gas density, 67
calculation of molar mass of gas, 65
and gas density, 66–67
molecular formula from, 66
relationship to mole, 158
size and empirical formula, 32*M*
Molar volume, 60
calculation of, 60
of selected gases, 60*t*
Mole, 29–32
Avogadro's number in, 30–31, 158
and balanced equation, 158–59
conversion from grams, 32
examples of, 31*f*
formulation of, 30–31
molar mass, 31–32
relationship to molar mass, 158
SI symbol, 30*M*, 158
Molecular biology, 623
Molecular formulas, 168
for ascorbic acid, 168
calculation from empirical formula, 66
of compounds, 4–5
determination by mass spectrometry, 168
nature of, 3, 5
Molecular mass, 30
determining with mass spectrometry, 534–35
molecular formula from, 66, 168
Molecular models, 106*B*
ball-and-stick models, 106*B*
computer models, 106B, 640–41*B*
space-filling models, 106*B*
stick models, 106*B*
Molecular orbitals, 208–9
bonding molecular orbital, 209*f*, 209
Molecular shape, 104–12
angular, 109
bent bond, 110
electron-pair arrangements, models for, 107*D*
linear, 107*D*, 110
molecular models, 106*B*
and multiple bonds, 110–11, 112
prediction of, 112
tetrahedron, 107*D*, 108*f*, 108–9
triangular, 107*D*, 110
triangular pyramidal, 109
with two central atoms, 111*f*, 111
and VSEPR (valence-shell electron-pair repulsion) model, 104–7, 112
Molecular solids, 390*t*, 390
Molecular speeds:
average speed, calculation of, 72
determination of, 71*f*
and diffusion, 70–72, 72*f*
mean free path, 72
speed of oxygen, 71*f*
and temperature, 70, 72*M*
Molecular structure, determination of, 536
Molecules, 1–2
biomolecules, 615
intermolecular forces, 73–74
interstellar molecules, 688–90, 689*t*
Lewis structures for, 98–100
models of, 1*f*
observation of, 2–3*B*
origin of, 688–90
vibrational energy, 536–37
Molecules, observation of:
gas chromatography, 541
infrared spectroscopy, 536–38
mass spectrometry, 534
nuclear magnetic resonance spectroscopy, 538–40
X-ray crystallography, 541
Moles, determining number in gas, 65
Monoclinic sulfur, 224, 225*f*
Monomers, 717, 718*t*
Monosaccharides, 627–28
structure of, 627*f*
Motion, Newton's law of, 56
Müller, K. Alex, 743
Multiple bonds:
carbon, 272–73
covalent bonds, 100
Multiple bonds (*cont.*)
Lewis symbols for molecules, 100
molecular shape with, 110–11, 112
Muscovite, 700
Myoglobin, 622, 624
heme group in, 624*f*
structure of, 622*f*, 624
Myristic acid, 524*t*, 525*t*

Naphthalene, 294
Natural gas, 283
Neon, Ne, 88
Neoprene, 724–25
Nernst, Walther, 477
Nernst equation, 477–80
and equilibrium constant, 478–79
in measurement of pH, 480
Net ionic equation, 147
Network solids, 390–91
covalent network solids, 390
ionic solids, 390–91
metallic solids, 391
three-dimensional, 390*f*
Neutralization, of acid, 144
Neutralization reaction, and enthalpy change, 308–9
Neutral solutions, 432
Neutron activation analysis, 164–65
Neutrons, 22
charge on, 22*t*
composition of, 22*M*
Newton, Isaac, 181
Newtons, 29*B*, 55
Nickel–cadmium cells, 482
Nitrates, 589
copper nitrate, 372–73
as meat preservative, 590
Nitration reaction, 293
Nitric acid, 244*M*, 588–90
as oxidizing agent, 590
production of, 588–89
Nitrites, 590
Nitrogen, N, 50–51, 87
characteristics of, 50
liquid nitrogen, 50*f*
oxoacids of, 588–90
reactivity of, 50
Nitrogen cycle, 708
Nitrogen dioxide, 585–86
free radicals of, 588
preparation of, 586, 587*D*
Nitrogen fixation, 50–51
and Haber process, 50
meaning of, 51
in soil, 51
Nitrogen monoxide, 4, 5, 584–85
free radicals of, 588
preparation of, 585, 586*D*
Nitrogen oxides:
and acid rain, 602, 605
dinitrogen monoxide, 588
dinitrogen tetraoxide, 585–86
and free radicals, 586–88
nitrogen dioxide, 585–86
nitrogen monoxide, 584–85
and ozone depletion, 599

Nitrogen oxides (*cont.*)
  properties of, 585*t*
Nitrogen triiodide, decomposition of, 318*D*
Nitroglycerin, 317*B*
Nitrous oxide. *See* Dinitrogen monoxide
Nobel, Alfred, 317*B*
Noble gases, 88, 97
  properties of, 88
  types of, 88
Nodes, 211
Nomenclature, meaning of, 277
Nonbonding orbitals, 210
Nonelectrolytes, 130
Nonmetal halides, 127
  formation of, 127
  molecular shape of, 127*f*
Nonmetals:
  halogen reactions with, 126–27
  p-block elements, 203
  and periodic table, 85, 86–88
  physical properties of, 85
Nonstoichiometric compound, 742
n-type semiconductors, 735
Nuclear chemistry:
  nuclear energy, 671–78
  radioactive decay rates, 662–66
  radioactivity, 653–61
  radioisotopes, 667–70
Nuclear energy, 671–78
  chemical aspects of, 676–77
  as energy source, 321
  fission, 672, 673–76
  fusion, 672, 677–78
  isotopic enrichment in production of, 676–77
  mass defect, 671
  nuclear binding energy, 671–73
  nuclear fuel in production of, 676
  production of, 672
  risks and benefits of, 678
Nuclear forces, nature of, 34, 653–54
Nuclear magnetic resonance spectroscopy, 538–40
  low-resolution/high-resolution spectra, 539*f*, 540*f*
  NMR spectrometer, 539*f*
  structure of organic molecules, 540
Nuclear reaction, 33
  compared to chemical reaction, 33
  definition of, 653
Nuclear reactors, 674–76, 675*f*
  components of, 674, 675
Nucleic acids, 632–35
  deoxyribonucleic acid (DNA), 632–35
  nitrogenous bases in, 632, 632*f*
  ribonucleic acid (RNA), 636–37
  structure of, 633*f*
Nucleophile, 504
Nucleophilic reagent, 504
Nucleophilic substitution, 504, 512, 523
Nucleosynthesis, 685–87
  process of, 685
Nucleotides, 529, 630
Nucleus of atom, 22, 34, 653–54
  metastability of, 656
  stability and instability of, 34, 653–54
Nuclide, representation of, 653*M*
Nutrition:
  carbohydrates, 644
  fats, 644–45
  minerals, 645, 647
  proteins, 645
  vitamins, 646–47
Nylon, 726–27
  synthesis of, 727*D*, 727*f*
  types of, 726

Observations, importance of, 62*B*
Oceans:
  composition of, 704*t*, 704–5
  maintenance of composition of, 704–5
Octet, 93
Octet rule, 93, 94, 97
  application of, 97
  exceptions to, 97
Oil of wintergreen, production of, 523
Oils, composition of, 525*t*
Oleic acid, 524*t*, 525*t*
Olivine, 697*f*, 697
Onnes, Kenneth, 742
Ores, minerals, 701
Organic chemistry, nature of study, 264, 499
Organic compounds, 163
  alcohols, 504–11
  aldehydes, 514–18
  aliphatic compounds, 290
  amides, 530
  amines, 528–29
  amino acids, 531–34
  aromatic compounds, 290
  carboxylic acids, 518–22
  combustion of, 166–67
  disulfides, 514
  esters, 522–27
  ethers, 511–12
  functional groups, 500–501
  haloalkanes (aklyl halides), 501–4
  hydrocarbons, 264, 500–501
  ketones, 514–18
  nature of, 166, 264
  numerous number of, 499–500
  study of, 264, 499
  thiols, 512–14
Organic compounds, structural information:
  gas chromatography, 541
  infrared spectroscopy, 536–38
  mass spectrometry, 534
  nuclear magnetic resonance spectroscopy, 538–40
  X-ray crystallography, 541
Orthoclase, 700
Orthorhombic sulfur, 224
Ostwald process, production of nitric acid, 588–89
Outgassing, 703
Oxalic acid, sources of, 521
Oxidation, 48–49
  of aldehydes, 517–18
  definition of, 235
  of ketones, 517–18
  process of, 48–49, 132–33
  and reduction, 49
Oxidation numbers, 232–35
Oxidation numbers (*cont.*)
  rules for assignment of, 232–35
  use of, 235
Oxidation–reduction (redox) reactions, 122, 131–38
  acids, 142
  alcohols, 510
  characteristics of, 132–33
  direction of, 472–73
  and electrochemical cell, 465–67
  electron transfer in, 149
  example of, 137
  halogens as oxidizing agents, 133–35
  metals, 378–79
  metal displacement reactions, 473*D*
  oxidation numbers, 232–35
  oxidizing agent in, 133
  reducing agent in, 133
  writing of, 133*M*
Oxidation states:
  chlorine, 250*t*
  copper, 371
  higher states, 374
  iron, 367
  metals, 365, 374, 378
  phosphorus, 241, 246*t*
  sulfur, 235–36*t*
Oxides, 47, 86*M*, 86, 354
  acidic oxides, 354
  aluminum oxide, 361
  basic oxides, 354
  chlorine, 252
  copper oxide, 372
  examples of, 46
  formation of, 46
  iron oxide, 368
  metal, 47*D*, 47–48, 351–52
  of nitrogen, 584–90
  phosphorus, 242–43
  production of, 354
  of sulfur, 227
Oxidizing agents:
  function of, 133
  halogen as, 133–35
  iron as, 370
  nitric acid, 590
  ozone as, 591
  sulfuric acid as, 230–32, 231*D*
Oxoacids, 227, 247–50
  naming of, 247–48
  of nitrogen, 588–90
  phosphoric acid, 243–44
  of phosphorus, 246*t*
  strengths of, 248–49
  structure of, 227
Oxoanions, 228
Oxygen, O, 46–49
  in atmosphere, 45
  and combustion, 47–48
  isotopes of, 24*f*
  molecular speed of, 71*f*
  obtaining from water, 46*D*, 46
  oxidation, 48–49
  oxides, 47
  reaction with iron, 368
  reactions with metals, 351–52
  reactions of metals with, 47*D*, 47–48

Oxygen, O (*cont.*)
  and reduction, 49
  test for, 48
Oxyhemoglobin, 371*B*
Ozone, 44, 591–92
  odor of, 591
  as oxidizing agent, 591
  and photochemical smog, 601–3
  production of, 590
  structure of, 590–91
Ozone depletion, 585, 594–99
  and chlorofluorocarbons, 596–99
  and nitrogen oxides, 599
  ozone hole, 597, 598*f*
  solutions to problem, 599
Ozone layer:
  formation in atmosphere, 595–96
  steady-state concentration of ozone, 595*f*

Palmitic acid, 524*t*, 525*t*
Paper chromatography, 16–17*D*
Paramagnetic compounds, 366
Parent ions, 534
Partial pressure, 68
  calculation of, 68–69
  Dalton's law of, 68
  and rate of chemical reaction, 120
Parts per billion, 173
Parts per million, 172
Pascal, 55
Pauli, Wolfgang, 206
Pauli exclusion principle, 206
Pauling, Linus, 101, 617
p-block elements, 203
Penicillin, 639
Pentane, 276
  structural isomers of, 276*f*
  structure of, 275*f*
Peptide bond, 617
Percent abundances, isotopes, 27
Percent ionization, calculation of, 428
Percent yield, 163
  calculation of, 163
Perchlorates, 252
Perchloric acid, 244*M*, 252
Periodic table of elements, 81–90
  development of, 82
  and electron configuration, 203–4*f*
  electronegativities of elements, 102
  example of, 83*f*
  Group I elements, 86
  Group II elements, 86
  Group III elements, 86–87
  Group IV elements, 87
  Group V elements, 87
  Group VI elements, 87
  Group VII elements, 88
  Group VIII elements, 88
  groups in, 83–84, 86–88
  metals, 84–85, 86
  nonmetals, 85, 86–88
  periods in, 83–84
  semimetals, 85, 86–88
  valence, 88–89
Periods, in periodic table, 83–84
Permanganate ions, 374*f*
Peroxides, 592
Peroxides (*cont.*)
  barium peroxide, 591
  and free radicals, 591
  hydrogen peroxide, 591
Perutz, Max, 623
Petrochemical industry, 499
Petrochemicals, 499
Petroleum, 283–84
  fractional distillation of, 283–84
  limited resources of, 319
  origin of, 702
  uses of, 284
Phase changes, 401–7
  boiling point, 405
  condensation, 402
  and entropy, 406–7
  evaporation, 402–3
  melting, 401
  vapor pressure, 403–5
Phenanthrene, 294
Phenol-formaldehyde, 729
  formation of, 729, 730*f*
Phenolphthalein, 444, 445*f*
Phenols, 168
  types of, 511*f*
  uses of, 511
Phenylalanine, 618*t*
pH:
  of acid solutions, 433
  of basic solutions, 433
  of blood, 439
  of bodily fluids, 439
  of buffer solutions, 441
  of common solutions, 447*D*
  in determination of ionization constant in acids and bases, 448–49
  of salt solution, 437
  in titrations, 447–48
  of water, 432
pH measures:
  application of, 448–49
  color changes in, 446*t*
  indicators, 444–46
  and Nernst equation, 480
  pH meter, 446, 447*D*
  pH scale, 431–33, 432*t*
  taste, 443
  titrations, 447–48
  values of common substances, 432*t*
Phosphates, as fertilizers, 244, 245*B*
Phosphides, 247
  formation of, 247
Phosphine, 247
  formation of, 247
Phosphoric acid, 243–44
  formation of, 244
Phosphorous acid, 247
Phosphorus, P, 87, 240–47
  allotropes (forms) of, 240–41
  halides, 247
  of matches, 242*B*
  oxidation states, 241
  oxidation states of, 241*t*
  oxides, 242–43
  oxoacids of, 246*t*
  phosphides, 247
  phosphine, 247
Phosphorus (*cont.*)
  phosphoric acid, 243–44
  phosphorous acid, 247
  polyphosphoric acids, 245–47
  properties of, 240
  reaction with oxygen, 48*t*
  reaction with water, 244*D*
  red phosphorus, 241
  sources of, 240
  white phosphorus, 240
Phosphorylated molecules, 630
Photochemical reactions, 187–89
  chlorine and hydrogen in, 188
  process of, 188–89
Photochemical smog, 599–602, 600*f*
  components of, 600–601*t*
  formation of, 600
  and ozone, 601–3
  solutions to problem of, 602
  toxic effects of, 601–2
Photodissociation, 592–93
Photoelectric effect, 185–87
Photoelectrons, 198
Photoelectron spectroscopy, 198*f*
  ionization energies from, 199*f*
Photons, 185
  calculation of energy of, 185
  photoelectric effect, 185–87
Photosynthesis, 44
  effects on biosphere, 707
  and modification of atmosphere, 703
Photovoltaics, 321
Physical change, meaning of, 14–15
Physical properties, of substances, 10
Physical states:
  bromine as example of change, 54*f*
  changes of, 53
  gases, 53
  liquids, 53
  phase changes, 401–7
  solids, 53
  and vaporization, 53
  See also individual states
Pig iron, 376
Planck, Max, 185*M*
Planck constant, 185
Planetismals, 691
Planets:
  characteristics of, 691*t*
  formation of, 691–92
Plasma:
  molecular state, 677
  state of matter, 690–91
Plastics. *See* Polymers
Plastic sulfur, 225–26, 226*D*
  molecular structure, 226*f*
  polymerization reaction, 225–26*D*
Plate tectonics theory, 693
Plato, 2*B*
p-n junctions, semiconductors, 740–41
Polar covalent bond, 104
Polarimeter, 533*f*, 534
Polarizability, 409, 410*t*
Polar stratospheric clouds, 598
Pollution:
  and acid rain, 602–5
  and greenhouse effect, 606

Pollution (*cont.*)
and ozone depletion, 594–99
photochemical smog, 599–602
pollutants, 584
process of, 583–84
Polonium, 87
Polyamides, 725–27
Polyatomic ions:
characteristics of, 145
Lewis structures of, 145–46
Polycyclic arenes, 294
Polyesters, 728–29
Polyethylene, 289, 717–20
polymerization of, 717–20
Polymer chemistry, nature of, 716
Polymerization reaction, 717–20
initiation of, 717
plastic sulfur, 225–26*D*
propagation steps in, 719
Polymers:
addition polymers, 717–24
condensation polymers, 725–29
copolymers, 724
cross-linked polymer, 719
phenol-formaldehyde, 729
polyamides, 725–27
polyethylene, 717–20
polyesters, 728–29
polymethylmethacrylate, 723
polyropylene, 720–21
polystyrene, 721–22
polytetrafluoroethylene, 723*B*, 723
polyvinylchloride, 721
rubber, 723–25
structure of, 225
thermoplastic polymers, 729
thermosetting polymers, 729
types of, 719*f*, 719–20
use in medicine, 722*B*
uses of, 716, 718*t*
Polymethylmethacrylate, 723
uses of, 722*B*
Polypeptides, 617
Polyphosphoric acids, 245–47
formation of, 245–46
Polyprotic acids, 436
Polypropylene, 720–21
Polysaccharides, 627, 628
Polystyrene, 721–22
formation of, 721
Polytetrafluoroethylene, 723*B*, 723
uses of, 723*B*
Polyvinylchloride, 721
production of, 721
p orbitals, 211
Positron, 655
Positron emission, in radioactive process, 655
Potassium, K, reactions with water, 353*D*
Potassium-argon dating, 664–65
Potassium-40 dating, 664
Potential energy, change in chemical reaction, 562*f*
Precipitation reactions, 122, 146–48
characteristics of, 147, 149
example of, 146*D*
examples of, 239
Precipitation reactions (*cont.*)
metals, 379
solubility rules, 147–48
Pressure:
atmospheric pressure, 55
and change in volume, 67–68
Dalton's law of partial pressure, 68
definition of, 55
and gases, 56–60
and gas-phase equilibria, 450–52
gas pressure, 55
partial pressure, 68–69
and volume, 61*f*, 67–68
Primary cells, batteries, 480, 481
Primary structure, proteins, 617*f*, 617
Principal valences, 88–89
Principles of Chemistry (Mendeleev), 82
Probability distribution, 195
Products, in chemical equations, 18, 30*M*
Proline, 618*t*
Promethium, 33*M*
Propane, 169, 274
structure of, 275*f*
Propanoic acid, 519
Propanone, 517
Propene, 286
polymerization of, 720–21
structure of, 286*f*
Propylene oxide, 512
Protactinium-234, 33
Proteins, 616–25
and amino acids, 616, 618–19
denatured protein, 625
fibrous proteins, 625
globular proteins, 625
molecular construction of, 616–17, 618–19*t*
nutritional requirements, 645
primary structure, 617*f*, 617
quarternary structure, 625
secondary structure, 617, 620–21*f*, 622
tertiary structure, 622, 624*f*
Protein synthesis, 635–37
and RNA, 636
Protons:
of atom, 21
charge on, 22*t*
composition of, 22*M*
p-type semiconductors, 736
Purines, and gout, 638–39
Pyroxenes, 698

Qualitative analysis, 157
Quantitative analysis, 157
Quantum mechanics, 181, 195–97
application of, 195, 197
Quantum number, 192–93
Quarks, 22*M*
Quarternary structure, proteins, 625
Quartz:
characteristics of, 696
crystals, 387*f*, 696*f*

Racemic mixture, 534
Rad, 661
Radiation:
α radiation, 659
Radiation (*cont.*)
β radiation, 659
gamma radiation, 659
health hazards of, 659, 660*t*
measurement of, 661
Radiation therapy, 670
Radioactive dating, 664–66
potassium-argon dating, 664–65
radiocarbon dating, 665–66
Radioactive decay processes
α-particle emission, 654–55
β-particle emission, 654–55
electron capture, 655
gamma-ray emission, 655–56
positron emission, 655
and stability, 656
Radioactive decay rates, 662–66
decay constant, 662
half-life, 662–64
and radioactive dating, 664–66
of radon-222, 662*f*
Radioactive decay reactions, equations, writing of, 657
Radioactive decay series, 656
uranium-238 example, 657*f*
Radioactive elements, transuranium elements, 670–71
Radioactivity, 33–34, 653–61
and Curie family, 660–61*B*
definition of, 33
measurement of, 658
nuclear forces, 653–54
and nuclear reaction, 33
nuclear stability/instability, 653–54
Radiocarbon dating, 665–66
Radioisotopes, artificial, 653, 667–70
in medicine, 669–70
in radiation therapy, 670
radiolabeling, 668–69
synthesis of, 667–68
Radiolabeling, 668–69
Radon, 88, 662*f*
Rate constant, 554
Rate laws, 555–57
first-order rate law, 555
integrated rate laws, 558–59
second-order rate law, 555
Reactants, 157–62
in chemical equations, 18, 31*M*
and chemical reactions, 120
effect on reaction rate, 554–61
limiting reactants, 160–62
Reaction intermediates, 557
Reaction kinetics, 550
Reaction mechanism, 556, 560–61
Reaction quotient, *Q*, 443
Reaction rate, 120–21
and activation energy, 562–63
Arrhenius equation, 565–68
average rate, 552
collision model, 563–65
definition of, 550
and enzymes, 572
and equilibrium constant expression, 560–61
half-life of first-order reaction, 559*f*, 559–60

Reaction rate (*cont.*)
  initial rate, 554
  instantaneous rate, 552–53
  integrated rate laws, 558
  measurement of, 551–52
  rate constant, 554
  rate-determining step, 556
  rate laws, 555–57
  rate-limiting step, 556
  and reactant concentration, 554–61
  reaction intermediates, 557
  reaction mechanism, 556, 560–61
  and temperature, 561–68
  transition state, 562–63
Reactor core, of nuclear reactor, 674
Redox reactions. *See* Oxidation–reduction reactions
Red phosphorus, 241
Reducing agent:
  function of, 133
  sulfur dioxide as, 237*D*
Reduction:
  of aldehydes, 517
  of copper oxide, 49*D*
  definition of, 235
  of ketones, 517
  and oxidation, 49
  and oxygen, 49
  process of, 49
Relative average atomic mass, 28, 29*B*
Rem, 661
Resonance structures, 292, 294–95
  and bond order, 295
Respiration, 44
Ribbon model, computer models, 640*B*
Ribonucleic acid (RNA), 636–37
  base pairing in, 636*f*
  messenger RNA (mRNA), 636–37
  synthesis of, 636–37
Rivers, acid rain effects, 603
RNA. *See* Ribonucleic acid (RNA)
Rocks, 693–94
  igneous rocks, 693–94
  metamorphic rocks, 694
  radioactive dating of, 664–65
  sedimentary rocks, 694
Rose extract, as indicator, 444*D*
Rubber, 723–25
  crude form, 723–24
  synthetic rubber, 724–25
  vulcanization of, 724, 725*f*
Rubidium-87 dating, 665
Rusting, 370, 483–84
  process of, 483–84
Rutherford, Ernest, 22–23*B*, 660

Salicylic acid:
  and ester formation, 523
  production of aspirin, 500*f*, 523
Salt bridge, 467
Salts, 355–58
  acidic salts, 437
  basic salts, 437
  calcium carbonate, 358
  and fireworks, 359*B*
  hydrated salt, 362
  neutral salts, 436–437
  pH, determination of, 437–38
  production of, 355
Salts (*cont.*)
  sodium carbonate, 356–58
Sanger, Frederick, 617
Saturated fatty acids:
  structure of, 525*f*
  types of, 524*t*
Saturated hydrocarbons, 288
s-block elements, 203
Scanning tunneling microscope, 2–3*B*
Scientific method, process of, 63*B*
Scientific notation, 21*M*
Scurvy, 646*t*, 647*f*
Secondary cells, batteries, 481–82
Secondary structure, proteins, 617, 620–21*f*, 622
Second law of thermodynamics, 326
Second-order rate law, 555
Sedimentary rocks, 694
Sediments, 694
Semiconductors, 734–41
  conductivity of, 734–735*f*
  doping, 735, 739*f*
  function of, 734
  compared to insulators, 737–38*f*
  integrated circuits, 741
  n-type semiconductors, 735
  parking-garage analogy, 738*f*
  p-n junctions, 740–41
  p-type semiconductors, 736
  silicon as, 735–40
  transistors, 741
  types of, 738*f*, 739
  valence and conduction bands, 737–38*f*
Semimetals, 85
  and periodic table, 85, 86–88
Separation of mixtures, 14–17
  chromatography, 16–17*D*
  distillation, 15*f*, 15
  filtration, 15*f*, 15
Serine, 619*t*
Shell model of atom, 90–92
  electron arrangement in, 90*f*, 90–91, 91*t*
  shell in, 90
  valence shell, 91
Shells:
  electron arrangement in, 202*t*
  of electrons, 200
Side chain, 275
Σ, sigma, meaning of, 313
Silica. *See* Silicon dioxide (silica)
Silicates, 697–99
  structure of, 697–99*f*, 698*f*
  types of, 697–99
Silicic acid, 697
  formation of, 697
Silicon, Si, 87, 694–701
  aluminosilicates, 699–701
  compared to carbon, 694
  purification of, 739–40
  as semiconductor, 735–40, 738*f*
  silicates, 697–99
  silicic acid, 697
  silicon dioxide (silica), 695–96
  surface of, 3*B*
Silicon carbide, structure of, 272*f*
Silicon chip, 741
Silicon dioxide (silica), 695–96
  as amorphous solid, 399, 400*f*
  empirical formula of, 6–7
Silicon dioxide (*cont.*)
  sources of, 696
  structure of, 6*f*, 394*f*, 695*f*
Silicon hydrides, 694
Silver, Ag:
  electrical conductivity, 349
  in electroplating, 492
Single bonds, 100
SI system, 21*M*
  prefixes used, 22*M*
Smalley, Richard, 267*B*
Soap, 525, 526–27*B*
  production of, 526–27*B*
Soapstone, 699
Sodalite cage, 731
Soda water, 269
Sodium, Na, 86*M*
  properties of, 344*D*
  reaction with water, 353*D*
Sodium bicarbonate, 357–58
Sodium carbonate, 356–58
  production of, 356
  uses of, 357–58
Sodium chloride, NaCl,
  crystals of, 96*f*, 129*f*
  electrolysis of, 136*f*, 485–86
  electron density distribution, 396*f*
  empirical formula of, 7
  impurities in, 11*f*
  production of, 485
  structure of, 6*f*, 95*f*, 352*f*, 398*f*
  uses of, 125
Sodium flouride, uses of, 125
Sodium hydroxide, preparation by electrolysis, 490–91
Sodium hypochlorite, properties of, 251*D*
Sodium nitrate, 590
Sodium nitrite, 590
Sodium-24 radiolabeling, 669
Solar energy, as energy source, 321
Solar nebula, 690
Solar system:
  asteroids, formation of, 692
  comets, formation of, 692
  interstellar molecules, 688–90, 689*t*
  meteorites, formation of, 692
  planets, origin of, 691–92
  stars, formation of, 686–87
Solar wind, 691
Solids:
  amorphous solids, 226, 399–400
  characteristics of, 53
  close-packed arrangement, 391–92
  crystals, 387–99
  energy levels in, 736–39
  ionic solids, 390–91
  lattice structure, 391
  molecular solids, 390*t*, 390
  network solids, 390*t*, 390
Solid solution, 12
Solubility rules, and precipitation reactions, 147–48
Solute, 169
Solutions, 169–75, 415–17
  aqueous solution, 12
  concentration of, 170
  concentration of trace constituents, 172–73
  enthalpy of, 416–17

Solutions (*cont.*)
  gaseous solution, 12
  homogeneous mixture as, 12
  miscible and immiscible, 415, 416*D*
  molarity of, 169–71
  solid solution, 12
  titrations, 173–74
Solvent, 169
s orbitals, 210–11
Space-filling model, computer models, 640*B*
Specific heat capacity, 307
Spectra, 184, 189–94
Spectroscopy:
  atomic absorption spectroscopy, 164
  mass spectrometer, 25, 26*B*, 534–35
Speed of light, formula for, 183
sp hybrid orbitals, 213*f*, 213, 214
sp^2 hybrid orbitals, 213, 214
sp^3 hybrid orbitals, 212*f*, 212, 214*M*
Spontaneous processes:
  and disorder, 323–26
  and entropy, 326
  nature of, 322*f*, 322, 323*f*
  spontaneous cooling, 324–25, 325*f*
Spontaneous reactions, 321–23, 323*D*
Stability, of atoms, 33–34
Staggered conformation, 280, 281*f*
Stainless steel, 377
Standard atmosphere, 35
Standard enthalpy of formation, ΔH°_f, 312*t*, 312–14
  examples of use, 313–14
Standard molar entropy, S° 327*D*, 327
  of water, 406*t*, 406
Standard reaction enthalpy, 306–7
Standard reaction entropy, 327–28
Standard reduction potentials, 470–72
  for half-reactions, 471*t*
Standard state, 306
Starch, 628
Stars:
  formation of, 686–87
  spectra of, 689*f*
  supernova, 687*f*, 688
States. *See* Physical states
Staudinger, Hermann, 716
Stearic acid, 524*t*, 525*t*
Steel, 350
  composition of, 346
  production of, 375–76
  types of, 376–77
Stern, Otto, 205*B*
Stoichiometry:
  empirical formula determination, 165–67
  of gas reactions, 168–69
  molecular formula determination, 168
  nature of, 157
  percent yield, 163
  reactants in, 157–62
  of reactions in solution, 169–74
  theoretical yield, 162–63
Stratosphere, 592
  characteristics of, 44
  ozone formation in, 595–96
Strong acids, 140, 141*t*
  types of, 244*M*
Strong bases, 143
Strontium-87 dating, 665
Structural formulas, 7–8
  characteristics of, 7–8
  examples of, 7*f*
  three-dimensional type, 7*f*, 8*t*
Structural isomers, 275–76
Styrene, production of, 721–22
Subshells, 200, 201–2
Substances, 9–11
  characteristics of, 9
  chemical properties of, 10
  examples of, 14*f*
  impure substance, 11
  mass percentage and purity, 11
  physical properties of, 10
Substituent groups, 277–78
Substitution reactions, 503–4
  alcohols, 509–10
Sucrose, structure of, 8*f*
Sulfates:
  copper sulfate, 372–73*f*
  iron sulfate, 368
Sulfides, 87, 236–39
  copper sulfide, 372
  hydrogen sulfide, 238–39
  iron sulfide, 368
  metal sulfides, 237, 239
Sulfite ion, 236
Sulfur, S, 87, 224–39
  allotropes (forms) of, 224–26
  crystals of, 225*f*
  halides, 239
  monoclinic sulfur, 224, 225*f*
  orthorhombic sulfur, 224
  oxidation–reduction reactions of, 232–35
  oxidation states of, 235–36, 236*t*
  oxides of, 227
  plastic sulfur, 225–26, 226*D*
  properties of, 224–26
  reaction with copper, 372
  reaction with oxygen, 48*t*
  sources of, 224, 239
  sulfides, 236–39
  sulfuric acid, 227–35
  sulfurous acid, 237
Sulfur cycle, 708
Sulfur dioxide:
  and acid rain, 602–5
  and coal, 320, 602, 605
  formation of, 227
  as reducing agent, 237*D*
Sulfuric acid, 227–35, 244*M*
  concentrated, reactions of, 229–30
  as dehydrating agent, 230, 231*D*
  formation of, 227
  industrial production of, 228
  as oxidizing agent, 230–32, 231*D*
  oxoacids, 227
  properties of, 228–29
  uses of, 228
Sulfurous acid, 237
Sun, elements in, 695*t*
Sunlight:
  spectrum of, 592*f*
  ultraviolet radiation, dangers of, 596
Superconductivity, 742–43
Supernova, 686, 687*f*
Surface catalyst, 570, 571*f*
Suspension, nature of, 12
Synthesis, empirical formula determined from, 165–66
Synthesis gas, 51, 505
Synthesis reactions, 122
  process of, 149–50

Talc, 699
Taste, as measure of acid/base, 443
Technetium, 33*M*
Technetium-99 radiolabeling, 669
Tellurium, 87
Temperature:
  and change in volume, 67–68
  and Charles' law, 57–58
  definition of, 64
  and electrical conductivity, 349*f*
  and entropy, 326*f*, 326–27, 329*f*
  and equilibrium constant, 452
  and gases, 56–60
  and gas-phase equilibria, 452–53
  and kinetic energy, 64
  meaning of, 303
  and molecular speed, 70, 72*M*
  and rate of chemical reaction, 120
  and reaction rate, 561–68
  and vapor pressure, 404–5*f*
  and volume, 67–68
Temperature measures:
  Celsius scale, 58
  Kelvin scale, 58
Termolecular reaction, 556
Tertiary structure, proteins, 622, 624*f*
Tetrachloromethane, 501, 503
Tetrahedron molecular shape, 107*D*, 108*f*, 108–9
Tetrahydrofuran, 512
Theoretical yield, 162–63
  problems in determination of, 162–63
Theories, formulation of, 62–63*B*
Thermionic effect, metals, 350
Thermite process, 375, 376*D*
Thermochemistry, nature of, 302
Thermodynamics, 302
  bond energies, 314–19
  enthalpy, 306–11
  entropy, 323–30
  first law of, 304–5
  Gibbs free energy, 331–34
  Hess's law, 310–11, 313*f*
  second law of, 326
  spontaneous reactions, 321–23
Thermoplastic polymers, 729
Thermosetting polymers, 729
Thiols, 512–14
  oxidation of, 514
  properties of, 513–14
Thomson, George, 197*B*
Thomson, J. J., 22*B*, 26*B*
Thorium-234, 33
Three Mile Island, 678
Threonine, 619*t*
Thymine, 632
Tin, 87
  as conductor, 738*f*, 739
Titrations, 173–74, 447–48
  acid–base titrations, 173*f*, 173–74, 174*f*, 447–48
  end point in, 447
  equivalence point in, 173, 448

Toluene, 293
Torricelli, Evangelista, 55*f*, 55
Trace minerals, 645, 647*t*
Transistors, 741
Transition metals, 203, 365–75
  complex ions, 367
  compounds of, 367–73
  magnetic properties, 366–67
  oxidation states, 365, 374
  reactions of, 367–73
Transition state, chemical reactions, 562–63
Transuranium elements, 670–71
  names of, 670, 671*t*
Triacylglycerols, 641–42
Triangular molecular shape, 110
Triangular pyramidal shape, 109
  molecules, 109
Trichloromethane, 501, 503
Triglycerides, 524
Triiodide ion, 135
Trimethylamine, 529*t*
Triols, 507
Triple bonds, 100
Triprotic acids, 244
Tritium, 25*M*
Trona, 356
Troposphere, 592
  characteristics of, 43–44
Tryptophan, 619*t*
Tyrosine, 511*f*, 619*t*

Uhlenbeck, George, 205*B*
Ultraviolet radiation, dangers of, 596
Unimolecular reactions, 556
  process of, 568
Unit cells, 391*f*, 391, 393
  atomic radius, 393
  dimension and density, 393
Universal indicator, 446
Universe:
  big-bang theory, 685
  See also Solar system:
Unsaturated fatty acids:
  structure of, 525*f*
  types of, 524*t*
Unsaturated hydrocarbons, 288
  testing for, 289*D*
Unshared pairs, 99
Uracil, 637
Uranium, U, and radioactivity, 33, 660*B*
Uranium-235, fission of, 673–74*f*
Uranium-238
  radioactive dating, 665
  radioactive decay series, 657*f*
Urea:
  and gout, 638–39
  production of, 530
Urushiol, 511*f*

Vacuum-flask calorimeter, 308*f*, 309
Valence, 88–89
  and electron configurations, 209–10
  electrons and core charge, 92–93
  of main group elements, 89*t*
  principal valences, 88–89
  in writing formulas, 89
Valence band, 737, 738*f*
Valence shell, 91
Valine, 619*t*
Vanadium, V, 346
Vaporization, 53
  molar enthalpy of, 403
Vapor pressure, 403–5
  build up of, 403–4
  and temperature, 404–5*f*
Vegetable fats, 523, 525*t*
Vibrational frequency, molecules, 536–37
Vinyl chloride, production of, 721
Visible spectrum, 184
Vitamins, 646–47*B*
  deficiencies, 646–47*B*
  water and fat soluble, 646*t*
Volcanoes, gases released by, 703
Volume:
  in Avogadro's law, 59–60
  dependence on temperature and pressure, 67–68
  molar volume, 60
  and pressure, 61*f*, 67–68
  and temperature, 67–68
Volumetric analysis, 157
Volumetric flask, preparation of solution, 171, 172*f*
VSEPR (valence-shell electron-pair repulsion) model, 104–7, 112
  prediction of molecular shape, 112
Vulcanization, of rubber, 724, 725*f*

Water, 411–14
  autoionization of, 431
  boiling point, 411–12
  density of, 411*f*, 411
  freezing, process of, 406–7
Water (*cont.*)
  freezing point, 411
  hard water, 357*B*
  and hydrogen bond, 412*f*, 413
  ice, 414
  obtaining oxygen from, 46*D*, 46
  pH of, 432
  as polar molecule, 408
  properties of, 10*D*, 143, 411–13
  reactions with metals, 352–54
  standard molar entropy of, 406*t*, 406
  state changes, 408–9, 411
  states of, 5
Water cycle, 708
Water of crystallization, 229, 362
Water softening, 731–32*f*
Water vapor, 5, 53
Watson, James, 623*B*
Wavelength, of electromagnetic radiation, 183
Wave numbers, molecules, 536–37*t*
Waves:
  diffraction, 197*B*
  interference, 196–97*B*
  in phase, 196*B*
  properties of, 183*f*
  speed of, 183
Weak acids, 140, 141*t*
Weak bases, 143
White phosphorus, 240
  reaction with oxygen, 243*D*
Work, and Gibbs free energy, 333–34

Xenon, Xe, 88
X-ray crystallography, 387, 388–89*B*, 541
  measurement of atoms and ions, 395–97
  requirements for, 541
Xylenes, 293

Zeolites, 701, 731–33
  as catalysts, 732–33
  as ion-exchange materials, 731–32
  structure of, 701*f*, 731*f*
Zinc, Zn:
  reaction with copper sulfate, 466*D*
Zinc sulfide, ZnS:
  structure of, 399*f*
Zone refining, 740

Photo Credits

All Demonstration photos: Tom Bochsler, Photography Ltd.

Chapter 1 **CO.1** Tom Bochsler, Photography Ltd. **Fig. 1.6** Tom Bochsler, Photography Ltd. **Fig. 1.7** Barry L. Runk/Grant Heilman Photography. **Fig. 1.17** Tom Bochsler, Photography Ltd. **Fig. 1.19** Bettmann. **Box 1.3** Van Bucher/Photo Researchers. **Box 1.1** Gerald Present, Ph.D./IBM. **Box 1.2** Bettmann. **Box 1.4** NASA. **MA p. 6** M. Thonig/Allstock. **MA p. 7** Paul Silverman/Fundamental Photographs.

Chapter 2 **CO.2** Stephen Frink/Stock Market. **Fig. 2.2a** Tom Bochsler, Photography Ltd. **Fig. 2.3** Tom Bochsler, Photography Ltd. **Fig. 2.4b** Tom Bochsler, Photography Ltd. **Fig. 2.9** Ned Gillette/Adventure Photo. **Box 2.1a** NASA. **Box 2.1b** BMW of North America, Inc. **Box 2.2** Butch Powell/Stockphotos, Inc. **Box 2.2b** Culver Pictures. **Box 2.3** Courtesy AIP Emilio Segre Visual Archives/Institut International de Physique Solvay.

Chapter 3 **CO.3** Ken Eward/Science Source/Photo Researchers. **Fig. 3.3** Tom Bochsler, Photography Ltd. **Fig. 3.4a** GE Research & Development Center. **Fig. 3.4c** Tom Bochsler, Photography Ltd. **Fig. 3.5** Tom Bochsler, Photography Ltd. **Fig. 3.7** Bancroft Library, University of California. **Fig. 3.9** Charles Falco/Science Source/Photo Researchers **Fig. 3.13a** Michael Freeman **Box 3.2a** Tom Bochsler, Photography Ltd. **Box 3.1** Historical Pictures Collection. **Box 3.3a-c** Tom Bochsler, Photography Ltd. **Box 3.3d** Michael J. McGlinchey and Luc Girard/McMaster University.

Chapter 4 **CO.4** Haramaty Mula/Phototake. **Fig. 4.2** Tom Bochsler, Photography Ltd. **Fig. 4.4** Charles M. Falco/Photo Researchers. **Fig. 4.5** Tom Bochsler, Photography Ltd. **Fig. 4.8** Tom Bochsler, Photography Ltd. **Fig. 4.9** Tom Bochsler, Photography Ltd. **MA p. 125** Zviki Eshet/Stock Market.

Chapter 5 **CO.5** Custom Medical Stock Photo. **Fig. 5.3a** Tom Bochsler, Photography Ltd. **Fig. 5.3b** Tom Bochsler, Photography Ltd. **Fig. 5.3c** Tom Bochsler, Photography Ltd. **Fig. 5.5a** Tom Bochsler, Photography Ltd. **Fig. 5.5b** Tom Bochsler, Photography Ltd. **Fig. 5.5c** Tom Bochsler, Photography Ltd. **Box 5.1a** Four By Five **Box 5.1b** Linda K. Moore/Rainbow.

Chapter 6 **CO.6** Pekka Parviainen/Science Photo Library/Photo Researchers. **Fig. 6.7a-d** Sargent-Welch Scientific Co. **Fig. 6.10** Sargent-Welch Scientific Co. **Fig. 6.11** Burndy Library. **Box 6.2** Dr. E. R. Degginger, FPSA. **Box 6.1** Science Photo Library/Photo Researchers. **MA p. 194** Tom Bochsler.

Chapter 7 **CO.7** Tom Bochsler, Photography Ltd. **Fig. 7.2a** Tom Bochsler, Photography Ltd. **Fig. 7.2b** Tom Bochsler, Photography Ltd. **Fig. 7.3b** Tom Bochsler, Photography Ltd. **Fig. 7.5** Tom Bochsler, Photography Ltd. **Fig. 7.6b** Tom Bochsler, Photography Ltd. **Fig. 7.7** Tom Bochsler, Photography Ltd. **Fig. 7.11b** Tom Bochsler, Photography Ltd. **Box 7.1a** Tom Bochsler, Photography Ltd. **Box 7.1b** Tom Bochsler, Photography Ltd. **Box 7.1c** Tom Bochsler, Photography Ltd. **Box 7.2** Courtesy AGWAY, Inc. **MA p. 224** George Loun/Visuals Unlimited **MA p. 224** Renee Lynn/Photo Researchers **MA p. 240** A. Kerstitch/Visuals Unlimited.

Chapter 8 **CO.8** Kenneth Eward/Science Source/Photo Researchers. **Fig. 8.1** Kristen Brochmann/Fundamental Photographs. **Fig. 8.5** Tom Bochsler, Photography Ltd. **Fig. 8.6a** Tom Bochsler, Photography Ltd. **Fig. 8.6b** Tom Bochsler, Photography Ltd. **Fig. 8.17b** Tom Bochsler, Photography Ltd. **Fig. 8.17c** Tom Bochsler, Photography Ltd. **Fig. 8.20b** Tom Bochsler, Photography Ltd. **Fig. 8.20c** Tom Bochsler, Photography Ltd. **Fig. 8.21** Tom Bochsler, Photography Ltd. **Box 8.1a** Clive Freeman/Biosym Technologies/Science Photo Library/Photo Researchers. **Box 8.1c** David McGlynn/FPG International. **MA p. 274** Tom Bochsler, Photography Ltd. **MA p. 274 (bottom)** David R. Frazier/Photo Researchers. **MA p. 294** Paul Silverman/Fundamental Photographs.

Chapter 9 **CO.9** Jim Olive/Peter Arnold. **Fig. 9.15** Science Photo Library/Photo Researchers. **Box 9.1** Bettmann. **MA p. 320** Galen Rowell. **MA p. 321** Grant Heilman/Grant Heilman Photography.

Chapter 10 **CO.10** Seth Joel/Woodfin Camp & Associates **Fig. 10.1** Tom Bochsler, Photography Ltd. **Fig. 10.7** Tom Bochsler, Photography Ltd. **Fig. 10.8** Science Photo Library/Photo Researchers. **Fig. 10.9** Tom Bochsler, Photography Ltd. **Fig. 10.10** Stephen Frisch/Stock Boston. **Fig. 10.15** Tom Bochsler, Photography Ltd. **Fig. 10.16** Tom Bochsler, Photography Ltd. **Fig. 10.18** Tom Bochsler, Photography Ltd. **Fig. 10.19a** Tom Bochsler, Photography Ltd. **Fig. 10.21** Bethlehem Steel. **Box 10.1** Robert Mathena/Fundamental Photographs. **Box 10.2** Milton Heiberg/Photo Researchers. **MA p. 344** Tom Bochsler, Photography Ltd. **MA p. 368** John Cancalosi/Perter Arnold.

Chapter 11 **CO.11** Craig Tuttle/Stock Market. **Fig. 11.2a** Runk/Schoenberger/Grant Heilman Photography. **Fig. 11.5c** Tom Bochsler, Photography Ltd. **Box 11.2a** William J. Leigh/McMaster University. **Box 11.1a** E. F. Smith Memorial Collection, Center for the History of Chemistry. **Box 11.1b** **Box 11.1c** The Royal Institution. **Box 11.1e** James King-Holmes/OCMS/Science Photo Library/Photo Researchers. **MA p. 401** Tom Bochsler, Photography Ltd. **MA p. 417** Tom Bochsler, Photography Ltd.

Chapter 12 **CO.12** Tom Bochsler, Photography Ltd. **Fig. 12.2a** Tom Bochsler, Photography Ltd. **Fig. 12.2b** Tom Bochsler, Photography Ltd. **Fig. 12.2c** Tom Bochsler, Photography Ltd. **Fig. 12.3** Tom Bochsler, Photography Ltd. **Box 12.1** Professors P. M. Motta and S. Correr/Science Photo Library/Photo Researchers. **p. 438** Tom Bochsler, Photography Ltd. **MA p. 446** Tom Bochsler, Photography Ltd. **MA**

Chapter 13 **CO.13** Tom Bochsler, Photography Ltd. **Fig. 13.1** Tom Bochsler, Photography Ltd. **Box 13.1** Science Photo Library/Photo Researchers. **Box 13.2** Stock Montage/Historical Picture Service. **MA p. 467** Tom Bochsler, Photography Ltd. **MA p. 480** P. Ceisel/Visuals Unlimited. **MA p. 490** Tom Bochsler, Photography Ltd. **MA p. 491** Tom Bochsler, Photography Ltd.

Chapter 14 **CO.14** Tom Bochsler, Photography Ltd. **Fig. 14.3** Bettmann. **Fig. 14.25** Varian Associates, Inc. **Fig. 14.28** Pete Saloutos/Stock Market. **Box 14.1** SIU/Peter Arnold. **MA p. 507, p. 508, MA p. 521, MA p. 523**, Tom Bochsler, Photography Ltd.

Chapter 15 **CO.15** Richard Megna/Fundamental Photographs **Fig. 15.12** Peter Arnold.

Chapter 16 **CO.16** Tom McHugh/Allstock. **Fig. 16.1** NASA. **Fig. 16.5** NASA. **Fig. 16.8a** Pete Saloutos/Stock Market. **Fig. 16.8b** Bill Ross/Allstock. **Fig. 16.10** John Shaw/Bruce Coleman. **Fig. 16.11** Richard Megna/Fundamental Photographs. **Fig. 16.12** Bochsler Photographic Associates. **Box 16.1** Anthony Howarth/Science Photo Library/Photo Researchers. **MA p. 592** Tom Bochsler, Photography Ltd. **MA p. 604** D. A. Humphreys.

Chapter 17 **CO.17** Dr. A Lesk, Laboratory of Molecular Biology/SPL/Photo Researchers. **Box 17.1a** Omikron/Photo Researchers. **Box 17.1b** Dr. A. Lesk, Laboratory of Molecular Biology/SPL/Photo Researchers. **Box. 17.1c** BBC. **Fig. 17.8** Prof. P. Motta/Dept. of Anatomy/University "La Sapienza", Rome/SPL/Photo Researchers. **Box 17.2a** Will & Deni McIntyre/Photo Researchers. **Box 17.3a** Geis/Gouaux/Science Source/Photo Researchers. **Box 17.3b** Ken Eward/Science Source/Photo Researchers. **Box 17.4** Biophoto Associates/Science Source/Photo Researchers.

Chapter 18 **CO.18** Tom Bochsler, Photography Ltd. **Fig. 18.6** Bettmann. **Fig. 18.7a** Visuals Unlimited. **Fig. 18.7b** Visuals Unlimited. **Box 18.1a** Granger Collection. **Box 18.1b** University of Pennsylvania, Smith Collection. **Box 18.2a.1** John D. Cunningham/Visuals Unlimited. **Box 18.2a.2** Dan McCoy/Rainbow. **Box 18.2b** Richard T. Nowitz. **Box 18.11b** Dr. Jeremy Burgess/Science Photo Library/Photo Researchers.

Chapter 19 **CO.19** Lightscapes/Stock Market. **Fig. 19.1** NASA. **Fig. 19.3** Royal Observatory, Edinburgh/Science Photo Library/Photo Researchers. **Fig. 19.4** Courtesy Department Library Services/American Museum of Natural History. **Fig. 19.8** Michael Dalton/Fundamental Photographs. **Fig. 19.9** Karl

Hartmann/Phototake. **Fig. 19.14** Clive Freeman, The Royal Institution/Science Photo Library/Photo Researchers. **Fig. 19.15** Visuals Unlimited. **Box 19.1** Prof. W. Alvarez/Science Photo Library/Photo Researchers. **MA p. 699** Tom Bochsler, Photography Ltd. **MA p. 700** Grant Heilman Photography. **MA p. 701** E. Spiegelhalt/Woodfin Camp & Assoc. **MA p. 709** Manfred Kage/Peter Arnold.

Chapter 20 **CO.20** Manfred Kage/Peter Arnold. **Fig. 20.2** Tom Bochsler, Photography Ltd. **Fig. 20.6** Dupont Photo. **Fig. 20.24** Charles M. Falco/Photo Researchers. **Box 20.1a** Visuals Unlimited. **Box 20.1b** Linda K. Moore/Rainbow. **Box 20.2** Dupont Photo. **MA pp. 717, 722, 724, 727, 728** Tom Bochsler, Photography Ltd.

The International System of Units (SI)

SI Base Units

Physical Quantity	Name	Symbol
Length	meter	m
Mass	kilogram	kg
Time	second	s
Electric current	ampere	A
Temperature	kelvin	K
Amount of substance	mole	mol

SI Derived Units

Physical Quantity	Name	Symbol	Definition
Volume	liter	L	$10^{-3} m^3$
Force	newton	N	$kg \cdot m \cdot s^{-2} = J \cdot m^{-1}$
Power	watt	W	$kg \cdot m^2 \cdot s^{-3} = J \cdot s^{-1}$
Pressure	pascal	Pa	$kg \cdot m^{-1} \cdot s^{-2} = N \cdot m^{-2} = J \cdot m^{-3}$
Energy, work, heat	joule	J	$kg \cdot m^2 \cdot s^{-2}$
Frequency	hertz	Hz	s^{-1}
Electric charge	coulomb	C	$A \cdot s$
Electric potential difference	volt	V	$kg \cdot m^2 \cdot s^{-3} \cdot A^{-1} = J \cdot A^{-1} \cdot s^{-1}$

Physical Constants

Constant	Symbol	Value
Atomic mass unit	u	$1.660\,54 \times 10^{-27}$ kg
Avogadro constant (Avogadro's number)	N_A	$6.022\,14 \times 10^{23}\ mol^{-1}$
Boltzmann constant	$k = R/N_A$	$1.380\,66 \times 10^{-23}\ J \cdot K^{-1}$
Elementary charge	e	$1.602\,18 \times 10^{-19}$ C
Faraday constant	$F = N_A e$	$9.648\,53 \times 10^{4}\ C \cdot mol^{-1}$
Gas constant	R	$8.314\,51\ J \cdot K^{-1} \cdot mol^{-1}$
		$0.082\,06\ L \cdot atm \cdot K^{-1} \cdot mol^{-1}$
Mass of an electron	m_e	$9.109\,39 \times 10^{-31}$ kg
		$5.485\,80 \times 10^{-4}$ u
Mass of a neutron	m_n	$1.674\,93 \times 10^{-27}$ kg
		1.008 66 u
Mass of a proton	m_p	$1.672\,62 \times 10^{-27}$ kg
		1.007 27 u
Planck constant	h	$6.626\,07 \times 10^{-34}\ J \cdot s$
Speed of light	c	$2.997\,9246 \times 10^{8}\ m \cdot s^{-1}$